DI042389

Growth in a Changing Environment

Growth in a Changing Environment

A HISTORY OF
STANDARD OIL COMPANY (NEW JERSEY)
EXXON CORPORATION
1950-1975

BENNETT H. WALL
Principal Author

Assisted by
C. GERALD CARPENTER
Who Drafted Several Chapters

and
GENE S. YEAGER
who assisted with the
Latin American Section

EARL N. HARBERT
Editorial Consultant

McGRAW-HILL BOOK COMPANY
New York   St. Louis   San Francisco   Auckland   Bogotá
Caracas   Colorado Springs   Hamburg   Lisbon
London   Madrid   Mexico   Milan   Montreal
New Delhi   Oklahoma City   Panama   Paris
San Juan   São Paulo   Singapore
Sydney   Tokyo   Toronto

Library of Congress Cataloging-in-Publication Data

Wall, Bennett H.
    Growth in a changing environment: a history of Standard Oil
Company (New Jersey), Exxon Corporation, 1950-1975/Bennett H.
Wall, principal author; assisted by C. Gerald Carpenter who drafted
several chapters and Gene S. Yeager who assisted with the Latin
American Section.
        p.   cm.
    Bibliography: p.
    Includes index.
    ISBN 0-07-067915-0
    1. Standard Oil Company — History.   2. Exxon Corporation — History.
3. Petroleum industry and trade — United States — History.
I. Carpenter, C. Gerald.   II. Yeager, Gene S.   III. Title.
HD9569.S82W35 1988
338.7'6223382'09 — dc19   88-5945
                                    CIP

1234567890   DOC/DOC   8921098

ISBN 0-07-067915-0

Contents

Introduction   v

Prologue   xiv

iv

Introduction

ACCORDING TO THE BEST INFORMATION available to me, the idea behind this volume originated in a conversation between J. Kenneth Jamieson, then Board chairman and chief executive officer of Exxon Corporation, and Herbert Longenecker, then President of Tulane University. The time may have been the fall of 1973 or perhaps, the winter of 1973-1974.

Precisely when Longenecker first approached me about taking the responsibility of principal author remains unclear. At the present time, I do not recall. But I do remember that, a few days later, Tulane Provost David R. Deener sought me out to promote this project as one that held special importance for the university. Even so, exactly why I accepted the combined Jamieson-Longenecker-Deener invitation to confront such an enormous task still represents something of a mystery—especially to me.

Certainly, the general idea of tackling a big topic always held appeal. But nothing so large as a 25-year history of Exxon!* In the end, perhaps my acceptance should be traced to unadulterated ego. At any rate, I agreed to take on the Exxon assignment.

To begin, I telephoned friend Ralph W. Hidy of the Harvard Business School and discussed particulars of the contract to be written for me. As always, Hidy provided wise counsel, based on his experience as author of the first volume of the existing three-volume history of the Standard Oil Company (New Jersey)—now Exxon. Then feeling well armed with sound advice, I awaited the results of my initial consent. The ball had been put into play, and it remained for the moment in Exxon's court.

---

*The three earlier volumes had been written by historians at the Business History Foundation. They are: Ralph W. Hidy and Muriel E. Hidy, *Pioneering in Big Business, 1882-1911: History of Standard Oil Company (New Jersey)*, (New York, 1955); George S. Gibb and Evelyn H. Knowlton, *The Resurgent Years, 1911-1927: History of Standard Oil Company (New Jersey)*, (New York, 1956); Henrietta M. Larson, Evelyn H. Knowlton, and Charles S. Popple, *New Horizons, 1927-1950: History of Standard Oil Company (New Jersey)*, (New York, 1971).

Sometime that winter Jamieson gave a luncheon for several Tulane folk, including my wife and me. After lunch, Paul E. Morgan, senior public affairs advisor at Exxon, discussed with me goals for a proposed fourth volume of the (Jersey) Exxon history: essentially, something readable; and an account that would focus largely on Exxon's people. At this time, I discovered that Morgan, too, had telephoned Ralph Hidy, to ask about my qualifications for the job. So, as middle man, Hidy played an indispensable role.

Months later, Dave Deener telephoned me at the Claremont Graduate School, where I held a visiting appointment at the time, to confirm that the Exxon Board had approved a grant to Tulane University for my services in writing a fourth volume of the company history. Now, the game had turned to my court; a staff had to be employed, and cooperation from Tulane's History Department secured. After due deliberation, in the manner of academe, that department approved my working on the book. Meanwhile, two young Ph.D.s—C. Gerald Carpenter and Gene S. Yeager—were employed, in part to assist my authorial efforts. At last, sometime in the fall of 1974, we began to work on project Exxon.

As Mrs. Rose B. Koppel established a project office at Tulane, we consulted Paul Morgan in New York and began interviewing people. Exxon assisted by placing notices announcing the project in several company publications, and by urging that employees talk freely with us. Yeager, Carpenter, and I found ourselves spending much time in New York City, especially in the Exxon Records Center and the company Secretary's Office. Not everything emerged from our search (notable among the missing were many records subpoenaed for the long U.S. civil suit of the 1950s). But the sheer volume of records available could not be regarded as less than overwhelming.

For each of us, a personal schedule filled rapidly, especially for me—a program of frequent travel combined with a demanding load of teaching. Yeager, who possessed a keen mind, soon proved himself a hard worker. When another opportunity appeared, he resigned after one year, having completed a rough sketch of what became the present Latin America section. Carpenter remained an essential part of the project for four and one-half years, during which time he drafted several chapters for later revision by the principal author. In fact, even after Carpenter secured a university teaching appointment, he continued to provide assistance. My debt to him is great, especially since, during his years writing about Exxon, Carpenter established himself as a promising business historian.

Both Carpenter and Yeager contributed importantly; and once they were gone, I struggled as principal author to keep on track. Here, it would be futile and merely time-consuming to detail the delays and frustrations encountered; certainly, I have never lacked for advice from an array of self-appointed historians. Of course, much still remains that could be added to the Exxon story. But time has set some arbitrary limits.

X   X   X

In New York, on October 29, 1981, while I was conversing with Howard W. Page, Emilio G. "Pete" Collado came by. The three of us began to talk about the Exxon book— then we drifted from that to the general subject of history. Collado, an omnivorous reader, and Page, clearly another book person, considered this writer's task for several minutes. Then, turning to leave, Collado remarked: "You know Howard, after it is all researched and written, when you read it, it will bear only a faint resemblance to what actually happened and who did what." With that said, he left, and Page and I grinned in mutual understanding.

Yet it remains my sincere hope that what is written here does resemble— and more than faintly—what actually *did* happen. But if it does not, then I am chiefly responsible. For among the many persons who tried to help me, like Page and Collado, none should be found at fault.

X   X   X

In fact, so many current and former employees of Exxon assisted the principal author and his two associates that the act of singling out some of them for special notice opens the way to leaving out others who helped as much.* All these people, moreover, appear in my memory more like friends and helpers than interviewees; they made themselves available to talk about the company and about the people who made it. Over many lunches and dinners, and often in private homes, these persons, their wives, and their guests cheerfully talked shop about Exxon as they knew it. Certainly, much of what they said became influential in shaping the final narrative, as each conversation added missing pieces. I thank every one.

Siro Vázquez once remarked that, after this book had been completed, his wife hoped for another to be written, based principally on interviews with the wives of the men interviewed for the first. Unwilling as I am to contemplate any sequel, I still would be remiss to omit my statement of gratitude to the Exxon women who did provide help: Willie Cleveland, Katharyn John-

---

*Such a listing does appear after the final chapter in this book.

son,* Betty McCoy, Nancy Morgan, Eileen Page, Suzanne Pratt, Kay Shepard, Ruth Turner, Mary R. Williams, Dorothy A. Windels, and several
others. All these ladies proved to be informed conversationalists, whose
abiding interest in the company showed years of thoughtful concern.

In countless hours of conversation, Edward F. Johnson and his wife
Katharyn taught me a great deal. Johnson, a broad-gauged, humanistic
lawyer, strongly supported the notion that the public had the right to know
about Jersey, and that the company had an obligation to inform the public.
He believed, moreover that as writer of the company history, I should have
unrestricted access to both records and persons, plus the freedom to use what
I found. Through the years, Johnson demonstrated an unusual ability to put
events and people into an enlarged context.

Howard Page and his wife, Eileen, also liked to talk about the company
and its problems. Through her work with British Petroleum Company, Ltd.,
Eileen Page got to know many of the industry's leaders. Page patiently (most
of the time) overlooked my vast ignorance of the Middle East, as he
explained how and why things happened, 1950-1975. Later, he read with
care several draft chapters and then, in long conversations, enlarged upon
his written critique. As principal author, I grew especially fond of Page, who
displayed a brilliant mind and an obvious genius. Certainly, it was no surprise to learn that a number of very successful, younger executives at Exxon
admired Page enough to make him their role model.

Smith D. "Sparky" Turner also made himself an indispensable resource
for the Middle East chapters printed here. Turner provided expert information regarding conferences, agreements, and especially production formulae. No matter how many drafts, never did I get the chronology of Middle
Eastern events accurate enough to satisfy him. By both letter and telephone,
Turner supplied sound advice and instant answers—many of which he alone
possessed.

Again, special credit for the materials in this volume should go to Wallace
E. Pratt, the legendary geologist who on several occasions offered me a
running commentary on Exxon, the oil industry, and the men who made it.
Today, his knowledge of the petroleum industry remains unrivalled, and he
is missed. For there can be no doubt that Pratt was an oil giant—an all-time
great, who enjoyed to the end of his long life an almost total recall of events.

*Mrs. Johnson once told M.J. "Jack" Rathbone the company should pay the expenses of wives to
enable them to accompany husbands who had to be away from home on business for a long
period of time. Rathbone implemented her suggestion.

Herbert Longenecker, who resigned as Tulane President in 1975, encouraged the principal author across the entire period of research and writing. David R. Deener did all that a person could to get this project started, and so long as he lived, his support remained invaluable to the principal author. Again, Ken Jamieson did everything he could to make certain this volume appeared: he generously granted interviews, read all or portions of chapters, checked interview transcriptions, and by letter and telephone supplied valued information.

M.A. Wright drew on his capacious memory of Exxon and its affiliates to detail key episodes about which company records preserved little. In long interviews, I easily came to understand how so engaging a person could have been marked early in his career for corporate leadership. Wright provided detailed information, read all or parts of several chapters, loaned rare books from his personal library, and made himself exceptionally helpful and friendly.

Paul E. Morgan deserves much praise; for in what must have been the most frustrating assignment of his career, he managed to remain even tempered. Morgan's knowledge of the Exxon empire is encyclopedic; at times, his memory proved more accurate than the printed Exxon word. In an important way, Morgan made this book happen. He reviewed the entire Latin American section after Gene Yeager left the project, using his personal knowledge of the Exxon operation to add flavor. Both he and wife Nancy deserve not only my thanks but also a commendation from Exxon.

Clifton C. Garvin, Jr. supported this project from the time it started until he retired as Exxon's chief executive officer in 1986. His help is appreciated.

George H. Freyermuth, whose magnificent performance as head of Public Relations at Jersey in the 1940s and early 1950s did much to reverse the public's low opinion of Jersey, merits a place here too. Freyermuth is a great narrator, and after listening to him, I knew why Director Robert T. Haslam chose him to explain the I.G. Farben affair. Another Public Affairs executive, William P. Headden, also cleared the way for this book. He believed in Exxon, and his stories often stimulated the author's curiosity about the company. Again, John O. Larson's enlightening conversation, plus his and Mrs. Larson's warm hospitality and easy acceptance of interruptions to their routine, made interviewing him a pleasure. In addition, he carefully critiqued several chapters of this book.

Ralph and Betty McCoy provided warm friendship and hospitality, enlivened by endless Jersey stories. McCoy arranged a number of special interviews, introduced me to oil people, and generally supported the project. William B. Cleveland thoughtfully prepared for interviews by first writing and sending a memoir of his years with the Jersey affiliate, Standard-Vacuum Oil Company, and later with Esso Eastern Inc. He and his wife, Willie, detoured to Athens, where he supplemented the memoir with additional information. Finally, he read several chapters and made suggestions.

Ed Holmer, President, Exxon Chemical Company, reviewed an early draft of the chapter on Chemicals, and found it wanting. Holmer then asked whether the author was willing to dig into records in Holmer's personal files. When the author responded positively, Holmer spent several weekends gathering stacks of material. Afterward, he explained some of the more technical documents, and before he finished, Holmer had read several versions of the Chemicals chapter. Equally ready to help with anything relating to chemicals was T.A. White of Exxon Chemical's Public Affairs Department. White prepared the glossary, the chemical profile, and the organizational chart—all to be found at the end of the Chemicals chapter. Both Holmer and White deserve much credit.

Zeb Mayhew, despite a serious hip problem, related fascinating stories concerning his years in production and as head of Esso Exploration Inc. He and his wife quickly made an interviewer feel at ease. Overall, Mayhew proved to be a storehouse of information about exploration and production. Like him, Harold W. Fisher and David A. Shepard always found time for interviews, and both furnished information unavailable elsewhere. Thoughtful men, they consistently underplayed their own roles in the achievements of the company. Yet their distinguished careers, documented in company records, clearly refute such modesty.

Lloyd W. Elliott provided sound information on Standard-Vacuum Oil Company and Esso Eastern. Regrettably, we were unable to interview him often. Robert H. Milbrath read several chapters and made suggestions that prevented grievous errors, especially in regard to events in Europe and concerning the corporate name change.

Emile E. Soubry and Wolf Greeven presented the author with authoritative accounts of marketing and of the problems they encountered at Jersey. Soubry also talked a good bit about how the Board of Directors functioned. Greeven, who played an important role in the name change, read that chapter and critiqued it for us, as he did with some European sections. Siro Vázquez patiently answered more questions in general conversation than we could bear to ask him in a polite interview. An elegant gentleman always, he proved to be both interesting and attentive to authorial needs.

One of the most illuminating interviews, in which both C. Gerald Carpenter and I participated, involved Hugh de N. Wynne, who, while riding the train from Princeton, carefully planned what to say. Wynne's mastery of minute detail about Libya proved fascinating; he held us spellbound. Later, he also read the chapter on Libya and suggested a number of improvements.

Over more than a decade, certain interviews naturally stand out. One of these, I conducted with George T. Piercy, but unfortunately it remains lost to this history because of a jammed tape recorder. Piercy remembered events dating from the Rathbone years (1950-1965) through those of Clifton C. Garvin, Jr. (1975-1986), in many of which he had played an important part.

Henry B. Wilson proved to be one of the most knowledgeable persons we encountered. He reached back in memory, to bring forth events, dates, and stories—most of which could not be found in company files. Certainly Wilson possessed a broad overview of the company's role in society, and he never seemed reluctant to give an opinion.

This author especially appreciated the time Exxon attorneys Frank X. Clair and David G. Gill spent explaining a number of court decisions involving the company. Concerning its name change, Clair proved particularly helpful. Gill, over at least ten years, furnished legal information on suits, hearings, cases, and decisions.

Cecil Morgan assisted the principal author in locating materials and finding persons who retained precise information. In a number of interviews, Morgan related novel insights and unique information. His presentation, moreover, amounted to a well-organized lecture. As a teacher, I could only admire.

Dorothy A. Windels of the Exxon Secretary's Office provided assistance that few others in the company could have rendered. She frequently had material verified, dates and names checked. She also contacted several annuitants with questions for which our research provided no answer. Her knowledge of the company proved invaluable to us. We owe much to her.

At Exxon Company, U.S.A., Hines and Rex Baker aided with interviews, and Hines Baker later corresponded with the principal author answering sticky questions. The Baker brothers knew the Humble-Exxon USA story well, and willingly shared it with me. Both men belonged to a vanishing breed—old school gentlemen.

Other former Humble executives also freely gave of their time and knowledge: Richard Gonzalez, Nelson Jones, L.T. Barrow, Morgan Davis, and Merrill Haas. Each of them made significant points concerning the Humble story. To them and especially to Carl E. Reistle, Jr., the author's debt is great. Reistle, in long conversations, answered many questions, and he explained in detail several of Humble's operations. Reistle proved to be superb. His easy relaxed manner, combined with a highly accurate memory, made it a pleasure to interview him. What good stories he told! And Leslie C. Rogers provided such friendly assistance as only a well-informed public affairs executive (Exxon USA) can supply: interesting conversations, introductions, addresses, telephone numbers, plus arranging for additional research.

X   X   X

All the authors are indebted to Rose Koppel, Elizabeth Varadan, and Sheree Dendy for managing the office, transcribing interviews, and arranging files. Ms. Koppel worked almost six years on Exxon, while Ms. Dendy has labored even longer. Without their fine grasp of all relevant materials, the principal author would certainly have erred more often. In particular, Ms. Dendy spent blinding hours checking the manuscript, the galleys, and the page proof—all without losing her sense of humor. Her alertness caught many mistakes; and, at times, her memory of names and events seemed truly limitless.

A word of praise is not sufficient for Dean Kay Orrill of Tulane University, through whose office for many years Exxon-related business passed. She has been ever kind, patient, and understanding.

In 1980, while I was still writing the first draft of this volume, arrangements were made to transfer the entire operation to the University of Georgia in Athens. Tulane had provided for me and my associates everything for which we asked, but changes naturally alter conditions. Two Georgia administrators, in particular, helped to make this move possible and congenial: Paul C. Nagel and Lester D. Stephens, who, as department head, somehow arranged office space for both the Exxon project and me after my formal retirement. To these gentlemen and to the entire University of Georgia History Department, my thanks are offered.

Again, I would be negligent if I did not express my gratitude to Matt Moran of Exxon, who gave expert advice on the format of this book. His help in organization and styling and his command of the publication process have made it better.

There are several persons who have been more than purveyors of information, helpers, or critics. They deserve an especial category of praise and appreciation.

X   X   X

Earl N. Harbert has served as editorial consultant throughout the entire project, sagely advising the principal author on such matters as organization, usage, and revision of the text. In general, he has filled the place of in-house critic, and he deserves credit for his constructive role.

Second on such a list is Henrietta M. Larson. In the early days of this Exxon project, I spent about a week in Northfield, Minnesota, where I daily talked with her about her work on the third volume of the Jersey story. She and her sister Norah proved to be most hospitable, gentle folk. Henrietta provided much information about Jersey in the 1950s, although *New Horizons* ended its coverage in 1950. This connection proved extremely helpful.

Later she read several chapters and offered suggestions about them. She also loaned me a raft of material to study—books, interviews, and pamphlets.

Viewed in perspective, I think it only fair to say that she possessed almost as much information about the Jersey Company as any person who ever lived—even Wallace Pratt and Ed Johnson. To have her available to help was simply wonderful.

Finally, my wife, Neva Wall, deserves my thanks for listening, without complaint, to endless oil talk. For almost thirteen years, she has endured my thinking aloud, as I tracked through the maze of material. That is a long time, especially since her husband has never worked for an oil company. Her patience and understanding make me grateful.

Prologue

Jersey: World War II to Mid-Century

THE DAY WAS JUNE 24, 1950. In Independence, Missouri, the President of the United States, Harry S. Truman, interrupted his visit to listen to his Secretary of State, Dean Acheson: "Mr. President, I have very serious news. The North Koreans have invaded South Korea."[1] Back in Washington the next day, President Truman approved a draft resolution to be offered to the United Nations Security Council. It stated that this armed attack on the Republic of South Korea constituted an "unprovoked act of aggression" by the North Koreans. Later, this phrase was softened to read "breach of the peace," and in that form the resolution was approved by the Council. Meanwhile, Americans everywhere heard news of a far away war—a war that very few suspected would play a major role in U.S. history. Yet the United States Congress agreed with the President; at all costs, the Communist North Korean invasion must be repelled and another world war averted.

Americans anxiously watched the news from Korea. The U.S. Defense Department [Defense] sent two additional divisions of American soldiers to bolster the battered forces of General Douglas MacArthur, and with their aid, the North Koreans were stopped near Pusan. While fighting went on, Defense accelerated procedures to fill out existing skeleton battalions and regiments with World War II veterans still in the reserves, as well as with draftees. President Truman, acting with the overwhelming approval of his diplomatic and military advisers, turned down an offer of assistance from Chiang Kai-shek, Chinese Nationalist President on Formosa. Any such move as the use of Chiang's troops, Americans feared, might lead to Red (Mainland) Chinese military action in support of North Korea. World peace seemed to hang by a thread.

A month passed, and new offers of support and assistance poured into the White House from all over the nation. Executive Committee minutes from a meeting of the Standard Oil Company (New Jersey) [Jersey] Board of Direc-

xiv

tors indicate how one such offer originated: "Mr. [Frank W.] Abrams stated that he planned to write President Truman pledging, on behalf of the Company, full support of the Government's position in actively resisting the forces which threaten the nation's security and world peace."[2] Yet these dry-as-dust minutes do not reveal either the extent or the flavor of the Board deliberations, although it may be significant that the characteristic phrase used to indicate merely routine acceptance—the Committee [or Board] "finds no objection to this"—does not appear. In any case, Abrams, chairman of the Jersey Board, on July 24, 1950, dictated a letter to President Truman. A very thoughtful man, Abrams wrote:

> My dear President Truman:
>
> Through its long experience in conducting a worldwide business the Standard Oil Company (N.J.) has had opportunity to observe the ills that flow from tyranny and the benefits that come from human liberty. As one American industrial concern, we want you to know that in today's test between these extremes in human values, we are wholeheartedly behind our Government's position.
>
> The management of this company is ready to devote the organization's facilities, skills and equipment to the full support of our Government against the forces that threaten the nation's security and world peace.[3]

Whether President Truman, busy with the mounting problems of the Far Eastern struggle, actually saw the letter will never be known. Obviously, however, alert members of his staff did; for within five days Abrams received this signed reply:

> THE WHITE HOUSE
> WASHINGTON   July 31, 1950
>
> Dear Mr. Abrams:
>
> I appreciated most highly your good letter of the twenty-sixth.
>
> I have always known that when the country is faced with diffi-

culties the men in charge of great business organizations are always ready to do whatever is necessary to meet the situation. I came in close contact with them in the Second World War and I am thoroughly convinced that all of them are patriots. You demonstrated that by your letter of the twenty-sixth.

Sincerely yours,
(signed) Harry S. Truman[4]

Truman's statement, "I am thoroughly convinced that all of them are patriots," probably raised eyebrows throughout the Jersey organization, where memories of less friendly government attitudes remained fresh. Abrams had George H. Freyermuth, head of Public Relations, circulate copies of the President's reply throughout the company, and many executives recalled 1942, when Truman had expressed an opposite conviction that the government and the company stood poles apart. In fact, the two letters together served to demonstrate a great change in both the corporation's relationship with Mr. Truman and the President's attitude toward the Standard Oil Company (New Jersey). Perhaps the politician in the White House did not appreciate this contrast; but many people at Jersey did.

X   X   X

1942 represented a year of trauma for most Americans. The Pearl Harbor attack of December, 1941, had brought a shocking national defeat, but the same Japanese action also served to unite a nation which, for almost a decade, had been badly divided over the question of support for Britain and France in their struggle against fascism. By 1942, all debate had ended, as Americans turned to the active pursuit of an enemy, fighting with guns, planes, ships, and in factories at home. While volunteers, angry over the Pearl Harbor debacle, rushed to enlist in the armed forces, an aroused civilian population demanded to know why their nation should find itself in such an unprepared state — and who should be held responsible. The news media announced a steady Japanese advance toward Australia and India, increased German submarine activity in the Atlantic, and American defeats everywhere; while Congress questioned leaders in business, labor, and government. In this national climate of irresponsible accusation and wrathful inquisition, Jersey was not spared.

Some years earlier, the United States Department of Justice [Justice] had already begun to investigate business relationships between American and German firms. A month before Pearl Harbor, Justice filed suit against Jersey and several of its affiliates, alleging violations of the Sherman Antitrust Act of 1890. Of course, the corporate relationship between Jersey and I.G. Farbenindustrie [I.G. Farben], one basis for the suit, had never been kept secret; in fact, when the two business concerns banded together for research purposes, Jersey fully informed both the U.S. Department of State [State] and the U.S. Justice Department. To a man, the Jersey Board knew that the United States had benefited from their patent exchanges with Farben through their Joint American Study Company [Jasco]. They could point with pride to their company's role in developing the 100 octane gasoline required for most military aircraft, toluene, and synthetic rubber—with patents, research data, and formulae that emerged from cooperative work with Farben. So, Jersey Company directors and executives expressed shock when Justice agents, led by Assistant Attorney General Thurman Arnold, began to subpoena Farben-related correspondence from Jersey.[5]

As the time for hearing the case approached, Edward F. Johnson, Jersey counsel, moved to weaken the suit: "Early in the discussions I took up the matter of the consent decree with the lawyers assigned by Arnold to handle the Justice Department's case, having first obtained the Board's consent."[6] The Jersey Board gathered in informal session at the Mayflower Hotel in Washington, D.C., to decide whether or not to settle the case. Some members, confident that their course would be judged moral, legal, and in the best interest of the United States, objected to any consent decree; they wanted to fight the case in court. Chairman Walter C. Teagle and Vice President Everit J. Sadler headed this group. But in a tense session, lasting beyond midnight, those favoring a settlement prevailed. Principally:

> to obtain vindication by trying the issues in court would involve considerable expense and months of time of the Officers and Directors and many employees of the Company and its subsidiaries; that their work is of greater importance to the stockholders than court vindication and that the Company ought not to remain in a position which the Department of Justice considers to be questionable.[7]

After the vote to accept the consent decree, Johnson and Vice President Ralph W. Gallagher left the hotel and made their way through the wartime blacked-out city to the Justice Department building. There, Arnold awaited their answer. The date was March 25, 1942.

X   X   X

While the Department of Justice moved against Jersey, the Senate Committee on National Defense, chaired by Senator Harry S. Truman, was busy collecting information on the role of American business in the nation's defense preparations.[*][8] This committee had been created for the purpose of investigating industrial production problems, and executives of the Jersey Company, already under fire from the Justice Department, expected to be called before it. They were unprepared, however, when the Truman Committee labeled Jersey as the principal villain in America's failure to develop adequate synthetic rubber production before 1941. In fact, in responding to questions from both committee members and counsel, Jersey officials proved quite vulnerable because few of them knew anything concrete about the details of Jersey's contract with I.G. Farben, or about their company's earlier history.[9] Walter Teagle—who knew a great deal about both—never was called to testify. Meanwhile, the ensuing publicity defamed the corporation in the eyes of a disturbed nation. In particular, the press quoted Senator Truman as stating that the Jersey Company had committeed treason, although he actually qualified that remark.[10] The accusation made headlines everywhere. Senior Jersey employees began to withdraw from contact with the public, and newer employees expressed shock and amazement. Morale in the giant organization plummeted.

Walter Teagle, a legendary figure in his own right, grew embittered over what he believed to be unfair treatment of the company by both the Congress and the press. He decided to retire on November 30, 1942. Vice President Sadler and Director Donald L. Harper planned to end their Jersey careers on the same day. Although Chief Executive Officer [CEO] William S. Farish already had been ill during the period between the Temporary National Economic Committee [TNEC] hearings in 1939 and the various congressional committee investigations in 1942, no one anticipated that on the day before Teagle, Sadler, and Harper severed their company connection, Farish would suddenly die. In fact, Farish's death at almost the same moment as the executive retirements served to strip the corporation of most of its senior management—at a time of combined national and corporate emergency.

*The Truman Committee began to conduct hearings on March 5, 1942.

Already the Jersey Board of Directors had determined that Gallagher, a "gas man," would replace Chairman Teagle; now, he also assumed the position of CEO, and thus a double share of new responsibilities. Gallagher, for years, had been a recognized leader in developing Jersey's gas transmission lines, as he played a major role in establishing safety standards for the industry. In 1937, he first joined the Jersey Board as vice president and member of the Executive Committee.[11] Now, for more than a month, the Jersey Board did not select a new president. They did, however, vote to elevate to the Board Frank W. Pierce, a labor relations expert, and Robert T. Haslam, general sales manager of the Standard Oil Company of New Jersey [Esso] and a one-time engineering professor at the Massachusetts Institute of Technology [MIT]. No one questioned the ability of these new appointees; yet the fact remained that Gallagher and the Jersey Board suddenly had to confront tasks and situations unparalleled in earlier company history. Today, the scope of their achievements should be measured against the enormity of the problems they faced.

In 1942, the Jersey Company faced a crisis greater than that created by the Supreme Court 1911 Dissolution Decree. For, in addition to its loss of internal morale, the company now began to pay a much larger price for the outside loss of public confidence, a loss that no one had anticipated. Declining sales documented that conclusion. Jersey was in trouble with its customers, and more than most of his fellow managers, Haslam had ideas about how to respond.[12] With the approval of the Board, he had letters mailed to both shareholders and customers explaining the company's position, and he placed copies in every service station that sold Jersey products. His direct approach impressed the Board of Directors, who turned to him for leadership in corporate public relations.

Gallagher, whose background in gas marketing led him to view the corporation differently from most directors, suggested that Haslam engage a professional public relations firm, and Haslam wisely chose the firm of Earl Newsom & Company. With the assistance of Newsom and his associates, Haslam sketched out a new company organization designed to improve contacts with what he called "The Public Mind." After the directors approved the plan, Haslam began to build a staff of selected Jersey employees and competent outsiders, who immediately started to prepare information for the use of Jersey executives called to testify before congressional committees. The new unit also worked to help company officials prepare for speeches and interviews. In February, 1943, the Jersey Board officially sanctioned a Public Relations Department; Haslam served as contact director, Vice President Wallace E. Pratt became alternate Board contact, and Newsom continued as an outside adviser. Afterwards, public relations continued to play a key role in the company's business.

X X X

Meanwhile, the war went on. In his volume, *Standard Oil Company (New Jersey) in World War II*, Charles S. Popple has detailed the part of Jersey and its employees in the allied struggle against the Axis powers [Germany, Japan, and Italy], but a few additional points require comment here.[13] The company lost half its tanker fleet to enemy submarines, while thousands of employees of Jersey and its affiliates either volunteered or were drafted into the armed services. Despite these losses, from quickly devised plans Jersey produced large quantities of toluene and 100 octane gasoline for the armed forces. Jersey scientists such as Eger V. Murphree were lent to priority wartime activities like the Manhattan Project, which ultimately produced the A-Bomb; and some executives became administrators for the national office of the Petroleum Administration for War [PAW], and for other wartime organizations. In the same years, Jersey's pioneer efforts in the development of synthetic rubber paid handsome dividends, when the nation was shut off from all supplies of the natural product. Yet, even though the war directly shaped the course of Jersey production, the Board anticipated peace by making plans for the return of employees in the armed services. Guarantees to them included seniority rights, supplementary income, and pay increases. By making these plans, naturally, the Board opened a serious study of the company's future.

X X X

Although wartime changes affected the Jersey Board significantly, the leadership of the company remained impressive—despite the death of Farish and the resignations of Teagle, Sadler, and Harper. Already, CEO Gallagher had served Jersey for more than forty years, and soon he established himself as a very wise entrepreneur. One Board member termed him "very sagacious," while another said, "he was a balance wheel in a very difficult period." Vice President Eugene Holman had been associated with Jersey and its affiliates for more than twenty-two years. During this time, he gained special recognition as an expert on oil production, and he served as president of the Creole Petroleum Corporation [Creole] which operated in Venezuela. Throughout his career, Holman proved that he was "tough minded," as one director later remarked, "a scarce asset," and a person who also had shown "the quality of mildness." Frank Abrams, who later became Board chairman, proved to be an oil man unfettered by any company constraints on original thought. Abrams, over many years of service, demonstrated great respect for learning and for creative ideas. In many ways, he was also an effective diplomat. His assistant, Courtney C. Brown, often commented on the intensity with which Abrams listened to those with whom he differed, as he sought to understand their opinions. More than most executives of his era, he advocated advanced ideas about the corporation's proper role in society.

During the war, the great geologist Wallace Pratt left the Jersey Board, but his influence lingered on for years. Pratt had joined the Board in 1937, after almost two decades with the Humble Oil & Refining Company [Humble].* A contemporary described him as "a person with a great soul. . . . He had a strong influence on the Board but was not a leader except through the impact of his personality." Certainly, he proved to be a shrewd judge of men, and he played a major role in shaping the Board's understanding of alien peoples and movements, and in developing a new attitude toward Jersey's corporate relationships with foreign governments and institutions. Although famed as a geologist, he was also a broad-gauged, well-informed industrial statesman, who held a world view. Another unusual man, Orville Harden, became the great negotiator for the Jersey Board. Yet he remained a very private person who, according to one story, turned down the post of chief executive when it became apparent that Teagle would resign. Harden always attended Board sessions well prepared; and his thinking encompassed the totality of the oil business— exploration, production, refining, marketing, and profits. Few who knew him questioned the popular assertion that Harden was the hardest working member of the Jersey Board.

Another well informed member, Robert Haslam, made his company reputation by establishing, in 1927, a research operation at Standard Oil Company of Louisiana [Louisiana] in Baton Rouge. Afterwards, Haslam never returned to his faculty position at MIT; instead, he became executive vice president and manager of the Standard Oil Development Company. He also served as manager of Esso Marketers** before moving on to the Jersey Board. After interviewing many Jersey employees in 1945, one observer wrote this concerning Haslam: "The technical men look on him as little

*Humble Oil & Refining Company became Exxon Company, U.S.A. [Exxon USA] on January 1, 1973.

**Established in 1933 to centralize marketing operations of Standard Oil Company of New Jersey, Colonial Beacon, Standard Oil Company of Pennsylvania, and Standard Oil Company of Louisiana.

short of a god. The executives give him credit for their [the?] fine morale among executives and management."[14] Highly articulate, Haslam possessed an uncanny ability to clarify and simplify answers to complex questions. He understood both the oil business and its people, and he developed a strong conviction that the public had a right to information.

Frank Pierce, who joined the Board on the same day as did Haslam, already had worked with Clarence J. Hicks, the famed labor relations expert, assisting Hicks in developing Jersey's successful employee programs. Pierce became the Board's spokesman in that area. As its specialty products man, the Board selected Frederick H. "Fritz" Bedford, Jr., who had proved himself expert on sales. Finally, Thomas C. McCobb, the ninth member of the Board, had served as company comptroller. Principally, he kept the Board informed about accounting needs and business systems management.[15]

Among these nine members of the re-constituted Jersey Board, at least four—Abrams, Holman, Harden, and Pratt— cast long shadows in American petroleum history. The Board itself, led for a time by Gallagher and then by Holman, made several significant contributions to Jersey's growth and development (and thus to the remarkable postwar economic history of the United States): 1) the institution of a new and more effective public relations program, one that better served both the public and the corporate interests; 2) the development of an effective personnel program, to recruit and retain key executives; 3) a readjustment of profit sharing with foreign host nations; 4) the mounting of an effective legal challenge to the Red Line Agreement of 1928 (which had constricted Middle East oil production and marketing for many years) and after reaching a settlement, the purchase of 30 percent of the Arabian American Oil Company [Aramco]; and 5) the establishment of an adequate financial department within the holding company (along with comparable departments in the affiliates).

Some of these pre-1950 Board initiatives were long overdue; the need for modern accounting practices, for example, already threatened disaster for the company when the Board instituted dramatic changes. Yet many of these contributions effectively shaped the fortunes of Jersey during the 1950s and after.

X   X   X

As a result of the adverse publicity growing from congressional charges that Jersey had hampered the American war effort, the Board of Directors and other concerned company officials took steps unprecedented in company history. When they acted to insure that their version of events reached a larger public, a new era emerged: "Public relations began at Jersey" on the day when Assistant Attorney [General] Thurman Arnold alleged that the company's patent agreement with the I.G. Farbenindustrie represented the basic reason for the national shortage of synthetic rubber—at a time when rubber was badly needed for the war effort.[16] John D. Rockefeller, Jr., deeply disturbed by the public furor over his family business and acutely distressed by personal attacks on his family, quietly demanded that the Jersey Board improve "the company's standing with the public." Haslam, as manager of Esso Marketers, added weight to this argument when he told the Board "that his organization could not deliver satisfactory sales unless something was done to improve Jersey's public image."[17]

In the past, Jersey had ignored the question of "public image" to follow a policy of silence, designed to prevent competitors from obtaining trade secrets. This failure to provide the public with information about the company's activities coincided with a general lack of understanding of the American economic system—especially the petroleum industry. Populistically inclined politicians had long used this public ignorance to personal advantage, and they often made political capital of Jersey's silence. In fact, Jersey's calculated reserve stemmed from management's belief that the best public relations meant simply putting high quality products into the marketplace. At least a few company officials regarded any formalized Public Relations Division as unnecessary, even wasteful. Again, Jersey had a long-standing tradition of refusing to respond to attacks on the corporation; it dated back to the early muckrake era, at the end of the nineteenth century. John D. Rockefeller, Sr. had simply chosen to ignore all charges against both the company and himself, and his successors at Jersey followed his practice. Against this background of reticence, the Board reacted to create its new department—Public Relations.

One-time Jersey Vice President and Director David A. Shepard pointed out in 1973: "The Jersey organization worked so hard and so effectively on needed fiscal and technological rearrangements that they seemed to have insufficient time left to pay adequate attention to the important field of public relations. They paid a heavy price for this omission."[18] Many questions concerning the business simply were not being answered. Gradually the directors realized that no major corporation could afford to isolate itself from the society in which it must function.[19]

In fact, no Public Relations Department existed at Jersey until February 17, 1943, when George Freyermuth joined his friend, Haslam, in this work. Freyermuth, a graduate of the University of California at Berkeley, had xxiv

studied fuels and gas engineering under Haslam at MIT, where he acquired an excellent grasp of the scientific and engineering aspects of the petroleum business. Freyermuth, moreover, was a very articulate speaker, who could explain the more technical aspects of products to company salesmen. At one time, Haslam briefed Freyermuth on the Jersey-I.G. Farben patents that had created so much controversy, and then arranged for him to speak on that subject to Jersey employees in Radio City Music Hall. His explanation proved so effective that, afterwards, the Board relieved Freyermuth of other duties and assigned him to work with Haslam on a permanent basis.

Yet a few Jersey directors did not approve of the new department.[20] These directors preferred to ignore the information problem and to occupy themselves instead with the war and other pressing business. Orville Harden, "a brain . . . thought in terms of cents per gallon and what the return on investment would be, and whether this was a suitable long-term investment or whether this [or that] government would survive," never really became involved in public relations.[21] Frank Pierce and Fritz Bedford also were not enthusiastic, and other directors could not be counted as advocates.[22] However, so long as Gallagher, Holman, Abrams, Pratt, and Haslam supported public relations fully, few others criticized what to earlier Boards probably would have amounted to a misuse of company funds.[23]

Unlike so many of his peers, Abrams showed no fear of the press; and until his retirement, he cheerfully accepted a role as the unofficial company spokesman. In fact, Abrams proved to be an exceptionally effective speaker—slight in stature, a good voice, and with a puckish sense of humor, he obviously enjoyed addressing audiences of all kinds. He also delighted in telephoning authors of articles about Jersey and inviting them to meals, where he would point out errors in their stories. Like Holman, Abrams understood the disadvantages of simple "image making." As he once remarked:

> It always seems rather sad to me that we of the industrial and business world deceive ourselves that we can make friends and influence people through such things as paid newspaper advertising, pamphlets, and billboards. Some of these may help under certain conditions, but when it becomes the main channel of our effort, I think it is almost an insult to the intelligence of the average reader.[24]

Abrams preferred a program of solid information—adequate to explain the Jersey Company to anyone who would listen.

Among the executives, Abrams, Haslam, and Pratt made scores of speeches, as did other directors and managers in affiliated companies. Although the burden was shared, it proved inevitable that Abrams and Haslam would be remembered for doing the most effective platform jobs. Public relations, to Haslam and his supporters, remained a difficult but still a doable task. Earlier, he had written to Earl Newsom "that the machinery for a modern corporation's contact with the Public Mind has traditionally been minimized in Standard's get-up and your first job may involve lifting the Company's management out of traditional conceptions."[25] Haslam saw his first problem as one of building an efficient, ongoing organization "to serve as a constructive force in the future." Finally, Newsom also accepted this view, and along with Haslam, he deserves credit for the largely improved public relations at Jersey.

Gallagher, Holman and Abrams now made it clear that the company and all its employees must operate on the premise that the public had a right to full information. As one employee phrased it, "We liked to say and we wanted to feel . . . that nothing that the Company did was in conflict with the public interest."[26] Then he went on to explain: "We had to get them [the Board] to realize that the press and the people representing the press weren't basically or necessarily antagonistic; that they didn't necessarily abuse information."[27] Soon, Earl Newsom brought in an organizational chart of a proposed Public Relations Department. When Freyermuth went to Haslam saying it would not work, Haslam replied, "Well, why don't you go ahead and organize it."[28] Freyermuth then became the key liaison between the Newsom group and Haslam and the Board.

Freyermuth's communications philosophy concentrated on selected audiences—educators, intellectuals, and other opinion leaders. Filling the boxes on his chart were the names of old Esso hands: K.E. Cook, David Davis, Reginald Sloan, N. D'Arcy Drake, W.M. Craig, Anne Adams, and William P. Headden. And despite a company prejudice against outsiders, Stewart Schackne, one of Newsom's staff, soon joined the group, along with another Newsom employee, Edward Stanley, an experienced newspaperman who had served in the Office of War Information [OWI]. Stanley, in turn, persuaded Freyermuth to hire two old friends, E.R. Sammis, who had begun his career as a playwright, and Roy Stryker, an outstanding figure in visual communications, and from *PM*, the influential New York tabloid, James W. Crayhon, a newspaperman formerly with the Associated Press.

Haslam suggested—and Freyermuth and his successors continued for many years—the institution of press dinners, at which outstanding speakers were invited to discuss some new or exciting venture of the company. Many

Jersey executives also attended these dinners, and they encouraged the reporters and editors present to ask questions and to discuss points of interest. At every opportunity, the guests were reminded that more information could be obtained from company headquarters.

Elmo Roper of the Roper Organization, Inc., later became consultant to the Board, for whom he undertook numerous opinion surveys. One of the first revealed that the morale of most employees seemed high. The survey of workers at Louisiana Standard in Baton Rouge may be termed typical. There, Roper found that more than 43.8 percent of the workers rated their company far above average in its policies affecting employees, while 26.6 percent believed it to be at least a little above average, and 76.2 percent of the employees believed management to be "generally fair and decent."[29] On one important point, however, the Roper survey did not gather clear information—what the employees thought of the parent company and of the controversy then surrounding it. As one of the employees had remarked about Jersey to an interviewer several days before the Roper interview, "It's the deadest thing in the world on publicity."[30] Another employee said to the same interviewer, "Their [the Jersey Company's] greatest mistake in overall policy is their negligence in presenting their own case," and he went on to say that in thirty-five years of work there, he had never been asked to do "a thing that was wrong."[31]

Edward Stanley suggested building an archive of war photographs and paintings. Roy Stryker then assembled numerous well known photographers, like Russell Lee and Gordon Parks, and turned them loose. Next, Stanley contracted to purchase paintings from sixteen nationally known artists, eight of whom recorded "oil industry activities on the home front while the rest worked in the theaters of war."[32]

In the fall of 1944, Carl Maas, a distinguished art editor and critic, joined Public Relations, where he worked to strengthen both the fine arts and the performing arts sections. As Freyermuth put it, we "wanted to identify oil with . . . people, . . . the human part—the workers and the families of workers and the fact that they were human beings just like everyone else."[33] Using notable workers with brush, pen, and camera, Maas and Freyermuth sought to replace the old public stereotype of oil rigs and gushers. They also believed that they could help artists survive when their livelihoods seemed threatened by public indifference. Maas later arranged exhibitions of new art, beginning in New York in January, 1946, and continuing until 1951. In all, four separate shows of about thirty-five paintings each were displayed at 125 university and museum galleries.

Within a few years, Jersey achieved wide recognition as a supporter of the arts—number one, Maas said—and other companies began to follow this lead.[34] Overall, the outstanding photographs of the Stryker team and the paintings by such artists as Thomas Hart Benton and Adolph Dehn gained acceptance from a segment of society that had previously condemned Jersey, while the company came to recognize that its size and importance in the community carried special obligations.

As early as the 1930s, Jersey had made and distributed films, although few had received any national attention. Indeed, "It is ironic that much of Jersey's reputation as a supporter of the Arts comes from its underwriting of Robert Flaherty's *Louisiana Story*, a film for which the company took no screen credits." After earlier documentaries on other subjects had earned an international reputation for Flaherty, Freyermuth commissioned him to do a film on oil, and then sent him to tour Pennsylvania, Oklahoma, Louisiana, and Texas seeking material. While crossing the Louisiana bayou country, Flaherty found the locale for his film, which he began shooting in 1944 and concluded in early 1948. The hope was that the film would show the oil industry as part of the "normal day-to-day productive life of the American people."[35] First shown in New York in 1948 (and soon thereafter in theatres throughout the United States and abroad) the *Louisiana Story* quickly became famous.[36] Its musical score won the Pulitzer Prize in 1949 for Virgil Thomson, an American composer. Even though Jersey received no film credits, inquiring reporters learned to call the Public Relations Department for information about the *Louisiana Story*, and Flaherty always gave the company full credit for supporting such a risky venture.

Meanwhile, everywhere in the company, there was a new emphasis in listening to, as much as telling, the Jersey story. Specialists poured in to discuss domestic and foreign problems with the Executive Committee, the Board, the Coordination Committee, and even with lower echelon officials. Discussion groups were inaugurated in Boston, New York, Philadelphia, and other cities where they continued activities for several years. For example, Milo Perkins of New Deal fame advised the company on international economics, while Arthur Newmyer & Associates gave advice about political developments in Washington—such things as pending legislation, commission actions, committee hearings, new appointments, and other activities that might affect the company's interests. Newmyer merely reported and helped to arrange meetings. No effort was made to lobby either the Congress or the administration on behalf of Jersey. In fact, Haslam and Freyermuth fought to divorce the Public Relations Department from every kind of activity that could be termed "lobbying."

X   X   X

One of the Board's most farsighted decisions was to commission the writing of a company history. Books usually outlive programs, and for almost four decades, this written history has remained unaffected by those structural changes that have occurred in the company. As Vice President David Shepard remembered, the decision to commission a company history crested on a feeling of "uneasiness engendered by the fact that in most U.S. libraries the only reference work on the oil industry was the very unfriendly book by Ida M. Tarbell, *The History of the Standard Oil Company*, first published as a book in 1904."[37] In any case, at Jersey six previous attempts to get a history underway had proved unsuccessful.[38]

Freyermuth could never remember whether he or Schackne had first read Henrietta M. Larson's article, "Danger in Business History," in the *Harvard Business Review* for the spring of 1944.[39] In that article, Dr. Larson pointed out the shortcomings of the Tarbell volumes. In June, Freyermuth wrote to Dr. Larson, and subsequently, he and Schackne went to Boston to meet her and Dr. N.S.B. Gras, both professors at the Harvard Graduate School of Business Administration. These conversations were reported to Holman, Abrams, Pratt, and Haslam, who encouraged full exploration of Larson's ideas.

Early in 1945, Dr. Larson—now on leave from the Harvard Business School—surveyed the records of Jersey and its principal affiliates, and conducted a number of interviews. Serious conversations soon began, and the Jersey leaders agreed that the company would gain by having a new and more complete history available to the public—on library shelves and in bookstores. As the only surviving member of the Executive Committee of that time, Wallace Pratt elaborated on this point in 1960, when he admitted that the Committee had realized that even they did not have full information about the company:

> Gradually it dawned on us that it would have been extremely helpful if we had been in a position to show our critics a comprehensive history of our long business career—in Germany and elsewhere—prepared by a competent, completely detached independent agency, given free access to all records and authorized in advance to publish its own findings.

These ideas helped to endorse the proposals of the Business School Faculty.[40]

On March 3, 1947, Freyermuth, accompanied by company Secretary Adrian C. Minton, Counsel Edward Johnson, G.S. Koch, head of the Tax Department, Schackne, and D'Arcy Drake joined the Executive Committee to discuss the proposal made by the Business History Foundation formed by Gras and his associates. The Committee decided that it would support the writing of a four-volume history of the company.[41]

When the first volume of the Jersey history, *Pioneering in Big Business, 1882-1911*, by Ralph and Muriel Hidy, appeared in 1955, it received uniform acclaim from reviewers. Since that date, it has been regarded as one of the most important contributions to the history of business in the United States. A second volume, *The Resurgent Years, 1911-1927*, written by George Gibb and Evelyn Knowlton, published in 1956, also received most favorable comments. Volume III, *New Horizons, 1927-1950* appeared in 1971, and it earned a new share of critical acclaim.[42] Overall, most reviewers have praised the Jersey Company for opening its files to researchers; here, Volume III has been singled out for its lucid presentation of vast quantities of such material. In fact, the company's decision to open its records for the entire history project marked a new era of confidence, and a rare moment in the entire history of American business. Looking back to 1947, it is now apparent that the company's authorization of a corporate history—written free from all company interference—was one of the most successful public relations ventures ever undertaken by Jersey.

While these selected features of public relations activity in no way purport to be comprehensive, this enumeration does indicate the dedicated approach taken by the Jersey Company officers and the Public Relations Department in quest of a modern definition of corporate responsibility. It also demonstrates the great freedom that Jersey management allowed their Public Relations staff in the effort to improve the company's relations with the world outside. As affiliate companies began to develop their own Public Relations Departments, most of them borrowed from the Jersey experience. But the fundamental ideas and the most impressive achievements first belonged to the parent company.

X   X   X

The second significant innovation made at Jersey between World War II and 1950, occurred because Gallagher found time to strengthen Jersey for participation in the war effort, as well as to position Jersey for its anticipated conversion to peacetime production. While doing both jobs in a quiet, unobtrusive manner, he focused the Board's attention on the necessity of rationalizing more formal and exacting procedures as parts of a coherent personnel development program. By war's end, the Board had approved launching a comprehensive plan which specifically designated certain measures and required the use of many "assistants" to department heads, managers, and other executives. In all, Gallagher persuaded the Board that, while the cost for such training might be high, the long-range benefits to Jersey would more than offset this cost.

Cecil Morgan, who had a long and distinguished career with Standard Oil Company of Louisiana as well as with Jersey, remembered the early days when the latter company operated without any formal personnel policy. In his words:

> Now as far as the executives are concerned, Baton Rouge was used pretty much as a training ground for many years. There were teams of members of the executive group who recruited new engineers . . . . it has to be recognized that the company was a company of engineers. . . . This team scouted the colleges all over the country and took the cream of those they could recruit.

Morgan himself served on a 1940s talent search committee, and he used as an example of what these committees found, M.A. Wright, at the time in the Jersey Producing Department. Speaking from notes made in preparation for the interview, Morgan remarked that Wright had been tagged early—and correctly—as a prospect for rapid advancement to management. Morgan continued:

> Mike was a very hard-headed, meticulous worker with all the proper motivations and at the same time, a sort of narrowness of purpose, concentration of purpose. . . . He came along very well and he acquired the education [for administration] because Mike was an extremely able person and extremely fine person.

Wright went on to become executive vice president and director of Jersey and later, chairman of Humble Oil & Refining Company, as well as an effective spokesman for the entire petroleum industry.[43]

Appropriately, it remained for Wright to articulate the key reason for Jersey's continuous development of top managers. Across its century of existence, Wright said, the decision to structure the Personnel Division of the Jersey Company represented a major contribution to American industry. He cautioned, however, that a crucial part of the program—yet an absolute necessity for any success—was company attitude. Wright went on:

> . . . it has to have a management that you know believes deeply
> that the development of people within the organization is impor-
> tant, and it has to be done on a planned basis, and you have to pick
> good people in the beginning. . . . You have to guide their experi-
> ence so that when they get to be, say the right age, why they're
> ready to step into really high level positions.

This plan, Gallagher, Abrams, and many others accepted and effectively supported.

Wright referred to the time when Jersey annually hired 500-1,000 persons, recruited principally from the top 15-25 percent of their college classes where, as Wright phrased it, "they either worked for grades or did not have to. . . . You need both kinds of people." Given ambition and intelligence, Jersey would provide sufficient training and experience—to insure that the company would "end up with the best organization relative to competition." John O. Larson, who served fifteen years as company secretary, agreed with Wright about the importance of the personnel program: "It's really the people that have been developed to run the company."[44]

In retrospect, it now appears that the importance of Gallagher's role in organizing a Personnel Department for the Jersey Company cannot be over-estimated. And when he succeeded Gallagher, Holman also pursued a search for talent throughout the corporation. Both men had grown familiar with the tradition of powerful directors and managers at Jersey; as leaders, they sought to emphasize the selection of key executives who understood the total business of the company, rather than promoting only functional specialists to top administrative positions. Moreover, Gallagher and Holman expected every company leader to participate continuously in this ongoing search. What had begun as a mere idea soon developed into a refined corporate hunt for outstanding candidates to replace every manager, up to the level of the Board.

X   X   X

A third major change at Jersey, instituted by Gallagher, Holman, and their fellow Board members, derived in part from their special effort to anticipate those social and economic changes expected to prevail in the post-World War II period. The chief result of these expectations became a more equitable sharing of the profits from crude oil production with host nations; for Jersey, this change began in Venezuela.

There, the Creole Petroleum Corporation, a Jersey affiliate, daily produced many thousands of barrels of oil for the United States and its allies.

Local pressure to increase Venezuela's share of the oil income grew markedly in 1935, when General Juan Vicente Gomez, the long-time Venezuelan dictator, died and General Eleazar Lopez Contreras succeeded him as President. Dr. Guillermo Zuloaga was named to represent Venezuela in discussions with Henry E. Linam, president of the Standard Oil Company of Venezuela [S.O.V.], an older company also owned by Jersey. These discussions aimed at defining a new profit sharing agreement, and they continued at intervals during the next two years. Afterwards, in 1941, General Isaias Medina Angarita succeeded Lopez Contreras as President. Medina, cognizant of acute nationalistic pressures within Venezuela, hoped to channel some of this enthusiasm for change into popular support for his new government. At the time, the war made foreign trade hazardous, and the Venezuelan economy suffered as a result. Then, unemployment caused the government to seek "to obtain a revision of the terms under which foreign oil producing companies were operating in that country; it proposed that the government's share of the profits be increased even under the old concessions issued under the petroleum law of 1922."[45] In fact, these government efforts to increase its share of the petroleum companies' profits received solid support from most well informed Venezuelans.

In New York, when the first hint of renegotiation surfaced, the Jersey Board expressed mixed opinions. Everit Sadler and his supporters demanded that Venezuela continue to honor existing concessions and contracts between the company and the government, while Abrams, Pratt, and others favored renegotiation. But America's political allies needed Venezuelan oil, and, before the participants could reach an impasse, the U.S. Department of State acted to produce a settlement between the oil companies and the Venezuelan government.

Wallace Pratt represented Jersey in Washington talks and, after reporting to the Board, moved on to Venezuela carrying full authority to act for the company. There, he engaged in lengthy meetings with Venezuelan officials, their consulting engineers, and executives of the two Jersey companies, S.O.V. and Creole. When Pratt returned to New York, he led the fight to persuade the Jersey Board to turn in old contracts for new grants, and to agree to other changes designed to split the profits more fairly. Pratt's position was valid historically, as he argued "that foreign operations in the future would survive not on the strength of contracts but on the basis of recognition by both sides of a mutuality of interest and on an equitable sharing of benefits."[46] As a geologist, Pratt might well have believed that vast quantities of oil remained to be found in Venezuela. In any case, he could see no alternative open to the company. Abrams and Holman agreed with Pratt. But Sadler fought any change to the end, despite his knowledge that this position would not receive the sanction of the United States government. Of Sadler, Pratt dryly remarked, "The only friend I had up north, Sadler, was usually on the opposite side from me."[47]

The 1943 agreement with Venezuela increased that country's share of oil profits by approximately 80 percent, and under a revised system of taxation provided that approximately 50 percent of net profit be returned to the host country. Dr. Guillermo Zuloaga, a participant in negotiations, declared: "The decision of the company [Jersey] to convert to the law of 1943 was a culmination . . . of a process of education of the executives which led to the victory of the new school over the old school of executives in Standard."[48]

Jersey actually made its decision to give up old concessions, to apply for new ones, and to accept additional taxes—partly because of the exigencies of the war—but also because most Board members realized the importance of friendly relations with Venezuela. Holman, Pratt, and Abrams remained genuinely convinced that a global business must recognize the legitimate requests of host nations.[49]

As a result of the change in Venezuela, Arthur T. Proudfit was selected to be Henry Linam's successor as president of the Standard Oil Company of Venezuela and vice president of the Creole Company. To Dr. Zuloaga, Proudfit represented the younger school of executives, one who had learned in Mexico* the dangers of a too rigid policy on contracts and concessions. Venezuelans came to have confidence in and respect for this Jersey representative. Once it was reached, the Venezuelan decision itself became the basis for subsequent alterations in Jersey contract and concession agreements, not only in Venezuela, but in other countries as well.[50]

Reviewing Jersey's course of action during the Venezuelan dispute, historian Henrietta Larson concluded in *New Horizons*, "The importance of the amicable settlement of the Venezuelan issue can hardly be overstressed."[51] In 1944, all of the Jersey Company's properties in Venezuela were merged into the Creole Petroleum Corporation. Then, because of the allies' continuing need for oil, Creole immediately undertook an expanded exploration and production program.

Meanwhile, the Venezuelan model agreements attracted the attention of oil-rich Middle Eastern countries, and the profit sharing rule soon became the guiding principle in Jersey's relations with host nations throughout the world. In Venezuela, most Jersey employees understood the rationale for their company's position; indeed, they gradually became convinced that

*Where in 1939, final settlement of concession (and other claims) was reached.

only the company's accommodation of Venezuelan demands managed to delay, for the moment, a drive toward full nationalization of oil property.[52] In retrospect, observers have estimated that renegotiation permitted Jersey to operate Creole for an additional twenty years. Given the circumstances in the world, 1943-1950, Jersey directors made the pragmatic decision.

X   X   X

The fourth significant move of the Jersey Board between World War II and 1950 came in the Middle East, where the company acted to acquire another major source of oil. In this extremely important development, Eugene Holman and Orville Harden played the most impressive roles, aided by Counsel Edward Johnson.

While today, oil men and government experts still do not agree on details, the fact remains that during World War II many representatives of both groups grew concerned about the adequacy of American domestic petroleum supplies for the postwar era. One obvious cause of this uneasy feeling was the overwhelming success of the allied war effort—to which the United States furnished almost seven billion barrels of oil. Jersey reacted to its perceived need to replenish supplies by pushing ahead with its search for new sources at home, and even more, by renewing interest in possible sources abroad.

Much earlier, in March, 1925, Jersey and several other U.S. companies had joined companies from Britain and France in the Turkish Petroleum Company, Limited [TPC].* All these parties signed a restrictive covenant termed the Red Line Agreement.**[53] After much haggling and negotiating, the State Department, promoting American participation in that area, approved the operation of American companies within the Red Line restrictions. Then, headed by Jersey and Socony-Vacuum Oil Company, Inc. [Socony-Vacuum],*** five (originally seven) American oil companies, anxious to play a role in Middle Eastern oil developments, created the Near East Development Company [NEDC] to bargain more effectively.[54] In July, 1928, the TPC, consisting of the Anglo-Persian Oil Company, Limited [APOC], the Royal Dutch-Shell Group [Shell], the Compagnie Française des Pétroles,

*TPC changed its name to Iraq Petroleum Company, Limited [IPC] in 1929.

**So called because of the area lined in red pencil wherein the partners (termed "Groups") would not participate in the manufacture or production of oil except through the company or its agents.

***Became Socony Mobil Oil Company [Socony Mobil] in 1955; in 1966, it became Mobil Oil Corporation [Mobil].

S.A. [CFP], and the Participations and Investments, Ltd. [P&I, owned by Calouste Sarkis Gulbenkian], agreed to admit the NEDC into their concern as a full partner. Each of the Groups received 23³/₄ percent participation, and Gulbenkian retained his 5 percent share. By 1934, three American companies had withdrawn from the NEDC. Jersey and Socony-Vacuum bought their shares; thus, each held rights to 11⁷/₈ percent of the Iraq Petroleum Company.

The world surplus of oil during the Great Depression of 1929-1939 worked to delay the search for petroleum in the Middle East, but as the war clouds of World War II rolled in, development began in earnest. At the outbreak of the war, Gulf Oil Corporation [Gulf], The Texas Company, and Standard Oil Company of California [Socal] had already brought in oil wells there, and many American and British concerns were interested in concessions.[55]

Minutes of the Jersey Board reveal how the search for additional Middle East oil and the problems resulting from the Red Line Agreement grew increasingly complex. The whole story is one of uncertainty about the government and about the role of the company before, during, and after the war. An astute person who knew the directors during this period characterized them collectively as "affected in manner similar to shell shock. . . . They were incapable of making decisions."[56] While this judgment may be a bit harsh, the minutes establish that the directors moved with more than their customary caution.

A 1944 report by the noted geologist Everette L. DeGolyer indicated that Saudi Arabia had one of the largest oil deposits in the world, and that oil concessions there could be developed quickly. Standard Oil Company of California and The Texas Company—already operating inside Saudi Arabia through their subsidiary, Aramco—boasted oil in quantity.* That oil could be loaded onto tankers in the Persian Gulf, although both military leaders and oil executives perceived that a pipeline to the Mediterranean would provide a much cheaper and more secure means of delivery. Meanwhile, other revenues of King Abdul Aziz Ibn Saud [Ibn Saud] dwindled sharply because of the exigencies of the war.[57] In September, 1944, the King addressed a letter to William A. Eddy, American Minister to Saudi Arabia, seeking to ascertain American intentions toward his country. That letter may well have stimulated Washington's interest in his oil-rich nation.

On June 22, 1945, at President Truman's suggestion, Fred M. Vinson, Director of the Office of Economic Stabilization, conferred with representatives from the Navy and War departments. The record of the conference

---

*Both companies serviced markets in the Eastern Hemisphere.

indicates that Truman (like Roosevelt earlier) understood the Saudi Arabian problem and that he kept congressional leaders informed.[58] Later, in August, 1945, James V. Forrestal, Secretary of the Navy, addressed a memorandum [which he never mailed] to James F. Byrnes, Secretary of State, in which Forrestal expressed a conviction that oil and its by-products represented "the foundation of the ability to fight a modern war," and that within twenty-five years the United States would face sharply declining oil reserves. As Forrestal wrote:

> I don't care which American company or companies develop the Arabian reserves, but I think most emphatically that it should be *American* [italics his]. So far as the five United States companies [The Texas Company, Standard of California, Gulf, Jersey, and Socony-Vacuum] are concerned, I think it would be possible to have them act in concert, although this is a matter which will probably have to be cleared with the Department of Justice.[59]

Finally, Forrestal pointed out that the Aramco concession constituted "one of the three big remaining oil puddles in the world."

Earlier, on March 3, 1941, Stuart Morgan of the Near East Development Company had written to Wallace Murray of the State Department, enclosing several legal opinions as to how the war affected the existing Red Line Agreement. One such opinion, provided by an authority on contracts, A. Andrewes Uthwatt of the British Royal Courts of Justice, stated that the existing contract had been frustrated "by supervening illegality" and, essentially, that the IPC partners could now ignore its restrictions and make whatever arrangements they desired without further reference to the agreement of 1928. Morgan also enclosed a letter from R.W. Sellers, then representing the Socony-Vacuum Oil Company in London, to H.F. Sheets, Jr., of Socony-Vacuum's New York office, declaring that the opinion "by Uthwatt was in the nature of a bombshell." Sellers went on to say that—if the Red Line Agreement had been invalidated by the war—then "I.P.C. and the Groups are all released from the terms" and could go on to do things that previously had been prohibited them. To Murray, Morgan wrote: "If Mr. Uthwatt's opinion . . . should hold good in law, it is feasible that . . . the Red Line be vitiated" and other arrangements in the Middle East could be made.[60] Ultimately, Jersey attorneys agreed with both Uthwatt and Morgan. Solicitors for CFP and Gulbenkian, however, argued vigorously that the 1928 Group agreement remained valid. Only a court test in London could decide this legal question.

X   X   X

In New York, meanwhile, important changes took place within the Jersey Board. In 1944, Holman became president and chief executive officer of the company. When Gallagher retired, Abrams replaced him as chairman. Among the membership, Wallace Pratt took early retirement, and John R. Suman, a seventeen-year veteran with Humble, replaced him in 1945. Stewart P. Coleman, a chemical engineer with the Humble Company since 1920, joined the Board in 1946.* Chester F. Smith, who had a long background in refining, was added to the Board in 1944. Already he had served as president of Jersey's domestic affiliate, Esso, for several years. The directors also selected Jersey Treasurer Jay E. Crane to join them (in 1944) along with (in 1945) Bushrod B. Howard, a shipping specialist who had distinguished himself as the manager of Jersey's maritime service.

Among all the new Board members, Suman might be termed the key selection. Like Holman, he had been hired by Pratt, who now delighted in Suman's promotion.[61] Ever direct and outspoken, Suman soon proved that he would hold his own in the rough-and-tumble discussions of the Board. His reputation had been built on his imaginative program for oil conservation and reservoir engineering at Humble, where he served as vice president in charge of production. Over many years, he proved invaluable to the Board, as will be shown later.[62]

X   X   X

As a general rule, the war and their companies' responsibilities held the attention of the Jersey Board until September, 1945. No special concern for the Middle East seemed justified since, prior to the war, the Red Line Agreement had presented no special burden for Jersey or Socony-Vacuum, because ample supplies of cheap petroleum products could be found in Venezuela and in the United States, and these could be transported to markets in Europe.[63] But as the Board turned to study postwar operations, the potential advantages of locating additional oil supplies closer to European markets became clear. Federal officials also had to convert from wartime priorities, of course; so they could not avoid some delay and confusion in the process of developing a peacetime attitude toward American corporations operating in the Middle East. At Jersey, a final decision regarding participation in Aramco was postponed, as the *New York Times* editorialized on February 23, 1944: "Fair as the leases may be under which Americans now

---

*Earlier, Coleman had headed Jersey's Economics Department, and the Coordination Committee.

operate in Saudi Arabia and the Bahrein Islands, it is clear that by entering the Near East for oil we also enter Near Eastern politics. This country therefore needs a foreign oil policy."[64] Almost a year later, the *Times* predicted that the Saudi Arabian area "may well prove to be [the] most interesting foreign area to the United States in this century."[65]

Whether or not they read the *Times*, the Jersey Board felt compelled to solve the problem of finding new oil by buying into Aramco. Holman teamed with both Harden and Coleman in planning strategy and in negotiating with Socal and The Texas Company. Meanwhile, other Jersey officials quietly reviewed government positions in Washington and London. To prevent misunderstanding, the Board sought to clear its every move abroad with U.S. government agencies. In January, 1945, Harden visited Secretary Harold Ickes and Ralph Davies of the Interior Department.[66] In August, the Executive Committee asked Harden to discuss the Red Line Agreement and Aramco with Assistant Secretary of State W. L. Clayton.[67] During the next several months, Harden again saw Clayton and Loy Henderson of the State Department to review the Group Agreement of 1928. On August 21, 1946, Jersey Counsel Edward Johnson told the Executive Committee that Messrs. D.N. Pritt and Gerald Gardiner, British counsel retained by Jersey's Montagu Piesse, had rendered an opinion in which Johnson concurred, "that the Group Agreement dated July 31, 1928, signed by direct and indirect participants in the Iraq Petroleum Company, Ltd., had been dissolved when France became an enemy of Great Britain during World War II." Johnson added that no subsequent action by the participants in IPC had caused the reinstatement of the Group Agreement. Already, Sir Valentine Holmes, representing the Socony-Vacuum Oil Company, had reached the same conclusion, as had Jersey lawyers.[68]

On December 9, Harden informed the Executive Committee that he had authorized the mailing of instructions previously drafted on October 3, 1946, to the effect that Jersey, Socony-Vacuum, and the Near East Development Company considered the Group Agreement, dated July 31, 1928, dissolved.[69] In the intervening weeks, Harden, Johnson, and Middle East Committee Chairman Woodfin L. "Woody" Butte continued to discuss IPC affairs with counsel retained by the other Groups.

While the legal work and the consultations went forward, numerous Jersey experts arrived in Saudi Arabia to study operations.

On December 13, 1946, Leonard F. McCollum, one of the ablest production specialists in the company, reported favorably on his trip to Iraq, Kuwait, and Arabia.[70] Then G.M. "Mose" Knebel, head of Jersey's Exploration Department, came back "with his eyes popping. . . . His advice to

Jersey always was, 'Go after the elephants,' and leave the little things alone. And man, this was a real elephant."[71] Such enthusiastic reports served to whet Jersey's appetite for holdings in the Middle East. Because of uncertainties about the legality of the Red Line Agreement, moreover, Holman managed to persuade representatives of the California and Texas companies to grant Jersey and Socony-Vacuum an option to buy a 40 percent share of Aramco. These two companies then split the 40 percent 30/10, with Jersey receiving the larger share—at the time, an appropriate division because Socony-Vacuum had virtually no markets west of Suez, while Jersey had important markets east of the canal and in Europe.

Shortly afterward, Compagnie Française des Pétroles brought suit in a British court to prevent the Near East Development Company partners from acquiring a share in Aramco, and to require them to act only through the Iraq Petroleum Company Group Agreement. Gulbenkian supported this position. In reply, Jersey Standard and Socony-Vacuum filed a "Statement of Claim" denying the allegations of CFP. Since none of the Groups in IPC desired to carry the case to court, continuous efforts were made to resolve the dispute. As Shepard put it, "Nobody, not even Gulbenkian, wanted his case tried."[72]

On December 26, 1946, the Executive Committee decided that it would be advisable to have a representative of Jersey visit Saudi King Ibn Saud to express the company's appreciation for the chance to share in the development of Aramco's petroleum concessions; but most importantly, to afford the King an opportunity to "express directly to an accredited company representative" any comments he cared to make concerning Jersey's action in acquiring a stock interest in Aramco.[73] Butte later wrote that the Committee consulted Harold Hoskins* about whom to send, and that Hoskins said, "Send a general." So, Suman headed the delegation and Butte went with him. Concerning the meeting between the King and the Jersey director, Butte recalled, "He [Suman] and Ibn Saud took to each other like a couple of old Indians and within fifteen minutes were joking and slapping each other's legs, and he [Suman] authorized us to send a cable that afternoon which would let us go ahead and exercise our purchase option on the Aramco shares."[74] On May 2, 1947, Suman reported to the Executive Committee that "he had a most satisfactory audience with King Ibn Saud, who expressed his approval of the proposed arrangement between the company" and Aramco, and that Ibn Saud gave every assurance that his government would continue to cooperate.[75]

In London, representatives of the four main Groups met in the Shell Transport and Trading Company Building to work out a new agreement,

*of Socony-Vacuum, Inc.

while Mr. Gulbenkian operated from his center, the Hotel Aviz in Lisbon. At issue were technicalities concerning oil liftings, production, and transportation facilities. So, Middle East Advisor Smith D. "Sparky" Turner procured a large sheet of paper and covered it with figures explaining how these various requirements would be met and priced. He then explained the technical aspects of the agreement to the senior executives of the four main companies. Butte, who was there, remarked, "Sparky . . . did an absolutely masterful job . . . so even those senior executives all understood it (and I tell you that's something)." Because of its size, Turner's paper became known as the "horse blanket."[76]

Despite the talks, drafts, counter-drafts, meetings in New York, London, and Lisbon, time passed without any firm agreement being reached. Gerald Gardiner indicated to Edward Johnson that—to protect Jersey's option to buy into Aramco—Gulbenkian's case should come to trial on November 1. This would allow for appeals, should they be needed. Since time was short, Shepard and Gulbenkian's solicitor hurried to Lisbon to persuade the P&I owner to accept the draft of the new Group agreement. Orville Harden, an unusually fine diplomat and a tough negotiator, also visited Lisbon, where he pacified (to a degree) the angry Gulbenkian. But, Gulbenkian still demanded every detail in writing.

Since each of the Groups held a different set of ideas, based on variously worded clauses of agreement on separate pieces of paper, no one appeared to know precisely *what* was being agreed to. So Butte borrowed a typewriter and worked all night to put the clauses together in a single document, which Austin P. Foster, General Counsel of Socony-Vacuum, labeled "The Bucket." It soon became "the single negotiating text."

On November 1, 1948, the trial opened in a court of the Queen's Bench. Later, Butte wrote, "I doubt that since the trial of Bardell V. Pickwick has there been such a muster of gentlemen in wigs who presented 'all that pleasing and that extensive variety of nose and whisker for which the Bar of England is so justly celebrated.' "[77] The lead barristers for all the companies requested a conference, and the judge granted a stay of twenty-four hours. In the meantime, according to Nubar Gulbenkian [Calouste Sarkis Gulbenkian's son], "Session by session, we worked gradually toward a basis for agreement." Representatives of the various Groups then met, seeking to resolve all outstanding points disputed.[78]

Additional clauses covering objections by Gulbenkian added fifteen or twenty pages to the agreements written earlier. Apparently, no one sorted them out; instead, they were all printed together. Not surprisingly, Gardiner referred to the entire agreement "as a fantastic farrago of nonsense," but in total effect these pages canceled the Red Line restriction and opened the way for Jersey to enter Aramco.

In New York, company committees studied various aspects of the problem, and public relations experts appeared before the Board to discuss the possible public reaction to Jersey's purchase of a share in Aramco.[79] Finally, Eugene Holman, Frank Abrams, and Orville Harden, on the way to discuss with their counterparts in the California Company certain actions of Aramco, flew to the West Texas ranch of Wallace Pratt, where for a day and a night the four outstanding oil men considered the purchase from all angles—with a growing conviction that they had made the right decision.[80] On November 18, 1948, Frank Abrams announced that he had received information from Orville Harden and Edward Johnson concerning successful negotiations just completed among the stockholders in the Iraq Petroleum Company, Limited, and the signing of a new agreement, dated November 3, 1948.[81] On December 2, Holman reported, "The purchase by the company of a 30 % interest in the Arabian-American Oil Company and the Trans-Arabian Pipeline Company had been consumated."[82] That Aramco purchase represented nothing less than a landmark decision in the history of the Standard Oil Company (New Jersey).

X   X   X

The last major change made during the war years and before 1950 developed out of Jersey's emerging need for executives trained in economics, international finance, and world monetary systems. As early as 1932-1933, Teagle's efforts to merge with the California Company had foundered principally because of the inadequacies of the Jersey accounting system. Teagle wrote at that time, "our past methods of accounting our published balance sheet and earning statement do not correctly reflect either the actual earnings or the book value of the company." Ordinarily the Jersey CEO would have moved to make swift and dramatic changes in house accounting and bookkeeping practices, but at that time a major depression engulfed the United States and the world. Teagle went elsewhere to help the New Dealers fight the recovery battle and since his associates Farish, Harden, and Sadler had opposed the merger, little change in accounting practices occurred.

The world monetary crisis persisted, and Jersey reached out to the Federal Reserve Bank of New York to hire Jay E. Crane, who had handled the Foreign Department of that institution. Crane served as assistant treasurer and treasurer before he was elevated to the Jersey Board in 1944. Unlike other

Board members, his hiring set a precedent for Jersey, which would continue to seek financial experts from outside the company. Crane understood the complex problems of international finance, and he adjusted easily to the changing role of corporate treasurer. In 1947, Crane persuaded Emilio G. Collado, who held a doctorate in economics from Harvard and a degree in engineering from MIT, to join Jersey's Treasurer's Department. Collado had been involved in international finance for many years, while he served in the U.S. Department of State and the U.S. Department of Treasury. At Jersey, he joined Leo D. Welch, an international banker of much experience. Not surprisingly, Crane, Welch, and Collado exercised great influence in Board meetings, especially in monetary matters. Together, they developed an effective financial organization that served the company well for many years.

X   X   X

As a series of corollaries to its increased concern with what the public thought of the company, the Jersey Board made a number of lesser decisions that also had long term effects. In March, 1947, Eugene Holman's assistant, Dr. Frank Surface, appeared before the Executive Committee to present a memorandum on "The Responsibilities of Business in the Field of Higher Education." In it, he asked whether business could not justify broader support of liberal arts education, as well as contributions to engineering and the sciences.[83] The Executive Committee studied the problem and on May 23, 1947, appointed a committee "to develop and recommend a policy and program which might be followed by the Company and its affiliates in the educational field, including a basis for financial assistance to colleges."[84]

In September, Abrams spoke on the "Crisis in Education" program of the Advertising Council, where his remarks on "The Stake of Business in American Education" received widespread notice, perhaps because of his statement that business would aid education if legal grounds could be found.[85] Many lawyers expressed reservations, and a spirited national discussion ensued. Finally, Jersey assumed a leading role in resolving the question of corporate aid to education. By 1949, the company had initiated action in New Jersey designed to find a way to aid higher education. While the last court decision would be delayed for several years, Jersey acted on its own to make direct grants to institutions for a variety of programs not related to technical or engineering problems or research. The entire discussion of such aid within the company—as well as this company action—supports the conclusion that the Jersey Board had begun to view business problems within the broadest context of social values.

X   X   X

Some lesser decisions (among other Board activities between 1942 and 1950) did prove significant in Jersey's development. Several of these involved gaining a measure of compensation for properties seized or destroyed during and after the war. Here, Holman, Johnson, and other executives spent seemingly endless hours presenting company claims before many different commissions and boards. Litigation concerning the patents seized by the Alien Property Custodian continued until 1948, when Federal Judge Charles E. Wyzanski, Jr. ordered most patents returned to Jersey.[86] Even today, many Jersey executives contend that the infant American petrochemical industry took on new life with that judicial decision.

For the petroleum industry, as well as for the entire United States, ominous notes were sounded between 1945 and 1950. At the end of the war, Interior Secretary Ickes sagely remarked that the "war for the most part had been fought with American oil." Not many ordinary Americans understood what he meant, but shortly thereafter, eleven major U.S. oil companies asked the government to support their global search for oil, because the future held the "definite probability of [the United States] becoming a net importer of oil." These alarming cries received additional dramatic emphasis from continuing news reports of Russian troop movements near the Iranian border. In 1946, as any possibility of recovery of Jersey's prewar Rumanian and Hungarian properties lessened, Jersey confronted an increasing problem of finding oil for European affiliates.[87] Naturally, the onset of the Cold War between Russia and the United States aggravated this problem. But the weather did not help either. The severe winter of 1948 produced oil shortages everywhere, and strikes by coal miners in both the United States and Europe placed an added supply burden on the oil companies.

Efforts by the United States to help rebuild Europe through aid programs seemed to hinge on providing oil. In 1948, Frank Abrams, then in London, remarked that the speed of European recovery depended on the rapid development of Middle Eastern oil. To assist in meeting this crisis, Jersey affiliate Anglo-American Oil Company, Limited, began a major addition to its Fawley [England] refinery in June, 1949. Now, the problem of moving oil became pressing, and Jersey increased its order for new tankers, originally purchased to replace those lost during the war.[88] Still, as the rift between the superpowers widened, predictions of depleted oil reserves continued to be heard. Pratt, then retired, headed a study group for the National Security Resources Board, which recommended that states require reduced oil production to conserve supplies. Their report termed the status of our proven oil reserve capacity alarming—"there is now no reserve productive capacity at all"—and they endorsed both the importation of oil from the Middle East and the preservation of oil in the United States for emergency use.[89] Yet few Americans believed these or other informed statements about oil scarcity. As new cars appeared and eager customers purchased them, domestic gasoline sales increased, further depleting domestic supplies. Despite the success of wells brought in offshore by Jersey's affiliate, Humble (and others), in 1948, pressures mounted to find new sources of supply.[90]

Already an "inward trickle" had begun; and as more imported oil arrived, new problems developed—especially international problems involving oil imports and (ultimately) oil quotas. The Jersey Board soon became a strategic force in politico-economic struggles around the world. As a beginning, Jersey's subsidiary operating in Venezuela, Creole Petroleum Corporation, supplied much oil to U.S. and Canadian affiliates; while a domestic affiliate—the Humble Company—actively sought to increase its production in the United States. Finally, their struggle pitted domestic versus foreign producers, and it engaged the entire industry.

As early as May, 1947, Jersey declared that it had no plans to import Middle Eastern oil; however, less than a year later, the company announced the arrival of oil from that region.[91] Events during that cold winter simply had forced the company to reverse its earlier decision. In early 1949, Governor Beaufort H. Jester of Texas fanned the flames of controversy by writing, and then making public, a letter to American oil companies requesting that oil imports be reduced because they were hurting "the oil industry in Texas." Holman, in reply, stated that Jersey had taken steps "to get production in line with demand," that Creole had cut its production sharply, and that Middle Eastern imports had been reduced.[92] By 1950, the division of opinion among U.S. oil companies had pushed the import question into the political arena.

X   X   X

The war record of Jersey was one of which the company and its employees could be proud; the drive that enabled it to reach great production heights and levels of efficiency continued when peace returned. Gallagher, Holman, Abrams, Pratt, Harden, and Haslam provided an unusual leadership record, one that proved outstanding however measured. They tackled major problems while at the same time achieving day-to-day successes; still they acted to provide the basis for a greater Jersey. Their vision enabled the company to move into the second half of the twentieth century in a far better position than anyone could have anticipated. Under their direction the company's public approval rating reached its zenith, their personnel program produced preeminent leaders, and their drive to locate and to acquire available oil reserves paid handsome dividends. The decision they made to renegotiate concession agreements with host nations proved enlightened. Too, the company's financial leaders became adept at managing income from the pound sterling countries after World War II, thereby strengthening the company financially. Yet no one of that generation could have foreseen the striking and dramatic changes that lay ahead; nor could anyone have known just what resources would be needed to meet the new challenges. A description of what happened to Jersey during 25 eventful years (1950-1975) comprises the remainder of this volume.

Growth in a Changing Environment

Chapter 1

The Corporation in 1950

A S IT ENTERED THE SECOND HALF of the twentieth century, Standard Oil Company (New Jersey) [Jersey] had clearly established its position as the leader of the international petroleum industry. Worldwide, Jersey and its affiliates accounted for almost 14 percent of total crude oil production outside the Communist sphere, 16 percent of the crude runs to refinery,[1] and 17 percent of the petroleum products sold. Net crude oil production for its consolidated companies (Western Hemisphere affiliates) averaged more than a million barrels daily in 1950, and the company's share of gross crude oil production by non-consolidated (Eastern Hemisphere) affiliates added another 172,000 barrels per day [b/d].[2] Product sales by consolidated affiliates totaled 1,492,000 b/d, and the worldwide average exceeded 1,700,000 barrels.[3] The magnitude of these operations matched their global scope. The 140 countries, colonies, and dependencies in which the company's more than 250 affiliates and subsidiaries operated spanned an area that accounted for over 90 percent of the non-Communist world's consumption of oil.[4] Outside of the Communist countries, Jersey played a significant role in every important oil region, both producing and consuming.[5]

Marshalling the resources of so immense an enterprise to compete successfully across such a diverse and vast demesne presented a serious administrative challenge. To meet it, the company needed an overall organization that would coordinate the operations of all its affiliates and yet retain the flexibility necessary to permit their adapting to local conditions. Fortunately, the need for this structure had been recognized nearly a quarter of a century before and, through a series of management changes, the company's leaders had been moving toward it.

The corporate structure of 1950 represented the substantially completed evolution of those early efforts to organize Jersey's worldwide interests on a basis that combined effective control and coordination with maximum flexibility. And, with only a few minor alterations, the 1950 organization

1

remained intact until the end of the decade. But the significance of the plan went beyond its specific features or the fact that it served the company well during an extended period of rapid change. It represented a statement of business philosophy, even though the Jersey executives who participated in the evolution of corporate organization did not consider themselves business philosophers; yet in their pragmatic efforts to adapt their company's structure to meet changing times, they articulated a very definite philosophy of management, one whose basic soundness is evidenced in its durability. Adjustments continued to be made as the company's needs and the environment in which it operated changed. Yet, while the form of the organization has altered substantially, the principles of management shaping the structure have remained essentially the same. Examination of Jersey's 1950 structure thus serves a dual function: it delineates the administration of the company during the decade of the 1950s and, at the same time, it reveals the fundamental tenets that have continued to define the management of both affiliates and the parent company.

Very clearly, decentralization of management was the most important of the principles underlying the 1950 organization, following naturally from the fact that over-centralization of management triggered the first steps in the process that led to alteration of the structure. In 1927, Jersey's president, Walter C. Teagle, became concerned about the growing administrative burden carried by the company's directors, all of whom served as officers and employees of the corporation. The rapid growth of the company's interests, as much as its geographic spread, generated increasingly complex problems that required greater attention to overall planning and coordination. Yet, because Jersey was in 1927 an operating company as well as a holding company, that same growth also forced its directors to devote more time to problems of day-to-day operations. Responsibility for operations often encouraged directors to identify so strongly with individual departments or single affiliates that the general interest of the company was sometimes neglected. Teagle clearly acknowledged the Board of Directors' primary responsibility for overall direction and results. With the assistance of Director Everit J. Sadler, he worked out an organizational scheme designed to permit management to concentrate on broad principles and strategies rather than details of operation. In essence, the plan aimed at decentralizing the management of actual operations while centralizing the functions of coordination, policy-making, and planning—a distinction urged on Teagle by Frank A. Howard, head of the company's Development Department.[6] The initial step, making Jersey exclusively a holding company, did not entirely solve the problem, but it did establish decentralization as one of the fundamental principles of overall administration. Later changes strengthened and refined the structure while keeping the same general lines so that, by 1950, the proper responsibilities of both parent and affiliate company managements had been well defined.

The 1950 organizational manual spelled out in specific terms the distinction between the centralized functions of the parent company and the decentralized responsibilities of its affiliates. As a holding company, Jersey depended upon the managers of its affiliates for administration and operations. Whether wholly or partially owned, many of these companies had become almost autonomous units, with separate boards of directors accountable for their performance and results. Closer to developments and more sensitive to local conditions, such managers ideally possessed a responsiveness and competitive flexibility that would have been impossible with centralized administration. Basically, the parent company provided "the centralized supervision . . . essential for protection of its affiliated interests."[7] While carefully avoiding any interference with affiliates' operating authority, the holding company supplied the broad oversight, planning, and policy-making necessary to take full advantage of the Jersey system's total resources. Using its wider perspective, the parent company's management sought to guide the affiliates in a manner that would insure that overall performance exceeded what they could achieve separately. Thus, decentralization provided a distribution of responsibility designed to obtain the greatest benefit from the particular strengths at each level of management.

In carrying out its responsibilities for coordination, policy-making, and planning, the parent company maintained a watchfulness over all matters affecting affiliates, especially those with a potential for broad or long-range effects on the company's business. Within this charge, Jersey's management concentrated on two areas: fiscal matters and personnel.

Jersey's equity interest in each of its affiliates would have furnished ample justification for its concern with their financial performance and with the pattern of their capital investments. However, under the system of centralized policy-making and planning, the parent company's involvement in these matters went even further. Affiliate balance sheets became fundamental tools for Jersey managers in their general supervision of the company's interests. In a similar way, evaluation of affiliate capital budgets served as a primary means of coordinating actions, making sure that all spending programs coincided with policies and plans designed to benefit the overall interest.

The role of the parent company as principal lender and financial advisor to its affiliates strengthened this advisory control. In part, control resulted from the parent company's tradition of self-capitalization. While the general petroleum industry had been marked by a high degree of self-financing, Standard Oil Company (New Jersey) ranked as an outstanding example of the practice. From 1946 through 1950, Jersey paid out in dividends only 34.8 percent of its $1.5 billion earnings. The remainder it reinvested. However, as banker to the affiliates, the parent company excelled. It not only loaned its own money at the lowest first class credit rates, it also borrowed money more cheaply than affiliates could, passing the funds on to them in low interest loans.[8] By thus strengthening the position of individual affiliates, this centralized management of the Jersey system's tremendous financial resources not only improved the coordination of the parent company's many interests but in doing so also enhanced their performance.

Although capital remained important, the necessity of providing capable managers proved even more fundamental to the long-term success of Jersey and its affiliates. The same might be said of any enterprise, but the decentralized nature of Jersey's management gave its directors special reason for considering it their "prime duty to see that the affiliated companies . . . [had] competent and efficient management." Thus the parent company's efforts "to judge and improve such managements"[9] warranted the highest priority. To carry out this important function, Jersey relied on a multifaceted system, characterized by the same balance of decentralization and centralization found in the overall management structure. A high degree of authority exercised by affiliate management resulted from this system. Because of this characteristic of the Jersey system, the everyday administration of the company's business played an important role in developing the abilities of its executives;[10] however, as in other areas of management, centralized policy-making and planning enhanced the effects of decentralization. As part of its general supervision of the company's interests, the parent company's top management held formal reviews of each affiliate's management prior to signing Jersey's proxies for their annual shareholders meetings.

A similar mixture of decentralization and centralization characterized efforts to provide the necessary supply of managerial talent for the future. With the encouragement and broad supervision of the parent company, each affiliate maintained its own comprehensive executive development program. In total effect, these programs served to supplement the efforts of Jersey's top management, who had the advantage of a more complete view of the company. Drawing on the opportunities available throughout the system, they could make arrangements for executives' training and experience rarely possible within any single affiliate; thus they helped create a pool of broadly prepared executives requisite for the future management of both the affiliates and the parent company.

The decentralization espoused by Teagle dominated the overall structure of 1950 and furnished the basis for Jersey's working relationship with its affiliates. The organization through which the parent company executed its centralized functions also revealed another principle of management, one that had been a tradition in the company even before Teagle's time: from its early years, the Standard Oil Company (New Jersey) had always relied heavily on group discussions prior to decision-making, and that feature, the 1950 organization attempted to preserve. During World War II, decision-making at Jersey had become more and more the work of a variety of committees,[11] giving new currency to the old joke that Jersey should be called the "Standard Meeting Company." However, few, if any, of the company's executives seriously questioned the traditional cooperative approach to problem-solving. In dealing with the scale of the company's operations and the many complicated issues involved, the need for group discussion and review before decision-making seemed irresistible. Frank Abrams, the company's chairman in 1950, explained part of the logic by noting that "it stands to reason that if you get five men together and one man is wrong, the mistake is going to be picked up. Or if one man has a good idea, the others will contribute to it and develop it."[12] Many of the problems the company faced simply extended beyond the capacity of any single individual, and these problems were increasing in number and magnitude. As the company grew and the world in which it operated became continually more complex, the need to meet problems with as much experience, judgment, and expertise as possible could only increase. The complexities of gathering and bringing together information about all aspects of the company's far-flung and diverse interests, integrating that information, and formulating general policies and plans, and, finally, coordinating the company-wide application of those guidelines dictated a cooperative approach.

The nature of group decision-making at Jersey can be better understood by looking at its operation.[13] In the 1950 organization, the application of this principle at Jersey could be seen at all levels, beginning with the company's most important group, its fourteen-member Board of Directors. The significance of the Board as a decision-making body could hardly be overstated, for under Jersey's administrative system, it also constituted the company's top management. Unlike most other large American corporations, Standard Oil Company (New Jersey) had a Board of Directors composed entirely of its own working executives, an "inside" Board. The company often drew criticism for this arrangement on the grounds that it did not fully protect shareholder interests, and company executives periodically reconsidered the wisdom of continuing the practice. However, the structure of the Jersey Board seemed so fundamental to the company's basic processes that its management rejected any change throughout the 1950s.[14]

Of course, the company's commitment to its "inside" Board encouraged a reliance on group decision-making. An examination of the Board's functions and its role in Jersey's decentralized system of management makes clear both the relationship between the two and the practical nature of group decision-making as practiced within Jersey.

The decentralized operating authority that characterized Jersey's relationship to its affiliates contrasted sharply with the way in which it distributed responsibility for centralized functions. In the structure of 1950, the Board was the single body with a coign of vantage that encompassed all aspects of the company's business. Thus, while Jersey's staff departments and committees played an important role in gathering and analyzing information, the parent company's key functions of general supervision, policy-making and planning centered to a remarkable extent in the Board and its committees.

To carry out these functions, Jersey's Board met regularly every week. Between these meetings, an Executive Committee met daily to exercise an almost full authority to act for the Board. While the Executive Committee had only five regular members, the other nine directors, designated as alternate members, usually attended its sessions. In effect, then, the Board convened almost daily.

In the area of executive compensation the Board delegated its authority to a Salary Committee composed of the five regular members of the Executive Committee. This group established salary levels for all executives below Board level. During 1950, the Board also began discussing the appointment of an Executive Development Committee to supervise the company's efforts in that area. By mid-1951 the Board decided to give the Salary Committee those added responsibilities, reconstituting it as the Compensation and Executive Development Committee [the COED Committee]. The committee established policy in this important area, administered Jersey's executive development program, and gave general supervision to the parallel programs of the affiliates. Additionally, it selected executives to fill top positions in both affiliates and the parent company, and it also guided the development of the most promising individuals. Although not formally involved, all directors contributed to the Board's execution of this function through their supervision of the company's operations.[15] The time devoted to the executive development program indicated the great importance attached to ensuring an adequate supply of capable managers for the company's rapidly expanding business.[16]

The obligation for executive development proved only one part of the Board's broad responsibility for supervision, policy-making, and planning; the totality of these functions necessitated and, in the opinion of many, justified Jersey's reliance on an "inside" Board. In identifying and studying the general conditions and problems facing company interests, the Board defined its area of concern broadly. To gather as much information as possible about the world in which the company operated, the Board drew on both affiliate managements and its own staff departments, augmenting these internal sources of information by employing consultants, and by inviting public leaders and outside experts to speak to the directors and certain other executives. Having established the need for a policy, the Board undertook a study of the issue through either a committee of its own members or a staff department or committee. Finally, after this preparation, the Board took up the matter, endeavoring to reach a consensus based on free participation of all of its members. Where one of the directors evinced serious doubts about projected plans or programs, this served to indicate the need for further study. While Jersey's Board could act swiftly when necessary, the company's decentralized form had been designed to place the authority for quick decisions on matters of lesser importance at a lower level and to enable the Board to deliberate carefully on broader needs.

In addition to establishing policy for areas of broad concern, the Board and the Executive Committee devoted much time to reviewing the performance and plans of the affiliates, insuring compliance with general policies and principles. Maintaining this supervision and coordination depended upon effective two-way communication. To provide this, Jersey employed a dual system. First, the parent company's management examined each affiliate's capital-spending plan and its overall performance and outlook in a series of formal reviews held annually. These sessions provided an important opportunity for the directors to gather basic information and to communicate with affiliate managers. They also used these sessions to appraise lesser executives.

The Executive Committee did not formally approve or reject affiliates' budgets. However, it did comment on and make recommendations on them. Each affiliate decided whether to accept these recommendations and suggestions, another "expression of Jersey's policy of decentralizing responsibility." Given the meshing of the total company interests and the Board's control of shares in the affiliate, generally they heeded the Executive Committee's advice. Yet, notable exceptions did occur.

The other advance in effective intercorporate communication came about during World War II. The plan for decentralizing Jersey and separating operations from the holding company developed sufficiently by 1943 to enable then-President Ralph W. Gallagher and his associates to spot several weaknesses in the structure. Too much authority was being exercised by the chief executive officer [CEO] and the Executive Committee. The congressional investigations of 1942 had indicated a gap in the information about operating companies. Effective liaison between the Board and the affiliated companies did not exist, at least at that time. Hence, the "over-all coordination" of the company's interests had been weakened. So had "the morale and unity of the Board of Directors."

Characteristically, the Jersey Board reached a decision only after making a broad study of the company's administration. This resulted in a significant and rather durable alteration; a "wide distribution of the administrative burden among the directors under a system of 'contact' assignments."[17] Except for the two highest officers, each director received responsibility as either principal or alternate contact director in some specific function(s) or area(s). In effect, the contact director became the official liaison between his particular assignment and the Board. The Gallagher refinement of the Teagle plan also continued to relieve the directors of routine duties in order to free them for "long-range planning, general policy and high-level control."

At the same time (1943) in order better to serve the affiliates, a system of functional departments evolved. These departments, headed by coordinators, "extended in principle" to all the affiliates. Each had responsibility "for the particular branch of operations" it represented. Their principal duties fell into three categories: first, they had to keep abreast of worldwide developments such as "demand, competition, technology and governmental relations that might affect either the company or their function." Secondly, these coordinators "constituted a continuous liaison of the parent with affiliates" and finally, "according to the *1950 Organization Manual*," they informed the Board and Executive Committee "of significant developments and opportunities outside the Jersey group" for their particular operating interest.

War-induced pressures slowed down the full development of the functional coordinating departments for several years. However, by 1950 they appeared to be operating effectively, and essentially as planned.[18] As the company's business grew in size and complexity, the role of the Board members as contact directors became crucial. Although they concentrated on those companies and departments for which they had primary responsibility, directors followed developments in their alternate areas as well. The use of dual contacts for each department and affiliate gave the system at least some of the advantages of group decision-making, but the degree of participation of alternate contact directors in the oversight process varied greatly. Nevertheless, the system of dual contacts proved important in maintaining the Board's supervision of the company's business, for it increased the likelihood that, whatever subject came up for discussion, at least one director with specific knowledge about the affiliate or department involved would be present.

Although the Board endeavored to have primary contact directors available for important discussions of affiliate or departmental matters, the nature of their duties sometimes made this difficult, for some of them spent as much as a third of their time visiting affiliates around the world, observing their operations and talking to their managers. Through these visits they gained a detailed understanding of the affiliate's business. Even more important, they observed and acquired much information about the type of personnel needed to continue providing both the affiliates and the parent company with managers for the future.

The contact directors, then, were an essential element in the successful operation of Jersey's decentralized system of management. They gathered much of the information necessary for the Board to execute its centralized functions and communicated policy to affiliates either directly or through Jersey staff departments. In doing so, the contact directors performed a crucial role in clarifying and maintaining the line between the affiliate's responsibility for operations and the parent company's overall responsibility for policy-making and planning.

The heavy burden placed on Jersey's directors made any addition of non-employee directors inconceivable without a fundamental change in the company's system of management. In the Jersey system of the 1950s, functions centralized in the parent company could be executed only by a Board of working executives like that already in existence. Despite extensive staff work, many questions that came to the Board, especially those involving investments, still required expert technical knowledge available only through training or long experience. Other matters called for a thorough understanding of the intricacies of all the interrelationships of the international petroleum industry. The Board and Executive Committee deliberations, and even more the contact work on which it relied, demanded the full-time attention of all directors.[19] The nature and extent of the Board's decision-making responsibilities and the processes through which it gathered and dispersed information meant that, although Jersey's directors had no administrative responsibilities for operations, they constituted, within the company's decentralized structure, a board of managers rather than a board of directors in the usual sense of the term.[20] Under those circumstances, the Board felt that the election of non-employee directors would only dilute its ability to function effectively.

The duties of Jersey's directorate and the role of the Board in the company's management naturally influenced the criteria for choosing directors. The nature of the Board's centralized functions dictated the selection of executives capable of viewing the company's interests broadly. At the same time, the Board's work also required a high level of expertise in several aspects of the company's business. Individual directors needed such knowledge to function effectively as contacts, but the all-encompassing responsibility of the Board functioning as a group also required it. Thus, the selection of executives to be directors depended upon both the qualities and abilities of the individuals involved and upon the possession of just those traits required to give the Board the mixture of talents needed to execute its collective functions.

The Jersey directors of 1950 clearly reflected these standards. They had, in varying degrees, understood the need for, approved of, and participated in the administrative changes instituted by Teagle and Gallagher. While moving toward their election to the Board they had observed its functioning and understood how it worked. As "ponderous, slow and meticulous" as the Board moved, they saw clearly the advantages of achieving consensus there. Nor were they inflexible in their application of company principles and policies; rather they carefully considered alternative solutions and new ideas. In the Jersey tradition, the rule was "Care was at a premium, haste at a discount."[21]

By 1950, Jersey directors no longer held a significant portion of shares in the corporation. In fact, only one member had family connections that might have influenced his selection to that body, and his living contemporaries now pronounce that notion false. Virtually all of the Board members had spent their careers working with the company or an affiliate. A number possessed overseas experience, and many had mastered one or more of the major functions of the company [producing (including exploration), refining, marketing, marine transport, and pipelines. In 1952, the last two were combined into the Transportation Department under a single coordinator]. Generally, they had graduated from college with degrees in engineering and some had acquired expertise through experience. A brief summary of the careers and traits of the five regular members of Jersey's Executive Committee of 1950 will illustrate this pattern.

President Eugene Holman fits the mold of the traditional American oilman and at the same time his career exemplifies the experience and qualities Jersey Boards looked for in their search for executive talent. In 1944, Holman replaced Gallagher as CEO of the corporation. By the mid-century mark, Holman had become, in the opinion of many persons, the world's outstanding oilman.

While growing up in Monahans, Texas, Holman trapped and hunted for food and skins. Moving from Simmons College [Texas] to the University of Texas, Holman acquired a Masters degree in geology. During World War I, he served in the Signal Corps. In 1919, Wallace E. Pratt, chief geologist of the Humble Oil & Refining Company [Humble] hired him and sent him to Shreveport, Louisiana, where in 1922, he became superintendent of the Arkansas-Louisiana division. Soon, Jersey Director Everit Sadler learned about him and moved to utilize his administrative potential. Holman served as officer of several affiliates in South America and in 1933, became the president of Creole Petroleum Corporation, Jersey's most important affiliate there. In 1936, he returned to Jersey as head of the Producing Department and in 1940, joined the Board. Holman thus came to head the company through the exploration and production side of the business. A key figure in the Jersey empire, he was elected president in 1944, at the age of 52.

A man of unimpeachable integrity—who always "saw a long way down the cotton row,"—his vast knowledge of the logistics of oil proved invaluable to Jersey during the sixteen years [1944-1960] he served as president and chairman. The breadth of his view and the soundness and practicality of his judgment gave his opinions weight in the Board's deliberations. More importantly his capacity for viewing questions objectively helped to establish the temper of the discussions. Holman listened with full respect for the opinions of others; yet he proved ready to take strong stands on matters to which he attached importance. When he did so, others recognized the strength of his arguments and usually accepted his position. As CEO he acted as a mediating influence and firmly refused to tolerate cliques, either in or out of the boardroom.

As an executive Holman's personal style suited well the decentralized nature of Jersey's management system. Predisposed to consider the broad picture, he resolutely avoided becoming enmeshed in details; he followed the rule "analyze, organize, delegate and supervise." An excellent judge of people, he utilized the practice of selecting the most appropriate person or persons for an assignment, explaining what he wanted and then letting them work. He listened to and carefully read their reports before making his decision. Delegation of authority, he viewed as a "good management tool" and "a means of testing and training executives."

When dealing with the managers of affiliates, Holman used an identical practice of delegating responsibility and authority. Although some situations required a different course of action, Holman encouraged contact directors and parent company staff departments not to write or telephone affiliate managers, but in all cases, to require a monthly letter on the subject of operations. That policy, regularly followed, coupled with the Board's annual reviews of affiliate budgets and operations, and the inspection tours made by contact directors and other executives, in his opinion, would both provide sufficient direction and permit the executives of affiliates to develop their managerial talents fully.

Fortunately for Jersey, Eugene Holman's foresight brought great rewards. By 1950, he had already directed the purchase of a 30 percent interest in the Arabian American Oil Company [Aramco] along with other significant moves. Too, he indicated a full awareness that social and political forces engendered and nurtured during and by World War II had altered the future of international business organizations such as Jersey. He understood the reasons for the global interest in economic and social improvements, and indicated sympathy for them. Years earlier, as president of Creole, he had displayed an understanding of the importance of listening to and answering the questions of Venezuelans about the company's role in that country. As chief executive of Jersey he continued to inject into discussions comparable questions about corporate social responsibility. Though he came from "the old school of hardheaded economists," he exhibited none of their characteristics in his human relationships. Because of an innate shyness that handicapped all his platform performances, Holman expressed himself best in small gatherings. Certainly he stood out in his conviction that history would render the ultimate judgment of success. Though his Jersey experience helped polish his performance, Holman always possessed the attributes of a born leader.[22]

The talents of Jersey Board Chairman Abrams in many ways complemented those of President Holman. Abrams ultimately became the most renowned humanist in the history of the Jersey Board, "as close to an intellectual as the Board ever had," one contemporary called him. Abrams and Holman made an excellent team. These top managers had helped to create a new view of the relationship between the public—stockholders, employees, consumers, governments, and other groups—and the company, and in communicating their view of Jersey's relationship with them. Abrams excelled in and enjoyed speaking on most any public occasion. Because for Holman, even shareholder meetings proved to be very uncomfortable affairs, as time passed Abrams gradually assumed the mantle of company spokesman.

Abrams graduated from Syracuse University in 1912, and immediately secured employment in a Jersey refinery. He worked his way up the refining ladder, and in 1933 became president of the Standard Oil Company of New Jersey (incorporated in Delaware [the Delaware Company]). In 1940, he joined the Jersey Board, served a brief term as vice president from 1944 to 1945 and, when Gallagher retired in 1946, Abrams became chairman of the Board. Among the earliest of the Jersey leaders to recognize that the continued existence and the growth of the company depended upon the consent of the public, and that general acceptance and approval of the corporation required it to assume certain obligations as well as privileges of citizenship, Abrams helped to broaden the definition of social responsibility he believed should guide Jersey. Each of the directors—and most of the company executives—assumed a role in explaining this public-corporate relationship. But here, Abrams excelled.

These two Jersey leaders were impressive as a matched team. Abrams stood erect, though not tall and he possessed a sort of elfin quality that, combined with his easy approach and cordial manners, led people to feel at ease when around him. Most persons whom he met trusted him—he simply did not look like a "Robber Baron." And while not an orator, his simple, direct manner of address did much to counteract the public stereotypes of big businessmen. Abrams genuinely believed in Jersey; he projected no superficial impressions of the company, polished no images and sought to define the company always in the totality of its actions. Therefore, he came to take a remarkable part in expanding the conception of corporate social responsibility, not only for Jersey but for all American business.

Eugene Holman typified, for many who knew him, the Texas-American oil executive. Where Abrams excelled in conversation, Holman ever remained painfully diffident. His auditors, in many cases, grew more uncomfortable as he talked. His most effective role proved to be that of listener and decision maker. Even though slightly stooped, he appeared to be taller than in fact he was. His tanned face and searching eyes bespoke the outdoorsman, and he did enjoy hunting big game and fishing for trout. A direct person, he said what he thought. His mind absorbed facts about oil and about Jersey and he possessed almost total recall of this information.

Some observers, after watching the Holman-Abrams team in action, came to wonder whether all giant corporations did not need two such top executives in order to function effectively amidst the mounting complexities of the second half of the twentieth century.[23] As a team they well exemplified the newly reorganized Board; each was a specialist, yet held the broad overall view.

Orville Harden, the third member of the Executive Committee, drew his special strength from an extraordinary grasp of the fundamentals of the company's business and from his understanding of how each fundamental interacted with the others. A person gifted with tremendous drive, he ever sought mastery over all problems. Harden began his career with Jersey as an office boy and spent many of his nights going to school at Columbia University. Gradually, he rose through the ranks. One of his early assignments made him a member of the old Manufacturing Committee. In 1925, he chaired the staff department of an early version of the Coordination Committee. In 1928, he joined that committee and held his post until 1934. In 1929, he became a Board member and in 1935, a Jersey vice president. On June 1, 1937, he began serving as vice chairman of the Executive Committee. His service as vice president of the company and vice chairman of the latter committee continued until he retired in 1953. For many years, he served as director responsible for the operations of all the Jersey companies in Latin America except those controlled by the International Petroleum Company, Limited [IPC].

Harden's pleasing manners and intuitive understanding helped make him the great negotiator for the company, a role that led to his responsibility for Jersey's interest in the Middle East. One of his colleagues termed him "the most important and influential man on the Board after the middle 1930s," and certainly few other members ever equalled his capacity for work. His long experience gave him a broad view that few in the company shared. One of his associates likened him to a computer. Harden, he said, could translate barrels of oil into refined products and price these; while at the same time assessing capital and transportation costs, taxes and labor, "he came up with profit figures." Characteristically, he prepared for Board and Executive Committee meetings by mastering all the detail of the topics on the agenda, anticipating questions that might be raised and considering every aspect of the subject. With such thorough preparation, his influence on the Board naturally followed.[24]

The fourth member of the Executive Committee, John R. Suman, held a college degree in mining engineering from the University of California at Berkeley. From 1912 on, he worked in Texas as a geologist, engineer, and executive. Wallace E. Pratt of Humble persuaded W. S. Farish, then that company's president, to hire Suman, and with only limited experience there he quickly became a director. Suman had authored one of the first books on the subject of petroleum production. He early joined several of his Humble associates in advocating oil conservation and especially in improving the science of petroleum engineering.

Suman's great value to the Jersey Board, to which he was elected in 1945, derived from his "knowing 'independent' producers and their problems and being able to discuss matters with them. Many difficult situations arose where the 'independents' viewed things differently than the 'majors.' Some of these differences were valid but many were based on suspicion of motives."[25] Given his wide experience in exploration, labor issues, and production engineering, Suman proved to be an invaluable member of the Executive Committee and the Board.

Jay E. Crane, the last member of the 1950 Executive Committee, joined the Board in 1944. Crane epitomized the Jersey answer to a conglomeration of financial problems brought on by World War II. While the directors responsible for all other areas of the company's operation universally came to the Board after long experience with Jersey or its affiliates, following World War II the financial specialists elevated to that body had acquired their expertise in outside financial institutions. When Crane became Jersey's assistant treasurer in 1935, he had spent twenty years with the Federal Reserve Bank of New York. There, he had charge of foreign operations. In his years at Jersey, he studied closely the company's operations abroad and the effects of the war crises upon them. In 1949, Crane became a member of the Executive Committee with contact responsibility for all the company's financial departments.

The remaining members of the Board showed similar preparation for their work. Each of the contact directors for the four principal functions—transportation, refining, marketing, and production—had developed into a recognized expert in his field, with extensive experience at affiliate and parent company levels. Bushrod B. Howard, contact director for transportation, graduated from the United States Naval Academy and, by the time he became a director of Jersey in 1945, he had spent a quarter century working with the company's foreign and domestic tanker operations. Chester F. Smith, contact director for refining, started his career in the company's Bayonne, New Jersey, refinery while still in college. Concentrating throughout his career on the management of Jersey's refining business, Smith's expertise in the field had been recognized by his election as vice president of the American Petroleum Institute's Refining Division. He also served as head of the Standard Oil Company of New Jersey before joining the Jersey Board.

The contact director for marketing, Emile E. Soubry, began his career with the company at an even more tender age. Starting as an office boy for Anglo-American Oil Company, Limited, Soubry held numerous sales posts there before becoming that company's chairman in 1939. As Jersey's Coordinator of Foreign Marketing from 1943 to 1949, Soubry broadened the base of experience upon which he drew as contact director for marketing and European affiliates. The other major function, producing, had as its contact director, John W. Brice, a geologist by training and an expert on production matters by virtue of a career spent with several of the company's major producing affiliates, as well as in the Producing Coordination Department of the parent company. Because the most important Latin American affiliates were predominantly producing companies, Brice also served as contact for that region.

Most Jersey directors had contact assignments based on their training and experience, although some "covered" areas for which their principal qualifications were enthusiasm or a willingness to take on the necessary work. By 1950, Frederick H. Bedford, Jr., had been Jersey's director in charge of specialty sales for almost twenty-five years. Stewart P. Coleman, with a Ph.D. in Economics from the Massachusetts Institute of Technology [MIT], had spent the first part of his career in Humble's Research Division before concentrating on the economics and coordination work of the parent company. In 1950 he shared contact responsibility for the company's Coordination and Economics Department with Smith, and also served as contact director for Standard Oil Development Company [later Esso Research and Engineering Company]. By 1950, Frank W. Pierce, an expert in industrial relations, had given up contact responsibility for that area, but remained contact director for Imperial Oil Limited, of Canada. Responsibility for employee relations and public relations had passed to Monroe Jackson "Jack" Rathbone, a chemical engineer who had risen from a position as draftsman at the Baton Rouge refinery to head first the Standard Oil Company of Louisiana and then the Standard Oil Company of New Jersey. Soon after the two merged to become Esso Standard Oil Company, in 1948, Rathbone joined the Jersey Board as contact director for that important refining and marketing affiliate. Henry H. Hewetson, a former president and chairman of Imperial, had begun his career as a refiner, but later worked in marketing. As a new director in 1950, Hewetson's contact duties were limited to serving as alternate for marketing for Imperial and Latin

America, but during the 1950s he became primary contact for a number of marketing and refining affiliates.

Robert T. Haslam, whom Hewetson replaced on the Board in June, 1950, serves as the primary example of a Jersey director whose career pattern shifted abruptly away from the area of his formal training. A professor of chemical engineering at MIT, Haslam joined the company in 1927 to establish the research program at Baton Rouge, but within a few years he became a leader in marketing. By the early 1940s, he was the guiding force in creating Jersey's first public relations program.

Together, combinations of specialists covered all elements of the company's business, and the Jersey Board appeared well equipped to deal with the great variety of matters coming under its purview. To insure that top management had the time necessary to concentrate on overall policy and planning, however, some mechanism had to be devised for handling the more routine of the parent company's centralized functions. This responsibility devolved upon the system of Jersey staff departments and committees, which served both parent company and affiliate managers. For the former, these departments provided the crucial staff work without which the directors could not have performed effectively either as individuals or as a decision-making group. In this role, the staff not only offered analyses and recommendations on affiliate problems and proposals, but also served as additional sources of information about affiliate operations and personnel, supplementing the contact work of the directors. Staff organizations also performed other essential functions.

While contact directors provided the principal line of direct communication between the Jersey Board and affiliate top managers on policy and matters of an extraordinary nature, the staff departments and committees furnished the primary link on all routine business, especially with the operating departments of affiliates. As suppliers of "counsel and advice" to affiliates,[26] they not only disseminated policy throughout the operating level, but also made available information of a quality and scope that individual companies could not match on their own. With their broad perspective, the staff groups could apprise affiliates not only of developments throughout the Jersey system, but also of industry-wide trends. In this way they facilitated the coordination of affiliate planning. The staff departments also placed on call for the benefit of affiliates not only the expertise of their own organizations, but that of the entire Jersey system. Finally, they encouraged communication among the affiliates, bringing specialists together at periodic group meetings for an exchange of ideas and information on technical and administrative subjects.

Since, in 1950, Jersey's supervision of affiliate operations followed pre-dominantly functional rather than corporate or geographical lines, the most important of its staff departments were the offices of the coordinators who oversaw the major functions: producing, refining, transportation and mar-keting. Although these departments and their functions were not entirely new, their form and their role in Jersey's decentralized system of manage-ment developed during World War II and in the immediate postwar period.

Simply, the demands of wartime operations had placed a premium on efficient use of the company's resources, and as top management looked toward reconversion, they recognized the continued importance of effective coordination. Initially, they foresaw a postwar period of declining demand and excess capacity but instead, markets all over the world began to expand rapidly. Yet, while their projections had been wrong, the steps they had taken to prepare for shrinking operations proved to be equally important in guiding the company through a period of rapid growth. In fact, that growth called for even greater strengthening of the company's capabilities for coor-dination. In the postwar years, functional coordinators, who had previously been responsible only for foreign operations, began to exert greater influ-ence over domestic activities as well. In 1947, the Board formalized this development by making their responsibilities global. At the same time, they expanded the scope and level of activities of these departments and greatly enlarged their capabilities for serving affiliates. Their role in guiding affili-ate planning through analysis of their capital budgets also increased, although authority for final decisions remained with affiliate managers.

While the coordination of affiliate planning and operations in their func-tion remained the primary duty of the coordinators, they also had another responsibility. The contact director for an affiliate could designate one of the functional coordination departments as a "source of contact and channel of communication" for the affiliate on a corporate basis. In this capacity, the coordinators improved affiliate communication with the contact directors, and through them, with the Board. By giving affiliates a single representa-tive in the central office, they also helped them deal with other staff depart-ments, especially where problems or proposals involved more than one function and where coordination of staff department actions became essen-tial.[27] Thus, functional orientation remained paramount, and affiliate op-erating departments dealt directly with their counterparts in the parent company.

The work of the functional coordinators was supplemented by a number of other parent company staff departments which served holding company and affiliate managements, both directly and through the coordinators. Some, such as the Medical, Insurance and Social Security, and Employee Relations Departments, worked primarily on coordinating the actions of their counterparts in the affiliates, although they also administered the parent company's programs in their fields. Other departments (Treasurer's, Comptroller's, Tax, Law, Budget, Secretary's, Public Relations, and Coordination and Economics) also advised and counseled affiliates, but their responsibilities for parent company operations took a substantially greater part of their efforts.

To deal with problems and proposals that cut across functional lines, Jersey had several staff committees. The most significant of these and, in fact, the most influential group in the company other than the Board and Executive Committee, was the Coordination Committee. The organizational chart published in 1950 conveyed graphically the importance of this body to Jersey's management system. Directly below the Board and surrounded by the principal staff departments, the committee occupied, in the schematic, a central location that symbolized the multifunctional role it had come to play since its inception during the 1920s.

Initially, the Coordination Committee had been charged only with overseeing the flow of crude oil, thus insuring the efficient use of refineries and an adequate supply of products for domestic markets at the lowest overall cost. After World War II, the growing importance and scale of foreign operations created a need to coordinate all operations on a worldwide basis, and the Coordination Committee's charge expanded. However, an even more important change had already begun to take effect.

Beginning during World War II, the committee became increasingly concerned with meeting future as well as existing needs. By 1950 this had combined with an expanded geographic scope to make "long-range planning on a worldwide basis," its primary function.[28] The committee's responsibility covered all aspects of the company's operations, including its growing chemical business, but throughout the decade the emphasis remained clearly on providing the capacity needed to meet the rapidly expanding demand for petroleum products throughout the world.[29]

The decentralized nature of the company's management and the realities of its ever-changing environment precluded the imposition of a rigid program for development; but to take advantage of its total resources, the company still needed a general plan for its growth, one based on the overall interest. The Coordination Committee provided a means both for developing such a plan and for measuring the actions of affiliates, to insure the best use of available resources. In the words of the organization manual, it assisted the parent company and the affiliate managements "in evaluating advantageous and alternative courses of commercial action, particularly on a long-range basis."[30]

Based on the company's needs, one important aspect of this process was the gathering, consolidation, and dissemination of information. Because of its overview, broader than that of any single affiliate or staff department, the committee operated from a unique position to establish the basic assumptions underlying the company's planning. With all of the affiliates gathering data based on those common principles, the committee could produce an accurate survey of overall future needs which greatly improved planning at both parent company and affiliate levels. However, the Coordination Committee's responsibilities with respect to the planning process clearly had become larger and more direct than ever before.

The Board had authorized the committee to approve additions to affiliate capital budgets within a limited range, but the committee's real influence rose from the fact that it studied and discussed practically every important project or proposal before it went to the Board for final consideration. Top management relied upon the committee as its "study and advisory group" to frame the issues and to establish the context within which the final decisions would be made. Without the Coordination Committee to do this groundwork, any effective functioning of the Jersey Board in managing the business would have been "impossible."[31]

The key to carrying out these responsibilities lay in the coordination of affiliate capital budgets which the committee supervised. It not only oversaw the functional budget analysis performed by the Coordination Departments but also analyzed affiliate budgets on a corporate basis.[32] Finally, before completing its recommendations to the Executive Committee, the Coordination Committee brought together executives from affiliates all over the world, to give the capital spending plans one last review. In a week of all-day sessions, executives scrutinized every aspect of proposed affiliate budgets. Involving as many as three hundred individuals, these annual conferences, often held at the Westchester Country Club outside New York, provided an opportunity for the exchange of ideas across functional and corporate lines and encouraged the participants to view their work within a broader context.

Along with other meetings held by the Coordination Committee and the various staff departments throughout the year, the Westchester gatherings played an important role in Jersey's executive development efforts. The oral presentations that constituted the bulk of these conferences served to give exposure to many promising young executives. Although not linked formally to the work of the COED Committee, this function of the Coordination Committee had been traditional since its early years and, as the company grew larger, became increasingly important.[33] The membership of the Coordination Committee reflected its varied functions and its importance in the company's structure. Two directors who served as chairman and vice chairman of the committee attested to the value of this committee to the Board. In 1950, Chester Smith held the top position, assisted by Stewart Coleman. When Smith retired in 1954, Coleman took over the chairmanship. The functional coordinators, the treasurer, and the head of the Budget Department represented the principal Jersey staff departments. Along with its other functions, the Coordination and Economics Department served as the committee's staff, and its manager also held membership on the committee. Finally, the New York-based affiliates—Standard Oil Development Company, Esso Shipping Company, Esso Export Corporation, and Esso Standard Oil Company—each had a representative.

This pattern of membership, which had emerged in the early postwar years, changed only slightly during the decade. The controller joined the committee in 1953, thus completing the representation of Jersey's financial staff departments, and reflecting the increasing emphasis on finance in the committee's work. In the following year, the Coordination and Economics Department underwent a reorganization that divided it into a General Economics Department and a Coordination and Petroleum Economics Department. The head of the Coordination and Petroleum Economics Department, which continued as the committee's staff office, and the chief economist then became members of the committee. Finally, in mid-decade, the committee added Jersey's Middle East advisor to its ranks, in recognition of the expanding role of that turbulent region in the company's worldwide supply picture.

In addition to its regular members, the Coordination Committee welcomed to its weekly meetings any executives of affiliates who happened to be visiting New York, a practice which helped broaden the understanding of these managers, and which provided yet another opportunity for the committee to get affiliate views.

In carrying out its work, the Coordination Committee relied heavily on subcommittees of its members. After 1952, it also had the benefit of a Coordination Committee for Europe. Composed of parent company representatives stationed in Europe along with managers of local affiliates this "study and advisory group" had the responsibility for keeping the Coordination Committee better informed about the overall problems and needs of the European affiliates and for helping those companies solve supply and other problems that involved more than one of them. Although the stated purpose of its establishment centered on "decentralizing the European coordination activities,"[34] in practice it could not really take over the responsibilities of the Coordination Committee. It did, to a degree, facilitate communication among affiliates in Europe; but the difficulties that led to its formation could not be so easily remedied, and the coordination of Jersey's numerous European interests remained an area of serious concern.[35] (The struggle to gain control of the Jersey subsidiaries in Europe is discussed in a separate chapter.)

While it would be virtually impossible to include in these pages a truly comprehensive survey of Jersey's affiliates at mid-century, their importance should not be underestimated. The great variety of their activities and interests, and the special and often unique influences of affiliate operations on the course of Jersey's history, added complexity to the development of the entire business. Like the management committees of the Board and parent company, and the staff departments organized to coordinate and direct overall management, each of the larger subsidiaries played a considerable individual role in doing the work of the whole company. By focusing on only a few of the major affiliates, this section attempts to provide a sample of this influence as well as some estimate of the results. Only by keeping the selectivity of this account firmly in mind can the reader appreciate the richness and excitement of the increasing part played in 1950 and after by the many affiliate companies that helped to increase the scope of Jersey operations around the world. Like other similar relationships, the ties between parent and the various offspring represented a vital and always reciprocal exchange of information, influence, and experience, as ideas and men moved both ways along the links between companies.

How the Jersey Board of 1950, with layers and slices of functional and department committees throughout the company, actually managed the holding company and its many affiliates while responding to internal stresses and externally accelerating pressures, cannot quickly or briefly be narrated. Governance of Jersey and its holdings continued to follow the predominantly functional patterns that grew out of the Teagle-Gallagher changes in the administrative structure. Some contact directorships were assigned on a regional basis and some recognition of regional distinctions permeated staff reports. For that time, these developments may be considered experimental. However, the totality of the problems the Jersey Board and the committees encountered (around the world) and sought to resolve may best be shown by a geographical approach. Certainly such an examination will depict the nature and scope of the company's business at mid-century, even though the discussion is necessarily limited to selected subsidiaries.

Among its affiliates, Jersey frequently made a distinction between its domestic and its foreign operations. The United States constituted the world's largest and most important petroleum market, and Jersey's domestic operations accounted for over half of its total investment worldwide, and over 40 percent of its profits. In 1950, they produced $173 million of the company's $408 million in consolidated net income.[36] Marketing domestically was circumscribed by the results of the 1911 dissolution of the Standard Oil Company, which had the effect of restricting Jersey's use of the name "Standard" and, by extension, the use of its trademark "Esso" to eighteen states in the East and South. The Esso Standard Oil Company, its principal refining and marketing affiliate in the United States, still had the largest sales of a Jersey affiliate—$1.5 billion in 1950. However, because of the low profitability of the refining and marketing functions, this income netted only $52 million for the company. Though the wholly owned Carter Oil Company operated a refinery and marketed in eight Western states, it remained largely a producing company; hence, it recorded a better net profit. Nearly all of its $20 million in profits from $173 million in sales during 1950 came from its producing operations in the Rocky Mountain and Mid-Continent areas.

The differential in profitability among diverse functions can be dramatically illustrated by the case of the Humble Oil & Refining Company, the Texas-based domestic affiliate in which Jersey then held a 72 percent equity interest. In addition to extensive marketing in its home state and a small amount in New Mexico, Humble's giant refinery at Baytown manufactured products not only for itself but for other Jersey affiliates and non-affiliated companies as well. A subsidiary, Humble Pipeline Company, also operated one of the largest pipeline systems among the Jersey affiliates. However, by far the most important of Humble's operations was its production of crude oil, much of which went to other affiliates. Its operations along the Southern rim of the United States, from Florida to California, made it the nation's largest crude oil producer. Chiefly from this source Humble netted the $129 million it realized on its $793 million gross income in 1950,[37] making it Jersey's most profitable domestic affiliate. Humble also stood out among Jersey's United States affiliates for its autonomy. Owing to the stringency of the Texas antitrust laws and the historic public animosity toward Standard Oil in the state, its management resolutely insisted that while the parent company owned over two-thirds of Humble's stock, it could exercise no greater control over its affiliate than that incidental to voting its shares at the annual shareholders' meetings. Between these meetings, Jersey offered advice and Humble kept it informed of developments, but Humble had a unique and powerfully independent position within the decentralized Jersey system.

While the production, refining, and marketing of petroleum products remained Jersey's principal business, the company also had other kinds of U.S. affiliates. In addition to Humble's pipeline system, Jersey had a direct interest in several pipeline companies including the Interstate Oil Pipe Line Company, which transported crude oil in Oklahoma, Louisiana, and Mississippi, and the Plantation Pipe Line Company, operator of a refined-products pipeline from Louisiana to North Carolina. In all, the company held a position, directly or through its affiliates, in almost a dozen domestic pipelines. It also owned a majority interest in the Interstate Natural Gas Company, Incorporated, which produced gas in Louisiana.

Along with these investments directly associated with its basic business, Jersey had an interest in other companies supporting some phase of its activities. Gilbert & Barker Manufacturing Company (wholly owned) manufactured and marketed oil burners and service station equipment, and Ethyl Gasoline Corporation (held jointly by Jersey and General Motors Corporation) produced and sold the principal anti-knock compound used by Jersey and other companies in the production of motor gasolines.

The Enjay Company, Inc., the sole marketer of chemical products manufactured by Esso Standard Oil, grew increasingly important in another area of operation. While the United States provided Enjay's principal market, it also did a substantial export business. This placed Enjay among a group of domestic affiliates whose operations extended far beyond the borders of the United States. Several of these companies served affiliates worldwide. For example, Esso Export Corporation served as a broker for all Jersey affiliates, facilitating sales of crude oil and products among them and, in some cases, handling sales to and from non-affiliated companies. Although Esso Export reported a small profit, its primary mission centered on expanding the volume of affiliate sales by removing impediments to interaffiliate transactions, especially problems associated with the currency exchange restrictions of the postwar period. In keeping with its role, Esso Export always sold to affiliates at the lowest market price, charging only a modest brokerage fee for its services.[38] The newly created Esso Shipping Company, which had taken over the functions of the Marine Department of Standard Oil Company (New Jersey), operated Jersey's United States-flag fleet. Finally, the Standard Oil Development Company served as Jersey's principal research and development organization. Through its own appropriations and a system of contracts, it served affiliates throughout the world.

Jersey owned 70 percent of the largest petroleum company in Canada, Imperial Oil Limited. Unlike Jersey's domestic affiliates, Imperial had integrated operations that spanned that entire nation. After bringing in its Leduc No. 2 well in Alberta, in 1947, the affiliate's attention focused on the rapidly developing crude oil operations in the Western Provinces. By 1950, it had begun to put large amounts of capital into upgrading and expanding the refining and transportation facilities through which this growing production would find its outlet. One result was the completion of the Interprovincial Pipe Line (Company) in which Imperial held almost a 50 percent interest. With Canada at that time increasing its petroleum consumption faster than any other nation, this first link between its burgeoning producing region and its major refining and marketing areas insured Imperial's continued rapid growth, as Canada moved toward self-sufficiency in petroleum.[39]

Jersey's Latin American operations illustrated again the advantage in profitability enjoyed by producing in comparison with the other functions. A variety of affiliates refined and marketed petroleum products throughout the Caribbean area, and in every country south of Mexico except Bolivia and Ecuador. In most of these countries refineries served only the relatively small local markets, although in 1950, Brazil, where law confined refining to its nationals, ranked as Jersey's fourth largest market after the United States, Canada, and Great Britain. Producing ventures constituted by far the most important of Jersey's Latin American operations, and the most important producer was Creole in Venezuela. Like Humble, Creole refined and marketed products for local markets, but its production of crude oil far overshadowed these functions. Jersey's 94 percent equity interest in this subsidiary yielded $156.7 million in profits in 1950, almost 40 percent of the parent company's worldwide total. In addition to direct crude oil sales abroad, much of Creole's production went to its new Amuay refinery and the giant Aruba, Netherlands West Indies [later Netherlands Antilles], refinery of Jersey's wholly owned Lago Oil & Transport Company, Limited, from which products moved to markets throughout the Americas and Western Europe. Although not on the same scale as Creole's, the International Petroleum Company, Limited's [headquartered in Canada] integrated operations in Peru, Colombia, and Venezuela also yielded substantial profits to the parent company, again almost entirely from crude oil production.[40]

Taken as a region, Europe (including North Africa) and the Eastern Mediterranean area provided Jersey's second largest market. In the aggregate, the affiliates there had $955 million in sales in 1950. But aside from half interests in small producing operations in the Netherlands and in West Germany, the company's holdings in the region consisted of a diverse assortment of refining and marketing affiliates. Because of their concentration in these functions, they produced as a group a net profit of only $13.5 million, of which Jersey's equity interest amounted to $10.5 million. Yet, the large volume of the market represented potentially large returns, especially as the European economy recovered from wartime and postwar dislocations and the per capita consumption of petroleum products continued its rapid increase. Moreover, despite the narrow profit margins that plagued refining and marketing operations, the substantial outlet for crude oil represented by these affiliates remained crucial to the success of the more profitable producing affiliates in Latin America and the Middle East.[41]

Because of its role in the company's operations and future plans, Europe continued to be a constant source of concern for the parent company's management. Operations there suffered not only from the difficulties inherent in the predominant functions but also from the prevalence of nationalistic restrictions that made coordination of affiliates in different countries nearly impossible. In several countries separate companies even carried out diverse functions with differing degrees of Jersey ownership. The resistance of many affiliate managers to what they considered unwarranted and indefensible interference added to the parent company's problems of administration. Substantial amounts of stock in such European affiliates as Standard Française des Pétroles, S.A., N.V. Nederlandse Aardolie Mij., and Esso Standard Italiana, S.p.A. belonged to minority shareholders, and a few were still managed by executives whose family ties to the enterprises predated Jersey's acquisition of their stock. But some resistance to direction from New York existed even where these conditions were absent.[42]

To supplement its communication with the European affiliates, Jersey relied on its shareholder representatives. Stationed in Europe and assigned to one or more affiliates, these men had responsibility for advising affiliate managers concerning Jersey policy. Although they had no line authority, they did provide a source of on-the-spot information that served to make affiliate actions consistent with Jersey policy and with overall company interest.

Shareholder representatives "served another useful purpose." European governments played an increasingly important role in the struggle to reconstruct the economy. One director who understood that situation wrote that sometimes these governments requested that affiliates "take some action which would have been very detrimental to Jersey's overall interests. Most of our companies in Europe were headed by nationals and it was not easy for them to protect Jersey's interests when it was considered . . . against the National Interests." He went on to state, "It was left to the Shareholders Representatives to convey Jersey's thoughts to the Government. In most cases our Shareholders Representatives were well known and respected by the Government Officials and at times assisted by the U.S. Ambassador."[43] This diplomatic mission of the shareholder representatives appeared to be accomplished without excessive friction. Nevertheless, despite their efforts, shareholder representatives provided a communications mechanism until only such time as some form of regional organization could be effected.

Another task of the shareholder representatives resulted from the 1942 Consent Decree, which still hung heavy over Jersey's operations after World War II. The Board, in 1942, had assigned a lawyer, Edward F. Johnson, to monitor the company's effort to abide by the decree. When hostilities concluded, shareholder representatives began to examine the trade practices of the Jersey affiliates in their assigned areas. Travelling to New York, they reported to Johnson whatever actions they had deemed questionable. He noted these remarks in a memorandum of the conversation. Further, the representatives included such information in their annual reports to the Board.[44]

Among the European affiliates, the largest in terms of sales was the Anglo-American Oil Company, Limited (renamed Esso Petroleum Company, Limited in 1951) which operated in the United Kingdom and Eire. Organized in 1888 by John D. Rockefeller, Anglo-American ranked as Jersey's oldest overseas affiliate. In 1950, the company continued the process of upgrading and modernizing its gasoline marketing facilities, in anticipation of rapid expansion in that portion of its business. At the same time, it had begun construction of what would, when completed in 1951, be Europe's largest refinery. The new £37.5 million plant at Fawley, England, signified a new trend in the European oil business. Expanding markets and the availability of Middle East crude oil made large-scale modern refineries in Europe economically feasible, and government pressures to conserve foreign exchange credits made their construction there politically prudent as well.[45]

Of the other European countries, the largest gross sales by Jersey affiliates came in France and French North Africa ($197 million in 1950), Italy ($144 million), and the Scandinavian countries ($135 million). West Germany, Belgium, and Holland also represented substantial refining and marketing operations.[46] Thus, Jersey affiliates operated in all of non-Communist Europe except Spain and Portugal.

From the end of World War II, Jersey had been shifting the source of supply for its European affiliates from the Western Hemisphere to the Middle East. Europe remained a major outlet for Venezuelan crude oil and refined products, but Jersey affiliates there obtained an increasingly large proportion of their supplies from two Middle East producing affiliates in which the parent company held a minority interest. Through its half ownership (with Socony-Vacuum Oil Company) of the Near East Development Corporation, Jersey had held a 11⁷/₈ percent equity interest in the Iraq Petroleum Company, Limited [IPC] since the late 1920s. After World War II, Jersey had added substantially to its Middle East reserves and production by purchasing a 30 percent interest in Aramco. Already, Jersey obtained ten times as much oil from Aramco as from IPC.[47] The recent completion of the Trans-Arabian Pipeline [Tapline] provided a link between Saudi Arabia and the Mediterranean port of Sidon, Lebanon. That vital connection with Europe encouraged the further rapid expansion of the Aramco producing areas and Europe's continuing shift to Middle Eastern supplies.

Although European markets took most of Jersey's Middle Eastern production, some went in the opposite direction, to markets east of Suez. In the Far East, Australasia, and East and South Africa, Jersey had one affiliate, Standard-Vacuum Oil Company [Stanvac], jointly owned with Socony-Vacuum. Created in the 1930s to couple Jersey's production in the East Indies with Socony-Vacuum's marketing operations in the region, Stanvac had always enjoyed relative autonomy. Jersey helped select its management, but participated only minimally in the direction of its investment.[48]

Stanvac served a vast geographical expanse, and most of its marketing area had very low per capita consumption of petroleum products. Nevertheless, it provided an outlet for crude produced in Sumatra and New Guinea, and for the products of Stanvac refineries at Sungei Gerong in Sumatra and near Melbourne, Australia, as well as two smaller Japanese refineries in which the affiliate held a majority interest.

Jersey's affiliates, then, covered the globe. Together with the parent company they formed a complicated business system designed to serve many individual markets efficiently, and to provide both for current needs and for future development in a manner that would benefit "the General Interest"[49] as well as its constituent parts.

Contact (and alternate) directors using the functional committees, the departments, and the Coordination Committee kept abreast of what happened in the various affiliates and in all the countries in which they operated. Personal contacts at budget sessions and functional committee meetings, usually involving all or most of the managers of affiliated companies, added to the directors' information about the problems encountered and the solutions devised or recommended by the managers. The Jersey Company, in effect, in 1950, operated as an "open" organization, bringing to bear on its vast empire the wealth of information and managerial talent possessed by the entire organization, for the benefit of even its least important subsidiary. While these interchanges occurred, the Jersey Board appraised executive talent and shifted promising employees within its corporate holdings, providing them with an opportunity to develop fully their managerial skills. "Who will take our places?" remained one of the most important concerns of the directors. Line departments reporting to the holding company sought to anticipate personnel problems and provide good working conditions from the Board members downward. Few major Jersey executives left the company and, in the opinion of some competitors, Jersey actually stockpiled managerial talent. By 1950 with its combination of decentralized operating responsibility and centralized policy-making and planning, Jersey's structure seemed to offer a mixture of flexibility and stability ideally suited to the period of rapid growth projected for the worldwide petroleum industry. Nevertheless, the changes accompanying that growth would sorely test the company's capacity to respond and adapt itself to a very different business environment.

Chapter 2

An Investment Strategy for the 1960s

AS STANDARD OIL COMPANY (NEW JERSEY) [Jersey] President, Monroe
J. Rathbone continued to be an aggressive hard-working chief executive, constantly firing memoranda about company plans and its operations at the Board of Directors, at the functional committees and at anyone who he thought needed to be involved in his version of the company's future. In the Exxon Corporation records, inadequate sources render impossible the detailing of how company executives responded to most of these memoranda. But a sufficient number of sources were located, occasioned by one Rathbone memorandum, to enable one episode in the company history to be portrayed—the struggle over investment strategy.

Implementation of the decision to diversify into non-petroleum related enterprises did not (before 1975) prove to be a brilliant idea or stroke. One Exxon executive termed it "an unmitigated disaster, costing billions, absorbing top management time, and tarnishing the Corporate image." The record from 1960-1975 supports the conclusion that in several instances the company has appeared, or been made to appear, naive in areas outside petroleum-related enterprises. The debate leading up to diversification of investments is detailed below.

X   X   X

Shortly before he retired in May, 1967, Jersey Director Cecil L. Burrill remarked to Board Chairman and Chief Executive M. L. "Mike" Haider, "The Board is afraid to diversify and at the same time is afraid not to."[1] Burrill also remembered that Haider agreed with him. Both men had participated in the characteristically—for the Board—long, sometimes heated discussions of a management reorganization proposal circulated by President Monroe J. "Jack" Rathbone on March 7, 1960, entitled "Memorandum on Organization: Investment Planning and Review."[2] In this document Rathbone proposed that the Board appraise its philosophy of

30

investment return and the procedures used in reviewing proposed investments of the company affiliates. To do this, he indicated that the entire management structure might have to be altered. The occasion for the request that the Board consider such changes resulted from studies beginning in 1958 of an ad hoc group that worked under Rathbone's orders to review the company's structure and investment policies and procedures, against the changes for the oil industry that he believed in the offing. Rathbone took the position, "look we are in love with barrels of crude, busting a gut to get the extra barrel—making no money—we have to figure a better way to get the stockholders more money."[3]

The Iranian Consortium Agreement had helped stabilize for the time that country's oil output, and in conjunction with the production from other countries, led to a world oil glut that presented marketing and other problems for all transnational oil companies. Succinctly put, Jersey for one, annually sold more oil for less profit. As the value of oil wavered, the question of earnings emerged. Some directors questioned that Jersey stock actually merited the term "growth." One of them later pointed out that the:

> setting for the serious consideration given diversification included a long-term financial outlook by David J. Jones, which showed Jersey facing a future of limited opportunity for investment in the oil and gas business and an increasing cash accumulation of large proportions. Some of the directors, particularly those with operational backgrounds, were quite uneasy with this forecast. However, not having the data or time to devote to our own analyses, no particular issue was made in formal sessions.[4]

Rathbone in providing the Board with his memorandum, sought to encourage study of this problem, and he successfully precipitated searching debate on the subject. Viewed from the vantage of the 1980s, some of the forecasts, as might be expected, did not prove to be accurate.

Responding to Rathbone's memorandum, the Board acted quickly. On March 17, 1960, a newly created Board Advisory Committee [BAC] assumed the review function of the now dissolved Coordination Committee and, with staff assistance from an equally new Investment Analysis Department [IAD] headed by Jones (formerly of the Treasurer's Department), instituted more rigid standards and a more comprehensive approach to investment proposals. To encourage the freer discussion required by these new standards, the BAC excluded many executives and affiliate representatives from its decision-making deliberations. Some directors and BAC members believed that discussing an area in which Jersey had not formally charted its policies and priorities would be so time-consuming that the energy of these persons would be used productively by their continuing normal activities. During the discussions, representatives of all Jersey departments answered questions and explained specific data, but the executive sessions were attended only by BAC members and company counsel.

This formal search for a new order for Jersey lasted more than two years, and the issues raised in those years provided material for subsequent debate across the next decade.

While the Board's action in establishing the BAC and the IAD implemented proposals made in the Rathbone memorandum, they fulfilled only a portion of the mandate included therein. He had also charged the BAC with formulating and recommending specific investment policies which, upon approval by the Board, would serve as guidelines for BAC evaluation of proposals and thus permit greater decentralization of that function. Up to this time while the company's investment activities had been guided by certain general assumptions, no formal statement of principles or policy existed. Thus the Rathbone report stimulated long overdue management consideration of the problem and also precipitated a fundamental reconsideration of Jersey's entire investment philosophy. In keeping with the Jersey tradition governing review and study of problems, careful staff research began immediately.

As head of the new Investment Analysis Department, Jones found himself at the center of this review of Jersey's investment program. Because of the IAD's position in the new structure, Jones was told that it had a "mission to study broadly the company's investment activity." As "an essential first step" in carrying out that directive he undertook to assay the many and varied factors that he believed would affect Jersey's future investment program. During the summer of 1960, he wrote for the Board's consideration a lengthy memorandum entitled, "Basic Assumptions in Preparation for Investment Proposals."[5] Known to Jones as the "Basic Assumptions Report" and to others as the "Jones Report" this document became the focal point for an almost unique episode in the history of Jersey's management. The company's top executives for the first time listed and evaluated the assumptions they used in making investment decisions. Developing logically from Rathbone's efforts to restructure and redirect Jersey's investment procedures, the Jones Report and the discussions it prompted marked a crucial step in the movement toward a more comprehensive and selective investment review process and Jersey's first formal, detailed investment policy. But the change did more than improve the way in which the parent company handled affiliate investment proposals: it ultimately altered the means of implementing some of the basic functions of the parent company.

Noting that the company had "embarked upon a new approach to its investment problems," Jones introduced his memorandum by pointing out that although the company made particular investments contingent upon special consideration in each case, all such decisions were subject to certain general considerations. In order to improve management's judgments about specific proposals, it needed a clear statement of the governing assumptions about the environment in which those investments took place. Jones stressed that in a "large, complex organization such as Jersey . . . the explicit recognition of key assumptions or judgements. . .[was] essential for effective communication, staff study, and management deliberations."[6] Implied, but of overreaching importance, was the idea that the process of articulating the company's first formal investment policy—as mandated in Rathbone's memorandum—could not proceed without such a statement of basic assumptions.

Since the purpose of his report was to guide Jersey rather than the general investor, Jones carefully tailored it to the company's specific outlook and needs. Balancing the relatively long lifespan of most of the company's investments against the difficulty of predicting the future with any degree of assurance, Jones chose a time horizon of eight to twelve years—the decade of the 1960s. Rather than survey the total business environment for that decade, Jones focused selectively on those areas and aspects more relevant to Jersey's investment decisions: "International Affairs, United States Economic and Business Environment, Technological Factors, Foreign and Domestic Petroleum Industry Environment, [and the] Jersey Organization."[7]

The Jones Report, however, went beyond assumptions about the investment environment, to suggest the policy directions indicated by those assumptions. While consciously avoiding any specific predictions of the future, the document offered instead "either a range of plausible future developments with some indication of the most likely future position, or . . . a set of eventualities which, if adopted as a basis for policy formulation, [would] . . . tend to offer the best chance of maximizing gain or minimizing loss." This approach about the future investment environment reflected Jones' belief that the company's basic policy should feature ' "fail-safe' characteristics" so that the company would run "the least risk of damage" if the principles formulated proved too optimistic and, yet, not forego the possibly greater benefits that might be obtained if events developed more favorably than assumed.[8]

The message was clear. The problems of worldwide excess capacity and weak prices then afflicting the petroleum industry could not reasonably be expected to correct themselves, at least through the 1960s. As a result, Jersey faced the prospect of continued decline in its return on investment, even for the foreign operations that had boosted overall performance during the preceding decade. Jones expressed particular concern that if the poor earnings performance of the late 1950s had not already cost Jersey its status as a "growth stock," the continued decline that seemed inevitable surely would. The outlook for continued poor performance thus threatened the company's ability to attract and keep "top grade management talent" and to finance future expansion on the most favorable terms. Jones believed that the company abounded in skilled technocratic managers; however, he sought to provide a place for the entrepreneurial types who had perhaps been overlooked by Jersey's management up to that time. He believed that political as well as economic trends would work to make the decade of the 1960s one of unprecedented uncertainty. Many of the company's most valuable assets lay in less developed countries where political instability exposed them to the risk of nationalization, but even in more stable, industrialized nations a worldwide trend toward statism threatened to restrict the company's control over its property.[9]

Given this realistic outlook—gloomy to some BAC members—Jones opined that while the company could do nothing at the time that would substantially improve its financial performance in the short term, it could and should begin immediately to make the alterations necessary for improving its return on investment. Specifically, as Rathbone had suggested, the company had to re-order fundamentally its approach: all new investments had to be evaluated primarily in terms of their effect on overall earning performance and their impact on the goal of regaining "growth stock" status.[10]

Noting the immense size of the company's existing investment in the petroleum business, Jones made it clear that nothing could alter the company's future commitment to that industry and that some new investments in that area would continue to be needed. Nevertheless, he called for a far more selective approach to investments there, beginning with "a selective abstention from plowing back . . . into accustomed lines of oil investment" the funds generated by the company's oil and gas business. As Jones put it, the company first had to "stop doing unsound things in the oil business."[11] For example, to him, increasing the number of service stations to market an ever-increasing number of barrels of oil without first analyzing the potential return on such outlets, broadened the investment base without a corresponding increase in profits. He believed that Jersey should not gauge creativity or entrepreneurship by the traditional oil business yardstick. While Jones specified the need to abstain from new investments in finding long-term crude oil supplies, his admonition against expanding capacity without careful consideration of financial prospects applied to all functions. In particular, he advised that domestic investments in the oil and gas business be approached cautiously. With over half the company's net investment already in the United States, any overall improvement in earnings performance would have to begin here. As a further step to safeguard earnings in an uncertain world, Jones also recommended that future investments be concentrated in those areas of the world with more favorable investment climates—North America, Western Europe, Australia, and Japan.[12]

Despite Jones' attention to investment selectivity in the petroleum industry, his "fail-safe" approach called for something more. Although more than half of his general conclusions focused on the prospects for further petroleum investments, clearly the striking element in the Jones Report was its advocacy of diversifying the company's investments. While he explicitly recognized the risks inherent in entering new and unfamiliar fields, Jones saw a possibly greater risk in confining "investment exclusively to a line of activity which may not have as much financial buoyancy as formerly."[13] Selectivity in petroleum investments could improve financial performance most if the funds thus conserved were invested in lines of business with better prospects for growth and monetary returns. However, Jersey's existing large investment base meant that any incremental gains in earnings from diversification could affect overall performance only slowly. As a result, Jones urged that company leaders recognize Jersey's late start and the need to begin steps toward diversification immediately.[14]

The Jones Report introduced diversification more as a strategy than as a program. As a result, it dealt with the subject primarily in terms of its outer limits and the company's general approach to it. Emphasizing both the importance of not being inhibited by Jersey's previous experience in the petroleum industry and the primacy of the performance criterion, Jones urged that no industry or field of investment be precluded from consideration. While he seemed to favor substantial diversification as soon as was feasible, Jones stressed that there should be neither preconceived goals nor limits placed on the process. However, he did not recommend that the company make "portfolio" investments; rather he urged that it participate in the management of its investments, ordinarily seeking 100 percent control of the voting stock where no serious obstacles or disadvantages prevented such a course.[15]

In approaching diversification, Jones called for clear recognition of the fact that Jersey was entering "a new branch of management activity," one which differed from existing patterns of investment planning in two fundamental ways.[16] First, rather than responding to demand for a single commodity, the company would have to survey a broad spectrum of activities and recognize or even create a demand that had not yet been formulated.

Second, responsibility for recognition of, and acting on, these opportunities would rest primarily with the parent company rather than its affiliates. Since improving the performance of the company's oil and gas investments would require the full attention of affiliate managers, diversification had to be the responsibility of the parent company. In addition, only the parent company had the freedom and capacity necessary to range widely beyond the company's existing fields of endeavor.[17] The parent company, Jones suggested, should be open to new ideas, new investments, and be in a position to monitor them.

Hence, entry into a new field of management activity raised a significant problem about how to proceed. Acknowledging that Jersey had a virtually unique "wealth of human talent," Jones qualified the value of that asset by noting that the nature of the company's business had made greater demands on skills of an administrative nature.[18] By contrast, diversification of the company's investments would demand greater levels of entrepreneurship and creative marketing than the company appeared to possess at that time. Nevertheless, he urged that the lack of such skills or those associated with a particular industry not be considered a barrier to entry into any new field. Such limitations could always be overcome through acquisition of managerial talent either through merger or recruitment. In the meantime, Jones urged the company to inventory the managerial talent that it could later use in its diversification effort.[19] While he offered solutions to the problem of the shortage of entrepreneurial talent, Jones recognized that the demand for that talent would always exceed the supply. Since the parent company would be primarily responsible for finding and creating new investment opportunities, he called for the movement from all the affiliates of entrepreneurially oriented managers to Jersey. At the same time, he suggested that such managers be concentrated in those geographical areas with the most favorable investment climates although he recognized that otherwise unfavorable climates might generate worthwhile opportunities for diversification.[20]

Organizationally, Jones predicted that diversification would dictate within the parent company a larger staff to find and develop new investment possibilities at the same time that the number of staff members devoted to the company's petroleum business was being reduced in the interest of lowering operating costs. This larger staff would be necessitated both by the need to survey developments in many fields, including research within and outside the company, and by the new need to generate investment proposals within the parent company itself.[21]

In the meantime, Jones' Investment Analysis Department would "press on with its studies to identify broad fields of endeavor which seem, on general grounds, to warrant closer investigation in the search for investment projects."[22] Assuming that these studies would "in the not too distant future" identify such fields with generally good prospects, Jones suggested turning the process of investigation over to "special reconnaissance groups" for "detailed and integrated studies of the marketing, technological, financial, and other related areas" upon which they would base specific investment proposals or, possibly, a recommendation that the field not be an area for investment. He also raised the possibility that these groups would become the "nucleus of the future management team" if the project were to result in a new, non-petroleum affiliate.[23]

In mid-September, 1960, the Executive Committee received copies of the Jones Report and discussed it briefly. President Rathbone then wrote to members of the Board and the Board Advisory Committee on Investments asking each individual for a written response. Although Rathbone welcomed the inclusion of any relevant comments on the subjects discussed in the report, he provided a format designed to focus attention on those portions of the report most directly pertinent to future policy decisions. He asked each executive to set down his comments, particularly dissenting ones, about the "Implications for Jersey" that Jones had drawn from his assumptions, the "Conclusion on Investment Climate in Foreign Areas," and the "Assumptions and Judgements as to the Nature of the Jersey Organization." In the case of Jones' "Conclusions," Rathbone asked for a ranking in terms of priority for attention. To facilitate prompt discussion of this important document, Rathbone requested that the responses be returned by October 7 so that the Executive Committee could take up the topic during the following week.[24]

Taken collectively, the responses to the Jones Report constitute a remarkable document. As a conscious and explicit statement of where each of these executives stood in 1960, this exercise provided an essential step toward the company's first full statement of its investment policy.

Various executives called the Jones Report "thoughtful," "thought provoking and timely," "impressive," "a major accomplishment," "comprehensive," "scholarly," and even "courageous."[25] Rathbone praised the report, terming it "realistic." While no one found the report acceptable without reservation, even some who raised serious objections to certain points stressed the significance of Jones' accomplishment.[26] Overall, the response might best be characterized as cautious approval.

While the exceptions taken by different individuals do not lend themselves to easy generalization, certain basic patterns did emerge. Few were willing to accept the level of pessimism they perceived in the Jones Report. Several respondents conceded that the 1960s would present greater political risks than the preceding decade, but a number of them took issue with that evaluation, citing, in one instance, a long list of political developments— domestic and foreign—that ranged from the Korean War to the Castro revolution in Cuba. Nevertheless, no one argued with Jones' assessment that improving the company's earnings performance in the coming decade would require extreme care and selectivity in its investment activity.

Jones' proposal of a "fail-safe" approach to investment policy brought responses that clearly indicated the feelings of this body of top executives about the general process and their role in it. While accepting the need for careful evaluation of investment plans, many objected to the concept of a "fail-safe" strategy because it seemed to imply that risks should be minimized. To them, risks always accompanied opportunities. The function of investment planning and evaluation was not to eliminate risk, but to assure that risk was balanced by commensurate potential for profits. To this group, the history of Jersey gave clear testimony to the wisdom of always being willing to take appropriate risks and to the dangers of a too conservative policy.[27] More specifically, they objected to the idea of strictly limiting future investments to those areas of the globe offering the most favorable climate. As several pointed out, investment climate and investment opportunity were far different things. Those areas with the most favorable investment climates were also the places where many other investors were most likely to be competing, while areas with less favorable investment climates might offer greater opportunities, especially considering the resources that might give Jersey an advantage.[28] Certainly, few of these executives would be willing to eliminate from consideration large areas within which very attractive specific opportunities might be found.

While Jones expressly qualified his apparent attempt to categorize major areas of possible investment fields, many of the executives, nevertheless, saw it as an important tendency of the entire report. They emphasized that while general guiding principles had to be established, nothing could replace the judgment of experienced executives. In fact, the same difficult conditions that made the formulation of such a policy statement necessary also made the intangible factors even more essential.[29]

The respondents were nearly unanimous in challenging Jones' assertion that Jersey might be short on entrepreneurial talent. While many admitted that concentration on the petroleum business had limited the full expression of that talent, few agreed that there was a serious deficit. Moreover, many stressed the fact that such talent had been demonstrated in the process of developing various petroleum ventures throughout the world.[30]

Response to the Jones Report ranged over the whole of its content and even beyond, but all the executives recognized and responded to the radical recommendation that lay at the heart of it. They reiterated Jones' firm declaration of the company's primary commitment to the petroleum industry and the vital importance of recognizing that fact and planning investment policy accordingly. With a single exception, all of those who ranked Jones' conclusions according to their priority for action placed the necessity of doing everything possible to improve the return of the company's oil investments first. A number of the executives who took this position insisted adamantly that Jones had badly underestimated the possibilities for profitable investments in the petroleum business during the coming decade. For them, the greater selectivity counseled by Jones, along with the continually expanding demand for petroleum products, would, before the end of the decade, permit improving returns from Jersey's primary business that would substantially enhance overall performance.[31]

Despite such strongly expressed reservations, only a handful of the executives could be described as essentially rejecting the concept of diversification. Among these, the most adamant were Directors Howard Page and Emile Soubry. From the perspective of almost a half century in the oil business, Soubry viewed the petroleum industry as still "one of the great growth industries." While he acknowledged a "temporary respite" in profitability due to overexpansion, he predicted that the company would return "very soon" to the conditions in which it would be "as much as Jersey could do to finance the enormous expenditures necessary to meet its obligations in the oil industry."[32]

While Soubry's faith in the future of the oil business could hardly have been stronger, Page, who later characterized himself as "the only one definitely, positively, absolutely against . . . [diversification]," clearly took the lead in resisting the new policy.[33] However, for him diversification posed a threat in another area, the "dilution of management and analytical talent." His position derived from a vision of the challenge facing the company that differed from that of the Jones Report to "keep [the] eye on the main chance." To Page the difficulties of the coming decade consisted of much more than slim profit margins and diminished earnings; rather, approaching investment policy in those terms would result in a dangerous misallocation of the company's human resources. In his view "new and continually changing conditions" in the foreign oil business had created "really fundamental problems" that even some of the company's top managers did not yet appreciate. The "old formulas" could no longer be relied upon, and the

problems had not even been studied thoroughly. Needless to say, the executive session in which Page presented his views proved to be very stimulating.
Even with all of the company's best talent concentrated on the foreign oil
business, Page warned that Jersey would "have difficulty in coping." Any
move that diverted managerial talent to new, unrelated ventures would, of
necessity, weaken what he saw as the already marginal ability to solve problems in the company's primary business. In a new twist on the acknowledged size of the existing business, Page pointed out that "a small error, due
to dilution of talent to other things, multiplied by our volume. . . . [could]
result in more loss than the profit of a dozen glamour industries." Moreover,
he argued, while Jersey's management might have some advantage over
others in the international oil business, there was no reason to suppose that it
had any superiority in finding and developing new "glamour" fields. In fact,
he pointed out, "the built-in bureaucracy" inevitable in such a large company would make the company fatally slow "in many fast moving, small
new industries" unfamiliar to Jersey management.[34]

To Page, the proper course appeared clear. With Jersey in the oil business
"for better or for worse," management's job was "to make the best of it."
Before the company considered entering new fields to improve its return on
investment, it should concentrate all available resources on preserving the
return from its existing business. Meeting that challenge had far more relevance to the long-range prosperity of Jersey than did the recovery of "growth
stock" status, and the fundamental and continuous changes in the complexity of the problems faced by the managers of international oil companies
would require the total commitment of managerial talent, a resource always
"in short supply."[35]

While Page's response to diversification made the greatest impact, several
other executives were only slightly less outspoken about the priority that
should be given to the company's oil business. Page had urged concentration
on Jersey's foreign interests, but M. A. "Mike" Wright, while pointing out
the importance of investments already made abroad and their tremendous
potential for generating future profits, also noted his "sincere belief" in the
possibility of high returns from carefully selected investments in the domestic petroleum industry. In both cases he stressed the need to consider investments in terms of their effects on the company's integrated operations and
the importance of that factor in choosing investments that would have the
maximum effect on the overall earnings performance. Although Wright
acknowledged the need to have the Investment Analysis Department seek
out possible areas for non-oil investment, he saw diversification as beginning logically with fields directly related to the company's oil business and

moving farther away only when those opportunities had been fully exploited.[36]

Siro Vázquez, Jersey's coordinator for producing activities, echoed Wright's statement of priorities and the opinion that the company had not yet exhausted "the opportunities for safe and profitable investments in the fields of energy and petroleum derivatives and related industries."[37]

Stewart P. Coleman, former head of the Coordination Committee, added a new note to Page's concern over the dilution of management talent. Where Page had stressed the overall ability of Jersey's management to deal with the complexity and rapid change of new conditions and problems, Coleman argued that the Investment Analysis Department itself should play a major role in helping the company improve the performance of its existing business by guiding the establishment of policies that would work toward the restoration of the profitability of the petroleum industry. Once the department had taken care of that primary obligation, Coleman had no objection to the proposed diversification studies.[38]

With the exception of Page and Soubry, all the Jersey executives responding to the Jones Report seemed willing to concede that the company had reached a point where it should look seriously at the possibility of entering new lines of business. Beyond that vague concession and their insistence that the oil and gas business receive first priority on capital and management resources, no consensus could be said to exist. Few, if any, appeared willing to share Jones' opinion that no field should be excluded from consideration. Most of the directors and executives who commented favorably on diversification indicated that any new field should have close ties with Jersey's current business. The positive response of several readers to Jones' mention of the fuel cell exemplified the tendency. Investment in that area offered the possibility of both profitable investment in a new field and an opportunity to increase the outlet for the company's petroleum products.[39]

The Jersey executives who endorsed diversification suggested a variety of areas for investigation, but by far the most enthusiastic and widespread support centered on what Director Wolf A. M. Greeven termed the "most obvious development," expanded investment in the field of chemicals.[40] In the discussion of diversification, chemicals occupied a unique position. It presented more than a tie with the existing business. Part of the current business, chemicals, at the same time, were also a new field. For more than forty years Jersey had been making and selling petrochemicals, and management clearly considered that part of the company's existing business. Within the limits of current operations, even Howard Page accepted that view. In fact, given their traditionally higher rate of return compared to oil investments and the possibility of increasing the outlet for crude oil, such extensions of Jersey's chemical interests promised to play a significant role in improving the overall performance.

But the appeal of chemicals as an area for diversification was not confined to expanding the business along the same lines. Quite apart from Jersey's search for new attractive investment areas, developing trends within the chemical industry exerted pressure on the company to extend the range of its chemical operations. Since World War II, the chemical companies that purchased the basic chemicals produced by Jersey had shown a tendency to integrate backward, filling increasing proportions of their own needs by manufacturing from refinery gases. As H. W. "Bud" Fisher, contact director for chemicals, pointed out, the continued success of the company's chemical business would require expansion into new areas.[41] Thus, the further development of Jersey's chemical business linked naturally with the new need for diversification. Greeven argued for investments that would ultimately have the company marketing branded products to their final consumers.[42] To Board Chairman Leo D. Welch, diversification in chemicals required no more than encouragement; the Jersey Board had only to make the company's chemical management aware of their new willingness to entertain proposals that in the past had been rejected as too far removed from established lines of business.[43]

For some of the strongest proponents of chemicals, the debate over diversification offered more than a chance to press for new investments in that field. It became the occasion for renewed efforts toward a long-sought-after goal. Marion W. Boyer, chairman of the BAC, and George T. Piercy, head of the Coordination and Petroleum Economics Department, joined Fisher in urging that Jersey's chemicals business be given separate management in order to make it more effective. Such a move would also largely remove new chemical investments from the oversight of executives primarily concerned with Jersey's oil business, and would also smooth the path to further diversification without unnecessarily diverting management's attention.[44] With the new initiative in Jersey's investment process, achievement of separate management for chemicals seemed to have a better chance than ever before.

Certainly the chemical field would be central to any diversification effort, but beyond that general proposition, nothing about the extent or direction had yet been agreed upon. The lack of consensus on the specifics extended to the plan of action suggested by Jones. While most of the respondents agreed that the parent company would play a different and larger role in the investment process, many objected to the implication that diversification would be highly centralized. Consistent with their belief that the company possessed considerable entrepreneurial talent, they maintained that affiliate managers constituted a vital worldwide network for identifying new investment possibilities. A cautious approach to organization also marked attitudes toward the role of the Investment Analysis Department. While most executives endorsed the IAD's preliminary investigation of new investment fields and the proposed inventory of executive talent, several noted that the building of a diversification staff—either within or outside the IAD—would be premature. They preferred reliance upon existing affiliate and parent company staff departments, at least until firmer plans for diversification had been established.[45]

In the meantime, the majority seemed to favor a course designed to prepare the way for diversification once a positive decision was made. In addition to the IAD studies and the inventory of executives, several respondents endorsed Jones' recommendation that the company pay closer attention to developments both inside and outside their own research labs as a means of identifying potential investments.[46]

Despite Rathbone's original intention to begin full discussion of the Jones Report early in October, the last written response was not submitted until November 30. Part of this delay derived from the fact that this report contained many ideas about which not all of the Board members were accustomed to thinking, and they demanded time to digest it and study all its ramifications. Ever the driver, Rathbone next instructed his executive assistant, C. H. "Chad" Carpenter, to analyze and catalog the comments as an aid to further discussion when the Executive Committee took up the "complicated and controversial" subject on December 14.[47]

As an introduction to Carpenter's presentation and a means of focusing discussion, Rathbone wrote a memorandum, briefly sketching the major points contained in the Jones Report. Noting the "undesirable outlook" for the oil business and the adverse effects of declining return on investment, Rathbone reduced Jones' recommendations for investment policy to three basic needs:

(a) to become more discriminating and selective in our investments in the oil business both as to area and function; and

(b) to aim at more diversification of investment, by acquiring more knowledge of good return investment opportunities outside the oil industry, and being prepared to invest a certain part of our funds available for investment in such "outside" investment opportunities as they become apparent; and

(c) to organize ourselves from a personnel and organizational standpoint to implement (a) and (b) more effectively.[48]

Carpenter's analysis showed that despite serious reservations in some cases, Jersey's top management agreed in general with "the major assumptions, evaluations, and conclusions [of the Jones Report] . . . as well as the relative order of importance of such conclusions. . . ."[49] Since the next step seemed to be "to synthesize and crystallize these conclusions into a broad Company investment policy," the Committee endorsed Rathbone's recommendation that a task force be established with responsibility in three general areas. First, the group would prepare a broad overall statement of Jersey investment policy for the 1960s against which it would review existing functional budget objective statements to insure that investment in existing lines would be consistent with the overall statement. Paralleling this, the group would also prepare a specific investment statement relating to outside investment policy, tying it to the broad statement of policy. Secondly, the task force's mandate called for recommendations as to appropriate changes in Jersey's organizational pattern and personnel that would help implement the investment policy statements. Finally, the task force was asked for recommendations on the presentation of these policy statements to affiliate managers and the role of affiliates in implementing the policies.[50]

The members of the task force, nominated by Rathbone with the concurrence of the Executive Committee, represented the full range of Jersey's operations both functionally and geographically. Marion W. Boyer, who had Rathbone's complete confidence, became the chairman and represented the BAC and Refining; Howard W. Page, the Middle East and Far East; Cecil L. Burrill, the IAD; M. A. Wright, the United States and Producing; William R. Stott, Europe and Marketing; Michael L. Haider, Latin America, Canada, and Producing; and Emilio G. Collado, Finance.[51] At its first meeting, on January 3, 1961, the special committee appointed David J. Jones its secretary.[52]

To facilitate its work, the committee divided itself into two sub-groups. The first, headed by Burrill and including Collado and Wright, undertook the drafting of a statement of investment policy and objectives. The second, chaired by Haider with Page and Stott as members, addressed the problems of organization posed by the new departure in policy, specifically in the areas of research, chemicals, and diversification. As head of the larger task force, Boyer attended most of the meetings of both parts and chaired the sessions of the full group.[53]

The special committee's original mandate called for it to report in the spring of 1961, but its work actually took much longer. In part, this resulted from other demands upon the time of the executives involved. Heavy travel schedules made meetings of the full task force difficult to arrange. In addition, because of the nature of the issues involved, the development of consensual policy statements required repeated and protracted discussions and many redrafts. The difficulties lay first in striking a balance between "undue generalization and unwise stress on particularities" and second in the choice of language to convey policy recommendations in a controversial area. Boyer willingly traded time to obtain a more coherent and cohesive report. Finally, at the end of September, 1961, he submitted the first fruits of the subcommittee's work.[54]

Entitled "Statements of Investment Policy," the report began where the Jones Report left off. The initial section announcing the "basic purposes" of the policy statements, made clear the impact of Jones' "Basic Assumptions" by stating that the investment outlook indicated a change from the investment policy followed by Jersey since 1945. Although more optimistic in some respects than Jones, Boyer's group generally accepted the same grim view of the future of the petroleum business. Proceeding from that assumption, the new report concentrated on establishing more specific policy recommendations to supplant the broad implications Jones had drawn from his analysis of future trends.[55]

Reflecting a sentiment evident in a number of the responses to the Jones Report, the introduction to the "Statements of Investment Policy" featured reaffirmation of both the freedom and the responsibility of the company executives to use their judgment and initiative in solving investment problems. The limitations of the policies received further expression in two statements stressing the need for continual reappraisal to keep the guidelines up-to-date.[56] Yet, despite this tendency to qualify carefully the applicability of the policy statements, the report made clear that, within the limitations set, the guidelines represented the Jersey Board's determination to direct affiliate investments along a path supporting a sharply defined strategy geared to overall corporate needs.

That strategy represented a combination of the ideas contained in the Jones Report, the response to it, and the efforts of the task force to shape the two into a workable policy that would balance investments and returns without sacrificing essential flexibility. Since the ultimate goal of the policy was "a position of unquestioned investment excellence" in the eyes of the shareholders and the investment community, the strategy called for investments promising higher than average returns with an emphasis on projects likely to yield those returns promptly.[57]

Higher returns, of course, implied greater risks. As the responses to the Jones Report affirmed, a "fail-safe" approach did not seem appropriate to Jersey's needs, and the new statement of policy stressed the need to insure that increased risks were justified by the promise of commensurate earnings. The effort to improve such performance would have to be accompanied by increased selectivity in the choice of investments, more definite performance standards, and fuller and more comprehensive evaluation of proposals.

Emphasis on the principle of balancing risk against return also indicated the determination to build flexibility into the new investment policy. Like the Jones Report, the task force statement declared that opportunities for higher returns could be found most readily in the developed parts of the world, North America, Western Europe, Australia, and Japan, although the new report reflected some skepticism about opportunities for foreign investors in Japan. At the same time, the policy recognized that higher projected returns would justify individual projects in areas having greater political and economic risks. Similarly, sufficiently high deferred returns could warrant compromising the preference for prompt returns. The desire for flexibility was also reflected in the policy's stipulation that political considerations, government regulations, or safety requirements could demand investments that would not ordinarily be considered if judged solely on the basis of their rate of return. Effect on overall integrated operations and the need to defend existing investments also qualified as valid reasons for making exceptions to the general emphasis on higher returns from new investments.

The policy statement's specific recommendations for Jersey's petroleum investments illustrated both the double emphasis on selectivity and flexibility. Conceding that the petroleum oversupply would continue to retard earnings for some years, the task force nevertheless projected "steady and substantial growth" in demand which, when combined with rigid cost controls and selective investment, would create opportunities for improved earnings performance. In marketing, refining, and transportation, the report urged emphasis on investments that would reduce operating costs and enhance overall performance and profits on an integrated basis. The task force also recommended attention to marketing investments aimed at finding and developing new uses for oil that would increase total demand. For the refining function, the guidelines acknowledged that other considerations had to be taken into account, but urged concentration on projects to cut costs, improve the performance of the integrated enterprise, or expand capacity in areas promising greater returns such as specialty products. Given the long-range outlook, the policy statement warned against expansion of capacity in advance of actual need. The special role of transportation in Jersey's integrated operations made a policy of focusing on overall effect especially pertinent to that function's investment decisions. While then ample worldwide crude tanker capacity seemed to preclude capital investment in that field, a different situation and the scale of Jersey's operations suggested investment in the transportation of specialty products, chemicals and gases, and expansion of the company's pipeline holdings.

The producing function presented a somewhat different problem from the rest of the petroleum business. The general condition of overcapacity certainly existed, but the implications for investment policy differed. Investments there involved greater uncertainty and, generally, required a longer time lag between the initial investment and profits. In addition, producing had less geographic flexibility. Nevertheless, it could make a contribution toward the objective of improved return by cost reductions and by concentrating a greater proportion of the investment funds on the "rapid exploitation" of existing properties.[58] New reserves would also be needed, but the policy statement urged selectivity with emphasis on areas accessible to markets and those that offered prospects for fairly rapid development. At the same time, the task force called for diversification of sources to reduce political risks. The probability of an extended period of oversupply of oil from conventional sources indicated little opportunity for development of tar sands or shale, and the policy statement simply called for keeping abreast of developments in those fields.

Although by far the greater part of "Statements of Investment Policy" centered on Jersey's petroleum business, the critical issue remained the question of diversification. Not surprisingly, the new report introduced the question of expanding the company's "field of enterprise" with a firm statement of commitment to the oil business as a field for both operations and investment:

> Jersey will continue to have its main business in the integrated, international petroleum field for the foreseeable future, and will mainly depend on additional petroleum investments and improved operations to raise its rate of return during the next few years.

Nevertheless, when the report asked "Is diversification required?" its answer was clearly in the affirmative.[59]

Recognizing that "an almost exclusive dependence upon one commodity" made the company vulnerable in light of an unsettled future, the task force recommended diversification on the basis of two benefits. First, investment in other fields offered "security of income" to reduce the risks of concentration in a single industry, in this case a very volatile one. Second, as Jones had stressed, investment in new lines could contribute to improving overall earnings performance. While recommending "no predetermined amount" of diversification, the report envisioned "new fields," ultimately contributing "an appreciable part" of the company's total revenues.[60] Realistically the company would have to rely on oil and gas until the 1970s after which time diversification would become profitable.

Despite recognition that investment choices turned on subjective evaluations, the policy statement stressed the importance of clearly established standards of estimated return which could be weighed against other investments in the same or different fields. This emphasis on opportunity costs was reflected also in two other guidelines. With overall return a prime concern, selective disinvestment became "an essential aspect of investment management" since it facilitated the allocation of investment funds to fields promising higher returns. The policy of selectivity received additional reinforcement from the statement's explicit approval of substantial capital accumulations where attractive investments with appropriately high returns were not available.[61]

Given its special position among the company's interests, the field of chemicals had a natural place at the center of any diversification strategy. To the task force the field offered "sufficient potential for profitable diversification to warrant increased effort to expand the manufacture and marketing of chemical products on a worldwide basis." To this initial broad endorsement, the committee quickly added a note of caution that echoed the tone of the entire report. Citing the recent entry of newcomers to the industry and the rapid expansion that had increased both risks and competitive pressure, the policy statement stressed the need for "discrimination and judgement in the selection of projects" with greater attention to "the comprehensive . . . study of consumer markets and technological trends." Nevertheless, the committee foresaw wide opportunities in the chemical field that justified a more aggressive approach.[62]

The task force's report established two investment objectives to guide Jersey's diversification in the chemicals field. First, it urged that the company "increase the rate of investment in the manufacturing and marketing of chemical products to the extent justified by attractive indicated project returns. . . ." Second, it proposed that an increasing proportion of such investment be "directed toward the production and sale of finished chemicals and consumer products" where economic considerations and the security of outlets for more basic chemicals warranted such investments. This emphasis on forward integration received further reinforcement from a recommendation that new or expanded investment in raw materials production be evaluated in terms of the prospects for adding more advanced intermediates, finished chemicals, or fabricated products. As a means of gaining knowhow and personnel to facilitate forward integration, the report also suggested acquisition of established producers or sales outlets.[63]

While the "Statements of Investment Policy" envisioned broadening the range of Jersey's chemical operations to a substantial degree, the strategy also incorporated certain clear limitations. In order to achieve and maintain a "broadly based competitive position," the report urged concentration on a limited number of fields with "a degree of association among themselves and with existing Jersey activities sufficient to provide substantial mutual support." To reduce the risks of entry into new fields as well as to enhance the integration of the entire chemical effort, it also stressed the importance of developing more than one product in any field.[64]

Although recognizing the necessity of judging any specific proposal on its own merits, the task force listed eight fields seemingly "compatible with Jersey's broad investment plans": plastics and resins, synthetic fibers, elastomers, fertilizers and agricultural chemicals, surface active agents (detergents), solvents and surface coatings, lube and fuel additives, and automotive specialties. All were recommended not only because of their relationship with each other and Jersey's existing business, but also because they provided reasonably large volume outlets and substantial potential for growth well-suited to Jersey's overall investment needs.

While the more venturesome investment guidelines applied to chemicals, they also included greater latitude concerning the geographical location of investments. The report recognized that the greatest number of suitable chemical investment opportunities were likely to be found in the industrialized nations, but it also cited India and certain Latin American countries as possible locations. Even in the rest of the world where serious political and economic barriers existed, the task force recognized that increased risks might be justified by "special situations" such as rapidly expanding demand for chemical fertilizers in underdeveloped nations.

Despite its "generally aggressive stance" on chemical expansion, the task force warned against "immature manufacturing investment." In direct contrast to the petroleum business, the greatest element of uncertainty in chemicals concerned markets rather than supply. Thus, successful expansion in the manufacture of chemicals demanded increased attention to market analysis and closer integration of chemical research with the marketing effort. In addition, the task force noted the need for a stronger worldwide supply and marketing organization.[65]

The investment policy guidelines for chemicals endorsed no specific project but did sketch out concretely and in detail the outline of a substantial diversification effort. The company's readiness to go forward in that area could be gauged by comparing the guidelines for chemicals with those for "other fields." The principles governing diversification remained the same in both cases, but where the discussion of chemicals offered specific products as examples, those principles remained abstract when applied to other fields. Although the "Statements of Investment Policy" presented a more detailed set of rules for selecting new fields for investment, they were hardly more specific than the Jones Report in suggesting what those fields might be.

As in the case of chemicals, the report stressed concentration in a limited number of fields with emphasis on better than average returns, potential for growth, and linkage with Jersey's existing business and resources. Even more than in chemicals, entry into a few selected fields seemed necessary for success because of the increased demands placed on senior executives by the novelty of new lines of business. The task force reasoned, moreover, that diversification across a broad range would lead to returns approximating the average for industrial concerns rather than the higher than average goals they sought.[66] Selectivity thus seemed to dictate a small number of fields.

Being selective, however, entailed more than attention to return on investment. Again the task force emphasized the importance of balancing diversification with integration, urging the choice fields that might serve "as growth nuclei for activities branching off from them and with a good chance of linking up with other [company] activities." This meant not only choosing areas in which Jersey could expect to have "particular advantages" but also activities that would improve overall returns by increasing "the utilization of existing skills, resources, and facilities." The latter tactic included not only creating new outlets for oil, gas, and petrochemicals but also taking advantage of the company's financial skills and resources and its technological knowhow in such areas as "large scale continuous processing of bulk materials over land and water, fluidized solids technology, [and] multiple product manufacturing." Finally, increased use of existing facilities encompassed such direct steps as utilizing marketing facilities and exploiting the existing worldwide network of affiliates.[67]

Among the existing resources most crucial to diversification was Jersey's substantial research program. The Boyer task force's "Statements of Investment Policy" concluded with a set of recommendations addressing the research needs of the company not only in its diversification effort but in its overall drive to improve its return on invested capital. In essence the new policy required both an increase in the research effort and a sharpening of its focus. Diversification obviously dictated a research program extending beyond the bounds of the company's existing interests, and the report called for greater awareness of, and interest in, diversification on the part of Jersey's research arm. Consistent with the rest of the policy statement, however, the guidelines for research stressed the need to proceed from more familiar lines of business, specifying the application of "basic research and creative engineering" in the areas of new energy sources, energy converters, and chemicals. But the proposed extension of the research effort went much farther. The task force called for expanded research in fields beyond Jersey's existing business. The breadth of the new investment effort also demanded increased attention to research developments outside the company's own laboratories, a worldwide effort to detect developments that might present diversified investment opportunities.[68]

If extending the research program in this way were not to absorb inordinate amounts of its resources, the company would need to concentrate at the same time on increasing the effectiveness of the effort. Thus, while insisting on sufficient support for petroleum research "to insure profitable competitive advantage," the task force called for increased emphasis on fields closer to the diversification effort.[69]

The drive to improve earnings performance, however, applied to new as well as existing fields of endeavor, as did the goal of more rapid commercialization of new research breakthroughs. To achieve this, the task force urged increased "emphasis on marketing orientation in research and engineering aspects in product fabrication."[70] That meant a shift toward concerns that formerly received limited emphasis by Jersey researchers: broad economic analysis, product application research, and market evaluation.

The new priority given to coordinating the company's research effort with its larger objectives, and especially its investment program, required changes in structure as well as emphasis. The necessity of closer liaison with the marketing and planning activities of both the parent company and the affiliates led the task force to recommend the establishment of additional foreign task force centers closer to those affiliates and markets, in order to supplement the work of Esso Research labs at Abingdon in England. Finally, to support the new role of research, the report endorsed the establishment of a Research Coordination Department in the parent company.

The draft "Statements of Investment Policy" submitted by Boyer in September, 1961, represented a consensus among all of the task force members except Howard Page, who maintained strong reservations that he preferred to present to the Executive Committee when it considered the policy statement.[71] Despite the general agreement on principles, however, the Executive Committee took more than a year to establish the final language of the policy statement and adopt it. The delay reflected both the difficulty of stating the new policy in terms appropriate to guide the decentralized company into new fields of investment and the problem of providing an organizational basis for the successful implementation of the policy. In both cases the length of the process indicated not so much indecision or extended intensive debate but rather the recognition that such a new departure demanded careful consideration.

For the most part the portions of the policy statement relating to the company's existing business required little discussion. In fact, diversification in the field of chemicals received endorsement in a unique form. To expedite the expanded investment in the field that had already begun, the Executive Committee authorized the drafting of a separate "Statement of Policy for Chemical Activities" to guide expansion in the field while the overall "Statements of Investment Policy" remained pending. Even before the Executive Committee took up the question of general diversification, it approved on March 23, 1962, a policy statement clearly designed to push Jersey forward in the field of chemicals. Although deferring "a more detailed outline of investment objectives" until the "Statements of Investment Policy" was approved, the chemical policy statement made clear the Board's determination "to intensify [Jersey's] . . . chemical effort by broadening its markets, expanding its range of products, and integrating further forward toward finished chemicals." Recognizing the demands of this new approach on the company's management, the policy statement pointed out that it had now become "necessary to regard chemical operations as a separate field of business from petroleum, and to place it in the charge of full-time specialized chemical managements. . . ."[72] For the parent company this meant the designation of a vice president to administer chemicals and the creation of a Chemical Coordination Department which would take over much of the parent company's responsibility for chemicals, including those previously held by Marketing Coordination, Refining Coordination, the Regional Coordinators' Offices and all other Jersey departments. In addition, the new department would assume the staff responsibilities for chemicals then held by Esso International Inc. Since separation of chemicals management in the operating affiliates and coordination of operations presented difficult and complex problems, H. W. Fisher, contact director, was elected vice president and was assigned the task of formulating detailed plans for corporate and operating organizations necessary to carry out successfully the new policy. The following year a new affiliate, Esso Chemical Company Inc. replaced the Chemical Coordination Department as the entity responsible for overseeing Jersey's expanding chemical operations. It also became the management unit for that company function.

In April and May 1962, the Executive Committee at last took up the more difficult portions of the "Statements of Investment Policy," those dealing with diversification and research and their organizational requirements. In mid-June, the Executive Committee reviewed its work of the past several months dealing with the investment policy and organizational changes proposed by the Boyer task force. With approval of revised plans of organization for research and diversification, the Committee essentially concluded its work on the organizational aspect of the new investment policy. Still to be considered, however, were further revisions of the "Statements of Investment Policy" and a policy statement covering "Investments in Association With Others."[73]

While the "Statements of Investment Policy" awaited final action, Jersey began implementing the organizational changes designed to tighten the relationship between its research arm and its operations and investment policy. The first step was the appointment of Eger V. Murphree as Jersey vice president, to serve as research policy coordinator. Long the guiding spirit of the company's research program as president of Esso Research and Engineering Company, Murphree continued in that post while taking on his new responsibilities. As research policy coordinator, he chaired a newly created Research Policy Committee which assumed responsibility for coordinating research policy on a worldwide basis. Emphasizing the global span of the committee's function as well as its mandate to tie research to operations, the membership included representatives from foreign and domestic operating affiliates as well as research companies and Jersey staff departments. Since policy coordination had formerly been a function of the Esso Research board, the new structure also called for changes in that body. Now responsible only for the coordination of research operations, the Esso Research board was streamlined and reduced to nine members all of whom were executives of that affiliate.[74] With these changes, the company's research organization was placed in a better position to support the new investment policy. Changes in the organization necessary for the policy of diversification, however, awaited the adoption of the policy statement.

By early 1963, the Jersey Executive Committee had at last approved a revised version of "Statements of Investment Policy" and began making plans for conveying the new policy to affiliate and parent company executives. Finally on June 3-4, the Board revealed in the first meeting ever of its executives from all over the world the investment plans that it had been formulating since the submission of the Jones Report in September of 1960.[75] At that meeting, Rathbone told the executives, "this is it."

Dissemination of the policy statement cleared the way for implementing organizational changes related to the new developments. To carry out its policy of investment in new fields, the Jersey Board had approved the creation of a New Investments Department in the parent company, and in September it began its work under the leadership of Donald O. Swan. Within a few months the staff department was supplemented by the creation of a new affiliate, Jersey Enterprises, Inc., which would operate the parent company's diversified ventures. At the same time as the establishment of the New Investments Department, the company also consolidated its investment planning functions. While the General Economics Department remained separate, the company moved toward better coordination of its investment policy by establishing a new Coordination and Planning Department, which merged the functions previously served by the Coordination and Petroleum Economics Department with those of the Investment Analysis Department. After studying the possibility of setting up a separate Long Range Planning Department, the Board apparently decided to include such functions under the new department as well. Jones became the first head of the new comprehensive planning unit. These changes at the staff level soon allowed a realignment of responsibilities among top management. In November, 1963, the Board approved a proposal to shift to the BAC and its staff much greater responsibility for affiliate capital budgets, raising the upper limit of the BAC's discretionary authority over budget additions from one million to five million dollars. At the same time, the Board moved to strengthen the policy-making functions of the Executive Committee by shifting more responsibility for overseeing operations to the Board and the contact directors.[76]

By the time the organization for diversification was put into place, the company had already taken important steps in some areas toward implementing the policy. The diversification begun in the late 1950s in chemicals continued. Forward integration was taking place in a number of products, but was symbolized by the company's entry into the manufacture of plastics. From its first commercial operations in polypropylene, the company ventured into the production of plastic film, plastic laminate for counter tops, and plastic fibers. Although these activities were later curtailed, they indicated the early commitment to forward integration as a means of diversification in chemicals.[77] The other major effort of the chemicals diversification drive of the early 1960s centered on the production of chemical fertilizers. In 1960 and 1961, the company entered into four joint ventures in fertilizer manufacture in Central America and the Caribbean area. In the latter years the company also made its first commitments for fertilizer investments outside that area. By 1965, Jersey had attained sole ownership of the Caribbean facilities and added a plant in Jamaica. In addition, Esso Chemical had new fertilizer manufacturing plants in the planning or construction stage in the Philippines, Malaysia, Pakistan, Lebanon, Greece, Spain, and the Netherlands.[78] In Canada, Imperial Oil Limited was also selling fertilizers as a preliminary step toward manufacturing them.

The discussion of diversification also coincided with a revitalizing effort to find new uses for petroleum as a way of improving profits in the company's main line of business. Since the creation of the Board Advisory Committee on Investments, it had a subcommittee which had taken over the work of the New Petroleum Uses Committee, established in 1956. In February, 1961, the Executive Committee approved organizational changes designed to improve and intensify the effectiveness of the effort by assigning a full-time executive to the project and placing him at the head of a New Uses Advisory Committee, which would replace the BAC subcommittee. While the new group would include some of the subcommittee members, it would be composed primarily of executives from affiliates and parent company staff departments including Humble Oil & Refining Company and its Enjay Chemical Company division, Esso Research, the Coordination and Petroleum Economics, Investment Analysis, and Marketing Coordination Departments.[79] New Uses Activities remained closely allied with the company's research effort, although from 1963 until 1965, it had separate contact directors. In the latter year Esso Research resumed primary responsibility for coordination of all of Jersey's research activities and absorbed New Uses Activities. This latter committee had covered a broad range of possibilities, but much of its work in the early 1960s centered on three areas: developing the technology for the use of oil in the manufacture of iron and other metals, investigating possible uses of oil in agriculture either as a basis for mulch or as a stabilizer of shifting sand in arid areas, and the use of petroleum in building blocks, although the latter was soon dropped.[80]

While chemicals remained the primary area of new investments and the drive to develop new uses for petroleum represented a significant effort to apply Jersey's research capabilities to the task of earnings improvement, much attention also focused on diversification beyond the scope of the existing business. Of course, even here the activities of the New Investment department meshed closely with those of Esso Research and the New Uses effort, and a clear distinction could not be made in some cases. In fact, the first investment of Jersey Enterprises involved just such a project. By the time the New Investments Department came into being, Esso Research had proceeded far enough in developing a process called Fluidized Iron Ore

Reduction (FIOR) to justify the construction of a pilot plant. With Imperial Oil Limited's refinery at Dartmouth, Nova Scotia, chosen as the site for the test unit, the parent company Board created Jersey Enterprises as a vehicle to finance this plant and the further development and commercialization of the process. During 1964 and 1965, Jersey studied investment in full-scale FIOR plants in India and Venezuela.[81] The link between the New Investments Department and Esso Research could be seen in another notable departure from Jersey's usual lines of business. In 1965, Jersey Enterprises entered into a joint venture with Nestlé Alimentana, S.A. to develop further an Esso Research process for the production of protein-rich food supplements from hydrocarbon-based feedstock and, if successful, to manufacture and market such products.[82]

Jersey Enterprises, however, did not remain merely an operating company for exploiting Esso Research technology. The New Investments Department wasted no time in putting the affiliate to use in its own diversified investment. On January 30, 1964, the parent company Board endorsed Jersey Enterprises' acquisition of American Cryogenics, Inc., a corporation "engaged in the production of cryogenic gases and related equipment." Although the business failed to expand as anticipated and Jersey subsequently disposed of the company in 1968,[83] it served to indicate the rapid implementation of the diversification policy.

As the Jones Report had anticipated, the diversification effort had to be centered in the parent company, but Jersey's decentralized structure also dictated an essential role for the operating affiliates as a network for both locating investment opportunities and for exploiting them. Among the operating companies Imperial Oil Limited moved most quickly. In fact, the Canadian affiliate had accommodated its organizational structure to the possibility of diversification even before the unveiling of Jersey's "Statements of Investment Policy."[84] By the time Jersey announced the new policy, its Canadian affiliate had already begun preliminary investigation into what would be its first field for diversified investment other than chemicals.

In 1964, Imperial confirmed its earlier judgment by forming Building Products of Canada Limited, a wholly owned affiliate which acquired the assets of a Canadian corporation engaged in manufacturing and marketing a broad range of items based primarily on forest raw materials.[85] At the same time, Jersey's New Investment Department undertook studies concerning investment possibilities in the field of forest products.[86]

As these examples indicate, the process of diversification in its early years amalgamated the efforts of staff and operating organizations throughout the Jersey enterprise—the parent company's New Investment Department and New Uses Activities, Esso Research, Esso Chemicals, and the operating affiliates. The new effort got off to so swift a start in several areas, that many of the early diversification ventures failed to develop as anticipated. With one major exception these early and varied efforts had little, if any, lasting effect on the company's operations. However, their history did illustrate that even after the most careful deliberations some lessons could only be learned through experience.

The single operating area in which the diversified investments of the early 1960s had an important and continuing effect was, of course, chemicals. Although the first rush of enthusiasm led to some major mistakes here as in other fields, there can be no doubt that the principal immediate effect of the first round of diversification was the impetus given to expanding and broadening Jersey's investment in chemicals. In fact, the two movements seem inseparable. Developments in Jersey's chemical business, combined with conditions in the oil industry, clearly had a part in Rathbone's decision to reappraise the whole investment process. The resulting policy, in turn, threw new weight behind the expansion of Jersey's chemical business, resulting in larger investments, investment in new fields, and separate management. Jersey's chemical business had been growing rapidly since the mid-1950s, but its emergence as a major element in the chemical industry can be clearly dated from this period and the diversification drive.

However, the significance of the developments that began with the formation of the Board Advisory Committee on Investments and culminated in the "Statements of Investment Policy" goes far beyond the effects on one field and even beyond the issue of diversifying Jersey's investment base. Diversification had broad and lasting implications, but they were only part of the significance of the reevaluation of Jersey's investment function—and not the major part. The adoption of "Statements of Investment Policy" had a much more fundamental import. It marked the first time the company had ever set down in a comprehensive formal document the guidelines to be followed in its future developments. As such, it represented a watershed in the company history because the policies adopted and the very act of issuing them in a policy statement altered the investing function and with it the relationship between the parent company and its affiliates. Under Rathbone's leadership the Board made these decisions on procedures to be followed in diversifying, ever aware of the continuous and cautionary admonitions of the Jersey legal department that purchases of companies in any industry most likely would be challenged by either or both the Federal Trade Commission or the U.S. Department of Justice.

Thus, this reappraisal of Jersey's investing process in the early 1960s brought the function to a new level of consciousness. After that examination and the adoption of an investment policy statement, the process of investment analysis took place in a different framework, one of clear-cut policies with goals stated rather than assumed. The end result was to promote a much more sophisticated analytical approach to the investment function. The new emphasis on selectivity in choosing investments in existing as well as new fields amplified the role of the parent company in the investment process. While this affected affiliates as well as the parent company, the most significant changes involved the latter where the BAC and the Coordination and Planning Department were at the center of the new developments. The later organizational changes made in November, 1963, were recognition of the improvements in investment analysis instituted by the BAC. Further revisions in the budget review process and the adoption of corporate and functional investment objectives in 1965 showed the evolution of that process to be a continuing effect of the company's new approach to its investment function.[87]

The changes arising from greater selectivity were intensified by the addition of another element—diversification. Without diminishing the priority of oil operations in the company's business, the decision to diversify Jersey's investment base caused a fundamental change in perspective that enhanced the effects of selectivity on the evolving functions of the parent company. Diversification implied a competition for capital resources that helped shape a new role for Jersey. Through the 1950s the company had remained the administrator of an oil empire; its function with respect to investments had been that of supplying necessary capital to finance the expansion of its affiliates in established lines of business. While for some years these actions of the Board represented changes in atmosphere rather than in substance, from the beginning they signalled all Jersey affiliates that the parent company would insist on financial results and that their activities would be searchingly reviewed. In essence Jersey announced that the Board of Directors represented the court of last appeal [non-employee members were added] and that while the policy of decentralization would continue to be in effect throughout the company, all affiliates would be held strictly accountable to the parent company. As a result, for the Jersey Board investment policy formation and investment analysis became its central functions in guiding the development of the integrated enterprise.

The evolution of the structure and functions of Exxon proved to be a continuous response to ever-changing conditions, many unanticipated by company executives and planners. In some cases, Exxon never implemented carefully planned alterations because new factors quickly made old plans obsolete. One example may be found in the proposed investment strategy of the 1960s. Based on the proposition that "the company's cash generation was going to be so large . . . [that] there would *not* be sufficient opportunities to invest in the *OIL* industry," the program hammered out in 1960-1963, did not become operative when that cash surplus did not appear. Now, it is easy to understand why the company did not pursue new diversification plans, in the face of such alternate investment opportunities as those appearing in Prudhoe Bay, the North Sea and the whole synthetic fuel industry. None of these could have been foreseen by Rathbone, Boyer, Jones, or other company executives. Hence, no major shift in investment policy occurred as a result of the study growing out of the Jones Report. What diversification did occur "represented a very small percentage of Exxon's capital investment program."[88]

Yet, as a discrete episode in the history of Exxon, this study of a new approach to investment policy was valuable and important. If the specific results had little economic impact, the exercise itself forced a reluctant review of management patterns in the postwar world; and, in the form of intellectual strength, gave the Board a more flexible approach to using the great powers they would need to deal effectively with the problems of the future.

Chapter 3

1960 Reorganization

"I'VE BEEN IN THE OIL BUSINESS for 38 years," said Monroe J. "Jack" Rathbone, the Standard Oil Company (New Jersey) [Jersey]'s new chief executive officer [CEO] in 1960, "and I've never seen anything like this. We've had oversupply before; we've had competition; we've had mean political situations—but never such a combination."[1] When Rathbone made his observation, the international petroleum industry faced a troubled future. The unprecedented array of economic and political problems afflicting it could not be easily resolved. Jersey, as the largest global petroleum company, experienced the full force of these adverse conditions. Though the 1950s had not been the most profitable years in industry history, the general expansion of the business during those years was spectacular. But now after a brief recession in 1958, both the company and the entire industry seemed to confront a very different set of circumstances.

Well might Rathbone express his puzzlement at the combined political-oil situation. In 1953, President Harry S. Truman had been succeeded by Dwight D. Eisenhower, and the relatively calm Secretary of State, Dean Acheson, by the more loquacious John Foster Dulles. But the Cold War and containment policies continued despite changes in the style with which they were applied. The death of Russian Premier, Joseph Stalin, in 1953, had served to muddy the diplomatic waters: on the one hand the Russians mounted a peace offensive while on the other their surrogates in North Korea, Vietnam, Iran and elsewhere actively propagandized and, in some cases, fought. Eisenhower confronted these and other problems. For the United States and Western powers, the armistice in Korea cooled one trouble spot on the globe, while problems in Southeast Asia bid fair to threaten world peace even more.

The settlement of the Iranian-British oil controversy, satisfactory to the West, did restore oil from the former country to world markets while in Iraq, anti-Western revolutionaries seized control of the government and threatened the oil supply from that source. The Iranian Petroleum Law of 1957, opened that country to other than consortium members; and Jersey and its consortium partners felt threatened by the actions there. Changes in the laws concerning participation in Iraq caused Jersey and other members of the Iraq Petroleum Company [IPC] to question the security of that area. Fidel Castro seized control of Cuba and immediately revealed his Russian connections. He nationalized the properties of foreign corporations [Jersey valued its loss at $72 million]. The spread of anti-American feeling and international communism led to outbreaks of mobs which jeopardized the life of Vice President Richard M. Nixon while he toured South American countries and caused President Eisenhower to cancel an anticipated and much desired visit to Japan. A new bloc appeared on the world power stage: that of the neutralists, which served to create problems for the two super powers, the United States and Russia. The neutralists consisted largely of what Westerners termed "emerging nations."

Venezuela, where Jersey's Creole Petroleum Corporation [Creole] had year after year returned handsome profits, raised taxes on oil from 50 percent of net profits (including royalties) to 60 percent. This action placed oil from that country at a disadvantage in competition with Arabian crude and slowed down sales. Libya came in, though, with a fine light crude oil much in demand and which could be shipped to Western Europe for about one-half the cost of shipping Persian Gulf oil to the same place. The U.S. applied import quotas on foreign oil, thereby limiting the domestic market.

For Jersey and other oil companies with foreign markets, sales volume increased while profits decreased. The plentiful supply of oil from the Middle East and Libya continued to depress prices while only the major oil companies sought to control production. Many companies began to consider placing limitations on their exploration budgets; Jersey did not! Adding to the confused oil marketing picture, Russia began to export oil to the West, at prices often lower than those in existing markets. Certainly the environment in which international oil companies operated changed drastically during the decade.

The Jersey Company's ability to adapt to a changing business and political environment had already been demonstrated through its survival for more than three quarters of a century, during which it experienced almost uninterrupted success. Yet, because of the company's size and complexity, changes in the plan or style of management usually came only gradually and piecemeal. The long-term consequences of any drastic alteration in so large and intricate a structure had never been fully explored; and fundamental changes ordinarily took place only after all other alternatives had been exhausted. Even in the period after World War II, when conditions changed radically, Jersey's basic structure remained largely unaffected. Only in the last half of the 1950s, when serious dislocations became inevitable, did Jersey's management begin to restructure some of the company's vital components.

The motive for reorganization arose from a shared realization that Jersey by adapting to changing conditions without altering its existing organization had produced problems for its management. After 1927, when President Walter C. Teagle first made the decision that Jersey should be a holding company only, the administration of the company's vast properties had been predicated upon the ideal of decentralized operating decisions and centralized policy-making and overall planning—in the company's terminology, a proper balance between the sacred principles of decentralization and coordination. The intense competition and more rapid pace of the petroleum business in the late 1950s made decentralized operating authority more important than ever, but the complexity of new problems called for a broader perspective than that available to any individual affiliate. Nor could corporate management, working in the existing coordinate structure, always supply the necessary breadth of view. With their functional orientation, Jersey's staff departments had no way of factoring into the decision-making process the non-technical considerations that had become critical to an increasing number of business decisions. Only the Board of Directors and the Executive Committee had overall responsibility, or full capacity, to view the company's business comprehensively. As a result, more and more problems were moving up the administrative pyramid to the Board for resolution.

As this tendency increased, a twofold danger could be foreseen. One, the increasing number of managers assigned to affiliate executive committees slowed the decision-making process. This altered the ability of the operating affiliates, and of the company as a whole, to quickly and effectively resolve urgent problems. Two, the burden of coordinating affiliate actions hindered the company's top management in executing their responsibilities especially in the area of policy-making and overall planning. Now, it was clear that to ignore these crucial functions would be as damaging to Jersey as the failure to coordinate the many individual solutions to short-term problems. Historically, these entrepreneurial functions had ever been the special responsibility of Jersey's Board members. They could not be delegated.

By the end of the 1950s, it had become obvious that improvement of Jersey's coordinating and integrating functions could no longer be postponed. Responsibility for operations had to be re-allocated so that top executives would be relieved of the routine duties that threatened to absorb all of their time.[2] In the United States, Jersey first responded by unifying its domestic operations in a new Humble Oil & Refining Company [Humble] (incorporated in Delaware). Beginning late in 1959, a series of mergers among the company's existing affiliates quickly created a corporate entity that was better designed to meet competitive conditions.

The desirability of unifying domestic operations had long been apparent to Jersey leaders. Close coordination of its affiliates in the United States, however, had always been stymied by the fierce independence of Jersey's principal domestic producing affiliate, the Humble Oil & Refining Company incorporated in Texas. From the time of its first investment in that company in 1919, Jersey had owned a majority of Humble's stock. By 1954 it held almost 88 percent of the stock.[3] Yet, Humble maintained a nearly complete autonomy.

In all cases where Jersey, as a holding company, owned less than 100 percent of the stock in an affiliate, it exercised control only through voting such stock at the annual meeting; but Jersey directly controlled wholly owned (100 percent) affiliates. Generally, the company used its stock to elect the directors (if any) of affiliates. However, where an affiliate operated well and responsibly, the Standard Oil Company (New Jersey) rarely interfered. This proved to be the case with Humble.

No formal written understanding existed between Jersey and Humble regarding the manner and extent of the control to be exercised by the former after it acquired a majority stock interest in the latter company. This policy existed from the beginning of the business relationship between the two, and it was based on two conditions: First, Humble possessed a group of talented, aggressive, and "successful young business men in whom the Jersey people had confidence and who they felt . . . with freedom of action . . . could accomplish more than if controlled from New York. The other reason was a legal one."

This legal arrangement actually had roots deep in the history of the oil industry in Texas. Beginning in the late 1890s, the Texas legislature passed a series of antitrust laws "even more severe than the Sherman Act." Among the principal targets of these Texas acts, the Standard Oil Company (New Jersey) held a prominent place, and action against the affiliates of the unpopular trust came quickly.[4] A suit filed by the even more disliked Waters-Pierce Oil Company in 1898[5] resulted in that company being perpetually enjoined from operating in Texas on the grounds that it had been part of the Standard Oil Trust. A newly organized (1900) Waters-Pierce Oil Company obtained a permit to do business in Texas, but again in 1906, the state sued, with the result that the Jersey Company paid a $1.6 million fine and Waters-Pierce forfeited its charter. Using individual representatives, the Standard Oil Company again attempted to conduct business in Texas by creating the Corsicana Refining Company and acquiring two other local oil companies. However, the state once again brought suit and forced the sale of these properties in 1909.[6]

About a year after the court-ordered sale of Corsicana and other companies, the properties merged into the new Magnolia Petroleum Company, 85 percent owned by John D. Archbold and Henry C. Folger, Jr., directors of Standard. Again the state sued, charging as defendants—in addition to the Magnolia Company, Archbold and Folger—the Jersey Company itself. The judgment, rendered in 1913, found the Standard Oil Company (New Jersey) guilty of violating the state's antitrust laws prior to 1909, and fined the company $500,000. Declaring this penalty to satisfy fully the state's demands relative to that period, the court further specified that Jersey had not violated the antitrust laws between 1909 and 1913. Despite this "clean bill of health," the company withdrew from Texas until 1919, when it bought over 50 percent of Humble's stock.[7]

When Jersey made this purchase, the Texas company's attorneys stressed the need for it to "exercise no further control over Humble than was involved in voting its stock at stockholders' meetings." Despite rigid adherence to that policy, the Attorney General of Texas filed an antitrust suit against Humble in 1923, seeking penalties that included a permanent injunction against Jersey doing business in Texas. The Attorney General charged that Jersey, "through its ownership of some 60 percent . . . of the stock of Humble, was doing business in Texas without a permit and was operating . . . in violation of the anti-trust laws of Texas." He further alleged that Jersey, having been convicted of violating these laws, could not do business there, except by complying with certain statutory procedures. Humble answered the allegation that it had been Jersey's agent in Texas by flatly declaring that Jersey's influence on its operations had been limited to the voting of its shares and that such action did not constitute control as defined by the antitrust laws. The court's decision in Humble's favor referred particularly to Jersey's practice of having Humble's regular proxy committee vote its stock, and to the fact that despite Jersey's equity interest in Humble, the latter company "continued to have substantially the same officers and directors, practically all Texans." The decision further declared that "These officers . . . [were] in complete control and management of the business . . . [and that] the testimony . . . [did] not disclose any effort on the part of . . . [Jersey] to exercise, control or manage appellee [Humble].' "[8]

The 1923 case solidified Humble's independence by demonstrating both the continuing danger of antitrust action against Jersey in Texas and the effectiveness of Humble's autonomy as a defense. State officials made one other attempt (1931) to prosecute Humble and Jersey under the antitrust laws, but the relationship between Jersey and Humble was not fully explored, due to a settlement of the case before trial. Still, the threat of political attack against the Standard Oil Company remained a constant influence on the Jersey-Humble relationship. By the mid-1950s, Humble's management considered its relationship with the public to be as satisfactory as at any earlier time in its history, and it attributed the favorable situation to its ability "to live down such odium as may have existed in the public mind against Standard."[9] Meanwhile, Humble's resistance to Jersey's control remained undiminished.

True, Humble recognized a "community of interest" with Jersey and tried to consider that company's overall operations in its decision-making,[10] and it always stood somewhat aloof from the functional coordination and capital budget review processes of the holding company; operating more like a coordinate than a subsidiary company, its relationship with Jersey remained confined to discussions and advice at the top executive level.

The importance of Humble in Jersey's domestic operations and the success of its management as well as the specter of the Texas antitrust laws had always suppressed any desire of the parent company to bring the affiliate under closer control. However, around 1955, pressure for such a change began to increase. Humble's advocacy of oil import quotas directly contravened the overall interests and the public stance of the parent company, although it may have served to highlight Jersey's lack of control of its affiliate. More importantly, as domestic competition tightened and other companies began to streamline their organizations,[11] Jersey's leaders saw as essential some basic change in the relationship that would help them maintain their share of the domestic market in the years ahead. Such improved coordination would require consolidation.

The achievement of that objective, in turn, meant a more complete exercise of Jersey's control over Humble. Toward that ultimate end and the immediate goal of qualifying to file consolidated tax returns, the parent company made a new effort in 1954 to increase its stock-ownership in the Texas affiliate, offering to exchange nine shares of Jersey stock for ten of Humble. As a result, Jersey's equity interest in the affiliate increased from less than 80 to more than 87 percent.[12] The following year, the Jersey Board made an offer to purchase up to 20,000 additional Humble shares and authorized the treasurer to acquire more stock, up to an aggregate total of 50,000 shares, when attractive opportunities presented themselves. In 1958, under a voluntary exchange offer, Jersey increased its portion of the shares in Humble from 88 percent to 98 percent.[13]

As the economic recession of 1958 further exacerbated the problems of the oil industry, the pressure to make Jersey's domestic operations more efficient increased. Before he retired early in 1957, Jersey's general counsel, Edward F. Johnson, had told the Board that an expanding theory of antitrust law posed a potential, if as yet "unsubstantial" threat to Jersey.[14] The decision of the United States Supreme Court substantiating the Justice Department's contention that E. I. du Pont de Nemours & Company's [Du Pont] ownership of stock in the General Motors Corporation had resulted in restraint of trade, meant that transactions between parent companies and partially owned subsidiaries could be considered as discriminating against outside firms.[15] As Johnson pointed out, the fact that Humble's large reserves of crude oil had been used substantially by Jersey might be construed by the court as depriving non-affiliated companies of the full opportunity of competing for those supplies.[16] His advice to the Board coincided with that from another highly placed attorney, that of Rex G. Baker, Humble's recently retired general counsel. Where Johnson remembered that he saw a potential threat to both Jersey and Humble from the decision in the Du Pont-General Motors case, Baker perceived a real "peril" to both companies. He, therefore, urged Jersey to become sole owner of Humble so that any transactions would be between the holding company and a wholly owned affiliate.[17]

This situation demanded change, but effective corporate reorganization had to be built on action, not speculation. A positive response required a responsible executive who recognized the situation, translated that recognition into a plan, and then convinced other executives of the need to implement it.[18] In this case, Chairman Eugene Holman assumed the role of leader.

As the first step toward gaining complete control of Humble, in 1957, Holman arranged the nomination and election, to the Jersey Board, of Humble's president, Hines H. Baker. Baker became a full-time executive of Jersey, serving as member of the Executive Committee and as vice president. The first such elevation of a Humble director since 1945, Baker's coming to New York automatically strengthened the liaison between Jersey and Humble and opened the way for consideration of a structural change in the corporate relationship. Soon after Baker joined the Board, Holman asked him to begin work on a plan by which the parent company could gain complete ownership and control of Humble. Baker related, "When I went up there, Gene Holman made a statement to me that he hoped I would work out a merger with Humble and later on the thing got up to real basic discussion." Rathbone, too, wanted Baker to undertake this assignment. So Baker made some notes which he drew on for information in the Board and Executive Committee meetings. Baker said:

> I set out to do it in a way that would accomplish wholesome relations between the Humble people and the Jersey people for all times. We wanted to be fair to Jersey and at the same time be fair to the Humble stockholders. . . . The ownership at the time that I was up in New York and at the time of the exchange finally in 1959, was something above 98 percent. . . . That [2 percent] prevented the absolute control of Humble by Jersey and involved all kinds of legal problems and the inability to operate the holding.

About that time the Supreme Court ruled in the DuPont-General Motors case and Rex Baker went to the Jersey executives expressing the necessity of getting 100 percent ownership of Humble. Hines Baker then devised the inter-corporate stock transactions necessary should Jersey acquire the missing 2 percent of the stock.[19]

In its efforts to overcome the reluctance of the minority shareholders to sell their Humble stock, Jersey could not have had a better representative than Hines Baker. Baker, a very sagacious lawyer, was partially cast in the mold of the legendary Texas oil man. He had known Holman for more than forty years and the two Texans worked well together. President Rathbone, too, respected the soft-voiced Baker and strongly backed Holman and Baker in company councils. The consensus of those directors associated with Jersey in 1957, is that only Hines Baker could have engineered this stock transfer at that time. While not a founder of Humble, he had joined that company in 1919, and had served as general counsel for many years before becoming president. In the fraternity of oil, he and his brother Rex were referred to as the "Baker Boys," or "Baker Brothers," and they achieved a remarkable record at Humble. During their years there, they gained an almost unrivalled reputation for personal and public integrity and the remaining minority holders of Humble stock knew them well. Thus, they believed in Hines Baker's sincerity when he stated that he wanted to be fair both to Jersey and to them.

Accomplishing this goal turned on the difficult task of convincing Humble's minority shareholders to part with their stock. Except for those held by Humble to cover executive stock-option plans, the outstanding shares were almost all in the hands of either employees or the families of the company's founders. The tremendous sense of pride and the loyalty these people felt toward Humble had long been regarded as the principal obstacle to merging the Texas affiliate into Jersey. Earlier offers to exchange Jersey stock for Humble shares had reduced the ranks of the holdouts, but in a number of cases family identification with the company still meant serious resistance to Jersey's proposed acquisition. However, competitive pressures and the threat of antitrust action simply left Jersey with no alternative.

Working with Jersey's Leo D. Welch, Hines Baker insisted on a voluntary exchange formula that would offer five Jersey shares for four of Humble's, arguing that any lower offer would fail to attract the shareholders who had refused to participate in the earlier exchange. Armed with this offer, Baker succeeded in convincing most of the remaining minority shareholders of the need for change and that the best interests of all would be served.[20]

With less than 2 percent of Humble stock outstanding, Jersey moved toward the goal of consolidating all domestic operations. Before he retired late in 1958, Hines Baker drafted a reorganization plan covering both the merger of Humble into Jersey and the unification of all domestic affiliates. On August 17, 1959, Rathbone—who had early become a strong proponent of the move—and other representatives of Jersey met with Humble executives (in Houston) to discuss Baker's plan. Basically, the proposal called for the creation of a new Humble Oil & Refining Company, to be organized in Delaware, after which the existing Humble (a Texas corporation) would, with the approval of its stockholders, transfer all its properties and operations to the new company in exchange for all of the stock of Humble (Delaware). Then, subject to stockholder approval, Humble (Texas) would merge with Jersey, leaving the latter company as the sole owner of Humble (Delaware). Finally, as soon as feasible, The Carter Oil Company [Carter], the Esso Standard Oil Company [Esso, Esso Standard], the Oklahoma Oil Company, the Pate Oil Company and lesser domestic affiliates would be merged into Humble (Delaware) which would survive as the operating company. Rathbone then discussed the plan with the principal executives of the other domestic affiliates who, according to his report, were all "strongly in favor of it."[21]

With management support assured, on September 1, 1959, Jersey and Humble announced the forthcoming consolidation of domestic operations. Wanting to consummate the plan as quickly as possible, Rathbone assigned Jersey representatives to work with Humble officials in developing a definitive merger proposal for submission to stockholders. Two days later, on September 4, the new Humble Company received its charter in Delaware, and on September 21, it announced the membership of its board of directors. As anticipated, Humble (Texas) had the largest representation, three of the seven members, while Carter and Esso Standard were each represented by two directors. The new president would be Morgan J. Davis, then president of Humble (Texas).[22]

On the first of December, after special stockholder meetings, the new Humble (Delaware) Company began operations. At year's end (1959), Esso Standard and Carter were incorporated into the new company, and during 1960, other domestic affiliates, Enjay Chemical, Pate Oil, Oklahoma Oil, and Globe Fuel Products, Incorporated, also merged into Humble. Pending the evolution of a more precise organizational plan, these companies operated as divisions of Humble under their existing names and with their own managements and boards of directors.[23] Forging them into a single operating organization remained a task for the future. Nevertheless, the mergers that had already taken place (exploration, producing, shipping, refining and manufacturing and marketing) now put all U.S. operations under a single management. This marked one of the most significant steps in Jersey's drive to improve its performance. Even with the interim organization, the benefits anticipated from this bold step quickly became realities.

The selection of the Humble top executives to manage the new structure permitted significant progress in the ongoing drive to reduce costs, especially by means of ending duplication of efforts. More importantly, the authority vested in the new Humble board made possible better coordination in all areas. With all responsibility for domestic operations centered in a body devoted exclusively to their coordination, and with lines of communication shortened and more clearly defined, many internal duplications, conflicts, and inconsistencies could be effectively avoided. Of equal significance to Jersey, decentralizing much of the responsibility for domestic operations served to free the time of the parent company's top executives, and enabled them to concentrate on managing the company's worldwide interests. Yet, despite the relief rendered by this domestic reorganization, the substantial burden of its overseas operations still rested on Jersey's directors, and this reduced the Board's ability to look ahead and plan grand strategy.

The scale of the company's domestic operations did make it feasible to delegate most of the responsibility for coordination of U.S. operations to the new affiliate. But for most of the rest of the world, effective management required a perspective that incorporated several national markets and, thus, several affiliates. Clearly, an improvement had to originate in the parent company, and involve significant structural alterations.

As in the case of the Humble merger, the planning and execution of such fundamental changes required leadership from a capable and influential executive. With Holman preparing to retire, his heir-apparent as CEO, President Rathbone, took up the mission of reorganization with alacrity.

Intellectually and temperamentally well suited to fulfill the role of creative innovator, in his earlier career Rathbone had honed all the skills necessary to accomplish this important task. To institute changes upon taking control of the company was natural, of course, but Rathbone's training and inclination dictated that those changes would be substantive, and that he would become known as "the reorganization man." Possessing "an engineer's mind"[24] and schooled in the refiner's world of narrow margins, Rathbone believed in the need "to seek out, test, develop and adopt new methods and new ideas . . . as part of a recurring process" aimed at improving operations and administration.[25] During almost six years as chief operating officer, his propensity to follow closely the development of staff recommendations[26] had given him a detailed knowledge of the entire decision-making process at Jersey, and reinforced his awareness of the importance of a sound organizational structure to the continued success of the business. He believed the development of a suitable organization to be an essential function of a chief executive,[27] and when he assumed that position in 1960, he naturally turned his attention to making those changes in structure that would better allocate responsibility for both current administration and the broader functions of policy-making and planning.

As president under Eugene Holman, Rathbone had already altered the company's organization by allowing the Executive Committee to concentrate on "policy determination, consideration of important current problems, and studying and discussing longer range problems." In 1958, he had authored a plan that established for the first time a clear-cut distinction between the functions of Executive Committee members and those of the remaining directors and, by extension, between those of the Board and the Executive Committee. "Substantial additional authority and responsibility" he had shifted to the contact directors (now with a few but important exceptions, just those Board members who were not regular members of the Executive Committee). Rathbone encouraged these executives to pass much of this new responsibility down to the staff-level coordinators and department heads, for their greater contact duties required increased time and attention. As a result, while they remained alternate members of the Executive Committee who could attend meetings without explicit restrictions, they clearly understood that they would attend only those meetings relating to one of their contact responsibilities or to some more general subject in which they had some particular interest.[28]

With the contact directors now carrying a greater portion of the administrative load formerly borne by the entire Board, the Executive Committee could be strengthened to handle enlarged responsibilities for overall planning and policy-making. With only a few exceptions, the regular members of the Committee, now designated as executive vice presidents, were relieved of their functional contact responsibilities. The Committee then increased the number of its members from five to six. The retirement of Director Jay E. Crane, almost a year earlier, had left the Board without a specialist in finance. But the new role of the Board required the membership of such a person, so a veteran financial expert, Leo D. Welch, was soon added. The function of the executive vice presidents as general executive aides to the president and to the chairman of the Board received full definition; and a rotation system under which the executive vice presidents could assume the duties of the two top officers in their absence was also established.[29]

These changes, coupled with minor modifications in the administration of the capital budgets of affiliates, relieved the Executive Committee and its individual members of many routine administrative chores, but still the Committee's responsibilities in this area remained substantial.[30] More importantly, the Board and Executive Committee continued to be the only bodies charged with developing an integrated plan for Jersey's global operations as well as for appraising the diverse influences that affected them. As he prepared to take command of Jersey, Rathbone sought an improved structure to serve this integrating function, one that shifted much of the responsibility to staff specialists, thereby freeing the top executives to consider and act on staff recommendations.

Operating conditions of 1960 were different from those of a few years earlier. For several years, Jersey's *Annual Reports* indicated fluctuations in net consolidated earnings after a high in 1957. It should be remembered that these in part reflected sharp, retroactive tax increases for Creole in Venezuela; however, Rathbone believed that a decline had occurred in real profits from other affiliates too. On the domestic scene, this could have been occasioned by the recession that began in late 1957, and lasted about two years. He believed changes in the organization and a rationalization of the company from the Board down were necessary. When he discussed the problem of reorganization, he used for his frame of reference an ideal balance between decentralization and coordination, "the cornerstones of Jersey's management theory."[31] In pursuit of this goal, the Board, led by Rathbone, effected major changes in two of the fundamental processes by which the parent company influenced the activities of its affiliates: investment planning and contact by staff and top executives.[32]

An important exception to this new approach was Imperial Oil Limited [Imperial] located in Canada. Jersey owned more than 70 percent of the stock in this affiliate and over the years (as in the case of Humble), had seldom interfered in its management. Imperial drew on Jersey's resources and expertise; in fact, a high degree of collaboration existed between the two companies from which both benefited. But in Canada after World War II, nationalistic political feelings dictated a climate of opposition to any domination of Canadian businesses by American firms. Imperial operated only within Canada; its managers understood the petroleum business; and its investment planning appeared to be both conservative and sound. Thus, Jersey decided that Imperial was already operating in accordance with the "new" policies.

As in the unification of the domestic affiliates, the goal in reorganizing the parent company's contact with its foreign subsidiaries was better coordination within a decentralized system. To achieve that end, Rathbone envisioned a fundamental reorientation of the contact structure. While functional coordination would remain an important part of the total process, a new regional structure shifted the primary emphasis to a corporate approach, allowing each affiliate to take a broader view of its business opportunities.

The need for this new regional-corporate orientation had been imposed by the changed conditions under which affiliates now operated. Before 1960, the parent company maintained communication with affiliates through both contact directors and staff departments. Under that system, both contact directors and functional coordinators tended to concentrate on the function that coincided with their own primary orientation and to give relatively little attention to other aspects of the affiliates' business. Thus, even with the provision for corporate contacts, this predominance of the functional orientation meant that "Jersey's top-level responsibility tended to blur . . . at the joints where the various functions dovetailed."[33] Now, however, rapid political and social change, as much as economic stringency, required a broader view of each affiliate's business, one that could best be taken from an overall perspective. Yet, even better integration on a corporate basis would not alone provide a sufficient response to such critical times.

The business difficulties facing the petroleum industry in 1960 showed no respect for national borders. Jersey obviously needed to coordinate the policies of its affiliates across broad geographic areas, such as the home countries of important regional political and economic arrangements like the

European Common Market. This coordination on a corporate-regional basis promised to improve the Jersey handling of current problems by establishing new lines of communication more appropriate to the complexity of those problems.

The idea of a regional approach came as no surprise to Jersey. By 1960, a variety of organizational structures had provided some regional coordination in several areas. The jointly owned Standard-Vacuum Oil Company had been given responsibility for all of Jersey's interests in the Far East after 1933. Esso Mediterranean, Inc., which was established in 1959 "to acquire and coordinate the activities" of ten small marketing affiliates in the Eastern Mediterranean area, North and West Africa, and the Iberian Peninsula had really been a forerunner of the new regionalism. This company also served as a regional coordinating company although only for a single function.[34]

Contact with the Middle East most nearly foreshadowed the regional structure adopted in 1960. As early as 1951, a Middle East advisor in New York provided staff support for contact directors concerned with that important area although responsibility was still divided among several directors. Two years later, Director Howard W. Page assumed contact responsibility for all Middle Eastern affiliates except marketing companies, leaving the region under a single contact director with a regional coordinating staff. However, supervision of jointly held producing, refining, and transportation ventures could not be equated with the task of coordinating majority or wholly owned and integrated affiliates engaged in several (or all) functions of the oil business.

The greatest problems of coordination and the greatest need for a regional approach existed in Europe and in Latin America. The Jersey Board first recognized this in 1950, when it began experimenting with contact assignments on an "area basis,"[35] making a single director responsible for all European affiliates, and assigning another as contact for all Latin American affiliates except the Creole Petroleum Corporation, the International Petroleum Company, Limited, and the Panama Transport Company. However, within a few years, business pressures forced the redistribution of contact assignments. By 1960, six different directors had been given contact assignments for Latin American affiliates and three others had served for European companies. The Board made no further effort to coordinate Latin American affiliates on a regional basis, but, beginning in 1952, a European Coordination Committee, chaired by Howard Page, attempted to provide a regional perspective for European affiliates, though on a purely advisory basis. Periodic meetings with executives of several European companies offered an opportunity for operating managers to share experiences and ideas about common problems and the occasion for contact directors to clarify Jersey policy;[36] but this structure had little practical effect on the coordination of European affiliates.

The parent company's General Economics Department, at the suggestion of the Executive Committee, also attempted a review of Europe as a region. In 1958, they established the position of European economist, to be resident in London. This person analyzed and reported on not only the economic activities and trends in each of the Western European countries, but also the "economic aspects of regional developments . . . such as the European Common Market and other supranational arrangements."[37] Despite this start, however, regional coordination of European operations remained an aggravating problem that intensified with the increasing competition of the late 1950s.

When Rathbone prepared to take over as chief executive officer, he identified a number of the company's main problems and policy concerns and assigned committees of directors to investigate and suggest solutions to each.[38] One such concern was the improvement of policy communications between the parent company and its European affiliates. By December, 1959, Director Emile E. Soubry had recommended the establishment of a corporate coordinator for Europe. This executive, who would be located in New York, would relieve the functional coordinators of their responsibilities as corporate contacts and also "serve as a channel of communication on corporate matters between the company, its representatives in Europe and the affiliated companies in that area." The shareholder representatives (preferably with a more appropriate title) would continue to function but would be reduced in number from five to four persons. To coordinate the flow of information relating to corporate policy and developments of a regional nature, the shareholder representative, based in London, would be designated Jersey's senior representative in Europe, in addition to his duties as liaison among affiliates in the United Kingdom, Eire, and the Scandinavian countries.[39]

During the same year, Director M. A. Wright proposed structural changes to facilitate intelligence gathering and coordinate "broad policy direction and action." He suggested the appointment of one primary and one alternate contact director for all Latin American affiliates. As a full-time assistant to these directors, he would select a Latin American executive assistant, preferably "a man of stature with broad experience in Latin America." This executive would chair a Latin American Committee, and be responsible to the contact director, "charged with centralizing and coordinating intelligence and action in the fields of politics, public relations, as well as broad corporate and operating policies and programs." While the members of the committee would remain administratively attached to parent company staff departments—Producing, Refining, Marketing, Coordination and Petroleum Economics, Public Relations, and Government Relations—they would concentrate on Latin American problems and "function actively in the Latin American Committee." Despite the innovations in his plan, Wright's approach to regional coordination stressed minimum alteration in the existing structure; he envisioned no change in the assignment of corporate contact responsibility to functional coordinators.[40]

Working closely with Secretary John O. Larson and his staff, Rathbone had been conducting his own study of the advantages of regional coordination. Drawing on the work of Soubry and Wright as well as his own ideas about the organizational needs of the company, he revised a tentative plan submitted by Larson, until he at last developed a structure suitable for the corporate-regional orientation he envisioned. In January, 1960, Rathbone presented the proposed plan to the Board. Stressing the intensely competitive environment, the growing complexity of the situation, and the trend toward regional economic blocs, he proposed a maximum decentralization of authority and responsibility, coupled with the application of "as much experience and judgment as possible" in meeting problems wherever they occurred. To satisfy these requirements, he outlined the new parent-affiliate relations.

Rathbone divided the company's operations into six geographic regions with "a contact director team" for each region, consisting of "two working contact directors and an advisory contact director" who would be a member of the Executive Committee. The regions denominated were: United States; Canada; Europe and the Mediterranean area and West Africa; Latin America; Middle East; and Far East and East and South Africa. In the case of the working contact directors, one would be designated as the primary contact for the region and the other as alternate, although Rathbone emphasized the necessity for maximum teamwork in order to provide affiliates with the best possible support and advice. In Europe, the necessity of accommodating the heavy workload dictated a division of affiliates; the plan essentially distinguished between those in the Common Market countries and those outside. The primary contact director for each group would be the alternate for the other, with a single advisory contact. Thus, a three-man team prevailed in both regions.

While the primary-alternate contact director provision merely applied the existing contact system to a regional-corporate structure, the advisory contact director marked a sharper change from past practice and became the key element in making the regional structure a decentralized one. This executive had "to cross check the judgment of the contact directors, to act as a sounding board, and to help decide how far the contact directors should go on their own initiative and what should be referred to the Executive Committee or the Board." In effect, he would head "a sort of regional executive committee" that would improve decision-making for the region by establishing shorter and clearer lines of communication. Thus relieving the Executive Committee of the necessity of making many less important decisions, the new structure served also to reinforce the distinction established in 1958 between regular members of the Executive Committee and the other directors. This further emphasized the broader policy-making functions of the Executive Committee.[41]

Next, to facilitate the work of these new contact director teams, Rathbone proposed the establishment of regional advisors or coordinators for each of the regions outside North America. Based on the model of the Middle East, these experienced senior executives with small staff departments would serve "as executive, administrative, and coordinating aides" to the contact director teams. Since the Middle East region already had a functioning regional advisor, and because the status of Stanvac in the Far East and South Africa regions remained unsettled in the wake of its court-ordered dissolution, the immediate effects of the new system essentially were confined to the European and Latin American regions.[42]

Rathbone's plan for a regional approach to Europe incorporated Soubry's idea of reducing the number of shareholder representatives. Rathbone wanted P. J. Anderson, who was based in London, recognized as the senior representative. While all the European shareholder representatives would communicate directly with either the European regional coordinator in New York or the contact directors for Europe, Anderson would be responsible for arranging regular meetings of the four shareholder representatives. These sessions would allow them to exchange information and adopt a uniform position on regional matters. In addition, Anderson would continue to serve as local contact for representatives of New York staff departments resident in London and as principal contact with other international oil companies based in London.[43] In the final plan adopted, the number of representatives was reduced to two and their title was changed to "Jersey representative," but their duties remained essentially the same as those spelled out in Rathbone's proposal.[44]

In New York, the European regional coordinator, in addition to advising and assisting the regional contact directors, would take direct responsibility (along with the shareholder representatives and European affiliates' managers) for handling "such matters as can be adequately and conveniently handled" without reference to the contact directors. Where other parent company staff departments were better able to deal with particular subjects, he would make referrals after which those departments would deal directly with the affiliates or shareholder representative. However, even in those cases, the regional coordinator had to be kept informed. Finally, the European regional coordinator received responsibility for maintaining information "on general political, economic or social developments in the European region. . . ."[45]

In Latin America, conditions in the separate countries differed, and the recognition of "a very definite common threat . . . of nationalism, anti-Yankeeism, underdevelopment, need for foreign capital but reaction against it, and an ever-present threat of communist infiltration" convinced Rathbone of the importance of a regional approach. Here, affiliate managements were more responsive to parent company influence. Except for the absence of Jersey representatives, the regional structure essentially duplicated that in Europe, and the duties of the Latin American regional coordinator closely paralleled those of his European counterpart.[46]

In February, the Board began discussions on the regionalization plan. Rathbone's research and preliminary work on the proposal had been thorough and he was adamant; so the Board made few changes.[47] In fact, the only departure from the regional-corporate contact system provided for coordination of eight exploration ventures through a centralized Producing Coordination Department.[48]

Even before the final endorsement of his regional structure, Rathbone had begun to focus on the other important field of Jersey's interest in which he hoped to improve the decision-making processes through structural change—investment planning and review. Rathbone decided the recommendations from a committee headed by John W. Brice did not go far enough, so he turned instead to the three director members of the Coordination Committee: Stewart P. Coleman, Marion W. Boyer, and Cecil L. Burrill.[49] After consulting with them and considering their recommendations, Rathbone prepared for all Board members, in early March, a confidential memorandum outlining a new structure through which the Board (if it approved the change) would exercise its responsibilities in the field of investment planning.

A need for improvement was evident. In practice as well as theory, Jersey acted as the central banker for its worldwide operations. It allocated investments so as to maximize earnings and yet be certain that funds for the most important projects remained available. Acting as a holding company, one of Jersey's central functions now became the careful supervision of affiliate investment programs, to guarantee that they would maintain the financial strength of the parent company and insure a satisfactory return to its shareholders. In the preceding years return on net worth had been declining rather steadily. By 1959, return on equity was barely half what it had been in 1950. Moreover, competition in the marketplace now threatened the ability to generate the necessary capital to meet long-term requirements. While anticipating no immediate financial crisis, Rathbone still saw a definite need for greater attention to the financial aspects of the capital investment program. And the increasing importance of political and social considerations to the company's activities argued for an even broader analysis of potential investments.

Under the existing structure and procedures, such considerations were limited to the Board and Executive Committee. Yet, to make the needed improvements in investment analysis without altering this structure would also impair the ability of those bodies to execute their policy-making function. Thus, Rathbone's proposal coupled broader investment analysis and planning with more decentralization of responsibility for those functions. The key to his plan was a new Board Advisory Committee on Investments [BAC] composed of four directors and eight other Jersey executives. In effect, this new committee would replace the existing Coordination Committee; one of the principal ongoing activities of the BAC would be the review and appraisal of investment projects and the formulation of recommendations to the Board on investment policies. To support the BAC in these efforts, Rathbone called for the formation of an Investment Analysis Department [IAD] as a "full-time service and study group" charged with analyzing "on a broad and continuing basis both existing and proposed investments." Since analysis of all the company's investments would be impossible for a staff group of the size envisioned, the department would concentrate "on those which . . . [seemed] to present unusual aspects or have far-reaching investment implications." The department's function clearly paralleled the investment review activities of the BAC, but its designation as a staff group to assist all directors as well as the BAC signified a new concern with broader analyses of investments at all levels, and a recognition of the importance of staff work in achieving that goal.[50]

Although the BAC assumed the primary function of the Coordination Committee, its establishment involved much more than merely organizational change, designed to improve performance of the same task. The reorganization of 1960 actually redefined the task itself. The Coordination Committee had, in recent years, paid increasing attention to the financial aspects of investment proposals, but its deliberations continued to be dominated by a highly traditional emphasis on technical feasibility and logistical impact. As in the early postwar years, the committee's main concern seemed to be balancing future capacity in the various functions with projected demand, thus maximizing the efficiency of Jersey's entire system. Taking projected demands as given, it concentrated on meeting demand in each market at the lowest possible overall cost.[51] The job of answering such questions as how to pay for the necessary increases in capacity remained primarily a function of the Board. As a result, financial considerations were not smoothly integrated into the overall investment planning process. For the same reasons, political and other non-economic factors—also in the Board's province—generally received scant attention as investment factors in Coordination Committee deliberations. Thus, under the existing structure these increasingly important considerations could not be integrated into the investment review process in its formative stages.

The BAC essentially reversed the emphasis of the Coordination Committee and, in the process, altered Jersey's approach to investment planning. While the new committee retained responsibility for insuring the technical balance of Jersey's system and its adequacy to meet future demand for oil, in practice much of this burden shifted to the staff departments, whose heads made up the non-director majority of the committee, and to the Worldwide Petroleum Supply Subcommittee consisting of representatives of the four functional coordinating departments, the Coordination and Petroleum Economics Department, Esso Export Corporation, Creole, and Humble.[52] Charged with considering "*all* factors . . . including political, economic, technical and financial factors" relating to investments, the BAC was designed to improve investment planning both by providing a comprehensive and thorough evaluation of proposals and by integrating all investment considerations at an early stage of capital budget planning. In essence, despite continued attention to logistical balance, the new arrangement stressed consideration of investments as investments throughout the process.[53]

Once taken, this new approach had a liberating effect on those areas of the company's business outside the traditional mainstream and carried even greater implication for future development within the mainstream. The new emphasis on comprehensive analysis and the greater attention to all financial considerations together promised to improve the company's return on the oil investments that constituted by far the largest portion of its total business.

At the same time, the process of applying to oil investments the more rigid financial standards that had traditionally applied only to "non-oil" projects improved the chances of the latter in the competition for capital resources. While management had no intention of wholesale diversification, this emphasis on investments as investments opened the way for new departures. As the largest and best established "non-oil" field, chemicals benefited most dramatically, but Rathbone also proposed diversification along a broader front, wherever new uses for petroleum might be found. Burrill, who joined the Board in 1959, was given charge of diversification as one of his assignments. After gaining some years of experience in that post, he came to believe that "the company was afraid to diversify and at the same time, was afraid not to. So that very little was done by way of diversification outside of chemicals."[54]

The appearance of the BAC made possible long-range, as well as immediate improvements in the crucial area of investment planning and review. The selective approach implicit in thorough evaluation of proposed investments, particularly the new emphasis on financial considerations, promised a more rational distribution of available capital funds, one that would continuously strengthen the company's overall position. Moreover, by delegating to the BAC authority broad enough to allow more definitive recommendations, the Board and Executive Committee had freed themselves to concentrate on only the most important questions, including policy formulation and strategic planning. However, in Rathbone's conception, the BAC should also have a specific role in helping in these tasks by developing for the Board's consideration, "specific investment policies."[55]

Some evolution of policies incorporating the BAC's comprehensive approach to investment planning now became a prerequisite for the reorientation of an overall policy based on technical feasibility and functional balance to one focusing on the broad corporate effect of new investments. Only by a detailed plan that stressed consideration of the total investment environment could the Board insure that, even in the early stages of capital budget development, affiliate analysis of potential investments would be compatible with the functions of the BAC and consistent with Jersey's overall interests. In creating this framework, the BAC itself offered the Board the best possible staff assistance.

To lead and chair the BAC in the execution of its new duties, Rathbone chose Marion Boyer, a veteran company executive, then serving as vice chairman of the Coordination Committee. Deputy Chairman Howard Page and Vice Chairman Cecil L. Burrill had both acquired experience as heads of Jersey's Coordination and Economics Department. For some time, Page had been closely identified with the company's Middle Eastern producing and refining operations. All three had served in the federal government: Boyer as general manager for the Atomic Energy Commission, Page as program director of the World War II Petroleum Administration for War, and Burrill in the same position in the Petroleum Administration for Defense during the Korean Conflict. Stewart Coleman, the last chairman of the Coordination Committee, remained a member of the BAC until his retirement in 1961. All four men had previous experience on the Coordination Committee.

While the BAC did expand the scope of the company's investment analysis and planning, it was not designed to perform all of the functions of the old Coordination Committee. With the demise of the latter, some of its most important responsibilities had to be shifted elsewhere in the corporate structure. Replacement of the Coordination Committee by the BAC, therefore, had measurable impact in areas far removed from the field of investment analysis and planning. The Coordination Committee under the leadership of Everit J. Sadler, for example, had become during the 1940s an important mechanism for educating affiliate and functional staff executives about the overall interests of the Jersey system. Large meetings conducted around the world, brought together technical and non-technical personnel, to exchange information and ideas. The largest of these sessions, the week-long annual budget review meeting, sometimes involved as many as three hundred participants. While Rathbone came to question the efficiency of these huge sessions, they clearly worked well educating company personnel and in promoting personal contacts that helped unify the giant corporation's far-flung operations. In addition, the meetings gave valuable experience and exposure to those young executives who made many of the presentations; the sessions were an important, if informal, part of the company's executive development process.[56] After the Coordination Committee disappeared, young promising managers continued to appear before the Executive Committee and the BAC during the annual review of operating results, in budget sessions, and as a part of planning meetings. So this significant function of the Coordination Committee continued. Too, some of the educational activities of the old Coordination Committee continued. For example, one participant wrote:

under Dave [David J.] Jones' leadership and my [Cecil L. Burrill] guidance, we found it necessary to educate many company officials, including the members of the Board on the mathematical theory of investment. It was impossible for us to proceed without looking at returns on investment using the discounted cash flow technique. And some rather elementary educational presentations

were made at as high a level as [the] Executive Committee in order
to prepare company executives for the use of this technique, which
had not been considered before that time.

Even so, the general responsibility for executive development necessarily
gravitated toward the affiliates, under the overall direction of the parent
company's Compensation and Executive Development Committee
[COED]. Increased reliance on these programs sharpened an existing interest in identifying the most promising young executives as quickly as possible.
Decentralization encouraged the development of the younger managers by
creating additional opportunities for them to demonstrate initiative and
sound judgment.[57]

Under Rathbone's guidance, the whole Jersey reorganization of 1960 took
effect with remarkable smoothness, especially in view of the magnitude of
the changes. Taken together, the BAC and regionalization (including the
Humble mergers) marked a basic reorientation of the parent company and
its relations with its affiliates. Yet, for all its innovation, the new structure
represented a projection of the same management philosophy that had
guided the company since the late 1920s. In continuing the effort to strike an
ideal balance between coordination and decentralization, the reorganization of 1960 actually managed to strengthen both. The organizational
changes of 1960, designed to achieve a combination of better support and
greater control, modified the decision-making processes of the company in
two ways: to make them more responsive to affiliate needs, lines of communication and authority had to be shortened and clarified; and, to provide
a better base for decision making, the quality and quantity of the information available had to be improved. The essential step in the reorganization
came in the shifting of responsibility for corporate action from the Board
and Executive Committee to the contact directors and to staff departments
of the parent company. The new regional organization, and especially the
actions of the BAC, reflected this change by superimposing a corporate orientation on an existing functional structure.

The shift also brought fundamental change in the coordinating function.
Staff departments proved better equipped than the directors to gather and
systematically assess information about the various factors affecting the
company's business. By dealing directly with affiliates, these departments
had an advantage in communicating a broader, better integrated view of
the company's overall interests to operating managers. As a result, decisions
made at the affiliate level could be based on a better appraisal of their ultimate impact. In addition, the fact that decisions came to be made within a

similar integrated framework helped to coordinate the process and to retain the benefits of the technical expertise.

Finally, the Jersey Board could delegate greater authority for operating decisions because its affiliates better understood company policies. Originally, decentralization had meant merely shifting responsibilities to the affiliates, but the reorganization of 1960 also placed the emphasis on delegating authority within the parent company. The result was clearer delineation of roles at all levels. Assigning greater responsibility to teams of contact directors sharpened the distinction between the Executive Committee and the other directors. The broad duties assigned the regional coordinators and the BAC emphasized the expanded central role of staff departments in coordinating affiliate activities, while more specifically defining the realm of the functional coordinators. Through the new channels of communication and information, problems, and recommendations could move from affiliate to Executive Committee, while each level of the structure filtered out just those matters it felt capable of handling, and passed the rest on to the next level. At the end of the process the Executive Committee would still decide the most important and difficult questions, but relieving it of "some of its 'less vital' work would give it 'more time to think in the broad policy and planning area,' " its most crucial function for the long-run success of the company.[58]

Early in April, 1960, implementation of the plan began. The BAC took over the Coordination Committee's old conference room as well as its responsibility for capital budget reviews. To encourage objectivity and frank discussion of projects, the committee adopted Howard Page's suggestion that all affiliate representatives leave after making their presentations, leaving the committee to deliberate "*in camera.*"[59] Since the committee met three times a week to work off a backlog of investment proposals, its meeting place became known as "the 'back' room."[60]

By the end of the month the Board had approved the final changes in the company's revised "Plan of Organization and Management Guide" and appointed three regional coordinators: M. W. Johnson for Latin America, Wolf Greeven for Europe, the Mediterranean area and West Africa, and Smith D. Turner for the Middle East.[61] When Rathbone assumed the position of chief executive officer on May 1, 1960, his post existed within an organizational structure of his own design.

Rathbone firmly believed these sweeping changes would end the continued drop in Jersey's net earnings that characterized the period, 1957-1960. By merging Carter Oil Company and others with Humble, he clearly

foresaw the economies possible through the elimination of separate exploration, refining, production research facilities and personnel. By November,
1960, Jersey had reduced the number of its employees worldwide from a
1957 high of 160,000 persons to 148,000. Too, the Jersey president anticipated nationwide marketing; first, under the Enco [Energy Company]
trademark to be used by Humble outside the eighteen states where it marketed in accordance with what they then believed to be part of the 1911
Supreme Court Dissolution Decree, and if that failed, something else.
Breaking away from the traditional "Esso" name mattered. Regarding
Enco, Rathbone said, "If we like the name, we'll probably use it all over the
country," and then he remarked, "You can't imagine what a hell of a job it is
to develop a new trademark." But Jersey had to find markets and approaching marketing on a national scale might help.

Perhaps more than any other change from the Jersey of the 1950s to the
one Rathbone perceived as operating in the 1960s, funds would now be made
available for the company's petrochemical development. He said, "Chemicals do not loom large in our budget in our over-all figures—but they will."
And the budget line for petrochemicals in 1961 carried almost twice as much
for them as did that of 1960. Jersey may not have diversified in the exact
sense of the word, but petrochemicals did benefit from what the company
called diversification. The new president believed in this outlet for company
investment. On one occasion Rathbone confidently stated, "This dashing
into petrochemicals is going to lose a lot of money for some companies
because petrochemicals are not a guaranteed gold mine. And it won't be us."
He told an interviewer, "Hell, I was a chemical engineer to begin with."
Rathbone proposed to find new uses for oil and gas through research programs.

Rathbone also foresaw the reorganization as giving field personnel authority to make more important decisions, thus shortening communication
lines. Decentralization of responsibility was accompanied by the new centralization of command. Thus, he believed, decisions would be made faster
and from this, much needed executive talent uncovered and developed.

It should be clearly understood that this reorganization of Jersey, once
started, proved to be part of an ongoing process as Rathbone and his successors as CEOs sought to keep the company abreast of the rapidly changing
world in which it operated. Jersey, after the Teagle reorganization of 1927,
had served as the central banker for its many subsidiaries, had examined the
budgets of affiliates and had supervised their investment programs as well.
In this process, the Board and the Executive Committee had reviewed all

proposals in the light of economic, political and social considerations as they viewed them.

The Rathbone plan of 1960 established the Board Advisory Committee to take over that assignment. The BAC, unlike the Coordination Committee earlier, devoted little time to the logistical impact and technical feasibility of the plans of affiliates; rather it formulated for the Board, recommendations on investment policies, especially those with far-reaching implications for the future. Too, the BAC concerned itself more with the advisability and profit possibilities in plans, than how to finance them. The Board had to provide solutions for the question of funds. The BAC sought to bring into focus on budgets and investments, both the short-term and long-range implications of all factors that could affect their decision. Rathbone devised the IAD as a resource-study group to augment the investigation and analysis of the BAC.

Rathbone foresaw the BAC and IAD working closely "with the all important Coordination and Petroleum Economics Department [CPE] which is concerned with the logistics of running Jersey." The five other functional departments, in turn, operated directly with the CPE unit. In reality, the plan called for a four-tiered operation with the assigned directors monitoring the first two phases: the affiliate budget and financial proposals first went through the functional departments where specialists examined them; they next went to the CPE group for additional review. There, substantive information relative to the project received full consideration. When the CPE finished its appraisal—the BAC study took place. The BAC had the power of decision. Here the involved directors gave opinions and answered questions. Then the BAC, *in camera*, again examined the proposals and gave final approval, subject to the entire Board agreeing on financing. Further, about mid-year the entire matter was reappraised to see "how well the basic assumptions and projections have held up."

Three companies were hardly affected by the 1960 reorganization of Jersey; Creole, Imperial Oil Limited and the Standard-Vacuum Oil Company [the latter in the process of being dissolved after April 13, 1960]. Each of these had assigned to review its operations, a primary contact director, an alternate contact director and an advisory contact director.[62]

Rathbone's plan of reorganization, though adopted and placed in effect in 1960, did not stop with changing the responsibilities of the Board, the Executive Committee and the Coordination Committee and shifting some of their burden to the BAC, the CPE and the IAD. The regional coordinators and the contact directors also worked with the new committees and departments. Theoretically, thus freed, the Board and the Executive Committee could devote more time to planning company strategy and formulating policy. In fact, for some time both bodies changed their meeting policies, to one of holding less substantive meetings. For several months, the Executive Committee did very little (when compared to previous years), while it awaited a clearer definition of its new role and function. However, in less than a year, Rathbone came back to the Board with additional ideas.

On January 18, 1961, the Executive Committee considered a policy guide, drafted by the Reports Review Committee, seeking "broad guidelines for the establishment by company departments of requirements for recurring reports from affiliates in a manner consistent with the company's objective of holding such requirements to an essential minimum." Rathbone believed this, too, would save the time of the various persons responsible for studying affiliates' reports—especially those of the directors personally concerned.

During the next year, the Board reviewed memoranda outlining the responsibilities of directors, the organizational pattern of the company, the duties of the Executive Committee and the Compensation and Executive Development Committee, "and certain officers and other executives in the overall organizational chart." On April 12, 1962, the Board recommended "that the memorandum and chart be revised" and that the discussion be continued at a later date. As a result, on July 5, 1962, the Board appointed a new vice president, Eger V. Murphree, the company's noted research scientist, to serve as research policy coordinator.

By 1963, the Executive Committee had prepared a draft covering investment policy, the writing of which occupied a large portion of that Committee's attention for about a month. The Committee then approved publication and distribution of the statement entitled "Organization for Investment in New Fields of Endeavor." Rathbone then called a conference in New York, to communicate the Executive Committee's views on that subject to company executives and those of affiliates. Shortly after that meeting, in view of the fact that the reorganization had not provided for a coordinator of chemical activities, the Board appointed the president of the Esso Chemical Company Inc., to the BAC.

Rathbone had appointed Board members Michael L. Haider, Marion W. Boyer and Cecil L. Burrill to a committee to review planning as conducted under the 1960 reorganization. The committee report had been reviewed and approved by the COED Committee before it was presented to the Board on June 13, 1963. It recommended the merger of the IAD and the CPE into a single department to be known as the Coordination and Planning Department [CPD]. In addition to the responsibilities assumed from the two departments it displaced, the new department should:

study and recommend long-range corporate objectives, coordinate the long-range plans and objectives of the company interests, and assist the Directors and other executives and staff groups of the Company as well as the Managements of affiliated companies, in analyzing and coordinating investment proposals and their implementation, especially those of a multifunctional nature.

The committee also recommended that the manager of the new department be added to the BAC and that Burrill be made primary contact director, Howard Page, alternate director and Boyer, Executive Committee advisory director. The Board approved this report. Within three months, the Executive Committee placed a ceiling of seventy-six on the number of persons that could be employed by the CPD.

It is obvious that the various efforts of Rathbone and the Jersey Board to keep the 1960 reorganization plan flexible and to make variations in the original schematic, where needed, continued for some years. However, Board minutes both illustrate and indicate the depth of the administrative revolution. Where earlier Board minutes had usually ended with "the Board finds no objection" to whatever was under consideration, after 1960, the minutes reveal that the BAC advised the Board of the actions taken. Further, it is apparent that on November 8, 1963, the Board empowered the BAC and its contact directors with full investment review authority. In 1962, the BAC had twelve members including four directors; by 1964, the number of directors on it had been reduced to two and the membership dropped to ten.

Rathbone had, indeed, restructured the company and in so doing, enabled its management to focus their attention on the strategy necessary to help resolve the problems developing throughout the world. One interviewer summed up Rathbone's achievement: "The Company has developed itself to the utmost point in flexibility, ready to roll with a punch today but always with a calculating eye on tomorrow."[63]

So comprehensive a structural change in Jersey organization had not occurred since Walter Teagle made it a holding company in 1927. Conceivably only so dynamic, thorough, and knowledgeable a leader as Rathbone could have achieved so many and such varied changes in the company. Time alone would tell whether this revamped structure would prove more flexible than its predecessor organization in meeting the difficult-to-anticipate problems of the future.

Chapter 4

Humble Oil & Refining Company— Exxon Company, U.S.A.

AMONG STANDARD OIL COMPANY'S (NEW JERSEY)[Jersey] affiliates, Humble Oil & Refining Company [Humble] long enjoyed a special relationship with its parent.

For Humble, the years from the end of World War II to 1950 represented a glorious era. In this affiliate's home state of Texas, demand for gasoline and other oil products surged beyond all expectations, as the Texas Railroad Commission allowed Humble wells to produce even more oil than they had pumped in the peak war-demand year of 1944. Like other companies, Humble pushed the search for new oil, even though exploration and production costs outgained prices—despite the removal of wartime price controls in 1946. In proven fields, wells generally ran deeper, while wildcatters actually broke drilling records for depth in their search for new pools. No longer did the issues of proration and conservation have the political impact of the prewar years, as most U.S. oil producing states moved to establish some form of production control.*

While restriction of production to market demand clearly operated to keep prices up, some Humble executives believed that production controls also stimulated the search for oil. Other Humble leaders viewed price increases as an invitation to large-scale importing of oil produced abroad, since production costs were much lower, especially for Middle Eastern oil. In addition, during the 1940s, the U.S. government—some departments at least—equated the development and control of these foreign oil sources with American national security, and they encouraged domestic companies to acquire interests in foreign producers.[1]

For the Humble Company and for its shareholders, the second half of the twentieth century actually began two years early, with the death of Board Chairman Harry Carothers Wiess in 1948. Wiess represented the last of the

*California excepted.

company founders to serve as chief executive officer [CEO]. He was succeeded by Hines H. Baker. Both before and after the death of Wiess, moreover, significant changes had to be made in the company bylaws—and these changes played an important role in shaping the history of Jersey's "special" affiliate.

X   X   X

Early in 1948, Humble shareholders approved an increase in the size of the board of directors, from eight to ten members. Then at the Annual Meeting in May, the stockholders elected Wiess as chairman and Hines Baker as president. The position of chairman had been vacant since 1941, when R. L. Blaffer retired. Alone as president, Wiess had ably carried the burden of executive leadership through the war years, acting to shape the postwar policies of the company. While he remained its principal officer, Baker became in reality Humble's CEO. When Wiess hand-picked Baker as his presidential replacement, his action showed that, between the two men, there existed mutual confidence and respect. In fact, given the unclear lines of authority at Humble's top level, only two such cooperative men could have administered the company jointly.

Other than Baker, the only person who might have become president was John R. Suman, but Wiess made it clear even before 1945 that he did not intend to advance Suman to that position. So Suman, an extraordinarily able petroleum engineer with a genius for gaining and maintaining the respect of independent oil men, left Humble in 1945 for the Standard Oil Company (New Jersey) where he served as a director and vice president. At that time, Wallace E. Pratt was preparing to resign from the Jersey Board, and he persuaded CEO Eugene Holman to bring Suman to New York, where he served the Jersey Board well.

In Houston, Hines Baker had joined the Humble Law Department in 1919; he became a director in 1937, and a vice president in 1941. Along the way, he achieved a national reputation as an expert on conservation. In fact, many oil men insist that Hines Baker (together with his brother, Rex G., also a lawyer) first laid out the principles of oil field conservation law. Hines took on the role of a fighter—in his gentle, scholarly way—for voluntary oil field unitization after observing the wasteful practices rampant in East Texas during the 1930s. Unitization, to him, meant lower and more uniform production costs as well as conservation of the oil. Backed by Humble chieftains, W. S. Farish, Blaffer and Wiess, Baker placed the Humble Company solidly behind voluntary—not federal—oil field unitization.

After becoming vice president of Humble in 1941, Hines Baker grew concerned about the company's relations with several of its constituencies, especially labor and the public. From afar he had watched the pillorying of his close friend Farish* during the Temporary National Economic Committee investigations of 1938-1941. Then, Baker witnessed the antitrust and patent crusade led by U.S. Assistant Attorney General Thurman Arnold. Reactions to Jersey's answers in the 1942 Senate investigations of Pearl Harbor did not console him, and like many others, Baker concluded that the rough and unmannerly treatment of Farish during his testimony was a major factor in his death from a severe heart attack some days later. For these and for other reasons, Hines Baker sought to improve public relations at Humble.

While Humble Company leaders had always been active in their communities throughout Texas, public relations now began to supply additional information to the world outside. In particular, the company advertised that it was "manned by Texans," which helped to make the Humble Company credible to its constituency at home and identifiable elsewhere. Baker believed that Humble employees should enjoy good public visibility because of their fine performance, and clearly many of them agreed. So Wiess, Baker, and other Humble executives decided to tell the Humble story to the world in a company biography. When Jersey suggested consulting the Business History Foundation, Inc., Humble responded positively. Wiess and the two Bakers met with foundation executives and agreed to make Humble "records available for research, to arrange interviews with its employees" and to assist in a number of other ways. The foundation was given "full freedom to determine what should be published." Viewed in retrospect, this "we have nothing to hide" attitude represented Hines Baker's goal for public relations in Humble—an ongoing reputation as a well-run company, honest and honorable in all its dealings. The resulting company history appeared in 1959 as the *History of Humble Oil & Refining Company*, written by Henrietta M. Larson and Kenneth Wiggins Porter.[2]

X   X   X

The corporate plan for shared executive responsibility—with Wiess and Baker cooperating fully and working as a team to operate—never received an adequate trial. Shortly after he was elected chairman, Wiess underwent surgery from which he never recovered. So after his death, the Humble bylaws underwent further change. This time, authority was concentrated in the office of the president. Baker also persuaded his friend, the able geologist

*Farish served as Jersey president and chief executive officer, 1937-1942.

and exploration expert, L. T. Barrow,* to replace Wiess as chairman, but now with a primary responsibility for "long-range plans and broad corporate policies."[3] Earlier, in 1948, shareholders had acknowledged increased importance to the company of exploration and production by filling two newly created positions on the board with Morgan J. Davis and Carl E. Reistle, Jr. Davis was an outstanding geologist and exploration man, and Reistle achieved a notable record as a producer. When Barrow succeeded Wiess, Davis became the sole director overseeing exploration.

Baker and Barrow considered their first major objective to be that of replenishing supplies—of finding new oil and gas to replace what Humble sold. But locating and producing new fields began to prove increasingly expensive, especially since some of the exploration now moved offshore, to the little known continental shelf of Louisiana and Texas. Despite large sums expended, Humble increased its reserves only moderately in 1948, while, as a whole, the oil industry did much better. Even so, the Humble Company *Annual Report* for 1948 called attention to the fact that the petroleum industry's proven reserves contributed "greatly to national security." Finding new oil seemed a patriotic duty.

Because demand for oil and products remained relatively constant in 1949, the Texas Railroad Commission reduced allowable production days by 17 percent. For Humble, with a number of fields flowing at "high maximum efficient rates of production," this curtailment amounted to a 25 percent cut from 1948. Humble executives felt the commission ruling on production to be "a disproportionate reduction" that adversely affected company operations and decreased income. Yet Humble failed to secure an exception from the commission. As the 1949 Humble *Annual Report* declared:

> Humble did not expect its operations or net income to continue at the high levels of 1948. On the other hand, it had not anticipated that it would bear such a disproportionate reduction in operations and income incident to the readjustment.

Of course the production cut in Texas created many serious problems for Humble's management. As a group, they took measures aimed at achieving greater economy, so a new pattern of supply gradually began to emerge. As the 1949 *Annual Report* concisely stated: "imports increased." Nothing more was said about the radically changing nature of the industry.

In fact, after World War II, Humble implemented new technology to increase the yield from proved fields, and better trained exploration crews

*Barrow retired in 1954, and J. A. Neath replaced him as chairman.

and improved equipment reduced the per foot drilling cost. Yet Humble exploration crews were forced to range farther afield in their search for petroleum. While as late as 1947, the company explored principally in Texas and Louisiana, by 1948, it had begun active onshore exploration in California—an occasion that marked the first appearance of a Jersey-affiliated company in that state since the famed U.S. Supreme Court Dissolution Order against the Standard Oil Company in 1911.

Now another Supreme Court decision interrupted Humble's offshore activities in Texas, Louisiana, and California, in June, 1947, when the court held that the U.S. government had "paramount rights" to all "lands, minerals and other things of value, lying seaward of the low water mark on the coast of California." When the federal government filed additional suits against Texas and Louisiana, the issue of tidelands oil entered American politics, where it remained confused for six years.

Some three years after the United States vs. California decision, the Supreme Court ruled in separate cases against Texas and Louisiana. Both decisions reaffirmed the principle that states did not own the three-mile marginal belt along their coasts, but that the federal government did. Humble executives and many others considered these three court decisions to be an outright extension of federal sovereignty in an area where state sovereignty had never been seriously questioned. They looked to the U.S. Congress for eventual relief; however, in the interim, they also sharply reduced offshore operations. At Humble, the leaders believed that the states would do a better job of handling leases along the continental shelf than would the federal government, even though royalties would have to be paid in any case.[4]

X   X   X

From 1948 to April 14, 1957, Hines Baker ran Humble; in 1957, Morgan Davis succeeded Baker as CEO. By December 31, 1959, the Humble Company had been completely absorbed by Jersey—to exist as the sole U.S. oil affiliate of the latter company. But, before that date, during nine years of the 1950s, Humble pursued its independent ways with very little direction from the parent holding company. After December, 1959, however, Jersey policies often shaped Humble's actions, although initially, as might be expected, not without some friction. Finally, the Humble years 1950-1975, provide two greatly differing stories: one for the years 1950-1959; another for 1960-1975. Here, we shall be concerned with both.

X   X   X

In 1950, the Humble Company represented only one of twenty companies in the United States and Canada in which Jersey held a "direct majority interest." Although Humble's gross revenues were only slightly more than half that of another Jersey owned company, Esso Standard Oil Company [Esso], the net income for Humble more than doubled that of Esso. In fact, the principal portion of Jersey's U.S. and Canadian profits came from just four companies—Humble, Esso, Imperial Oil Limited [of Canada], and The Carter Oil Company. Alone, Humble produced $129 million of the $231 million total, or 56 percent, and approximately 32 percent of Jersey's world-wide net profit of $408 million.

While these profits indicated excellent organization and management, Humble also provided other advantages. Much of the industry's profit derived from crude oil, and Humble represented the single largest producer of crude in North America; indeed, in the Western Hemisphere only Creole Petroleum Corporation [Creole] of Venezuela produced more crude. Humble produced more than 110 million barrels of crude in 1950, while it claimed reserves (so far as they could be judged) sufficient to last some twenty-two years at the current rate of production. [Known reserves for the entire industry were estimated to last only sixteen years.] By 1951, Humble had 10,866 producing wells, most of them flowing wells; however, the number of wells produced by pumping or gas-lift methods was increasing annually, as crews drilled or produced on 2.4 million acres. Unoperated leases amounted to an additional 12.7 million acres.* Humble stood tall as a vast enterprise indeed.

In exploration, the company became excited over the discovery of California's Castaic Junction field in 1950. Subsequent drilling in 1951 seemed to confirm a major field, and Humble moved toward full scale operation in California. By 1953, after the U.S. Congress had restored full control of the tidelands to the states, the company also shared (with another company as operator) in California's West Newport Beach field—another important discovery. In all, 1950-1952 set records for Humble's exploratory drilling, but then in 1953 such drilling dropped off, although a considerable amount of new natural gas and oil was discovered onshore.

By this time, Humble exploration crews had branched out to drill in Mississippi, Alabama, Louisiana, New Mexico, Arizona, California, and Florida; while the company made plans to drill in Oregon and Washington. Late in 1953, Humble began to invest in leases offshore of Louisiana, Texas, and California. Morgan Davis once explained how the geologists sold this idea of purchasing offshore leases to the Humble board:

*For relevant statistics, see the table at the end of this chapter.

> We got together and we took the Louisiana map with the oil fields
> and the gas fields spotted on it, and we just turned it over into the
> continental shelf outside. We already knew that there were domes
> out there, but we didn't know how many. . . . That's what sold the
> thing to the board. We said, 'Here's the way it looks on the land,
> and all you have to do is turn over here and visualize . . . the
> continental shelf which was once land.'

For years, Humble poured money into offshore leases but without achieving much production. Yet Davis, David Carlton and other geologists, and geophysicists insisted that big oil provinces lay offshore. So the geophysicists continued to amass information, and the company kept on acquiring leases. Only in the late 1950s, however, did production show sizeable increases, and then Humble's investments in leases offshore were shown to be prudent.

Overall, the search for oil in 1954 proved quite successful. In that year and in 1955, Humble continued exploratory drilling in the Neches field in East Texas, which showed great potential. Offshore Louisiana, Humble drillers located oil and a major sulphur deposit off Grand Isle. In 1956, exploration activity again increased, and nearly one-third of the new wells resulted in discoveries of either oil or gas. At the end of the year 1958, thirty-eight geophysical crews of all types—one more than in 1957—were in the field. In 1957, drillers proved to be only moderately successful, and in 1958, the company again reduced that operation. So from 1950 to 1958, Humble's Exploration Department remained strong and active. The sharp fluctuations in the numbers of wells drilled and in exploration activity often reflected forces other than the suitable prospects for drilling.

Of course many factors affected Humble and its leaders, and modified, in turn, the exploration plans made by Barrow and Davis. Here, we shall mention only a few: In 1950-1954, the Korean War increased demand for certain oil products, which caused the Texas Railroad Commission to revise upward the number of days wells could produce. When the war ended, demand dropped, and the Texas body decreased allowable production days. Again, the fight in Congress over reducing the oil depletion allowance from 27.5 to 20 percent (or less) served to foster another set of executive headaches and to raise concern about exploration, especially since increases in U.S. oil imports threatened Humble's position. Also the 1956 decision by President Dwight D. Eisenhower to veto the Harris-Fulbright Bill that would have deregulated natural gas (that is, remove gas from price control by the Federal Power Commission) led Humble to reconsider its exploration program for natural gas.

In the same years, Humble researchers developed the jet drilling bit, which reduced drilling costs. Drillers also began to use small-diameter bits, and this slim-hole drilling usually reduced costs by as much as 15 percent; while in some wells compressed air or gas came to be used instead of drilling mud. Many other technical and technological improvements helped also; yet the total outlays for exploration continued to increase as wells went deeper into harder rock formations.[5]

During the Suez Crisis of 1956-1957, Humble supplied eight million barrels of crude oil to Jersey affiliates in Europe. In fact, the Humble management stood ready to supply however much might be needed on a long-range basis. As the 1956 *Annual Report* remarked:

> The closing of the Suez Canal from November 1956 to March 1957 emphasized the hazards involved in dependence on oil from the Middle East and the importance of larger proved domestic reserves. It meant greater demand for domestic oil, not only immediately but also for the long run, in order to avoid risk of increasing dependence by the United States on foreign oil.[6]

Humble had grown to world stature.

X   X   X

Between 1951 and 1959, Humble consistently added to its reserves of both oil and gas. Several fields of consequence were located in East Texas and West Texas, but also in Louisiana, both onshore and offshore, as well as in California. By 1956, production of crude oil had topped the 1948 figure, rising to 371,000 barrels per day [b/d]; and the next year (1957), Humble production—with Suez closed—rose to 460,000 b/d. During the last half of that year, the Texas Railroad Commission cut back production from sixteen to twelve days per month, in the absence of any foreign emergency. Humble, in turn, pointed out that such operating conditions raised unit costs and handicapped the search for new reserves. Already, Humble had begun to use pipelines instead of barges to move its growing offshore production from wells off the Louisiana coast, which started production in 1957.

In 1958, Humble started to build what the Jersey *Annual Report* termed the world's largest gas cycling and processing system, near the famed King Ranch in South Texas. The Humble *Annual Report* added that the 30-inch transmission line would be completed in 1959, when the processing plant itself would be constructed. When complete, it would process 800 million cubic feet of natural gas daily, and recover approximately 28,000 barrels of liquid products during the same period.

Everywhere, demand for natural gas kept soaring; yet Humble explorers located more than an ample reserve supply. The King Ranch surplus, for example, was stored in several depleted oil fields. Humble researchers and production experts also kept experimenting with gas reinjection to aid in oil recovery. By 1957-1958, they had become expert in this practice.

Although the actual number of wells drilled varied from year to year [see table at the end of this chapter], finding new reserves inevitably added high costs to company operations. As Humble's *Annual Report* for 1953 remarked:

> Maintenance of adequate reserve capacity, while desirable in the national interest, adds materially to the oil industry's investment and costs. . . . It can no longer concentrate its attention primarily upon the necessity for expanding capacity. The need increases for intense effort toward control and reduction of costs.

Humble sought to reduce costs by more alert and efficient management, the wider spacing of wells, and many technological improvements. Also very important, more production resulted from old fields as they were unitized. In fact, demand for crude oil and refined products affected production more than it did exploration. When gasoline sales dropped, the refinery cut back production, and Humble bought less crude from independent producers. [See table at the end of this chapter.] Still, Humble executives knew that they faced a rough assignment in predicting demand in a changeable market. The alternative strategy of storing oil above ground remained an expensive price for failure.

X   X   X

Despite the high priorities on exploration and production, Humble executives especially prized the Baytown refinery. Always, they insisted that this model plant operate smoothly. Their quest for efficiency led them, for example, to close a small refinery in San Antonio in 1950, and to undertake a three-year modernization program at Baytown. In 1951, Humble dedicated a new Baytown refining research center—research, needed to advance continuing operations. Company executives also secured government approval for an exploration-production research center to be built in Houston. Humble continued to operate the butyl rubber and butadiene plants for the U.S.

government as it had during World War II. Largely because of the Korean War, all of those plants operated at capacity in 1951; but when synthetic rubber demand fell off in 1952, production was cut back. Elsewhere, the Humble Company persevered in its efforts to minimize water contamination through filtration and special treating units. It also started negotiations "with a chemical company to handle acid sludges as a means of further reducing air contamination."

During most of the 1953-1958 period, demand for refined products fluctuated as a result of the Suez Crisis, government controls and pressures, imported oil, and other variables. In this environment, important decisions worked to alter the nature of the old company. For example, Humble purchased the butyl rubber and butadiene plants in 1955, for $29 million, just in time to catch a market boom. During that year and 1956, Humble's plants produced at record levels. Earlier, the 1950 Humble *Annual Report*, on a note of triumph had stated: "Butyl rubber is superior to natural rubber in the manufacture of inner tubes from the standpoint of cost as well as quality. Practically all inner tubes are now made of butyl rubber." Now, that faith in butyl appeared to be vindicated; however, in 1954, the introduction of tubeless tires sharply reduced the demand for butyl rubber. Yet, by 1956-1957, new uses actually increased demand. Again, butadiene quickly became important in the manufacture of polymers, while its use in synthetic rubber manufacture also increased. Encouraged by these developments, Humble also built the world's largest benzene plant, as the company strengthened its commitment to petrochemicals.

X   X   X

Along its entire product line, marketing supported Humble's operations, and vigorous efforts to increase sales were made throughout the 1950-1959 period, under the the leadership of Harry W. Ferguson. One of Ferguson's best moves was to build bulk stations large enough to have space later for central warehouses and to acquire sites large enough for pipeline products terminals. For each complex, he judiciously selected a location.

Consumer sales often depended on product quality; so Esso Extra gasoline, which hit the market late in 1950, filled the needs of the ever-increasing, higher compression automobile engines. It continued to be an exceptionally popular name brand for many years, but Golden Esso Extra, a companion brand that followed in 1956, did not enjoy such sustained popularity. Obviously, the travelling public did not feel the need for a third grade of motor gasoline.

Always, the Humble Company prided itself on its attractive service stations. Old stations were torn down and new ones erected as part of a continuous upgrading process. Periodically, all service stations were reconditioned, new equipment was added, and display space refurbished. The customer deserved the newest and best.

Through conferences, training sessions, supervisory visits, award programs, and similar controls, Humble sought to improve customer relations. So completely involved did company executives become in the sales of gasoline and products that in 1958, they developed an eight-week training program principally for beginning operators. By 1956, a dealer council had also emerged, to meet with company managers and to review the retail program. As a result of these measures, once Humble became the leading gasoline marketer in Texas (1951), it continued to increase its percentage of that market. One reason for this steady success can be found in the credit policy: "Humble-Matic" charge plates were mailed to all credit customers in 1952; five years later, after a tremendous increase in users, Humble issued modern plastic credit cards. Again, Humble became the first oil company in Texas to adopt this easy way to transact business.

Aside from customer service stations, other sales divisions of Humble handled natural gas, crude oil, and petrochemicals. While, from the beginning, crude oil provided the major source of income, gas, and petrochemical sales increased almost annually during the 1950s.[7] Humble's business could never be thought of as static.

X   X   X

When the 1958 *Annual Report* used the word "People" to head the section on persons who worked for the company, it caught the flavor of the Humble Company before Jersey acquired 100 percent of it in 1959. Humble resembled a large family, both in informality and in its lack of job stratification. Persons who worked there for even a short time recalled that special environment, even long after it had disappeared. No wonder the *Annual Report* so often commented on the "high morale" that characterized Humble operations. Its employees knew the pre-1959 company as something more than a mere employer.

When the Korean War began in 1950, a number of Humble employees entered military service—under terms of the same generous leave policy that had been in force during World War II, which prescribed regular raises and guarantees of reemployment. At home, on-the-job accident and death reports were heard by the corporate Board, which usually commissioned a full investigation and often acted on its findings. In this and many other ways, Humble showed itself to be concerned for the welfare of its people.

In the company view across more than forty years of existence, Humble's problem of securing a reasonably well-educated labor force was never solved; but the situation did improve, although in some respects education in the South and Southwest developed only slowly. So, while Humble might have its pick of college-trained engineers and scientists, it also had to run a variety of training programs to help less well-educated employees. Thus, principally because of the Comprehensive Driver Training Program at Humble plants, auto accidents in 1950 were reduced 17 percent against the 1949 total. A new overall safety record in the company's history was achieved in 1951; there were only 3.2 disabling injuries per million man-hours. This earned Humble a number of national and regional safety awards, and the company went on to even better plant safety. In 1954, no fatal accident occurred, and in 1957, the natural gas operations established a new industrial record of 3.2 million man-hours without a disabling accident. This record was made possible by management's willingness to spend money on safety.

Mobile medical units circulated among field crews—giving physical examinations, dispensing pills, and offering advice. Company infirmaries at Baytown and Houston received constant attention. For example, in 1958 the medical staff became concerned with the effects of plant noise—"sound pollution"—on the hearing of Humble employees. Further reports indicated that excessive noise could be unproductive and dangerous, so Humble immediately took steps to reduce noise levels in its facilities. Again, when Humble researchers began to use radioactive materials, the company issued clear instructions concerning how they would be handled, and supervisors enforced these rules. In the nine years before Jersey merged all the U.S. affiliates into Humble, medical insurance programs there were greatly improved, as the company assumed an increased share of the costs.

Years earlier, Humble had put into place a retirement program, termed the Annuity and Thrift Plan. President Wiess believed that Humble "was privileged" to be able to pay its employees well, so the company Annuity and Thrift Plan guaranteed $2.00 (annually) to go with each $3.00 the employee contributed. Death benefits were also generous. As inflation affected retirees' income, the Humble board approved special contributions to enlarge the fund. Almost 98 percent of Humble employees belonged to this program in 1950, and by 1958, slightly more than 99 percent had joined. Too, the company's group insurance program also compared favorably with any in the industry, as did Humble's vacation policy, and other employee benefits and privileges.[8] All together, the Humble Company prided itself on retaining those people it hired. "Employee turnover was low, as usual," the 1954 *Annual Report* stated. This saved the company money.

X   X   X

In 1957, Roper Associates conducted a series of interviews for the Jersey Company. The Roper report quoted an unnamed labor leader: "The labor policy of the Standard Oil Company . . . is very bad . . . ," and went on to criticize the entire industry, of which both Jersey and Humble were members:

> When I said that the oil industry's labor policy was bad, I was thinking of the fact that they have encouraged company unions, and fought legitimate unions. They're very tough to get collective bargaining with. Many workers belong to the Oil Workers International, but the oil management has been successful in splitting up the workers.[9]

In truth, whatever Humble Oil did regarding national and international unions between 1950 and 1958, the fact remains that in elections called by the National Labor Relations Board, the company unions won. For example, in the 1956 election held at the request of the Metal Trades Council,* Humble employees voted to retain the Baytown Employees' Federation as their bargaining agent.

Why this pattern proved to be general remains an open question; but some facts seem relevant. In the Roper Survey of fifty-one "opinion shapers" ranging from U.S. Senators to newspaper editors to labor leaders, a number of the respondents commented favorably about the constructive labor policies of the Jersey companies—especially on high wage levels and a generous retirement system. In fact, some leaders of independent unions came to feel frustrated because Humble (like other Jersey affiliates) had "given so much to the workers for so long that these unions have never been able to make themselves as necessary to the workers as other unions." One union leader remarked, "Even before we start bargaining a lot of our battles are won."[10]

Cecil Morgan, who had a long and distinguished career with Standard Oil Company of Louisiana [Louisiana], and Jersey, took this view of why the affiliates could pay better wages:

*Member, American Federation of Labor-Congress of Industrial Organizations Affiliates.

The company's wage policy was such that they were just a little bit ahead of everything else in the area. One reason why the company could do that . . . the oil industry has a ratio of labor to physical equipment that is completely different than the automobile industry [for example]. Fewer employees operate a greater amount of equipment in the oil industry than in most any other industry.

Morgan went on to point out that the many technical advances in the industry required labor to acquire new skills and thus become eligible for better pay. Morgan believed Jersey and its affiliates planned the career program for labor almost as thoroughly as they designed the one for executive development. He emphasized that the company's health care, medical, and hospitalization plans also made working for Jersey affiliates attractive. These all served to offset efforts of outside labor unions to persuade employees to join their organization.

Some years later, Carl E. Reistle, Jr., who headed Humble, recalled that after the Teamsters* had won an election at Bayway, New Jersey, he told the plant manager:

. . . Why don't you ask your top supervisors to keep abreast of the thinking of your rank and file employees? . . . Your union representative is usually a pretty good guy if you treat him right. Make his union job one of which he can be proud. If you just brush him off, you don't help him. You should remember that his problem is your problem.

Then he said, "I don't care what union we have. We can get along with any union."[11] While some leaders denounced Humble as paternalistic, others complained that it coopted top union laborers for management training programs.

What labor leaders believed to be a preemptive move on the part of management designed to stifle labor organization, was usually part of the famed selection process operating throughout the Jersey system, the Compensation and Executive Development [COED] Committee. An outgrowth of the World War II years, when Ralph W. Gallagher headed Jersey and constantly urged all supervisory forces to look for and report on talented persons, under Eugene Holman it became regularized as an open-ended search for management talent. Hines Baker liked the idea; he and Wiess installed it at Humble,

---

*International Brotherhood of Teamsters.

where the 1955 *Annual Report* detailed: "Humble continued its Executive Development Program, including appraisal of management personnel, special intra-Company training assignments and participation in university courses in management."

Throughout Humble, supervisors and administrators on all levels held conversations with employees in their units at regular intervals and discussed with them their job performances. As Reistle put it, "Of course we have the records to go back and . . . to review men. We do this very carefully each year with all the ones who have any potential." For some employees, they recommended better jobs; and for some few undoubtedly, they recommended outright release. But written reports from the lower levels moved up to the next, ultimately to the division head, and on to top management. Talented employees received quick attention.[12]

The COED Committee served many purposes. It gave Humble employees the feeling that the company could be flexible and that some better position could be reached. It served to bolster morale and to stimulate performance. People on many levels become friends with potential company leaders.

Yet just how well the COED Committee system actually worked at Humble is not precisely clear. As Morgan Davis recalled:

> I'm sure we didn't formalize to the extent that Jersey did. We did along toward the latter part of my service. We spent a lot of time on it. I think we realized . . . that . . . one of our prime responsibilities was to furnish somebody to take our places. . . . We took it very seriously.

And as another observer remembers:

> One of the weaknesses of the old Humble Company was that they did a damn poor job of developing people. . . . They ran their business on a completely functional line. The exploration people never crossed over to the producing people, and vice versa. They only came together at the top.

Hindsight allows for no easy consensus.

X   X   X

Reistle, who rose through producing to become the Humble Company's CEO, remained for years one of the company's most popular executives, though he remembered that many of the employees would remark when they saw him approach, "Here comes Mr. No," because even though he listened to them, he frequently turned down their proposals. Reistle said on many occasions: "I always felt that the most valuable asset we had was our people." He thought the oil industry traded on the importance of the individual, and on teamwork. Regarding union representatives, he said, "I always listened carefully to their complaints." Reistle attended many personnel committee meetings, especially those where unions were represented. He usually took with him what he termed "bright young men" who kept the minutes. Many employees "couldn't express themselves very well," but the carefully drawn minutes of the meetings did not indicate that. This practice helped the union representatives to show their constituents that they had been fairly represented.

Humble's leaders constantly reminded managers, from foremen up, that they spoke for Humble: "You listen and you make the decision." As Executive Vice President, Reistle told them:

> Now look, when a guy brings in a grievance, or you have a problem [which you can't answer], you talk to them and say, 'Now I'd like to think this over. This is something that I ought to study before I give you the answer.' Then, you should contact the next in command and discuss the problem with him. Be sure that you and your management associates are in accord on the proper answer so that there will be no problem if the answer is challenged and appealed to higher management.

This decentralization of authority and responsibility helped develop good management, and it also gave all supervisory employees a feeling of involvement in company decisions.

Humble leaders, in turn, believed in their system and in the company: "The reason I'm working for Humble is because it's the best damn job I can get," Reistle repeatedly told employees. And he wanted employees to believe it also. Good morale meant high productivity. M. A. Wright, who for eight years served as Jersey's contact director for Humble, also told employees and associates at Humble that they had the "best job in the world." Other executives made the same statement.

From its founding, Humble testified to its dedication to integrity; its executives wanted no sharp trades. As Wright recalls, "We had plenty of money, and we had more opportunity than we could really take care of, and we basically felt like we were moral and honest people." Once, when asked about peculation in the company, Wright laughed and asked the questioner whether he knew Hines and Rex Baker, and Wallace Pratt. When answered "yes," he remarked that they were almost as puritanical in their moral codes as Farish and Wiess had been. Carl Reistle also preached on this text:

> We want the reputation that we are fair with our suppliers, our customers, and the people from whom we buy or lease land. We should remember that we will in all probability deal with them many times in the future. If we follow this practice, they will respect us and treat us fairly.

In his explanation, Reistle pointed out that Humble had to be a leader; "if it was the law, we obeyed it." Humble executives wanted to avoid any public criticism of their company.[13]

X　　X　　X

Several Humble executives viewed the Jersey-New York operation as much too centralized, even though they accepted (and even liked) the holding company contact directors. This difference in operating styles naturally produced some resistance to "headquarters," which the tortured legal history of Jersey in Texas (featuring repeated efforts by that state to bar Jersey from operating there) did nothing to relieve. In fact, a few Jersey executives in New York accused Humble leaders of using this ongoing threat of legal action by the state of Texas to keep the Jersey Company out of Humble affairs.[14] Yet despite some difficulties from time to time, no real friction appeared to alter the relationship between the two companies.

Humble, from the beginning, had excellent leadership on virtually every level. Jersey's legendary CEO, Walter C. Teagle,* did not tolerate meddling in Humble affairs; for his own replacement at Jersey, he chose Humble CEO W. S. Farish.** So later, when Humble-trained Eugene Holman became Jersey CEO, on June 12, 1944, Humble executives felt certain that their independence would be continued, as it had while both Teagle and Farish headed Jersey. Holman "was very strong in resisting . . . [those] who desired to encroach on the day-to-day operations of Humble," but not everyone at Jersey shared this view. Mike Wright remembered that the tradition of independence at Humble continued even after Jersey acquired 100 percent of the stock. He stated:

*November 15, 1917-June 1, 1937.

**Farish became Jersey CEO on June 1, 1937.

Almost until I went there [1966], the relationship of the two organizations was distant. The president of Humble would go see the
Jersey president and discuss what to spend on capital investment in
the next year. . . . Jersey exercised very little influence.

When Wright moved from Jersey to head Humble, he notes: "one of the
problems I had was that New York wanted to get too deep into the operation
of Humble Oil & Refining Company, really staff people, not top executives."[15] Although the two companies continuously exchanged top-level
leaders, even that caused some ripples at Humble, where many persons
considered it rank desertion to leave for Jersey.[16]

Yet when men of stature—Coleman,* Farish, Pratt, Suman, and
Holman—moved on to Jersey, their reputations increased rather than diminished. Of course scores of other Humble employees also gained national
and international visibility while working for Jersey, especially on the functional levels of exploration, producing, refining, and pipelines. Thus, when
in 1957 the holding company decided to consolidate the U.S. domestic companies, a large number of persons at the new Humble Company had already
worked with representatives of Jersey or other affiliates.

The greatest intangible strength of the company proved to be an attitude
that characterized personal relationships during Humble's first forty-two
years. While Jersey played banker and advisor, Humble grew in its own way,
under the leadership of Farish, Pratt, Blaffer, Wiess, Baker, and others. In
fact, so swiftly did Humble grow that the system of personnel stratification,
so characteristic of large corporations, never could catch up. Thus, executives from the founding group easily called by name many of the "rough-
necks," roustabouts, and other workmen scattered throughout the
company—and they were answered in kind. When an employee aired a
complaint, he usually knew that a founder would hear it, and that some
action would be taken. This easy familiarity, part Southern, part Western,
characterized the unusually good company morale during Humble's first
four decades.

Henry B. Wilson, a Humble man who moved on to head Jersey's Public
Affairs Department, once remarked: "I just don't know of a board that was
as close to its personnel. . . . There really was a terrific esprit de corps in
that company." Wright also pointed out that, because of its independent
operation, Humble employees felt farther removed from Jersey than did the
employees of other affiliates. During the eight years he served as contact

*Steward P. Coleman served as a Jersey director from January 1, 1946 to March 31, 1961; vice
  president from June 2, 1955 to March 31, 1961.

director for Humble, Wright recalls, "there was very little relationship between the two organizations except that a very high level of friendship existed." Budget review remained perfunctory. Dr. Richard J. Gonzalez, a
one-time Humble director, remembered a good many things about company personnel before the merger. "Top management," he said, "still had a
lot of friends in the ranks of the working corps." He continued:

> . . . even as the company mushroomed very quickly . . . it was
> still interesting to go down to the Baytown Refinery and see your
> executives walking around the yard and talking with people, and
> knowing something about their families.

In discussing the employees of Humble, Gonzalez said that the stock buying
plan for employees followed right up to the 1959 merger, achieved a benefit
no one really could have expected. "Fairly modestly paid people ended up
accumulating very sizeable fortunes," and this gave them a special interest
in the company and helped create high morale. Wilson, who became pretty
thoroughly indoctrinated in the Humble mystique remarked about this: "I
have never in my life seen a morale or a feeling that you couldn't take off
your coat and fight any sonofabitch coming along who said a goddamn
word about Humble. . . . That Humble thing was something."[17]

X   X   X

Wright, in a long career with Exxon,* became familiar with many people
in the company's Exploration and Producing Divisions, including Wallace
E. Pratt, about whom he said: "To the explorers, to the exploration men, and
to the geologists, Wallace Pratt was something of a god. They all looked up to
him. And other oil men revered him almost as much as they did."[18] After
Pratt retired from Jersey, on July 1, 1945, his influence could still be felt at
Jersey and Humble.** For example, in the 1950s, two of Pratt's close friends
were at the helm of the Jersey Company—Board Chairman and CEO
Eugene Holman and President Monroe J. Rathbone. Many years earlier,
Pratt, who had hired Holman, recognized his potential for managing.
Rathbone, who in those early years headed Standard Oil Company of Loui-
siana, first knew Pratt as Humble's chief geologist and later as vice presi-
dent. The two spent many hours discussing oil reserves and the importance

---

*Standard Oil Company (New Jersey) became the Exxon Corporation on November 1, 1972.

**Pratt stepped down as vice president of Jersey on January 31, 1945, but at Eugene Holman's
request, he remained on the Board of Directors. John Suman later replaced him.

of continuously adding to those assets. Rathbone, classed by other oilmen as a refiner, eventually acted on these ideas. In fact, as long as Pratt remained alive, he continued to advise top executives of both Jersey and Humble.

In 1955, while in France, Pratt reflected on the corporate relationship between the two companies, Jersey and Humble. He knew that in the years he served as Jersey vice president and director,* this matter had often concerned their respective boards. Yet the problem had not been solved. Finally, he decided to telephone Rathbone and propose a plan that might enable Jersey to acquire all the Humble stock—one that could be worked out without the legal complications that many persons anticipated.

While this went on, Jay E. Crane, treasurer,** sought to secure more Humble stock for Jersey. Leo D. Welch, who succeeded Crane as treasurer,*** followed the same practice. In 1954, Jersey also mounted a major effort to acquire Humble stock, and when that drive ended, some four years later, the holding company owned more than 90 percent of the shares of the affiliate. Still, attorneys for both companies believed Jersey should have complete control—100 percent ownership. In part, Jersey's fear of federal antitrust actions may have stimulated this suggestion, but there were also other considerations. Jersey Secretary John O. Larson**** described one of them:

> As long as you had 15 percent [or so] of Humble stock owned by some of the members of the founding families, it was a pretty hard position that the directors and top management were in. People who owned 10 or 15 percent were telling them [Humble] what to do and Jersey was telling them what to do. You had to get rid of that conflict.[19]

Before he retired, Pratt's good friend, Jersey Chief Counsel Edward F. Johnson, also became deeply concerned about the possible antitrust aspects of the relationship between the two companies. Johnson believed that the corporate division of Jersey affiliates in the U.S. offered a target for antitrust attack—as part of an expanding theory of law that held commerce to be illegally restrained "within and by a corporate family." Other Board members would have been surprised to learn that Johnson believed that, in the Jersey corporate setup, "The weakest unit is Humble," which, because of its

*1937 to 1945.

**1938-1944; Director, 1944-1957.

***1944-1954; Director, 1953-1963; Chairman of the Board, 1960-1963.

****1955-1969.

relationship with Jersey, "is being restrained from expanding into areas into which, . . . it would normally be expected to enter." Justice lawyers might also contend that Humble's crude reserves were being utilized essentially for Jersey's benefit, thereby depriving other companies of the opportunity to compete for them. Prophetically, Johnson added that bringing about a consolidation of the Jersey domestic interests [affiliates] into one corporation would have advantages, and he suggested that the move to do so should begin immediately.

Meanwhile, in Houston, Rex G. Baker, who had become Humble general counsel in 1943,* studied the corporate relationship independently. Baker also foresaw legal problems, and he remarked that the only way to avoid them was to have Jersey buy 100 percent of Humble. Pratt knew about the concern of both attorneys when, in 1955, he proposed his own solution to the problem.[20]

X   X   X

Hines H. Baker had come from the first group of Humble executives to be hired by the company founders.** After Baker, Pratt believed that Morgan Davis, who had been a Humble director in charge of exploration since May, 1948, should advance to leadership. Historians of Humble have described his work before 1948 as demonstrating "high general ability, good judgement, and capacity for management."[21] So to Pratt, Davis seemed the ideal person to lead a new generation of Humble chief executives.

Holman, Rathbone, and Pratt all knew that Humble shareholders trusted Hines Baker implicitly; that neither his personal nor public integrity had ever been questioned. From France, Pratt telephoned Rathbone that day in 1955, and suggested a "power play": ask Baker to come to New York and help Leo D. Welch and Treasurer Emilio G. "Pete" Collado*** acquire all remaining Humble stock. Then, Davis could be free to run the 100 percent owned Humble Company. This move would also open the way to CEO for Director Carl Reistle. L. T. Barrow, who headed exploration for thirteen years after Holman left, and who also supervised that function as director, vice president, and chairman, was one of the persons in whom Pratt had great confidence. Barrow retired in 1954. Reistle himself once remarked, "I don't think Mr. Pratt had any strong feeling about me one way or the other.

---

*Director, vice president, and general counsel.

**The Humble Company was incorporated in 1917.

***1954-1960; director, 1960-1975; vice president, 1962-1975.

But, he was very high on Morgan and very high on Barrow. So he wanted Slim [Barrow] and Morgan to become involved in the senior management of Humble. . . . It was my lot to fall in with that deal." Pratt obviously considered all three men top management talent. Reminiscing about those men, John Larson, Jersey's secretary, said: "As I recall, Wallace Pratt was the individual . . . who kept pressing the Jersey Board to move Hines Baker up so that they could get . . . Morgan Davis before it was too late." Larson went on to state that Jersey desired a much closer working relationship with Humble, and it needed Baker "to get the benefits of his influence and also [because] he would be a unifying force."[22] When he returned from France in 1956, Pratt stopped by the Jersey offices in New York and talked to Rathbone and Holman. Obviously, they agreed that the shift in management might benefit everyone concerned and especially the relationship between the two companies.

At the Annual Meeting of the Humble shareholders on April 29, 1957, Baker resigned, and Davis was elected president of Humble. When Pratt heard the news, he wrote to Rathbone:

> Humble's annual meeting and the resulting changes in the official family remind me of the incident which led me to telephone you from Europe some two years ago and of my later conversation with you and Gene in his office about a year ago.
> It appears to me that events have developed to a point where the solution we all planned and hoped we might bring about at that time is now a fact. I am gratified because I am confident the changes serve the best interests of the men involved. Humble's new president and Jersey's new Board member are now both in position to render a larger service and I think it likely that each of them will come to a greater satisfaction in his new capacity.[23]

The Jersey shareholders met on May 22, 1957, and elected Hines H. Baker director, vice president, and Executive Committee member, as had been arranged. Baker had already moved to New York.

In the meantime, and independent of this move at Jersey, another decision served to draw attention to the Jersey-Humble relationship. In fact, it made the elevation of Hines Baker to the Jersey Board of Directors resemble an act of genius. This was the decision of the U.S. Supreme Court in the U.S. vs. E. I. du Pont de Nemours & Company [DuPont] case. The court heard this case on November 14-15, 1956, and rendered its decision on June 3, 1957.

The 1917-1919 purchase by DuPont of 23 percent of the stock in the General Motors Corporation provided the basis for this suit. The court rendered an opinion that this stock interest enabled General Motors to obtain an illegal preference over competitors in automobile finishes and fabrics from DuPont. Thus, the court stated that this DuPont partial ownership of General Motors could be used to substantially lessen competition; it could tend to create a "monopoly" in a line of commerce. The court claimed that the DuPont-General Motors stock arrangement violated provisions of the Clayton Act of 1914, as amended in 1949.

One of the lesser reasons for public attention to the DuPont case was occasioned by divisions in the court that issued the decision. Only four justices affirmed the lower court decision, while two dissented. Three justices recused themselves. Interestingly enough, Robert A. Nitschke, who had headed the U.S. Justice Department team that prepared the criminal—later civil—cartel suit against Jersey and other oil companies in the early 1950s, had since switched over to the corporate side. He was among the lawyers whom DuPont (General Motors) retained for this trial.

In Houston, Humble General Counsel and Director Rex Baker heard enough about the DuPont case to realize that the principles involved might well be said to parallel the Humble-Jersey relationship.[24] Rex Baker thought of a Humble Company 100 percent owned by Jersey. He said: "I originated the idea but then it was just in order to meet an antitrust situation." He explained the situation this way:

> Jersey operating through Humble was a perilous thing because they had other subsidiaries of Jersey with which Humble was doing business. . . . Jersey then was leaving itself liable to the charge that it was favoring their subsidiaries in their business transactions, to the detriment of other potential customers, and therefore were guilty of discrimination and violating the antitrust laws. . . .

After he received the full text, Baker pored over the General Motors decision for several days. Then on June 7, 1957, Rex wrote a letter to brother Hines Baker in New York, telling him that Jersey should move expeditiously to acquire the remaining 2 percent of Humble's stock. Otherwise, it probably could anticipate a suit comparable to that which Justice had won against General Motors.[25]

X   X   X

During his eighteen month stint as Jersey director, Hines Baker worked closely with Holman, Rathbone, and Welch. When he left New York at the end of December, 1958, he had completed a merger plan in which Jersey offered five shares of its stock for four shares of Texas Humble stock. On September 2, 1959, the Jersey Board of Directors announced its decision to consolidate all "producing, refining, marketing and marine activities of affiliated companies in the United States in order to achieve nationwide operation under unified management."* To achieve this, the company proposed to follow the Baker plan, which called for the creation of a new Humble Oil & Refining Company in Delaware [new Humble] as a wholly owned subsidiary of Humble [Texas]. The latter would then transfer all its assets and operations to new Humble—if the Texas Humble shareholders [98 plus percent Jersey] approved the merger on November 25, 1959. Then, Texas Humble would merge with Jersey, which latter company would also acquire the obligations of Texas Humble. The plan also called for affiliate boards to take all proper legal steps, so that on December 31, 1959, Texas Humble would cease to exist.

The so-called Baker plan represented largely his personal ideas. But Baker also listened to Rathbone, who operated on an entirely different theory from Holman's. All three men moved cautiously, involving not only corporation counsel but also affiliate heads. As Morgan Davis recalled:

> I was commuting regularly to New York, and I took an apartment up there in the old Ambassador. . . . And we started working this thing out. Well in the fall of '59, we put it together in preliminary shape and invited the presidents of these different affiliates, like Bill [William] Naden [Esso Standard] and John [W.] Brice of Carter Oil Company and Duke [Harold W.] Haight [to participate].

When completed, the Baker plan outlined the broad features of the reorganization, while the details were left to be worked out by the board of directors of the new Humble Company after it was created.[26]

On January 15, 1958, Rathbone had taken the unusual step of writing a two and one-half page New Year's letter to "Presidents of Affiliated Companies." The CEO liked numbers and he furnished them; but first, he pointed out that:

> The preservation of this important competitive tool—the financial strength of every member of the Jersey family—is the problem

*For more about this merger, see the chapter, "1960 Reorganization."

which we believe merits our primary concern and that of the management of every affiliate. . . .

It now appears that our worldwide major product sales for 1957 were about 150,000 b/d less than we estimated they would be in the spring of the year and . . . prices realized have been lower than estimates at that time.

Rathbone predicted that 1958 would not be a good year; the oil business, he said, was in a very competitive situation, and he asked all affiliates to review critically their operating costs. Finally, Rathbone expressed his conviction that all the affiliates and Jersey could do a better job with fewer people. All together, he formulated ideas about establishing regional units for Jersey affiliates—to superimpose a control-advisory structure between them and the parent company. Humble, in his plans, would become the U.S. regional company.

On December 1, 1959, the Jersey shareholders met and approved the entire package. Texas Humble shareholders already had voted in favor of the change on November 25, and Humble Oil & Refining Company [Texas] ceased to exist on December 31, 1959.[27] Morgan J. Davis became the first president of the new Humble Company.

When the Humble Company announced the merger, it also stated that Esso Standard Oil Company, New York; The Carter Oil Company, Tulsa; Oklahoma Oil Company, Chicago; and Pate Oil Company, Milwaukee,* soon would be merged with it, and that Humble would have its headquarters in Houston. A news story pointed out that in addition to the potential for nationwide operations, the reorganized company would provide greater opportunity for the development of personnel, "greater flexibility in making investments" and more opportunities for growth and expansion. This, both Jersey and Humble executives expected, would better enable the new Humble to meet existing and expected competition.

X   X   X

The properties pulled together for the new Humble Company were vast and well established: Carter Oil had 2,700 employees and assets of $193 million; Esso Standard Oil had 25,000 employees and assets of $1 billion; Humble Oil had 16,800 employees and assets of $1.5 billion; while Pate Oil

---

*Globe Fuel Products Co., Chicago, also merged with Humble in 1960.

and Oklahoma Oil jointly had 1,800 employees and 600 service stations.*
The totals proved quite impressive:

Employees	46,300
Payroll	$337 million
Service Stations	22,000 (37 states)
Other Outlets	8,000
Crude Production	430,000 b/d
Refinery Runs	1 million b/d
Assets	$3 billion
Domestic Reserves of Oil & Natural Gas	Largest of any U.S. Company
Oil & Gas Wells (Active)	17,850
Pipeline Capacity	700,000 b/d Crude and Products
Tanker Transportation Capacity	350,000 b/d[28]

Nationwide operations provided one ostensible goal for the new firm, but the possibility of increased efficiency also spurred Jersey on. For example, Jersey affiliates Humble, Carter, Esso, and the smaller companies had all fought for a share of the market by adding outlets. As one writer pointed out, that represented "one of the grosser misallocations of land, labor and capital. . . ." D. Woodson Ramsey, who directed Humble's marketing program, described the problem he faced:

> The industry has more damn gasoline stations than we have customers—the oil industry problem is to find some way to deliver gasoline to the consumer other than by having service stations on every corner. It is too expensive.

Duplicate plants and facilities could be eliminated, and such functions as exploration and production could easily be combined. Refinery runs would be better planned and transportation better scheduled. For the planners, sufficient reasons existed to create the regional oil giant. Now, the problem appeared to be one of getting the entire operation to work efficiently. As Davis said, "Our goal is to achieve half of Jersey's earnings, and they have the same ambition for us."

Jersey attorneys immediately took the precaution of clearing all copyrights and patents, so that operations could begin on a national scale. They

---

*Assets not given for these two.

also protected for Jersey and for Humble the existing titles, patents, and copyrights of all the merged companies, in case they should be needed for later use. In fact, even after the shareholders had approved the merger, Frank X. Clair, one of Jersey's copyright and patent lawyers, attended a session of the Executive Committee where he stated a conviction that: "The merger of Esso Standard into Humble as now proposed would weaken the legal bases upon which 'Esso' and 'Standard Oil' were protected." The Committee approved the steps being taken to protect "Standard Oil" for trademark purposes and to cover Esso Incorporated. They also agreed that after merging with Humble, Esso Standard should be known as Esso Standard Oil, a division of Humble Oil & Refining Company.

In the year of the merger (1959), Humble and Esso both had near-record results, while Carter had the best performance in its 82 year history. In fact, in December, 1959, the last month in which Carter operated as a separate company, crude oil production reached an all-time high for a single month. Obviously, such a combination of companies would comprise a stronger company than the sum of its units. Adding to the breadth of the new Humble was Enjay Company, Inc., which it took over on May 31, 1960, giving it ranking among the top ten suppliers of petrochemicals in the United States.* As a division of Humble, it expanded rapidly.

X   X   X

"Five months ago, the shareholders of Standard Oil Company (New Jersey) and the old Humble Oil & Refining Company gave their approval to one of the most far-reaching and significant reorganization plans of either company," *The Humble Way* stated in 1960. And although some of the employees of the merged companies failed to see them, the advantages appeared obvious. Both President Rathbone of Jersey and President Davis of Humble sought to be reassuring: "We're not going to need any less people. We're going to need more," Rathbone said, but that did not prove to be the case.** Davis extended himself even further, promising few transfers.

The obvious reason for these errors was that, in creating a single U.S. company, there had not been time to rationalize the new structure. Instead, the merger happened so fast that it jolted the entire petroleum industry. One Humble executive recalls:

---

*Also discussed in two additional chapters, "The Chemical Story" and "Reorganization of 1960."

**See table at end of this chapter.

We had a board, and a headquarters, and a name—and that was it. We didn't have any of it planned out in advance. We didn't know who was going to fill the jobs—we just sat around a table and asked: 'Who will these men be?'[29]

In fact, the new Humble board of directors included H. W. "Duke" Haight, who had once run the Creole Petroleum Corporation; E. Duer Reeves, director and vice president; John W. Brice, executive vice president; Carl E. Reistle, Jr., executive vice president; D. W. Ramsey, Jr., vice president; H. W. Ferguson, vice president; William Naden, executive vice president; and Morgan J. Davis, president. John Brice had served as Jersey director before becoming president of The Carter Oil Company.[30] In all, this first Humble board included three executives from the old Humble, two from Esso Standard, and two from Carter Oil. Esso Standard and Carter joined Humble on January 1, 1960; Pate Oil Company and Oklahoma Oil Company were added later that year. As Hines Baker explained his reasons for wanting a single board that included representatives of the merged companies:

That way, you would get a common board who could bring these things together and in a short while reorganize it so that it [the new Humble] could do a real job. But in the meantime, the pride of each organization and the feeling of achievement . . . would be kept.

In retrospect, the period of transition, 1960-1962, lacked the unitary and relatively smooth operation that characterized most Jersey reorganizations, but no other had been so extensive. One contemporary source warned: "It will be months, perhaps years, before the ultimate organization structure will be completed." And years later, Davis remarked: "This [the new organization] was understood by all concerned to be an interim type of organization," as he went on to describe some of the things that the new Humble executives did and felt. Since a lot of Humble people believed Jersey to be too centralized, "We decided to play a waiting game. So we made a lot of charts."

Jersey Contact Director Wright, heavily involved in the merger, became Humble board chairman and CEO on June 1, 1966. His opinion of the changes, as could be expected, differed from those of Baker and Davis:

You know when we merged all of these entities in the United States under one company . . . in some ways I think it's almost harder than companies without the same background. You'd think when you put [together] Carter Oil Company, Humble Oil & Refining Company, Interstate Pipeline, and the tankers, and all these various energies, that since they were all subsidiaries of Standard (New Jersey), that it would be like one happy family. It didn't turn out to be quite that way.

Wright believed that part of the problem grew out of the old Humble's tradition of independence—Humble pretty much ran itself while other affiliates did not. This created some problems between the staff and men in the field.

J. Kenneth Jamieson, who joined Humble on January 1, 1961, as director and vice president, recalls: "I was really, in a sense, somewhat removed from the argument as a complete outsider. . . . There were a helluva lot of personality conflicts . . . between the various groups of people." Most of the consequential struggles, Jamieson believed, resulted from differing perspectives: Esso Standard men traditionally emphasized refining and marketing, while Humble gave priority to exploration and production. The Humble Supply Division seemed very weak, so Jamieson moved many of his best managers into it. Clifton C. Garvin, Jr. and Donald O. Swan both achieved impressive records there.

The key factor, however, was that the merged companies were all completely owned by Jersey. While many executives can remember incidents where conflict occurred, Jersey CEO Rathbone* made the plan work. As Davis remembered:

> Rathbone was on our side and understood our problems. And when they'd run to him with their stories, well, he'd tell them, 'Well now, we've got to get this straightened out; we're going to make one big company out of this thing, we want everybody to jump in and put his shoulder to the wheel.' Jack, of course, had such tremendous respect from everybody in the organization that was a lot of help.

Of course, some did retire or resign.

Despite Morgan Davis's attempting generously to shift much of the credit for getting the new Humble to succeed, personally he deserves more praise for its success than he claimed; or as senior Texans termed it, "for bridling that horse." The merger brought to his desk major problems in rapid succession. Davis proved a fine administrator in a near impossible position.

*Monroe J. Rathbone become CEO on May 1, 1960.

Humble's new management plan concentrated power in a nine-man board of directors. The next lower level, Coordination and Planning Management, included both directors and functional vice presidents. Finally, the Advisory and Service Management group covered Controllers, Tax, Law, Secretary's, Employee Relations, Treasury, Public Affairs and Economics and Planning.

Humble now represented eight units: four regions, three divisions, and one subsidiary.* Regarding this new organization, Davis remarked: "The regional type organization will develop men who are familiar with more than one phase of our operations before they enter company-wide administrative positions." Yet the plan also emphasized the old Jersey principle of decentralization: "Each operations officer is accountable to executive management for all assigned activities in his geographic region or operations division and has responsibility for competent and profitable conduct of the Company's business, including the efficient use and development of manpower."

As one indication of the new Humble's intention to become a nationwide marketer, in the fall of 1960, the company opened service stations in Ohio, Oklahoma, and California. Since the courts barred it from using the name "Esso" in some states, Humble accelerated its search for a name that could be used everywhere. "Our objective is to come to one brand—which one it will be, and when, is another question," said D. W. Ramsey, Humble's director of marketing programs. The company also continued to purchase marketing properties in states in which it had not previously operated.

X   X   X

Jersey President Rathbone studied the Humble situation with care. Realizing that after Davis and Reistle, the Humble talent pool for top management would be almost empty, he engineered some changes; so Jamieson, president and director of International Petroleum Company, Ltd. [IPC], came to Humble as director, member of the executive committee, and vice president on January 1, 1961.

This story of Jamieson's move from IPC to Humble bears retelling. As Carl Reistle remembers, "Well, Jamieson wasn't my pick at all to start with. He was an unknown quality." Ahead for Humble lay the retirement of both

*Regional offices were located in Houston (southwest region); New Orleans (southeast region); New York (eastern region); and Tulsa (central region).

The three divisions were: Manufacturing, Marine, Enjay Chemical Company; the subsidiary was: Humble Pipe Line Company.

Davis and Reistle. Then, as the latter put it, "What are you going to do next? Jamieson was obviously a man of great ability . . . so we brought Jamieson in." Reistle stated that as originally planned, Jamieson would succeed him as president and CEO. This appeared to be a common assumption throughout the petroleum industry. After Jamieson joined the Humble board, Davis surveyed that situation and made a decision:

> Along in '61, I believe it was, I suggested to Jack Rathbone that he make me chairman of the board and put Carl Reistle in as president, because it was getting pretty close to retirement for me and I didn't see any reason why Carl shouldn't pick up the reins and get all the experience he could. They did that. Carl was fine.

In September, the Humble board elected Davis chairman of the board and CEO. Reistle moved up to president.

Other employees provided a more difficult problem for the managers of Humble. In 1957, there were 19,604 employees; by 1958, the staff had been reduced to 17,600, as Rathbone suggested in his January letter. The *Annual Report* for 1958 reported:

> After careful analysis of the problem, in which consideration was given to the welfare of all employees concerned, a special voluntary retirement program was offered to employees 55 years of age or older. It included a partial absorption by the Company of the annuity discount caused by early retirement.

In view of the fact that Jersey had made no secret of its effort to acquire 100 percent control of Humble, it is highly probable that Esso Standard, Carter, and other affiliates reduced their work forces in anticipation of the forthcoming merger, and because of pressures from Jersey.[31]

With Jamieson aboard and following Jersey's lead, the Humble board began to discuss personnel with the executives of the three divisions. Here, the burden of conferring with Humble managers fell principally on Reistle, who has described the creation of the new Humble era—an era marked by reductions of land exploration in favor of offshore (continental shelf) drilling. Morgan Davis opposed this move, but now Humble belonged to Jersey.[32] As Reistle recalls:

> As we began to change our exploration policies, we found that it was impossible to run 100 or more company rigs at a profit. This

gave us a surplus of supervisory personnel and forced us to eliminate a number of jobs. From then on, we used contractors.

The changes disturbed him greatly.

After the merger, in fact, Jersey's interest in the new Humble increased.
The parent company fed engineers into supply and refining as it sought to
develop more profit for those functions. Since Jersey needed to improve its
financial situation, Humble provided an obvious opportunity for doing so.
The efficiency of many company functions drew attention. Reistle remarked that it became obvious that the "whole method of running producing operations as well as the refinery operations" needed to be changed.

Consolidated Humble had 42,000 employees, but management believed
that the company could operate efficiently with 30,000. Apparently, Carter,
under John Brice, suffered most severely. Reistle pronounced Carter's work
force a great resource: "Gosh, some of our best men we got out of Carter . .
some damn good men. . . ." But Carter did have a large staff, and it
showed very little production. Even though management's plan appeared
fair, some persons objected bitterly to being "forced out."[33]

In the early 1960s, Rathbone anticipated problems over the employment
of minorities. He sent for Davis and told him about an assignment—"a kind
of chore," but ". . . I'm going to get you to do it." He then explained that the
reports on the employment of blacks that he had received did not satisfy
him. To Davis, Rathbone said, "I want you to go around to every unit and see
what the situation is. I want to know how many black workers we have. I
want to know what they're doing. I want to know how many supervisors
you've got among them." Then the Jersey CEO told Davis to do whatever
was necessary to get the information—since all the units were now part of
Humble.

After travelling throughout the entire Humble network of regions, divisions, and the subsidiary, Davis commented: "I found two or three black
janitors at the other two refineries [Everett and Bayonne]. And no black
unions. . . . I went to Pittsburgh where we had a grease plant . . . I guess
they had forty blacks working there." Davis met with these men and learned
that they had their own union and were not interested in joining or merging
with a white union.

The Baton Rouge plant employed a good many black men; they, too, had
their own union. Davis found that, like the men in Pittsburgh, they didn't
want "any part of the white union. . . . Baytown . . . was a carbon copy of
what I'd found in Baton Rouge," he added. Finally, Davis took his report to
Rathbone in New York, who read it, and according to Davis, almost had a

stroke. The news that, except for Pittsburgh, Jersey employed blacks only in the South, Rathbone found hard to believe.

As might have been expected, Rathbone soon took action. By the time he left office on February 28, 1965, minority employment had increased throughout the Humble Company. Special training programs had been developed for workers in need of such assistance. In 1968, some six years after Morgan Davis first surveyed minority employment by Humble, Roy Wilkins, head of the National Association For the Advancement of Colored People [NAACP], personally presented to Humble the Award of Merit for recruiting, training, and employing minorities. The citation also mentioned Humble's displays of Afro-American painting. By 1968, nearly 30 percent of all new Humble employees came from minority groups (or about 600 in all). Many of them—more than two-thirds—were in white collar or professional categories. Even in retirement, Rathbone must have been pleased by the sharp reversal in Humble recruiting practices.

X   X   X

Jamieson became executive vice president of Humble on January 1, 1962. Later that year, John Brice, president of the Carter Division and director of Humble, took early retirement; and in the following April, Director William Naden also took early retirement. When Davis retired on November 1, 1963, he had (according to *Business Week*): "capped his career with a rare management experience: The transformation of a middle-sized, regional oil producing company into a national integrated oil giant that ranks among the top 10 of all U.S. companies both in revenues . . . and in earnings. . . ." Humble had become the largest domestic oil company, and only its parent, Jersey, and Socony Mobil Oil Company [Socony Mobil] of the international oil companies were larger.

A number of executives interviewed believed Davis to be a good administrator, though a bit stiff and authoritarian. Holding company executives resented the fact that he concentrated authority in his own hands—he would handle all contact with Jersey. Some of the Jersey Board members wanted Contact Director Wright to put a stop to that system; but Davis was a close friend of Rathbone—and Rathbone tolerated policy deviations now and then.

There are numerous letters indicating this lack of clear communication lines. In early August, 1961, Jersey Vice President William R. Stott wrote Davis that Jersey had established a Material Coordination Division within the Comptroller's Department. Shortly thereafter, Jersey indicated that it would like executives to visit all Humble refineries concerning material operations. Davis replied:

> You are probably conscious, as I am, of the considerable number of visits that are made to Humble facilities by representatives of Jersey and its affiliates. While I realize, of course, that for the most part these visits are necessary and desirable, nevertheless we feel some responsibility for minimizing wherever possible the time and effort that is required to arrange for visits of this character.

Davis then directed Stott to contact Harry W. Ferguson, Humble's director in charge of refining—"an appropriate counterpart."

After Davis's retirement, Reistle became board chairman and CEO and Jamieson became president. The changes seemed to confirm Rathbone's plan when he brought Jamieson to Humble.

X X X

Naturally, the surge of activity occasioned by the consolidation of Jersey's domestic affiliates drew attention away from other important operations. Humble's moves toward national marketing have already been considered; so far, however, nothing has been mentioned about the significance of Jersey-U.S. Department of Justice relationships in stimulating that development.[34]

For many years, Standard Oil Company (Kentucky) [Kyso] represented the largest contract customer of Standard Oil Company of Louisiana, and this relationship continued until the late 1940s. Then, in a corporate reorganization, Jersey merged Louisiana into Standard Oil Company of New Jersey, incorporated in Delaware [The Delaware Company],* and that company began to fill the Kentucky contract. When Humble became the sole domestic oil company affiliated with Jersey in 1960, the supply contract shifted again.

Terms of the contracts between the Jersey affiliates and Kyso are not available, but it is known that the Kentucky Company negotiated so well that the prices paid for oil and products were sometimes lower than those Louisiana charged Jersey affiliates. In fact, Jersey President Rathbone often discussed the Kentucky contract. Since all the Jersey affiliates involved produced the principal portion of their oil in Louisiana and Texas, the proximity of the Kentucky Company's marketing area made it an ideal sales

*The Delaware Company became Esso Standard Oil Company in 1947.

territory for them. Perhaps this fact influenced those companies when they came to discuss price with Kyso; in any case, the question of buying that company frequently entered into Jersey and Humble deliberations. As Carl Reistle recalls, "Kentucky was a wonderful outlet for the Jersey refining area, and for Standard of Louisiana, and for Esso Standard. Well, they said they couldn't buy it."

Rathbone always opposed the purchase, perhaps because he, like so many Jersey executives of that era, was overly concerned with possible antitrust actions against Jersey by the U.S. Department of Justice. Both Reistle and M. A. Wright agreed that this antitrust phobia often subtly influenced decisions by the Jersey Board. As Wright once remarked: "I think that the employees of Jersey were the most antitrust sensitive of [those of] any organization in the United States. . . . That was the first concern of most everything we did." When Rathbone continued to oppose the purchase of Kyso, Wright remembers:

> We talked about the Standard [Jersey]-Kentucky deal from time to time. . . . A fellow named Smith [W. C.] was head of Standard (Kentucky) and he and Rathbone were good friends. And whenever we'd talk about it, why Jack would usually end up by saying, 'You know Smitty and I are good friends and he's promised me that if they ever would consider selling, why they'll sell it to us, and for us not to press on them,' and so when the Standard of California bought them, Rathbone was real upset. He thought he had been double-crossed by Smith.[35]

But there is no telling what might have happened had Humble attempted to purchase the Kentucky Company. However, a reasonable conclusion is that, over time, events demonstrated that the course pursued by Rathbone was wise—given the treatment generally accorded Jersey by the Justice Department.

In the last several years of President Dwight D. Eisenhower's administration, the Antitrust Division of Justice received a transfusion of energy when William P. Rogers became Attorney General. A burst of antitrust activity resulted, and some attorneys in the Antitrust Division became intrigued by the terms of the Humble-Kyso contract, and they became convinced that the contract violated the 1942 Supreme Court Consent Decree which forbade noncompetitive practices violative of U.S. antitrust laws. They also believed that the contract violated the 1911 Dissolution Order of the Supreme Court.

Jersey and Humble, aware of this interest, became concerned over the possibility of another antitrust suit, and they instituted changes with the hope of halting the Justice investigation.[36] But on December 2, 1958, Justice filed an antitrust suit against Kentucky Standard, Humble Oil, and Jersey—charging that the six-year contract made in 1956 ignored the 1911 decision because Kyso agreed to buy more than 80 percent of its oil and products from Humble. Smaller companies had not competed in the bidding. Justice also alleged in the complaint that Humble agreed to keep out of Kentucky Standard's marketing area. In fact, had Humble made the effort to market in that territory, it is probable that any U.S. court therein would have found Humble guilty of marketing in the area reserved for Kentucky after the 1911 decree.[37] Although the amount of money involved in the contract is not known, Kyso's more than 8,500 retail outlets in 1957 sold 718 million gallons of Humble gasoline, while they merchandised more than 1.5 million gallons of crude refined into other products.

As has been indicated, Jersey frequently bought small oil companies—generally marketing operations—and it acquired these properties all over the United States.[38] In May, 1960, in an exception to that trend, Wright, Jersey contact director for Humble, ignored the civil antitrust suit pending over the Kentucky contract and brought to the Executive Committee a recommendation that Humble buy the Monterey Oil Company, whose management had approached Humble with an offer to sell. Monterey engaged primarily in producing oil and gas in California and Texas. Now, Monterey proposed to secure clearance for this sale from the Justice Department, and if that proved unsuccessful, Humble would have an option to withdraw its offer. The Jersey Executive Committee gave its approval to the transaction, conditional upon Justice viewing it favorably, and the Committee approved a ceiling price of $83 million for the properties. In January, 1961, Justice apparently supplied assurances that Humble and Jersey found satisfactory, for the Monterey deal went through at a price of $81 million.[39]

After President John F. Kennedy was inaugurated in January, 1961, he appointed his brother, Robert F. Kennedy, to be Attorney General. As antitrust activity in the Department of Justice again increased, the case involving Humble, Kyso, and Jersey took on new interest. Justice directed its special attention to the smallest firm, Kyso. At the same time, Standard Oil Company of California [Socal] also signalled intentions to become a national company. Already, it had production in the Gulf Coast states, for which the market areas of Kentucky Standard would provide ideal outlets. Talks began between Socal and Kyso executives, with Justice monitoring them, and certainly both Jersey and Humble became aware of these negotiations. Some individuals in the

Jersey Company have even stated that the president of Kentucky Standard again gave assurances to Rathbone that Kyso would never sell itself to Socal. Attorney General Kennedy approved the sale of that company to Socal. By June, Justice attorneys had drawn up a consent decree, which Kentucky Standard signed on June 6, 1961, one day after Vice President Stott had discussed these matters with the Jersey Executive Committee. Finally, Humble and Jersey decided not to appeal; they signed the consent decree on July 7, 1961. Under its terms (after July 1, 1966), neither Socal nor Kyso could buy any oil products from a Jersey Standard concern for the five state market served by Standard of Kentucky. For good measure, Justice also added six other states to the proscribed marketing area. Attorney General Kennedy advanced as his reason, "We have agreed to this judgement because we believe it will restore competition." The policy of 1961 did not last.[40]

X   X   X

Humble leaders, like Jersey leaders, realized that by marketing several brands they were confusing the buying public and hurting sales.* Beginning in 1960, Humble introduced a new name in a "few localized test Markets"— "ENCO" for Energy Company. President Davis remembers: "We spent a good deal of money setting up the word "ENCO. . . . We came up with the idea of being the energy company of the United States." As Reistle recalls: "We wanted to go to a new name across the country." In the discussions that went on with Jersey men, he learned that, "In those days, doing away with Esso was like doing away with God!"

When the tests proved successful, on May 12, 1961, Humble completed the change to Enco in twenty-one states, essentially in its southwestern and central regions. In nineteen of the twenty-one states, Humble products also carried the Enco name, while in Texas and Ohio, Humble products were sold under the Humble name. On the East Coast, the company continued to market its traditional Esso products. But on the façade of all Esso stations, "Humble" was prominently displayed. The company also ran large advertisements emphasizing the letters "EN" and "CO" in "Energy Company of America," which were always underlined. These often used photographs of great energy attractions, such as hurricanes, tornadoes, and Niagara Falls. Davis termed this "the biggest advertising and promotion campaign in its

---

*Few persons complained about this more than Rathbone. But he could never bring himself to act on his conviction. Simply, he liked both names, "Jersey" and "Esso," and didn't want either to disappear.

[Humble's] history," as Humble marketed two brands in 45 states for more than eleven years. Davis said, "We thought it [Enco] went fine."*

Predictably, Humble immediately made the move to establish alternate outlets for its products in the former Kentucky Standard territory and Jersey backed the move. Within six weeks, the Jersey Executive Committee "found no objection" to Humble's constructing 192 service stations to market its Enco brand gasoline. And in August, the Jersey Executive Committee approved an increase in the Humble budget of $18 million for service station construction. Humble and Jersey next began testing legal rights to the Esso name, and inexorably (as is described elsewhere) this led to changing the name of both companies.[41]

X   X   X

Whatever the executives of Humble and Jersey thought of the Justice Department's decision to permit California Standard to purchase the Kentucky Company, neither the affiliate nor the holding company changed any plans to expand Humble operations. As has been mentioned, Humble exploration crews began to search for oil on land in California in 1948, and in 1958, the company purchased half interest in some leases off the continental shelf there. Humble also began to plan a refinery that it needed for the West Coast market. Obviously, the leaders of Jersey and Humble reasoned that if Justice believed strongly enough that competition needed to be restored in the Southeast in the old Kentucky marketing territory, then Justice would not object if Humble moved to develop West Coast interests. If competition needed stimulating in the East, then in the West, where Socal dominated the business, the same principle should apply—or so they reasoned.

Both Jersey and Humble continued to purchase properties in the Western United States—and elsewhere.[42] While they generally bought smaller companies, it became evident that the easiest way to gain a stronger position in the West would be through the acquisition of an established company—one with production and refining. Even without major production, Humble had 200 stations in California, marketing under the Enco brand; yet they had only 1.5 percent share of the five-state Far West market. A portion of these sales came from Carter Oil, which still marketed under its brand names through some 600 service stations (even though Carter had been merged with Humble).

*On July 1, 1972, the Enco name was discontinued; the Jersey Company became Exxon Corporation in November, 1972, and a few weeks later, Humble became Exxon Company, U.S.A. [Exxon USA].

In the fall of 1963, Davis and Reistle informed the Jersey Executive Committee that Humble had been approached by a "non-affiliated U.S. company as to its interest in acquiring a portion of that company's assets." Humble investigated the proposal, and the two men indicated that it planned to pursue the offer. The Committee expressed no objection, but it wanted the full Board to consider the information before terms and conditions of the sale had been reached. Buying another oil company of the size outlined by Davis and Reistle required careful study.

The Humble spokesmen identified the "non-affiliated company" as Tidewater Oil Company [Tidewater], owned by the J. Paul Getty interests. Buying the West Coast assets of Tidewater would, in the opinion of many oil men, facilitate the entry of Humble on the West Coast market, since Tidewater had 3,900 service stations that could market Enco gasoline and Humble products. For Humble, entering the California market had not proved to be a speedy process. Even obtaining choice locations for service stations had become quite difficult.

On the negative side, across several years, Tidewater had spent little effort and money on marketing, and its West Coast Division was deep in debt. Tidewater did have considerable production, but its marketing needed revitalization. Since J. Paul Getty had begun to emphasize East Coast sales in order to utilize both his extensive Middle East reserves and his new modern refinery in Delaware, he had agreed to sell his Western properties. On the surface, this proved to be fortunate for Humble.

Carl Reistle became board chairman and CEO of Humble in November, 1963, and to him the Tidewater deal presented an opportunity to move decisively into the West Coast market. But the deal had to be thoroughly reviewed, so Clifton Garvin was sent to California to investigate the Tidewater offer. As Jamieson remembers, "I guess I was the lead on the Tidewater thing there. . . . I did quite a lot of work on it. Of course Garvin was working with me on it and I was working under Reistle." While Garvin and Jamieson checked, the Jersey Executive Committee agreed that the price to be paid for Tidewater should not exceed $330 million.[43]

Soon Garvin began to question the possible benefit to Humble and Jersey of acquiring Tidewater. His reports to Reistle and Jamieson reflected this opinion. Jamieson tended to agree with Garvin, while Reistle continued to press for the merger. Moreover, by purchasing Tidewater, they would be following the Justice Department line of promoting competition in the West, just as the California Company's purchase of the Kentucky Company served that purpose in the Southeast.

On November, 22, 1963, Vice President Wright discussed the features of the Tidewater purchase with the Jersey Executive Committee. Tidewater wanted to sell its marketing assets in eight Western states,* plus its 135,000 b/d refinery at Avon, California. Five foreign flag tankers either in operation, under contract, or construction were to be included, as well as two United States flag vessels. Tidewater officials were to initiate talks with Justice to see whether it had any objection to the transaction. In case the government took an unfavorable view of the sale, Humble would cancel its offer. Wright said the deal was to be concluded on March 31, 1964, and the Jersey Executive Committee "expressed no objection" to the Humble plans. Then, the Committee recessed for lunch, and learned that President Kennedy had been shot in Dallas.[44]

On the same day, the Gettys, headed by George, met with Humble executives to go over final details before the lawyers completed the necessary contracts and other papers. Reistle, Jamieson, and Attorney Nelson Jones represented Humble, and they met George Getty and his group at the offices of the law firm representing Tidewater. Shortly after the negotiations started, the two sides differed on the agreed-upon price by about $100,000. Reistle recalled: "So here they were arguing and everybody had been talking and discussing it, and they all just got mad . . . about the thing and you could see that the deal was beginning to get on everybody's nerves." Reistle then turned to Getty and said that neither he nor anyone else could fix a precise value for the property. Whether the $100,000 was added on or subtracted would not make that much difference. Then, Reistle reached in his pocket and pulled out a silver dollar saying: "I'll match you for that $100,000. If I win, the price remains as it is stated. If I lose, we will pay $100,000 more than that." Getty agreed and they flipped coins. Reistle won! He recalls, "From then on, everybody was easy again and so we made the deal." They agreed to meet later in Delaware to verify and sign the papers; then that group went outside and learned that President Kennedy had been assassinated.

Five days later, November 27, 1963, Humble and Tidewater issued a press release announcing the sale of Tidewater's Western properties. So far the Justice Department had not indicated that it contemplated any action. As the Humble press release noted, Tidewater would continue to produce oil and gas in the West, Mid-Continent, and Southern areas, while it would refine and market petroleum in the East only. Neither leases nor producing wells were included in the properties being sold, in order to forestall possible

*They were: Washington, Oregon, California, Idaho, Utah, Nevada, Arizona, and Hawaii.

Justice objections to shutting off the sale of Tidewater crude to other companies.[45]

Through the sale, Tidewater would obtain money with which to retire its debt. This would enable it to expand its marketing, i.e., be more competitive in the East. Humble would gain marketing facilities, plus a West Coast refinery which it needed. To many observers, it appeared that the transaction would benefit both companies and increase competition on both sides of the continent. Still, a letter from the Antitrust Division of the Department of Justice, stating that it objected to the sale, would void the sale, as the contract stated.

Executives and attorneys for the two companies met in Wilmington, Delaware on March 31, 1964, to conclude the transaction. Then, Jersey Attorney Thomas E. Monaghan notified Justice that the purchase agreement had been signed.[46]

Accustomed as were business leaders—and especially those of the Jersey Company and its affiliates—to the inconsistent interpretation and application of the antitrust statutes by the Justice Department, as well as to the Federal Trade Commission's ever varying enforcement of its many rulings, they still may not have been prepared for the Justice move against Humble and Jersey in the Tidewater case. Businessmen generally had assumed that when Lyndon B. Johnson became President, relations between government and business might settle down, to be less fractious than they had been during the last years of the Eisenhower administration and under Kennedy. Yet major businesses began to complain over what Chairman Frederick R. Kappel of American Telephone and Telegraph Company termed "an overdose of investigation." Companies everywhere were being investigated, and many of them indicted for alleged civil and criminal antitrust violations.

For Humble and Jersey executives, however, this 180 degree turnaround could not have been altogether unexpected. In the 1950s, Jersey Counsel Edward Johnson and Humble Counsel Rex Baker had told their respective boards of directors that they believed significant changes in the interpretation and application of the Sherman and Clayton Acts were in the offing. Too, both attorneys had warned that renewed activity could be expected from the Federal Trade Commission. And, when Tidewater and Humble signed the papers to complete the merger, antitrust activity was expanding.

Justice finally moved against Tidewater and Humble during the week of April 12, 1964, acting under Section 7 of the Clayton Antitrust Act. As *Business Week* described it, the basis for the new suit was indeed a novel interpretation of antitrust theory: corporate giants should grow from within, rather than by merger. In its complaint, Justice reached afield to ask that Humble divest itself of Monterey Oil Company, which Humble had acquired more than three years earlier, in February, 1961.* And Justice added a count that might well have crippled Humble's planned expansion, by requesting:

> That defendants Jersey and Humble be enjoined and restrained from acquiring, directly or indirectly, any stock, financial or other interest in, or any of the business or assets of, [sic] any person engaged in the United States in the production, transportation, refining or marketing phases of the petroleum industry.

These charges proved to be extensive and inclusive. Justice alleged that if the deal were allowed to stand, it would reduce competition, and eliminate Tidewater as a source of supply for Humble competitors and for cut-rate gasoline dealers. Justice asked for a preliminary injunction to stop the transaction.

Here, it should be noted that even before Justice moved, Jamieson, Garvin, and some other Humble executives had begun to question the long-range value of Tidewater to Humble. Jamieson remembers: "It came to the point where the Justice Department said they'd challenge it, and we just dropped it, for which I was very happy." Jamieson also noted that the potential for another antitrust suit did prove to be a factor in that decision. Then he added: "The more we got to looking at the property, [the more] I thought it got worse and worse looking. . . . I was quite happy when we decided to drop the thing." Garvin, too, proved to be delighted with the decision of the Justice Department to stop the sale of Tidewater to Humble. Years later he remarked that the company was "lucky" that Justice did move.[47]

Tidewater officials announced that the company would "resist the suit" since they believed the "transaction would clearly foster competition." Humble, more cautious, awaited study of the complaint by its attorneys.

Before the April 30, 1964 purchase deadline, Humble and Jersey had

*At that time such an action was not unusual. In 1960, Tidewater and Standard Oil Company of California had purchased the Honolulu Oil Corporation. Justice sought a temporary injunction to prevent the acquisition, but failed to obtain it. After the transaction was completed, Justice filed a new suit against California Standard and Tidewater. In May, 1964, the case was still in court and neither company then knew whether it could keep the assets it had purchased. *New York Times*, May 2, 1964.

In July, 1966, Phillips Petroleum Company purchased almost the same properties which Humble had bought and returned. In November, 1976, Federal Judge Warren F. Ferguson of Los Angeles ordered Phillips to divest itself of Tidewater holdings. *Wall Street Journal*, November 14, 1976.

decided to cancel the Tidewater purchase agreement. The official reason for the decision was the Justice contention that it would violate antitrust laws. Humble said that, even if it won, the lengthy litigation would delay the acquisition. The court dismissed the Justice case against Humble without comment because neither Humble nor Jersey had filed an answer to the complaint. Justice generously voided the charge that Humble had violated the antitrust laws when, in 1961, it purchased Monterey Oil Company. More important than either of these actions, however, was the fact that Justice no longer demanded that Humble and Jersey be barred from purchasing other oil companies or properties.

At the time Carl Reistle received the news that Justice planned to sue to stop the Tidewater transaction, he, Jamieson, and the Humble executives conferred with Rathbone, Wright, and others at Jersey. They agreed that fighting the case in court might jeopardize all future acquisitions.[48] For Jersey Vice President Wright, the Tidewater retreat changed the course of Humble Oil. Wright believed that decision not to fight the case became a cardinal Humble principle. He described for an interviewer what happened: "Tom Monaghan, who was the single [general] counsel in New York, was keeping the Justice Department advised that we were negotiating . . . so he went down to tell them that we had reached an agreement . . . and these people did not say anything one way or the other. But the next day . . . the Justice Department filed a complaint." Wright went on: "They blocked it. . . . Well, it was an interpretation, I guess, of antitrust." For him, the Justice suit did not involve competitive practices. Had the Tidewater-Humble sale been effectuated, the combined companies would have had only an 8.5 percent share of the West Coast market, while six other companies would have had more. As Wright said: "We suspected it was just a case of bigness." Justice believed Humble, backed by Jersey, could "go on a 'grassroot' basis and do it ourselves."

When sometime later, Justice intervened to stop another proposed merger involving Humble, the principle was confirmed. In Wright's words, "we were just barred because of size, . . . and so we concluded with those two cases tying in together that the basic problem was one of size and so we didn't try to acquire anything else." After that, according to Wright, "We just adopted a rule of our own that we wouldn't buy any oil property from anybody who was producing over 10,000 b/d, which is not much." For several years afterward, this principle guided Humble, which limited purchases to "a few wells here and there and that was the kind of rule we ran our business by out here then. . . ."[49]

The collapse of the Tidewater deal left Humble as it had been late in 1963.

At that time, either by construction or purchase, Humble had increased the number of its service stations to more than 1,000 in the eight-state area where Tidewater operated. Yet, the company still needed a refinery, and studies were started to determine where to build it. At the same time, the inconsistency in the Justice Department's application of antitrust statutes did not go unnoticed by executives of other oil companies. One of them remarked:

> Humble apparently was willing to pay a premium to get into the West Coast market. There aren't many others who could raise that kind of money, and the Justice Dep't. would probably jump on those. The point is, what's going to happen to these companies who for whatever reason want to sell out?[50]

[The answer to that question came twenty years later.]

X   X   X

Reistle remembers that "After the Tidewater deal fell through, we decided to build a refinery. . . . Our marketing people felt that we ought to . . . handle a lot of the Jersey crude that had to be imported." In March, 1965, Humble hired Bechtel Inc. to locate a refinery site, which it did. Reistle remembers that the place recommended was a swampy area at Moss Landing near Castroville, a few miles north of Monterey and Carmel, and that a great portion of the 450 acre property involved had been devoted to farming artichokes: "Well, we might as well have stuck our finger in a beehive because all these people didn't want any industry there. . . . They could see that a modern refinery . . . would upset their economic system. And then . . . [they believed] that we would destroy the ecology." Humble paid $3 million for the artichoke farm, and legal action began. Humble ultimately won a favorable decision, but one so hinged with restrictions that it would have to obtain approval for every change in the refinery plans.[51]

After a study of these restrictions, in June, 1966, Humble decided to relocate its refinery on some surplus U.S. Navy property it had acquired at Benecia, north of San Francisco. The new choice proved to be fortunate. When the Humble Company announced that it planned to build a refinery at Benecia, the release stated that capacity would be 70,000 b/d, and it estimated the cost at $135 million.* Construction began in 1966, and plans

---

*Costs rose to $169 million by 1968, Executive Committee Minutes, January 3, 1968, Secretary's Office, Exxon.

called for completion in 1969.

Even before construction started, Humble moved cautiously. Various spokesmen discussed plans with the director of Parks and Recreation for the State of California, the Sierra Club, and the Save San Francisco Bay Association, to ascertain their objections and to ask for suggestions. The company announced its intention of building a model plant from the environmental standpoint. Bay water currents, air currents, plant life, soil, and even sound levels were charted by experts. Normal smokestack height was to be increased 115 feet, to place it above any wind downdraft. Mufflers and resonators were planned to stifle noise. Plans called for air (instead of water) to be used in the cooling processes. Various environmental groups made quite a number of suggestions, and changes cost the company $10 million more than the original amount budgeted. Yet Humble spokesmen pronounced them very worthwhile in view of the community goodwill gained, as most environmental groups praised the company for its continuous efforts to maintain the quality of the environment.[52]

X   X   X

Back home in Houston by 1964, a new marketing era had dawned for Humble; it began with the birth of the "Put A Tiger In Your Tank" campaign to push Esso Extra, a premium grade gasoline. Behind the rapid success of this worldwide advertising campaign lies a complicated tale—one which demonstrates Jersey's vitality as a modern corporation. In the beginning, a single affiliate, the Esso Petroleum Company, Limited [Esso U.K.] used a rather wild looking tiger to advertise Esso Extra in England, 1953-1957. Though pleased with the results, Esso U.K. temporarily retired the fearsome beast from duty as a corporate symbol.

Then, in 1959, Oklahoma Oil Company, based in Chicago, sought ways to increase sales of its premium gasoline. They turned the problem over to an advertising agency, which passed it, in turn, to a young trainee, Emery Thomas Smyth. Motivational research associates told Smyth that people sought power when they bought gas. He thought of the tiger and next of the "Put A Tiger In Your Tank" slogan. So he drew the beast, which—despite his efforts—still fell short of looking amiable. But Oklahoma liked it and used it for television advertising in the Northwest.

Meanwhile, Jersey's merger of its domestic affiliates occurred in 1960, and President Morgan Davis went to Chicago to see the Oklahoma executives. While there, they showed him slides of their advertising program featuring the tiger with the soon-to-be famous slogan. Davis said:

This tiger was a very fierce looking creature. . . . It wasn't a character that would make you feel very friendly. After seeing a few [slides] . . . I thought it was pretty corny, and I said, 'Just cut it out. I just don't believe it's any good.'

In the reorganization, W. W. Bryan, who had been marketing vice president for Carter, joined the Humble team as director and vice president. Earlier, he had borrowed the tiger from Oklahoma and used it in marketing campaigns in the West. When the time came for product advertising, Bryan remembered the tiger: "Let's take a look at the old tiger program that somebody killed [Morgan Davis]. . . ." When they dug up the sketches, Bryan said: "This is not a very nice looking tiger. But let one of my artists see if he can fix it up." Then he put the Humble Advertising Department and the McCann Erickson Advertising Agency to work on the project, and the "friendly" tiger was born.

Bryan developed the tiger's personality. The beast, in his opinion, appeared as a good-humored, friendly, helpful animal—and very effective. Apparently, virtually everyone else thought so too, as the tiger went on to become one of the most famous symbols in advertising history. While it promoted Esso Extra, sales of other grades of gasoline also improved wherever the tiger appeared. The tiger program went well in foreign countries too—even those where a tiger had earlier been used in advertising, such as England.[53]

X   X   X

Within three years, Humble again sought to expand its West Coast marketing by negotiating with Socal for the purchase of its Signal Oil Co. Division. Signal had about 1,500 service stations and 150 bulk storage facilities. Justice reacted to the news of the proposed sale by indicating that it had no plans "at this time" to enter a challenge; essentially, this meant that Justice saw no violation of the antitrust laws if the transaction went through. Since Signal belonged to Socal, the latter would be weakened in the competition for markets, while Humble could gain the 3 percent held by Signal. With the acquisition, Humble would increase its share of the West Coast market from about 2 percent to more than 5 percent. The price Humble finally agreed to pay was slightly more than $100 million.[54]

X   X   X

Naturally, Humble's expansion to the West Coast included both market-ing and exploration. The company hoped to locate oil in sufficient quanti-ties for the new Benecia refinery, but through the 1950s, California strikes proved insufficient. So Humble continued to depend on supplies from the Middle East. Meanwhile, geologists and geophysicists pushed the search northward. Precisely when Humble and Jersey became interested in the prospect of finding oil in Alaska is not clear, but the two companies certainly knew about Alaskan exploration prospects as early as the end of World War I. All the oil companies took note of President Warren G. Harding's action in 1923, when he set aside a large portion of the North Slope (near Prudhoe Bay) between the Brooks Range and the Arctic Ocean, as Naval Petroleum Reserve No. 4 [Pet Four]. Harding also pointed out that there were "large seepages of petroleum along the Arctic Coast of Alaska and conditions favor-able to the occurrence of valuable petroleum fields," as he established a 23 million acre reserve to provide future supplies of oil.

If the Jersey Company—or any of its affiliates—joined in the rush to secure Alaskan claims shortly after that event, the authors of the second volume of the *History of Standard Oil Company (New Jersey): The Resur-gent Years, 1911-1927* did not record that fact.[55] What is known, however, is that many companies—Jersey may have been one—did secure leases on land adjacent to Reserve No. 4. Their subsequent failure to make required im-provements on the land led to revocation of these leases. Of course, reasons for this lack of development by oil companies are easy to find: foul weather, a transportation problem, and especially the low price of other oil. For Jersey and for Humble, then, exploration went forward in warmer, more accessible climes.

In fact, between Harding's action and World War II, little exploration activity disturbed the Alaskan Arctic wilderness. During that war, the U.S. became concerned about the volume of oil from domestic sources used to supply its allies; so in 1944, the U.S. Navy authorized drilling in Reserve No. 4. This exploratory drilling led to an announcement that a field of 100 million barrels had been located; but even so, due to the high cost of drilling and the obvious transportation problem, all drilling ended when the navy left in 1953.

Before this, in 1944, Wallace E. Pratt, one-time Humble chief geologist and vice president, and at that time, Jersey director and vice president, published an article in *Harper's* magazine, "Oil Fields in the Arctic." Here, Pratt declared that the lands that bordered "the Polar Sea—the Northern shores of Siberia, Alaska and Canada—are marked by a series of conspicu-ous seepages of oil." Geologists, he pointed out, estimated that in the Western Hemisphere alone, about one-half million square miles remained to be explored—an important prospective area, all of it above the sixtieth parallel. He also noted that "Point Barrow oil contains a valuable lubricating fraction and is low in sulphur." Later, Pratt claimed no special credit for pointing to Prudhoe Bay as a likely oil prospect: "Hell's Bells. Both the surface and subsurface evidence has been there all along for anyone to see. No question about it. We were bound to make a strike eventually."[56]

In the late 1950s, Humble became seriously interested in Alaskan oil, partially because it recognized a threat to other sources of supply after Egypt's Colonel Gamal Abdul Nasser closed the Suez Canal. Everyone knew that an inadequate supply of oil could paralyze the economy of the West. The Humble *Annual Report* for 1956 called attention to "the importance of large proved domestic reserves."[57] One-time Humble Geologist, J. Ben Carsey, like Wallace Pratt, always believed oil would be found in Alaska. In 1954, he had sent Geologists Dean Morgridge and Alex Osanik to Alaska to make aerial surveys. In 1955, they published a report outlining that territory's oil potential. Carl Reistle, Humble's director in charge of producing, recalls the company's response: "At that time, both [Morgan] Davis and I were intrigued with Alaska. But we made a bad choice there. We selected the tertiary out on the Aleutian Islands [actually the Alaska Peninsula] and made a partnership arrangement with Shell [Royal Dutch-Shell Group] to get part of that. . . ."[58]

Humble also increased reconnaissance work in Alaska. While the drilling proceeded, it made aerial magnetic surveys over some 10,500 square miles. The first Alaskan well, Bear Creek Unit 1, in the Kanatak Area, was scheduled to go down 12,500 feet, under the extreme conditions detailed in the 1957 Humble *Annual Report:* "The remote location, severe cold, and high winds have presented unusual difficulties." John Loftis, exploration manager of Humble, went to Alaska in 1958. He remembers: "We established an exploration office in Anchorage. Of course, you could only work there in the field in the summer time."

Drilling continued in the Humble-Shell well during 1958. It was finally abandoned in March, 1959, after it had reached a depth of 14,300 feet. Humble's share of the well cost more than $7 million, making the prospects in Alaska seem bleak. In Loftis' words: "And then we sort of gave up the ghost, [and] brought that office out of there." Reistle expressed the decision to leave Alaska more succinctly:

> Morgan [Davis] was chairman of the board . . . he was a fine exploration man. . . . When we tested the area which we thought

was the most desirable and the easiest, we had mounted quite an exploration effort up there. . . . We had a lot of information [about territory] north of the Aleutians . . . but after we drilled that dry hole, Morgan said, '. . . Let's get the hell out of here!' So . . . we pulled . . . completely out of Alaska.

Loftis recalls that Richfield Oil Corporation [Richfield] brought in a field in the Kenai Peninsula, where Humble applied for additional leases. But before the U.S. Department of the Interior could rule on that application, conservationists in what Alaskans termed "the lower forty-eight" states mounted such vehement opposition that Secretary Fred A. Seaton soon "stopped the issuance of leases and closed down existing oil operations."

Although the shows of gas and oil in the Humble-Shell well did not prove to be of commercial importance, Humble, acting alone, did locate a significant deposit of iron ore. This lease, the company held.[59] In fact, Humble maintained an interest in Alaskan oil long after its active search for crude had ended. But to continue a search for reserves equal to or greater than the amount of crude oil and gas produced proved increasingly expensive, so in 1959, Humble closed the Anchorage office.

A year later, J. R. Jackson, who was handling the assignment of geologists in the Los Angeles area, and who also believed that oil would be found in Alaska, went to John Loftis, general manager of Humble's Minerals Department, and asked that Humble try again there. When Loftis agreed, Jackson sent Morgridge and land man Robert J. Walker to Alaska to make more aerial surveys. The North Slope area proved attractive and Jackson began to shuttle back and forth, Los Angeles to Houston, in an effort to persuade the Humble board to invest heavily in that area. Merrill W. Haas, a long-time distinguished Carter Company geologist who became Humble vice president for exploration in 1960, blamed the board's reluctance to invest in Alaska on the Bear Creek well disaster—the most expensive dry hole in Humble history.

To some of the geologists and production men, the indecision represented a form of management paralysis. Perhaps they overlooked the fact that gradually the Jersey system of long study and full reports was replacing the entrepreneurial style decision making that previously characterized Humble Oil. So Morgridge and others made additional studies for the board. Up to December, 1963, Haas and Jackson believed Humble could have gathered the information that would have enabled it to obtain almost 100 percent of the leases on the North Slope. But in the end not enough funds were provided for full exploration by the seismic crews.[60]

Meanwhile, by 1964, the Richfield Oil Corporation had acquired leases to 800,000 acres in the Prudhoe Bay area, 200 miles east of Point Barrow. Richfield had done a good job of "geologizing" the structure. Jackson and Haas knew this and after the latter assured the board that expenditures would be limited to $20 million, a merger of Alaskan interests with Richfield was approved.

In Los Angeles, Jackson initiated talks with Rollin P. Eckis, Richfield president. Soon they made a deal; Humble paid Richfield $1.5 million toward its 1964 exploration budget plus $900,000 for 50 percent interest in its leases. Humble also agreed to spend $3 million on seismic and other exploration; afterwards, the two companies would be equal partners with Richfield serving as operator. Carl Reistle described these events simply: "We got re-enthused on Alaska, so our exploration people, under Tom Barrow [?], went to Richfield and made a deal with them for $3 million to get half of their North Slope properties." The principals agreed that additional money would be needed for exploration and geophysical work. M. A. Wright, Jersey vice president and contact director for Humble, said that Jersey approved the arrangement because, "Richfield had good exploration people."[61]

The Humble-Richfield agreement of 1964 specified that the two companies would share equally in all "exploration and development north of the Brooks Range and east of Pet Four." Both partners agreed to a buy-out plan; neither company would sell its Alaskan interests without first offering them to the other. At the October, 1964 lease sale, the new partnership acquired considerable acreage, showing revived faith in Alaskan oil.

Humble influenced Richfield to end exploration in the foothills of the Prudhoe Bay acreage where the land was owned by the U.S. government and move the seismic crew over to state acreage. Ultimately, exploration there provided the information that led to the Prudhoe Bay discovery.

Alaska announced in October, 1964, that in December there would be a competitive lease sale on lands west of Prudhoe Bay. Quickly Humble's Bill Smith coordinated geophysical operations "covering the more prospective leads." In the bidding, however, British Petroleum Company, Ltd. [BP]* and Sinclair Oil Corporation [Sinclair] secured many more tracts than did Richfield-Humble.

Regardless of whether Humble or Richfield took the leading role in the Alaskan bidding, more important is the fact that both made substantial sums available for the joint exploration account. Each agreed to bid as much

*Before 1954, Anglo-Iranian Oil Company, Limited.

as $15 million for the leases and to put up $25 million for wildcats—for a total of $40 million. When he decided to invest that sum, the Richfield chairman ignored his board, and Humble had problems securing Jersey's approval. Richfield-Humble utilized the time between the December, 1964 sale and the announced July 14, 1965 sale to explore further the remaining Prudhoe Bay acreage. On these findings it proposed to bid. Humble wanted to increase the money available for bidding, with each partner adding $1 million. Under the agreement, the partners had to match funds. But Richfield had already encountered problems with the U.S. Department of Justice, and an Antitrust Division lawyer had suggested a merger with Atlantic Refining Company [Atlantic]. Certainly before July, 1965, it did not appear likely that Richfield would sink another $1 million on North Slope leases. And they did not match the Humble offer.

The result proved disastrous for the Humble-Richfield bidding; Haas remembered that had the bids included the extra $2 million, they:

> would have given us still another opportunity to come away with most of the field. In fact, we would have won every tract except one had the board [Humble] not required us to reduce our bids on the flanks.

Even then, the Richfield-Humble combination won on the largest number of tracts, 69 of the 104 on which it bid. British Petroleum Company obtained most of what was sold. As Wright told an interviewer: "Tom [Barrow, senior vice president of Humble] was smart enough a person to buy some acreage which turned out to be the Prudhoe Bay structure and Arco [Richfield] went along with it."[62]

Already planning a full operation, Richfield pioneered the use of a Caterpillar train, loaded with freight, that ran from near Fairbanks to the Arctic Slope. This surface route stayed open about seven months of every year. So, early in 1965, the company shipped drill rigs, prefabricated camps, pipe, and other supplies to Fairbanks and chartered air freighters to haul the stores to the exploration camp.

X   X   X

As Humble and Richfield increased their interest in the Alaskan North Slope, another oil concern also became very active here. Expelled from Iran in 1951, BP decided to adopt "a policy of world-wide diversification in exploration, refining and marketing." When it received favorable reports on

Alaska's North Slope, BP joined with Sinclair in acquiring leases. Wildcatters for the combine began to drill, and while prospects seemed good, they found neither oil nor gas in commercial quantities.

Beset by financial problems as well as the antitrust suit, Richfield merged with Atlantic Refining Company on September 16, 1965, to form the Atlantic Richfield Company [Arco].[63] Before that time, the Richfield-Humble committee had decided to locate its first wildcat on acreage leased from the U.S. In February, 1966, Arco and Humble began moving drilling equipment and material to that location named Susie No. 1. Ten months later, drilling was suspended there after reaching a record 13,517 feet and after a cost of $4.5 million. Arco and Humble knew that Sinclair and BP had drilled at least six dry holes on the North Slope during the three preceding years. Charles S. Jones, formerly chief executive officer of Richfield, commented that with Susie No. 1 dry, plus all the other dry holes, oil prospects on the North Slope looked pretty grim. Humble again influenced Arco to move on to acreage leased from the state, so crews moved the drill rig to Prudhoe Bay State No. 1 on the Prudhoe structure.

On December 26, 1967, a loud, vibrating noise attracted forty oil men to the wildcat well being drilled. Humble Geologist Gil Mull was quoted:

> It was about 30 below zero and there was a thirty knot wind blowing. We could hear the roar of natural gas like four jumbo jets flying right overhead. A flare from a two-inch pipe shot at least 30 feet straight into the wind. It was a mighty encouraging sign that something big was down below.

There was! As weather permitted, drilling continued. On March 13, 1968, the great Arctic pool was discovered—the fabulous Prudhoe Bay field— which another Humble geologist described as "a mean, nasty, unforgiving place to work." After that first well came in, Arco borrowed a nearby BP rig and set it up about seven miles southeast of Prudhoe Bay No. 1. There, by June, 1968, the drill bit had penetrated a 400 foot oil column, adding confirmation that the great oil field, an "elephant" in oil field argot, did exist.*

Humble and Arco employed the leading petroleum geology firm, DeGolyer & MacNaughton of Dallas, to estimate the size of the discovery. Their studies showed that the pool contained five to ten billion barrels of oil—"one of the largest petroleum accumulations known to the world today." The

*By 1985, cumulative oil production totalled 800 million oil equivalent barrels. For Exxon, in 1985, Prudhoe contributed 331,500 equivalent b/d.

estimates of other geologists ran much higher; in general, all agreed that the Prudhoe Bay field contained at least one and one-half times as much oil as the East Texas field, which up to that time had been the largest ever found on the North American continent. By now, oil company expenditures in Alaska totalled more than $1 billion.

William D. Smith, who assessed the potential of the strike for the *New York Times*, wrote: "This discovery . . . will open the door to a vast and fabulous play stretching from Alaska across the Canadian Arctic." He went on to point out that the oil possessed such low gravity that the pour point might enable it to be moved through an unheated pipeline. While everyone agreed that the oil had very little sulphur, some persons insisted that it compared favorably with the earliest Pennsylvania oil of the 1860s. And an Eskimo claimed that, occasionally, chunks of frozen oil had been carved from the ice and then burned as it thawed. Some enthusiasts went even further, to declare that Alaskan crude oil would prove so pure that it could be used in motors without being refined.[64]

In light of the subsequent battle over the ecology and environment that took six years to settle, it should be noted that Humble—for years an advocate of oil field unitization—Arco, and BP made numerous operating agreements covering any oil found. Each field would be operated as a unit, and all waste would be disposed of properly. These agreements prorated drilling costs and guaranteed maximum oil extraction.

For Humble and Arco the Prudhoe Bay discovery presented problems, many of them welcome results of success. Regarding one problem, Humble Chairman Wright* recalled that after the second well was drilled: "Bob Anderson [Robert O. Anderson, Arco chairman] and I went to London to see British Petroleum. They had acreage on the flank, and so we went over and talked to them . . . about buying their acreage. . . . They wouldn't go for that," even though BP was not drilling on their leases at that time. Next, Humble and Arco pressed BP to begin drilling, as a way of defining the area of the strike: "We were ready to start moving toward building a pipeline, and we were going to get it done in three years. This was in '68," said Wright. To BP, the two men offered the loan of "a rig or two" if that would speed up their test program.

BP had long sought access to the U.S. market. Now, the Alaskan strike provided a means to that end, for BP held almost 54 percent of the Prudhoe Bay oil field. BP next sold the bulk of its Alaskan production to Standard Oil Company (Ohio) [Sohio] for 25 percent of Sohio shares, plus additional

*Wright became chairman and CEO of Humble Oil & Refining Company on May 1, 1966.

stock (correlated to Alaskan production after 1972). Thereafter, BP's interest in Sohio rose to 54 percent. Meanwhile, the Jersey Board and Executive Committee kept informed about the Prudhoe Bay strike, and about the debate going on at Humble and in New York over how to move Alaskan oil to refineries. Even though many oil experts insisted that "several years of exploration" must elapse before this determination could be made, Humble and Arco proceeded rapidly toward a decision. In fact, they seem to have decided that they might need both tankers and a pipeline. For, on June 6, 1968, Humble and Arco applied to the U.S. Department of the Interior for a permit to build a pipeline. From that date, apparently, everyone assumed that a pipeline would be constructed, though many people also hoped for the success of tanker trials in the Northwest Passage.

Already it seemed clear that Humble, Arco, and BP together owned most of the Prudhoe Bay field, so their key executives formed a study group concerned with how best to move the North Slope oil to market. Later, the three firms organized an unincorporated company—Trans Alaska Pipeline System [Taps]—to plan a pipeline, and in the next two years, Taps spent $100 million. In August, 1968, Humble, Arco, and BP employed Pipe Line Technologists, Inc. of Houston to study the feasibility of a pipeline from the North Slope to some Alaskan port. These pipeline experts produced a cost estimate between $800 million and $1 billion. Other possible routes were also considered. The idea motivating Humble, BP, and Arco remained constant, however; they wanted an 800 mile pipeline to a year-round port on the Alaskan coast. From that point, tankers would move the crude oil to refineries such as Humble's Benecia, north of San Francisco, or through the Panama Canal to others in Texas, Louisiana, or on the East Coast.[65]

On December 4, 1968, the Jersey Executive Committee listened to Humble declare that it had "evaluated a number of possible alternatives for transporting Alaskan North Slope crude oil to market and had concluded tentatively that the possibility of moving ice-breaking tankers through Arctic waters (the Northwest Passage) appears to have substantial economic advantages over other alternatives, providing such movement is feasible." The Committee approved Humble's chartering (with associated companies) for two years the S.S. "Manhattan." If a test run through the Northwest Passage proved successful, the "Manhattan" would establish a shorter route from Alaska to the thickly populated U.S. East Coast. The distance from Barrow to Philadelphia was approximately 4,500 miles, with ice covering about one-half of it.

The "Manhattan" itself had been built for Stavros Niarchos in 1962; it was a 115,000 deadweight ton tanker, 940 feet long. A Massachusetts Institute of

Technology [MIT] team, with assistance from the U.S. Coast Guard, designed an ice-breaking prow for the ship, which was cut into four parts and then rebuilt for strength. The rebuilt hull measured more than 1,000 feet and the tonnage stood at 150,000. Now the largest U.S. vessel afloat, the "Manhattan" could no longer be termed a tanker; instead, it had become an icebreaker-oceanographic research vessel, with a crew of 179 men and one woman. In addition to sailors, scientists, reporters, and marine engineers, the lone woman aboard, Helen D. Bentley, used her influence as President Richard M. Nixon's nominee to chair the Federal Maritime Commission as a ticket to passage.[66]

Already inflation had increased cost estimates for the pipeline to $1 billion. Even so, on January 31, 1969, the Jersey Executive Committee expressed no objection to Humble's negotiating an agreement with Arco and BP for the "construction and operation of a joint-interest 48-inch crude oil pipeline" to move North Slope oil to Valdez, and for terminal facilities and necessary service tankers. On the same day, the Executive Committee reintroduced the question of using tankers—which had been thoroughly aired on December 4, 1968. Since then, estimates of the cost of this test also had to be revised. Finally, the Executive Committee declared that no further additions to Humble's budget would be made until the exact degree of participation by "non-affiliated companies" could be ascertained.[67]

While both plans for moving Alaskan oil were being discussed, Humble and "associated companies" arranged to purchase 500,000 tons of 48-inch pipe from Japanese steel manufacturing firms for a total cost of $392 million, of which Humble paid 25 percent. At the time, no U.S. steel companies fabricated pipe of that dimension, although Bethlehem Steel Company (and others) offered to gear up to make it. But no American company could move fast enough to deliver such pipe in September, 1969, when construction was to begin. In contrast to later prices for pipe, as Wright said, "we got it real cheap."[68] For joint company exploration, Atlantic Richfield operated the drill rigs at Prudhoe Bay. But back in the lower forty-eight states, Arco appeared to be headed toward a collision with Humble. On June 5, 1969, the Humble Company sued Arco, seeking a court order that would require Arco to sell one-half the land it had acquired in its merger with Sinclair Oil Corporation on March 4. Arco announced its intention to fight in court, since it had acquired the land as part of the Sinclair purchase. Humble contended that the 1964 agreement with Richfield gave the former the option to buy at the cost to Richfield half of any North Slope interests which Richfield acquired in any merger (such as that with Sinclair); and that the latter's merger with Atlantic did not alter that agreement. Humble asked for a court judgment validating its right to purchase an "equal undivided share in oil and gas leases recently acquired by Atlantic-Richfield in the contract area." The court continued the case until December, 1969, while the two companies reviewed developments in their North Slope field.

Finally, the two companies reached a settlement: Humble was to direct operations in most of the jointly-held North Slope areas, and Arco agreed to sell Humble half interest in two tracts of land on the North Slope (totalling 5,000 acres) plus one-third interest in 11,500 acres in the Santa Barbara Channel (near where Humble discovered an oil field in 1969). For this settlement, Humble paid Arco $2.75 million cash.[69]

X X X

In May, 1969, Jersey shareholders held their Annual Meeting in San Francisco, where Chairman Michael L. Haider reported on earnings and pronounced both the oil import system and the oil depletion allowance vital to national security. President J. Kenneth Jamieson* added his assessment of the Alaskan oil strike: "In the United States, Alaska has become, almost overnight, potentially the most important new oil province—for both our company and the industry." While no one guaranteed it, everyone seemed to believe that by 1972 Alaskan crude would be available to refine.

Later in the summer, Humble produced numbers for the public, as Wright remarked:

> We expect that Humble's investment in the North Slope will approach $400 million before we produce and sell our first barrel of Alaskan oil in 1972. That figure includes our share of expenditures for developing the Prudhoe Bay discovery and the Trans-Alaska pipeline, and our attempt to open the Northwest Passage to commercial transit with an ice-breaking experiment this summer with . . . the 'Manhattan.'

Jack F. Bennett, head of Humble's Supply Division, supported this cost estimate by noting that, in 1969, it cost $142 per foot to drill a well in Alaska, against $13 per foot in the lower forty-eight; while the cost of maintaining a rig ran $18,000 per day in the Arctic and $3,000 in West Texas. In summary, Bennett pronounced Alaska's North Slope "the most hostile environment in the world."[70]

Meanwhile, vigorous opposition to pipeline construction had begun to

---

*Jamieson became Jersey president on March 1, 1965.

develop, and on August 31, 1969, conservationists flocked to Fairbanks to rally at the University of Alaska. This meeting was organized by U.S. Secretary of the Interior Walter J. Hickel, himself an Alaskan. Representatives of the oil companies, the state of Alaska, and the U.S. Interior Department also attended, as did scores of corporate lobbyists. More than anything else, this gathering sent a signal to oil companies warning them that the pipeline might not be constructed as planned, due to opposition from both native claimants to the land and conservationists in Alaska and the lower forty-eight states. Years later, Wright recalled, "We really didn't understand what had to be done in the Arctic. . . . We really didn't realize the tundra problem." The main contest between conservation-preservation groups and the oil companies was not without political overtones; it focused on the question of how much time would be required before all needed environmental safeguards could be determined and construction authorized.

As the press in the lower forty-eight states began to receive both information and misinformation about the Prudhoe Bay strike, a debate developed concerning just what oil would mean for Alaska. The *New York Times*, on November 10, 1969, advocated a cautious approach to pipeline problems, and it approved the deliberate procedures then being used by Congress. But an ill-informed editorial writer attacked Alaska Governor Keith Miller, claiming that he should oppose Humble, Arco, and BP "for the greedy haste with which they are prepared to endanger a vast territory—the land, its people and its wildlife—for the sake of a quick and enormously profitable return on their investment." Other editorial writers remained equally uninformed about what was happening at the scene.

On September 10, 1969, Alaska sold 450,000 acres of leases on the North Slope for $900 million, plus a royalty of 12.5 percent of the gross value of any oil discovered. Representatives of oil companies had swarmed to the sale, to pay for these leases an average price of more than $2,000 per acre, while Alaska ignored protests from conservationist groups as well as from native claimants to the land.[71]

On balance it now seems clear that, despite some errors and accidents, Humble Oil had a fine record in the field of conservation and respect for the environment. Certainly Humble was ahead of the public in acknowledging the problem. Daniel Chasan, the author of *Klondike '70* and no friend of the oil companies, visited the North Slope in 1970 and wrote down his impressions. He credits Humble and Arco with running a good Alaskan operation, and remarks that he searched in vain for the damages to the environment that preservationists had predicted. Again, Mary Clay Berry, in her book, *The Alaska Pipeline*, notes that the Arco Prudhoe Bay installations represented "the showplace of the North Slope, [they] look more like a park than an oilfield." Both Humble and Arco sought expertise and advice from the Naval Arctic Research Laboratory in Barrow, but other companies were not so careful. Humble and Arco even went so far as to support research on grasses that would grow during the brief Arctic summer, to replace damaged tundra. Finally, this proved to be a long, expensive process, which delivered only limited success.[72]

Quite early it became evident that Arco wanted the pipeline more than did Humble. Before 1974 and the assertion of power by the Organization of Petroleum Exporting Countries [OPEC], Humble seemed less enthusiastic about the immediate construction of a pipeline. A much smaller company than the other two, Arco needed to get oil to its markets, in order to recoup its tremendous investments in Alaska. As for BP, the merger contract with Standard Oil Company (Ohio) provided stock incentives; so BP's share in Standard of Ohio would automatically increase as Prudhoe Bay oil came on stream. With clear self-interest in view, both partners accused Humble of "dragging its feet" on the pipeline. When asked to explain, Humble Chairman Wright responded, "We aren't as eager as they." Then he added, "But it isn't a question of being eager, just the desire for a more economical, efficient way." As things turned out, time proved of no great importance; instead, other factors intervened in the pipeline controversy.

X   X   X

During 1969, the question of ownership—title to Alaskan land—merged into the uproar over the pipeline location and the clamor over environment and ecology. Also, it had become evident that Taps had not been properly organized. Five companies had joined the original three, and each company exercised one vote. But nothing emerged from its meetings, which have been described as "organized chaos." So the eight owners chartered a Delaware corporation, Alyeska Pipeline Service Company, Inc., in which Humble had 25.52 percent of the stock, while Arco and BP each had 28.1 percent, and the other five companies held the remaining shares. Now, Alyeska was to be responsible for the design and construction of the pipeline, and it began to operate in August, 1970. Humble's Edward L. Patton, who had supervised the construction and operation of the recently erected Benecia refinery, was named president and chief executive officer. On their part, Jersey directors also learned firsthand about Alaskan problems, for on June 23, 1970, they met in Fairbanks, where they thoroughly reviewed North Slope developments, including the pipeline.[73]

Meanwhile, in June, 1969, Humble President Charles F. Jones* told the New York Overseas Press Club that the voyage of the "Manhattan" was a test to see if the water route to Prudhoe Bay could be used all year. Two months later, on August 24, the tanker left the Chester, Pennsylvania shipyard on that trip. Already, Humble had poured $36 million into renovation, while Arco and BP had added $2 million each.[74]

By mid-September, the "Manhattan" had reached Alaskan waters, and it began to return. In Canada, discussions of possible pollution by the ship appeared in the newspapers and then in the legislature. By February 20, 1970, these questions were causing even the prime minister to doubt the wisdom of a second voyage (which Chairman Wright and Humble were already planning). Humble, in turn, accepted Canadian claims on pollution and agreed to post a performance bond covering spillage and other possible accidents.

The "Manhattan" left port again on April 3, 1970; the second voyage was designed to test conditions in the ice and related areas. The ship made no attempt to go through the Northwest Passage; yet the voyage lasted more than six weeks.[75]

Still, Humble and its associates continued to explore pipeline routes as well. During the summer of 1970, a survey for a line to Chicago was completed, but when cost estimates topped $3 billion, the idea did not survive. Then, on October 21, Humble announced the suspension of studies on ice-breaking tankers, stating that though their use "to transport crude oil from Alaska's North Slope to United States' markets is commercially feasible . . . the pipeline transportation appears to have an economic edge at present." Actually, the principal reason for dropping tankers appears to have been the great cost of constructing a loading facility in waters off the Prudhoe Bay field—estimated at $200 million. But Humble added other reasons too; under the Jones Act of 1920, vessels engaged in domestic commerce had to be built in U.S. shipyards and manned by U.S. crews. This, Humble determined, would double the cost of ship operations. Still, company Chairman Wright said, "We now know that icebreaking tanker transportation is a workable alternative, and this gives us much greater flexibility in meeting future transportation needs."

Of course, Humble and other Alaskan companies had to wait many months before construction of the pipeline could begin; meanwhile, world-wide inflation accelerated, and the cost of the pipeline kept mounting. Thinking back on that period, Wright told an interviewer:

*Charles F. Jones succeeded Jamieson as president of Humble on September 10, 1964.

We concluded that the pipeline would be cheaper, but the interesting thing was when we ran the Esso "Manhattan" up there and back, we thought we were talking about tankers that would cost about $30 million a piece and by the time we ran this ship up there and back, we found out we would have had to have double hulls, and we would have had to have more power and so forth, why the price at that point jumped up to $100 million at least. Therefore it didn't look like it was economic. But in the meantime, we were using $1,003 million for the cost of the pipeline and that escalated up to $8 billion. So we might have done better if we'd gone to tankers.

Wright may have a point. Certainly any clear answer to the pipeline vs. tanker question became an early casualty of inflation.[76]

While Humble and its associated firms analyzed the data regarding the relative merits of the tanker vs. pipelines, yet another consortium emerged to test still another line—this one for gas transmission, from the North Slope and the MacKenzie Delta to Chicago and Ontario. This group included Trans-Canada Pipe Line Company and a number of U.S. natural gas companies. Humble provided them with information and suggestions.

Humble created a Northwest Project Study Group to draw together information and ideas for this pipeline. Jersey and Humble soon had still another committee studying the gas transmission line problem; this one they termed Non-Affiliated Gas Arctic Systems Study Group. The Jersey Executive Committee suggested that the two combine, in view of the fact that yet another committee (ad hoc) in which Jersey's Gas Department would take a prominent role, remained to be created. This ad hoc committee would implement all recommendations if Humble constructed the pipeline.

For some time, the consortium surveyed routes by which to move natural gas. Humble and several other oil companies expressed interest in a pipeline to Valdez (Gulf of Alaska), where they proposed to liquefy the gas for shipment. But as drilling in the Beaufort Basin proceeded, the combined companies studied a plan to bring the gas to Southern Canada and the U.S. Midwest through a 2,500 mile pipeline to North Dakota; then by existing lines to Chicago, Sarnia, and Detroit. Two of Jersey's affiliates, Imperial Oil Limited, and Humble, participated in these surveys, but in the end, little resulted from them.[77]

In fact, the pipeline story quickly became one so intricate that it defied logic and reason. Had the Arab nations not used the oil boycott against the United States (and several other Western powers) on October 16, 1973, it is entirely possible that politics might have prevented any congressional action on the line—since coincidental to the North Slope strike, Jersey drillers found other oil. Two important fields were discovered in Australia in 1967, and the Brent field was located by an Esso-Shell team in the North Sea in 1972. This latter field was conservatively estimated to contain one billion barrels of oil.

What made the North Slope different, and a focal point for national attention, was the feeling that it represented the last unspoiled wilderness belonging to the United States. In addition, of course, the Prudhoe Bay field did have much oil.

Primarily a result of the "Torrey Canyon" accident, the American environmental-ecological movement boomed in the 1970s. This ship ran aground off the coast of England on March 18, 1967, spilling 120,000 tons of crude into the ocean.* Later in January, 1969, the movement assumed larger political significance when a Union Oil Company of California well (A-21) off Santa Barbara erupted, spewing its oil and endangering popular beaches. Despite the well being successfully capped, oil and gas in significant quantities continued to leak to the surface, and damage to the shorelines brought out environmentalists.

On January 1, 1970, President Nixon signed the National Environmental Policy Act of 1969 [NEPA], one of the provisions of which required involved federal agencies to prepare environmental impact statements in advance of all final recommendations. These statements were to be made public.

Humble, BP, and Arco, on February 1, 1970, applied to the U.S. Department of the Interior for a permit to make all necessary geological and engineering studies of the pipeline route. The study application was approved, but a second application** for a permit to build a pipeline was stalled by Secretary Hickel's inability to find answers to questions posed by the House and Senate Committees on the Interior, and the failure of Taps to answer Hickel's questions satisfactorily. Meanwhile, other oil companies were converging on the North Slope, where much damage was being done, especially to woodlands and tundra. During the months between the Prudhoe Bay strike (March, 1968) and the pipeline application (June, 1969), Northeastern Alaska little resembled a controlled lease area. Wildcatters were numerous; in fact, for some months, the Fairbanks Airport ranked first in the world in freight shipments. There were no reports available on drilling—every company operated under rigid security rules. During that time, Humble oil scout Charlie Guion achieved considerable notoriety by going too close to a

*The "Torrey Canyon" was owned by the Union Oil Company of California.

**Made under the Mineral Leasing Act of 1920.

Gulf Oil Corporation-BP drill rig in a helicopter, looking for "information." His action led to a judicial injunction limiting how close helicopters might approach Gulf drill rigs.[78]

This feverish activity attracted special attention from environmental-ecological groups, especially after the oil spills. Their numbers grew rapidly, and Alaskans had time to make a more effective presentation of their claims. While the pipeline application remained stalled in the Interior Department, on April 13, 1970, Judge George A. Hart, Jr. of the U.S. District Court issued a preliminary injunction at the request of environmental groups. This enjoined Interior Secretary Hickel from issuing a permit for construction of the Alaskan pipeline across any public lands, until a final determination could be made on a permanent injunction. During the hearing, the Interior Department did concede some danger of oil spills.

For the next three years, this struggle over the Alaskan pipeline raged on, not only in Alaska but in the rest of the Continental U.S. as well. State and congressional committees listened to hours of expert testimony; commissions studied the tundra, the Eskimo, and the fauna; several organizations with differing objectives emerged to represent the Eskimos; politicians wavered from side to side as they became better informed. Throughout this period, coalitions among interested organizations shifted constantly. The National Environmental Policy Act of 1969 required an environmental impact statement from the Interior Department, and that, in turn, required more time to prepare. Generally, the state of Alaska and its officials wanted the line built, and built quickly; while the native Eskimos also supported building the line, but only with strong guarantees. For the oil companies, as well as for most conservation-preservation groups, the entire struggle represented a painful learning experience. By 1970-1971, when Alyeska took over pipeline development, recent plans called for 422 miles of line, some of which had to be put on steel structures ("stilts") to protect the permafrost and the routes used by migrating moose and caribou.

Oil companies faced a dilemma: no crude could be moved until a pipeline route was approved and the pipeline constructed. Humble, backed by Jersey, held the position that Wright had outlined earlier—in favor of the most economical and safest way to get the oil out. But Humble's partners, Arco, BP, and now the other five companies in Alyeska, pressed to get the injunction lifted and the permit issued because, as smaller companies, they needed a faster return on their North Slope investments.

On February 3, 1971, Alyeska announced that it would start construction during the summer—if it could get the injunction lifted. David R. Brower, president of Friends of the Earth, attacked the pipeline idea in a *New York*

Times article entitled, "Who Needs the Alaska Pipeline?" And as though he were responding to Brower's question, Interior Secretary Rogers C. B. Morton* repeatedly assured the nation that "any decision is months away," thus serving to confuse the situation further.

When the District of Columbia Court of Appeals reaffirmed, on February 9, 1973, the constitutionality of the Mineral Leasing Act of 1920, it declared that the Interior Department could not grant more than the fifty-four feet of right-of-way provided for in that act. The U.S. Supreme Court later refused to review the decision of the Appeals Court. Soon, supporters of the pipeline, with the assistance of President Nixon and Secretary Morton, pressured Congress to act immediately to exempt the line from the provisions of both NEPA and the 1920 Mineral Leasing Act; and a bill was introduced.[79] Now the pipeline issue created a heated debate in Congress, even though many questions remained unanswered.

Still, the bill to allow construction of an Alaskan Pipeline did pass; it became law on November 16, 1973. While excitement attending the oil boycott of the U.S. by Arab nations hastened the bill's progress, in fact all the final signing really meant was that construction could begin. By now the projected cost of the pipeline had increased to $5 billion.

The pipe itself had been coated against rust and stored at Prudhoe Bay, Fairbanks, and Valdez, at a cost of $5 million per month. Alyeska moved quickly to increase its office force, from fifty to three-hundred and fifty persons. In Fairbanks, Burgess Construction Co. merged with Houston Pipeline Co., and both companies began to accept employment applications. Ultimately, however, the major portion of the pipeline contract went to such giant companies as Brown & Root, the Bechtel Corporation, and three other companies which created a building consortium called Arctic Constructors, Inc. This group paid high wages to laborers in return for the unions' no-strike pledge and other guarantees. By 1976, employees of all these companies numbered more than 22,000.

While this hiring went on, the Alaskan legislature in special session removed numerous provisions in state laws to which the oil companies objected. But the legislature also increased the mineral severance tax from nine cents to 27 cents per barrel of crude. Alaska also geared its tax rate on oil to the crude wholesale price index, so that estimates of the 20 mill property tax indicated it would raise $14 million in 1974, and $100 million in 1975. One version read: "The pipeline and its support facilities will double the existing tax base of this state of 302,000 people." Overall estimates indicated that Alaska would receive an annual income of $500 million when the pipeline

*Replaced Walter J. Hickel in 1971.

flowed at its full capacity of two million b/d.

Alyeska obtained a state deed to 900 acres of land at Valdez (for $10 million) and began to erect wharves and stockpile equipment and materials. While the pipeline construction itself would represent a premier engineering feat, another difficult problem would be presented by the service road that ran alongside the line. Both the road and the pipeline would have to cross 800 rivers and streams. Each bridge required a separate permit; in fact, Alyeska had to secure almost 1,100 permits of all types from such governing agencies as the Corps of Engineers. Almost everyone agreed that the combined construction time could not be less than three years, and it might be considerably more.[80]

Reports from wildcatters drilling on the North Slope estimated the most recent proven reserves at some 25 billion barrels. During the early 1970s, the U.S. Navy permitted some wildcatting in Pet Four. As a result of this drilling, the navy estimated reserves of between 14 and 15 billion barrels of oil in that single reserve. The U.S. Geological Survey placed Pet Four reserves between a low of ten and a high of 33 billion barrels.[81]

Actual construction on the pipeline began during the early part of 1974, when the now defeated environmentalists adopted a "watch" program. J. Kenneth Jamieson, Exxon's chief executive, in an effort to placate all such groups, remarked: "They have pushed us into doing things that we probably should have done anyway." A stronger statement emerged from former Governor and U.S. Interior Secretary Hickel: "If the pipeline had been built in 1969 or 1970, it would not only have been an environmental disaster, but an engineering disaster."[82] By June, construction of the service road parallel to the pipeline route had begun, and men, machinery, and materials poured into Southern Alaskan ports. In fact, the work progressed rapidly, largely because, as one observer remarked, "The pipeline traded money for time."

The first Prudhoe Bay oil entered the pipeline on June 20, 1977. Already four thousand welding irregularities had been corrected, and engineers had checked the above-the-ground line in every possible place. No spills occurred. A month later, crude entered the storage tanks at Valdez, somewhat behind the early predictions. While, for Alaska, the sudden wealth from taxes created new problems, for the lower forty-eight states, Alaskan oil arrived at a propitious time. They welcomed the flow which, at capacity, would supply 7 percent of their total needs. For them, the project was a success.

For Exxon Company, U.S.A. [formerly Humble],* the search for Alaskan oil had only begun. Along with Arco, Exxon USA already owned about 40

*Humble's name was changed in January, 1973. For further details, see the chapter "The Company Changes Names—Jersey to Exxon."

percent of the oil and, alone, Exxon USA owned 37 percent of the natural gas in the Prudhoe Bay field. Figures on precisely how much Exxon USA would eventually obtain would become available only many years later; however, it is reasonable to assume that the figure would approximate the two billion barrels of crude oil predicted in the beginning.

X　X　X

All together, Humble Oil & Refining Company bought and sold millions of acres of land across the years, and their tracts varied in size and in the quantity of oil and gas beneath the surface. But, it is very likely that, next to the King Ranch deal, the most important single acquisition ever made by the corporation was the West Ranch, which had tremendous effect on the development of the Humble Company, of Houston, of the Southwest, and even of Rice University. This purchase was handled by L. T. Barrow, an executive who knew the resources of West Ranch as well as anyone could. Yet, even Barrow did not foresee the full significance of Humble's actions so close to home.

The West Ranch was located in Texas, twenty-two miles south of Houston. Two oil fields, Clear Creek and Friendswood, were developed on it; even so, as late as the 1950s it also remained a cattle ranch, with an abundance of wild birds and animals, and a stand of fine timber. A chain of lakes cut through, the largest of which was Clear Lake. Negotiations over the West Ranch property had proved to be memorable.[83] H. P. Pressler, Jr., a Humble attorney, worked behind the scenes, handling the legal issues, while Barrow represented Humble. Virtually every day Barrow went to J. M. West's office—for almost five months. West customarily wore a large Stetson hat indoors and out. Reistle, then in charge of production for Humble, once related to an interviewer Barrow's version of his dealings with West: "So Slim [Barrow] reported to his associates: 'We've practically got it made.' " But this was not the exact case.

Reistle went on: "Barrow said, '[I] spent all morning talking to Mr. West, and he sits there with his hat pulled down over his eyes so I can't see him, and he'll sit there for fifteen minutes and not say a word. Then he says, "Slim, will you go over just what you'll give me?" ' Slim said, 'I'd go over it. Then he'd just sit there and be quiet.' " In time, however, West agreed to Humble's terms: $8,500,000 plus oil and gas royalties ranging from three-sixteenths to three-eighths plus other considerations. Pressler then drew up a contract, and Barrow's success as negotiator became part of the Humble legend. From the time Humble purchased West Ranch in 1938 until the 1950s, the situation at the ranch remained idyllic.

X   X   X

In 1914, a Ship Channel was cut through Buffalo Bayou and Galveston Bay. This encouraged growth in the area, as did the oil and gas discoveries. Every decade, in fact, Houston's population doubled; and after World War II, industry flocked to the city and spread along the banks of the Ship Channel, which had been deepened to serve ocean-going ships. The ready availability of feedstocks from Humble's nearby Baytown plant, like the development of water transportation, helped to attract new industries. Small towns, such as Friendswood—a Quaker settlement—began to grow significantly, leading their residents to express concern about water and air pollution. In 1950, the Humble Company capital budget included $465,000, for a long-range program to improve the quality of those refinery effluents entering the Houston Ship Channel. Some years later, state and federal legislation forced other companies along the channel to follow the Humble lead.

Humble became concerned that the population sprawl might reach West Ranch, which did not yet border on the Ship Channel. In due time, the Houston Chamber of Commerce told Hines Baker and the Humble board that the expansion from Galveston Bay might require some of the West acreage.[84] In fact, the sheer size of West Ranch seemed to guarantee new controversy, so, during the 1950s, the Humble board of directors discussed several proposals for using the surface portion of the West Ranch property. Some board members favored developing an industrial complex, while others believed that the surface land should be sold. These and other alternatives were still being debated when, in 1957, Hines Baker moved on to Jersey, leaving Morgan Davis behind as Humble's president.

In 1958, the Humble board gave the West's big stone house and a private lake to Rice Institute,* along with more than 22 acres of land. Rice President William V. Houston proposed to use the house as headquarters for a geology laboratory, and it did prove to be a valuable property. George Brown, head of the Brown and Root contracting firm, and a friend to many Humble executives, as well as the Chairman of the Board of Governors of Rice Institute, played a major role in this transaction. While at Rice, Brown had been a roommate of Albert Thomas, one of two representatives from Houston in the U.S. Congress. Thomas, also a close friend of several of the Humble

*The Rice Institute became Rice University in 1960.

board members, knew about the West house gift, which later provided the precedent for a much larger gift from Humble Oil to Rice.[85]

Another group of Houstonians formed a syndicate to buy the remaining surface rights to the West Ranch. All of the persons involved in the syndicate were Reistle's personal friends, but, he continued to oppose the sale.* In his words:

> We fished together. We hunted together. We'd travelled around the world together. . . . Well, I didn't want to sell the damn thing . . . I think part of it was that as a producer it is always a problem to have a [different] surface owner. You've always got that [problem]; you've broken a pipeline; you've killed a cow; you've put a road where you shouldn't have put it and you've always got these problems. . . . So we had it for quite a few years, until 1960. We could begin to see the developments . . . and so we hired the Lehman Brothers—a group to make a study of how we could develop that property. It ended up that they offered us $30 million for it. But after careful consideration, this offer was turned down.

While Humble executives concerned themselves about the reorganization of Jersey's domestic companies, the low profits on petroleum products and the possible sale of West Ranch, on October 4, 1957, the entire Western World was jolted by the success of Russia's Sputnik I. The Space Age had arrived.

In July, 1958, the U.S. Congress passed the National Space Act, for which Senator Lyndon B. Johnson was the principal sponsor. This bill provided for unitary control of the space program under a National Aeronautics and Space Administration [NASA]. The purpose was two-fold: to end the feud among the military services, each seeking to control the space program;** and second, to create a single coordinated, decision-making body to guide the U.S. program, which was to be "devoted to peaceful purpose for the benefit of all mankind."[86] In reality, the image of the U.S. as the dominant global power had been seriously damaged. Only swift and vigorous action could counter this blow to its national pride and reputation. NASA was created to provide that action and to alter world opinion.

After passage of this pioneer space legislation, Lyndon Johnson became

*By this time, Reistle had become president of Humble.

**Werner von Braun and his army group of experts, composed mostly of ex-German scientists, working at Redstone Arsenal near Huntsville, Alabama, did put a 30.9 pound sphere, Explorer I, into orbit on January 31, 1958. This event occurred after the Russians had orbited Sputnik II.

chairman of the Senate Aeronautics and Space Committee, and (perhaps by coincidence) Albert Thomas chaired the Independent Offices Subcommittee of the House Ways and Means Committee, which handled the portion of the budget that fixed funds for NASA. From 1958 to 1960, while the U.S. space program encountered repeated failures, Johnson and Thomas applied heavy political pressure to secure the space headquarters for Houston. In 1960, Johnson ran for Vice President, so the main burden of political maneuvering fell to Thomas, who naturally worked closely with Brown.[87] When the first NASA administrator resigned, James E. Webb, an experienced corporate administrator and a person with excellent political connections, replaced him. At that time, Webb was serving as director and assistant to the president of Kerr-McGee Oil Company, and he was a friend of many Humble executives. So, he sought allies for NASA among the politically powerful oil interests. As late as June, 1961, Webb wrote that locating NASA in the Southwest might bring strong support "from one of our largest industries— the oil industry," and he wanted to "generate this interest in a practical manner." Certainly, Johnson, Thomas, and Brown could not have been unaware of Webb's desires in this respect.[88]

On January 20, 1961, John F. Kennedy and Lyndon B. Johnson were inaugurated as President and Vice President of the United States. In April, 1961, with Kennedy's approval, Johnson pushed through Congress a bill designating the Vice President as head of the National Aeronautics and Space Council [NASC]. This move provided Johnson with yet another powerful position to use in influencing the location of the Manned Spacecraft Center [MSC]. As chairperson of NASC, he appointed Brown as a civilian member of that body.

Kenneth S. Pitzer, a nationally known scientist, succeeded William V. Houston as president of Rice in June, 1961. Pitzer also had important connections; and he had chaired the General Advisory Committee of the Atomic Energy Commission. As a part of his plans for Rice, Pitzer announced that there would be "substantial development[s]" in several areas, including space science. All of the main actors in the NASA drama had taken their places.

Humble Oil also entered the NASA selection process, as on September 20, 1961, Jersey Executive Vice President M. A. Wright told the Jersey Executive Committee that Humble had donated 1,000 acres of West Ranch land to NASA*—to serve as the site of a manned spacecraft center. Humble representatives, Wright said, would soon come to the Committee with proposals

*The minutes make no reference to Rice University. Wright has noted that the Committee should not be blamed for any such omission.

for the development of the remaining West property, some of which had already been given to Rice University, with the "proviso" that the institution would offer the land to NASA as a site for MSC. Although Humble executives had discussed donating this land to Rice earlier, and while some evidence reveals that Johnson, Thomas, Brown, and Webb had discussed Rice donating land for NASA, Humble did not move until the fall of 1961.

Some years later, Reistle (at the time, vice president) described what went on at Humble headquarters during that period:

> Morgan Davis was chief executive officer at that time and he came in and he said, 'Carl, Mr. Brown was talking to me last night and he said that we ought to talk to Albert Thomas.' Well, Albert Thomas says, 'There's a good chance for me to get NASA for Houston and for Texas if you fellows would give a piece of that West Ranch to Rice Institute and, in turn, Rice Institute would give it to the government to build NASA.' Well, this is just what we were looking for. We could visualize then that we could have a real development there. So, as I recall, we originally gave them 900 [1,000] acres, and we later made some additional acreage available through Rice Institute.[89]

About three months after the inauguration of Kennedy and Johnson, Thomas and Webb met with Johnson in Thomas' office, and property owned by Rice was discussed as a MSC site. A memorandum written by Webb clearly establishes that fact.

Humble followed its 1,000 acre donation for Rice with an additional gift of 600 acres. This, the oil company specified, could be sold to the United States government. Rice later received $1,400,000 for the 600 acre property. The land transfer to Rice, which, in turn, gave and sold it to NASA, resolved the 1950s debate over the disposition of West Ranch, but only temporarily. On September 19, 1961, James Webb announced that Houston met all of the criteria for the MSC including a body of land bordered by water—the Houston Ship Channel. NASA began to plan for the center.

X   X   X

Reistle, who as board member knew the company executives pretty well, has stated that, in the 1950s, Humble really had no one trained to develop real estate. So, Humble's management spent a good deal of time discussing with engineers and real estate corporations the prospects and possibilities for the remaining tracts of West Ranch. Finally, Humble selected three developers with outstanding reputations, including Del E. Webb, head of the corporation bearing that name. Webb came to Houston with a draft of a development plan, and Morgan Davis, representing Humble, met him in negotiations. By this time, the NASA selection had been announced. With that decision, all peace and quiet disappeared on West Ranch. Humble had purchased a tract of land to make the property accessible to water. When excavated and dredged, this would enable industries located there to use Galveston Bay and the Ship Channel. The tract was named Bayport.

On January 18, 1962, Webb and Humble joined forces to develop 15,000 acres of the West Ranch property. By this time, Humble had given land to the Port of Houston, and the Port Authority had began work to develop a port suitable for ocean-going vessels at Bayport. On a strip of land purchased by Humble in 1960, the Port Authority dredged a channel and a turning basin. These improvements were financed by bonds (which Humble bought) to be paid off from port income.

The Humble-Webb agreement led to the Friendswood Development Company; and while Webb supplied management, Humble detailed executives to work with it, especially after the corporation bought 15,000 acres of land from Humble for $1,200 per acre. A clause in the contract covered necessary funding: Humble would lend, or cause funds to be lent, to the venture. All key decisions were to be jointly made. Friendswood soon began development of the residential part of a new community to be called Clear Lake City; first, however, Friendswood built roads and sidewalks, installed sewer lines, checked drainage in the area, and had the electric, telephone, and gas companies run lines. By 1966, more than $30 million had been spent; and while houses were selling well, current income could not support such an outlay. Webb had just bought another property in the West, and he agreed to sell his interest in Friendswood Development to Humble. Originally, Humble had lacked sufficient knowledge of real estate. But, as Reistle said, "Del Webb taught us a great deal." Soon, Humble had thirty builders working for Friendswood Development, under the direction of John Turner.[90]

Once underway, Friendswood Development proved that it could sell property, as Bayport became attractive to business—especially chemical industries that could use the easily available feedstocks from Humble's Baytown refinery, across the Ship Channel. Soon, a grid of pipelines carried such feedstocks.[91]

X   X   X

On October 10, 1966, W. W. Bryan and John Turner went to the Jersey Executive Committee with a proposal concerning the expansion of Friendswood Development Company. M. A. Wright, the new Humble CEO, had worked out an arrangement with Robert Kleberg, chief executive of the King Ranch: Friendswood Development and King Ranch Incorporated would purchase 50,000 acres of land some twenty-five miles north of Houston, for joint development. The Jersey Executive Committee approved Humble's spending its half share of costs on the property. Behind this arrangement, stood several important details. While Wright also knew Kleberg quite well, Reistle and Kleberg had been close friends for many years. So it was natural for the latter to come to Reistle and tell him, "Carl, I've got a chance to buy the Foster Lumber Company. They've got 52,000 acres of land, and I can buy it for $15 million." The two men discussed timber prospects on this land, which lay in two parcels, and also the possibility of a big gravel deposit. Reistle admitted that he knew little about developing real estate, and asked, "Why don't you go to Mike Wright and offer him half interest for $7.5 million?" And Wright had not yet been Humble CEO for one full year when Kleberg came to him with the same proposal. Feeling a bit uncertain because Friendswood had not shown much of a profit, Wright stalled while he sent two experts to the property to cruise the timber. Finally, Humble and Kleberg (King Ranch) agreed to become partners, and Humble ran the operation. Within a few years, sales of gravel and timber had paid almost the entire cost of the Foster land. So Humble continued to purchase land both there and in the Friendswood area.[92]

X   X   X

Meanwhile, Humble continued to donate land to institutions and colleges in and around Houston. In September, 1969, Friendswood gave the University of Houston 128 acres for a graduate center; and Red Bluff Development Company, created to hold West Ranch land after Humble bought out Del Webb, added 335 acres on which the university could develop two additional undergraduate campuses—a total of 463 acres valued at $7.3 million.

In 1969, the Friendswood Development Company announced three new real estate developments: Kingwood, on property north of Houston where lots became available for sale in 1970; Woodlake, located on 250 acres that Humble purchased in Houston; and Rollingbrook, near the Baytown refinery, an apartment and commercial complex. Eventually, Trailwood—the first village in Kingwood—won special recognition for its greenbelt system from the American Society of Landscape Architects.

By the 1970s, Friendswood had begun to yield good income for Humble, and sales continued to be quite brisk. Once Kingwood began full operation, the company income from the properties rose to more than $30 million. Despite the heavy capital outlay in both developments, there appeared to be little doubt that the real estate business would profit Exxon Company, U.S.A.*

Given the growth of Houston, the development of NASA, and the development of the Gulf area in oil and petrochemicals, West Ranch stands out as one of Humble's great purchases. The residential and urban development continued to be orderly, as Humble sought to encourage all property owners to protect the land and the water. For industries locating in Bayport, Humble operated a waste disposal plant, but the company also set rigid standards for waste as a part of the agreement to sell the land. On the West Ranch many trees were left to screen residential areas from industrial areas, and all business establishments sat well back from roads, creating a proper landscape. Humble continuously monitored the air for pollution, but the company hoped the various towns would accept that task eventually. In fact, the entire residential-urban development revealed careful and well-coordinated planning. By 1975, both Clear Lake City and Bayport had proved to be successful, and Kingwood later joined the list.[93] John Turner became recognized across the United States as a master of both commercial and residential real estate development.

X   X   X

In 1947, the Humble Company decided to make more systematic charitable contributions. A company foundation, created at this time, began to donate two-thirds of the money Humble and its affiliated companies had budgeted for giving. Later, one-third of the total designated for the contributions program went to the Esso Education Foundation, created after 1953. By 1972, the Humble Company Foundation had made 97 grants, principally to public and private educational institutions across the U.S.A. Most of these contributions were earmarked for specific academic departments; generally, the grants also stipulated that the funds be used to improve the quality of teaching and research or to assist outstanding students. Humble also followed the practice of encouraging employees to give to their alma maters by matching contributions.

At intervals, Humble also made gifts of property to educational institutions.

*Note the name change (as previously explained).

In 1964, it gave the University of Tulsa the $5 million building that had once housed facilities of The Carter Oil Company, and after the 1960 consolidation, the Humble Company Research Center. Tulsa used this structure for its College of Petroleum Sciences. The University of Oklahoma, in 1965, received a $600,000 building at Leonard, which became its Earth Sciences Laboratory. As has been mentioned, while developing Humble's West Ranch property, Friendswood Development Company donated land to the University of Houston, and the Humble Company made a direct acreage gift to the Rice Institute, as a part of the plan to secure NASA for Houston.

As Humble's program of giving developed, it increased in both scope and amount. Contributions supported mobile home Information Centers for disadvantaged areas, which served to provide counselling about educational opportunities as well as to distribute information and literature. The foundation sponsored college leadership institutes; "Earn and Learn" programs in stenographic and office skills; supported scholarship programs for outstanding black students; made regular donations to symphony and other musical organizations; and in general supported the arts. While by no means unique among business organizations, the Humble Company Foundation—later Exxon USA Foundation—had its own Board of Trustees, and it deliberately sought to provide financial assistance to interesting and important projects.

After the 1972 name change, the foundation increased its support of the fine arts: a Shakespeare Festival in Houston; the New York Philharmonic; the Midland-Odessa (Texas) Symphony and Chorale, and many other projects. It also provided funds for a project entitled, "Recording the Roots of Navajo Culture" designed "to find and preserve through photography these records of a vanished culture." Almost no one has doubted the value of this exercise in corporate giving.

X   X   X

In all, the Humble experience, 1950-1975, included much hard work and a number of important changes that served to dramatize the company's enlarging role in American national life.

On July 25, 1959, six weeks before the big merger, Morgan Davis announced that Humble proposed to build a new building, a million square foot office structure, 44 stories tall, covering one full city block. When completed, it would serve as headquarters for the entire operation and be the tallest building in the U.S. west of the Mississippi River. Ground was broken for the new structure on February 16, 1960, and plans allowed two years for construction. In fact, it was slightly more than two years when, on September 4, 1962, small groups of Humble employees began work in their new quarters. All employees settled in by March, 1963, and in that month and April, the Humble Company held open house to introduce its $32 million building—truly a showpiece.

Yet Humble was just another big company when Hurricane Carla struck the Louisiana Coast, on September 11-14, 1961. Morgan Davis wrote to Wright on September 29, to report the damage. Already, the company had contributed $25,000 to the Baytown community for assistance. Now, up-to-the-minute reports indicated that physical losses to company property reached "somewhere around $2,500,000." As Davis correctly judged: "All in all, I suppose this is a very small loss when the size of the hurricane and the area of operations affected are considered."

During the first five years after the merger, Humble profits doubled, while the work force dropped by 25 percent. Jamieson recalls that the "consolidation of facilities was dramatic." This included three uneconomic refineries, although Rathbone, the old refiner, protested against eliminating any of them; however, Reistle gathered the numbers and Rathbone studied them one weekend. Then, he handed the report to Reistle and told him to sell.

By 1966, Humble had 966 service stations in the West, using principally the Enco brand. Competition in this market had become fierce, especially with Union Oil, as every oil company pushed for a larger market share.

Humble's natural gas sales increased annually; as early as 1963, when sales increased 13 percent over the previous year,* the average daily sale amounted to 2.4 billion cubic feet. Humble actively explored for natural gas; it bought some fields outright and acquired others through merger. The company also acquired pipelines and extended those it owned, as the gas business became a very profitable part of total operations.[94]

$$X \quad X \quad X$$

As the merged company began to function effectively, significant changes began to be made at Humble. First, a number of the directors from its first board took early retirement and left the business scene. Then, Jamieson, who had been elected president on November 1, 1963, resigned (on September 10, 1964) to become executive vice president and director of Jersey. Charles F. Jones replaced him as president.

As Jamieson remembers, he had little advance warning that the holding company might want him; after all, while he had worked for several Jersey affiliates, he had never worked directly for the parent company. And natu-

*In my source, no total figure is given.

rally, he liked Houston and Humble. Reistle explained Jamieson's move this way: "He had to go up there. We originally had it planned that he [Jamieson] would succeed me and he wanted to. He very bitterly accepted the other job . . . he didn't want to go." Some years after he had retired and returned to Houston to live, Jamieson was asked whether he had any plans to go to New York in 1964. His negative answer came quickly; he had no such plans: "I was quite happy here, but I had long ago made the decision that if you don't go with the system, you might as well run the local grocery store down there." Of course, the system also had its reasons.

In Houston, Humble announced a complete reorganization of its field operations, on November 11, 1965. The four regions were abandoned in favor of a functional organization, which consisted of seven marketing regions and six exploration-production divisions. Each functional unit reported to a new vice president in Houston, and refining operations remained unchanged, as did the Marine Department.

Then, in February, 1966, Jersey announced that at Humble Wright would succeed Reistle when he retired. For eight years, Wright had served as contact director for Humble, and he liked most of the people there. Still, as he pointed out: "I was the first non-Texan that had headed the company down here." Reistle, he said, was born in Colorado, but he had worked for the company so long (thirty-two years) that he considered himself a Texan. And on June 1, 1966, Reistle retired.

Wright proved to be an excellent choice to head the Humble division of Jersey. He had worked for Jersey or its affiliates fifteen years before he became a director in 1958, and he had acquired global experience in production. Wright's public activities also proved to be of tremendous value to both Humble and Jersey. As executive vice president of Jersey, Wright became concerned over the failure of Jersey Company executives to seek out opportunities for public statements about the company. So he became one of the leading spokesmen for the industry, who utilized all company resources to help make his addresses—later published—into media events. With an engaging frankness, Wright answered questions and thus added to his reputation. Soon he began to testify on energy related matters to committees of both houses of the U.S. Congress.

Wright proved to be a major positive force in the effort to get accurate information to the public during the years 1966-1976, when the entire petroleum industry suffered in public opinion. Humble, Jersey, and the industry all benefited from Wright's decision to speak out for them.[95]

X   X   X

When oil spills, smog, and other indications of air and water pollution began to spur public criticism and as the medical profession first pointed to hazardous chemicals in the air, water, and food as probable sources of cancer and other diseases, oil companies assumed the unenviable role of public villains. In a speech to the Houston Chamber of Commerce, given less than six months after he became CEO of Humble, Wright stated that some businesses recognized that a fair share of the pollution derived from industrial practices. He predicted that unless business moved to clean up the air and water, the people would force action through their duly elected representatives. "The final responsibility of effective business leadership," Wright said, "is to educate the public as to what can be done and how much it will cost."

This pollution problem had been created by people, by cities, and by industries. But as had so often been the case since the 1920s, Jersey received first billing in notoriety. Of course, even today, it remains difficult to assign precise degrees of responsibility for pollution, due to intense emotion that has surrounded the entire issue. Certainly, Humble possessed an excellent record in oil field conservation, and executives such as W. S. Farish, Wallace Pratt, John Suman, Hines Baker, and Rex Baker had demonstrated for years that their company sought to obtain the maximum oil available from all wells without polluting or causing damage to the land. As has been mentioned, in 1950, Humble initiated a pioneer study of pollution in the Houston Ship Channel. When Baytown citizens protested the increasing contamination of Galveston Bay, Humble called on the State of Texas to monitor pollution there, and acted to tighten its control over refinery effluents.

Yet pollutants spread from the Ship Channel to the tidelands and estuaries. Efforts to clean up both air and water pollution in the 1950s stalled when court decisions made prosecution of guilty corporations more difficult. As late as 1967, federal authorities rated the Ship Channel the "worst pollution problem in the state." Humble repeatedly sought to control pollutants from its plants. Refineries obviously played a major role in polluting the air, but despite Humble's efforts to control its own emissions, some problems remained. Isolated efforts to comply with regulations and laws designed to end pollution did not achieve much judicial sanction before 1975.

Jersey and Humble both used committees to study pollution problems, and both realized that the vagaries of public opinion could directly affect sales. When Jersey changed leaders, however, public relations—in this case, public information—suffered. Where Holman believed in an ongoing, large-scale public relations effort—one of the great public relations operations in corporate history—Rathbone regarded such formalized public relations as something of lesser value; he believed in a different approach. Thus, personnel changes and cuts in funding, as well as experimentation with different courses, together served to alter the Jersey public relations operations, until the company could no longer maintain the high public approval rating it had held in the early 1950s. Then, the consolidation of all domestic affiliates into Humble served to divert much attention in both Houston and New York from public-connected problems. Reorganization simply demanded too much time.[96]

Once Humble had achieved a semblance of unity, the company began to attack the pollution problem. By now, the great national debate on auto emissions had begun, and cities such as New York were sufficiently concerned about smog to demand that low-sulphur oil be burned in power-generating plants that supplied the city's electricity. In January, 1962, Jersey established a Heavy Fuel Oil Council, to survey all the affiliates and determine how much oil with low sulphur they had available or could locate. Director Harold W. Fisher headed this council; and he periodically reported on its activities to the Jersey Executive Committee. On January 13, 1967, the Committee told Fisher to have the council intensify its efforts to locate low sulphur oil; and to pull all Jersey groups together for that purpose. On January 24, 1967, the Committee set up a task force to aid the council, and then discharged its advisory committee on low sulphur fuel.

The head of Humble's Planning Department, A. A. Draeger, addressed a letter to "Members of the Board" on the subject of the use of lead in gasoline, on May 28, 1968. He estimated that the cost to Humble of removing all the lead would be $500 million by 1980. He indicated that his figures were estimates; then, he went on to detail the cost of removing lead by degrees, a slower and seemingly less expensive process, estimated at 1.2-2.0 cents per gallon. Draeger suggested that Humble adopt the slower procedure to permit company experts and others to do more research on motors and unleaded gasoline.[97]

In 1970, perhaps following the advice of its Planning Department, Humble introduced "Big Plus" gasoline, which featured a low lead content. At the same time, additional "Alert" service stations, which featured self-service and low prices, opened to the public, and new car washes and car care centers in selected locations soon followed. The latter facilities relied on an important development from Esso Research and Engineering Company, a diagnostic-repair tune-up procedure (preventative maintenance) for automobiles which enabled mechanics to locate quickly engine malfunctions that could increase exhaust pollution. Despite some advantages for motorists, these centers never proved popular, and they were gradually closed. In larger ways as well, the mission of Humble expanded to match the times. In 1962, Morgan Davis urged the Jersey Executive Committee to let Humble adopt the policy, "under which it would participate in all aspects of the energy business in which opportunities to invest profitably and advantageously can be identified. . . ." The Committee agreed to consider the matter further, and on October 15, it approved a statement:

Humble Oil & Refining Company is in the energy business and it is the policy of the company to participate in all aspects of this business in which opportunities to invest profitably and advantageously can be identified.

Obviously, from the first part of this declaration, Humble borrowed its famous nickname, "The Energy Company."

Jersey moved on its own to encourage affiliates in searching for minerals of all types (not gas and oil alone), including uranium. In 1966, studies were made of new business opportunities in the field of nuclear processing. Humble itself had for some years resisted pressures to buy oil shale properties; but in 1967, the company began to add shale to its holdings. Next, in 1973, Exxon began coal gasification, at first on a limited scale, but this processing activity soon expanded. At the same time, some plans formulated in 1966 for guiding Humble's mineral research necessarily had to be altered, largely because the company lacked the specially trained managers required. In 1968, Humble-owned Carter Oil Company, using the name of another Humble company—Monterey—created the "Monterey Coal Company" to deal exclusively in coal operations. The new company had offices in Carlinville, Illinois as well as in Houston. In September, 1968, Monterey bid on a coal contract to supply a Commonwealth Edison generating plant that furnished electric power to Chicago. Carter had leased 30,000 acres of land in

Southwest Illinois, and Humble claimed that Monterey could supply two or two and one-half million tons of coal annually. Monterey won the contract and scheduled production to begin in 1970.

In 1974, Carter Oil Company created another subsidiary, The Carter Mining Company, to be headquartered at Gillette, Wyoming, to develop the Rawhide Mine. This coal mine was projected to have a capacity of five million tons annually, and plans called for it to open in July, 1976, with an environmentally responsible mode of operation—one that would establish "that modern surface mining is compatible with environmental goals."[98]

In magazine articles, Jersey and Humble spokesmen pointed to the ongoing need to protect the environment. They freely admitted that the oil industry had shared in creating the problem, and now they wished to call public attention to company efforts in establishing proper industrial practices, and adequate state and federal regulations. Exxon, in short, was signalling a readiness to meet its obligations to both the environment and the public. Exxon USA contributed money to support university scientists engaged in pollution studies. Jamieson preached that "total energy supply and demand [must] be recognized as a single problem." Haider, in his call for cleaner air and water, reminded the public: "The opportunity is still ours for a reasoned and effective response to the challenge of contamination." Wright, among others, constantly promoted the use of improved technology and careful work as pollution deterrents. Simply, as one person recalls the message, at Humble, "Careless technology can no longer be afforded."

In California the Humble Company purchased leases offshore during 1953—after Congress had restored control of the tidelands to the states. In 1967, real interest in offshore California leases became evident, as the Jersey Executive Committee approved Humble's plan to bid on tracts being offered near the Santa Barbara Channel. The Humble bid of $21 million won considerable acreage, and in February, 1968, Humble bought 47 tracts west of Santa Barbara for $218 million. Then in 1969, as part of the settlement with Arco, Humble acquired a one-third interest in 11,500 acres in the Santa Barbara Channel area.

In September, 1968, Humble applied to the County Supervisors of Santa Barbara for a permit to build a natural gas and crude oil processing plant—at a site that first had to be rezoned. Such rezoning required a public referendum, and environmental groups (including a new Citizens Committee) rushed to oppose the Humble petition. In the November election, the Citizens Committee won, thereby defeating, in effect, the entire Humble proposal.

When Alaska's Walter J. Hickel became Secretary of the Interior, he reversed (April 12, 1969) previous rulings and announced that oil drilling would be permitted in the Santa Barbara Channel at five locations. Immediately, there was a great public outcry against his decision. Humble continued to drill, and in July, 1969, discovered oil in commercial quantities on two of its leases. But the Humble decision to produce this oil again aroused a new uproar from environmentalists, including a group of survivors from the old Citizens Committee, now called "GOO" for "Get Oil Out." Certainly, the Santa Barbara Channel continued to represent a hot political issue for some years. Humble, like other companies that sought to operate in the vicinity, did work hard to find some means of minimizing damage to water fowl, fish, beaches, and water.[99] Yet despite these efforts, in 1970, Humble and a number of other companies were accused of "unsafe operations of oil wells" in the Gulf of Mexico, as the U.S. Department of Justice filed 150 misdemeanor counts. The central question concerned the use of storm chokes on the wells; and, while Humble denied all charges, it paid a fine of $300,000 rather than pursue the case through expensive litigation.

Then, in early December, 1970, a mysterious explosion occurred at the Humble plant at Linden, New Jersey; and several people were injured. Immediately, the area buzzed with rumors of sabotage. Humble investigated the near-tragedy for three months, and concluded on March 5, 1971, that an over-heated reactor caused the explosion "rather than sabotage." Of course, in so vast an enterprise, some incidents such as these were bound to occur; nevertheless, when they did, a portion of the public proved all too ready to view them as evidence of careless management. Public relations always represented a big job, never to be finished. For example, Jersey CEO Jamieson insisted on an annual joint review with the U.S. Coast Guard covering all problems that related to Humble (and other Exxon affiliates), especially those concerned with its Marine Department. As the Coast Guard has testified, these surveys revealed surprisingly few incidents. Humble quietly went on installing pollution control devices at its refineries, and training all personnel in the current procedures for keeping the environment free of pollution.[100] Certainly, the company always felt that it was doing its part as a good citizen of Texas, the United States, and the world.

X   X   X

As chairman and CEO, Wright gradually turned the Humble Company into a Jersey company. Of course, it took him some time to change the board of directors, but as he pointed out, there really wasn't any necessity for

Humble to have a board when Jersey owned it all. So, after he became chairman in 1966, Wright began to have some new charts made. They helped him, as did his vast experience with oil exploration and production, and even more with the men at Humble. A Management Committee did much of the work of merging Humble into Jersey. And within two years, Wright could point out to Cliff Garvin that, while the *Fortune* "500" did not include Humble, it nevertheless ranked eleventh among other oil companies in sales, and fifth in net income. Altogether, Wright proved to be an excellent choice to run Humble [Exxon USA].[101]

After reorganization, the Humble Company itself became a lean giant; and with the passage of time, it joined other American companies in changing from a regional to a national business organization. Yet even to the present, Humble has retained much of the special flavor that distinguished it in the beginning.

HUMBLE OIL & REFINING COMPANY
FINANCIAL REPORT

(Million Dollars)

| | December 31 | | | | | | Expenditures for Drilling | | |
Year	Total Assets	Net Fixed Assets	Shareholders' Investment	Net Income	Dividends	Capital Expenditures	Total Including Dry Holes	Dry Holes	Payroll and Other Compensation (a)
1950	935.5	679.6	779.6	129.4	71.9	110.5	69.4	23.9	100.3
1951	1,059.9	770.6	865.2	169.5	80.9	158.8	105.6	29.3	106.0
1952	1,105.7	854.4	928.5	145.3	81.8	160.3	118.3	43.1	119.3
1953	1,186.3	915.8	1,008.6	164.3	81.8	154.2	100.8	37.4	127.7
1954	1,245.1	981.8	1,072.9	146.3	81.7	161.2	108.6	35.0	132.5
1955	1,323.2	1,092.5	1,163.9	175.0	82.8	210.4	109.7	38.0	138.4
1956	1,433.0	1,166.8	1,257.6	179.0	86.1	183.9	127.5	48.1	151.2
1957	1,524.8	1,270.4	1,338.1	175.9	96.9	235.4	146.1	61.8	167.4
1958	1,568.6	1,288.0	1,390.0	136.5	100.8	154.9	78.7	35.6	161.0

(a) Includes employee benefits.

Source: Humble Oil & Refining Company, *Annual Reports*, 1956-1958.

HUMBLE OIL & REFINING COMPANY
OPERATING DATA

| | | Refinery | | Number of Wells | | | Unoperated | |
Year	Net Production (b) (Mil. Bbls.)	Crude Runs (Mil. Bbls.)	Trunk Line Deliveries (Mil. Bbls.)	Producing at Dec. 31	Completed During Year Producers	Dry Holes	Acreage Under Lease at Dec. 31 (Thous. Acres)	Number of Employees at Dec. 31
1950	110.7	81.0	217.2	10,294	669	147	13,470	17,653
1951	136.6	90.0	270.8	10,866	893	240	13,171	18,112
1952	138.6	92.1	262.0	11,333	808	294	15,363	18,800
1953	143.0	89.3	240.2	11,967	660	238	14,390	18,821
1954	132.5	83.4	218.7	12,551	798	176	13,555	18,740
1955	142.3	90.4	233.4	12,832	665	192	14,347	18,729
1956	148.4	95.8	256.1	13,145	692	213	15,701	19,001
1957	143.9	86.0	246.5	13,457	648	219	16,260	19,604
1958	121.2	83.2	213.1	13,538	410	147	14,280	17,600

(b) Crude oil, condensate, and liquids from natural gas processing plants.

Source: Humble Oil & Refining Company, *Annual Reports*, 1956-1958.

Chapter 5

The Chemical Story

1918-1960

THE HISTORY OF the Exxon Chemical Company, from modest beginnings to a complete company in 1963, can only be compared to a child learning to walk; two steps forward, one backward; sudden starts and quicker stops; ups and downs; crawling before standing. For those individuals who helped chemicals grow and for Exxon. Corporation, this evolution represents an exciting chapter in the holding company story. It also typifies the triumph of ideas in the marketplace.

When World War I began, oil companies had no market for gases resulting from then-current refining practices. Generally the ethylene and propylene derived therefrom served as fuel for boilers, or the refineries flared these gases. The increased demand for an acetone solvent led Standard Oil Company (New Jersey) [Jersey] to begin construction of an isopropyl alcohol plant at Bayway, New Jersey, utilizing propylene gas. From this gas, they produced isopropyl alcohol, which could then be converted to acetone. This resulting distillate would serve as the wanted solvent.

The Jersey alcohol plant began production in 1920. By this time the market price for alcohol had dropped to almost $1.50 per gallon, and Jersey's production cost came to about $3.00 per gallon. Characteristically, the corporation continued to search for a way to lower production costs as a means of recovering its expenditures.[1]

After the war, President Walter C. Teagle of Jersey sent a team of engineers to check on the damages to Jersey property in Germany and France. They also sought information about the German war-accelerated processes of manufacturing oil and rubber from coal. Teagle's effort to obtain this information intensified in the face of a threatened oil shortage, while demand for gasoline increased. Too, Teagle desired to improve efficiency in Jersey's production and refining. Given his background as a chemical engineer, it was natural for him to initiate several actions that not only secured Jersey's market position, but helped to shape the company's future in chemicals.

Thus Jersey's petrochemical company evolved largely from an interest in hydrogenation and the investment in the isopropyl alcohol plant first, and the equally early interest in synthetic rubber second. The story begins with Teagle. With an assist from Board Chairman A. C. Bedford, in 1918 he brought to Jersey E. M. Clark, superintendent of the Standard Oil Company (Indiana) Wood River refinery. Clark richly merited the description as "a perceptive administrator as well as inventor" whose "inventive abilities were marked, and he was uninhibited in his scientific attitudes."[2]

Clark began to analyze the Jersey operation, especially that of Bayway, and to supervise applied research there. He continued to use as his legal consultant Frank A. Howard, a distinguished inventor as well as a junior partner in a prestigious Chicago firm specializing in patent law. When Teagle read some of the Clark-Howard correspondence, he immediately invited the latter to draft a plan for a Jersey research organization. Ultimately in 1919, Teagle and Clark persuaded Howard to direct this new venture. Howard never returned to the law; rather, working with Clark and encouraged by Teagle, he soon brought into the company a number of distinguished scientists and engineers. [In selecting these men, he used the advice of Robert A. Millikan, later a Nobel Prize winner.] Many persons who could not be persuaded to join Jersey were retained as consultants, among them Ira Remsen, a leading chemist of that era (president emeritus of Johns Hopkins University) and Dr. Warren K. "Doc" Lewis, head of the Chemical Engineering Department at the Massachusetts Institute of Technology [MIT]. Lewis proved invaluable as a friend of Jersey who encouraged bright MIT graduates to join the company. The Standard Development Department resulting from the efforts of these and other persons became, in October, 1927, The Standard Oil Development Company [SOD].

In these years between the creation of the Development Department and that of a separate chemical corporation, the name of Frank A. Howard was preeminent. Aside from Board members, he probably contributed more to Jersey, and certainly more to Jersey chemicals, than any other person in the history of the corporation. In 1927, acting on the advice of Professor Lewis, he "borrowed" Professor Robert T. Haslam of MIT to head an additional research center at Baton Rouge. The intermittent problems of oil shortage and glut, plus a rubber supply fluctuating in price and quantity confronted the Jersey Board. They also encountered daily such engineering riddles as automobile motor-knock, and the wasteful refining process used in making gasoline. Attempts to solve these problems began. Change did not occur without friction, however, for the eager young scientists and engineers employed in Howard's Development Department soon "shook up" the more experienced and conservative Jersey managers. Refiners especially sought to resist change and to follow tested practices.

Clark (and Teagle) provided support for Howard and his associates. Clark, with new ideas about refining and chemicals, but no practicing diplomat, agitated old Jersey hands as no one had for years. Soon he joined the Board, and then, with his backing and that of Teagle, Howard's organization began to modernize most of the company's technical operations.

Howard relentlessly drove toward broadening the research area for his department. But except for Teagle, Clark, Everit Sadler, and Heinrich Riedemann, many directors, like the lesser managers, remained unconvinced by his proposals and those from his expanding department. These men possessed a very limited vision of the future of the chemicals-from-oil business. Jersey research engineers and chemists readily learned and ever understood the difficulty of relating to marketers, refiners and producers in high places

At Jersey, the tradition of marketing died slowly. Years later, Osgood V. "Otz" Tracy, then president of the Jersey affiliate, the Enjay Company, Inc. [Enjay], recounted an incident illustrating this communications problem. Tracy remembered that the Jersey Coordination Committee came to Baton Rouge "to get a picture of what the future might hold for petrochemicals." Eger Murphree, the famous Jersey scientist, spoke for about two hours, describing the future of chemicals derived from petroleum. When the speaker finished, a committee member remarked, "The only word I understood throughout the whole thing was alcohol."[3]

During these exciting years for chemists everywhere, Howard and Teagle first contacted Badische Anilin und Soda Fabrik, a large German chemical firm, a part of I. G. Farbenindustrie [I. G. Farben]. In 1926 they visited a Badische plant; in 1927 and 1929, Jersey signed contracts with I. G. outlining the methods to be used in developing the hydrogenation process in the United States. Subsequently, as is well detailed in George S. Gibb and Evelyn H. Knowlton's *The Resurgent Years*, the two companies negotiated other contracts and agreements. Some persons contend that the American petrochemical industry stemmed largely from the implementation of these agreements. Certainly Jersey, for one, acquired invaluable information from the Germans and in return gave little—because its small, young Development Department had little to offer. Thus in 1927, in one giant step, Jersey jumped from very limited applied research into the most advanced pure research in the world. In that same year, Jersey consultants Lewis and Haslam persuaded fifteen MIT faculty members, graduate students and graduates in chemical engineering to join Jersey's Standard Oil Company of Louisiana's [Standard of Louisiana] Baton Rouge refinery as a special task force; their assignment, "to help pull the nation from the brink of energy disaster by finding a way to make synthetic gasoline before dwindling oil reserves run out." Company old timers referred to this group as "the 15 virgins" because none of them had any "practical oil industry experience." One year later Haslam brought down George H. Freyermuth, a chemical engineer with a master's degree from MIT in fuels technology. The earlier fifteen graduates in Baton Rouge referred to him as "the sixteenth virgin." [4] One of these "virgins," Marion W. Boyer, remembered his shock when, on his first day in Baton Rouge, Haslam turned over to him the task of overseeing the construction of the hydrogenation plant.

The frantic work in Baton Rouge was halted in October, 1930, when Columbus M. "Dad" Joiner brought in his wildcat well in East Texas. The oil market quickly became saturated, and oil prices dropped to ten cents per barrel. "Even in those days Howard had synthetic rubber on his mind," one of the Baton Rouge chemical engineers recalled; he "had a very fertile and active mind, keeping his sights on future needs." [5] I. G. Farben scientists had experimented with a process of developing rubber from butadiene, and Howard wanted Jersey to share in this development. [6] Teagle, too, saw the great advantage the company and the nation would derive from an indigenous source of rubber. So in 1930 Jersey joined I. G. Farben in creating a Joint American Study Company known as Jasco, Incorporated. Jasco received from I. G. Farben numerous patents, while Jersey had naught to contribute but money.

In the decade 1920-1930, Jersey made great strides toward modernizing the company, and no function benefited from these changes more than chemicals. From its Farben agreements, Jersey secured along with other scientific data, much information on the chemical processes that ultimately led company researchers to synthetic rubber and 100 octane gasoline, both of inestimable value to the allies in World War II. In 1942, in the spotlight of the war, Jersey drew open condemnation and accusations of lack of patriotism for its relations and agreements with I. G. Farben. Judging from the testimony given before then-current congressional committees, and the sensational news stories, one would assume that Teagle, Clark and Howard possessed the secret of producing synthetic rubber from oil long before the Farben chemists did. In fact, Jersey's interest in this product did date to World War I—but the Germans had more advanced technology.

During the depression (1929-1934), research proceeded rapidly in Baton Rouge. Howard, Haslam, MIT Professor R. P. Russell, and their team continued to push toward new and cheaper chemical compounds. They concentrated their efforts on improving the process of making toluene,[7] and later developing on a commercially feasible scale butyl rubber, which Jersey scientists W. J. Sparks and R. M. Thomas had invented. During these years Jersey recruited the brilliant Eger Murphree from a subsidiary of the Allied Chemical and Dye Corporation, to direct research leading to new chemical compounds. Among those who came with him was the able Otz Tracy, termed a "real entrepreneur" by one who knew him well.[8] By 1938, the Jersey research group had concentrated in three major research areas: the production of butadiene, the oxidation of wax, and a nitrogen project. Though quite significant later, no one of these had as yet proved economically practicable, leading one participant to state, "I do not think there had been any more unsuccessful ventures ever proposed by the Jersey Company."[9] Still, by 1939, the development group moved Jersey through the process of steam cracking liquid products of refineries to make the ethylene required to produce very profitable ethyl chloride.[10] The Baton Rouge steam cracking plant had opened for Jersey the source of olefins.[11] These researches and discoveries provided the basis for oil-chemical industry developments over the next forty years, and they proved especially important during World War II.

The early war years offered time to reassess not only Jersey's relationship with the public but all other aspects of its operation as well. By 1942 the company's top management had begun the difficult process of surveying the changes likely to be wrought by the war; they sought to anticipate problems of conversion and to chart a course for the postwar period—at best an arduous task. War demands affected some areas of the company's business more than others, especially Jersey's operations in the field of petrochemicals—or oil-chemicals, the term then applied to them. The need for Jersey to reconsider its role in the industry derived both from broader international developments and from the company's particular situation. The United States had only recently entered the conflict when Jersey's management began studying the company's future participation in the oil-chemicals field. The necessity of addressing basic policy questions quickly became apparent.

With the normal sources of natural rubber and several other products interrupted, war mobilization created demands for basic organic chemicals that, because of the magnitude of supplies needed, would have to come from petroleum. High octane aviation fuel and synthetic rubber ranked high on this list. So the Teagle-Clark-Howard-inspired improvements in refinery technology helped make it possible to utilize more of these by-products in chemical operations.[12] The military demanded higher quality petroleum products. At the same time, the oil-chemical industry's demand for raw materials, particularly the chemically reactive olefin gases, continued to expand. Thus, by the mid-1940s, most of the raw materials required by the oil-chemicals business—principally olefin gases—had to be made specifically for that purpose from selected feed stocks.[13] The war also produced more specific concerns related to the company's chemical business. The dissolution of the company's cooperative research agreement with I. G. Farben and the Alien Property Custodian's seizure of the patents held by the two companies through Jasco, magnified and dramatized the uncertainty about the future direction of the company's chemical effort. The conflict also changed fundamentally future prospects for synthetic rubber, an area in which the company had pioneered. Although the government now controlled the rubber industry, Jersey leaders assumed its eventual return to private control. Just what that industry would be like after the war, and what role Jersey would play in it, became pressing issues as the company's management took up the broader question of how to proceed in the oil-chemicals field.

Within the company a debate developed, and two distinct issues emerged for the consideration of the Jersey Board. The first was "the basic problem of the policy of the company in relation to the maintenance and expansion of this [its chemical] business." The second question was how to "organize, or reorganize," its chemical operations. Howard, now president of the SOD, and universally recognized as Jersey's guiding figure in the chemical field, led this discussion. As a point of departure, in 1943, Howard noted Jersey's historical policy not to enter any other field, "outside the oil industry, save only to the extent that there is some special business relation between the company's oil business and the non-oil business in question." While this general policy had been sufficient to cover such developments as the company's participation in the Ethyl Corporation and the Standard Alcohol Company and its production of petroleum additives, Howard contended that the company needed a more flexible policy as it moved farther into the development of chemicals tied to the oil business only by the raw materials from which they were made. He believed that the decade of cooperation between Jersey's technical staff and that of I. G. Farben had resulted "in greatly widening the horizon of the company with respect to possible new oil-chemical developments" and by making Jersey's operations in the oil-chemical business "in scale and variety much greater than that of any other oil company in the world." Characterizing Jersey as one of "only a very few oil [or chemical] companies" with "anything like an up-to-date picture of the immediate commercial possibilities [of petrochemicals] . . . ," and noting that it also possessed a technical and manufacturing organization capable of expanding the business, Howard argued for a policy statement clarifying the relationship of the oil-chemical business to the company's oil business; one that would guide the expansion of the former.[14]

Howard also favored Jersey's developing its own chemical business. Although the risks were greater, this would permit a more complete dovetailing of chemical processes with refinery operations, while at the same time placing fewer restraints on the future development of the refineries' oil business. They, in turn, would benefit from the impetus given by chemical processing to closely associated refining processes. By complementing the company's oil business, its chemical interests could thus be allowed to expand as rapidly as economics dictated.

In summary, Howard argued that "fundamental economics and practical experience both favor the maintenance and normal expansion of our business, of converting oil raw materials into primary chemical products." The potential profits of the oil-chemical business and its tendency to make the associated refinery a more intensive manufacturing enterprise—thus more profitable—justified a policy of investing necessary capital, time, and effort in its expansion. Finally, he pointed out that if the oil industry failed to expand the business, chemical companies would be forced to do so, and they would preempt the field. Jersey had to act quickly.[15]

In pressing for this expansion of the oil-chemical business Howard understood the relations between that function and other operations of the company. While such expansion went beyond the production of refinery gases, he generally limited it to those processes most closely related to refinery operations: the conversion of refinery gases to primary chemical raw materials. Thus during World War II, even Jersey's chief proponent of petrochemicals did not openly advocate integration forward into the realm of the chemical companies.

Perhaps Howard erred in this approach to the Jersey Board, for Clark, Sadler and Riedemann no longer sat in such councils to back his proposals. Howard realized—as did no other person—that Jersey research engineers and scientists constituted one of the ablest, most tightly bonded teams in the field of petroleum chemistry. Among the oil companies no one disputed their primacy.[16] And he wanted the Jersey Board to utilize the experience, morale and know-how of this group. Persons who remember the war years point out that Haslam, who went on the Board in 1942, immediately received the task of improving Jersey's public relations, as well as that of serving as the contact director for development. With his attention divided, and with the Board composed largely of refiners and producers, they contend that great opportunities for Jersey chemicals, then and later, did not receive the consideration they deserved, despite Howard's pleading. Further, it is true that Howard came out of the wartime congressional hearings a much less influential figure. So conceivably, the changes in personnel on the Jersey Board and wartime stringencies did dilute the influence of the leading advocates of the oil-chemical business at the time.

One participant commented on those years somewhat wryly. He pointed out that expenditures on chemical research and development were kept low, and that research on butyl rubber absorbed the largest portion of the budget. Jersey also needed capital for rebuilding its European refineries, and "top management had not yet evolved a definite chemical policy for the future." He added, "To us in the chemical field, this was very short-sighted in view of the increased activities of our competitors, both chemical companies and oil companies. . . . In 1950, the Chemical Products Department sold our plant [Buna-N] to U.S. Rubber. This was a blow to our chemists' morale. . . ."[17] If this is a fair statement, then the Jersey Company's failure to capitalize on the available skills and training of its chemists and technicians represented a great loss. [It should be noted, however, that while Jersey disposed of the Buna-N (butadiene acrylonitrile) plant, it continued to fund research in the butyl rubber process.]

One other loss occurred during the war and in the five years following. Perhaps because of self-engendered pressure resulting from the public censure Jersey received from congressional committee hearings, the company strained every resource to develop high quality synthetic rubber from isobutylene as well as butadiene. It is true that, due to the "preoccupation by Jersey's chemical people with synthetic rubber, the very heart and soul of the rapidly growing world petrochemical industry was missed, namely thermoplastics."[18] However, such mistakes were not unusual in the hurry to develop the Jersey chemical business.

The Jersey Board certainly moved more slowly than usual in formulating a clear policy concerning petrochemicals, and this despite Frank Howard's moderate approach and limited goals for that operating division. Eugene Holman, who would soon be elected president of Jersey, expressed some lingering apprehension about the effect of diverting to chemical manufacture refinery gases that could also be valuable in the processing of petroleum products,[19] but further study confirmed the complementary nature of the two lines of business. Thus petrochemicals could no longer be considered an outlet for by-products. Oil industry giants such as Jersey were in a superior position to develop the business for a number of reasons. Well situated to select and supply the most satisfactory processing stocks, oil companies enjoyed a technological advantage in processing normal petroleum products used to make chemical raw materials, and they possessed better means of absorbing the by-products from chemical processes. The use of olefins by both refineries and chemical plants would make possible optimum use of these gases. [20] More importantly, the technological similarity of the two fields would strengthen both.

Howard's views on the direction of chemicals expansion seemed to be "accepted by the Jersey organization as a whole,"[21] but that acceptance did not result in a definite statement of policy on the subject. The company's top management did, however, take concrete steps toward solving what Howard considered a secondary problem, the question of how to organize the company's chemical business.

Consistent with its position as a "stepchild" of the petroleum industry,[22] petrochemicals had never found a satisfactory, permanent home in the Jersey organization. Despite some earlier efforts to find a rational corporate basis for directing the company's chemical business, petrochemicals continued to be handled by a variety of affiliates, with no central policy-making structure. In addition to Jersey's equity interest in Ethyl Corporation, in varying degrees oil-chemical operations had been conducted since the 1920s by Standard Oil Company of New Jersey [incorporated in Delaware]—hereafter cited as Esso Standard—Standard Oil Company of Louisiana, Humble Oil & Refining Company [Humble], Imperial Oil Limited [Imperial] in Canada, and Standard Alcohol Company. Moreover, several joint ventures, with affiliates and other chemical companies, had been launched in Europe before the war. Sales were made by affiliates "in part directly to outside customers, in part to and through one another and in part through Stanco Distributors, Inc." which latter firm specialized in marketing Jersey's lube oil additives and synthetic rubber. Howard provided such coordination as existed among these companies. Next H. W. "Bud" Fisher of Esso Standard [after 1945, director], with a staff of engineers, chemists and salesmen had partial administrative responsibility for oil-chemicals, while Dr. M. B. Hopkins, president of Standard Alcohol Company, served as a general staff executive for the business. R. P. Russell, a director of Ethyl Corporation and executive vice president of SOD, also had some direct responsibility. On the Jersey Board, Frank Abrams, F. H. Bedford, Jr., and Haslam all had important general responsibility for either the manufacture or sale of oil-chemical products.[23]

Recognizing the need for some "organization . . . in which would be vested the sole responsibility for recommending to the Board what Jersey's position should be with respect to chemical and potential chemical products," in 1943 Orville Harden took the lead in establishing a permanent Chemical Policy Committee. Under the chairmanship of Howard, this committee would be responsible for promoting the development of the company's chemical business by working with affiliates and the parent company's Coordination Committee and by making recommendations to the Board.[24] In addition to Howard, the Policy Committee included Fisher as vice chairman, Director Chester F. Smith, Russell, M. J. "Jack" Rathbone with Hopkins as secretary.

At the same time, Fisher received operating responsibility for sales and promotion of the chemical business.[25] At first these functions were divided between Stanco Distributors, Inc. and Standard Alcohol, but in 1947 they were centralized at last in the newly created Enjay Company, Inc., which had responsibility for both domestic and export chemical sales.[26]

The development of Jersey's chemical business continued to be slow throughout the early postwar years despite these changes. The Chemical Policy Committee and Enjay gave the company's chemicals effort some central direction and a degree of independent management. The election of Fisher as director of Esso Standard, in 1945, and his subsequent appointment to the parent company's Coordination Committee[27] indicated the improved status of chemicals within the company. However, no major expansion occurred. As Howard noted, research in petrochemicals increased, despite the Board's maintenance of a tight rein on capital expenditures.[28] But the lower priority given oil-chemical investments eroded Jersey's leadership position in chemicals among oil companies in the field; by 1948 other companies had rapidly closed the gap, as the industry expanded to supply large new markets for plastics, synthetic fibers, and synthetic detergents.[29]

In 1945 Frank Howard stepped down as vice president. His contribution to Jersey, especially to chemicals, across more than twenty-five years can be documented. What caused his early retirement is not known; conceivably he sought to escape the weight of obloquy associated with his prewar role in dealing with I. G. Farbenindustrie.[30] Jersey retained him as consultant for several years. But with Howard gone and Haslam taking early retirement in 1950, the Jersey Company lost the principal advocates of chemicals among its top executives. Nevertheless, as a consultant, Howard continued to stimulate discussion and action on the Board and to make petrochemicals a larger share in the company's business.

X   X   X

On May 12, 1950, Howard wrote a broad-ranging, strong and, in places, prescient letter to Board Chairman Abrams and President Holman. This interesting document resulted largely from a desire on the part of the management of the Ethyl Corporation to expand into the oil-chemicals field. Further, Howard, aware that the expansion of the Jersey chemical business had not kept pace with the mushrooming growth of that industry, deemed the Jersey position in that field precarious. He conceded that the heavy capital requirements and other "demands of the basic branches of the oil industry have left little time or money for new ventures" in oil-chemicals, but he tagged as the principal reason for the loss of position, "the lack of recognized leadership in the holding company for this branch of the business." Recognizing that this might be "a matter of opinion," Howard explained that:

> the missing factor needed to make the separated fragments of our oil-chemical business into an effective competitive unit, which can hold its position in this business, is detailed knowledge, initiative and follow-through centered in one individual at the top. That means an officer or director of the holding company charged with the positive duty of acting as a leader for this business.

This person's leadership, Howard said, "must be exercised in a manner consistent" with the company's management structure, "and the policy control of the Jersey Board." Howard emphasized that such leadership could build and maintain morale and "an esprit de corps among the [then] three groups of chemical research men . . . and the Enjay group." Failure of the Board to provide such an executive meant to the chemists that "they are isolated stepchildren of a great company which has no real interest in them, or their relatively trifling work, which lies off the main path of the company." New research led to new and cheaper basic chemicals, and these contributed to "national growth, employment, defense and prosperity." In this exploratory field, Howard contended, Jersey had "the most able and experienced unit in the oil-chemical field" and in his opinion, the company "had an obligation to the nation, as well as to its own stockholders to develop and supply materials from oil needed for growth of the chemical industry." He pointed out as examples Jersey's leadership in developing "unique concentrations of special raw materials such as olefines [sic] and aromatics" and that in these chemicals they had the "longest experience, the best research and production talent." The Jersey relationship with the Ethyl Corporation, he concluded, would benefit also from a top leader "familiar with

Jersey's chemical business."[31]

The Howard letter created quite a stir among company executives. Rathbone, who had just joined the Board after years of outstanding service in the Louisiana Company, appeared by training and experience to fit the Howard model of a Board-level champion.[32] Abrams turned to Rathbone for advice and a study of the Howard proposal on July 10, 1950. He indicated that the Howard letter had been discussed at lunch that day and that the group "believed that Frank's suggestion of separate Board representation of the Chemical business was not in keeping with our principles of organization, and the same end result could be secured by other means." Abrams admitted that the company had "no clear policy understanding . . . concerning the extent to which it is prepared to go in promoting the Oil Chemical business." He called for a thorough review of the oil-chemical business at Jersey.[33]

With his boundless energy plus his dedication to hard work, Rathbone began a six months' assessment of the problem. The fact that he had only joined the Board in late 1949 may have led him to ponder the phrasing of his report to Abrams. Certainly his sympathies lay with the chemical people, but the Board would not have been receptive to outright support of chemicals. He realized that, in a practical sense, little could be achieved for himself or the company by taking so strong a position favoring them as had Haslam. On January 10, 1951, he replied to Abrams and at the same time responded to Howard's earlier letter. Rathbone noted that Howard's comments "are well taken but I think his suggested solution is not the most effective which can be devised." The new director then launched into a review of the oil-chemical business. First, he pointed out that for Esso Standard's chemical activities—the principal Jersey firm engaged in the business—between 1945 and 1949, "the average return on gross investment (after taxes) was 15.8%," a much better return than that from the oil business. For 1950, he projected for that company a return of 24 percent on the original investment. Then he stated that in the same five-year period $9,000,000 had been reinvested in new chemical facilities; "Thus with net earnings . . . the oil-chemical business has provided funds amounting to $33,000,000, of which $9,000,000 was reinvested in the chemical business and $24,000,000 was made available for dividends or investment in the petroleum business." To Rathbone the business rated a strong plus, for it provided additional capital to "the petroleum end of the business." He next by figures and charts established that, "In every case, except Du Pont, the [various leading] chemical companies plowed back a much greater proportion of earnings than did our Chemical Products Department." His figures established Jersey's leadership among oil companies in the chemical business, with a ranking of eleventh among all chemical companies. Further, he said, "It will be noted that Jersey's chemical products return is above average for the 14 chemical companies listed."

When Rathbone reached that section of his report, it would appear that he had established the validity of the main points in Howard's letter; the value to Jersey of the chemical business, the need to expand, and the efficiency of the Chemical Division employees. But then he advocated a course that across the years proved costly to Jersey, one that led after twenty-two years additional experience to a complete reversal of policy. He asserted that "It is no appreciable effort for our operators, our mechanics, our technical people or our managers to understand and engage in oil-chemical operations as compared to our regular run of petroleum operations." While that may have been correct in 1951, hindsight must judge the notion in error. He went on to state that the oil-chemicals and the oil business dove-tailed smoothly, and "The profits of the Chemical Products Department are always additive to those which could be secured if its products were sold as petroleum products instead of chemical products." Once again, viewed from the coign of vantage of thirty years hindsight, the hard-hitting Rathbone made a judgmental error. He described research in chemistry "as so closely allied to oil research that they would have to carry it on even if we were not in the chemical business at all." Jersey chemical research, he placed in four categories: a) alcohols and derivatives, b) petroleum additives, c) synthetic rubber and allied products, and d) chemical raw materials for other industries. These indicated to Rathbone the diversity of the business. He recommended that the affiliates establish chemical businesses but when they desired to do so, they should receive advice and encouragement from Esso Standard's Chemical Products Department. He advocated that 50 percent of the net earnings of chemicals be returned for investment as needed.

Rathbone agreed that postwar capital demands "to expand the basic oil business" had been a major factor in the Jersey failure to push chemicals more heartily, but he perceived as an equally important reason uncertainty among Board members as to whether an oil company should be in the chemical business at all. Company affiliates entering the chemical business did present some problems both of policy and management; hence Jersey required some guidelines for chemicals. After a few statements appeasing those Board members who would relegate chemicals to a minor position in the company, Rathbone indicated where his own opinion lay. As executive and later president of Standard of Louisiana he had loomed large on the scene when Haslam, Russell and their MIT engineers began the research that had led to the development of the chemical business for the company. Himself a chemical engineer by training, he understood what they were doing.

Sagaciously, Rathbone concluded with the recommendation that oil-chemicals should be separated from the oil business, that it no longer should be "played down," and that chemical operators had not previously been given the "standing" valuable for morale at Jersey. He pointed out that "year after year the chemical products management have put up [for approval by the Board] projects estimated to yield very high financial returns, unsuccessfully." He advocated more freedom of operation for chemicals, and at the same time greater accountability to the holding company.[34]

Copies of the Rathbone letter went to a list of directors and other executives, among them Eger Murphree, then president of Standard Oil Development Company, who responded to the lengthy Rathbone review on January 22, 1951. While Murphree agreed with a good portion of the report, he took issue sharply with one point—a long-time bone of argument by the chemical people—that of having the Chemical Products management review the nature and scope of the "research program in the oil-chemical field contemplated by the Development Company each year and which is to be charged to the Chemical Products Department, with authority to refuse to accept charges for any part of the program they are unwilling to finance." Murphree stated that under existing accounting procedures, development and chemicals agreed upon such charges. He urged that this working arrangement be changed. The difficulties that emerged had to do with the uncertainty of knowing, at the time research went on, whether either oil or chemicals or both would utilize the process developed. Too, some chemists believed that patent income was unfairly divided. To Murphree, "The distribution we make in the cost of research between the oil field and the oil-chemical field seems to us to be quite artificial." If procedural change proved necessary, then he advocated that "work on oil-chemicals" be "treated the same as research projects in the oil field."[35]

Howard "read with interest and appreciation Rathbone's report." He called on Abrams to help Rathbone "become a recognized leader of this chemical business throughout the Jersey interests." The only point on which Howard differed substantially with Rathbone concerned the proposed transfer of the Enjay Company to Esso Standard.[36]

Like the earlier proposals of Howard, Rathbone's suggested policy statement would endorse the desire of the company to continue expansion in the oil-chemical field, subject to a number of conditions. Most of these restrictions Eugene Holman had sketched out earlier.[37] They simply described the parameters within which the business had already developed.

Jersey's Executive Committee discussed the Rathbone review of the petrochemical business and his proposed policy for over a year. The Committee, after adding several restrictions governing expansion, finally adopted a "General Statement with Respect to Petroleum-Chemical Operations of Jersey Affiliated Companies" in March, 1952.[38] The 1952 policy statement did, for the first time, present a definite framework within which expansion could be considered. But it did not convince oil-chemicals personnel that their position in the company had changed. The statement still limited expansion. Despite these flaws, it came closer than had any previous policy announcement to a recognition of oil-chemicals as, in Frank Howard's words, "a permanent and altogether desirable branch of the oil business."[39]

If the policy Rathbone put forth for oil-chemicals gained acceptance, his suggested organization did not. Enjay did not merge with Esso Standard at that time, although its sales effort continued to be closely supervised by the Esso Standard Chemical Products Department. The Refining Coordination [Refcor] Department of the holding company rather than the Chemical Products Department of its affiliate assumed the function of advising and assisting affiliates on potential oil-chemical projects.[40] This arrangement proved more conducive to coordinating development throughout the Jersey system, but by placing oil-chemicals within the functional structure of the parent company, Jersey forfeited the advantages of a separate organization specializing in the coordination of all aspects of the business.

Proponents within Jersey continued to think that expansion of the company's oil-chemical business was being unnecessarily retarded. While the "100 per cent oil atmosphere" of the 1940s had dispersed somewhat, discrimination in the competition for capital investment funds remained a concern.[41] Presentations for investment funds to the parent company's Refining Coordination Department had to be supported by a more thorough analysis, establishing a substantially larger rate of return on capital invested than did proposed oil investments. Defenders of oil-chemicals argued that they were being asked to promise both minimum risk and maximum return on the same investment,[42] in effect a double standard, clearly exceeding the policy statement's requirement "that proposed capital investments by affiliates in this field should meet the usual tests of good business judgment and satisfactory financing arrangements."[43] Although few of the projects they developed failed to be approved, they still felt that some highly attractive investments could be found among those rejected.[44]

Despite these apparent handicaps, development of the company's oil-chemical business did accelerate under the new policy. With the added stimulus of the Korean War mobilization, sales increased from about $75 million in 1950 to almost $100 million three years later.[45] With a return on investment substantially higher than that from the company's oil business, this growth encouraged the approval of larger capital outlays. Still, Jersey slipped from eleventh to twelfth rank among U.S. chemical companies and, more disturbing, other oil companies, especially Shell [Shell Oil Company], threatened its leadership in the oil-chemical field.[46] Experience had clarified the appropriate province of the oil companies within the chemical industry, and Jersey management had no intention of relinquishing their company's historical claim to it. However, only part of the story stems from corporate pride, for conditions in the oil business continued to make further oil-chemicals growth increasingly attractive. After a temporary softening of the chemicals market in the first half of 1954, sales by Enjay resumed their upward trend, with 1955 sales promising to exceed the previous year by 50 percent.[47] In Europe, recovery and the end of currency restrictions, combined after 1954 with the renewed availability of Iranian oil, greatly enhanced the attractiveness of oil-chemical investment abroad.

In 1955, the role of chemicals in the Jersey structure underwent fresh analysis and evaluation. The Executive Committee requested that a special committee of the Coordination Department study the policy of Jersey and its affiliates concerning petrochemicals. Charles J. Hedlund chaired this committee. When they reported on May 24, 1955, the Executive Committee approved the recommendations made and in accepting them, indicated that chemicals from petroleum had "come of age in its own right." They added that Jersey should make a "more aggressive effort" in that field. The Executive Committee also exhibited concern that the company had virtually no foreign chemical activities "while Shell was firmly entrenched throughout Europe and was becoming active in other areas." They obviously expected action from all the affiliates engaged in manufacturing and marketing petrochemicals.

The 1955 statement of Jersey's chemical policy, while in places echoing the earlier 1952 guidelines, did include some changes. But the new one clearly reflected the company's enhanced respect for chemicals at Jersey. It provided a more positive approach governing chemical expansion and greater emphasis on foreign developments. Petrochemicals, to the Coordination Committee, had now become "a rapidly growing and profitable field of business throughout the world" rather than the "relatively new and rapidly expanding field" described in the 1952 statement. The report urged affiliates to "move aggressively to make the most of investment opportunities." The new importance of chemicals is nowhere better reflected than in the statement that expansion might require development of "additional managerial and staff personnel" to handle the "petroleum-chemical business" rather than diverting the time of the existing managers and staff from the basic petroleum business. Affiliates were encouraged to present proposals.

The report also suggested organizational changes "to take cognizance of the growing importance of petroleum-chemical activities . . . " and to stimulate their expansion worldwide. While Esso Standard and Enjay would continue to have principal responsibility for chemical operations in the United States, a chemical coordinator would be added to the Refining Coordination Department of the parent company. Charged with coordinating chemical operations worldwide, this person would be responsible specifically "for providing the stimulation and leadership needed for the further development of such activities abroad." The report anticipated changes in Marketing Coordination [Marcor] but deferred them until the spring of 1956, when the experience gained in the interval could be assayed.[48]

Directors Marion W. Boyer and Stewart Coleman with Dr. Harry G. Burks, Jr. of the Refining Coordination Department received the task of selecting the new chemical coordinator. After discussing possible candidates with Board members and the Coordination Committee, they unanimously agreed that "Otz Tracy was the ideal man for the job" and they "obtained the Executive Committee's approval to approach him."[49] However, Tracy refused the position and they chose Robert M. Jackson of the Chemical Products Department of Esso Standard Oil Company to fill the post.[50]

X   X   X

The world of plastics began to emerge. Low-density polyethylene became available to chemical firms in 1955; Jersey's domestic chemical plants quickly moved to produce the newest thermoplastic, polypropylene. "This product was poorly handled and to this day has not been very successful," one observer commented.[51]

Inner tubes had been for years the largest outlet for butyl rubber, but the emergence of the tubeless tire in the 1954 model car represented a serious threat to this market. Though a major supplier of butadiene for the manufacture of the general purpose styrene-butadiene rubber [SBR] used in all tires, Jersey had continued to concentrate its main effort in finding new uses for butyl. Its scientists had invented butyl, and company pride remained a factor in this decision. One chemist likened the effect of the tubeless tire on butyl to that of the boll weevil on cotton. "Making butyl just for inner tubes was like raising cotton used to be in the South—a one-crop proposition. Along came the boll weevil and it turned out that everybody was better off when we had to diversify."[52] Despite this note of optimism, Jersey's butyl sales by 1958 had dropped to 52,241 tons, a decline of 40 percent in three years. Butyl simply lacked the wear qualities necessary for general purpose tires.[53]

But by 1958 the research efforts of Jersey scientists began to provide new uses for butyl and to improve on the old ones. The effort to develop an all-butyl tire had failed. The real push to restoring the health of Jersey's synthetic rubber business lay elsewhere. From the laboratories came new applications for butyl in related specialty rubbers, especially chlorinated butyl rubber [chlorobutyl]; also an improved truck tire inner tube, and other uses. In this recovery the export market, especially Europe, where Enjay had been promoting butyl since World War II, played an important role. By 1959 butyl sales exceeded 105,000 tons, more than double those of 1958, and two large additions to manufacturing capacity were underway.[54]

The most dramatic increase in Jersey's chemical business during this period came in 1955 when Tracy, president of Enjay, finally succeeded in arranging for the Esso Standard Oil Company and the Humble Company to purchase from the U.S. Rubber Producing Facilities Disposal Commission the butyl rubber plants the affiliates had built and operated for the government during and following World War II, which the company had long desired but had not been able to persuade that commission to sell. [55] This acquisition substantially enlarged the petrochemicals business of both companies. Coinciding with the new statement of policy on petrochemicals expansion, the purchase of butyl plants precipitated and induced the organizational changes necessary to permit full exploitation of these properties. To help the affiliates develop and serve the markets for their production, technical service laboratories were established within Enjay. The Jersey Executive Committee in October, 1955 approved transfer of its equity interest in Enjay to Esso Standard. This action strengthened the company's principal petrochemical marketing affiliate in the United States.[56]

As foreign affiliates began marketing petrochemicals, they created not only outlets for their own production but also additional markets for American-produced chemicals and a broader base of support for research and development expenditures. However, the main impetus for expanding foreign chemical operations derived from the needs of the foreign affiliates themselves.

With the restoration of Iranian oil supplies to world markets late in 1954, a chronic surplus of crude oil producing capacity plagued the international oil industry. Combined with a decelerating growth rate of demand in the major markets and a surplus international refining capacity, this situation engendered an intense competition that exerted downward pressure on Jersey's profits. While the United States remained somewhat insulated from the direct effects, these factors hit hard in Western Europe, where the surplus crude had its greatest impact.[57] In addition to the extremely competitive situation in fuels, other forces peculiar to Europe increased the need to respond to the 1955 call for chemicals expansion.

European refineries, often built to satisfy host-country foreign exchange requirements or to appease burgeoning nationalism rather than based on demand or the company's economic criteria, had particular need for the supplementary return on chemical investments to make them profitable or at least to cut losses.[58] The fact that most European affiliates had only refining and marketing functions, and not the more profitable producing operations, further intensified their interest in petrochemicals profits. The Western European market became increasingly attractive after currency exchange restrictions began to be reduced, and it grew more so as economic unification under the Common Market became a reality. Actually, foreign affiliates in some of the larger markets had begun to develop petrochemical investment opportunities even before the new policy statement of 1955,[59] but the parent company's top management envisioned a much broader effort, with participation by company interests worldwide within five years.[60]

The new initiative had not proceeded very far, however, when it encountered organizational problems. Humble and Esso Standard had organizations that allowed coordinated development of their chemical interests without interference from their petroleum business; but foreign affiliates undergoing rapid increases in their petrochemical investments had no such inter-affiliate structure. The relatively small size of these companies and especially of their chemical operations compounded the problem.[61] While the parent company had established no firm policy, it quickly suggested a pragmatic solution calling for only minimal changes in organization. Jersey advised those integrated affiliates beginning or expanding chemical manufacturing operations to distribute responsibilities among existing functions, and to appoint a chemical coordinator with duties generally confined to the evaluation of existing chemical operations and the promotion of new ones. For marketing affiliates with no manufacturing operations the existing organization handled the expanding chemical business.[62]

The parent company offered new levels of support to both types of affiliates. The chemical coordinator within Jersey's Refining Coordination Department, given responsibility for overseeing all company petrochemical interests, had particular responsibility for "providing the stimulation and leadership needed for the further development of such activities abroad."[63] In addition to promoting new foreign chemical production, this executive also helped arrange marketing assistance for foreign affiliates on several levels. Enjay and the Marketing Coordination Department took the lead here, providing expert marketing advice. Esso Export Corporation [Esso Export], the company that handled sales among all Jersey affiliates, had a special chemical group that gave supply information and assistance. Marketing technical service came from Chemical Liaison in Europe, operating out of England in a manner comparable to the role of the Enjay Laboratories staff in the United States.

Despite this structure for support, foreign affiliates continued to encounter difficulties as they entered or expanded in the petrochemical field. Basically, chemical production closely approximated ordinary refinery operations, while chemical marketing differed fundamentally from oil product marketing. Oil marketers handled relatively standardized products. Chemical salesmen, on the other hand, constantly had to provide technical service to customers, including tailoring products to their special needs. This difference necessitated not only a new approach on the part of manufacturers, but it required also a corps of experienced specialists in all functions. Only time could eliminate this deficiency.[64] The same remedy could be applied for technical problems associated with the start-up of new activities and for entry into new markets.

The relatively small size of the chemical markets outside of the United States remained a chronic problem even in Europe, where the potential for economic unification might have augured relief. Despite potentially large markets there, the chemical business in Europe remained in the hands of a number of affiliates, each primarily restricted to a single national market. This fragmentation increased the difficulty in coordinating the European market for the overall benefit of Jersey. Small units did not justify a separate chemical organization in every affiliate, and without integration, development of petrochemicals could not proceed.

Nevertheless, substantial expansion in petrochemical investment outside the United States did begin to take place after 1955. The first steps toward expanding petrochemical manufacturing by foreign affiliates occurred in those companies with large-scale refineries already engaged in some chemical activities. As it began expanding its Fawley refinery from 150,000 to

200,000 barrels per day [b/d] capacity, Esso Petroleum Company, Limited, incorporated a new steam cracker for the manufacture of ethylene and butadiene to supply the nearby plants of non-affiliated companies making plastics and synthetic rubber.[65] Soon other affiliates entered the petrochemical field. By 1957, approved capital expenditures for petrochemical projects had risen to a level of $88 million, and a major portion went overseas. Imperial Oil Limited also entered the petrochemical field, bringing on stream at Sarnia a plant to produce alkylate feedstock for the manufacture of synthetic detergents, the first such facility of its type in Canada.[66]

Certainly the affiliates' interest in petrochemicals intensified. J. Kenneth Jamieson, then serving as one of Imperial Oil Limited's vice presidents, remembered, "I was really looking after that end of the business. . . . We decided to start moving into the chemical business at Imperial," based on how well Jersey had done in petrochemicals. Imperial's formal entrance into the field dated from a butyl plant constructed during World War II. Jamieson added, "the petrochemical industry in Canada started to grow, and it was the logical thing for Imperial to get in it. We hired a fellow from the old Esso Standard Company [Clay Beamer] to come up and start a chemical end. Actually, I hired him, and convinced him to come up." Regarding the policy statement's invitation to affiliates to present proposals, Jamieson recalled the time he and others at Imperial came down to make a budget presentation for chemicals before a scheduled meeting of the Jersey Coordination Committee. Because the coordinators could not attend, they sent deputies. Jamieson said, "So we made our presentation . . . then said, 'as soon as you fellows approve this . . . ' They said, 'Well, hell, we can't approve anything.' I said, 'What am I doing here wasting my time for then?' " When the parent company sent a "big long memorandum . . . in essence saying that you should not establish a separate chemical marketing organization, that it should be handled by the conventional marketers . . . I was the contact [director at Imperial] for the chemical business, and I just threw the damn thing in the wastebasket." And when Frank Howard learned about these activities at Jersey, he wrote to Rathbone, "how happy I am with the present world-wide aggressive program of the Jersey interests in petrochemicals."[67]

By 1959, more than one third of the company's total petrochemical investment lay outside the United States, and investment plans proposed to continue the shift to foreign countries. With major additions coming on stream in France and Germany, and construction of Australia's first petrochemical installation underway by the Standard-Vacuum Oil Company [Stanvac][68]—which also began discussions with the government of Japan about a large petrochemical venture in that country—the scope of Jersey's chemical interest was becoming truly global.[69] From a negligible portion in 1956, foreign sales in 1959 represented 36 percent of Jersey's total chemical sales of $256 million.[70]

Despite this rapid growth, foreign chemical operations were not yet playing the broadened role anticipated from them in the 1955 decision to expand petrochemical operations globally. The overall success of that development depended not only upon European and Latin American affiliates selling the products of their own new chemical facilities, but also on maximizing their imports of petrochemicals produced by Esso Standard, Humble and Imperial. Full exploitation of the company's advantage in large-scale processing and its research and development expenditures depended upon international markets. Enjay and the Marketing Coordination Department of the parent company began taking steps to increase these outlets in 1958, but made only slow progress compared to the accelerating growth of international chemical markets.[71] Nevertheless, the expansion of the company's chemical business worldwide was sufficient to intensify the problem of coordinating sales. While this was one of the consequences of the company's expansion during the 1950s, a more fundamental consideration became the new shape of Jersey's multinational chemicals business.

The bulk of these sales came from the United States, where Enjay increased its market share to maintain the twelfth rank among U.S. chemical companies.[72] Yet, despite Enjay's continued dominance, the appropriate perspective for viewing Jersey's chemical interests became more and more international. In 1959, foreign petrochemical projects accounted for more than a third of the company's $300 million total investment in the field; they would represent, moreover, an even greater portion of the additional $100 million outlay expected in the following two years.[73] But the rapid growth of the company's foreign chemical operations simultaneously with the broadening of Enjay's business, brought with it more than a prospect of participating in the world's faster growing markets. Expansion along such a broad and varied front also posed a challenge that Jersey's chemical management would have to face in the 1960s.

1960-1967

By 1960, Jersey's chemical sales had tripled since 1953, and the company projected continued and vigorous growth. Under the impetus of the 1955 decision to build a global chemical operation, foreign petrochemical sales promised to increase even more, as the bulk of new capital investment continued to take place outside the United States.[74] The success of the European

Economic Community [EEC] augured well for these investments.

To a large extent the anticipated expansion of petrochemicals was based on the continuation of the forces responsible for the previous rapid growth. The crude oil surplus continued in the 1960s to generate strong competitive pressures and to reduce profit margins in petroleum product sales. Although the chemical industry had begun to experience a slight decline in profitability, it still remained far more lucrative than the petroleum industry. Attractive opportunities at home and abroad dictated Jersey's pattern of continued investment in new chemical facilities.

The tendency of traditional chemical companies to integrate backward, manufacturing their own feedstocks from refinery gases, threatened the position of most basic chemical suppliers. Some producers became concerned that, unless they integrated forward to become competing producers of finished products, they would be squeezed out of the industry.[75]

At Jersey, administrative changes served to promote further growth in the petrochemical field and to enhance the company position in chemicals. Eugene Holman, a great oil man who "never really got excited about chemicals," retired as Jersey's chief executive officer [CEO] in 1960, bringing to the helm as president, Rathbone, a refining specialist with a longtime interest in petrochemicals, who had become a strong believer in their future. Too, as the author of the company's first policy statement with definite guidelines for petrochemical development, Rathbone, in 1952, had helped pave the way for the expansion of the 1950s. Now, as Jersey's CEO, he enthusiastically pushed expansion in chemicals as an important part of the company's overall strategy for meeting the difficult conditions of the 1960s. Crude oil surplus capacity and the decelerating rise in demand for petroleum products appeared to dictate a major program of diversification. While a higher return on invested capital and greater long-range stability and growth remained primary concerns,[76] the prospect of new, profitable outlets for crude oil had an additional appeal.

When Rathbone became CEO of Jersey, *The Lamp* in an understatement described him "as a congenial, energetic man . . . with a prodigious capacity for work." The account pointed out one of his management techniques: "It is not unusual, on Monday mornings, for him to parcel out sheaves of yellow paper covered with ideas about company affairs that he has jotted down over the weekend."[77] The same issue which announced his new appointment also carried "The President's Letter," in which Rathbone painted a glowing future for the petrochemical industry and concluded with the statement, "our company expects to participate fully in this dynamic growth." At the shareholders meeting in 1960, he told those present that despite "overcapacity and overproduction" new products would help maintain income; polypropylene, he said, "has many qualities and properties superior to those of polyethylene." He continued to express unquestioning approval of chemicals in the *Annual Report* for that year.[78]

Perhaps by coincidence, in that same year, 1960, Rathbone set in motion the Board study that examined company investment practices. Even before he advanced to command Jersey, the Board Advisory Committee on Investments [BAC] had assumed the review function of the now abolished Coordination Committee, with its staff work being done by the newly created Investment Analysis Department [IAD]. From the spring of 1960 until June, 1962, when the final report, "Statements of Investment Policy" appeared, the Board considered this new approach to the business of investing. They assumed that in the life of the company the time had arrived when it behooved them to search outside the traditional petroleum business for fresh opportunities. Further, they expected to have a large cash surplus to back this new departure. Harold W. Fisher, contact director for chemicals, pointed out that for chemicals to continue to prosper, it would be necessary to expand into new areas, and to integrate forward. This required more investments. In fact, chemicals emerged from these deliberations in a strong position, with new guidelines for the future, a "Statement of Policy for Chemical Activities" [March 23, 1962].

One basic assumption of this new investment strategy, proposed in 1960 by David J. Jones, echoed the 1951 report on "Jersey's Oil-Chemical Business" made by then Director Rathbone.[79] According to this view, successful managers from the petroleum and from the petrochemical operations could easily move from one to the other. While a sort of high-flung notion at the time, it remained current for several years before the company reversed its position.

So beginning in 1962 under Rathbone's leadership, the new word became "diversify" and the Jersey Board took the obvious step of expanding the company's chemical operations as rapidly as possible, both in older lines and in new areas.[80] Rathbone remained convinced that Jersey would not make mistakes and that chemical profits would figure significantly in Jersey's overall performance.[81] Both the BAC and the new IAD[82] promised to remove the last vestiges of the anti-chemical feeling that had once pervaded the Coordination Committee and constrained petrochemical growth. The decentralization of investment responsibility that occurred as part of the same reorganization gave affiliate managers a mandate to develop promising petrochemical investments. They quickly responded to this change.

Under the new policy the growth of Jersey's petrochemical operations continued, with emphasis on integrating forward to become competitive with the chemical companies.[83] This movement involved not only a "stepped-up marketing drive"[84] but also a steady increase in "prescription marketing," tailoring products for specific customers' needs, considered a distinguishing mark of a true "chemical" operation. Enjay Company had already advanced quite far in this direction, backed by technical service and product development laboratories at Linden, New Jersey; Baton Rouge, Louisiana; and Baytown, Texas. Esso Research and Engineering Company's [Esso Research] laboratories at Abingdon, England, provided similar services for marketing affiliates in Europe, the Mediterranean area, and North and West Africa. As the company prepared to move into more sophisticated chemical products, Esso Research's " 'mature' chemistry and chemical engineering organization" represented an important asset.[85]

While the major part of new chemical investments were outside the United States, Enjay, headed by John E. Wood, III, continued in the early 1960s to be, "for all practical purposes . . . Jersey's 'chemical company.' "[86] Rising sales volumes led to an expansion in facilities for the production of paraxylene and ethylene that made Jersey the world's largest "merchant" producer of the latter chemical. But industry-wide overcapacity in basic petrochemicals had introduced a period of weak prices. While Enjay tried to improve the profitability of this important area of its business, it concentrated greater efforts on expansion elsewhere.

Part of the response to low prices for basic petrochemicals centered on expanding other established lines of business, especially synthetic rubber and solvents like ethyl alcohol, isopropyl alcohol [IPA], and methyl ethyl ketone [MEK].[87] Expansion in synthetic rubber proved especially indicative of the growth of the company's business during this period. Despite the failure of the butyl tire, the combination of newly developed uses and increased export sales kept demand for butyl high and led to continued expansion of productive capacity. At the same time, Enjay concentrated on the development and commercialization of new specialty rubbers. In 1960 they had introduced chlorobutyl rubber and butyl latex, and after several years' research they developed ethylene-propylene rubber [EPR] with outstanding resistance to ozone and chemicals. The Enjay Company commercialized this rubber, thereby demonstrating the increasing sophistication of the development process and the new relationship between Enjay and Esso Research.[88]

Research and development played an even greater role in the other major element of chemicals growth during the early 1960s, the drive to integrate Jersey's chemical operations forward toward the consumer. The most evident symbol of that drive was the new polypropylene plant at Baytown, Texas.

Although the company had begun research, looking toward entry into the plastics field as early as 1953, it started commercial operations only at the end of 1959. Based on a continuous manufacturing process perfected by Jersey engineers, the 40 million pound per year facility marked a major step along the path Rathbone envisioned for Jersey's chemical interests. With the market for this versatile plastic expected to reach one billion pounds per year by 1970, the decision to market a plastic suitable for use by fabricators signaled Jersey's determination both to adjust to changes in the chemical industry and "to rise to the challenge presented by the global glut in crude."[89] The $15 million spent in perfecting the process assisted the later commercial development of other plastics, and the company wasted no time in taking advantage of it.[90] In 1961, the company introduced a butadiene-styrene copolymer with numerous industrial surface coating applications. However, this venture ultimately proved unsuccessful; it was divested. While Enjay Laboratories' excellent polymer applications facilities made possible a rapid expansion of sales for the company's versatile polypropylene, continued work by Esso Research produced new varieties as well as new chemicals used in plastics manufacturing.[91]

Jersey's integration forward in chemicals took on a new dimension in 1961 when Enjay, with Humble as its new parent company [The Enjay Company, Inc. merged into Humble Oil & Refining Company and became Enjay Chemical Company on May 31, 1960], moved into the actual manufacture of fabricated plastic products. A joint company formed with J. P. Stevens & Co., Inc., acquired the assets of National Plastic Products Company, Inc. In addition to continuing that firm's primary operations of manufacturing "saran and polypropylene monofilaments and a melamine laminate known as Nevamar," the new company began work on the application of research already done by Esso Research and Stevens on the manufacture of textile fibers from polypropylene.[92] Enjay also obtained Humble's consent to acquire the assets of Extrudo Film Corporation and began expanding its facilities for the production of polypropylene film for packaging. At the same time, it prepared for even greater expansion in polyethylene film, Extrudo's principal product. Also Enjay, about this time, began preliminary work on high-density polyethylene, looking toward the fabrication of plastic containers;[93] yet the product never developed. The initial enthusiasm for diversification and for chemicals went far, but whatever may be said for these projects and joint enterprises, in the end Jersey unloaded the polypropylene fiber and Nevamar laminate businesses.

Still, these developments demonstrated the substantial progress of Enjay during the early 1960s toward integrating and expanding Jersey's domestic chemical operations. To achieve the rapid growth envisioned for the company's overall chemical enterprise, however, would have required concentration on the areas with fastest growing markets, and these were outside the United States.[94] While Enjay remained for several years the keystone of Jersey's chemical business, the pattern of growing sales and capital budgets attested to the importance of foreign operations in the company's strategy for future development.

Thus, by the early 1960s, Jersey had a solid foundation on which to build its global chemical enterprise. Its industrial chemical operations overseas, though not large, ranked well with those of other American companies. Overall, by the end of 1961, Jersey's affiliates outside the United States had twenty-five chemical plants on stream in Canada, Europe, and Australia, with an equal number under design or construction in these and other areas.[95] While overseas expansion at the beginning of the decade presented a variegated pattern reflecting differences in local conditions, the scope of the entire program demonstrated the direction of Jersey's worldwide expansion in chemicals.

In Canada and Europe, affiliates stressed rapid growth in the basic chemical industry. Most American firms moving into these areas left basic chemical production to established local companies, choosing instead to concentrate on more sophisticated products. In contrast, Jersey affiliates, especially in Europe, took advantage of their favorable feedstock position to become important suppliers of olefins and aromatics to the expanding industry abroad. Construction of new large installations in Sweden and the Netherlands and expansion of facilities in England, France, West Germany, and Canada evinced the importance of this business. Jersey realized that the success of the EEC, and the almost certainty that Great Britain would join that organization, had reduced the risks customary to such expansion. Most of the growth in the early 1960s centered on basic chemicals, solvents, and synthetic rubber. [96] Affiliates in the United Kingdom and France took leading places in the specialty rubber industry of Europe. Building on sales previously based on imports, the French company increased its equity interest in Société du Caoutchouc Butyl [SOCABU], the only butyl manufacturer on the continent at that time. Later the British affiliate constructed another butyl plant.[97]

In the Far East, where Jersey and Socony Mobil Oil Company under court order slowly worked out a division of their jointly owned affiliate, the Standard-Vacuum Oil Company, a similar pattern emerged. There, in the companies and properties given to Jersey, the tendency toward forward integration proved somewhat stronger than in European and Mediterranean countries. In Australia, the affiliate brought on stream a basic chemical manufacturing plant in 1961, and in a joint operation with an American rubber company, entered the styrene-butadiene synthetic rubber field. In Japan, even preliminary discussion of a basic petrochemical installation at Kawasaki included the possibility of manufacturing polypropylene and other end-use products at a later date.[98]

Rapid expansion along the broad front of the chemical industry worldwide placed "heavy pressure" on Jersey's chemical organization. [99] Although the quest for simultaneous growth in a variety of areas was based on substantial resources—a global chain of affiliates, a well-financed, large-scale research program, considerable financial support, flexible, energetic management—observers questioned whether any oil company could adapt to the intricacies of the international chemical business and achieve results proportionate to those of specialized chemical companies. In the case of Jersey, the lack of a unified organization for coordinating chemical operations abroad, where most of its and the industry's growth was taking place, raised a particular problem.[100]

Unlike most major chemical producers with large overseas operations, except for a minor role played in coordinating chemical exports, Jersey had no separate international division or subsidiary to manage exports and foreign production. Nor did it have a separate worldwide marketing entity. Instead, its organizational structure for international chemical operations in the early 1960s consisted of a parallel organization tacked on piecemeal to parts of its oil business. Affiliates with chemical operations managed them through a variety of structures, depending upon the size and character of their business.[101]

To the extent that Jersey could be said to have any coordination responsibility for all chemical interests, it resided in Fisher, the Board contact director for chemicals, and in a small staff group within the Refining Coordination Department (although other functional coordinators and contact directors continued to have some responsibilities in the field). Admitting it was "a little bit illogical" to have a Chemical Advisory Group overseeing marketing and other functions from within Refining Coordination, Fisher maintained that after serious consideration the decision had been made to keep it there because manufacturing seemed to be the essential function of the chemical business. Even sales, with an emphasis on technical service and "prescription marketing," had a close relationship to manufacturing. Thus, within a structure for coordinating an oil company, the Refining Coordination Department did offer the most suitable location for supervising the expanding chemical complex.[102]

The Chemical Advisory Group, while nominally responsible for coordinating all chemical development, devoted from 80 to 90 percent of its efforts to the area outside the United States. Headed by Robert M. Jackson, the group felt that its principal task was the promotion of new chemical projects. While ideas might come from an affiliate or some other department of the parent company, the chemical staff group originated most proposals and it was responsible for drawing relevant data from various parts of the Jersey organization and coordinating the development of the project and its presentation to the BAC. However, under Jersey's philosophy of decentralized management responsibility, the final decision to present a project rested with the affiliate itself.[103] Jackson's method of operations reflected both Jersey's decentralization and the absence of a hierarchical structure for chemicals development. He spent much of his time visiting some twenty-five affiliates, meeting officers and manufacturing technologists informally, trying to improve communication on chemical matters, and encouraging them to develop their own projects whenever possible. Some time later, Fisher described Jackson's style, "we had a fellow named Robert Jackson and [he] was a promoter of great energy. And he went around the world literally a number of times just looking for opportunities for us to get in the chemical business."[104]

Although new projects constituted the principal concern of the chemicals group in the Refining Coordination Department, Jackson's staff also had responsibilities for overseeing continuing chemical operations. In this, their duty overlapped that of chemicals groups within the Marketing Coordination Department and Esso International, Inc. To cope with this structure and facilitate development of the chemical business, representatives of all these groups met every two or three months with the chemical coordinators of each of the affiliates.[105] Through these meetings and other informal contacts, the damaging effects of overlapping responsibilities and a confusing organization chart could be ameliorated; however, the strain exerted by accelerating development and the cumulative burden of ever-larger operations dictated a fundamental reconsideration of the chemical organization's structure. The need for a single, unified organization for chemicals coordination within the parent company soon became obvious, but the inability of foreign affiliates—especially those with small chemical operations—to develop adequate specialized organizations to handle their own rapidly growing chemical business represented an even more basic weakness.[106] Jersey's decentralized structure required strength at both levels.

The need to tighten and revamp the structure of the company's worldwide chemical operations led to consideration of various alternative organizational designs, but for those divisions with the broadest perspective—

functional, geographical, and temporal— a mere realignment of the chemicals organization would be insufficient. If reorganization were to achieve its ultimate goal of maximum chemicals growth based on sound investments and good management, it would have to proceed from a fundamental change in Jersey's approach to the chemical business.[107]Although proponents within the company had succeeded over the years in raising the status of petrochemical development, it remained, especially at the affiliate level, an important and profitable—more or less subservient—sideline to the main effort. Given Jersey's relative size, this hardly could have been otherwise. Yet its absolute magnitude, its growing importance within the world chemical industry, its integration forward into a greater range of more sophisticated products on a global scale, and its increasing direct competition with specialized chemical companies all argued for the separation of Jersey's chemical business from its oil business, at least to the extent that this could be done without losing the advantages inherent in its relationship with the parent company.

In 1960, CEO Rathbone asked his administrative assistant, George T. Piercy, to investigate Jersey's chemical business, formulate recommendations and report back to the Executive Committee. Piercy toured the various chemical affiliates in Europe and America as well as the plants of Jersey's major competitors. He discussed with Jersey personnel what they thought should be done to improve both the operations and earnings of their plants. What he found is partially reflected in the notes he took of the various conversations. He began:

> as mentioned by Mike Haider I interviewed [persons at] most of the larger affiliates. It was quite startling to find that the organizations were considerably different. Some were following the parent company's directive of 1955 . . . one indicated that they had filed this communication in the waste basket as they already had a different organization and so did the U.S. affiliate.

Some executives wanted to change aspects of their operation but Jersey directives, they indicated, limited such changes. Piercy found morale generally low and he noted, "all affiliates except one thought that the chemical effort should be grouped together." The lack of a clear delineation of the role and responsibility of Esso Export and the Marketing Coordination Department served to confuse managers. On this point "criticism of the Parent Company was loud and forthright." Some interviewees indicated that directives from Esso Export and Marketing Coordination "didn't help much."

When one of these arrived, the Refining Coordination Department usually wanted it restated to explain limits of responsibility.

In Europe, Piercy found that no study had been made of operations or organization. European affiliate executives unanimously favored a coordinating group there. Piercy commented, "our two men there then were nothing more than a relay station."

Generally Piercy found few grounds to criticize what the chemical affiliates had accomplished but he sensed frustration everywhere. Reflecting, he addressed to himself a question, "One does wonder why our chemical effort is as small as it is?" Then he added, "Jersey was the first or one of the first in petrochemicals—yet our effort is small, particularly outside the U.S." Then he listed the percentage of the gross revenue derived from chemicals in the total for several competing companies and for Jersey:

	% gross from chemicals
Phillips [Petroleum Company]	15.4
Shell [International Chemical Company, Ltd.]	6.5
California [Standard Oil Company of California]	8.2
Jersey	4.8

Piercy wrote in his notes: "I observed a lack of stimulation from above" and that all the managers he interviewed "thought competitors were well advanced of Jersey." On marketing chemicals he "observed [that the] Enjay organization is really not product oriented." A great portion of what he heard, noted and reported had been repeated to Jersey directors over and over by advocates of a greater role for chemicals. The Piercy report indicated that "one might consider the performance [of chemicals] extraordinary" considering the manpower devoted to that portion of the Jersey operation. Too, worldwide investment opportunities had come not because of vision or investigation but because the company's far-flung operations had concentrated in the raw material phase of the business. The assessor doubted that without other and newer products such opportunities would long appear. Hence, in his opinion, the existing organizational structure would "probably not be sufficient to continue to generate investment opportunities" as it had been doing.

"Jersey," Piercy said, "should separate its chemicals from oil and adopt a product line or market approach." After briefly outlining the courses open to Jersey, Piercy recommended that "the ownership and operation of the chemical manufacturing facilities . . . be placed directly under the chemical entity" or separate company. All chemical functions of Marcor, Refcor and Esso Export could be combined under one group with total responsibility for stimulating, developing, promoting and selling chemicals. He would add to this new department long range corporate planning plus full responsibility for "Jersey's success in chemicals." This body would require a larger staff. While some of his recommendations indicated that he had not time to rationalize the full impact of his first recommendations, the statement provided an insightful view of the future of chemicals in the Jersey organization.

In the Jersey chemical structure Piercy foresaw the need for three divisions at the top, one each for the United States, Europe, and "Rest of the World." To advise the chemical company, he added a Board Advisory Committee [BAC] for chemicals, to operate almost as did the comparable body for Jersey proper. Unlike the Jersey BAC, busy with oil, members of the BAC for chemicals could "develop a wide and intimate knowledge of chemicals." Piercy concluded, "But it is the recommendation of this study that a separate corporation be set up—wholly owned by Jersey and operating parallel to Humble." Chemicals would then receive the attention they deserved, would be more manageable and the personnel would have better morale. Further, chemical managerial talent could be sought and developed for that purpose and organization, alone. According to Piercy, the arguments pro and con for this change had "been made many times in the past." Now he believed only the question of timing the change to be important.

Piercy turned over the report to Rathbone who indicated his approval.[108] Then the chairman appointed a special committee of the Board composed largely of the younger members to review chemicals.

Piercy appeared before this committee and despite the artistry of his presentation, no hasty action ensued. The Executive Committee, also, continued to study the recommendations. Later when the Piercy report came before the Board, several members objected to implementing some or all of them. William R. Stott, for one, fought the separation of chemical marketing from that of the general oil business; others disliked the separate company idea.

Piercy always believed that he received the assignment to review chemicals because Rathbone knew that he would soon be leaving for Imperial. He did move on to Canada before the squabble ended. As Rathbone remembered and Piercy learned, Jersey moved deliberately to free chemicals. But the Executive Committee did finally act and the Esso [later Exxon] Chemical Company Inc. of 1963, to some extent, resembled the company proposed by Piercy two years earlier. However, not until 1966 did the organization change to the extent he had visualized.

On March 23, 1962, the Executive Committee accepted most of the Piercy recommendations and at the same time approved a statement of "Company Policy for Chemical Activities" designed to achieve the necessary separation upon which reorganization would be based. Noting the company's intention "to intensify its chemical effort by broadening its markets, expanding its range of products, and integrating further forward toward finished chemicals," the policy statement declared the necessity of regarding "chemical operations as a separate field of business from petroleum" and of placing it "in the charge of full-time specialized chemical managements in operating affiliates." While recognizing that local conditions might require continued joint operations "for some time to come," the policy called for separation "as fully and rapidly as practicable. . . ." Rathbone, who now virtually ran the Executive Committee, had called for this separation in his January, 1951 letter to Frank Abrams.

The Board went beyond several of the Piercy suggestions. Piercy had pointed out that, because of possible tax, legal, and other problems, the holding company should notify the foreign affiliates of their intention to follow the same structural reorganization there at a later time. This, he made clear, would enable those companies to begin the study and planning necessary to accomplish this; however, the Board moved to integrate all chemical operations at the same time.[109]

To implement its decision, the Committee approved the designation of a vice president for Chemicals "to assume overall leadership of the chemical effort." Fisher, the contact director for chemical activities, assumed this new title and began developing recommendations as to the form of corporate and operating organizations for the implementation worldwide of the new policy. To assist Fisher, the Committee authorized the formation of a new Chemical Coordination Department; this combined the chemical responsibilities previously held by the Marketing Coordination, Refining Coordination, and Regional Coordinators Offices, as well as the staff responsibilities for chemicals vested in Esso International Inc. To head this department established in June, 1962, Rathbone selected Harold G. Mangelsdorf. Mangelsdorf had served on one of the CEO's pet projects, the Advisory Committee On Human Relations, from 1956 to 1958, at which time he became Deputy Coordinator of Refining Coordination.

Fisher, who along with Rathbone, Tracy and others had long endorsed the separation of chemicals from the oil business, had a problem with Mangelsdorf. He later stated that he "had a lot of trouble selling this idea to Mangelsdorf" and ultimately they "had a knock-down drag-out argument about this business of splitting off the chemical business." Obviously Mangelsdorf accepted the Board decision.[110]

As Fisher and Mangelsdorf began the difficult task of splitting off the chemical operations of overseas affiliates, Humble's management proposed to begin implementation of the new policy domestically by strengthening Enjay Chemical Company's control over its chemical business. While Enjay would remain a division of Humble until the success of the separation could be evaluated, direct responsibility for marketing all chemical products and for the manufacture of all chemicals at sites other than Humble refineries would be given it. In addition, it would furnish guidance to Humble's Manufacturing Department and Esso Research with respect to Humble's chemical business. Henceforth, Enjay would work directly with the parent company's Chemical Coordination Department to assure integration with Jersey's overall chemical interests. [111] Separating the domestic half of Jersey's chemical business presented only minor problems. The Enjay Chemical Company's long history and large market had already allowed it to develop a solid organization of chemical specialists. Certainly it was the most sophisticated and most nearly self-sufficient among Jersey chemical entities.

A year later, on April 29, 1963, Jersey announced the formation of a new company, Esso Chemical Company Inc. [Esso Chemical], "to expand the worldwide chemical business." Sales of chemicals by Jersey Standard affiliates had "amounted to $342 million in 1962," representing an increase of 65 percent over a five year period. Designed to handle "international movements of chemicals," to advise and assist the "40 Jersey affiliates" in the business, this new company clearly reflected the increasingly high opinion held of chemicals in the holding company. To head the new company, the Board appointed Mangelsdorf, then serving as Coordinator Chemicals Coordination. Robert M. Jackson, head of the Chemical Advisory Group, became executive vice president. The reason given for the creation of the new company was "another move in Jersey's intensified drive to broaden markets for its chemical products, to expand the range of those products and ensure the most efficient results from the company's substantial investments in chemical operations." Esso Chemical would also assist in the formulation of research programs; and "furnish technical sales service and laboratory assistance to affiliates and customers." Jersey Assistant Treasurer, E. A. Herberich, joined Esso Chemical as director, along with others from Jersey's chemical affiliates. Four days earlier (April 25, 1963), by circular letter, Jersey had authorized the establishment of a regional office in Europe designed to provide marketing, sales, technical and manufacturing assistance to affiliates and to assist them in obtaining supplies. [112]

In their new positions, Mangelsdorf and Jackson became even more influential proponents of further and rapid chemicals diversification. Not all Board members or even lesser executives evidenced complete assurance that the course being pursued by the Esso Chemical executives redounded to

Jersey's benefit.

In early November, 1963, Jersey Director Wolf Greeven sought more information about an aromatics plant in Rotterdam upon which construction had begun in 1962. Conceivably, since Greeven served as contact director for marketing in Europe, this inquiry could have resulted from the Board's desire to review some of the projected plants there. In any event, Greeven received a letter stating:

> The pressure to build an aromatic solvents plant in the Netherlands began early in 1957 when Bob Jackson of Refcor [Refining Coordination] started to build his chemical coordinator nucleus. The first correspondence I was able to find was a letter . . . written by Herb Minich, *one of the disciples*, in which he states that . . . it was the consensus of opinion [of the European Coordinators] that Jersey should examine the opportunities for marketing and manufacturing aromatic solvents and aromatic chemicals in Europe.[113]

Whether Greeven and the Board used this and other information to delay construction on the Rotterdam plant is not known. Work on that facility did shut down for an interval and recommenced later. The plant went on stream in 1964.

At about this time (1963) the Jersey diversification program, with its principal emphasis on chemicals, began to go sour. The Executive Committee and the Board, with Rathbone in the chair and with Director and Executive Committee member Stott seconding his lead, began to approve capital expenditures without the usual thorough analysis and study, as though riding the crest of a wave. For years a Jersey tradition had been "go slowly," but not at this time. A saturated worldwide petroleum market and dwindling profits forced the Board to look at new markets for chemicals, despite indications of erosion of profits in that business also. Study of the Board and Executive Committee minutes establishes that all manner of projects and proposals came before it, and most of them received approval. Rathbone sensed that pressure to approve almost everything had mounted when, in October, 1963, he wrote the full Board:

> The Executive Committee, on Tuesday, reviewed a Fertilizer Plant for Cyprus and an Ethylene/Polyethylene Project for Colombia. It so happened this was the first time that I and some others on the Committee had *even* heard of these two projects. While this is not a criticism, it raises the question as *to the desirability of giving the* Executive Committee *some* advance notice when work is being

initiated on projects in new geographical areas, new functional areas or of any very unusual nature.

He added that this covered all subjects brought to the attention of the Board and the Executive Committee. Only members who were properly informed could give sound advice on project planning prior to the formal presentation. He concluded with the suggestion—from him it amounted to an order—that the "Contact Directors try to be sure that the Committee is informed about such projects when it appears appropriate."[114]

On October 25, 1963, Esso Chemical organized Esso Chemical S.A., with offices in Brussels, as the European regional headquarters "for the overall benefit of Jersey and our affiliates' chemical business in Europe." For some reason, no evidence exists concerning the failure of Esso Chemical to establish a regional office in Latin America or in the Far East; however, this failure could have been occasioned by the acceptance of the fact that the European Economic Community had been clearly established, and Europe was, of course, the largest overseas market.[115] The "Chemicals Financial and Operating Review" presented to the Jersey Executive Committee in April, 1964, pointed out that "75% of our manufacturing investment in Europe is in chemical raw materials" and that the "absence of exclusivity and the susceptibility for major chemical companies to integrate backwards to produce their own raw materials" exercised downward price pressure on these products. Uncharacteristically for Jersey and its affiliates, the review recommended joint ventures in forward integration. Jersey had always entered such arrangements only as a last resort, whatever the reasons advanced. But from the report it is clear that though those managing Esso Chemical sensed the need for qualified, experienced and highly trained personnel on all levels, they moved ahead without them. Plans called for rapid expansion in the new business lines of plastics, fibers and surface coatings as well as continued investment in older areas.[116]

Esso Chemical's investment budget rose from an average $53 million in 1961, 1962 and 1963 to $257.1 million for 1964. The November 2, 1964, chemical budget presentation for Executive Committee review stated, "It will be Jersey's largest final chemical budget." The review attached no great significance to "failure of market outlets to develop as expected."

The desire to expand into fertilizer production represented one of the boldest ventures in the history of Jersey. Most of these heavy investments occurred in the newer nations, then termed "developing countries." So far as the records show, there is no clear reason for this. The Food and Agriculture Organization of the United Nations proclaimed a war against hunger in the 1950s, and the World Bank called on Western powers to aid emerging nations through programs designed to increase food production. The media took up this cry with, in fact, a neo-Malthusian tone to most of its coverage. Some transnational corporations, Jersey among them, came to believe that the way to attack the problem lay at the base—crop yields. With their know-how and organization, these corporations felt that they could do something profitable in the economic field that would also achieve a positive social good—so they built fertilizer plants. The spring, 1963 edition of *The Lamp* featured an article, "The Earth Can Feed Us All." After listing the countries and locations of the Jersey fertilizer plants abroad, the author then explained "why international oil companies can enter the chemical fertilizer business, at the right time and place, both with profit to themselves and with benefit to the countries where they operate."[117] While few informed persons would quarrel with anything in the article, it did not explain how fertilizer plants or the use of their products could overcome the downstream obstacles in marketing, preservation, and distribution. Several other well-written and equally enthusiastic articles on fertilizer and food appeared in *The Lamp* during the next three years, most of them with the theme "more fertilizer means more food." Fertilizer and time would solve the problem of hunger; Jersey would help supply the fertilizer. So the tremendous potential demand for chemical fertilizers and the basic role of food production in the development of emerging nations created for Jersey an ideal overlapping of the needs of the corporation and those of the host countries.[118]

Although Stanvac had been supplying refinery gases to non-affiliated fertilizer manufacturers since the mid-1950s, not until the end of the decade did any Jersey affiliate propose to enter the field directly. In 1960, International Petroleum Company, Limited [IPC] launched the first such program with construction of a plant adjacent to the Cartagena refinery in Colombia, built to manufacture ammonia for conversion to ammonium nitrate fertilizer at facilities in several Latin American countries. With the encouragement of Jersey's Executive Committee, other affiliates in the Caribbean area soon followed suit. By the end of 1961, ownership or management of all of these projects had been transferred to IPC whose interests now included ammonia plants at the Aruba refinery as well as Cartagena and fertilizer manufacturing facilities at Cartagena, Aruba and in Costa Rica and El Salvador. The fertilizer plants had all been developed as joint ventures, but during the year the Executive Committee decided that to insure flexibility in future development, IPC should avoid any new joint projects with partners.[119]

At the same meeting in which it voted for a more independent approach to expansion in fertilizers, the Executive Committee also attempted to blueprint future developments. Discussion of Esso International Inc.'s assuming responsibility for supply and marketing arrangements for the fertilizer projects associated with the Cartagena ammonia plant pointed to the parent company's increased interest in the new line of business. By approving the formation of a small staff group within the Refining Coordination Department, organized to explore prospective investment opportunities and to coordinate the activities of affiliated companies in the field, the Executive Committee actually encouraged other affiliates to enter the fertilizer business.[120] Within the year, Jersey's interests had made new commitments for joint ventures in ammonia and chemical fertilizers that marked the company's first such projects outside Latin America.[121]

Based on a 1963 study, Jersey sought to enter the fertilizer field in the United States. In 1963, the company began negotiations for the purchase of the American Agricultural Chemical Company [Agrico], hoping to integrate backwards and obtain a supply of phosphate, a basic ingredient needed in fertilizer. Director Fisher handled these matters skillfully but still the effort failed. On recommendation of Esso Chemical, the Executive Committee next authorized new negotiations for the purchase of the Virginia-Carolina Chemical Corporation [VCC]. This approach also failed. Next, the Committee endorsed an effort to buy the Potash Company of America [PCA], and Mangelsdorf pursued this plan. However in 1966, the United States District Court in Newark, New Jersey, enjoined the company from buying PCA. After discussion on the motion of member Emilio G. "Pete" Collado the Committee decided not to appeal this decision, and the setback virtually ended Jersey's efforts in the United States to move laterally through purchase of fertilizer plants or to integrate backwards by purchasing suppliers of needed ingredients.[122] So the attempt to develop a fertilizer business in the U.S. failed.

Joint ventures abroad became common in 1964, and the company rushed to get into plastics, plastics fabrication, fertilizer and other businesses wherever it could. Management's philosophy is illustrated by the following: "a final point is that in attempting to enter profitable new fields, such as plastic and plastics fabrication, it is important to seize attractive opportunities, including acquisitions, as soon as they are identified."[123] This statement contravened most Jersey precedents, for normally such acquisitions would have been preceded by thorough analysis of markets, technology, and potential labor supply. This latter portion of such studies came to be increasingly important as the political pressures of the host nations demanded increased employment of national personnel.

Esso Chemical's five year plan prepared in 1965, called for expansion of the capital invested in fertilizer plants from $119 million in 1964 to $613 million in 1970, with a growth in total sales during the same years from $605 million in 1964 to $2,300 million in 1970. At the same time estimated profits would increase at an average annual rate of more than 25 percent, rising from $23,600,000 to $255,000,000. The illogical nature of this plan became apparent during the review by the Executive Committee:

> The projection for gross plant [for the overall company] called for an increase of from $80,000,000 to $2,725,000,000 . . . and necessary manpower would jump from 14,431 to 58,800 persons. At this time Esso Chemical profits had dropped to 8.4 per cent on capitalization.[124]

The same presentation to the Executive Committee mentioned favorably the Copet [L. Compania Petroquimica Industrial y Comercial S.A.] acquisition, "The Copet know-how—technology, business and marketing—is outstanding and forms the cornerstone of our expansion [in fibers]."[125] The proposal for 1966 chemical investments carried the statement, "Jersey's interest in Copet in Argentina provides a nucleus of business and spinning know-how on which to base future growth."[126]

By 1965 several indications of trouble signalled the Jersey Board that chemical investments had run ahead of the company's ability to manage effectively. Significant statements indicating this may be found in presentations such as that to the BAC in 1965; here, the company had "a very small (and unprofitable) position in conventional thermoplastics, as shown by . . . a $9,500,000 loss on a year-end investment of $45,000,000."[127]

Too, "net income did not parallel in sales"; hence, "the return on chemical net worth in 1965 was just under 4 percent." The forecast for 1966 indicated about the same return. While expected losses occurred in start-up plants, there were other problems; for example the "costs of building an organization in Europe to make possible the substantial growth we are planning."[128]

The National Plastic Products Company, Inc., owned equally by Jersey and the J. P. Stevens & Co., Inc., had never really functioned well: "The . . . venture in the U.S. has not been satisfactory and urgent action is required to resolve this," ran one report. The Executive Committee finally approved the purchase by Humble of the Stevens share,[129] giving Jersey complete control of its VECTRA polypropylene textile fibers.

As the various Esso Chemical affiliates grew, the company moved toward forward integration in a number of plants. By this means, management sought to make their units more competitive with those of other major chemical companies. This effort to enter the marketplace by producing finished products received the approval of Jersey's Executive Committee. However, the Board approved only a modest beginning toward its implementation, advising Esso Chemical to seek future appropriations through normal investment review procedures.[130]

The first step under this directive came in 1964, when Esso Aktiengesellschaft [Esso A.G.], Jersey's German affiliate, acquired Bisterfeld & Stolting, a West German corporation engaged principally in the fabrication of brake linings. Then Esso Chemical S.A. purchased several Danish companies that manufactured and sold plastic packaging. Esso Italiana entered the field through further participation in Rumianca S.p.A.'s Sardinian complex. Esso Petroleum Company, Limited (England) also moved to enter the plastics fabrication field. All of these "ill-conceived" programs were later abandoned. They occurred despite Jersey's traditional emphasis on cautious investment. One observer stated that such actions indicated a "lack of analysis of the marketplace."[131]

During the next two years, forward integration in the European chemical operations proceeded without major interruption. In those years Jersey entered the chemical industry in Spain and Greece. In both cases, the chemical plants developed as part of larger industrial projects, and the justification for investment rested heavily on a desire to promote the petroleum side of the business. To encourage the development of higher value industrialization within their countries, host governments made approval of refinery permits contingent upon specified levels of related petrochemical investments. Late in 1965, Jersey built chemical facilities at the Northern Greece Development Company's industrial complex at Thessaloniki. In Spain, Jersey gave authorization for a refinery and a caprolactam plant at Castellon de la Plana. They also built a plant to manufacture nylon fibers at Sargosa. The Spanish fibers project based in part on Copet technology, despite glowing promise, never really flourished.[132] Both the fibers and caprolactam plants were ultimately divested. Outside Europe, Canada, Japan and Australia became the only foreign markets in which company interests dictated a policy of building basic petrochemicals capacity and of integrating forward.

Having begun with no petrochemical investments in 1957, Imperial Oil Limited had, by 1964, entered a period of very rapid expansion. Its plants completed or under construction during the next two years included one for polyvinyl chloride, and additional capacity for basic petrochemicals (ethylene, benzene, toluene, and xylene). Further evidence of this trend could be seen in the new polyvinyl chloride applications laboratory at Sarnia, built in 1964, and in the acquisition of Building Products of Canada Limited, a firm engaged in the manufacture and sale of a range of fabricated products for the construction industry based primarily on forest products raw materials.[133]

Most of Jersey's petrochemical business in the Far East, previously served by the Standard-Vacuum Oil Company, consisted of fertilizer projects and the sale of imported chemicals, but joint projects in Australia and Japan continued to develop along a broad front. Although these markets remained substantially smaller than those of Europe and the United States, developments followed a similar pattern. While expanding both the range and capacity for the production of chemical raw materials, affiliates in Japan set up a technical service laboratory to aid in marketing butyl rubber, and completed a plant to manufacture a special alcohol for plasticizers used in flexible vinyl. In Australia, forward integration stressed synthetic rubber [SBR].[134]

As the balance of Jersey's chemical investments shifted in response to faster growing markets abroad, the domestic affiliate, Enjay, continued to serve as a model for the company's other chemical interests. Taking advantage of Esso Research's ongoing program in a broad range of product lines and of its own increasing sophistication in manufacturing and marketing, Enjay continued its efforts to move closer to the final consumer, while simultaneously expanding the raw materials base upon which its business rested. Despite tightening competition in the field, Enjay responded in 1964-1965 to long-term growth trends by increasing capacity for manufacturing ethylene and butadiene, and by adding new facilities for the production of ethylbenzene. Also, Esso Research developed new processes for deriving commercially useful specialty acids from petroleum, and for separating normal paraffins from refinery streams.[135]

Esso Research's efforts led to new products in the fields of solvents, surface coatings and additives, but Enjay's search for additional products continued to focus on synthetic fibers, plastics, and synthetic rubbers. The affiliate found new applications for butyl and its other specialty rubbers, while earlier acquisitions provided a base for the extension of the end-products business. The Extrudo Film Corporation set up its third polyethylene film plant in 1963, and Humble, now with complete control of National Plastic Products Company, prepared to support further development in the plastics fabrication field with a new plastics coloring laboratory, a manufacturing plant at Lake Zurich, Illinois, and expansion of its Nevamar product line in the decorative laminates field.[136] Five years after beginning production of its first plastic, polypropylene, Enjay participated in the industry in a variety of ways, ranging from supplying others with raw materials to the manufacturing of fibers and plastic film. Although this represented only a relatively small part of Enjay's business, Jersey continued to be optimistic about products made from polypropylene.

In the two years after Esso Chemical Company Inc. was created in 1963, it succeeded in promoting the rapid growth of Jersey's foreign chemical business. Through the establishment of Esso Chemical S.A., it also helped coordinate the management of chemical activities in Europe, the Mediterranean area, and Africa.[137] Yet the mandate to provide separate specialized management of Jersey's chemical interests remained largely unfulfilled.

If Esso Chemical were to achieve unified global management of Jersey's chemical interests, it would necessarily have to play the leading role in a new drive to separate chemical and oil activities. However, by early 1965, Jersey's top management had begun to consider even broader changes in the structure of its chemicals business. In January, the management of Humble Oil & Refining Company proposed new steps to consolidate even further the control of all domestic chemical operations under Enjay Chemical Company, which would continue as a division of Humble, based in Houston.[138] While the move would have streamlined Enjay's operations somewhat, advocates of a more independent chemicals program feared that the new structure, although it might improve the management of domestic chemical operations, would also result in the subordination of Enjay's interests to Humble's main concern—petroleum. While acknowledging the practical difficulties inherent in any attempt to separate physically refinery and chemical operations, Humble proposed to avoid a situation in which chemical management would "be frustrated and defeated by disinterest on the part of refinery management." Early plans developed in New York called for the president of Enjay (presumably the subsidiary responsible for marketing and coordination of domestic chemical operations) to join the Humble board of directors, thus assuring the stature of the affiliate, assuming the selection of a candidate of real capabilities. Further, their proposal called on Humble's board to confirm an interest in and a "responsibility for, the success of the U.S. chemical business" as well as to agree that any chemical investment opportunities it considered "not interesting" could be processed through "the Jersey chemical review procedure, as of possible interest in the over-all chemical investment program." Humble's manufacturing specialists would have to confirm their responsiveness "to the commercial needs of the chemical business."[139]

Despite these efforts by Humble to maintain supervision of Jersey's domestic chemicals business, the reorganization of the company's chemical interests had already begun to move in another direction. In March, 1965, the parent company's top management reviewed plans "to consolidate in Esso Chemical by year-end 1967 the over-all management of the chemical interests of the company and its affiliates."[140] By that fall, chemical proponents in New York, led by Enjay's President Clifton C. Garvin, Jr., had wrung from Humble an agreement to transfer its Enjay Chemical Division along with $200 million in assets to the parent company, for reassignment to Esso Chemical Company in 1966.[141]

The decision to make Enjay a part of Esso Chemical was without question a most important step in strengthening and unifying the company's worldwide chemical business. However, it was only one part of a renewed effort to complete the separation of chemicals management worldwide mandated in 1962. Of course, the prospects for success varied from region to region. In some parts of the world physical separation of operations presented only minor obstacles, but independent management would have to be built almost from the ground up. In other areas, involved corporate relationships prevented easy division of chemical and oil management.

Understandably, Europe and the Mediterranean area presented the most complex problem. There, Jersey interests participated in a great variety of chemical operations through numerous corporate vehicles. The already-established Esso Chemical S.A. was coordinating regional chemical operations, but difficulties in segregating the chemical business of affiliates in many different circumstances had to be overcome. Projects developed outside of the regular affiliate system, such as the Greek chemical complex, the Lebanon fertilizer plant, and the company's interest in Amoniaco in Spain seemed readily transferable to Esso Chemical. Affiliates involved only in marketing operations created no real problem; continuing their existing corporate arrangement appeared to be the logical course.[142] But the chemical operations of integrated European affiliates presented greater difficulties.

Early in 1965, plans for the future development of Jersey's European chemical interests called for only modest steps toward separate management. Recognizing the difficulties of dividing most physical operations, the program stressed the strengthening of the coordination and stimulation of functions of Esso Chemical and its regional headquarters in Brussels. Plans included consolidating market research in Esso and centralizing European technical service laboratory work at Brussels; however, only the chemicals plant at Stenungsund, Sweden, seemed appropriate for ownership and management by Esso Chemical.[143]

In response to the increasing concern over the performance of chemical investments and the need for stronger management, this cautious outlook soon gave way to a more ambitious program. Later in the year, plans called for the consolidation of Jersey's chemical investments under Esso Chemical including the separation of the European chemical interests. For most of the Western European countries, new subsidiaries of Esso Chemical would own and manage new investments in each country. Where possible, the new companies would also own—or at least manage—existing chemical operations. While the separation of the chemical business within Humble seemed to indicate the feasibility of such a step elsewhere, the special difficulties of the European situation soon became apparent when Esso Chemical tried to implement the new policy there. Executives of European affiliates accepted the need for separate chemical management, but by August serious obstacles had begun to appear.

The enthusiasm of the Jersey Board for fertilizer expansion met no such frustration. From the completion of its first four fertilizer plants in Latin America in 1963, the company proceeded with its program to make itself one of the world's major fertilizer producers "at one leap."[144] Esso Chemical's forward integration into the manufacture and sale of fertilizers and extension of its investments to new areas of the world became a symbol for the Jersey interests' whole program of "broadening . . . [their] worldwide chemical business, adding new product lines, and . . . strengthening . . . [their] position as manufacturers of finished chemical products."[145] Responding to heralds of a "green revolution" in the developing world, Jersey sought to take the lead "in creating fertilizer capacity where it . . . [was] needed most." With demand projected to continue increasing at 10 percent each year, the developing world market in chemical fertilizers, with its greater ease of entry, fitted perfectly with Jersey's chemical strategy of growth and diversification. Between 1963 and 1965, both the planned capital outlay and the total capacity of company fertilizer projects more than doubled. By the latter date, total capacity of existing or planned facilities had risen to over one million tons of ammonia and 1.8 million tons of finished fertilizers, and the projected capital cost stood at $200 million.[146]

In the Caribbean area, where Jersey's first fertilizer plants went on stream in 1963, expansion continued. By the time production had begun at facilities in Colombia, Aruba, Costa Rica, and El Salvador, Esso Standard Oil, S. A. [Esso Caribbean] had already moved to acquire a Puerto Rican firm engaged in blending and marketing fertilizers. Additional construction of a fertilizer plant, begun in Jamaica in 1964, increased total capacity in the area to 650,000 metric tons. From the beginning company affiliates in the countries of production and elsewhere sold fertilizers made by Jersey interests through existing marketing organizations.[147]

Jersey normally looked upon joint ventures with a somewhat jaundiced eye, participating in such arrangements only where necessary to gain entry to a new area of industry. Company interests usually insisted upon majority ownership or, at least, management control. Hence, Esso Chemical's decision to acquire International Development and Investment Company's equity in Aruba Chemical Industries N.V., Abonos Colombianos, S.A., and Fertica, S.A. in 1964, appeared natural.[148] Having gained experience in the manufacture and sales of fertilizers, the Jersey organizations sought flexibility both for the operation of its existing now-integrated fertilizer business and for anticipated expansion.

As the first fertilizer plants came on stream in the Caribbean area, expansion in other parts of the world had already begun. In the Far East, where rising population and low crop yields made the potential impact of increased fertilizer use most promising, fertilizer facilities reached the planning or construction stage by 1965 in the Philippines, Malaysia, and Pakistan. The Philippines project, in particular, showed the experience gained in the Caribbean. Planning from the beginning a fully integrated operation with a complete line of fertilizers, Esso Standard Fertilizer and Agricultural Chemical Company Inc. (Philippines) actually began agricultural extension work to develop the market for its Engro brand of fertilizers more than a year before it completed the Bataan plant.[149]

In Europe and the Mediterranean area, the partially owned Amoniaco Español S.A. made the greatest progress in the fertilizer field. Its 310,000 metric ton plant at Malaga, Spain, was Jersey's main fertilizer facility outside the Caribbean area to go on stream before 1966. The continuing favorable investment climate in that country led Esso Mediterranean to raise its equity interest in Amoniaco from 50 percent to 75 percent in 1963, and to propose in 1965 an even larger ammonia and fertilizer plant at Tarragona. As the ammonia plant associated with the Greek industrial complex at Thessaloniki neared completion, Esso Mediterranean's partially owned Lebanon Fertilizer Company began construction of a fertilizer manufacturing plant designed to provide an outlet for part of the ammonia.[150] In South Africa, Jersey acquired Triomf Fertilizer and Chemical Industries, Ltd. in 1965.

After earlier deferring action, Jersey's Executive Committee, in 1963, approved its first fertilizer project for Western Europe, a 100,000 ton ammonia plant to be built at Rotterdam. Later studies led to a proposal for an additional facility in France, but Jersey's directors chose instead to take advantage of Common Market economic unification and enlarge the Rotterdam installation, increasing its ammonia capacity to 300,000 tons per year and adding large urea and nitric acid plants.[151] Imperial Oil Limited also entered the fertilizer field at this time, marketing products from a nonaffiliated manufacturer as a preliminary step to building its own plant, which on completion in 1969 at Redwater, represented a major investment for Imperial. Along with the Rotterdam project, this Canadian development represented the movement of Jersey fertilizer into the agricultural regions of developed nations.[152]

With the rapid expansion in the mid-1960s, Jersey's chemical interests seemed to be well on the way toward the role envisioned for them at the beginning of the decade. By 1965, Jersey's chemical sales worldwide had reached $600 million, making the company the fourteenth largest chemical producer in the world,[153] and Esso Chemical could boast an annual growth rate in sales of 13 percent in each of the previous two years. More important to Jersey, this expansion showed promise of increasing chemical contributions to total profits, the basic goal of the chemical expansion program. In addition, an analysis of the growth trends indicated that company chemical interests were following the course charted by managers seeking long-range expansion.

Most of the growth in sales during 1964-1965 derived from well-established lines of business, and the United States remained Jersey's largest market for chemicals. However, the fastest growing markets for chemicals lay elsewhere, and Jersey's chemical interests moved toward them. Foreign sales continued to increase more rapidly than the domestic market, and new product lines—especially fertilizers—grew faster than basic chemicals, solvents, additives, and synthetic rubbers. Investment trends also reflected this vision of Esso Chemical's future. Although representing only 30 percent of the company's chemical business, these new product lines accounted for almost half the capital budget in chemicals for 1965.[154] Foreign investments also continued to grow more rapidly than foreign sales. By the middle of the decade over half of Jersey's chemical investment lay outside the United States. Already a significant factor in the worldwide industry, Jersey's interests promised to become even more important in the future.

Certainly, the conditions that had inspired the rapid expansion in chemicals—high return and potential future growth compared to petroleum investments—continued to prevail.[155] Thus, even as Jersey's Board noted a "somewhat less certain" investment climate at the end of 1964, and called for "particularly critical review" of additional capital investments, it approved financial assistance to Esso Chemical Company, and gave that relatively small affiliate the largest amount of direct equity capital and the second highest level of overall assistance among all affiliates.[156] In the following year, the Executive Committee, reviewing the program of investments planned for 1966 and a proposed five-year chemical investments program, once again noted the importance of carefully reviewing new investments while affirming "its previously expressed view that a continued orderly expansion of chemical activities . . . would be desirable."[157] Based on its expansion program in the first half of the 1960s, management regarded "the broad field of chemicals as a prime area of expansion."[158]

Yet, even with this rapid growth and management's conviction that chemicals would "be an increasingly important contributor to Jersey's profits in the future," the outlook at mid-decade could best be described as one of optimism. Despite rapidly expanding sales, Jersey's chemical investments remained less profitable than the chemical industry's average, and chemicals had not yet begun to make the contribution to overall earnings anticipated from the continued large capital outlays. While some of this failure could be attributed to the normal—if not fully anticipated—costs of entering and developing new markets,[159] the fact remained that rapid expansion in many areas of an increasingly competitive industry called for the most effective management possible. As more of the company's chemical investments reached the operational stage in the second half of the decade, providing the necessary expertise would become increasingly critical to the success of Jersey's chemical interests.

The profit on chemicals in 1964 fell short of the anticipated $40 million by more than $16 million; it was less than the 1963 profit of $33 million. This shortfall certainly came as a disappointment both to the Esso Chemical managers and to the Jersey Board, for sales had increased almost $90 million. Reasons advanced for the poor showing included: less recovery "than had been estimated" from "fertilizer affiliates"; "price weakness is occurring in these products [general chemicals] because lower cost products are becoming available from larger, more efficient plants"; "drop in both volume and price of methyl ethyl ketone following the entry of Celanese and Sinclair into the market"; "competitors are becoming more aggressive"; "a product quality disadvantage compared with our major competitor, du-Pont [sic]"; "severe operational and mechanical problems"; and other similar statements. Still, chemical executives remained optimistic: "we have staked out our problems and have programs underway to solve them."[160] Despite the firm belief of Mangelsdorf and his associates that chemicals would soon increase both market position and profits, several members of the Jersey Board asked for a complete review of the entire operation.

Rathbone moved to start such an evaluation by asking Jersey's newest executive vice president, J. Kenneth Jamieson, to review the company's entire research program. Jamieson had moved up to the Board from Humble, where he served as president, and he had broad experience in the company. Appraising the research operation became his first task as a Board member.

Jamieson talked to George Piercy, who had appraised chemicals for Rathbone and the Board in 1961, about the problems confronting that unit. Further, he visited the plants of several competing companies to see how they operated. Insofar as chemicals figured in the Jamieson report, he emphasized the need for an independent chemical company and especially the need to attract qualified personnel to be left there and not rotated into the oil side of operations. Further, he questioned whether the expansion of the late 1950s and early 1960s had been consistent with traditional Jersey policies. To him, much growth of "hardware" had been achieved with too little analysis of potential markets. Yet, the Jamieson report led to no major change in the structure of the Esso Chemical Company.

X   X   X

In July, 1965, the BAC approved an investment plan for the functional 1966 budget. In the Executive Committee presentation on October 29, 1965, the Esso Chemical Investment Planning Department head, R. E. Bittner, pointed out that recycling with affiliates "to fix up [investment] plans" had begun. "The Long Range" plans had been developed along product lines: "Additives, Raw Materials, Solvents and Elastomers . . . the 'other' category." Earnings did not keep pace with sales growth.[161]

Meanwhile, the Board moved to straighten out its chemical affiliate. Mangelsdorf and Jackson retired, and this opened the Esso Chemical Company presidency for Enjay President Garvin. On February 17, 1966, in a letter, President Garvin announced to all employees a virtually complete change in senior management personnel as well as a new organizational structure. In 1964, Garvin had become executive assistant to the president, Michael L. Haider, and he continued in that position when Haider replaced Rathbone as chairman and CEO. On March 1, 1965, he left that post to become president of Enjay Chemical Company and from there went to Esso Chemical at the end of the year. In his letter Garvin stated that for several months "careful study has been given to the question of restructuring the Company's organization." This study provided the basis for the changes being made. Four "strong regional operating affiliates, each of whom will be fully responsible and accountable" would be established. Because of the complexity of the business, the New York office, Garvin added, would play an important role in the overall conduct of each business line. He went on to say "Esso Chemical Company will be organized to manage and staffed to handle essentially all of its business, including staff and auxiliary services."

President Garvin, Executive Vice President Donald O. Swan and four senior vice presidents, "all resident in New York," constituted the board of directors. The president of each regional affiliate and the senior vice presidents reported directly to Garvin and Swan. The four senior vice presidents functioned as line executives concerned with worldwide problems. Selected for these posts were, R. E. Bittner, E. A. Herberich, E. C. Holmer and C. F. Van Berg. The new structure called for a Management Committee to be established to "review all proposed project investments, except where approval has been delegated to the regional operating organizations." The Management Committee which met three times a week conducted these reviews.

In the headquarters company the four new product management departments became "(1) Agricultural Chemicals, (2) Chemical Raw Materials and Industrial Chemicals, (3) Plastics and Fibers, (4) Elastomers, Additives, Resins, Solvents and Surface Coatings." The reorganized company incorporated many other changes based on the experience of the Esso Chemical organization, modifications of the Piercy report and ideas of the BAC.[162]

In the regional structure, the lines of authority were clearly drawn; "Local chemical affiliates within each geographic area will report to the regional operating company." The latter company would provide, "in varying degree[s], . . . the regional operating affiliate . . . [with] guidance and assistance on project initiation and development activities, and other functions." In this manner, the interest of overall regional effectiveness could be preserved.

This regional structure facilitated interaffiliate cooperation and allowed the maximum benefit to be derived from overall operations. Esso Chemical had recognized this need in the creation of Esso Chemical S.A. in Europe. By 1967, Esso Eastern Chemicals Inc., with headquarters in New York [became Essochem Eastern Ltd. in 1971 and moved to Hong Kong] and Esso Chemical Inter-America Inc., with headquarters in Coral Gables, had joined Esso Chemical S.A. and Enjay Chemical Company to complete substantially the regional structure. In addition, the Canadian organization functioned simultaneously as a fifth entity. In this regional structure the Esso Chemical organization showed its Jersey heritage most clearly. A strong tradition of decentralized management, combined with the need to integrate better its increasingly sophisticated foreign operations, had led Jersey's management to move toward a regional orientation as early as 1960. Local managers could now respond to local situations, and they could more effectively utilize the advantages of Jersey resources outside their immediate control.[163]

As the Esso Chemical organization took shape, its operations expanded rapidly. Transfer of all domestic chemical operations out of Humble to Enjay Chemical occurred in mid-1966,[164] and, while financial and tax considerations continued to block the transfer of existing operations in Europe, the company established separate chemical affiliates there, in order to undertake new projects and to assume management—ultimately ownership—of existing facilities when such a step became feasible.[165] In other regions, where chemical and oil operations were not so closely tied, separation came more easily. By mid-1968, Esso Chemical owned most of its own manufacturing facilities, and the forging of a worldwide management structure to direct them had been essentially completed.[166]

The three-tiered corporate structure, with its functional and regional organizations that overlapped the authority of product line vice presidents appeared unwieldy. But to the management of Esso Chemical, it represented "the only workable method of attaining its objectives," not only splitting off chemical operations but also meeting a more important, continuing responsibility "to maintain a truly international chemical business while striking a delicate balance between growth and profits."[167] Although the quantitative effects remained difficult to assess, Esso Chemical executives saw this organization as yielding greater integration and better coordination of their global enterprise, along with offering a greater emphasis on planning. Worldwide supply-demand charts were drawn up for the first time, important problems identified, and actions to solve them began.[168]

Garvin and Swan attacked the problem of supplying the managerial talent necessary to make the new system work. Without a sufficient number of chemical specialists to staff it, any organization would be useless; in fact, many experts considered the wealth of experienced specialists to be the principal advantage enjoyed by those chemical companies competing with the affiliates of other oil companies. Now, Garvin and Swan found themselves in conflict with one of the basic tenets in Jersey's management philosophy: placing a high value on breadth of experience, the company has traditionally developed its executives by shifting them into and out of a variety of functions and companies. This practice had insured that the management of every chemical affiliate was composed largely of "oil men." While continuing to rely heavily on recruits from Jersey petroleum interests, Garvin also began trying to retain more of them within his organization, by selling the idea of a career in chemicals. Without abandoning his belief in the value of broad experience, he sought to strike a balance at Esso Chemical between that principle and the competitive necessity of specialized knowledge,[169] by separating the chemical business without isolating it from the best resources of the parent company.

As Esso Chemical proceeded to staff its new organization, expansion of the worldwide business continued apace. Meanwhile, investment objectives of Jersey, stressing increased investment in chemicals, remained essentially unchanged, and the research effort in chemicals intensified.[170] As Jersey's chemical sales accelerated in 1966, capital investments in all areas of business were designed to keep pace. New facilities for established product lines reflected the anticipation of continuing growth worldwide for this foundation upon which the company's chemical business still rested.[171] In part, this investment represented further efforts to integrate forward in new product lines. The potential for future growth in these fields—especially fibers, plastics, and agricultural chemicals—also prompted large direct investments.

Expansion in plastics continued to attract a portion of Jersey's capital funds. In the United States, the favorable outlook for Enjay's polypropylene business justified additional capacity at Baytown in 1968. In the same year, the company's first low-density polyethylene [LDPE] plastics plant was completed at Baton Rouge to support the marketing effort underway, and Esso Chemical also obtained a license to build other LDPE plants throughout the rest of the world. Enjay also took steps toward this long-range goal of moving further in the vinyl plastics field by entering the plasticizer and phthalic anhydride businesses, integrating forward from its higher olefins, oxo alcohols and orthoxylene raw materials.[172] Imperial had moved into the polyvinyl chloride market in 1966. Now it also integrated forward into fabricated plastic products, buying a major interest in Poli-Twine Corp. Ltd., a company engaged in manufacturing and marketing polypropylene twine and rope in Canada.[173]

Agricultural chemicals continued to expand, although the direction of their development began to change after 1966. As fertilizer projects inaugurated earlier came on stream, Esso Chemical affiliates took over their management and with it the task of developing markets for their output. However, serious problems began to appear as management's attention and additional funds were required to cover overruns, necessary modifications, and heavy start-up costs.[174] While investment objectives for 1967 continued to stress agricultural chemicals as a likely field for investment in developing areas,[175] actual investments in fertilizers shifted to more developed countries: South Africa, the Netherlands, Great Britain, and Canada.[176] Although more difficult to enter, the established markets of these nations promised greater opportunities for Jersey affiliates to undertake the kinds of large-scale operations to which they were accustomed. The geographic distribution of Jersey's chemical investments after 1966 continued to demonstrate the commitment to building a truly global organization. Thus, the exceptional growth of Imperial's chemical business, once it entered the field in 1957, continued as it prepared to manufacture fertilizers that would supply the market it had already developed for the Engro brand.[177] The area of operations for Esso Eastern Chemicals provided a broad range of investment opportunities. As fertilizer plants were completed in the Philippines and Malaysia, chemicals development occurred in Japan and Australia, the two markets in the region with substantial growth potential in non-agricultural chemicals. Subsequently, Australian and Japanese affiliates added to their petrochemical investments.[178]

The new teams at Esso Chemical had hardly begun to function when early in 1966 Senior Vice President Herberich, in a finance and operations presentation to the Executive Committee, pointed out that the estimated profits for 1966 had been revised downward from $56.4 million to $27.3 million and that net income did not parallel growth in sales. Projected capital expenditures for 1966 had risen from the 1965 figure of $78.8 million to $239.3 million.[179]

One member of the new Esso Chemical board remembered:

> The period from March . . . to October of 1966 was a nightmare for the new management. Event after event piled on the other and were invariably negative. These included occurrence of investment overruns on major projects, indicating lack of management control . . . unusual costs indicating lack of management information and control. The European regional operations were particularly worrisome as indicated by a visit by the entire six-man team to Brussels.

When the various regions' estimated 1967 investment appropriations came in, they totalled over $400 million. Then, "The Management Committee went into closed session now aware that the foreign operations of Esso Chemical Company were out of control."[180] The minutes of that meeting do not reveal the intensity with which members studied and discussed possible courses to follow, but they do show that for the first time company management slashed the proposed 1967 investment appropriation back—to "a level of $150 million." In that meeting the new managers asserted control of the company and served notice that, instead of "expansion for expansion's sake," the new guiding philosophy would be "controlled expansion."[181]

For President Garvin, Executive Vice President Swan and Senior Vice Presidents Bittner, Herberich, Holmer and Van Berg, 1966 became a succession of "surprises." As they delved into the previous six years' operations, they made all possible efforts to re-assert control, and increase efficiency; the new management especially sought to implement effectively many large projects already funded in 1965 and 1966. The largest share of the reduced 1967 investment plan of $150 million went to the United States, and the only major project commissioned outside of North America was in France. It quickly became evident that the new group of managers would attempt to resolve problems by directly cutting back non-profitable operations, to maintain "a more manageable rate of capital expenditures." Still the proposal erred on the optimistic side, as they stated, "Beginning in 1968, a build-up in rate of appropriations should be recommenced, reaching a level around $300 million per year late in the period." But a major slump in the global chemical industry sales [1966-1970] so compounded the internal problems for Esso Chemical management that "significant growth of capital appropriations" did not recommence until 1974.[182]

As the role of the chemical organization began to change in 1966, the first concern of the new company leaders became the design and establishment of an organized structure that would allow it to meet its "primary objectives[:] . . . to maximize profits and create a sound basis and framework for continued substantial growth."[183] The further separation of Jersey's chemical business from its oil operations represented an integral part of this restructuring, but in turn it magnified the problems that the new organization would have to address. As separation of chemicals proceeded, not only would a greater burden of management fall on Esso Chemical; the task of coordinating chemical growth with the program of oil and gas interests would also become both greater and more complex.[184] Yet this job would be essential if chemical development were to benefit fully from the resources of the parent company and make the maximum contribution overall. Part of the problem resulted from the reluctance of the petroleum-based Accounting Divisions that also handled much chemical accounting to provide cost information in the detail required by chemical management. Full and reliable information on true profit or loss was difficult to obtain. Garvin and his crew first had to establish new management information systems and secure control of all their accounting methods before accomplishing anything more.[185]

In 1966, the investment policy of the company was changed. The Esso Chemical board made the decision to curtail sharply its investment program, and to refocus the worldwide chemical organization on profit improvement of the existing business until such time as prospects for achieving a good return were demonstrated. This may very well have been the most significant decision made by the new management team.[186]

The profit improvement program aimed at turning around the total Esso Chemical operations called for "categorization of a large number of separate, discrete businesses" into three classes: those to be improved, those either to be improved or sold off and those to be sold off. The divestment list began with only two operations but as investigation proceeded, it grew rapidly.

Ed Holmer related that from the fateful (for chemicals) September 28, 1966 meeting of the six-member Esso Chemical board until 1972 "the name of the game was increasing profits and increasing return on investment as opposed to justifying new 'hardware.' "[187]

Jersey Board Chairman M. L. Haider reported to the shareholders on May 17, 1967, that most of Esso Chemical's investments had been "designed to lay the basis for future growth and were not expected to result in high immediate earnings," that they had "not yet developed to the point where they are yielding an acceptable profit rate."

In 1951, Director Rathbone had recommended that "oil-chemicals" be separated from the oil business, and he must have smiled when he heard Chairman Haider say, "Several years ago, when we decided to go into the chemical field on a large scale, we concluded that the chemical business was quite distinct from the oil business. We therefore set up a separate corporation known as Esso Chemical Company."[188] More than fifteen hectic years had passed.

The problem of balancing growth with profits plagued Jersey's entire chemical effort after 1966. However, the principal area of concern was fertilizers, the prime source of optimism for chemicals only a few years earlier. In that field all the vexations of the overall chemical effort were magnified. As Esso Chemical affiliates brought new fertilizer plants on stream, the industry worldwide faced a prolonged period of overcapacity, especially for the nitrogenous fertilizers in which Jersey had specialized. In its enthusiasm for expansion in fertilizers the company had not only failed to recognize that capacity more than equalled demand, but actually had made its plight worse by failing to anticipate those technological changes about to revolutionize the fertilizer industry. Between the time of Jersey's decision to build small plants to serve local markets and the completion of those facilities, the optimum scale of such installations was multiplied several times by new technology. As a result, the operating costs of Esso Chemical's new fertilizer plants in the Caribbean area and Lebanon prevented them from competing successfully against imported products from large-scale plants unless they had very substantial tariff protection, which governments refused to provide at that time.[189] Construction delays, cost overruns, engineering modifications, and operating problems in some of the new plants further undercut Esso Chemical's ability to wring profits from tight markets.

The Garvin-Swan team assumed that improvements would follow naturally from the separation of chemical from oil activities, and also establishment of a unified organization headed by Esso Chemical Company. As they extended and strengthened the regional system of management, chemicals managers sought to concentrate on areas where better integration and efficiency would yield greater returns from existing business, due to reduced costs and larger sales.[190] The increasingly disappointing returns from fertilizer investments sparked a special emphasis on market improvement in developing areas. Realizing that overcapacity in fertilizers resulted from insufficient demand rather than lack of need, fertilizer affiliates in 1967 increased their market development efforts through enlarged agricultural extension programs in the Far East and Central and South America. Backed by agronomists and technicians furnished by affiliates, independent local dealers attempted to build demand for the products of their Esso Agroservice Centers with demonstration projects and educational programs.[191] In conjunction with this comprehensive approach, Esso Chemical continued to expand its agricultural chemicals product line, particularly in pesticides (although it did little direct marketing of them).[192] However, the practical rate of expansion remained limited. Combined with technical problems and high operating costs, relatively slow growth held down returns from fertilizer investments.

By 1967, experience insured that Esso Chemical Company "had no interest in entering the agricultural business in another underdeveloped country. . . ." Jersey management was cautious about any fertilizer project or any entry into new geographic areas or lines of business.[193] Nevertheless, the chemical affiliate's strategy for improving profitability recognized a continuing need for further selective chemical investments, both as a source of new profits and as a means of enhancing the performance of existing facilities. Despite the worldwide overcapacity plaguing the chemical industry, Jersey's long-term strategy still called for growth in its chemicals business. The task of Esso Chemical was to stabilize development by making sure that new investments would contribute to the goal of a better balance between profits and growth.

Some Exxon employees still refer to "the fertilizer binge" as a phrase to describe the boom years in the Jersey Company chemical organizations. Available materials do not supply enough information to enable one to piece the entire story together. It is clear, however, that worldwide over-production of oil and surplus refining capacity combined to make inevitable a search by large oil companies for petroleum-related business that would increase net profits. For Jersey in the early 1960s, this situation happened to interlock with reports showing that the company's cash flow would continue to increase, and with the Jones proposal concerning a new investment policy. Even though profits in Jersey chemicals wavered in the 1955-1960 period, the adoption of the new policy appears to have—at least for a time—superseded usual Jersey demands for careful analysis before investment and for avoid-ance of joint investments. [This, the Jones Report did not recommend.] Certainly no meaningful portent of trouble appeared to create doubts among the Jersey Board, for a highly venturesome plan was presented, with the goal of securing at least 10 percent of the plastics business in Europe. However, this grand strategy was never implemented. It became clear to the new top management team that the plan was unrealistic and, accordingly, it was dropped.

A Jersey director who participated in the discussions regarding chemicals in those years wrote later that chemical diversification took place much too quickly, without adequate study of markets and, in the case of fertilizers, also of the difficulties of manufacturing; "The Colombian and Central American projects were prime examples of bad investments." A search lo-cated no specific business with a growth rate better than that of the oil industry, so the company turned to chemical investments. Only chemical diversification was left to be vigorously pursued.[194]

Another Board member who listened to the discussions concerning invest-ments and chemicals remembered that:

> All their forecasts showed that there was going to be an over-whelmingly large cash flow. And they felt that they should use that cash flow to diversify. . . . I think the main protagonist of that philosophy was Jack Rathbone, that . . . we could manage any-thing. . . . They were going to have all this tremendous cash flow to get rid of . . . all we had to do was to build plants . . . and completely neglected the fact that the prime rebuttal was market-ing . . . we built a lot of plants, including the fertilizer plants and made a lot of pretty bad errors . . . we were going to get in forest products, God knows what all. . . . The thing they overlooked is:

first of all, the farmers [in underdeveloped nations] don't know
how to use the fertilizer; secondly, they've got no way of financing
it; and third, if they do increase their yield, there's no way to market their crop.

He continued by saying that for all the fertilizer plants to be profitable,
the whole infrastructure of agriculture in those nations would have to be
rebuilt, and later remarked that Stott must have "mesmerized Rathbone."

Possibly the explanation from the director, Harold W. Fisher, who had
almost continuously worked with some phase of chemicals at Jersey deserves
notice. He pointed out the:

> forecast of surplus cash . . . never materialized. [But] This [pros-
> pect] also put a lot of zip into the chemical business. Leo Welch,
> when he was chairman . . . was very keen to get into the chemical
> business. We were under a lot of pressure to expand the chemical
> business rapidly. 'Jack' Rathbone caught this fever too. I don't
> know if maybe he had not originated it. But anyway he and Leo
> were both enthusiastic about this.

When Fisher returned from England and the Middle East in 1959, he
talked to Rathbone:

> Jack had sort of given me my sailing orders on the thing. He said,
> 'Now look! We want to get as big as DuPont, as fast as we can in the
> Chemical Business!' [Fisher had reservations about how fast this
> could be done.] But this was the theory and this was the Board's
> decision, and I am not saying that I disagreed with it. . . . We
> went absolutely crazy to get into these darn things. We tried to buy
> the Potash Company of America and the Department of Justice
> wouldn't let us do that. It was the best thing that the Department
> of Justice ever did for us, in my opinion. . . . Bill Stott said, 'Bud,
> the damned Esso salesmen down there [South America], they can
> sell anything—fertilizer and what not!' Well, I had my doubts
> about it. But we went ahead with it anyway.[195]

While there is little doubt that the fertilizer episode of chemicals represents the most expensive error committed in that development, even today
some persons contend that had divestment of the chemical plants not occurred, Jersey would have recouped more than its expenditures. But the
final decision to present a project rested first with the affiliate and after the
proposal had received proper review from various committees, next with

the contact director. By that time, in the normal course of decision making, approval by the Executive Committee was generally assured. However, there can be no doubt that Rathbone's strong support of chemicals affected, and perhaps influenced, some Board members. The strong-minded Rathbone, in furthering investment in that area, chose Mangelsdorf and Jackson to help execute his plans—to have chosen opponents of his program for chemicals would have been little short of the ridiculous. As is the case so often in institutions, blame for lack of success is given the few, while many share in successes.

X   X   X

1967-1975

On November 4, 1966, executives of the Esso Chemical Company made their 1967 budget proposal to the Investment Advisory Committee [IAC]. Ed Holmer presented the plan. Referring to the expansion in plant investment installed over the two previous years, he said that it represented a rate "of 27% per year. We have concluded that this is too fast. This is particularly apparent in Europe." The short term financial outlook appeared gloomy. He went on: "Plant operations are ragged, even in mature locations." Marketing had gone slowly. After advancing several explanations for these circumstances, he stated: "Hence a sharp change in plan is proposed, a change which is designed to gain proper control of the business and to place our growth on a more systematic basis." Holmer then revealed details of Esso Chemical's plan to reduce the level of 1967 appropriations to $150 million.

The results, in his opinion, would be far from certain. Some prime opportunities for investment would be lost while large contracts for industrial chemicals would certainly go to others. But, he pointed out, "It simply means that we are people-limited and, particularly, management limited." In response to the budget presentation, the Investment Advisory Committee commented: "IAC was very pleased . . . and felt that the plans for corrective action and the reduced investment program for 1967 were entirely appropriate."[196]

In the Accomplishments and Objectives [A & O] Review several weeks later, Don Swan pointed out the only thing the executives knew "with complete certainty. . . . We have been trying to do too many things too fast."[197] But even cutbacks on investments and personnel did not improve profits for some time. The chemical industry suffered from overproduction and fluctuating prices.

Tariff negotiations conducted at Geneva, Switzerland in June, 1967, found the United States agreeing to reduce duties on chemical imports by an average of 43 percent. Other nations reduced tariffs on chemicals an average of 26 percent. For the chemical company executives, this action presented another problem "difficult to assess."

The international trend toward foreign governments participating in chemicals also vexed the officials because it "raises the prospect of competing for sales in a climate where political considerations could outweigh economic ones." In the same IAC presentation, Swan provided an example of the latter, "The Greek PVC operation which in effect is supporting ethylene and VCM plants of uneconomic sizes, installed as a part of the refining complex obligation." There were others.[198]

Later in December, 1967, Holmer, reporting to the Jersey Executive Committee, informed them that the chemical management had been too optimistic, "by last spring it had . . . become apparent that several years, rather than twelve to eighteen months, would be required to gain proper control of the business outside the United States."[199] Work continued to bring chemicals under control.

Management under a single global structure made possible better planning and closer integration of the worldwide chemical investment program. To improve its execution of this function and to reduce the problem of cost overruns on most projects, Esso Chemical, in concert with Esso Research and Engineering, instituted in 1967 improved planning and estimating techniques, designed to provide more accurate cost and performance forecasts in the early stages of investment study.[200] However, the full benefits of these changes would not be reflected in the company's profit and loss statements for several years. In the meantime, Esso Chemical completed the major expansion program it had approved several years before, enlarging its physical plant by 70 percent from 1967 through 1969.[201] The resulting demand for capital funds temporarily increased the burden on both the parent company and chemical cash flow.[202] Certainly with the focus on improving efficiency in existing lines of business, the expansion program did not offer the shortest route to higher profits within the context of the long-term growth strategy.

In 1968, R. A. Winslow became president of the troubled Esso Chemical S.A. region, replacing J. F. Wright who retired. Garvin, after two fruitful years with Esso Chemical, went to Jersey as director, member of the Executive Committee and executive vice president. Don Swan then assumed the presidency of Esso Chemical and Ed Holmer replaced Swan as executive vice president. Archie L. Monroe, controller of Esso Chemical, replaced

Senior Vice President Herberich who joined the parent company. Most of these changes came after the Garvin team had worked out the principal schematic on how to organize the company.

The management group at Esso Chemical began a major program of divestment that continued for several years. Declining prices and overproduction of basic chemicals everywhere partially influenced this; some plants had managerial problems, and the lag of one to two years between planning and starting construction sent building costs up. With one stroke, top management eliminated the once-glittering European plastics plan. They dropped major investment plans for both plastics and olefins at Antwerp, as well as projects to acquire plastics fabrication companies and the building of resin plants elsewhere. A major Cologne steam cracker plant appropriation was cancelled. The rescue of the European operation occurred at the proper moment and all available human resources became necessary to complete construction of the eight major plants coming onstream in 1967 through 1969.[203] Busy as all the chemical executives obviously were, nevertheless, they found time to tackle internal matters of great concern, though of lower priorities in the company. In 1968, they installed a safety program for employees, entitled the Risk Management Program. Properly executed, they believed that such a program would reduce accidents as well as lead to increased production.

In 1968, chemicals management found time to draw up strict guidelines for planning new investments. These included such cautionary phrases as, "Relate Investment Rate to Management Resources"; "Restrict Major Investments to Stable Locations Where Large Plants Are Feasible"; "Avoid Isolated Opportunistic 'One of a Kind' Investments." These provided a measure of safety for the future—certainly a strong reminder.[204]

Beginning in 1969, Esso Chemical decided to abandon its once ambitious plans for expanding in the fibers business, to divest its existing operations, and to concentrate its energies elsewhere. Don Swan emphasized this point in an April letter to Garvin, "We cannot afford the Fibers business. Considering the scale of operations, technical expertise, and marketing capabilities required to be successful, we have concluded that our resources can be better employed in other chemical activities and we should plan to withdraw from the business."[205] This turned out to be an excellent decision. The worldwide fibers industry soon became depressed; even the leaders experienced very poor profitability throughout the entire next decade.

In 1969, Esso Chemical decided to divest an additional $92.2 million of agricultural and fiber assets, including plants in Central America, the Philippines, South Africa, Spain, Lebanon, Jamaica, Copet in Argentina and

Enjay fibers.[206] The weeding out process continued. Regarding this program, Ed Holmer wrote, "Beginning in 1968, a major, deliberate, carefully-planned program of divesting [all] major operations . . . [began] and lasted until 1971. A total of $500 million of gross plant was divested." Senior Vice President Van Berg and, subsequently, Monroe "spearheaded" the complex divestment program.

Esso Chemical in 1969 had passed the low point for its profits, although its performance remained unsatisfactory. More importantly, the problem of profitability in chemicals had become, by that date, part of a more general concern with Jersey's overall performance. As the expansion program of the previous years neared completion, the chemical affiliate reduced sharply its new investments and continued to shift them to more profitable product lines.[207] Forward integration remained an option in chemicals growth strategy, but management's attitude toward it had become much more passive,[208] in line with the stress on consolidation. The managers of Jersey's chemical enterprise had already fashioned the essential elements of the strategy that would ultimately restore the balance between profits and growth. In this formulation, the establishment of a global chemical company had been an indispensable step. In the implementation of the strategy, the new organization would be equally important.

The decline in Jersey's consolidated earnings put the problem of chemicals profitability in a new light. Since taking responsibility for the company's chemical business in 1966, Esso Chemical had made great progress toward the integration of its worldwide chemical empire. Through organizational changes and a major capital expansion program, it had worked to create a global chemical company capable of deriving maximum benefit from Jersey's total investment. Yet, despite these efforts, the profit and loss statements remained disappointing. When the parent company's overall profits slipped, the pressure for more rapid improvement in chemicals performance increased. Jersey President M. M. Brisco could claim with some justification that 1969 had been the "turn-around year for chemicals."[209] Although the predicted larger contributions to consolidated earnings still lay several years in the future, the program of divestment begun in 1968 marked the final extension of the consolidation strategy forged by Esso Chemical. After almost a decade of rapid expansion in chemicals, Jersey was on the verge of reaping its rewards. In the years since 1966, it had charted and begun to execute the strategy necessary to achieve its goals, but the organization still needed refinement and improvement.

In the spring of 1970, a group headed by Holmer, Esso Chemical Company's executive vice president, completed a major study of the company's organization. The study was motivated by strong feelings that the decision-making processes within chemicals had become unnecessarily cumbersome and time-consuming. There appeared to be excessive checks and balances built into the system of management.

As a result of this study, major changes were made. The company greatly increased the power of worldwide product line management. Instead of having groups of product lines reporting to four vice presidents, separate product line vice presidents were appointed, with each given greatly enhanced authority over the planning and conduct of his worldwide business. The "group" vice president positions were eliminated. Simultaneously the company eliminated large functional departments involving manufacturing and marketing on the basis that all such activities should be the direct responsibility of the product lines. These changes were made not only in headquarters but also in all of the regional organizations. As a result, excessive checks and balances of the earlier organizational structure ended.

Moreover, the study defined the dynamics of the management system. A so-called "two-boss" concept became thoroughly ingrained throughout the worldwide organization. They established worldwide product line business plans as the cohesive element which made the system work. The concept required that the worldwide product line vice presidents and the presidents of the regions would agree on all major plans or changes in plans before these would have any status in the organization. [Charts indicating these changes appear on page 244.]

From this point on, the chemical organization became a true "matrix" organization with management responsibilities shared equally between the product line and regional managements. Further, each executive became fully accountable for the financial performance of his business segment.[210]

During the course of this study, another major issue arose involving the effectiveness of the research and engineering effort provided to chemicals by Esso Research and Engineering Company. The study group saw clearly that the size and complexity of the chemical company had grown to the point where it required direct control over its research and engineering functions. Accordingly, it recommended that research, development, and engineering activities be moved from Esso Research and Engineering Company to Esso Chemical Company. After much deliberation and selling of the concept to the parent company, the decision to make this step was made in April, 1972. Technology became the last major activity to be separated from petroleum; this step essentially completed the separation process begun nearly ten years before—much as Rathbone had visualized and recommended about twenty years earlier.

In 1968 and 1969, the chemical company handled well a number of difficult "start-up" problems in widely separated places. Generally, as plants
came onstream, lower profits ensued.[211] Profits for 1968 reflected this. So in
December, 1968, when Ed Holmer circulated the abbreviated versions of
presentations he and Swan had made to the Operations Review Committee,
he pointed out that the 1966 cutback in investment would really be felt first
in 1969, and that "Major improvement in book profit is firmly anticipated."[212] He and Swan believed that the bottom of the decline in profits
had been reached. Not satisfied with projected profit levels, Swan wrote:

> We have arrived at certain decisive actions aimed at improving the
> future structure and profitability of our business. These include
> selective divestments, definition of the business lines and geo-
> graphical areas in which future investments will be emphasized,
> and conversely, de-emphasized, and major reorganizational steps
> in certain of our operating regions.

In 1968 chemicals lost $15.8 million of after-tax income; however, in the
United States profits reached $57.2 million before taxes. All foreign regions
reported losses.[213]

In fact, chemicals reached the low point in 1968. It took time to unload
unprofitable operations. In 1968, the company established a reserve for
anticipated losses from such divestment, and went about the task of persuading the Jersey Board to accept the plan. Esso Chemical net income, after
divestment and taxes in 1969, amounted to $26.1 million and in 1970, $47
million. The projected net income for 1971 came to $65 million. Swan wrote
Garvin, "We are still not satisfied with these projections and have embarked
on a planning program focussed on the period through 1973 to enable us to
realize by 1975 and if possible sooner, truly acceptable levels of profitability."[214] Holmer several days later, in presenting the chemical company corporate outlook to the Jersey Management Committee, stated "that very
substantial profit increases should be obtained between now and 1975."[215]

In 1970, chemical management became involved in the Jersey Company's
name change plans. Jersey Board Chairman and CEO Jamieson, terming
the name change an "investment in the future" pressed to have the name,
Exxon, apply to all domestic petroleum and related operations. On November 25, 1970, responding to Jamieson's pressure, the Board reconstituted the
name change steering group and the chemical company designated Senior
Vice President Archie L. Monroe to serve on it.[216] In July, 1972, after further
study by the steering group and the Management Committee, the Board

announced that they would ask for shareholder approval of the new name. When approval came at a meeting in October, 1972, the change became official. On January 1, 1973, Esso Chemical Company Inc. became Exxon Chemical Company [Exxon Chemical], a division of Exxon Corporation, and the Enjay Chemical Company became Exxon Chemical Company U.S.A., an operating division of Exxon Chemical Company.[217]

By 1972 Esso Chemical could see the results of its efforts. As the world economy began to catch up with capacity, worldwide sales for the company increased almost 17 percent to $1.258 billion, exclusive of interaffiliate sales.[218] As the company divested its plants, keeping those that could be properly managed, profits increased.[219]

Results in 1973 were even more gratifying. Now the renamed Exxon Chemical Company saw another 22 percent increase in its revenues. Moreover, improving prices and the continuing drive for greater efficiency resulted in quadrupled earnings and a return on capital employed of 18 percent. Chemical earnings accounted for 8.3 percent of Exxon's consolidated total, more than twice the 3.4 percent of the year before. Net income to Exxon reached $202 million. Forecasts indicated that this latter would reach $300 million in 1974. As it turned out, this profit forecast was greatly exceeded.[220]

All geographic areas and product lines shared in the improvement, but the results of foreign operations were especially notable. After years of low earnings abroad, the company's foreign chemical business actually became more profitable than that of its U.S. operating division, Exxon Chemical Company U.S.A., yielding a 20 percent return on capital employed. As sales and capital investments continued to rise more rapidly abroad, Exxon prepared at last to draw the full benefits from its global organization in chemicals. Better integration and worldwide planning had clearly strengthened the performance of the total enterprise, while the decentralized management enhanced flexibility.

The responsibility given regional and local managers for meeting the conditions of their diverse markets promised additional benefits. In 1972, Esso Chemical introduced in the United States a line of Escorez resins with applications in the rubber, tape and coatings industries. Originally commercialized by Esso Chemical affiliates in Europe, Escorez resins were manufactured in France and shipped to the United States. Although a relatively small part of the company's business, its importation symbolized both the potential benefits to be derived from the variety of the company's operations and the truly global nature of its organization.[221]

In April, 1972, as a result of the 1970 organization study, it was decided that research, development, and engineering should be moved from the Esso Research and Engineering Company to Esso Chemical. During the remainder of the year, a major study was conducted to determine how these critically important activities should be integrated into the Exxon Chemical organization. On February 16, 1973, the completed Technology Study became available to the Exxon Chemical Management Committee. The group, headed by Holmer, made seven recommendations. They recommended the establishment of an integrated chemical research and process engineering technology organization for each product line, answering directly to the appropriate product line vice president. They also recommended a separate central, "highly expert, cross-product line world-wide chemicals engineering organization" to be located at Florham Park, New Jersey. A third recommendation called for adjusting "the balance of centrally-based technical resources between Europe and the United States." The report also recommended that product line technology planning should be carried out in considerably greater depth than in the past. Additionally, it provided for the appointment of a worldwide technology vice president to maintain a broad overview of the chemical company's entire technical effort. It concluded that operating excellence should be improved by "beefing up" the number and effectiveness of plant technical personnel. Finally the institution of a program for career development of "technically oriented personnel" completed the list. Overall implementation of the Technology Study provided the chemical company with a technology organization much more responsive to business needs.[222]

1973 represented the first really good profit year at Exxon Chemical; moreover, the higher profitability rate continued through 1974. The company had achieved the objective of "being among the best in the chemical industry." Too, the program of putting together greater stress on safety [the Risk Management Program], thereby placing much more emphasis on manufacturing excellence, paid handsome dividends. Safety performance improved "to a level of 1.7 disabling injuries per million man hours." For that, the company received the Award of Honor from The National Safety Council.[223] Obviously, manufacturing efficiency also improved.

The boom in demand for chemicals in 1973 had helped to sustain prices and to push profits up. However, demand exceeded supply capability about mid-year, and toward the end of the year feedstock shortages "further aggravated the situation." Swan wrote Howard C. Kauffmann in the Financial and Operating [F & O] Report on March 29, "In this climate, the 'manufacturing excellence' program begun in 1971 and heavily emphasized in 1972 and 1973 paid off handsomely." Simply put, Exxon Chemical had produced more from existing equipment. The coming on stream of a major addition to the Baton Rouge olefins plant helped increase profits.[224]

While the year 1973 may have been a profitable one, a major problem did arise—the energy crunch. In the fall of that year (October 16), the Organization of Petroleum Exporting Countries [OPEC] almost doubled the per barrel price of oil. J. K. Jamieson, Exxon's Board chairman and chief executive officer, in the *Annual Report* for that year pointed out, "The immediate causes were deeply rooted in unresolved political problems in the Middle East. What was different this time was that the consumer, particularly the American consumer, felt the effects."[225] During the year, Exxon Chemical had, in fact, experienced shortages and had been forced in turn to allocate "supplies of some of its products to customers." Tight petroleum supplies made it likely that the situation would deteriorate further before it improved. Even the Federal Energy Office requirement, imposed during the last quarter of 1973 (and early in 1974) that refineries produce 5 percent more gasoline seriously affected the chemical business.[226]

Environmental concerns, and changes in technology in 1973, called for new developments and opened new opportunities in several product lines. More sophisticated technologies throughout the oil industry and the creation of new national oil companies resulted in larger markets for additives and chemical specialties outside the Exxon system, and increased the competitive pressure on research activities in those fields. New technologies and new regulations designed to protect the environment also brought rapid changes in the solvents business. Exxon Chemical had to expand its facilities in that product line to maintain competitive flexibility. Pressure on Exxon refiners to provide low-sulphur fuel oil also increased the company's supply of sulphur and necessitated greater marketing efforts by the agricultural chemical product line.[227]

To maintain Exxon Chemical's strong position in chemical raw materials and synthetic rubbers required additional measures. Plant expansion and construction of new facilities in France had already strengthened Exxon's competitive position in butyl rubber, as well as making it the only company with ethylene-propylene rubber facilities in both hemispheres. To continue leadership worldwide would require further increases in capacity. A strategy for chemical raw materials called not only for expansion to meet the growth of markets in various parts of the world, but also for continuing efforts to increase Exxon Chemical's own consumption of olefins. Again, the recent rapid growth of plastics sales had made expansion in that field highly attractive as a means of upgrading the company's product line.[228]

Low-density polyethylene and polypropylene continued to be the mainstay of Exxon Chemical's plastics business. The company also maintained two small polyvinyl chloride operations in Canada and Greece, but its involvement with the vinyl plastic industry was increasing through its role as a leading supplier of plasticizers produced by its chemical intermediates product line.[229] In the strategy formulated at the end of 1973, supplying the vinyl industry with plasticizers would involve expanding in the United States and abroad, especially in Europe and in the Far East.[230]

The worldwide energy crisis increased the pressure on Exxon Chemical management, just as it placed new burdens on all petroleum companies and their affiliates. The rationing of supplies to customers could only be a stopgap arrangement. The Exxon Corporation, because of its management and planning, suffered less than did its competitors. In view of the problems created by the Arab oil embargo, planning assumed even greater importance.[231]

In this unsettled environment, Swan, on May 1, 1974, wrote Exxon Director and Senior Vice President Kauffmann outlining chemical's "principal objectives and strategies." Swan pointed out that these could not be considered new for "they have all been operative for a considerable period." The first and basic objective he considered to be that of maintaining "profitability at levels equivalent to the best of competition." Exxon, Swan said, had the right, "to expect the management of its chemical business to produce results comparable to the best managed companies in the chemical industry." He added that when chemicals met that standard, good returns on investment would follow.

As for strategies, Swan wrote that the company aimed to control investment rates to such levels as could be managed effectively by the chemical organization. Along with management resources, chemicals would appraise "engineering resources [available] to implement new projects." This had to be done without impairing existing operations and levels of return. To him, "Growth has been and will be viewed as a dependent variable, subordinate to maintaining acceptable current returns."

Chemicals, according to Swan, would continuously seek to improve relations and understanding with Exxon's petroleum management to insure the quantity and quality of fuels and feedstocks necessary for its operations. This area, given the uncertainties existing in the world supplies of petroleum, remained extremely critical to their other objectives. He added that investments had to be consolidated in "9 or 10 stable, industrialized countries" while the company continued to market worldwide. Productivity improvements would be sought as superior to "grass roots investments." Management resources would be concentrated in the existing lines of business—

"no step-outs to entirely new product lines are envisioned"—while the company continued to push those "relatively specialized and proprietary business segments" likely "to enjoy exceptionally high returns."

Swan concluded by analyzing in short paragraphs the nine product lines then managed by Exxon Chemical, indicating areas of possible divestment; some where "bottle-necks" existed and others where growth might be expected. As Swan indicated, Exxon Chemical already operated on these guidelines. Thus, to the Exxon Board, this statement represented a basic schematic for chemicals for the next decade—one that had evolved after years of corporate debate and planning—and one that proved to be sound.[232]

In *ChemReport* for 1974, Swan attributed the success obtained in "A Most Unusual Year" largely to "the timing and combination of our programs for manufacturing and marketing excellence." Operating earnings doubled [after taxes, but excluding interest expense, foreign exchange gain/loss and minority share of earnings, all net of related income taxes] with $456 million reported from a revenue of $3.3 billion. Exxon Chemical now ranked among the five most profitable worldwide chemical companies. Swan added, "if our sales were considered separately from Exxon Corporation's, we would rank in the top 50 on the *Fortune* '500' list." Production at the chemical company had reached capacity in 1973; however, the 6 percent increase above capacity in 1974 provided proof of the overall productivity of the organization. The petroleum crisis in the early part of the year had cost "some capacity" but Exxon Chemical had engineered around this "with the cooperation and flexibility of [the] Exxon petroleum organizations."[233]

For Exxon Chemical, 1974, despite its problems, was a good year. However, on April 1 of that year, Executive Vice President Holmer left the company to become an executive vice president of Esso Middle East, charged with developing industrialization projects in countries rich in hydrocarbon resources, for example, Saudi Arabia. He was replaced by R. A. Winslow who had served as president of the two largest chemical regions, Europe and the United States. Holmer had chaired several of the key committees whose reports, when implemented, had helped bring the chemical company to its high position in 1974. Certainly he had been a stabilizing factor during the eight years in which he served. With Swan and Garvin, he had led the fight to get the Jersey Board to approve the divestment of the unprofitable chemical companies. It is obvious that his departure came as a result of the Exxon Corporation's Compensation and Executive Development Committee's desire to move him upward in the managerial network.[234]

Worldwide recession affected Exxon Chemical Company in 1975. However, its performance improved during the latter part of the year, and its return on capital employed was at a competitive 14 percent level for the year. Decline in demand for all product lines resulted in a loss of 10 percent in sales and 7 percent in revenue, as compared to 1974. Meanwhile plant expansion doubled over the previous year. European plants that grew included solvents facilities at Antwerp, a steam cracker expansion at Cologne, and plastics facilities at Meerhout (Belgium). Completion of a major fertilizer plant addition in Redwater (Alberta) also took place, and construction continued on a solvents plant in Singapore. Major additions were started in the United States, at Baton Rouge, Bayway, and Summerville (South Carolina).[235] With supply and costs of petroleum and other fossil fuels now a major uncertainty globally, the chemical industry faced an unsure future. Yet the impressive achievements of this Exxon affiliate stand forth in the record. Exxon Chemical has come a long way since Frank Howard referred to it as the Jersey "step-child."

Glossary of Chemical Terms

Acetone . . . . Oxygenated solvent derived from isopropyl alcohol.

Acrylonitrile . . . . Vinyl cyanide chemical intermediate. See Buna-N.

Additives . . . . Chemicals added in small quantities to improve qualities of fuels and lubricants.

Aromatics . . . . Liquid hydrocarbons having a closed-ring structure. Most important are benzene, toluene and xylenes which, like the lower olefins, are basic petrochemical building blocks.

Benzene . . . . See Aromatics.

Buna-N . . . . Synthetic rubber made by the copolymerization of butadiene and acrylonitrile.

Butadiene . . . . A diolefin used in the manufacture of synthetic rubber.

Butyl Rubber . . . . Synthetic rubber made by the copolymerization of isobutylene and isoprene olefins.

Caprolactam . . . . Chemical used in the manufacture of nylon fiber.

Chemical Intermediates . . . A wide range of chemicals that are used in the production of industrial and consumer products.

Chemical Raw Materials . . . See Petrochemicals.

Copolymerization . . . . Reaction of two or more molecules yielding a complex molecule, e.g., synthetic rubber.

Elastomers. . . . .   Products that exhibit elastomeric properties, e.g., synthetic rubbers.

Ethylbenzene . . . .   Chemical derived from ethylene and benzene.

Ethylene . . . .   One of the important olefin petrochemical building blocks produced by the steam cracking process.

Feedstock. . . . .   A term that describes crude oil, its fractions or other raw material that enters any process equipment.

High-Density Polyethylene
   (HDPE) . . . .   Polyolefin thermoplastic made by polymerizing ethylene. Greater rigidity and higher softening temperatures are characteristics that distinguish HDPE from LDPE.

Hydrocarbon. . . . .   Molecule containing hydrogen and carbon.

Hydrogenation . . . .   Addition of hydrogen to a molecule.

Isobutylene . . . .   One of the important olefin petrochemical building blocks produced by the steam cracking process.

Isopropyl Alcohol . . . .   Also Isopropanol . . . also "Rubbing Alcohol" . . . Solvent derived from propylene.

Low-Density Polyethylene
   (LDPE). . . . .   Polyolefin thermoplastic made by polymerizing ethylene. Highly branched chain structure.

Methyl Ethyl Ketone
   (MEK) . . . .   Oxygenated solvent.

Naphtha . . . . Hydrocarbon fraction derived from the distillation of crude oil.

Normal Paraffins . . . . Saturated high molecular weight hydrocarbons.

Olefins . . . . Unsaturated hydrocarbons that include ethylene, propylene and butylenes derived from the steam cracking process.

Oxo Alcohol . . . . Higher alcohols derived from higher olefins (C_6-C_{12} range) by the "Oxo" process.

Paramins® . . . . Exxon Corporation trademark for family of additive products for improving qualities of fuels and lubricants.

Petrochemicals . . . . Ethylene, propylene, butylenes and butadiene derived from the steam cracking process; also aromatic hydrocarbons—benzene, toluene, and xylenes (BTX). They are basic chemical raw materials used in the manufacture of a wide range of products.

Plasticizers . . . . Chemical products for making vinyl products soft and pliable.

Plastics . . . . Polyethylene, polypropylene, polyvinyl chloride (PVC) and other polymers which can be formed into desired shapes under heat and pressure.

Polyethylene . . . . See low-density and high-density polyethylene.

Polymer . . . . High molecular weight molecule, e.g., polyethylene, formed by the linking together of thousands of the original molecules, e.g., ethylene.

Polymerization . . . . Reaction of the same molecule yielding a new compound, e.g., reaction of propylene to form polypropylene.

Polyvinyl Chloride (PVC) .. A thermoplastic polymer derived by polymerizing vinyl chloride monomer.

Polypropylene . . . . Polyolefin thermoplastic made by polymerizing propylene.

Propylene . . . . One of the important olefin petrochemical building blocks produced by the steam cracking process.

Refinery Gases . . . . Mixture of lower paraffinic and olefinic hydrocarbons that served initially as the source of olefins, e.g., propylene.

Resins. . . . . Amorphous solid, semi-solid mixtures of hydrocarbons having special properties, e.g., adhesion.

Solvents . . . . Hydrocarbon fractions derived from crude oil for use in surface coatings, polishes, inks, adhesives and other industrial processing. The oxygenated solvents are made from olefins for use in surface coatings and as chemical intermediates for a wide range of products.

Steam Cracking . . . . The basic commercial process for producing ethylene, propylene, butylenes and butadiene using heat and steam to break up molecules.

Thermoplastic. . . . . Plastic that softens under heat.

Toluene . . . . See Aromatics.

Urea. An organic fertilizer.

VCM . . . . Vinyl chloride monomer. See polyvinyl chloride.

Exxon Chemical Profile in 1975

WORLD HEADQUARTERS

Exxon Chemical Company . . . . . New York

REGIONAL HEADQUARTERS

Exxon Chemical Company
U.S.A. . . . . Houston, Texas
Essochem Europe Inc. . . . . Brussels, Belgium
Esso Chemical Inter-America
Inc. Coral Gables, Florida
Essochem Eastern Ltd. . . . . Hong Kong
Esso Chemical Canada—
Division of Imperial Oil
Limited. . . . . Toronto, Canada

FINANCIAL DATA
(Millions of Dollars)

Sales Revenue. $3,043
Earnings from Operations 199
Return on Capital Employed
(percent). 14.0

REVENUE BY PRODUCT GROUP
(Millions of Dollars)

Agricultural Chemicals. . . . . $ 297
Primary Petrochemicals 1,027
Polymers and Fabricated
Products   619
Solvents and Specialty Chemi-
cals 1,100

SALES BY REGION
(Percent)

United States of America 38.9
Europe, Africa and Middle East . 38.9
Canada. 11.0
Latin America 5.9
Far East 5.3

RANKING AMONG CHEMICAL COMPANIES

Worldwide 13th
U.S.-Owned. 5th

EMPLOYEES

Worldwide 18,000

ESSO CHEMICAL COMPANY*
AND
PRINCIPAL AFFILIATED ORGANIZATIONS

Organization Prior to 1971

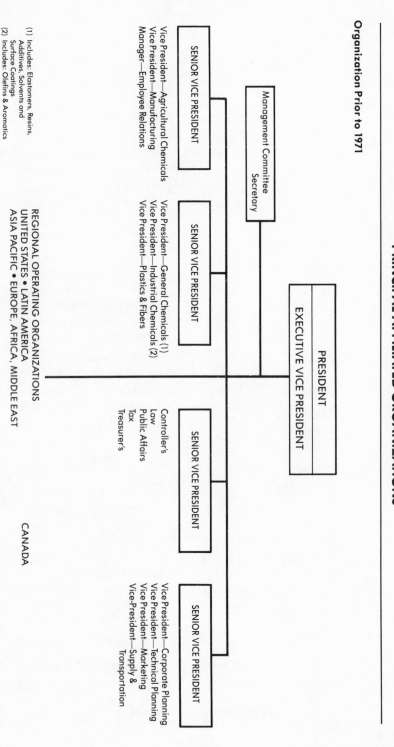

PRESIDENT

EXECUTIVE VICE PRESIDENT

Management Committee Secretary

SENIOR VICE PRESIDENT

Vice President—Agricultural Chemicals
Vice President—Manufacturing
Manager—Employee Relations

SENIOR VICE PRESIDENT

Vice President—General Chemicals (1)
Vice President—Industrial Chemicals (2)
Vice President—Plastics & Fibers

SENIOR VICE PRESIDENT

Controller's
Law
Public Affairs
Tax
Treasurer's

SENIOR VICE PRESIDENT

Vice President—Corporate Planning
Vice President—Technical Planning
Vice President—Marketing
Vice-President—Supply & Transportation

REGIONAL OPERATING ORGANIZATIONS
UNITED STATES • LATIN AMERICA
ASIA PACIFIC • EUROPE, AFRICA, MIDDLE EAST

CANADA

(1) Includes: Elastomers, Resins, Additives, Solvents and Surface Coatings
(2) Includes: Olefins & Aromatics

244

Organization After 1971

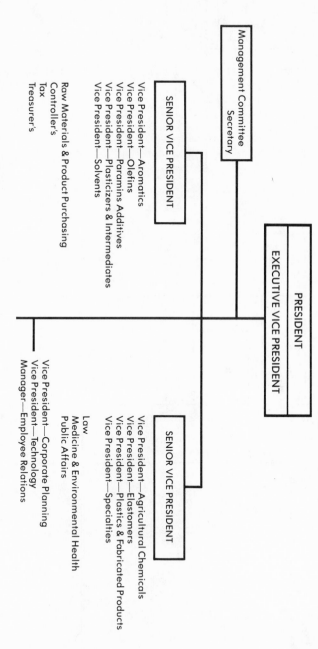

PRESIDENT

EXECUTIVE VICE PRESIDENT

Management Committee
Secretary

SENIOR VICE PRESIDENT

Vice President—Aromatics
Vice President—Olefins
Vice President—Paramins Additives
Vice President—Plasticizers & Intermediates
Vice President—Solvents

Raw Materials & Product Purchasing
Controller's
Tax
Treasurer's

SENIOR VICE PRESIDENT

Vice President—Agricultural Chemicals
Vice President—Elastomers
Vice President—Plastics & Fabricated Products
Vice President—Specialties

Law
Medicine & Environmental Health
Public Affairs

Vice President—Corporate Planning
Vice President—Technology
Manager—Employee Relations

REGIONAL OPERATING ORGANIZATIONS
UNITED STATES & LATIN AMERICA
ASIA PACIFIC • EUROPE, AFRICA, MIDDLE EAST

CANADA

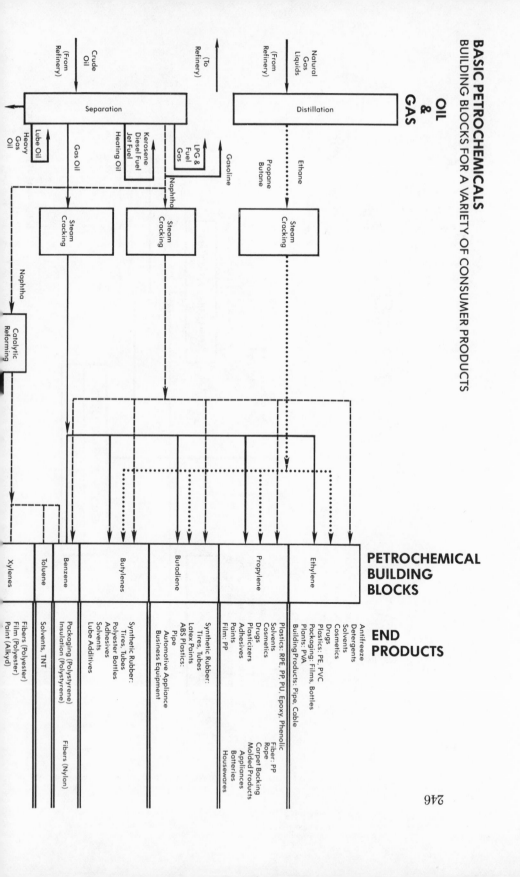

BASIC PETROCHEMICALS
BUILDING BLOCKS FOR A VARIETY OF CONSUMER PRODUCTS

Chapter 6

Beginnings in Europe

FOR EXXON* IN EUROPE the period 1950 to 1975 was one of remarkable expansion in every aspect of the oil business. Post-World War II recovery there was made possible by an abundance of relatively low-cost imported oil which supplemented coal and other indigenous energy sources. As recovery took place at a progressive rate, the demand for oil and oil products also increased in similar fashion, and this required backup facilities—tankers, pipelines, storage, refineries and marketing outlets.

Exxon's traditional role in Europe had been marketing oriented, but now the company had to switch gears and provide expertise in the other industry functions. This meant an influx of technical people into its manpower and management ranks, as well as a coordination effort to ensure that all the new pieces would fit together. David A. Shepard (later Jersey director and vice president) had been chairman of Anglo-American Oil Company, Limited [became Esso Petroleum Company, Limited in 1951] from 1945 to 1949, and other Americans had served with European companies from time-to-time. Now, however, Jersey regularly assigned Americans to the management, including the boards, of European companies. There was, in fact, a considerable exchange of personnel in both directions, to provide training and experience for many executives.

With important oil and gas discoveries at Parentis, France (1954), Groningen, Netherlands (1959), and in the North Sea (1969), another element—production know-how—also came into play. With the installation of producing facilities in the 1960s and 1970s, Europe was to become an upstream operation of great importance to both the Western European governments (especially in terms of money supply and exchange balance) and Exxon affiliate companies operating in the host countries. For example, in 1975, Europe accounted for 15 percent of the Exxon total worldwide natural gas production, 45 percent of Exxon worldwide natural gas sales, and 36 percent of Exxon worldwide refinery runs.

*Standard Oil Company (New Jersey) became the Exxon Corporation on November 1, 1972.

Equally important during these years was the trend toward diversification in Exxon's overall operation, especially in chemicals. Many of the local affiliate companies experimented with non-petroleum marketing-related businesses, such as motels and steam generation, in an effort to increase their sales volumes. (The story of chemicals became far more complex, and it is in chapter 5 of this volume.)

During these twenty-five years, European operations provided Exxon with a booming market—but also with problems of poor profitability and low returns on investment in the marketing and refining operations. These factors demanded special attention and required Exxon to improve coordination among its affiliates, and finally to establish a more formal European regional organization.

All of these elements of Exxon's European experience are detailed in the following sections.

Chapter 7

Exploration and Production in Europe

BEFORE WORLD WAR II, Jersey had production in Hungary and Rumania, but when both of these countries fell into Soviet Russia's sphere of influence after the war, the puppet governments nationalized Standard Oil Company (New Jersey) [Jersey] holdings without compensation.[1] In fact, much of Europe's oil before, during, and immediately after the war came from the United States and Venezuela, so the change did not amount to a radical shift in suppliers. For example, oil from the United States for the allied cause in Europe amounted to almost seven billion barrels.

Following World War II, as soon as Jersey had reorganized its Western European affiliates, a number of them began a fresh search for oil, either with their own companies or in joint venture companies created for that special purpose. Then as rapidly as Europe's use of oil for energy increased, the effort to locate indigenous sources of supply took on additional momentum. Again, after 1950, the several Middle East crises served to dramatize and emphasize the precarious nature of Europe's dependence on oil sources, especially in the Middle Eastern region. Each new crisis increased interest in exploration for oil in some closer place. Ultimately new discoveries of oil occurred in the Parentis field in France, in West Germany, and in the Netherlands on the continent.

The Parentis field in Southwestern France near Bordeaux was found in 1954, by Esso Standard Société Anonyme Française [ESSAF]. This company had secured a permit from the government of France allowing exploration of a large acreage, thereby making it the first such foreign firm to receive a French concession. [Since that company was partially owned by French investors, this satisfied the French law requiring some national participation.] M. A. Wright, at that time Jersey's coordinator of worldwide producing activities, recalls this exploration: "It was mostly in a lake, a shallow lake, and . . . we sent some fellows from Venezuela over there . . . because they'd worked [Lake] Maracaibo . . . to work on this development. . . ." Directed by James A. Clark, renowned Jersey production chief, and using techniques developed in Venezuela, the drill crews brought in the Parentis field. Production soon reached 50,000 barrels per day [b/d] to make the French field the largest producer in Western Europe.[2] Many Europeans became optimistic about more oil being found, but increased government restrictions served to dull the interest of major oil companies in exploring further in France, and hope dwindled for another major commercial find.

In Italy, Esso Standard Italiana [Esso Italy or ESI] sought for some years to secure exploration rights in the favorably regarded Po Valley but, in this case (fortunately for Jersey since no oil was found), Italian law discriminated in favor of Ente Nazionale Idrocarburi [ENI], the government company. Jersey sought to persuade the U.S. Department of State [State] to intercede in Italy as a way of opening the Po Valley to all oil companies but to no avail. Despite continued interest and intermittent efforts, ESI did not obtain a permit to explore there.[3]

Even with the lack of great success in finding quantities of oil in Europe, exploration continued. Finally, in 1959, drillers in the province of Groningen in the Netherlands struck dry gas.* Delineation drilling then began, and the results proved not only important to Jersey but were of major significance to the development of the industry in Europe.

X X X

Some years earlier, Jersey and Royal Dutch-Shell Group [Shell] had reached an understanding: when the former exercised its exploration rights in Cuba and the latter used similar rights in Holland, they would share what was found as 50-50 partners.** In fact, based on the geophysical evidence available before World War II, neither company had much reason to believe that significant oil or gas existed in either place. But the exigencies of the second World War led Jersey to accelerate its search for oil and gas, and this exploration included Cuba, which proved unsuccessful.

Somewhat later, just after the war, Shell started its search for oil in Holland, where it found gas and oil in small quantities. A geologist on Paul

---

*Dry gas is that not associated with crude oil. Note also that during World War II, oil had been located at Schoonebeck (1943).

**There are many versions told of this arrangement. Wright's version is given here. Zeb Mayhew, also of Jersey's Producing Department at that time, remembered the Wright version as the one he heard. He verified the Wright story.

Ruedemann's staff in the Jersey Producing Coordination Department learned about these discoveries and remembered the agreement concerning the Jersey-Shell venture in Cuba and Holland. He told Ruedemann who then went to Wright, at that time Jersey's deputy coordinator of production, and reported that "When we traded off for them [Shell] to come into Cuba, I [we] got a Letter [of Agreement] from them saying that if they ever explored for oil in Holland, we would have the right to participate in it 50-50." Wright then asked him to locate the letter as a step toward establishing the terms of Shell's promise.

Producing Coordination found the letter and gave it to Wright, who then contacted Shell. Shell searched its files and found a copy which confirmed the agreement. As a result, in 1947, Jersey and Shell formed a joint exploration and producing company in Holland, N.V. Nederlandse Aardolie Maatschappij [NAM]. Wright later described this cooperation by Shell as one of the "greatest ethical" demonstrations he had ever witnessed. He said that Shell records had been moved several times during five years of war, had been bombed in one location, and that no one at Jersey would have questioned it if Shell had pled inability to locate any copy of the understanding.[4]

NAM's principal purpose at that time was to explore for and produce crude oil. The natural gas found did present a problem, but the company decided against either reinjecting or flaring it; rather, "Shell on behalf of the partners agreed to make such gas available to the Dutch government [gas distributing company] at a very nominal price." Gas alone was not satisfactory, however; so NAM wildcatters continued the search for oil and drilled two hundred dry holes during the next thirteen years.

Then on August 14, 1959, at Slochteren Well No. 1 in the province of Groningen, at 10,000 feet down, it happened—not oil—but dry gas that had been trapped underneath a salt bed in "gently sloping dome-shaped" Rotliegende sandstone in the Permian era, some 220 million years earlier. From the first it appeared to be an important find. NAM operators then sought to have the Netherlands Inspector of Mines come and examine the discovery, for that country issued production licenses only after he had certified the existence of a commercial deposit. Once that license had been secured, delineation drilling could proceed.

Jersey's Zeb Mayhew, who had served as producing coordination assistant to Wright and who at that time held a top position in that department under Siro Vázquez, recalled that when Groningen came in, for some months, the NAM partners gave it minimum publicity. The companies needed time to analyze potential markets and develop production plans. This involved study of operations from the wells to the customers. Later, news of the Groningen discovery created excitement in the offices of many oil companies, all of them anxious to locate gas or oil closer to or in Europe.

Some years later, May 18, 1966, President J. Kenneth Jamieson speaking to Jersey shareholders remarked, "The Groningen discovery in 1960* excited a good deal of speculation concerning the possibility of oil or gas fields under the North Sea." But in 1959 and 1960, international legal rules adequate for settling questions concerning continental shelf areas did not exist. The 1958 Continental Shelf Convention Agreement did not take effect until June, 1964. So during the years intervening, the governments of the United Kingdom, the Netherlands, Norway, and Denmark established boundaries and "negotiated bilateral agreements defining specific areas to be under each nation's jurisdiction," which became codified later. When West Germany protested to the International Court of Justice, procedural rules enunciated by that body enabled the U.K., the Netherlands, and Norway to adjust their claims so as to provide West Germany with a North Sea area.

Meanwhile this discovery of gas created an entirely different situation for NAM. For when the Netherlands government learned something of "the magnitude and significance of the find," it quickly appreciated the fact "that exploitation could not be on the same basis" as the earlier NAM discoveries. Deep well drilling required large outlays of capital as would pipelines and market development for the quantity of gas believed to have been discovered. Locating this natural gas in a commercial quantity excited the Dutch government, while at the same time development of the gas posed monumental problems for both it and for the NAM partners, who of course were delighted by the find. C. R. Smit of Esso Nederland N.V., Wright, and Marketing Contact Directors Wolf Greeven and W. R. Stott (joined by Shell executives) all began to explore licensing rights, and Jersey also alerted affiliates in neighboring countries to look for possible markets. The Groningen field enjoyed an excellent location, one that could provide comparatively easy access to markets in Germany, Belgium, France, Luxembourg, the United Kingdom, and even Italy. Though all parties expected delays before the gas could be marketed, few of them anticipated that it would be in the late 1960s before large quantities would begin to move to these countries.

The original Jersey position in negotiations with the government of the Netherlands was that NAM should receive two-thirds interest in the joint gas company they planned to create (with Shell) to exploit the field, and the

---

*Jersey publications variously give the Groningen discovery date as 1959 and 1960.

state-owned company, Staatsmijnen (States Mines), one-third. On May 16, 1961, Greeven wrote Stott that negotiations were deadlocked and that "the production and export issues are still at stake." He went on to say, "It looks as though the resulting deadlock may require further discussions with Shell at a high level and possible discussions with the government directly." Robert H. Milbrath, president of Esso International Inc., met with Shell and States Mines personnel early in June in an effort to resolve some of the points at issue. Milbrath's group "focused on the premium nature of natural gas as a fuel" and attempted to convince the Dutch leaders that:

> rather than allow the gas to replace coal and residual fuel oil . . . it would be very much in the interest of the Dutch economy to permit its export to European neighbors as a premium fuel earning attractive levels of foreign exchange.

Shell and Jersey, of course, could provide the capital, the personnel, and the technology to develop the transportation and marketing; thus all the partners would benefit. And in the end, the Dutch government did accept this proposal.

But reaching full agreement consumed time. The parties continued to meet and cablegrams to and from New York flowed freely. Representatives of Shell and Jersey met twice during September, 1961, and later that month, Smit summarized the situation in a memorandum: "The feeling is that some compromising here and there might be necessary and even advisable." He then raised several questions, the principal one being whether in the company to be created to develop the new field, Jersey and Shell would accept only a 50 percent joint equity, with the Dutch government holding the other 50 percent—for "political reasons." Under this "concept the State would have automatically a 50% say in management, but could not override the partner"—a position Milbrath earlier had indicated might solve the dilemma. This would permit Jersey and Shell each the freedom to manage its share of the investment, which had long been a Jersey objective in all joint enterprise ventures. Too, Dutch government involvement in the company's business outside the Netherlands presented additional important economic and legal problems.

On October 12, 1961, Greeven reviewed the Groningen gas situation with the Jersey Executive Committee. As he advised the Committee, "recent developments suggest that it may be necessary to provide for a somewhat greater ownership interest on the part of States Mines." Greeven, Jersey's principal negotiator, informed the Committee that he had authorized an agreement with Shell and the Dutch government calling for States Mines to have 50 percent equity interest and NAM 50 percent with respect to transportation and sales within the Netherlands. The Committee approved his action.

It should be noted that in these proposals to provide States Mines with an equity interest in the company then under consideration, States Mines was to share in the exploration and producing investments, as well as in the costs and profits associated with the newly discovered gas reserves.

The Dutch government next indicated a desire to create a domestic profit center in Holland, built solely with revenues from exported gas. When Greeven heard this, he reacted strongly against it. In a cabled message he stated that Jersey would not agree to have all the profit from the gas operation siphoned into the Netherlands. Speaking for Jersey, he said that as a minimum the "border sales price will have to be competitive and permit commercial profit for foreign importer [sic] adequate to compensate for sales effort and/or investments."

In February, 1962, Milbrath who had chaired a "special study group" gave a lengthy report on Groningen gas, including an estimate of the magnitude of the reserves, a summary of legal problems encountered (and expected) with the Netherlands government as a consequence of developing the field, and economic, political and transportation problems his group believed might arise in the future. The Committee continued to hear reports and to give its advice and opinion to the contact directors, the regional coordinator for Europe, and others involved in arranging for the gas to be produced and marketed.

Meanwhile, NAM exploration crews continued to work, and in July, 1961, the Committee learned from new geophysical data "that a substantial portion of the area underlying the North Sea may offer prospects for discovering large volumes of hydrocarbons, particularly natural gas." Jersey and Shell had expressed interest in some 15 million acres off the Dutch Coast and about 22 million acres off the coast of Great Britain. Jersey moved swiftly to have applications for licenses submitted in both countries, while at the same time sounding out Shell regarding sharing 50-50 in exploration costs and in producing any oil or gas found. Their partnership continued with their U.K. affiliates during the development of the North Sea area.

The negotiations among Jersey, Shell, and representatives of the Dutch government dragged on into 1962. Finally, on September 19, Director Stott secured Executive Committee approval for a NAM affiliate development company, N.V. Nederlandse Gasunie [Gasunie], in which Jersey and Shell each would have 25 percent equity interest, States Mines 40 percent, and the

Netherlands government 10 percent. To transport and market the gas, the Committee approved two months later, two additional companies: Esso Holland Inc., to "hold the Company's economic interest in the gas produced," and International Gas Transport Maatschappij to "develop projects involving the transmission outside the Netherlands of gas sold for export," including pipelines.

NAM continued exploratory drilling to delineate the extent of the Groningen field, and Jersey awaited the results. Thus, the company *Annual Reports* for 1959, 1960, and 1961 made no reference to this discovery, while that of 1962 made only a passing reference to the Netherlands; however, it did mention that geophysical tests in the North Sea had started. Generally the company has relied on careful testing and evaluation before publicizing its various strikes, or their extent. Jersey has never been noted for exaggerated reporting of oil and gas discoveries. So no reference to the Groningen strike was made.

The *Annual Report* in 1960 did emphasize Jersey's success in selling the idea of central heating with oil as the fuel. "Jersey's affiliates were among the first to recognize the potential of the European home heating market," it stated, and then went on to say that "home heating sales [in Europe] have more than doubled over the last three years." So the affiliates continued to encourage heating oil sales even as Groningen and other gas was being readied for the European market. These companies already possessed the equipment to handle oil, their refineries were geared to manufacture it, and the consumer price remained relatively constant.

Then in 1963, Jersey and Shell finally reached an accord with the Dutch government. Jersey arranged with the Humble Oil & Refining Company [Humble] for Dutch Gasunie and States Mines personnel to visit Humble gas facilities in Houston, Texas, and to meet managers of major U.S. gas pipeline companies. The visitors sought to learn more about the problems connected with "municipal distribution, appliance manufacture and gas merchandising and promotion."

Later, on December, 1963, the Jersey Executive Committee was told that Groningen gas reserves approximated 45 trillion cubic feet.* The Dutch government ruled that 18 trillion cubic feet should be reserved for local distribution, while the remainder could be exported. The announced reserve placed the Groningen field "among the largest in the world." The same account stated that the marketable volume of gas would remain small until 1965.[5] Later tests indicated that the field extended into a portion of

---

*The SONJ *Annual Report* for 1963 gave the figure as 39 trillion cubic feet.

Germany where Jersey and Shell exploration and producing affiliates already held concession rights.

A year later (1964), Jersey in its *Report* pronounced natural gas "a bright energy prospect in Europe," as it pointed out that Groningen gas alone possessed three times as much potential energy as did all Europe's known oil resources. The strategic location of the field in a "highly populated and industrialized area" provided a special advantage for distributing the gas. Work on a pipeline had already begun; some homes and plants had converted to natural gas; and already the Dutch were using 100 million cubic feet daily. Groningen gas for them meant an increasing amount of foreign exchange, hence, an improved economy and a better standard of living. It quickly moved their country toward becoming a self-sufficing energy state. A new era for natural gas was transforming the European market, and Jersey was determined to play an important role in it. Gasunie reported 700 miles of pipeline already laid in 1965. Of equal importance, 40 percent of the customers for natural gas in the Netherlands had converted from coke-oven gas to natural gas, while NAM marketers had concluded agreements for gas sales in Belgium, Germany, and France. The Jersey *Annual Report* for 1966 also noted that work on a major trunkline system for Groningen gas had been started and then stated with the same caution which usually accompanied such remarks, Jersey "anticipated that expanding gas sales from this source will have an increasing positive effect on company earnings."

Groningen gas sales did rise, despite a continuing decline in the market price for heating oils, which began in 1965. Exports rose from 14 million cubic feet per day in 1966 to 111 million in 1967. Although Jersey considered natural gas a relatively unknown energy source in the European market, the company nevertheless noted that "affiliates' revenues and earnings from gas sales now are beginning to reach impressive levels." Later, in 1968, some Jersey gas from offshore fields in Great Britain became available for marketing. Jersey and Shell by this time had located two major deposits of gas; one they termed the Leman Bank—the largest gas field found in the North Sea for many years, and the other, the Indefatigable field. In 1968, the share of Jersey from the two amounted to 13 million cubic feet per day.

Exploration by NAM drillers continued. Jersey also sent teams to explore the Dutch offshore area and when, in 1964, Great Britain ratified the United Nations Continental Shelf Convention* other companies joined in explora-

*The Convention required 22 signatures before it could take effect. The NAM discovery at Groningen, followed by other discoveries in Germany, increased interest in the potential of the North Sea for drilling both oil and gas. Thus, pressure from oil companies helped to encourage the British to sign. The Convention was drawn up in 1958.

tion of the Channel area. Zeb Mayhew remembered that activity. He said:

> We did some fiddling around there in the early and middle '60s, drilling some real shallow wells, gas wells in some peoples' back-yards, little tiny wells . . . that produced enough gas to cook a few meals on and then we gradually got a sort of greedy eye on the North Sea. . . .

The Jersey geologists and seismograph experts came to believe the English Channel sea beds dipped from both sides down towards the center with faults along the middle. They also believed that this structure extended north into the North Sea. So Mayhew went to talk about exploration licenses with John Angus Beckett, Under Secretary in Britain's Ministry of Power, and head of its Petroleum Division. He liked Beckett's attitude. Beckett posed a question to Mayhew, "Why should the British government put their money into foreign soil when they have a bunch of crazy companies just dying to do it?"

Mayhew, pleased with the answer, began to calculate the area for which he thought Jersey should apply. Then he went back to the British Minister of Power and "applied for the whole cockeyed North Sea area." He said: "We so shocked the minister . . . that he said, 'Well, let them have it if they want it. They're crazy enough, they must know something is there, nobody else has asked for it.'" Then the minister asked Beckett, "What about BP [British Petroleum Company, Ltd.]?"

Mayhew remembered that the Jersey move startled BP and Shell: "BP got caught with its socks completely down. Well the next morning BP and Shell were on our doorsteps." Mayhew added that both company exploration units "had been laughing at us for drilling these little gas wells." So the three companies began to negotiate. Then BP broke off. Jersey and Shell then entered a 50-50 partnership to cover the North Sea area. Finally the Jersey-Shell exploration and producing joint venture included the United Kingdom and Dutch portions of the North Sea. Mayhew said: "We were looking for oil. Of course we found it."

As a Shetland Islander once told Jersey Public Affairs Writer Sandford Brown, "Nothing you do is ever easy when you're working on the North Sea." The laconic speaker was guilty of understatement, for while seismic and other surveys could be made there with relative ease, North Sea drilling required new skills and technology in the oil industry. In fact, two of the reasons the search for oil and gas in the North Sea occurred were because the price of, and demand for energy had risen enough to make the gamble worthwhile. Here, history again became the stirring saga of men struggling against raw nature. Then too, as drillers, divers, and geologists searched the North Sea, they felt the ever present danger of the active mines left behind after World War II. But Jersey (Esso Petroleum Company, Limited) and Shell U.K. took the gamble and secured licenses to explore; so their organizations began to search despite long odds.

BP secured exploration licenses some time after those awarded the Jersey-Shell company, and it brought in a gas field in the North Sea in 1965. The Jersey-Shell company brought in the next gas field in that area—off the British Coast—in 1966.

Though small by U.S. standards of consumption, gas sales soon began to add significantly to the income of Jersey affiliates in Western Europe. New pipelines fanned out to Brussels, Frankfort, Hamburg, and throughout the Netherlands, while Paris was reached via a Belgian trunkline. In Amsterdam, the two power plants that supplied electricity to the city converted to natural gas and much of the smog which had long plagued the city disappeared. Here and elsewhere, gas helped industry to meet the new regulations setting plant waste emission standards. Sales of German gas also increased rapidly. Everywhere in Europe, marketers worked to secure contracts to supply fuel to plant and apartment complexes. Gas sales rose and so did sales of heating oil. Even so, natural gas only supplied 6 percent of Europe's energy requirements in 1970.

By that time, Jersey-Shell explorers had moved up the North Sea, to begin drilling off the Scottish Coast. They continued to look for oil but without much luck on the British side, at least for a few years. The first great oil find occurred in Norwegian waters by the Phillips Petroleum Company in 1969—the Ekofisk field—followed in 1970 by West Ekofisk. In that year, Jersey and Shell located the Auk (Josephine) field off Scotland. Experts estimated that Auk could begin producing in 1974. The Brent field, the largest yet discovered in British waters, was located in August, 1972. There, the drilling platform was established in one of the deepest sections of the North Sea, in 475 feet of water. It had been built to withstand waves 100 feet in height and gusts of wind up to 150 miles per hour. The Brent structure, the most expensive at the time,* required 24,000 tons of steel. This strike occurred 100 miles northwest of the Shetland Islands and after five weeks of exploratory drilling the Anglo-Shell-Esso combine reported tests indicated a production rate of 300,000 b/d could be expected, with a confirmed reserve of three billion barrels. The same statement gave 1976 as the earliest possible

---

*E.g., the fifty-three story Brent platforms cost $390 million each in 1975.

production date. The first oil produced would be loaded on tankers, weather permitting, while the task of laying a pipeline to the Shetlands went on.

Meanwhile, delineation drilling continued in the Cormorant field, again with little success. In this latter field, however, drillers did find oil in 1973. The partners fully expected to find as much oil there as they had located in the Brent field. Estimates of North Sea proven and probable reserves ranged from nine to twenty billion barrels in 1973.

While Jersey-Shell partners poured capital into North Sea exploration, they also continued to emphasize gas marketing. Jersey-interests' share in 1970 rose to almost a billion cubic feet of Groningen gas per day, plus 357 million cubic feet per day from Germany, and 170 million from offshore United Kingdom.

Even the slowing-down of the European economy in 1971 did not affect gas sales. In that year, Jersey's share was more than two billion cubic feet per day from Groningen, German, and British fields, with added benefits from the higher market price it brought. Apparently as one observer commented, "potential customers in Europe are ready and eager for natural gas."

The oil crisis of 1973 added to the fervor with which companies sought to acquire production from the North Sea area. By that time, natural gas was supplying the European market with the equivalent of almost a million barrels of oil per day. Jersey sales continued to increase, reaching 3.8 billion cubic feet per day by 1975. So earnings from all gas sales added substantially to Exxon [Standard Oil Company (New Jersey) became Exxon Corporation on November 1, 1972] earnings.

In that year, Jersey possessed 50 percent equity in the Dunlin, Auk, Brent, and Cormorant oil fields in the North Sea, plus the interest in Groningen and other gas fields in Holland. Jersey held a 10 percent interest in the largest Norwegian oil field, in which the government also owned 50 percent, and with seventeen other companies, had begun development of a North Sea crude oil transportation and delivery system, expected to become operative in 1977. The joint owners of this line expected to deliver one million barrels of oil per day from their North Sea sources. Quantities of wet gas had been located in several oil fields, such as Brent, in which Jersey had interests. Meanwhile, the price for oil quadrupled from 1970-1975, while that for gas increased sharply. The Netherlands had gas to export, while Germany still imported more than half the gas it needed. Norway enjoyed the fortunate position of energy self-sufficiency, but in 1975, the United Kingdom produced only about one-half its requirements. Of course, all of this complex pursuit of local energy sources began in a sandstone formation NAM drillers located after what appeared to be thirteen years of fruitless search. In the decade that followed much was accomplished.

Yet for Jersey, as for other oil companies, with billions invested in Europe and the North Sea, problems still remained. Host governments continued to seek an increased share of the income from their natural resources, once production had begun. Thus, the high profit levels generated by the increasing prices of Groningen gas led to revisions of the contracts and revenue agreements between the NAM partners and the Netherlands government in 1971 and again in 1972. In Britain in 1975, impending changes in the scale of royalties and corporate income taxes (to take effect in 1975) increased the government's share to about 75 percent. In Norway, legislation passed parliament that enabled that government's share to reach 75 percent also. As a result of those and other governmental demands, North Sea exploration slowed down, though the actual production of oil and gas remained constant.[6]

X X X

By itself, the full story of exploration for oil and gas in Europe between 1950 and 1975 would take volumes to relate. As in most cases where oil or gas has been found, a large element of chance affected the Groningen discovery. That field, approximately twenty miles square, was under a geological dome that flattened out in all directions. In those upper rock layers, only small quantities of gas were ever found. But after the strike, subsequent drilling in other Permian rock further away led to significant new finds of both gas and oil. This field alone increased Europe's energy supply from sources close at hand at a very critical time in European history. The fact that it stimulated offshore exploration along the English Channel area, and in the Channel, adds greatly to the significance of that find. Later discoveries of oil and gas in the North Sea occurred largely because of what exploration experts learned at Groningen and in the Channel. And when in 1973, the Organization of Petroleum Exporting Countries [OPEC] gained control of petroleum prices and demonstrated its new power by unilaterally increasing the price of oil, North Sea oil production became quite profitable. And if Exxon and other participating oil companies profited by these discoveries, so did the governments of Western Europe and the peoples of the Free World. These gas and oil finds helped the West to maintain its economic stability in a time of uncertainty, especially during the years of the petroleum revolution following 1970.

Chapter 8

Jersey Refining in Europe, 1950-1975

WORLD WAR II AND THE YEARS immediately afterward marked a new beginning for the Exxon Corporation* in Europe. Exxon's affiliates rebuilt refineries, in most cases from total ruins, and then immediately began to enlarge and remodel them in response to the changing consumer market and rapid advances in technology. In Europe, between 1950 and 1975, oil displaced coal as the primary energy source. War damage to the coal mines, plus the new availability of low cost Middle East oil brought the liquid fuel to every nation, even though supply disruptions unsettled the European markets.

These changes had lasting effects on Exxon affiliates. They required an almost complete transformation of refining operations in Europe. By the time of the oil crisis of 1973-1974—after some thirty years of post-war change—these European affiliates operated a technically sophisticated and truly massive refining complex on their continent—refineries vastly different from those of the prewar years. This new empire of modern refining, Exxon had built on the bare ruins left by modern history's most destructive European war.

1945-1950

In the war's immediate aftermath, widespread destruction and severe dislocations characterized the entire European economy; as American Secretary of State George C. Marshall warned in 1947, "the entire fabric" of Europe's economy verged on dissolution.[1] When representatives of the Standard Oil Company (New Jersey) entered Continental Europe to reconnoiter in 1945, they found seven of the ten Jersey affiliates' refineries destroyed—refineries that had represented more than 50 percent of total

---

*Standard Oil Company (New Jersey) [Jersey] became Exxon Corporation [Exxon] on November 1, 1972.

European capacity. Again, entire units of Standard Française des Pétroles, S.A.'s [Esso France] Port Jerome refinery had been removed piece by piece from France by the Germans, who shipped them for reconstruction elsewhere.

Human destruction was great as well. Affiliate work forces had been scattered by the war, and the fate of many key persons remained a mystery. And even the three refineries that remained intact—Fawley in England, Atlas (Antwerp) in Belgium, and Kalundborg (asphalt plant) in Denmark—could not be fully utilized because they lacked crude oil for processing.[2]

In the hectic final six months of 1945, Jersey officials boldly confronted decisions about the future of the company's European business and especially about the role of refining in that future. Given the dim prospects for restoring integrated operations, should the company rebuild its refineries? If so, what scale of operation and what level of technical sophistication would be best for each affiliate and for itself? What portion of Europe's product needs should be met by local refineries? What share should be imported? For everyone responsible, the simmering European political environment that surrounded the economic dislocation served largely to complicate every effort to find answers to these questions. In addition, all Europe faced serious financial problems centering on the obvious shortage of the foreign currency required for economic recovery. Yet, despite so much uncertainty about the future—including unpredictable government policies on dollar exchange and repatriation of profits—Jersey's Board made every necessary decision with good speed. Finally, it was decided the parent company would construct an entirely new refining industry to serve its European affiliates, and do so, moreover, in record time.

This decision meant more than simply beginning a large scale construction program under difficult and uncertain conditions; it also entailed a major shift in traditional worldwide patterns of oil flow. Before the war, most refining occurred near the source of production, and since there was little production in Western Europe, Jersey's affiliates there actually had few refineries. These few included several small crude topping plants plus two facilities limited to the production of asphalt and related by-products. With a total crude refining volume of only 60,000 barrels per day [b/d], Jersey's European refineries could supply a mere fraction of total regional demand. Two-thirds of the petroleum products marketed in Europe came from the Western Hemisphere.[3] Postwar conditions, however, dictated a fundamental change.

First, because of the shortage of foreign exchange, Jersey faced the large scale problem not only of supplying the petroleum products necessary for economic recovery, but of doing so at the lowest possible cost in foreign currencies. In every country, this consideration influenced the company's decisions about both the source of crude and the pattern of refinery construction. Using foreign exchange calculations of that time, Middle Eastern oil was considered to have a higher pound sterling value than crude from the Western Hemisphere, Jersey's traditional source for European markets. So Middle Eastern oil began to supplant Western Hemisphere crude in those markets.

This move fitted perfectly with the Jersey worldwide supply strategy, simultaneously freeing Western Hemisphere capacity for use within the Americas and permitting expansion of the outlet for crude oil from the company's growing Middle Eastern production. At that time, however, Jersey did not own sufficient refining capacity in the Middle East with which to meet Europe's increasing demand for products. Again, the solution to this problem neatly coincided with the general concern for protecting Europe's foreign exchange balances. Since two dollars worth of crude oil would yield products worth approximately three dollars, by shifting the refining of most products to Europe, Jersey could take a major step toward conserving scarce exchange currencies.[4] But so fundamental an alteration in the world's oil map also necessitated major changes in both the refining and transportation functions.

To begin implementing its decision, Jersey arranged for the immediate delivery of the crude oil needed to restore operations at the few usable European refineries, while it began planning the rebuilding of other facilities. By 1946, over half of its prewar European refining capacity had again become operative, and in the following year that proportion reached almost three-fourths.[5] Of course, Jersey's ultimate goal represented something larger than the rehabilitation of prewar facilities; for the company realized that it would need refining capacity several times greater and equipment much more technologically advanced than that of the old refineries if its European affiliates were to meet their own postwar needs. The most ambitious single project undertaken by Jersey at that time centered on the expansion of the Fawley refinery, on the southern coast of England. There, an increase in crude capacity from 18,000 b/d to 110,000 b/d made the refinery the largest in Europe and gave it alone almost twice the total capacity of all Jersey's European prewar refineries together.

Jersey's emphasis on saving foreign exchange and on capitalizing the pound sterling value of Middle Eastern oil soon proved critical to the economy of Great Britain. So a special series of consultations between Anglo-American Oil Company, Limited and the British Labour government began in 1945. Fears that the government might nationalize the petroleum industry made negotiations more difficult, but the two parties finally reached agreement in 1949. Yet, because of the emergency need to begin the Fawley project, Jersey took the extraordinary step of making firm purchasing commitments even before the final agreement. As a result, construction proceeded at a record pace, and a new Fawley refinery went on stream in late June, 1951, six months ahead of schedule. It represented sufficient capacity to meet almost all of Great Britain's, Denmark's, and Ireland's demands. Moreover, the refining of Middle East crude at this plant helped largely to alter Britain's foreign exchange balances and hastened the end of British gasoline rationing. In addition, Fawley included technical facilities that stimulated the growth of a new petrochemical industry in Britain.[6] The overall result of Fawley's rapid reconstruction and of the general growth of Britain's petroleum refining industry provided a general stimulus to its economic recovery.

Clearly Jersey could have opened the door to a plethora of problems had it acted in too much haste or with too little care in reconstructing the Fawley or other plants or in building completely new ones. Thus steps were taken to beautify the plants' surroundings wherever Jersey undertook construction, and Fawley proved the first. In 1951, Anglo-American had hundreds of trees planted to screen the refinery; then after some land-leveling efforts, the entire area was grass seeded. And for the final pastoral touch, sheep were purchased to keep down the grass and a shepherd hired to watch them.

These bucolic touches hid the fact that in more far seeing moves, the company had expended large sums for pollution control devices, effluent treatment, control of chimney-stack discharges, smell and noise control and in desulphurization units. All of these innovative ideas came more than a decade before the environmentalist movement really got underway and at least a decade before state or national legislatures began to react to the problem of global pollution.

On the continent, meanwhile, a new refining industry grew from the ashes of existing plants and from the purchase of other facilities. Equipment from France's Port Jerome refinery had to be tracked down and repatriated from the Central European refineries where the Nazis had moved it. Port Jerome resumed operations in 1946, and six years later, the capacity had been doubled to 50,000 b/d. In combination with a tiny 600 b/d refinery at La Mailleraye, Port Jerome allowed Esso France to supply much of France's needs, using Middle East crude. In occupied Germany, the Ebano asphalt plant in Hamburg restarted in 1947, and over the next four years was altered to a general refinery enlarged to three times its original capacity.[7] Elsewhere, Jersey was soon able to restart refineries in Vallø, Norway; Antwerp, Belgium; Teleajen, Rumania; and the Kalundborg asphalt plant in Denmark. [Rumania nationalized Teleajen in 1948.]

The restoration of the Italian refineries proved more difficult. Because of the political uncertainty surrounding the future of its San Sabba refinery in Trieste, Standard-Italo Americana Petroli-Società per Azioni [SIAP], a Jersey affiliate in Italy, began to seek alternate refining capacity. In 1947, the affiliate reached an agreement in principle with Azienda Nazionale Idrogenazione Combustibili [ANIC], an Italian state-owned company, to share in the restoration and modernization of two ANIC refineries. Three years later, the two companies set up a joint company, Stanic-Industria Petrolifera Società per Azioni [STANIC], to buy and operate ANIC's Bari and Leghorn refineries. Under SIAP's management, Bari was modernized and Leghorn was expanded to 21,500 b/d, giving Jersey twice its prewar capacity in Italy. These moves to increase Italy's refining capacity promised to reduce by $4 million per year that nation's foreign exchange balance of trade against the importation of finished products. In addition, the arrangement had the blessing of the U.S. Economic Cooperation Administration [ECA].[8]

Jersey's European affiliates by 1948 surpassed their prewar crude refining capacity, and two years later, without the use of Teleajen, they exceeded that capacity by more than 60 percent, with 100,000 b/d of capacity. Moreover, by 1950, three-fourths of Jersey's supply of crude oil for Europe came from the Middle East.[9] But even such increases in refinery volume could not keep pace with Europe's rapidly increasing demand for petroleum products; more development would be required.

X X X

1950-1960

The factors that dictated rapid rebuilding and expansion after the war continued to operate during the 1950s, but they were joined by new forces that intensified the need for additional European refining capacity. The same considerations of national interest—especially the scarcity of foreign exchange—led Jersey to set a goal of self-sufficiency in the European market, which the unprecedented growth of demand there made difficult to achieve. Before the war, oil in bulk had been used primarily for bunkering ocean-going ships, but with its increased use for heat, power, transportation, and farming, it quickly gained a larger share of the European energy market. Extensive war damage to coal mines and outdated extraction methods caused their restoration to be slow and expensive, so coal provided little postwar competition in most of Western Europe. Furthermore, the growth of Europe's petrochemical industry also began to have a modest effect on oil supplies. In 1950, Europe's total demand for petroleum of 1,300,000 b/d already dwarfed the 800,000 b/d of 1938, and at that time the shift to oil was just beginning.[10] The Organization for European Economic Cooperation [OEEC] estimated, in 1955, that petroleum would increase its share of the energy market in Europe from 20 percent to 25 percent within a decade.* In fact, that prediction grossly underestimated the shift; by 1960, oil supplied 32 percent of Europe's energy. Thus, when combined with the rapidly expanding economy, the change to oil guaranteed a continually growing market for new refining capacity. For example, in 1953, officials in Great Britain predicted a doubling of the demand for oil by 1963.[11]

For Jersey this growth pattern, while straining the company's ability to respond, corresponded well with other overall corporate needs. By the 1950s, the Middle East had become an important source not only for crude oil but also for profits. The continued rapid expansion of Aramco [Arabian American Oil Company]—in which Jersey possessed a 30 percent share—plus supplies of oil purchased from Anglo-Iranian Oil Company, Limited,** and after 1954, the availability of oil derived from participation in the Iranian Consortium, required that the company sell ever-increasing volumes of Middle East oil. So Jersey's basic strategy in Europe came to center on achieving "the maximum outlet for its crude," and the company geared its marketing, refining, and transportation functions primarily to that all-important goal.[12] Where spot surpluses of certain products arose, affiliates were encouraged to search out buyers rather than to eliminate such a surplus through any reduction of crude runs in the refinery involved. Refinery cutbacks were the least acceptable alternative in the interests of Jersey.[13]

Caught between the marketers' unrelenting drive to increase sales and the seemingly unlimited sources of supply in the Middle East, Jersey's refiners worked diligently to keep refining capacity even with a growing demand. This important effort received special attention not only from Jersey's European affiliates but also from both the Coordination Committee for Europe and the parent company's Coordination Committee. The latter conducted special studies of European refining needs in 1952 and 1955, and continued to emphasize European considerations in a 1956 study of worldwide supply and refining needs.[14] All this time, the European growth rate remained

---

*OEEC became the Organization for Economic Cooperation and Development [OECD] in 1961.

**Anglo-Iranian became British Petroleum Company, Ltd. [BP] in 1954.

phenomenal. When the new Fawley refinery went on stream, in 1951, Jersey claimed a crude refining capacity of 200,000 b/d in Europe. That figure rose to 280,000 b/d by 1953, which was more than four times the total of European affiliates' prewar capacity. Total capacity reached 550,000 b/d in 1958, with an additional 286,000 b/d of capacity scheduled for completion by the end of 1960. Although Europe still imported some products from the Western Hemisphere, dependence on that traditional source steadily declined.[15]

This remarkable growth in total refining capacity came both through expansion of existing refineries and construction of new units. In the early 1950s, Jersey's Italian affiliate added two new refineries; Augusta with a crude capacity of 22,500 b/d and Trecate (70 percent owned) with 20,000 b/d capacity. Then in response to increasing industrial demands and foreign exchange pressures, the company approved construction of a 25,000 b/d refinery at Antwerp. They designed this plant to make the Benelux* countries self-sufficient in refining, and as an indication of the trend toward greater technological sophistication in European refining, the plans included catalytic cracking facilities. In the late 1950s, additional refineries were added to Jersey's growing European plant; in 1958, at Cologne, West Germany, an 87,000 b/d refinery incorporating a powerformer and hydrofiners went on stream. This new refinery also included integrated petrochemical facilities.

The following year a 46,000 b/d refinery was completed at Bordeaux, to refine crude oil from the French affiliate's field at Parentis, which had started producing in 1954, being the first field in Europe, outside of Soviet Russia and its satellites, to yield oil in commercial quantities. Esso Teoranta (Ireland) brought on stream a 56,000 b/d refinery at Whitegate, in which the affiliate held a 40 percent interest. With its powerformer and hydrofiner, this new facility helped to make the company self-sufficient in the Eire market.[16]

In addition to these new refineries, company affiliates added new capacity and more advanced equipment to existing facilities. Even before the Fawley refinery went on stream in 1951, the company had already begun to consider expanding it, and in 1957, a new 75,000 b/d atmospheric pipestill began operating. This raised the refinery's crude capacity to 185,000 b/d. Fawley also added a hydroformer and a powerformer and expanded its catalytic cracking capacity, thereby increasing its ability to produce both higher value light products and higher octane gasoline, shortages of which became a problem throughout Europe. Although no other European refinery could match

*Belgium, Netherlands, and Luxembourg.

Fawley's overall size in the 1950s, several did expand dramatically. Esso Aktiengesellschaft's [Esso A.G. or Esso Germany] Ebano refinery at Hamburg grew to 40,000 b/d crude capacity in 1951 and to 54,000 b/d by 1959. The addition of fluid catalytic cracking facilities and a hydroformer also increased the ability of that refinery to convert a greater proportion of its crude oil to higher grade products. STANIC's Bari and Leghorn refineries in Italy were expanded in 1956 and 1957 to 41,000 b/d and 42,000 b/d respectively. At the same time, they added facilities necessary for an increased production of high octane gasoline. Even Jersey's newest European refineries were enlarged during the decade—Antwerp to 40,000 b/d in 1958 and Trecate to 26,000 b/d with added catalytic cracking capacity. Although the Port Jerome refinery did not increase in crude capacity, it did add fluid catalytic cracking facilities, a reformer, and a powerformer to upgrade its conversion capacity.[17]

In addition to changes aimed at expanding and improving petroleum product production, several Jersey installations added facilities that brought them into the petrochemical business. This entry allowed them to use refining capacity more effectively and to increase crude oil outlets. Fawley led the way by constructing three new chemical units in 1958, but Esso France also began to participate by supplying ethylene and butadiene for a synthetic rubber plant. The new Cologne refinery joined the trend by incorporating chemical facilities in its original design.[18]

Of course this expansion could not occur without creating problems. Foreign exchange considerations continued to figure prominently in most plans for refinery expansion, with Jersey always trying to hold dollar investments to a minimum as well as seeking host government permission to use foreign exchange for the purchase of all necessary crude oil supplies, equipment, services not available in Europe, and government guarantees of its right to repatriate dividends and loan payments.[19] And Jersey, despite its substantial efforts, continually had difficulty keeping up with Europe's growing demand. The 1952 study of refining needs noted that even with planned expansions at Fawley and elsewhere, the Scandinavian affiliates would still need an additional 94,000 b/d of crude refining capacity by 1960.

Meanwhile, problems encountered in the product mix generated by Europe's refineries in processing Middle East crude proved even more disturbing. The fundamental difficulty was that the European demand for middle distillates, especially fuel oil, remained out of proportion to the demand for gasoline. Compared to the United States, Europe used significantly less gasoline than fuel oil. As a result, while European affiliates were expected to be self-sufficient in gasoline by 1960, middle distillates remained in short supply. Larger quantities had to be imported each year. The special study of European refining needs made in 1955 addressed this problem, by suggesting several possible ways either to increase outlets for gasoline or to change the product mix in favor of middle distillates. Finally, this study of the problem of adequate fuel oil was to prove of little value to Jersey, for in 1959, a jointly owned company* of Jersey and Royal Dutch-Shell Group [Shell] brought in what ultimately was established as a tremendous dry gas field in the Netherlands province of Groningen. Exploitation of this discovery assumed major importance in the 1960s. This "find" enabled Jersey marketers to push natural gas as an alternate energy source for heating. But the shortage of fuel oil did continue through the 1950s. As a result, Jersey suffered from the increasing tendency of European governments, especially Italy, to involve themselves in the petroleum business. Although the company considered all of the activities of the Ente Nazionale Idrocarburi [ENI] a threat to private companies, refining offered a special concern because of its joint operation of STANIC with one of Italy's government companies. By the mid-1950s, Jersey sought to extricate itself from STANIC, by buying out ANIC, but the company failed to do so at that time.[20]

X   X   X

1960-1965

During the 1960s Europe continued both its economic upsurge and the shift to petroleum as the predominant energy source. We have noted that this shift to petroleum required Jersey's refiners to respond to that rapidly expanding market. In fact, as the European Common Market moved toward greater economic integration, that stimulus combined with the on-going shift to petroleum products** to double the demand for oil in Europe during a decade. Again, as in the 1950s, Jersey met the challenge through the construction of new refineries and the expansion of existing ones, always with the fundamental goal of moving the maximum amount of Middle East crude oil. Now, Jersey sought to expand these European outlets even more rapidly by moving into new markets. The pressure to do so increased when the company discovered new supplies of oil in Libya in 1959. As crude oil from that country quickly became available, Jersey pressed to find markets for it that would not displace the still-growing supplies from traditional

*N.V. Nederlandse Aardolie Maatschappij [NAM].

**As has been noted, the price of petroleum remained low throughout this decade.

sources.[21] Finally, the early 1960s signalled a major change in the company's chemical business, which had become closely entwined with its refining function. In 1963, the parent company separated the chemical business from that of oil.

During the first half of the 1960s, the European affiliates continued to add new large-scale refineries in response to the phenomenal growth of their markets. Jersey affiliates brought on stream three new facilities in 1960: these raised the company's total crude refining capacity to 800,000 b/d. This amount equalled prewar Europe's total demand for petroleum. The 40,000 b/d Slagen refinery near Oslo, Norway, came on stream to fill part of the refining shortage of Scandinavian affiliates.[22] Milford Haven, Esso's second refinery in Great Britain, added 100,000 b/d of crude capacity.[23] The new 100,000 b/d Rotterdam refinery signified not only a rapid growth of the economy of the Benelux countries but also the overall growth of European demand and the increasing integration of Europe's economy. Original plans for the Rotterdam refinery called for a 40-50,000 b/d facility, but the final design reflected a decision to supply part of the Swedish market as well as much of the Benelux demand. All three of these new refineries were constructed with powerformer capacity, an indication of the advanced technology required by the European market.[24] In 1962 and 1963, three more refineries were added: in Germany, the Karlsruhe refinery with two hydrofiners, a powerformer, vacuum pipestill, and thermal cracker—eventually reaching a running level of 68,000 b/d of crude. A second German refinery, Ingolstadt, gave Esso Germany an additional 74,000 b/d of crude capacity as well as more hydrofining capacity. In both cases the selection of equipment gave particular attention to the special characteristics of Libyan oil with the goal of using "the maximum amount of Libyan crude oil practical [sic]."[25] Esso International Inc. purchased the 24,000 b/d Tidewater Oil Company refinery at Kalundborg (Denmark) and made plans to more than double its capacity. In 1965, Esso Standard S.A.F. [ESSAF] began to operate a new 60,000 b/d conversion refinery at Fos. In addition to a powerformer and a fluid catalytic cracking unit, this installation had other facilities that allowed an even greater yield of high grade products from each barrel of crude oil. The construction of this type of refinery emphasized the growth of the market that made self-sufficiency in more products economically feasible. It also included the new technical equipment which Jersey then began to use in its other European refineries.[26]

In the Benelux countries, both the Antwerp and Rotterdam refineries expanded by adding new units. Antwerp increased its crude capacity from 60,000 b/d to 100,000 b/d while adding a vacuum pipestill plus solvents production facilities. At the same time, Rotterdam increased its crude capacity to 153,000 b/d, expanded its powerformer capacity, and added an aromatics plant.[27] Esso Standard Italy [ESI] closed its aging San Sabba refinery in Trieste, and ended its partnership in STANIC. Too, it purchased California-Texas Oil Company, Ltd.'s remaining interest in the Trecate refinery in 1962, and undertook an expansion of that facility from 26,000 b/d to 117,000 b/d crude capacity by adding a hydrofiner and two powerformers. When the Trecate expansion came on stream in 1964, the affiliate also added a new 4,500 b/d lube plant at Augusta, thus allowing production of several products in Italy that had previously been imported. The rate of refinery expansion in Italy actually led to a shortage of skilled refiners—a situation not unusual in such boom periods.[28]

While no other European affiliate matched ESI in its rate of refining expansion, others did experience substantial growth. The new Milford Haven refinery increased its crude capacity to 130,000 b/d in 1964, and its powerformer capacity to 22,500 b/d the following year. Slagen refinery also grew quickly, jumping to 58,000 b/d capacity in the same year. In Germany, a second atmospheric pipestill raised crude capacity at Cologne to 119,000 b/d while the addition of a 17,000 b/d vacuum pipestill permitted better utilization of crude oil. Ingolstadt added an asphalt plant, and even the older Hamburg refinery added a lube oil hydrofiner and, through debottlenecking, raised its crude refining capacity to 65,000 b/d. Although it did not increase its crude capacity, the Port Jerome refinery added de-waxing capacity.[29] This change permitted the processing of greater quantities of Libyan crude.

The rapid expansion of Jersey's European refining plant and capacity compelled the extension of Esso Research and Engineering Company's [ERE] operations to Europe. Although never intended to handle all the problems encountered regarding refining or refinery construction, having ERE personnel available for consultation proved beneficial.[30]

This phenomenal growth of European refineries could not utilize all the crude oil produced and available from Jersey's worldwide sources before 1970. So the company sought to expand in parts of Europe where it had previously had no significant marketing operations, and to capitalize on the advantages of refining closer to markets. During the early 1960s the company offered to construct refineries as well as other industrial facilities in order to enter the markets of Greece and Spain. It also considered the building of a refinery in Portugal, but that plan never materialized.[31] Jersey's investments in Spain also included a variety of facilities, but the principal attraction there for the company was entry into a new market that would guarantee an outlet for the plentiful supply of Eastern Hemisphere crude oil. Even a decision to increase the portion of the market reserved for the government from 20 percent to 45 percent failed to discourage Jersey.[32]

Yet the actions of European governments proved a continuing and growing source of concern for Jersey's affiliates on that continent. The company found itself competing with the Italian state company, ENI, especially in North Africa. On the continent, the government of France gave preferential treatment to Union Generale de Pétrole [U.G.P.], a state-owned company, in the granting of refining licenses. Jersey argued in vain that such actions violated the spirit of the Treaty of Rome that, in theory, governed the transition period of the Common Market.[33] Yet, with petroleum and gas gaining an ever increasing share of the European energy market, Jersey could only expect greater government intervention in the industry either through public corporations or regulation.

X X X

1965-1969

During the last half of the 1960s, Jersey continued to increase its European refining capacity dramatically, still in response to a sustained growth of the European market. Now, however, this expansion did not take the form of building new refineries. Instead, the company's refining function benefited from the formation of Esso Europe Inc. in 1966, because the advent of the regional company allowed better integration of all existing refineries—an important factor in the development of a greater variety of products, and in anticipating emerging demands in the area of pollution control and for more efficient operation.

The largest of Jersey's new refinery ventures during this period came in France, where the aging of the Port Jerome refinery called for additional distillation and cracking capacity in the eastern part of the country. Esso France's plans for a new 60,000 b/d refinery were undercut, however, by the granting of a government refining license to Compagnie Française de Raffinage [CFR], an affiliate of the state-owned Compagnie Française des Pétroles [CFP]. As a result, Esso France became a 40 percent partner in a 90,000 b/d refinery near Metz (with CFP and Elf Union, another government-controlled corporation).[34]

Jersey affiliates also entered partnerships to run refineries in Germany and Switzerland. Elwerath A.G., a company in which Esso Germany owned an equal share with Deutsch Shell A.G., purchased a 38 percent interest in Gewerkschaft Erdol-Raffinerie Deurag-Nerag in 1968. Esso

Standard (Switzerland) [Esso Switzerland] joined with seven partners to set up Raffinerie du Sud-Ouest S.A., which purchased a refinery near Collombey, Switzerland, in 1966.[35]

These refineries gave Jersey affiliates some additional crude refining capacity, but these gains proved modest when compared with those coming from the expansion of already existing facilities. The largest single project boosted Esso-Nederland N.V. [Esso Netherlands] Rotterdam refinery's crude capacity from 153,000 b/d to 338,000 b/d and added a second powerformer with 20,700 b/d capacity. This more than 100 percent increase in capacity indicated the growth of the European market, its greater integration under the Common Market, and Jersey's increased ability to take advantage of that integration by means of Esso Europe. The dedication of part of the powerformer capacity to the production of feedstocks for Esso Netherlands' Rotterdam chemical plant reflected the growing importance of Jersey's chemical business in the refining function. Finally, the inclusion of a 76 ton per day desulphurization plant as part of the project reflected the company's response to the growing environmental concerns.[36]

Although the Rotterdam expansion dwarfed all other Jersey refinery projects, expansions at the Augusta, Fawley, Port Jerome, and Karlsruhe refineries and smaller scale additions at several other locations showed the continued pressure to keep up with Europe's growing demand. ESI, in 1965, brought on stream at Augusta a new 100,000 b/d atmospheric pipestill which it increased in 1968 to 112,000 b/d capacity. In the intervening years, a catalytic cracking unit and an alkylation unit were added, thus implementing the refinery's capacity for more complex processing.[37] Fawley added 105,000 b/d of crude refining capacity in 1966, followed by a second powerformer and a variety of new chemical and computer facilities. The modernization and expansion of the Port Jerome refinery the same year improved the profitability of that plant while the expansion of its dewaxing facilities further increased its ability to use crude oil from Libya—an important consideration for Jersey's European refiners both because of the need to find markets for growing supplies of crude oil and because of the increasing pressure to reduce the sulphur content of fuel products.[38] Esso Germany's Karlsruhe refinery also experienced major expansion; in addition to a second atmospheric pipestill there that raised crude refining capacity from 100,000 b/d to 170,000 b/d, the addition included a second powerformer, four hydrofiners, and other facilities that permitted higher conversion of crude oil into products.[39]

Although Esso Denmark's Kalundborg refinery was the only other Jersey refinery to add substantial new crude refining capacity during the period before 1969 (increasing from 30,000 b/d to 72,000 b/d), a variety of projects greatly increased the capacity of other refineries to produce higher octane gasoline and a growing list of more complex products.[40] Slagen in Norway, and Whitegate in Eire, both added powerformer capacity, and at the Trecate plant in Italy, a new hydrofiner came on stream. The Ingolstadt refinery in Germany increased its hydrofining capacity and added a catalytic cracking unit and second vacuum pipestill, while the Cologne refinery added capacity for the production of asphalt and chemical feedstocks. At Hamburg, a major new lube and grease plant capable of producing hundreds of grades of lubricants for the domestic and export market went on stream as part of a larger effort to increase the production of lubricants for the European market.[41]

The increasing European demand required that Jersey's affiliates make investments in response to the lessons of the 1967 Middle East crisis that closed the Suez Canal. In the wake of the crisis several European nations enacted legislation imposing compulsory storage requirements on companies. Furthermore, the closing of the canal accelerated the shift to supertankers, which required larger berthing facilities near several refineries.[42]

While responding to changes in its existing markets, Jersey also brought on stream new refineries in areas it wished to enter, but these new ventures were not without their difficulties. The 55,000 b/d refinery at Thessaloniki, Greece, financed and built by Jersey under the terms of the agreement between Thomas A. Pappas and the Greek government, went on stream in 1966. Jersey later moved to acquire title to not only the refinery but also to the rest of the hydrocarbon complex that included ammonia and petrochemical plants. Two years later, the refinery finally became profitable, but low product prices in Greece and higher transportation costs resulting from the closing of the Suez Canal made even marginal profitability difficult to achieve. Moreover, while Jersey's plans had been based upon sharing the growth of the Greek market only with the state-owned refinery, the Greek government appeared ready to approve a third refinery, to be financed by Aristotle Onassis, that would undercut those plans. The unsatisfactory performance of the rest of the Greek industrial complex compounded these problems. Neither the ammonia plant nor the petrochemical facilities appeared likely to make a profit for some time, while a steel plant in the same complex absorbed increasing amounts of capital thereby accumulating debts that would serve to make profitable operation difficult. In the meantime Jersey hoped to increase the capacity of the refinery to 64,400 b/d through debottlenecking.[43]

Jersey also encountered difficulties with its Spanish refining investment.

When the refinery at Castellon de la Plana came on stream in 1967, its failure to generate a profit accentuated Jersey's consistent desire to gain full equity ownership through purchasing the shares of the Spanish partners. The Spanish government thwarted these plans, however, by adopting the Refining Decree of March, 1968, which limited to 40 percent foreign ownership in refineries. Thus Jersey had to reduce its interest in Esso Standard Española, S.A. [Esso Spain] refining to that level, and this action in turn complicated plans for future development because the local partners would have to provide the 60 percent additional equity capital.[44]

By the end of the 1960s, Jersey had doubled its European refining capacity from 800,000 b/d to more than 2,000,000 b/d (app.), broadened its product refining and substantially updated the technology of its entire operation. While it expected Europe's shift to petroleum to decelerate in the coming decade, the company did anticipate a continuation of some trends from the past. But the 1970s brought a dramatic and unexpected break with previous postwar experiences.

X X X

1969-1972

The growth of demand for petroleum products and the consequent increase in Jersey's refining capacity continued to mark the European refining function through 1972. During the 1960s, oil's share of the European energy market had already doubled and stood at 60 percent at the end of the decade. Although the percentage increase was expected to be much less during the 1970s, the overall growth of the market continued the pressure to expand.[45] Refiners, however, faced other considerations besides keeping capacity up to demand. Many of these problems arose as a result of Europe's increasing concern with environmental issues and changes in the demand for different products in the European market.

As the 1970s began, Jersey's refiners were still responding to rapidly increasing European demand, adding substantial new refining capacity each year with the expectation that capacity would have to double during the decade. In fact, Jersey's Management Committee learned in 1970 that because of its previous planning tools, the company had "traditionally underforecast demand" in Europe.[46] In response, the committee approved the use of "a trend line based on historic actual growth" for planning future refinery capacity additions in Europe. On the basis of this plan, the company considered adding another 350,000 to 400,000 b/d of crude refining capacity in

1974, in addition to the 750,000 b/d increase already planned for 1971-1973.[47]

The largest refinery expansion projects of 1969-1972 were major overhauls of the Hamburg and Trecate refineries. In addition to 125,000 b/d of crude refining capacity, the German refinery got a new vacuum distillation unit, a powerformer, hydrofiners, a lube vacuum pipestill and phenolfiner, and other facilities including a central control house with a process computer. A new pipestill at Trecate brought crude refining capacity there to 222,000 b/d in 1972, while other units improved the ability of the refinery to make better use of crude oil.[48]

In addition to these larger projects, the Slagen, Fawley, and Collombey refineries all added substantial new capacity, while debottlenecking operations at Fos, Thessaloniki, and Augusta created additional capacity at those sites. Catalytic cracking facilities at Augusta and Fawley were expanded, and Fawley added new chemical facilities. Finally, Port Jerome joined Fawley and Hamburg among the European refineries with computerized controls, thereby increasing the efficiency of operation.[49]

Despite the continued rapid growth of demand for petroleum products in Europe, the development of an environmental movement there began to place additional burdens on the company's refiners. Concerns about air and water quality were not new, and the company had long been taking actions to guard against pollution. Esso Petroleum Company, Limited, as early as 1964, had participated in the formation of a joint industry-government group to establish codes of safe practice for pipelines and storage facilities. The new refineries already had some facilities for reducing air and water pollution, and comparable investments had been made in the older plants. For example, the Karlsruhe, West Germany refinery already easily satisfied the most stringent government standards in Europe, cleaning the water it used with mechanical skimmers, sedimentation, aeration, and biologically active sludge.[50]

While the prospect of increasingly rigid governmental standards would require greater investments that might not be easily recovered in the competitive European market, the imposition of limits on the sulphur content of fuel oil, and the lead content of gasoline sold there proved to be of even greater concern in the early 1970s. Low-lead gasoline could be achieved by additional plant equipment; this, however, did not place Jersey at a competitive disadvantage since the same standards applied equally to all refiners. But sulphur standards for products presented a greater problem. The company already had desulphurization equipment installed at Rotterdam, but that would not be sufficient in case of the widespread adoption of tighter sulphur standards for fuel oil. This meant that the company would have either to add expensive desulphurization capacity at other refineries or increase its reliance on low-sulphur crude oil. Since Jersey's European affiliates remained largely dependent on crude oil from the Middle East, this placed it at a relative disadvantage compared to some of its competitors in Europe. Thus, the addition of desulphurization capacity would not be a financially feasible, nor an easy solution. The company would either have to find new sources of low-sulphur crude by 1975 or face a loss of market position in Europe.[51]

The sulphur problem pointed to a more fundamental problem concerning fuel oil. The company forecast a downward trend in total fuel oil sales in Europe in a study made in 1969. For the refiners, this created an unprecedented situation. After decades of trying to handle maximum quantities of fuel oil, they now had to plan new investments to increase yield through better conversion of crude oil. Otherwise, the company had to consider reducing its market share of certain products in Europe. Thus, environmental and economic considerations in combination threatened to reverse the direction of Jersey policy in Europe since World War II; no longer could the company automatically assume that any investment could be justified so long as it increased the outlet for crude oil.

It should also be noted that in the 1960s and early 1970s the availability and accessibility of Groningen gas—at rates competitive with heating oil—led to the expansion of that energy source, and to a corresponding decline in the demand for heating oil. The movement to protect the world environment also developed rapidly. Though Jersey had anticipated this for some time and had moved to provide cleaner fuels even before the governments passed legislation requiring them to do so, more change became necessary. So gas provided the quick and one obvious solution to the problem. But as the gas supplanted oil, major alterations in affiliate refineries also became necessary, as well as changes in the volume and quality of crude oil. Groningen gas increased profits as rapidly as pipelines could be laid and contracts secured. Inevitably—certainly by 1969—given the quantity of gas already discovered and brought on stream, several affiliate refineries for all practical use were too large. Thus a different and plentiful energy source began to supplant heating oil in the 1960s, just as oil more than a decade earlier had replaced coal.

Jersey began considering "optimum marketing strategies" in 1970, for Great Britain, Switzerland, and Germany, strategies that would reduce its market share in some products and lead to some disinvestment.[52] The oil embargo of 1973 interrupted this planning and completely disrupted the world of oil. It led to new strategies for European refining.

X   X   X

1973-1975

Jersey's European refining plant continued to expand until 1973, when the Organization of Arab Petroleum Exporting Countries' [OAPEC] oil embargo and the subsequent price increases completely altered the world of oil. Those sudden changes, in conjunction with existing concerns about Jersey's fuel oil business and the profitability of some areas of the company's European operations, led to a retrenchment in total refining capacity and a new emphasis on so-called "Prescription refining," in place of the wide-open, large-volume refining of previous years.

As we have seen, those expansions in crude refining capacity that came on stream after 1972 reflected earlier expectations that European demand would continue its steady growth. New distillation capacity at Milford Haven in the United Kingdom raised that refinery's crude capacity to 305,000 b/d. A new atmospheric pipestill at Fos helped raise the French refinery's capacity from 70,000 b/d to 186,000 b/d.[53] After 1973, however, while additional crude refining capacity was added, the European affiliates took the unprecedented step of actually reducing capacity at several refineries. The largest of these projects occurred at the giant Rotterdam refinery, where a 153,000 b/d pipestill was "mothballed."[54] The company also moved to dispose of its interest in the Castellon de la Plana refinery in Spain, not because of excess capacity but due to disagreement with the Spanish shareholders about the future course of development for the joint venture plus a desire to use Jersey crude oil and capital in other more profitable ways.[55]

Although Jersey's European affiliates ceased to expand crude refining capacity, large-scale investments in their refineries did not end. In some cases they added more advanced equipment to enhance their ability to perform the type of refining required by new market conditions. Such additions were part of the 1973 expansion at Fos, and similar investments were made at Fawley, Hamburg, Cologne, Augusta, and Port Jerome between 1973 and 1975. In addition, several refineries added facilities to meet the new lead and sulphur standards required for gasoline and fuel oil. Karlsruhe brought on stream a new low-lead gasoline project in 1974, and Ingolstadt followed the next year. Both of these German refineries made increasing investments in equipment designed to reduce the refineries' impact on air and water quality in the areas in which they operated.[56]

By 1975, there were twenty-one refineries operated by Jersey's European affiliates, and the great majority of each of these had more crude capacity than the combined total production of all Jersey's European refineries before World War II; several of them, in fact, possessed five times as much. In terms of technology, the change was equally great, not only in the production of chemical feedstocks but in the manufacturing of basic fuels, lube oils, and other petroleum products as well. Yet, in one sense, the refiners of 1975 were like those of 1950: fundamental changes still confronted their world, and as refiners they knew that they had to adjust to a rapidly changing environment.

Chapter 9

The Development of Esso Europe

THE FIRST MAJOR STEP toward a regionalized managerial system in Europe occurred in 1960, when Standard Oil Company (New Jersey) [Jersey] began to reshape its own structure. Six years later, the creation of regional affiliates, including Esso Europe Inc. [Esso Europe] denoted the end of major reforms in the parent company's operations. Long before this time, however, Europe had presented special problems for management, and after 1950, it was clear that the administrative structure in Europe had to be altered. As Michael L. Haider, one-time Board chairman, observed, "You always have problems when you're in somebody else's country." And besides the many countries in Europe, there were more than one hundred Jersey affiliates. Twice each year those European executives involved in financial matters appeared in New York for budget reviews. Company Secretary John O. Larson remarked, concerning these and other such arrangements, "Our overseas operations were pretty fragmented."

At the time, however, the Board viewed Europe as merely one disappointing example of declining financial returns on investment, a case study which had inspired Jersey President Monroe Jackson Rathbone to undertake a thorough examination of the entire structure through which the parent company guided its subsidiaries.[1] Although the company did not report operational statistics on a regional basis, the significance of the European performance can be understood after a review of the hemispheric results reported in the company's financial statement. Jersey's net assets in the Eastern Hemisphere grew from $539 million in 1955 to $3,576 million in 1963,* while net earnings from that source increased only from $171 million in 1955 to $355 million in 1963.[2] Europe not only represented the major share of these investments; it also contained the fastest growing oil product outlets. The reduced product pricing arising from excess producing capac-

---

*Beginning in 1964, the company switched from a net assets basis to capital employed.

ity, however, served only to accentuate the historic problems associated with Jersey's efforts to coordinate operations throughout the European region.

Across the years, the varied markets and diverse affiliates of Jersey in Europe had virtually defied any exercise of more than the loosest control by the parent company. Some of the affiliates had operated for almost as long as had the parent company; and of these, several were already well established at the time they first became part of the Jersey system. Distance and diversity also tended to guarantee substantial autonomy to each of the European affiliates. For decades, the families who founded the businesses continued to operate them with only the slightest acknowledgement of Jersey's ownership.[3] And, of course, the same factors that hampered Jersey's exercise of control also tended to discourage anything more than limited, formal communications among the European companies. Cooperation was simply nonexistent.

Even before World War II, Jersey executives were aware of the European problem; conceivably some of them envisioned a long range solution. But Europe, between the two world wars, did not appear to be the best place to initiate sweeping administrative changes, which, in any case, would have disappeared in the raging fires of Hitler's territorial ambitions. Again, since only in Hungary and Rumania did Jersey have affiliates with functions other than marketing, efficiency did not seem to require a regional approach to management.

After World War II, and for some years, the Jersey affiliates needed direct contact with the parent company to help with such problems as rebuilding and restoring plants, securing loans for operations and advice on the proper technical equipment to install. Also, national legislation designed to regulate foreign exchange created financial problems. The money shortage led Jersey to finance construction of new refineries and the expansion of established ones in many nations that for years had imported all or most of their product requirements. Yet, with immediate postwar markets still limited, making those facilities economically profitable demanded more frequent interaction among European affiliates. By 1960, European operations had become more complex, as a number of affiliates added oil and gas production and chemical manufacturing to their operations. With Marshall Plan aid, Europe had shifted away from coal to oil and natural gas as its primary fuels, thus accentuating the already rapid expansion of the European market. Finally, the economic unification which culminated in the founding of the European Economic Community [EEC] in 1957,* opened up possibilities of substantial

*Became effective, 1958.

economies of scale—provided that Jersey could improve the coordination of its European operations.

Certainly by 1960, Europe's real and potential importance to Jersey's overall operations had been well established. At about the same time, the world's seemingly chronic glut of crude oil made the profit level generated there unacceptable—significant increases in sales volume with a decreasing rate of return.[4] These circumstances required greater structural efficiency and improved management if Jersey were to continue to operate there at all.

Haider, who as chief executive officer [CEO], drove the regionalization plan through the Board, has described the situation as it existed before the 1960 restructuring and reorganization: "There was a contact director; in fact, there were two contact directors. Again you had a sort of divided responsibility, no real overall supervision." Jersey's functional coordinators held responsibility for overseeing the operations of all European affiliates, the assignment of coordinators coincided with each affiliate's most important function, and each coordinator reported directly to the contact director for that general area. In addition, the parent company's Coordination Committee guided each affiliate's capital investment program. Clearly, the functional bias built into this system tended to prevent a regional perspective from developing at Jersey's headquarters.[5]

Supplementing the role of the functional coordinators and the Coordination Committee were two other administrative structures, which evolved in response to special situations during the 1940s and 1950s. The first of these, shareholders representatives, had first been sent to Europe at the conclusion of World War II, to assist in rehabilitating Jersey's affiliates there.[6] Wartime destruction, coupled with the lack of normal plant repair and maintenance, plus the special needs of reconversion, all created problems that economic dislocations of the immediate postwar period served to intensify. Initially, European governments concerned themselves with domestic problems as they passed laws which hampered the resumption of normal business. Under conditions such as these, it was important for Jersey to place representatives on the spot—representatives empowered to deal with both national and supranational government agencies, as well as with affiliate managers. Based throughout Western Europe and assigned responsibility for assisting managers and for interpreting the parent company's policy to them, as well as for representing Jersey's interests to official agencies, these shareholders representatives played an important role in reviving the fortunes of Jersey's European affiliates.

From the beginning, however, the position of shareholder representatives was anomalous and poorly defined, especially since their functions remained entirely advisory. Affiliate managements still remained free to report directly to contact directors and coordinators in New York, and, since few political situations called specifically for the participation of the parent company's representative, the handling of governmental relations could easily become a source of friction between the shareholders representatives and local boards.* Thus, while shareholders representatives did serve as a valuable channel of communication at times, they did not significantly improve the coordination of European operations, either through Jersey or by means of their cooperative efforts, since there was no regular consultation among them.[7]

As the need for coordination within Europe increased during the 1950s, the parent company added a second mechanism, the European Coordination Committee [ECC]. Composed of heads of Jersey's largest European affiliates, the committee resembled the London Council of the 1930s and 1940s.[8] Once formed, the ECC met periodically to discuss the overall regional outlook. This created a series of opportunities for executives to share views, discuss mutual problems, and attempt to resolve the differences that inevitably arose as affiliates became increasingly interdependent. Yet, while the meetings did enhance regional perspectives, they failed to mitigate the intense nationalism that flourished after World War II. Nor did they lessen the intense rivalry that had always characterized the actions of the various companies. Lacking authority over the affiliates, the ECC remained relatively ineffective in the area of regional coordination.

One partial solution to the weakness of both the shareholder representatives and the ECC had been represented by periodic efforts of some of Jersey's departments to improve coordination by stationing a representative in Europe. For example, the Jersey Economics Department established the office of European Economist, where a European-based specialist functioned to gather information on a regional basis and transmit it to both the home office and the European affiliates.[9] Despite the presence of such specialists, however, the lack of any specifically assigned functions for them served to dilute their impact in Europe.

In 1959, Emile E. Soubry, a Jersey director with more than forty years experience in European operations, conducted a thorough study of the region's needs. His report called for better coordination, and it became the basis for the changes made in Europe during 1960.[10] First, the number of shareholder representatives was reduced from five to three, and their title

---

*Harold W. Fisher, shareholder representative in England, for example, found his *official* relations with Esso Petroleum Company, Limited quite difficult.

was changed to "Jersey representatives." Next, Paul J. Anderson, Jersey's London-based representative, was designated as the parent company's senior executive in Europe. While these executives continued to operate with considerable independence, Anderson not only coordinated their efforts, but he also brought them together periodically to exchange information and opinions, in the hope that they would reach a more uniform stance on regional matters. In addition, he served as local contact for representatives of New York staff departments resident in London, and as principal contact with other international oil companies based there.[11]

CEO M.J. Rathbone began to reorganize the entire Jersey Company in 1960. Secretary John O. Larson, who helped supply the details for many of Rathbone's plans, believes that Rathbone had no superior "from the standpoint of having interest in the organization structure and the way the company was managed." Rathbone saw that problems existed on continents other than Europe, and he realized that the domestic operations should not be exempted from his proposed reforms. With Eugene Holman and Hines H. Baker, in 1958, he set in motion a plan for a changed and unified domestic affiliate because he knew that any reordering of Caribbean, Central, and South American affiliates would require time to be effected. Rathbone, with very limited foreign experience, listened to Soubry, William R. Stott, and others on the Board. And to the Soubry study of 1959, Rathbone added in 1960, a new detail based upon the existing contact director system, but designed to enhance the corporate and regional focus of the improved structure. In place of the existing plan, based on functional contact directors, Rathbone's proposal featured an informal "regional executive committee" for Europe composed of three directors. He divided the European affiliates into two groups, essentially on the basis of membership in the Common Market. A single director would become the primary contact for one group and an alternate contact for the other. The second director would take the reverse role, giving the affiliates the advantages of both closer attention and consultation from two experienced and well-informed executives. Finally, a member of Jersey's Executive Committee would be designated as advisory contact for the entire region. Working as a team, all three would coordinate the operations of European affiliates, with the advisory contact reviewing the judgment of the two directors, and helping to decide the extent to which the contact directors should act on their own initiative or yield to the Executive Committee or Board.[12]

At the center of Rathbone's arrangement was a new staff office suggested by Soubry, Regional Coordinator for Europe. Appointed to fill the new post was Wolf Greeven, who worked with all of those who had interests in Jersey's

European operations. Acting as an advisor and assistant to the regional contact directors, he passed on to them those items that in his judgment required their attention. After consultation with the Jersey representatives, Greeven made all the decisions that did not require attention from the directors. Whenever appropriate, he made referrals to Jersey staff departments, and they, in turn, kept him informed of their actions. Since his own staff remained small, Greeven depended on the functional staff departments to help him fulfill the regional coordinator's largest responsibility for maintaining intelligence on political, social, and economic developments throughout the region.[13]

Even with this refined administrative structure, European operations still failed to improve significantly. Although more and better information reached New York, actual coordination of European operations lagged. Most European affiliates remained as independent as before. Although the Jersey representatives now occupied a better position to develop a regional perspective, they lacked the authority to press their views on affiliate managers. Worse yet, the regional coordinator and the contact director team in New York also lacked any real authority to act directly on affiliates. In fact, in some ways, the changes had made the whole system more cumbersome.[14]

In 1963, Jersey's top management again sought to develop an effective solution to the European problem. They assessed several alternatives, including a much stronger regional management system proposed by Haider, the parent company's new president.[15] Traditional procedures proved too inhibiting, however, and the Board chose to pass over Haider's proposal and instead to modify the structure adopted in 1960. The new plan called for expanded cooperation among the European managers and strengthening ties with New York. First, the creation of a European Council, composed of the chief executives of all European affiliates, occurred. This body mirrored the ECC except for its broader membership and different leadership. Although the group exercised no direct control over individual affiliates, it did provide a forum for exchanging views and broadening the outlook of affiliate managers in Europe. In focus and function, it could be compared with its predecessor, but Jersey's increasingly regional approach, and the creation of the regional coordinator's office, together promised to improve its effectiveness.

To strengthen New York's voice on the European Council (and thus strengthen the council as coordinator of regional operations) the Jersey Board assigned David A. Shepard, giving him the title of European Vice President.[16] A long-time director, and executive vice president, Shepard was: "to guide, on behalf of the Board of Directors and in cooperation with the contact directors ... the coordination of Company policies with respect to Great Britain, countries associated in the EEC and other countries in Western Europe as well as the Middle East," and to provide direction to company staff in those countries—a broad assignment, to say the least. Shepard also represented, when necessary, the company's interests to local governments, agencies of the European Economic Community, and "other national and supranational organizations."[17] His appointment marked the first time that Jersey had stationed a director in Europe since before World War II.[18]

Now the company appended this structure to the regional coordination organization adopted in 1960, thus creating the most elaborate European coordination structure in Jersey history. But while this new approach did render some improvement, it did not provide a solution to the fundamental and chronic problems affecting European operations.[19] In fact, the new complexity added to the burdens placed on New York staff departments and on European contact directors.[20]

More fundamental alterations in Jersey's organizational structure required other changes in company leadership. Haider, as Board member and as Jersey president since April 1, 1963 had suggested additional new European arrangements to Rathbone, but without success. As he put it, "I came to the conclusion—and this was discussed with Rathbone before he retired but he wasn't very enthusiastic about it—that we set up a management group in London and [have] that management group tied in with New York." For Haider, it seemed simply, "getting related things into one package." As Secretary Larson noted, Haider worked very closely with study committees: "Now Haider was the man on the Board who bought the concept [Esso Europe] and convinced the Executive Committee that we ought to go this route... ." Then Haider became CEO and Board chairman on March 1, 1965. Larson had for years attended Board and Executive Committee meetings where the problems of the European affiliates had been discussed, and he knew Haider's views well. He recalled:

> A committee* was formed at my recommendation, of which I served as chairman, to determine what could be done to get a little tighter control over our foreign operations and reduce the demands on the Jersey parent company providing guidance on matters that they shouldn't be concerned with ... . That committee ... had Luke Finlay, Wolf Greeven and a couple of others [on it] and we

---

*Larson pointed out that they drew on a staff called Organization Planning (termed at times a department) in the company Secretary's Department for assistance in this study. The group had originally been created to aid any of the contact directors who needed special studies made.

made a number of recommendations, one of which was to bring all the European companies into one unit and have it managed and treat that as an affiliate that would report to New York.

Later, the Executive Committee approved the Larson committee report, and Haider then appointed another committee, with Greeven as chairman, to work out a general plan for achieving these results.

J. Kenneth Jamieson, who had replaced Haider as Jersey president, was also a strong believer in decentralization and regionalization. So he strongly supported Haider's plan for control of the European affiliates. Howard Page, Jersey's first vice president for a region—the Middle East—also quickly recognized the advantages of the new structure. So did Nicholas J. Campbell, Jr. who became a Board member on May 19, 1965. Campbell, who had Latin American experience, had represented Jersey with Page and Lloyd W. Elliott in dividing Standard-Vacuum Oil Company [Stanvac] properties, and he had also served as president of the Japanese affiliate, Esso Standard Sekiyu. Traditionally, Jersey CEOs in the last analysis shaped Board decisions, and Haider, who gave every appearance of being easy going, nevertheless, possessed a strong will. The ultimate decision was largely Haider's, but he recognized the tactical advantages of both full support from the Board and further study of full details in making the change.[21]

In May, 1965, Haider called an informal meeting attended by Jamieson; Marion Boyer, chairman of the Board Advisory Committee on Investments; Shepard, European vice president; Stott, contact director for Europe and for marketing; M.A. Wright, Jersey vice president; and Larson, secretary of Jersey. Focusing on the problem of "deciding on the form and responsibilities of the headquarters organization and the manner in which Jersey should look after its investments in affiliated companies," the participants appraised the existing structure and philosophy. They concluded that while Jersey's well-established principles of management remained valid and "should be continued in any reorganization plan," they also believed the existing system of affiliate management to be "an inconsistent and unbalanced pattern of organization" in which the parent company in some cases dealt directly with affiliates, and in others worked through regional management groups. Affiliates varied greatly in the degree of their autonomy; moreover, the level of parent company supervision and involvement bore no clear relation to the size or importance of the affiliate.[22]

In general, the group discussed "three broad alternative systems of management": (1) uniform reliance on local operating managements and the elimination of "intermediate holding or management affiliates"; (2) the establishment of a single international subsidiary to manage all of Jersey's foreign operations, except Imperial Oil Limited [Imperial], the Middle East minority interest companies, and chemical operations; and (3) reliance on "strong regional management groups responsible for overseeing all affiliates within a region." Based as much as possible within their regions, these new groups would be responsible for "practically all the coordination work" currently done in New York, and this would permit delegation of "broad responsibility and authority" to regional managers with a corresponding reduction in the size of the New York staff.[23]

The six regions designated at the meeting paralleled the existing organizational structure: the United States, Canada, Latin America (excluding Venezuela), Europe, the Far East, and Africa. Because Jersey owned only a minority position in the companies holding the concessions, the Middle Eastern organization was to be left intact. Both Humble Oil & Refining Company [Humble] and Imperial already fit the pattern designed. In fact, the "degree of management responsibility, authority and autonomy ... enjoyed" by those affiliates was accepted as the standard to be applied throughout the new structure.[24] That decision clearly indicated the groups' intention to reform rather than simply realign responsibilities on a regional basis. Here, reorientation of Far East and African operations offered few problems. By breaking off Esso Standard Eastern's [Esso Eastern] African operations, and adding them to Esso Mediterranean, Inc.'s [Esso Med] responsibilities, a regional structure could be easily realized.[25] Since effective regional management groups already operated in those areas, the only major change required a strengthening of the coordination function there. The regionalization of Latin American operations also proceeded smoothly, building on progress made since 1960.

Although the proposed change would substantially strengthen regional management in all areas outside North America and the Middle East, the creation of a regional management group for Europe quickly became the principal focus of the discussion. While agreeing that a single group would ultimately be desirable and endorsing active efforts to develop such plans, some participants felt that then-current political and economic conditions dictated a stage-by-stage progress toward that goal. As a beginning, a feasibility study of a single management group for Scandinavian affiliates reported progress within weeks of the meeting, and other studies of possible regional management groups for Holland and Belgium, Germany and Austria, and the Common Market were planned.[26] Unexpected developments, however, quickly disrupted this cautious program.

On May 19, 1965, Haider reported the meeting to Jersey's Board and asked directors for their suggestions or recommendations on the subject of "the over-all organizational structure" of the company.[27] In mid-June, Larson prepared a memorandum outlining the results of the May meeting, to which he attached a draft of Jersey's organizational chart as it would appear with the proposed regional management groups.[28] At the end of the month, Haider presented a more detailed report to the Board, highlighting Larson's memorandum and noting that the proposed arrangement would also permit a substantial change in the size, nature, and functions of the headquarters staff. Because the regional management groups would be taking responsibility for many of the coordination functions then performed in New York, headquarters staff could be reduced to "a relatively small number of widely experienced headquarters' executives," who would be able to concentrate on such limited functions "as over-all planning and coordination, executive development, and financial planning and coordination." Thus, these changes would not only eliminate the regional coordination offices in New York, but would also reduce functional coordination staffs. After discussion, the Board agreed in principle with the Haider proposal and endorsed plans of the Compensation and Executive Development [COED] Committee to establish an organizational study committee and other special study groups. The membership of these groups was selected from among the directors and other New York personnel, who were instructed to develop plans for the implementation of the proposal and to decide on its timing.[29]

To lead this special mission, Haider named Marion Boyer, who would serve as chairman of the special Organization Study Committee. A.O. Savage, Jersey's controller, and Director Campbell completed the group. In August, Boyer called Campbell back from vacation to get started on the task. As their first step, they discussed the decision to adopt regional management with managers of the parent company's staff department and New York-based special organizations, such as Esso International Inc.[30] The study committee found that a majority of these groups favored a change. Having completed that stage of the study, the committee next began trying to gauge the reactions of European affiliate managers.[31] Their positive response to the idea of stronger regional coordination by an organization based in Europe led the committee to draft final recommendations for the COED Committee to take to the Jersey Board.[32]

On November 3, the Organization Study Committee presented the COED Committee with a progress report confirming "the feasibility and desirability of a regionalized organization for the Company's oil and gas operations, except for its Middle East interests, Libya, and Creole" and outlining proposed regional structures for Europe, Africa (except Libya), the Far East, and Latin America (except the Creole and Lago Petroleum Corporations).

The study group then went into greater detail regarding implementation of the proposed European organization, noting that details would be given further consideration by the new regional management once it was constituted. After discussion of the proposal, the COED Committee approved the progress report for presentation to the Board, noting however, that final consideration of the European proposal should be deferred until Haider could discuss the study committee's conclusions with the chief executives of European affiliates. The COED Committee also decided that "arrangement should be made for study and definition of the structure, functions, and staffing implications for the parent company" concurrent with the on-going study of regional organization.[33]

Later in the month, Haider visited Europe with Jamieson, Stott, and M.W. Johnson, the regional coordinator for Europe. Meeting with the chief executives of all the European affiliates in Stresa, Italy, Haider understood that they supported approval of the shifting of coordination functions to Europe; nevertheless, he was apprehensive about their reactions to a proposed structure under which they would lose their direct contact with New York, as well as some of their traditional autonomy.[34] What happened at the meeting is best described in Haider's own words:

> I had a long agenda of a lot of mutual problems for Europe, etc., which was mainly just something to kind of break the ice and talk about until we got to the main dish. And so finally, we got to the part of the program where I talked about organization of the European operations. Serge Scheer, who was head of the French company, put up his hand and said, 'Mike, I'd like to speak to that.' And I thought the British and French were the ones that were really going to fight us. So when Serge put his hand up, I thought, 'Well, here it comes.' In effect what he said was, 'What the hell took you so long?'

The Esso United Kingdom's [Esso U.K.] chief executive, Sir Hugh C. Tett, echoed Scheer's response, commenting, "This is a helluva good idea. What took Jersey so long to wake up to this?" The other affiliate spokesmen joined Scheer and Sir Tett in voicing their approval of the new plan. Only Vincenzo Cazzaniga, the head of Esso Standard Italiana-Societa per Azioni, expressed serious reservations about regionalizing European operations. Haider later related, "What the European affiliates themselves recognized is that there was a lot of interlacing of their affairs with other affiliates. And there was nobody to arbitrate or to give them guidance." Esso Europe would provide a place where they could thresh out their problems without involving the parent company in blow-by-blow details.

Initially, a few European affiliate executives groused a bit because, under the new company, there would no longer be required trips to New York each year. One or two Jersey directors, including one who voted for the change, wondered aloud whether the direct budget presentation made by affiliate executives on such trips had not provided the Executive Committee and the Board with a fine opportunity to assess the affiliate executives' management potential and to form a first hand opinion on personnel. But these mild objections amounted to little at the time.[35]

Having received assurance that the affiliates would support the new arrangements, the Jersey management now moved quickly to implement its reorganization plan. At the end of November, Haider reported to the COED Committee and the Board on the results of the Stresa meeting, and the committee gave final approval to a "fully regionalized organization."[36] To head the new European regional organization, the committee chose Campbell, in part because he was both a member of the Organization Study Committee and Jersey's newest director.[37] Moreover, his experience in the Stanvac dissolution and his term as president of Esso Sekiyu had provided him with the ideal background for leading a new regional company. Campbell resigned from the Board to accept this new post on May 18, 1966. The committee next chose Robert Milbrath, president of Esso International, Inc., to be executive vice president; as senior vice presidents, they appointed M.W. Johnson, the regional coordinator for Europe, and H.T. Cruikshank, an American serving as a managing director of Esso U.K. For the other directors of the new group, the committee selected the heads of three of the largest European affiliates: Scheer, Tett, and Cazzaniga. These officers were instructed to form themselves into a new study group, to work out details of the new organization, and to select a staff for it. At the same time, the COED Committee recommended a shift of Esso Eastern's African operations to Esso Med and a change of the latter's name to Esso Africa, Inc. [Esso Africa]. Finally, the committee noted that implementation of all these changes should be coordinated as fully as possible.[38]

By mid-December, 1965, the Organization Study Committee had turned its attention to the implications of regionalization on the "future composition of the Company's headquarters organization, including special purpose affiliates in the New York City area." The new officers of Esso Europe continued to design their own organization, while the regional contact directors for Africa, the Far East, and Latin America studied organization needs for their regions, in order to make further recommendations to the COED Committee.[39]

By February, 1966, Jersey appeared ready to announce all the changes.[40]

Campbell and his fellow directors had already begun selecting the new staff and moving them to London. Although some executives necessarily came from New York, the two main sources were European employees of affiliates in the region, and Americans based in Europe as representatives of Jersey's functional coordination and other staff departments (or of special functional companies based in New York). The top managers aimed at selecting as many Europeans as possible for the new company, and in the end, only about one-third of those chosen were Americans.[41]

By September 1, 1966, the new regional management organization for Europe had taken control. On that date, all correspondence between the affiliates and New York began to be re-routed to London, including the 1967 European affiliates' budgets, which were already in the final stages of approval. Despite the magnitude of the change, the transition proved remarkably smooth. Nevertheless, Esso Europe still had a major task ahead in trying to work out its relationship with each of the affiliates, as well as the changes in the affiliates' relationships with the parent company.[42]

Solutions to those problems hinged on the success of the new regional organization's efforts in coordinating European operations. And that, in turn, depended upon getting the cooperation of the European affiliate managers. To achieve that end, Esso Europe added two additional European directors, so that a majority of the board of directors would be Europeans. In addition, the regional company brought together the chief executive officers of all the European affiliates several times a year, as a means both of improving intraregional communication and enhancing the personal relationships among them. One of the early and remarkable results of the new organization was that easier relationships did develop. European executives, who had traditionally treated each other as competitors, began to view Jersey's European operations as a cooperative effort. The increased opportunity for resolving conflicts among these companies contributed to the change, as did the improved coordination through the offices of Esso Europe.[43]

Despite their quick approval of the coordination group in principle, several of the affiliate chief executives remained skeptical when the organization was established. Their initial support had been based upon the assumption that they, through the new organization, would exercise greater control over their own coordination. In essence, they visualized a strengthening of the European Council. Nevertheless, once the new group began to function, their irritation lessened and several became staunch converts to the change.[44]

For Jersey, the advantages of this major structural renovation quickly became evident, and the improved coordination of all European operations proved to be most important among these. Before, the efforts of Jersey's functional coordination departments to keep up with European developments had always foundered on the problem of maintaining uniform supervision of diverse operations. The addition of a regional coordinator corrected this situation, but also placed an additional level of management between the affiliates and the parent company staff departments—without adding to the ability of the regional coordinators to oversee operations within their region. Now, the establishment of Esso Europe, especially in the context of worldwide regionalization, provided both the advantages of the regional coordinator, and the necessary staff to follow adequately local developments. Where Jersey's functional staff coordinators had been able to visit small affiliates only every two or three years, the functional coordinators of Esso Europe could make the same contacts several times each year, thus providing an improved familiarity with affiliate problems and means of easy consultation.[45] Moreover, the new organization permitted the coordination of the various functions within the region, a goal that neither Jersey's functional coordinators nor its regional coordinators had been equipped to achieve.

Other advantages to the change could not have been clearly anticipated. For example, one was the greater capacity of the regional management group to use and develop the pool of executive talent available within the European affiliates. So long as the European affiliates perceived themselves to be separate entities, the careers of executives had been limited almost entirely to affiliates in their own nations. Those working for smaller affiliates had very limited opportunities to develop their managerial potential. By bringing together executives of different nationalities, Esso Europe not only created new opportunities for the most talented managers, but the new company also helped to break down feelings of national separateness among the affiliates, thus opening the way for the exchange of promising executives among the affiliates themselves. In addition, the new structure helped Jersey to gauge the individual potential of its European managers. As executives of Esso Europe, they could be given broader responsibilities, while regional and parent company top managers could become more familiar with their capacities. The most promising—those slated for the top positions in their national companies—would be fully evaluated and, if appropriate, assigned to the parent company for additional experience.[46]

For the individual, assignment to both Esso Europe and the Jersey parent offered fresh opportunities to expand career horizons. Finally, this improvement in Jersey's European managerial system had a "ripple effect" going far beyond the company itself. The success of the European Economic Community came to depend heavily upon its ability to break through barriers similar to those that had handicapped Jersey operations in Europe before the creation of the company's regional organization in London. Hence, as Esso Europe (and other comparable organizations) helped inspire a spirit of cooperation across national lines, each gain also helped the EEC.

The creation of Esso Europe also cleared the way for better coordination of research operations. Through the initiative of the regional managers, the established affiliate research organizations agreed to rationalize their areas of specialization. The new company substantially reduced duplication in research efforts that had been characteristic of an earlier time.[47]

One other unanticipated benefit of regionalization came from the greater decentralization permitted by the new system. All previous structures for coordination had placed a heavy burden on the contact directors. As planned, even the regional coordinators had been envisioned largely as executive assistants to the parent company's working directors. While contact responsibilities continued under the new structure, much of that burden shifted to the regional managements because of their locations and because they were better equipped to make decisions. As a result, the employee directors could devote more of their attention and time to policy making and to long-range planning. This shift increased the capacity of the Jersey Board to carry out these functions, which, in turn, transformed the role of the Board within an increasingly decentralized system. Then too, this alteration in the Board's role permitted another significant change, one that Haider simultaneously pursued—the addition of Jersey's first non-employee directors to the Board. Where before, even a full Board of working directors had been hard pressed to maintain their contact responsibilities, the regional structure afforded better coordination with a reduced number of contact directors. Although this possibility had not been foreseen in planning for regionalization, Campbell later noted that the changes in the parent company Board "would not have been possible without regionalization."[48]

Yet, despite these impressive changes, the first measure of Esso Europe's success was in its primary mission, the coordination of European operations. The changes in management went smoothly, but the task of resolving the long-standing problems of European operations would be a much more difficult task, especially when external developments made the business environment in the region even more unfavorable.

Chapter 10

Marketing in Europe

THE STANDARD OIL COMPANY (NEW JERSEY) [Jersey] established a number of marketing affiliates in Europe during the last quarter of the nineteenth century. For the next sixty years, Jersey's primary concern in Europe was to provide these outlets with an adequate supply of crude oil and refined petroleum products. In the early years, however, most profits came not from marketing but from upstream operations. So as Jersey's European strategy gradually developed, it proved to be simple and direct: sell the maximum volume of product and make adequate space for the quantity of crude the company acquired in both hemispheres.

This story of how and where Jersey secured plentiful supplies of oil and products to sell has been described elsewhere in this volume. It is sufficient to note here that the company obtained interests in several large sources of oil during the post-World War II period through joint agreements with other oil majors and with national governments, as well as by means of long-term purchase contracts. Inevitably the task of disposing of oil and refined products fell upon the Marketing Department, and the job the marketers did for more than two decades was simply tremendous. In no place did their efforts show better results than in Europe. As one observer characterized the Jersey records in those years (1950-1973): "Brilliant expansion of product sales made it possible to profit by buying any really cheap crude [oil] available to meet the increase [in demand]." Jersey marketers met every challenge and positioned their company to take every advantage.

Some American visionaries of the post-World War II era welcomed the development of the great oil producing areas abroad. They believed that supplies from these fields would relieve the strain on U.S. resources, which could then be largely preserved for future crises. At the same time the new sources of oil would provide adequate supplies of petroleum for the restoration of Europe. Certainly such political leaders as Harold L. Ickes, James V. Forrestal (and others) wanted Europe to recover quickly from the ravages of war, but they also feared a future shortage of American domestic

295

oil as an all-too-likely possibility, since they knew that U.S. oil had been used generously in both world wars. Therefore, these observers encouraged Jersey and other U.S. companies to step up their search for oil abroad and to buy equity shares in those companies that had already found it.[1] To complete European recovery, they believed, the production and use of this foreign oil would require all the developmental and marketing skills which Jersey and its peers possessed. As early as 1950, one U.S. Department of State spokesman declared: "Use of Middle East oil conserves Western Hemisphere resources which are vital to the Allied nations in an emergency"—a point well remembered.[2]

Yet Europe after the war comprised a number of independent nations, with widely varying marketing regulations. This market proved complex. In addition, Jersey counsel and shareholder representatives continuously had to monitor affiliate compliance with the terms of the 1942 Consent Decree (explained elsewhere) and to be certain that no affiliate entered into cartel agreements, even though some governments, through market allocation arrangements, effectively cartelized all marketing.[3] Supervision became more efficient, however, after Jersey brought all its European companies together under one regional administrative unit—Esso Europe Inc.—in 1966.

Certainly Europe provided the obvious market for Middle East oil, and by 1950 Europe appeared well on the road to recovery. Market competition (where allowed by the governments) quickly heated up, as Europe rapidly turned from a coal-based economy to one based on oil. By mid-century, the Middle East supplied 75 percent of Europe's oil requirements.

Yet many difficulties still confronted the marketing organizations of Jersey's European affiliates after 1950: two problems, in particular, assumed major significance; the inadequate supply of foreign exchange, and a shortage of products. The first of these had forced virtually all countries in Europe to hedge domestic trade with currency restrictions and rationing designed to keep dollars or dollar exchange at home. This policy, in fact, served also to preserve home markets for national businesses at the expense of American firms.

Jersey's management anticipated that chaotic financial circumstances would accompany the ending of World War II, with its expected recovery of world trade. In essence, as is described in the third volume of the Jersey history, *New Horizons*, the company operated in two differing worlds of international finance; one (led by the U.S.) in which currency could be freely converted into dollars [hard money] and another (led by Great Britain and the Bank of England) in which members were dominated by a single source of foreign payments [the sterling area]. The member states in each group varied from time-to-time. Additionally, most of the Western European nations sought to conserve their gold and silver reserves; hence, they were often short of dollar exchange. Still despite this division in the world of international finance, many involved governments sought to aid world economic recovery and to revive world trade.[4]

In anticipation of postwar financial difficulties, Jersey in 1944, elected Jay E. Crane to the Board of Directors. Crane, an expert in international banking, had joined the company in 1935, as assistant treasurer. When he became treasurer, he persuaded the Board to strengthen his department so as to have sufficient personnel to advise on currency exchange problems. Crane, in 1947, helped convince Emilio G. [Pete] Collado to leave government service and become Jersey's foreign exchange manager. Collado had been at the Bretton Woods Conference of 1944, which established the International Monetary Fund, and had served as delegate to other significant economic conferences. He and his staff advantageously positioned Jersey resources for the time when inflation and serious financial stringencies struck Western Europe in 1948-1949. With David J. Jones, also of the Treasurer's Department, he worked out a system for using sterling to purchase needed supplies and goods in different sterling area nations, and these agreements stimulated marketing.[5] Jersey supplied the extra crude oil for sterling and used the exchange to buy needed materials. But the exchange problem continued to have to be negotiated on an affiliate-by-affiliate basis.[6]

Nor was this the only problem. Before Jersey would undertake to rebuild and improve its important Fawley (England) refinery—much needed to meet product demand in that country—the company requested some assurance that the new Labour government did not intend to nationalize the oil business, even though that government had compensated other property owners who underwent nationalization. Jersey insisted on a firm answer to the nationalization question, and finally sufficient guarantees were made to satisfy its Board. Renovation of the Fawley facility soon began.

Even while juggling foreign exchange currencies, Jersey also loaned the Anglo-American Oil Company, Limited* $100,000,000 to enlarge the refinery at Fawley. This giant plant went on stream in 1951, and by 1952, it had almost reached its full capacity of some 130,000 barrels per day [b/d], an amount believed to be sufficient to supply most of the still-developing

---

*In 1951, Anglo-American Oil Company, Limited became Esso Petroleum Company, Limited [Esso Petroleum].

demand in the British, Danish, and Irish markets. Soon construction began on a pipeline from Fawley to London, followed by similar pipelines to other major market areas—which together resolved in large part the delivery problem. Yet, so rapidly did the sales of Esso Petroleum increase that in 1957 alone, it acquired 150 additional retail service stations. Not surprisingly, new demands for products soon led to the construction of another large refinery at Milford Haven, Wales, in 1958-1960. By enlarging the refinery at Fawley, Jersey had signalled its departure from prewar marketing strategy: now the company determined to shift refining capacity from producing to consuming countries so as to supply the market more effectively. As a result, the greater quantity of products available in Great Britain and Ireland after 1951 permitted additional consumption without a substantial increase in the drain on foreign exchange, which led in turn to additional market expansion.

Meanwhile, it also became obvious that the strategy of supplying Denmark from the Fawley refinery introduced problems other than those of transportation. In 1954, the Jersey Executive Committee voted to hold in abeyance plans to add additional equipment at Fawley, "until the extent to which Esso Denmark [Dansk Esso Aktieselskab] will continue to draw product supplies from Fawley is determined." Of course, while the Fawley improvements waited, sales volumes in the Scandinavian market continued to increase. Soon D. L. Wright, Jersey Scandinavian shareholder representative, initiated discussions with the Norwegian government concerning "company interests"—building a refinery there to supply "a portion of the product requirements of affiliated companies, and perhaps others, in that country, Sweden and Denmark." For some time, Wright tried to get Norway to allow refinery equipment to enter duty free, and he sought agreement with that government concerning the proposed wholly-owned affiliates repaying Jersey with dollar exchange. Finally on November 29, 1956, after Wright reported partial success in achieving these objectives, the Executive Committee agreed to send the Norwegian government a letter of intent and to accept one in return; and construction on the Slagen (Norway) refinery began.

Looking back, it is obvious that Scandinavian sales developed so rapidly that, even with products from Slagen, Esso Denmark required still more. Jersey marketing coordination experts, aware that the company had a dominant position as the largest marketer in Denmark, were also concerned about selling off expected surpluses from Jersey's Middle East production; so they began to consider building a refinery in Denmark. Instead, in 1962, the company had an opportunity to buy Dansk Veedol, a Tidewater Oil Company (24,000 b/d capacity) plant at Kalundborg, a plan which Jersey's Marketing Coordination Department [Marcor] and Esso International, Inc. both endorsed after looking "at the purchase on a broad business basis." They also recommended that the refinery capacity be increased to process 40,000 b/d "of Aramco [Arabian American Oil Company] crude." Esso International arranged the purchase of this plant on the grounds that it anticipated continued increases in sales in the area.

Together these three examples—Fawley, Slagen, and Kalundborg (also discussed elsewhere)—illustrate the effective coordination that existed in many of Jersey's functional departments. The three show just how well the Marketing Coordination Department could operate in bringing together those affiliates needing products with the others seeking additional outlets for crude oil. Each of these cases furnishes a typical example of feverish activity in marketing Jersey products; while elsewhere in Europe, Jersey's refinery construction and expansion program moved almost as swiftly to keep up with increasing product demand.[7]

In postwar West Germany even before marketing could begin, former employees were located. While the installations of Deutsch-Amerikanische Petroleum-Gesellschaft [DAPG] had been virtually wiped out during the war, the staff, in large part, had survived and were anxious to begin business as soon as new equipment and new products could be supplied.[8] Of course, DAPG in war-torn West Germany did not prove to be the only Jersey affiliate short of products to sell.

With Marshall Plan aid, the European recovery quickly began; yet years elapsed before Jersey affiliates secured all the proper products and equipment. All the while Jersey continued sending its own experts to Europe, who consulted on operations, and then reported back to the home company. These experts from Marketing passed back information on problems successively to Contact Directors Emile E. Soubry, Peter T. Lamont, William R. Stott, Wolf Greeven, Robert H. Milbrath, and Howard C. Kauffmann. More than ten years after the war had ended, for example, Stott received a letter stating that marketing in Austria still remained unsatisfactory. A large number of unsolved problems remained, but more important, some Austrian marketers had actually been trained for other jobs. The Austrian affiliate (Esso Standard (Austria) A.G.) continued to use war-surplus tank trucks, "practically rebuilt by company personnel in their spare time," and no long-range plan for service stations existed, while the company presently operated only three. Postwar Austria retained price controls on virtually all products, and the affiliate could not borrow capital, which would have to be sent in. Specific recommendations made to Stott included a request for a copy of the Jersey accounting manual in German, and for the loan of an American executive familiar enough with the Jersey control system to install it in the Austrian marketing program.[9]

Another observer, J. C. Anderson, wrote A. O. Savage, the company controller, that the French affiliate did not appear to be price-competitive with dealers in neighboring countries. He added that the Jersey affiliates in France also had grave personnel problems that prevented effective reorganization; only resignation or retirement seemed to reduce the number of employees. Other observers commented on the French government's circumscribing all business operations with excessive rules and statutes. Later, several writers observed that many of the French regulations skirted the European Economic Community [EEC] Treaty (1957), violated its spirit, and served in practice to make petroleum a French state monopoly.[10]

When in 1953, the new refinery for Belgium, the Netherlands, and Luxembourg (Benelux)—at Antwerp—went on stream, Gerhard Geyer of Esso Aktiengesellschaft [Esso A.G. or Esso Germany], fearing that it might supply motor fuel to Southern Germany, wrote to Lamont inquiring about such competition. Jersey, Lamont said, always coordinated such operations with the local marketing affiliate, so as to avoid disturbing its market.[11] Afterwards, in fact, the German company under Geyer's direction began to develop rapidly; machine accounting procedures were adopted, some personnel were upgraded, and Esso A.G. hired fresh new university-trained engineers and technicians. Overall by the middle 1950s, the German company had achieved a significant position among affiliates in Europe.[12]

Italy, on the other hand remained an unpromising place for additional Jersey investment until long after the war. Italian government regulations continued to favor Ente Nazionale Idrocarburi [ENI]—the government owned company—at the expense of the Jersey affiliate, despite the latter's effort to gain help from the U.S. State Department in encouraging Italy to allow freer competition. As Anderson lamented after a tour of Italy around 1956, "I do not . . . believe that I have a reasonably clear understanding of the political situation . . . ." But perhaps no one did, for Italian prices in the 1950s remained as unstable as the government. By 1961, another observer could report, "Substantial improvement in market prices is necessary if private oil companies are to continue to operate in the Italian market."[13]

In general, Jersey encouraged all of its European affiliates to pursue a policy of "solus situs," a securing of exclusivity for its products on a service station forecourt. In the British Isles and in Ireland, this policy proved difficult to implement; some service station operators habitually stocked petroleum products manufactured by as many as six competing companies. Here,

Jersey attacked the problem through its affiliates, Esso Petroleum Company, Limited and Esso Teoranta [Irish-American Oil Company]. Fortunately, in Contact Director Soubry, Jersey possessed an English-born, self-made, and very articulate marketing specialist with long experience in England. He was able to persuade British and Irish marketing units to begin practicing exclusivity, and thanks to unrelenting pressure from the parent company as much as to improving demand for Jersey products, the "solus" practice soon became standard.[14] A concurrent expansion of Esso brand advertising naturally assisted in product marketing. Later on, the development of the European Economic Community (or Common Market) dovetailed beautifully with this brand-recognition program, and the Jersey image gained in customer recognition on the continent as well.

As Europe recovered from the war, Jersey moved to establish bulk sales to railroads, power companies, and other large and rapidly developing industries previously committed to coal as an energy source. The Organization For European Economic Cooperation [OEEC] predicted in 1955 that, within ten years, petroleum would own 25 percent of the total energy market—a level much underestimated in fact, as Europeans moved even faster to use oil for heating. Also, in addition to seeking conversion of well-established industries, Jersey marketers sought out newer growth industries, such as airlines, and guaranteed to provide them with efficient fuel service. Esso Teoranta installed a hydrant system to service planes at Shannon Airport in 1955, utilizing the most advanced technology in the world at that time.[15]

As the European energy market altered, Jersey affiliates on the spot explored every possible avenue for increasing sales. To exploit the growing mechanization of agriculture, they published and distributed farm magazines, worked out joint demonstrations with agricultural agents, conducted forums and classes, and even sponsored such competitive events as the World Plowing Match. European affiliates also acted to encourage private travel, after careful studies showed that it would increase product sales. First, the French company established a touring service headquartered in Paris, in 1953, and other affiliates gradually followed this example. In 1962, the *Annual Report* indicated that more than 5,000 Esso stations in Europe cooperated in the "Esso Touring Service" network. These stations distributed local maps and tour information without charge, and sold other highway maps and tour guides at nominal prices.

Most European affiliates concentrated for several years on reestablishing and improving previously existing services. Service stations received constant upgrading, full-service stations replaced existing one-pump outlets, building designs were improved—all efforts centered on giving better service to customers and on lowering operating costs. Everywhere the *ESSO* name and the blue oval, red-lettered trademark replaced the various local symbols that had served to confuse customers and thus to prevent effective brand identification. This multiplicity of changes cost a great deal of money, and despite the traditional corporate autonomy of affiliates in the decentralized structure of Jersey's management, most of these funds could not be obtained within Europe. So the parent company increased its loans to European affiliates. As the Executive Committee minutes show, such capital outlays usually received prompt approval. Frequently the Committee later allocated additional sums to cover cost overruns. Jersey's financial aid program paid new dividends as most affiliates quickly increased profits.[16]

Improvement in the fortunes of the affiliates went hand in hand with a rise in the standard of living for most Western Europeans. The various U.S. and international aid and reconstruction programs encouraged many changes, including conversion to new methods of production and distribution, a shift from the use of animal power to machine power, higher wage levels, and improved educational programs. European recovery, as a whole, emerged as an achievement little short of miraculous. By 1955, the purchasing power of the average European had increased to between 10 and 15 percent above the highest prewar level. European specialists came to the United States to study production methods, and when they adopted them, Americans flocked to Europe to supervise the installation of new machinery and techniques.[17] Cooperation of all kinds translated into the postwar European boom.

All of these alterations in the European market placed additional strains on the parent company, as Jersey executives sought to adapt to the enlarged regional market that was replacing a collection of smaller national markets. Now, Jersey's Marketing Coordination Department had to find ways to help regional marketers improve sales, to regularize procedures concerning supply and distribution, and in general to derive full benefit from a coordinated program. In 1952, a series of biennial marketing conferences was inaugurated; at these, Jersey officers exchanged information with European executives. Then, in alternate years, Marcor teams held smaller seminars and conferences, which served to resolve misunderstandings and to emphasize common goals and principles. Newly tested procedures could be explained in such groups, and the advantages of product changes made clear. Marcor experts also brought to such meetings new ideas about organization, personnel, tax, and accounting procedures, showing why these would be more efficient and economical. Marketing Department executives emphasized more efficient product distribution systems as well as the advantages of Jersey's brand recognition program. Experience in dealing with the problems growing out of postwar Europe's resurrection proved valuable to Jersey executives. The shareholders elected a number of them to the Board, where they continued to advise on European affairs. Spending much of their time in Europe consulting with executives of the various affiliates, they acquired more knowledge of the local operation and a better understanding of its problems—to the overall benefit of Jersey.

As Europe approached a unitary market, affiliates necessarily had to fit into an intricately balanced vertical operation, which Jersey restructured for them over time. In 1957, for example, Lamont told Coen R. Smit of Esso-Nederland N.V. about a planned meeting between Jersey executives and the heads of the European affiliates, to discuss "what improvements could be made." Earlier, Smit had indicated that such a meeting would be very helpful, but Lamont wanted to be certain that the persons who attended would confine their conversation largely to policy matters, "consultation on any broad subject involving either organization or broad policy." He included in this range, of course, industry and economic changes and, more important, Lamont saw this gathering as an opportunity "to assure that close coordination exists in the thinking of our top men in Europe and our top men in Jersey." As a result of such meetings, many of the activities of marketers in Europe reflected closely coordinated Jersey programs, now mounted on an international scale. The competitive struggles to secure goodly shares of the European heating oil and jet fuel markets provide ready examples. Thus, when in 1957, Jersey decided to appoint a New Uses Committee, it asked the European affiliates to appoint similar groups to ferret out new ideas for using petroleum products.[18]

Again, the development of refineries in consuming areas encouraged a policy of coordinated strategy. Each decision to build or enlarge a particular refinery had to be based on full consideration of the size, composition, potential, and competitiveness of its contemplated market. European Contact Director Lamont referred to these needs when he wrote Esso A.G.'s Geyer in 1954, concerning the policy of Jersey "to obtain the maximum outlet for its crude and to obtain the maximum employment of its refinery facilities."[19] Any lack of sound regional planning increased the staff work of each affiliate involved, which in turn presented problems for Jersey's marketing (and other) coordinators and for the Executive Committee. Ultimately, the necessity of carefully coordinated planning for every major investment provided one of the basic reasons for the creation of Esso Europe Inc., in 1966.

As part of its general duties, the London office of Esso Export Corporation [Esso Export] handled sales to non-affiliated brokers. Yet as a matter of policy, Esso Export never charged non-affiliate brokers less than the price that affiliated companies paid. This consideration for affiliates helped the parent company to integrate the European affiliates successfully into a recognizable "worldwide company."[20]

Another excellent example of this close coordination was provided by the Suez Crisis of 1956-1957. When Egypt blocked the canal and Syria cut the pipelines to the Mediterranean, chaos—especially in petroleum marketing—might well have ensued. Instead, reassured by Jersey leaders, various affiliates pooled their talents and resources, estimated their oil needs, distributed products on an almost unified basis, and carried on their business during the crisis with only minor interruptions of service to their customers. As a historic challenge, Suez brought home to the entire Jersey Company the advantages of operating cooperatively, and of coordinating their efforts at all times.[21]

X X X

Before 1950, almost 90 percent of the oil shipped to Europe was financed by the Economic Cooperation Administration [ECA]; and in such countries as Great Britain, more than one-fourth of the total dollar deficit was produced by shipments of oil from American companies. By 1960, however, these circumstances had largely altered. As widespread recovery continued, Middle East oil fields produced larger and larger amounts of crude. Jersey exploration crews discovered the first oil field in Libya in 1957, and by 1959, that country provided oil in commercial quantities, reaching 670,000 b/d in 1968 and 1969 (Jersey's share). The European supply picture changed rapidly as exploration and production in new geographic areas proved successful.

Of course, the plentiful new supply of oil for the marketers was itself a tribute to Jersey's careful management. At the same time, however, the vigorous expansion of the European market during the decade of the 1960s also caught many oil executives off guard and led Jersey strategists to confess, by 1970, that they had "traditionally underforecast demand" there, as they moved to adopt a more flexible set of criteria to control future estimates.[22] In fact, petroleum and product consumption increased at a rate more than twice that of overall energy demand, which, based on Europe's economic growth, had averaged an increase of 5 percent per year. In 1970, Europe's consumption of petroleum reached 12 million barrels per day, or 60 percent of the continent's total energy needs. Some observers gave as a reason the Treaty of Rome, signed in 1957, which had added impetus to the economic integration of Europe. Yet overall, it was clear that the plentiful supply of oil provided cheap fuel for additional and largely unanticipated growth.

The low price of oil acted to increase Europe's petroleum dependency because the cost advantage of oil over coal as an energy source continued unabated. Marketing competition for sales actually increased, as Libyan production mounted and as a number of large U.S. independents with concessions there sought European markets to replace U.S. markets, which were not available because of the mandatory oil import controls put in effect by the U.S government in March, 1959. Overall, the growth in productive capacity in the Middle East, coupled with Russia's entrance into European markets, created a huge oil glut, and this in turn reduced profit margins. In the *Annual Report* for 1960, Jersey pronounced the Soviet Union "a serious competitor" for European markets.

Yet even earlier, Russian competition posed special problems for Jersey's European affiliates, if not for the parent company. For example, Director Soubry reported to the Executive Committee on October 8, 1953, that the Soviet Union sought to make a state-to-state agreement with Denmark to sell (or barter) oil. Soubry explained that this oil came from Jersey's one-time Rumanian company, which had been expropriated and for which no compensation had been received. Then on June 25, 1954, Director Lamont told the Executive Committee that a treaty being negotiated between Denmark and the Soviet Union would allow the former to purchase a large quantity of petroleum products from the latter (or from countries it controlled). Esso Denmark had been asked by the Danish government to prepare to market these Soviet products. Finally, Esso Denmark refused to distribute Russian oil products. Jersey strongly supported the action of its affiliate, citing the fact that Russian (or Russian bloc) oil represented the use of Jersey oil and properties without compensation by Hungary and Rumania.[23]

As Europe recovered, Jersey's European affiliates began to confront a backlash from some of the political controversies that had become associated with Middle East concession negotiations. Esso Standard (Switzerland) ran into this situation in 1961, in a national market that consumed only one-half of one percent of the world's total petroleum product. Don J. Thompson of

Esso Switzerland wrote the Regional Coordinator for Europe, M. W. Johnson, to explain why the company had not been able to settle its income tax for 1958:

> The authorities continue to point to the large profits of the 'Mother Company' and insist that part of this relates to operations in Switzerland and that we have paid artificially high prices for products in order to contribute to the Jersey profit. With an economic boom in Switzerland they cannot comprehend why everyone is making profits except the oil industry.

Jersey's Swiss affiliate fought the charge because, as Thompson pointed out, for over sixty-seven years it had returned a profit in all except eight. When Berne authorities learned that *another* major oil company had discounted oil to an independent company, they apparently assumed that *all* majors sold to their affiliates at high prices, and then lowered them for sales to independent companies. Further, cut-rate Russian oil products had by now begun to appear in Switzerland—"Soviet oil offered in free world markets at artificially low prices motivated by the Russian Government's economic and political objectives," and not based on the objectives of sound business policy. These finished products for the Swiss market had been refined in Italy and Germany, and sold to independents "at extremely low prices"; the independents, in turn, sold them as bargains and made quick profits.

Some weeks after writing the letter, Thompson went to New York to see Contact Director Wolf Greeven. Meanwhile, Esso Switzerland had refused to accept a Swiss government proposal to juggle its income figures so as to make them show a profit. While in New York, Thompson prepared a paper to explain to the Swiss government just why the company had made no profit in 1958. He declared that both Jersey and Esso Switzerland sought only "to earn profits sufficient to provide reasonable returns" on their large capital investments in that country.[24]

What appears unusual in this instance is that the Swiss government was insisting that the affiliate conspired with its American parent company to transfer profits as a way of evading Swiss taxes. As usual, Jersey responded by forming a committee to study the problem, which was not considered in isolation. Jersey records show that there were other cases of friction with various European government agencies, usually because of comparable misunderstandings, especially in West Germany, and Italy.

Geyer of Esso A.G. wrote a letter to Greeven in 1961 and sent a copy to Director Stott (who then shared responsibility for marketing), in which he explained the West German Parliament's concerns in passing a new import tax that affected petroleum products. Geyer and his associates met with parliamentary and government representatives, finding that "again and again attention was called to the disproportion existing between the growing profits of the international oil companies and the P/L [profit-loss] statements of their German affiliates." Some of these persons alleged to Geyer that parent company billing for supplies "appears to be not in line with free market conditions, resulting in the fact that profits are shifted to foreign countries while payment of the German income tax is evaded."

Geyer suspected that the coal interest had pushed for an increase in the heating and fuel oil tax, and for quota regulation of oil imports. At the same time, rebating and discounting to affiliates and independents by some oil majors helped to strengthen the coal industry's position. Now, due to the serious nature of the problem, Geyer advised that Jersey at least adjust the price of heating oil.[25]

In Italy, in 1961, the director of the Jersey affiliate, Esso Standard Italiana [Esso Italy or ESI], Dr. Vincenzo Cazzaniga, told Greeven that the Italian Parliament had become interested in the matter of ESI's financial losses. Cazzaniga added: "It is to be expected that in investigating the balance sheets still to be settled, the competent authorities will examine with particular attention the question of crude prices." Later, with Jersey's assistance, he drafted a White Paper which supplied charts and other information on Middle East oil pricing, oil company profits, and the importance to Italy of stable oil prices. But as he predicted earlier, in Italy the issue of major oil company profits remained very much alive. In 1967, Cazzaniga sent Johnson (then a director of Esso Europe) portions of a report on the Italian oil situation that emphasized the alleged diversion of profits by parent companies; the "monotonous regularity" of losses by affiliates of the major oil companies, while the balance sheets of those companies were "characterized by very high profits." The report raised a question: What are the interests that keep the affiliate companies in Italy—and in the many other European countries—in the areas of refining and distributing? Then the report postulated an answer: the "forced division of the profits in the various phases," so as to concentrate all profit in production. While persons familiar with the Italian petroleum situation might well think that Enrico Mattei of Ente Nazionale Idrocarburi, in his campaign against the international oil majors, had successfully indoctrinated many Italians with his opinions, the fact remains that his views were commonly held in Europe.

These examples point up the fact that by the late 1950s and early 1960s, Jersey's European affiliates encountered problems elsewhere than in direct marketing. Profit margins began to decline, even as governmental pressures on the company actually increased. Given Jersey's tremendous expansion, based on increased product sales volumes and accompanied by large capital outlays, some affiliates—despite extraordinary efforts—still could not sell enough product to return a reasonable profit. Ultimately, this problem along with others prompted Jersey President Monroe J. Rathbone to begin the rationalization of the entire management structure in Europe and elsewhere, a program that his successors continued.[26]

X X X

As a general rule, the European affiliates sought to increase gasoline sales as the best way of maintaining satisfactory profits for Jersey and themselves. The rapid growth in heating oil sales—although welcomed—posed more complex problems for refiners. In fact, when Libyan crude oil became available, it proved to be especially adaptable to gasoline production and more difficult to convert to heating oil. So beginning in 1964, Jersey moved to stimulate gasoline sales by developing a "New Look" program for its service stations. Also in 1964, its consumer advertising began to employ one of the world's most successful advertising symbols, the famed Esso Tiger. First used by Esso Petroleum in England, the tiger soon migrated to the United States, where the "Put a Tiger in your Tank" program created something of a sensation. Later, in Continental Europe, the tiger proved equally popular with retail customers. Everywhere, he served as a powerful stimulant to sales and to Jersey's comprehensive brand identification program.[27]

During 1962, Wolf Greeven, regional coordinator for Europe, appointed a committee (which included representatives of many European affiliates) to study and evaluate the suggestion that the company's marketing interests could be stimulated by building "a network of travel stations located on the principal highways of western Europe." Passenger car registrations increased annually, and American tourists swarmed on the highways. Esso Sweden opened a small hotel adjacent to a full service station as an experimental unit in 1963, and the success of this Laxå motel encouraged the marketers. The Esso touring service quickly perceived the potential advantage of motels, as did the marketers of automobile products. Esso Denmark moved next, and on a larger scale. On February 19, 1963, the Jersey Executive Committee heard its proposal to embark on a program of building "apartment and office buildings, motels and restaurants at some twenty-nine locations in Denmark," adjacent to service stations. The Executive Committee "found" no objection and approved funds for construction. A week later, Esso Germany unveiled its own proposal to build a motor truck servicing terminal near Hamburg, with parking facilities for such vehicles, and a motel-restaurant combination. Again, funds for the project were voted.

H. T. Cruikshank, deputy coordinator of Jersey's worldwide marketing operations, told the Executive Committee on August 13, 1963 that the motel-restaurant committee report had indicated new opportunities for opening these units in many places, and that Swedish, Danish, German, and Italian affiliates had already begun construction of eight of them. Cruikshank explained that the affiliates planned to lease the motel-restaurant combinations to management firms; while regular dealers would run the service stations. In all, the Committee approved new proposals for eight other affiliates to build eighteen additional motel-restaurant facilities, fifteen of these to be constructed adjoining existing (or previously budgeted) retail outlets. Later, in 1964, Jersey's Executive Committee authorized a new company, Motel Services, Inc.* to manage these motel-restaurant combinations. At that time, only four had been completed—three in Sweden and one in Italy—while construction continued on six others. The *Annual Report* for 1965 stated that fourteen others would be finished by 1967.

More than 1,400,000 Americans visited Europe in 1965, and travel services optimistically predicted that the number would reach 4,500,000 in ten years. As the leading gasoline marketer in Europe, Jersey shared in the profits from this travel; gasoline sales increased from 805,000 b/d in 1960 to 1,215,000 b/d in 1965. While "making an attractive return on investment" may well have provided the primary, perhaps the only necessary, motivation for entering the motel-restaurant business, Jersey also saw it as an important addition to its well established touring service.

On May 19, 1966, the Executive Committee reviewed the status of the entire hotel program in Europe. Thirty motels had been planned, or were then under construction, while seven already had begun to serve the travelling public. The Committee reiterated that units ranging up to 100 rooms served company purposes best when built in conjunction with service stations. It also urged the affiliates not to build larger high rise structures in downtown areas. While endorsing further expansion, the Committee cautioned that growth should be orderly, with not more than ten new units opening each year.

Many of the executives who supported this pioneering diversification from Jersey's traditional petroleum business continued to be enthusiastic

*This became Esso Motor Hotels Inc. in 1966.

about it. In 1967, Cruikshank wrote ESI's Cazzaniga: "The simple facts remain that the European Motel Hotel Programme has given every indication of success with a vast untapped potential remaining that should prove even more successful with the experience that has been gained." Cazzaniga obviously seemed reluctant about this new business, so Cruikshank reassured him: "The Motel Hotel Programme is not intended to be a substitute for the Retail Development Plan but rather is supplementary to it . . . no other company, petroleum or hotel, has advanced to the great extent that Esso has throughout Europe. . . ." Both Jersey and Esso Europe Inc., Cruikshank concluded, remained convinced that the program would stimulate additional sales.[28]

By 1966, the effort to lure more of the travelling public to buy gasoline at the Esso oval sign had expanded in many directions. Snack bars were added at all of the motels, and Esso Motor Hotels Inc. developed car parking areas, "car parks," for tourists. All of these "parks" carried Esso products, and most marketers felt confident that they helped to sell gasoline. In 1968 alone, twenty-nine motor hotels opened, with nine others still scheduled for completion. All told, as the *Annual Report* stated, a total of thirty-eight in nine countries.

Greeven encouraged Motel Services, Inc. and its subsidiaries not to wait for choice locations to become available, but instead to search for them constantly. He suggested that all motel plans include "very ample drinking and eating facilities." Jersey built no cheap houses; rather, it made every effort to have the properties accord with scenic settings. (In Hanover, Germany, the winner of an architectural contest designed the motel.) Multilingual managers ran the motels. Snack bars provided food for those in a hurry, while each house had a dining room serving dishes indigenous to the particular country as well as international cuisine. Guests could confirm reservations at other Esso motels while domiciled in any of them. The service stations, fully equipped, provided not only gasoline and other products but also repair services and maintenance work. Most of them offered tow-truck service to handle breakdowns and accidents. While Jersey and its affiliates never did realize the profits earlier projected, increased gasoline and product sales indicated that the motor hotels had helped business. And while operating motels, Jersey did all facets of the business properly.

In 1968, the Executive Committee approved building a 153 room resort motel-restaurant on the island of Djerba, off the southern coast of Tunisia. So the chain continued to grow. Seven new motels opened in 1971, bringing the total number to forty-nine. In that year, the *Annual Report* credited them with "another year of profitable operation."

Up to this time, the motel strategy had been developed by Esso Europe with Jersey's approval. Any final evaluation of success or failure seems largely problematical; for even today the determination of how much such investments affected product sales remains difficult, and measuring the good will created by providing travellers with American-style housing must be deemed equally baffling. Originally, the entire venture had been motivated by the company's desire to increase product sales, and it seems evident from the continued expansion of the motel-restaurant chain that, up to 1971, Jersey believed, at the least, that the profits and other benefits derived from providing such services were worthwhile. In 1972, however, Exxon Corporation [Exxon; Standard Oil Company (New Jersey) became Exxon on November 1, 1972] reversed its earlier support, and sold or leased out twenty-seven units, retaining only thirty-two units (mostly in Sweden). The company pointed out that, even though the motel-restaurant chain—which had spanned the European continent—remained profitable, it nevertheless still involved "escalating amounts of capital and management time." In short, Exxon decided that henceforth it would concentrate in "the corporation's primary fields of oil, gas, and chemicals."[29]

X X X

After the first Suez Crisis, most European governments became very concerned about their increasing dependence on oil. They evidenced this concern in a variety of ways: some nations passed laws regulating and restricting petroleum marketing; others, like Italy and France, determined to promote their own national oil companies, and thus disagreed strongly with Common Market policies by adopting legislation that limited the operations of all privately owned concerns; and a few other nations passed laws designed to favor coal, and to slow down its displacement by oil as their fundamental source of energy. The 1959 Jersey *Annual Report* acknowledged this type of action: "Large surpluses of coal led certain European governments to take steps to give their coal industries artificial competitive strength." Elsewhere, government price controls hampered marketing from time-to-time, and some few countries mandated reserve storage.[30] Prodded by their governments, the state-owned companies of France and Italy soon began to explore the possibility of finding new concessions or forming joint companies in such producing countries as Iran and Iraq. All of these actions invited constant surveillance from Jersey managers and specialists, who acted to monitor marketing costs and profit margins.

Jersey contact directors for marketing, among others involved in that function in Europe, responded to these government actions by studying reports incessantly, by constant travel and personal contact, and by conducting meetings on timely topics with affiliate committees or their spokesmen. For example, in April, 1963, Director Stott and six other Jersey officials attended the "18th Scandinavian Committee Meeting," where they first listened to reports on supply problems. W. D. Bayles, Jersey's political expert in Europe, explained some problems connected with Common Market developments—the French had objected to many of its policies, and no common energy policy for Europe appeared to be likely in the near future. Bayles identified as the principal barriers to such a policy: "The level of European coal production; import duties on petroleum products; special treatment of petroleum imports from preferred areas; petroleum imports from the Soviet bloc; refinery protection" and security stockpiling. Others at the meeting reported on the particular aspects of these problems that affected the Scandinavian countries, including pending legislation.[31] While such meetings may have proved helpful in alerting affiliate managers to potential difficulties, and while they may have helped to erase nationalistic tendencies that could hamper concerted action, they also proved both time consuming and expensive.

After the Arab-Israeli War of 1967, the attitude of European governments toward international petroleum companies changed markedly, despite the fact that, because of the companies' joint effort under the sponsorship of the United States (previously discussed), Europe actually suffered little economic damage during that war. But the later boycott of oil moving from most of the Arab states to the United States, Great Britain, Holland, and West Germany dramatized what *could* happen. Petroleum, principally from the Middle East, supplied Western Europe with almost 60 percent (in 1966, 8,576,000 b/d) of its total energy needs. During the Suez Crisis of 1956-1957, Jersey had been able to draw on Western Hemisphere supplies. In 1967, however, even the appearance of "Supertankers" did not reassure the heads of European states. For one thing, tanker rates had risen sharply, while consumption in Europe had also risen precipitously. In Europe, the short duration of the 1967 war had allowed the Organization for Economic Cooperation and Development [OECD] to limit its activities to fact-finding, for which purpose it created a special committee, the International Industry Advisory Body [IIAB]. This committee found that, while product stocks had fallen below desired levels, a building-up process had begun. It estimated that the desirable quota would be reached by January 1, 1968. In Great Britain, stocks available for another emergency already averaged 83 days' supply for all products. British concern over the problem centered largely on the quantity and availability of future oil supplies during emergencies.

John Angus Beckett, Under Secretary in Britain's Ministry of Power and head of its Petroleum Division, planned a visit to New York in September, 1967 to discuss these problems before going on to more such talks with Venezuelan officials. Esso Europe's President Nicholas J. Campbell suggested that Jersey send him as much information on Venezuela as possible, in order that he might be able to brief Beckett beforehand.[32] Some two years later, Jersey Vice President Milbrath, in a letter to Esso Europe's Vice President Cruikshank, described a meeting with Beckett and others on the same subject. Essentially the British officials sought Jersey's views "on the security of supply in the face of heavy dependence on Arab oil." The Ministry of Power regarded the Esso Petroleum inventories as very low. Milbrath pointed out that Jersey continued the search for oil in order both to diversify its sources of supply and to limit its dependence on any one source; and that the company's "flexible European supply arrangements" would help it meet any such difficulty.[33]

Governmental policies, cheap Russian oil, and price cutting by those large independent companies with new Libyan oil to sell added to the seemingly ever-increasing producing capacity in Middle East and Mediterranean concessions, to drive prices down. Then a new factor increased the complexity of the marketing problem in the late 1960s: Groningen gas. A joint Jersey-Royal Dutch-Shell Group [Shell] company exploration crew brought in a vast field of dry gas in the Netherlands province of Groningen in 1959, and as Jersey began to market its share of the gas, it naturally competed with heating oil. Gradually the plentiful supply of this gas forced Jersey to renovate its refineries as the cleaner fuel, gas, took over a larger share of the market and forced a decline in heating oil sales. As a result, Jersey affiliates' profits tumbled. Only Esso France could show a profit in each year of the 1960s, and that company still did not have sufficient income to finance all needed improvements. Jersey officials repeatedly discussed outstanding problems with executives of European affiliates, and advised them about what to do to make their operations more efficient. Yet the great difficulty remained: too much oil and products were being sold with too small a profit margin. And by the 1960s, with most of Jersey's refineries using Middle East crude, with its high sulphur content, public concern over environmental damage in Europe was mounting. This also served to increase the sales of Groningen gas.

Taken as a whole, Jersey's European affiliates actively supported

Conservation of Clean Air and Water in Western Europe [CONCAWE], a Hague-based body that advised the Council of Europe on pollution matters. Jersey joined other major oil companies in voluntarily agreeing to stop disposing of oil and contaminating waste in the ocean in 1964. Product costs necessarily reflected the changes in environmental standards; even simple changes, such as required labelling and packaging alterations, proved expensive. Legislation imposing new standards usually demanded additional capital outlays, and not always were the affiliates able to increase prices to amortize them. Jersey possessed the know-how to meet different standards for fuel oil; but as Milbrath reported to CEO Michael L. Haider in 1969, it seemed "hard to visualize how increased costs due to desulfurization investments could be recovered in the marketplace."[34]

X X X

Elsewhere in Europe, Jersey's marketing organization by the late 1950s continued its work of increasing the volume of total sales, while at the same time maintaining satisfactory profit levels. An especially attractive market was large-scale entry into those European states where marketing activities had previously been limited—Greece, Spain, and Portugal. Again, the Libyan discovery required additional markets, and these same countries seemed to be obvious outlets for some of that oil. In 1959, Jersey incorporated a new affiliate, Esso Mediterranean, Inc. "to coordinate activities of affiliates in the rapidly expanding markets of the eastern Mediterranean, North and West Africa and the Iberian Peninsula." This company moved beyond marketing when it considered manufacturing petrochemicals in Spain in 1963. A year earlier, Jersey took steps to establish an industrial and marketing presence in Greece. In association with Thomas A. Pappas, a prominent Boston businessman of Greek origin, an agreement was concluded with the Greek government calling for the construction of a 55,000 b/d refinery and of a petrochemical complex. Plans were also made for the possible construction of a steel plant. The agreement with the Greek government was "validated by decree in December 1962." As part of the agreement, Jersey received a ten-year (1966-1976) crude oil supply contract for amounts of up to 2 million tons per year.

The association with Pappas arose out of the latter's desire to encourage U.S. investment in Greece to help develop the economy of his mother country, and Jersey's desire to extend its activities to Greece. Thus it was that in 1963, Jersey began to provide "financial, engineering and managerial assistance to an industrial development project in Thessaloniki, Greece." That same year, Jersey's Executive Committee also approved a plan to establish 200 service stations in Greece over the following ten years and to construct a marketing terminal and an industrial complex near Athens. In 1964, a Jersey affiliate began to market gasoline in Greece and in 1965, Jersey incorporated Esso Standard (Hellas) Commercial & Industrial Company A.E. to handle the marketing effort.

Paralleling development by Jersey of its refining, petrochemical and marketing interests, Pappas moved forward with plans for the construction of a steel manufacturing plant with interest free loans of $12.25 million provided by Jersey to the Pappas interests.[35]

As a result of changes in the Greek government and the coming to power of political leaders who had been critical of the agreement concluded with the government in 1962, Jersey and Pappas were required to resume negotiations with several successor Greek governments. These political complications resulted in delays in the construction of the plants in Thessaloniki and prompted a thorough review by Jersey's management. Jersey's newly established Board Advisory Committee on Investments [BAC] decided that the expectations for early profitability of the Thessaloniki plants had been overly optimistic. Nevertheless, Jersey authorized continuation of the negotiations with the Greek government. In the end, the refinery and the petrochemical plants were completed in 1966, and later, Jersey acquired formal ownership of all the shares of the refinery and petrochemical companies which had been funded by the company.[36]

Spain offered a different kind of opportunity for Jersey. In 1962, a company study indicated that "economic conditions in Spain are favorable at the present time. . . . The climate for investing in Spain is good" and that Jersey should capitalize on this opportunity for expanding its marketing despite the fact that a virtual monopoly controlled by the government existed there. The Esso Mediterranean, Inc. board of directors studied this memorandum and then advised against entering Spain in 1962, but they did suggest that communication with the Spanish government be continued at six-month intervals, and that the company bid on Spanish crude oil contracts.

A year later, in April, 1963, the Jersey Executive Committee reviewed Director Stott's report advocating participation with other private investors in a large petroleum complex in Spain—refinery, fertilizer plants, petrochemical manufacturing facilities, tourist facilities, and other developments—totalling $240,000,000. Stott, who negotiated with Spain, indicated to the Committee that further action would require "detailed discussion and investigation." R. B. Carder, Jersey regional assistant to

Stott, told Greeven in August, 1963, that he had informed Jersey officials in Spain that they were authorized to form a new company (a Spanish law required that 25 percent of the company be owned by Spanish shareholders). Carder continued, "The Spanish shareholders mentioned are all well known figures in Spain and apparently we have had extensive dealings with all of them." The Banco Español de Credito [Banesto] also would take shares. Carder, in a subsequent memorandum to Stott, advised caution, especially regarding refineries; if all the proposals "be approved, there would be considerable surplus capacity in 1966." Regarding petrochemical manufacturing, given the government project, plus another joint private industry proposal, "it is certainly questionable that the Spanish market will justify two such ventures." Finally, other asphalt plants due to come on stream "will take care of the Spanish requirements for some time to come." Further information on proposed plants, Carder had not found "readily available."

Still negotiations continued. In 1965, Esso Mediterranean, Inc. turned over all operations in Spain to a new company, Esso Standard Española. Then, in the following year another corporation, Esso Iberia Inc., took over consulting (and other services), to Jersey affiliates in Spain and Portugal. Jersey next created another affiliate, Esso Petroleos Españoles [EPE], to handle its Spanish petroleum business. EPE built a refinery at Castellon de la Plana in which it held a 40 percent equity share. Esso Europe planned to supply it with naphtha and liquefied petroleum gas [LPG] after refinery operations began in 1967. As Campbell, then president of Esso Europe, wrote Jersey CEO Haider in February, "The situation beyond 1967 is undetermined. . . . We can expect that the government will call upon us to contribute to the domestic market in increasing amounts." The partners in the joint company who held a 25 percent equity share, according to Campbell, became dissatisfied with the projected rate of return during the start-up years, and now they wanted to reduce their holdings. One reason they desired to sell was that government restrictions on tanker transportation of product—"pirate prices"—Campbell termed them, would limit profits. He found the Spanish partners unenthusiastic about investing in other contemplated plants. In April, the Executive Committee authorized Campbell to begin negotiating for the purchase of the Banesto shares in EPE.

Unanticipated political changes occurred in Spain in 1967-1968, and in February, 1968, the new government began to review proposed legislation that would restrict the petroleum business even more. Robert T. Bonn, head of the Spanish affiliate, cabled Greeven that these "clearly represents [sic] retroactive change in rules of game which were in effect when Esso Petroleos was formed. . . . [it] is only refinery to which [the] change applies." The growing balance between world demand and productive capacity, followed by demand outstripping supply, caused the Spanish situation to deteriorate rapidly thereafter. Early in 1974, *ABC*, Spain's largest and most influential newspaper, in a pro-Organization of Petroleum Exporting Countries [OPEC] editorial, called for nationalization of all property owned by the major national oil companies. Fifteen months later, Clifton C. Garvin, Jr., Exxon president, authorized Campbell to negotiate the sale of EPE and of its sister company in Spain, Productos Quimicos Esso [PQE]. Garvin stated that this decision had been reached after thorough study, and despite the fact that "current and projected future profitability of both companies is acceptable by normal Exxon standards." Exxon, Garvin pointed out, believed that it could put "to better use the crude, capital, technical and management resources currently committed" to those two companies. Finally, in 1975, Exxon divested the principal part of its Spanish operation, retaining only a marketing unit dealing largely in lubricating oils, aviation gas, and specialty products.[37]

To the west, the Jersey company, Esso Portugal, had developed a small market in jet fuels in Portugal by 1955, though it held no land marketing permit. Through contracts with the Portuguese government, it had assisted Esso Exploration Guine Inc. in obtaining a concession in Portuguese Guinea. Jersey later surrendered that concession in 1961, but the company also sought to continue working through Esso Guine. In 1962, Jersey proposed to build a refinery at Oporto, Portugal, but changes in the government ministry slowed the final decision until June, 1963, when the projected refinery was turned down. In the discussion that followed, the Portuguese Minister of Economy advanced no substantial reason for the negative decision. He did state that "he would like to see us expand our [Jersey's] activities in Portugal, that it was his intention to grant us a distribution quota for the Portuguese market . . . since at the moment he could not see any 'political' objection." In fact, the minister pressed Jersey to build a refinery and oil bunkers in the Cape Verde Islands, but Clark R. Egeler of Marcor indicated that economic analysis showed clearly that a refinery investment without a market outlet in Portugal remained "unattractive." This analysis caused Esso Portugal to review its investment program, and the fact that the minister indicated his intention to grant Esso Portugal a marketing quota, Egeler said, "could possibly provide the opening into the Portuguese market which we have been seeking for some years."

Carder sent a copy of the Egeler letter to Director Greeven, who wrote across the page before he returned it, "Portuguese black bean soup." At this point, negotiations seemed to be stalled, although Jersey went ahead to request a marketing quota. Subsequently F. B. de Castro Neves, general manager of Esso Portugal, wrote Egeler that the minister would not grant an import license until he could be assured that Esso Portugal intended to build a terminal. The Castro letter ended Jersey's effort to secure a share of the Portuguese oil business, except for a limited marketing operation.[38]

Each of these three European ventures received the usual deliberate and careful study that accompanied Jersey moves. Yet it should be obvious that the company did not derive the expected benefits from them. Few, if any, Jersey executives anticipated the many changes that took place in the world of oil during the 1960s and 1970s; yet, given the available facts, Jersey's decisions to move into Greece, Spain and Portugal must now be regarded as sound. As world events developed, however, the company was lucky to escape from the aftermath of these three decisions without greater losses.

X X X

As a whole, total profits from European affiliates improved in 1967, even though the Middle East war occasioned higher freight (and other) charges, raising overall costs. Jersey emphasized its natural gas, gasoline, and heating oil sales in an effort to keep profits up. In 1968, Esso Europe assigned a unit value to each service station and then began a program to make each one a more efficient sales unit. Marginal outlets were simply closed; and modern, large-volume stations replaced them, located nearer to newly developing suburban areas and interstate highways. Some stations introduced self-service, and some included car care centers. The variety of Esso brand products on the shelves increased—to include tires among other necessities—and each new product was carefully designed to attract customers and stimulate sales. Seven car care centers opened in the Scandinavian countries. Elsewhere, the marketing of natural gas began "to reach impressive levels."

This emphasis on quality development continued in 1970, as affiliates also "persevered in their efforts at greater selectivity." In that year fuel oil sales increased 15 percent, and distillate sales rose by 11 percent. Meanwhile, the continued blockage of the Suez Canal increased tanker transportation charges, and the oil producing countries imposed higher taxes—so product prices went up sharply.[39] At the same time, environmental restrictions on refining and petroleum products also added additional charges against the narrowing profit margin.

For oil companies, 1971 brought its own share of marketing problems everywhere, and Europe proved to be no exception. Although European consumption of petroleum products declined, reflecting a general slowdown in economic activity, 1971 sales about equalled those for 1970. OPEC moved closer to dominating world production. Member states continued to raise prices on oil, which in turn brought more cautious consumption. Actually, reserve stocks increased late in the year, when demand sharply declined. People continued to travel, however, and European gasoline sales rose 4.5 percent for the full year.

Its *Annual Report* for 1972 pointed out that Exxon "continued efforts to enhance downstream return by improving marketing profitability in the face of continuing cost inflation." One way to achieve this cost efficiency was to eliminate uneconomic retail outlets: in 1971, the company closed 294 stations in Europe; in 1972, it shut down 576. Meanwhile, the Marketing division applied more rigid tests before determining to open new sales units, in promising areas, as Jersey built fewer stations in 1972 than in 1971. Yet overall, the European market made a larger contribution to Exxon earnings in 1972 than it had in the previous year, due largely to strong growth in natural gas and chemical product sales.

The oil producing states moved again to increase per barrel prices during 1973, and late in the year, OAPEC [Organization of Arab Petroleum Exporting Countries] imposed production cutbacks (as well as embargoes on supplies to Holland and other countries). Despite these obstacles, Esso Europe's sales rose 4.3 percent, because of economic growth in the region and the company's success in competing for traditional markets. In all, 1973 proved to be a record-shattering year, even though, during the last several months, the Arab production cutbacks forced the company to allocate supplies to its customers.

This allocation of oil continued in 1974, while some of the producing countries lifted their embargoes against the U.S. (and several other nations) and increased production. Several OPEC members continued to increase charges for their oil, and higher energy prices helped to produce an economic slowdown, which in turn brought a decline in sales. So product inventories rose at the same time that governmental price controls on many products forced company marketers to spend many hours in a special effort to make these programs work. In general, however, sales declined worldwide. This downturn continued in 1975, when European sales dropped 10 percent, with "gasoline the only product not sharing in the decline." Profits also declined partly because of the price controls in many countries and despite the closing of marginal retail units.[40]

Overall then during the late 1960s and afterward, although Jersey (and later Esso Europe) worked hard at marketing, profits and the affiliates' share of the market continued to demand concentrated attention as they fluctuated. The record is confusing, although now it seems clear that Jersey's loss in market share went not to other long established major international competitors, but rather to newer companies that entered the European market with newly found Libyan or Algerian or Nigerian oil, obtained at low prices and without extensive capital investments. By the middle 1960s, chronically poor profits led Jersey to commission new studies to determine "optimum marketing strategies" for Great Britain, Germany, and Switzerland. These joint exercises called for affiliates to have "the option of sharing in the industry growth pattern," together with the alternative of deciding not to participate in such growth. Again, Jersey wanted a full consideration of the total business of each affiliate—whatever strategy might lead to improved profitability, even a program of divesting business properties. Jersey planners no longer could content themselves with the general theory that Europe provided simply an outlet for products. Now Europe represented a target for potential profitable investment, something much more than a market.[41]

This new strategy continued to define affiliate activities as the company entered the 1970s. Particular examples, such as the sale of motor hotel properties and of certain investments in Greece and Spain have already been discussed. As much as any specific decisions, these moves symbolized Jersey's changed pattern of business philosophy, and the shortage of oil after 1973 dramatically accelerated this general trend toward selective investment.

X X X

Despite healthy profits in 1973, the year the producing states first successfully restricted production, the economic outlook for Exxon (and for other international oil companies) appeared unsettled. The company confronted the situation directly. Exxon had evolved a new regional administrative structure during the 1960s; the revamped regional units responded to the oil shortage with a "feet to the fire" philosophy that had become a tradition at the parent company. Uneconomic units were quickly reorganized or sold; personnel were reduced in number; and a general belt tightening ensued. Marketers in Europe no longer had to conceive ingenuous plans for increasing sales volume; now they sought the most profitable way to dispose of severely limited supplies of petroleum products. Meanwhile, since the public failed or refused to understand the new complexities of the oil problem, they loudly insisted that the international oil companies were the real villains in the world drama of petroleum.

In general, these and other problems led the Jersey Company to adopt an overall strategy aimed at greater autonomy in the field and more centralized operations—in a word, regionalization (discussed elsewhere in this volume). Of course, Europe had always been a particularly difficult area in which to exercise proper management controls, and during the post-World War II period, the management task was further complicated by the activities of the Common Market. So by the time Jersey formally moved to create a unified European regional structure—Esso Europe Inc.—in 1966, some European managers had already begun informal cooperation across national boundaries. In short, by 1966, European management was ready for the Jersey move.

But even so, merely by imposing a new regional level of executive management, Jersey did not quickly solve all the familiar problems of doing business in Europe. Nevertheless, in 1967, J. Kenneth Jamieson, Jersey's president, after a trip through Scandinavia, foresaw fair prospects for the new plan: "The regional organization is maturing into an effective instrument for improving operations in Europe," he wrote. "The manner in which Esso Europe is serving as a vehicle for executive development, . . ." also seemed impressive. This talent search, always ongoing at Jersey, had earlier been limited in each of the national—hence smaller—European affiliates. Now, because of the change, Jersey's integration of all its European affiliates into "overall forecasted programs" proved much easier than expected. At the bottom line, Jersey's success in reaching better profit levels owed much to regionalization.

In the early years of Esso Europe, however, new lines of authority had to be created between Jersey's established inter-regional and functional companies and the new regional group. For example, in 1967, Director Siro Vázquez agreed to settle a dispute between Esso Europe and Esso Exploration Inc. First, he pointed out that while neither company received exactly what it wanted, the agreement "preserves the security of exploration information which is essential to us"; Esso Europe would receive the necessary general information regarding exploration matters, and the other company would "obtain Esso Europe's guidance on all exploration questions which may involve future marketing problems, development commitments and European policy angles." As Vázquez demonstrated, effective coordination in Europe required that such distinctions be made.

For three quarters of a century, European statesmen as well as affiliate executives had become accustomed to dealing with Jersey leaders on a personal basis, and many statesmen initially disliked the intrusion of the new regional management of Esso Europe between national governments and the executives of the parent company. Britain's Angus Beckett registered such disapproval to Robert Milbrath in 1969. Beckett insisted that his relations with Jersey had been "placed somewhat in jeopardy" by regionalization, that "on some of the broader international oil matters they [the British Cabinet] had felt greater satisfaction in discussing these matters with the appropriate people in Jersey rather than with Jersey's representatives in Esso Europe." Milbrath tactfully replied that the Esso Europe managers could "open any of our resources" to Beckett and privately suggested that all Jersey directors who visited Great Britain and other countries should pay courtesy calls on such prominent government officials as Beckett.[42]

More important, despite all the difficulties, President Campbell and his Jersey associates did manage to draw the European affiliates closer together. Financial profits achieved after 1966 represented only one form of their success, which also had to be defined to include the new spirit of cooperation that Esso Europe engendered. Certainly, the reorganization of business activities in Europe played an important part in helping Jersey [Exxon] get through the next troubled decade in world oil affairs.

X X X

Chapter 11

Jersey and Russian Oil in the West

THE PROBLEM OF RUSSIAN OIL as it emerged after World War II was not a new one for Standard Oil Company (New Jersey) [Jersey]. The company had marketed products in Russia even before 1900; and after World War I, Jersey moved to purchase properties near Baku (from the Nobel interests)—which deeply entangled the company in Russian oil affairs. After the first world war, Russia actively marketed oil from expropriated properties, which served to complicate further its relations with the American company. Over almost all of this time, Jersey sought to realize something more from its multi-million dollar Russian investment, either through compensation for lost property or from specially priced oil that Jersey could sell. In its effort to solve this continuing dilemma, the company worked very closely with the U.S. Department of State [State]. But clearly, Jersey had made a poor investment in Russia. Oil properties seized in the Bolshevik Revolution were never returned nor was adequate compensation provided.

In fact, after the 1920s, little formal contact between Russia and Jersey occurred—until the end of World War II, when a new complexity emerged; such problems as Jersey and other international oil companies had with Russia now derived principally from properties seized by Russian satellite states.

After World War II, Joseph V. Stalin continued to serve as Premier of the Union of Soviet Socialist Republics [U.S.S.R.] until his death in 1953. Under his control, that country developed a "coal-steel-based economy" geared to a policy of economic isolation. This Russian policy largely answered the demands of the Cold War, and of the containment philosophy that characterized the foreign policy of the United States and its Western allies. Yet Soviet coal-steel economics proved expensive, especially in the case of railroads, which required large amounts of capital and labor. After Stalin's death, and several years of political struggle, Nikita S. Khrushchev emerged to become Premier of Russia in 1958, a position he held until 1964. In that year, Aleksei N. Kosygin became Premier, and Leonid Brezhnev rose to first secretary of the Communist party.

Khrushchev radically altered the nature of Russia's economic development. Some observers have described this change as a "Soviet Economic Offensive" because it once again introduced into world—and especially European—markets such recently unavailable items as Russian crude oil, and also because the new Soviet trade (and aid) arrangements involved both barter and credits. Already, several Eastern European Communist states had defied Soviet preferences and acted on their own to expand trade with the West. Further, during the five years between the death of Stalin and Khrushchev's takeover, several revolts in these satellite nations had helped to create pressures within the Soviet system that forced Russian leaders to revise their economic policies. As a result of this relaxing of barriers, trade between Soviet bloc nations and the outside world more than doubled in value between 1953 and 1959.

At the twenty-first party Congress, in 1958, Khrushchev announced new goals. These included increased production of oil and gas, and more emphasis on lighter industry—a shift away from coal-steel toward more sophisticated products. This step, the Kremlin believed, would free manpower and capital and enable the Russians to develop such new industries as petrochemicals. It would also mean trading with non-Communist countries for many products needed in the Communist world, including food and consumer goods.

The Russian satellite states applauded this policy reversal. In general they hoped to further their own economic development through such trade, and they welcomed Russia's overtures to the West. Among the many raw or semi-processed products Russia possessed, for which ready markets existed or could be created, was crude oil, in part because Russian refining had lagged behind crude production.

In retrospect, the fearful overreaction of the Western nations to Russian oil sales appears to have been largely baseless. Economists and analysts at Jersey made several in-house studies which showed that if Russia, across the years ahead, sought to satisfy its own anticipated consumer demands, its then-known reserves of oil could barely supply the needs of the Soviet world. But in the United States, an illogical fear of every Russian move had developed, and it became habitual during the Cold War; American paranoia over oil sales now seems to have belonged to that pattern of misconception.

In the U.S., some international oil companies assumed that the Russian purpose was to drive down the price of oil in already over-supplied markets, thereby injuring Western industry. Of course, Russian competition could create production problems for the majors, especially in Middle Eastern countries, where consistent pressure for annual production increases existed.

At the same time, producing nations would be injured by decreases in per barrel prices. And each new consumer state gained for Soviet oil would mean another nation tied to Soviet supplies, thus making its economy subject to Russian influences.

Perhaps this sudden reversal of Russian economic policy, at that time, could not have been viewed in any other way. Yet most Western fears proved groundless, as later events have demonstrated. Among the international oil companies, such as Jersey, the sudden appearance of Russian oil in the market, together with a realization that much of it sold at very low prices—far below the usual levels—made "political oil" a cause for nervous fear.[1] In fact, Jersey reacted in several ways: One of these grew from the earlier post-World War II property losses, at the hands of Russia and of her satellite nations—by expropriation. These losses, while not large when measured by Jersey's totals, nevertheless represented considerable sums of money. In response to company claims for compensation, moreover, little action from any of the host nations had been noted. So Jersey's pride now hurt. Already, Chief Executive Officer [CEO] Eugene Holman, with the backing of the full Board of Directors, had established a policy that virtually forbade any type of operation in or with a country that had expropriated company property or nationalized concessions. For many years this principle endured, and when the company finally abandoned it, political pressures from the United States government played a major role in the change.

Jersey entered legal claims against both Hungary and Rumania in August, 1955, when the United States Congress passed Public Law 285, under which the company filed for approximately $68 million and $60 million respectively. [It is interesting to note that while Rumania paid $11,300,000 of the award allowed Jersey (about $31,700,000) by 1975, Hungary paid less than a million dollars (of $28,000,000 awarded).] As a matter of good business practice, furthermore, even when economic studies indicated that the company could make a profit, Jersey refused to buy oil from countries that had expropriated its assets. In fact, Jersey leaders had no reason to trust Communist intentions, and many reasons to question them. So the Cold War policy of the West conformed nicely to Jersey's views. In 1960, and later after, the company suspected that Soviet Russia intended, largely through its crude oil export policy, to injure or even destroy Jersey and other major oil companies as forces in international trade.

How many other American business leaders agreed with Jersey's attitude of prohibiting trading with the U.S.S.R. and its satellite states remains uncertain. But clearly, as Soviet interest in trade increased, a considerable segment of Western corporate opinion began to endorse the search for new markets in the Communist world. No so at Jersey! At conferences held in Petersburg, Virginia, in June, 1958, and Tarrytown, New York, in April, 1959, Jersey leaders discussed with their peers in the U.S. business world the particulars of Russian "economic warfare" and of its impact on Western trade. Subsequently, in a note to David Rockefeller, Jersey Vice President Leo D. Welch echoed these conversations: "Unless we insist upon respect for patents, copyrights, contracts, etc. before trading starts they will never be honored." Then, he went on to say that there was considerable "fuzziness" surrounding Soviet credits for consumer goods. Welch believed that the West should "only give ground if the Soviet is willing to give up its cold war, subversions, Berlin demands, Germany partition, etc., etc." He added, "Our weapons are not too many and we should make the most diligent use of those which we have, and one is credit."

Shortly after economists from Jersey Coordination and Planning Department had completed an appraisal of Russian crude oil in existing (and future) world markets, in the spring of 1960, George T. Piercy, the manager of that department, toured the Soviet Union as a member of an American oil industry delegation. This group visited most Russian producing and refining areas, seeking to answer the overriding question: "What does the Soviet drive to increase production mean to the world petroleum industry?" While Piercy found no clear answer, in Baku, one of the oldest oil producing areas, he did observe "a fluid catalytic cracking operation that duplicated a process patented by Esso Research and Engineering Company. Our hosts did not explain how the design had been acquired, but they were eager to know what improvements had been made in later models." Piercy finally decided that this Russian plant represented "a carbon copy" of the first one that Jersey had constructed at Baton Rouge, in 1942, right down to a sign reading "Model One." The Russians had changed "absolutely nothing." Since neither the United States government nor Jersey ever furnished plans for this type refinery, the Russians had obviously duplicated the original facility without receiving the permission of its owner. Nor was this incident unique.

In a "Statement of Position On Oil Trade With The Soviet Bloc" dated June 26, 1963, Jersey clarified its position:

> Jersey is convinced the U.S.S.R. had singled out the oil industry for attack because of its importance to the economic strength of the free world, because its size and efficiency make it a major symbol of successful private business operation, and because it provides a major link between nations with a high degree of industrial development and those that are less developed.

This position paper noted that references in an "authoritative Soviet publication" had singled out the petroleum industry in the West as a basic target for attack, and identified oil as the foundation of Western capitalism. Further, the Jersey statement repeated a fact well known in the West: Soviet trade policy always had political, and often even military, implications.

Later in 1963, the *Petroleum Intelligence Weekly* (October 7) announced that an engineering consortium headed by a French concern, and including Dutch, German and British firms, was drawing plans for construction of a 240,000 barrel per day [b/d] refinery in Russia. Jersey scrutinized this story carefully, and sent it to experts for comment. The American company, like other international oil majors, insisted on tight security in agreements with European construction firms—especially with regard to equipment and processes and, in most cases, complete refinery designs. Jersey's general interests seemed at risk, but finally the company's economists and production analysts found most interesting the statement that "Cat[alytic] cracker and alkylation units which the consortium cannot supply because they are based on U.S. patents will be built by the Russians themselves." Jersey's experts doubted that any single European firm possessed enough information to build an advanced type of refinery, but they also agreed that a combination of European firms could do the job. So the Russian situation attracted even more attention in company offices.

On October 22, 1963, the President of Esso Research and Engineering Company, W. C. Asbury, wrote to Jersey Director Harold W. Fisher regarding Russian refinery plans. Asbury recalled that some refineries had been built in Russia during World War II, with designs based on the Houdry [Eugene P. Houdry] process for cracking oil. To Fisher, he suggested that a picture of a refinery in Stalingrad resembled one built by the M. W. Kellogg Company in Yugoslavia just after World War II; and that, more recently, the Russians had visited modern refinery units in Cuba, nationalized after Fidel Castro's 1959 revolution. And Jersey itself, under an agreement with the Iraq government, had built a refinery in Baghdad before the revolution of 1959 took place there. In Asbury's opinion, then, the Russians did possess sufficient information.

Several days later, Fisher learned from another source that, after the Iraq revolution, the Russians placed as many as forty advisors at the Baghdad refinery; they "had access to all the plans, specifications and operating manuals for the units," but still there was no catalytic cracker at that plant. About Cuba, stories differed. Some versions of oil lore had the Russians dismantling units and then shipping them to Russia. The best information seemed to establish that the Russians had built some catalytic crackers and catalytic reformers, but: "They do not have the technical know-how to build modern low cost units and to operate them efficiently . . . they do not know how to produce good catalyst for these plants." Ultimately, the U.S.S.R. confirmed these suggestions of deficiency by seeking to buy a modern refinery from the West. But Jersey at least proved adamant in its opposition to every such proposal.

Finally, the Jersey Board Executive Committee endorsed a policy statement (issued January 15, 1964) that, in essence, would guide the Jersey Company for almost a decade. Entitled "Soviet Bloc General Trade and Oil Trade: A Review and Recommended Jersey Policies," it succinctly stated three reasons why Jersey did not encourage freer trade with the U.S.S.R. and its satellite nations: First, Jersey believed that the Soviet bloc sought to achieve world domination through economic warfare. Second, the U.S., acting as an example for its allies, should take a "more restrictive stand . . . in order to exercise moral suasion on them." Since Jersey represented one important company in an industry adversely affected by Soviet bloc trade, it should remain as guarded as the U.S. government, "and perhaps more so," in order to be effective in "enunciating a restrictive policy with our government." Third, even non-restrictive trade would put pressure on the Soviet bloc "to export oil in order to gain foreign exchange to pay for its imports." This statement went on to exclude such trade as equipment used for peaceful purposes, drugs, consumer goods, etc.—unless such trade involved bartering or long-term credits. At that time, the Jersey Company affiliates' trade with Russia amounted to less than $200,000 per year.

However, later in March, 1964, Jersey's Government Relations Department Manager, Luke W. Finlay, wrote Director Wolf Greeven to indicate that, while the thaw in U.S. and U.S.S.R. diplomatic relations had affected Jersey only slightly, other U.S. businesses seemed to have revised their Cold War attitudes. Finlay appeared shocked by the suddenness of this change in stance as announced by certain businessmen at the Business International Roundtable discussion held in Washington, often "to the effect that somebody is going to trade with Russia and the United States ought to jump into the contest, and pick up a share of the trade for itself, rather than leaving it to all the others." Jersey, on the other hand, stuck to its course, despite what other businesses did and despite the "on again, off again" diplomacy of the United States government.[2]

X   X   X

Jersey began to be aware of Soviet exports to Western Europe as early as 1953, when shipments of Russian crude oil to the West averaged 35,000 b/d, but some two years before the U.S.S.R. became a net exporter. That the matter continued to interest Jersey executives as well as its affiliates is evidenced by a draft of a letter of April 2, 1954, to all [European] affiliates, by European Contact Director Peter T. Lamont. He pointed out that the oil sold or bartered on a government-to-government basis would lead the purchaser to require the industry to process it. This, Lamont continued, was "a matter of basic principle, in that such requirement is tantamount to expropriation in whole or in part of our investments to the extent used for the benefit of the state to the exclusion of the rights of the owners." With minor alterations, his statement summarized the Jersey position over the next several years.

In fact, Jersey's leaders correctly anticipated the economic problem. For example, while visiting affiliates in Scandinavia in September of 1953, Director Chester F. Smith reported an increased availability of Russian oil products at prices below the world market. He noted that a customer of Jersey's Danish affiliate, Dansk Esso Aktieselskab [Esso Denmark], had already begun to purchase Russian fuel oil, and that the Soviet Union sought to include petroleum supplies in a pending bilateral trade agreement, to be signed with Denmark in July, 1954.[3] In December, 1953, C. L. Bauer of the Coordination and Economics Department prepared a study of the "more aggressive attempts" underway "to expand the oil trade from behind the Iron Curtain to free world areas." Basing his analysis largely on trade agreements signed during the last half of the year, Bauer estimated Communist exports to Western Europe at 63,900 b/d of crude oil and products, excluding the prospective sales to Denmark.[4] Though the amounts involved were relatively small, Bauer warned of "potential competitive pressures in the period ahead." Even assuming that the Russians could dispose of 100,000 b/d in 1954—unlikely, Bauer noted—the Soviet share would only represent 1.5 percent of total foreign demands. Yet even this small quantity "could cause serious competitive impacts in some countries."[5]

His understanding of history, particularly of the historical pattern established by the Soviet Union after 1917, when it re-entered the European oil market after the revolution, worked to magnify Bauer's apprehension. He knew that, after having withdrawn from world markets for several years due to internal domestic instability, the Soviets re-entered during the 1920s and quickly captured 14.5 percent of foreign demand by the early 1930s. In that history, Bauer saw a parallel to the new Russian drive to expand exports and regain Western markets after World War II. Earlier, when "the Russians invaded the oil markets of Europe to recoup their lost position," Bauer warned "price considerations affecting direct sales were not important in Soviet negotiations." Now, Russian strategy seemed once again to ignore economics in favor of political motives.

Jersey's concern, however, involved more than the impact of lower Soviet prices on the already soft European market; greater attention centered on the questions of reserves—the quantities of Russian oil remained unknown—and of the means by which this oil would be introduced into Europe. Jersey's affiliate in Denmark had already been asked by the Danish government "to investigate the possibility of importing and distributing Russian source products," an arrangement Dansk Esso considered objectionable for a number of reasons: First, this arrangement would place the affiliate in the dangerous position of handling the same Rumanian oil against which the parent company had filed an outstanding claim.[6] More importantly, Russian oil would displace Middle East and Western Hemisphere crude, for which at that time Jersey needed every available marketing outlet, and upon the continued use of which it had made substantial investments in Europe. The most fundamental problem, however, derived from the role of the Danish government in the transaction. Generally, the Jersey Board objected in principle to direct state-to-state bargaining (instead of company-to-company or company-to-state transactions). In this case, the company's choice was limited either to cooperating, or if it refused, to seeing the Danish government enter the oil business on its own. In either case, any sizeable influx of Soviet oil would disrupt the market and alter the role of private oil companies.[7]

For without question, Jersey recognized substantial forces that could make Russian oil an enduring factor in the European oil business. Western European governments would continue to be attracted both by the bargain prices and by barter arrangements possible only under bilateral trade agreements with the Soviets. With foreign exchange still scarce in many Western nations, any chance to trade or barter products directly appeared attractive. This feature, in fact, appealed strongly to Denmark, with its increasing food surpluses and decreasing sterling exchange; but the same advantage also appealed, in greater or lesser degree, to all European governments except, perhaps, Great Britain and the Netherlands.[8] Oil, as a trading commodity, could be handled easily, without substantial capital investment or drawn-out negotiations.[9]

Dansk Esso, acting in accord with Jersey policy and with the support of the American Embassy in Copenhagen, refused to handle Soviet oil, but the

Russian threat of 1953 soon became a reality. In May, 1954, an official of the Danish State Railways informed Dansk Esso that Denmark would be importing Soviet oil and that the State Railways, as its distributor, would cartelize the marketing. Plus, private oil companies were expected to handle Russian products on the basis of their proportionate share of the national market. Otherwise, the alternative would be the creation of a state-owned oil company. At the same time, the Danish government warned of possible adverse public reaction directed against private companies.[10]

The railways action produced a sharp difference of opinion between the Danish managers of Dansk Esso and Jersey's representative in Copenhagen, D. L. Wright. Erik Frandsen, vice chairman of the Dansk Esso board of directors, flew to New York to acquaint the Jersey Board with the arguments for accepting the Russian products, as well as to provide justification for keeping the Danish government out of the oil business. Proportionate acceptance, he argued, would preserve Dansk Esso's large volume business with the Danish State Railways—presumably the principal recipient of the Russian products should the government be forced to enter the business.[11] Wright countered by stressing the negative effects that any acceptance of Russian oil would have, not only on Dansk Esso, but upon other Jersey affiliates in Scandinavia and throughout Europe. Since the Danish government could not force private oil companies to handle the oil, he urged Dansk Esso to continue to resist, and he assured the company of continuing support from the United States Department of State.[12]

A meeting of Jersey executives with State Department officials [including Robert H. S. Eakens, chief of the Petroleum Policy Staff], in June, 1954, however, produced a less satisfactory understanding. For the United States government representatives made clear their reluctance to encourage any categorical refusal to handle Soviet oil. While generally supporting the company position, the State Department carefully noted the growing attraction of such trade from the point of view of the Danish government. Further, it recommended that Dansk Esso base its final decision on its "best commercial judgment."[13] Although the American Embassy in Copenhagen continued to protest any move to force private oil companies to distribute Russian oil, the State Department at home took the position that it could not prevent Denmark from importing the oil. The question of whether to distribute it or not remained with individual oil companies.[14] As a final result, Jersey "reluctantly" recommended that Dansk Esso make its own decision, on the basis of the Danish national interest. Assuming that the affiliate would take the Russian products, the parent company then offered guidelines for the novel transaction. These suggested proportioning the Soviet oil among all operating oil companies; and that Dansk Esso purchase Soviet oil through the Danish government.[15] It further advised Dansk Esso not to take any initiative in acceding to the government's request; rather, Jersey recommended that any action be carried out in concert with other Danish companies.[16]

At a meeting of representatives from all Danish oil companies, on August 2, 1954, Dansk Esso followed this advice, joining with its competitors in rejecting the government plan for proportionate distribution. The Danish government, as expected, cancelled Dansk Esso's contract to supply the State Railways.[17] But later, when Russian-Danish trade negotiations bogged down, the company was invited to renew the old supply agreement on existing terms.[18]

As things happened, the failure of the Russian and Danish governments to conclude an oil agreement for 1955 saved Jersey temporarily, but the threat of Soviet oil displacing some portion the company ordinarily sold to its affiliates in Europe did not end. By the end of 1959, in fact, the U.S.S.R. had replaced Venezuela as the world's second largest oil producer, and it had rapidly increased exports as well. Russia's 300,000 b/d (exported to areas outside the Soviet bloc) largely went to Western Europe.[19] By 1960, the Soviets had captured 8 percent of the European market for crude oil, and they continued to increase that share by reducing prices.[20] Naturally, the Russians could set prices at any level, since they remained free to ignore cost considerations in pursuit of increased market share. The Soviet oil commissar, speaking at the Second Arab Congress, provided unwelcome news for Jersey's marketing specialists and Board of Directors when he announced the Soviet intention of regaining its pre-World War II share (almost 19 percent) of the world export market.

X   X   X

In 1959, after Fidel Castro took control of Cuba, he moved to force Cuban refineries belonging to Jersey, Texaco Inc.,* and Canadian Shell, Ltd., to use Russian crude oil. Soviet Minister Anastas Mikoyan, on his visit to Cuba in February, 1960, arranged to barter ten million barrels—half a year's supply—for five million tons of sugar. Later, a Cuban economic mission arranged to buy additional Soviet oil, at a price more than 30 percent below the cost of the crude from Venezuela—then used in all three refineries. As a final response to Castro, the refineries and their owners simply closed and sent their foreign personnel home. Cuba then nationalized the property of

*Prior to 1959, known as The Texas Company.

the two U.S. concerns, including Jersey's $35 million refining plant. Next, Russian engineers appeared to adjust all machinery and equipment for processing low-grade Russian crude oil.

For Jersey, this nationalization represented the culmination, rather than the introduction, of serious problems with Castro's Cuba. Earlier, for some months, his government had refused to allow the company to convert Cuban pesos into the dollars needed to pay for oil. And the entire oil problem represented a mere complication in the larger negotiations between governments on the subject of Cuban sugar. Robert Anderson and others at the United States Department of State had convinced themselves that Cuba would not dare to nationalize oil refineries, and so they advised U.S. companies to reject any use of Russian oil.

X   X   X

Elsewhere, the potential effects of Russian oil continued to bother Jersey. M. W. Johnson, regional coordinator of European marketing activities, visited three Scandinavian countries and Belgium in the fall of 1959, and returned to report what he had found. Russian actions in Europe, he found difficult to understand. Their price on September 28, 1959, was 25 percent below the market price, or $1.50 per barrel f.o.b. Johnson went on to question any Russian interest in who hauled the oil, or concern about what happened to it after the sale. Here, political motive seems invisible, but the system of barter, Johnson found more certainly political (even though still confusing): thus, Sweden bartered timber for Russian oil; while, in addition to oil, Belgium purchased timber from Russia. As much as possible, Jersey kept abreast of worldwide developments through periodic studies made by its Coordination and Petroleum Economics Department.[21] Yet the seriousness of its concern increased with the passage of time and the growth of Soviet exports and export potential.

Another departmental study of the Russian Seven Year Plan, taking effect in 1959, showed that the U.S.S.R.'s annual investment in its oil industry would more than double by 1966. If planning goals could be met, crude oil pipelines would bring oil westward from the expanding Ural-Volga fields— and increase by almost 100 percent the amount available for sale in Western Europe. These goals, Jersey specialists believed, were attainable.[22] Thus, the prospect of even greater (Soviet) penetration of markets appeared stronger than ever, at a time when surplus productive capacity worldwide seemed to guarantee continuing pressure on pricing structures.[23] While some Jersey executives still spoke of the Soviet Union's motive as, "mainly to obtain a political advantage in the cold war," others believed that "the basic reasons" were "primarily economic." They predicted that Russian exports would rise to between 750,000 to 1,000,000 b/d by 1965.[24] In fact, this view must now be regarded as realistic, even though Jersey specialists, in 1960, pared their earlier estimates of Soviet oil exports. At that time, they sharply questioned the prevalent Western assumption that Soviet strategy was designed to disrupt world markets. Instead, their careful study indicated that an absence of taxes and royalties plus cheap transportation charges from the Black Sea, permitted Soviet crude to be marketed at very low prices.[25]

Jersey, in turn, could meet this competition, these experts felt, by price discounting east of Suez; but that action seemed unlikely to be successful in Europe. Moreover, the possibility of trading for necessary commodities scarce in the Soviet Union might make it advantageous for the Russians to sell crude oil below production cost, especially as a way to win an increasing share of the market.[26]

X   X   X

Overall, Jersey analysis portrayed crude oil trading as a means used by the Soviets to attain Communist goals of economic growth and self-sufficiency, while at the same time noting that Soviet leaders recognized trade as "an instrument of power and influence and do not hesitate to exploit such activity for political advantage." In the first year of the Khrushchev Seven Year Plan, the Soviet oil industry actually exceeded its projected targets of crude oil and natural gas production; nevertheless, Jersey experts questioned whether the remaining goals for the next six years could be met, due to known shortages of equipment. In addition to direct price competition, the study also considered the option of Jersey's purchasing Soviet exports as an alternative to investing new capital to expand Middle Eastern crude production. While low prices for the Soviet oil might make such a step profitable, "the intangible, but very real, political damage that Jersey would incur by substituting Soviet for Middle East supplies" appeared to overshadow "any economic incentive."[27] Jersey's apparent competitive advantage in the Far East raised another possibility—selling oil to Communist China, North Korea, and North Vietnam, as a way of offsetting Soviet exports to the West, and possibly weakening relations between those countries and the Soviet Union. This option seemed to be foreclosed, however, by United States government policy against trade with China, and also by the lack of sufficient convertible currency in those countries.[28] Finally, the study considered whether Jersey and the oil companies should ask the United States government to intercede with North Atlantic Treaty Organization [NATO] countries, requesting them to set quotas that would limit "their imports to some maximum percentage of their domestic demand." With Soviet imports filling 6 percent of the current NATO needs, however, a 10 percent quota would have substantially affected only Italy, while reducing total consumption of Soviet oil by no more than 35,000 b/d from the projected 1961 level of 190,000 b/d.[29]

In July, 1960, Jersey's Executive Committee considered the problem of Soviet crude oil exports and their impact on Western markets. Then, it seemed to some directors that Russian price competition might drive a wedge between that country and other producers, such as Venezuela and the Middle East. No Jersey policy statement emerged, but the directors had a chance to consider the Soviet problem, as well as some possible solutions.[30] After the Executive Committee adjourned, Jersey President and Chief Executive Officer Monroe J. Rathbone created "a special 'Task Force' Committee of Directors" to give the problem "regular attention on a comprehensive basis." He chose Emile E. Soubry, Jersey's contact director for Europe, to head that committee; other members were William R. Stott, Harold W. Fisher, and Emilio G. Collado. Rathbone charged them with studying all aspects of the Russian oil situation, developing recommendations, and reporting to the Executive Committee on a regular basis. Among the "countermeasures" he asked them to consider were both steps that might be taken by private companies, and those that required government action.[31]

The committee, in turn, appointed a Staff Committee, chaired by Rathbone's assistant, George T. Piercy, and charged it with gathering data and suggesting possible solutions.[32] The Staff Committee also cooperated with those affiliate managers in Europe who sought to come to terms with the Soviet threat. Later (May, 1961), when Wolf Greeven, a marketing specialist, joined the Board, the monitoring of Soviet oil exports and of the Free World response became one of his new assignments; and Executive Committee minutes reveal that he frequently conferred with Common Market representatives and with the U.S. Department of State. Yet the Committee minutes do not indicate that any policy emerged as a result of Greeven's reports.[33]

During the second half of 1960, however, the mission of the Soubry committee became more critical. In August, the worldwide oversupply of crude oil capacity, aggravated by Russian exports, finally caused British Petroleum and Jersey to announce a reduction of around 6 percent in the price posted for crude oil. By this time, the Russian share of the European market had reached 8 percent; and larger quantities of Soviet oil seemed destined for that market because of continuing pipeline construction to the West. Naturally, the United States and all the major oil companies were concerned. Too, in the Middle East and Libya, new sources of oil had been discovered, and production begun; in fact, the excessive producing capacity there, coupled with Russian oil, provided the impetus for the posted price reduction that encouraged several Middle Eastern nations and Venezuela to form the Organization of Petroleum Exporting Countries[OPEC]. Obviously, that organization, in turn, added to the problems of the companies.[34]

Another equally serious threat arose later in the fall of 1960, when Ente Nazionale Idrocarburi [ENI], the Italian state oil company, negotiated a new agreement with the Soviet Union to replace one due to expire at the end of that year. In almost every feature, that compact alarmed Jersey. It showed that Italy would increase its dependence on Soviet oil faster than had any other Western European nation—as ENI's imports would satisfy 21 percent of all Italian demand by 1965. The price, $1 per barrel f.o.b. Black Sea ports, represented the lowest price yet quoted for Soviet oil; and, as Jersey executives quickly realized, less than half the equivalent posted price for comparable oil from Kuwait. Price undercutting by the U.S.S.R. thus threatened either to drive all private oil companies out of Italy, or to force further reductions in Persian Gulf posted prices, upon which Middle Eastern governments relied.[35] When Jersey committees analyzed such information as Vincenzo Cazzaniga and his associates at Esso Standard Italiana Società per Azioni [ESI] could provide, they concluded that the Russian oil contract was an "attractive deal" for ENI and for Italy—one from which ENI, at least, would make money.

Almost as disturbing to Jersey was the list of goods Azienda Generali Italiana Petroli [AGIP] (ENI's marketing arm) promised to deliver to the Soviet Union under the barter arrangement provisions of this contract. In addition to synthetic rubber, these included 40″ diameter steel pipe and other pipeline equipment. This material would be used to increase the volume of Russian oil available in Europe, since the U.S.S.R. expected to continue building pipelines to the West.[36] Already, in 1959, the United States government had refused to permit such export of strategic materials to the Soviet Union. But, the Russians obtained a sufficient quantity from their European customers [Italy, West Germany, and England].[37] Italy's interest in trading with the Soviet Union came as no surprise to Jersey executives, who knew that ENI's head, Enrico Mattei, had always been a severe critic of the international oil companies. His aggressive challenge to the majors seemed to represent a larger trend that favored state-owned oil companies and state-to-state trading. Mattei simply understood that he could use low cost Soviet oil to further his larger objectives. For example, there were reports that the Russians had offered an additional 40 million metric tons of crude oil over an unspecified period, "for sale worldwide without any restrictions as to destination"—which meant that Mattei could sell Russian crude to other European nations, thus making ENI "a prime instrument for ... [Russian] penetration of European markets."[38]

Jersey President Rathbone, a firm anti-Communist and an avowed supporter of the Cold War, made an address before the American Petroleum Institute [API] in November, 1960. In it he criticized the Italian agreement with the Russians at length. He characterized Soviet oil as "a political thing" aimed at "sowing dissension" and exerting a "disruptive effect on the whole NATO area," and he called on the United States government to make official diplomatic representations protesting the agreement.[39] That government's failure to respond only strengthened Italian belief that the protests of the oil companies amounted only to desperate attempts at saving their posted price system and their alleged excessive profits.[40]

In this troubled atmosphere, the Soubry committee presented its report, "Strategy for Meeting the Threat of Soviet Oil," during February, 1961. Noting that Russian exports to the West were already "causing serious problems," the report envisioned an "even more serious" threat "in its growth potential." Assuming that the pipeline then under construction by the European state members of the Council For Mutual Economic Assistance [COMECON] would be completed by 1964 or 1965, and that it would eventually connect with the Russian pipeline, the committee estimated a potential for 750,000 b/d in exports to the non-Communist world through 1965, with a sharp rise to as much as 1,000,000 b/d after that date. Scandinavia, Italy, Germany, and Japan appeared to be "the major targets." In contrast to earlier Jersey assessments, the Soubry report placed greater emphasis on Russian political motivation; oil exports were "being used to obtain strategic materials, to strengthen the Soviet economy, and to expand Communist economic and political penetration" of the West. Although the export trade might be "rooted in certain economic requirements," the report termed it "part of the whole Soviet campaign to promote their interests in the world at the expense of ours."[41]

In its recommendations, also, Soubry's committee emphasized the political nature of the oil export problem. First, it considered and rejected price competition to limit Soviet exports. Here, the Italian-Soviet agreement had established that the Russians would sell at even lower prices when "necessary to move oil and secure critical goods." Nevertheless, the report recommended immediate steps to decrease the Soviet price advantage, both by trimming Jersey's operating costs and prices as much as possible and by getting "producing countries to share the burden of lowering prices. ..." This diplomatic effort was to be accomplished in part by pointing out to the governments of oil producing countries the fact "that Soviet oil [was] ... backing out their oil—" meaning that only so much could be marketed. So far, those governments had been reluctant to accept this fact.[42]

Since there might "be no bottom solution" alone was unrealistic. Thus, the Soubry report advocated an "appeal for government action to limit the inflow of Soviet oil," especially in Europe. The objective was to hold Soviet imports "to no more than its present percentage share of demand in a given country," and the diplomatic appeal should stress the "serious economic, political and strategic threat" posed by too heavy dependence on politically unreliable Soviet supplies, as well as the danger of alienating "traditional suppliers" in the Middle East and Venezuela, possibly leading to "a producing cartel or expropriation." Soubry noted a tendency of Russian trade to concentrate on "top-bracket strategic materials" that "strengthen the economic and political potential of the Soviet bloc" and offered a more general argument against its expansion: In return for limiting Soviet oil import, Western governments might expect "some indication that private companies would move toward lower crude prices."[43] With this advantage in view, the report acknowledged the "critical [need] to bring the problem to the attention of the U.S. government and to get its understanding and support." At the same time, affiliates in Western Europe would consult with their governments and urge limiting Soviet imports "to the lowest possible level." Finally, a similar approach would be made to supranational groups, like the Common Market, the European Free Trade Association, and NATO.[44]

Since some increase in Soviet exports appeared inevitable, the Soubry report recommended trying "to minimize its impact" by letting it flow selectively into areas where Jersey interests would suffer the least damage and where the Russians would realize the fewest benefits. Essentially, this meant keeping the U.S.S.R. out of "profitable and growing markets" in Europe and Japan. Company policy, would, of course, "have to be tempered by any Western political or strategic considerations" that might arise and by consideration of the impact on Jersey's "long-run volume, prices and profits."[45] In conclusion, Soubry underscored the role of the U.S. government:

> It is critical, therefore, that the U.S. government should understand the problem of Soviet oil, and be willing to give its cooperation and support. Then it would be much easier to approach the governments of Western Europe and to work with the oil producing countries in a sustained effort to meet the Soviet threat.[46]

X   X   X

On the basis of Soubry's recommendations, Jersey increased its efforts to gain a better hearing and increased cooperation from various governments that shared an interest in containing the spread of Soviet oil.[47] First, in March, 1961, Rathbone sent to Allen W. Dulles, Director of the United States Central Intelligence Agency [CIA], a copy of the committee's final report. In an accompanying letter, Rathbone observed "that the only real hope for dealing with this problem effectively ... was some sort of collective action by Free World governments which would limit or eliminate the flow of Russian oil to Free World markets."[48] Dulles' response, while sympathetic, served largely to increase the company's concern; as he indicated, that, according to his intelligence sources, exports of Soviet oil to the Free World would reach 1,000,000 b/d *by* 1965, rather than after that date.[49] Outside the U.S., on February 17, 1961, Jersey representatives met in Paris with executives of its European affiliates, to discuss the evolution of a Common Market energy policy—subject of an upcoming meeting of European Economic Ministers. Here, the company's concern centered on the fact that, when the economic union became "fully operative," energy imported by any member would then assume the status of Common Market energy; it would flow freely within the six nations. Directing particular attention to the problem inherent in Italy's large (and increasing) imports of Soviet oil, the Jersey meeting agreed to focus action on the Common Market Commission's "limited treaty powers" and on "prior inter-governmental consultations with respect to trade agreements." As a "second line of defense" the Jersey executives pointed to a provision in the Treaty of Rome that allowed member nations to "adopt protective measures in their own interest" during the transition period. These devices for protecting markets all required governmental action; so Jersey urged affiliate managers to seek a "voice, vis-a-vis their governments" both directly and through leadership in oil industry associations.

Specifically, managers were encouraged to contact their countries' Economic Ministers, Foreign Ministers, and Defense Ministers, stressing, in the last instance, "the political and security dangers involved in the hope that the Ministers of Defense might prevail upon NATO to take energetic action vis-a-vis Italy." The "question of placing pipeline material back on the strategic list of items barred in trade with the U.S.S.R. [was to] ... be raised directly with NATO and with Defense Ministers of member nations." Finally, W. D. Bayles, Jersey's roving European representative, was delegated to contact the Commission of the Common Market.[50] In a subsequent memorandum, Common Market affiliates received a substantial summary of the dangers represented by the Russian pipeline system into Eastern Europe. It reported that Enrico Mattei of ENI had suggested that Italy might build a pipeline from the Italian border to connect with the Russian line at Bratislava, Czechoslovakia, only 200 miles distant. With this pipeline in operation, the Italian company could flood the Common Market with products made from Russian crude. So, in Jersey's view, completion of the Russian pipeline system represented the last real obstacle to a full-scale Soviet invasion of the European oil market. After the pipeline began to operate:

> the limits on Soviet ability to export oil will ... largely be confined to whatever rate they set for exploration and production. ... Production could be increased and demands could be sharply limited if it were decided to use greater oil exports as a means of waging more intense economic and political warfare.[51]

By the summer of 1961, Jersey had placed in the hands of the United States government, several European governments, and such supranational organizations as NATO, the European Common Market, and the Organization for European Economic Cooperation [OEEC] critical material detailing the expansion of Soviet oil pipelines and exports, and concerning the Russian threat to the security of the West.[52] Then in July, 1961, Rathbone made a direct approach to Secretary of State Dean Rusk, who listened as Jersey's chief executive reported on recent conversations with government and business leaders in Scandinavia and Great Britain. During their talk, Rathbone also summarized for Rusk the contents of a Jersey memorandum on "The Soviet Oil Offensive," written by George Piercy.[53] Briefly, Piercy had forecast Soviet control of 11-12 percent of Western European demand by 1965, and warned that Soviet oil not only had "security implications" for any nation that became too dependent on unreliable supplies, but also that Soviet pricing might well be politically based. Again, allied nations that proposed to barter goods for Soviet oil might well gear their industry to Soviet needs only to see their markets disappear as soon as Russia had achieved its goal of economic self-sufficiency. Finally, Europe's increasing reliance on Soviet oil endangered traditional relationships with suppliers in the Middle East and Venezuela, where reduced royalty payments and tax income would guarantee "internal unrest."[54] None of these possibilities could have pleased Rusk.

Having made a case for action, Rathbone went on to explain that private oil companies alone could do little to combat Russia's "progressive threat to the security of the United States and to the Free World." An initiative by government—in particular, by the government of the United States—already was badly needed. In his conversations abroad, he had found general dissatifaction over "the absence of any definite position or policy statement by the U.S. Government on the question of Russian oil." So Rathbone proposed that the United States take the lead in establishing a united policy among its Western allies. That policy would limit Soviet oil imports to existing levels, or even seek some reduction (in certain cases, such as Italy, Sweden, and Greece, where Russian oil was already being purchased in sizeable quantities). Rathbone also urged new diplomatic efforts to channel existing oil imports away from "more critical areas" like industry and transportation, and to shift allied exports to Russia to consumer goods and away from producer goods. Regarding Soviet exports to less developed neutralists or uncommitted nations, he advocated some coordination of foreign aid with private trade, as a way of blunting Soviet efforts "to spread their economic-political influence." Certainly, alternatives to purchases of Russian oil could be arranged through collaborative efforts of the U.S. government and private U.S. companies. Finally, to help Jersey and other majors in dealing with non-Communist producing nations, Rathbone asked for government assistance in convincing the Middle Eastern and Venezuelan governments that Soviet oil posed a threat to them. Ultimately, and on the largest possible scale, he hoped that all producing governments would use their influence with the governments of consuming countries to stem the flow of Soviet oil.[55]

Already, by the time of Rathbone's talk with Rusk, the public had begun to understand the security threat posed by Soviet oil. Press accounts commonly referred to Russian oil exports as "one of the most potent weapons" used for "the economic and political penetration of the free world."[56] Rathbone himself had attacked such use of trade as a weapon against the Free World in a speech given at the dedication of a refinery in Slagen, Norway.[57] Then, a few weeks later, on June 20, 1961, Minority Leader Everett M. Dirksen echoed Rathbone in the U.S. Senate, proposing that the United States government actively warn its European allies against the dangers of Russian oil.[58] On the following day, Gordon W. Reed, chairman of the Texas Gulf Producing Company, told a meeting of the American-Arab Association that the United States should take the lead in establishing an economic equivalent of NATO, by setting quotas to limit and control Soviet dumping.[59] Again, Rathbone spoke out, calling for "a working partnership between business and labor and government" to meet Cold War demands "too vast for government or business corporations to resolve alone."[60] Among industry groups, the Independent Petroleum Producers Association of America [IPPA] warned of the dangers posed by the Soviet oil offensive, and called for government action to block the "Soviet plan for world domination."[61] In his own speech to the American Petroleum Institute, Senator A. S. Monroney advocated a full-scale partnership of government and industry, to curtail the flow of Western capital equipment and technology that was allowing the Soviet Union to export even more oil.

All these concerns of both American political leaders and oil executives now appear to have been justified. The historian Mira Wilkins has pronounced "Soviet oil ... a threatening aspect of East-West trade" during the 1950s and early 1960s. By 1960, Russia had regained second place—behind the United States—in oil production and for several years its exports of crude had increased annually. As Wilkins points out: "By 1961 Soviet oil supplied 35 per cent of Greek demand, 22 per cent of Italian demand, 21 per cent of Austrian, and 19 per cent of Swedish demand." Russia also supplied all of Iceland's petroleum and products, Finland with 80 percent of the oil it needed, and Japan with 200 million barrels annually. Quite properly, Western oil companies and their leaders were alarmed; they had made tremendous investments in Europe and the Middle East after World War II, and now they saw all earlier plans and prospects put in jeopardy by the unsuspected moves of Russia.[62]

X   X   X

In June, 1961, President John F. Kennedy and Soviet leader Khrushchev held their famous confrontation in Vienna. Obviously, this event helped to convince the President that a Russian threat to U.S. security, based at least in part on political oil, could not be easily dismissed. While Kennedy still had the politics of petroleum on his mind, he talked to a prominent lawyer and public figure, John J. McCloy.* At the time, McCloy served as advisor to the President on arms control and disarmament. Yet he also worked as a lawyer, serving as counsel to Jersey and four other large U.S. oil companies, as well as to both the Royal Dutch-Shell Group [Shell] and British Petroleum Company, Ltd. [BP]. Kennedy may have been aware of this special government-oil relationship embodied in McCloy, for their discussion considered Soviet penetration of European oil markets and the possible dependence of some European nations on that oil. Kennedy seemed convinced that American security was endangered by both the use of Russian oil as a political weapon

---

*Of Milbank, Tweed, Hadley, and McCloy of New York. Already McCloy had served as Assistant Secretary of War; President of the World Bank; and U.S. Military Governor and High Commissioner for Germany among other posts.

and the growth of Russian power in the Middle East. McCloy later explained why Kennedy proved receptive:

> It was a combination of fear that perhaps Western Europe or the free world would become too dependent upon Soviet oil. There were some moves at that time being made that sort of pointed in that direction, together with a general emergency of an infiltration or pressures in that part of the world as the British began to move out of it. So this was the political background that I think provoked the national interest.

J. Kenneth Jamieson, who joined the Jersey Board on September 10, 1964, has identified this shared sense of crisis over Russian oil as the beginning of the movement toward oil company unity. Later, he told an interviewer:

> The McCloy Committee* really dates back to the movement of Russian oil to Western Europe. I guess they (the international companies) were concerned that the Russians would just dump oil in there and wreck the whole damn market. So the McCloy Committee sort of evolved from concern with that problem and as the Middle East became more and more critical for the companies, the committee moved into that area.

As an attorney, McCloy knew that large U.S. oil companies feared antitrust suits, so he mentioned to the President that some government assurances might be necessary if private companies agreed to act together in a common national interest:

> I said that if at that time any action seemed to be necessary, I thought it would be desirable to talk to the Department of Justice [Justice] about it. He [President Kennedy] asked me if I had talked to them about it, and I said no, I-hadn't.
>
> To the best of my recollection, he right then and there made an arrangement for me to see Mr. Robert Kennedy [Attorney General], and I did talk with him.

---

*The McCloy Committee referred to the union of oil companies formed to bargain with OPEC in 1971. McCloy served as attorney for all of the signers of the "Message to OPEC," and he secured the Business Review Letter authorizing this union.

During the next ten years, in fact, McCloy discussed with each successive Attorney General* the problems caused by the spate of Russian oil and the Middle East situation—always in the light of possible joint action by the oil companies. In time, the growing demand for oil worldwide, coupled with declining surplus producing capacity served to erase the danger from Russian oil. But the shrewd attorney never forgot that the route to some union of the oil companies had been opened (if not used) in 1961.** He and President Kennedy did discuss OPEC, and the President appeared to agree that, at some future date, such a bonding of the oil companies would be necessary.

X   X   X

While pressures favoring U.S. government actions to curb Soviet oil exports increased during 1961, the Energy Committee of the Common Market moved ahead by developing a voluntary plan designed to restrict Soviet imports into the community to 250,000 b/d—approximately, the existing level. Italy, however, adamantly refused to accept this quota system, even though other members exerted pressure. Meanwhile, subcommittees of the Organization for Economic Cooperation and Development and of NATO also took up the problem.[63] Jersey noted these developments, while still attempting to persuade the United States government that some action was necessary to support its European allies.[64]

Within the company, Rathbone appointed a second Task Force of the Board to serve as "a new focal point for information and responsibility for the Russian oil problem." Soubry had retired. Rathbone chose Greeven to serve as chairman, and added David A. Shepard to the committee, specifying as well that Piercy continue as a member of the Staff Committee. In addition to conducting studies and developing recommendations for action, Rathbone charged the second task force to act as a clearinghouse "for all Jersey and industry information . . . and actions, . . . [to] keep apprised of the forecasts for the future" and current sales and prices, to organize "appropriate visits and presentations to oil industry organizations, governments, supranational bodies, and other interested groups," and to follow up and evaluate recommendations and actions.[65] More specifically, the new committee soon took on a wide variety of tasks, including supplying information to affiliates, contacting American Embassies and consulates, proposing a policy statement to API, and determining Jersey's own policy on the use of

*Nicholas Katzenbach, 1964-1967; Ramsey Clark, 1967-1969; John N. Mitchell, 1969-1972.

**See the chapter "Confrontation in Tehran and Tripoli, 1971."

quotas as a means of restricting Russian oil imports. At its second meeting, the committee expanded its membership to include representatives from the regional coordinators' offices, Esso International Inc., and the Government Relations, Public Relations, Producing Coordination, and Coordination and Petroleum Economics Departments.[66] In November, 1961, this enlarged group put aside other business to concentrate on developing a general policy statement for Jersey. Working from Greeven's draft memorandum, the committee formulated a "Statement of Position on the Threat of Communist Trade," which it submitted to Jersey's Executive Committee just after Christmas, 1961.[67]

Unlike the Soubry report, the "Statement" contained no references to commercial steps that might be taken by the oil industry, or, for that matter, to any action that private industry might initiate. Instead, it proceeded "from the premise that closely knit and scrupulously implemented economic coordination must be created, and that the United States must take the lead in this endeavor." By creating "economic and commercial coordination" to "parallel our existing military alliances," the West could establish a trading system that included "as many of the non-committed countries of the Free World as possible." Ideally, this would achieve maximum benefits for each nation, and minimize the damage inherent in trade with the Communist bloc. To achieve that end, the study recommended policies to control strictly Russian access to strategic materials and technology.[68]

The Jersey policy statement called attention specifically to the problem of Soviet oil imports, and then suggested that an agreement among non-Communist nations "would reserve to Free World sources a given [minimum] percentage of their oil import requirements (say 95%) on the premise that the security of these supplies is assured by long-term continuity of off-take." Such a position could be sustained by both the strategic advantage of holding Soviet imports to no more than 5 percent in any country, and the "politically and morally acceptable justification ... [based on] the stated objective of the industrialized countries to help the 'underdeveloped' nations to improve their economic and social status." In addition, the Jersey report recommended invoking the "anti-dumping clause of GATT [General Agreement on Tariff and Trade] and the Common Market" and a NATO boycott of oil products from refineries processing Soviet crude. Further, to keep the Soviets from using trade to strengthen their threat to oil markets, the policy statement called for an immediate proscription from trade by Sweden, Japan, and each of the NATO nations of "all items destined to complete the Soviet pipeline systems and satellite refineries... ." These same nations would also be asked to enforce stricter controls over their technology, so that no vital secrets would reach the Russians. Finally, with respect to "underdeveloped import countries," the statement cited a provision in the United States "Act for International Development" of 1961, which discouraged recipient nations from becoming dependent for fuels on sources "inherently hostile to free countries"; and it called on government agencies to see that the fuels incorporated in development plans derive from non-Communist sources.[69]

After this comprehensive "Statement" had been reviewed by the Jersey Executive Committee, it was distributed to the Secretaries of State, Defense, Commerce, Interior, and Treasury and to the Central Intelligence Agency. (Jersey also planned to distribute it to members of Congress.) Copies also reached oil industry groups, such as the (now inactive) Soviet Oil Committee of API, and the National Petroleum Council's Committee On the Impact of Oil Exports From the Soviet Union, whose working subcommittee was still headed by Piercy. Jersey representatives in Europe received copies with instructions to seek an endorsement from other major oil companies. Executives of Jersey's foreign affiliates found their copies accompanied by a request to keep them confidential until notified to release the text.[70] One way or another, Jersey made certain that its corporate attitude toward the U.S.S.R. would have a voice.

X   X   X

Meanwhile, by 1962, the pressures on governments—of the U.S. and of other nations—were beginning to show measureable results. In January, the U.S. Interior Department asked the National Petroleum Council to prepare a study of Soviet oil exports; and later in that year, the Joint Economic Committee of Congress commissioned a broader study, covering the subject of East-West trade. Both studies emphasized the dangers posed by Russia's use of "trade as an ideological weapon." Elsewhere, Senator Kenneth Keating of New York conducted hearings of the Senate Internal Security Subcommittee on the Soviet oil offensive. Relying on a commissioned study by Dr. Halford Hoskins, Keating charged publicly that Soviet motives were political and designed only to gain leverage over NATO policy.[71] Now, the American Maritime Association added its voice to the call for government action—twice demanding a government-enforced boycott against the owners of tankers carrying Soviet oil. Given the extremely high surplus tanker capacity of the time, however, such a move would have had little practical effect. Instead of refusing charters to the Russians, desperate ship owners, feeling the effects of rock-bottom spot-charter rates, continued to provide the Soviet Union with easy access to European deep-water ports.[72]

In Europe, national governments also felt compelled to attempt some coordination of their policies on Soviet oil imports. In March, 1962, the Parliamentary Assembly of the Common Market agreed to a draft policy that included a stability of supply clause; it was approved by all member nations except Italy. In September, when the Common Market Council of Ministers met to discuss an energy policy for their countries, a quota on Russian oil—probably 10 percent—was widely rumored even though, again, the Italians remained opposed. In the end, however, the Council of Ministers failed to agree, as other nations joined Italy in finding the proposal flawed. Some objected to the treatment of coal, and others wished to postpone action until the question of British membership in the Common Market had been settled.[73]

Meanwhile, public attention had shifted to NATO's discussions of the Soviet oil threat, as completion of the Soviet's COMECON pipeline, scheduled for 1964, seemed destined to remove, at an early date, any real constraint on Russian oil exports. Even so, any rapid completion of the line clearly depended upon shipments of large diameter pipe and other pipeline equipment from the West. Of course, the United States had consistently refused to remove such pipe from its "positive list" of strategic materials banned from trade with the Soviet Union; but Sweden, Japan, and even some NATO members—especially Italy, West Germany, and Great Britain—continued to supply a major proportion of the 40-inch pipe necessary to complete the COMECON line. When NATO did meet to discuss the Soviet oil threat, therefore, the United States took the lead in seeking to bar such pipe from East-West trade.[74]

In the Senate, Jersey's George Piercy previewed the National Petroleum Council's report before Keating's Internal Security Subcommittee. Piercy emphasized that the oil industry was the principal target in the Soviet economic offensive against the West, and he reiterated that the capability of the Russians to export more than one million barrels of oil per day to the Free World largely depended upon whether or not the West took some form of united action to control this flow. Senator Keating, presiding at the hearing, called on the United States government to bring pressure on its allies to halt the shipment of all strategic materials to the Soviets.[75]

Before the end of 1962, the American government did, in fact, enjoy some success in applying pressure within NATO. Despite resistance from Great Britain, Italy, and perhaps other members, NATO voted in November to declare large diameter steel pipe a strategic material with vital military significance; such pipe was then added to the list of items banned from export to the Soviet bloc. Although existing commitments were exempted, West Germany and Great Britain soon cancelled contracts for exporting the pipe, leaving Italy as the only NATO member to continue sending such pipe to Russia. NATO also requested that Japan refrain from such export, but Japan refused to do so. Nevertheless, the NATO action sharply curtailed the supply of pipe that could be used to complete the COMECON line, and it quickly fell behind schedule. Thus, NATO gained credit for the first united action that negatively affected Soviet oil exports. A construction delay, however, did not bring the Soviet threat to an end.[76]

Curiously and without explanation, during 1962, the rate of Soviet exports to Common Market nations unexpectedly declined. At the time, this decline was viewed as temporary, or perhaps even deliberate. Experts in Soviet behavior predicted a continued Russian drive to capture 18-19 percent of the Western European market—the same "historic" share claimed by the U.S.S.R. on the basis of its pre-World War II trade. Moreover, in oil the Soviets now seemed interested in integrating forward, by refining their own crude in the West.[77] In fact, the growth rate of Soviet oil exports did pick up again in 1963, although, Russia's share of an enlarging market remained in the 8-9 percent range—the level the U.S.S.R. had first attained in 1961. Jersey kept close watch and even moved to direct action, as a defense against new Russian moves. In March, the company announced a massive oil deal between Esso International and ENI, involving 80,000,000 barrels of crude oil, to be delivered over a period of five years at a price rumored to be 20 percent below the posted price in Libya, the source of most of the oil involved. Given ENI's special status as the largest non-Communist purchaser of Soviet oil, as well as its history of antagonism toward the major oil companies, this agreement seemed to mark a major breakthrough. For Jersey would replace the Russians as chief supplier to the Italian government. In its terms, this pact proved that the American company could be flexible in defense of its commercial interests. In addition to the substantial price concession, as Jersey President Michael L. Haider pointed out, the company would provide Italian-nationalized firms with some business whenever they were competitive with other suppliers. But he added, "We are not accepting payment in equipment." Although Jersey insisted that the agreement with ENI was simply a commercial transaction, some observers believed that the United States government also played a role. Again, speculation developed over whether this unprecedented contract had been made possible by the death of Mattei, in an airplane crash in 1962, but finally both ENI and Jersey denied this rumor, stating that negotiations had begun before that accident. Without any question, however, Jersey interpreted its Italian connection as a major setback for the Soviet Union.[78] Yet this view proved premature and perhaps even naive. For later in the same year ENI announced a six-year pact to purchase 175,000,000 barrels (app.) of Russian crude—more than twice the amount covered by the agreement with Jersey. Despite its magnitude, however, the new arrangement actually led to a decrease in the Soviet share of Italian oil imports, from 30.9 percent in 1963 to 29 percent in 1969.[79] The market, of course, kept growing. Throughout 1963, Jersey maintained its vigilance, especially when the Russians sought to use Western technology to upgrade Soviet refineries, or even to buy new refineries from the West. Company executives saw in these proposals new evidence of a continuing danger, and they surely felt relieved when in January, 1964, the U.S. Commerce Department finally rejected a Soviet bid to purchase an American refinery.[80] By the end of 1963, however, public opinion in the United States had moved toward a more liberal position on general trade with the Soviet bloc, especially with the Eastern European satellites. Jersey met this changing national atmosphere with deliberation, as the company prepared a new study on general trade with the Soviet bloc—including a re-examination of Jersey's policy on the oil trade. In essence, the company stood firm by rejecting any liberalization of U.S. trade restrictions, because it believed that Soviet goals had not changed. Economic integration within the Soviet bloc made it impossible to separate trade with the satellites from direct trade with Russia; so Jersey saw no reason to change American policy on general trade or to revise the company's very strict guidelines covering exchanges with the Soviet bloc.[81]

By consciously refusing to modify its stand, Jersey placed itself outside of an evolutionary trend in international economics. During the winter of 1964, pressure on the United States government to liberalize its trade policy—especially from Great Britain—began to build anew. By March, 1964, many American business leaders had joined politicians in demanding that President Lyndon B. Johnson relax restrictions on American trade with the Communist bloc, and thus bring the United States into alignment with its allies. After hearings by the Senate Foreign Relations Committee, its Chairman, J. William Fulbright, called for increased trade to be used as an instrument of detente with the Soviet Union. In April, the United States Chamber of Commerce appealed for looser trade restrictions, so that American businessmen could share in the lucrative trade that allied nations seemed all too eager to monopolize, despite political pressures from Washington.[82]

Meanwhile, Jersey continued as one of the lonely advocates of tight restrictions on Soviet trade, especially in sophisticated equipment and technology that might assist the Soviets in competition for Western markets. After completion of the COMECON pipeline, for example, the company turned its attention to a new threat—Russian natural gas—which Jersey predicted could "become a prime economic tool employed by the Soviets in the cold war." This gas could either be sold directly to the West or used as a substitute fuel to free additional petroleum for export. In either case, its industrial development depended even more than oil on an extensive pipeline system—which added an obvious urgency to the old need for maintaining the NATO ban on the sale of large diameter pipe.[83] So despite some relaxation of restrictions on NATO members' trade with the Soviet bloc during 1964, large diameter pipe remained proscribed.[84]

In fact, Soviet efforts to increase oil exports to Europe did continue, as new steps were taken toward refining and direct marketing through Nafta Limited [Great Britain] and Nafta S. A. [Belgium]. By 1970, the Soviets had purchased an interest in a Belgian refinery that would finally make possible the refining of petroleum products in the West—adding another outlet for Soviet crude oil.[85] At the same time, the U.S.S.R. also stepped up efforts to export natural gas directly to Western Europe, reaching agreements by 1969 to supply Austria, Italy, and West Germany. This Soviet expansion was assisted by the lifting, in November, 1966, of the NATO ban on large diameter pipe, and indeed much of early payment for gas took the form of such pipe.[86] Of course, such natural gas imports were even more vulnerable than oil to a sudden Soviet action, and again, Jersey felt apprehensive, especially since Russian gas threatened full development of Jersey's interest in natural gas from its Groningen (and related) fields. Yet this time the company failed to launch any trade defense comparable to its earlier campaign against Soviet oil. By 1966, Jersey seemed resigned to accept a Russian presence as a fact of commercial competition.[87]

Finally, this changed Jersey response proved to be wise. For, although Soviet oil exports continued to increase in absolute terms, its share of the Western European market stayed consistent at about the 8 percent level. Growing domestic demands actually led the Soviet Union to urge its European satellites to seek more of their oil supplies from the Middle East. As for natural gas, the failure of Soviet supplies to reach planned levels simply accentuated the oil problem at home.[88] In addition, a change in Soviet internal oil pricing, designed to reflect real exploration costs, resulted in a doubling of the price in 1967, and with that increase, the Soviet capacity to undercut prices elsewhere in Europe, quickly disappeared.[89]

Equally important, Western attitudes toward Soviet oil imports changed markedly over time. While the Russians continued to use trade as a means of acquiring hard currency, necessary to buy goods scarce in the Soviet economy, after 1964, they shifted the focus of their interest—away from oil and toward acquiring wheat and building up their chemical fertilizer capacity.[90] By 1964, prescient observers noted that demand for petroleum within the Soviet bloc had increased so much that even the completion of the pipeline system to East Europe was "not expected to present a major threat to the oil industry of the West."[91] But Jersey remained less than sanguine about this prospect, although by the late 1960s, the company's attitude could be accurately described as watchful but "unworried."[92] In 1969, when the Soviet Minister of the Oil Extracting Industry, Valentin D. Shashin, predicted that Russian exports to the West "would not continue to rise significantly in the future because of growing domestic requirements,"[93] he merely confirmed what most oil men believed: that the Soviet oil offensive had "largely run its course."[94] In fact, the growth rate of Soviet oil exports to non-Communist Europe actually declined from 1969 through 1972—only to pick up rapidly with the Middle East War and oil embargo in 1973.[95] Yet this was the new world of oil. Confidence had replaced fear, and Soviet expansion no longer generated the excitement of the earlier years, perhaps because of the emerging events in the Middle East.

Finally, then, between 1953 and 1974, Jersey experienced failure in its effort to influence a public policy that would deal effectively with the threat of Russian oil. Of course, corporate self-interest played a part here; but Jersey's opposition to Soviet commercial expansion also represented something much larger than merely an extension of earlier demands for compensation, or a blind response to competition in the market. In fact, after 1954, Jersey (and other international oil companies) were forced to enter a new world of oil—and to play an unrehearsed role in a scene of complex economic and, even more, political significance. If, as many Jersey executives contend, the company had been "encouraged" by the U.S. government to enter and act in Iran and elsewhere in the Middle East, no student of modern business should be surprised. Politics everywhere exerted their powerful influences on markets. And neither Iran nor Russia can conveniently be regarded as cases to be studied in isolation from the rest. For, when Iran shut down its production, only to reopen and expand, the surges in oil volume were felt throughout the world. Finally, to accommodate Iranian production without disaster, cutbacks had to be made elsewhere, and new problems emerged for all oil companies. Irrespective of the quantity, the publicity attendant sales of Russian oil, and the ever-looming threat of increased volumes becoming available in European and some other markets, dramatized for the Persian Gulf states and several other producing states the need for some union for the influencing of crude oil prices. This led directly to the formation of OPEC and to the constant increase of its influence, despite its slow beginnings.

In the end, however, Russian crude oil merely complicated, for a time, an already complex and untidy marketing picture. On one hand, the national governments, especially those of the U.S. and the U.S.S.R.—but also governments in Europe and the Middle East—planned and experimented with the basic tools of business and economics. And on the other hand, private companies like Jersey tried to continue their old practices of doing business while they sought explanations of the new rules. In this confusion, Jersey found itself caught. Russia and Soviet oil became a target for corporate fears and a cause of corporate frustrations. The real reasons for Jersey's anguish lay deeper than an East-West rivalry. This episode in Jersey history was a symbolic prelude to the crisis of 1969-1974.

Chapter 12

Imperial Oil Limited

I<small>N</small> THE *ANNUAL REPORT TO THE SHAREHOLDERS* for 1950 G.L. Stewart, president of Imperial Oil Limited [Imperial], stated: "Activities of Imperial Oil Limited kept pace with the expanding [Canadian] economy in 1950 and again reached record levels." Forty-four percent more crude oil had been produced than in 1949, the volume of refinery runs had increased by 15 percent, and total product sales by 13 percent. Imperial operated eight refineries and was building a ninth; the company had eight marketing divisions spread across Canada. Stewart went on to detail plans for new tankers, more pipelines, and for refineries to be enlarged or constructed. Throughout the report, he made clear that all Imperial executives viewed the future optimistically.

In conclusion Stewart noted that during the previous few years Canada had "compressed extraordinary growth," especially in the areas of manufacturing and natural resource development. The Canadian standard of living had improved, the work force had been enlarged, more farms had become mechanized—the list could be extended—and Canadian oil had been a significant factor in all of these changes.[1]

The positive tone of this mid-century report reflected the booming post-World War II Canadian economy. And the story of Imperial's role in Canada's growth and development after 1950 properly begins before that date, with what has been described as one of the most important economic events in the entire history of Canada—the discovery by Imperial drillers, on February 13, 1947, of the Leduc oil field [Leduc is 17 miles south and west of Edmonton, Alberta]. Rarely can an historic turning point in a nation's economic history, or in a national industry, be fixed so accurately as is the case with Imperial's Leduc discovery. Here was the first event in the long chain that would make Canada a major oil-producing country; and it would also turn Alberta into Canada's "richest resource province." No other single event in Imperial's history so changed its operations.[2]

At the time of the Leduc discovery Henry H. Hewetson, who was later to join the Board of Standard Oil Company (New Jersey) [Jersey], was chief executive officer [CEO] of Imperial. For him and the Imperial board of directors the Leduc discovery provided the gratifying climax of years of exploration in Western Canada. But Leduc also brought a major problem—connecting the new field to established markets.[3] To solve that problem, a pipeline was required. After careful study, Imperial asked the Canadian government to incorporate the Interprovincial Pipe Line Company [Interprovincial], which would build a pipeline from Edmonton, Alberta to Regina in the neighboring province of Saskatchewan.

When this request originated, there was no federal legislation regarding pipelines, and the government decided that before an incorporation of Interprovincial could be approved, a pipelines act was needed. Parliament passed the Pipe Lines Act of Canada on April 29, 1949. On the following day the Interprovincial Pipe Line Company was incorporated (along with four other pipeline companies). On June 10, following public hearings, the Board of Transport Commissioners authorized Interprovincial to build its line to Regina.

The application for the incorporation of Interprovincial marked the beginning of a continuing program to develop markets for Western Canadian crude oil which occupied Imperial for the next two decades. This program included not only a sponsorship of pipeline systems to move the oil, and the building of new Imperial refineries; it also provided supply guarantees to other companies to encourage their use of Western Canadian crude. As one example the nearby city of Edmonton, an obvious as well as a natural market for Leduc crude, had no refinery. So Imperial quickly acquired from the United States government a small World War II-built refinery at Whitehorse in the Yukon Territory. It was dismantled and shipped to Edmonton, where a crew reassembled it, and it went on stream in July of 1948—just 17 months after the discovery of Leduc. Next, to expand the Edmonton market, Imperial entered into a crude-oil supply agreement that enabled McColl-Frontenac Oil Company Limited [now Texaco Canada Limited] to build its own refinery at Edmonton.

By the summer of 1949, major new discoveries had been made, and it seemed probable that Western Canada would be able to produce enough oil not only to supply the refinery needs of the three Prairie provinces but additional crude for other refineries as well. Accordingly, Interprovincial sought permission to extend the original pipeline to Superior, Wisconsin, where crude could be loaded on tankers for shipment, principally to the Imperial refinery at Sarnia, Ontario. On its way to Superior the line could also serve refineries adjacent to the route within the United States.

On September 12, the Canadian Board of Transport Commissioners approved this proposed extension. That decision resulted in a parliamentary debate thereby publicizing the entire matter. This debate, thereby proved to be an omen of later government actions that increasingly inhibited operations and decisions of the oil industry there. To build the extension, a choice had to be made between an all-Canadian route and that ultimately chosen, the trans-American route to Superior. Some members of the Canadian Parliament opposed the latter route using nationalistic arguments. Canadian Minister for Trade and Commerce C.D. Howe challenged these objections, demanding to know what could be wrong with selling oil to the United States. But an *ad hoc* committee, contending that a Canadian birthright was being sold for gold, joined the Nationalists in opposition, using in their platform the slogan, "Put Canada First." Despite the fact that the route chosen had been selected because of lower construction costs ($10 million approximately), fierce lobbying continued. Howe pointed out that the higher cost of the Canadian route would penalize Alberta producers, who would receive a lower price for their oil. He and his fellow Liberal party members stood firm, and the Transport Board decision was upheld.[4] In this debate Imperial and its partners had confronted a relatively new but very important force—that of a resurgent Canadian nationalism. Its power would surface again on the issue of new pipelines, and nationalism would increase in significance across the years.

First, the pipeline extension also had to clear legal hurdles in the United States. Most important, a separate company was required to operate the U.S. section of the line. In 1949, Interprovincial established a U.S. subsidiary, Lakehead Pipe Line Company, Incorporated [Lakehead] which provided the corporate structure that managed affairs on the American side of the border.

The actual building of the 1,120 miles of pipeline from Edmonton to Superior began in April of 1950. Construction was completed in a record-breaking time of 150 days. Imperial took a major business risk when it sponsored Interprovincial, and it succeeded. The company felt confident that Western Canada would soon become a major oil-producing area, one large enough to justify the cost of this pipeline. At the time the extension was first proposed however, no person could have known for certain. Imperial had taken the gamble, not only by assuming a one-third ownership but also by guaranteeing the line's throughput. As W.O. Twaits, who served as the chief executive officer of Imperial from 1960 to 1973, said in a 1974 speech: "In effect, we mortgaged the company."

While Interprovincial pushed ahead with construction of the pipeline, two of three new large Imperial tankers under order to haul the crude oil from Superior to Sarnia "were being readied." Construction had also begun on a new refinery at Winnipeg, scheduled to be completed in June, 1951. All together, the availability of the pipeline plus its own refineries would enable the company to supply customers throughout the entire Prairie area with petroleum products refined from western crude oil. This area included that portion of Eastern Manitoba previously supplied with shipments from Ontario.[5] In April, 1951, the first tanker, the *Leduc*, reached Sarnia.[6]

Meanwhile, to help finance its heavy postwar program, which included not only the previously mentioned building of new transport and refining facilities but also continuing exploration, field development, expansion and modernization of existing refineries and modernization of marketing operations, Imperial sold its interests in three subsidiaries—Irving Oil Limited, the Royalite Oil Company, Limited and International Petroleum Company, Limited [IPC]. These moves caused some complications for management. Although the decision to sell Irving and Royalite apparently did not entail serious debate, the sale of IPC proved another matter. It was reached only after careful consideration of many other alternatives. W.O. Twaits recalled the circumstances this way:

> At that time we owned an 80-percent interest in Irving Oil Ltd., which we sold, and we sold off our interest in the old Royalite Company. ... Most important of all, we decided to sell off our subsidiary, International Petroleum. We were selling our own control over Latin American oil to gamble on how much oil we could get out of Western Canada. And I can tell you, we sweated over this for a long time.[7]

But finally Imperial did sell IPC to Jersey, and then it sold an additional $24 million in debentures. Together, these two actions raised more than $100 million which, along with retained earnings, made it possible for Imperial to carry out a heavy capital program in that period. As time went on additional financing had to be found.

Overall, Imperial's continuing exploration success—Redwater, discovered in 1948 and Golden Spike in 1949—served to focus the attention of the international oil community on the potential of Western Canada. Every major oil company in the United States, and many elsewhere, plus many important independents, joined the rush for acreage in Alberta and other Canadian provinces. In 1949, the oil companies spent more than $100 million on exploration in Western Canada alone.[8]

In 1952, Michael L. Haider, at the time deputy coordinator of oil-producing activities for Jersey, was elected president of the American Institute of Mining and Metallurgical Engineers [AIME]. Earlier he had worked at Imperial, where he became a vice president, so he knew the Canadian oil story well. Speaking before an AIME group shortly after his election as president, he summarized the dramatic developments in the history of Imperial and the Canadian oil industry since the Leduc discovery. Haider stated that by the end of 1952, the petroleum industry would have spent one billion dollars to develop Canadian oil resources; 3,000 producing wells had been drilled; refining capacity tripled; a 2,000-mile pipeline-tanker system had been constructed for transporting western crude to Ontario. All this, he termed "a typical oil story in its magnitude." While Canada in 1947 had imported more than 90 percent of its oil from the United States and the Caribbean area, Haider said that estimated production in 1952 would supply "almost 40% of the total country's requirements." And Canada's demand for petroleum and oil products had increased almost twice as fast as comparable demand in the United States: 82 percent vs. 43 percent.[9]

Success in exploration spurred the development of new markets for Western Canadian crude production. Imperial joined with several other companies in a consortium to incorporate the Trans Mountain Oil Pipe Line Company [Trans Mountain] to serve the West Coast refineries in the Vancouver, B.C., and Puget Sound, Washington, areas. By the terms of the agreement, Imperial subscribed to 8 2/3 percent of the common stock and assumed 50 percent of the deficiency guarantee, under which payment would be made of interest and principal on borrowings should they not be met from revenue generated by the proposed line.

Before June, 1950, when the Korean War began, virtually all of the crude oil and products used in the Pacific Northwest area arrived by tanker. Thus, the construction of Trans Mountain received a high priority from the governments of the U.S. and Canada. Accordingly, on March 31, 1951, Trans Mountain was incorporated by a special act of Parliament, and on December 10, the Board of Transport Commissioners heard the company's application for a permit to build the line. Three days after this hearing, a permit was issued. The actual construction of this pipeline proved to be a test of men, technology and engineering skill. In final form it extended 718 miles and crossed 72 rivers and streams; along the route the winter snowfall reached a depth of more than 50 feet in mountain valleys. All work had to be accomplished at altitudes ranging from 1,125 feet to 3,920 feet above sea level. The job of building the Trans Mountain line began in 1952, and as many as 2,500 men at a time worked on the line. Finally in the late fall of 1953, oil from Edmonton reached Vancouver through the line. Subsequently, Trans Mountain built an additional line to the Puget Sound area, to take care of customers there. Also, Imperial increased the capacity of its Vancouver refinery from 12,000 to 22,500 barrels per day [b/d] by 1953 and to 36,800 b/d by 1973. In British Columbia, refining capacity grew to 150,000 b/d by 1975.

As a way of developing even larger eastern markets for western oil, in 1952 Imperial contracted with British American Oil Company Limited [later Gulf Oil Canada Limited] to supply three million barrels of crude each year to that company's plant near Toronto. This reduced in that approximate quantity crude previously imported from Illinois. In the same year, the company also contracted to supply five and one-half million barrels per year for ten years to the Sarnia refinery of Canadian Oil Companies Limited. In both cases, Imperial guaranteed prices competitive with the cost of oil from Illinois.[10]

As the crude oil production capability of Western Canada continued to grow, it became apparent that more efficient facilities than the pipeline-tanker combination would be required to move oil to Ontario. In 1953 Interprovincial extended its line to Sarnia, and in 1956, the line advanced to the outskirts of Toronto; in 1962, a branch line was built to Buffalo, New York and spur lines went out to St. Paul, Minnesota, and Toledo, Ohio.

The year 1968 saw another major extension of the Interprovincial system. It had become clear that even greater pipeline capacity would be needed to serve the expanding demand for crude oil in Ontario and the Great Lakes area of the United States. The most practical way to do this seemed to be to build a new line to Chicago and then on to Sarnia.[11] Unlike the earlier proposal to build a line to Superior, the Chicago extension of Interprovincial did not become a matter of controversy in Canadian politics. But some opposition did emerge, this time from the Independent Petroleum Producers Association of America [IPPA]. In fact, they put enough pressure on the U.S. government to block the projected line for some months, using objections based on an overall program of opposition to imported oil. Finally, late in 1967, the two governments reached an agreement on the issue, and the U.S. issued a construction permit in January, 1968. The Superior-Chicago line was completed in that year, and in 1969, Interprovincial completed the Chicago to Sarnia pipeline.[12]

Between 1950 and 1975, Imperial's petroleum exploration program concentrated company efforts on oil-prospective, rather than gas-prospective, areas. As a result Imperial's production of natural gas, a large part of which came from plants that separated gas from oil in the field production process, remained secondary to its output of oil. Because of this the company concentrated on developing new markets for crude oil, while potential natural gas markets were left for others to develop. However, Imperial's gas production gradually became an important part of its upstream operations; and between 1951 and 1975, its natural gas sales increased from 20 million cubic feet per day to 412 million cubic feet per day.[13] At the same time, the company also substantially increased its output of liquefied petroleum gas [LPG]. Business in all important lines increased far more than had been foreseen in 1950.

Imperial President and Chief Executive Officer J.R. White formally announced in 1955, that the company had decided to enter the petrochemical field. He stated that the company would build a detergent alkylate plant in Sarnia. By the following year, petrochemical plans had advanced to the point where the *Annual Report*, for the first time, devoted a separate section to this phase of business operations. The new section in the report called attention to construction of a large plant, "the first big venture of any oil company in the Canadian petrochemical field," being built to provide "large quantities of basic chemicals" for other processing companies. To be completed in 1958, it would produce ethylene, propylene, butylenes, butadiene, aromatic distillates and tar. Much of the push behind Imperial's movement into the rapidly growing field of petrochemicals came from J.K. Jamieson, who had joined the Imperial board in 1952. Jamieson saw to it that Imperial invested in this area before other competing oil companies made their moves; his election in 1953, to vice president with responsibility for chemicals, served to confirm the momentum of Imperial's move.[14]

In particular, Imperial created a Chemical Products Department consisting of three men. Clay Beamer, who headed it, had been brought in from Jersey's U.S. organization by Jamieson.[15] Beamer toured Imperial's plants, interviewing chemists and trying to persuade them to join his new department. As the Chemical Products Department expanded, its influence grew. The next half-dozen years saw additional plant construction and other efforts that paralleled the development of the petrochemical industry in the United States. During the early 1960s, Canadian agriculture, especially wheat producers, began to expand their acreage every year. With this attractive market in view, Imperial decided to enter the fertilizer business.

In 1964, Imperial began to sell Engro brand fertilizer [manufactured for Imperial at that time by another company] through company agents in several provinces.[16] By 1967, the company had gained 12 percent of the "total prairie fertilizer market." It also had built more than 400 warehouses across the plains, and could announce "the start of construction of a fertilizer complex at Redwater, near Edmonton," estimated to cost $50 million. That complex was to go on stream in 1969.[17] By that time, however, Imperial's entry into fertilizers had proved no more rewarding than had Jersey's. Yet despite a continued decline in farm prices, Imperial held on to its fertilizer plants. When in 1972, prices of farm products and sales of fertilizer began to improve, they reached such levels that the company moved to increase the capacity of the Redwater plants. By 1975 their capacity was 700,000 tons annually, and Imperial's position seemed stable.[18]

In 1964, Imperial acquired Building Products Limited [Building Products] largely because it presented an opportunity for the large-scale use of "oil industry products, especially asphalt and fuel." Building Products also manufactured plastic pipe, light reflectors and siding for housing—all of which had some market potential.

In 1969, Imperial announced the creation of a new division, Esso Chemical Canada [Essochem], to be given responsibility for the company's entire chemical business. Another reason advanced for its creation was "to point up the fact that in this field it is backed by the marketing and technological strengths of [Jersey's] Esso Chemical Co. [mpany] Inc. [orporated] and its world-wide affiliates."[19] Again, the reorganization appeared by the end of the 1970s to have been largely successful. However, it is important to remember that during the late 1960s and before the Arab oil embargo of 1973, Imperial's chemical division operated under a serious handicap: production costs of U.S. Gulf Coast chemicals had fallen so low that they could be shipped to Canada, pay the 10 percent duty, and still undersell Canadian chemical products. But the Arab oil embargo of 1973 and the oil price increases dictated by the Organization of Petroleum Exporting Countries [OPEC] triggered a strong market for chemicals. In these years, Essochem had the advantage of being able to secure most of its feedstocks below world prices from Canadian sources. In 1974, Essochem sales reached $318 million. This represented an increase of $120 million over 1973.[20] In 1975, the sales figure grew another $2 million. Chemicals at Imperial had come of age, as these record figures showed.[21]

X   X   X

By 1968, virtually all pipelines in which Imperial possessed an interest had been automated. Computer-assisted consoles controlled the pipelines, scheduled the flow of oil, performed some accounting and handled most problems associated with engineering. By this time, Imperial shared in pipeline systems from Maine to Montreal and from Toronto to Vancouver, and gathering lines collected its oil and gas from widespread fields. Imperial's exploration activities ranged from offshore in the Atlantic to British Columbia, and from the Canadian-U.S. border to the Arctic. The establishment of a petroleum engineering laboratory and research center in Calgary, Alberta, in 1954, foreshadowed another important development in the company history.[22] Together with three other companies in 1959, Imperial also joined in a research and testing program leading to the extraction of liquid hydrocarbons from the Athabasca oil sands of Northern Alberta, the world's largest known petroleum deposit.[23]

With the completion of the Interprovincial line between Edmonton and Sarnia, Imperial saw no further need for its Cygnet, Ohio line, previously used to bring oil from the U.S. So in 1953, it sold this line to U.S. interests.

From the outset, the question of oil and gas exports had been decided by the Canadian government. Initially, the decision lay with Alberta as the major producing province, which determined whether or not it would permit oil or gas to be moved beyond its borders. Here, the criterion was years of supply; in the immediate post-Leduc period, Alberta stipulated that it had to be assured of 50 years' supply of natural gas for its own citizens before it would allow any to be exported from the province. Of course, Interprovincial pipeline movement of oil and gas, and their export to other countries, fell under the jurisdiction of the federal government, and gas exports to the United States also required the approval of the U.S. Federal Power Commission.

The question of oil and gas exports first emerged as a serious political issue in 1958. At that time the views of oil men differed sharply from those of the public and of some government agencies. Thus:

> The situation made it imperative that the Government of Canada should inform itself as to whether permitting the sale of oil and gas outside the country would . . . as industry claimed, provide the necessary incentive to discover new reserves out of all proportion to those required by the country.[24]

The Canadian election of 1957 resulted in the defeat of the Liberal party, after 22 years of control, and the victory of the Conservatives. In that year, the government appointed a Royal Commission on Energy and selected to head it, a Toronto industrialist, Henry Borden [the Borden Commission, it was generally called]. In 1958, this commission began to conduct hearings. Until that time, "organized, sustained efforts to market Canadian oil and gas were practically non-existent except by individual members of the industry."[25] Now, the Borden Commission sought to gather the basic information it needed to develop a national policy on exporting gas and oil.

Based on its findings, the commission recommended that the Canadian government issue permits regulating on the basis of the rate of discovery of additional reserves the quantities of gas and oil to be exported. The commission also recommended that the government create an Energy Board, a measure generally approved by the Canadian oil industry. These recommendations proved to be "the stabilizing influence that has ensured industry that if it finds oil and gas reserves beyond those that will be required by Canada over a reasonable period of time, the export of such products will be permitted. That does not, ... guarantee markets," but it did provide for the industry "a fair idea of what avenues for expansion are open for the future."[26]

The U.S. oil quota system imposed after 1957, the reopening of Suez, and the return of Iranian production to world markets, all had served to complicate problems for oil companies everywhere. In Western Canada, producers had a surfeit of oil; while in Eastern Canada, refiners were being undercut by offshore oil products, available at "distress" prices. The shut-in production in Western Canada stemming from the world crude supply situation actually posed another problem for the Borden Commission. In Western Canada, a lobby created by independents and intermediate oil companies, a group becoming more politically powerful, began to push for a government-backed extension of the Interprovincial pipeline from Toronto to refineries in Montreal. They contended that this would end that heavily populated area's dependence on oil imported through the Portland, Maine pipeline to Montreal—largely South American and Middle Eastern oil. The American press and many persons appeared to believe that most major transnational oil companies opposed the "Montreal pipeline concept." At that time, the oil majors had large investments in the Middle East. From fields there, oil could be moved by tanker [supertankers developed after the Suez Crisis of 1956 - 1957] to Portland and by pipeline to Montreal for "a fraction of the cost of western Canadian crude."[27] Thus, in the late 1950s—a time of low profits for the oil companies generally—economics made it more attractive to move Middle East oil to Montreal than it would have been to deliver oil from Alberta to Montreal through the contemplated pipeline. Further, the U.S. government expressed concern over the effect on the Venezuelan government, if its oil markets were further reduced.

Imperial did not consider Montreal an economically advantageous market for Canadian crude. The company's position is accurately reflected in an editorial from the *Imperial Oil Review:*

> If Canadian crude oil could be made available at Montreal at competitive prices, this company would use it and displace imported crude oil from any source whatever. But until every effort

has been made to widen the more attractive markets for Canadian oil, a Montreal pipeline would be poor business judgement for everyone concerned—the producers, refiners, consumers and tax-payers.[28]

Imperial found particularly galling certain allegations that its objections to the Toronto-Montreal pipeline derived from Jersey's influence. Imperial not only prided itself on its Canadian nature—it also knew the benefits that had accrued through its long affiliation with Jersey. [29] In the last analysis, the Borden Commission did not recommend that Interprovincial build the Toronto-Montreal pipeline; instead it recommended that Western Canada producers "push for markets in other areas (for example, the Chicago and Northern U.S. refiners) before it asked the government to investigate any further the rather questionable economics of a big-inch oil pipeline stretching all the way from Alberta to Quebec."[30]

From the recommendations of the Borden Commission came the creation of a National Energy Board [NEB] in 1959. Two years later, a policy was established under which oil and gas exports were to be permitted when a specified surplus above Canadian needs existed. The National Oil Policy [NOP] set voluntary production targets for several years ahead. Implicit in these limited controls was a serious threat to the oil majors: the Canadian government could open (via pipeline) the Montreal market to Western Canada producers at any time. The National Oil Policy also declared, in effect, that all refineries west of the Ottawa Valley (in essence, all refineries in Ontario and west) and refineries in those areas of the United States that could be reached economically would be, in the future, the natural market for Western Canadian crude oil. The same announcement stated that only Canadian crude was to be used in refineries west of the Ottawa Valley and that efforts should be made to enlarge export markets.

The major international oil companies appeared able to cope with a gradual erosion of oil prices during the 1950s and 1960s. A basic reason for the decline of prices was that producing capacity increased faster than demand. One cause was the continuous granting of concessions by Middle Eastern and Mediterranean governments to intermediate-sized international oil companies, plus pressures on earlier concession holders to increase production by those same governments which wanted more income. In 1957, U.S. President Dwight D. Eisenhower invoked a program of voluntary import controls, largely because of pressure from the politically powerful IPPA and the coal producers. When voluntary controls failed, he instituted in 1959, a mandatory control of imports, a program that seemed to work for a time. Under this program, Canadian exports to the U.S. continued without special license.

X   X   X

During the 1950s, Imperial sales of all products increased, but its share of the Canadian market declined; in 1950, the company sold 64 million barrels of products, while in 1959, it sold 107 million barrels; during the same period, however, its share of the total market dropped from 50 percent to 35 percent. In the decade, existing competitors expanded their operations, and additional major oil companies moved into Canada, bringing generous funding and excellent marketing organizations. The advent of these companies increased the already keen competition in the Canadian petroleum industry. Imperial had planned carefully to meet this competition. Starting immediately after the war, the company consistently spent much of its capital to expand and modernize its existing refineries. For example, between the time it went on stream in 1948, and 1950, the capacity of the Edmonton refinery was tripled. In all, the heavy financial requirements for refinery modernization and expansion, added to extraordinary capital needs demanded by continuing exploration, field development, and crude transport facilities, left little money aside to intensify marketing operations during the 1950s. What capital expenditures could be made for marketing, Imperial directed toward improving the efficiency of its product distribution system.

Imperial marketers were not so sanguine as to believe that—even with plentiful funds—the company could possibly have maintained the 50 percent share of the market it had held in 1950. Still, they could not help but feel that a meager capital allocation penalized them in the contest for markets. As one example, while the total number of Canadian service stations increased by more than 4,000 during the 1950s, in 1959 there were nearly 700 fewer Imperial outlets than there had been in 1950. By the end of the decade, however, the marketers' belief that the company's spending policies had handicapped them proved incorrect.

It became increasingly clear in the late 1950s that Imperial had entered an era of brutal competition, one that eventually extended into the 1970s. Price wars occurred in major marketing areas. In these years, the correctness of earlier capital allocation became apparent. As things turned out, the earlier basic capital expenditures on physical plant construction and improvement, rather than on expansion of market outlets, placed the company in a sound position later to meet the challenge of sterner competition.

Imperial backed its marketers and refiners with the largest research program in the Canadian industry. Of course, it also had access to the research of Standard Oil Development Company [SOD; after 1956, Esso Research and Engineering Company] and to research at companies in the worldwide network of the Jersey Company. This left Imperial researchers free to specialize in areas in which they had expertise and to seek solutions for problems peculiar to the Canadian industry. The company had pioneered in research leading to the removal of impurities from lubricating oils and the development of longer-lasting oils and greases. One of the processes that Imperial developed—phenol extraction—came to be used by 40 percent of the world's refineries.[31]

As well as pioneering in new processes, the personnel at Sarnia gained a reputation for solving problems in lube oil applications. In 1951, for example, responding to a plea from General Motors of Canada Limited [GM] for help in combatting excessive wear, bearing corrosion, and excessive carbon deposits in their locomotive diesel engines, Imperial's Sarnia research group provided GM's engineers with an lubricating oil they had been working on for two years in anticipation of such problems. GM engineers proclaimed that it had solved their problem. Imperial researchers also developed a new additive, the first of its kind, which inhibited wax crystals from growing in heating oils made from Canadian crude, thus retarding solidification and allowing the oil to flow at lower temperatures. Soon, the use of this additive for oil in cold climates spread across the globe, and for some years Imperial remained the sole supplier. By the 1960s, the total research program of Imperial's two units at Sarnia and Calgary more than equalled the total research and technical service programs of all other oil companies in Canada. As part of its research effort, Imperial continuously sought new uses for petroleum. In 1960, it participated in tests at the Dominion Foundry and Steel Ltd. [Dofasco] plant in Hamilton, Ontario, which demonstrated that the use of fuel oil could reduce costs and increase the capacity of plants engaged in iron ore smelting.

A special section of the *Annual Report* in 1961, explained Imperial's full and varied research program: "certain members of the staff are immersed in long-term work ... [while] most of the staff are occupied with more pressing shorter-term work."[32]

In October, 1970, Imperial introduced " 'Esso 2000' a low-lead gasoline designed for 1971 model cars" and for some 20 percent of earlier year models. The company announced that this gasoline would be "a first step toward supplying fuels needed to meet future automobile emission standards." [33] Once again, the product had been changed to meet the new demands of a changing time. One sign of change was the company's increasing concern with the environment. Imperial environmental coordinators and associated personnel composed an important segment of the company, a group of employees large enough to undertake a major environmental survey in 1970, with the intention of eliminating as much "in-plant" pollution as possible.[34]

Imperial had been far ahead of the general public in its concern for the environment and the extent to which company operations might cause pollution. As early as the 1930s, an Imperial chemist, Alexander McRae, had cruised the St. Clair River near Sarnia in *The Juicy Scoopy*, a converted lifeboat, taking samples of the water. Based on his laboratory tests, the company recognized the problems and soon began to alter its industrial processes so as to reduce the discharge of waste products into the water. Later, in 1953, Dr. R.K. Stratford, head of the company's Research Department, persuaded the province of Ontario "to undertake a continuing cooperative study of environmental quality in Sarnia's Chemical Valley." This study monitored both air and water quality, and the lessons learned and solutions applied there became guideposts for other refining and chemical producing plants owned by Imperial.[35]

By 1963, Imperial had ordered its ships to retain food wastes and other garbage for later disposal ashore [Canada ordered all vessels to follow this practice in 1970]. From an early date engineers and scientists throughout the company had worked to prevent noxious odors from escaping the plants and waste from polluting nearby land and water. So Imperial's 1970 survey of the environment simply followed its traditional practice.

During the late 1960s, Imperial also took a leading role in the establishment of cooperative action groups to protect the environment. In 1970, along with other companies, it assisted in the formation of two major industry associations—the Arctic Petroleum Operators Association and the Petroleum Association for the Conservation of the Canadian Environment [PACE]. To complement their efforts, other local and regional organizations were formed in such major operational centers as Sarnia and Montreal. As the press release that announced the establishment of PACE said:

> The Association is designed to serve as a pacesetting function within the industry. It will facilitate the exchange of technical information on pollution prevention, foster environmental and ecological research and develop joint industry programs such as spill prevention and cleanup. It will provide the main point of contact with government, industry and other groups interested in preserving the environment.[36]

In turn, the Arctic Petroleum Operators Association was established to study ways of protecting the Arctic and sub-Arctic environment and to advance solutions for problems encountered there. With more than 50 years experience in petroleum exploration and production in the sub-Arctic, Imperial appropriately took the lead in the formation of this cooperative group.

In all, Imperial emerged as a pioneer in corporate efforts to leave the environment as it had been found. The company made significant technological contributions to the industry, and these directly assisted in the preservation of Arctic ecology. Achievements were made in new drilling methods designed to minimize possible environmental damage.[37]

X   X   X

In the *Annual Report* for 1950, Imperial pointed with pride to a long history of freedom from labor disputes. The company attributed this "in large part to the loyalty and enthusiasm of the employees and also to the careful attention that has been given to employee relations." As an early leader in creative employee relations, the company had organized a joint council system which provided an excellent climate for employee-employer cooperation.

This council system instituted in 1918, grew out of studies initiated in the United States by John D. Rockefeller, and carried out by, among others, W.L. Mackenzie King, later to become prime minister of Canada. Use of the councils permitted amicable resolution of problems involving working conditions at the local level, while the councils themselves served as a channel through which employee concerns and opinions could be expressed on matters that were regional or national in scope, such as wages and benefits.

By 1950, there were 90 joint councils, representing many of the 12,500 employees in Imperial.[38] Although the number of councils did not increase significantly between 1950 and 1975—consolidations in some phases of company operations offset expansion in others—the function of the councils continued to evolve. Council committees encouraged by the company, acted to study and recommend ways in which the quality of working life at Imperial could be improved. Employee Relations Manager R.A. Wilson described this program this way: "In this development the foreman has less of a role and the employees in his group more of a role in deciding what's to be done. It's a more democratic system."[39]

In the Winnipeg refinery, for example, at the suggestion of the elected representatives of the employees, Imperial introduced an experiment in compressing the work week. The results were so positive that the three-day or four-day work week spread to all other Imperial refineries except Ioco, which does not have an industrial council and whose wage-earning employees are members of the Oil, Chemical and Atomic Workers Union. Another development that sprang from the council system of study and recommendation was flex time. Here, office employees, while putting in a full working day and not disrupting essential routines, decide individually when their work day will begin.

In 1950, and in virtually every year thereafter the *Annual Report* reflected the company's continuous concern with its employee benefit plans. Thrift plans, pension and annuity plans, group insurance, hospital and surgical benefit programs, all received constant attention. The company acted to upgrade and broaden benefits for all employees including those who had retired.[40] By 1955, Imperial paid accident, sickness, and survivor insurance for its employees and shared with them the cost of group life, hospital and surgical insurance and the annuity and thrift plans. Ninety-nine percent of all employees participated in the two latter plans. Among other possibilities, thrift plan contributions could be used to purchase Imperial stock.[41] In 1958 and 1966, the insurance and health care programs were integrated with the corresponding systems of the provinces and the nation. In its continuing study of employee problems and relations, Board members and other major executives visited all major company operations and sought to meet employees and their families.[42]

Imperial's own Compensation and Executive Development [COED] Committee became effective in the years following World War II as a resource for providing leadership in the times ahead. Functioning on the same basis as the comparable Jersey committee, Imperial's executive appraisal system sought to identify those employees with high potential for leadership and to make every effort, through placement and training, to enable them to realize their potential to the fullest.

Other company training programs enabled many employees to improve the skills needed to operate the increasingly complex machinery and equipment used in Imperial's ever changing plants. Experiments began at Sarnia in 1956 with Joint Industrial Councils, which led to more intensive employee training. Employees who sought to promote their personal development through job-related courses also received assistance. Additionally, Imperial maintained programs of scholarships, fellowships and research grants. In 1962, it installed the Imperial Oil Higher Education Award program to aid children or wards of employees and annuitants. For eligible high school graduates, the company paid four years' tuition at a Canadian university or other institution of higher learning. Sixty-two students enrolled under the plan in that first year.[43]

As a consistent goal, Imperial had sought to lead the Canadian oil industry in wages and benefits. Keen competition made this role difficult; however, by the 1970s, the employee benefit package taken as a unit—wages, pensions, health benefits, holidays, etc.—enabled the company to rate well with about 90 percent of its competition.

In 1947, Imperial opened a Medical Department, and after that time the company continued to lead the way in providing on-the-job health services for its personnel. Between that date and 1975, the Medical Department grew from one doctor and one nurse to seven physicians, two industrial hygienists and 17 nurses available in a network of health centers across the country.

X   X   X

From 1916 to 1957, Imperial maintained its corporate headquarters at 56 Church Street in Toronto. Office space at that address was limited and operations had become cramped even before World War II. W.G. Charlton, who joined the company as an office boy at Church Street in 1935, recalled that a large number of departments and companies were crammed into one building. Soon some departments and affiliates had to move to nearby buildings which Imperial gradually acquired. Finally, in 1950, Imperial's managers grew impatient with make-shift solutions and appointed a three-man committee to select a site and determine what would be required for maximum efficiency in a new building. That committee selected a site on St. Clair Avenue, three miles from the old location and, at that time, some distance from Toronto's main business district. Once again Imperial in moving to its new location led the way; in the following years other businesses moved into that area.

The committee had selected a spectacular site. Standing on an escarpment, the new building could command a view of Toronto unmatched by any other structure at the time. However, this same cliff posed excavation problems. Actually, these delayed construction of the new headquarters at 111 St. Clair Avenue West until 1957, when the contractor finally turned the building over to the company.

X   X   X

Imperial, as the best-known and largest oil company in Canada, has been regarded from the founding to the present as a prime source of information on the oil business for the public, the news media, and the government. In 1944, the company gave regular staff status to the public relations activities that had earlier been conducted by Assistant to the President, W.F. Prendergast.[44] Two years later the Public Relations Department had 28 employees. A primary factor in this rapid growth "was that in Canada there is no national equivalent of the American Petroleum Institute [API]." The Canadian Petroleum Association [CPA] concerned itself only with upstream functions—exploration, production and pipelines—and at the time its interests centered in Western Canada. Other sources of information were required. After Leduc, the Imperial Public Relations Department found it necessary to decentralize operations by establishing field offices at major operating centers. Here, field representatives acted as one-man, one-secretary regional Public Relations Departments.[45]

During the 1950s, when many oil majors and their affiliates moved to have their histories written by trained scholars and agreed to open company archives to them, Jersey secured clearance from Imperial for researchers to use Jersey-related records. One scholar commissioned by Imperial wrote a history of that company which was never published. President W.O. Twaits saw very little value and considerable expense in publishing such a history.[46] Jersey and some of its affiliates, like Humble Oil & Refining Company [Humble], conceived the publication of corporate histories as a way of setting the public record straight. Obviously Imperial felt otherwise.

Yet at about the same time, in the late 1950s, Imperial's management became increasingly aware that the company needed to take more positive stands on public issues in Canada, in order to gain a fair hearing. Many citizens believed that the major oil companies had reaped fantastic profits from excessive product prices and the exploitation of publicly owned natural resources. These mistaken notions required some form of correction. One way in which the company could attack erroneous beliefs of the public was by stating its position in the *Imperial Oil Review* and the *Imperial Oil Limited Annual Report*. The *Review* also carried general interest articles and, on occasion, specific editorials written to express Imperial's view on subjects of special concern such as oil and gas prorationing, wage and hour controls, Arctic and sub-Arctic pipelines and their effect on the environment.[47]

Meanwhile, in the pages of the *Annual Report*, an interesting change took place. Until 1957, these reports described only internal company developments. From that time on, however, especially in presidential messages, attention was given to a variety of external factors that influenced company operations. These influences included Canadian government attitudes and actions. Later, beginning in 1973, a special section on government actions and the implications for the company of these actions, appeared in the *Annual Report*. Thus, between 1950 and 1975, Imperial developed its yearly statements from a quiet voice on purely business matters into an outspoken source of commentary on government actions in which the company and its employees had a vital interest.[48]

Another important change in the company's relations with the public derived from alterations in the climate of opinion in which Imperial operated, especially changing public expectations concerning business as a social entity. The concept of social responsibility was not new to Imperial—early in its history the company had demonstrated an appreciation of the fact that Canadians had to be regarded as something more than just customers. By the late 1920s, the company had begun to sponsor *The Imperial Hour of Fine Music*, a weekly series of hour-long concerts of classical music, broadcast live over a national network of radio stations stretching from Montreal to Vancouver. For many years, Imperial also sponsored the highly popular program "Hockey Night in Canada," on the Canadian Broadcasting Corporation network. The company's advertising carried corporate messages on subjects such as profits and prices, as well as product and service information. More recently, some Canadian Nationalists objected to the corporate commercials on the grounds that these messages often sought to influence political debate on oil policy.

In other ways, especially through its donations program, Imperial consistently supported worthwhile projects in education, health and welfare, arts, and sports. While the company realized that donations had a low profile, it also felt that good taste required an absence of publicity. It had become apparent as early as the 1940s that "the other side of Imperial," its conduct as a corporate citizen in the local and national community, had become important as a yardstick by which Canadians would judge Imperial's overall performance. So public relations began to reflect this fact; "other side" projects were undertaken, and they began to attract a higher level of public attention.

Imperial, along with other Canadian corporations, also had a long-standing policy on political contributions. Like a good corporate citizen, the company contributed to the country's political machinery, much as a private citizen contributes to the political party of his choice. In 1975, at the [U.S. Senator Frank] Church Committee Hearings on Multinational Corporations, the fact emerged that Imperial Oil Limited had, without objection from Exxon Corporation [formerly Jersey], made annual contributions to Canadian political parties. As soon as this information appeared in the newspapers, Imperial made it clear that its contributions violated no law, and that the company did not support political philosophies that would be detrimental to the company. During the hearings, Exxon stated that contributions by its Canadian affiliate averaged $234,000 per year over the preceding five years.[49]

These revelations caused an uproar in both Canada and the United States. Needless to say, because the Imperial contribution story surfaced during the investigation of Exxon (and other multinationals), both Canadian Nationalists and other citizens who favored nationalizing the oil industry saw evidence of an abuse of resources, and of the evils of foreign capital in Canadian markets. Certainly, the hearings did nothing to help Imperial Oil Limited or any other company involved. In the summer of 1975, Imperial made the decision to cease making political contributions in the future.

X   X   X

As a general rule, the Leduc discovery acted as a spur to change within Imperial, and overall, the process of adapting to post-Leduc pressures and problems proved smooth: the company changed but did not completely reorganize either its management or staff structure during the years 1950-1975. Most of the changes involved rethinking the administrative structure and later, the development or alteration of departments and functions that could respond to the more plentiful supply of gas and oil that Leduc and later discoveries guaranteed.

By 1960, then, Imperial's operations had become "increasingly dispersed" and "complex" as well as being subject to "rapidly changing technical and economic conditions." So the Imperial board of directors, after considerable study, undertook a reorganization that would "concentrate the maximum authority and accountability for operating performance in the field," leaving the central headquarters office to concentrate on overall control and long-range planning, for both personnel and facilities. In this reorganization the (now defunct) General Operations Council [GOC] more than any other executive group, "assumed a major responsibility" for planning coordination and investment screening.

As approved, the plan reorganized the executive function (the collective responsibility at board level) and changed the board contact system. The board as a whole was expected to "be familiar with people and conditions in all areas of operation." President Twaits pointed out that the Imperial communication system "cannot substitute for direct contact." Under the new plan the executive function would be concerned with three broad areas of responsibility: a) Operating Activities; b) Administrative Activities; and c) Regional Policy. Under area one, the plan called for grouping all revenue producers into four classifications: Exploration and Production, Chemicals, and Petroleum Products, with a fourth group open for later expansion. Administrative Activities was divided into such groups as Finance, Employee Relations, Research and Technical Development. As the plan stated, problems in this area "frequently originate at the field level, but usually acquire corporate importance and ultimately [require] corporate policy decision." Thus, the Administrative Activities category essentially dealt "with staff or administrative activities within which lie some of our [Imperial's] most difficult problems." Regional Policy, the third area of defined responsibility, acknowledged that there were certain "specific political and economic considerations which affect corporate decisions." Here the governments of the provinces and of the nation were recognized as persisting influences on corporate decisions that would compose the particulars of Regional Policy.

The reorganization plan allowed for no departmentalization at the executive level. The board and Executive Committee together continued to monitor departmental operations and to facilitate compromise where overlapping existed between areas or departments. To accomplish this, the plan defined director contacts: a) Administrative Control Contacts—day-to-day relations with the assigned departments, and b) General Policy Contact—concerned with group performances on a broad basis. Overall the plan established a revised contact director system adapted to the needs of a vertical oil company. When reorganization went into effect, board members would be freer to study and plan for policy decisions. While assignments did vary, all board members, "individually and collectively has [had] a single responsibility for the company as a whole," and departmental managers were encouraged to consult with any director. Twaits concluded his announcement of the forthcoming change by pointing out that, in an organization like Imperial, "it is essential that all information and experience be brought to bear on a problem with the least interference from actual or presumed organization barriers." He sought freedom of action for both top executives and field managers; they must be allowed to enhance policy decisions. "Easy access" throughout the company would serve to improve all functions.[50] The reorganization plan went into effect during the summer of 1963.

X   X   X

From 1950 to 1959 Imperial's capital outlay and exploration expenditures totalled just over a billion dollars. To finance this massive outlay, the company reinvested some $230 million in earnings, went to the money market several times to borrow, and sold off some properties, as previously mentioned. In 1957, Imperial President J.R. White told a meeting of the Canadian Petroleum Association:

> Imperial had some $30 million in cash at the end of the War [World War II], and felt quite comfortable about its financial position. In no time that $30 million went out the window, and was followed by our shareholdings in International Petroleum Co. Ltd., in Royalite Oil Co. Ltd. and in some other outside companies. Beyond this we had to borrow $100 million and issue more than four million shares to raise additional funds.[51]

Imperial's earnings during the decade increased every year except 1958 when, reflecting growing competition in the product market and a decline in crude oil production resulting from reduced exports, profits dropped sharply. Earnings as a percentage of shareholders' investment during the decade had ranged from 13.4 percent in 1955, to 8.3 percent in 1958; dividends paid during the period totaled $283 million, 55 percent of earnings after taxes. By the end of 1959, the book value of shareholders' investment had risen to $630 million, compared with $256.7 million in 1950.[52] It should be noted that in the low performance year, 1958, an agreement was made between the United States and Canada concerning export and import policies. Whether the Canadian-United States delegations ever implemented that pact is not known. Still, the text of the "so-called agreement, which is all the [U.S.] State Department or the Canadian Embassy in Washington are aware of," reads:

> The Canadian and United States Governments have given consideration to situations where their export policies and laws of the two countries may not be in complete harmony. It has been agreed that in these cases there will be full consultations between the two governments with a view to finding through appropriate procedures satisfactory solutions to concrete problems if they arise.[53]

Imperial would henceforth consider one more political variable, like the decision in 1957, by U.S. President Eisenhower to invoke the voluntary Oil Import Program, which affected Canadian exports.

Imperial began to shift the emphasis in its exploration program in the early 1960s. In 1961, it let drop exploration permits of more than one million acres in the Atlantic provinces, and by 1967, it had brought all exploration activity on the eastern mainland of Canada to a close. This move did not mark the end of the search for oil in Eastern Canada, however; for soon the company began to explore the "frontiers," which included the Atlantic Ocean offshore.

In Imperial's assessment, the Southern Basin—which extended from Southwest Manitoba across half of Saskatchewan and two-thirds of Alberta, on into Northeastern British Columbia and the southern part of the Yukon and Northwest Territories—had become a mature area in terms of oil exploration. Later events proved this assessment to be correct. Even the Rainbow-Zama area of Northern Alberta, which in 1965 gave indications of being a major discovery, turned out to contain reserves much smaller than those of the major fields discovered in the previous two decades. Other promising fields contained largely gas, not oil; while their size and distance from existing transport systems, combined with the prevailing low wellhead prices, meant that many of the new-found fields simply did not warrant development.

By 1965, on the average, one well for every 10 square miles had been drilled in Alberta. Yet other vast sedimentary basins in Canada had barely been touched.[54] So Imperial shifted exploration to those other basins—offshore in the Atlantic, the Arctic Islands, and Canada's Western Arctic. At about the same time, the company increased its efforts to develop synthetic crude oil. Already 10 million acres of exploration permits had been acquired in the Western Arctic, as well as a half interest in 40 million acres in the area of the Grand Banks, southeast of Newfoundland. Seismic surveys and drilling began in both areas. As might be expected in the initial exploration phases of such large areas, the initial drilling turned up many dry holes. Overall results by 1965, however, were sufficiently encouraging to warrant the continuation of exploration.[55]

In 1968, Humble Oil & Refining Company and Atlantic Richfield Company [ARCO] drillers brought in the Prudhoe Bay oil field. This field, the largest single strike ever made in North America, was estimated to contain nearly 10 billion barrels of recoverable oil reserves. Some geologists estimated that amount to be only a small fraction of the quantity that would ultimately be found.

It would be an understatement to say that the Prudhoe Bay discovery created great excitement in Toronto and New York. J.A. Armstrong, who became Imperial's president in 1970, described the effect in an address in Toronto: "During 1968, while a phenomenon called Trudeaumania was sweeping Canada, a phenomenon called Prudhoemania was sweeping the oil industry in North America." He went on to point out that the Canadian (and Imperial) interest in the Prudhoe discovery was magnified by the fact that the Alaska North Slope and an adjoining Arctic coastal area might have similar geological histories. "To put this another way," he said, "there was every reason to hope that, in the Canadian Arctic, oil discoveries similar to Prudhoe might be found."[56] For Imperial, exploration retained its special excitement, not untouched by disappointment.

The company's optimism over the prospects of the Western Arctic, for example, was first reinforced in January, 1970, when Imperial drillers made an oil discovery at Atkinson Point on the Beaufort Sea. Subsequent drilling proved it not a major strike, however; yet to Imperial, "the fact that geological results obtained from the drilling confirm that the Arctic is a highly prospective region for crude oil and natural gas" tempered the disappointment.[57]

By 1975 it had become apparent that the area was still highly prospective—but now for gas rather than oil. Approximately two hundred million barrels of oil had already been discovered. Had these been in the Southern Basin, they would have made a substantial contribution to the Canadian oil stream, but the difficulties of moving Arctic oil to market meant that they must lie untapped. Tanker movement at the time appeared out of the question, and in Imperial's assessment, reserves of at least one billion barrels would be required to justify the expenditures needed to develop the fields and to build a pipeline that could carry the crude to Southern markets.[58]

About one-third of Imperial's exploration holdings in the Western Arctic lay offshore in the shallow waters of the Beaufort Sea near the Mackenzie Delta, where in 1970, the company's Arctic search for oil was extended offshore. In 1971, Imperial discovered a major gas field in the Delta—delineation drilling was to show that it contained more than three trillion cubic feet. Additional gas discoveries in the area by other companies brought the total to more than five trillion cubic feet. These reserves, by themselves, were not large enough to justify the very high cost of development and connection to market, but the gas reserves of Prudhoe, some 300 miles to the west, were also awaiting movement to market. A pipeline that could carry both Prudhoe and Beaufort gas seemed to be an economic proposition, and in June of 1972, Imperial and other companies formed the Gas-Arctic Northwest Project Study Group to examine the feasibility of building a pipeline from Alaska and the Mackenzie Delta to southern markets. The consortium that formed the study group—by the end of the year it numbered 25 member companies—eventually established two companies. One, Canadian Arctic Gas Study Limited [CAGS], would complete the feasibility study. The other, Canadian Arctic Gas Pipe Line Limited [CAGPL], would make the permit application to build the line if the studies proved its feasibility.

In itself the story of this whole project—planning for the movement of Arctic natural gas—provides us with a case history of how the Canadian political climate for petroleum development had changed between the immediate post-Leduc years and the 1970s. Interprovincial Pipe Line Limited was incorporated on April 30, 1949, and eighteen months later it began operations. The Gas-Arctic Study Group was formed in mid-1972. At the end of 1975—42 months later—the movement of Arctic gas still remained under study by Canadian government agencies. No operations were underway and nothing was being earned. Yet by that date members of the consortium, either on their own or through their participation in CAGPL, had spent more than $100 million on engineering, environmental feasibility and preliminary design studies, plus the costs of regulatory hearings. In all, three different groups proposed to build Arctic gas pipelines. One would carry Prudhoe gas only as far as the port of Valdez, on the Alaskan Coast, where it would be liquefied for shipment to U.S. markets; a second group would build a relatively small-diameter line to carry only Mackenzie Delta gas. CAGPL, the third group, intended to move both Alaskan and Canadian gas. In the opinion of Imperial and other member companies, this latter would provide the most efficient and economic way to transport gas from both areas, and it would also provide a maximum overall benefit to Canada.

The Valdez pipeline fell under the sole jurisdiction of the United States, while the two joint U.S./Canadian proposals soon became a "hot potato" for Canada's federal government. From the first, various special interest groups generated heated public debate. Some groups exhibited genuine concern over the effect of a line on the delicate Arctic ecology and on the life-styles of the native inhabitants in that region. Other pressure groups saw in the proposals a political lever with which they could move the federal government to accede to special demands. Following a time-honored Canadian tradition, the federal government met the situation by establishing a commission—this one empowered to consider and recommend to the Minister of Indian Affairs and Northern Development the terms and conditions of any right-of-way that would allow passage through the Northwest and Yukon Territories, taking into consideration the environmental and socio-economic impact of the proposed lines. Then hearings began on March 3, 1975, and lasted for 21 months. These were followed by National Energy Board hearings, and Imperial filed a number of briefs for both the hearings of the commission and of the NEB.

(In December of 1976, Imperial Board Chairman and CEO Armstrong pointed out that the nature of the pipeline decision could give a clear signal that considerations other than the domestic supply situation were dictating the development of Canada's petroleum resources.[59] His words proved prophetic. The recommendation of the commission, along with various political considerations, led the government to deny the CAGPL proposal and to approve a route different from that proposed originally by either of the Canadian applicants. The delays resulting from this decision meant that, by the end of the decade, the prospects of moving the gas reserves of Prudhoe and the Mackenzie Delta were little, if any, better than they had been in 1970, when Imperial first started studying the feasibility of Arctic pipelines.)

While the necessary pipeline studies took place and the proper governmental bodies held more hearings, Imperial moved ahead with a major exploration program in the Beaufort Basin. At the same time, the company also expanded its search offshore in the Atlantic and began new exploration in the Arctic Islands. In 1970, Imperial acquired 21 million acres off the coast of Labrador, and by the end of 1971, its wholly-owned exploration permits in the Atlantic totalled some 46 million acres. In that year its associate in the Grand Banks venture, Amoco Canada Petroleum Company Limited, also began drilling from one semi-submersible rig, which was joined by another semi-submersible in 1972. In all, ten offshore wells were drilled that year. Two locations indicated the presence of hydrocarbons, but tests indicated that the sites were non-commercial. Imperial entered into a farm-out agreement early in 1973, which reduced its interest in the Grand Banks acreage from one-half to one-third, in return for $50 million of exploration drilling by Amoco, Skelly Oil of Canada Ltd. and Chevron Standard Limited. Under the terms of this farm-out, Imperial would not be required to make exploration expenditures during the two-year period of the agreement. By the end of 1974, no commercial shows of either oil or gas had been found and drilling was suspended.

Elsewhere, Imperial started seismic exploration of its large holdings off the Labrador Coast in 1974, and continued these in 1975. Water depths in places exceeded 3,000 feet. In 1975, the company also began supporting research to develop drilling and production technology for deep-water ventures. This research would enable the company to drill in the area in the 1980s. In the Arctic Islands, about 700 miles from the North Pole, Imperial began seismic surveys and drilling under a farm-in agreement with Panarctic Oils Ltd., which continued until the end of 1974. (In 1976, the company entered into a joint venture with the Arctic Islands Exploration Group on a major farm-in program for the offshore area near these islands. Imperial had a 35 percent interest in that venture.)

While stressing exploration of the frontiers, Imperial still managed to maintain, at a relatively low-level, an exploration program in the Southern Basin. By 1975, higher prices for oil and natural gas, coupled with tax and royalty changes and government measures to encourage drilling, helped bring about an increased level of industry exploration in Alberta and British Columbia. In 1975, the company acquired additional leases in Alberta, ran seismic surveys in the Alberta foothills, and participated in the drilling of 26 wells in the two provinces.

(By the early 1980s, and with the benefit of hindsight, many observers considered that Imperial's emphatic shift in exploration to the frontiers had brought little return but disappointment. As early as 1975, the company had discovered major gas reserves in the Beaufort Basin, and Imperial undertook an all-out effort to connect these fields to market, but by the early 1980s, the Beaufort program had failed to add one cubic foot to the company's marketable reserves. Again, some oil had been found, but not nearly enough to overcome the barriers of distance. Offshore in the Atlantic, the search had produced nothing except dry holes. Nevertheless, Imperial executives, like other oil company officials, realized fully that exploration is a long-term operation whose rewards often take many years to achieve.)

X   X   X

The establishment of a Canadian National Oil Policy in 1961 (described earlier), and new production regulations in Alberta (1964), significantly increased Imperial's revenues. Primarily because of the ruling on the natural markets for Western Canadian crude, the volume of production of that crude nearly doubled between 1960 and 1969—from 519,000 b/d to more than 1,100,000 b/d.[60] Two rulings of the Alberta Oil and Gas Conservation Board in 1964 proved particularly significant for Imperial—one affected the crude oil prorationing formula and the other the spacing of wells. An earlier prorationing formula in Alberta had guaranteed each producing well a production rate sufficient to repay costs and to provide a return on investment. Because the new formula allowed greater production from fields with large potential, Imperial received a larger share of market allocation. The well-spacing policy required that fewer wells be drilled or operated to produce oil from any given field. This ruling served to lower development and operating costs. Altogether, Imperial's crude oil and natural gas liquid production in 1969—at 179,000 b/d—nearly doubled that of 1960. Natural gas production increased from 119 million cubic feet per day to 390 million cubic feet per day during the same period.[61]

Canada, by 1969, had become self-sufficient in oil supply. While still importing some oil, the nation possessed domestic production, part of which went to export markets, that more than equalled its total consumption. Canada was "the only industrialized country in the non-Communist world to achieve that happy position."[62] Nineteen sixty-nine also marked the peak year for both Canada's and Imperial's reserves of conventional crude oil. Because of the maturation of the Southern Basin and the lack of exploration success in the frontiers, reserves afterward began a slow decline. In 1975, Imperial's reserves at 1.1 billion barrels represented a drop of more than 600 million barrels since 1969, while total Canadian reserves in the same period dropped 2.5 billion barrels.

X   X   X

In 1959, the company had joined with three other corporations—Cities Service Athabasca Inc., Royalite Oil Co. [now Gulf Canada Limited] and Richfield Oil Company [later Atlantic Richfield Co.]—seeking to resolve the problem of producing oil from the Athabasca oil sands. In 1962, this consortium applied to the Alberta government for permission to build a 100,000 b/d plant. Because Alberta crude oil then had difficulty holding its markets in competition with cheaper offshore oil, the provincial government denied the request. Two years later, the consortium formed Syncrude Canada Ltd. [Snycrude] and, in 1968 and again in 1969, applied for a permit from the Alberta government to build an 80,000 b/d plant.[63] Finally the 1969 application was approved, and in 1971 approval was received to increase plant capacity to 125,000 b/d as inflation and other factors made the projected 80,000 b/d plant uneconomical. Although earlier estimates had placed the total cost of the plant at $600 million, rapidly escalating construction expenses soon pushed the estimated cost to first one and then two billion dollars. At that point, the Atlantic Richfield Oil Company withdrew from further participation. Actual expenditures by this time exceeded $1 million per day, so a new partner or partners had to be found, and quickly. The search produced additional partners—the Canadian federal government and the provincial governments of Ontario and Alberta, and the new complex went on stream in 1978, and by 1979, Imperial had invested $529 million in Syncrude.

Solving the Athabasca sands riddle did not alone occupy the attention of Imperial researchers. Between 1964 and 1975, the company had spent more than $35 million on experiments designed to produce oil from another oil-sand deposit at Cold Lake in Alberta. The Alberta Energy Resources Conservation Board had estimated that the Cold Lake area contained 164 billion barrels of oil. Even if only a small fraction of this could be recovered, that amount would make an important addition to Canadian oil production. But again, the problems remained stubborn. Unlike the Athabasca sands, those at Cold Lake were too deeply buried to respond to surface mining techniques. Engineers found that a process using steam pumped into the formation to release oil particles would create a flow to the well bore where oil could be pumped to the surface. In September, 1975, Imperial Senior Vice President D.C. Lougheed told an audience at Cold Lake that the company had finally solved the technological problems and now felt that it could produce heavy oil from the formations.[64]

X   X   X

Beginning in 1965, Imperial's business activities took on a new look as the company began to prospect for uranium and other minerals. Coal and other mineral deposits had been located earlier, [65] but no such finds had ever received sustained attention. Now in 1966, Imperial (in conjunction with three other companies) discovered large reserves of columbium. These deposits were located near James Bay in Northern Ontario. Soon Imperial crews became active in eight Canadian provinces and the Yukon Territory. By 1972, after several years of extensive geological studies, Imperial began to acquire coal leases, and in 1973, field crews located a deposit of lead and zinc in the Gays River district of Nova Scotia. By 1975, there were more discoveries, as promising finds of uranium, base metals, and precious metals had been located. In all, Imperial Oil Limited had become one of the leading minerals exploration companies in Canada.

X   X   X

Between 1960 and 1975, Imperial continued to modernize and expand its plants and its own transport facilities as well as to participate with other companies in expansion of the Canadian pipeline system. Additional lines were added to connect new-found fields to established trunk systems, and product lines were built from refineries to major marketing terminals. In all of its pipelines, Imperial installed modern automatic controls to increase overall efficiency. On the waters Imperial substantially increased the capacity of its existing fleet and built new vessels for use on the Great Lakes and in the Coastal Service. Older ships were sold or retired. The prohibitive costs of maintaining an ocean-going fleet forced Imperial to rely on chartered vessels to move offshore crude.

Thus, the expertise in transportation and supply that Imperial had demonstrated throughout its history now played a major role in helping Canada through the oil supply crisis of 1973-1974, which stemmed from the Arab oil boycott. During that crisis, Imperial shipped about eight million barrels of Western Canadian crude to refineries in Eastern Canada, using both the St. Lawrence Seaway and the Panama Canal routes. [In addition to this direct movement, Exxon's pooling arrangements for supplying crude also helped secure alternative supplies for East Coast refineries.]

By 1960, Imperial had finished its postwar refinery expansion and modernization program, and management could shift its attention to reducing operational costs and increasing the yield of more valuable products. From 1960 to 1965, capital spending on refinery projects averaged slightly more than $6 million per year, compared with an average of $26 million per year [excluding chemical facilities] for the decade of the 1950s. During the remainder of the 1960 decade, capital expenditures for refining increased substantially, averaging $28.7 million per year. A fluid coking complex at Sarnia, started in 1965, accounted for the bulk of the increase. Its on-line computer, the first to operate in Canada, made that refinery the most sophisticated in the nation.* In 1971, the company announced plans to build a new refinery (Strathcona) with a computer-operated, totally automated system on the site of the old one in Edmonton. It would supply products to terminals in Calgary, Regina, and Winnipeg by pipeline. In the same year, Imperial also announced projected additions to facilities at Sarnia, Ontario, Dartmouth, Nova Scotia and Montreal East, Quebec. Although construction of the new Strathcona complex was hampered by a shortage of skilled labor, the plant went on stream in 1975.[66] With its start up, Imperial refineries in Calgary, Regina and Winnipeg were closed.

Imperial's marketing in the 1960-1975 period proved to be full of innovative ideas and experimental thinking. In the retail gasoline field, activities reflected the public's new purchasing patterns, occasioned by rapid urban growth, establishment of shopping plazas, extension of trunk highway systems, and changes in consumer attitudes and expectations. Imperial responded to the changes by adding to its merchandise lines and by redesigning its service stations to make them more attractive. The broadening of its service stations to include a line of Voyageur restaurants, takeout

---

*The new units proved "tempermental," thereby contributing to a decline in earnings, leading to the first drop in profits since 1958.

food facilities and laundromats occurred about the same time. Around 1964, selected automotive service centers began to offer attractive lines of merchandise as well as diagnostic automobile clinics and extensive repair services. Imperial improved its services in the growing pleasure boat market with special gasoline and motor oil and blending pumps. In the domestic heat market, the company moved aggressively to meet competition, and, in 1963, offered its Esso heat customers complete burner service, "all for the price of the oil." Less obviously, the drive to improve Imperial's market position included upgrading storage depots, warehouses, truck transport and equipment. This program extended across Canada.[67]

Imperial's Canadian marketing efforts, during the 1960s, received a boost from the famous Esso Tiger. This friendly tiger first appeared in Jersey marketing on May 22, 1964, and in February, 1965, W.R. Robinson already had written to Jersey Director Wolf Greeven to outline the Canadian situation in regard to Imperial's utilizing the tiger program. The main problem he found to be:

> the use of a slogan similar to that employed by Humble by Proctor & Gamble for its product 'Tide.' Their variation, although not used uniformly, is 'The Tiger in Tide.' The acceptance of this program in Canada was Imperial's principal reason for not adopting the Humble Tiger Program this year. There is also the problem which we discussed at our meeting regarding any legal objections which Proctor & Gamble might have to our using the 'Put a Tiger in Your Tank' slogan in Canada.[68]

Ultimately Imperial and Proctor & Gamble quietly resolved all legal questions and the tiger started his advertising career in Canada. This whimsical, but obviously powerful animal, quickly captured the imagination of young and old alike; he boosted sales to a point where competing companies felt compelled to fight back with anti-tiger campaigns. Eventually, however, the tiger ran his course and faded from the scene, leaving behind a remarkable history as one of the most successful advertising symbols ever devised.

Even so, while product and service promotions played a large role in marketing during the 1960s, the dominant factor in the entire downstream phase of the business remained price competition. Surplus refinery capacity, both domestic and foreign, stemming from surplus offshore crude capacity, resulted in products made available at distress prices. This triggered price wars. As early as 1960, for example, the average price the company received for a gallon of gasoline sold to large-volume (commercial) customers was 14 percent lower than it had been in 1955. The 1960s also saw a large increase in the number of "Private Branders"—outlets that relied on gasoline sales almost exclusively for revenue and profits, and that sold gasoline at low pump prices to attract the motoring customer. For Imperial the competitive situation was further exacerbated by "leakage" of low-cost products across the National Oil Policy line. Transgressions of this type reached the point where President Armstrong declared:

> The National Oil Policy is not law—it relies on the voluntary participation of the oil companies involved. So far as crude oil is concerned, it is working—refineries west of the Ottawa Valley are supplied from Canadian fields. The policy is not, however, working as well so far as products are concerned. There is a gradual increase in the volume of products moving westward across the NOP line, either from refineries in Quebec or overseas. We estimate that in 1969 product movement totaled some 65,000 barrels per day. Of this volume, 40,000 barrels per day was true leakage or, in other words, backed out overhead products which could logically be supplied by Ontario refineries.[69]

During this competitive period Imperial continued to work to improve relations with its dealers and agents. By 1970, while the company sales had increased 33 percent during the decade to 381,000 b/d, the monetary return on capital employed in the petroleum refining-marketing phase dropped to an unsatisfactory 5.7 percent. Clearly, price could be singled out as the dominant consideration in customer decisions about where to purchase gasoline. As a result, Imperial began to open self-serve stations that gave it the means and flexibility to meet these competitive conditions. The first self-serve station went into operation in 1970.

Throughout the decade, the company's research effort continued. At Sarnia, a new three-story glass-walled wing to an older building was added to the research center, and by 1970, the staff there numbered over 200.

Overall between 1950 and the end of 1975, Imperial's capital expenditures totalled some $3.8 billion. During that period, shareholders' equity grew from $257 million to more than $1.5 billion and the company's long-term debt in 1975 stood at $342 million.

X   X   X

In an examination of Imperial's history from 1950 to 1975 two facts stand out. First, the company continued to manifest the pioneer spirit that characterized its decisions since the 1880s. All the evidence confirms a continuing existence of this innovative, adaptive, creative thinking that had made the company successful in the nineteenth century. Second, the growing influence of government on almost all decisions became more apparent year by year, as the business environment grew more controlled. By 1970, the common denominator in investment decisions throughout the industry was government attitudes and actions. Much earlier in the days after Leduc, provincial and federal legislation and rulings had appeared to be concerned largely with upstream operations—the exploration, production and transmission— of oil and gas. By the 1960s, however, the attention of governmental bodies was also directed to the downstream aspects of the petroleum business—refining and marketing. In that decade alone, various agencies of provincial governments conducted five separate studies of industry marketing practices. Only one of these reviews found serious fault with the industry, the report of the Alberta Gasoline Marketing Inquiry Committee. This document proved "highly critical of the oil industry's dealer relations."[70] Yet when Imperial studied that report, it found such gross errors of fact that, in every case where the company "could be identified, the report was either wrong, or incomplete or both."[71]

The actions of OPEC in 1973 triggered governmental responses that had a profound effect on the oil industry. In that year, controls were placed on the export of crude and some products, and an export tax on oil was introduced. This tax limited the well-head price of Canadian crude oil to levels well below the international prices. Other legislation provided for the allocation of petroleum products at the wholesale level and the rationing of such products at the retail level, if such action became necessary. In 1973 and 1974, the oil producing provinces raised their royalty rates, and on top of this, in 1974, the federal government announced that provincial royalties could not be deducted when calculating income for federal tax purposes. W.O. Twaits likened these actions to those of "two hogs rooting under a blanket for the same acorn." The exploration for oil declined. In 1974, the federal government's "voluntary" freeze on prices of petroleum products also added to the burden of the oil companies. By 1975, the federal government had begun to phase out exports of crude oil.

Naturally, the first half of the decade of the 1970s proved to be a period of increasing preoccupation at Imperial with government and public relations. Its manager of Public Affairs, R.E. Landry, remarked, "There is nothing new in government involvement in the business process in Canada. While government, as Mr. [Pierre] Trudeau has reminded us, may have no place in the bedrooms of the nation, there has always been ample evidence of its presence in our boardrooms." However, Landry went on to say that despite the precedents, a growing number of Canadians felt that something had gone badly wrong with their country's concept of a mixed economy.

Imperial Chairman and President and CEO, J.A. Armstrong phrased this view in terms more meaningful for the oil industry when he reported to shareholders in 1975:

> In retrospect, how could we have anticipated that in less than two weeks following the 1974 Annual Meeting we would be faced with a situation where federal, as well as provincial, policies would have switched from being venture-oriented to revenue-oriented? No one anticipated an intensification of the federal/provincial struggle over resource revenue sharing with ensuing tax/royalty legislation that would jeopardize Canada's future supply of domestic crude and natural gas.[72]

By the 1970s, it seemed clear that the long-standing Canadian-U.S. cooperative program, built on mutual restraint and self interest, needed significant repair. As might be expected, during the third quarter of the twentieth century, latent Canadian nationalism had appeared from time-to-time. At their strongest, these feelings complicated the role of American investors in Canadian enterprises and especially of those business leaders who managed the enterprises in which they invested. Most of these American individuals and firms had entered Canada expecting to follow the rules and regulations applying to them, and a number of Americans had become Canadian citizens. The companies spent substantial sums helping to develop Canadian resources, but during this period, they grew restive as they began to anticipate possible discrimination in nationalistic advocacy of programs that called for denial of employment and even expropriation. The 1973 oil crisis fueled the flames of this sentiment.[73]

The troubled 1960s and 1970s for the oil industry served to stimulate not only Canadian Nationalists but those who feared domination by the United States. In 1969, Canadian Prime Minister Trudeau told the National Press Club in Washington, D.C., that living next to the United States was in some ways like sleeping with an elephant, "no matter how friendly and even-tempered is the beast one is affected by every twitch and grunt."[74] His statement had particular relevance for the oil industry, at that time 60 percent controlled by U.S. companies. For Imperial, in particular, Canadian criticism of foreign investment proved a distressing problem. As the largest oil company in Canada, and the acknowledged leader of the national petroleum industry, Imperial suffered many attacks. Time after time, the company had proudly announced its Canadian flavor, and these statements were sincere. Yet Imperial also realized that benefits had accrued to it and to Canada through affiliation with Exxon. Overall, Imperial had gained much from access to Exxon's worldwide experience and expertise. When Imperial went to the market for funds, Jersey provided a share so long as the nature of the financing permitted. And personnel from Jersey affiliates—though relatively few in number—played a big role in helping the Canadians develop exploration and production skills, after the Leduc find, just as Imperial personnel gained valuable worldwide experience through service abroad with other Jersey affiliates. Perhaps, most important of all, in Jersey, Imperial had a major shareholder that understood the oil business, one that had the patience to sacrifice short-term rewards for longer-term benefits. As Chairman Twaits described this benefit in 1974, in his final address to the Imperial shareholders:

> Then, as now, we had a major shareholder with world-wide expertise in the energy field and a philosophy that could, and still does, reconcile its interests with the aspirations of Canadians. Indeed it was the support and encouragement of our major shareholder during those long dry years that kept Imperial exploring in the West—a point worth recalling when foreign investors are being criticized.[75]

Chapter 13

Latin America

FOR EXXON CORPORATION, LATIN AMERICA has traditionally been a strong marketing territory, providing a sizeable percentage of total corporate profits, year in and year out. During the 1950s and 1960s, however, the largest profit centers were in producing areas: Venezuela and Perú, with Venezuela assuming a major role in the total corporate picture.

The area is a vast conglomeration of independent nations with various forms of government, with people of diverse ethnic backgrounds, and with obvious differences in geographical make-up and socio-economic development. Political instability, which has been an historical fact of life there, seemed to reach a crescendo during those two decades. Here, one primary challenge confronted the corporation: how to continue to operate in a volatile environment that became more and more hostile to foreign private investment.

ARGENTINA

When Juan Perón was overthrown in 1955, the Argentine nation found itself economically if not spiritually bankrupt. Esso Argentina had managed to survive during the Perón years under the guidance of H. A. "Ham" Metzger, despite deteriorating economic conditions and tight controls on dollar payments. Ham Metzger had served in Colombia and Bolivia, the latter during the days of nationalization of the company there, and he had become skilled in business diplomacy. The nationalistic atmosphere in Argentina, especially with respect to foreign oil companies, did not abate after 1955, and the government oil corporation, Yacimientos Petrolíferos Fiscales [YPF], received wide responsibility for oil matters from one government after another. While some efforts were made to bring private oil companies back into exploration and producing ventures, the vacillating policies of these different governments, including that of Juan Perón when he returned to power in 1973, and that of his wife after his death, left little in the upstream for development. Through the years, Esso operated refining facilities, tanker transport, and marketing, generally on a profitable basis. On December 6, 1973, Victor E. Samuelson, director of Esso Argentina and manager of Esso's Campana operations, was kidnapped by Marxist guerrillas. He was released after five months of incarceration upon payment of a ransom. His abduction occurred during a period of extreme violence between the left and the right sectors of the political parties which forced most expatriates to leave the country.[1]

Esso Argentina provided a "proving ground" for a number of Exxon executives, including Ham Metzger, who ended his career as a vice president and director of Creole Petroleum Corporation in their New York office. Others were D. C. Dunaway, B. G. Tchalidy, William E. Patrick, Hugh Wynne, James Dean, Robert Dolph, Charles Wheeler, Theodore Wieber, Juan Yañes, and P. J. Kinnear.

ARUBA

The island of Aruba, part of the Netherlands Antilles in the Caribbean near the northern coast of Venezuela, is the site of Exxon's Lago refinery. This plant began operations in 1929 as a unit of Standard Oil Company of Indiana, which sold it to Jersey Standard in 1932.[2] At that time it processed 110,000 barrels per day [b/d] of crude oil from Venezuela, and by 1939, crude runs had reached 230,000 b/d. Because Aruba provided an important source of supply for the allies during World War II, a German submarine shelled it. After the war the Aruba plant continued to expand, with investments amounting to over $60 million during the decade of 1949 to 1959 and another $64 million from 1959 to 1967. From 1968 to the early 1970s, investments in plant equipment jumped to $260 million, mostly for fuel oil desulfurization facilities to meet U.S. environmental standards. By 1975, Lago operated a complete and flexible fuel products refinery, processing crudes largely from Venezuela. With a refining capacity of 500,000 barrels per day, its facilities included crude and vacuum distillation, visbreaking, alkylation, naphtha fractionation, hydrogen plants and hydrofining/go-fining for desulphurization of naphthas, kerosene, gas oils, sulphur recovery, and a sulphuric acid manufacturing plant.

The original manager of Lago, Lloyd G. Smith, for whom an avenue is named on the island, was succeeded by John J. Horigan, and then Odis Mingus, Walter A. "Bud" Murray, J. M. Ballenger and R. L. Trusty.

BRAZIL

By 1950, Jersey's affiliate in Brazil—the then Standard Oil Company of Brazil—already had forty years of operating history. During this period,

Esso had become Brazil's major petroleum marketer with a nationwide distribution network. Imports made up virtually all of the supplies, and Brazil became Jersey's largest outlet for Esso products in Latin America.

The decade of the 1950s saw a number of important events affecting Brazil's petroleum industry in general and Esso's operations in particular. At that time, the political environment was dominated by a rising nationalistic fervor which carried over into the next decade. A military junta overthrew the dictator Getulio Vargas, in 1954, and two years later established a democratic system of government. This lasted until the João Goulart regime was overturned in 1964, when the military once more stepped in.

During the Vargas government, an oil monopoly law passed in 1953 led to the formation of the state oil company, Petrobrás.[3] Under the slogan, "O Petróleo é nosso," [The petroleum is ours] Petrobrás assumed 100 percent control over Brazil's limited crude production and exploration activities. Further, with the aim of making Brazil self-sufficient in petroleum products, Petrobrás moved aggressively into the refining sector, to the exclusion of private interests, including the multinational companies. Petrobrás' efforts largely succeeded in reaching the goal of product self-sufficiency through purchase and expansion of essentially all the existing refineries, and the country ceased to be a major importer of oil products by the end of the decade. Meanwhile, the Brazilian government's National Petroleum Council continued to regulate the petroleum industry, especially in terms of controlling prices and margins.

With the growth of Petrobrás, the nature of Esso's operations changed from that of an importer of Jersey products to that of a distributor of locally refined products. However, Esso continued to be the dominant marketer, since Petrobrás did not become a major factor in the market place until the 1970s. Jersey also continued to supply a portion of Brazil's crude imports.

In spite of the increasingly difficult political environment, Esso moved to diversify its operations during the latter part of the 1950s and the early 1960s. In 1956, a lube blending plant was inaugurated in Rio de Janeiro, under the name of Solutec.[4] A second blending plant followed in 1960, and in 1963, Iretama was incorporated as a separate affiliate to handle Esso's growing chemical business in Brazil.[5] By that time, Esso itself had undergone a change in name and corporate structure, in compliance with a presidential decree requiring foreign oil companies to be locally incorporated. The new name adopted, Esso Brasileira de Petróleo S. A., remains in use today.

The 1960s were marked by profound political change in Brazil. A rapidly deteriorating economy, coupled with widespread political unrest, helped to trigger the overthrow of the leftist-leaning government of João Goulart in a bloodless military coup in 1964. The political and business environment for foreign capital and particularly the international oil companies changed dramatically almost overnight. Just prior to the military takeover, hostility to foreign oil companies had reached the point where expropriation of Esso's operations was imminent. However, the new military government quickly announced the cancellation of all plans to expropriate. Indeed, it welcomed foreign capital as essential to the country's economic development.

The economic policies of the military government soon began to show positive results, as Brazil entered a period of 12 percent-plus annual economic growth—commonly referred to as the Brazilian economic miracle—which extended into 1973. Petroleum consumption grew at a similar rate, and the industry went through a phase of rapid expansion.

Until 1970, the state company, Petrobrás, remained a relatively minor factor in the market. Then Petrobrás, aided by a government order requiring all state companies to purchase Petrobrás products, began to penetrate the market more aggressively at the expense of private marketers, including Esso. [6] By 1975, Petrobrás had gained one third of the market, replacing Esso as the largest petroleum marketer in Brazil. Yet, despite the Petrobrás entry, Esso sales continued to grow, reaching 130,000 b/d in 1975, and Esso Brasileira retained its position as Exxon's largest marketing company in Latin America.

During most of the 1950s, the company was under the management direction of M. W. "Johnny" Johnson, who was succeeded by C. J. Griffin, G. W. "Bill" Potts, and Lionel J. Bourgeois, Jr., all long-time Exxon employees.

CHILE

One of the most stable democracies in Latin America during the nineteenth and most of the twentieth century, Chile provided a favorable market for Jersey. Although foreign companies were not allowed to explore for oil or refine it, Esso Chile carried out a full marketing operation and, in addition, owned part of an important pipeline. In 1970, a Marxist government, under Salvador Allende, although receiving only a third of the national vote, assumed power.[7] Social and economic reforms, including speed-up of nationalization of foreign-owned mining companies, created a number of internal and external pressures on the government, and, in 1973, a military junta seized control of the government. Throughout this turmoil, Esso Chile continued operations and maintained its business position.

Management of Esso Chile through most of the 1950s and 1960s was handled by G. S. Warner and Armando Chellew, a Chilean national. The staff, with few exceptions, was composed of Chilean nationals. Chellew retired in 1966, to be replaced by Hobart "Hobe" Gardiner, an American with long experience in Latin America. He stayed until the Allende years, when Leo Swinderman replaced him. Others with experience in Chile were L. S. Ridgeway, R. B. Carder, P. J. Kinnear, J. F. Campbell and J. E. Glover.

COLOMBIA

Exxon's interest in Colombia dates back to 1917, when the Tropical Oil Company began development of the De Mares oil field.[8] Later, the company built a refinery at Barrancabermeja. In 1951, these reverted to the government, by agreement, and the government's oil company, ECOPETROL, drew up another agreement with INTERCOL, a new Jersey affiliate there, to help operate the oil field and the refinery. The arrangement to operate the oil field soon ended, for the workers had been so well trained by Tropical that they required no extended assistance. The agreement to manage and expand the refinery lasted, however, until 1961.

Meanwhile, the sales organization of Tropical became the basis for the formation of Esso Colombiana, a marketing organization handling the distribution of fuels, lubricants, and accessories. Also, in 1957, INTERCOL put a new refinery on stream at Cartagena on the northern coast, and in 1974 this was sold to ECOPETROL.[9] But Exxon affiliates continued to run a fully integrated operation in Colombia: exploration in the eastern plains, the Magdalena Valley, the Caribbean Coast, and other areas; production of oil and gas in Provincia and Bonanza; refining in La Dorada; transportation by river barges and pipelines; and marketing. At the end of 1975, a contract was signed with the government to explore and possibly develop large coal deposits in the Guajira Peninsula.[10]

During the 1960s, the company also helped to develop a fertilizer industry, building plants near the Cartagena refinery. These were sold to local shareholders in 1972, when Exxon decided to withdraw from the production of agricultural chemicals in most parts of the world.

A number of Exxon executives gained valuable experience in Colombia, where the affiliates were subsidiaries of International Petroleum Company, Ltd. [IPC], a Canadian corporation which had moved its headquarters to Coral Gables, Florida, in the early 1950s. These included L. P. Maier, M. A. "Mike" Wright; and Milo M. "Mike" Brisco and Howard C. Kauffmann, both of whom later became presidents of the Exxon Corporation; J. K. Oldfield, C. F. "Charlie" Van Berg, O. C. Wheeler, E. C. Borrego, F. J. "Espy" Espinosa, W. J. "Bill" Nutt and R. Castro were also involved.

THE CARIBBEAN AND CENTRAL AMERICA

Exxon's interests in the Caribbean and Central America underwent a number of corporate changes over the years. The old West India Company became Esso Standard Oil S.A., a company incorporated in Panama with headquarters in Cuba. This organization conducted marketing operations throughout the area until the Cuban revolution in 1959. Shortly thereafter, a new company, called Esso Standard Oil S.A. Ltd. and commonly known as Essosa, was incorporated in the Bahamas. All assets and liabilities of Esso Standard Oil S.A., except those in Cuba and Guatemala, were transferred to Essosa. Corporate headquarters were moved to Coral Gables, Florida.

In 1962, Essosa personnel in Coral Gables were transferred to IPC, the Exxon affiliate which operated in Colombia, Ecuador, Perú, and in Venezuela through its undivided interest of 25 percent in the Mene Grande Oil Company.[11] Further, in the reorganization Essosa was divided into an Eastern and a Western region with staff support from IPC. In 1964, the two divisions were reconsolidated into an organization known as CCA [Caribbean-Central America]. This consolidation lasted until the formation of Esso Inter-America in 1966, with management divided into three groups— one located in Puerto Rico covering the Caribbean, one in El Salvador covering Central America, and one in Jamaica which also handled the Bahamas and Bermuda. In 1970, the groups located in Puerto Rico and Jamaica were combined and returned to Coral Gables.[12] The Central American group remained in El Salvador and was called Centam.

Although this area of Latin America had been much like many other areas in terms of political turmoil and growing nationalism, the tactical problem of supplying and servicing more than thirty separate political entities proved a major challenge. In times of shortages of supply and rising prices, there was, of course, the additional problem of dealing with each separate government to obtain relief from controlled prices, especially on gasoline. Nationalism played a crucial role in the supply picture, since each country felt that it should have its own refinery within its borders. In response to this nationalistic feeling (refineries like a national airline were a status symbol) and under encouragement from local governments, a number of small refineries were constructed by the industry. For example, in the early 1960s, Jersey built three of these in the area: one in Jamaica, one in El Salvador, and another in Nicaragua. Negotiations continued for the construction of several others, in such places as the Dominican Republic and the Bahamas, but these never came to fruition for Jersey.

With respect to the Caribbean and Central America in this period, political turmoil was particularly evident in the Dominican Republic, Haiti,

Panama, Guatemala, and, of course, Cuba. As a result of the Castro take-over there, all company assets were nationalized, involving a loss to Jersey as certified by the Foreign Claims Settlement Commission of $71.6 million. This included a refinery and marketing facilities as well as the disruption of the headquarters' operation for the entire area. Some of the Cuban employ-ees moved to Coral Gables.

In the Caribbean and Central America almost all local operations were managed by non-Americans, in part because over the years Jersey had care-fully selected and trained nationals in each area. People like A. W. "Ar-chie" Spillett and Rey Canez from Haiti, J. P. "Joey" Proudfoot from Trini-dad, Harold Cole from Barbados, Manuel Guardia and E. A. "Chebo" Morales from Panama, Danilo Lacayo from Nicaragua, Nigel Ince, J. C. "Jackie" Thwaites from Jamaica, V. I. "Vince" Diego from Cuba, John Carey from Bermuda and many, many others remained the backbone of a widely diversified and geographically complex component of the Exxon sys-tem.

ESSO INTER-AMERICA

In 1966, all of Jersey's interests in Latin America except Creole and Lago were placed under a new company, Esso Inter-America.[13] In effect, this move not only merged IPC and Essosa, but it brought all other companies under centralized management control—Argentina, Brazil, Chile, Para-guay, and Uruguay. Once again, Jersey followed the familiar pattern of regionalizing operations as it was doing in Europe and in the Far East. Esso Inter-America's first president was Howard Kauffmann, who had spent much of his career with IPC.

Esso Inter-America, based in Coral Gables and known as EIA, provided an efficient organizational structure, with a president and a few directors to handle area and functional responsibilities. When Howard Kauffmann left EIA to become executive vice president of Esso Europe in 1968, he was replaced by James F. Dean, who had served with Creole and who had been president of the Argentine affiliate. By 1969, EIA had crude production of 93 mb/d of crude, 165 mb/d of refinery runs, 370 mb/d of market sales, and a manpower table of 8,246 employees.[14] Geographically, the company cov-ered a territory from Bermuda to Tierra del Fuego, a north/south distance of some 6,100 miles, and from east to west a distance of 4,100 miles, from Recife in Brazil to Guatemala City. Within the territory, not counting Cuba and Mexico, there were more than 220 million people under thirty-five different flags and speaking five different languages. Business in the region was car-ried out in twenty-six different currencies through thirty-nine separate corporate affiliates. Only 1.3 percent of the employees (110 people) in the region were North Americans, with another 228 working in the headquarters office in Coral Gables.[15]

In 1971, Jim Dean moved to Esso International Inc., and Leo Lowry, after a long career with Creole, became president of EIA. He guided the company into the mid-1970s. In 1973, EIA assumed managerial responsibility for Aruba and in 1976 for Venezuela, thus placing the entire Latin American area under its wing. At this point, the company still had crude production of 93 mb/d of crude oil, 322 mb/d of refinery runs, 337 mb/d of market sales, and 7,991 employees.[16]

X   X   X

The Exxon experience in Perú, through the International Petroleum Company, Limited, and in Venezuela, in large part through the operations of the Creole Petroleum Corporation, merits more complete discussion: IPC-Perú because of the unusual political circumstances affecting its activities, and Creole because for many years it was Exxon's largest source of crude production and a major profit center. A review of these two companies follows.

Creole, the Breadwinner

JERSEY'S MAJOR AFFILIATE in Venezuela, Creole Petroleum Corporation, was the principal source of crude production and the company's single most profitable operation until the mid-1960s. By 1950, Creole represented more than thirty years of continuous expansion and development. It had become an industry showcase in Latin America, if not the world.

Jersey's earliest efforts in Venezuela had been undertaken by the Standard Oil Company of Venezuela [SOV], an affiliate based in the eastern zone of that country. Throughout the early and middle 1920s, this company conducted extensive exploration under extremely difficult conditions without discovering commercial deposits of oil. Ultimately, Jersey Director E. J. Sadler determined that the company should secure a foothold in the Maracaibo basin of Western Venezuela, which competitors were already developing with good result. In 1928, Jersey acquired majority ownership of the small Creole Syndicate, incorporated in Delaware by speculators who bought and traded leases in Venezuela. While not basically an operating company, Creole Syndicate had formed a partnership with the Venezuelan Gulf Oil Company, which had discovered oil in the shallow water "kilometer strip" concessions along the eastern shore of Lake Maracaibo. Shortly after the acquisition, Jersey turned the old Syndicate into a holding company, Creole Petroleum Corporation, with SOV as its affiliate.[1]

Ironically, SOV brought in its first commercial producer in the East, Moneb 1 at Quiriquire, only a few weeks after Jersey had purchased Creole and its valuable concessions. Yet, while the eastern zone remained attractive, Jersey focused attention on the now-booming western state of Zulia and its Maracaibo basin. Here, Creole's major problem was that competing companies already held the promising concessions at Lake Maracaibo. Facing this situation, Jersey determined to strengthen its position. An opportunity

396

arose in 1932, and the company seized it by purchasing the Lago Petroleum Corporation, an affiliate of Standard Oil Company of Indiana. Thus, Jersey laid the foundation for one of its most significant overseas operations.

Standard of Indiana's willingness to sell resulted from overcapacity, declining prices in the United States, and general concern about the Depression. Furthermore, Indiana suffered a lack of foreign markets to which it could divert Venezuelan oil if threatened import quotas were imposed in the United States. In April, 1932, Jersey and Indiana completed an arrangement under which Jersey would exchange some $135 million in securities and cash for Lago and several other Indiana affiliates, principally in Mexico and South America. For Jersey, the acquisition proved valuable in several respects: Lago brought Jersey production of some 88,000 b/d, estimated reserves of over 500 million barrels, highly important concession areas, and a pool of talented men who were experienced in Lake Maracaibo operations. Also with the purchase, the company acquired two other Indiana affiliates, Lago Oil and Transport Company, Limited, and Lago Shipping Company, Limited, which would be valuable to operations in Venezuela. The first of these companies had just completed construction of a 110,000 b/d refinery on the Island of Aruba, some fifteen miles off the Venezuelan coast. The second operated a fleet of twenty-one shallow draft "lake tankers" which transported oil from Lake Maracaibo to the Aruba refinery. Together, the three new companies provided the essential production, refining, and transportation components that were required to develop an integrated operation in Western Venezuela.[2]

The sudden expansion of Jersey operations in Venezuela created organizational problems which would not be resolved until almost two decades after the Lago purchase. Jersey appointed Eugene Holman as president of both Creole and Lago. Standard Oil of Venezuela retained its separate identity. In 1937, SOV and Lago's responsibilities were divided on a clear geographical basis, SOV operating in the East with headquarters in Maturín, and Lago based at Maracaibo in the West. At the same time, the top managements of Jersey affiliates were relocated to the Venezuelan capital, Caracas.

Jersey had clearly begun to streamline its Venezuelan activities by the 1940s. World War II brought new problems which the cumbersome multi-company structure could not meet effectively. By 1943, the government of Venezuela suggested that the vestiges of the pre-1932 companies be eliminated. Jersey consequently began to merge its Venezuelan interests and by 1944 had accomplished the task. Under a formula accepted by Creole stockholders in August, 1943, Creole acquired all of Lago's interests, except the Aruba refinery and the lake tanker fleet. SOV was liquidated and all

Venezuelan operations were consolidated under Creole. A Jersey affiliate, IPC, held a 25 percent interest in another Venezuelan oil producer, the Mene Grande Oil Company. Henry E. Linam of SOV became Creole's president, and Arthur T. Proudfit of Lago was named Creole's general manager.

Despite the reorganization, difficulties remained. SOV and Lago personnel had different backgrounds in operations, and they often found it hard to cooperate effectively. SOV employees had developed a strong sense of camaraderie in the isolated jungle camps of the East. They had become accustomed to a wide range of company-supported services which were not needed in the more cosmopolitan Maracaibo area. Because conditions in the East were generally harsher and oil deposits less abundant, SOV personnel often felt themselves more "macho" than their counterparts in the West. Lago employees, for their part, argued that drilling over water in Lake Maracaibo had nothing in common with technically routine land-based operations in the East. These attitudes began to change as employees were transferred between the two regions.[3]

World War II created a number of other challenges for Jersey and the Creole Company. Initially, German submarine activity in the Atlantic and material shortages caused a drilling slowdown, but by 1943 the allied war effort began to tap Venezuelan reserves for increasing volumes of petroleum. Between 1943 and 1948, total output increased by an average of 165,000 b/d, and by the latter date, the country was producing 1,340,000 b/d, two and one-half times the prewar average. For Creole, the war sparked a boom that would be sustained through the late 1950s. An exploration and development program permitted the company to increase its Venezuelan production substantially and its estimated reserves to some two billion barrels by 1945.[4]

More significantly, perhaps, the war accelerated important political currents. Nations throughout the Caribbean joined the allied cause and supported the war effort, largely as suppliers of petroleum, hemp, quinine, and rubber. Their citizens pondered such democratic pronouncements as the "Four Freedoms" and the Atlantic Charter. No exception to this rule, Venezuela became the scene of fervent political activity, led by a party of young liberal reformers, Acción Democrática [AD], which hoped to replace traditional dictatorial government with a more democratic system.

After the death in 1935 of long-time dictator General Juan Vicente Gómez, the new Venezuelan government began to demand that foreign oil companies increase their contribution to national development. Meeting this problem created a sharp division in the Jersey ranks in the years from 1941 to 1943. One faction, consisting of SOV President Henry E. Linam and

Thomas R. Armstrong of the New York Production Department, urged that Jersey stand fast, that compromise would encourage other nations to act in similar fashion. Another group, headed by Jersey Vice President Wallace E. Pratt and Creole's Proudfit, argued for flexibility regarding concession agreements. Pratt had earlier proposed this position unsuccessfully in Mexico, and he felt that Mexican nationalization of foreign oil operations vindicated his approach. Ultimately, Pratt's argument prevailed, and Linam and Armstrong resigned from the company. The new contracts which the companies and Venezuela concluded in 1943 resulted in a combination of royalties and income taxes accruing to Venezuela that would equal 50 percent of profits. For their part, the companies received forty-year extensions of existing concessions as well as access to additional acreage. The new direction in company-government relations had far-reaching implications. Other firms operating in Venezuela adapted to the new arrangement, and in time the "50-50 principle" provided a model for oil development in other areas throughout the world.[5]

The policy of conciliation and adaptation proved a wise one: the reform-minded Acción Democrática continued to gain strength and by 1945 assumed full control. AD's public pronouncements on the oil industry were clear, and many observers credited the 1943 decision with creating a general climate of cooperation. Another reason for the overall amicable relations between Creole and the new government was the presence of Arthur Proudfit, who became company president in January, 1944. Proudfit had worked in Mexico as a driller for Edward Doheny's Huasteca Company, and, in 1932, he joined Lago as assistant superintendent at Lagunillas on Lake Maracaibo. An astute judge of people, Proudfit developed an easy rapport with Venezuelans. He perceived the significance of the growing strength of the political parties in the legislature and concluded that, for Creole to have a continuing role in Venezuela, it was imperative to "prove that we were necessary implements to [Venezuela's] progress."[6] With the assistance of Creole's Henry Pelkey and Jersey's K. E. "Ted" Cook, the company established a Public Relations Department in late 1945. To demonstrate the mutuality of interests which Proudfit thought so important, however, required cooperation with the Venezuelan nation across a broad front. Venezuelans had long complained that the oil industry created dislocations in their economy, drawing workers from agriculture and other industries, leaving the nation dependent upon imports from other countries. This pattern became especially pronounced during World War II, when oil revenues soared while imported foodstuffs and manufactured goods became less available. Observing the situation, AD leaders, most notably Petroleum and Hydrocarbons Minister Juan Pablo Pérez Alfonso, began to urge that Venezuela reinvest or "sow" its oil revenues by developing agricultural and other industries, which would provide an economic base after the petroleum reserves had been depleted. The government felt that there was need for the oil companies to participate more actively in general national development. One opportunity to do so occurred with a visit by Nelson Rockefeller to Venezuela and other Latin American nations, in 1946. Rockefeller inaugurated a scientific program of food production, distribution, refrigeration, and practical education. The vehicle created to implement this program in Venezuela was the Venezuelan Basic Economy Corporation. Creole quickly became an important sponsor of the program, providing some \$7 million over a period of several years.[7]

Creole's interpretation of Venezuelan political developments proved essentially sound; although Acción Democrática did not make the radical moves toward nationalization they had previously threatened, they did insist that the oil companies adjust to a series of important changes. The AD's stated objectives during the late 1940s included: increasing taxes, direct government exploitation of oil, reinvestment of profits in the domestic economy, and higher salaries for oil industry workers.[8] The latter was particularly important because the AD party had a broad base of labor support.

One of the new problems which Creole faced between 1945 and 1948 was increased militancy among its Venezuelan laborers. In 1946, oil field workers organized for the first time in an industry-wide federation of some forty unions, FEDEPETROL. In June of that year, the new organization secured contracts which insured workers two weeks vacation per year and full pay for weekends and other days of rest. In 1948, labor negotiations entered a new phase, when FEDEPETROL negotiated a uniform contract with the entire petroleum industry. This new departure, the three-year "collective contract," was to be the basic bridge between management and labor until 1975. Creole based its position in the 1948 negotiations upon a belief that amicable, enduring, labor relations were preferable to short-term considerations. Creole negotiators offered the union unexpectedly generous benefits. In the contract which these negotiations produced, the workers received a three week vacation, increased salaries, and a guarantee that commissary prices of certain basic foodstuffs would not be increased with inflation. Creole, for its part, insisted that there be no encroachment on management prerogatives and that negotiations would be conducted directly with the unions, rather than through the government, as had been the case.

The 1948 contract did improve labor relations, but, paradoxically, it created friction between Creole and other companies. As the largest oil producer in Venezuela, it always wielded great influence in collective labor negotiations. Creole, however, was willing to grant a much larger settlement than smaller companies, many of which felt Creole's generosity unnecessary.[9]

The most important problem in government-company relations in the postwar period centered on company income taxes and royalty payments. The formula established in 1943 required the companies to pay a 16$\frac{2}{3}$ percent royalty plus an income tax and a graduated surtax. Although this formula initially gave the government some 50 percent of profits, by 1945 production and prices had increased spectacularly, and payments to the government, although larger, no longer equalled the company's profits. On December 31, 1945, the government imposed an additional surtax on production. Proudfit argued that the new tax would have the effect of nullifying the 1943 agreements.

Rómulo Betancourt, leader of Acción Democrática, rejected Proudfit's position, claiming that the 50-50 principle was the essence of the 1943 accord. Creole consequently paid an $18.7 million surcharge.[10]

In 1946, the company's earnings once again exceeded those of the government, which moved to rectify the imbalance. In December of that year, Proudfit once more protested such measures, arguing that the government had assured him in 1945 that it would not impose further surcharges. A new, mutually acceptable agreement was clearly needed, and the two parties began another round of negotiations toward that end. Ultimately, in 1948, Creole bent to the inevitable; it accepted the 50-50 principle as a law. Under the new arrangement, Creole would pay the standard rate of 2.5 percent tax plus royalties and a surtax graduated to 26 percent. If this formula did not result in a 50-50 settlement in any given year, Creole would pay the difference. In addition, Minister Pérez Alfonso insisted that Creole make good its deficits for the years 1946-1947 by investing in public works.

The adoption of the 50-50 rule in Venezuela was seen throughout the world as a precedent-setting effort to develop host country economies through increased participation in the profits of foreign enterprise. The principle was later adopted in other oil-producing countries. Creole and Jersey actually found the new step a positive one in many respects. "Fifty-fifty" was a convenient yardstick—a simple expression of the mutual interests which companies and host countries shared, and seemed to represent a new direction in international enterprise.[11]

Ironically, after Creole had successfully weathered the difficult transition to democracy in Venezuela, a military coup in 1948 toppled the government of Rómulo Gallegos and returned the nation to rule by military dictatorship. The new regime would control Venezuela for a full decade, largely under the direction of General Marcos Pérez Jiménez. From his exile in Cuba, Gallegos charged that the oil companies had supported his overthrow. Creole and other companies immediately denied the allegation, and no evidence was ever presented to support his charges. However, for some Venezuelans these assertions served to identify the oil companies with dictatorship. This view was also encouraged by the fact that the Pérez Jiménez government softened many of the AD's social and labor policies, and generally gave the oil companies greater latitude in their Venezuelan operations.[12] In fact, although Creole could not express itself publicly on the change of government, it apparently found the new developments disappointing. Arthur Proudfit had, of course, maintained contact with all the parties in Venezuela and was careful to use the phrase "Venezuelan Nation" rather than "party." Yet, he did feel that democracy represented the wave of the future in Venezuela.[13]

CREOLE IN 1950

By 1950, Creole had emerged as one of the most vital links in the Jersey circuit. As one journalist observed, "Jersey would undoubtedly exist without Creole, but it would exist neither at its present profit level, nor as the integrated worldwide production-refining-sales empire it is today."[14] Creole's principal value was as a producer of crude. Its production had soared from some 100,000 b/d in 1941 to 668,000 b/d in 1950, making it the world's largest single petroleum producer. It supplied approximately one-half Jersey's crude, approximately 300,000 b/d more than Humble Oil & Refining Company. Creole held some one-half of Venezuela's 4.4 billion barrels in proven reserves in 1949 and produced approximately 45 percent of Venezuelan crude. Creole's significance for the Jersey Company, however, is better reflected in its profit statements. In 1949 it generated almost one-half of Jersey's net income, although it accounted for only one-sixth of the company's total investment. One of the world's lowest cost/unit producers, it was paying out excellent annual returns on invested capital at mid-century.[15]

Creole, by 1950, had enjoyed generally favorable relations with its Venezuelan hosts. Arthur Proudfit delegated nearly all management responsibilities to his executive vice president, Harold W. "Duke" Haight, and other board members; he devoted his own efforts to improving relations with the

Venezuelan people and government. For Proudfit, Creole's major priority was coming to terms with its environment. He observed in a 1949 policy statement that:

> Our main job . . . is not just to tell the government what we are doing; it is to convince the public at large that we are not a big octopus, that we are putting back into Venezuela a great deal more wealth than we take out, that we are an asset to the culture, education, and general welfare of the country. This is best done in direct, daily business and social contacts with the Venezuelan people. How well we succeed depends on the time, experience, and inclination of our own staff members to improve these contacts. The company cannot excuse the attitude of any Americans on the staff, however bent on their own details of work, who are remiss in these important public relations. Creole policy is to measure Spanish-language fluency and ability to get along with the Venezuelans as a basis for promoting members of its expatriate staff; the company is also devoting its resources of selection, training, and technical education to placing Venezuelans in every staff position it possibly can. We feel that if American industry abroad can demonstrate in these and other ways that it fairly shares its benefits with the people who serve as its host, then any other problems will work out pretty naturally. The oil industry has gone a long way on that path here in Venezuela.[16]

Creole's effort to "Venezuelanize" its operation had many aspects. The company, for example, stepped up its teaching of Spanish to expatriate employees by starting an intensive Spanish language school after World War II. Inaugurated by Creole's George Culp, the program was later operated by professional teachers. It included six hours of language training per day and one hour of instruction in Venezuelan culture and history. One observer described it as a "grueling modification of the Berlitz method with an oil-field twist."[17]

Perhaps one of Creole's outstanding achievements by mid-century occurred in the area of employee relations. In the words of one observer, the company engaged "in a greater variety of cradle-to-grave activities than any non-governmental enterprise on native U.S. soil."[18] By 1949, it had provided ten thousand houses for its workers and supplied them with free utilities. It operated twenty-two commissaries which sold many staple items at cost. Its schools annually educated over 5,500 children and provided home economics courses for wives of company employees.[19] Creole had excelled early in health care, a practical necessity for a company whose workers often found themselves quartered in isolated jungles and lake camps. In 1947 and

1948, it cooperated with the Venezuelan Ministry of Health to virtually eliminate malaria throughout the nation. By 1947, the company operated ten hospitals with 387 beds and was planning a large new central hospital at Maracaibo, which was later built.[20]

Creole's innovations in employee relations had far-reaching implications; for example, it was the first company to employ large numbers of women, a radical turn in that relatively conservative society. However, families soon grew willing to entrust their daughters to Creole, and by 1949, the company employed some 1,500 women.[21]

Venezuelanization had advanced to such a point by 1950 that 92 percent of all company employees were nationals. But this was not simply done. Rather, it required that great attention be given to training and employee development. As early as 1938, Creole had given scholarships for professional and technical training, and its educational program had expanded steadily. By 1950, it was training some 3,200 employees. In addition, it had begun requiring that every expatriate employee train one of his subordinates—preferably Venezuelan—to perform his job.[22] Creole employees were also well paid. In 1950, Venezuelan workers averaged some $6,638 per annum in wage and benefits which included the costs of camps, schools and hospitals—compensation somewhat greater than that which Humble paid its workers in Texas.[23]

AMUAY—THE WORLD'S LARGEST REFINERY

World War II, as has been noted, touched off a major program of expansion in the Venezuelan petroleum industry—an expansion in which Creole set the pace. This burst of activity was sustained through the 1950s, with a brief period of retrenchment in 1949 and 1950. The single most important development of the boom period for Creole was the construction of Amuay refinery on Venezuela's Paraguaná Peninsula.

Creole's original decision to construct the new refinery grew out of several circumstances. Most of its crude oil was refined in Aruba, and as crude production increased dramatically, the Venezuelan government felt that any additional refining capacity should be built in Venezuela as part of its "sow the petroleum" program. Creole's refinery on the San Juan River at Caripito had a 60,000 b/d capacity, which principally served the eastern division. Its small refinery at La Salina on Lake Maracaibo was well situated to process crude output, but it could not be enlarged because of its location. Also, Lake Maracaibo's outlet to the Caribbean was partially blocked by sandbars, and the installation could be reached only by shallow draft tankers. Nor was the aging "lake tanker" fleet particularly suited to serve the

Maracaibo fields after World War II. Too, the company and the government had reached an agreement during the war that Lago's Aruba refinery would not be further expanded, and that preference would be given to refinery projects on the mainland.[24]

During the war, Creole made the decision to construct a new Venezuelan refinery with deep water access. In 1945, the company received permission to launch such a project at Turiamo Bay, a point which had the advantage of being located midway between the eastern and western production zones, but which would require expensive pipelines over difficult terrain. Before the Venezuelan authorities could approve the project, however, a new government assumed control, and its petroleum minister balked at the proposal. The new government regarded Turiamo as an essential site for naval and commercial port facilities and urged that the company shift its attention to the underdeveloped Paraguaná Peninsula.[25]

In 1946, the company began the enormous task of building a refinery at Amuay. Perhaps the major problems which the project posed grew from the hostile physical environment. The Amuay site, located on a desert shoreline far from the traditional areas of settlement and activity in Venezuela, was to be one of the world's first "grass roots" refineries, built without the support of an adjacent established community. Paraguaná's few inhabitants made meager livings as fishermen or goat herders. Creole set up temporary headquarters in "Villa Zorca," an adobe hut, until rows of quonset huts could be constructed. The company contracted out much of the construction work, which brought in great numbers of Venezuelan and American workers. Yet sustaining this effort in desolate Paraguaná proved difficult. Water was imported by tanker from the United States or Panama at great cost. Semi-wild goats wandered over the construction site and frequently ate the labels from packages. One supervisor complained that a goat ate note paper from a clipboard that he had put down while reading gauges.[26]

When Creole chose the Amuay site, it committed itself to construct the two major pipelines which would connect the refinery to sources of water and crude oil. The crude pipeline was to extend from a pumping station at Ulé on Lake Maracaibo, some 232 kilometers from Amuay. To pump over 300,000 b/d of crude to Amuay for processing or loading directly onto large tankers required that a specially designed large diameter pipeline be constructed. The most difficult phase of the project involved the stretching of a corrosion-resistant pipeline over a fifteen-mile-wide bay. In order to accomplish this task, Creole engineers floated sections of pipe across the bay, welded them together, and gradually submerged the sections. One observer noted that the operation suggested "the effort by a palsied giant to thread a needle held by a giantess with St. Vitus Dance."[27] The project suffered several setbacks when heavy seas snapped large sections of pipe before it could be properly anchored on the sea floor. By December, 1948, however, Creole completed the work and began pumping crude to Amuay for loading on tankers.

Securing a reliable supply of fresh water proved to be quite as difficult as linking the refinery to crude oil sources. For the first several years of Creole's presence at Amuay, Jersey dispatched newly-constructed tankers on their maiden voyages to Venezuela loaded with fresh water for Amuay. The cost of this expedient was calculated at over fifty cents per barrel. In 1948, however, Creole joined forces with the Venezuelan National Institute of Sanitary Works and with Shell Oil Company, which also was building a refinery on Paraguaná. The three partners drilled wells and built a reservoir at Siburua on the mainland. The Siburua pipeline followed a narrow neck of land which connected Paraguaná to the mainland. It was completed in August, 1950, and supplied water to satisfy the needs of the two refineries and several towns, which by then had a combined population of 140,000 inhabitants. The advent of fresh water on Paraguaná visibly altered the landscape in the immediate vicinity of the refinery. Creole beautified its residential and working areas with tropical plants, and housewives began to nurse luxuriant gardens.[28]

The Amuay refinery went on stream on January 3, 1950, operated by veteran engineers who were transferred from other Jersey refineries and the La Salina refinery, which closed in December, 1949. Amuay was one of the first great refineries to be constructed by American oil companies in crude producing countries. It was also unique in that it gave the Jersey Company a flexible new source of heavy fuel oil for the American Eastern Seaboard market. Amuay, in fact, was a pioneer in the "swing refinery" concept, storing vast quantities of fuel oil during the summer months and releasing it in periods of high demand.

A project of major proportions, the construction of Amuay put the Creole Company under considerable financial strain and dictated considerable readjustment. Originally conceived as a 60,000 b/d refinery which would cost some $60 to $65 million to construct, by 1948 Creole was contemplating a 100,000 b/d installation. In order to achieve the greater output, however, the company had to effect a broad range of economies in construction and eliminate several of the facilities it had planned at Amuay. But the postwar period was one of rapid inflation, and construction materials were in especially short supply. By 1949, the estimated cost of a 100,000 b/d refinery with residential and other support facilities had risen to some $200 million. Accordingly, in 1949, the company returned to its prior plan for a 60,000 b/d refinery, although even the smaller facility now cost Creole over $150 million to construct.[29] However, Creole executives charged with this project had the foresight to retain the general layout and provide the infrastructure that would later permit expansion several fold.

Amuay would ultimately become one of the cornerstones of Creole's operation, but its costly construction required that radical economies be made throughout the company. In 1949 and 1950, Creole underwent a period of retrenchment during which many employees were laid off. In 1949, Creole reduced its capital expenditures and, in addition, cut back sharply on inventories and on general operating expenditures.[30]

It was also during this readjustment phase that the Jersey Board chose to increase its equity in Creole. In 1949, Creole's minority ownership accounted for approximately 7 percent of its total shares. In that year, Jersey set about increasing its share to over 95 percent. The purpose of this campaign was to consolidate Creole and Jersey federal tax returns. In addition, the move seemed to reflect the increasingly central position of Creole in Jersey's plans.[31]

The 1950s: Social and Technological Advances in the World's Largest Overseas Firm

In the early 1950s, Creole faced two unprecedented challenges: the start-up of the Amuay refinery and the simultaneous decline of the eastern production zone. These problems required liberal doses of both technical and sociological imagination, as the company applied all of its resources to meeting the new demands.

Like many oil companies operating in remote, inhospitable areas, Creole had worked for decades to develop comprehensive programs of employee welfare. Where no adequate housing, sanitary systems, hospitals, or food supplies were available, the company provided them for its workers. By the early 1950s, the Creole board was beginning to see the disadvantages of such a system. First, liberal company support seemed to make workers overly dependent upon Creole. In many remote camps, company maintenance men replaced lightbulbs in employee housing, and housewives purchased all the necessities of life in commissaries. Second, many of Creole's eastern zone camps were producing diminishing quantities of petroleum. To shut down production in such areas would mean stranding large populations of dependent employees. Third, camp life often created a gulf between company employees and the local population. Oil camps, typically, were surrounded by chain-link fences, outside of which large squatter communities developed. In 1951 and 1952, Creole conducted a study which demonstrated that company towns and surrounding fences were major sources of resentment by non-employees.[32] Creole's effort to reverse the tendency toward company enclaves became known as "Community Integration," an innovative step for a company so long committed to paternalism. On its own initiative, the firm began seeking ways of encouraging workers to rely upon the community for education, medicine, transportation, food, and housing.

Creole's test project for the community integration concept was Judibana, a new residential community built to house workers at the Amuay refinery. Creole designed Judibana as an "open community." Planners laid out the town so that one lot of every four would be reserved for a non-Creole employee. Some houses were to be built by the company for rental to staff employees who were frequently shifted from one assignment to another. The majority, however, were to be constructed or purchased by the employees themselves. Supermarkets, banks, drug stores, travel agencies, motels, theaters, and restaurants were also to be owned by individual entrepreneurs rather than by the company.[33]

In order to implement community integrations, the Creole board, led by Duke Haight, sought means of encouraging home ownership. The company built model homes and urged workers to construct similar housing, using local contractors. The company also began a low interest loan program so that workers could afford their own houses. It also launched a "Realistic Rental" policy whereby rent in company housing would be raised to approximate community levels. Employee paychecks were increased proportionately. The effect was, once again, to eliminate the barrier between company town and surrounding community. In Judibana and the Bolívar Coastal Field [BCF], along the Lake Maracaibo producing area, the plan was especially attractive: workers could, in relatively painless fashion, earn equity in a home which they would own at the age of retirement. In the eastern division, Creole had more difficulty in effecting a transition to employee-owned homes because it was not an area of expanding activity. Here, Creole cut prices for existing houses drastically and, as a last resort, often made outright gifts of company housing.[34]

Community integration was a long-term success, but it did encounter considerable resistance from unexpected quarters. Many Venezuelans, long accustomed to company paternalism, balked at the prospect of having to supply their own needs. Venezuelan legislation, furthermore, had for many decades been designed to make the petroleum industry provide housing and medical treatment. In some cases where oil companies had supplied the essential services—telephone, gas, and electric power—the government would not accept the task for itself, even when offered intact transmission systems as outright gifts. Despite Creole's best efforts, camp towns and company paternalism continued in some degree. Creole persisted in the effort to eliminate them. A second major open community, Tamare, just south of La Salinas, brought many of the Judibana experiments to the BCF in the mid-1950s. In most areas, however, company commissaries and other signs of the old system remained.[35]

The company integration plan also involved transferring schools, hospitals, and other services to local agencies. Fortunately, this coincided with the desire of the Venezuelan government to "sow the petroleum" in infrastructure and development by requiring local contractors to build and maintain many of the services peripheral to the oil industry. Creole encouraged national entrepreneurship and capital formation. Similarly, the company gave preference to Venezuelan products wherever possible. Even before the company had completed the 165 bed hospital at Maracaibo in 1950, it determined to lease it to an order of missionary nuns, reserving a percentage of the beds for Creole employees and dependents. Gradually the company integrated its schools with local institutions as well. The effect, once again, was to build local capability in providing services.[36]

Technological Development in the 1950s

With the construction of Amuay, Creole had created the basis for an effective production, transportation, and logistical system. Most of Creole's production operations now centered in the BCF, while the bulk of its refining was concentrated on the Paraguaná Peninsula. The two poles of the Creole operation were intimately linked through pipeline and by modern communications. It remained only to perfect, augment, and modernize this basic operation.

The Ulé-Amuay pipeline eliminated the need for small "lake tankers" which had formerly carried crude out of Lake Maracaibo. Approximately the size of Lake Erie, this inland sea was connected to the Gulf of Venezuela and the Caribbean Sea by a shallow passage, the Tablazo Strait. Navigation through the Strait was blocked by sandbars and by difficult currents. Ordinarily, the Strait had twelve feet of water, but during the dry season only vessels drawing ten feet or less could navigate it—and then by following a crude path of tree branches and palm fronds stuck in the sand by passing seamen. The typical lake tanker was a small vessel of about 6,000 tons. Such a vessel ordinarily stood in line to pass out of Maracaibo with its load of crude. Upon reaching Aruba, it pumped its cargo into a larger tanker, the *George Henry*, permanently stationed at the wharf to provide storage.

Although Creole now had direct pipeline connections with a deep water terminal at Amuay, it nevertheless entertained hopes of improving sea access through the BCF. In 1951, when the Venezuelan government announced plans to dredge the Maracaibo Channel, it requested that Creole subscribe $10 million in bonds—approximately one-third of the projected total cost. This project involved dredging a twenty-one mile channel to a depth of thirty-five feet and the construction of breakwaters to protect the channel from the sea. Understandably, the project proved very difficult, and the government found it necessary to seek additional funds. When the channel was completed in 1956, Creole and the Creole thrift fund, an employee savings program, had purchased approximately $22 million in bonds and, in addition, the company loaned substantial amounts as well. Creole's funding amounted to one-half of the project's total cost.[37] To take advantage of improved access to Maracaibo, Creole began a major expansion of the industrial-logistical complex at La Salina in the early 1950s. With the aid of the Frederick-Harris Consulting Firm and George McCammon, Creole's chief engineer, Creole planned a deep water terminal to be built on a man-made island approximately one mile from the eastern shore of Lake Maracaibo. Construction resulted in an island storage and loading facility capable of handling four large tankers simultaneously. On shore, the company built a 100 acre industrial zone to service the entire lake operation. This complex included a bulk storage facility for cement, with a capacity for over 100,000 sacks, a huge pile casting yard, two large dry docks, and a series of smaller warehouses that contained over 45,000 articles by 1974. In addition, the complex included the largest private launch repair shop in the world and, by 1957, an immense central workshop which met the needs of the entire BCF.[38]

To serve the new deep water terminals at Amuay and La Salina, Creole commissioned a fleet of four large tankers. By 1960, all four, the "Esso Amuay," the "Esso Caripito," the "Esso Maracaibo," and the "Esso Caracas," plied the waters between the BCF, Amuay, and Aruba. In the same year, the company retired the last lake tanker, the "Esso Mara." The new 32,000 dead weight tons [DWT] air conditioned tankers were specifically designed for Venezuelan-Aruban service and sailed under the Venezuelan flag with national crews. These vessels had permanent ballast and special large pumps so they could unload quickly. On a typical voyage, such a ship took on a partial load at La Salina, sailed out of the lake and topped off its cargo at Amuay for the final run to Aruba. Although much larger than their predecessors, the new ships were also capable of negotiating the winding San Juan River to the Caripito refinery in Eastern Venezuela. Creole's transportation system had become infinitely more flexible and efficient with the addition of tankers and the dredging of the new channel. In a world of constantly increasing tanker capacities, however, terminal facilities had to be constantly modified. By 1963, much larger ships, such as the "Esso Manhattan" (160,000 DWT) were paying visits to Amuay. The deep water harbor at that site had to be dredged repeatedly.[39]

Unquestionably, Creole's major technical advances during the 1950s were responses to the peculiar problems of oil production in the Maracaibo Basin. Petroleum production in the BCF, of course, meant drilling offshore. Technology for such drilling was in an elementary stage in 1950; companies like Creole found it necessary to develop even the most fundamental techniques themselves. In addition, the lacustrine environment of the BCF imposed an entire range of exotic and persistent problems. Lake Maracaibo was swept by fierce "chubasco" storms between July and October. Lake water itself contained large amounts of free oxygen, and its salinity varied sharply from dry season to wet season. With the completion of the Maracaibo Channels, salinity began to increase. The combination of these factors caused an unprotected pipeline, exposed to lake water, to be eaten away by corrosion within four years. Compounding this problem, Maracaibo was the home of various species of the teredo, a marine boring worm, which destroyed virtually any protective coating applied to prevent corrosion. Similarly, "bear hair" and other aquatic vegetation clogged pumps and filters. Early on, Creole created a "corrosion group" which conducted experiments at La Salina with the aid of the Clapp Marine Biological Laboratory. Company engineers coated pipelines with asphalt, but found it just as subject to teredo damage as asphalt impregnated wood had been. Maracaibo teredos also proved too much for rubber-coated metal—the first recorded teredo damage to rubber. Enamel and plastic paints and fiberglass coatings, combined with cathodic charges, provided increased protection by the late 1950s. It was not until the 1960s, however, that Creole discovered plastic coatings capable of warding off both teredos and corrosion. Even then, Creole's vast network of underwater pipes was subject to occasional leaks, and remote pipeline monitoring devices were needed to maintain the system's integrity.[40]

Creole's most impressive engineering achievements were in offshore drilling. Such work had at first been accomplished on "cribbing"—tree trunks piled log-cabin fashion in shallow water to provide drilling platforms. Subsequent generations of BCF drillers worked from platforms mounted on wooden piles and concrete pylons. By 1927, Lago began to drill from seventeen pilot "termite platforms." Gradually, the surface of Lake Maracaibo became dotted with rows of large permanent platforms resting on rows of pylons. In the 1950s, Creole engineers began to realize that large, costly platforms could be eliminated if the weight of the derrick and drilling equipment could be borne by a movable barge or ship. Between 1955 and 1958, company engineers had put into operation a floating rig mover or "crab" with the aid of the Sverdrup and Parcel Company of St. Louis, and Higgins, Inc. of New Orleans. This device was a self-propelled marine fork lift, which moved essential drilling equipment to a small platform and derrick.[41]

In the mid-1950s, Creole received additional concessions in the center of Lake Maracaibo. The ambitious drilling program in the deep water of the new areas made it imperative that the company streamline its drilling techniques. Creole and the Central University of Venezuela combined forces to solve the problem. The solution was a cantilever drilling barge, a large, floating power plant containing a derrick which jutted out over the water. Such a device could anchor alongside a small three pile platform, drill a well, and move off to another site. The stationary platform bore little weight and could go into production without the need of a derrick. The first such cantilever rig went into operation in 1963, despite critics' fears that the ungainly vessel would overturn. The device proved highly successful, cutting down rig time substantially. Its success convinced Creole engineers of the need to apply the cantilever concept to the repair and maintenance of old wells. The derricks of such wells could be removed to accommodate the cantilever barge, of course, but many older platforms themselves were too high above the water for the new rig. The answer was a "high profile" cantilever barge, which could serve platforms up to eighteen feet in height— virtually every Creole platform on the lake. The new technique cut overtime work substantially, and gradually the neat rows of derricks marching across the face of the lake began to be replaced by modest slabs of concrete set on a few pylons. By the 1970s, the art of over-water drilling had been perfected to the extent that production platforms could be mounted on "monopiles," single concrete tubes with hollow cores through which drillers could run their pipe and tools.[42]

One of the outstanding technical innovations credited to Creole is the large capacity over-water gas conservation plant. By the early 1950s, gas conservation had become a matter of increasing concern for Creole and Venezuelans alike. Gas injection was hardly new to petroleum production men, but the over-water aspect of the Maracaibo project created special problems. First, the fields to be served were extensive and produced immense volumes of natural gas. Second, ordinary reciprocating power plants and compressors would have weakened the pilings upon which a BCF pressure maintenance plant rested. With the aid of Ingersoll-Rand and Cooper-Bessemer, Creole experts (notably Federico Baptista and William Glendenning) began to explore centrifugal compression and turbine power generation. The first installation of a gas conservation plant, Tía Juana 1, was dedicated on January 15, 1955, after four years of study and eighteen months of construction. It boasted ten 6,000 horsepower gas turbines capable of injecting 137 million cubic feet of gas per day. It burned 17 million cubic feet daily. Unlike Maracaibo production platforms, "T.J.1" was a manned installation. It contained instrument shops, maintenance shops, six offices, shower, and dining room. It stood fourteen feet above the water, on piles which had been driven by 200 ton weights.[43] The second platform, Tía Juana Conservation Plant 2, for example, measured 231 by 440 feet and stood on hundreds of piles in 85 feet of water. It was supplied by twelve flow stations and had four huge cranes mounted on its surface.

Creole engineers concluded that four major conservation plants were sufficient to serve the core area of the company's lake operations. This amounted to an investment of over $100 million. By the late 1960s, the need for pressurization in outlying areas led Creole to the "miniplant" concept, and eventually a total of twelve prefabricated, remotely controlled miniplants were located strategically around the lake. Though undeniably important, the total impact of the pressurization program is difficult to assess. At the outset, reservoir pressures had been dipping and production was declining. After the plants began to function, the higher pressures augmented production greatly. Furthermore, the plants removed the spectacle of gas flares along the skyline of Maracaibo, thereby eliminating a major source of public resentment. Ultimately, more than 80 percent of Creole's gas was either reinjected, employed in gas lift production, reduced to distillates, or piped to Venezuelan homes and industry.[44]

Overall, Creole's most notable technical accomplishments can be seen in its responses to peculiar Venezuelan problems. The company excelled in large scale production and logistical operations, the transport of large volumes of crude, the injection of large volumes of gas into reservoirs, pumping of heavy crude, and the organization of huge marine operations.

One principal area of Creole experimentation was its navy of crew and work boats. Transport of personnel to and from work sites on the lake required round trips of one-and-a-half to three-and-a-half hours per day. Construction and maintenance tasks demanded vessels capable of bearing immense loads of equipment. From an early fleet of wooden-hulled sailing vessels, Creole developed by 1959 an aquatic transportation section staffed by 1,400 employees. In 1955, Creole operated 162 personnel launches of thirty-two to fifty feet in length, 128 barges and work boats, and twelve large tugs. It also maintained a maritime dispatching office at Tía Juana equipped with the latest radio equipment. This center coordinated movement all over the lake. Each of the three BCF districts, La Salina, Tía Juana, and Lagunillas, operated round-the-clock facilities. In 1952, the personnel of these units scraped and painted 4.2 million square feet of hull and deck. Creole truly operated one of the world's great fleets.[45] In fact, it was the world's largest small craft fleet.

Creole production methods generally kept pace of its other technical advances, but once again the company's performance was strongly influenced by its vast cushion of crude reserves. One area in which Creole excelled was the application of field unitization techniques. Although the Venezuelan government began urging conservation in the early 1960s, the usual incentives for unitization were not particularly present in Venezuela. Most Venezuelan production came from huge, continuous concession blocks. Only some 20 percent of Creole production was credited to competitive areas, and in several instances, Creole production men controlled entire reservoirs by themselves. Creole's rule, generally, was that whenever secondary recovery methods began to be applied to a reservoir, unitization should be considered. The most ambitious such effort was made in Southwest Bachaquero in 1959 and 1960, and that reservoir remained the largest example of unitization in the world until Prudhoe Bay, Alaska, began production some seventeen years later.[46]

Closely akin to Creole's push for unitization were its attempts to space its lake wells at optimum intervals. Moving from the shore in La Salina or Tía Juana toward the center of Lake Maracaibo, a boat passenger would notice that newer, deep water wells were spaced much more widely than wells nearer the shore. Ultimately, Creole settled upon a 77 acre per well spacing scheme in mid-lake. The engineers placed wells with such accuracy that sighting down a row of derricks, an observer could clearly see the curvature of the earth.[47]

Evidence of the skill and adaptability of the Creole engineers and production people was brought home in October 1962, at the time of the Cuban Missile Crisis. Saboteurs thought to be Cuban agents one Saturday night about midnight dynamited several key Creole installations, including transformer stations, knocking out more than half of Creole's crude oil production. That same night an all-out effort was mobilized that brought some production back on stream within 36 hours and it was essentially all recovered within 60 days.

At Amuay, Creole's constant innovations gradually gave shape to the seasonal storage concept. The Amuay refinery had been designed in the late 1940s as a bulk producer of heavy fuel oil for power plants and heavy industry, destined for the sharply seasonal market of the Northeast United States. Amuay consequently required vast storage capacity, so that oil refined in the summer months could be ready for shipment for the winter season starting in October. Creole Directors Guillermo Zuloaga and Siro Vázquez conceived a solution to this storage problem. With the aid of T. W. Lamb, a Massachusetts Institute of Technology consultant, they and other Creole engineers applied earthdam technology to the construction of huge open storage pits. Such pits, much less expensive than tank farms, were possible to construct at Amuay because of the nature of the soil. The open pits were dug and sealed with a native clay which made them impermeable, and when full, the only notable problem encountered was the propensity of Venezuelan pelicans to land in them, mistaking the oil for water. This problem resolved itself within a few years, as the birds gradually learned to be wary.

Amuay's open pit storage expanded easily as the refinery increased its capacity. Ultimately, three pits were capable of containing 29 million barrels of petroleum products, and additional steel tank farms gave the refinery a total storage capacity of 46 million barrels. The dimensions of the earth tanks were impressive. One such tank, planned in 1955, was to contain 3.5 million barrels, measured forty feet in depth, and had a surface measuring over fifteen acres. Before construction began, however, the Creole board increased this reservoir's capacity to eight million barrels. The same pit was expanded in 1962 to a final capacity of 11.5 million barrels.[48]

Creole made a wide variety of other technical advances in the 1950s. For example, oil field communications first conducted over AM radio bands were easily accessible to the public. Wives of Creole workers sometimes monitored the AM frequencies and were able to announce promotions, pay raises, and similar intelligence before news reached the individuals concerned. Surprisingly, in the beginning Creole did encounter resistance as it sought to improve communications during the 1950s. Particularly in the eastern zone, the bosses of isolated camps resented the possible loss of autonomy that they believed improved communications would bring. Creole consequently lagged behind Mene Grande and other companies in the communications field.

In 1948, Creole and Mene Grande completed a direct dialing system with four substations, the largest direct dialing system in the world at the time.

By the mid-1950s, Creole engineers began retiring AM radio equipment and replacing it with FM to handle long distance communications. Gradually, old lead-sheathed wires, subject to fluctuations in temperature and condensation, were replaced by modern radio equipment and huge antennas. Ultimately, it became possible for a supervisor to communicate directly from his launch in the center of Lake Maracaibo with an advisor in Caracas, Houston or New York.[49]

Under the direction of Durward Wilkes, manager of Creole's Refining Department, the Amuay refinery in particular became the site of numerous technological advances, despite its continuing role as a bulk producer and the absence of a substantial local market for more exotic products. The refinery's throughput capacity increased steadily, as demonstrated below:

Year	Barrels per Day (b/d)
1952	75,000
1955	145,000
1960	355,000
1970	450,000
1975	630,000

By 1972, Creole had added a fifth distillation plant (a 180,000 b/d unit) to its complex at Amuay, which also boasted a huge 200,000 b/d capacity desulphurization plant built between 1968 and 1972 to meet U.S. government environmental regulations for low sulphur content fuel oil. By the late 1960s, Amuay was producing lube oils, avgas, gasoline, jet fuels, kerosene, diesel, asphalt, and, of course, heavy fuel oils. Its hydroformer, installed by future Amuay manager, Renato Urdaneta, was the only such unit in the world for many years.[50]

Creole engineers and workers also developed an entire range of lesser technical improvements. For example, these men developed a simple device which permitted oil drums to be turned and stored on their sides, thus eliminating the corrosion produced by upright storage. Creole technicians also invented new displacement devices to measure volumes of oil pumped onto ships.

In the BCF, Creole constructed sophisticated electrical systems based in La Salina and nearby Punta Gorda. From a simple steam-operated generation system in 1933, Creole built a system of ten gas-operated generating units which produced ninety megawatts by the mid-1970s. These units, three powered by gas turbines and seven steam-driven, were surrounded for security purposes by a twenty foot wall topped with mesh wire at Punta Gorda. The electrical system that radiated from the plant served flow stations and other installations as much as one hundred kilometers distant.[51] Also, two generators, mounted on barges, were available for emergencies and could be towed to any point of the lake where they might be required.

The 1950s were indisputably prosperous years for Creole and for other oil companies in Venezuela. National oil production soared, in part because of the increased markets created by the Korean War, the rebuilding of Europe, the nationalization of Iranian oil, and the closing of the Suez Canal. Also, until the late 1950s, foreign oil companies in Venezuela found themselves working in a less volatile, less openly antagonistic political climate than had been the case in the previous decade. These favorable conditions had clear implications for Creole. The company's earnings—whether measured as a ratio of investment or of production—remained exceptionally high. Now, however, Creole was operating on an infinitely larger scale than it had in the 1940s; hence, its profits were much larger. In 1954, for example, the company's net profit after taxes represented a 24 percent return on shareholders net equity, and it had $1 billion in total assets. It operated 3,000 producing wells itself and shared in the production of 716 others. It employed 14,000 workers. In 1954, Creole alone produced 48 percent of Jersey's net income.

In 1954, Creole capped its series of successes with the construction of a modern new headquarters in Caracas. A ten story, earthquake-resistant building, this edifice housed 1,400 workers and fourteen departments. This structure soon became a visible symbol of the new Creole Company.[52] Yet, its success, like the prosperity of the other petroleum companies in Venezuela, was not unqualified. For one thing, there were ominous hints; some critics charged that the Venezuelan boom was artificial and would slow as soon as world conditions stabilized. Others observed that all of the traditional problems associated with underdevelopment—uneven distribution of wealth, a stagnant agrarian sector, and moribund domestic industry—were actually being exacerbated by the boom in petroleum.[53] Still others charged that material progress outpaced social progress in Venezuela. In a more subtle but equally important sense, the sheer dimensions of Creole's operations and the weight of its influence—its very success—created suspicion among Venezuelans. Friends and detractors alike often referred to the company as "Mama Creole," while some spoke of the president of Creole as Venezuela's vice president.

Creole's integrated operations proved to be a training ground for people who went on to important positions in other Exxon operations. They include Bill Cleveland, Jim Dean, Hal Siegele, Dr. Siro Vázquez and Hugh Wynne in Production; Zeb Mayhew, Bill Wallis and Dr. Amos Salvador in

Geology; Woody Butte, Jack Clarke, Nick Campbell, Martin Jones, John Keffer, Dick Lombard and Hub O'Malley in Law; Charlie Leet and Robert Menk in Refining; Al Bruggemeyer, Ivan Cunningham, Martin King and Robert Mays in Finance; Hal Wright in Employee Relations and Wolf Greeven in Marketing.

Others spent many years within Creole and made important contributions to its success. Among these were Willie Franks, Phil Wolcott and Dr. Guillermo Zuloaga in Geology; Don Bancroft, Dr. Federico Baptista, Jim Barnett, Lawrence Cade, Bob Eeds, Lawrence Eldridge, Jay Gamble, Nicanor García, Bill Glendenning, Herbert (Shorty) Hegglund, Ray Ingram, Joe Stagg and Morris Pitman in Production; Tom Burnham, Sam Mathis, George McCammon, Maurice Wilkinson and Paul Widner in Engineering; Jack Creamer and Dr. Carlos Lander in Employee Relations; Oliver Knight, Dr. Guillermo Rodríguez Eraso, Dr. Ernesto Sugar and Henry Winter in Coordination and Planning; Harry Jarvis, Neal Griffin and Durward Wilkes in Refining and Ham Metzger, Dan Stines and Ray Winkler in Export Sales.

In addition the company was ably served by outside legal counsel, Dr. René Lepervanche Parparcen through the 1950s and 1960s. He was a former president of the Organization of American States [OAS] and served as legal consultant to the company from 1945 until his death in 1968.

Social Responsibility

By the mid-1950s, Creole directors had begun to confront the need for the company to become more actively involved in the social and economic development of Venezuela. That the host country required infusions of know-how, imagination, and capital for development was obvious. Creole seemed to have many of these requisites itself, and its almost embarrassing profit record gave company directors a special sense of responsibility for overall development of Venezuela.

This movement toward more active social and economic involvement proved as much a turnabout for the company as the shift from camp towns to open communities had been. Creole's traditional role had been largely that of a nonparticipant, except in the immediate area of oil production. The company provided inexpensive gasoline and service to the domestic market—often at a net loss—and medical and social services to some nonworkers in addition to its employees. Since the early 1940s, Creole had also given college scholarships to Venezuelans. Overall, however, the company made an effort to present a low profile to the Venezuelan public. True, it did publish an in-house magazine, *Nosotros*, and in 1939, it began a slick format, monthly magazine, *El Farol*, which published articles by Venezuelan scholars and writers and circulated throughout the country. Similarly, in 1943, it launched a radio musical show in Caracas, "Variedades Esso," and in 1952, began a radio news show and later a television news program, "El Observador." But in general, Creole tried to remain as inconspicuous as possible and had little public exposure in the areas of social and economic development.[54]

On October 16, 1956, the Creole board took a bold step, creating the Creole Foundation, the first such organization operated by an American corporation outside the United States. In time, the new institution came to epitomize imaginative approaches to social and economic development and it was widely emulated. Although many individuals gave direction to the foundation over the years, the concept itself was shaped by an economist, Joe Slater, and Hal Wright, a member of the Creole board, with strong backing from Dr. Siro Vázquez and Dr. Guillermo Zuloaga, members of the Creole board in the mid-1950s. These men worked to document the need for a foundation and to prove to a tight-fisted board that the idea was feasible and would yield positive results. Within a few months after the institution's creation, Slater resigned to take a post with the Ford Foundation.[55]

Although from the outset the Creole Foundation was dedicated to the development of human resources, the unique rationale upon which it operated evolved gradually over a period of several years. Slater's original idea was refined by his successor, foundation President Alfredo Anzola, and by long time managing director, George Hall, who came to Creole from the International Institute of Education. Perhaps the foundation could best be described as "non-operating"; it coordinated, originated, and funded projects but did not actually administer them directly. This philosophy grew from a belief that Creole should encourage "institution building" by channeling aid through new or existing indigenous organizations and those institutions with which Venezuelans could help themselves.[56] The relationship of the Creole Foundation with the Creole Corporation was an arm's-length one: the foundation, constituted as an independent, non-profit organization, applied to Creole each year for operating funds. Its charter provided for Creole's funding up to $1,530,000 per year for the first five years.

The foundation itself was directed by a three man board of Creole directors. In addition, there was an advisory committee of five outsiders, which included three Venezuelans, Pedro Grases, Arturo Uslar Pietri, and Agusto Márques Canizales, and John Camp from the Venezuelan Basic Economy Corporation and George Hill, a rural sociologist. In this way, the foundation was set up so that its activities could be reviewed to screen out potentially embarrassing or sensitive projects. In practice, the foundation always operated with a very small staff and minimal overhead expenses, while the Venezuelan Advisory Board added a creative touch to the foundation's activities. One of its members, Pedro Grases, a distinguished Venezuelan scholar who had taught in many of the world's great universities, was also active in a variety of philanthropic and cultural organizations.[57] Most of Creole Foundation's initial efforts were in areas in which the company already had some experience—education and culture. Then, the foundation began its own program of scholarships for studies in American and domestic universities. It supported national ballet troupes, museums and, in 1958, helped sponsor a visit by the New York Philharmonic Orchestra under the direction of Leonard Bernstein. It also made a special attempt to improve higher education, in what Anzola spoke of as a conscious effort to create an "intellectual elite" that could use its acquired talents for national betterment. The foundation subsidized university exchange seminars, set up laboratories at major universities, and awarded a $10,000 prize for the most important scientific achievement in any given year on a Venezuelan subject. The foundation contributed to schools of forestry and soil mechanics, and aided university libraries. It also gave some support to non-Venezuelan institutions, among them the Smithsonian Institution, the Hispanic Foundation, the Academy of American Franciscan History, and the Experiment in International Living.[58]

Higher education remained a major Creole Foundation concern throughout its history. The foundation scholarship plan became its best known and most prestigious program, probably because of the great number of recipients who later became distinguished in public life and business. Creole scholarships were awarded on the basis of need, academic record, and according to the field of study. Anzola and his associates felt strongly that Venezuela required special expertise in engineering and the sciences. The need for petroleum engineers, for example, was felt acutely, but the typical Venezuelan student studied law, medicine, or social sciences instead. Consequently, science and engineering students were given preference in the annual fellowship competition. By 1964, the foundation was awarding $250,000 in scholarships annually. Of the 404 scholarships awarded between 1957 and 1964, 194 were given in engineering and agronomy, 51 in physics, mathematics, biology, and chemistry, while the remaining grants were distributed among the humanities, social sciences, and other fields. Fellowships for advanced work in foreign universities amounted to less than 10 percent of the total grants given. The foundation also added other programs which complemented the scholarship plan. With other foundations, it participated in EDUCREDITO, a revolving fund created in 1964 to lend money to students. In the late 1960s, it inaugurated a program of scholarships in nautical and marine studies at the National Merchant Marine School. The Creole Company itself maintained its ambitious program of education for employees' children. The Creole Foundation also supported this program by providing small collateral grants to United States universities, in proportion to the number of company dependents enrolled.

In addition to its participation in the relatively familiar field of higher education, the Creole Foundation moved to broaden its interests and to include practical training at all levels. The directors were especially concerned about the plight of the urban poor of Caracas. There, migration from rural areas had greatly increased since World War II, and squatters' makeshift barrios began to appear in ravines and on mountain sides throughout the city. By the late 1950s, approximately one-quarter of the Caracas population was housed in such "belts of misery," where up to 60 percent of the inhabitants were children. The need to provide basic educational opportunity for these children became a principal concern of the foundation, which concluded early that barrio youngsters required prevocational training that would equip them for the sorts of jobs available in a rapidly modernizing society. In 1957, the foundation's officers persuaded four primary schools to offer prevocational courses for fourth through sixth grade boys. Despite the scarcity of manual arts instructors, barrio youngsters took to the program readily, and their school attendance improved dramatically.

Once the prevocational experiment had become firmly established, the Creole Foundation turned it over to the Ministry of Education and to a religious order. By the late 1950s, the foundation had begun to extend the prevocational approach, to reach rural youngsters. The foundation provided financial support to a variety of organizations, all of which were willing to experiment in rural education. It supported an agronomy school run by the Salesian Order in Valencia, which taught children to care for ducks, rabbits, and chickens, and underwrote the "Samaritans of the Fields," a religious group that sent teams of roving educators into rural villages. It supported a rural normal school, "El Macaro," which prepared school teachers who would be stationed throughout rural Venezuela. It also joined forces with the University of the Andes in an attack on adult illiteracy in the State of Mérida. Another joint project, this time in cooperation with the Central University and the University of Florida, was designed to support research at the University's Maracay Agronomy School.[59]

The constant purpose of the foundation was to help break down inertia and build self-supporting institutions for future development. One example was Fé y Alegría, a program resembling the prevocational plan, but much broader in scope. Fé y Alegría was launched by Fr. José María Vélaz, a Chilean Jesuit, in 1954, and heavily supported by Venezuelan businessman Gustavo Vollmer. Vélaz's original plan was to teach literacy and provide a stable, healthy environment for barrio children in a period when 80 percent of Venezuelan youngsters did not complete the third grade. By the mid-1960s, the staff of Fé y Alegría was housed in a six story building. It employed some 500 teachers and operated slum schools throughout Venezuela. Financial support was generated through an elaborate system of committees that staged raffles and other fund raising events.[60]

Although most foundation activity was heavily pragmatic, the organization did become involved in several scientific undertakings. In 1958, the La Salle Natural Science Foundation established a research station on Margarita Island to study marine resources and the needs of native fishermen. Venezuela has 2,200 miles of coastline and in 1958, some 200,000 Venezuelans depended directly upon fishing for a livelihood. With Creole Foundation support, La Salle began a school in which modern fishing techniques were taught.

Another Creole Foundation venture in applied biology was the Llanos Study. The Llanos is a vast, forbidding plain in Eastern Venezuela. A land of extremes, the Llanos is alternately too wet or too dry to be suitable for efficient cultivation or ranching. In addition, a hard, impermeable rock-like material often lies just beneath the surface, preventing crop growth and drainage. On the Llanos, the Creole Foundation cooperated with the Venezuelan Society for Natural Sciences. The research station which they established at Calabozo experimented in new grass varieties and better agricultural techniques.[61]

Other activities of the Creole Foundation are too numerous to discuss in great detail and can only be summarized here. One program, conceived by Eugenio Mendoza, a Venezuelan philanthropist, the Dividendo Voluntario para la Comunidad, received a boost from a foundation-sponsored international seminar of executives in 1963. Dividendo ultimately became a federation of almost five hundred private enterprises that had recognized their obligations to Venezuelan society. Participating firms pledged 2 percent to 5 percent of profits for social action, and Dividendo served as a clearinghouse for projects worthy of their assistance. The Creole Foundation also provided encouragement to seminars in which Venezuelan businessmen could study the future prospects of their country and develop strategies to cope with anticipated problems.[62] The Creole Foundation sponsored an almost endless variety of programs in the field of education. It supported the Inter-

American University Association and made possible one-month visits by undergraduates to the United States. It funded student residence halls and meeting centers operated by the Opus Dei Organization. The foundation assisted in funding an institute of business and public administration studies, organized by Dr. Carlos Lander, a long-time Creole director, and a school for the deaf and mute. It also aided Junior Achievement programs in Venezuela and participated in the founding of several institutions of higher education. In 1961, the Creole Company turned over the facilities of its Jusepín camp to the administration of a new Eastern Venezuela university, Universidad de Oriente. The Jusepín camp was equipped with a variety of large buildings, 660 houses, water, gas, and sanitary facilities. It thus became a home for schools of petroleum engineering, veterinary science, zoology, geological engineering, agriculture and mining.

On balance, the most important contribution of the Creole Foundation was its unique institution-building effort. Unlike traditional philanthropic organizations, it acted through great numbers of other organizations and institutions. Certainly both the Creole Foundation and the Creole Corporation did occasionally make direct gifts to the poor and suffering, such as victims of the 1967 earthquake. Yet the key to the foundation's excellent image was its keen sense for evaluating worthy projects and institutions and its patient nurture of selected projects, once underway. Over 50 percent of the foundation grants resulted from outside initiative. In sum, Creole Foundation programming was both pragmatic and experimental. To the casual observer, foundation activities may seem to be characterized by a certain randomness, but upon closer examination clear patterns do emerge. The foundation, of course, showed a consistent dedication to social and economic development. This interest was expressed in various ways as political and social conditions in Venezuela changed. During the first few years of its existence, the foundation concentrated upon higher education and prevocational training. Later on, it launched community organization drives in the barrios and in rural areas. It also helped the relatively moderate Andrés Bello University to establish a social sciences school. By the mid-1960s, the foundation had begun an attempt to demonstrate the potential of corporate funding and know-how for social and economic development. During this phase it supported Dividendo, Fé y Alegría, and similar agencies.[63]

The Creole Corporation launched another project in the early 1960s, when Harry Jarvis was president of the company. It was designed to help the badly depressed economy of Venezuela at that time. This was the Creole Investment Company [CIC], capitalized at $10 million. The rationale behind the formation of CIC was that the nation needed an injection of capital to turn around the depressed economy. A show of interest and support by Creole, with cash on the barrelhead, not only would perk up the economy with injected capital, but would also provide a better morale, and rekindle local faith and confidence in the future of the private sector.

Creole's policy was to invest in new as well as existing businesses that required capital, but on a less than 50 percent basis only. Once an investor, CIC provided professional managerial talent on a part-time basis, often in the form of taking active participation on the boards of directors of the companies. From 1961, when CIC started, until 1963, it invested $5.4 million in twenty-two companies with activities in a wide range of endeavors, such as cattle and sheep ranching, textile, brick, paper and fiberboard manufacturing, seed corn and starch processing, truck farming, candy making, and mushroom growing.

Investments were supervised by CIC's General Manager Frank Amador, a bilingual native of Las Cruces, New Mexico, who had worked for Creole in Caracas for seventeen years, heading in turn its research, economics, budget, and tax departments. In the first two years of operation, he and his six-man staff evaluated more than 350 proposals from Venezuelan businessmen. The $5.4 million of investments mobilized more than $21 million in other shareholder investments and long-term loans.

The effort proved to be a helpful stimulus, and many of the enterprises prospered. In some cases, CIC was required to take more than a minority position, to try to keep the business afloat. On balance, the fact that Creole demonstrated its belief in the future of the country perhaps proved more rewarding than the pragmatic results. CIC began liquidating its interests when it was decided that CIC had served its purpose and should be phased out. By the time of nationalization, in 1975, all investments had been liquidated.

THE LAST YEARS

The revolution that overthrew Pérez Jiménez in January, 1958, marked a turning point in Venezuela for Creole and the oil industry.

During the 1950s, Creole enjoyed favorable relations with the Venezuelan people. As has been pointed out, Arthur Proudfit, experienced in dealing with various governments in the country, steered a steady course for the company during the increasingly autocratic regime of Pérez Jiménez, often a delicate matter. But the company kept to its work of producing, refining, and shipping oil and pointedly stayed away from the political aspects of the national scene.

When Proudfit moved to the Jersey Company as a director in 1956, Duke

Haight became president of Creole; he maintained the same kinds of policies which Proudfit had espoused. Haight, a long-time international operator with oil field experience in Mexico and Egypt, increased the emphasis Creole had always placed on acting as a responsible "guest" in a host country. He stressed language training for all expatriates, technical training, and management development for Venezuelan employees, elimination of camps, wherever possible, by integrating them into the local community, and increased social responsibility, as evidenced by the creation of the Creole Foundation.

The 1958 revolution, basically a reaction of the Venezuelan people to years of repression of their civil liberties under Pérez Jiménez, was not aimed at North Americans nor at their companies. The fact that oil operations were not disrupted nor U.S. citizens molested in any way attests to the more fundamental causes of the revolt.

However, during 1958, the provisional ruling Junta called for general elections, and the ensuing scramble by a number of parties and politicians led to the awakening of some dormant anti-American attitudes, and a number of office seekers fanned the embers of this feeling to further their particular objectives and arouse their special constituencies.

In December, 1958, the then president of the provisional Junta, Dr. Edgar Sanábria, proclaimed a higher tax on profits, retroactive to the first of the year. This affected the entire economy, particularly the oil companies, since it abrogated the 50-50 principle, raising the government's take to about 65 percent in the case of Creole, whose tax payments to the government on 1958 earnings increased by some $92 million.

The new tax was announced just as Duke Haight was leaving Venezuela for Christmas vacation, and it caught him by surprise. He prepared a straightforward presentation of the facts as he knew them, and, against the advice of Directors Leo Lowry and Robert Mays, and Public Relations Manager Everett Bauman, issued the statement before departing the country:

> The new income tax law decreed by the Provisional Government of Venezuela on Friday, December 19, is a hard blow at the oil industry of Venezuela, and in my opinion has drastically altered the climate for foreign investment in this country.
>
> By this action, Venezuela becomes the first country in the world to break the so-called 50-50 principle of equal shares in the fruits of industry, completely disregarding acquired rights and ignoring the moral if not legal obligation to negotiate this rupture with the interested parties.
>
> Some other oil producing countries have recently concluded

certain oil agreements which depart from 50-50 but in no case have existing concessions or fiscal agreements been modified. Action was taken in spite of promises made to the oil industry here by the President of the Government Junta in a meeting at Miraflores Palace on February 5, 1958 that no changes affecting the oil industry would be made by the Provisional Government and that existing laws would be adhered to.

On that occasion the President requested cooperation of the oil industry and requested the head representatives of oil companies to assist in establishing confidence abroad for continued investment in this country under the new government.

At that meeting I had the honor to be selected to reply to the President of the Government Junta on behalf of the Venezuela oil industry and thank him and the government for the fine sentiments expressed. I also stated my conviction that the government could count on enthusiastic support and cooperation of industry because of mutual interests involved and the long record of harmonious relations in discussion, negotiation and settlement of mutual problems.

Only last month at a meeting in the office of the Minister of Mines and Petroleum, the present Minister reiterated that the same petroleum policies previously in effect would be continued. The action of Friday night also disregards repeated statements made by the constitutional president-elect during the recent political campaign that any change in petroleum policy in Venezuela would be negotiated with the Venezuelan petroleum industry as a commercial matter.

This action therefore comes as a great surprise and is viewed with alarm because it raises great doubts as to future actions.

During 1956 and 1957 the oil industry acting in good faith and in belief that 50-50 would not be unilaterally broken paid approximately $686 million to the federal treasury for new concessions. It seems possible that some of these companies will now feel that they have been defrauded.

The 50-50 agreement which has now been ruptured derives from negotiations between government and the oil industry in 1943, during which time many long-standing problems were settled by the enactment of the 1943 Petroleum Law and the income tax law. The principle of equal division of the fruits of the oil industry was established at that time.

In the general settlement which was then worked out, the old petroleum concessions were continued under the new petroleum law and given a new life of forty years—an extremely important quid pro quo since many of these concessions were near the end of their original life.

During the three year period from 1945-48 further negotiations were carried out with the oil industry with the object of assuring that the nation obtained an equal share of the oil income. These negotiations resulted in the modification of the income tax law and the 50-50 principle became fully effective and guaranteed.

It is logical to think that agreement reached by free negotiation cannot be unilaterally terminated without the destruction of confidence and understanding.

The present unilateral retroactive action provides no quid pro quo and reduces incentive to further investment in the oil industry in this country.

It must be remembered that new taxes are additional costs for industry. This increase in cost comes at a time when there is great surplus producing capacity. Our principal business is to export oil abroad and in our foreign markets we must compete with many other producers. We are now placed at great competitive disadvantage. We will undoubtedly lose markets and this will reduce our sales income. Inevitably this will lead to reduction in investment.

New capital investment is of greatest importance in the oil industry. Much has been said here about the "excessive" profits of the oil industry. Such statements do not take into account the fact that a large part of these profits is reinvested each year in Venezuela to provide producing capacity for the industry. Even a larger part of profits must be invested abroad to provide marketing facilities for Venezuelan oil.

Finally, it is highly uncertain whether the government will obtain with the additional income tax a larger share of the oil income than it would receive under the 50-50 agreement. Only time will tell. Among other things, it will depend on the amount available for reinvestment and the climate which exists for investment in this country.

Venezuela's Minister of Mines took little time to refute Haight's statement, proclaiming the action to be "exclusively a matter of national sovereignty," and the government let it be known that Haight would no longer be welcome in the country.

The election resulted in a victory for Rómulo Betancourt and the Acción Democrática party, and they governed the country for the next five years. To replace Duke Haight with someone who could work harmoniously with the new government posed a dilemma for the company, and there was really only one viable solution: Arthur Proudfit resigned from the Jersey Board and returned to Venezuela in early 1959, once again as Creole's president. He set to work to build a solid relationship with the government and to restore employee morale, which had been badly shaken. He did his job well and retired in 1961. Harry Jarvis, another long-time Jersey employee with wide Latin American experience, replaced him. More Venezuelan employees were added to the Creole board. Also moving up the executive ladder were men with backgrounds in finance. Leo Lowry, who for some years had been a director and member of the executive committee of Creole, became executive vice president. Robert Mays was named a member of the Creole board. Mays, who had brought with him to Creole considerable experience in financial management worked to establish at Creole, with Lowry's support, one of the most effective financial control systems not only in Venezuela but throughout the oil industry. Introducing the company to full and effective use of computer technology proved no easy task. However, with largely Venezuelan employees, most of whom had less than a high school education, a modern computerized information system was developed. Under Duke Haight especially, both the engineering and operating people in Creole had been taught to be fully accountable for the results of their expenditures. The collaboration between the financial departments and the operators in emphasizing the "bottom line" was a major contribution to Creole's success. As a result, even when the companies in Venezuela were later nationalized, the Venezuelans' respect for Creole's management efficiency and technological and financial capabilities was evident. Mays later became controller of Exxon.

Leo Lowry replaced Jarvis as president in December, 1964, only to face demands by the government for even greater participation in earnings. This was particularly manifested by large income tax claims for previous years, alleging that the company's export sales prices had been inadequate. Pressure for outright nationalization of the industry also increased. For example, the government decreed that one-third of the Venezuelan domestic market for oil products should be ceded to the national oil company.

The expanded social programs of the Acción Democrática government had put strains on the federal budget, and there was increasing anti-American sentiment. Government demands accelerated after Betancourt completed his five-year term and was succeeded by Raúl Leoni, also of the Acción Democrática party. After lengthy negotiations, agreement was reached with the government to apply tax reference values rather than actual realizations for income tax purposes. This, together with tax rate changes, brought Venezuela's share for 1967 to 73 percent.

In 1969, Rafael Caldera of the Social Christian party became President of

Venezuela in a peaceful transfer of authority, but there was no let-up in the policy of seeking a larger "take" for the nation. New tax laws and arbitrary reference values for oil tax purposes were imposed, resulting in a government take in excess of 80 percent. These actions, in effect, broke the tax reference price agreement signed in 1967 covering a five-year period. The government also decided to reserve the natural gas industry to the state exclusively.

Lowry left Creole in 1971 to become president of Esso Inter-America, turning over the management reins to Executive Vice President Robert N. Dolph. The new executive vice president was Guillermo Rodríguez Eraso, a Venezuelan. He and Dolph had worked closely together during the late 1950s in Creole's economic planning. Their first major crisis was the Hydrocarbons Reversion Law, adopted by the Venezuelan Congress, which drastically changed the rules under which assets would be transferred to the state by the oil companies, upon expiration of the concessions. In addition, the law required deposits by oil companies into a special guaranty fund, to assure transfer of all physical assets in satisfactory condition. In 1973, another law was passed which allowed the government oil company, Corporación Venezolana del Petróleo (CVP), to take over all oil marketing operations within Venezuela. Late in the same year, the Acción Democrática party regained control of the government by winning the fourth peaceful election since the fall of Pérez Jiménez. The new president, Carlos Andrés Pérez, who was inaugurated in March, 1974, stated that he would try to speed up reversion of the concessions. In August, 1975, Congress enacted the law which nationalized the oil companies. It became effective on December 31, 1975.

Throughout this period, Creole continued to operate in its usual manner—maintaining its policy of "Venezuelanization," refinery expansion, investments needed for effective operations, and other efforts to retain Creole as an asset to the nation. The company cooperated with the government, and nationalization of Creole's operations was achieved through agreements which called for a payment of $507 million for the assets taken by the government. Separate agreements were signed for a four-year technical assistance contract and for future oil purchases.[64] Of the total compensation of $507 million, $72 million was to be paid in cash and the remainder in 6 percent bonds maturing over a five-year period. Of the total amount of the bonds, $206 million had already been deposited in the guaranty fund.*

Nationalization was the culmination of a gradual encroachment by the government on the oil industry that started with the 50-50 agreement of 1943 and grew more forceful in 1958 and later, when the government's take surpassed 90 percent. Thus, from 1957, when Creole had its peak net income of $397 million, until its last year of operation in 1975, when it netted $111 million, there was a marked erosion of profits even though production showed sizeable increases for most years.

A review of Creole's history of productivity, leadership, and its continuing attempts to enhance the Venezuelan economic and social atmosphere naturally raises the question: Why was the company nationalized?

There is little doubt that Creole through the years, for the most part, conducted its business in a statesmanlike manner. It settled problems by negotiations, generally acceding to the wishes of the national government. Its labor relations were good, and training of Venezuelan personnel excellent.

Perhaps one answer lies in the nationalistic feeling that, during the 1950s and 1960s, enveloped Venezuela as well as other countries in Latin America and in the rest of the world. In a column in *The Journal of Commerce*, May 24, 1951, Petroleum Editor Wanda M. Jablonski, who later founded the widely respected *Petroleum Intelligence Weekly*, wrote this:

> A little before the whole Iranian crisis exploded, an American company operating an important concession in another foreign country decided it was time to check up on its own local "public relations."
>
> So the company had the equivalent of a Gallup poll made to see how it stood with the local population.
>
> Four main questions were asked. The results ran something like this:
>
> 1. Do you think the prices we charge you for oil products locally are fair or too high? 80% replied "very fair."
>
> 2. Do you think we have conducted ourselves in a manner to make us welcome in your country? 85% replied "yes, we like you."
>
> 3. Do you think the profits we split with your Government are fair and equitable? 85% replied "very fair."
>
> 4. Do you think we should be nationalized? 98% replied "yes."

If, for Creole, the handwriting was on the wall in 1951, then the fact that the company survived for another quarter century may be testimony to superior management. There is no doubt that its painstaking efforts to continue to operate under ever more difficult conditions paid off in terms of returns to the shareholders.

Problems in Perú

WHEREAS JERSEY AFFILIATES in Venezuela and Colombia enjoyed relatively cordial relationships with host governments and local populations, the experience of IPC in Perú was one of intermittent acrimony and bitterness. In 1914, Walter Clark Teagle, then president of the Imperial Oil Limited, created IPC to hold the foreign properties of Imperial. Ten years later, IPC found itself embroiled in a legal dispute with the Peruvian government over the question of the ownership of a producing field known as La Brea y Pariñas, on the northern Pacific Coast of Perú. This never-resolved controversy endured until the expropriation of the company's holdings in 1968-1969. IPC's problems in Perú turned on a long-standing dispute concerning subsoil rights—a dispute which the company inherited when it began leasing properties from the London and Pacific Petroleum Company in 1914.

For centuries, Spanish law had vested in the state title to all subsoil resources. Individuals or companies could obtain temporary licenses or concessions to exploit minerals, but could never receive permanent title to the deposits. However, in 1826, Simón Bolívar's fledgling government, in order to pay off the national debt, broke with this tradition by selling to private parties its rights to various properties, including the pitch mines on the La Brea y Pariñas estate. This exception to traditional law provided the basis for continuing conflict between government and private enterprise. Subsequent regimes saw this reversal of traditional subsoil law as an attack on national sovereignty and attempted to get the property restored to the nation.

A long series of owners transferred La Brea y Pariñas subsoil rights with each sale of the property. In 1863, after a British engineer struck oil just north of the estate, the Peruvian government began seeking some basis for regulating and taxing the unique property. In 1888, the government concluded that, due to the nature of the case, standard mineral codes would not apply. Accordingly, the government ruled the La Brea y Pariñas estate liable

431

for only nominal taxes and indicated that the owners would not have to meet the minimum exploration and development requirements demanded by national mineral legislation. When IPC leased the property from British oilmen in 1914, total taxes each year amounted to $150. Taxed under the then-existing uniform mineral codes of Perú, it would have paid about $6 million.[1]

The British lessors filed petitions with the Peruvian government asking to submit the title controversy to a Peruvian court. Perú agreed to this request, and the court promptly ruled the IPC claims invalid. This led to intervention by the British government, and in 1921-1922 the United Kingdom and Perú agreed to submit the question to arbitration by an international tribunal. The resulting decision, known as the "Laudo Arbitral" of 1922, based on an earlier agreement by both governments, provided that La Brea y Pariñas would be regarded as 41,614 separate mineral claims or concessions. Taxes would then be levied at the rate of thirty soles per year on those parcels that actually produced petroleum, while unproductive parcels would pay a nominal tax of one sol per year. In addition, the award stipulated that the owners would pay a lump sum of $1 million to satisfy all claims for alleged back taxes. Further, under the Laudo Arbitral, Perú acknowledged the claims at La Brea as private property.

Although the Laudo represented a sincere and impartial attempt to regularize the status of the La Brea y Pariñas claims, some Peruvians noted that it left subsoil rights intact and that it continued to provide for taxing the holdings at an abnormally low rate. However, the Peruvian government observed its provisions and began taxing the property on the new basis.[2]

The three decades following the Laudo were prosperous ones for IPC, notwithstanding periodic outbursts of Peruvian nationalism. Shortly after receiving the judgment of the International Tribunal, in 1924, Imperial purchased La Brea y Pariñas from William Keswick, heir of the London and Pacific Petroleum Company's founder. Throughout the following years, the company proceeded to develop its Peruvian properties and to operate its refinery at Talara. IPC was a low-cost producer. The fields, conveniently located near the sea, made the export of most of the production to foreign markets easy. In addition to property taxes, IPC paid export taxes on the crude oil and income taxes to the Peruvian government which amounted to substantial sums of money.

While the company maintained profitable operations through the late 1940s, government policy prevented it from investing in the exploration and development of new fields. Throughout those years, IPC requested the government to permit it to explore the promising montaña region on the eastern slope of the Andes as well as the Sechura Desert, a coastal area adjacent to La Brea y Pariñas.[3] While this met with the approval of the executive branch, so much opposition to granting the request developed in the Peruvian Congress that the company shelved the idea. However, in 1952, under a new law, concessions were granted to IPC and other companies in Sechura as well as in jungle areas and on acreage south of Lima. The companies found no promising oil deposits and soon withdrew.

IPC established new methods of secondary recovery and rational field management at La Brea y Pariñas. An old marginal field, La Brea presented company experts with something of a geological puzzle. Located on a belt of heavy faulting, its geological formations were so crosshatched with fractures that seismography and other modern exploration methods proved difficult.[4] Consequently, IPC employed conventional as well as innovative techniques to probe every corner of its properties. In developing proven reserves, the company quickly adopted the latest techniques. Texts on petroleum engineering regularly cited the IPC production operation as a model of efficiency. It was one of the first firms in the world to experiment with gas reinjection to force oil to the surface. As early as 1943, it had injected natural gas for conservation purposes, and in 1950 it began injecting liquid petroleum gas [LPG] to increase its oil recovery. Between 1953 and 1957, it invested $64 million in such secondary recovery projects. By the 1960s, secondary recovery measures provided approximately 30 percent of all oil lifted from La Brea y Pariñas. Hence, production in the old fields remained quite low.[5]

Inability to increase production continued to be the most important effect of International's confinement to the La Brea y Pariñas holdings. Crude production reached the yearly maximum of 17.6 million barrels in 1936, slumped to 14.2 million barrels by 1939, and remained at about 6 million barrels per year throughout the 1950s. By 1968, La Brea y Pariñas produced only 5 million barrels annually despite the extensive secondary recovery measures.[6]

The Peruvian government's policy of restricting company activities proved the major reason for International's failure to increase production. When World War II erupted in Europe, the Peruvian regime froze most consumer prices to curb inflation. After the war, it deregulated the prices of all products except petroleum and its derivatives. As a result, the company's return on products marketed in Perú eroded rapidly, for creeping inflation yearly reduced the real value of Peruvian currency. Even in the 1950s, Peruvians were paying about seven cents per gallon for gasoline—probably the lowest price in the world at that time.

A long-standing requirement of the Peruvian government provided that International supply domestic demand before it exported any petroleum. While this hampered the return on investment during the 1930s and 1940s, the restriction did not impose any particular hardship upon the company. However, between 1938 and 1958, national consumption increased 400 percent while prices and production levels remained relatively unchanged. Thus, IPC, the only significant producer in Perú, found itself obligated to divert oil from the high profit export market to low profit domestic outlets, a quite dramatic shift. Whereas in the 1930s, International exported some 70 percent of its liftings, by 1957, domestic consumers absorbed 75 percent of the total product of the company. In the early 1960s, Perú became a net petroleum importer.[7]

Understandably, IPC Presidents Michael Haider (1954-1959) and J. K. Jamieson (1959-1961) sought to break away from restrictions that served to cripple the company's operations. Efforts to reverse these government policies produced some limited success, partially based on a growing realization among Peruvians that the company could not fully supply domestic markets under the existing price control system. In 1952, the government raised the price for regular gasoline and the price of premium to 11¢ per gallon.[8] Another slight increase in the price of gasoline occurred in 1954. But in both instances the company argued that the increases failed to keep pace with inflation and other costs. In 1952, legislation established a new petroleum code which guaranteed the government 50 percent participation in profits and authorized a depletion allowance for oil companies. While this did not apply to La Brea y Pariñas, which continued to operate under the 1922 Laudo, export taxes and income taxes, as noted above, provided revenue to the Peruvian government equal to or greater than the amount that the new law would have generated.

Beginning in the late 1940s, Standard Oil Company (New Jersey), the principal owner of Imperial, began tightening its control over IPC. That company had continued to employ many of the executive personnel of the old London and Pacific Petroleum Company when Imperial had purchased it during the third decade of the twentieth century. The British operators had a reputation for conservative, paternalistic labor policies; they viewed local public opinion rather cavalierly. These attitudes had been readily assumed, in turn, by Imperial's largely Canadian personnel, who replaced the British. Now, executives at Imperial and Jersey both believed such policies and attitudes to be a serious liability for a company confronting rising Peruvian nationalism. They began to centralize control over the affiliate, with the hope of changing policies and attitudes. This development coincided with the growing need of Imperial to finance exploration and development activity in Western Canada, where a new oil field had been brought in. In 1948, Jersey began acquiring IPC stock from Imperial through an exchange of three Jersey shares for each twenty shares of IPC. By late August, 1948, Jersey held over 50 percent of IPC stock and by February, 1949, had increased its equity in the company to 82.5 percent.[9] Shortly thereafter, IPC and Jersey decided to move the IPC offices to the Miami, Florida area, where management would be closer to operations in South America.[10]

Having failed to reverse the steady deterioration of its position through increased prices or by obtaining additional concessions, IPC now began moving to enhance its image among Peruvians. This effort had two principal aspects: improved labor conditions and increased attention to social and economic development. IPC proved to be a comparative neophyte in these fields, unlike the Creole Petroleum Corporation, which had achieved progressive changes in both community and labor relations. To meet this competition, IPC tried to make many more improvements in much less time.

Largely because of the legacy of its British and Canadian management, IPC had long been regarded as the "most conservative of the Jersey companies in Latin America."[11] Throughout the 1930s and 1940s, most of its Peruvian laborers were the sons of former company workers who had been born on the La Brea y Pariñas estate. The company did make early progress through the hiring and promoting of Peruvian nationals, to fill technical and managerial positions. By 1939, the company retained only two hundred expatriate workers out of a total of five thousand employees in Perú.

In almost every other respect, however, working and living conditions around the International refinery complex could be termed inadequate. When IPC first acquired La Brea y Pariñas it found "unspeakable" conditions prevailing. Bubonic fever and smallpox were rife, and neither fresh vegetables nor treated water was available to workers or their families. While IPC took measures to rectify some of the conditions that occasioned these problems, both Imperial and Jersey, in the 1940s, came to realize that the company still lagged behind other affiliates in most aspects relating to personnel and working conditions. In 1945, a three-man team, appointed by the parent companies, visited La Brea and, after thorough study, filed a highly critical report recommending a broad program of reform. The report noted that among the lowest paid workers 47 percent lived in houses without electricity or running water and that for most workers there were no private baths or toilet facilities. Skilled and unskilled workers alike inhabited old wooden six- or eight-family barracks. The visiting team concluded that IPC should completely raze the city of Talara and other installations, and rebuild them at an estimated cost of $11 million.

IPC management largely resisted implementing the recommendations made in the report until Jersey acquired controlling interest in the company, in the late 1940s. When Jersey had achieved that control, it selected one of the report's authors, H. A. "Red" Grimes, a former Creole manager, to be the new IPC general manager in Perú. Grimes immediately launched the recommended program of broad reform. Between 1952 and 1954, new brick houses, schools, recreation facilities, and a Talara shopping mall were constructed. Under Grimes' leadership, the company revolutionized health care for its employees. Massive programs in immunization, tuberculosis, and venereal disease control quickly began and were continued. The number of hospital beds per capita doubled. So effective were these changes that by the mid-1960s, Talara had one of the world's lowest mortality rates from infectious diseases.[12]

IPC, then, had made dramatic, albeit tardy reforms in Perú. Nevertheless, some sincere and some politically motivated Peruvians continued to resent the company's presence in their country. Satisfying such groups proved difficult. Their members argued that Talara, a desert oasis almost six hundred miles north of Lima and an example of a company enclave, actually enjoyed closer ties with other countries than with the host country itself. Paradoxically, the company's concessions to its laborers alienated Perú's conservative establishment. By 1964, IPC's basic wage was almost double that prevalent in the Lima metropolitan area. Domestic employers also claimed that IPC created dissatisfaction and pressures for higher wages among their workers.[13]

An additional source of antagonism developed when IPC acquired interest in another Peruvian company, the Compania Petrolera Lobitos. Lobitos, the wholly owned subsidiary of Lobitos Oil-Fields, Ltd., a London holding company, operated "Concesiones Lima," a producing district adjacent to the La Brea y Pariñas estate, while it utilized IPC to refine and market its crude. In 1957, International acquired a 50 percent undivided interest in the Lima Concessions. As a result of this action, IPC controlled 90 percent of the Peruvian petroleum output, rather than the 65 percent it had owned before that date. IPC critics interpreted the acquisition as a movement toward monopoly, and their indictment of IPC grew correspondingly stronger.[14] But government approval of the acquisition reflected the expectation of greater activity in the Lobitos area under the new operators, which in fact resulted in more production.

Hence, in the late 1950s, IPC found itself on an increasingly shifting, if not yet treacherous base in Perú. Its financial position became nearly untenable as it shifted to supply unremunerative domestic markets. In addition, Nationalists and left-wing reformers continued their harsh criticism of the company. In 1958, IPC responded to these economic and political conditions by cutting the company drilling program, curtailing investments, and by declaring surplus more than one thousand Peruvian employees, who remained on the payroll without working until called back to work later on. To publicize these changes, the company launched a public relations effort, taking out advertisements in Lima newspapers and appealing to the public through a variety of communication channels.[15]

IPC's dilemma posed a problem for the Peruvian President, Manuel Prado, who needed to assure a continuing supply of oil for his country. Following conversations with IPC executives, and with the advice of his new Prime Minister and Minister of Finance, Pedro Beltrán, who, as owner and editor of the newspaper *La Prensa*, had been one of the President's major critics, Prado concluded that oil prices would have to be increased. This would insure such a continuing supply, while helping to improve the atmosphere for foreign investment, and, although not a minor consideration, increase the government's own revenue through resulting income taxes. Thus, price increases were decreed in the latter part of 1959, and these were structured so as to avoid widespread public reaction. For example, the prices of regular gasoline and kerosene were raised only slightly.

Now faced with a politically effective solution to the oil problem, opponents of Prado revived the almost forgotten argument about the 1922 Laudo award, seeing this as a rallying point for Leftists and Supernationalists. They had the support of *El Comercio*, a leading newspaper owned by the Miro Quesada family, who for a number of reasons opposed both Prado and their competitor Prime Minister Beltrán. *El Comercio* waged a very telling propaganda campaign, which resulted in increased pressure and, indeed, a number of bills in Congress aimed at nationalizing IPC.

Meanwhile, Prado began negotiating an agreement with IPC under which the company would guarantee a 50 percent participation in all profits to the government, as well as promise to reinvest a percentage of its earnings in Perú. Finally, both parties agreed that the government would draft a bill for the "adaption" of La Brea y Pariñas to the uniform petroleum law of 1952.

Responding to the outcry in July, 1960, Prado named Prime Minister Beltrán to head a commission to investigate the entire IPC question. The commission concluded that the subsoil should never have been alienated, that the 1922 Laudo was "defective," but that nationalization should be a last resort.

In September, 1960, the Prado government submitted new legislation, based upon Beltrán's report, which provided for "progressive nationalization" of La Brea y Pariñas. IPC would pay $3.5 million in cash and surrender subsoil rights to all areas in which it operated. The appearance of this bill produced a new storm of heated rhetoric from both sides to the issue. IPC management, concerned about this latest tempest, borrowed two members of Jersey's staff, Paul E. Morgan and G. W. "Bill" Potts, to review the company's public relations in Perú. Morgan and Potts spent two weeks in Perú during May, 1961. While there, they interviewed persons ranging from political party leaders to workers in Talara, Lima and Arequipa. They found the company's attitude to be both defensive and defeatist. The ghosts of British, Canadian and American imperialistic practices prior to World War II still survived every effort to portray the company as concerned and enlightened. After writing a report, Morgan and Potts made a presentation to the IPC board, one so effective that Ken Jamieson, obviously pleased, stated, "Who do you guys think you are? Huntley and Brinkley,* the way you gave that report." IPC Vice President Brisco remarked that the oral report and recommendations on this touchy problem were the best he had ever heard, and in a letter to Jersey President Haider, he moved that the Morgan-Potts team be sent to check the operation of other affiliates.[16]

Evidently the continuous efforts of IPC (and Jersey) to alter its reputation proved unrewarding. The Peruvian military became concerned about the need to be identified with nationalistic feelings, and proclaimed that the Laudo should be declared null and void. In June, 1962, ten days before the end of his legal term in office, Prado was ousted by leaders of the armed forces. The new government called for elections. The APRA party, which for years had been involved in opposition to the military, won, and the government annulled the results. Another election resulted in the selection of Fernando Belaunde Terry, and he was inaugurated President in July, 1963.

Belaunde, a young reformer, belonged to the Acción Popular party, which had advocated immediate expropriation of IPC properties in 1960. Upon taking office, however, he steered a much more flexible course. In his inaugural address he promised to negotiate a mutually acceptable relationship with the IPC within ninety days. The company refrained from public discussion of the issue while, during the next two weeks, amicable negotiations were conducted. IPC proposed to release its subsoil title in exchange for exclusive rights to explore and develop 247,000 acres of oil property for thirty-five years. The company would be taxed on a Venezuelan-inspired

*David Brinkley and Chet Huntley composed a popular team of NBC television news reporters at that time.

formula, paying a 12^1/$_2$ percent royalty in crude oil, as well as a 2 percent ad valorem tax. In addition, the company would make certain cash payments to the government immediately. This package proved more favorable to the government than that offered under the 1952 law.

Once again, Belaunde operated under heavy pressure, just as had his predecessor. In particular, *El Comercio* as well as the extreme Left attacked his conciliatory oil policy. Apparently responding to these pressures, on October 17, 1963, the Belaunde government submitted an unexpectedly harsh counterproposal. This formula included a revocation of the 1922 Laudo and payment of $50 million in "back taxes." The government also proposed a 12^1/$_2$ percent production royalty, a 33^1/$_3$ percent export tax, an annual surface tax of $250,000, and an income tax rising gradually from 60 percent to 70 percent. The government package also included a twenty-year contract and mandatory reinvestment of 50 percent of company profits. According to company calculations, this formula would result in deficit operations. When IPC refused to accept the proposal, Belaunde became increasingly menacing, recommending that the Congress either impose his formula or nationalize company property. At this point, IPC took the offensive. In a joint statement released by Brisco, who had become IPC president in the United States, and General Manager Lorne Smith in Perú, the company denounced the new government plan as confiscatory, and reiterated that the company owed no back taxes. This further outraged Belaunde, who "admonished" Smith and announced that the company statements were "false and tendentious affirmations."[17]

On October 30, Smith issued a public statement that he had been unjustly admonished for stating the truth. He also ran full-page advertisements in Lima newspapers stating IPC's interpretation of the case. In addition, International employees in Lima held a luncheon at which they honored Smith and released a public letter supporting the company position.[18]

At this juncture, the U.S. State Department began to figure prominently in the IPC case. Under President John F. Kennedy, that department and its representatives in Lima had merely urged that the conflicting parties resolve their differences in mutually satisfactory fashion. When Lyndon B. Johnson entered the White House, however, he named a fellow Texan, Thomas Mann, as Assistant Secretary of State for Latin America. Belaunde viewed this development with suspicion. In fact, the department did suspend all foreign aid to Perú late in the year, citing only "red tape" and delay in explaining the action but never specifically mentioning IPC. Although IPC apparently had no role in the decision to curb aid, Belaunde regarded the action as company-inspired; in turn, he became more suspicious of IPC.[19]

By early 1964, IPC and Belaunde had reached an impasse. Congress shelved the President's bill and passed new legislation authorizing him to "make suitable arrangements to resolve pending matters." With this, a new cycle of negotiations began. Brisco and, on occasion, Haider, participated in the new round of discussion, but the principal negotiations were conducted by Fernando Espinosa, now IPC general manager. Espinosa, Cuban-born, urbane and tactful, was ideally suited to mediate International's problems in Perú. He had already helped negotiate the solution to a potential crisis for IPC in Colombia. Espinosa quickly gained the President's confidence and ultimately held some sixty conferences with Belaunde. By late July, 1964, the two men reached an unofficial agreement. The terms on which they agreed provided that IPC exchange its subsoil rights for a cancellation of alleged back taxes. A combination of taxes would result in approximately 65 percent government participation in profits. The company would also receive twenty-five-year operating and producing concessions on its oil properties and its refinery.[20]

Once again Belaunde's attempt to settle the dispute met with sharp opposition from *El Comercio* and from politicians opposed to him. Espinosa served as a mediator between Belaunde and some of his political antagonists, but the criticisms eventually had the effect of strangling the negotiations. Moreover, Belaunde had come to believe that Jersey would convince the State Department to resume foreign aid payments if he stalled discussions with IPC, and he made resumption of such payments a condition for settlement of the oil affair.

When in December, 1964, Belaunde finally submitted a revised proposal to the company, it was far different from the one that he had drafted with Espinosa. The Belaunde plan left the La Brea y Pariñas subsoil rights intact but he did not waive back taxes as had the previous agreement. Also, the new plan required 90 percent government participation in profits. Company operations in Perú would be directed by a five-man commission to which the government would appoint three members. Predictably, IPC expressed "surprise and consternation" and rejected the new proposal.[21]

By 1964 and 1965, IPC's hopes for a satisfactory settlement with Belaunde had been dashed. The company found itself struggling against an increasingly bellicose Peruvian nationalism. Furthermore, IPC's major historical defense against nationalism—Perú's dependence on it for oil supplies—was now rapidly disappearing. In 1957, IPC produced 95 percent of all Peruvian oil on its Lobitos and La Brea y Pariñas properties. By 1968, IPC only accounted for 30 percent of all Peruvian production. One competitor, the Belco Corporation, equalled IPC in crude production. Simultaneously, IPC had lost its former preeminence in refining. In 1960, IPC processed 95 percent of all Peruvian crude oil, but by 1967 Chevron [Standard Oil Company of California] was running a 10,000 b/d unit, and the Peruvian state company had nearly completed a 20,000 b/d refinery. This trend had two major effects: it reduced the government's reluctance to nationalize, and it confirmed the belief of some Peruvians that IPC had intentionally limited exploration and production efforts in the past.[22]

During the years from 1965 to 1968, the IPC affair became a cause célèbre as well as the subject of a growing volume of polemical nationalistic literature, even though considerable support remained for IPC as an operator—if not owner—within the country.[23]

Paradoxically, the erratic actions of the U.S. government probably contributed to the clamor for expropriation. During 1964 and 1965, Perú received virtually no U.S. foreign aid. Assistant Secretary of State Mann "deobligated" previous commitments to Perú and made no additional infusions, presumably hoping to force a settlement of the IPC affair. In late 1965, Lincoln Gordon replaced Mann in the State Department. Gordon quickly concluded that the policy had yielded no positive results, and he restored aid until 1967, when another consideration became more important. American policy by the late 1960s was based upon a premise that Latin American countries should spend their defense budgets on counter-insurgency equipment rather than on high performance aircraft. In 1967, the Peruvian military sought to purchase supersonic jet fighters in the United States. When they were refused, the Peruvians bought Mirage V aircraft from France. In retaliation, the State Department once again curtailed aid to Perú. This episode antagonized many Peruvians, who became convinced that the State Department had acted at the behest of IPC . The impression was so widely accepted that the U.S. Ambassador to Perú, John Wesley Jones, was dubbed "Mr. IPC " in a Lima periodical.[24]

Pressure upon Belaunde to expropriate IPC properties became intense from this point, for IPC opponents, including technocrats who envisioned government operation of the oil industry, attempted to drive a wedge between the company and other firms operating in Perú, arguing that the case was "unique" because of the subsoil question and because of alleged bribery by the company in the early twentieth century.[25] On July 26, 1967, Belaunde found it expedient to sign a nationalization law which had originated in the Peruvian Congress. This permitted the company to "temporarily continue operation" while it appealed the law in Peruvian courts. Company appeals proved ineffective; the courts ruled that IPC would have to return over $144 million in profit earned "illegally." This figure represented the government's estimate of all IPC profits for the past fifteen years, the period covered by the Peruvian statute of limitations. On November 25, the government demanded an additional payment of 26 million soles ($1 million) in "complementary taxes."

While the company continued its appeal, two other factors apparently saved it from immediate expropriation: first, the state petroleum agency, Empresa Petrolera Fiscal (EPF), seemed reluctant to assume responsibility for the company's complex operations; second, Ambassador Jones began to take a more active role in urging Espinosa and Belaunde to resolve their differences. Finally, on July 25, 1968, IPC accepted Belaunde's proposal that it waive its subsoil rights in exchange for a quitclaim on alleged debts, and operate its facilities on a service contract basis. But the two men could not agree so easily on the details of the settlement, specifically the price at which the IPC-operated refinery would accept La Brea y Pariñas oil now owned by the government.

In August, 1968, Espinosa finally agreed to the Belaunde plan, apparently after Jones had convinced company management that no better alternative existed. On August 13, the agreement was signed in Lima. The President, some cabinet members, and the presidents of both legislative chambers immediately flew to Talara as a political gesture. The agreement provided that the EPF would sell La Brea y Pariñas oil to the IPC refinery at $1.97 per barrel and that the government would take over all subsoil rights.

For several weeks there was almost general rejoicing, but it was short-lived. Part of the agreement concerned the relationship between IPC and the government oil company, EPF. The document developed separately for this purpose did not specify the prices of some IPC services to EPF, nor did it establish the date of transfer of properties to the government. Moreover, less than a month after signing the act, the president of EPF alleged that the last page of the document, on which he had pencilled the dollar prices EPF expected to receive from IPC, had been suppressed or destroyed. The Belaunde government and IPC both stated that the "missing page 11" had, in fact, never been a functional part of the document. This question, never resolved satisfactorily, provided a focus for the Peruvian press and politicians, resulted in increased pressures upon Belaunde to confiscate IPC holdings outright, and offered an opportunity for military intervention. On October 3, 1968, the military, under the command of General Juan Velasco Alvarado, seized upon the IPC case as a pretext to stage a coup against Belaunde. The new military government immediately revoked the agreement and on October 9, deployed one thousand troops to occupy the producing fields, the refinery, and the facilities at La Brea y Pariñas. In January,

1969, Velasco took over the remainder of IPC properties, including its offices in Lima.[26]

After the expropriation, the principal remaining questions involved the level of compensation to the company and the possibility of U.S. government action. Jersey did regard the expropriation as an "abrogation" of the Act of Talara, but apparently realized that the action of the Peruvian government was irrevocable.[27] IPC claimed that it had lost approximately $108 million in property in Perú. The Peruvian government, however, filed a counterclaim of some $690 million which it felt represented the value of all crude oil "illegally" lifted since 1924, when IPC had become the owner of La Brea y Pariñas.

At first, the U.S. State Department approached the expropriation problem cautiously. On November 8, Ambassador Jones issued a statement that the United States might make representations on the company's behalf or impose sanctions if Perú did not compensate adequately for property seized. At the same time, Jones recognized Perú's sovereign right to expropriate. The Velasco government pronounced the expropriation an internal affair and, on November 27, began to marshal public opinion against possible U.S. action. Velasco sent agents throughout Latin America, attempting to tie the matter to the broader issue of inter-American "economic aggression."[28]

By late 1968, Peruvian press releases suggested that the United States would reduce Perú's sugar quota and invoke the amendment [1962] of Senator Burke Hickenlooper that prohibited foreign aid to countries that nationalized or excessively taxed U.S. foreign corporate property. Hickenlooper himself argued that the amendment to the Foreign Assistance Act was mandatory and that the government would have to invoke it against Perú. Apparently in response to that threat, the Peruvian government agreed to participate in direct conversations with American officials. Richard Nixon appointed a special emissary, John Irwin, to negotiate the matter. Irwin held inconclusive conversations with Peruvian representatives throughout the spring of 1969, in Lima and in Washington. Shortly thereafter, Secretary of State William Rogers announced that action under the Hickenlooper Amendment would be "deferred" until IPC had pursued its case through the Peruvian appellate process.[29]

That the American government never took conclusive action in the IPC matter can be traced largely to then-current inter-American political considerations. The expropriation of IPC occurred about the time President Nixon had planned to dispatch Governor Nelson A. Rockefeller on a goodwill mission through Latin America. Shortly before August 9, the final deadline for action under the Hickenlooper Amendment, the Peruvian government denied IPC's final appeals. Velasco contended that IPC had never legally owned the La Brea y Pariñas estate and stopped payment on a check that he had written ostensibly to compensate the company. The U.S. State Department chose not to react and apparently closed the case.[30]

IPC's experience in Perú stands in sharp contrast to the history of other Jersey affiliates in Latin America. When it first acquired La Brea y Pariñas, IPC acquired two important legacies: a conservative British social and labor policy, and a unique title to the subsoil. Both served IPC well during the early years of its operation, but after World War II each became a distinct liability. Despite the company's sincere and continuous efforts to alter its image in Perú, IPC became a symbol of everything Peruvian Nationalists disliked about foreign enterprise in their country. Nationalism thus became the principal consideration, and by the 1960s, many Peruvians would not accept any solution short of getting back the subsoil rights.[31] Whether or not IPC could have forestalled expropriation by moving more quickly to reverse its position is problematical. President Belaunde did express the opinion, after the confiscation, that the company would not have been expelled from Perú had it released its hold on the subsoil in the early 1960s,[32] ignoring the fact that the company had repeatedly offered to do precisely that.

In February, 1974, Ambassador James R. Greene was sent as a special emissary by President Richard Nixon to negotiate with Perú on all problems that had developed since 1971, including, of course, the IPC issue. The ensuing conference resulted in an agreement under which Perú made a cash payment to the U.S. government of $75 million and released frozen funds of a similar amount. The U.S. government took the position that it was entitled to determine the distribution of funds to U.S. companies as it saw fit. Esso Standard [Inter-America] Inc., now owner of 99.45 percent of IPC, received $23 million.[33]

Chapter 14

Jersey in the Middle East

IN THE YEARS BETWEEN the U.S. Supreme Court ordered dissolution of the old Standard Oil Company (New Jersey) [Jersey] in 1911, and 1950, the Jersey Company had always been crude oil short. The year after the dissolution, Jersey had product sales of 125,000 barrels per day [b/d], but production of only 10 percent of the quantity to support these sales. From that time until 1950, Jersey production did not supply the quantity of oil and products it sold—it remained crude-hungry—nor did the company in all those intervening years ever pioneer any important crude discoveries, or production.

Initially, Jersey could satisfy its market demands with crude purchases in the United States. The desire for a guaranteed supply, however, led it, after World War I, to purchase the majority interest in the Humble Oil & Refining Company. Later in the next decade, with the backing of the United States government, Jersey acquired a 11⅞ percent equity in the Iraq Petroleum Company [IPC].* Shortly thereafter it also acquired the controlling interest in the Lago Petroleum Corporation in Venezuela.** All of these companies later developed spectacularly, to provide Jersey with a large portion of the crude it needed.

Like most major oil companies (and most Western governments), Jersey did not anticipate the demand for petroleum that came after World War II. This judgmental error, plus U.S. government concern over a possible oil shortage, provided the company with additional reasons to seek crude. It also became evident during the war that a large proportion of the world's unutilized oil resources were located in the Middle East. Jersey had traditionally supplied its European markets with production from the United States and Venezuela, but the phenomenal increase in demand in the former country also caused the company to look for supplies in the Middle

---

*Throughout this section, IPC is used to refer to Iraq Petroleum Company and not to International Petroleum Company.

**The story of Jersey's acquisition of a share in these last two companies is told elsewhere in this volume.

East. Jersey President, Eugene Holman, seized the opportunity in December, 1946, agreeing in principle to acquire a 30 percent share in Arabian American Oil Company [Aramco]. At the same time, Jersey Vice President Orville Harden announced another agreement in principle to purchase from Anglo-Iranian Oil Company, Limited [AIOC—after 1954, British Petroleum Company, Ltd., BP], 800 million barrels of Iranian (and Kuwaiti) oil over a twenty year period. Both pacts received confirmation later and oil obtained under these agreements largely went to supply Jersey's European markets.

Still the maw of Jersey markets remained unappeased—the quantities of oil received from all affiliated companies proved inadequate to keep up with expanding sales. The 1954 acquisition of a 7 percent interest in Iranian production, however, did help ease the Jersey supply problem. In 1959, Howard W. Page reached an agreement with BP for 215 million barrels of oil over twelve years. In 1966, just before the Harden contract expired, Smith D. Turner negotiated a new one (to replace both the one ending, and Page's 1959 contract), enabling Jersey to purchase 2.1 billion barrels of BP supplied oil from Iran, Kuwait and Nigeria over a fifteen year period. [This contract was terminated in 1979, when 81 percent (app.) of the volume had been delivered.]

In addition to the above-mentioned purchases and contracts, Jersey rather consistently purchased large volumes of Aramco crude from its partners. These volumes were in excess of those actually lifted over the company's 30 percent share. And on occasion, Jersey purchased oil from other sources.

The common rationale behind all this crude acquisition by Jersey was the aggressive expansion of product sales. It acquired any really cheap crude oil available to meet the increased demand. Conversely, some producers had more reserves than they could utilize, while others possessed a goodly supply of crude and needed capital for expansion. Too, where concession contracts specified minimum liftings, some producers needed another outlet. This, Jersey gladly provided.

Through actions enumerated above, Jersey acquired a position of being able to replace, from the Middle East, dwindling supplies for Europe previously obtained from the Western Hemisphere, and, eventually, to ship Mid East crude to the U.S.—even to Texas. [The table at the end of chapter 15 gives figures on the results of this program, from 1952-1975, and serves to illustrate the build-up of Jersey's Middle Eastern supply position. Note that the 1975 volume was almost eight times that in 1952.]

In 1950, the Iranian revolution occurred, bringing to power the Nationalist, Mohammed Mossadegh. Under his leadership, Iran unilaterally nationalized the local oil industry, and cancelled the AIOC concessions. In the resulting snarl, much needed Iranian oil virtually disappeared from world markets. Finally, in 1954, at the request of the United States government, Jersey and other companies entered an Iranian oil producing consortium that eventually brought a twenty-year solution to the political turmoil in that country.

The irony of the situation, meanwhile, stemmed from conflicting actions within the U.S. government which, through the Department of Justice, had long been accusing Jersey as well as other oil majors of various antitrust violations—taking specific legal actions in 1942 and 1952. Conversely, the exigencies of the Korean War, the Iranian revolution, the several Suez crises and Arab-Israeli confrontations, required that America supervise an organized and orderly approach to the task of providing oil supplies where and when they were needed. Thus, the U.S. government requested that the oil companies—on several of these occasions—join together to do the job. The resulting combinations of oil majors, the Departments of State and Defense believed to be absolutely necessary for purposes of national security while the Department of Justice continued its skeptical course. Hence, it appeared that some government departments worked at cross purposes with others. For Jersey, at least, while the needs of national security, and world peace proved imperative, the shadow of the antitrust enforcers remained ever present.

The following sections provide only part of a complex history, one that reflects the complexity of an age that defies any full summary. Here are some of the details of this tangled story of oil and war, national and international politics, and legal actions, together with an account of their effect on Standard Oil Company (New Jersey).

Chapter 15

Jersey, Antitrust, Iran and Stanvac

IN MARCH, 1942, the exigencies of World War II led the United States Department of Justice [Justice] and the Standard Oil Company (New Jersey) [Jersey] to agree to resolve the pending civil antitrust action through the signing of a consent decree. The decree provided that "without trial or adjudication of any issue of fact or law herein and without admission by any party in respect of any such issue . . . ," Jersey should release certain patents which Justice contended had been used to create a patent "cartel." The Jersey Company also pleaded *nolo contendere* to charges in a companion antitrust action that it had violated the antitrust laws of the United States, and it paid a number of small fines to settle that suit. In so doing the company and its executives in no way admitted any violation of any law.

When in 1952, the Justice Department instituted a grand jury proceeding to investigate possible violations of the Sherman Antitrust Act of 1890 and other antitrust laws by Jersey and other international oil companies, Jersey, confident that it had adhered to the agreed settlement, appeared ready to contest this action. But again the national interest, as judged by the Truman and Eisenhower administrations, led to a change from possible criminal action to a civil suit in 1953, and ultimately to another consent decree in 1960.

Between the first decree and the 1952 investigation of Jersey and other companies, ten years elapsed. Company executives claimed that they had for these ten years followed the 1942 Decree to the fullest, in letter and in spirit; that every affiliate had been warned against violating its provisions; and that Edward F. Johnson, the company's general counsel, had monitored compliance with the decree each year. The 1952 proceedings seemed to some Jersey executives merely a new form of harassment, for despite all of the company's precautions, a revival of government legal action had ensued. These men loudly protested their innocence of all wrongdoing, and they firmly believed what they said. The Justice Department in the 1953 civil suit contended that the company had continued to violate the antitrust statutes during and after World War II. It further alleged that Jersey, in concert with other major oil companies, had "cartelized" operations abroad, in clear violation of these statutes. *

Jersey's directors and other executives believed that the Consent Decree of 1942, negotiated with the U.S. Department of Justice in the I. G. Farbenindustrie [I. G. Farben] case, had settled for the time all "cartel" and antitrust charges. With that episode resolved, Jersey turned its attention to the production of war materials for use by the United States and her allies during the second World War. Certainly the wartime record of the company offers substantive evidence that it did strain all facilities and resources to promote the Allied Victory.[1] Nevertheless, at least one astute observer later remarked that there really had been no sound basis for assuming that antitrust pressures would lessen; rather, at that time [1942] he assumed that the interest of Justice in Jersey's operations would continue after the war. His opinion proved to be prophetic.[2]

In fact, new antitrust and "cartel" issues began to arise from the decree. Attorneys in the Antitrust Division of Justice sought to document the operation of the so-called Achnacarry [AS IS] Agreement of 1928 and of the cooperative working arrangement, the Draft Memorandum of Principles [DMOP], supposedly drawn up by the world's largest oil companies. Too, the requirements of war had created a demand for expert oil men acquainted with the nuances and niceties of the international oil trade. This need brought to Washington, and to the Petroleum Administration for War [PAW] many oil company executives. But antitrust lawyers, congressmen, and others were suspicious of petroleum experts; they believed that the PAW representatives from oil companies wished to use the wartime grant of practical immunity from antitrust laws to gain postwar advantages for their parent companies, especially in competition with the independents.

In December, 1943, Attorney General Francis Biddle wrote to Assistant Attorney General Wendell Berge, inquiring about possible violations of antitrust laws found in the various directives published by PAW. Berge, in reply, noted that these orders depended on joint action, but that individual sanctions such as the withholding of transportation permits could be used to enforce them. He indicated his belief that "patterns of both Government and private operations are being crystallized which may be urged as a basis

---

*In 1978, Professor Burton I. Kaufman in his monograph, *The Oil Cartel Case* (Westport, Connecticut, 1978) examined the 1952 case. He obtained considerable information through the Freedom of Information Act. He found no conclusive evidence of oil power dictating national policy; rather he observed that the interests of the United States, as viewed by its national administration, and the interests of the international oil companies frequently intermeshed; both pursued similar objectives but for different purposes.

for a permanent system of business . . . the maintenance of a complete status quo in the petroleum industry."[3] Thus, the petroleum administrator issued varying directives, and members of Congress and others questioned whether these directives did not sustain the prewar "cartel" arrangements. Judging by internal departmental memoranda, lawyers at Justice continued to monitor—and properly so—all such combinations effected during the war under the PAW tent, involved in the production, refining, marketing, and transportation of oil. It is quite evident, too, that some of these Justice attorneys had not approved the issuance of the 1942 Consent Decree, and that they expected to develop another major antitrust suit involving Jersey.[4]

Edward F. Johnson, a Jersey attorney (general counsel, 1945), and other company counsel continued to explore with Justice and with the staff of the Alien Property Custodian the questions growing out of the I. G. Farben case. Johnson indicated to the Assistant to the Attorney General, James P. McGranery, that Jersey would like to see all these issues resolved "in the best interest of the American public." He added, "The welfare of our company obviously depends upon the welfare of our country."[5] Yet, after digging into the records, some Justice attorneys were anxious to begin a new grand jury investigation of Jersey.[6]

Surviving correspondence indicates that the existing International Aviation Associates [Intava] may have played the largest role in persuading the Justice Department that Jersey and the Socony-Vacuum Oil Company [Socony] had violated both the 1911 Supreme Court Decree and the antitrust statutes by acting together to form Intava in 1936.[7] In January, 1946, Frank W. Gaines, Jr. of the Justice Department Antitrust Division wrote to Herbert A. Berman of that department concerning the Intava investigation:

> It may be when all things are considered, . . . a reasonable arrangement and that the intentions of the parent corporations are good. . . . However, . . . both of the parents have had unpleasant collisions with the anti-trust laws in the past; [and] . . . they are guided by expert counsel.[8]

Aside from Intava, the search for evidence against Jersey also went forward. On September 18, 1946, Gaines wrote to Robert A. Nitschke, another Justice attorney, about a letter found in the files subpoenaed from Jersey in 1941, "It seems to me that this letter may be the opening wedge to a full investigation of the celebrated 'AS IS' agreement."[9] On October 4, 1946, Gaines wrote another memorandum to Nitschke stating that "from the information at hand it appears conclusively" that Jersey participated fully in the "cartel" created by the Achnacarry and subsequent agreements, that some of these "cartels" extended throughout the war and "are being re-established." He related to Nitschke that antitrust lawyers derived their information "primarily from three sources: (1) the Standard-I. G. investigation . . . ; (2) the current Intava investigation; and (3) statements and documents given to us in confidence by a former employee and minor official of some of the foreign subsidiaries of Jersey." Gaines explained that Jersey Assistant Counsel Thomas W. Palmer had advised him that any "cartels" in which Jersey had an interest at the time of the consent decree "were local in character and were immediately dissolved" but that in view of the other material he had found, Palmer's statement could "hardly be relied upon." Gaines recommended to Nitschke that a staff be selected to explore the matter fully "because of the strong evidence of a conspiracy already at hand."[10]

On the same day, W. B. Watson Snyder, also of the Justice Department, wrote to Nitschke supporting the Gaines recommendation and citing as one reason for his support an opinion that the agreements under which Justice exempted certain actions of PAW committees during the war had been violated. This "amounted" to an "official rubber stamping" of the prewar "cartel" arrangements.[11]

Gaines and Watson Snyder found a receptive listener in Berge. In November, 1946, Berge wrote a memorandum for Attorney General Tom C. Clark, which reviewed the material gathered by his colleagues. He concluded with the recommendation that "the assistance of a federal grand jury for further investigation of the above described matters is desired." He also asked for authority to submit the information collected on Jersey and the other major oil companies to such a body impaneled in the U.S. Court for the Southern District of New York (or some other district).[12]

Jersey Attorney Palmer responded to a request from Gaines in November, 1946, stating that Jersey did not have (nor did they have in 1942) copies of the agreements desired. He added that wartime controls in foreign countries had ended all voluntary trade arrangements in which Jersey subsidiaries may have been involved. These agreements had not been revived, said Palmer.[13] Subsequently, Jersey Director R. T. Haslam wrote a memorandum to Attorney General Clark covering the points of their discussion of December 6, 1946. In it he reviewed the petroleum situation in Latin America and Europe, country by country, pointing out all of the trade controls deriving from various governments' regulations. Some person at Justice noted on the front of this letter, "Watson Snyder, Does this indicate any violation of Federal law?"[14] Clearly, the Jersey Company sought to supply information to the Attorney General and his staff as a means of forestalling another antitrust suit. Records indicate that other major oil companies also attempted to follow the same strategy. Concerning The Texas Company,* Gaines wrote Nitschke, "Obviously they would be unwilling to give us a straight story voluntarily since they are in the same cartel with Standard, et al."[15]

Perhaps the continued correspondence and visits with antitrust lawyers influenced Jersey's decision to break up Intava, for the company could not have been unaware of the antitrust lawyers' interest in its operations. On March 3, 1947, the company announced the termination of the joint arrangement with Socony that had created Intava and indicated that it would buy out its partner. The dissolution was completed by 1948.[16]

However, another move led by the Jersey Company to have the Red Line Agreement dissolved [a July, 1928 arrangement specifically made in connection with Jersey's entry into the Iraq Petroleum Company] and to purchase (with Socony) shares in the Arabian American Oil Company [Aramco, covered elsewhere in this volume] soon came to the attention of the Justice Department. In 1946, Jersey "agreed in principle to purchase vast amounts of oil from the Anglo-Iranian Oil Company and to join AIOC in building a pipeline from the Persian Gulf to the Mediterranean"; however, the agreement was modified later, and the line was never built. Later that year, Jersey and Socony announced a potential agreement with Aramco in which the former two companies would purchase "thirty percent and ten percent, respectively, in the third."[17] While Jersey had kept the Department of State and the Department of the Interior informed of all negotiations, the projected acquisition was not formally called to the attention of Justice until late in 1946. Some attorneys there resented the delay. Watson Snyder wrote a memorandum to Berge on January 8, 1947, indicating that Creekmore Fath, Middle East advisor to Department of Interior Secretary Julius A. Krug, had suggested that Justice "ask Mr. [Edward F.] Johnson" to submit copies of all the contracts entered into by the international companies regarding Middle Eastern oil. At the same time, Watson Snyder remarked that on December 10, 1946, he had been informed that the oil companies had given "some information" to the Attorney General, probably requesting some form of clearance. But he did not know whether any response had been made.[18]

Berge, on January 10, 1947, exchanged memoranda with Watson Snyder. The former indicated that he knew the State Department approved of the Middle Eastern oil transactions. Watson Snyder reported to Berge that he

---

*Renamed Texaco Inc. [Texaco] in 1959.

had called Assistant Attorney General H. Graham Morison, who insisted that no approval "had been given by that office." Morison indicated his opinion that the State Department "is making a serious mistake in backing these oil agreements." Then, he told Watson Snyder "that this Department should not involve itself by asking for information from the companies involved because he felt we were in a better position to later prosecute if we did not have all the information in advance."

Watson Snyder told Berge that he disagreed with Morison because "most of us entertain the view that the new Red Line Agreements are . . . carrying out an existing world cartel between these companies." The law, as he interpreted it, forbade such combinations in both foreign and domestic commerce, and the "Standard Oil Company [meaning Jersey] is already engaged in broadcasting press releases intended to give the impression that complete governmental approval has been given to their activities." Watson Snyder then enclosed a draft letter "to Mr. Edward F. Johnson asking him to come to Washington and discuss the matter with us."[19]

A few days later, Watson Snyder examined the State Department records on the entry of American companies into the Iraq Petroleum Company and the Red Line Agreement. He made a memorandum concerning the salient points. Basing his judgment on this document, he opposed the impending Aramco arrangement even though he had not yet examined the contracts or discussed the matter with State Department officials.[20]

Harry T. Klein, president of The Texas Company, sent to Attorney General Clark copies of the Aramco-Jersey-Socony agreements on March 12, 1947. Watson Snyder and Gaines studied them carefully. The former, in a longer memorandum, pointed out that in his view the various subscription and offtake agreements constituted part of "a chain of evidence which shows a violation of the decree of 1911." He cited Section 6 of that document, which forbade the old Standard Oil companies in combination from making "express or implied agreements" concerning the "price or terms of purchase" or sale "of petroleum or its products in interstate or international commerce." Next, he recounted the various combinations Jersey had entered—in his opinion in violation of this decree. In his list, he included the Standard-Vacuum Oil Company [Stanvac], a jointly owned Jersey-Socony venture in the Far East. After evaluating the reasons offered in support of these several combinations, he confronted the question of Aramco: "now the Jersey, California [Standard Oil Company of California (Socal)], Socony, Aramco deal is being explained away on the ground that California and the Texas Company are not financially able to build pipelines in the Middle East." Hence, the latter two companies required the financial resources of Jersey and Socony "to aid them in fully producing Saudi Arabia and moving the production to market." Watson Snyder viewed the prized Aramco concession, owned by California and Texas, as being in competition with Jersey-Socony joint operations:

> The claim has been made, and the Honorable Thurman Arnold [formerly Assistant Attorney General] has stated that he had evidence to prove it, that Standard Oil Company of California directors are compelled to carry out the bidding of Rockefeller and Jersey on the threat that the latter [?] can muster enough proxies to throw out such directors on their failure to comply. The control of Socony and Standard of Indiana [sic] by Rockefeller interests is a matter of public record and the evidence from Mr. Arnold would explain why California was forced to take Jersey and Socony in as partners. Both California and Texas are very rich companies and it is not conceivable that they needed finances from other sources to carry on their operations in the Middle East.[21]

Just how many Justice Department attorneys agreed with Watson Snyder and Arnold (if quoted accurately), is not known. But in that branch of the government there were others who agreed with the view that the Rockefeller interests—using Jersey—manipulated the entire Aramco transaction.[22]

Nitschke reviewed all the evidence for Morison on January 31, 1947. He admitted that he did not know "positively" whether local "cartels" had been "revived since December 31, 1945, the date on which their 'clearance' from PAW expired." Then, as a consequence of "Standard and Shell . . . conducting meetings in regard to the AS IS situation," he deduced that "it is logical to assume that the component parts . . . are occupying their old places in the overall conspiracy." He believed that the new Middle Eastern arrangement had been arrived at through "prior understandings," and joined with his colleagues in recommending a grand jury investigation. One of the readers, [probably Morison], scribbled hard questions all over this memorandum: Why had Justice not been able to get copies of either the AS IS or DMOP Agreements? What "considerable information" had been uncovered indicating more exactly the nature of the conspiracy? How do we know the unsigned "so-called draft" memorandum was treated as a "binding agreement" by the Royal Dutch-Shell Group [Shell] and Jersey? The writer marked passages of the memorandum with the phrase "lawful," and comparable words, indicating that Nitschke had failed to convince Morison that a "cartel" existed. In one paragraph, Nitschke had described how Jersey President Walter C. Teagle had played host to Sir Henri Deterding of Shell and how the two had travelled across the United States in Teagle's private railway car. The questioner commented, "Surely we haven't come to point of challenging American hospitality?" There can be no doubt that all the activity, research and discussion in the Antitrust Division of Justice still failed to convince Morison of the need to impanel a grand jury at that time.[23]

Whether Assistant Attorney General Morison's dismissal of the evidence gathered by the attorneys of the Antitrust Division disturbed or influenced them in their effort to develop a new antitrust or "cartel" case is not known. But Gaines, Watson Snyder, and others at Justice continued their search. In the winter of 1946-1947, the Swedish government investigated the "Oil Cartel" in that country and found some evidence of the DMOP's existence *prior to 1939*. The Swedish report included alleged facts about market allocation, price fixing, and similar non-competitive controls. Gaines secured a copy of part of the report and passed it on to Nitschke. He followed this some weeks later with additional information. Then he raised the question concerning "the wisdom of allowing our larger oil companies to form partnerships controlling most of the important sources of oil production in foreign countries." It is not clear whether Jersey knew of the interest of Justice in the Swedish "cartel," but on August 8, 1947, Director Robert T. Haslam wrote to Attorney General Tom C. Clark:

> Regardless of what might have happened in Sweden prior to the War ... I can state to you emphatically that, to the best of my knowledge and the knowledge of my associates here in New York ... there is no cartel in oil products in Sweden in which our affiliates participate.

Haslam appended to his letter a product sales table covering the years 1938-1946. Apparently this combination served to end discussion of the Swedish "cartel" in the Justice Department.[24]

During the remainder of 1947 and early 1948, the attorneys in the Antitrust Division of the Justice Department continued to be interested in Jersey documents acquired in earlier investigations. They also followed with special interest various news stories concerning the impending Red Line court case [a Jersey effort to have the 1928 agreement declared invalid] and the activities of Aramco. Attorney General Clark also considered whether the Middle East arrangements being effected by the oil companies violated antitrust statutes. In the spring of 1948, he asked Watson Snyder, Gaines, and George P. Comer, to give him their opinions on those subjects "from the point of view of their particular work." This, they did by June 7.

Comer in his "Economic Report on Middle East Petroleum," clarified several issues regarding oil pricing that had confused both Congress and the public. He stated, "Whether an antitrust proceeding against Aramco . . . should be pursued at this time in view of the great need for Middle East oil in world markets, is a policy question on which the Economic Section of the Antitrust Division is not called upon to express an opinion." He did suggest that Jersey and Socony be required to divest their 40 percent holding in Aramco, though he pointed out that the "public interest" might be served by allowing them to help finance the pipeline. To him, the new interlocking of the old Standard Oil companies in Aramco provided opportunities for price fixing, for circumventing the various decrees against Jersey, for discrimination, and for other violations of law.

Gaines reviewed the "cartel" story up to 1948. He admitted that the evidence they had developed "is not entirely conclusive," but he also thought that a grand jury could resolve "some of the details which are as yet unknown." However, he backed away from his earlier position that Jersey (and the other companies) had continuously violated court decrees and the antitrust laws; a grand jury investigation at that particular time, he said, "would be a matter of inter-departmental policy."

In his response to Clark, Watson Snyder traced the history of Middle Eastern oil. He found the "off-take" agreements of the Aramco partners to be a direct violation of "antitrust principles," but he made no recommendation concerning a possible grand jury investigation.[25]

Meanwhile Jersey sought to arrange a settlement of the Red Line case outside the British courts. Working out the details with the other principals to the agreement consumed considerable time, and not until November 3, 1948, did they finish the task. Then, litigation quickly ended, and in December, 1948, Jersey and Socony consummated their stock purchase with Aramco. With this transaction, Jersey acquired additional significant production in the Eastern Hemisphere. Newspaper stories related details about these Middle Eastern involvements, and the memoranda circulated among Justice attorneys indicated their continuing interest in these Aramco oil agreements.

Several of these documents provided the basis for Rudolfo A. Correa's memorandum of January 10, 1949, addressed to Assistant Attorney General Herbert A. Bergson. Correa summarized the Iraq Petroleum Company's operation and pointed out, "These arrangements under this agreement may also present a violation of Section 6 of the Standard Oil decree (1911)." However, he "recommended that no opinion be given with respect to the agreements and that no action be taken concerning possible violations."

This, he believed proper in view of the activities of "the State Department, Department of Interior, Department of National Defense and National Security Resources Board, with respect to international oil."[26]

Jersey Vice President and Director Orville Harden, Arthur Newmyer of Arthur Newmyer & Associates [Jersey's government relations advisor in Washington], and an Aramco attorney, on March 10, 1949, delivered to Attorney General Clark the schedules, agreements, and contracts (with copies of cablegrams attached) entered into by the Aramco partners. Clark asked Watson Snyder to analyze these documents, and the latter went over them carefully. He concluded that no price competition outside the United States and the Soviet Union could be expected in the future when "oil produced from those vast reserves [the Middle East] will be the controlling factor." But once again, Watson Snyder failed to suggest that a grand jury be called.[27]

In June, 1949, Newmyer arranged for Harden and Charles L. Harding, a director of Socony, to confer with several of the Justice attorneys. Bergson, Manuel Gorman, and Watson Snyder represented Justice. Harden, who knew the full history of IPC as well as any person then living, explained the story to the listeners and accounted for the noted Red Line Agreement. The purpose of the Harden-Harding visit, however, was to assure Justice that the "arrangements between the owners of Aramco, did not involve the fixing of 'world oil prices' " and that competition would result in a reasonable price everywhere. Watson Snyder found their arguments "not very conclusive."[28]

All of this investigatory activity at Justice had its counterpart ferment in the Federal Trade Commission [FTC]. Possessing statutory authority to move against persons, partnerships or corporations when it had "reason to believe" them in violation of the laws prohibiting unfair methods of competition, price fixing and other such practices, the FTC had investigated the oil industry several times. In 1944, the commission had decided to conduct a long-range investigation of international "cartels," including what it termed "the most important"—oil. But the FTC did not complete this study. However, the publicity attending the purchase by Jersey and Socony of shares in Aramco stimulated the commission to revive this probe. On December 2, 1949, the commissioners of the FTC adopted a supplemental resolution specifically calling for an investigation of the alleged petroleum "cartel," especially any restrictive agreements made among U.S. based companies and those of "other nations in connection with foreign operations and with international trade in petroleum and petroleum products and of the relationship of such agreements to domestic trade in and pricing of the American petroleum industry." Commission subpoenas went out to the involved oil companies, including Jersey, in January, 1950.[29] The FTC commissioners designated John M. Blair, assistant chief economist of the FTC, to direct research. Later, he claimed to have edited and arranged the *Staff Report*.[30]

After a few days, Justice and the FTC staff, who had been communicating from time to time, joined forces. On February 8, 1950, six persons interested in Jersey's (and other oil companies') Middle Eastern activities met to "discuss international oil practices as such practices might effect [sic] the domestic oil market." Three of these, headed by Holmes Baldridge, special assistant to the Attorney General, represented Justice, and Corwin Edwards led the three FTC representatives. They sought to resolve any potential conflict "between the oil programs of the two agencies." The Justice attorneys indicated that they would pose no objection to an FTC economic study of the major oil companies. However, should the *FTC Report*, when completed, indicate a necessity for legal action, then before making any formal move, the two agencies would again discuss the matter. The Justice spokesmen told those of the FTC that "in view of our extensive oil program, covering a period of ten years, and which was not yet completed," they would object to the FTC instituting any legal process "in the domestic oil field."[31]

The FTC decision to study the oil "cartel" jarred the entire industry. Commenting on this element of surprise, The *New York Times* "Topics of the Day in Wall Street" column pointed out that most of the records sought had already been made public and that the practices complained of either had ended before World War II or had been going on for twenty-five years.[32]

The FTC staff investigation became a matter of great concern to the Justice Department. Their searches of Jersey (and other company) files in 1941-1942 and in 1946-1947 had established for many attorneys there what they believed to be sufficient documentation of activities violative of the antitrust laws. Up to this time, Assistant Attorney General Morison had rejected all recommendations from the Antitrust Division lawyers to move forward in the case because he felt that they had not uncovered sufficient evidence to warrant grand jury proceedings. One of these presentations, he refused to press because it "completely ignores the fact that distribution of oil both foreign and domestic is in fact a quasi-public utility business . . . and is inately [sic] by nature of a monopolistic character."[33]

Now the explosive mixture of fact and rumor surrounding an *FTC Staff Report* demanded general attention. Jersey's Chief Executive Officer [CEO], Eugene Holman, who had made no reference to the then-current FTC investigation in his statement to the shareholders in 1950,[34] responded to the fast spreading stories and gossip. He said that no one in the company had yet seen any such report. Then he reiterated his often-made statement that the Standard Oil Company (New Jersey) belonged to no oil "cartel."[35]

X X X

The Korean War, beginning in June, 1950, represented to many Americans a political extension of the Cold War in an area with which few were familiar. Believing that already their generous aid programs had placed Western Europe on the road to recovery, Americans were more concerned about the success of the North Atlantic Treaty Organization [NATO, to which General Dwight D. Eisenhower had just been appointed as Commander] than about the war in Korea. Nevertheless, in bipartisan fashion they backed President Harry S. Truman's decision to check what he termed "Soviet aggression" in Korea. Yet the task proved difficult, for after the defeat of Japan and Germany in 1945, the great American industrial-military war machine had been dismantled. Gone too were the powerful and sometimes overlapping war agencies, among them the Petroleum Administration for War. The committee that replaced it had only advisory powers to the military; it soon became evident to the administration and to Congress that the situation required new combinations of experts who could guide the nation's wartime effort.

In September, 1950, Congress passed the Defense Production Act, thereby providing a basis for mobilizing oil and other industries. This act authorized the creation of a Foreign Petroleum Supply Committee [FPSC] that included several subcommittees. Bruce K. Brown, president of the PAN-AM Southern Corporation, who had served as assistant deputy administrator of PAW during World War II, was appointed assistant deputy administrator of the Petroleum Administration for Defense [PAD] in the Interior Department. Utilizing his many contacts to bring key oil executives back into service, he set about creating an administrative structure that would help regulate oil production and refining, in the face of the anticipated shortage of jet fuel. PAD committees directed petroleum allocations of all domestic and some foreign oil supplies. But late in 1950, the Korean War changed shape. The entry of Chinese Communist forces broadened the international complexion of the struggle, for at the time, many political leaders and their advisors considered Red China to be politically yoked to Soviet Russia. This view led to an increased concern about adequate petroleum supplies for Western Europe and for U.S. forces stationed there. Then in April, 1951, Iran nationalized oil operations in that country. The Iranian action added to the U.S. government's concern over the adequacy of oil supplies in both Europe and

Korea. Once again the government called on the oil companies to provide expert assistance.

The Interior Department, on June 25, 1951, acting under the authority conveyed by section 4(a) of the Defense Production Act, drafted the protocols of a Voluntary Agreement Relating to the Supply of Petroleum to Friendly Foreign Nations, and on June 29, Interior Secretary and Director of the Petroleum Administration for Defense Oscar L. Chapman invited Jersey to become a member of the Foreign Petroleum Supply Committee.

As Defense Secretary and Director of the Office of Defense Mobilization [ODM], Charles E. Wilson also wrote to Jersey (June 26, 1951) requesting it to join with seventeen other companies in acting under the Voluntary Agreement. This Wilson letter explained that the Attorney General had approved the agreement and cleared it with the Federal Trade Commission "as contributing to the national defense under Section 708(c) of the Defense Production Act of 1950." The Jersey Executive Committee discussed these requests on June 29, and on July 2, 1951, Eugene Holman accepted the invitation for Jersey. Shortly thereafter, Chapman and Bruce Brown persuaded Jersey Vice Presidents Stewart P. Coleman and Emile E. Soubry, to serve as chairman and member respectively of the FPSC. J. A. Cogan and C. L. Burrill, members of Jersey committees dealing with petroleum economics, also joined the FPSC.

Jersey Director Peter T. Lamont secured approval of a letter from heads of several Jersey committees and sent copies of it to all foreign affiliates. He sought to reassure the executives of European affiliates that "Jersey does not anticipate there will be any difficulty in meeting your requirements as presently indicated, except for residual fuel oil." The shareholder representatives were instructed to monitor all supply problems. Lamont pointed out:

> You can appreciate that the U.S Government is vitally concerned in this matter because of the effect which an oil shortage would have upon defense activities as well as the economic welfare of the U.S. and other friendly countries, particularly including Western Europe.

The oil company representatives did their work well, but not before charges had appeared in the press that the Voluntary Agreement merely sanctioned the "cartel" arrangements. Interior Secretary Chapman heatedly replied to some of these allegations in writing to Manly Fleishman, administrator of the Defense Production Administration. Fleishman had received a letter from Attorney General J. Howard McGrath asking for all the documentary material available about the activities of oil companies involved in the FPSC. Chapman stated that he had no material "documentary or otherwise" that would substantiate those charges; that "In sponsoring the voluntary program, I met and dealt with the oil companies which were in a position to help, namely, American oil companies engaged in foreign operations." He added, "I had no more knowledge of the moral character of their prior operations than I am sure that you . . . and the Attorney General had when all of us approved the Voluntary Agreement, Plan of Action No. I." Chapman pointed out "of my own personal knowledge . . . the American companies necessarily involved did not feel that they could safely meet together, even under the direction of their Government . . . except under such assurances as it provided in Section 708 of the Defense Production Act of 1950." He went on, "They had to assure themselves that such cooperative actions as might be undertaken, under Government direction, would not subject them later to prosecution under our antitrust laws." He knew that the companies could quickly obtain all the information necessary for estimating the amount of oil needed in Europe and in the Korean War. In fact they had already done this at the request of the government. He said the Departments of State, Defense and the PAD had sought to prevent the disruption of the economic life and defense efforts of the nations involved; "There was no other way," he stated. Chapman heartily endorsed the work of the FPSC.

The Jersey employees continued to work for the FPSC until the termination of the Voluntary Agreement on April 30, 1953. Some years later Coleman described what had happened: "When the Iranian crisis arose, President Truman caused the creation of the Foreign Petroleum Supply Committee to provide a coordinated effort to replace from other sources the supplies which had previously come from the [now] nationalized oil operations in Iran." Coleman added that this cooperation provided the only way oil could be delivered to friendly European countries and the U.S. military forces stationed there; the surge of anti-American feeling in Europe required remedial action. Dismissing the "cartel" arguments advanced by many Representatives and Senators, he pointed out that the FPSC had advisory powers only.[36]

Meanwhile, Blair and his associates at the FTC completed the *Report to the Federal Trade Commission . . . on The International Petroleum Cartel* and turned it over to the commissioners in October, 1951. As so often had happened, news and gossip about this Staff Report circulated throughout Washington and among the oil companies. President Truman's watchdog committee immediately classified it as a top security document.

Two months after the FTC staff had finished writing, the Select Committee on Small Business of both Houses of Congress published and distributed a report on *The Third World Petroleum Congress*, written by Elmer Patman [one-time lobbyist for the Superior Oil Company], who had been appointed by the committee to investigate oil "cartel" activities.[37] In his report Patman charged that a "cartel" threatened to destroy the industry. Jersey again responded that it belonged to no "cartel."[38] However, the allegations in the Patman report served to stimulate popular discussion of an oil "cartel" and especially of Jersey's role. In late 1951, the *FTC Staff Report*, still not available to the oil companies nor to the public, was sent to those government departments concerned with international business and trade and with foreign affairs. In March, 1952, Watson Snyder wrote Morison, "I hear that at least oral representations against publication have been made by both PAD and State." The Intelligence Advisory Committee met in special session to consider the consequences of the release of the *FTC Staff Report*. At this session the discussion centered around whether U.S. foreign relations would be damaged by the release of the study, and whether it would furnish "the Russians with excellent propaganda material." During the discussion, Central Intelligence Agency [CIA] Director General Walter Bedell Smith stated that "a clique in the Federal Trade Commission . . . are urging that this report in toto be released." The committee agreed on this estimate of the result of publication of the report:

> Official publication of this report would greatly assist Soviet propaganda, would further the achievement of Soviet objectives throughout the world and hinder the achievement of U.S. foreign policy objectives, particularly in the Near and Middle East, and would otherwise tend to injure U.S. foreign relations and strategic interests.

They believed that even selective publication would "only lessen the degree of harm that would result" from any publication. Impressed, the President, for a few days more, refused to allow the report to be made public.[39]

On June 5, 1952, he again blocked the release of the no-longer-secret report. Truman wrote to Senator Thomas C. Hennings, Jr., of Missouri, denying the latter's request to have the report made public. As the President explained, he "concurred with the views of his advisors that the public disclosure at this time of the contents" of the FTC study "might adversely affect our foreign relations and hence would not be in the national interest."[40] Later, Hennings pointed out that Truman had acted on the advice of the Departments of State and Defense.[41]

Yet, both the House and the Senate Judiciary committees continued to seek access. Finally, however, President Truman acceded first to the request of the Senate Select Committee on Small Business that the report be made public. Once they had obtained it, the committee made it a part of the report of its Subcommittee on Monopoly and had the document printed.[42] The *FTC Staff Report* was made available to the public on August 22, 1952. The various allegations in the report, and most of the documents it included contained relatively little new or startling information. Instead, the editors had filled it with details about the opening of the Middle East to U.S. oil companies, about the IPC, the first Red Line Agreement, the AS IS and DMOP Agreements, and Aramco.

Justice, now quite certain that with the *FTC Staff Report* and its own findings it possessed all materials requisite for a grand jury probe, requested that President Truman authorize full-scale court action against the "cartel." The President was caught. He had always advocated the expansion of American business abroad, including oil. Still, as a professed free trader, he objected to international "cartels" and approved of breaking them up. Yet to approve the request of Justice, and to move against alleged oil company collusion in producing, transporting, and refining petroleum while permitting these companies to operate in the same fashion under the Defense Production Act of 1950 (as amended in 1951 and 1952)[43] appeared to the President to be totally inconsistent and antithetical; certainly, the incongruous nature of the situation was not lost on him or his advisors.[44] Then too, events in Iran had created an oil crisis for Great Britain, and the United States sought to guarantee an adequate oil supply for friendly nations. So for the moment, the beleaguered President delayed. But the Department of Justice, now led by Morison, persisted.

On June 23, 1952, Truman yielded to the pressure by signing a memorandum to the Secretaries of State, Defense, Commerce, and to the Federal Trade Commission, stating that he had asked the Attorney General "to institute appropriate legal proceedings with respect to the International Oil Cartel." He requested that those addressed cooperate with Justice "in gathering the evidence required."[45] Attorney General James P. McGranery assigned a top level Justice lawyer, Leonard J. Emmerglick, "one of the Department's most experienced and seasoned attorneys,"[46] to handle the case. Emmerglick had grand jury subpoenas served upon twenty-one large oil companies, including Jersey, requiring them to produce specified documents by September 3, 1952. This action aimed at establishing criminal conspiracy—a charge which if successfully proved could have carried jail terms for those found guilty.

On July 9, President Holman wrote to Jersey shareholders and employees concerning a "secret" FTC *Report* on the alleged "international oil cartel." He pointed out that the report still had not been made available to the company, but that the press obviously had obtained copies. Holman again emphasized that Jersey belonged to no "cartel,"[47] international or other.

On July 17 and 24, 1952, the Jersey Board discussed the *FTC Staff Report* which by this time, although still a security document, had been obtained by the press. Also, on August 12, the Board sought from General Counsel Johnson and his associates an explanation concerning just what material should be produced in response to the Justice Department subpoena. Twelve department heads (or their representatives) attended this session along with six directors. On September 4, after discussing the "cartel" investigation, the Board appointed a "top-drawer" committee to study and to make recommendations affecting company policies.[48]

On September 19, President Holman and Chairman Frank W. Abrams signed another letter to all company shareholders and employees. It stated that a subpoena had been served requiring the company to produce a number of documents dating back to 1928, and that company attorneys had filed a motion to have so broad a subpoena withdrawn or modified.[49] The company executives strongly defended Jersey against both the *FTC Staff Report* and the Justice Department's allegations, concluding their remarks with the statement that "Neither the FTC staff report nor the Grand Jury investigation confronts us with any specific charges of illegal action of any kind."[50]

Holman next wrote a "Letter to Heads of Affiliated Companies" dated September 26, 1952, indicating that a copy of the *FTC Staff Report* had been received by Jersey. He enclosed a copy of the September 19 shareholder-employee letter and attacked the report as largely inferential. He ended with a denial that Jersey belonged to any "cartel."[51]

There is no question that the *FTC Staff Report* and the commencement of the grand jury investigation of the oil companies created both domestic and foreign problems for the PAD, for the oil companies, and for the United States government. The allegation in the report, that the Voluntary Agreement [see FN 43] maintained and sanctioned "cartel" agreements among the major oil companies, ultimately created problems for the PAD staff. As early as January, 1952, representatives of Justice and the FTC had tangled with PAD representatives. Watson Snyder described to Morison what had happened:

The writer and Mr. Corwin Edwards of the Federal Trade Commission made very blunt remarks at these conferences to the effect that approval under the Defense Agreement was in reality amounting to no more than a waiver of the Anti-Trust laws as to the foreign oil cartel arrangements already in existence for the last 23 years. PAD representatives in a very threatening manner took the position that PAD and the State Department desired that everything be done to meet the serious shortages due to the closing down of Iran.

After this meeting, the Attorney General called on the administrator of the Defense Production Administration to keep Justice informed regarding oil company activities under the Defense Agreement. In May, 1952, Bruce Brown wrote Oscar L. Chapman, interior secretary and administrator of the PAD:

> The situation continues to 'stew' in the newspapers. The FTC, either through ignorance or possibly through design, seems to be taking the position that the only objection to the publication of this report as an official Government document is that it would impair the Iranian situation. Nothing could be further from the truth. The Iranian situation is probably past repair. What the publication of the report *would* do is mess up the whole Middle Eastern oil situation in all countries.[52]

Brown and his associates believed that they had enough to do as they tried to alleviate what appeared to be an ongoing shortage of aviation gasoline, especially after the AIOC Abadan refinery closed. But now, they also had to furnish copies of all "voluntary" agreements made, plus minutes of all committee sessions to the Defense Production Administrator, for eventual review by the FTC and Justice.

On September 11, 1952, the Jersey Executive Committee received a report from one subsidiary, the Creole Petroleum Corporation, that its relations with the Venezuelan government had been adversely affected by the "cartel" investigation and by the publication of the *FTC Staff Report*.[53] Representatives of other oil companies pointed out that all the publicity about the concurrent suit and report had provided the Soviets with good propaganda, and filled the Arabs with "suspicion and distrust" of the American oil companies. Public opinion surveys financed by the oil companies had established these facts.[54]

The oil companies wanted an extension of time to comply with the grand jury subpoena; they considered the stipulated date, September 3, 1952, unreasonable, especially in view of the quantity of documents that the Justice Department demanded. Thurman Arnold, the famed "trustbuster" and the one-time Jersey antagonist, now representing the Sinclair Oil & Refining Company, stated that the search of his company's records would cost $300,000. He believed the subpoena to be unreasonable. Hugh Cox and Gerhard Gesell, representing Jersey, agreed. In addition, they told Emmerglick that they needed "time to orient themselves with the industry practices." Emmerglick asked the court for a thirty-day extension and noted "these three companies [Jersey, Socony, and Socal] are commonly related in that the Rockefeller interests are the dominant stockholders."

Compliance with the subpoena presented problems of time and expense, and it also raised, as in the case of Aramco (or other jointly owned companies) new questions about which company had to produce what records.[55] Further, the subpoena included the AIOC, the Royal Dutch-Shell Group and other foreign based companies. The attorneys for those companies argued that their records would be considered "classified" by those countries and indicated that they might litigate the subpoenas. A legal imbroglio ensued. Foreign nations began to issue rulings to companies domiciled within their territories forbidding compliance with the American subpoena.[56]

At a conference in December, 1952, with the representatives of the Departments of State and Justice concerning the question of documents involving foreign nations, Arthur Dean of Sullivan & Cromwell, counsel for Jersey, pointed out that in some Middle Eastern countries plus Argentina and Venezuela, "the situation was extremely delicate and dangerous," and that a request to withdraw documents might "result in cancellation of concessions or permits to do business." Jersey's Howard W. Page, he stated, had recently surveyed "the feeling against foreigners" in the Middle East and observed political demonstrations against "foreign business interests." Jersey, Dean added, did "not engage in business in those countries"; rather, its subsidiaries incorporated there did. Dean said that Jersey preferred that all such requests for company records be made by the Department of State, which later did provide assistance.[57]

X   X   X

As the Justice Department attorneys labored trying to perfect the "cartel" investigation, the *FTC Staff Report* attracted its own share of attention from the advisors to President Truman and, especially, in the Departments of State and Defense. At that moment, both departments were vitally concerned with events in the Middle East, an area which, in the President's opinion, was extremely important to Western defense plans—especially Iran. On one occasion, Mr. Truman placed his finger on Iran on a globe and stated, "Here is where they will start trouble if we aren't careful . . . if we just stand by, they'll move into Iran and they'll take over the whole Middle East."[58] Of course, there were problems for the United States and the major oil companies in other countries as well and, while the President gave his first attention to Korea, the situation in Iran rapidly deteriorated.

The Iranian Majlis* passed a nationalization law on April 30, 1951. [An explanation of this legislative action appears later.] When Shah Muhammad Riza Pahlevi assented to the law, it took effect on May 1. The British protested this unilateral action with no promise of compensation, and the United States, already engaged in Korea, pressed for a settlement. Europe needed Iranian oil, and the high octane fuel from the Abadan refinery figured in plans to defend the free world.

For the United States, Iran had become an important battleground in the Cold War. Most American foreign policy experts believed that—should the Russian influence over that country increase—friendly control of the Middle Eastern oil fields and pipelines upon which the Western World had come to depend would be jeopardized. Historically, the Russians had shown special interest, for Russian troops, unlike those of Great Britain and the United States, did not leave Iran at the conclusion of World War II[59] but only much later. Meanwhile, the occupation of Iran by the Great Powers during World War II actually intensified "resentment against foreign influence" on the part of some Iranians. Later, the attention of the Western powers and of the major oil companies grew increasingly fixed on Iran, as petroleum supplies began to loom larger in Western security plans.

Howard Page, who was in the Middle East much of the time during the Iranian crisis, came to believe that nationalism in Iran differed from "so-called Arab nationalism or Pan Arabism. Iran has been a political entity—a real country for many centuries," he said, adding that nationalism there was confined to its boundaries. Forcing the Russians out of Azerbaijan after World War II, he considered an early manifestation of this Iranian nationalism, and the effort to get rid of the British influence another. Thus, the AIOC became another victim of it—concession terms had not been important. Nothing that the AIOC did would have mattered. Page regarded the Iranian affair as a drama of emerging nationalism—more anti-British with anti-foreign overtones.

---

*The lower house of the Iranian Parliament.

In 1949, the AIOC in a supplemental agreement increased Iranian royalties from four to six shillings per ton and offered a 23 million pound payment on credits accrued. The Shah and many members of the Majlis objected to these terms. In February, 1950, a newly elected Majlis assembled. Within four months, General Ali Razmara became Premier. "Nationalization now" became the cry of street agitators and the mobs who followed them. Razmara sought minor concessions from the AIOC. Despite pressure from the United States, the company would not yield.

One of the influences on this Iranian position derived from Venezuela. In 1943, Venezuelan Oil Minister [Mines and Hydrocarbons] Juan Pablo Pérez Alfonso had engineered the inclusion of a clause in the new income tax law that, when combined with the royalties, supposedly established the government as a 50-50 profit-sharing partner of the oil companies. For a time this law actually slowed down the trend toward nationalization of oil in Venezuela. But soon the Venezuelans had their oil concession agreements and tax laws translated into Arabic. Next, they sent a "diplomatic mission" to spread the word to their competitors in the Middle Eastern countries. Officials in Iran and Saudi Arabia refused to see them, but in Iraq, they did confer with officials in both Baghdad and Basrah.[60] In this and other ways, many ideas about what constituted 50-50 income-sharing spread through the world, along with the news that Venezuela was receiving income tax plus royalties from the companies established there.[61] In most cases, royalties did not change; the increased income for the host nation came from added income taxes on the company profits. Whether prompted by knowledge of the agreements with Venezuela or not, the Saudis and Iraqis certainly sought additional income from both Aramco and IPC.

Jersey's interest in the Iranian crisis stemmed from the fact that Anglo-Iranian Oil Company had been supplying it with crude from that country under the contract negotiated by Director Orville Harden in December, 1946.[62] This oil went to European markets, largely.[63] When in 1951, AIOC could no longer fill the contract from Iranian sources, that company increased its Kuwait production to keep Jersey supplied with crude oil. Another pressing concern with the Iranian situation derived from the Cold War policy of the United States, which guaranteed friendly nations an adequate supply of oil. Obviously, what happened in Iran was of vital concern to Jersey, for anything that threatened its supplies of oil from that part of the world directly threatened its European sales.

X   X   X

On May 14, 1951, Jersey Director Bushrod B. Howard attended a meeting called by George McGhee, Assistant Secretary of State, where executives of the leading oil companies expressed their views on the Iranian-British controversy. Howard returned to New York and drafted a letter to McGhee, which was read to Jersey's Executive Committee on May 17, 1951. The Executive Committee approved the letter "subject to review by Judge Manley O. Hudson, outside Counsel" [Hudson was a noted international lawyer]. When Hudson approved the letter, Howard sent it to McGhee. Speaking now for the company, Howard saw the Iranian seizure of AIOC properties as a threat to foreign aid and national security, and thus a "direct challenge to the United States economic foreign policy." As he went on, "Even more serious repercussions would follow if the Iranian doctrine of 'nationalization' were to be adopted throughout the Middle East and in other areas where strategic resources of immediate importance to the Western World are being developed under contracts between governments and foreign investors." Finally, Howard urged that the U.S. government take immediate steps to spell out its official position. This request represented a company position, for, in fact, the Jersey Executive Committee had debated the Iranian oil situation on both May 15 and May 17, when the minutes show: "Iranian Oil Situation Discussion." On May 21, the Jersey Assistant Secretary, John O. Larson, circulated copies of the Howard letter to the entire Board.

Meanwhile, the Department of State in a lengthy press release dated May 18, 1951 announced an official U.S. position on the British-Iranian controversy. It "stressed to the governments of both countries" the need to solve the dispute in a friendly way, through negotiation. Great Britain had been advised that the U.S. believed "that arrangements should be worked out with the Iranians which give recognition to Iran's expressed desire for greater control over and benefits from the development of its petroleum resources." The U.S. had, at the same time, pointed out "the serious effects of unilateral cancellation of clear contractual relationships," and that "legitimate objectives" could be achieved through friendly negotiations. This statement also made it clear that U.S. oil companies "which would be able to conduct operations" on a scale sufficient to supplant AIOC, would not "be willing to undertake operations in that country" because of Iran's seizure of that company's properties. Finally, the release emphasized the great interest of the U.S. in the "continued independence and territorial integrity of Iran." Earl Newsom, advisor to Jersey's Public Relations Department, concerned that President Eugene Holman and the Board should receive the latest information, sent to them copies of a realistic article entitled, "Oil

Agreements With Underdeveloped Nations." Fred W. Palmer, the author, pointed out that the weak Iranian government had been a pawn of "the Nationalist idea (backed by strong Moslem forces seeking to return to old ways)." Oil income, he said, did not always go to the people "but to the governments—where sometimes it is squandered." When it disappeared, the foreign oil company became the scapegoat. Palmer advocated offering new contracts "to underdeveloped countries," with provisions for a portion of the income to be set aside for projects of benefit to all the people; a real program of financing education for young students of the host country; provision for facilities for educational activities; and other programs. His proposal actually hinged on the first point—new contracts—signed before pressures built up to alter old ones. President M. J. Rathbone circulated the memorandum to the Board and to other key officials, as an "interesting suggestion." George H. Freyermuth, public relations manager, stimulated further conversation regarding Iran and comparable countries in an analysis of the memorandum.[64]

Already, as has been noted earlier, executives of Jersey, acting upon the request of the U.S. government, had joined representatives of other oil companies to insure that any possible loss of Iranian oil would not injure our Western allies or handicap our military efforts in the Korean War. This cooperation continued for two years.

X   X   X

When the Iranian crisis began, however, Jersey confronted several other pressing problems in the Middle East. As early as 1948, David A. Shepard, then serving as Jersey's representative in Great Britain and a director of IPC, had written concerning Iraq, "time is running against us" because of "upward revisions of oil royalties elsewhere in the Middle East" and because Iraqi knowledge of "important amounts of oil in the Concession might injure IPC's bargaining position." As Shepard predicted, trouble lay ahead. In fact, IPC narrowly escaped nationalization in 1949.[65] Obviously, as Shepard phrased it, "The day . . . has passed since it is . . . possible to hold a rich concession without seeking actively to bring it, so quickly as is practicable, into efficient production." Iraq continued to press for "share participation."[66] Woodfin L. Butte, on Jersey's Near East Committee, wrote at about the same time, "If we wish the IPC Group to continue to control the entire petroleum resources of Iraq, [we must] strive to have the Iraq Government remain always well content with the situation." He said they "would stand alone . . . if one of those goals were not an increased revenue."[67]

In October, 1949, Page, who had been serving as executive assistant to the chief executive, Jersey's President Holman, replaced Shepard as IPC director and as company representative to Great Britain. Page toured the Middle East, and upon his return to London, sent a letter to the other IPC Group representatives, asking them to review a memorandum he had prepared. He pointed out that across the years, Jersey leaders had put more money into IPC "than they had taken out by a wide margin." Now, after carefully analyzing the entire IPC operation, he suggested that the Groups take a more active role in its management, saying, "The Groups have, in total, the greatest experience and best authorities on oil problems in the world. This source of consulting advice should be put at the disposal of the IPC Management as an aid to them." [68] Many years later, Page, in an understatement, described what he had done:

> I was instrumental in getting IPC to sit down . . . and to make some plans in advance as to what we might do. . . . [To think about and discuss what the IPC response would be to hypothetical but quite probable demands by the Iraq Government, such as why can't we have a 50/50 agreement as Venezuela has and is being considered for Saudi Arabia.] This was bitterly opposed by some other people who simply said, 'That's not the way to do it. You wait until they ask for something. . . . ' I said that . . . wasn't the way I thought it should be handled.[69]

On January 26, 1950, Page wrote to Jersey Director John W. Brice, relating details of a session with IPC Group representatives in London, from which he had emerged very pessimistic. He said prophetically that despite progress in the "past year, IPC will never be a good operating company because of its fantastic set-up."[70] Part of the problem lay in the nature of the Groups, especially in the one owned by C. S. Gulbenkian [Participations and Investments, Ltd.] and the Compagnie Française des Pétroles [CFP]. Gulbenkian, in a letter to Jersey's Orville Harden, complained about friction among the Groups, adding, "We are wasting millions."[71]

Page's position with the Group representatives led to his being added to a negotiating team that conducted discussions with the Iraq government in 1951 and 1952. On February 3, 1952, the IPC Group of companies and Iraq signed an agreement providing for "the equal sharing of profits."[72] This agreement was altered later by the unilateral withdrawal by the revolutionary government of all non-producing areas and by some minor changes in terms, but the actual producing operations were controlled within the framework of this agreement for nearly two decades, in itself, no small achievement.[73]

Aramco, in which Jersey owned a 30 percent share, also faced difficulties after 1948. The Saudis knew about the Venezuelan 50-50 agreements with the oil companies, and they constantly pressed Aramco's management for more money.[74] Each time the Saudi demands quickly became known to the U.S. Department of State. Aramco management discussed the royalty and tax problem with the U.S. Treasury Department. The result: Treasury official George A. Eddy was dispatched to Saudi Arabia to discuss money and income with representatives of that government. Eddy recommended an income tax system. He left, and a Washington tax lawyer, John F. Greaney arrived to help draft a Saudi income tax law.[75] Before the Saudis and Aramco signed the tax agreement, King Abdul Aziz Ibn Saud issued two royal proclamations. The first of these, more general in nature, dated November 4, 1950, levied a tax of 20 percent of net profit, with the objective of arriving at 50-50 profit-sharing by combining income tax and royalty payments. [Under this decree, Saudi income did not reach 50-50, and the rate soon changed.] On December 27, 1950, Ibn Saud promulgated a second decree; this altered the earlier one by establishing graduated tax rates ranging up to 40 percent of the company's profits. On December 30, 1950, the two parties signed a new contract[76] to accommodate the changes to the original concession agreement. They also endorsed some minor changes.[77] In June, 1950, Aramco announced this new profit-sharing agreement with Saudi Arabia. As U.S. Secretary of State Dean Acheson wryly commented, "Aramco had avoided [a struggle over its concession] by . . . graciously granting what it no longer had the power to withhold."[78] When news of this agreement reached Iran, its government moved to take a more rigid position in negotiations with the Anglo-Iranian Oil Company.[79]

The Shah had appointed General Ali Razmara as Premier in June, 1950. He and the Shah diligently planned developments for Iran and pushed other reforms that they could finance. They waited for expected loans from the American government and the International Bank for Reconstruction and Development before instituting massive economic, educational and social programs—The Iranian Seven Year Plan. These loans, Iran expected to repay from the royalties to be received from the AIOC. But in September, 1950, the Iranian government learned that the anticipated loans would not be forthcoming. In turn, the rejection stirred the Majlis to demand increased royalties from the oil company. Almost overnight, anti-American and anti-foreign outbursts increased in number and seriousness. Meanwhile, Russia stepped up her military activities along the Iranian border at the same time that she began negotiating for a new trade agreement. In October, 1950, a tumultuous anti-American outburst took place in the Majlis. Most American experts, hired to plan and supervise the economic and technical reforms of the Seven Year Plan, departed for home. In December, newspapers reported that the Iranian Army had released ten leaders of the pro-Russian Tudeh party, a political group outlawed since February 5, 1949.

In the winter of 1950-1951, the National Front party of Dr. Mohammed Mossadegh gained power by demanding that the Anglo-Iranian Oil Company be nationalized. Mossadegh, an elderly, long-time member of the Majlis, had for years been a proponent of nationalizing the company. Shrewd, clever, and a knowing practitioner of all the political arts from the lowest demagoguery to high statesmanship, in June, 1950, he became chairman of a special committee to study an arrangement first proposed by the AIOC in 1949. Mossadegh believed that, "It is better to be independent and produce only one ton of oil a year, than to produce 32 million tons and be a slave to Britain." So his committee reported "against the agreement on the ground that it did not safeguard Iranian rights and interests." After accepting this report, the Majlis entrusted the same committee with the task of charting a course for government relations with AIOC.[80] Now, Mossadegh carefully cultivated the xenophobic attitudes of his followers. He recommended to the committee that they approve nationalizing the local oil industry and cancelling the AIOC concessions. The committee asked General Razmara whether such action was practicable, and he replied negatively. Four days later, a religious fanatic assassinated him. The Majlis then approved the nationalization of AIOC, and cancellation of its oil concessions effective when the Shah signed the act, on May 1, 1951. Meanwhile, the NATO Commander, General Dwight Eisenhower, then in Paris, and struggling to pull the Western nations together, noted in his diary, "Lord knows what we'd do without Iranian oil." Certainly European economic recovery hinged to a considerable degree on that single energy source.[81] The new Iranian Premier, Hussein Ala, rejected British protests and questioned the right of the British government to intervene in a matter between Iran and a private company. Then, he resigned when the expropriation bills passed, and the Shah turned to Mossadegh to form a new government. On May 8, 1951, the still-outlawed Tudeh party held major rallies and also presented its list of demands to the Majlis. From all appearances, a revolution had taken place.

Nationalization of the AIOC had occurred unilaterally, contrary to provisions in the concession agreement. Nor did it seem to matter that, somewhat belatedly, the company indicated its willingness to arbitrate the matter and to apply the 50-50 principle in a new agreement. The British took the case to the International Court of Justice at the Hague in late May, 1951. The court enjoined both parties to preserve the *status quo* while it considered whether to hear the case. Iran then withdrew as a "signatory of the Court convention," and refused to recognize the court's jurisdiction.[82] Early in June, President Truman in letters to Mossadegh and to the British Prime Minister Clement Attlee urged renewed negotiation. Next Iran created the National Iranian Oil Company [NIOC] and gave British employees of AIOC a week to decide whether they would work for the new company. The British reacted by ordering all their tankers away from the great Abadan refinery and by sending home other personnel. The situation day-by-day worsened as Secretary of State Dean Acheson let the British know that the United States would not support armed intervention. Then, Mossadegh agreed to receive Averell Harriman as President Truman's special envoy. On July 15, 1951, Harriman arrived in Iran in the midst of a Tudeh-sponsored demonstration that soon turned into a riot. However, this visit to Tehran, and later to London, did quiet talk of hostilities, and Harriman managed to secure Mossadegh's commitment to diplomatic negotiations.

Next, the British sent to Iran a mission headed by Richard R. Stokes, Lord Privy Seal. The group arrived on the day that AIOC closed the Abadan refinery, which for the next several years refined only enough oil needed for domestic consumption, with a token amount for export. Briefly, the Stokes mission failed to find any hope of settlement, and it went back to London at the end of August. Harriman himself returned to Washington on August 22, 1951. Against American advice, the British filed a condemnatory resolution with the U.N. Security Council, and Mossadegh came to New York to argue Iran's case before that body. As experts had predicted, the British resolution failed. On September 27, 1951, Iranian soldiers took over the Abadan refinery, and all remaining British AIOC employees left. To complicate matters further, in the general election in Britain, the Labour party went down to defeat, and the Conservative party headed by Winston Churchill and Anthony Eden now took over to deal with the Iranian problem. Meanwhile, on July 5, 1951, the International Court ruled that it lacked jurisdiction in the matter, and since all diplomatic negotiations appeared to promise little hope of resolving the dispute, compensation for the expropriated property replaced political or legal control as the major issue.

Mossadegh had returned to Iran, where general elections took place on December 18, 1951. After the election, he resigned his position as Premier, but within a few weeks the Shah had to reappoint him. During that interval, the British consulates were closed.

As could have been anticipated, all these changes battered the Iranian economy for more than one reason. As Secretary of State Acheson wrote, "Mossadegh was aided by the unusual and persistent stupidity of the company and the British Government in their management of the affair." Then he added, "never had so few lost so much so stupidly and so fast."[83] Of course, other Americans—State Department and oil company executives—freely criticized both the AIOC and the British government. However, Jersey's Page, who was in the Middle East when the trouble erupted, after long reflection wrote, "After a careful study of the events, I expressed the opinion that while I and other Americans would have handled the situation differently, I believed that, in the circumstances, neither I nor anyone else could have prevented the nationalization."[84] The forces of nationalism unleashed by World War II simply had proved too powerful to be checked. Page, well informed on such matters, went on to become the Middle Eastern specialist on the Jersey Board of Directors. Finally, he believed that "Mossadegh didn't want to make an agreement with any foreigner. I couldn't have made an agreement with him; I wouldn't have even attempted it."[85]

The Iranians and Mossadegh quickly learned that no sizeable export market existed for their oil. Once the AIOC technicians and staff had left Abadan (October, 1951),[86] the British government advertised widely, promising that all persons or countries buying Iranian oil would be sued. Most American companies likened purchasing such oil to buying stolen property. Several of the oil majors that had concessions and agreements with Middle Eastern countries adopted the attitude, "There but for the grace of God go I." Those companies disliked the idea of fattening themselves on the remains of AIOC; they questioned whether any stable government could be established in Iran and, because of that doubt, asked whether an agreement would last long enough to be worth making. Of course, if any American company such as Jersey had purchased Iranian oil, it could have faced expensive lawsuits. Virtually all of the American companies, like Jersey, had properties in Britain; these could have been attached or seized as an award in an unfavorable legal decision there.

The Korean struggle had preempted the services of independent shippers, and other than those belonging to AIOC, a scarcity of tankers operated against Iranian efforts to sell oil. Jersey, like other companies capable of handling and marketing Iranian oil, knew also that it still faced grand jury action and possible prosecution for criminal violations of the U.S. antitrust statutes. Page wrote concerning this, "They [the companies] therefore gave no consideration to joint action and in any case had no economic interest in the purchase of Iranian oil as they had their own low cost sources." They "could easily obtain oil at a cost much less than Iranian prices" by increasing production in these other properties in the Middle East and Venezuela. Page noted, "at that time none of the Arab states cooperated with Iran by restricting their own production." The American companies made more money by producing their own oil.[87]

The conclusion is inescapable that none of the majors bought Iranian oil. Page, years later, explained this: "we weren't buying it before [nationalization] and we weren't buying it afterwards." Despite oft-printed statements to the contrary, the oil majors did not conspire to boycott NIOC oil. They did not buy because of the above stated reasons. Secretary of State Acheson pointed out, "we . . . quietly discouraged American private purchases of oil in Iran. . . . The inevitable litigation could cause irritation." More pointedly, on December 6, 1952, the State Department released a news bulletin stating that persons who purchased oil in Iran did so on their own appraisal of the legal risks involved.[88] Soon, Acheson became convinced that both Iran and Britain "were pressing their luck [positions] to the point of suicide" and moved to stimulate negotiations. When Churchill and Eden visited Washington in January, 1952, Acheson used the occasion to press his point that the stalemate "offered little promise for British interests in Iran and considerable danger for ours elsewhere in the Middle East."[89] The United States suspended military aid to Iran in the same month, and allowed for only token economic assistance. And Britain, in a stronger fashion, forced a tanker loaded with Iranian oil into the port of Aden. Later, a British court upheld the AIOC claim to the cargo. So the stalemate continued.

X   X   X

In June, 1952, the Democratic party, in convention in Chicago, nominated Adlai E. Stevenson and Senator John J. Sparkman as candidates for President and Vice President respectively. Perhaps this had influenced President Truman to turn over to the Senate Committee on Small Business the declassified *FTC Staff Report*. Sparkman announced that his committee had the report on the oil "cartel," and when reporters queried Jersey Chairman Frank Abrams about it, he replied, "I want to repeat categorically that Standard Oil Company (New Jersey) is not a party to any international oil cartel and, in fact, doubt that such a cartel exists."[90] However, a few days later, the Department of Justice served a subpoena requiring Jersey to produce documents and, possibly, executives to testify at some later time.

As if the worry of the antitrust suit were not enough, Jersey executives also began to receive reports of the growing hostility of an ever larger number of Arabs to both the United States and the company. On July 7, 1952, the Executive Committee listened to an outside consultant's report on the subject of "American Interests in Arab Lands." The Committee then asked Director David Shepard to discuss with Arthur Newmyer, George H. Freyermuth and Thomas E. Monaghan, "the question of what, if anything could be done to stop, or possibly even reverse, the rising resentment in Arab lands of the United States Government's policy with respect to their countries." The Committee also requested Vice President Orville Harden to obtain the opinion of the Aramco executives on that question.

Before reports could be prepared suggesting how relations between Arab nations and the United States might be improved, the Executive Committee learned that on August 22, 1952, the date of the *FTC Staff Report*, a federal complaint had been filed against the company (and on Esso Export Corporation) "for recovery of alleged overcharges on sales of Arabian crude oil" in transactions involving two agencies working to promote European recovery, which the U.S. government supported: the Mutual Security Agency [MSA] and the Economic Cooperation Administration [ECA]. The Justice Department alleged that Jersey had overcharged these agencies almost $32,000,000. These claims received widespread publicity.* While discussing these charges, the Executive Committee also considered how to implement company policies "regarding relations with the public, stockholders and employees as affected by" the *FTC Staff Report*, the "cartel" investigation and this latest complaint. Interestingly, within eight days the Creole Petroleum Corporation reported that its relations with the Venezuelan government had been "affected adversely" by the same three developments. Several days later the Committee approved the use of letters to shareholders, employees and to heads of affiliated companies, presenting the Jersey position on these two cases and on the *FTC Staff Report*.

On October 11, 1952, the subject of how to settle the Iranian question and

---

*These claims, which are the only ones relating to the "cartel" case, that were brought to judicial determination, were dismissed in 1957, after trial before U.S. District Judge Thomas F. Murphy in the Southern District of New York, and the U.S. Court of Appeals for the Second Circuit unanimously affirmed the dismissal in 1959. [*United States vs. Standard Oil Company of California, et al.*, 155 F. Supp. 121 (S. D. N. Y. 1957), Aff'd, 270 F 2d 50 (2d Cir. 1959). The Socal-Texas-Caltex case was chosen as the test case, and the others were discontinued after the Second Circuit opinion.] The Court of Appeals stated (270 F 2d at 59) ". . . the conclusion is clear, as succinctly summarized by the trial court, 'that the prices financed by the ECA [Economic Cooperation Administration] were in fact the lowest competitive market prices.' " Thus of the charges made by the FTC and by the Antitrust Division with respect to alleged "cartel" activities, those that came to trial were dismissed and the oil companies' position vindicated by the courts.

move the oil led to another U.S. government interdepartmental memorandum. The point raised by the State Department with the Attorney General concerned whether a group of American oil companies could combine to organize a company to purchase oil from the NIOC, keeping within the framework of the antitrust statutes. Apparently this idea grew out of the Truman-Acheson conversations. Justice advised that such action alone would constitute "unreasonable restraint upon our foreign commerce" and would "be illegal under the antitrust laws."[91] Justice then sought to outline a procedure by which State Department objectives "might be attained" without coming into conflict with the "national policy of free competition."[92]

Just before the 1952 presidential elections, President Truman "wished aloud . . . that the Anglo-Iranian dispute could be settled before his term expired." Both he and Secretary of State Acheson believed that Iran would never make a new concession contract with the British, and that the British would never seek such an agreement without compensation for the one unilaterally ended by Iran. Both countries would dislike an independent American solution, they believed. Acheson understood that such a solution "would require the cooperation of the major American oil companies, who alone, aside from Anglo Iranian, had the tankers to move the oil in the volume necessary."

On November 7, 1952, Acheson discussed with the President whether his authority under the Defense Materials Procurement Act could permit several American oil companies to buy and sell Iranian oil, provided Iran could be helped by a large loan. The President liked the contemplated plan. The Department of Justice opposed the oil companies meeting with the State Department; President Truman overruled Justice.

On December 4, representatives of Jersey, Socony, The Standard Oil Company of California, The Texas Company and Gulf Oil Corporation [Gulf] met with members of the State Department to discuss a plan to move "an additional ten million tons of oil per year." Those present agreed that any settlement made should not injure the existing 50-50 agreements then prevalent in the Middle East; that AIOC had to be compensated; and that the governments concerned had to approve any plan agreed upon. None of the oil company representatives indicated "any particular desire to move this oil"; instead, they emphasized that grave problems would arise if they cut back production elsewhere in the Middle East to accommodate Iranian oil. At another meeting on December 9, with the same persons present, the State Department representatives emphasized that Iran would never "agree to a concession." The oil executives believed it possible to peddle the Iranian oil in world markets. On that basis, Paul H. Nitze, director of the State

Department Planning Staff, flew to London, where he reported on this to the British.[93] The British and Mossadegh tentatively agreed that the plan could be the basis for negotiation. Then on January 29, 1953, Mossadegh denounced the plan.

Of course, these actions attracted considerable publicity. An editorial in *Fortune* entitled "That Sinister Oil Cartel," questioned whether successful prosecution of the big oil companies by the Justice Department, along the lines of the *FTC Staff Report*, would result in lower prices for petroleum products. Remarking, "It is hard to see where Antitrust is going or what it can do," the editorial went on:

> Even if the government trust busters did not have the stimulus of a presidential election, they could hardly be expected to overlook the foreign operations of American oil companies. When in recent times have they been presented with material that lent itself to such easy and dramatic conclusions?

On November 9, 1952, the *New York Times* commented on the antitrust investigation: "If the American companies have done something wrong or illegal in the developments of these vast oil resources abroad, from which they now are obtaining some 2,500,000 barrels daily of crude oil, the feeling is that some settlement should be worked out without litigation which might require years."[94] During the next several days, the Acheson-Eden talks regarding the Iranian crisis again focused attention on the oil companies and on Iran.[95] *New York Times* pundit Arthur Krock, in his column "In the Nation," pointed out that the British and Dutch governments had refused to allow oil companies incorporated in those countries to honor the subpoenas issued in the antitrust investigation on the ground that to do so would constitute an invasion of their sovereignty. This, he said, put the Department of Justice "up a tree." Krock wrote that the suit stemmed from the *FTC Staff Report* "sparked by Commissioner John Carson, generally rated as a crusading 'liberal' doctrinaire," and that one reader had described it as "leftish as a quarter-to-nine." He pointed out that the agreements with the Arab nations "were not frowned on in that period (1940-1945) by the United States legal guardians of the anti-trust and anti-cartel laws." Terming this case a "vote-getting stratagem," he concluded, "This policy is also highly approved in the Soviet World. It buttresses the Communist charge that American business men are crooks, especially (Moscow tells the Arab peoples) American oil men."[96]

Secretary of State Acheson believed that he had the answer to the question: Would the oil companies cooperate? He next tackled an even larger question. "Would the Department of Justice, which was already proceeding against them for alleged violation of the antitrust laws in their Middle East operations, allow them to cooperate if they wished to?"[97] In early January, Acheson arranged for meetings of the Justice, State, and Defense Departments, and the Chairman of the Joint Chiefs, General Omar S. Bradley. Attorney General McGranery and representatives of the department's Antitrust Division attended. According to Acheson, these antitrust representatives put on quite a show:

> . . . they wanted no truck with the mammon of unrighteousness, were a good deal more certain of what would be permissible under the law than the Supreme Court up to that time had shown itself to be, and had no hesitation in disagreeing with me on foreign policy aims and with [Robert Morse] Lovett [Secretary of the Defense Department] and Bradley on the military risks and consequences of an absorption of Iran in the Soviet system.[98]

In addition, Acheson believed that the same antitrust lawyers leaked confidential information about these meetings to the media.

The British began to thaw. Top level discussions continued between them and the United States as they sought to refine a proposal that the Iranians might accept. The next development occurred on January 6, 1953, when the Departments of State, Defense, Interior [PAD], and Justice filed a report on "National Security Problems Concerning Free World Petroleum Demands and Potential Supplies" with the National Security Council [NSC]. Their joint statement asserted that any trial of the oil companies "on criminal charges would be harmful to critical American foreign policy objectives."

In the opinion of three of these departments, criminal investigations had already seriously impaired the prestige of the United States and the oil companies in the Middle East and elsewhere, which could lead to heightened international problems. The CIA also filed a classified statement on the probable consequences of pressing the cartel investigation. Further, Defense and State recommended that the President instruct the Attorney General to drop the criminal investigation and to file a civil action instead. The Justice Department urged just the opposite, arguing that the alleged "cartel" provided a serious threat to "our national security." The Justice statement emphasized "even-handed administration" of the law and conscientious enforcement of the Sherman Act. In turn, the National Security Council supported the recommendations of State and Defense.[99]

After receiving the recommendation of the NSC, Truman acted swiftly. On January 12, 1953, he wrote to the Attorney General, "I am of the opinion that the interest of national security might best be served at this time by resolving the important questions of law and policy involved in that investigation in the context of civil litigation rather than in the context of a criminal proceeding." He went on to say that this decision would apply "only if the companies involved agreed to the production of documentary material which the grand jury subpoena directed them to produce." Truman later told Emmerglick that based on General Bradley's assurance that the national security called for the change, he had decided to drop the criminal investigation. The President added that he wished the civil action "vigorously prosecuted."[100] Conceivably, Truman did this precisely to insure that the Antitrust Division would understand what he and others believed was the threat to national security. The Truman order of January 12, changing the "cartel" investigation from "the context of a criminal proceeding" to "the context of civil litigation," did not result in any further action before the end of his presidency.

Truman's decision did cause an uproar in the offices of the Federal Trade Commission and the Justice Department. FTC Commissioner and Acting Chairman Stephen J. Spingarn angrily said, "It would appear that honest and sincere men in the Government have fallen victim to the worldwide propaganda campaign conducted by the major oil companies since the publication of our report."[101] The President made his decision on January 12, 1953, and one day later Attorney General James P. McGranery decided to meet with the attorneys of the five companies to tell them that the criminal investigation would be dropped and replaced by a civil suit, provided they "agreed to produce . . . all the documentary evidence required of them under existing grand jury subpoenas."

Ed Johnson, Jersey's general counsel, and outside attorney Arthur Dean of Sullivan & Cromwell, represented Jersey at this session, which quickly became explosive. First, Emmerglick outlined the Justice plan to implement the President's offer. Next, as Ed Johnson described it, "and lo and behold, McGranery came through the door striding like a peasant at market time. . . . He came to a halt. I believe he actually clicked his heels . . . he started the conversation in this fashion: 'I came to tell you, not to negotiate with you.' " He told the thirty-five attorneys present to sit down. Then he announced that the proposal he offered "was 'cold turkey' and that the companies could take it or leave it." The proposal of the Justice Department required all the oil companies to pledge not to move to have the suit amended or dismissed until such time as the specifics of the government's case could be fleshed out from the documents subpoenaed. The civil charge at the time had only skeleton form. This feature of the arrangement agitated the lawyers who considered that agreeing to the proposal would permit Justice to "fish" in their companies' files. Johnson stated that when Mc-Granery finished, "Art Dean stood up . . . all of us stood up and [Dean] said, 'Thank you, Mr. Attorney General,' and started out and everybody started with him." Dean stopped a moment to tell McGranery that so far as the Standard Oil Company (New Jersey) "was concerned, the offer was unacceptable."

A number of attorneys then repaired to the headquarters of Newmyer & Associates, where Dean discussed the affair with representatives of the press. Johnson said Dean told them the whole story, "Art was straightforward." Dean termed the proposal "outrageous blackmail," adding that he found the action of McGranery "discourteous and insulting."

Prior to Johnson's trip to Washington, Frank Abrams had cautioned him to be sure that "we should do nothing to affect our (Jersey's) very good position with the government." As Johnson remembers, "Abrams sent for me the next morning when he had seen the big headlines in the paper. . . . Abrams wanted to know from me if they (outside counsel) had been told and instructed not to antagonize the government representatives." Abrams appeared angry as he questioned Johnson, who replied, "Mr. Dean expressed his opinion and went out. And I must agree that I'm on Mr. Dean's side. He did what I believe to be the right thing." They then sent for Dean. "So Abrams started to talk to Dean, leading up to show his displeasure of what had taken place. But before he had gotten very far, Art said, 'Now Mr. Abrams . . . if you would like for me to withdraw from this case, I will do so right now.' . . . Abrams would have none of that. We all had a good laugh by then."[102]

Dean had indicated that no compromise agreement appeared possible, pointing out that certain documents involving the security of the United States should not be made public.[103] This point always irritated the Antitrust Division lawyers or, as Emmerglick phrased it, "The National security argument is usually made in one context or another."[104] Dean had added another question at the session with Emmerglick, but not then answered—whether the incoming administration had agreed to honor the agreement then being proposed. Spingarn, proud of the *FTC Staff Report*, remarked bitterly that the whole affair provided "further dramatic proof that democratic governments are but the puppets of monopoly." The oil companies refused to accept the government proposal, and the court continued the case until January 28, 1953, at which time they were to produce "documents located in the U.S. which are the subject of an outstanding subpoena." The court took this action to give the incoming Eisenhower administration opportunity to consider what action to take. The outgoing and incoming advisors to the Department of State then drafted a memorandum stating that they believed the January 28 date allowed too little time for the new administration to study the case fully. They suggested that the court approve a "stand-still" agreement for a period of sixty to ninety days for that purpose.

In the meantime, the Secretary of Interior, Oscar L. Chapman, sharply questioned some of the Justice Department's statements. He wrote Attorney General McGranery a long letter taking issue with "inaccuracies in statements to the Press of McGranery's Special Assistant." These had to do with first, the allegation that the "international oil cartel" had caused fewer wells to be drilled in the U.S.; hence, reducing the U.S. reserve capacity, and second, that the "cartel" had caused the existing "world-wide shortages of aviation gasoline." Chapman quoted the statements made to support the allegations. Further, Chapman charged that when Bruce Brown, of PAD, had pointed out that both the charges were inaccurate, the Acting Assistant Attorney General had reaffirmed them. Chapman produced figures showing that there had been a steady increase in the number of wells drilled each year and said that only lack of pipe prevented additional drilling. He also charged that the same person had told the court that he proposed to free the Middle East of "cartel" control, so that competition for oil there would resume. Chapman asked McGranery how more oil from that part of the world, with resulting lower prices, would encourage more drilling in the U.S.

The Interior Secretary also challenged the statement that the "cartel" had caused the scarcity of aviation gasoline; instead, he asserted that in May, 1952, one-third of the refining capacity in the U.S. had been shut down by strikes. He went on to praise the oil company record. Then he asked McGranery to correct the "erroneous impressions" left with the court. For that, there was too little time remaining in the Truman administration, perhaps, but the letter does illustrate the cross purposes that appear to have characterized relations among government departments and agencies.[105]

X   X   X

In 1953, the Republican party, General Eisenhower, and especially his Secretary of State, John Foster Dulles, came into power after having freely belittled the Truman-Acheson foreign policy during the campaign. However, when given the responsibility for international policy, Eisenhower and

Dulles actually changed their earlier views. Dulles ultimately came to endorse the policy of containment of communism as a proper Cold War strategy even more strongly than had his predecessor, Dean Acheson; but as one writer remarked, "with less patience." In February, 1953, Dulles visited six Western European nations plus Great Britain to promote the formation of a stronger defense commitment. France had refused to join the European Defense Community [EDC] because it considered that too few restrictions had been placed on Germany. Elsewhere, Russian Premier Joseph Stalin died in March, 1953. Whether this affected the Dulles decision to accede to the French demands is not known, but when he did, both France and Germany joined EDC in September, 1954. [The Russians, of course, countered with the Warsaw Pact in May, 1955.]

In the Far East, President Eisenhower played the leading role in negotiating a truce which replaced the Korean War. The United States continued to recognize the Republic of China on Formosa as the *de jure* government of that nation, with Chiang Kai-shek as its head. In 1953, the French forces in Indochina began the retreat that led to their ultimate 1954 defeat by the North Vietnamese Communist forces of Ho Chi Minh. The Geneva Accords of July, 1954, divided Indochina into Laos, Cambodia, and split Vietnam into two sections, North and South, along the seventeenth parallel. Both Eisenhower and Dulles disliked this arrangement, and the United States refused to sign the documents. In September, Dulles announced a new pact in the making, which he believed would check the spread of Asian communism there and stabilize the existing governments. Termed the Southeast Asia Treaty Organization [SEATO], the new agreement provided a defense mechanism to preserve the status quo.

Even while Dulles labored to erect these barriers against the spread of communism in early 1953, the Middle East, and especially Iran, loomed large on his horizon. He considered this area even more important to the United States than the Far East. With the Iranian situation degenerating from confusion to chaos, Dulles perceived that new action by the United States would be necessary. First, he made plans to tour that area. Iranian Premier Mossadegh urged him to include that country in his itinerary and as late as April 6, Dulles considered such a visit likely. For whatever reasons, he did not go there on his Middle East journey of three weeks—probably because the Iranians had reversed their position on negotiations in late January. Dulles appeared a bit baffled by the fact that while in the Middle East, he wished to discuss with his peers in the governments of those countries he visited, the United States' fear of Russian communism, while the Arab diplomats wanted to talk only about their fear of Israel. Many of the Arabs believed that Israel, backed by the U.S., would simply expand. Dulles gathered that, for them, the issue of Palestine was focal. Upon returning to the U.S., he announced a proposal to enter into a series of mutual assistance (defense) pacts across the Middle East. Turkey, Iran and Afghanistan would become the first line of defense against Soviet expansion toward the Persian Gulf and on into Asia. Dulles labeled these states the "Northern Tier." To erect such a wall of containment, however, Iran would have to be saved for the West.

X   X   X

In one of the first Eisenhower cabinet meetings (January 23, 1953), Attorney General Herbert Brownell, Jr. indicated that he proposed to ask for a delay in the "cartel" proceeding because of its controversial nature.[106] President Truman's order of January 12, 1953, had not been implemented, but the oil companies continued to press Justice on the issue of what papers and documents it desired. Jersey's Eugene Holman telephoned Secretary of the Treasury George M. Humphrey on April 7, indicating that he had been in touch "with the four men who are interested in doing whatever they can," and Humphrey reported this to Dulles, who said Brownell should be in "on it." Dulles asked Humphrey to get in touch with Brownell and "do spadework" before they met again [Holman had met with Dulles and Humphrey on April 1, 1953]. Dulles added that these conversations "should not be mentioned at the working level."[107] Holman was genuinely concerned that even a civil suit would create significant problems in the Middle East for Jersey and other oil companies. On April 8, 1953, the National Security Council met and agreed that Brownell should ask the court to terminate grand jury action and to institute a civil action under the antitrust laws. They also approved continuation of the screening committee, with the view of preventing public disclosure of evidence involving national security. Dulles did not oppose this course provided "the proceedings are conducted with due regard for all matters affecting the national security" but he did state[108] his opinion that the civil suit would injure U.S. relations with Middle Eastern countries.

On April 20, 1953, Dulles and Brownell discussed the NSC meeting on the "cartel" case.[109] The following day the government instituted civil action and at the government's request, the court ended the grand jury investigation.

In asking the court to make this decision, Justice Attorney Emmerglick pointed out that his request was based "upon the considered judgement of two Presidents, two Secretaries of State or their principal representatives, two Secretaries of Defense, and in addition, the Chairman of the Joint Chiefs of Staff, the Central Intelligence Agency, and a number of present and former cabinet members."[110]

X   X   X

As the Iranian situation steadily deteriorated into a power struggle between Mossadegh and the Shah, all order disappeared and mob rule took its place. The Shah went on vacation, and then fled to Iraq, and later to Rome [where CIA Director Allen W. Dulles visited him]. President Eisenhower, Secretary Dulles and many others became convinced that Iran would fall to the Soviets, for all reports indicated that Mossadegh daily moved closer to the Russian-backed Tudeh party. In the United States, Eisenhower searched for a plan that would solve the Iranian problem and still protect the interests of the free world. Even before he assumed the office, in fact, British intelligence had approached the CIA concerning the possibility of installing some government other than that of Mossadegh in Iran.[111] Under the code name, "Ajax," the intelligence agencies of both countries had interviewed the Shah and evaluated the Iranian situation. Everything, it appeared, must hinge on what might happen when the Shah was restored to his throne. On the narrower oil question, Eisenhower decided to follow up the Truman-Acheson plan by having American oil companies and the AIOC jointly move all cargo. But before they would agree to become involved, the American companies sought some guarantee that their combined action would not be treated as a violation of the antitrust statutes.

Meanwhile, on May 27, 1953, Jersey stockholders met. As a group, they revealed concern over the many allegations, charges, indictments, and actions directed at the company. However, shareholder Judge Louis Goldstein, in a speech of some length, defended management and moved the adoption of a resolution affirming "their faith and confidence in the loyalty, integrity, experience and ability of the directors and officers" of the company. Goldstein expressed satisfaction that they had conducted the affairs of the company and its business in accordance with the laws of the United States of America. Much to the gratification of Holman and the other executives, the motion passed with near unanimity.[112]

The "Ajax" plan proceeded. In July, Secretary Dulles telephoned his brother, CIA Director Allen W., to ask about "the other matter" [the CIA-British Intelligence backed revolution?] which the latter had not mentioned at the NSC meeting. Allen replied that "it was cleared directly with the

President and it is still alive." He indicated that "the young man [the Shah?] may pull out at the last minute, he is an unaccountable character but the sister [Princess Ashraf?] has agreed to go" [back to Tehran?]. On August 18, 1953, a revolution began in Tehran, and the Shah returned on August 22. He appointed General Fazlollah Zahedi, Premier. On August 25, Secretary of State Dulles announced that the U.S. "had a miraculous second chance [in Iran] and shouldn't go over the same old cycle again."[113] But the return of the Shah to the throne of Iran and the appointment of Zahedi as Premier, did not bring an end to the pressure on the U.S. and Britain. Iran was still bankrupt and the Eisenhower administration believed a full resumption of oil exports to be essential if Iran was to be saved from the Russians.

On the advice and with the cooperation of Attorney General Brownell, the complaint in the "oil cartel" proceeding did not include "the British, Dutch and other foreign companies as parties defendant in the action."[114] This omission pleased the State Department and disturbed lawyers in the Antitrust Division of the Justice Department. However, it did indicate a new degree of cooperation among top officials in those two departments.

On the last day of August, 1953, Dulles and Brownell agreed on the language the latter would introduce in the "cartel" case meeting of the NSC:

> Agreed, with the concurrence of the Attorney General, to recom-
> mend to the President that in the interests of national security in
> view of the Iranian situation, the Attorney General be requested to
> conduct proceedings in the so-called oil cartel suit now being car-
> ried on (NSC Action No. 766-A) with due regard for their effect on
> foreign relations.[115]

The NSC approved this statement.

The name of Herbert Hoover, Jr., a consulting oil engineer who had ad-
vised the Venezuelan government in negotiations with Jersey (and other companies)[116] in 1942-1943, quickly surfaced, as he was viewed as an individual capable of negotiating some form of working agreement between, first, the British and the Iranians; and second, the NIOC and the oil compan-
ies, now termed "the consortium." Everyone appeared to agree that the AIOC could not be revived as a predominantly British operation; on the other hand, the companies involved all had agreements with countries in the area, and any accord reached had to be comparable with arrangements in those countries.

Dulles checked with Harold Stassen, Thomas E. Dewey and the State Department's legal advisor, Herman Phleger, all of whom endorsed Hoover, who indicated his willingness to serve.[117] Next, Eisenhower appointed him petroleum advisor to Secretary Dulles.[118] Hoover then talked to Stanley N. Barnes, Assistant Attorney General, about the legality of the consortium and about problems he anticipated in the negotiations. At this time, Justice sought to persuade Hoover (and the State Department) to use American companies "not presently in the Middle East" rather than oil majors. However, neither the State Department nor the Defense Department liked the idea.[119]

Beginning negotiations, Hoover arrived in Tehran on October 17, 1953, and aided by Ambassador Loy Henderson, entered discussions with a committee of four Iranians appointed by Premier Zahedi. These talks stalled on a question of priority, whether Iran should first resume diplomatic relations with Britain or whether a solution to the oil problem should be found first. Britain desired the former. The U.S. backed Britain, and on December 5, 1953, diplomatic relations between the two countries resumed. Next, Hoover summarized for the Iranians just what information was needed before a new oil policy could be developed. He kept London informed of these conversations, and before long he began to shuttle among Tehran, London, and Washington.[120]

Hoover had requested that Justice advise him about how to establish a consortium "with as little likelihood as is possible of prosecution under our antitrust laws."[121] The oil companies desired strong assurances of non-prosecution, in exchange for participation. When the AIOC proposed a conference to be held in December, the companies sought guarantees from State and Justice against indictment for conspiring under the Sherman Antitrust Act. Jersey's Vice President, Orville Harden, wrote the State Department whether they "could see any objection to Jersey representatives participating in the conference." He added, "from the strictly commercial viewpoint, our company has no particular interest in entering such a group but we are very conscious of the large national security interest involved." Justice replied that Jersey representatives could attend only if Hoover appointed them as advisors.[122] The oil companies then met and indicated their "readiness to form a provisional consortium for the solution of the Iranian oil dispute."[123]

Leonard Emmerglick, however, believed that the oil companies requested this meeting as "a possible attempt to secure that kind of approval from the Attorney General which would ultimately make it impossible to go on with the cartel case." He opposed any grant of immunity to the consortium. The State Department did not concur; they pressed Justice instead to decide whether the antitrust laws would apply to the proposed consortium.[124] Again the inconsistent position of the U.S. government and of the oil companies assumed great importance. How could the oil companies agree to participate in a consortium established under governmental auspices to do that for which they were being sued for allegedly doing on their own? How could the government ask them to do in one place what it held illegal in another? Brownell said that the suit would continue—but in a way to "avoid undesirable publicity." Eugene Holman, Jersey's chief executive, apparently spoke for all the oil companies in sessions with Dulles, Brownell, Humphrey and others[125] as he posed these and similar questions.

Justice attorneys were successful in getting Brownell to point out, in a memorandum to the NSC, that while he and others sought a legal solution to the consortium problem, U.S. companies had stalled the case by refusing to produce the documents requested. He wrote, "no hazard to national or company interests will result from disclosure of the facts to the primary guardian of national security, the Government itself." He blamed the slow-down on the oil companies alone.[126]

Even before Jersey, Socony, Socal or Texas could begin to act with the consortium, these partners in Aramco had first to explain to King Ibn Saud that they "would not be able to increase our liftings . . . for awhile in Aramco and . . . there might be chaos in the area if we didn't [go in]." They pointed out that "we were doing it as a political matter at the request of our government." King Saud pleased both the State Department and the Aramco partners when he replied that they should, in his opinion, go ahead. He added, "In no case should you lift more than you are obligated to lift to satisfy the requirements of doing that job."[127]

The American oil companies refused to participate in the consortium as first proposed until they had received adequate assurances that it did not violate antitrust laws. Page explained, "I can guarantee you we wouldn't have gone into the consortium [on a waiver of the antitrust laws] . . . they know a waiver lasts just as long as the political party was in there [power], if that long."[128] The NSC had already decided that "the enforcement of the Antitrust laws of the United States against the Western oil companies operating in the Near East may be deemed secondary to the national security." Now, the Justice Department had to review the legality of the consortium.[129] On January 14, 1954, the NSC, the secretary of the Treasury, and the director, Bureau of the Budget, met and advised the Attorney General that "the security interests of the United States require that United States petroleum companies participate" in the consortium.[130] Brownell responded on January 21, that if the proposed consortium, as developed, followed exactly the statement sent him, then it "would not in itself constitute a violation of antitrust laws."[131] The oil companies had made their case well. A *New York*

Times petroleum expert noted that the U.S. government, anxious to resolve the Iranian problem, had applied pressure on the oil companies encouraging them to join.[132] Finally, Howard Page best expressed the oil companies' position when he said, "This consortium was set up at the request of the U.S. Government and the British Government. It was not our idea at all . . . none of the companies."[133]

Nevertheless, the American companies plus AIOC, Shell and Compagnie Française des Pétroles resumed talks in February, 1954. They decided how they would divide the oil, clearing each step with the three governments involved.[134] Then, the Iranian government invited the consortium to send a team to begin negotiations.

By this time the Western European powers, especially Great Britain, had become anathema to most Iranians. This made the appointment of an AIOC official to head the consortium team unacceptable. Since five of the companies involved were American, logically one of their representatives could be selected to lead the negotiating team in attempts to resolve the dispute. For this task, the companies chose Jersey's representative, Orville Harden [Harden had retired as vice president in December, 1953]. A brilliant man, Harden had established his mastery of the world oil trade over more than thirty years of service to Jersey, during which he had earned a well-deserved reputation for skillful negotiating. Middle Eastern leaders liked him for his courtesy and gracious manner, and trusted his integrity. Some observers stated that Harden possessed an unerring instinct for understanding the positions of all those with whom he dealt. Thus, with these qualities and with his mastery of seemingly limitless detail, he could negotiate from a position of strength.

All members of the Harden team came from Jersey, AIOC and Shell. Harold "Bill" Snow of AIOC and John Loudon of Shell headed the delegations from those companies. They brought with them outside tax counsel. Harden's Jersey team included Smith D. "Sparky" Turner, an expert who had acquired invaluable experience handling Middle Eastern problems through long service there and in London; Thomas E. Monaghan, Jersey counsel; and from London, Phyllis Carrier.

Other American companies apparently resented their lack of representation on the negotiating team. For some time, they maintained delegations in Beirut, on a stand-by basis, perhaps as emergency resource groups. But Harden never called them in.

When he and his associates arrived in Tehran, they found Torkild Rieber, serving as consultant to the Iranians. Rieber, then president of the Barber Asphalt Company, had at one time been chief executive of The Texas Company. Later Howard Page came to believe that Rieber had been sent to Iran by the U.S. government "to try to steer Iran into a proper frame of mind for future negotiations." Page did learn that Rieber had been outspoken with the Iranians about the absolute necessity of reaching some agreement with the oil companies.

The Harden team located its headquarters in a Tehran Army officers club, vacated especially for them. There "amid luxurious Persian carpets they froze, and were almost killed with kindness. The most famous chef in Iran was brought in . . . vodka flowed like water, caviar was heaped out like hash," Turner recalled. The team could not endure such richness. Herbert Hoover, Jr. diplomatically moved Harden to the Darband Hotel and gradually, the rest of the team moved there also. Formal meetings with the Iranians took place in the White Palace. When the Harden team conferred in private, Ambassador Loy Henderson made available for their use the library-conference room of the American Embassy.

Discussions began in Tehran, in April, 1954. Compensation to British Petroleum Company for their shares of the remaining AIOC properties which would be retained and controlled by the national oil company, such as a small inland refinery and marketing facilities for the whole country, were matters for BP alone. On the larger issue, the normal procedure would let the consortium pay Iran for the properties they would operate, and Iran then pay BP for what had been expropriated. However, if this precedent had been followed, Iran would have lost face. So, before leaving London, the consortium members agreed to pay compensation directly to BP. During one early morning meeting of the Harden group, shocking news arrived; the Majlis had passed new legislation prohibiting further grants of oil concessions, precisely the devices that the consortium had expected to use as tools.

Somewhat frustrated by this turn of events, the team debated the alternatives for endless hours. One possibility, that of making the consortium an agent of NIOC, did not receive enthusiastic support, for both the Iranians and the Harden team feared the results of such a step. But finally, this proved to be the only way out of the dilemma. So over long periods of time, representatives sought "appropriate wording, safeguards and qualifications" which would serve to protect and guarantee all interests of both groups.

Dr. Ali Amini, the Iranian Finance Minister and head of the Iranian negotiators, impressed the Harden group with his sheer ability. But he was trapped between his desire to save his country from ruin—oil exports had to be resumed—and his fear that if he agreed to a contract unacceptable to the Majlis, both he and the Shah would be expelled. Fuad Rouhani, the legal advisor to Amini [later with of the Organization of Petroleum Exporting

Countries—OPEC], also made a favorable impression in the minds of both the Americans and British.

The pace of the discussion was not hectic. Religious holidays slowed down the negotiations while providing the Harden team members an opportunity to rest and tour a bit. When the groups came to the bargaining table, however, no agreement could be concluded. The difficulty lay in defining the rights and obligations of each side under the agency concept. Harden decided to return to London with his group, in order that they might confer with their principals. As indications of both good will and of their intention to return, they left several team members in Tehran—and referred to them as "hostages." Iranian demands also had discouraged Rieber, who joined the Harden group at the Tehran airport. He seemed to consider the task of arriving at an agreement with the Iranians impossible.[135]

As the Harden team left, the Iranians handed them a new document containing that country's current demands. This became known as the "Airport Memorandum," and these demands discouraged all of the oil company representatives. Harden returned to London a sick man, and he could not go back to Iran. At a meeting called in London to assess the discouraging situation, Howard Page, who had been elected to the Jersey Board on January 1, 1954, attended as Jersey's Middle East contact director. The consensus of those attending this meeting was that "the companies had been given a hopeless task and the matter should be dropped." But representatives of the British and American governments pressed for another effort. Page appeared to be the only person at the meeting who believed that such a negotiation "was not an impossible task and that several approaches were worth trying." He told the group that Jersey would be willing to participate in such an effort, and he was selected to head this team despite the fact that the two other negotiators, John Loudon of Shell and Harold E. Snow of BP, were senior to him.

Some sources have indicated that Enrico Mattei, who had achieved the grouping of all the state owned Italian companies into the Ente Nazionale Idrocarburi [ENI] and who had cooperated with AIOC, now sought a share in the consortium. To do this, he would have had to approach first the governments of Great Britain, the U.S. and Iran. Had they approved, next he would have had to approach the participating oil companies. Page, however, doubts that Mattei approached any of these; and had he done so, Page further doubts that any concerned government "would have attempted to force the oil companies to accept him as the U.S. government did in the case of the Independents (strictly American) who also had no existing non-U.S.A. production to replace with Iranian production." Certainly, this author has seen no persuasive evidence that Mattei did seek entrance into the consortium.[136]

The Page mission arrived in Tehran on June 20, 1954, and started seven-day-week negotiations that lasted until August 5. When this Western oil delegation arrived in Iran, they began meeting in the American Embassy, which Page described as "a . . . place, hot, cramped, . . . in the basement." The Page group did not desire to use the embassy for their work; however, Hoover insisted that it was the proper place to negotiate, for it had been "swept" for bugs. Page remembered that "Hoover . . . was scared stiff of the Russians" and that he brought in an expert on "bugging" [listening to conversations] to demonstrate to the Page group just how easy it was. The expert pointed out that the sound of running water could disguise conversations. Page then used the expert's words to avoid using the embassy. He said, "Fine. I know just the place," and then he suggested that the sessions be moved to a "villa in Tadrich" that had once belonged to AIOC.[137] There "water spilled from the swimming pool to a fish pool and then cascaded down to another pool where they washed dishes." So the party moved to the villa, where the official negotiations took place, and even "the expert couldn't say it [the place Page desired] was not good."

The negotiators established an office in the villa garden, and the secretaries worked in the building. Tables placed directly above the cascading water served to destroy the usefulness of any "bugging" devices, whether of Russian or other origin. The Western delegation lived in the Darband Hotel, a hostelry with a lovely view of the mountains but as some remembered, inferior food.[138]

Dr. Ali Amini, still heading the Iranian negotiators, immediately suggested, and Page agreed, "that neither team should be allowed to indulge in recriminations. . . . The discipline worked." They began discussions where the previous talks had ended with new proposals being fully introduced.

Torkild Rieber was not present to advise the Iranians. Possibly he did not return from London because he still considered the Page mission hopeless. In another change, the Page team, unlike the one that preceded it, encountered no mobs; "they [the Iranians] were really quite calm and law-abiding." However, Page and Dr. Amini did have personal bodyguards. Page named his "Dick Tracy." In a personal way, Page liked Amini, who he remarked:

> had the best mind of anyone I had met before or since. While he knew initially nothing about the oil business . . . , I would say that when the negotiations ended there probably were not more than ten men in the world who understood the workings of the

international oil business as well as he did.

In the case of an impasse in the discussions, the two jointly reconciled the points in dispute.

Page and Amini, with the aid of their associates,[139] finished the document termed "Heads of Agreement" and on August 5, Page left for Holland, where he dropped off copies to be signed by the European consortium members. He went on to New York, where the American companies signed, and from there to England and Tehran. There, he and Dr. Amini began work on a final agreement but the drafting proved to be difficult, since much of it had to be done by lawyers. Unexpectedly, a cable arrived from the consortium group in London ordering an end to all negotiations, stating that a point of law was involved. Page "was dumfounded," for the cable also stated the group would no longer be bound by the "Heads of Agreement." He quickly chartered a plane and moved everyone in his group out of Iran except for his wife and secretary. He left them behind because he was not ready to make "an irreversible break." Before leaving Iran, however, Page cabled London and called for a meeting of all the lawyers involved. This meeting took place when he arrived in London on September 1. The lawyer who had raised the point that stalled the negotiations presented the point at this meeting. He then walked out, refusing to discuss or argue it. Some of the other lawyers presented their views and then those present decided to vote on the validity of the point in question. The vote was unanimous against the interpretation that had caused the break in the negotiations. Then the company representatives decided that the drafting of the final agreement should continue. By this time they were being flooded with demands by Iran that were outside the Heads of Agreement, occurring exactly as Page had predicted when the group insisted that he inform the Iranians that the consortium would no longer support the Heads.

Accordingly, Page stated flatly that he would not return to Tehran unless the group broadened his terms of reference to include consideration, at least, of the most reasonable of the new Iranian demands. The company representatives granted Page's request, and then he and a few advisors (mostly lawyers) returned to Tehran on September 7. The moment belonged to Page.

Once the negotiations resumed, Page and Amini soon reached an agreement. The resulting document had to be printed, verified line-by-line, and agreed to by the principal negotiators. While all this took place, new Iranian demands required more negotiation, which led to changes on the printed copies. Finally, the negotiators signed at 3 a.m., on September 17, when they sent the final version off by chartered plane to be approved by the company representatives. After the companies approved the draft consortium agreement, it then became necessary to derive from it an exact Persian version in the Farsi language. This proved to be quite difficult to accomplish and Page required that an expert in that tongue be sent out from England to advise and check on each provision in the agreement. This, too, had to be signed by the negotiators and by the companies before being submitted to the Majlis for ratification and to the Shah for his signature.

Each company in the consortium had to establish its own trading company in Iran to buy and sell the oil produced under the agreement. The procedure was simple and routine, so it was deferred until after the Majlis had ratified the agreement. The lawyers submitted the required forms but nothing happened. By appealing to Dr. Amini for help, Page managed to get the incorporation certificates at the last minute so that oil could start moving into tankers as soon as the Shah signed the agreement. Page learned some six months later that his edict "no gifts to any government employee," issued to his team on the first day, occasioned this delay. Porters in the ministry where companies incorporated, had refused to carry the papers around to be signed because they expected tips [their principal source of income], and none had been forthcoming.[140]

On September 17, 1954, Page and Dr. Amini signed the consortium agreement. Then, it went to all of the involved companies for their endorsement and signatures, to the Majlis for ratification and finally to the Shah for his approval. Until that happened, no oil could flow. This took time. The debate in the Majlis was long and bitter because some members questioned whether the new agreement conflicted with the March, 1951 nationalization law.

The appointment of a strong Nationalist, Hasan Sadr, to head the joint Majlis-Senate oil committee that considered the draft agreement insured careful review. Dr. Amini, Fuad Rouhani and other oil experts testified before this legislative body. Its examination of the consortium agreement took almost a month, and when the work was completed, the Majlis ratified the pact on October 21, 1954. Next, the Iranian Senate approved on October 28, and the Shah signed the agreement on the following day. Amini and Page flew to Abadan to make speeches on October 30, 1954, while oil flowed into the waiting tankers as they had planned. By November 4, Page was back in New York, having achieved what many had believed to be impossible.

It should be noted that not all U.S. companies wanted to sign the agreement. Several companies stubbornly refused to agree, "They were forced in . . . and . . . did not want to go in."[141]Finally, all of the five U.S. oil majors did sign and join with the three foreign based companies, to form the original consortium.* As one historian of British Petroleum wrote, "Thus it seemed, honour would be served all around."[142]

Of course, AIOC did not like being booted out of Iran. But given the circumstances, it had no real choice, and it accepted the compensation paid by the other consortium members. Regarding that price, Sir Francis Hopwood of Shell remarked, "This must be a very fair price, as it was agreed after long negotiations between an unwilling seller and unwilling buyer." No more accurate evaluation is possible.

In return, President Eisenhower and Secretary of State Dulles pronounced the accord a new source of strength for the free world. Then, the State Department arranged to satisfy, at least to a degree, the complaints of Justice and the independent oil companies regarding the failure to include oil companies that had not yet demonstrated much interest in the Middle East.

Yet the consortium agreements themselves did provide that, during the first six months after the effective date, the five American participants *might transfer* up to one-eighth of their interest [total 5 percent] "to one or more other established American oil companies," with the proviso that neither the British nor Iranian governments objected. Further, the transferee had to become a party to all the agreements and had to be "financially responsible." When the six months open period for such transfers ended, no participant could transfer "all or a part of his Participation Interest" to other parties for at least five years.

In the various agreements, nowhere was any requirement imposed on the "participants to take in additional companies." In fact, as Kenneth R. Harkins and W. B. Watson Snyder wrote to Assistant Attorney General Barnes, "no method has been provided by which the State Department could require the five American companies . . . to admit additional companies." Justice, they said, had never been informed that such an agreement existed. On the letter, however, Barnes noted, "not officially, but H. Hoover has so informed me orally." When the five majors indicated their willingness "to transfer at cost a portion of their participation," the State Department requested that interested independents apply for shares. All together, these smaller oil companies applied for 36 percent of the total consortium take. On or around April 15, 1955, the Iricon Agency Ltd. was organized under the laws of Great Britain, to hold the interest in the Iranian Consortium released by the five American majors. Commenting on this development,

*See list, footnote 134.

Page remarked in jest, "Because people were always yakking about it, we had better put some independents in there."[143] He added an opinion that since these companies had few outlets abroad, they could not move the volume of oil necessary to fund the Iranian government. Certainly, Jersey and the other companies had no intention of voluntarily yielding European or other markets to the independents. At a later time, Page noted that some of the companies who received a portion of the 5 percent allocated by the majors could not handle their share of the oil.[144] In 1955, however, each of the five majors assigned 1 percent of its share to the Iricon Agency, made up of eight independent American oil companies.*[145] The State Department selected those companies to participate, and each paid a proportionate share of the compensation due British Petroleum.[146]

With the consortium agreements signed, Jersey now had a 7 percent participation in Iran, from which country it had been excluded earlier. The Iranian shutdown of 1951-54 had led to increased production in Kuwait, Saudi Arabia, Iraq, and especially in Venezuela, where it reached record levels. With the large productive capacity in other countries, the resulting surplus created larger problems for Jersey and the entire industry.[147] Surely Jersey had economic motives for entering the agreement, but it did not enter to obtain additional production. Had the Russians moved into Iran and dumped oil on world markets, Jersey would have been seriously hurt. Had Jersey refused to participate, this too would have damaged the company's prospects. Contrariwise, had the Iranian nationalization worked, other producing countries would have been emboldened to expropriate foreign properties. All these possibilities had been thoroughly discussed at Jersey. In 1974, twenty years later, Howard Page told the Senate Subcommittee investigating multinational corporations that "up to this time, probably no, we haven't gotten our money back." But he believed that in the long run, Jersey would recover and even profit. As for the 1954 decision, politics rather than profits account for Jersey's participation.[148]

If nothing else, the Iranian crisis established clearly that in this case the United States government used American business firms to further its Cold War containment policies. The two presidential administrations involved acted for political as well as economic considerations. National security, in their view, must take precedence over the desires of FTC and Justice Department lawyers. Finally, the post-1954 success of the consortium arrangement appeared to justify the position of the National Security Council. After finishing his terms as President, Eisenhower pronounced settlement of the

*See list, footnote 134.

Iranian problem one of the major achievements of his entire administration. And the consortium itself, with only minor modifications, endured for some twenty years.

X   X   X

Many Americans believed that the Justice Department lacked any authority to grant immunity for statutory infringements—since only Congress could change the law.[149] Given the arrangements presented to the Jersey Board of Directors, however, no alternative to membership in the consortium appeared feasible. Many persons believed that the Defense Production Act of 1950 (with later changes) had provided that the national security take precedence over the antitrust statutes. Earlier, and in the civil complaint filed in connection with the "dismissal of the grand jury inquiry and the vacating of the existing subpoenas," the Justice Department moved that Jersey and other companies be perpetually enjoined from a long list of activities alleged to be in restraint of trade in foreign and domestic commerce. At that time, Frank Abrams, speaking for Jersey, had pronounced these charges false and based "on interpretations of business actions which give an entirely wrong impression of the actual relationships between the companies involved." He expressed confidence that Jersey had broken no antitrust laws and once again stated that the company belonged to no "cartel."[150] Other members of the Board believed that when the criminal investigation was dropped, and replaced by civil proceedings, the company's best interests had not been served. Jersey's outside attorney, Arthur Dean, stated this view in a conference with Justice Department officials. Jersey, he said, preferred not:

> to be placed in any bargaining position with our own government. We don't want any of these ministers or any of the heads of these foreign governments to say that we were afraid to stand trial on the criminal side of court. Speaking only for Jersey as a pure litigating matter outside the security interest of the United States, we would prefer to be tried.[151]

Since the NSC had already made its decision, the question of criminal trial appeared moot. Yet Dean believed strongly that Jersey would win the case.[152]

On the other side of the court, the Antitrust Division of the Justice Department believed that a civil case, properly prosecuted, "still held a promise of significant relief." Late in 1953, seven Justice attorneys, using "court discovery procedures," began to reexamine the Jersey (and other companies') files. In all, they secured more than one hundred thousand documents and letters, and shipped these materials to Washington, where other staff members sought to fit them together with other documents, to form a comprehensive base of evidence. Federal readers included both lawyers and economists. One interdepartmental group, the Government Screening Committee, including members from the State, Justice and Defense Departments and the CIA, pored over them seeking "possible security sensitivity." As Antitrust Attorney Barbara J. Svedberg phrased it, government officials "read documents month after month" and "in some cases they were extremely scrambled." Another attorney noted that "Jersey appears to have an administrative policy of making 10 or 12 copies of every incoming document. At times as high as six copies of the same document will be scattered all through the Jersey deliveries." It took four Justice attorneys almost thirty months to examine 43,930 Jersey documents.[153] Ultimately, Justice selected about two thousand papers from all the defendants for trial purposes.[154] At the same time, the Justice Department pressed the court to require all the American companies to furnish additional material regarding their relationships with foreign-based oil companies. Too, Justice wanted to see the correspondence and documents in the possession of the Department of the Interior, acquired while it operated the Petroleum Administration For War (1941-1945) and the Petroleum Administration For Defense (1951-1953).

X X X

In 1955 and part of 1956, Jersey, through its attorneys, continued to insist that it had no interest in settling the case with a consent decree. Justice antitrust lawyers, now headed by Wilbur L. Fugate, continued to press the oil companies to supply documents from foreign subsidiaries. When it became evident that—without further legal action—these would not be forthcoming, Justice began to prepare for trial. Again on June 19, 1956, Jersey attorneys still headed by Dean, conferred with the Justice Department lawyers, and repeated arguments concerning the sensitivity of certain papers. Dean made the point that there were some documents, "disclosure of which would be dangerous or embarrassing to Jersey for other than antitrust reasons." [These papers, Justice was invited to see and take notes on.] According to one source, Jersey (and the other defendant companies) "admitted that it participated fully in the original Achnacarry and other related AS IS agreements up until 1938," when it ordered all such agreements ended. To this,

Justice replied that Jersey had produced no evidence that these agreements were ended. Indeed, Justice went further and alleged that Jersey and its peers had cartelized operations during World War II and afterwards. Dean remarked that he had "all kinds of documents to show we didn't do those things." Finally, after a full discussion, he insisted that Jersey remained uninterested in a consent decree.

Meanwhile, in 1956, the NSC further refined the case by recommending that the government not seek relief through challenging joint production, refining, storage and transportation ventures among the defendant U.S. oil companies, or to other such operations with CFP, British Petroleum, and Shell. Justice followed this recommendation; however, Emmerglick and his associates believed these NSC decisions "cut the very heart out of the cartel case."[155]

During the Suez Crisis of 1956-1957, the "cartel" case did, in fact, "stand-still." Recognizing that an oil shortage might occur in Europe—an event disastrous to both our military and our allies—the Eisenhower administration granted immunity (for a specific period only) from antitrust laws to all defendants in the case. For the involved companies, this marked the second time in five years that they had been asked by their government to do exactly that for which the same government at the same time was prosecuting them. And again, the oil companies, under the leadership of Jersey's Coleman, did come together to form the Middle East Emergency Committee [MEEC] to advise on supplying oil to Europe.

The patent unfairness of the government's prosecuting the oil companies on one hand while working with them on the other, bothered many officials. So Justice moved to obtain consent decree agreements from one or two of the companies, in the hope that those remaining would agree to negotiate settlements.

Though earlier, Jersey had consistently refused a consent decree settlement, the NSC decision of 1956 so narrowed the case that only joint marketing operations remained subject to litigation. Justice attorneys had one such joint marketing company in mind—the Standard-Vacuum Oil Company, created in 1933. At that time, this corporation, jointly owned by Jersey and Socony, complemented the interests of both companies in the Far East and parts of Africa; Jersey had production and refining but few markets, while Socony had markets but lacked oil. Together, the companies set aside a large part of the world for Stanvac in which they agreed not to compete directly. As one guarantee, the two top Stanvac executives reported only to the chief executive officers of the two parent companies, and not to department heads. Just before, and increasingly after World War II, the Stanvac arrangement began to bind the owners, who had come to see such an organization as unwieldy and perhaps, unworkable. This complicated Stanvac operations. Jersey Director Lloyd W. Elliott, who spent eighteen years in the Far East, most of it with Stanvac, listed several reasons that had caused both parent companies to consider dissolving the affiliate; but to him, meddling by business executives proved the major problem. He remarked, "That's what was making it more and more difficult all the time to continue the Standard Vacuum."[156]

Before Justice began to eye Stanvac as a way to salvage something from the antitrust suit, Jersey and Socony had seriously explored the possibility of ending their joint ownership—for business reasons. Concerning Jersey's position in Stanvac, another important Jersey official remarked:

> This difficulty should be cleared up. A division of the Stan-Vac properties between the parents seems to be a likely way to bring about the desired result, if Justice can be persuaded to go along. If this step should be taken apart from the litigation, it might not be necessary to consult Justice . . . the importance of clearing up this legal tangle at some appropriate time should not be overlooked.[157]

Jersey's interest crested when their counsel reported that "the U.S. Government looked on this action as a basis for settling as to Jersey" the civil antitrust suit begun in 1953.[158]

Negotiations leading to a consent decree in 1957 opened on May 15, when Director David A. Shepard and Counsel Thomas E. Monaghan met with a number of Justice lawyers to discuss the possibility of settling the case. Stanvac remained the issue, and Shepard and Monaghan indicated that both Jersey and Socony Mobil [Socony-Vacuum became Socony Mobil in 1955] would have to accept the various forms of relief agreed to. Having achieved some general understanding, they decided to approach Socony on these matters after terms had been defined.

After hearing the Justice proposal, Jersey decided to reverse its earlier position, in order to end the long legal battle. Dean stated that Jersey "entered into such negotiations with the greatest of reluctance and only because officials of their foreign subsidiaries had represented that a trial of this case might greatly injure Jersey interests abroad." The negotiations proceeded, and in September, 1957, Assistant Attorney General Victor R. Hansen sent Monaghan a copy of a consent decree signed by the government and by Jersey. Jersey was allowed seventy-five days to persuade Socony to sign the document.

Two months later, Monaghan and Shepard returned to the Justice Department, where the latter stated "that the Jersey Board had been very much surprised at Socony's refusal to consent to separating Stanvac's marketing assets." Numerous conferences had taken place between spokesmen for the two companies. Now Shepard told the Justice attorneys that Jersey believed that Socony "had come around to the point where it was willing to discuss the actual breakup of Stanvac."[159] Howard Page, who participated in these Board discussions, believed that Socony refused "to go along" because "they were making more money out of Stanvac than Jersey, and they thought the government did not have a case against them or Jersey." Socony "had no intention of accepting a consent decree." The Board concurred in this opinion, but CEO Eugene Holman decided that the time had come to settle the case and to get on with the business of selling oil.[160]

While consent decree negotiations were consuming time, the pressures to act were mounting. Senators Joseph C. O'Mahoney and Estes Kefauver suggested that the Justice Department try the "cartel" case, rather than settle it by consent decree. Socony and Socal indicated to Justice and to the public that they would fight the case with every legal tactic at their command. Gulf, like Jersey, appeared willing to negotiate a consent decree. So did The Texas Company, at least at times. Meanwhile, the oil companies continued their resistance to making additional documents available, while Justice attorneys accused all of them of keeping key records overseas.

Jersey Counsel Monaghan, sometimes accompanied by Director Shepard or Eugene Holman, appeared endlessly at conference after conference with Fugate and other Justice attorneys. On November 13, 1959, Monaghan told Fugate that Jersey had worked out a plan to split up Stanvac and complete the move by December, 1960, if Jersey could "obtain a consent decree." Fugate refused to promise a settlement until he received the final court order on providing new documents. That order came a few days later. It required the defendants to produce "domestic and foreign documents in their files" and in those of their subsidiaries. Fugate noted that these would help in the case "against Socal and Texaco." On December 11, 1959, Fugate told Monaghan that the government would deal with Jersey alone, provided it had the "written agreement of Socony to the breakup of Stanvac."

Jersey apparently produced all the required papers, including the Socony agreement. Late in that year (1959), Elliott and Page reported to the Board on Stanvac dissolution problems. Socony now appeared ready to sign a "Statement of Intent" with Jersey regarding the breakup of Stanvac. On April 7, 1960, Elliott, Page, Norton Belknap, and Nicholas J. Campbell, Jr., associate general counsel, presented the Board with a proposed reorganization agreement between Jersey and Socony, which the directors approved.[161] The Board also authorized execution of the agreement on behalf of the company. They appointed Monroe J. Rathbone, Campbell, and G. D. MacConnachie to represent the company at the formal signing of the papers. This took place in Jersey City at 10:00 a.m. on April 13, 1960, at the offices of the Corporation Trust Company.[162] On November 14, the U.S. District Court entered a Final Judgment for Jersey ordering the dissolution of Stanvac within three years. The intent of the decree, as stated, provided "so far as practicable, [to] permit Socony and Jersey to compete with each other in marketing petroleum products in the area formerly served by Stanvac."[163] The decree also enjoined Jersey from "any future agreements or combinations by the companies to fix prices, divide markets or allocate production with any competitors in the world oil market." In accepting this decree, Jersey admitted to no violation of the law.[164] With reference to the decree (as well as those against Gulf and, later, Texaco), the Justice Department stated that "the main objectives of the suit have been realized" and "would go far to correct competitive conditions in United States foreign trade in petroleum and its products."

To represent Jersey in the final negotiations, the Board appointed Elliott, Page, Belknap, and Campbell, who had already spent months working on the problems of dissolution with executives of Stanvac and Socony. Campbell said:

> Finally, we were able to persuade Mobil [Socony] to sit down with Jersey to negotiate the Stanvac split-up. . . . I was deeply involved in that. We went for days either in Shorty Elliott's office or the old Chatham Hotel. We worked day and night. We had tense moments.

They agreed on a plan to split up the company, everywhere except in Japan, where a division, they believed, would take five years to arrange. But Justice insisted that Stanvac interests in Japan be dissolved more quickly; so the negotiations resumed. According to Campbell:

> At one point, the division of assets got to be such a critical point that they actually cut a carpet right down the middle of the room. We almost had a crisis because of two or three bicycles in the Osaka branch. . . . It was protracted and . . . a great tribute to Shorty Elliott and Paul Keyser who was their negotiator [Socony Mobil] . . . we finally got the thing worked out.

From the start of the "cartel" suit Campbell had believed that Justice had a poor case and that they sought "some way to settle . . . because they were convinced that they could not try their case and win it."[165] Holman kept pressing for a division of assets, but delays continued. Page has stated, "The thing broke down completely to start with . . . and I went over and Gene Holman talked to me about it." Socony Mobil representatives continued to tell Page that the government:

> has no case and we [Page and Elliott] said we don't think they have a case either but they are holding up a lot of stuff. In some cases we couldn't do the things we wanted to do because they would be illegal. So we wanted the breakup. Stanvac did not. They [Socony Mobil] were getting the best of the deal. They were not worried about antitrust.

Early in the negotiations with Socony, Holman told Page that it appeared that the "thing [breakup] had come to a complete impasse, and that he was . . . at his wits end to find how the hell to do anything." During the conversation, Holman recalled Socony saying, "They wouldn't do it for anything less than fifty million dollars" plus the sharing of properties. As Page remembers:

> So I thought about that for awhile and I thought of a scheme. And that was to offer them a piece of property in Venezuela which had not yet been developed, which if it contained the oil the geologists said, might be well worth fifty million dollars. But I never . . . believed geologists and I knew that they [Socony Mobil] wanted very badly to get in Venezuela. . . . So we made this proposal to them. That brought them in.

Jersey worked out a transaction with the owner of the property, the Creole Petroleum Corporation, in which that company "definitely came out ahead."[166] Jersey then pressed a final settlement with Justice while also pushing Socony Mobil toward a final agreement on the breakup of Stanvac. But even after Socony Mobil accepted the Venezuelan property, as Page suggested, working out details took three more years.

As it happened, the dissolution of Stanvac coincided with Jersey's own earlier plans. An increase in demand for refined products "together with the increased crude production and refining capacity in the areas [Far East] has convinced the companies that it will now be in the best interests of all concerned to serve the public as separate organizations." The position of

Stanvac's parent companies in the Far East no longer could be complementary. [The agreement did permit Stanvac to continue producing in Indonesia.][167]

At the time of the dissolution, 1959, Stanvac had just passed the billion dollar mark in sales, and it claimed assets of almost $900 million.[168] Although headquarters were in White Plains, New York, Stanvac had 37,000 employees in more than forty foreign countries. Understandably then, the mechanics of reorganization proved difficult but pragmatic. Jersey ever sought talent, and as Page pointed out, "The most difficult thing, I think, was the sorting out the people. We had to be sure we knew about the one we chose." In dividing the Japanese personnel, Page and Elliott agreed that they wanted a young man, M. Yashiro, who was outranked by about fifteen persons. Nevertheless, they selected him as Jersey's first choice, and his later success with the parent company vindicated their judgment.

Jersey wound up acquiring sole title to the existing marketing operations of Stanvac in Pakistan, India, Goa, Ceylon, Burma, Thailand, Malaya, Singapore, South Vietnam, Cambodia, Laos, Luzon Island in the Philippines, Uganda, Tanganyika, Kenya, the Malagasy Republic (Madagascar) in East Africa, and to one marketing affiliate in Australia. In addition, Jersey received one-half the assets of Stanvac marketing in Japan, plus one-half interest in one refinery and one-fourth interest in another. It also received majority stock interest in refining companies in the Philippines and in India.[169] Additionally, Jersey acquired the Stanvac headquarters building in White Plains, New York, along with the title to the name, Standard-Vacuum Oil Company. The Stanvac name was replaced by Esso Standard Eastern, Inc.

The Stanvac dissolution did not entirely satisfy Justice, but it did open up the Far East for Jersey. Even so, the unfortunate publicity attending the long "cartel" suit, filled with charges and denials, served to render even less secure from host nations the properties of all American oil companies in the Middle East. Reflecting on this point and drawing on his years of experience in dealing with that section of the world, Howard Page remarked:

> that [the cartel suit] stated very specifically that we were a cartel . . . that there were law suits. . . . It was no good telling anybody else, 'Well, it's the kind of thing that goes on in the States.' They [foreigners] said, 'obviously you're a cartel. . . . Your own government says so. What the hell do you expect us to expect.' Now, where that really hurt was that once [oil] prices started dropping . . . they said to us, 'Look, what . . . are you doing to us? You're a

cartel and you let the prices drop.' So we would try to explain that
we were not a cartel. And they said, 'Don't be silly. We got the
report [your government's] that says so, and you're just purposely
getting together to knock the prices down just to hurt us. It doesn't
make any sense, otherwise you'd roll the prices up, any good cartel
does that.' And we just couldn't convince these guys there wasn't
some dirty work afoot.[170]

Foreign governments kept themselves informed about the course of American law. As early as May, 1953, the Saudi Arabian government had secured
a full translation of the crucial *FTC Staff Report* that shaped the entire
"cartel" case.[171] From that date on, Middle Eastern heads of state applied
mounting pressure on the companies to increase the income from oil. And,
the political trend toward nationalization kept growing. Worldwide, the 50-
50 arrangement of oil companies with Venezuela increased pressure among
all host nations everywhere who demanded an increased share of profits.
The prospects of such arrangements plus the *FTC Report* might have precipitated a world crisis for the oil companies in the 1950s, had not the Iranian
fiasco sharply amplified a stark warning against mindless expropriation and
nationalization of oil concessions. Certainly, as the Departments of State,
Defense, and Interior reminded us in 1953, "the motives of any foreign
enterprise" were still suspect in the Middle East, and the very success of the
oil companies "invites envy and dislike." This joint document warned,
"There is, of course, no way of preventing adverse reactions arising from the
FTC Report, because it has been published," and went on to state that the
United States was waging a "constant struggle to dispel the concept, very
much alive, not only among communists but also in non-communist circles
of the free world, that capitalism is synonymous with predatory exploitation." [172] Now, from the vantage point of more than thirty years, it appears
that Truman did err in declassifying the *FTC Report*, and that Justice
wrongly pursued vigorously pushing the antitrust suit when other options
were open. Yet, even today, it is still not clear to what extent American
antitrust laws may apply to American business operations in some foreign
countries. What the Jersey story does provide is a classic example of confusion; a catalogue of the problems that must be expected when a transnational business is subjected to differing national policies emanating from
various government agencies that were themselves pointing forward toward
different objectives.*

---

*It is interesting to note that in 1981, Justice withdrew the antitrust case entirely, advancing as a
reason the lack of evidence.

X   X   X

In retrospect for Jersey, the hectic years, 1950-1954, appeared to have been linked by problems with which the company could grapple, but never really solve. The *FTC Report* plus the ensuing antitrust suit both represented situations of the type Jersey had encountered previously and had been able to settle. But, no one at Jersey could have envisioned in 1950, that the company would become involved as a major participant in the settlement of the Iranian crisis in 1954. In the beginning, these incidents were so separate as to be disconnected—certainly not hinged. And had any executive possessed such foresight, he probably would not have guessed that the company would join other oil majors, under governmental auspices, to become an arm of the Defense Department during the Korean War, while at the same time a party to court proceedings brought by the Justice Department for combining to restrain trade. Finally, the balancing of the antitrust settlement with Stanvac's dissolution could not have been anticipated. In fact, however, as other sections of this study point out, these hectic years served to show that as American business became truly international, necessarily its base broadened and it had to become increasingly involved with questions of both national security and American diplomacy. The activities of Jersey from this time forward take place in a public arena, under the glare of the media. For such exposure to worldwide publicity, no parallel existed in earlier company experience, as transnational corporations in the last half of the twentieth century became truly global concerns.

TABLE 1.

GROSS EXXON CRUDE LIFTINGS FROM
MIDDLE EAST ARRANGEMENTS
(Thousands of barrels daily)

	1975	1970	1965	1960	1955	1952
Crude Oil Offtake Under						
Equity Arrangements (1)						
Saudi Arabia	848	1,079	611	365	278	248
Iraq, Qatar and Abu Dhabi	89	258	201	225	89	50
Libya	104	631	520	–	–	–
Total Middle East and Libya	1,041	1,968	1,332	490	367	298
Crude Oil Offtake Under Special						
Arrangements (2)						
Saudi Arabia (Govt. Participation						
Crude Buyback) (i)	1,142	–	–	–	–	–
Iran (NIOC Purchase) (ii)	241	245	128	67	21	–
Purchases under other special						
arrangements (iii)	344	663	220	164	100	53
Total Middle East	1,727	908	348	231	121	53

(1)   Includes 100 % of Exxon and majority owned affiliate production and Exxon ownership percentage of the production of companies owned 50 % or less.

(2)   Includes:

   (i)   Buyback of parts of the governments' share of production under terms of participation agreements effective January 1, 1973, and as amended therafter.

   (ii)   Offtake from Iran from March 21, 1973, under terms of the purchase-sale agreement with the National Oil Co. (included in gross production prior to that date).

   (iii)   Major purchases from BP (Kuwait) and others under long term special arrangements.

N.B. This table used by permission of Exxon Corporation. See *letter,* J. P. O'Halloran (Esso Middle East) to George H. Lewis (Public Affairs), September 1, 1982, (copy), Files, C.B.H.S.

Chapter 16

Esso Eastern, 1961-1975

I N 1960, AFTER EIGHT YEARS of almost ceaseless negotiating, the Standard Oil Company (New Jersey) [Jersey] agreed to a consent decree offered by the United States Department of Justice [Justice] "as a basis for settling, as to Jersey, a civil anti-trust suit with respect to foreign marketing operations which had been pending against it and four other oil companies since 1952."[1] Part of the decree required that the Standard-Vacuum Oil Company, Inc. [Stanvac], owned in equal shares by Jersey and Socony Mobil Oil Company [Socony Mobil]*, be dissolved.

During the years 1953-1960, Jersey Director David A. Shepard and Counsel Thomas E. Monaghan spent countless hours in discussion with Socony Mobil executives before the latter would agree to dissolve Stanvac, and even more time discussing the case, Stanvac, and the consent decree with attorneys at the Justice Department. Jersey Vice Presidents Lloyd W. Elliott and Howard W. Page, aided by Associate Counsel Nicholas J. Campbell, Jr., and Norton Belknap of the Petroleum Economics Department, then worked interminable hours with their Socony Mobil counterparts in the effort to divide fairly the properties of Stanvac. Elliott and Page even had to go to Japan to negotiate with Stanvac's Japanese partners, who had a 50 percent interest in the company's refineries there. Finally, both Justice and the Federal District Court for Southern New York approved the settlement, which was intended to restore marketing competition between Jersey and Socony Mobil in the Far East.[2]

Except for its headquarters in White Plains, New York, Stanvac—a company with a billion dollars in sales in 1959—had all its operations east of Suez. By the terms of the consent decree, operating properties in more than thirty countries were to go to Jersey—from the east coast of Africa across South and Southeast Asia south to Australia and north to the Philippines and

---

*In 1955, the Socony-Vacuum Oil Company, Inc. became Socony Mobil Oil Company. In 1966, the latter became the Mobil Oil Corporation [Mobil].

Japan.* Although agreement to reorganize Stanvac was reached on April 13, 1960, the court did not render its initial judgment until November 14. Working out the complex details associated with this transaction, however, went on for more than a year. As the Jersey *Annual Report* for 1961 noted, progress had been made "moving through the intricate details attendant upon the division of a corporate entity of such size and importance."[3]

Jersey Regional Coordinator for the Far East H. W. McCobb anticipating that a switch-over to a new company might create problems settling company accounts with Stanvac late in 1961, sent a letter to all Jersey affiliates urging them to "adopt any measures necessary to insure that billings from your company to Standard-Vacuum are mailed" before December 31.[4] Stanvac planned to end joint operations by March 30, 1962, and at that time, a new affiliate would take over Jersey's share of the Stanvac properties. Employees referred to March 30, 1962 as " 'E-Day'—Emancipation Day." Jersey had created to take over its share of the Stanvac properties a new company, Esso Standard Eastern, Inc. [Esso Eastern or ESE].**

At an administrative level just below the Board of Directors and company counsel, many executives worked on committees to arrange the details of this transfer. Training programs for all former Stanvac employees had to be organized; tons of leaflets explaining the change had to be printed in dozens of languages, and old operating manuals had to be rewritten. Because Socony Mobil formerly handled all Stanvac marketing operations, hundreds of people now had to become familiar with the qualities and uses of Esso products, even to such details as lube specifications. The old system had existed for thirty years; thus for many employees the change to Esso amounted to starting anew.

Of course a major public relations advertising campaign accompanied the shift in brand identification. So among the many changes that were under way wherever Jersey sold its goods, arrangements had to be made to remove the Stanvac advertising symbol—the flying red horse—and to replace it with the red, white, and blue Esso oval, until that time little known in the Far East. As one participant said, "Esso was not a brand in our part of the world."

Earlier, the Elliott-Page committee had worked out an agreement with Socony Mobil making it certain that personnel from either company who joined Stanvac in the previous five years would return to the company they left. The remainder of the employees were assigned to whichever company (Jersey or Socony Mobil) received the Stanvac property (or was to supervise

---

*Socony Mobil acquired 100 percent interest in many Stanvac affiliates, the largest of which were located in Australia, New Zealand, Japan, the Philippines, Southern and Eastern Africa, and Hong Kong.

**Formed in 1961

it). Overall, for an operation of such magnitude, remarkably little cross-recruiting occurred. In fact, the breakup of so large an enterprise had few precedents, so Jersey, Socony Mobil, and Stanvac executives charted a new path or procedure with every decision they made in accordance with their instructions, "Do the best you can."[5]

Jersey leaders had tried to interest Socony Mobil in dividing Stanvac even before Justice filed the 1952 antitrust suit. After World War II, the original reasons for the joint company virtually disappeared; at the beginning Jersey had access to crude oil and possessed refinery facilities but needed markets, particularly in the Far East, while Socony Mobil had the markets but needed crude and refined products. In the 1950s, largely because of the leadership of Eugene Holman, Jersey grew more rapidly than did Socony Mobil, and especially was this true of its marketing organization. Jersey's sales leaders became very aggressive, and sought to become highly competitive in all parts of the world except in the Stanvac territory. There, crude oil sales constantly increased, but neither Jersey (nor Socony Mobil) could enter marketing contracts for crude because by agreement they had turned the Far East over to Stanvac.

The management systems of the Stanvac partners were unlike; Socony Mobil was highly centralized, while Jersey permitted its regional managers considerable autonomy—and decentralized its operations to a high degree. Not surprisingly, now and then friction developed between the two owning companies "concerning the prerogatives of Stanvac management." Jersey had developed a remarkable executive talent search through its Compensation and Executive Development [COED] Committee, in which system all affiliates participated. Because of the vastness of the territory and the many countries in which Stanvac operated, executive development opportunities were limited; expatriate Stanvac employees generally remained within a particular region and with the same national affiliate for many years. Again, surveys by Jersey functional specialists indicated what appeared to be untapped marketing areas, conclusions with which Stanvac management did not always agree. Since the most senior management of Stanvac reported directly (and only) to the chief executive officers [CEO] of the two parent companies, some Jersey officials felt that Stanvac did not welcome new ideas or even suggestions for change from outside. Elliott, who had spent eighteen years with Stanvac, indicated his own awareness of this problem, but it remained unsolved, along with others.

From the management standpoint, the Stanvac organization had become top-heavy, with five geographical regions and a regional vice president for each. These executives had broad authority over the conduct of the business including specialized functional matters in their specific areas; orders from headquarters functional groups to a field office cleared through them. Since the regional vice presidents travelled a great deal, their assistants left in charge sometimes vetoed orders of veteran functional officials, and unpleasant feelings resulted. Finally, at the top of its administrative edifice, Stanvac utilized a sixteen-person board of directors, who appeared "to work in splendid isolation."

For twenty-nine years, Stanvac had been a profitable operation for Jersey and Socony Mobil. Given the areas it served and the many problems of all types it confronted, the organization had done well. Yet the company's administrative structure with its five geographical divisions clearly failed to allow a unifying force or spirit to develop, or to permit executives to develop as they should have.

When Jersey men began to join Stanvac, they found it difficult to dent the general miasma that afflicted that company. A one-time Stanvac executive wrote: "Head office controls were so tight that it was impossible for the latter [field offices] to do a creditable job, and it was obvious that there was tremendous waste motion and lost time in Headquarters." So experienced Jersey executives enthusiastically tackled the problem of dividing the old company and establishing a new one to manage its 50 percent share. In fact, the sale of the White Plains headquarters, which became Jersey's in the settlement, to International Business Machines [IBM] Corporation in December, 1961 [for $10.5 million], could be interpreted as Jersey's desire to break with Stanvac's past as much as a chance to realize cash from an asset.[6]

With the division of Stanvac and the creation of a new company to service that territory, Jersey achieved a greater degree of flexibility in operations in a vast Eastern Hemisphere area where energy use was increasing more rapidly than in most other parts of the world. Certainly, from the moment it appeared, the wholly owned company was more attuned to the Jersey management system than its predecessor ever had been.

The region included quite a number of countries which, prior to World War II, had been colonies of Western powers or Japan. Some of these had gained independence through peaceful processes [India, Pakistan, Singapore, Malaysia]. Others had been forced to seek freedom through revolution or civil war [Indonesia, Indochina]. Nationalism—or nationalistic fervor—quite naturally continued to be one of the strongest forces influencing internal and external political and economic policies in all of these states. Both Stanvac and Esso Eastern felt its force. Shortages of foreign exchange and capital in those countries as well as the comparable scarcity of both entrepreneurial and managerial talent demanded careful attention by Esso Eastern, and the frequent and rapidly changing political situations presented another factor to be considered.

Prior to the dissolution of Stanvac, it had become apparent that Socony Mobil's major interest lay in the more politically stable countries, including especially Japan. The market there had proven to be a dynamic one of high growth. In the splitting of Stanvac's assets, Socony Mobil received either all or a large share of the business in these more stable countries. Jersey was determined to gain a good share of the business in those areas also, and it believed that some of the other developing countries had considerable growth potential.

At the time of the dissolution, Stanvac claimed assets of more than $900 million. For its share, Jersey acquired sole title to Stanvac marketing operations in Pakistan, India, Goa, Ceylon, Burma, Thailand, Malaya,* South Vietnam, Cambodia, Laos, Luzon Island in the Philippines, Uganda, Tanganyika, Kenya, the Malagasy Republic (Madagascar) in East Africa, as well as to one marketing affiliate in Australia.

In Japan, the Jersey Company received one-half the assets of Stanvac's marketing companies, plus one-half interest in one refining company, and one-quarter interest in another. The sensitive and involved nature of both Japanese government and management relations had to be accommodated by the new joint owners. Stanvac had held a 50 percent interest in Toa Nenryo Kogyo K.K. [Tonen], a large, efficient, and progressive refining company; Tonen in turn owned a petrochemical company in which Stanvac indirectly had a 50 percent interest.** The other 50 percent of Tonen belonged to several banks and some private investors (including the founder, Mr. Nakahara, and his family). Under longstanding agreements, Stanvac had the right and obligation to supply essentially all Tonen's crude oil requirements and to buy all its refined products. These arrangements along with the 50 percent equity were divided equally between Jersey [Esso Eastern] and Socony Mobil in the settlement. The Japanese operating management continued, along with the two American companies, to retain individually proportional representation on Tonen's board of directors. Future relations of the two companies required that Stanvac's reorganization not inhibit Tonen's growth.

The name, Standard-Vacuum Oil Company, also became Jersey's.*** All together more than one billion people lived in the territory served by the properties that Jersey acquired in the settlement. More important, under the decree both Jersey and Socony Mobil could develop and operate properties and companies anywhere in the old Stanvac territory.

Prior to World War II and immediately afterward, the Stanvac region's petroleum business (except in Japan) was characterized by generally stable market conditions. Stanvac provided an almost typical example of foreign

---

*The names "Malaya" and "Malaysia" used in this section refer to the federation of states in the Malay Peninsula, plus Singapore, Sarawak and Sabah, created in 1964 (Singapore became completely independent in 1965).

**Jersey received a 25 percent equity share in Tonen.

***Jersey acquired 100 percent voting ownership on March 30, 1962, and on that date changed the name to Esso Standard Eastern, Inc.

owned companies. Primarily a marketer, it had distributed and sold refined products imported from a limited number of export refining centers—the Persian Gulf, the United States, and Indonesia, which also had the only oil production of any consequence in the entire region. The advent of the refinery located near the markets, along with increased competition from newly discovered crude oil also seeking markets, changed that. Local affiliate and headquarter managers then confronted alternatives: multiple supply sources, surplus product outlets, varying and increasing government interest and participation in petroleum decisions.

Meanwhile, the pressure grew to identify, recruit and train local talent, looking ahead toward the day when the majority of staff and management would be nationals available to replace expatriates. This proved to be a continuous process. Esso Eastern became renowned for the international business and technical training it provided, and its personnel, most of whom knew English, became targets for recruitment by new foreign businesses as well as by local companies and governments.

Jersey leaders in ESE well knew that one legacy of the old Stanvac Company could not be sloughed off; that was the vastness of the company territory. Distance had created problems for Stanvac and it would for Esso Standard Eastern. From Australia to the latter's African territory, measured in miles, one-fourth the distance around the world. Persian Gulf crude oil from Jersey's affiliates or associates had to be transported many thousands of miles to Japanese refineries and other places. The travel time involved in inspection of company operations proved to be a factor that required regular consideration, even with the improvements in air travel. The distance among ESE affiliates and the travel time involved had to be accommodated in the management structure; executives who travelled needed excellent staff work from persons capable of accepting responsibility and willing to make decisions. These problems persisted despite technical and technological improvements in travel and communication.

When the decree was entered, the Jersey Company quickly established the administrative machinery necessary to pull the properties it received together.

At the top, Esso Standard Eastern's board consisted of seven employees, six of whom had varying terms of service with Stanvac:

> J. V. Pickering, President and Chief Executive Officer
> Rufus T. Burton, Executive Vice President and Director
> William B. Cleveland, Executive Vice President and Director
> Harold Midtbo, Director and Chief Financial Officer
> Willem Holst, Director and Economist
> Charles Leet, Director and Chief Manufacturing Officer
> Walter F. Spath, Director and Head Marketer.[7]

The one board member who had not previously been with Stanvac, Spath, had achieved a good reputation as marketing director for Esso Mediterranean, Inc.

Elliott played a major role in selecting these executives. Jersey's contact director for marketing, William R. Stott, had visited portions of the old Stanvac territory several times. He was aware that Stanvac operated virtually as an independent company. Several other Jersey directors were familiar with various countries in Stanvac's domain. But other than Elliott, no major Jersey executive knew precisely what their company had obtained in the settlement, and Elliott had been a Jersey director, based in New York for more than a decade. A real familiarity with Stanvac operations seems largely to have been absent. In fact, at Jersey headquarters, there were few of the typical studies or reports on Stanvac that had long been standard for other affiliates.

To obtain more complete information about this part of its Eastern Hemisphere operation, the Jersey Board called on one of its newest members, Wolf Greeven, to tour the area and report when he returned. Greeven had acquired much experience in marketing and, in 1962, he replaced Stott as contact director for that function. During 1962 and 1963, Greeven made four long trips to the Esso Standard Eastern territory: on the first of these he went to Japan; on the second trip he visited operations in Southeast Asia— Hong Kong, the Philippines, Cambodia, Laos, Vietnam, Malaysia, and Singapore; the third trip took him to India, Sri Lanka, Pakistan, Kenya, Uganda, Tanzania, and Madagascar; and on the last tour, he went to Australia and New Zealand. On all of these trips, Greeven was accompanied by one or more ESE executives, men to a large degree familiar with the area being visited: Cleveland, Pickering, Burton, D. V. Schworer, Frank Delong or Jack Wexler. The visitors made notes on what they saw and heard, and later discussed operations in each country. Then Greeven wrote reports on each visit and each country, frequently with assistance from those who had made the tour with him. These reports are almost models of their kind; obviously they filled the information gaps in Jersey headquarters—they are clear, crisp, acute analyses. Many of the judgments and predictions made in them were fulfilled in time.

Among the most interesting of these reports is that for trip number two, which covered Southeast Asian operations. Greeven, joined by ESE's Burton, and Wexler, the company planning manager (with extensive experience in Indonesia), began a four-week survey on June 2, 1962, by visiting Hong Kong. There, Esso Standard Eastern had a branch office selling lubricants, specialties and chemicals; it also represented a base for possible future expansion. At the time, the Royal Dutch-Shell Group [Shell] dominated the Hong Kong market, but the director noted that "The two major power [electric] companies currently supplying Hong Kong are not inclined to increase their investment in power generation." The certain need for large and continuing increments of electric power might provide a means by which Esso Eastern could enter the fuel oil market there.

In the Philippines, the Greeven group found "productivity too low," and that "top level management, group decisions, and policy discussions are not as well organized as we should like to see them." A shortage of "top management people" handicapped the operation. Concerning a fertilizer project then being developed there, Greeven wrote, "unless land reform or, . . . widespread financial credits are made available by the government, it is to be feared that the estimated consumption of fertilizers will not be achieved." This judgment revealed remarkable foresight.

From Luzon, the Jersey-Esso Eastern party went to South Vietnam. In his conclusions about that country, Greeven noted that its civil war would be of long duration and that many years would pass before any real political control of outlying areas could be established. Esso operations would have to continue on a wartime footing.

In Cambodia, the travellers noted probable "internal communist subversion" but that "Prince Sihanouk is in complete control and is universally liked and respected." The Greeven party did not recommend much expansion in Cambodia for Esso Standard Eastern. They did say that ESE should maintain and, if possible, improve its marketing position. But Cambodian reservations were based on the opinion that Sihanouk "believes in eventual communist domination of the area and that he hopes to be able to insure a favored position for his country in the new order." The group also stopped in Thailand. There, Greeven offered advice on several related problems, including sales personnel turnover, service station programs, and terminal facilities. While in Thailand, Greeven also noted that operations in Laos had been profitable largely because of the military situation—Esso Eastern, he wrote, "had the lion's share."

Malaya, the team found politically stable and economically "one of the bright spots in Southeast Asia." There, ESE had specialty and gasoline marketing, liquid petroleum gas marketing, and even well developed plans for a refinery underway. Terminal and bunker facilities in Singapore were believed to be inadequate, and to remedy this deficiency Greeven suggested a thorough logistical review. The Malaya survey also included Sarawak, Brunei and British North Borneo. The trip ended on June 28, 1962.

On September 17, 1962, Jack Wexler presented the report of the Greeven trip to the Jersey Executive Committee and answered questions about the recommendations. The Committee was assured that ESE had taken steps to implement some of the recommendations. For Jersey Board members, Esso Standard Eastern directors and executives, and for both companies' functional departments, this survey provided a rich source of information—as did others like it. While in no way so refined a document as such reports later became, it provided a useful summary of the total situation in each country as the Greeven party found it: political, economic, energy consumption and potential, the marketing picture—plus conclusions and recommendations.[8]

Meanwhile, Walter Spath had decided to circulate to Elliott, McCobb, and all the Esso Eastern personnel a full account of the changeover from Stanvac to Esso in Thailand, "not because the changeover was the best program, the biggest, or the most spectacular of the lot, for it was neither." Rather he chose it "because it so well typifies the full degree of coordination, cooperation, attention to detail, personnel application and infectious enthusiasm with which our field people . . . performed this highly successful accomplishment."[9] For so extensive an operation, Spath believed that all units had performed exceptionally well and they deserved commendation.

The Esso Standard Eastern board of directors took charge of their company in the same aggressive fashion that characterized the operation of most Jersey affiliates. Information flowed into their new headquarters at 15 West 51st Street in New York. The work of the parent company from the spring of 1962 until the end of the year received special notice in the Jersey *Annual Report*. It said, "Special emphasis was placed on improving standards and efficiency of distribution, establishing our own terminal and storage facilities, and assuring growth in the important centers where the company sells oil products." Regarding the contribution of the employees the *Report* stated, "Their success in gaining recognition for the Esso oval was shared by the loyal group of independent dealers and agents representing Esso Standard Eastern." ESE added such operations to those previously conducted by Stanvac as "marketing asphalt in Australia, . . . selling cargoes of diesel fuel and gasoline in South Africa," and also laid the groundwork for chemical sales. ESE sought to enter the aviation gasoline business in Hong Kong and across the Pacific islands.[10]

The question of a refinery in Korea first surfaced just after the court-ordered dissolution of Stanvac. In June, 1962, Cleveland wrote Elliott and McCobb an "Aide Memoire: South Korean Refinery," in which he stated that South Korea in January, 1962, had invited international tenders for "preliminary engineering and technical surveys for a petroleum refinery" and that ESE had sent in a veteran crew to accomplish that mission. It had been instructed to gather economic data, to study potential plant sites and to obtain all other information necessary for a feasibility study or a "detailed refinery proposal." Their analysis showed that South Korea required a refinery with 32,000 barrels per day [b/d] capacity, capable of refining light Arabian crude and virgin naphtha. Construction of this plant would result in net savings to that nation of $5 million per year in foreign exchange.

In the 1950s and 1960s, Jersey possessed a good supply of crude oil. For example, by 1962, production in Libya had reached substantial levels. So Esso Standard Eastern and other Jersey affiliates looked for places to sell oil;

and one basic reason to build a refinery in Korea "would be the opportunity to obtain an outlet for crude oil." ESE questioned that Korea would give exclusive rights to supply the crude. Even had exclusive supply rights been obtained there would have been a problem because Korea lacked the foreign exchange with which to pay for oil. There were other problems.

As the ESE team and Jersey experts studied the refinery question, they had to consider another problem—that of largess. Information then current in the Far East indicated that the government of Korea expected generous gifts would be made by firms seeking permits to operate there. Jersey policy specifically forbade such transactions under any circumstances.

The ESE board decided to present the South Korean refinery study to the Jersey Board for discussion. This, Cleveland did in a report covering both negative and positive aspects of the proposed refinery. When he finished, Jersey CEO Monroe J. Rathbone rose and after pointing out reasons for not going into Korea, he concluded, "No."

Cleveland then wrote the Korean government in Seoul stating for ESE in equally terse terms: "Our company is not prepared at this time to proceed with the project. As a token of our thanks, we are herewith presenting them [you] with the entire project for whatever use you wish to make of it."

ESE did not enter refining in Korea, though from time-to-time proposals to share in existing plants or to build new ones went to the Jersey Board. In 1968, the Executive Committee "expressed no objection" to ESE's exploring a 50 percent equity purchase in a new lubricating oil and fuel products refinery at Pusan; however, the Committee so hinged its approval with conditions that it probably killed the proposal. ESE did not build, nor did it share in, a refinery in Korea.

[The epilogue to this story justified Rathbone's judgment. On May 16, 1965, the Subcommittee on Multinational Corporations of the U.S. Senate Committee on Foreign Relations heard testimony that another oil company which did build a refinery in South Korea in the early 1960s, had been "forced" to contribute $4 million to the "ruling party" there. The gift of this money violated a South Korean statute, and the solicitor was an important Korean. Testimony also indicated that particular company's foreign operations suffered because of this and other such incidents. Of course, publicity such as this helped damage the reputation of American business abroad.][11]

X X X

Elsewhere, ESE management encountered major problems in negotiating with the governments of India, Ceylon, and Indonesia. Conditions in all three countries also made it difficult for Western oil companies to operate effectively in them. The Soviet Union had moved to assist both India and Ceylon to locate cheap supplies of crude oil. As in Europe, the usual factors upon which prices had to be based were largely ignored by the Russians—there were no taxes or royalties to be paid, no dividends to the company shareholders—this was political oil. But these two governments viewed Soviet oil simply as cheap oil, and both exerted unceasing pressure on ESE and other Western companies to match the Russian price.* An additional complication in India was that the government insisted that all companies search for oil. For ESE, operating conditions rapidly grew worse instead of better.

In India, the ESE board encountered problems that led it to hold numerous discussions on policy and procedures with the U.S. Department of State [State]. The *New York Times* expressed the point of view of many Indians in May, 1961, in a dispatch from New Delhi: "There has been an undertone of suspicion in some Indian circles that Western companies with large sales organizations here such as Stanvac really did not want to find Indian oil." The Indians believed the companies made more money importing Persian Gulf oil than they would have from oil found in India. So the government had begun to deal with lesser companies in matters of exploration and to propose building its own refineries with Russian and Rumanian assistance. The Indian government approved a French company's bid for an exploration permit, while it denied a similar request from Stanvac. The shortage of foreign exchange in India ranked first among problems there, but finding an answer to the question of competition from Russian oil in the markets proved no easier for Western companies to resolve in India than in Europe.[12]

In late September, 1962, the ESE board had a position paper prepared on India, an amended version of which it sent to the U.S. State Department and to the Agency for International Development [AID]. This paper summarized the familiar-for-Jersey problem of competing with Soviet oil on either a barter or monetary basis. The Indian government had stalled the building of a lubricating oil refinery, ignored a proposal to build a steamcracker to manufacture feedstocks for a chemical plant, delayed requests for exploration permits and, in general, vacillated, while its dependence on Communist oil grew. Esso Eastern did not intend to meet Russian prices, and it objected to the use of AID funds to finance such capital developments as refineries to process Russian oil. The Indian government demanded not only the 50-50 division of profits but "equal participation in the financing and management of any such venture" as exploration. When Esso Eastern's Burton offered 50 percent equity subject to ESE retaining management control, India turned down the offer.[13] Plans to enlarge the Bombay refinery encountered similar difficulties.

Dan E. Stines, assistant to Jersey Board Chairman Leo D. Welch, succinctly summarized the Indian situation in the spring of 1962. He pointed out to Welch that marketing there had not been profitable, that the Indian

*See the chapter entitled "Jersey and Russian Oil in the West."

government wanted greater equity in a combined refinery and marketing operation, and that the "main reason for staying in India is that it is too important a country not to be represented in." Stines recommended giving some thought to limiting investments there "to those that can be financed out of profits after paying at least a nominal dividend on our investment."[14] The "Indian Problem," as ESE referred to it, continued to trouble the best minds in Jersey as well.

In Ceylon, ESE encountered the Russian oil threat again. In December, 1961, Ceylon made a barter agreement with the Soviet Union, trading agricultural products for oil, principally because this arrangement improved Ceylon's foreign exchange position.[15] A few months later, the government of Ceylon signed another barter agreement with the United Arab Republic [UAR]. Jersey observers reported that the price was 25 percent under world market prices and low enough to be competitive with Russian prices. The UAR traded Egyptian cotton for Russian oil and then shipped the oil to Ceylon. The Ceylon Petroleum Corporation also contracted for oil from Rumania at a discount of 20 percent below the Abadan (Iran) price.

Jersey's Chief Executive Officer, Monroe J. Rathbone, took to the speechmakers platform in 1961 and 1962, in an effort to influence the U.S. Congress to take action regarding the Russian oil threat. Then, delegates to the American Petroleum Institute meeting in 1961, heard Oklahoma's Senator A. S. "Mike" Monroney attack Soviet oil as a deliberate attempt to disrupt Western oil markets. A congressional study group listened to Jersey Director Emilio G. [Pete] Collado testify that the amount of Russian oil sold in the West in 1961 had reached 600,000 barrels per day and that the flood had just begun. But from newly created nations suspicious of Western companies, the message continued to be, either refine Soviet oil and sell the products therefrom or match the price.[16] ESE and other companies refused to do either.

In Indonesia during August, 1961, Sukarno announced that henceforth that country would take 60 percent of the profits from the operations of all foreign companies. Previously, a 50-50 profit sharing arrangement had been in effect, and because Indonesia had considerable production, the shift represented a major change for foreign companies. In those Sukarno years, Westerners living in Indonesia endured much. Virtually all foreign publications were banned for long periods of time, "radio reception was poor," and there was "no real continuing communication with the outside world."[17] The turnover among expatriate personnel reached high levels. But in the producing fields the working force consisted largely of Indonesians, and a high percentage of other employees came from the national population, all trained by Stanvac (or its peer companies).

The U.S. government possessed no leverage that could be used to persuade Sukarno to consider any other course toward outsiders. U.S. assistance there consisted largely of foodstuffs, and the amount of AID funds had declined sharply.[18] Sukarno threatened several times to expropriate foreign oil company properties. Meanwhile the Indonesian economy suffered.

Thus, during the time Jersey and Socony Mobil were dissolving Stanvac, India established price controls for oil and Ceylon expropriated without compensation Esso Standard Eastern service stations and some other properties. Then, Ceylon permitted Russian tankers to unload, without charge, oil into ESE facilities. In Indonesia, the threat of expropriation hung over ESE and other companies which had investments there totalling over one billion dollars. For Esso Eastern, ending operations in those three countries became ever more likely.

X X X

At the beginning of its existence, Esso Standard Eastern had little choice except to continue with the same management techniques that had confounded operations at Stanvac. Instead of vice presidents, however, ESE designated each top executive as "coordinator." One of the devices Jersey used to break away gradually from the Stanvac tradition was to change names for the various positions and later to shuffle positions, bringing the organization closer to the pattern of Jersey. Greeven saw this clearly and discussed it with M. W. Johnson, regional coordinator for Europe. Johnson then wrote to him stating "your proposals follow pretty closely the classic Jersey pattern which has stood the test of actual practice." Regarding titles for heads of field organizations, Johnson said, "the use of them might have some psychological effect in assisting to change a way of thinking on the part of people who are used to taking detailed instructions from some distant point."

In this effort to break with the Stanvac past and to diffuse a new spirit in ESE, Greeven had the enthusiastic support of both Cleveland and Spath. They were able to persuade the ESE board to gradually whittle down the power of the coordinators until they became advisory only. The ESE board then agreed that all field managers should report to either Cleveland or Burton—the executive vice presidents. President Pickering served as an alternate for either of those two. This move elevated the status of managers of the various national affiliates and improved their morale. And while the timing may have been coincidental, the ESE board conferred additional authority and responsibility on all their managers just as Rathbone was implementing his own decentralization ideas at Jersey.

The old Stanvac antipathies as well as some regional jealousies then began to disappear as the Jersey operational procedures gained force. At the same time, the careful selection of personnel during the Stanvac breakup began to pay dividends. Old Stanvac hands and young recruits began to see the advantages of sharing information by means of frank discussions and careful reporting.[19] And such veteran Stanvac men as Robert E. Anderson,* who had moved around much of the Far East, now began to move up the administrative ladder.

In October, 1962, Spath, ESE contact director for marketing, wrote to Greeven: "Each month now I receive the first (delivered by hand) sales report and analyze it promptly upon receipt. Discussions are held with our functional and regional people where corrective action is needed, and communication with the field is not deferred." On this occasion he reviewed for Greeven ESE sales activities in the Philippines, Malaya, Calcutta, Kuala Lumpur, East Africa, and Pakistan.[20] Again, to stay current with field operations, ESE top management travelled widely, and they discussed operating problems on the spot as well as in the office. To train new managers, they installed the Jersey COED Committee system, now composed of ESE board members, and appropriately called the "Management Development Committee." Its goal was to identify persons who exhibited exceptional potential before their tenth year with the company, as well as those who might become problems. Each general manager (formerly titled Stanvac field manager) reviewed his group of employees annually with the Management Development Committee.

Following the Jersey system of frank discussion when in conference, the ESE board meeting opened to departmental managers this question for discussion: "Why Regional Offices?" In the following exchange, the executives with little or no Stanvac experience reached overwhelming agreement that the old divisions had outlived their value to administrators. So ESE soon dropped the divisional system it had inherited from Stanvac.[21]

Soon this critical study of administrative procedures and forms began to show results. Volume sales went up 5 percent in 1963 for the entire territory, while in Japan, they rose more than 20 percent. Two new refineries in which the company had interests went on stream that year; one in Port Dickson, Malaysia, and the other in Adelaide, Australia. ESE finally reached an agreement with the Indian government to build a lubricating oil refinery near Bombay. The fertilizer plant program also appeared to be moving on schedule. In fact, the only entirely negative event of the year occurred in Ceylon, where the government expropriated the remainder of ESE's property without offering compensation.** The Hickenlooper Amendment was then invoked and all U.S. aid to Ceylon stopped.[22]

One problem consistently encountered in ESE operations stemmed from

---

*Later secretary of the Exxon Corporation [Exxon], November 1, 1972-September 30, 1977.

**Firm protests by the U.S. government as well as by Jersey later led to Ceylon's paying compensation for the expropriated property.

the general desire of all the governments with which the company dealt to enjoy equity participation in all affiliates incorporated to operate locally. Jersey directors and company executives had encountered this issue in almost every area where they operated. In long conversations they had explored it. Early in 1963, ESE's Burton wrote McCobb summarizing both the extent of such participation in the Far Eastern region and the agreements ESE had made concerning this thorny matter. Burton pointed out that "we do not contemplate equity participation when another effective solution is available. . . . We would not offer such participation except in cases of clearly demonstrated necessity." He believed that "many government officials and other influential individuals . . . very possibly see this issue in an emotional light and exaggerate the economic advantages that their countries will obtain from local participation," and he concluded by remarking that ESE would try to arm field management with the best information available on the advantages of 100 percent ownership.

The accompanying exposition of ESE's attitude regarding minority participation indicated some areas of potential conflict. It pointed out that the "newly emerging" countries in Asia and Africa viewed "foreign-owned and controlled business as a survivor of Colonialism." Since petroleum operations were too vital to be entrusted to foreign hands, nationally owned enterprises had to become large enough to prevent Socialist or Communist takeovers. National pressures to control affiliate companies increased each year, and ESE and Jersey could not stem the trend toward more government participation.[23]

By 1964, Esso Standard Eastern had begun to show even better returns. Sales by volume increased 11 percent; a new refinery and another enlarged one in Japan provided products for that booming market. Construction had begun on a fertilizer and an ammonia plant in the Philippines and an ammonia plant in Malaysia, while the volume of chemical sales rose by 50 percent over those of 1963. As later events proved, two items in the Jersey *Annual Report* for 1964, did not receive the attention they deserved; one referred to a joint venture with China Light & Power Company Limited [CL & P], and the other to a new company exploring for oil and gas in Australia.[24]

X X X

At the time of the Stanvac dissolution, Jersey received only a small marketing unit in Hong Kong. As the Greeven-Wexler report noted, Shell had 68 percent of the Hong Kong market, "including about 3.6 million barrels [of crude oil] per year to the power plants." The authors also found that the two existing private electric companies in Hong Kong appeared reluctant to expand to keep up with the growing demand. Yet the city needed a third generating plant, which ESE might have the opportunity to build, thus generating power from its ample supply of crude oil, selling the power to one of the two companies. Esso Eastern considered this idea as a possible way of entering the Hong Kong market.

In 1964, when ESE established a conventional marketing organization in Hong Kong, the local sales group told Cleveland on one of his visits that only by joining a major industrial customer and securing that business could ESE increase its share of the market. The ESE salesmen wanted to get the fuel oil business of CL & P, a company owned by an investment group headed by the Kadoorie family. Cleveland's investigation indicated that while CL & P needed to expand, it was cash-short at the time. Further, it had experienced some difficulties with the government. So he quickly seized the opportunity by requesting action when he reported to the New York office. All ESE executives knew that Shell could match any offer made; also they realized that the management structure of CL & P was quite complex and that both the Hong Kong and the British governments would have to approve any proposal. Nevertheless, the group at ESE New York headquarters proceeded to develop a long range capacity and financial plan for submission to the power company and to both of the concerned governments. All of this necessarily had to be done in almost complete secrecy.

Charles Leet, who had a reputation for being both sagacious and thorough, headed the small group designated to prepare the proposal. Chosen to assist him was a very experienced financial man, T. H. Tonnessen, Esso Standard Eastern's treasurer, plus Richard Marriner, head of the ESE Economic Coordination Department, and Mac Shivers, a refinery expert. This group began to work in Hong Kong with China Light & Power executives. ESE and Jersey executives joined them for brief intervals. Usually they met with Lawrence Kadoorie,* whose enthusiasm for the project grew daily.

The ultimate agreement proposed that ESE purchase a 60 percent equity in a new power manufacturing company, that it sign a long-term contract for fuel oil, and that the company organize and install under government sanction a "scheme of control," which in effect would allow electric power tariffs to provide reasonable rates of return and govern the distribution of profits to the various shareholders. The agreement also delineated the conditions that would prevail on employment of the so-called development fund. These agreements satisfied the CL & P owners.[25] When both groups had signed the document, ESE automatically gained a considerable share of the Hong Kong fuel oil business and managed to shock the Shell marketing organization in the process.

The 1966 Jersey *Annual Report* noted that the "affiliate began supplying all the fuel needed to operate the existing generators of China Light & Power

---

*Lawrence Kadoorie's wife, Miriam, attended most of these sessions.

Company under a long-term contract." Even as this happened, construction continued on new power generating facilities to be tied into the CL & P distribution system.[26]

X X X

The very different story of the discovery of oil in Australia combines all the elements of high drama with the fortuitous circumstances that often accompany oil strikes. Earlier in the twentieth century, Australia could claim little oil production, but in 1960, several promising finds were made. Broken Hill Proprietary Company Limited [BHP], Australia's largest industrial organization—really a conglomerate—then became concerned that "oil fever" would lead to encroachment on its extensive coal lands. The company, therefore, decided to take out exploration permits in the Sydney Basin as a means of protecting some of its properties. The BHP management decided to look for the oil, but before doing so, they sought to hire as consulting geologist, Lewis Weeks. Weeks had spent more than thirty years in the Jersey organization when he retired as the company's chief geologist. Even in retirement Weeks continued to maintain a high degree of professional visibility; at this time he served as president of the American Association of Petroleum Geologists. BHP sent an executive to New York to employ him. Weeks then went to see Jersey Executive Vice President and Director M. A. Wright. For years Jersey had a policy that, in general, discouraged its annuitants from working for other companies where information gained as a Jersey employee might be used. And earlier when another oil company had sought to employ Weeks as consulting geologist, Wright had told him, "we would look . . . with a cold eye on you going to work using all this information that you got while you were with us." But when Weeks came back to discuss the BHP offer, Wright said: "Well, we aren't particularly interested in Australia and I don't see anything wrong with you going over and doing this with these people."

In conversation with the BHP representative, Weeks bluntly told him the company would waste money looking for oil in the Sydney Basin. Despite this forthright assertion by the geologist, BHP did persuade him to come down to evaluate its basin property. Once there, Weeks asked another executive whether BHP really wanted to find oil. Assured that it did, Weeks went to Melbourne, where the company headquarters overlooked the Bass Strait between Australia and Tasmania. When asked where he would look for and expect to find oil, Weeks pointed to the "area of his interest," the tidelands and the sea, and told BHP Chief General Manager Sir Ian McLennan that he was certain oil would be found there. One day later, BHP applied for and later it received exploration permits covering 64,000 square miles in the Bass Strait—an area noted for strong winds and high seas.

Now, with acreage to explore, BHP needed expert assistance. In 1961, it created the Haematite Explorations Proprietary Ltd. [Haematite]* to handle its oil and gas affairs. Weeks advised BHP to do the required seismic work itself and to employ as consultants, Lyman Reed and Malcolm C. Baker, the first a Stanvac retiree and the second a retired Humble Oil & Refining Company [Humble] geophysicist. Both were employed, and geophysical work began.

Meanwhile, Jersey's Producing Coordination Department sent three men to Australia on June 1 to establish a "listening post." Under the Stanvac settlement, Jersey had complete freedom to explore in Australia. The listening team, headed by geologist Al J. Caan, kept informed about all the geological surveys being made. Haematite itself invited oil companies to study the survey results.

Next, BHP put its information into a report which representatives of several oil companies, including Jersey, studied. Since Australia had no laws concerning oil and gas, BHP also hired a New York lawyer to advise them on contracts and agreements. Then, in September, 1963, Weeks appeared in the offices of Jersey, his arms loaded with maps showing the first results of the Bass Strait surveys (Haematite completed the survey several weeks later). Weeks also shared the maps with other oil companies (some of the evidence was unfavorable). On Weeks' advice, BHP refused to consider any partner until Haematite had finished the surveys. But in Jersey's Producing Coordination Department, W. E. Wallis urged its head, Siro Vázquez, to act immediately. Vázquez went to Jersey's Board Advisory Committee on Investments [BAC], and on February 14, 1964, it approved Esso Exploration Australia Inc.'s** request to make a farm-in agreement to explore two blocks of the BHP area. Jersey was now ready for BHP's next move.

BHP then began to discuss with Weeks and the other consultants the specific terms and conditions under which it might admit a partner to form a joint company. By that time some twenty-one companies had become interested in the Bass Strait oil prospects. ESE President Cleveland praised the BHP attitude and approach:

> BHP went seeking a full partner. With no knowledge or experience in this art, they adopted an extremely sound attitude and opened their information to the larger companies of Britain and the United States. . . . They wanted a full 50% partner to explore for, find and develop the hydrocarbon deposits they had become convinced lay under Australian waters.

Earlier Weeks had suggested that BHP employ another Jersey annuitant, James A. Clark, to advise it on establishing Haematite and to explain the

*Later changed to Haematite Petroleum Proprietary Ltd.
**Later, Esso Exploration and Production Australia Inc. [EEPA].

Executives of Jersey/Exxon
1950-1975

*...tives pictured on this page held
...both of the two top posts in the
...ration—chairman and
...ent—during the years from 1950
...5.*

...rank W. Abrams

Milo M. Brisco

Clifton C. Garvin, Jr.

...ichael L. Haider

Eugene Holman

J. Kenneth Jamieson

...ard C. Kauffmann

Monroe J. Rathborne

Leo D. Welch

Vice Presidents
1950-1975

Paul J. Anderson

Executives below and on the next two pages served as directors between 1950 and 1975. With their colleagues shown on the preceding page, they formed Exxon's top management during those years.

Marion W. Boyer

John W. Brice

Cecil L. Burrill

Donald M. Cox

Jay E. Crane

Lloyd W. Elliott

Robert T. Haslam

Henry H. Hewetson

Bushrod B. Howard

Hines H. Baker

Thomas D. Barrow

Frederick H. Bedford, Jr.

Nicholas J. Campbell, Jr.

Stewart P. Coleman

Emilio G. Collado

Harold W. Fisher

Wolf Greeven

Orville Harden

Peter T. Lamont

Robert H. Milbrath

Howard W. Page

Frank W. Pierce

George T. Piercy

Arthur T. Proudfit

David A. Shepard

Chester F. Smith

Emile E. Soubry

Leroy D. Stinebower

William R. Stott

John R. Suman

Siro Vázquez

John R. White

Myron A. Wright

mployees in Saudi Arabia
arts of a drilling rig from an
o instructor. Aramco's training
s have turned out hundreds of
Saudis who now work the rigs in
ountry's oil fields. Gathering
bove are in Libya's Zelten oil
iscovered by Exxon in 1959 and
ed by the company until it sold
rests to the Libyan government
thdrew from that country in
desert traveler (right) stops to
an Aramco rig drilling in
rabia's Ghawar field, the
largest oil field.

Exxon helped build the 800-mile Trans Alaska Pipeline System (TAPS) connecting the huge Prudhoe Bay oil field on Alaska's North Slope to a warm water shipping terminal at Valdez (map). At left, below, a young bull moose pauses beneath the pipeline. Before the TAPS project began, a specially reinforced tanker named the "Manhattan" made an epic voyage to test the feasibility of moving oil all the way by ship. The "Manhattan" made it through the ice pack (right), but the difficulties of the trip helped establish the need for a pipeline.

BARROW

PRUDHOE BAY

P.S. 1

P.S. 2

P.S. 3   RANGE

BROOKS   P.S. 4

NOME

P.S. 5

P.S. 6

P.S. 7   FAIRBANKS

P.S. 8

P.S. 9

ALASKA   RANGE

P.S. 10

P.S. 11   WHITEHORSE

HAINES
JUNCTION

ANCHORAGE   P.S. 12

VALDEZ   SKAGWAY

TERMINAL   CORDOVA

WHITTIER   HAINES   JUNEAU

SEWARD   PRINCE WILLIAM
SOUND

ALEUTIAN RANGE

GULF OF ALASKA

Indicates sections of pipe
installed as of April 1976

MANHATTAN

Built in 1959, Exxon's refinery at Milford Haven, Wales (top) had an unusual degree of process control automation. The "Put A Tiger In Your Tank" campaign of the 1960s won advertising industry awards and boosted gasoline sales. In the 1970s, Exxon researchers used a pilot plant to develop "fluid bed combustion" technology for reducing air pollutants.

Esso tank truck cuts across the roadless grassland in this 1964 scene in Tanzania (above). Israeli pontoon boat patrolling Suez Canal just after 1973 war (right) approaches merchant ships stranded here when canal was closed during 1967. The 250,000-ton Very Large Crude Carrier "Esso Scotia" (below) offloads crude oil into a smaller tanker.

oys on the King Ranch in Texas
e) herd cattle within a few
gs of Exxon's natural gas treating
one of the world's largest plants
ind. Semi-submersible drilling
right is shown drilling on the
n field, one of several major oil
developed by Exxon and Shell in
oint venture in the U.K. North
eer (below) browse near an
producing well at Avery Island,
ana.

The Arab oil embargo of 1973 triggered a gasoline shortage which changed Americans' feelings about a product they were used to taking for granted. All across the country, automobiles waiting to fuel up formed long lines at service stations, drivers' tempers frayed and harried dealers had to dole out supplies based on odd or even-numbered license plates. The shortage was temporary—but the worldwide "energy crisis" was just beginning.

Esso is changing its name to Exxon

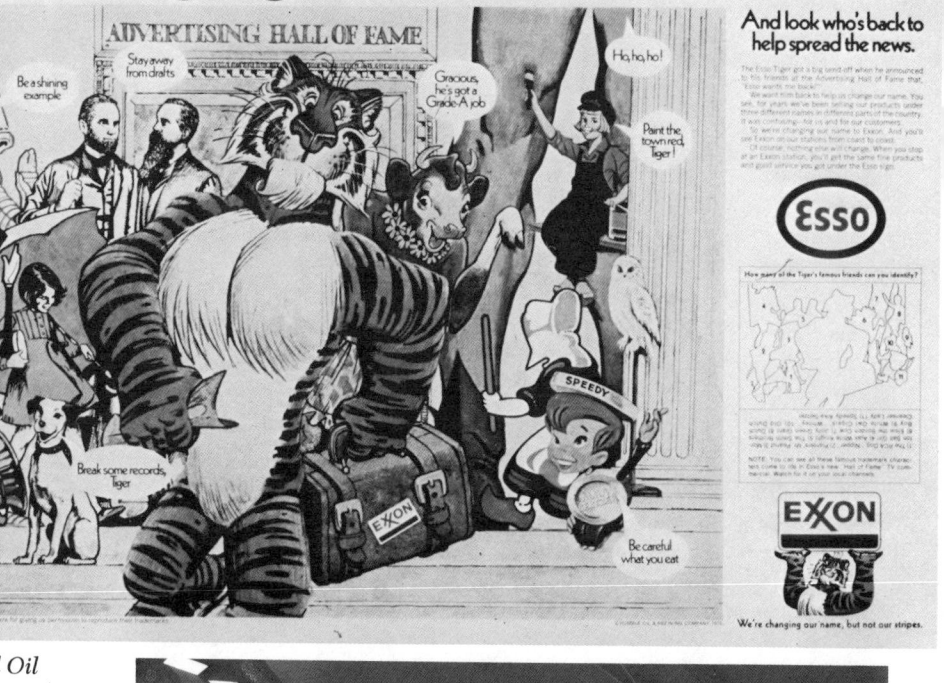

72, Standard Oil ›any (New Jersey) ;ed its name to Exxon ›ration and adopted xxon name for cts sold in U.S. ts. The Exxon Tiger i above with other ers of the "Advertis- all of Fame") was d from retirement to ›read the news.

Employees gathered at the New York Hilton (above) to hear the reasons for the name change spelled out by chairman J.K. Jamieson, (left) shown with new company logos.

Headquarters buildings for the company and its affiliates over the years have included (from top) the Caracas headquarters of Creole Petroleum Corp., Exxon's Venezuelan affiliate; the head office of Imperial Oil, Limited, the principal Canadian affiliate, at 111 St. Clair Avenue West in Toronto; and the corporate headquarters building at 1251 Avenue of the Americas in New York City.

various alternatives in so doing. BHP hired Clark, a veteran producing expert who had years of experience in the technical aspects of oil-finding and in administering producing companies. In fact, when Wright learned that Clark, with whom he had often worked, was going to Australia, he telephoned him in Houston and said, "If you have anything good over there, Jersey is interested."

On February 18, 1964, Vázquez cabled Caan and J. Stewart McClendon, a Jersey attorney then in Australia, that if negotiations with BHP appeared possible, Wright would be available to participate in them. When Clark learned that the BHP board of directors would make a decision on February 19, he wrote Wright to come immediately. Sure enough BHP sent its proposal to companies that had representatives in Australia—including ESE. It wanted an answer by April 1, one that included a schedule of future work.

The BHP proposal "triggered immediate action in New York." ESE cabled its Sydney office that Wright would arrive there on March 3, and would go immediately to Melbourne. Clark later said that Wright's visit influenced the BHP decision to select Jersey as its partner. Wright came with full authority to negotiate for Jersey.

For a week, Wright discussed the joint company and the oil business with Broken Hill executives. They knew little about Jersey, but the expertise of all the ex-Jersey men whom they had employed, as well as Wright's own vast knowledge of the industry, impressed them. He told them Jersey had the men, the resources, and the technology to do the job. Wright then made the Jersey offer to BHP: Jersey would drill one wildcat in each of five BHP blocks on Jersey's own account; should any or all of them become producers, BHP could then decide whether to take one-eighth royalty override [then customary] or put up its own money to become a 50-50 partner with Jersey. He advised the BHP executives to do the latter because it would be more profitable for them; also he felt that Jersey should have an Australian partner. Esso Exploration Inc.,* Wright said, would locate the oil if any existed in the Bass Strait. Wright then returned to New York.

No other company made such an offer. Shell had been informed (along with others) that BHP would listen to offers for farm-in drilling, but its Australian office routed the messages through regular channels to London. There Shell's headquarters staff learned of the opportunity too late. What happened in the other companies regarding the BHP offer is not known. Wright hazarded as a reason for so little general interest the fact that the BHP announcement went to marketing companies in Australia and not to company geologists. In any event, Jersey's initiative demonstrated serious intention to make an Australian connection.

Jersey drafted a complete proposal for a relationship with BHP and sent it to Melbourne. McLennan and his associates studied it and on April 17, 1964,

*Esso Exploration Inc. was created April 1, 1964 to operate Jersey's worldwide exploration ventures. Australia was its first.

Sir Ian cabled BHP's acceptance to Wright. BHP requested Wright to come to Australia for additional discussions leading to a legal agreement. Jersey replied that Zeb Mayhew, head of Esso Exploration Inc., attorney J. S. McClendon, and producing expert W. J. Barnett were on the way and that Wright would quickly follow them. Mayhew was to head the Jersey negotiating team.

The Jersey men learned very quickly that the Australians desired action; in fact, if BHP concluded a trade with Jersey, one of the conditions would be that drilling had to start before January 1, 1965. Wright and Mayhew both recalled that particular demand presented problems since neither of them knew where an idle drill ship could be found. Finally they located the Glomar III in the Gulf of Mexico. Wright said concerning the agreement, "One commitment I had to make was we'd have a rig running by January 1."

By the time Wright signed the agreement in May, the *New York Times* carried a story stating that Jersey had an option to drill on an acreage of 12,000 square miles in the Bass Strait. Soon the drill ship, Glomar III, now under contract to Jersey, left the Gulf of Mexico and headed for Australia— really a small fleet, consisting of a supply ship, followed by the Glomar III and then two crew boats. They headed straight for the first lease, not bothering to land in Australia because of complications attendant the labor unions there [the drilling took place in international waters]. Wright said, "We just barely got that thing turning over there by January 1." The vessel arrived in Australia on December 16, 1964, and began drilling operations on December 27.

The elements posed one of the major problems the drill crews encountered. Hard winds "blew all the time and there's no weather report [to go by] except in Tasmania and that part of the [weather] station was in Antarctica." Weather reports had to be made up daily. At times information from the Cape of Good Hope also helped in the preparation of the necessary daily weather reports.

On February 18, 1965, the bit struck natural gas. The next two wildcats drilled also hit gas in commercial quantities as well as some oil. Mayhew had returned to New York and moved on to Thailand; there he first heard about the gas strikes. Interestingly enough Mayhew remembered that the Esso Exploration crews—largely former Creole Petroleum Corporation [Creole] hands—had begun to train Australians as soon as they began drilling. He smiled as he remembered, "The Aussies thought we were miracle workers."

Subsequent delineation drilling provided the geologists with enough information for them to estimate that the fields contained more than four billion barrels of oil plus gas. Exploration continued. In 1967, two important oil fields were located. Production experts moved in. Crude oil production offshore commenced in 1969, and gradually increased to around 400,000 b/d by the mid-1970s. Later Broken Hill took bids on farm-in drilling on another lease. Several of Jersey's competitors offered cash for drilling rights. But Wright told Mayhew to offer instead of cash a line of credit for $75 million at nominal interest.* Jersey kept around a billion dollars in its liquid assets, as Wright said, "to go for things like this, I guess." This line of credit allowed BHP to participate in development of the earlier discoveries and to become a 50 percent working interest owner.

Years later Wright clearly recalled with obvious pride the loyalty of those annuitants who assisted in making the BHP alliance. He knew them all personally and apparently they knew each other. Wright said, "And had it not been for them, we wouldn't have ever obtained this Australian production. It was a real great experience."

Late in 1966, ownership of Esso Exploration and Production Australia Inc. was transferred from Esso Exploration Inc. to Esso Standard Eastern, which company handled Australian operations for Jersey. Until the share transfer took place, Esso Standard Oil Australia Ltd. [ESOA] had served as contractor for the exploration company. John H. Hamlin, who had managed the latter company, moved over to ESOA as deputy managing director in which position he actually ran it. Despite properties and persons involved, no major difficulties occurred during the transaction.

Overall, the partnership between BHP and Jersey proved to be a fine working arrangement. Since the BHP people had no experience working with oil, Jersey arranged with the Humble Oil & Refining Company to provide information to them. Many BHP executives visited Humble to learn oil techniques.

The joint company paid the Australian government 10 percent royalty plus taxes. And Louis Weeks received a 2¹/₂ percent override for his special contribution.[27]

X X X

From 1965 to 1970, Esso Standard Eastern, Inc.'s growth continued. Pickering became board chairman and Cleveland, CEO and president in 1966. Earlier, Burton had moved to London to head the office there, and about the same time Holst took early retirement. Sometime before Cleveland took over, Elliott talked with him and in the frank Jersey manner told him that, despite the progress being made, the Board believed more rapid ESE growth possible. After Cleveland reflected on the conversation, he decided to make some changes by reorganizing the company's management starting at the top. In doing this, he fitted some of his own ideas into standard Jersey practices. Since he disliked large boards, he never moved to replace Holst, and while Burton remained a director of ESE, he rarely came from London

---

*Both Wright and Mayhew agree on the $75 million figure.

to attend the board meetings in New York. Jersey, with its decentralized management, endorsed Cleveland's plan to run ESE with four directors (including himself). In 1969, Cleveland had the company name changed to Esso Eastern Inc. [EEI].

In 1965, the Jersey Board began its program designed to divide the company operationally into regions. It created Esso Africa to take over the territory of Esso Mediterranean, Inc., and added to that ESE's East and South African countries (acquired in the Stanvac settlement). ESE now covered Southeast Asia, India, Pakistan, Indonesia, Japan, Australia, New Zealand, Hong Kong, and the Philippines. In that year, ESE reported "14 per cent volume growth for the area as a whole." Regarding the Bass Strait discoveries, Jersey's *Annual Report* stated that studies had begun to determine the most effective way to market the oil, the best manner to move it to potential markets, and the best possible price.

In other Esso Standard Eastern territory, the 1965 *Annual Report* called attention to the settlement reached with Ceylon for the properties expropriated in 1962 and 1963, and to the increases being made, or projected, in refining capacity in virtually every place where refineries were located. In Malaysia and Pakistan, Jersey had begun construction of chemical plants.[28]

By 1966, the Vietnam War had begun to affect Esso Eastern's operations; the company helped supply "greatly increased quantities of aviation and motor fuels and heavy fuel oil for U.S. and allied military operations." Since production in Indonesia remained at about the same level year after year and ESE's sales increased annually, much of this additional product came from Jersey's Persian Gulf sources supplied by Esso International Inc. Thefts of entire truckloads—59 by 1968—and pilferage at times made deliveries difficult, and near the battlefront (and in many towns) damage and losses occurred due to actions by the Viet Cong. Luke W. Finlay and his Jersey security staff investigated the major losses and filed reports on them, as did the local affiliates.[29]

The discovery of other Bass Strait oil fields led ESE once again to study Australian production methods, distribution, and prices. Jersey's Producing Coordination Department assisted in this large scale assessment. The study concluded that the new oil discovered would be on stream in 1970. Meanwhile, in other sections of the ESE marketing area, sales continued to increase. Esso Exploration began drilling operations offshore New Zealand and near Malaysia, while the search for oil in central Sumatra continued. The Arab-Israeli War (1967) for a time reduced the available supplies of Persian Gulf oil for ESE's refineries, but despite that, growth continued.

H. W. Coxon of Jersey's Marketing Coordination Department [Marcor], toured Esso Eastern's marketing territory in April and May, 1968, in company with Ralph McCoy, a former European shareholder representative, and Walter Spath. They visited seven countries and attended an ESE marketing conference in Bangkok. Later in a letter to Spath, Coxon pointed out that while Marcor sought to help the affiliates improve their performances, he considered of more importance "how we can plan and prepare for the years ahead." To him, "it is in this area that Marcor must concentrate in gearing itself to assist the Regions." Coxon went on to suggest states in which ESE affiliates should seek to improve their share of the market; "it seems essential that our retail market studies . . . and land programs be pursued aggressively." He pointed out the need for national studies, "the kinds of outlets required in Japan are not necessarily the same as in Australia." Then, he explained how the Jersey system in marketing—and most everything else—operated: "The region role might be stated as one of strategic planning and direction while the affiliate is responsible for tactical planning, implementation and control."

Coxon criticized the Bangkok conference agenda and recommended that general topics be discussed, rather than limiting the agenda to stewardship reports. Yet he found the morale of the ESE marketing organization very good, especially in the Philippines and Malaysia. In Australia, he observed, morale was at a low point because of indecision over which basic marketing area to attack first—automotive, industrial, or farm.

This report indicates precisely how Jersey's functional departments interacted with regional affiliates. The parent company expected its executives to study their assigned specialties and to render reports honestly, though as diplomatically as possible.[30] Whether related or not, ESE had already become interested in its own marketing problems and had commissioned a study dealing with "dissemination of basic, predominantly qualitative information on the 'aftermarket' as a stimulus to innovation in the marketplace." Marcor considered the study a fine one, "intended to start the flow of creative thinking in affiliate management." So it circulated the report to the heads of other regional affiliates.[31]

The 50-50 partnership of Jersey and BHP in Esso Standard Oil Australia Ltd. operated with very little friction. In one case, however, differing governmental regulations and business practices created a slight problem. John Hamlin explained this: "As we developed our gas reserves and pursued active negotiations for gas contracts we found it necessary to publish reserves [figures] which would substantiate our negotiating position and at the same time satisfy BHP." Because the gas pool was located offshore, Hamlin said that ESOA could not have the usual "stepout and delineation wells on which to base our judgement." So some numbers proved high. The Jersey policy as explained by Cleveland had always been "ultra conservative on giving any well or test information." That also applied to oil.

Hamlin said ESOA had a running battle with the press and both state and federal governments, who wanted the reserve position "in respect of what they consider their extremely delicate relations and responsibilities to their shareholders and all the Australian stock exchanges." Thus on the one hand, ESOA responded to the Jersey practice of obtaining reserve estimates slowly, of seeking outside geological opinion and estimates; while on the other hand, the Australian government pressed for numbers more rapidly estimated. The government in this particular case also sought these figures, for it knew that if the ESOA reserve statement proved high enough, the entire Australian refining industry would have to begin to change to accommodate the Bass Strait oil instead of that from the Persian Gulf then being used. Hamlin said that the pressure to get reserve figures represented "a running battle."[32]

Under the reorganization agreement of 1960, the Indonesian oil fields continued to be operated jointly by a Socony Mobil and Jersey company— P. T. Stanvac Indonesia. In 1961, Jersey's share of crude oil amounted to 36,000 b/d. The company sought to obtain permits to explore for additional oil and in 1963, the government issued licenses allowing it to search for oil and gas in Central and Northern Sumatra.

Generally, Indonesian production fluctuated until the 1970s. The Northern Sumatra search for oil proved fruitless. Negotiations to sell the joint company, P. T. Stanvac Indonesia, began in 1965, but could not be concluded. In 1969, however, Indonesia bought the Palembang refinery for "about six million barrels of low-sulfur fuel oil of which Jersey's one-half share will be shipped to Hong Kong." Exploration continued in Sumatra for some years.

For EEI, 1969 proved to be a banner year; sales advanced; Australian crude oil production began; the new 720,000 kilowatt power plant in Hong Kong came on stream, in which the company [Esso Eastern] had a 60 percent equity share, proportionate representation on management matters, and rights to supply fuel oil at competitive prices.

In all its region, the general outlook seemed good. Drilling continued offshore Malaysia, and reports indicated likely oil and gas finds. The Singapore refinery was near completion. Cleveland had pared down EEI and its affiliate personnel without any loss in company productivity, while individual output increased.

By 1970, both gas and oil production in Australia had risen sharply, with sales of liquefied petroleum gas [LPG] beginning to petrochemical plants and to Japan. Refinery capacity everywhere either needed increasing or had been increased or studies had begun leading to increasing capacity. Singapore's 81,000 b/d refinery came on stream that year, while another refinery nearly as large was underway at Okinawa. Sales for the entire region advanced again, this time by 9 percent over those of 1969. Raw percentage increases over previous or other given years often lead to inexact conclusions. Thus in 1971, crude oil and product sales advanced again. The table in Appendix 2 illustrates the continued expansion achieved by Esso Eastern, as well as the ongoing development of oil and gas production by that company and its affiliates.*

It seems obvious that many factors combined to provide this growth. Some of them were: the Exxon management system (previously referred to) that led to improved morale in Esso Eastern; the enthusiasm with which Cleveland and his fellows attacked the marketing situation; the gradual development of thorough and complete operational report systems in which top executives such as Cleveland, Leet, Spath, and later, Maurice E. J. O'Loughlin, all participated as did Jack Wexler, and Richard Barham. Distance and time differentials made these reports increasingly important to management, both in the EEI office and at Exxon headquarters. In time, the reports developed into significant political and economic analyses of much value to the company. (See Appendix 2, page 902)

Another factor (also previously mentioned) in the success of Esso Eastern is that it began operations when the demand for energy in its territory was in a period of growth. Development within the company plus direct assistance from Jersey (Exxon) enabled it to locate considerable amounts of crude oil and national gas, all of which helped EEI during the period of the oil embargo and the price increases.

X X X

From 1970 to 1975, Esso Eastern continued to encourage the development of its Australian properties. On the other hand, in 1968, on Cleveland's suggestion, Jersey reviewed the outlook for marketing in nearby New Zealand and decided against it. As Jersey CEO M. L. Haider explained, "Our marketers attempted to obtain a respectable share of the New Zealand market with little success and at a sizeable annual loss."[33] In the 1970s, however, Australia provided a fine market and with domestic oil production increasing, Jersey wanted the Australian government to maintain the tariff of seventy-five cents per barrel.[34] The money collected on imports had been used to stimulate exploration for oil and gas, the latter now located in quantity. Meanwhile, Jersey's Refining Coordination Department [Refcor] indicated that Esso Australia needed more refining capacity—so it sought to expand the Altona refinery. As F. O. Dennstedt of Refcor told Jersey President M. M. Brisco and Director H. W. Fisher, "We feel that Esso Eastern's proposal is the most realistic solution to meet the problems at hand," though he found the projected return "is somewhat overstated." He assured the two executives that, if they approved, he would take the Esso Eastern proposal to the Investment Advisory Committee [IAC].[35]

The interest of Esso Eastern (and Jersey) in tax incentives provided for exploration in Australia served to add additional momentum to their search for oil and gas. The Jersey *Annual Report* for 1971 did not understate this accelerated exploration program when it said, "The company deliberately has fashioned an exploration program of considerable geographical diversity." Subsequently in 1973, Exxon* approved Esso Eastern affiliate EEPA's entering negotiations with the Australian government for rights to explore 80,000 square miles in deep water offshore the Exmouth Plateau in Western Australia. The Exxon Management Committee also approved EEPA entering a "production-sharing contract with the Government under which the Government would have a 51% interest and EEPA a 49% interest in the hydrocarbons produced."** Additionally in 1973, Esso Eastern sought approval of a long-range plan to acquire shale or coal reserves for a projected synthetic oil industry. The study indicated that a "crude shortfall will probably [continue to] exist in Australia in the mid-to-late 1980s."[36] Hardly a month elapsed before this proposal of Esso Eastern received approval.

Yet, within six months after visiting Esso Eastern in Sydney, Exxon's Producing Coordination Department drew a grim picture of the future of hydrocarbon exploration in Australia. The report praised the high caliber of the employees of Esso Exploration and Production Australia Inc. and pronounced the "overall technical quality" of the projects "excellent." But "the exploration outlook" at that time they pronounced "not bright." Since the opportunities were limited, the Producing Department questioned the use of much manpower in exploration off Western Australia and elsewhere. They also agreed with EEPA's judgment that gas might be located in some Australian basins and urged that these be reassessed. Of course these experts could not foresee the revolution that took place in world oil prices after the Arab-Israeli War of 1973.[37]

Esso Eastern undertook an Australian Strategy Study in 1974, to ascertain whether it should increase its marketing operation. Some executives in New York believed that the study would lead to the conclusion "that Australia is a viable marketing operation, that they can make a good 'two figure' return, and become cash self sufficient in the downstream." This, they believed could be done only "after considerable restructuring."[38]

Japan continued to be the most rapidly growing market in Esso Eastern territory in the 1970s. The 1972 Exxon *Annual Report* noted that Japan was "a principal and attractive market despite a high degree of government regulation." To meet the surging product demand, in 1969, Esso Eastern's 25 percent owned affiliate, Tonen, had increased the Kawasaki refinery from

---

*On November 1, 1972, Standard Oil Company (New Jersey) became Exxon Corporation.

**The Exxon Corporation normally did not enter into agreements with governments.

60,000 b/d to 180,000 b/d. By 1970, refinery runs in Japan totalled 321,000 b/d. Esso Exploration, in a joint venture with Japanese companies during 1971, began the search for oil and gas off Honshu, the principal Japanese island. Some years later, they discovered a gas field off the East Coast.[39]

In many of the fast developing countries of the Far East, national pride received a sharp boost when Esso Eastern (or another oil company) could be persuaded to build a new refinery. To some degree, this political pressure to locate refineries close to the consumer accorded with Jersey's post-World War II strategy.

The refinery on Okinawa is an example of this type operation. This island, the largest in the Ryukyu chain, had been occupied since World War II by the United States and governed by the U.S. Civil Administrator of the Ryu- kyu Islands. For eighteen years, however, the California-Texas Oil Com- pany, Ltd. [Caltex] supplied most of the petroleum product needs for Okinawa and the other islands. Nevertheless, when in 1970 the Ryukyu Electric Power Company [REPC] asked for bids on a refinery and the needed fuel oil, Esso Eastern entered one, as a way of securing markets there and gaining additional entrance into Japan when the territory ultimately re- verted to that country. Esso Eastern received the contract but the U.S. gov- ernment could not, of course, guarantee that Japan would honor the fuel oil contract after the reversion took place. So Jersey CEO J. Kenneth Jamieson wrote to U.S. Under Secretary of Defense, David Packard, on May 22, 1970, and strongly urged that "the U.S. negotiators [on the reversion] make a sincere effort to persuade the Japanese Government to accept the fuel supply contract along with the other REPC assets in whatever arrangement is fi- nally made." The fuel oil price, Jamieson noted, was "extremely favorable" and Esso Eastern proposed "to invest some $60 million" to fulfill its contract. Construction of the 72,000 b/d refinery proceeded rapidly, and the 1971 Jersey *Annual Report* noted that it was virtually completed. In May, 1972, when Okinawa reverted to Japan, two Japanese companies acquired a 50 percent equity interest in the already started-up refinery. Esso Eastern kept the fuel oil contract.[40]

The 1973 oil embargo by OPEC affected Esso Eastern operations as it did those of Esso Europe Inc. and Exxon USA.* In Japan, the government insti- tuted quite effective "voluntary" controls to limit consumption. Cleveland reported to Director Nicholas J. Campbell, Jr. in December that these mea- sures included a 10 percent reduction in petroleum and electrical use by large industrial establishments, lower motor vehicle speed limits, reduced outdoor lighting, and various other steps that decreased fuel needs. Cleve- land also told Campbell that if the crisis continued, he expected the govern- ment to enact oil conservation laws.[41]

---

*Humble Oil & Refining Company became Exxon Company, U.S.A. on January 1, 1973.

X X X

At the time Jersey and Esso Eastern became interested in exploring off-shore near Malaysia, in the South China Sea, Thomas D. Barrow was executive vice president of Esso Exploration. He had mentioned to Director Wright, who was then involved in negotiations with Broken Hill Proprietary Company Limited in Australia [described above], two or three places he wanted to explore, including Malaysia. Wright recalled, "Tom had been out there a month when I went out." They went to Kuala Lumpur to see the Prime Minister, Tungku Abdul Rahman, at his residence. The men "spent an hour talking business, and who we were, and what we did, and what we wanted to do. . . . We were willing to commit ourselves for quite a bit of money for exploration." Wright and Barrow liked Abdul Rahman. Wright remembers, "He was fascinated. We were the first oil people that he'd ever talked to. . . . Finally, he said . . . he thought we ought to work it out and that he'd put us in touch with his Minister of Mines."

Subsequently, Wright and Barrow met with the Minister of Mines, and negotiations on contract terms began. Esso Eastern and Esso Exploration sent out specialists to work things out, but it took many months to arrange details.[42] Finally, in 1965, Esso Eastern secured a license from the state of Sabah (North Borneo) covering five and one-half million acres, and exploration began later that year.

Esso Exploration drillers encountered shows of gas in 1970; evaluation began immediately. In 1973, further offshore drilling near Sabah located oil in commercial quantities—"each of the three wells . . . tested individually at more than 2,000 barrels of oil a day." While the gas discovered tested out a highly significant strike, the estimated quantity of oil was even better. "In world terms, it's not a huge amount of production," but "pretty significant as far as the country goes [and] . . . as far as Esso Eastern goes," as Jack Wexler described it. Several years later, drilling 150 miles off the East Coast of Malaysia, Esso Exploration made and confirmed a significant oil discovery. These coupled with the more modest reserves of Sabah required full study as production teams moved in.

Additional exploration and delineation drilling in 1973 and 1974 led to significant shows of gas, and Esso Eastern created a task force to review all Malaysian prospects. Developing a separate Malaysian exploration and producing organization received consideration too. Esso Eastern's eight-year search for oil in Malaysia finally succeeded. The company immediately secured additional licensing permits for exploration, while at the same time it assessed the potential of the discovery area. Two Sabah wells each reached 4,000 b/d production in 1974. Then the familiar pattern of dealing with producing nations occurred; in Malaysia, exploration and production stopped until a new agreement with the government could be reached;

however, the Sabah field continued to operate under a temporary arrangement.*

Though the effort of Tungku Abdul Rahman to create a state of Malaysia that included Singapore failed, that prosperous city became an independent nation. In fact, by 1970, Singapore showed signs of becoming the refining center for Southeast Asia. Esso Eastern's new refinery with 81,000 b/d capacity went on stream that year, giving the company a total refining capacity in its region of 413,000 b/d (in 1970). Immediately EEI began alterations to increase the Singapore refinery's capacity to 230,000 b/d, which would make it the company's largest in crude processing in Asia. Later in 1973, the company added a lube plant to its Singapore operation.[43]

Elsewhere, both marketing in Luzon and the Bataan refinery continued to create problems for Esso Eastern's management. Red tape—changing governmental restrictions—that actually slowed down payment by the refinery for crude—adversely affected profits and kept the company debt high.[44] Still, for some years, management sought and secured additional capital funds to expand refinery capacity. However, in 1972, the Exxon *Annual Report* singled out the Philippine operation for attention: "In the Philippines, restrictive regulations on prices and prospects of mandatory local ownership make the viability of the marketing and refining business in that country increasingly uncertain." The following year, Esso Eastern sold its entire Philippine operation (marketing and refining) to a local government oil company.[45]

Cleveland stated that, behind the sale, lay hard work, forethought, planning and "luck." He knew that this first peeling off of Exxon investments in the Far East could create some second guessing in world petroleum circles, "Consequently we prepared our case as thoroughly as we could." He added, "In effect, we were proposing to sell our assets for something around $40 million and to have the purchaser assume our debt." When Cleveland went to the Exxon Board with this proposal, President Clifton C. Garvin, Jr. asked just how he would approach the Philippine President. Cleveland told him he would have no game plan, but would use judgment as the conversation developed. The Exxon Board approved the divestment, and Cleveland then secured an appointment with the Philippine chief executive, President Ferdinand E. Marcos.

When Cleveland arrived in Manila, two business agents for persons left unnamed, called on him and offered to purchase the properties. Next, Cleveland saw the President. As he recalls, *"I got lucky!"* After he finished his sales talk, the President conferred with two of his most able ministers— "who were subsequently named to represent the Government in the negotia-

---

*In 1977, Esso Exploration brought in a major field offshore the East Coast of Malaysia. Exxon's share of production by 1980, amounted to 77,900 b/d, the maximum allowed under the depletion policy adopted by Malaysia that year.

tions which followed." That was the lucky break according to Cleveland—to be able to deal with the most informed and able representatives of the Philippine government. Within two weeks they consummated the sale for several million dollars more than the Board-authorized sale price.*[46]

This Philippine transaction marked the first sale of Far Eastern property by Esso Eastern. Elsewhere, in its territory, the company had divested other properties. In some places, however, it never really had a secure foothold. As Cleveland remembers concerning New Zealand, "It was never possible to re-establish ourselves [Esso Eastern] without making an investment disproportionately high for the possible benefits we might eventually attain."[47] As the oil shutdown developed in the 1970s and as the problems with producing countries increased, Esso Eastern (like other Exxon companies), found it difficult to convince consuming country governments of the reasons that made price adjustments necessary. In turn, some of the consuming countries exerted new pressure affecting the "profitability of refining and marketing operations."

In Esso Eastern's South Asian territory, in India, Jersey had begun to sell oil as early as 1882. Ninety-two years later, in 1974, the Indian government announced that it had purchased all Exxon assets there. The announcement proved premature, but accurate in the sense that negotiations were then underway that later resulted in the sale. Esso Eastern had encountered ever-changing problems in India. In the days of Standard-Vacuum Oil Company, India had announced (1948) that all petroleum developments would be confined to the public sector. From that time on, India taxed the privately owned petroleum industry more heavily than any other country in the region and controlled both prices and profits. It also regulated the volume of refinery production.

When Jersey created Esso Standard Eastern in 1961, executives of the latter company had found the Indian business climate a bit difficult, as did all foreign companies. The government demanded more than 50 percent equity interest in new refineries and lube plants. Then when an American company agreed to build a refinery on that basis, the Indian Ministry of Mines and Fuels informed Esso Eastern's J. W. Pickering that it thought that agreement should "set a pattern for future refinery negotiations" and that the Parliament would expect other contracts to be even more favorable to India.

The situation did not improve. In 1966, George P. Case, general manager of the India Division of ESE, gave Indian Prime Minister Indira Gandhi position papers on "Esso In India" and "Fertilizer Investments in India." When Ms. Gandhi indicated that India needed fertilizer "very badly and

---

*Wexler later remarked that Cleveland read every report available on the Philippines and that he knew the operation thoroughly when he decided to ask the Jersey Board to let him sell it.

most urgently," Case replied: "I pointed out that reasonable treatment and reasonable return on our petroleum investment might be helpful in assisting any future decision as far as Esso is concerned." The two papers stressed the need for fair competition and "impartiality in administering petroleum policy." Plant construction costs in India were very high, taxes equally so, and there was a question as to whether foreign investors were really welcome in India.

Nevertheless Jersey and Esso Eastern persisted in their effort to gain a share of the Indian market. A lube plant jointly owned with India came on stream in 1970. Instead of marking improvement in company-government relations, this plant led to more problems. So on March 13, 1970, the company offered to sell an equity in its Indian properties to the government. Esso Eastern received no substantive reply to this offer. [It should be noted that delay—in a Western sense—characterized all such negotiations.]

Two years passed without major alterations in the attitude of India toward foreign firms operating there. So on October 5, 1972,* the company made another proposal: A joint venture with the government's share up to 74 percent equity, or it could buy all of Esso Eastern's equity if a joint venture was not acceptable. In time, India indicated interest and talks began. These conversations provided the basis for the January 16, 1974 news release by the Indian government that it would buy 74 percent of EEI.**

That it took most of three years to work out an agreement was no doubt caused by frequent changes in the Indian cabinet—there were six Ministers of Petroleum and Chemicals during the period. Finally, the sale was concluded. The Indian government purchased "74 percent equity interest in the company's marketing, distribution and refining interests," and had an option to purchase the remaining 26 percent later.

Perhaps the 1974 Esso Eastern headquarters wire report to D. L. Snook of Exxon's Public Affairs Department expresses more precisely the reasons why Esso Eastern left India. The unknown writer said:

> From the beginning of its political independence in 1974, the Indian government has not only been unconvinced that two centuries of foreign presence contributed positively to the country, but it has also viewed certain foreign interests as obstacles to implementation of the government's announced desire to control industries of basic and strategic importance. Additionally, an attitude that foreign business ventures had inherent political affiliation with their home-country governments was developed throughout the colonial period, and that view is still in evidence.

*The Indian announcement gave October 3 as the date.

**This was the EEI proposal. A company needed a 74 percent concurrence of shareholders in order to pass certain special resolutions.

Certainly those views and opinions were not confined to India; rather Jersey [Exxon] correspondence indicates that company representatives encountered them rather consistently in what came to be termed Third World Nations.

In similar fashion, EEI sold to the government of Pakistan all marketing and refining operations in that nation and it sold its marketing assets in Cambodia and Laos to private investors in 1974. The collapse of the government of South Vietnam in 1975, cost Esso Eastern and Exxon some $30 million with interest.[48]

Some of these sales could have been influenced by the oil crisis of 1973, but most of them resulted instead from long and careful study by both Exxon and Esso Eastern. Cleveland served as that company's chief executive officer during the time in which virtually all of the divestment decisions were made, and he presented many of the proposals to the Exxon Executive Committee or to the Investment Advisory Committee. As one long-time Esso Eastern associate remarked of these decisions, "It was quite courageous [then] to constrict the size of the company."

Shortly after he became CEO, Cleveland and his associates installed an in-company report system, and from the affiliate and staff reports they perfected the criteria that enabled them to develop "rankings of the various countries according to investment climates."[49] Six years later, Esso Eastern no longer had either refining or marketing interests in those countries it had classified in the second investment level.

In most of those countries, subtle pressures of longer duration than any oil crisis first led Cleveland and his staff to consider divesting properties. While dormant on occasion, these underlying emotions and feelings surfaced in almost every crisis situation. Thus, in many new nations, foreign-owned companies served as reminders of the colonial past, and led their leaders to assert their conviction that the government alone should own all oil companies. In countries where dictators emerged, ownership of refineries and other plants became symbols of political status; too, friends and relatives could be placed on payrolls without arousing much opposition.

In such countries as India, the government by law and by regulations, favored the state company, thereby creating a business climate hostile to foreign oil concerns.* Many of the international oil companies were American and thus identified with U.S. policies. Many Asian (and other) nations took exception to at least certain trends in U.S. foreign policy, and often associated these policies with Americans in their countries. Some Asian leaders could perceive no benefits deriving from foreign investments in their countries across the years. Even so, Exxon prided itself (and its affiliates) on

---

*After 1959, India welcomed Russian technical assistance in the effort to locate oil. Of course, they continued to purchase Soviet oil.

being flexible. In 1973, the year of the energy crisis, the *Annual Report* stated:

> Exxon grew by living with economic and political change and by absorbing the stresses and strains necessary to balance the world's varying needs for a dependable flow of oil products with supplies available from a multiplicity of sources.

Esso Eastern and its managers thus were acting in a manner traditional for all Exxon executives when they made the decisions to abandon operations in these particular countries.[50]

X X X

Despite the divestments during its first decade, Esso Eastern's sales and profits grew impressively. Japan, Australia, Malaysia, Hong Kong and Singapore developed as rich marketing areas, and the output of EEI refineries increased annually. The oil and gas found in Australia and Malaysia proved of great value to the company as the world supply constricted. Old oil from its earlier source, Indonesia, also helped to meet company needs, but there, despite the efforts of P. T. Stanvac Indonesia, no new fields were found.

After 1962, a strong company bond began to mold Esso Eastern employees into a cooperative unit—a feeling rarely enjoyed by those who had worked for the predecessor company. Several company officers who watched this develop actually termed Esso Eastern a "discrete unit." Esso Eastern early tested the regional administrative structure later adopted by Jersey. The direct manner in which Pickering, Cleveland, Executive Vice President Robert Anderson (1967), and others moved to grasp full control of operations gave momentum to company feeling. Employees saw that the Esso Eastern management team relied on reports, read them carefully, and gave proper credit to individuals when the occasion warranted. More to the point, they saw streamlining of operations influenced by these reports: Public Relations and Government Relations became Public Affairs; Manufacturing and Product Movement were merged into Logistics; and the Treasurer's and Economics Departments became Finance and Planning. Staff reductions accompanied such consolidations. An important change placed authority for most major decisions in an Executive Committee, correspondingly reducing that function of the Esso Eastern board. Afterwards the board usually met annually, principally to ratify actions of the Executive Committee and to elect officers.

The company developed a training center in Singapore. Employees gathered there to take courses in a wide range of oil-related subjects—"from

instrument maintenance to management techniques." The joint partners
received invitations to take advantage of these sessions, and many of them
sent delegates.

Esso Eastern provided a wide range of consultants to advise managers on
situations or anticipated problems. The EEI home office paid special attention to these particular sources of information. The end product of this
training, in the opinion of many who served the company, was to keep the
organization flexible and receptive to new ideas.[51]

X X X

Cleveland had begun his search for executive talent as soon as he became
Esso Eastern CEO in 1966. Since the general managers reported to either
Cleveland or to the executive vice president, the person in the latter position
acquired much of the experience necessary to equip him for the top position.
After the Stanvac breakup, the executive offices of EEI had become institu-
tionalized rather than personalized. This made for a smoother operation
and it also made easier the shifting of personnel. And when Cleveland retired in 1974, Vice President Maurice E. J. O'Loughlin succeeded him.
O'Loughlin had served with Imperial Oil Limited, Exxon's Canadian affiliate and in 1965, he became manufacturing coordinator for Esso Eastern.
From that position he rose to director, vice president and on to the presidency.

Few who knew him would question that Cleveland had a special flair—a
style of his own as chief manager of Esso Eastern. Once given the authority
to decide, he made decisions; and once they were made, he acted on them.
Indeed throughout its organization Jersey tolerated few executives who
honed their administrative skills through procrastination. But Cleveland
surely belonged to the "do it now" group. Illustrative of this fact is the
history of the relocation of EEI's corporate headquarters from New York to
Houston.

In 1969, at a Jersey Board meeting, Cleveland questioned the proposed
cost of rental space in the parent company's new building, then being constructed in New York City. Director George T. Piercy then asked him why, if
the cost seemed too high, Esso Eastern proposed to remain there. Until this
moment Cleveland had not understood that the location of his headquarters
was a decision made by the affiliate, rather than the parent company. But
the Board assured him it would endorse any economical alternatives. By two
o'clock that afternoon, he had called a special meeting of EEI executives,
where they made a decision to secure an outside planning consultant. Some
six weeks later, the EEI management rejected the consultant's recommendations and employed a second person, who helped them narrow the choice of
possible headquarters locations to Houston or Dallas. Then, Cleveland,

Anderson, and Employee Relations Manager Bruce Weidenbacker chose Houston, where Exxon Company, U.S.A., Exxon Production Research Company, and Exxon Exploration Inc. already had their headquarters, as did scores of other oil companies. Even better, Exxon USA owned a prime location on the (then) outskirts of Houston, which was available for purchase. Exxon Corporation in New York approved the site, and as Cleveland wrote, "Then we made a really smart move, I think." They hired an outstanding architect who studied company needs and submitted a building plan, which EEI followed in detail. Their "smart move" in short was "not letting the in-house 'amateur architects' get in the way of the experts." The results proved the wisdom of the Cleveland decision.

More than half of all the New York office employees agreed to move to Houston. Those remaining stayed with Exxon in New York or transferred to affiliates. The move itself was staggered in three waves, one week apart. Within two weeks those already transferred had taken control of all operations, and at the end of three weeks everyone was in his or her Houston office. In nine months, Cleveland and his associates had decided to move, picked a location, engaged a contractor who built 100,000 square feet of office space, and made the move. After studying the result, it is obvious that "The Houston office of Esso Eastern Inc. is still one of Exxon's most attractive and functional office buildings."[52]

X X X

Although statistical comparisons without supporting data are often misleading and because there is little information on Esso Eastern's relative position in Jersey's total production, refining, and marketing for the years before EEI utilized its report system, it is difficult to trace EEI's growth with accuracy. Nevertheless, by the time Cleveland retired as president, better information had become readily available. Two years after he became CEO, Far Eastern Contact Director Brisco wrote to Cleveland concerning the "excellent presentation" he had made to the Operations Review Committee [ORC]. Brisco stated: "We agree with your proposal that it would be more appropriate, for next year's outlook, to present your financial results on a 'look through' basis." Then he added, "Jersey is considering similar changes in its financial reporting requirements." The ORC expressed pleasure that EEI had "developed rankings of the various countries in your area according to investment climates," and Brisco indicated that other Jersey departments had begun to study that procedure. Esso Eastern, as has been stated above, continued to develop its report system.[53]

During the height of the oil crisis, Exxon International Company suspended credit in the Esso Eastern area from September, 1973, through August, 1974, and orders went out to all company affiliates informing them of this policy change. On November 6, 1974, some weeks after Exxon International resumed the credit program, David Rockefeller's jet plane with a party of fourteen persons arrived in Bangkok, Thailand. Credit restoration did not matter in this case for neither Mr. Rockefeller nor his crew had an international credit card with them. [This fact, the press and storytellers overlooked.] The pilot approached Esso Thailand's airport crew for refueling and they refused to honor his request. Shell then supplied the fuel without any hesitation.

O'Loughlin received a full report stating that the incident had "amused" the Rockefeller party, especially John Louden, head of Shell, who was in the group. Exxon's Jamieson and David Rockefeller were already good friends, and C. O. Peyton, president of Exxon International, gathered the facts about the incident for the Exxon CEO. Peyton reported to Jamieson that "the Esso attendant was correct in refusing to refuel a private jet without either a proper credit card, or without having received prior clearance through normal company channels. . . ."[54]

X X X

Beginning in 1974, Esso Eastern, along with other Exxon affiliates, began to feel the pinch then closing in on the oil world. Its own affiliate, Esso Exploration and Production Australia Inc., on advice from Exxon, cut its exploration effort by one-third. The Investment Advisory Committee called on Esso Eastern to reduce its 1975 budget by $40,000,000. Some major replanning and restructuring of the budget proved to be necessary: the IAC requested a reappraisal of Malaysian gas and oil delineation drilling; asked that "mineral spending" be phased down and kept at the 1974 level; suggested that the Australian government be persuaded to allow Esso Eastern to allocate preferentially "our own crude to our refinery needs," and recommended several other budget alterations for 1975. IAC also recommended that the Australian coal program approved by the Exxon Producing Department in February, 1974, be either deferred or terminated. While budgetary considerations may have offered part of the reason for the latter recommendation, the slight "possibility of our [Esso Eastern] obtaining thermal coal exploitation permits in Australia" provided the primary explanation.[55] Along with regional offices, Esso Eastern received notification to "proceed with the development of a restructured overall 1975-78 [budget] plan for submission to New York by January 31, [1975]." This communication marked the beginning of an austerity program for Esso Eastern—and for other Exxon affiliates.

Overall, however, Esso Eastern entered the last quarter of the twentieth century with the fine record acquired during its brief existence. While not monumental discoveries, important sources of oil and gas had been found in

Australia and Malaysia. Exploration underway later than 1975 led to additional discoveries in both nations. Such markets as Japan, Australia, Malaysia, and Hong Kong had expanded beyond anyone's imagining at the time of the Stanvac dissolution. The success of the Hong Kong power plant—soon to be enlarged—had added credibility to the judgment of engineers, analysts, and planners.

Esso Eastern executives had become part of the Exxon COED network and could expect promotion to positions outside the national companies to which they were first assigned—when they developed as expected. Above all, Esso Eastern employees exhibited high morale—pride in themselves, their work, and their organization. They were fit and ready in the field and in management. By 1975, they justified the belief of Eugene Holman and J. Kenneth Jamieson that decentralization in the company would work—find good people, give them authority and responsibility and then watch them perform. Esso Eastern had provided a case history of the Exxon management philosophy.

Chapter 17

Jersey and the Suez Crisis

THE STORY OF THE FIRST SUEZ CRISIS, in 1956-57, further illustrates the confusion and lack of consistency in American governmental policies toward the transnational oil companies. This matter has been introduced in a prior chapter, covering Jersey, Antitrust, Iran and Stanvac.

Earlier, after World War II, Suez Canal traffic had begun its steady annual increase, with a new high of 14,555 ships passing through in 1955. These ships carried 1.5 million barrels of oil per day [b/d], of which 1.2 million barrels were destined for Western Europe. This oil represented two-thirds of all Western European requirements as well as two-thirds of the canal's total commerce, and provided for Egypt, a major source of income. In Europe, post-World War II recovery depended a great deal on a constant supply of Middle Eastern oil. The British government controlled the canal, and among the ruins of their great empire, they maintained bases along the canal as well as in Iraq and Jordan. Generally, the British encouraged Arab unity, while at the same time seeking to retain the remnants of empire.

The Egyptians themselves closely watched the struggle of the Iranians to rid themselves of British influence. In November, 1951, Iranian Premier Mossadegh, while visiting Egypt, expressed a view with which few Egyptians (or Arabs) would have quarreled: "All decisions concerning the Suez Canal or Iranian oil belong to the sovereign Egyptian and Iranian governments. Egypt and Iran share the same hopes and the same sufferings in all phases of their struggle." No clear, consistent policy or set of principles appeared to guide Western actions in the Middle East at that time.

In 1952, a clique of Egyptian Army officers, led by Colonels Mohammed Naguib and Gamal Abdul Nasser, deposed King Farouk and took control of the country. Nasser soon emerged as the principal figure in this military junta and as the strong man of Egyptian politics. By 1954, his stature in Egypt and the Arab world had become legend because he successfully concluded a treaty with Great Britain, that led to the withdrawal of British troops from the Canal Zone. Meanwhile, parallel Iranian success in achieving the consortium agreement in 1954, also ended British dominance there. These British setbacks soon led Nasser to set his sights on total control of the vital canal. His foreign policy included a more rigid opposition to Israel and some encouragement of armed raids by the fedayeen into Israeli territory. Israel replied to these in 1955, with an attack on Gaza, that easily defeated the Egyptians. When Nasser's request for additional American arms did not receive prompt action, he arranged for help from Russia via Czechoslovakia, creating yet another issue in the Cold War. As further warning, Egypt proclaimed that if the United States continued to supply arms to Israel, "the whole Arab world" would assume the United States had chosen sides in that Middle Eastern struggle. No neutrality seemed possible.

As a result of Nasser's attitude, the United States Department of State began to study the problems that would be created by a possible Suez closure. First among these was how to provide enough oil for American troops in Europe and for North Atlantic Treaty Organization [NATO] allies. During the Korean War, a similar oil and gas shortage had been countered under the Defense Production Act of 1950, which permitted the creation of a Foreign Petroleum Supply Committee [FPSC] which Standard Oil Company (New Jersey) [Jersey] Director Stewart P. Coleman had headed. By 1956, the 1950 act had been amended several times, as had the Voluntary Agreement Relating to Foreign Petroleum Supplies, under which the eighteen companies had banded together in 1950.

In February, 1956, the State Department began new discussions with representatives of the international oil companies, concerning the revised Voluntary Agreement, the probable basis for any joint action that might be required. Coleman attended one of these sessions, where he was given a copy of the revised agreement. He sent the document to Jersey Counsel Thomas E. Monaghan for review.

Monaghan, in his analysis, found the agreement to be full of holes. He pronounced it virtually unworkable, and pointed out that non-members of the FPSC abroad would have to be approached through the various U.S. Embassies, with no certainty of either a response or of accurate information in case of an answer. Unlike the situation under the earlier Voluntary Agreement, FPSC members could secure production and refining statistics only from their subsidiaries. The new agreement prevented the disclosure of any information gathered "to anyone other than the Administrator or Director." This prohibited the chairman from releasing the figures to his own committee. Monaghan found other clauses confusing, and pointed out that, for the agreement to be operable, several sections required change. What bothered the counselor, as well as the unworkability of the plan, was "how much the immunity from the antitrust laws will really be worth." He pointed out that in such cases as that of Arabian American Oil Company [Aramco], where any discussion would readily provide a relatively accurate estimate of Saudi Arabian production:

> it will be difficult to tell at the time whether a particular discussion is covered by the immunity. The Department of Justice might at some latter [sic] time contend that an individual company's estimate might have been deduced from an estimate with regard to a particular country, because that company had a large share of business in that country.

In sum, then Jersey wanted no part of any such agreement, and Coleman certainly presented Monaghan's objections to the Interior Department. A number of changes were made in the Voluntary Agreement when, and after, President Eisenhower authorized the creation of the emergency oil exchange. Also, since Secretary of State John Foster Dulles moved so vigorously in the entire Suez affair, it is reasonable to assume that he must have ordered his subordinates and aides to correct any flaws detected in the Voluntary Agreement.

Dulles, supporting the British, agreed to help build the Aswan Dam by promising a direct loan and by supporting Egypt's application to secure funds from the International Bank for Reconstruction and Development [World Bank]. Dulles believed that the American-British [-French] alliance was the keystone of peace in the world. At the same time, however, domestic political problems, plus inconsistent behavior on the part of Nasser, had led many powerful U.S. politicians, including President Eisenhower, to question the advisability of the loan. The arms agreement with a Soviet satellite state, Czechoslovakia, marked the beginning of Russian influence in Egypt. Nasser, angry over the delay in securing financial aid for the Aswan Dam, permitted Egypt to draw closer to Russia, especially as a source of munitions and loans. Russia thus jumped over the "Northern Tier" to secure a foothold in the "Southern Tier" of Arab states. This marked the first act in one of the greatest tragedies of modern diplomacy. At that moment, Nasser refused to go along; instead, he continued to bait the Western powers, using Radio Cairo to encourage the North African rebels against France and to attack British influence in what had been its Middle Eastern empire. Nasser boasted of supporting Syrian attacks on Israel, and he used every opportunity to denigrate the U.S., Britain and France. With all this anti-Western activity, he combined flirtatious behavior toward Russia and Soviet bloc nations. Acting independently, to the great horror of Dulles, he led Egypt to recognize Red China. Some Cairo press reports—perhaps planted—stated that Russia had offered Egypt a gigantic low-interest loan to be used to build the Aswan Dam. When he heard these reports, Secretary Dulles exploded, and he continued to burn when he learned that Nasser planned to visit Moscow in July, 1956.

At a meeting of the National Security Council [NSC] on July 13, 1956, President Eisenhower received a full briefing on the entire situation. Afterward, he instructed Secretary Dulles to withdraw the American offer of a loan, and Dulles, with great fanfare, did so. [From that time, Dulles has received credit for this action; however, the decision was the President's.] The President's action overruled the advice of several career diplomats, who argued that withdrawal of the Aswan offer might encourage Egyptian seizure of the Suez Canal. They proved prophetic. Nasser did seize the canal on July 26, 1956. President Eisenhower then ordered the creation of a Middle East Emergency Committee [MEEC] and charged it with finding ways to supply Western Europe with oil if Egypt blocked the canal.

Up to this point it would appear that a number of alternatives, properly explored, might have prevented the ensuing debacle. But in truth, no one of them was wholly agreeable to the U.S. The action taken by Nasser made him the strong man of the Middle East, not only for Egyptians but also for many Moslems in the Mediterranean area and elsewhere. Certainly his dramatic actions placed extreme pressure on the governments of Britain and France, the countries where most of the stock in the Canal Company was held. Their citizens had financed construction of the "big ditch" in the first place. Further, a large share of the petroleum consumed in those two countries came through the canal. Accordingly, the seizure of Suez created a practical crisis for both governments.

Jersey executives also recognized the value of the canal as vital to world oil transport as well as to the economic health of Western Europe, where the company had large marketing operations. In accord with traditional procedures, Jersey moved quickly to begin studies of the now complex canal situation and of the larger problem of oil supply.

Meanwhile in Washington, on September 26, 1956, Secretary Dulles held a press conference and made the famous remark that the United States had no intention of shooting its way through the canal. He also took the initiative in forming the Suez Canal Users Association, a group which met several times without arriving at any solution to the problem. The task became harder as vigorous applause from his fellow Arabs led Nasser to issue ever stronger statements about his designs. All this time he was encouraging more raids into Israel, and more than anything else, the forays exposed the weakness of all the governments involved—Egypt, Great Britain, France and the United States—revealing them to be guilty of poor planning and even worse communication. Of the group, France appears to have followed the most consistent pattern; however, when the government of Israel moved onto the scene, it acted more decisively.

In the U.S. a national election was approaching. Some observers have used the interests of politics to explain the apparent failure of President Eisenhower and Secretary Dulles to have intelligence and other reports properly evaluated as a basis for action. At the least, it is certain that the Central Intelligence Agency [CIA] as well as several highly placed foreign officials forwarded important information that, carefully analyzed, might have helped redeem the situation. Yet as one writer has commented, "The French and the British governments certainly did not, however, keep the United States informed of what plans they had." Concerning Great Britain, former Secretary of State Dean Acheson later went so far as to write, "It is fair to go further and say that, at that stage, its conduct was deceitful." Whatever responsibility Secretary Dulles had in this affair, it is true that he consistently opposed any use of force to open the canal, and that repeatedly, he made his position known to Britain and France.

As the crisis heightened, Britain and France alerted their army, naval, and air units. When Nasser blockaded the Gulf of Aqaba, where Israel had its Red Sea port, that country mobilized its border forces. Then, on October 29, 1956, Israeli troops, moving swiftly in what Premier David Ben Gurion proclaimed a defensive war, defeated the Egyptian Army in the Sinai Peninsula (and, within four days, moved on to Suez). Britain and France issued an ultimatum to Egypt and to Israel demanding that the fighting end within twelve hours, that both sides retreat ten miles, and that Egypt accept the temporary occupation of Suez by Anglo-French forces. Should the belligerents fail to respond to this ultimatum, troops of the United Kingdom and France would take steps to secure compliance. But Egypt ignored this dispatch. Fighting continued, and the British and the French moved troops into key positions along the canal. On October 31, Nasser ordered a number of vessels sunk as a means of blocking ship passage through the canal. President Eisenhower several times telephoned the British Prime Minister, Anthony Eden, and in strong terms denounced the British and French movement of troops into the Canal Zone. [Later, Eisenhower demanded that the British and French withdraw.]

At the United Nations [U.N.] the General Assembly passed a cease-fire resolution, which the belligerents largely ignored. The United States, in supporting that resolution, had for the first time in that body, voted against its allies (Britain and France) on a major issue. Yet this action should not have surprised either of those countries, for as early as October 2, Secretary Dulles "foreshadowed the later American policy of firmly disassociating itself from the Colonial interest of its European Allies" when he stated, "The Suez Canal ... is not an area where we [the United States, Britain and France] are bound together by treaty. ... There are ... [also] other problems where our approach is not always identical." But the moralistic stance of the United States in this affair seemed largely inconsistent with its often expressed desire to check the spread of Russian communism throughout the world. The Western Alliance, as well as the less tangible forces of Western influence in Mediterranean and Middle Eastern nations, were severely undermined. Most harmful of all, perhaps, many British and French statesmen and even more friendly people in allied countries felt themselves betrayed, victimized by the U.N. and the U.S., as they were forced to withdraw from Egypt. The Israelis, too, shared these feelings. And Nasser, who up to this time had feared that his government might be overthrown, now emerged a clear winner, with impressive stature and influence in the Arab World.[1]

What of Jersey and its interests in the Middle East during these hectic months? As early as April 10, 1956, the contact director for that region, Howard W. Page, reported to the Executive Committee that he had completed a preliminary review of the "prospective" Suez Canal situation. Following his remarks, the Committee commissioned a comprehensive study of Jersey's ability to move petroleum westward from the Middle East—without the use of the Suez Canal. This assignment called for review of the various alternative methods of fulfilling oil contracts with Jersey's European affiliates. In practical terms, a comparative analysis of capital outlay and operating costs for very large vessels, as contrasted with large diameter pipelines, was needed. The Committee also sought some assessment of the potential political risk factors involved. To prepare this complex study, the Wallace R. Finney Company was retained.

Page next reported to the Jersey Executive Committee on July 31, 1956; Finney had recommended construction of a pipeline from the head of the Persian Gulf, through Iraq and Turkey, to a Mediterranean seaport in the latter country, to be built at an estimated cost of $546 million. When completed, this 1,000 mile pipeline would have a capacity of approximately 850,000 b/d. The time required for construction, Finney estimated, would be four years.[2]

At the time, a four-year solution could not apply; events in the Middle East simply were moving too rapidly. European consumers were already anxious about their future oil supply. On August 9, 1956, two weeks after Nasser nationalized the canal, H. G. Burks of Jersey's Refining Coordination Department faced a question from Serge Scheer, president of the French Esso Standard Company, who asked what measures "could be envisioned to replace Persian Gulf crude if Suez Canal were closed?" Burks replied that "If Suez Canal were closed, change in routing of world oil supplies would be necessary, drawing on Western Hemisphere for oil to replace Middle East by shorter tanker route."[3] During this period of accelerating crisis, Jersey executives continued to assure the managers of their European affiliates that the company would make every effort to deliver adequate supplies. One of those managers wrote: "It is for us Europeans, a very great relief to know that thanks to the American help, our situation would not be imperilled if the Suez Canal were to be closed."[4] Characteristically, many Jersey executives had anticipated the problem and planned specific measures to solve it. They concluded that the United States would have to supply 900,000 b/d from its domestic reserves and that Venezuelan production would have to be increased by 100,000 b/d. Using these figures, they assumed the continued operation of the Middle East pipelines.

It has been noted that until the Suez Crisis began, the United States had enjoyed cordial postwar relations with Britain and France; now, the disagreements among wartime allies changed the climate for international operations and served to make both the U.S. government and the American oil companies "targets of suspicion and criticism."[5] Some experts even contended that Jersey's proposal to weaken Nasser's position by supplying Western Europe with oil from other sources would offend the Middle Eastern oil producing nations, and perhaps endanger the company's richest oil concessions, located in those countries. Yet, as early as August, 1956, the Oil and Gas Division of the U.S. Department of the Interior had begun closed door discussions with oil company representatives, to consider what to do in case Nasser closed the canal. Some meetings took place in London, which representatives of U.S. oil companies attended as observers. Discussion of "Anglo-U.S. cooperation on oil matters," however, was "taboo." President Eisenhower assumed responsibility for these sessions. He thought of the Suez Canal as the world's greatest "public utility" and the government-sponsored meetings paved the way for the Middle East Emergency Committee, formation of which, he had personally authorized.[6]

In early August, Arthur S. Flemming, director of Defense Mobilization, invited Jersey to assist in developing a contingency plan for moving oil supplies to Europe, later to be known as the "oil lift." Flemming pointed out that under the amended Defense Mobilization Act of 1955, the Justice Department had again approved the oil companies' banding together in the interest of national defense. [Not all the oil companies invited to participate desired to do so. The Gulf Oil Corporation later asserted that it was forced to join by the State Department.][7] Jersey Vice President Stewart P. Coleman, in reply, accepted for the company; he agreed to name a Jersey representative for the newly created MEEC.[8] On August 21, Assistant Secretary of the Interior Felix E. Wormser appointed Coleman chairman of that committee. [Earlier, during the Iranian crisis, he had chaired the Foreign Petroleum Supply Committee.] A day later Wormser requested the committee act swiftly, "to initiate such surveys and investigations and obtain, analyze and keep current such information and data as are necessary to estimate the petroleum requirements" of the involved nations, and "to prepare estimates on the availability of such requirements" in case of "a stoppage of transportation ... through the Suez Canal" or any of the Eastern Mediterranean pipelines.[9]

Throughout the correspondence between Coleman (and others at Jersey) and various government officials runs a theme of uncertainty; these men constantly sought assurances that the U.S. Department of Justice approved of the company's participation in the government's plan. Yet the old civil antitrust suit against Jersey remained unsettled; and again, Jersey and other oil companies found themselves in the position of being asked to do that for which they had already been brought to court. On August 15, 1956, Coleman wrote Flemming:

> We have taken note of the statements in your mentioned letter that the Attorney General has approved your request after consultation with respect thereto between his representatives, representatives of the Chairman of the Federal Trade Commission and your representatives, pursuant to Section 708 of the Defense Production Act of 1950 as amended ... and find it to be in the public interest as contributing to the national defense.[10]

When MEEC first met, on August 24, it worked on a plan of action that required approval from the Attorney General. In 1957, Coleman described that plan and how it operated. His version indicates the extent to which antitrust laws actually inhibited effective action by the oil companies, even during a serious international crisis, one that held the potential of developing into a threat to U.S. national security. The MEEC, Coleman wrote, "has no power to direct any company, whether a participant in the Plan or a nonparticipant, to take any action, nor has the Committee the power to request any company to take any action." It could not allocate available supplies. This could be done only by various government organizations. Voluntary committees, composed almost entirely of oil company personnel, studied transportation and marketing, analyzed production, and compiled a fund of statistical information, to be used by the concerned governmental agencies.

Company transportation experts had already informed Jersey executives that rerouting tankers around Africa would reduce oil deliveries to Europe by more than one-half, and that "virtually no additional tankers [were] available." In many European countries, the closure of Suez combined with the first signs of winter created great concern. The failure of the United States to state clearly just what it proposed to do in the Suez Crisis rekindled quiescent anti-American feelings in Europe. As the emergency became more acute, so did this sentiment.[11] Meanwhile, at various conferences and meetings, executives and government officials presented their estimates of the petroleum requirements for each affected country; they also sought to pool all available information on the best routes of supply for each country that could expect a shortage—largely those in Western Europe.[12] Within days, the MEEC moved to coordinate its plans with those of the British Oil Supply Advisory Committee, operating from Toronto, Canada. Jersey's Worldwide Supply Committee, anticipating the information needed, had compiled for its affiliates the data necessary for defining each company's role in this crisis. The Humble Oil & Refining Company, for example, reported "spare producing capacity of 1,250,000" b/d with "spare pipeline capacity about 550,000" b/d.[13]

MEEC believed that a liaison committee should be formed in London, and Coleman recommended H. W. Fisher to represent Jersey on that committee. Fisher had replaced Page as Jersey's chief London representative and as a director of the Iraq Petroleum Company [IPC].[14] Coleman wrote to Fisher that MEEC had "the same representation of American Oil companies, operating abroad as did the Foreign Petroleum Supply Committee, which dealt with the Iranian situation." As chairman of the latter committee, he pointed out that things were different now because "under the Iranian Crisis the European situation was fairly simple, in that Iranian supplies had been going only to British companies, this time all of Europe and a number of companies of different nationality will be affected."[15] In addition to Fisher on the London committee, other Jersey executives served on seven subcommittees of the MEEC.

Overall, MEEC confronted a very difficult task. Once at work, it had to establish liaison with such major oil companies as the Royal Dutch-Shell

Group [Shell], British Petroleum Company, Ltd. [BP], and Compagnie Française des Pétroles, S.A. [CFP], and solicit data about their operations. Even while planning, the committee sought to stimulate rapid development of crude production outside the Middle East, to chart routes to have such oil delivered, and expand port and tanker loading facilities. When Coleman attempted to resign as chairman on September 7, because of a recurring health problem, H. A. Stewart, director of the Oil and Gas Division of the Interior Department, persuaded him to withdraw his letter of resignation.[16] Jersey, tapping its vast resources of top executives, borrowed Imperial Oil Limited's J. A. Cogan, and assigned him to relieve Coleman of some of his workload. Cogan had served for several years as head of Jersey's Coordination and Petroleum Economics Department before going to Imperial; now, he moved to New York and quickly proved invaluable as the hard working expert who could provide concise summaries of the reams of statistics deluging the committee. Coleman concentrated on the world oil situation and after reviewing a number of projections regarding Western Europe's requirements for the approaching winter, he made the prophetic statement that these needs plus the Suez closures would result in a "serious upward pressure on crude prices."[17]

As the situation in Egypt became more confused, C. R. Smit of Esso Nederland, N. V. informed Jersey in a confidential letter that the Dutch government had established an organization to supervise gas rationing, and that "the necessary coupons for gasoline and other products have already been printed."[18] Then, questions arose regarding the use of information acquired by the MEEC, and Jersey's Associate General Counsel, Nicholas J. Campbell, pointed out that the Justice authorization did not permit "the *exchange of information* with others."[19] Similar to procedures in earlier emergencies, each participating oil company helped finance the committee operations, including the salaries of employees working for MEEC.[20]

Fisher insisted that Jersey keep its European affiliates informed about company plans, and he suggested that the shareholder representatives convey that information. Associate Counsel Campbell approved this idea.[21] Cogan then supplied Fisher with details of the MEEC actions; in turn, he relayed them to the shareholder representatives. Cogan also wrote to Fisher that if the pipelines went out, rationing of petroleum products would mean that the burden of supply would fall on Venezuela and Gulf Coast producers.

On September 28, 1956, the magazine *U.S. News and World Report* featured an interview with Jersey President Monroe J. Rathbone entitled, "Will Suez Force Gas Rationing?" Rathbone doubted the probability of gas ing because he did not believe that both the pipelines to the Mediterranean and the Suez Canal would be closed at the same time. But should both be shut down, then, in his opinion, rationing must occur. Questioning that the Arabs wanted to stop the flow of oil, he correctly predicted economic dislocation until the pattern of existing tanker hauls could be rearranged, and until larger tankers could be built.[22]

In October, Coleman drafted for Interior Assistant Secretary Wormser a report on the work of the MEEC. As a group, that committee viewed production, refining, and transportation of oil in such an emergency to be problems with which the industry could cope. However, another problem, they believed, lay beyond the scope of any combination of companies; it would require instead direct governmental action. That problem was compensating for the economic impact of the Suez shutdown on all of the oil-short nations. As Coleman wrote, "The Committee stated its view that these economic problems were of such significance that they would require for their solution action by the United States Government, as well as, and in conjunction with, action by various foreign governments." The oil executives pointed out that crude oil might have to be "moved by abnormal and expensive means." In conclusion, Coleman reiterated that the committee felt capable of carrying out any requests made by government spokesmen.[23]

On October 24, 1956, Jersey President Rathbone wrote Director David Shepard questioning the Free World's capacity "to construct 60,000 ton tankers" and asking whether steel plate, turbine gears, and boilers for these could be obtained. Rathbone believed that construction of such vessels would depend on the President's conviction that American national defense actually required their construction.[24] Jersey, in Rathbone's view, had more than a passing interest in building larger tankers. He went on to describe the Middle East as an increasingly troubled part of the world, where a number of developments could cause Egypt to shut down the canal, thus escalating charter fees for tankers. Even earlier, Jersey had determined to protect itself by ordering construction of new and larger tankers at the same time it continued a policy of large-scale chartering.

Meanwhile, Coleman's health continued to decline during late September and October, and Cogan and Arthur G. May, the executive secretary of the MEEC, picked up the burden, gathering information and coordinating plans with the Oil Emergency London Advisory Committee [OELAC]. Up to this time [November 1, 1956] and for some weeks thereafter, the oil company executives working for the MEEC could not be certain that anything they were doing would be useful. While Jersey and each of the other companies acted individually to supply its European markets as efficiently as possible, the MEEC sanctions did not yet apply. Because they could not act in concert—antitrust laws specifically forbade collusion in production, refining, transporting, and marketing—[also, the government's civil suit against the five U.S. majors had not been settled] considerable amounts of oil from the Persian Gulf came to the United States via the Cape of Good Hope, and some arrived from the Aramco pipeline. Under cooperative agreement this oil could have been exchanged with companies serving large markets in Europe, and the extra haul eliminated. Then another exchange could have been made for Gulf Coast or Venezuelan oil, with a company normally hauling from the Persian Gulf. But, in fact, all MEEC activity up to November 30, amounted only to so much "wheel-spinning," as oil companies continued to operate just as they had before Suez closed.

The United Nations General Assembly ordered a cease-fire in the Canal Zone in early November, 1956. On November 6, the British and French agreed to observe its terms, but they refused to leave the Suez Canal. In order to persuade Great Britain and France to withdraw, President Eisenhower continued to use oil power to bring pressure on them, and in a less direct way, on Israel. At the same time, his strategy aggravated the problem of supplying oil for non-Communist Europe.

Many petroleum consumers in Western Europe, as well as government officials there, believed that American companies had influenced the U.S. government and encouraged oil diplomacy so that the companies could consolidate their Middle Eastern positions. Certainly, the public on both sides of the Atlantic grew increasingly skeptical of multinational oil companies during this period; and European criticism of the U.S. became more vocal.

The oil companies themselves sought to have the U.S. government activate MEEC, but without success. As one oil company observer pointed out, "the psychological effect of the MEEC not being revitalized" was becoming increasingly serious and might lead to controls and regulation, which would hamper the industry later. He suggested some American action that would avoid leaving the impression in Western Europe "that they have been let down at a critical time" by the oil industry and by the U.S., "because of political feeling over the Middle East." Another oil company employee reported that despite the urging of the industry, "the administration simply refused to reactivate it [MEEC]." He added, "There is a strong feeling in Europe ... that the U.S. government was determined to punish Britain and France for their military action in Egypt ... and was in the process, punishing other countries to an equal or greater extent."[25]

In the meantime, Nasser sympathizers had bombed pipelines and pumping stations in Iraq and Syria, and on November 3, oil ceased to flow through the IPC pipelines. The political rift widened; even Saudi Arabia, a long-time friend, broke off diplomatic relations with Britain and refused to permit the loading of tankers bound for that country [or France]. In Kuwait, pro-Nasser workmen bombed pipelines and committed other acts of sabotage, forcing the entire oil operation to shut down. Every company associated with the British reaped a share of the general hostility directed against Britain by Arab nations and Russian satellites. With the Suez Canal presently useless, the IPC pipeline closed, and Kuwait production down, the winter looked grim to many Europeans.

X X X

At home in 1956, the United States still had considerable reserve capacity. Jersey itself had not imported any Middle East oil to the U.S. for more than a year. While a half dozen companies in aggregate owned most of the Middle Eastern producing concessions, at least fifty companies actually imported oil from that region. This lower priced oil kept prices down, and Oklahoma and Louisiana cut allowable production by 15 percent in May, 1956, and again in June. Both price and conservation provided reasons for limiting production. Also complicating this producing problem was the fact that the transnational companies operated at near peak production in the Middle East because governments in those countries demanded revenue and because the unsettled political conditions there encouraged the companies to amortize their investments as rapidly as possible. Any reduction would injure the relations of the United States with the host governments. Still, the long haul around Africa slowed down deliveries of Middle Eastern oil to Europe. Thus, by necessity, production there declined, and U.S. and Venezuelan production had to take up the slack. Since oil in these locations was now thousands of miles closer to Europe, Jersey and other companies increased production. [The January 1, 1957, *New York Times* reported that the Creole Petroleum Corporation had increased production in 1956 by 100,000 b/d "to help out in the situation."]

Many producers in the United States, led by Jersey's affiliate, Humble, now sought to obtain an increase in allowable production in Texas, where most of their reserves lay. The Texas Railroad Commission, dominated by Colonel Ernest O. Thompson and entrusted with control of allowable production, refused, however, to permit any increase. At the moment Nasser blocked the canal, the commission was permitting wells to produce only fifteen days each month. Other majors joined Humble in requesting an increase in allowables, which Thompson and the commission refused to grant. When Great Britain complained, Thompson replied, "We have already shipped her many barrels of crude, but we only get criticism for not going all out at her bidding. England apparently still looks on us as a province or dominion." His actions reflected the influence of the small independent producers, who before they would increase production, demanded an increase in the price of the crude they sold to major integrated companies. Humble actually did increase its purchases from West Texas producers by 10,000 b/d.[26]

Hindsight now supports the belief that President Eisenhower and Secretary Dulles overreacted to the Russian threat in the Middle East. Yet, substantial evidence of Russian ties with the Nasser government did exist in 1956; in fact, Premier Khrushchev had assured Nasser that Russia would support his drive for a United Arab Republic. Events in the Balkans, however, rendered Khrushchev's promise empty for that time. In October, after much rioting, Hungary revolted against her Communist master and drove the Russian guard out of Budapest. The Soviet leaders realized that the Western powers were diverted by the Suez Crisis and probably would pay little attention to events in the Balkans. So in early November, Russian troops marched into Hungary and, within a few days, brutally stamped out the rebellion, causing almost 200,000 Hungarians to flee the country. While the U.N. and the U.S. searched for a compromise that would persuade the British, French, and Israelis to withdraw from the Suez Canal, the Russian government, fresh from its Hungarian adventure, proposed to the United States that forcible intervention would solve the problem. Eisenhower rejected the offer, and Khrushchev indicated that "volunteers" from Russia might be sent to aid the Egyptians. Although Eisenhower's firmness discouraged such action, the canal remained closed.

Since many key decisions were made at the highest levels of diplomacy, and these determinations were not always well explained, the American oil companies often received a bad press, both in the United States and in Western Europe. Here again, the tight restraints that the Justice Department imposed on the MEEC increased the problem; although they were well aware of the negative response among the general public, the committee could do little about it for some time.[27] However, Fred A. Seaton, American Secretary of the Interior, sought to allay public apprehension about a possible oil shortage by stating, on November 10, that he did not foresee any rationing of gasoline. He said that the United States had an oversupply of oil and could easily supply Europe with 850,000 b/d. Seaton's remarks came at a time when the United States was importing more oil than was any other

Western nation.[28]

On November 11, 1956, the *New York Times* oil writer, J. H. Carmical, in a long article, summarized the effects of the Suez Crisis and the oil situation as he saw it. Venezuela, he pointed out, had become the Western Hemisphere's only major oil exporter. Since World War II, the U.S. role had reversed itself to become one of dependence on foreign oil supplies. Carmical noted that European recovery over the previous decade had been based on the availability of oil, and that if oil was not "made available to Europe, a large part of the billions spent by the United States on recovery there would simply go down the drain." He perceived the disparity between the amount of oil presently available to Western Europe through the Aramco Tapline and the amount needed as 1,750,000 b/d—a total that must be transported around Africa or from the U.S. and Venezuela. He continued: "This is the task that the oil companies now have assumed individually. If their activities should be coordinated as provided in a plan previously worked out with the United States Government, the job would be easier." But Carmical did not indicate—other than saying, "This plan ... may be revived"—whether he believed the MEEC would begin functioning soon, or at all.

On his part, Secretary Seaton indicated that no attempt had ever been made to put MEEC plans into effect. In fact, he confirmed that a meeting of the committee scheduled on the day Britain and France moved into Suez, had been cancelled and that MEEC "has no plans to meet again."[29] However, Wormser wrote Coleman on November 20, stating that the interruption to oil traffic from the Middle East had resulted in "grave and immediate impairment of their [Western European countries'] economies and [created] adverse effects upon the wellbeing of their peoples." Pointing out that "extensive redeployment and maximum use of the world's tanker fleet will be necessary to alleviate the shortages," Wormser concluded by saying that MEEC would again meet in a few days, to consider new plans to coordinate oil shipments.[30]

Obviously the several participating government departments shared no understandings nor goals. The oil companies served largely as pawns in a larger game. While Dulles at the State Department sought to keep MEEC inactive as one means of forcing the hands of Britain and France in Suez, Seaton and his Interior Department aides pressed the committee to continue the work of pooling of all information requisite for transporting oil to Western Europe. And the Justice Department added to the confusion, intermittently acting as if any cooperation of oil companies might be actionable under American antitrust laws. The stalemate might have continued except that, on November 26, the Organization for European Economic Cooperation [OEEC] offered a new plan for replacing MEEC. Under this plan, the responsibility for securing information on oil stocks, available storage, contemplated measures to restrict consumption, import-export figures, pipelines, and other transportation rested entirely on the governments of member states, who would afterward involve oil companies as they saw fit. Jersey's Associate Counsel Campbell reacted strongly against the proposal:

> It should be clearly understood replacement MEEC removes all authority for U.S. companies to take cooperative action with competitors, since antitrust immunity is based only on plan of action of MEEC which in turn has been approved by Attorney General under Defense Production Act. Therefore, existence and operation MEEC essential from a legal viewpoint.

He added that the proposed change would block, rather than open, the way for American companies to participate.[31] Finally, the proposal died for lack of support. Whether it ever reached the U.S. Office of Defense Mobilization [ODM] is not known.[32]

Throughout the Suez Crisis, Jersey executives carefully considered the opinions and advice of Fisher in London, as well as of the managers in their European affiliates. In addition, these executives received accurate information about the actions of European governments, other oil companies, and the OEEC from W. D. Bayles. Bayles, an assistant to the Jersey Government Relations Counsel, had been located in Europe for more than a year and one-half, and he usually reported directly to W. R. Carlisle of the Law Department.

In explaining Bayles' role, Coleman suggested the company "requires representation before many foreign governments and requires knowledge on the part of executives as to what the policies of the foreign governments may be."[33] President Rathbone himself described how Bayles worked when reporting "what was going on." According to Rathbone, he travels "around through Europe, discusses with the management of each one of our affiliated companies in Europe what their knowledge is of the situation in that particular country and he also establishes contacts with the secretariat or the executive group of the supranational bodies themselves." Jersey used the information supplied by Bayles to verify reports received from the managers of affiliates. When asked whether the correct inference would be that the company "operates a foreign service of its own," the unabashed Rathbone replied that indeed Bayles did provide that service.[34] In any case, his activities proved invaluable to Jersey during the first Suez affair.

On November 30, Coleman received a set of amendments to the original plan of action (dated August 10, 1956), and on the same day, Arthur Newmyer informed the company that President Eisenhower had decided to call on MEEC to help resolve the oil supply problem. Eisenhower had received assurances from the British and the French that they would withdraw from Suez. Three days later, Flemming wrote Coleman that the Attorney General had approved the amended plan and requested that Coleman once again represent Jersey.[35] In an internal memorandum written by George H. Freyermuth of Esso Export Corporation, titled "Industry and Jersey Supplies, European Area, as affected by the Suez Crisis," Jersey executives received the information that only "full use of tankers" by the companies in MEEC could forestall the crisis; "the movement of Middle East crude to the U.S.A. ... can be entirely replaced by U.S. Gulf/U.S. East Coast movements." Freyermuth pointed out that 800,000 b/d could be exported from the Gulf/Caribbean area to replace Middle Eastern crude on the East Coast, with the balance sent to Europe. By these arrangements, Jersey affiliates in Europe would receive 76 percent of their requirements.[36]

Sometime later, in 1957, Rathbone explained to a congressional committee how the oil "lift" worked. With the cooperation of many companies, Middle Eastern oil consigned to the United States was routed to refiners and marketers in Europe; while Jersey [and other producers], in exchange, supplied other companies with oil from the Gulf/Caribbean area, for use in their U.S. refineries. This exchange saved hauls of thousands of miles for the tankers.[37] Detailed planning required the full consent of the Justice Department and complete cooperation among the various European committees and MEEC.[38] Still, Jersey (among many companies) was forced to reduce its U.S. inventories, which reached a dangerously low level. Throughout the period, company representatives attended meetings of the European committees, acting as observers only.[39] One Board member, Peter T. Lamont, cabled Fisher that Jersey supported the U.S. government's position that any committee would be unworkable which included "representatives [of] other than international supply companies."[40] But the evidence indicates that a number of European governments sought to add to the committees' representatives of the consuming nations. The oil companies successfully forestalled this by insisting that such persons remain advisers and consultants, rather than becoming committee members.

Once Coleman and MEEC set to work in December to "reduce petroleum shortages in friendly foreign nations," they achieved considerable success. Finally, Wormser sent a telegram to Coleman to say that the plans effected on December 7, would terminate on or before March 31, 1957.[41] While

Coleman, Cogan, Fisher, and others worked to implement the overall MEEC programs, Associate General Counsel Campbell reminded them that they could switch oil and cargoes only with the participating companies: "If a transaction does not fall within these items [specified in the approved schedules], the benefits of the antitrust immunity cannot be claimed." He added that each transaction "should be so identified as to indicate clearly that it was entered into pursuant to a particular item or items of an effective schedule."[42]

As the oil exchange, which later took the designation oil "lift," became operative, Bayles reported that Western Europeans were increasingly skeptical about the role of the United States: "European public and government opinion is undoubtedly badly shaken with regard to reliability of the U.S. attitude." He pointed out that the U.S. support of Egypt's position in the United Nations (Egypt already had Russian backing) had caused some Europeans to recall two earlier occasions—Yalta and Potsdam—when, they believed, the United States had "sided with Russia on practically every issue. Europe has also had its confidence shaken ... from the moment it [the crisis] began, the oil companies were unable to give any assurances." [The public did not know that Dulles, on orders from Eisenhower, had suspended all operations of MEEC in the interests of U.S. foreign policy.]

Bayles observed that the companies had pledged to provide adequate oil for the expanding European economy, and their sudden apparent failure—regardless of who caused the crisis—could only damage their public reputations throughout Europe. The abrupt closing of the Suez Canal had made Europeans aware of just how much they depended on Middle Eastern oil. He added that in several European countries, government executives and the public already had begun to consider nationalizing oil companies, at the same time actively debating whether further conversion to oil really served to stimulate the economy. Some leaders had already begun to press for quick development of alternate supplies of energy. He ventured an opinion that the oil industry must take the initiative in allaying the "anxiety so prevalent in Europe," though he also admitted that just what they could do "is difficult to say." Finally, he urged Jersey to request from the U.S. government a clear statement that it "is striving to bring about a constructive, permanent solution of the Middle East problem." This would reassure those Europeans who questioned continued American support of "Colonel Nasser."[43]

Before the end of the year 1956, the Jersey representatives at MEEC came to appreciate the value of Bayles' analysis. They knew, moreover, that the number of persons in the United States who agreed with critical European opinion of the companies was increasing rapidly: "If the oil companies are as

competent as they say they are, then they should have been able to do something." Oddly, perhaps characteristically, this general criticism grew in volume at just the time when the MEEC companies were beginning to
cooperate to ease the shortage. But most critics placed blame for the November inactivity squarely on oil companies rather than the U.S. government,
where it more properly belonged.

As the drama of the oil exchange heightened, new controversy enveloped
MEEC, especially over the procedure used to select the particular companies
that were participating in this exchange program. Some temporary shortages of petroleum products in various parts of the United States triggered a
public debate. Certainly, neither the government nor the companies had
provided enough clear information to the public. In Congress, one representative after another rose to ask questions about MEEC. Anticipating a public
outcry, Jersey leaders emphasized their voluntary cooperation with
European-based committees, and full accounting. As Campbell wrote C. J.
Hedlund, chairman of a MEEC subcommittee, "It is more important that
company records be kept with utmost care." Next, Campbell told Fisher,
"Regardless of what requirements are set by the U.S. Government, it is most
important that each affiliated company maintains complete records of any
and all action taken in the emergency. ... Each action or transaction
[should be] ... readily identified as having been undertaken pursuant to
specific directives, requests or programs."[44]

The resulting "Oil For Western Europe" program really swung into high
gear in December, 1956. Freyermuth reported to the Esso Export Corporation, on December 18, that at a review of the "supply picture," agreement
had been reached to reduce previous estimates of affiliate needs for the first
quarter of 1957 by three million barrels. Freyermuth also expressed an opinion that "any general plan for pooling" or for distribution "would inevitably
be so complicated and so involved with political consequences that it could
not possibly get into effective operation in time to be of much help in this
emergency. ... Jersey's facilities and abilities should be used in Jersey's own
business." The affiliates faced "general price rises throughout Europe" because Shell had added an oil surcharge of approximately one dollar per
barrel, which its affiliates passed on to consumers. Jersey, Freyermuth held,
"would not want to put itself in a position of being vulnerable to the criticism that it is deliberately capitalizing on the current situation to its own
advantage." The affiliates, he wrote, "are well aware and genuinely appre-
ciative of the outstanding job which is being done by all Jersey groups."[45]

Here, Fisher too expressed concern: "It is probably too early to assess the
impact on the oil industry of the present crisis but there can be little doubt

that it will be substantial. The coal people will certainly be much more vocal for one thing." He believed the oil-consuming countries in Europe would try to develop greater unity outside of the OEEC which, in his opinion, was "still the most free-enterprise minded of all the inter-governmental organizations, by a wide margin." *Platt's Oilgram* of December 6 had suggested the creation of a panel of experts from involved countries, to act as consultants, and Fisher thought it important to have "well qualified U.S. *oil*" men selected.[46]

Meanwhile, MEEC committees had begun to develop new supply estimates for the first quarter of 1957. The oil situation seemed to improve, due first to the unusually mild weather in Europe, and then to the progress in eliminating "the Persian Gulf to the U.S. East Coast movements,"—made, as Cogan wrote Fisher, despite the inability of "Tide Water Associated Oil Co., Cities Service [Oil Company], and Gulf" to end their traffic.[47] True, the companies did prove unable to supply to each Western European refinery just the particular grade of crude oil it required; yet, overall the oil exchange appeared to work well. On Christmas Eve, Fisher wrote to Jersey Director P. T. Lamont, stating that "things have been moving so rapidly that it has been a little difficult to keep people posted." Since MEEC began its balancing act, "These supply people are undoubtedly very much pleased with the present situation. ... The whole Jersey organization has certainly done a magnificent job." During the same time, some European countries, France, for example, pooled all supplies of "crude and products, equalizing refinery runs ... in accordance with what they call 'solidarity.' "[48]

To the present researcher, however, the entire combination of MEEC with its various committees, the OEEC, OELAC, the Organization of Europe Economic Cooperation Petroleum Advisory Group [OPEG], and their subcommittees amounts to a miracle made of sheer confusion. The structure exceeds all hope of comprehension when the innumerable political agencies, committees, and persons representing various nations and groups are added to the basic design. One is left to doubt whether such a conglomeration could possibly function to any effect. Small wonder then that the general public became confused by, and sharply critical of, what was taking place. Some writers have even erroneously argued that the oil lift did not supply Europe adequately.[49] The truth remains quite different.

Yet a growing recognition of public concern led R. S. Fowler of the Interior Department's Oil and Gas Division (Director of the Voluntary Agreement) to ask fourteen public relations experts from participating companies to attend an MEEC meeting on December 28. The agenda for this session included a discussion of how to improve "the dissemination of information regarding operations and activities of the Committee."[50] Coleman noted on the agenda that a "possible unfavorable reaction to AM. [American] companies" might be the result of MEEC activities, and that possible "stimulation of govt. controls" could occur. Certainly, he was worried.

Jersey's William P. Headden attended this gathering and reported what took place to Stewart Schackne, manager of the Public Relations Department: Arthur Flemming stated that the oil companies had publicized and could continue to publicize their own activities, but that his office [ODM] would supply all information about MEEC activities. A State Department spokesman then pointed out that the strong anti-American feeling noted in Europe in November and December "was now, in his opinion, quiescent" but that any new European oil shortage would stir "latent [anti-oil company] sentiment." Companies could then expect the various governments to set oil reserve storage quotas, or the nations might press the companies "to develop sources of oil in more secure political areas than at present." A United States Information Agency [USIA] representative stated that his organization already had helped to disseminate widely information on the work of the committee and the participating companies, but USIA needed more information from each company. Then, Wormser emphasized that the principal problem lay in the "fairly widespread impression abroad that nothing had been done by American companies" before the activation of the MEEC. Early in the discussion it became evident that the U.S. government must move to counter the notion popular in Europe that the United States sought "to get a commercial advantage with regard to Persian Gulf Oil."[51]

Fisher, reacting to reports from executives of Jersey affiliates in Europe cabled Coleman on December 13, suggesting greater efforts on the part of Jersey and the other companies: "While we fully realize that you are taking all possible steps to improve Western Hemisphere supplies, we nevertheless feel that situation for oil industry and particularly for Jersey in Europe potentially calamitous." He recommended that Jersey use "all means of providing Western Hemisphere crude for export" to Europe. In reply, Coleman stated that he agreed with Fisher, and that improvement would be noted within a week.[52] Certainly Bayles had warned Jersey officials in time, but they had been unable to act rapidly enough to offset criticism of the efforts being made to supply Europe with oil. However, Jersey did send Schackne and Theodor Swanson of the Public Relations Department to join Lamont at the January 10 meeting of the Coordination Committee for Europe.[53]

In the 1956 *Annual Report* to Jersey stockholders, Eugene Holman and Rathbone stated that the closing of the Suez Canal and the interruption of oil flow in the IPC pipeline through Syria meant that, to meet consumer requirements, the company needed an all-out wartime effort. "Jersey's management had anticipated the possible closing of the Suez Canal," and they had "prepared plans" which were quickly implemented when Nasser seized the canal. At that time, Jersey affiliates were shipping 270,000 b/d through Suez and they received 80,000 b/d from the IPC pipeline system. By the end of the year, all "oil companies supplying Europe" faced the problem of the canal closure and the IPC pipeline shutdown. In this crisis Holman wrote, Jersey "accounted for about 40 per cent of the total emergency shipments to Europe from sources in the Western Hemisphere." Proudly, he pointed out that "This oil lift for Europe was effected without creating a shortage of petroleum products in the United States or in any other Jersey markets." This success, he attributed to "our previously prepared plans" and to the great flexibility of Jersey operations, plus our "diversified sources of supply."

The report also included some interesting figures for the stockholders: Increased demand led to Jersey ships' transporting 615,000,000 barrels of crude that year. To handle this, the affiliates used "their 108 owned tankers, many special service vessels, and numerous coastal craft and inland waterway barges." The company also chartered tankers. But the Jersey Board had found it inadvisable to be dependent on chartered tankers; four new company owned tankers had arrived, and sixty-two more vessels had been ordered, eighteen in the 46,000 ton class. Finally, the report told shareholders that Jersey's participation in MEEC had been requested by the U.S. government, and that this committee had managed to keep supplies of most petroleum products in Europe close to normal levels through December.[54]

On January 4, 1957, the *New York Times* carried the news that crude oil prices had increased 12 percent and that the Humble Company had been the first to raise its price. Hines H. Baker, then president of Humble, explained that the increase resulted from a sharp rise in demand and that, when Humble could neither secure additional production from its Texas reserves nor obtain the quantities needed at then-current prices, he had authorized the price increase. The Independent Petroleum Producers Association of America [IPPA] agreed with Baker: oil prices had to be raised.[55] But many members of Congress objected. Senators Estes Kefauver of Tennessee and Joseph C. O'Mahoney of Wyoming, among others, denounced the price increase as a raid on the pocketbooks of both the public and the U.S. government. They urged the Justice Department to impanel a grand jury to investigate the major oil companies for possible price-fixing agreements. Despite these complications, the oil lift continued. Yet it was clear that public complaints on both sides of the Atlantic had added a new dimension to what was originally a problem of supply logistics. Even before the New Year (1957) began, Arthur Krock in his *New York Times* column praised the oil companies, pointing out that the American oil companies had not, as Europeans thought, dragged their feet in November, when MEEC failed to operate. James Crayhon of the Jersey Public Relations Department sent a copy of the Krock column to Director Shepard with this remark: "The fact that information dribbles out in this fashion after all that the industry has done also seems to emphasize the need for more news out of the MEEC." Then he added, "We [Public Relations] still urge—and the [Arthur] Newmyer people concur—that shortly after the first of the year the MEEC hold a press conference ... *Mr. Coleman should participate in the press conference*, what the reporters need is a meat and potatoes session."[56] Crayhon simply believed that Jersey and other companies had performed well but that few people knew about it.

When January supplies of oil in Europe dropped to 75 percent of requirement, the public outcry led Jersey to press for greater action back home. While the corporation strained its resources to accommodate demand, Senator O'Mahoney took action. On January 8, 1957, Newmyer reported that the Senator planned to hold a hearing on "the Administration's Program to provide oil for fuel hungry Europe" before the Senate Antitrust and Monopoly Subcommittee, which he chaired. O'Mahoney contended that the U.S. government—not a "segment of the oil industry,"—should conduct the oil program. He claimed that his earlier protests about the failure to include representatives of independent refiners and producers to serve on MEEC had been ignored, and that the government had allowed transnational companies "to dominate world traffic in the production and distribution of oil and oil products." His determination to act through his committee resulted directly from the decision made by Humble to increase the price paid for crude oil. This decision, O'Mahoney concluded, would lead to increased prices rather than increased production. Instead of a few giants, "All American producers of oil products should be permitted to participate in this program," if production were to meet world needs.[57]

Further complications developed as some European countries authorized their United Nations representatives to obtain information about MEEC operations and about the roles of participating companies. France sent a delegation headed by a Socialist Deputy, Jacques Piette, to the U.S. with a charge to examine the program. F. O. Canfield cabled Shepard from Paris that Piette should see Coleman and Cogan; he recommended also that Piette be invited to Humble's Baytown plant.

Bayles wrote a seven-page report to Jersey executives on January 11, 1957. In it he warned that the major world political crisis was occurring, and that shifts in the balance of power had already taken place. "We can certainly assume," he said, "that the conditions for future operations in the Middle East will be different, as well as the climate in which we shall be doing business." Looking to the longer future, he saw two major problems: "(1) How to reduce Europe's dependence on oil; (2) How to achieve greater security in petroleum supply." For him, coal (with a major upgrading of facilities from the mine to the consumer) would provide the alternate to scarce supplies of oil—until nuclear power developed. The second problem might well be unsolvable for a time—greater security in oil supplies would be difficult to arrange. Any confidence in the Middle Eastern sources of supply, he predicted, would come slowly.[58]

On January 15, Wormser finally appointed an Information Subcommittee of the MEEC. Obviously, this came late, in fact too late to stave off the public furor over the oil shortage.[59] Certainly, if Cogan's report to Fisher accurately reported the amounts of oil sent to Europe, oil companies had adapted well to the interruption in their normal delivery patterns caused by the blockage of Suez. Cogan did write, however, that "the conclusion [had been reached] ... that every effort should be made to further reduce gasoline yields and imports in favor of heavier products." He coupled this remark with an announcement that a meeting of the "Information Subcommittee" would be held as soon as "the red tape of appointments and acceptances" could be concluded. "This would appear important in view of the continuing confusion on publicity on both sides of the water,"[60] he added.

At this moment, the British government, in a public statement, aggravated the situation by reacting to a telegraphic claim from the Texas Independent Producers and Royalty Owners Association [TIPRO] that European refineries actually produced less gasoline from crude oil than did those in the United States. Britain stated: "The nature of our problem may not be fully appreciated. Fuel oil is the greatest need because without it ... grave damage would be done to our economy." The British, under a rigid rationing program, preferred that available tankers bring in principally the heavier crudes from which they refined heating oil.[61] Cogan wrote Fisher that Humble had been "very active in endeavoring to purchase added crude" but that not as much as they required had been located. He also pointed out that European refineries expected to reduce their gasoline production only 1 and 4/10 percent in January; hence, he agreed with the British that "every effort should be made to further reduce gasoline yields and imports in favor of heavier products."[62] What other Jersey executives thought of the British announcement is not known. However, in January, the OEEC-OPEG Committees sought to resolve shortages among member nations through a study of each country's shortages, and by equalizing these from a common reserve, called a "kitty."[63]

MEEC also dealt with the oil supply problem in the countries near Suez— Egypt, Israel, Syria, Lebanon, and Jordan. Their shortages could be more readily replenished, although some deficiencies still occurred in those nations, due to less accurate information on reserves in storage than could be obtained from European states: "unidentifiable quantities of crude" moved to those countries.[64] In a few cases, persons and companies buying crude refused to divert the oil, and some port facilities could not accommodate the tankers normally used in the Persian Gulf hauls. Coleman decided to review the situation from London and Paris. He and Fowler, along with concerned members of MEEC flew to London on February 4, 1957.[65] Meanwhile, Secretary of State Dulles did not quiet critics in Britain and France when he stated at a press conference that he found the United States' principle of nonaggression difficult to apply when friends (Britain, France, Israel) appeared to be the aggressors.

In early February, Thomas P. Ronan, writing for the *New York Times* from London, pointed out that Europe might yet avoid a severe oil crisis. Oil receipts had increased, and he rated Britain, West Germany, Portugal, Belgium, the Netherlands, and Luxembourg relatively well off, with Sweden, France, Italy, and Denmark "more seriously affected." Interestingly, a few days later, Texas Railroad Commission Chairman Thompson told the House Committee on Interstate and Foreign Commerce that, "he did not believe there was a crisis in oil resulting from the closing of the Suez Canal." Shortly thereafter, the well informed *New York Times* petroleum expert, Carmical, explained that, given the progress in clearing the canal, together with the possibility of opening the IPC pipeline even for limited oil flow, the chief burden of the crisis had now shifted from MEEC to the diplomats.[66]

The Senate Subcommittees on the Judiciary and Insular Affairs voted to hold joint hearings on the oil lift and related matters, beginning on February 5, 1957.[67] Senator O'Mahoney chaired the Subcommittee of the Judiciary and Senator Kefauver that on Insular Affairs. O'Mahoney effectively prejudged the joint hearings when he opened them with the remark that MEEC, composed of representatives of fifteen giant corporations, did not constitute "a proper organization to represent the oil industry of the United States."[68] He stated that Congress had the duty of carefully scrutinizing MEEC, which included companies already "sued by the Federal Government for allegedly participating in foreign cartels." O'Mahoney and other members expressed their concern for independent producers and refiners, and expressed indignation over the oil price increase triggered by Humble on January 4. As Senator Matthew M. Neely of West Virginia said in his opening remarks, "The oil lobby appears to be more powerful than the President, the Congress and the people. ... The issue before us is whether the major oil companies have become a supergovernment above the law."[69]

X X X

The record of the committee hearings is difficult to evaluate. Clearly, the committee and its staff sought to prove that Jersey officials had been working on a plan to alter the sources of oil supplies and to reroute shipments at least four months before the Suez Crisis erupted—as if foresight alone constituted evidence of guilt. Had the committee staff bothered to read the 1956 *Annual Report* of the Jersey Company, they could have learned the truth. Despite repeated efforts, the committee did not demonstrate that the Attorney General had "reluctantly approved the plan of action" later adopted.[70] When the committee examined Coleman, Baker, and Rathbone, its members failed to comprehend the degree of decentralization in Jersey operations, especially the fact that Jersey allowed the Humble Company almost complete independence. Baker and Rathbone attempted to explain this situation, but in vain. Questions from committee members centered on who had authorized the price increase. Again, Baker repeated his statement: "The immediate cause of this price change is the sharp increase in demand for domestic crude oil, particularly in Texas, that has developed since the flow of Middle East crude oil to Europe and the East Coast was drastically curtailed by recent events." He justified the price increase by pointing out that wages had risen and drilling costs had skyrocketed since World War II. At the same time, steel and iron, and thus the unit costs of the ever-deeper wells, had gone up—while other costs had followed. But over a three and one-half year period crude oil prices had remained relatively constant. Baker related that Humble had requested the Texas Railroad Commission to increase allowable production, only to be turned down; that the smaller companies purchasing oil did not want or need more crude; and that Humble had sought to purchase oil from other sources, but without success. Only after that failure did he become convinced that a slight price increase would enable the company to acquire additional oil. So, on December 26, 1956, he had called Jersey President Rathbone in Baton Rouge, then flown to that city, where he went to the Rathbone farm. There, Baker informed the Jersey president that he believed a higher price paid to independent producers would enable Humble to obtain more oil. Rathbone indicated that he believed Baker's reasoning to be sound, and Baker returned to Houston. Finally, Baker stated before the committee that he had held no further discussions with Rathbone on that subject at any later time.

Rathbone went on to testify that "The Jersey company did not suggest that [price increase] to the Humble company. The Humble company arrived at that decision."[71] He did agree with Senator Alexander P. Wiley that the price boost had occurred at a bad time, especially in terms of Jersey's relations with the public. As Senator Kefauver indicated in his questioning, he believed that Jersey executives regularly attended Humble Board meetings, and so vigorously did Senator O'Mahoney pursue the question of Jersey's control over Humble that he later retracted some of his most extreme statements: "I was in error," he said, with reference to Humble's taking orders from Jersey.[72] The committee also made much of the fact that Jersey's Treasurer, Emilio G. Collado (along with T. H. Tonnessen also of the Treasurer's Department), had made a study projecting future oil prices and their effect on company income, which on December 12, 1956, concluded that a price rise of 25 cents per barrel would mean for Humble and Jersey increases in income of $46 million and (worldwide) $100 million respectively. Rathbone testified before the committee that Jersey statisticians and economists constantly reviewed the world situation and sought to anticipate all future problems, including price changes. But still some members of the committee appeared unable to grasp the fact that Jersey customarily conducted studies regarding affiliate income, production, marketing, pipelines, and other functional operations in order to estimate the profits of these various affiliates. Rathbone told Kefauver that Humble knew nothing of the Collado-Tonnessen report. Yet, except for Senator Everett M. Dirksen, the committee members seemed skeptical. Senator Kefauver challenged the statements by Jersey and Humble executives that no substantial price increases had occurred during the previous four years. He also asserted that both companies had diverted Venezuelan crude from Europe to the East Coast. Rathbone argued that the companies' figures on oil prices and profits were more accurate than the statistics provided by the Federal Trade Commission [FTC] which the Senator had used. Seth MacDonald from the FTC then made a statement that appeared to support the Rathbone [Coleman and Baker] position: In figuring profits, "they [FTC] took into account the total returns, the total profits, the total earnings of Creole [Jersey's 95 percent owned Venezuelan subsidiary] and not the 95 percent which Standard did."[73]

The members of the two committees attacked federal officials and oil company executives over the failure of the United States government to block Humble's unilateral increase in the price of crude oil. Interior Secretary Seaton, in rebuttal, quoted from a letter that U.S. Attorney General Herbert Brownell had written in response to Seaton's inquiry about whether a plan could legally be drawn that would incorporate a Voluntary Agreement on price. Brownell pointed out that price fixing violated the antitrust laws, and that he could give "no authority for a stabilization of prices." Further, the Voluntary Agreement did not mention prices, and MEEC had only advisory and recommendatory powers to the federal government.[74] Obviously, Brownell's ambivalent attitude on antitrust exemptions had influenced the thinking of those Interior officials who supervised the oil exchange program. Finally, Senator Kefauver forced Wormser to admit that while the government regarded regulation of oil prices as generally contrary to the antitrust laws, at the same time it had exempted them from those specific provisions of the law dealing with joint transportation.

The written records of these hearings leave the reader confused as to which group actually provided the larger target for criticism—oil companies or the Eisenhower administration. Certainly, the brutal treatment of some governmental officials signalled desire on the part of some committee members to enhance their public reputations and make political hay. These members challenged the legality of MEEC, and succeeded in establishing that some MEEC practices were at least highly questionable, and possibly illegal. Nor could combined efforts of oil company executives and minority members of the committee, led by Senator Dirksen, effectively counter the majority's findings.

On one point all agreed—government officials, oil company executives, and committee members; MEEC had been authorized and organized at the request of the federal government and on the assumption that the Suez Crisis amounted to a world emergency. Thus, though some of the senators were guilty of extreme remarks—Senator O'Mahoney, for example, said, "The building of these great international commercial empires rests upon the blindness of Congress to what the technological and scientific developments of our time have brought about"[75]—the committee made no recommendations of a substantive nature concerning changes in the antitrust laws. Essentially, the members were satisfied to hear allegations that oil companies had violated the agreement with the Attorney General that had provided the basis for the actions of MEEC. Toward the conclusion of the hearings, Senator O'Mahoney wryly remarked, "I say they [the oil companies] know what is going on before anybody in that country [none specified] knows, and before we in Congress know." Even so, he did not allege that the companies controlled the government.[76] He pointed out that the American states could not regulate the "commercial empires operating in foreign affairs," and added, while Rathbone sat on the stand, "The result of that is that there has been developed by the Standard Oil of New Jersey, as you have testified here and as your exhibits have shown, the establishment of an international state department, so to speak."[77]

In his own testimony, Rathbone emphasized that Jersey attempted to maintain good relations with all governments, as part of its effort to act in a legal and responsible manner; he denied however, that this interest could be construed as establishing "an international state department." Jersey, he said, simply wanted to preserve the free enterprise system. Coleman and Rathbone both insisted that the oil companies had done the hard job of supplying the oil-short European nations with oil more efficiently and more quickly than any government agency could have. To the oil companies, the problem was from the first "a purely physical supply job."[78] When the Suez Crisis began, the demand for tankers increased, and hauling rates shot up to previously unheard of levels. As Baker pointed out, this too, had influenced Humble's decision to raise the price of crude oil. Rathbone and Coleman stated that there had been no government effort to keep tanker rates down.[79] Certainly, independent tanker owners always charged all they could get. Stories of tanker charges of a million dollars (and more) for a single cargo of oil, carried from the Persian Gulf, around Africa, to Europe did have a basis in fact.[80] The oil companies, like governments and consumers, could be made victims as well as villains. So much emerged from the records. Finally, one interesting sidelight on the hearings deserves retelling. Just before the conclusion, committee members accepted as an exhibit a letter dated March 21, 1957, from one John J. Collins of the Jersey Standard Tanker Officers Association to Senators O'Mahoney and Kefauver.

Collins wrote that, judging from the testimony before the committee and the questions asked by its members, the "oil industry in general and the major oil companies in particular are being placed in the light practically of despoilers of our Nation and has been singled out as an industry that has followed a policy of the public be damned." He pointed out that the officers of Jersey's tanker fleet had not lost a day's work during the crisis. Collins wrote that he had never been a Jersey employee, that he had never met a Jersey director, and that, to his knowledge and belief, the company had never been guilty of any wrongful conduct in regard to his association. Jersey tanker crews, said Collins, received top pay, and Jersey sought good employees. Regarding the Joint Subcommittee's attempt to refute the Baker-Rathbone statement that there had been no increase in the price of oil for four years, Collins added that he could personally testify that in each of those years Jersey's seagoing employees had received improved wages and benefits, thus increasing the company's costs. He questioned whether the committee's pillorying of oil company executives served any useful purpose.[81]

All the time that these hearings occupied Jersey executives (and executives from other companies), MEEC continued to hold meetings. On February 11, the group assembled in New York, to hear reports on oil deliveries to Europe. In Paris, on the same day, representatives from the U.S. government and from MEEC met with the OEEC oil committee and OPEG. Stewart Coleman attended. Participants there learned that the Suez Canal might be open within a few weeks for limited tanker passage, though no one could forecast "the political handicaps which may arise as to the use of that Canal."[82] Coleman explained that "the most important change ... is the diversion of tankers formerly used in the Middle East/Europe and Middle East/North America Trade to move oil direct[ly] to Northern Europe from the United States Gulf Coast and the Caribbean," and that the "essential needs for Europe will continue to be met." He refused to say that there would be no additional increase in oil prices but did express an opinion that, in case price changes occurred, they would be based "on the relative qualities of crude and the relative distances from the consuming markets."[83]

Cogan reported to Fisher about MEEC activities on February 15. He sent with his letter a statement by Rathbone, made at a press conference on February 13. What the Jersey president said in defense of the company had so irritated Senator O'Mahoney that he expressed chagrin over "the fact that the president of Standard Oil calls a press conference to explain what he hasn't yet explained to his [O'Mahoney's] committee."[84] But as Cogan told Fisher, "oil company testimony" had not yet been scheduled, as of February 15.[85]

On February 20, Edward G. Moline of the State Department reported to Coleman on the condition of Suez. The canal, he said, now employed the same pilots and technical staff as before, the communications system had been only slightly damaged, and tugs were now becoming available. He believed that the authority intended to move convoys more slowly than "had previously been the practice" and with no night transit.[86] Obviously, MEEC and U.S. oil companies required accurate information on the canal situation; at that time, any passage through it would speed up European-bound cargoes, shorten tanker hauls, and save money. Further, once in general use, the shorter route would virtually eliminate the work of MEEC.

But in February, the European oil shortage intensified. Seaton wrote Coleman on February 25, stating that the data available indicated that the diversion of oil from Venezuela to the East Coast and Canada had not "been substantial up to the present date," and that the oil saved from reduction in American refinery runs had not been sent to Europe. He requested that MEEC consider new plans for diverting this oil to Europe.[87] The committee apparently acted on this request, for, on March 1, the Department of Interior indicated that Gulf Coast oil shipments had increased by 40,000 barrels daily.[88] On March 6, 1957, the *Financial Times* reported gasoline supplies in Britain so high that extra storage tanks had to be found.[89] While MEEC certainly deserved a major share of the credit for this increase, strict controls in the consuming countries, some rationing of fuels, and mild weather all helped as well to ensure the adequacy of oil supplies in Western Europe during the winter of 1956-1957.

On March 10, Defense Mobilizer Flemming "jumped the gun" by stating that in anticipating the resumption of the normal flow of oil from the Middle East to Europe, he would now attempt to resolve the oil import quota dilemma. Middle Eastern nations, he believed, would probably press to increase production "above the pre-crisis level to compensate for the cutbacks and Venezuela naturally desires to maintain production at the present record level." On the other hand, "independent producers ... are expected to press harder than before for limitation." The reconciliation of these divergent interests, he called a "most difficult" problem. Prior to the Suez Crisis, a hearing had been held in Washington "to determine whether excessive imports threatened the domestic industry and thus the national defense." Flemming delayed any decision on the matter when the Suez Crisis began, and several months elapsed before he could tackle the import question.[90]

In March, word came also that the Iraq Petroleum Company's pipeline to the Mediterranean would soon re-open, bringing the possibility of "early improvement in the Suez situation." Cogan resigned on March 7, to return to Imperial Oil Limited.[91] Rathbone, Coleman, and Baker finished testifying before the O'Mahoney-Kefauver committees. Cogan, writing a "last note" to Fisher, optimistically expressed the belief that "Mr. Coleman was ... completely successful in dispelling the confusion and misconceptions" about MEEC.[92] Meanwhile, MEEC and its subcommittees went on meeting in March. Wormser, addressing the National Petroleum Council on March 7, reviewed the work of MEEC oil lift programs and expressed one government official's "appreciation for the efficient and effective operations carried out by the companies making up" that committee.[93] Interior Department figures, made public the following day, indicated that the U.S. companies had sent 29 million barrels of oil and 27.7 million barrels in products of Gulf Coast petroleum to Europe between November, 1956 and

February 27, 1957. What the normal amount would have been is not known, but departmental figures showed increases for virtually every week during that period.

Suddenly, news of an oil discovery, a gusher in the Netherlands, on March 26, diverted, at least momentarily, the attention of the Atlantic-Middle East community. The Dutch-chartered N. V. Nederlandse Aardolie Maatschappij Petroleum Company, jointly owned by Jersey and Royal Dutch-Shell, brought in a new field at Wassenaar, near The Hague. Because of the crisis, this strike probably received more attention in the press than it warranted; but it did serve to encourage the companies to continue exploring for gas and oil.[94] Meanwhile, MEEC continued to function, although the situation that had brought it into being slowly improved. The *Financial Times* reported that gasoline was arriving in England at "over 90 per cent of the normal pre-Suez rate," while rationing had reduced consumption to only "60 to 65 per cent" of the gasoline available. In March, oil supplies all over Europe reached 85 percent of normal.[95]

As things quieted, Coleman wrote J. R. White, president of Imperial Oil Limited, thanking that company for loaning Cogan to MEEC; he also sent a note to Cogan saying that working together had been "really like old times."[96]

On February 19 and 20, 1957, Rathbone appeared before the Committee on Foreign and Domestic Commerce of the United States House of Representatives. His comments were reproduced and sent to company shareholders. He began, "There were not enough tankships in the world to bring around the Cape of Good Hope the same amount of Persian Gulf oil that the area had been receiving." However, when Egypt seized Suez in July, Jersey established a company committee to study what might be necessary if Egypt closed Suez and the pipelines ceased functioning, or both. Because of this planning, when both events occurred, Jersey did three things: First, the company rerouted all owned and chartered tankers; second, it began an extensive search in the Western Hemisphere for crude oil and products for Europe; and third, it used abnormal inland transportation [especially oil from Wyoming] to get more U.S. crude to the tidewater, for shipment to Europe.

Rathbone went on to say that Jersey had moved to Europe, by the end of November, 3,495,000 barrels of U.S. Gulf crude and 1,245,000 barrels of Caribbean crude. The company had also shipped 1,500,000 barrels of distillates and heavy fuel oil. The crude oil shipped represented 55 percent of the amount Jersey normally sent through Suez to Europe. He said, "No alternative would have been as effective as promptly as the Middle East Emergency

Committee." Because of the work done by that committee, "92 percent of European requirements were supplied," and he believed that more than 80 percent would be supplied during the first quarter of 1957. The American oil industry had done a good job, he remarked, and after addressing other matters, he ended his testimony.

On March 17, Senator John Carroll of Colorado and Coleman appeared together on the American Broadcasting Company's [ABC] television network. John Secondari, chief of the ABC Washington news bureau, served as moderator. The show, termed "Open Hearing," proved anything but open. Carroll again made the point that four months before Nasser seized the Suez Canal, Jersey, anticipating difficulty, had made detailed plans for operating in the emergency. He added that discussions had taken place in that interval among officials of the State Department and the oil companies. Carroll went on to charge that the government had gone with its "hat in its hand and asked these folks to come—to take care of the supply of oil in the breakdown."

Coleman replied sharply to the Carroll charge that "giant oil cartels" largely influenced foreign policy with the statement that he thought "the Sherman antitrust law the most wonderful law that was ever placed on the books of our Congress." He added that if Europe had such laws there would be few Socialists, and went on to deny that Jersey belonged to any cartel. Yet the "hat in hand" remark of Carroll about the MEEC brought Coleman's strongest rebuttal:

> We need this job like we need a hole in the head ... you as Senator shouldn't use an expression like that ... under complete government supervision, and as consultants, we were able to accomplish this job ... I worked like a dog. ... All the men with me ... have done a magnificent job in meeting this crisis. We've averted the collapse of Europe. And at the request of our Government, we met this [crisis]. ... We have not failed.

Coleman pointed out that Senator Kefauver held "a complete misconception of the function and the objective ... in the American oil industry." His own preference, Coleman stated, was for private capital to aid in developing foreign countries, rather than having the U.S. government accept this responsibility.

Senator Carroll then argued that, while he did commend the oil companies, he also objected to "the power that they exercise as companies" and pointed to the income from the Humble price increase for crude oil (a precedent followed by other companies) as reason enough for the government to move against the transnationals. He insinuated that because the companies made such vast profits from their Middle Eastern properties, they should absorb all additional charges resulting from the crisis. Congress should "knock off some of the tax benefits to these Middle East operations." At this point, the discussion with Coleman seemed to repeat the old argument: How could Jersey own 87 percent of the Humble Company and not control it? Coleman commented on a seeming "utter unawareness of the situation" on the part of Senator Carroll. He pointed out that oil producing companies had for many years known that as the U.S. oil supply dwindled, they must begin to pay higher prices for it. Unfortunately, the effect of a domestic shortage made itself felt at the time of the Suez Crisis.[97]

Several days later, Coleman wrote to the *Journal of Commerce*, pointing to errors in their reporting of his testimony before the Senate Subcommittee. Despite all the conversation on the record, the notion still persisted that because Jersey (and other companies) had studied what could be done prior to the Suez Canal closing, "all you had to do is push a button and everything was all right." Events had proved the truth to be different. Planning has its pitfalls![98]

As the March 31 deadline approached with the canal still not open, Wormser secured approval for an extension of the Joint Agreement expiration date until May 31, 1957.[99] While reports indicated that the canal might be opened for traffic earlier than scheduled, the British and U.S. governments "advised ship owners against use of the Canal," as did ship owners associations in a number of countries.[100] By April 10, L. A. Astley-Bell of OEEC Petroleum Emergency Group reported the new figures on supplies, deliveries, and consumption. He cautioned "that no kind of statistical exercise such as this can properly assess the problem which the sudden opening of the Canal at full capacity will present," and warned that such an upsurge would tax storage capacity. Astley-Bell indicated that the satisfactory oil supply and demand position had been achieved "only because of the abnormal steps taken to increase oil supplies from the Western Hemisphere and through the tanker savings accomplished by the industry and made possible by the industry committee machinery."[101] The records indicate that, despite the fact that oil became the lever to force the British and French from Suez, the oil companies acted jointly to avert the possible economic collapse of some Western European nations. As one Jersey man phrased it, "We did a helluva job and received damn little credit."

X X X

While in the winter of 1956-1957, the oil exchange proved an effective response to the Suez Crisis, other problems remained unsolved, casting deep shadows on Jersey's future. Both Coleman and Rathbone devoted much thought to possible solutions for future emergencies that might provoke as much trouble for the company as Suez had caused.[102]

Even before the Middle East crisis, Rathbone had ever been a thoughtful leader, and he remained intrigued with Page's idea of running a huge pipeline from Saudi Arabia through Iraq and Turkey, to the Mediterranean. The Saudi Arabia to Turkey line had also been suggested by Finney as the long-term solution to the Suez problem. According to Jersey Board and Executive Committee Minutes, during late 1956 and early in 1957, Rathbone, the man of vision, was conferring with the Department of State, other oil companies, and building contractors about this proposed large-diameter pipeline. On January 1, 1957, the State Department urged him to go to London for a conference of major foreign oil companies. At the same time, the department assured Rathbone that, while they expected no difficulty in obtaining satisfactory treaty protection for the proposed pipeline, they probably would meet some resistance in obtaining legal guarantees regarding oil rights in the Middle East.

Rathbone did go to London. Upon returning, he announced that the new pipeline would be built in the Middle East, passing through Turkey, and that it would be jointly financed by eight of the oil companies already engaged in Middle Eastern production. The estimated capacity would be 850,000 to 1,000,000 b/d, and the cost, $800 million. Rathbone further predicted that it would take three or four years to construct the line. He joined Secretary of State Dulles in saying that the facility could possibly have "international status," while Dulles suggested that such status was just as important "as it would be for the Suez Canal."[103] C. Douglas Dillon, Under Secretary of State, gave Rathbone a copy of the British "Report on Long-Term Requirements for the Transport of Oil from the Middle East," and the Jersey president returned the document to the Under Secretary along with an array of informed suggestions and opinions. With remarkable prescience, Rathbone foresaw the coming of the superships:

> We note the considerable emphasis placed on the use of large tankers around the Cape [of Good Hope]. We do not believe that 60,000 DWT [dead weight tons] tankers would meet the costs of movement through Suez unless returning via the Canal in ballast, but estimate ones of 80,000 DWT or larger would come close, particularly if there is an increase in Canal tolls.

Returning to the possibility of a pipeline, he wrote, "However, the largest tankers would show a less attractive return on investment than pipelines." He correctly observed that, "Use of such tankers would be seriously limited by water depth and docking facilities." Then he went on to discuss expertly the disadvantages of relying on superships. He disagreed sharply with the British conclusion that another pipeline would not prove feasible: he thought, to the contrary, that one should be built, especially if "treaty protection is obtainable." He even suggested an "international corridor between Israel and Egypt," with the pipeline running through it. While he admitted that such a territory would be difficult to establish, in Rathbone's opinion, this "international corridor" could be the solution to "this very troublesome Israeli-Egyptian conflict as well as to the Suez Canal problem as a separate issue."[104]

What happened to the Jersey plan for an additional pipeline is not known. Conceivably, the burgeoning world surplus of oil that appeared after resolution of the Suez Crisis dissipated interest in construction, or the ease with which the IPC line had been immobilized, made the company cautious.

Meanwhile, the MEEC, meeting on April 12, drafted a resolution addressed to Wormser, stating that in the light of reports of that committee and its subcommittees:

> as well as data furnished by the OEEC Petroleum Emergency Group, the petroleum supply position for the second quarter 1957 in Western Europe and the other friendly foreign areas affected by the substantial Middle East petroleum transport stoppage is so greatly improved ... that in the opinion of the Middle East Emergency Committee further implementation ... to alleviate petroleum shortages ... is not necessary during the second quarter, whether or not there is effective use of the Suez Canal during this period.[105]

The last moments of the crisis had come. Jersey employee Charles M. Furcht returned to the company after serving seven months as secretary of the Supply and Distribution Subcommittees;[106] others followed as the pressures on MEEC subsided. On May 5, Fisher cabled Campbell that OPEG could become inactive on that day, except for "completing certain historical information," but that the "mechanism will be maintained for the time being in the event it is needed to cope with some new crisis." He believed everything remaining could be formally "wound up" in three months. Several days later, Fisher again cabled Campbell that OPEG would be dissolved on June 30.[107]

Soon, congratulatory messages began to be received and sent; obviously the crisis was over.[108] On May 14, the British government announced an end to the gasoline rationing that had begun on December 17, 1956; and directed "British shipping to use the Suez Canal." These instructions precipitated intense political debate in Parliament because many observers considered this decision a capitulation to Egypt's terms for use of the canal. Strangely, at least in the view of oilmen, anti-Americanism also became rampant throughout much of the world.[109] The French government, however, continued to bar its ships from using the canal, as France defied Nasser. On May 14, Seaton announced the return to their companies of his twelve petroleum advisers: "In view of the satisfactory oil situation in Europe, it is now possible to terminate your appointment as a consultant,"[110] he wrote. Although the crisis had ended, the MEEC met on May 23, to discuss oil supplies and distribution in Europe. Obviously, disentangling the complex knots of committee structure took time. Seaton next wrote Coleman, terminating on May 31, the schedules under which exchanges of petroleum and tanker routings had been made.[111] This virtually ended the Voluntary Agreement and the resulting oil lift.

X X X

On June 5, 1957, Senator O'Mahoney invited Coleman [with Rathbone and Baker] to attend a public presentation by the staff of the Antitrust and Monopoly Subcommittee. On the agenda was an analysis of the records and results of the study of the "emergency oil life [sic] program, the price increase and related matters."[112] After listening to a lengthy recital, Coleman called the subcommittee's report "politically inspired."

As late as July, Campbell, Bayles, and Carlisle, all of Jersey, unofficially discussed with European officials who had helped handle the Suez Crisis, the ominous possibility of another canal shutdown. They agreed "that there is a sufficient 'threat of emergency' to begin to do something."[113] Antitrust statutes, however, prohibited their meeting with officials of other oil companies. Assistant Jersey Counsel Thomas E. Monaghan, on July 24, advised Paul Anderson, Frank Canfield, Luke Finlay, Hugh McFaddin, Ralph Bolton, Bayles, and C. A. Larson—all Jersey executives then in Europe—against such conversations:

> If you are approached by any European *group* such as one connected with OEEC with a request to participate in joint consideration or action on planning to meet an oil emergency arising out of

events in the Middle East, please do not accept or participate in any way without first getting in touch with us and obtaining our views.[114]

A federal grand jury had begun to investigate possible collusion in the crude oil price increase of January 3, an increase that had led to higher cost for all petroleum products, and Jersey continued to discourage any activities by company employees that might raise questions. Despite the action of the Justice Department, Interior Secretary Seaton had circulated to the petroleum executives a draft plan for future emergencies. Regarding this plan, Monaghan pointed out that he saw no objection to prior consideration of its provisions if "a top-level approach to the appropriate U.S. Government officials could produce a blueprint consistent with those provisions."[115]

In the 1957 *Annual Report*, Holman and Rathbone told Jersey stockholders that when the "difficulties [Suez] were overcome toward the middle of the year, the more than ample supplies of oil, which then became available ... were not absorbed as had been anticipated." Despite the last half-year leveling off in the worldwide demand for oil, profits "were the highest ever achieved by Jersey Standard." The report also noted that Jersey affiliates now had sixty-nine tankers under construction.[116]

Jersey executives continued to consider the problem of supplying their European affiliates with Middle Eastern oil. The large diameter pipeline to Turkey that Rathbone so desired did not materialize. The company's Worldwide Supply Committee often considered the prospect of another interruption to the normal flow of oil markets and certainly Jersey leaders did well to consider and reconsider every possible threat to Europe's oil supplies. Obviously, after Suez, the Middle East would never be the same. Only slightly more than a year later, a revolution occurred in Iraq, where, on July 14, 1957, troops led by a pro-Nasser General, Abdul Karim Kassem [Qasim], seized and killed King Faisal II, his family, and the former strongman, Nuri-al-Said-Pasha. General Qasim replaced Nuri-Pasha, who had been considered pro-Western, and formed a new government.

When this happened, Secretary Seaton quickly reactivated the Foreign Petroleum Supply Committee for the purpose of drafting a proposal to create a new MEEC, if the revolutionaries threatened Western Europe's oil supplies. Already, the Iraq revolution had begun to threaten IPC concessions. Negotiations between IPC and the government began and they would continue for several years. Meanwhile, the Iraq revolution had caused the president of Lebanon to fear for the security of his own country. As a precaution, he asked President Eisenhower to send troops to Beirut, and

Eisenhower complied with his request. At the same time, British troops entered Jordan uninvited. This show of force may have influenced Qasim; at least, he chose not to expropriate the IPC properties but instead to reaffirm the importance of oil revenues to his government and to his country.[117]

During this period, the FPSC met to consider how a new MEEC might operate. This time they determined to include representatives of more oil companies, to have a government employee serve as permanent chairman, and to work more closely with other federal agencies. They also decided to eliminate the cooperative plan for oil exchanges among companies, the key arrangement that had enabled MEEC to meet Europe's needs during Suez. What would replace it was never made clear.

X X X

Not until three years after the Suez Crisis had ended, did the oil companies receive a final legal judgment on the alleged conspiracy to raise crude oil prices which had occurred during that emergency. For some months, Acting Assistant Attorney General Robert A. Bicks had been aggressively pursuing corporations through investigations and indictments—so ambitious was his program, in fact, that a writer in *Fortune* stated that some business leaders believed Bicks sought to mold the "structure of U.S. business" to his "social, economic and political ideas." When in January, 1957, Humble announced the price increase of 35 to 45 cents per barrel (depending on grade and area) and other corporations immediately followed this lead, Bicks and his aides believed that the companies had entered into a conspiracy and violated antitrust statutes. While the O'Mahoney Committee questioned some oil executives, Federal Bureau of Investigation [FBI] agents interrogated others. Finally, Justice acted to impanel a grand jury in Alexandria, Virginia, and moved to subpoena evidence against those oil companies that had raised prices. In all, the grand jury subpoenaed more than one million documents from company files. Over a year later, it returned an indictment against twenty-nine companies, including Jersey and subsidiaries, for conspiring to raise, fix, and regulate the prices of crude oil and automotive gasoline. At the request of the defendants, the Alexandria court allowed a change of venue to the Tulsa court of Judge Royce H. Savage.

Before trial proceedings, both parties agreed to dispense with a jury and to submit briefs, in advance, spelling out the issues. "Judge Savage streamlined procedures and achieved agreement on so many facts" that only six hundred documents were admitted as evidence.

Trial began on February 1, 1960, and on February 2, Judge Savage denied a motion for acquittal. The Justice Department then notified Holman, Rathbone, and other Jersey executives, as well as executives from other oil companies, to be ready to testify. Holman did take the stand. He stated that Rathbone had discussed with him, late in December, 1956, the conversation with Baker, and that he (Holman) had discussed the prospective Humble price increase with the Jersey Executive Committee, but that "no action had been taken on the proposed price increase."

On February 13, Judge Savage issued an oral opinion on the case. He summed up the government's position: the economic conditions existing in December, 1956, and January, 1957, did not exert pressure for a general increase in crude oil prices; at the time, there was no shortage of automotive gasoline; and the closing of Suez did not substantially reduce stocks of crude oil. Then, he stated, "I don't agree with that first premise." The judge went on to say that the evidence presented indicated that exploration and production costs had increased greatly, and that even before Suez, some companies had considered "possible price increases." Humble, he said, had taken the lead "in announcing this price increase" but as he pointed out, "there was terrific economic pressure on Humble" to do so. He found economic justification for the price decisions made by each of the defendants, and he pointed out that Continental [Continental Oil Company] had been studying the economic situation for a year. That company had been "looking for the opportune moment to increase the price of crude." So when Humble announced the price increase for crude oil, Continental followed on the same day. Judge Savage concluded:

> It is my judgment that the evidence in the case does not rise above the level of suspicion. ... I think I should go further and say that after giving consideration to all of this evidence I have an absolute conviction personally that the defendants are not guilty of the charge made in this case. ...
>
> I have a firm conviction, upon the basis of this record, that there was not an unlawful agreement entered into by these defendants to increase the price of crude or products prices.

The judge next gave his opinion on "the alleged illegality of price agreements among members of the same corporate family." He recognized that some justification existed in law for the position of the government; however, in the absence of a controlling decision, he offered his "view that the mere approval by a parent corporation of the price schedules and pricing policies inaugurated and fixed by the subsidiary corporation does not constitute a *per se* violation of the Sherman Act." Jersey and the twenty-eight other oil companies stood acquitted. The legal argument of Hugh B. Cox, a Washington, D.C. attorney retained by Jersey, apparently influenced the outcome, for he argued that Jersey's parent-affiliate relations represented only the "common, historic, familiar and lawful way of doing business." With the Savage decision, the second Middle East crisis was finally put to rest.[118]

Bicks, however, believed that the government had gained something important, even in defeat. As he remarked, "The oil companies lost their case when the grand jury returned the original indictment. ... Whenever anyone drives into a filling station for a gallon of gasoline, they are bound to think, 'those price fixers!' "[119]

In a larger historical way, the first Suez Crisis marked the end of dominant British and French influence in the Middle East. Their replacement, the United States, only reluctantly became a great power there, but reluctant or not, in the years since Suez, the U.S. has taken the single largest role. If the West and the chief Western nations were the losers at Suez, and later in Iraq, the emerging power of the Middle Eastern states may be dated from that time, and, to the present, cannot be denied. For since Suez, the recurring crises in the Middle East have continued to shake all Western nations. As if to acknowledge the change in the world, Jersey continued its Worldwide Supply Committee for several years, together with some key subcommittees.[120] Surely, the lesson of Suez has become a permanent part of a Jersey education.

What Western governments learned remains more difficult to summarize. Certainly, Suez served as a lesson, for it showed for the first time that many Western nations (and some oil companies) had become dependent—perhaps too largely dependent—on Middle Eastern oil. On a smaller scale, oil companies realized they did need to depend for transportation on the Suez Canal, so the alternative, supertankers became a reality. On their part, Middle Eastern countries learned that the United States could not always be counted on to support its allies. In all, the twenty years following Suez proved to be filled with political ferment in the Middle East, with the oil producing nations ever ready to copy the examples before them, by applying both subtle and naked pressure against oil companies. Their purpose remained constant: to increase income and assume greater control over their natural resources.

Chapter 18

The Middle East, 1957-1970—Recipe for the End of the World

W HEN THE SIGNED CONSORTIUM AGREEMENTS were implemented by the companies, Iranian oil once again began to appear on world markets. The contracts dated October 29, 1954, with some modifications, lasted for more than twenty years. But neither that original agreement nor any later one made before 1975, with either Iran or any major oil producing country, operated without almost continuous dispute and, in many cases, open friction between the participating countries and the companies. Most concession contracts and agreements had to be altered at irregular intervals—whenever producing governments pressed companies for a greater share of oil income. To Howard W. Page, director and Jersey's chief Middle East negotiator (for more than a decade, beginning with the consortium agreements), these contract modifications constituted for Standard Oil Company (New Jersey) [Jersey] and other oil majors "a very slow acceptance of the facts of life which were constantly changing."

One reason for the strained relations between the producing countries and the oil companies was the intense nationalism and Pan Arabism that flourished in the Middle East. One writer has pointed out that the pervasive Nationalist revolutions went hand-in-hand with the "revolution of rising expectations," and that the latter fueled the revolutionary tendencies of the former. Even before the consortium agreement had received official approval from either the companies or Iran, a United States Department of State [State] representative noted: "To think that the Arab masses will learn a lesson from the Iranian debacle may be unrealistic." The intense nationalism in the Middle East heightened pressures on company spokesmen who negotiated every concession; and, in turn, it dictated to government officials an intense desire to gain every possible point.

By itself, the Iranian revolution served to stimulate widespread interest in both nationalism and the Middle East. These were matters of vital concern to Jersey. Board Chairman and Chief Executive Officer [CEO] Eugene Holman directed that company officers should be kept informed about the social, economic and political trends in the various countries throughout the world where Jersey had investments; including of course the Middle East. He wanted imaginative programs instituted to improve the company's relations with people in all host nations. Already, the company had an Advisory Committee on Human Relations, which sought to assess the impact of nationalistic forces and to recommend policies that should be pursued in response to these pressures. Jersey's consulting economist in the international field was the one-time New Dealer, Milo Perkins. In 1956, Monroe J. Rathbone (who became president in 1954) and Holman asked Perkins to suggest how the company should cope with the "forces of rising nationalism." Perkins responded with a brief memorandum in which he emphasized the proper selection and training of expatriate personnel, good relations with all governments, utilizing one "strategist" in human relations in each affiliate, and improving the company's relations with other extractive industries at home and abroad. Rathbone circulated the Perkins memorandum to the Board and to the Advisory Committee on Human Relations. Then, measures were taken to follow the recommendations as part of an on-going process.[1]

Between 1954 and 1975, under the guise of Pan Arabism and nationalism, major governmental changes occurred in Iraq and Egypt. Jersey's Howard Page once remarked, "Pan Arabism ... is a groping for a union of Arabs in some form or other free of any domination or pervading influence from outside." Then he added, "'Colonialism, Imperialism and Exploitation,' even though they no longer exist [as they once did] are still meaningful terms to most Arabs." To them, *any* agreements with foreign nations or companies seemed to establish the fact of continuing colonialism. So while these powerful undercurrents operated, within two decades three separate Arab-Israeli wars began and ended, all of which resulted in Arab defeats. These losses injured national pride in the defeated states; while the continued support of Israel seemed to offer further evidence of U.S. imperialism. The Israeli successes also weakened the influence of all moderate Arab governments.

Herbert Hoover, Jr., former Under Secretary of State, may have been absolutely accurate when he told a joint session of the U.S. Senate Foreign Relations Committee and the Committee on Armed Services, in 1957, that the consortium arrangements had preserved Iran for the West, and thus strengthened our "Northern Tier" of defense, as Secretary of State John Foster Dulles once termed the countries bordering Russia in that part of the world. But the Arab-Israeli War (the Suez Canal Crisis, 1956) offered the

Soviets an excuse to move into the "Southern Tier," where they continued to exert political influence, with varying degrees of success.

In 1953, Richard Funkhouser, a petroleum advisor in the U.S. Department of State, concluded a talk on the Middle East at the Army War College by stating: "A fixed and rigid position out there is in my opinion a fatal position. It is only by changing with changing circumstances can stability exist in the dynamic Middle East if we are to hold the oil, the area and the people." Three years later, based on the advantage of considerable experience, Funkhouser made a prescient remark in an internal memorandum for the State Department: "Concession contracts ... probably cannot be protected by guns or governments but only by enlightened thought, understanding and compromise." Jersey, in Middle East Contact Director Page, fortunately had an individual blessed with the two first-named qualities, and a willingness to attempt the third. He never viewed concession contracts as rigid documents, impervious to changing conditions and situations; rather, he considered such agreements valid so long as the mutual interests of all the contracting parties required that they continue. When the mutual interests of the parties came into conflict, negotiation for change should ensue. A negotiator with great stamina, Page owned the mental flexibility necessary for such meetings, along with a goodly share of imagination and vision. He carefully planned his every move, and he always had an alternate proposition to make when negotiations became deadlocked, or when his original suggestion failed to win approval. Following traditional company policy, he (and his successors) received great freedom to act during negotiations. He always remembered that just before he left for Tehran in 1954, he discussed his ideas about Iranian oil with other Board members. CEO Holman told him, "Howard, you know what has to be done. I don't think you're going to make any big mistakes. So you just go ahead and do the best you can. ... We will find out what's going on through regular channels. Don't worry about having to report." Holman concluded by telling Page that, should he need more authority, to ask for it "through regular channels."

Page had spent much time in London, where many companies operating in the Middle East had their headquarters, and he frequently travelled to the producing countries. He became known to, and was respected by, many government officials in that part of the world.

Jersey also possessed another competent student of Middle Eastern affairs in Smith D. "Sparky" Turner, the regional coordinator for the area. Turner had played an important role during the negotiations that ended the Red Line Agreement in 1946-1948, and he had also been a member of the Orville

Harden team sent to Iran in early 1954. Except for one negotiating session after that mission failed, Turner spent most of his time "urging Jersey's views on our partners and on the operating companies like IPC [Iraq Petroleum Company, Limited] and Aramco [Arabian American Oil Company]." Too, he represented Jersey in sessions with other oil company executives "as, for example, they, from time to time, developed and agreed upon revised complicated formulae for crude oil rights and payments for overlifting crude oil beyond their share."

Turner and Page made an excellent team; both understood the sophisticated formulae that appeared basic to the concession contracts; both were drivers, for whom work was a constant; and both were "naturals" for their positions. Many of Page's ideas and prospective plans were discussed with Turner, who could quickly grasp them, and just as quickly, bluntly state his opinion of their value in a given situation.

Harold W. Fisher, Jersey's first representative on the Iranian Oil Participants, Ltd., who also served a term as Jersey's IPC director and later as co-managing director—a full-time IPC position—during the time Page served as contact director responsible for the Middle East, was also a knowledgeable person with broad experience that proved valuable. Director David A. Shepard, too, had served as director of IPC for several years; he, also, had participated with the Jersey team that helped break the Red Line Agreement and thus, paved the way for the company to join Aramco. Shepard, who had lived for a number of years in London, acquired considerable information about the Middle East. These men—qualified, hard-working, and apparently tireless—gave Jersey first rank among the oil companies who sent diplomats to the Middle East.[2]

X X X

In 1960, five oil producing nations [Iran, Iraq, Saudi Arabia, Venezuela, Kuwait] formed the Organization of Petroleum Exporting Countries [OPEC]. From small beginnings, this federation grew to have great power by the third Arab-Israeli War in 1973. Since such non-Arab states as Venezuela and Iran played significant roles in the emergence of OPEC, Pan Arabism, though undeniably a factor, cannot fairly be called the prime reason for the creation of OPEC. After 1960, nevertheless, all international oil companies had to reckon with this completely different political and economic force, one that grew in influence while it sought consensus among the conflicting interests of both original members and newer producing states such as Libya.

The story of Jersey and Middle Eastern oil during these troubled years serves as a case study of the increasing power of producing nations; as the emphasis on oil shifted from sales of a commodity to its use as an economic, and ultimately, a political weapon. Further, as the number of U.S. firms operating abroad increased, and their business activities expanded, American government policy became more concerned with multinational actions. These firms, in turn (and this proved especially true of Jersey), began to consider the foreign policy implications of every major decision they made. Jersey's directors, for example, had to act as trustees for their shareholders as well as deal internationally with an increasing productive capacity in host nations, all of whom pressed constantly for more production and income. Therefore, the 1950s and 1960s were periods of plentiful oil and low profits.

The Western economies, geared to the rich Middle Eastern fields, greatly reduced the use of other energy sources such as coal, because of the low prices and ample supply of oil. Economies expanded rapidly, building on this base of a low cost, relatively clean fuel. Ironically, while productive capacity began to out-race demand (with resultant price drops), Jersey and other major international oil companies continued to be involved in legal battles and U.S. Congressional hearings, in which they were charged with "cartelizing" production and prices. Hence, these two decades were filled with controversy and with events about which there is, as yet, little agreement. As Jersey Board Chairman J. Kenneth Jamieson remarked to a *Wall Street Journal* reporter on August 28, 1969, "Tension in the Middle East seems to be getting worse and worse, and we view this as a most critical, dangerous area." A few months later, the *New York Times* quoted an unnamed oil company Middle East expert: "Viet Nam is just a brushfire compared to what can happen out there. Russia and the United States are facing each other head on in the Middle East. ... It's the perfect recipe for the end of the world and you better believe it."

Later, a Jersey Middle East executive, after reflecting on that troubled third quarter of the century (1950-1975), wrote that those years were the "one[s] in which the wheels came off—not four at once—but one at a time, ending with a situation little like that at the start."[3]

Iran, 1957-1970

The Iranian Consortium Agreement of 1954, called for the eight original members to incorporate two operating companies in the Netherlands and then to register them in Iran. The board of directors of each was to consist of seven members, two of whom the National Iranian Oil Company [NIOC] would nominate. Thus, during the last stages of the negotiations between Howard Page and Dr. Ali Amini, the Western consortium members arranged for the incorporation of Iranian Oil Exploration and Producing Co. and Iranian Oil Refining Co. At times, they represented the consortium members in dealing with NIOC.[4] To represent their collective interests, the original participants also incorporated a joint holding company, Iranian Oil Participants, Ltd., in which they owned all the shares. [Harold Fisher, Jersey's shareholder representative in London, was the first appointed to represent that company.] Then, each of these oil majors turned over to a subsidiary wholly-owned and incorporated for that purpose, the purchasing of Iranian oil. Jersey termed its affiliate Esso Trading Company of Iran [Esso Iran].[5]

Oil production in Iran during the last two months of 1954 pleased NIOC executives as well as Shah Muhammad Riza Pahlevi. In that year and during 1955, the volume of oil produced exceeded the minimum quantities specified in the agreement. Iran also received additional income of $10,000,000 from a supplementary loan by the U.S. government.

As time passed, however, the two Iranian companies that produced and refined the oil began to experience shortages of skilled labor. This occurred despite the fact that the Iranians took great pride in the knowledge that during the shutdown, local companies using Iranian labor had "managed through their own efforts" to supply all the oil needed for domestic consumption. The new consortium-created companies had pledged "to minimize the employment of foreign personnel."[6] In 1955, in an effort to live up to this agreement, the partners began to provide training in the Western Hemisphere for Iranian nationals employed by the two operating companies. Also in 1955, Jersey and its partners each began to train five Iranians per year for managerial positions. This program continued for several years.[7]

Originally Jersey and the other seven major companies sought to rotate their own personnel into the operating companies on a two year basis, but in April, 1956, Director Shepard reported to the Jersey Executive Committee that "it seems unlikely, at least for the next several years, that the Iranian consortium's operating companies will be in a position to fill with regular employees of their own the 300 or so key positions in their organization." He moved that the company extend its employees' assignments in Iran for an additional two years. The Executive Committee approved his recommendation.[8] A number of years passed before enough Iranians could be trained to operate the two new companies effectively, and Jersey continued to supply more personnel for the top positions [such as Abadan refinery manager] than its pro-rata share required.

X   X   X

Yet, even while Jersey and other international companies sought to re-introduce Iranian oil to world markets, the uproar created by government antitrust suits against American oil companies continued to create head-lines. Many newspaper stories alleged that the companies were receiving sympathetic treatment from the Eisenhower administration. Details of inter-company agreements did not get released at this time; however, several published versions of the Consortium-Iranian Agreement proved to be quite accurate. The October, 1954, issue of *Fortune* magazine carried many of the details, but it said very little regarding the arrangements effected among the companies.

Chairman Emmanuel Celler of the U.S. House of Representatives Com-mittee on the Judiciary became interested in the agreements. He also con-cerned himself with a related matter: how Attorney General Herbert Brownell came to give an advisory opinion to the National Security Council [NSC] that permitted American oil companies to combine with European companies to move Iranian oil. Testifying before the Celler Committee in May, 1955, Assistant Attorney General Judge Stanley N. Barnes stated that "the consortium agreement action, as I understand, was never subject to the scrutiny of the Antitrust Division." Since the opinion of the Attorney Gen-eral and the consortium agreement had been classified as "Secret," Barnes would not comment on that particular action. Kenneth R. Harkins, the committee co-counsel (at one time active in the Justice Department Anti-trust Division and the "cartel" investigation), volunteered that the Brownell opinion created a kind of "executive type of immunity."[9]

Before the hearing ended, several things became clear: first, widespread interest already existed in the exact terms of the Iranian Consortium Agree-ments; second, the oil majors had not selected the nine independent Ameri-can companies which combined as Iricon Agency Ltd. [Iricon] in the consortium; and third, the detailed terms of agreements among the oil com-panies would not be made public. These hearings also ranged afield to in-clude the civil antitrust suit against Jersey and other companies. Regarding the Brownell opinion, Barnes in his testimony repeatedly stated that "no immunity has been granted any oil company."[10] Continually beset with the problems of the world oil markets, with productive capacity increasing in oil producing countries, and under fire at home in a civil antitrust suit, while at the same time experiencing declining net income, Jersey and associated oil companies must have noted some of the congressional testimony with bewil-derment.

Chairman Celler introduced and entered in the hearing record a pam-phlet Jersey had issued as a supplement to "Standard Oil Company (New

Jersey) and Middle East Oil Production."[11] This document stated categorically that Jersey's participation in Iran began as a result of the United States' desire that Jersey participate, and that Jersey did so under "an opinion ... rendered by the Justice Department that participation in the proposed consortium would not violate the United States antitrust laws."[12] The pamphlet also related some details of internal agreements among the Iranian Consortium partners.

This publication and the *Fortune* article led the committee to request copies of all agreements regarding Iranian oil. Secretary of State Dulles refused to accede to this request on the ground that "I believe it would not be in the interest of the United States to divulge their contents at this time."[13] The committee (and, perhaps, the general public) failed to see why such information, already in the hands of scores of persons not cleared for U.S. security concerning classified information, should still be kept secret. When pressed, Dulles invoked the time-honored phrase that their publication "would adversely affect the foreign relations of the United States."[14]

The Celler Committee was in session to study then current antitrust problems; however, any fairminded person examining the reports of and documents placed before that body had to conclude that Jersey and the oil companies received a disproportionate amount of attention. The committee listened to former Justice Department antitrust counsel, Leonard J. Emmerglick's version of how Jersey and the indicted oil companies had stalled the government's efforts to secure documents needed for the earlier grand jury investigation. Co-counsel Harkins reopened old Jersey wounds when he introduced for the record (while Emmerglick was on the stand), a copy of an article from the *Washington Star* of April 5, 1942, that carried former Assistant Attorney General Thurman Arnold's allegations of the company's pre-World War II collusion with I.G. Farbenindustrie. Then, the committee placed in the record, certain documents regarding the oil industry mentioned in the Barnes testimony; also included were the government portion of the Iranian oil Consortium Agreement, and a number of documents and letters relating to the companies still engaged in the civil antitrust suit at home. This literature, as a whole, establishes an indisputable fact: By 1955, the American petroleum industry had become a highly visible target of U.S. congressional committees, as well as of the antitrust agencies of the government.

X   X   X

Under the terms of the consortium agreement of October 29, 1954, the companies had three months to start-up production; then during the remainder of 1955, they were required to produce 300,000 barrels per day [b/d]; in 1956, 450,000 b/d and in 1957, 600,000 b/d.[15] Soon, it became obvious that, despite an emergency U.S. grant of $45 million in 1953, immediately after Premier Mossadegh was overthrown [Iran already had received $23 million in U.S. aid during that year], plus $85 million in 1954, and $76 million in 1955, even with additional revenues from oil, the Iranian government would not have enough funds by 1955 to carry out its functions.

When the Shah returned to power, he required maximum oil revenues for his extensive development plans. The governor of the Central Bank of Iran, Abolghassam Ebtehaj, worked with a committee to draft an overall program. In its final form, the plan covered the years 1955-1960, and it called for a total investment of around $500 million dollars. But the estimated income from oil, plus loans from the International Bank for Reconstruction and Development [World Bank] and aid from the United States did not provide sufficient funds for this project. So Iran quickly began to press the consortium members to increase production, hoping that this move would also increase oil revenues.[16]

In January, 1956, Page told the Jersey Executive Committee that Iran had asked the consortium members to increase production by 200,000 b/d over the agreed upon base of "475,000 b/d" for that year. In the discussion that followed, Page stated that "due to present equipment limitations in Iran," the maximum increase in the volume of oil that could be made available for export would be around 80,000 b/d. Hearing this, the Committee agreed that when the companies met, Harold Fisher should indicate that Jersey favored an increase in liftings of 50,000 b/d but no more than the 80,000 b/d Page had stated could be made available. Subsequently, Page reported that the consortium partners had agreed to increase liftings 44,000 b/d.[17]

Since the consortium did not increase production to the levels Iran desired, the NIOC soon began to deal with individual companies. The reason the NIOC dealt with one company at a time derived partially from U.S. antitrust laws. Attorney General Brownell's ruling of January 21, 1954, which permitted the consortium to be formed, covered only that single combination of companies. No other agreements or understandings—among or between any of the U.S. participants—were covered. So in 1956, Iran asked Jersey to bid on petroleum concessions in offshore areas.

By investigating the possibility of acquiring additional oil concessions in that country, the directors were following a traditional Jersey course in strategic planning—of seeking to add reserves to the company's holdings even in periods of plentiful oil capacity. In this case, however, Iran indicated that any rights to concessions in offshore areas would be granted on terms different than those given onshore to the consortium. The NIOC would require a 50 percent operating interest in offshore properties, and the money for this share would come from bonus payments, rentals and like charges after oil in commercial quantities had been discovered.

In responding to Iran's requests, the Executive Committee asked Page to tell Fisher, who was going to Tehran, to continue discussions about the possible acquisition of offshore properties, and to make arrangements with Iran for reconaissance seismic surveys of likely areas. Fisher found the government representatives adamant in their position that the proposed offer represented the only possible basis for talks. So in August, 1956, he did not discuss exploration with them.[18] The Jersey Executive Committee then decided to let this particular opportunity go by.

Iran, however, persisted in its effort to obtain bids from Jersey for oil rights outside the consortium concession area. [Apparently no oil company made an offer under the Iranian terms; probably due to the continued expansion of productive capacity in other countries, plus the potential difficulty in marketing additional Iranian crude.] Eventually the government indicated its readiness to deal for a smaller percentage as the NIOC operating interest. Iranian officials asked Jersey "what principles it would regard as providing a satisfactory basis for oil rights in the areas in question?" In the Committee discussion on the subject, Page, M.A. Wright, Jersey's coordinator of producing activities, and Counsel Thomas Monaghan pointed to a forthcoming meeting of the company's foreign affiliates and proposed that the ramifications of this question of acquiring acreage with government participation be explored with them for possible effects on relations between those companies and their host governments. The Committee agreed that this course should be followed, but it cautioned that the company had ever been reluctant to accept government participation in such ventures unless most of the potential production could be absorbed locally. Too, where Jersey had assumed the obligation to dispose of another participant's share, this had always been considered a contribution of considerable value because of the company's heavy investments in downstream facilities. Jersey disliked such business combinations because it believed that when a government entity strengthened its direct access to world markets, the possibility of nationalization or unilateral contract abrogation by that government increased. In exploring such agreements, the Committee counseled those who did negotiate that among Jersey's important assets were "technical skill, organization and 'know-how.' "[19]

The Jersey Company continued to send its upstream experts to Iran to investigate likely concession sites. Several directors, plus Wright, representatives of four major Jersey affiliates and selected members of the Producing Coordination Department concluded after one visit to Iran that, under the government's proposal, Iran would receive considerably more than 50 percent of the net operating revenue; hence, the acceptance of that proposal would endanger the 50-50 agreements then operative throughout other oil producing countries. The Committee did, however, authorize Howard Page to negotiate an agreement with slight variations from the 50-50 principle.[20]

Page, Wright and a geologist, Zeb Mayhew, in company with other personnel then went to Iran in January, 1957, and talked to representatives of the government and the NIOC. Wright reported to the Executive Committee on January 24, that under the 50-50 proposal, the Iranians would actually receive 62 1/2 percent of the operating profit, and that their strictures regarding the structure of the joint company would tie up management and make it impossible for Jersey to exert sufficient control of the venture. Next, Dawson Priestman, a producing economics specialist, explained to the Committee that the 62 1/2 percent figure that Iran would receive derived from existing oil prices; should crude oil prices increase or operating costs drop below anticipated levels, the government's share of profit would go up correspondingly. Wright recommended that the Iranian proposal be rejected, which the committee did; and it also decided not to initiate a counter-proposal. Page was to be informed of the decision when he returned to New York, and the matter could be aired again if he disagreed. When he came back, Page also advised against accepting the Iranian proposal and against making a counter-offer. The Iranian demands remained a barrier to negotiations. Throughout 1958, and 1959, the Executive Committee deliberated on them, but it became obvious that the company would not meet NIOC demands.[21]

Jersey also discussed the possibility of acquiring offshore acreage from Kuwait at the same time that these Iranian prospects were considered. In 1957, Jersey brought into its discussions representatives of "another private oil company." Director Page, Fisher, and other Jersey representatives discussed these possibilities with the Jersey Executive Committee quite often in 1956 and 1957, but no final decision was reached.[22]

Whether the Iranians also conducted comparable negotiations with consortium members other than Jersey is not known, nor is it recorded whether Jersey informed its consortium partners about the talks with NIOC. The surviving records do not reveal the names of the private oil companies that Jersey brought into discussions about offshore properties in Kuwait, late in 1957, except for the British Petroleum Company, Ltd. [BP]. But the evidence makes it clear that the Iranians conducted negotiations with several other companies not in the consortium.

In July, 1957, the Iranian Parliament passed a law which served to give all NIOC proposals additional stature and which significantly increased that company's power. This parliamentary act permitted NIOC to determine exploration areas outside those of the consortium, including some offshore, and to enter joint ventures with foreign companies to explore and develop such areas. It is quite possible that the purpose of this statute was to strengthen NIOC's relative position, so that it could receive offers from companies not in the consortium.

Jersey expressed interest in obtaining a concession under the 1957 law. Page, accompanied by two lawyers and a geologist, went to Iran to "study" the law and to talk with authorities. He indicated in an interview that the company would pay a bonus for a concession. Jersey would offer a 50-50 profit-sharing contract to NIOC. Probably Page found one provision of the 1957 Petroleum Act so novel or unusual that it influenced his decision at that time. The law did not allow exploration permits to be granted before bids were made—a pig in a poke situation. This served to limit the technical information available to prospective concessionaires and greatly increased the odds against obtaining a paying concession. The bidders had to agree to underwrite all exploration costs, which could be recovered only when they located oil in commercial quantities; the same procedure applied to recovery of the prepaid first year annual rental. Page returned to New York and related his findings to the Executive Committee.

Years later, while reflecting on the Jersey effort to secure an Iranian offshore concession, Page remarked that, "the areas we went for were the poorest areas of the bunch, it turned out." He added that:

> there wasn't any real pressure on me to make a deal from New York, other than if I could get it for a low price. .... But we couldn't. Or not for conditions that we were willing to take at that time. You know 75/25 [profit-sharing], things like that. .... It wasn't my idea to go in there. So I think it was an idea of, 'Well, we ought to show them we're active around the place.'[23]

Page was correct. Many of the Iranian offshore concessions did not prove out.

On August 3, 1957, another oil company executive, Enrico Mattei, manager of Italy's Ente Nazionale Idrocarburi [ENI] concluded an agreement between that company's subsidiary, Azeinde Generale Italiana Petroli [AGIP], and NIOC. Mattei had long fought the international majors and he had loudly protested the failure of the United States, Great Britain and Iran to include ENI in the 1954 consortium. AGIP and NIOC created a jointly owned company to explore the concession. This partnership (and others formed later) paid the Iranian government 50 percent of the profits as taxes plus 50 percent as an investor in the concession development group. This has been commonly reported as a 75/25 percent profit-sharing agreement. Howard Page contends that to be incorrect. He stated, "The Italian agreement ... actually involved a 50/50 agreement with the National Iranian Oil Company, which included liability of half the development costs. This group [the developing group] in turn has a 50/50 concession with the Iranian Government. Iran does, through this device receive 75 percent of the profit from the concession." The oil majors viewed this move by ENI and Iran with somewhat angry incredulity and prophesied that the latter would profit very little from the arrangement. They pointed to the fact that Iran would contribute nothing to the venture until the discovery of oil in commercial quantities, and even then only an insignificant amount of money. Subsequently in June, 1958, NIOC and Pan American Petroleum Corporation [Pan AM—a subsidiary of Standard Oil Company (Indiana)] reached agreement on a concession. They, too, established a jointly owned company.

Oil was discovered in commercial quantities on the concessions of both joint companies; however, as the consortium partners had predicted, Iran rarely realized from any of these concessions (or those granted later) half as much income per barrel of crude oil as it received per barrel from that produced by the consortium. Iran, however, continued this program and later concluded other similar arrangements such as contractual-service agreements. These also failed to attract the larger foreign companies, forcing NIOC to continue emphasizing joint ventures.[24]

During the first Suez Crisis, 1956-1957, Iranian production followed, in general, the production rate specified by the Consortium Agreement of 1954. 1957 represented the last year for which the consortium had established basic production quotas, but throughout the term, the Iranians continually pressed for production in excess of the annual increase for other Middle East producing states. This supplementary increase, Page and the other negotiators could not contract for; they feared that other Middle Eastern countries (especially Saudi Arabia) would demand in turn higher offtakes. Instead, the consortium agreed to provide Iran an annual growth rate in crude oil production at least on the level of the Middle Eastern average. In fact, Iranian production over the next nine years increased 152 percent as against the overall Middle Eastern increase [by the same companies] of 136 percent. In Iran, production increased over a period of sixteen years at an annual rate of 14.2 percent.[25]

X X X

The entrance of Soviet Russia into the world oil market in the late 1950s served to complicate largely both supply and marketing problems. The increased Soviet oil sales to Western nations had a special purpose; they opened the way for the acquisition of the modern technology and industrial equipment much needed to convert Russian industry and make it competitive with the West.

Meanwhile, the problem of world over-productive capacity of oil became more complex when the U.S. resorted to a system of import quotas, first voluntary in 1957, and later mandatory in 1959. Soon, producing areas such as Nigeria, Libya, and Algeria were adding productive capacity and driving prices lower, especially since the latter two countries were much closer to market areas than the Persian Gulf states. Overall, production in these new producing nations came largely at the expense of the longer established Middle Eastern producers and Venezuela.

All of these actions and events created additional pressure on the consortium partners and exacerbated their difficulties in accommodating the continuous Iranian requests for increased liftings. Secretary of State (1961-1969) Dean Rusk, who saw the Shah on numerous occasions, once described the manner in which the Shah prepared his budget as quite uncomplicated. The Shah first decided for what programs he needed money and how much. When he arrived at that figure, he then called in the oil companies and told them how much they had to contribute to the budget total—contribute through taxes and royalties on their oil liftings. In that manner, irrespective of the world price for oil, the Iranian budget became tightly locked into oil production. Rusk also pointed out that after the Shah made his decision, all the ministers and officials followed his line; they echoed the Shah's remarks. And of course, other producing countries also requested the same increases. Meanwhile, the actual market price of crude oil fell everywhere.

Certainly the failure of major oil companies to increase production when requested to do so by the older producing nations intensified those nations' interest in creating some collective organization to serve as a unified bargaining agent. Early leaders in this movement were Saudi Arabia and Venezuela, from which two countries Jersey secured a large portion of its crude oil. As if to further complicate that picture, the U.S. government, then certain that politics rather than economics dictated the price of the cheaper Russian oil, consistently warned American oil companies—even as late as the 1960s—that they should placate, whenever possible, the Middle Eastern producer nations and thus serve American national interests. These warnings were based on a genuine fear in Washington that the political use of Soviet oil might undermine the existing stability of several countries in the Middle East—just what they believed to be the Soviets' plan.[26]

The growth of European demand for oil and products after World War II (discussed elsewhere in this volume) proved phenomenal. Jersey itself held a market position of almost 18 percent in Europe during the late 1950s. Most of this oil originated in the Middle East, where the company's equity position remained insufficient to supply the market. By purchasing 30 percent of Aramco [also discussed above] and by entering the 1946-1947 purchase contract with BP [at the time, Anglo-Iranian Oil Company, Limited] for 800 million barrels of oil over a twenty year period, Jersey helped to alleviate the shortage in Europe earlier. Later, in 1959, Page negotiated a contract with BP for an additional quantity of 215 million barrels (app.). When the first mentioned contract expired in 1966, Turner negotiated a new one with BP for more than 2 billion barrels of crude oil from Kuwait, Iran, and Nigeria over a fifteen year period. Jersey also consistently purchased oil from its Aramco partners; in fact, from almost anyone who would sell crude cheaply. Altogether, this amounted to a sizable ocean of oil, which crude-hungry Jersey required to service its expanding markets in Europe.[27]

The great volume of crude production available worldwide in the late 1950s had an effect on the price. Chiefly, the market price dropped below the posted price.* Then, the resulting decline in the selling price occasioned by the increase in the volume sold, served to decrease profits—as more crude oil reached the markets, it drove the price down.

Aware that this widening gap between posted prices and market prices had depressed the company's net profit margins, Jersey officials debated their next move. Rathbone had replaced Holman as CEO on May 1, 1960, and now he pointed with concern to the fact that the company continued to sell a greater volume of oil for less profit. The new CEO believed that an obvious way to increase profits would be to bring the posted price of Middle Eastern oil in line with the market price. So in July, 1960, in a Board meeting, a Rathbone proposal to reduce the posted price led to one of those memorable sessions for which the Jersey directorate had long been famous. Board members did not object to the necessity of some lowering of the posted

---

*The posted price originally represented the open market price at which crude was available to buyers generally. In the late 1950s, as new fields opened up and crude production swelled, discounting off the posted price began, and a gap developed and widened between posted prices and actual market prices. Since the host governments' taxes and royalty income were figured on the basis of the posted price, they objected to having that price lowered to reflect actual market levels.

price; yet, it is certain that the proposals debated were not phrased in simple, uncomplicated terms. Rathbone, who presided, possessed a wealth of experience; however, virtually all of it had been gained in the company's domestic operations. Several Board members who participated in the discussion, later questioned whether or not he fully comprehended the problems of dealing with Middle Eastern countries, or what might happen if the Board approved his proposal.

After Iranian oil had returned to world markets (1954), Middle Eastern Contact Director Page often pondered the effect on producing nations of a drop in oil prices that would force a closer alignment of the posted price with the actual market price. So on several occasions, he asked the Jersey Board to consider some revision of the agreements, "some sort of easing" in anticipation of that day. The Board, he said, "told me not to be silly, that the governments hadn't asked for it. Never give them anything they don't ask for. ... Well, the time came."

Earlier, in 1959, British Petroleum had reduced Middle East crude postings by 18 cents per barrel, taking the producing nations by surprise. In turn, they protested very strongly against it. The Suez Crisis had occasioned higher prices and increased demand; hence, it seemed natural that a drop in the posted price would be made when the shortage of oil disappeared and when the canal reopened in 1957. But the producing countries had budgeted anticipated income on the high posted price and higher volume. So no drop occurred. Too, postings in the Western Hemisphere, especially in Venezuela, proved to be quite flexible between 1957 and 1959.

Page had warned Rathbone against lowering the Middle East posted price in 1960. He knew that the governments would be more concerned about the effect on their revenues than any other consideration. Turner also strongly backed Page in opposing any posted price cut.

Page did not oppose the move on principle, but he argued vigorously against making it at that particular time. He "felt that before we did it we should make an agreement with the Gov'ts concerned that we would pick up 1/2 the loss to them or some similar agreement." Yet, all of the directors present understood that the posted price no longer reflected the true market price. Sales in the true market price range lessened the company's profit margin, for taxes were collected on the posted price. Page believed he knew how host governments would react to any pricing action. Further, he had no desire to create for Aramco problems with the Saudi Arabians, from whose country Jersey obtained far more crude than from any other source in the Middle East. For him:

The whole question was: were we going to force the governments to live up to the wording of their agreements and absorb 50 percent of the loss [under the 50-50 profit sharing agreements] resulting from lower real market prices which would bring their 'take' to less than when they signed the agreements.[28]

Precisely what happened at that Board meeting will probably forever remain unclear—surviving directors (and John O. Larson, the company secretary) are not absolutely certain they know who voted which way or for what. And the memories of other persons active at that time, who may have acquired portions of information about what took place, cannot be taken literally, for they did not attend and their testimony is not admissible. Everyone agrees, however, that the directors earnestly debated whatever position they supported (or opposed), that the issues were never simple, but always very complex. Page remembered that his proposal to delay action until agreement could be reached with the producing governments went down to defeat on the ground that it would break the 50-50 agreements. To him and several others, the maintenance of "a posted price above the market price" had already accomplished that. Obviously, the Board majority did not accept his view. Nor did they support Page's effort to delay action. A motion to cut Jersey's prices may well have passed.

One reliable version of what happened claims that the directors did back the Page motion; the cut would have to be made, but only after some agreement had been reached with the producing countries. This did not please Rathbone. So he decided that Jersey could cut the price without effecting any agreements with the producing nations. He then initiated the price-cutting action. No matter what did happen in the Jersey Board meeting, on August 9, 1960, one Jersey affiliate, Esso Export Corporation, announced reductions of from four to fourteen cents per barrel, in posted Middle East crude prices. Several other major oil companies quickly brought their prices into line with Jersey's.

Page, in an effort to soften the financial blow on the producing nations, sounded out representatives of several governments, proposing an arrangement whereby they would take only one-fourth of the loss and the companies would absorb the remainder. Just how the governments responded is not known, but Page quickly learned that "none of the companies would buy it." So this tactic failed. Fifteen years later, he described for the U.S. Senate Subcommittee on Foreign Relations what happened: "Exxon felt their posted price should reflect ... the market prices" and in 1960, dropped that price. "This," he went on "of course, affected the take of the governments, and that is when all hell broke loose, you might say and OPEC was formed." Turner also indicated something of the reaction. "I, too, had spoken against it, but was in Baghdad when it was announced and (to exaggerate) was glad to get out alive."[29]

Up to that time, the Shah had been a steady ally of the United States. To that country, he owed his restoration to the Iranian throne, and further, his relations with his Arab neighbors had not been good. Years later, American Secretary of State Henry Kissinger defined the relationship between the Shah and the United States during these years: "The Shah, a shrewd analyst of foreign policy, based Iran's security not on maneuvering ambiguously among the superpowers, but on alliances with America." But ally or not, the 1960 action of the oil companies certainly angered the Shah. As he said, "If the companies which produce our oil do not come and discuss the market situation with us, but cut our revenues without opening their books to us, what are we to think? The question is that we should not be left with the feeling that everything is being hidden from us." He went on to note that host governments had to base their budgets and development plans on oil revenues and that "as owners of the oil we must be treated as equal partners. The companies must realize that this is for their own good."[30]

At least one other reaction, that of Abdullah Al-Tariki, then Director General of Petroleum and Mineral Affairs for Saudi Arabia, closely paralleled that of the Shah. Tariki and Venezuela's Dr. Juan Pablo Pérez Alfonso, Minister of Mines and Hydrocarbons, had for some months in 1960 been advocating in conversations with representatives from Iraq, Kuwait and Iran the establishment of a Petroleum Affairs Commission, with whom they all could consult. Jersey's price cut, on August 9, 1960, led quickly to a meeting of these five key oil exporting nations in Baghdad, on September 10, 1960. From that meeting, OPEC emerged.

There is evidence to indicate that, largely to placate Iran, OPEC entered "the world stage on a note of moderation."[31] Certainly, Iran played a leading role. As a Muslim (though not Arab) nation, joined with four other major oil producing countries, Iran expected that OPEC would function as a collective bargaining agent for Iranian oil. In fact, for a number of years Iran served as the "bell camel" of OPEC, since no major international oil company had a direct agreement with OPEC, and all of them were in the Iranian Consortium. When it came time for the companies to negotiate, Page described how this happened:

> We said we won't negotiate with OPEC. We never heard of it. We
> have agreements with individual governments. So they [OPEC]

said, 'Well, alright, you start with Iran and remember whatever you agree in one place, you got to agree everywhere else.' ... And so we are kidding ourselves a little bit on this and it was a ... fact that we knew damn well we were negotiating with OPEC.

On October 10, 1960, the Jersey Executive Committee heard research reports on the Middle East, especially concerning government relations, concession terms and competitive trends. The discussion of these reports included "particular reference to potential problems which may be encountered by the private petroleum industry as a result of the activities" of OPEC. The Committee assigned several directors to pursue questions raised by the discussions with U.S. government authorities, and with the governments of other countries where Jersey operated. The Committee sought information about, and reactions to, OPEC before developing "a definitive Company policy position and plan of action."[32] OPEC, at this time, was less than a month old; yet the Jersey Executive Committee already took seriously the objectives announced at the Baghdad meeting.[33]

Of course, in many foreign nations there was no real comprehension of United States antitrust laws. Thus, it is unlikely that most OPEC leaders ever realized that, if the American oil companies joined together in negotiating production quotas and prices with OPEC, the Justice Department could charge them with violating the antitrust laws. [Some years later, in 1971, American companies had to deal with just that situation.] One OPEC representative, however, Dr. Pérez Alfonso of Venezuela, a lawyer by training, did possess some knowledge of U.S. law.

In September, 1961, Pérez Alfonso came to Washington and arranged a conference with representatives of the Departments of State and Justice. His ostensible purpose in requesting this meeting was to secure the reactions of those two departments to a proposed law in Venezuela which would fix export prices for oil. Both Justice and State Department spokesmen assured him that antitrust laws would apply only if the privately owned oil companies made price agreements concerning commodities imported by or exported to the U.S.A. At the same time, they advised him that any country could control its own export prices.

Pérez Alfonso explained that OPEC was planning to control both production and prices by bringing "pressure on the international oil companies *to do this under the supervision of OPEC.*" Such action by the companies seemed to violate U.S. antitrust laws, as Pérez was told. What he learned from Justice and State obviously affected the future course of OPEC, for that group finally did achieve control of both production and posted prices—and Pérez Alfonso played an important role in charting the course.[34]

X X X

The first OPEC pressure felt by the operating companies came at the fourth conference meeting in Geneva, in June, 1962. Among the important recommendations made there were moves to end the marketing expense contribution from OPEC members and to have royalty payments charged to operating expenses.

Iran began to negotiate with the consortium members to reduce its 1/2 percent contribution for marketing [amounting to 1 1/2 cents (app.) per barrel]. They reached agreement to have this reduced by two-thirds, and this arrangement quickly became standard for the other concession agreements. It remained basic until 1971.

Generally, in a 50-50 profit-sharing agreement, the producing country's share (including a royalty of 12 1/2 percent of posted price, as a minimum) was 50 percent of the profit, and the oil company's share, the remaining 50 percent. Now, if the producing countries could get the companies to agree that the royalty payment constituted an expense, the host governments would then receive 50 percent of the profits after this deduction, thereby increasing their total income.[35]

The oil companies balked at this OPEC notion when Iran presented it. Page led their attack on it. He drafted a counter plan which, he told the Jersey Executive Committee, he had been unable to get the consortium members to accept. He went on to explain a formula that would have provided payments slightly in excess of those under the then-prevailing system. The Committee approved his plan, even to opening of the records of Esso Iran to government representatives.[36]

While Page headed for Iran with a counter proposal, the consortium partners arranged to meet with Iranian officials in Paris during October, 1962, to take up the question of royalty payments. Jersey's Turner, who had been nominated by the Socony Mobil Oil Company* representative, made the consortium's presentation of the royalty matter. Finance Minister, Dr. Behnia, led the Iranian delegation. Since Behnia spoke no English, Fuad Rouhani (then also secretary general of OPEC) served as interpreter and tactical head. Turner made "a lengthy presentation and gave the Iranians a supporting memo[randum] and charts." He directed all of his argument to the point that the fundamental contract reached in 1954 was for 50-50 profit-sharing, and that the royalty provisions therein were not fundamental, but merely part of the mechanism used to achieve this 50-50 agreement. Turner wrote, "no concessions to the Iranian's demands were made at this meeting."[37]

*Renamed the Mobil Oil Corporation in 1966.

Later in 1962, Page again discussed petroleum affairs with government officials in Iran. He knew that Jersey in-house economic analyses and studies had established that Soviet oil could be delivered to Europe more cheaply than crude from the Persian Gulf, since neither taxes nor royalties had any bearing on the Soviet price structure. One version has Page using this realistic approach in an effort to convince the Iranians that the Soviet Union, Venezuela and Libya all possessed economic advantages in competing for European markets. The Russians made no secret of their desire to increase exports of crude oil, nor of the planned lengthening of their pipelines. When they completed these improvements, they would pose a real threat to Iran's oil.[38]

Meanwhile, OPEC had set up operations in Geneva, and sought to push for the royalty proposal with a flurry of memoranda that detailed the decline in world crude oil prices and the continuing erosion of purchasing power among the oil producing nations caused by worldwide inflation. The oil majors answered in kind with statements calling attention to marketing conditions and problems, and particularly emphasizing the widening gap between the posted price of oil and the market price. This disparity had led Rathbone to cut the posted price in the first place, thus earning for him in some circles abroad, the title "Father of OPEC."

Though the Paris negotiations with Iran fell through in 1962, Page continued to push for retention of the existing system. OPEC had chosen Iran and Saudi Arabia to handle these negotiations, because four of the major oil companies operated in those two states and because those two governments were more conservative than some others. But the negotiations led nowhere, since both the Iranian Consortium participants and Aramco strongly objected to considering royalty payments as operational expenses, according to the OPEC formula.

In 1963, discussions between Jersey and NIOC concerning an offshore concession resumed. Again however, the restrictions and procedures established by NIOC proved a formidable barrier to an agreement. Despite the fact that the territory being opened for bids lay in the Persian Gulf—the most promising non-consortium area—Jersey continued to balk. Finally on May 20, the Executive Committee advised "that it would seem prudent to seek to so qualify" as a bidder. It added a caveat, however, that the company would not bid unless other major sources of supply in that area were threatened. The Committee also suggested that the appropriate officials in Saudi Arabia should be told of the company's present thinking in this regard—and if necessary, officials in Iran as well.[39]

OPEC, now enlarged in membership, planned to meet in Riyadh, Saudi

Arabia during November, 1963, and the Jersey Executive Committee knew that the question of royalties would again be on the agenda. By this time, Page had developed fully his stop-gap plan as a response to the demand for full royalty charging against the expense of operating. First, he recommended that the Executive Committee endorse his proposal to have the consortium appoint a committee to negotiate with Iran, with the view of "determining terms and conditions" upon which the matter might be settled. The Committee approved his resolution. It next approved the Page proposal of discounts from posted prices, which they knew "would fall short by some 3 1/2¢ per barrel of offsetting the increase in the Government's revenue which would result from expensing royalty payments."[40] Meanwhile the OPEC member states delegated Secretary General Fuad Rouhani to negotiate for all of them. The Iranian Consortium then appointed a committee consisting of Page, J. M. Pattinson of BP and G. L. Parkhurst of Standard Oil Company of California [Socal] to represent it, and informed Rouhani that the committee could deal only with Iranian matters. Page and the two other members of the consortium negotiating committee met with Rouhani in London, but the meeting settled nothing. Then, the consortium members met and adopted the Jersey [Page] proposals as the basis for subsequent negotiations with Iran.[41]

In the OPEC meeting in Riyadh, in December, 1963, the question of where royalties should be placed in the company accounts received full consideration. Before that meeting, the oil companies had revised their offer to Iran, in order to eliminate several objectionable proposals. Now, the OPEC members could reach no decision on just what course to pursue should the companies persist in their offer. So OPEC stalled. Finally, just before the July, 1964 OPEC conference, a formula was developed that appeared to have a good chance of resolving the problem for a time. This three year offer—largely Page's plan—called for the royalties to be charged against expenses with an allowance off the posted price of 8 1/2 percent in 1964, 7 1/2 percent in 1965, and 6 1/2 percent in 1966. During 1966, a new agreement, based on the market price of oil, could be reached.

This proposal did not secure the approval of the ministers from all the oil producing nations, but Iran agreed to accept the Page proposal in principle. In order for it to cover 1964, the new contract had to be signed before January 25, 1965. At the seventh OPEC conference in Djakarta, late in November, 1964, the tentative Iranian agreement to accept the offer caused a serious rift among OPEC members—while Iran indicated that it would sign the proposed agreement, Iraq and Venezuela led other states in opposing it. Having settled nothing, the Djakarta OPEC meeting adjourned. Pressed by

Iran, Saudi Arabia and other members, the secretary general of OPEC called for another meeting to take place in late December, 1964.

Again, Page was selected to head the consortium negotiating team, and again they met with the Iranian representatives. Page explained his proposal once more. He assured the Iranians that to increase their take more than three and one-half cents per barrel in 1964, "would push up the cost and therefore the price of residual fuel [oil] on which the companies were making little, if any, profit." In both Europe and the United States, he pointed out, "The coal industry was willing and able to take large segments of the residual fuel market." His clinching argument was that "The Gov't's would lose 75¢ per barrel for every barrel of residual business lost as it would lock out crude outlet barrel for barrel." Later, he dryly remarked, "They finally got the point."

Page displayed some figures to show how the scheme worked:

(Settlement January, 1965 retroactive to January 1, 1964)

Posted Price (per bbl.)	$1.76	
	–0.15	(8.5% discount 1964; 7.5%, 1965; 6.5%, 1966)
Actual Sales Price	1.56	
Costs	0.26	(excluding government royalty) [= $1.30 total]
Royalty	0.22	(at 12 1/2%)
TOTAL =	$1.13	(posted price less 8.5% discount, less costs, less royalty)
Under 50-50 agreement	$1.13 ÷ 2 = $0.565	
For government add royalty	0.22	Percent
Total government take per barrel	$0.785	60.4
Total company take per barrel	0.515	39.6
TOTAL	$1.300	100.0

Page was led to advance his compromise solution by reasoning from a belief that "we could agree to the eventual 11¢ [increase in government take]" because "we expected our cost to go down at least 1¢ per year with increasing volume." He added, "It did." The resulting agreement with Iran soon became the controlling formula in most other Middle Eastern countries as well.

However, not all of the executives of the involved companies reacted enthusiastically to the Page plan, even though they agreed to it. Many of those persons still viewed contracts as sacrosanct, inflexible and subject to no alterations. Page understood this attitude, and he once remarked to an interviewer, "what you have to do is in effect negotiate with your partners all the time, day and night. They're always fighting." After this particular negotiated settlement, Page learned that the president of another oil company had gone to Rathbone and suggested, "You ought to fire that SOB Page who's ruining the oil companies."[42]

When OPEC met for the sixth time, in Geneva during July, 1964, this compromise offer of the companies received tentative approval, subject to minor changes. Iraq finally rejected it. Subsequently, OPEC removed from its agenda the question of how to bookkeep royalty payments. Then the companies and the oil producing states negotiated separate agreements.

Yet, Iran and OPEC had failed to move the posted price back to its pre-August, 1960 level, although Jersey (and several other companies) did add four cents to it, in order to balance with the price of other companies. Informally, the companies let it be known that before any future changes took place in posted prices, they would discuss these with the oil exporting countries.[43]

Any lowering of the marketing contribution appreciably increased Iran's income, as did the 1965 agreement in lieu of royalty expensing. Nevertheless, Iran continued to press the consortium to increase its liftings to a position higher than that of any Arab state. The Shah's development programs depended on increases in revenues. Too, national pride had been injured when Iran lost its position as the ranking oil producing state in the Middle East in 1951-1954. Jersey gave much consideration to Saudi Arabia, which in that period replaced Iran as the leading producer. Additionally, Jersey and the three other American major oil companies in Aramco,* together responsible for only a 28 percent share in the consortium (7 percent each), failed to see why they should support an increase in production of the more expensive Iranian oil when they had other sources of oil available. Too, given the surplus then existing in Middle Eastern productive capacity, plus the increased volume of cheap Russian oil, and the cost of adding additional capacity in Iran, Jersey knew full well that increases in production in Iran would require reductions in offtake elsewhere.

There is no question that at this time the Shah moved more aggressively

---

*The four U.S. firms that owned Aramco were Jersey, Socony Mobil Oil Company, Texaco, Inc., [before 1959, The Texas Company], and Standard Oil Company of California.

into Iranian government-consortium negotiations. When he did this, two other related actions took place: first, the interest of the British Foreign Office and of the United States Department of State in the Iranian Consortium negotiations significantly increased; and second, the Iranians became more and more convinced that the Aramco partners sought to keep production down in their country while making larger production increases in Saudi Arabia, thereby discriminating against them. In fact, the Shah and Iranian ministers became obsessed with Aramco and oil lifting in Saudi Arabia.

Early in 1966, George T. Piercy, then serving as Jersey's Middle East representative in London, visited U.S. Ambassador to Iran Armin H. Meyer, in Tehran. Meyer expressed concern about the "possible effect on Iranian production of the contractual arrangements between the Members [of the consortium] in comparison with those in other countries." Piercy later related to Jersey Middle East Regional Coordinator Smith Turner details of this discussion. Turner then wrote to Meyer, pointing out that (so far as he knew) these arrangements were confidential; certainly he did not believe the Iranians knew about them. Turner explained the system:

> In advance, for each year, each Member suggests a total production program and the highest figure that is supported by 70% of the shareholding is adopted. Each Member then has the right to take its shareholding percent of this total on payment (apart from payments to Iran) of only costs of about 25¢/barrel—say 15¢ for operation and 10¢ to BP (this latter 10¢ was agreed in 1954, to continue until 8.5 billion barrels had been exported, to compensate BP for assets given up in Iran). There are penalties if the Member does not take 75% of such share. Should it take more than its share, it would get only its shareholding percent of such excess at cost and buy the remainder at posted price from the other Members or, as has happened, purchase at whatever price it could from other Members who would not otherwise lift their full share (this has occurred at prices less than the half-way point between market price and cost).[44]

The Shah requested that the U.S. government approach the consortium members, obviously hoping that this would encourage the companies to accede to Iran's request for an increased offtake of crude oil. He made the same request to Great Britain. Shortly after conversations between the two foreign offices, a Near Eastern Affairs officer of the U.S. State Department talked to Jersey Vice President Howard Page and Socony Mobil's Henry

Moses, also a vice president. The department spokesman noted in his report, "Increase in Consortium Offtake," that Page "is acting as focal point for at least ARAMCO parents in Consortium." Page and Moses said that neither the Iranian government nor the British companies had a "leg [to] stand on in their criticism [of the] American members' performance." They also disputed the Iranian and British claim that the "respective offtake arrangements favor Saudi Arabia over Iran." Acting U.S. Secretary of State, George W. Ball, telegraphed a report of this meeting to the U.S. Embassy in Iran. When he reached the above mentioned point, he added, "We agree with Page this sensitive subject should not become part of argument with Iranians." But it did! Page and Moses successfully made the point that the increase in Iranian production had been higher than the average for Middle East producing states, and that this satisfied the consortium obligation in that particular; they added that the Aramco partners in the consortium had offered to increase their share of 1966 liftings from Iran "if other members will try to do same." BP and Shell had not replied.[45]

To increase further Iranian pressure, Premier Amir Abbas Hoveida asked the managing director of the consortium operating company, J.A. Warder, whether the "Shah's message was getting through" and whether "key company officials" should not meet with the Shah on this question. Later Hoveida told Warder that the Shah "suggested such meetings be held in Europe" during his forthcoming visit.

The consortium members planned to meet in London on April 27 to set the consortium's 1966 oil production program for the year [July 1, 1966 — June 30, 1967]. Page and Moses told the State Department representative that since the consortium had agreed to keep production of crude at the Middle Eastern average, they would probably aim for a 10-11 percent increase. Later, Warder indicated that he thought a 12 percent increase necessary; however, he told Premier Hoveida that any larger increase would require three years' planning. While theoretically, increases could be made, he said, these depended on variables such as weather, shifts in scheduled programming, etc., over which they had no control. When the U.S. ambassador talked to Hoveida, the latter mentioned 17.5 percent as the increase desired by Iran and that, while it might not be attainable in 1966, it would be "essential in coming years if economic and military goals [are] to be reached."[46]

Meanwhile, the Shah had also taken advantage of changes in Russian foreign policy. In the summer of 1965, he visited Russia for ten days, during which time he worked out a general plan for economic and technical cooperation with that country. The Iranian Parliament approved this agreement in February, 1966, and shortly thereafter, Iran began to construct a gas pipeline to Russia, while at the same time with Soviet credit, it began to erect "a steel mill complex" and machine tool plants. In the end this improvement in relations among Iran, Russia and Russian satellite nations served for a time at least to prevent a total impasse in consortium-Iranian relations.

While the Shah continued to demand more production from the consortium in order to finance his development programs, in the late 1950s he also developed a stronger interest in the security of Iran. He felt keenly that the United States had not supported its allies during the Suez Crisis, nor had the West supported Pakistan in its conflict with India. Too, he saw the actions of the relatively unchecked United Arab Republic [UAR]* leader, Gamal Abdul Nasser, as a threat to peace in the oil-rich Persian Gulf area. In several interviews and addresses in 1966, the Shah made the point that despite treaties, Iran must look after its own interests. To properly safeguard his country's independence, arms were needed. So the Shah's interest in building up Iran's defenses increased.[47]

In Iran itself, pressures on the oil companies mounted as the Shah (among others) talked pointedly about the power of the State to control its resources—loading speech after speech with anti-foreign rhetoric. Even had the consortium members agreed to increase the offtake to the level the Shah desired, the problem would have been only partially solved; for oil markets fluctuated from spongy to soft, and the additional oil could only damage that market.

X X X

At Jersey during the 1950s and 1960s, in keeping with traditional practice, a significant personnel movement continued as Page, Turner and the Compensation and Executive Development [COED] Committee sought diligently to choose persons who could fit into their Middle East regional organization, and then, to provide them with the background of experience necessary. It is conceivable that the committe chose too well, for other divisions of the company often claimed the outstanding talent even before it was seasoned to the demands of Page or Turner. In all, many persons served, and continued to serve the company exceptionally well, but usually not with Middle East. A few deserve particular attention for their contributions:

Paul J. Anderson, in 1948-1950, head of Esso Standard (Near East), Inc., a

---

*Egypt and Syria united until 1961; then Egypt alone until 1971.

portion of that time doubled as Jersey Middle East representative in London. In New York in 1950, he worked as chairman of the Jersey Middle East Committee. Returning to London in 1952, he joined Iraq Petroleum Company, Limited as director. Then in 1957, he became Middle East representative, from which position he was elected to the Jersey Board of Directors. His influence, however, diminished when in 1962, he announced his early retirement, to be effective in 1965.

Charles M. Boyer moved into the New York office of Middle East advisor in 1957. In 1960, he received a special assignment to work with Page and Lloyd W. "Shorty" Elliott on the court-ordered Standard-Vaccum Oil Company [Stanvac] breakup. He returned to Middle East in London in 1961, first as assistant Middle East representative and then as the Middle East representative in London upon the retirement of W.L. Butte. Page and Turner were both impressed with Boyer so they recommended that he become Turner's sucessor, when the latter had to retire in 1967. In 1965, Boyer traded jobs with George Piercy and served in the New York office as deputy Middle East coordinator.

Jack G. Clarke, who served in Creole Petroleum Corporation, transferred in 1959 to the Jersey Law Department, where he became the Middle East legal contact attorney. In 1965, he joined the London Middle East and a year later, he was promoted to Middle East representative. He returned to New York as assistant general counsel in 1968. There he remained in the Law Department, still assisting in Middle East negotiations until 1972, when he became general counsel for Exxon. He next joined Esso Europe as executive vice president and director, and in 1975, he was elected senior vice president and director of the Exxon Corporation.

Piercy, who became manager of the Coordination and Petroleum Economics Department of Jersey in 1960, and who served as executive assistant to CEO Rathbone for four years, went to Imperial Oil Limited [Imperial] in 1964, as a director. In 1965, he returned to New York as deputy Middle East regional coordinator, and later that year exchanged positions with Boyer. In May, 1966, he was elected to the Jersey Board of Directors, with the Middle East contact assignment. That post, he held for fourteen years.

Page in 1960, was elected vice president and as such, he continued to be the primary Jersey representative for that region; however, when he became executive vice president in 1966, although he maintained his interest in (and perhaps influence over) Jersey's Middle East strategy, he did not carry so heavy a burden. Page retired in 1970.

When Turner retired, Charles J. Hedlund succeeded him as Middle East regional coordinator. Hedlund, who had considerable line experience, outranked Boyer in the company structure. But at the time he could claim only a limited Middle Eastern background.[48]

Meanwhile in Jersey headquarters, on March 1, 1965, the Board elected President Michael L. Haider chairman and CEO, to replace Rathbone, who at that time ended his career of forty-four years with the company. J Kenneth Jamieson, who had been president of the Humble Oil & Refining Company [Humble] and who had moved to the holding company as executive vice president in September, 1964, now assumed the position vacated by Haider. Although Haider and Jamieson had gained significant experience with Imperial and International Petroleum Company, Limited, they both had only limited familiarity with Exxon's European affiliates and those in the Middle East.

As for Page and Turner, neither had a large Middle East staff—Turner once said, "I never had more than about 8 men, including my London group"—and some of the most talented individuals who worked with them were sent off to other departments where they became top executives. So in the critical years, 1970-1975, Middle Eastern experts at Jersey were scarce. Whether such experts could have performed as well, or better in those years will ever remain an unanswered question. The writer judges that, despite excellent planning and the equally sound selection of personnel, nothing would have prevented most of the changes that occurred in producing country-company relations in the Middle East; however, it is conceivable that some events could have been postponed to a later time.

X X X

As things happened the consortium partners agreed to meet during late April, 1966, in an effort to reach agreement on the percentage increase in their Iranian crude oil production. The Shah also indicated that he wanted to meet with the "key oil company officials" in early July, which for the American companies presented many problems including some with their own government.

Following their usual practice, Jersey and the other U.S. companies sought advice and clearance from the Departments of State and Justice before they met with the Shah. They asked whether their attendance at this meeting would in any way be construed as a violation of antitrust statutes. It is also possible that the increasing concern in the foreign offices of the U.S. (and Great Britain) over the Iranian Consortium oil lift question provided an additional reason for the oil companies to seek such advice. After all, the five American majors had become part of the consortium at the insistence of the U.S. government; and, keeping Iran in the Cold War alliance against Russia had been vitally important to the U.S. Yet the companies could not go so far as to meet every Iranian government request without jeopardizing their oil supplies elsewhere. If Iran received the desired 17 percent crude offtake increase, problems would follow. As Jamieson once phrased it, "you had the problem in the Middle East of walking a tightrope between Iran and Saudi Arabia. Each ... of them wanted, ... to have 100 percent of the growth of oil demand."[49]

The State Department referred the question of participating in these conferences to Justice, where Assistant Attorney General Donald F. Turner asked Wilbur L. Fugate, then serving as chief of the Foreign Commerce Section, for an opinion. In reply, Fugate indicated that the companies were then in a bind:

> At present there is an over abundance of oil which is a problem to producing countries. Each country, of course, wishes to continue a high rate of production at a high price. They have insisted that companies with concessions pay them royalties (in the form of taxes), on the basis of 'posted' prices although the companies assert that they actually must sell at a less price. As to production, the companies having concessions in several countries are in the position of having each insisting that there be no cut-backs.

After pointing out that Attorney General Brownell had approved the consortium agreements in a letter to President Dwight Eisenhower, dated January 1, 1954, Fugate said, "I see no new antitrust aspects of the meeting with the Shah." He viewed the meeting as within the outlines of the original agreement, which had already been approved. In his opinion, "it is rather late now to question it." Then he concluded, "it is probable, however, that the companies belatedly have seized upon the antitrust laws as a possible defense against the insistence of the Iranian Government on actions they dislike."[50]

This approval by the Justice Department cleared the way for executives of Jersey, and its U.S. partners in the consortium, to meet with the Shah in July, but in Iran, not in Europe. The Shah had specified that he wanted to meet with the top officers; however, "This was not a very popular idea with most of them and many, in any case, had prior engagements." From the major oil companies, only one CEO, John Louden of Shell, attended. Howard Page, who had been elected an executive vice president of Jersey in May, 1966, represented that company. Others of lesser or equivalent rank to Page represented their companies. Page described the meeting in this way: "The purpose of the meeting was not to negotiate. The Shah wanted to make clear to them what his demands would be and to put the fear of God into them in case they considered not accepting [them]."

Page had travelled to Lebanon to attend an American University of Beirut trustee's meeting. He left Beirut on July 4, and arrived in Tehran at midnight. On July 6, the consortium team lunched with Premier Abbas Hoveida and later that day, they saw other ranking Iranian officials, including Dr. Manuchehr Eghbal, head of NIOC. On July 7, the Shah received them and made his position clear. He indicated that unless the consortium met Iran's demands for an increased offtake, it would turn to the East [Soviet Union and its satellite states] for imports. Iran, he said, found the cost of military equipment and farm machinery from the West too high. Reports indicated that the Shah made these statements calmly and that they could not be construed as threats.[51] (An estimated value of these Iranian imports amounted to $700 million annually.) No final decision resulted from this conference with the Shah. Instead the parties agreed to have their consortium representatives meet again in September, for some decision on the issues.

The Near Eastern Affairs Division of State became busy. It arranged a meeting with the British Foreign Office, to take place two months later. Immediately, the preparation of briefing papers for these talks began. In late October, senior oil company representatives called on Ambassador Raymond A. Hare, to discuss these Iranian demands and to ascertain "possible motives for the pressure and demands put upon the Consortium by the Iranian Government." Hare called in five other State Department officials who were familiar with the problem. Jersey Director Piercy initiated the conversation by reviewing for Hare the rabid Iranian press campaign that had started in March, and that contained numerous errors concerning the consortium's record in meeting contractual obligations. Piercy pointed out that Iran sought to regain its historical position as the leading oil producer in the region, basing their demand on two specific factors; population and Iran's strategic position. Then Piercy added yet another reason which, he believed undergirded the Iranian arguments. He pointed out that the Shah had indicated in March and "again during his meeting with the member companies chairmen in July" that if Iran did not get the desired increase, it would turn to Russia for imports. Piercy finally told State Department representatives that the 17 percent figure "was an impossibility" and could not be met.

Other oil company spokesmen supported the Piercy views. As the discussions continued, it became apparent that Iran's earlier commitment to economic development had provided the primary cause for the pressure on the consortium, but at the time there were additional factors—arms and the desire to establish NIOC as a marketer independent of the consortium. Several oil men expressed their conviction that the Shah had become more concerned with "the financial requirements on the military side."

The oil men expressed interest in what Kermit Roosevelt of the Central Intelligence Agency [CIA], a close friend of the Shah, might have learned when the two met during the previous week. They also indicated that they wanted the State Department to do its best through channels to dissuade the Shah from "rash action," and, to "discreetly probe Iranian intentions." Piercy, in summing up his conversation, questioned whether the crisis originated with the Shah; and if so, whether he might be misinformed.[52]

State Department memoranda and telegrams concerned with the forthcoming session of the consortium and Iranian government officials continued to explore potential problems. The State Department found the Page-Moses position on yearly percentage increases to be correct; nothing in the agreement specified an increased rate of production. The agreement simply stated "that the Consortium policy shall be to ensure that its total production reasonably reflects 'the trend of supply and demand for Middle East crude oil.' "[53]

Following this meeting, the State Department had a full briefing paper prepared for talks with the British Foreign Office. State felt that it could only advise the American companies, and that negotiations between Iran and the consortium should settle the issues in controversy. Such an end might include revision of the 1954 agreement. Walter Levy, the noted oil consultant, told Under Secretary of State Eugene V. Rostow that the restricted secret arrangement in the participants agreement "would be political dynamite in the hands of the Iranians." State took the position that it could not believe the arrangement was not already known, but it did not press the point at that time.[54]

The consortium members began their meeting with NIOC officials in September, 1966, in London. There, on October 24, Dr. Eghbal, NIOC chairman, demanded a 17 percent increase in production. After deliberating, the consortium companies asked for a month in which to consider the request. The Iranians indicated that they would return for an answer on November 21.[55]

While all these consortium negotiations occupied the attention of many oil companies and of some Iranians, Chairman Eghbal concluded an exploration agreement between NIOC and the French company, Enterprise de Recherches et d'Activités Petrolières [ERAP]. In this new, perhaps revolutionary agreement, ERAP assumed all the exploration costs for certain areas in the Gulf. If and when oil was discovered, it agreed to buy back a large share [35 to 45 percent]. Both Eghbal and Prime Minister Hoveida flaunted this contract in loud speeches. Yet Page, Piercy, and Turner were sure that the contract was bad for NIOC.[56]

Meanwhile, the consortium oil company CEOs held a meeting in New York on November 14. Jersey's Haider attended, and the next day he called in Page, Piercy, and Turner. Haider related that the chiefs "were really scared." They had requested that Page head a team of three [J.M. Pattinson of BP and Socal's Parkhurst] to go to Iran later that month to meet with the Shah and NIOC officials. Haider had agreed to ask Page to do that, and Page inquired "what their fall back position was." Haider replied that they had none. Page agreed to go, but (as was his wont prior to such negotiations) "only if the chiefs agreed to some reasonable terms of reference." With Piercy and Turner, he then drafted proposals for the CEOs' meeting the next day. Page attended this meeting, where he "obtained reasonable terms of reference including my [his] idea regarding the Iron Curtain countries." [At that time, the U.S. government restricted the oil companies' marketing with nations in the Soviet system.] He saw Iranian sales to them as a way to placate the NIOC demand for consortium oil at cost—which they would otherwise have sold on world markets.

Page left for London on November 19. There he met with his two associates from BP and Socal and they flew to Tehran on November 21. When newspaper reporters in Iran sought to determine why these men had come to Tehran, one of the negotiators termed it, a "purely social" visit, while Page replied that he had "nothing" in his briefcase. Later a historian observed, "he evidently had something under his hat." After the expected round of formal calls on dignitaries and important officials, on November 23, they met the Iranian team and the Shah. Within three days, the two groups reached agreement, and on the 27th, they again met with the Shah. Page returned to London, where he briefed the executives of the consortium participants' holding company. One day later, he arrived in New York, where at 3 P.M., he met with Jersey's executives and Middle East specialists, and at 6 P.M. in the Mobil Board room, he explained the agreement to representatives of all the American oil companies in the consortium, never once complaining about how tired he was.

The American press did not cover these negotiations, but by December 12, the *New York Times* reported that an arrangement had been made, and on the 15th, the parties issued a statement. They had reached what amounted to a new agreement, one designed to placate Iran without dislocating world markets. Under the 1954 Consortium Agreements, the member companies had not been required to release any of the concession area before 1979, and then only 20 percent of it. Now, they agreed to release 25 percent of the area outside of the producing fields, after reaching mutual agreement on which part of the consortium concessions to give up. [Page stated that the consortium did not want to explore in the forseeable future the area given up.] They further agreed that the consortium would sell NIOC up to 20 million tons [140 million barrels] of crude oil during the next five years, with the specified proviso that this be disposed of in Iron Curtain countries. As Iran sold oil there, production would increase up to the specified figure. This, Iran could do because the recent changes in Soviet policy [previously discussed] had made trade with such countries as Rumania possible. This clause served to provide Iran with an oil outlet. Finally, the parties agreed to an increase in Iranian production of approximately 13-14 percent.[57]

It appeared that a satisfactory solution had been reached. Iran came away from the bargaining table with the satisfaction of knowing that their oil income would be increased to provide new funds for their economy. Meanwhile, with the additional acreage available, NIOC could push its joint venture program; while with the contract oil, plus that from jointly owned companies, Iran could now go aggressively after markets. This, it began to do. Under a 1965 agreement with Rumania, Iran had contracted to export $100 million worth of oil over a ten-year period. Now, under the revised consortium agreement, Iran could easily supply this oil, plus those amounts needed to fulfill subsequent trade agreements.

Nevertheless, the agreement pleased neither the Iranian officials nor the Shah. Secretary of State Rusk credited the Shah with many fine qualities: a zest for hard work, a capacity to absorb details, a dedication to his people and his country, while at the same time, being interested in the world and its problems. He was a well informed person. Yet Rusk pointed out that inflation constantly increased the Shah's projected development costs, and that this accounted for some of the incessant demands he placed on the consortium to increase production and revenues. But inflationary pressures were not the only reason for his pressing for even higher production. The Shah, Rusk said, really wanted to make Iran a great military power—an expensive dream and one that Rusk doubted the Shah could have achieved had there been no inflation. In Rusk's opinion, the Shah went too far, and many of his reforms, although much applauded in the West, actually promoted cabals among reactionary clericals and other groups, and helped produce the revolution against him.

Another part of the problem that the consortium encountered in dealing with the Shah derived from his personality, as well as the subservience accorded him by Iranians. He firmly believed in the future role of Iran as the key state of the Middle East; he desired to restore his nation to the glory and glamour it had twenty-five centuries earlier. Though only the second member of his family to rule Iran, the Shah felt the traditional grandeur of previous rulers on his shoulders.

One person who met in negotiating sessions with him, summed up the Shah's problem in this way:

> His ego just completely ran away with itself, compounded by the fact that he built up his royal image all the time. ... The whole system was so raw. I saw poor old Hoveida ... prime minister who was a bright fellow, smart, able ... when he came into the presence of the Shah, he'd kneel down and kiss his hand.
>
> Well, you know what kind of advice you're going to get from somebody that kisses your hand.[58]

Whether the Shah became enchanted with the military trappings he sought or whether he became disenchanted with the stand of the consortium on production is not really known; his attitude, however, suggested that he remained unconvinced that the consortium could not lift more Iranian oil. The agreement of December, 1966 took effect, but the basic issues remained unresolved. Events of 1967 temporarily eased the Iranian production oil lift problem. But this partial increase during and after the Six Day Arab-Israeli War, instead of being beneficial, moved Iran to become more dependent on oil revenue from the consortium.

X X X

On February 1, 1958, Egypt and Syria merged into the UAR while Yemen joined the new state a few days later. Some other Arab nations signed mutual defense pacts with UAR that year, thereby enhancing its prestige. This new state provided a broader stage for Egypt's Premier Nasser, who hoped to spread his revolution; so he immediately began to export elements of it, including armed troops.

Meanwhile, after the Suez Crisis of 1956-1957, Arab states appeared to divide into factions. Although Secretary of State Dulles sought to involve them in the Cold War anti-Communist defensive alliances of the United States, some of them rejected this attempt on the grounds that their greatest fear was not the Soviets but colonialism. To these latter states, the U.S.

anti-Communist activity supplied a convenient cloak for further imperialistic involvement in the Middle East; while the U.S.'s continued support of Israel provided adequate proof of these colonial designs. As has been stated earlier, this reaction confounded Dulles (and several of his successors).

In November, 1956, during the first Suez Crisis, a United Nations Expeditionary Force [UNEF] had been stationed in Egyptian territory at Sharm El Sheikh overlooking the Straits of Tiran, in Sinai, and in the Gaza Strip. Further, in 1957, President Eisenhower assured the Israelis that:

> they [the U.S.] considered the Gulf of Aqaba to be an international waterway legally open to the ships of all nations; and that they were prepared to exercise the American right of free navigation through the Straits of Tiran and to join with other nations to secure general recognition of these rights.[59]

Israel yielded to pressure from the United States and the United Nations [U.N.], but it warned that if the Egyptians closed the Gulf of Aqaba to Israeli ships, Israel would act according to its own interests. President Eisenhower believed that the Israelis had a right of access to the sea.

For ten troubled years, the UNEF did serve its purpose; no large-scale military action between Israel and its Arab neighbors took place. Then, in 1967, the situation worsened between Egypt, armed by Russia, and Israel, first armed by France and later, the United States. Across those years, raids into Israel by Palestinian guerillas, encouraged by Syria and Egypt, brought swift and savage retaliation. Relations between the two countries grew heated. So did the danger of war. As U.S. President Lyndon B. Johnson put it: "The danger implicit in every border incident in the Middle East was not war between Israelis and Arabs but an ultimate confrontation between the Soviet Union and the United States and its NATO allies." Egypt's relations with the U.S. deteriorated. In March, 1967, Soviet Foreign Minister Andrei Gromyko visited Cairo; subsequently, Russia warned Israel not to engage in war with Arab states. Syria persuaded Russia of an impending Israeli attack, despite the lack of conclusive evidence. This threat exacerbated the crisis situation fueled by the flow of radical Arab talk.

In May, 1967, Nasser asked the UNEF to leave Egyptian soil. Some unverified reports of Israeli troop movements made headlines in the press and the movements of Egyptian troops were shown on television. Nasser's prestige among the Arab nations declined because he failed to punish Israel for its attacks on the Palestinian guerrilla camps. This pressure forced him to take a foolhardy stance. On May 23, the Israelis learned that he planned to blockade the Straits of Tiran. President Johnson announced that the United States still considered the Gulf of Aqaba an international waterway, but Israel demanded that the United States, Great Britain and France honor their 1957 pledge regarding the neutrality of the Straits and, at the same time, honor the U.N. resolution of March, 1957, expressing the same view. The French ignored the request. Great Britain and the U.S. sought strong support from the U.N. Failing in this, they proposed to open the Straits with a naval squadron composed of vessels of several nations—the Red Sea regatta. But Nasser's action left Israel little choice. It mobilized; then Egypt mobilized amidst unanimous pledges of support from all (but one) of the Arab nations.[60]

Russia, from which country had come part of the intelligence about Israel that inflamed Nasser, now feared that it had set the stage for the final confrontation. So the Russians changed the tone of their diplomatic messages.

Nasser, now a victim of rash moves and flammatory rhetoric, also faltered. But the Israelis, relatively certain that the U.S. would not directly intervene, struck Egyptian air bases on June 5, 1967. In six days, they defeated the Egyptians, Syrians and Jordanians, plus assorted troop units from other Arab nations. And during those days, the unanimously passed U.N. Security Council ceasefire resolution was accepted by first Egypt, then Israel. Portending the future, within two weeks, Soviet arms shipments made their way to Egypt.

Nasser closed the Suez Canal on the second day of the fighting, and later ordered it blocked. [Suez remained closed until 1974.] On June 10, the United States declared an oil emergency because of the crisis. The Departments of Defense and State and the Office of Emergency Planning all agreed that the major problem lay in proper routing of the tanker fleets. The Interior Department moved to activate the Foreign Petroleum Supply Committee [FPSC], which had performed well in the crises created by the Korean War, the Iranian nationalization of the BP concession, and the first Suez Crisis. Twenty-one oil companies responded to Interior's call, an appeal which went out on the fifth day of the war, June 10, 1967. However, because of the actions of the Arab states supporting Egypt, FPSC continued for nearly three months beyond the duration of that struggle. Jersey's representative on the committee was George Piercy. The FPSC reported a plan of action on June 20, and after making the rounds of departments and bureaus, this proposal received final approval on June 30.

The problem first seemed—in fact, could have been—far greater than those earlier FPSCs had confronted; in addition to supplies for the Western allies, the United States had become bogged down in the Vietnam War. While the closing of Suez did not seriously affect the supply problem there, the actions of the Arab producing nations after the war began certainly did.

During the first week of June, 1967, the representatives of eleven Arab states met in Baghdad, where they discussed the critical Arab-Israeli conflict. They decided to use oil as a political weapon against the backers of any nation that directly or indirectly aggressed against any Arab country. Saudi Arabia, Kuwait and Iraq then stopped exporting oil. Other states followed their lead; so the FPSC activated an Emergency Petroleum Supply Committee [EPSC], on which Piercy also served.

Oil for the West now had to be routed around Africa; however, the first Suez Crisis, with its temporary canal closing, had led the oil companies to order larger tankers, termed "Supertankers." Since by now oil represented Europe's principal energy source, these supertankers proved invaluable [some of these had capacities of 200,000 deadweight tons, as opposed to the standard 45,000 ton vessels of 1956]. Using these ships, the major international oil companies effectively supplied their European customers. At the same time, however, tales of millions of dollars made on a single voyage kept the media and the public supplied with sensational gossip.

The measures of EPSC and the actions of the governments of the United States and Great Britain destroyed the effectiveness of the Arab embargo. The Saudi Oil Minister, Sheik Ahmed Zaki Yamani, sought to persuade representatives of other Arab states that their economies were being injured by the embargo. Failing in this effort, the Saudis permitted Aramco to resume exports, but only after the company assured them in writing that they would not ship oil to Great Britain and the United States—in their view, Israel's principal supporters. The other nations followed this example, except for Iraq and Libya. Late in July, Iraq authorized IPC to ship oil anywhere except those two named countries plus West Germany, Rhodesia and South Africa.

Obviously Iran felt bound only loosely to the Arab members of OPEC for it increased production, and loaded tankers for world markets. Too, Venezuela also increased production. With two of the principal oil consuming countries embargoed, the real losers, as Yamani had predicted, were the Arab producing states—their revenues dwindled—since less oil moved equalled less revenue.

On August 29, 1967, the Arab leaders met at Khartoum. There, they rejected a proposal by Iraq to ban oil shipments; instead, they voted to use oil as a positive weapon to strengthen the economies of those Arab states most affected by "aggression." The oil ministers realized that the continuation of the embargo against the United States and other major consuming countries injured Arab national economies as drastically as it did those of the pro-scribed states. Within a few weeks, the Arab states lifted the embargo. Several months later (January, 1968), the leading Arab exporting nations created the Organization of Arab Petroleum Exporting Countries [OAPEC].

Zuhayr M. Mikdashi, in his book on *The Community of Oil Exporting Countries*, states that the fundamental purpose of OAPEC was the integra-tion of the members' national economies into a single economy—a long-range view, if correct. At the time, however, Saudi Oil Minister Yamani viewed the purpose of OAPEC as an effort "to maximize the benefits derived by its members from their natural resources." He wanted oil to be used as a weapon to protect the economies and independence of the producing na-tions. It should be noted that Iraq did not join at that time.[61]

The Six Day War, in reality, gave Iran and Venezuela three months' pro-duction governed by no previous agreement. Afterwards, the Shah rarely failed to remind U.S. negotiators that he did not join the boycott and, on occasion, to use this evidence of goodwill as a lever against the consortium in his effort to obtain more production—the Shah used the phrase the "grave political risks" he had taken.

As for the Arab nations, the Six Day War caused them to become exceed-ingly angry. The defeat inflamed the masses against most foreigners and against Americans in particular. It also promoted the rise of Arab radical-ism. The basic lesson remained, as Jordanian Crown Prince El Hassan Bin Talal has written, "After the 1967 War, other Arab governments learned— and what a costly lesson—what we had known for almost two decades: Israel was to be an enduring reality of the Middle East. ...."[62]

In late June, 1967, Onnie P. Lattu, director of the Department of the Interior Oil and Gas Division, wrote Jersey President Jamieson to express appreciation for the assistance of Jersey specialists in the "first few critical days" when they sought to respond to "the emergency conditions created by events in the Middle East." Lattu had found them "extremely helpful in providing information not otherwise available to us." Three years later, that division of Interior completed a report on the 1967 Petroleum Emergency and sent a copy to Jersey for its suggestions. R.H. Herman sent it to Execu-tive Vice President Page with several wry comments: "The report ... con-tains nothing that is new to Jersey or significant for the future. ... In fact, the whole report implies that the U.S. Emergency Petroleum Supply Com-mittee [EPSC] was a significant factor whereas in actual fact, it was the international oil companies that handled the situation." Herman pointed out, however, that had the interruption to crude shipping from the Persian

Gulf continued, then action by that committee and other government agencies would "have been helpful." He suggested that Jersey continue to encourage the use of such committees "to improve the governmental mechanism to obtain faster response in the case of need." The report was also sent to Jamieson noting that the key points made were that uncertainties regarding Suez and the various pipelines from the Persian Gulf states made it mandatory to maintain "a large, flexible tanker fleet" and to diversify sources. Thus, Interior reported on the 1967 crisis which proved of such short duration that the FPSC never really geared for action.

The consortium problems with Iran continued to increase during the last part of 1967. An interesting addition to NIOC's usual arguments was that the government of Iran had made an error in calculating oil income for the "Fourth Development Plan period but [the] Shah could not admit error"; hence, the companies should help Iran overcome the problem created by that mistake. While negotiating, NIOC executives met with consortium representatives in London and Tehran. The oil companies held fast to the position that they could not reward Iran for cooperating with the U.S. and Great Britain during the recent crisis, nor could they increase production on a "need" [population] basis.

In mid-November, 1967, Gulf, Mobil and Jersey representatives met with an Assistant Secretary of State (and Kermit Roosevelt) to review the "mounting pressures" on the consortium. It is obvious from the report of this meeting that the State Department had received from the British Foreign Office information indicating that the American major companies were "being sticky on increasing offtake in contrast to British companies." This, the U.S. oil companies disputed. One question surfaced regarding the internal oil lifting arrangements of the consortium. The State spokesmen reiterated their view that the Iranians knew about this, but they agreed not to discuss it with Iran. The Shah had demanded a 20 percent increase in offtake, but the company spokesmen indicated this would be impossible to achieve; they stated that when they gave Iran the figures on estimated offlifting for 1968, and some indication of that for 1969, they did not believe the "Iranians will be happy." Finally, the oil company representatives agreed to send to the State Department a précis of their views for later use by Under Secretary of State For Political Affairs, W. Averell Harriman.[63]

The members of the U.S. Embassy staff in Tehran most concerned with the negotiations discussed the seeming deadlock over oil lifting there, and in early December, 1967, they sent their views to the Near Eastern Affairs Iranian Office in the State Department. They believed that "the chances of long term stability *in Iran are as good and possibly better than those in other* *producing countries and states of the Persian Gulf"* and that the political risks there were *"perhaps less than elsewhere in the Middle East."* This view, they arrived at despite the uncertainties that might arise when the Shah "passes from the scene." They also thought that he would not take less than a 12 percent per year increase in oil production over the next five years. Clarke of Jersey, along with a Texaco executive, told the staff that the consortium would give Iran increases as great as the Middle East average over the next several years; this, they estimated at 7-8 percent. The embassy group doubted that the Shah would accept such a low figure, and further stated that if Aramco's increase rose above such an average, the consortium members "will be faced with trouble in Iran." A point the embassy staff thought worthy of mention was the increase in Iranian production occasioned by the IPC* and Arab-Israeli crises. New production estimates, they felt, should consider this. Finally, they deplored the discussions of the overlift arrangements in the trade papers and Iranian press. The desire of the consortium members to protect their competitive positions, in their opinion, "in a broader context may have an adverse effect on US/Iranian relations."

While the embassy staff in Iran appraised the situation there, the U.S. Embassy in London forwarded to Secretary of State Dean Rusk a summary of the talks conducted in Tehran, among a consortium team, NIOC officials, Hoveida, and the Shah, which occurred on November 29 and 30, 1967.

Clarke again had represented Jersey. Hoveida, the report stated, "made no threats to the Consortium," although on one occasion he came very near. The principal argument centered around the secret participants' agreement and whether the members would benefit from changing it. Dr. Reza Fallah, an Iranian much respected by Jersey officials, had incorrectly concluded that all but one member would benefit from a change. Fallah told the consortium team that at the OPEC conference in Vienna (from which he had just returned) he had worked for moderation—a gambit designed to convince them of Iran's good motives. Since an impasse existed, Iran agreed to continue talks with the consortium team on December 7 in Tehran.

Shortly thereafter, a consortium negotiating team visited Iran, where the arguments of both sides continued. This time, however, the Shah brought up for discussion the actions of the consortium regarding possible Iranian investments, especially in petrochemicals. A State Department representative went to Iran and reported that after talking to Iranian government and oil company officials, he "was gratified that [the] consortium was prepared to

---

*Iraq and IPC had engaged in controversy over oil production for several years. This matter is covered later in this section.

make any increase over this year's ... increase." An Iranian spokesman told him they awaited the consortium answer.[64]

This report led Secretary of State Rusk to send a telegram to the American Embassy in London, in which he stated that if the discussions dragged on much longer, they might be aired in public, to the disadvantage of both parties. Rusk went on to say that the Iranian ambassador believed that spokesmen for his government had not fully grasped the oil company argument: that increasing the quantity of oil lifted "might not result in more oil revenues which are the objective on which GOI [Government of Iran] should concentrate." The ambassador realized that more oil production would depress prices, thereby decreasing government revenues. Rusk also learned that only Jersey, of the five American majors in the consortium, supported the move of Iricon and Shell to change the secret internal overlift agreement. The reason advanced by British Petroleum Company and the other four American majors for opposing such a change was that they had access to Kuwait or Aramco oil "where basic costs are lower."* A meeting of high level consortium company executives with Iranian officials had been scheduled for December 14, 1967 in London, and the Saudi Arabian government had asked that they be permitted to attend and, at the same time, conduct negotiations with Aramco. The reactions of the consortium spokesmen proved unenthusiastic because of the difficulty in conducting joint negotiations. Rusk's informant believed they would agree to the meetings on a simultaneous, rather than joint, basis.[65]

The day before the London meeting opened, Jersey and Mobil representatives again met with State Department officials. Charles Boyer, Deputy Coordinator For the Middle East, attended from Jersey. The group reviewed consortium production plans for 1968-1969. Again, State spokesmen pointed out that the overlift arrangement "has become an important issue with the Iranians" and they recommended replacing this arrangement with one more nearly like that Aramco had with its shareholders—"an irritant out of proportion to its true importance would be removed." It was pointed out that a basic change in overlift arrangements would enable the small U.S. companies in Iricon to get cheap oil without having put up more than a small fraction of the consortium's total investment. After further discussion, it was agreed that the world oil situation would not permit "any marked increases in offtake by Iricon or anyone else." The companies informed State that future meetings in Iran on December 18, and in London on January 3, 1968, would provide fresh opportunities to appraise Iranian demands.[66]

*So did Jersey.

On December 11, Under Secretary Rostow received a report from the U.S. Embassy in London that substantial agreement had been reached between the consortium and NIOC. Iran would receive an 11 percent increase in offtake in 1967, and numbers had been offered for 1968. Then, the two parties began to haggle over the adequacy of the level proposed for 1968. Again, the Iranians expressed the view that the overlift provisions discriminated against them, despite the fact that the consortium had liberalized this agreement. One State Department expert expressed his conviction that the Iranians knew about the Aramco overlift arrangement and wanted a system "at least as favorable" for themselves. Obviously, the dispute occurred because the consortium spokesman could not convince the Iranians of the validity of consortium positions—or the Iranians refused to accept them—a breakdown of communication and understanding. The consortium spokesmen contended that the revised overlift arrangement in Iran bettered that of Aramco, while the Iranians held the opposite to be true.

The consortium members continued to meet in London, attempting to effect agreement on their final offer. The Iranians simply waited.

A State Middle East expert recommended that State urge the consortium to discuss the overlift provision openly, since in his opinion Iran already knew its provisions; by doing this, "a troublesome issue would have been eliminated." The Iranians could then no longer claim discrimination. Next, State should advise the U.S. consortium members to be generous to the Iranians in the next consortium proposal; and if that failed, then the State Department should take "a more active role in the talks." Otherwise, this expert said, if Iran did not reach the same level of exports and same "absolute annual increase" as that accorded Saudi Arabia, "there will be serious problems in Iran, ... by 1973."

Indicative of the heat generated by these discussions was the Shah's statement to the consortium team that if no satisfactory agreement could be effected, he would take the matter to the Iranian people and that he feared no repetition of the Mossadegh affair. About this matter, he appeared determined and unyielding.[67]

The U.S. Embassy in London could not obtain any details of the December 14-15, 1967 meeting of the consortium. It did inform State that the parties were near agreement; however, the consortium wanted to see the Shah in January. Piercy had represented Jersey in London and had been selected to go to Tehran. Prior to those sessions, only Jersey had advocated early change in the overlift arrangements; now, it is apparent that other companies joined in contending that, irrespective of all other considerations, the time had come to demonstrate to the Shah that the consortium did not discriminate against Iran. Nevertheless, no solution was reached.[68]

By early 1968, Libyan production had changed the Iranian situation. Secretary Rusk notified the U.S. Embassy in Tehran that the five major U.S. international oil companies would probably not be able to meet the estimate of 9% consortium offtake that year. No increase above that could be made. By that time, a U.S. Embassy official had learned about the revised consortium overlift arrangement, a description of which he sent to James E. Akins of the Fuels and Energy Office at State. The writer called it "the Jersey formula" and termed it "almost diabolical," because of the early date on which participants had to determine their annual total offtake—no one could know in September, he said.

In his reply, Akins refuted that charge as incorrect. As he saw it, the Jersey plan would work well—certainly it could not be called "a gimmick to avoid permitting companies to overlift." Akins went on; "The Shah will learn of the move; if CFP [Company Française des Pétroles] or Shell or one of the Iricon companies doesn't tell him, someone will tell Walter Levy who will pass the word on." Akins did not understand why the American companies insisted that the revised offtake agreement be as complicated as the original; a simpler one "would have avoided the Iranian conclusion that there was a deliberate attempt to deceive them." Akins said State had briefed Levy extensively before he left to visit the Shah, and the department hoped he could persuade the ruler not "to move unilaterally against the Consortium." The Shah, said Akins, appeared obsessed with Aramco. But Akins did not believe that the Shah would reduce his demands.[69]

The Iranian developments had alarmed the State Department. Secretary Rusk called attention to this in a telegram: "Dept. increasingly disturbed by reports from Tehran, London and Paris that Aramco parents are being blamed for frustrating Shah's wishes for increased off-take and income. ...." These reports placed the other involved companies in the position of being willing to give in to the Shah's demands. Rusk noted, "we fear that high visibility of Aramco may lead Iranians to concentrate their irritation on its parent companies. But we also fear that French and British ... may be encouraging this tendency by trying to shift blame to Americans when there is none to shift."[70]

Finally, the negotiating parties reached an agreement; the consortium agreed to increase production by 10 percent and to advance Iran, against 1969 revenues, the difference between its take and the $1 billion the Iranians had demanded—needed for their Fourth National Development Plan.[71]

During the preceding ten years, the consortium had developed and expanded Kharg Island into the largest single crude oil export terminal, built numerous pipelines, booster stations, added facilities for recovering natural gas, and made many improvements in terminal facilities. While the 1967 Arab-Israeli War had made some improvements necessary, many were designed for future use. Further, under the agreement negotiated by the Page team in November, 1966, which provided NIOC with oil to sell to nations behind the Iron Curtain, much oil had been lifted. In the four years after that agreement, the consortium delivered to NIOC for such sales almost 48 million barrels.[72]

On March 28, 1968, Jersey's President Jamieson and Director Piercy attended a meeting with Under Secretary Rostow, State Department staff members, and representatives of several oil companies. Rostow explained State's position as one of non-interference in commercial matters; however, it now viewed the Iranian situation as one with political implications, "the oil companies must be concerned with the national interest as well as their own interests." He then reviewed the Middle Eastern political situation, which he described as "deteriorating." Russian influence had increased and the Arabs hoped to gain more power with Soviet assistance. In the view of the United States, according to Rostow, Iran had become the strongest state in the region and "it is very important to the U.S. in maintaining influence." The situation could result in an oil boycott. This, Iran did not desire; however, Arab pressures on it were great, and a boycott would enable Iran to strike back at the U.S. companies for not allowing it more production.

The State Department had discussed the situation with Iran. NIOC wanted a "5 year forecast of offtakes" from the companies, but State thought a two-year forecast would do. State had also involved the British government in these discussions.

The major oil company spokesmen indicated their view that preferential treatment for Iran would lead to reprisals from other producing states. The company spokesmen pointed out that the Shah's demands continued to be unreasonable. The French [CFP], they said, wanted oil at cost "without having a real market requirement for it." In the opinion of the major company representatives, oil-short companies sought to crack established markets with cheaper crude oil. Rostow quickly recognized the problem of achieving consensus in the consortium. He recommended that all companies agree to avoid open confrontation with the Shah.[73]

In July, 1968, at the sixteenth OPEC conference in Vienna, the members adopted a "Declaratory Statement of Petroleum Policy," based largely on a United Nations resolution approved in November, 1966. Essentially, the OPEC version of the declaration affirmed the right of permanent sovereignty of each country over the exploration and development of its natural resources. It established four objectives for members in the exercise of this sovereignty: first, all concession contracts should provide for payment of taxes on posted prices; second, the right of the government to recapture high earnings from outside companies or capital through additional taxes, after a reasonable time, should be guaranteed; third, concession acreage should be studied and relinquishment programs instituted, with the particular government a principal in determining the location of the acreage being given up; and finally, government participation in the ownership of existing and future concessions should be negotiated.

The host government would, under this resolution, fix the posted price which, the resolution recommended, should be correlated to price indices of goods traded internationally. It should also be noted that the OPEC policy declaration urged that operators be required to utilize the best conservation practices (and technology) in operating concessions.

While no immediate changes occurred as a result of the adoption of this resolution, events after 1970 accelerated the move toward bargaining for some of these goals—as the world petroleum market shifted from one operating for buyers to one for sellers. The 1968 document, which came after some years of study and many meetings by the OPEC Board of Governors and its permanent staff, in effect set forth an elaborate and quite detailed program for the ultimate control of all facets of each country's oil industry by its government. [By 1975, this program had been implemented.] Later in 1968, the OPEC Secretariat issued a policy statement on conservation practices, based largely on regulations in force in Canadian, Venezuelan and U.S. oil fields. When confirmed in November, 1968, at the seventeenth OPEC conference, it provided guidelines for the member states.[74]

It is apparent that during 1968, NIOC and the Iranian government began to press companies other than those in the consortium for increased offtake—Iran became interested in the total oil production in that country. [U.S. State Department officials had noted earlier that the Shah talked in these terms, while Eghbal and others referred only to consortium production.] In November, 1968, State received notice that Pan American would not invest heavily in an oil field in their concession until major negotiations had been concluded. ENI, too, had reservations about the oil share being demanded by Iran. Robert I. Dowell, the petroleum officer of the U.S. Embassy in Tehran, pointed out that continued "NIOC brinksmanship could possibly boomerang." In his view, Iran had the "'bread and butter' income of the Consortium and is as a result perhaps less anxious to settle with the smaller companies on anything but terms favorable to NIOC even if this means delaying production."[75] Obviously the OPEC resolution did serve to stimulate NIOC to press the smaller companies in a way similar to that used in negotiating with the consortium. For the time, at least, the results proved less satisfactory.

Early in 1969, State learned that some of the Iricon members questioned whether the major oil companies had been "straight" with the Shah. They advocated a different approach. Too, the Iricon representative believed that his Imperial Majesty deserved more consideration; Iricon knew that during his most recent negotiation with the consortium, the Shah had stated that he was negotiating "to obtain a special U.S. import quota for oil to be stockpiled in old salt mines, etc." Both State and Iricon doubted that these discussions would lead to a contract, due to the discovery of a major oil field in Alaska (1968).[76]

From 1960-1970, while Iran continued to be the "bell camel" of the OPEC nations, new forces arose to alter fundamentally the relationship between producing countries and the oil companies. The governments of those countries (except Venezuela) saw output levels as budgetary determinants, while the oil majors continued to see leapfrogging production demands as a threat to the economic viability of all production operations. Before 1970, the debate over production between these two parties grew even more intense. OPEC had started as an institution designed to achieve unity among the producing countries, as well as help them secure the maximum benefit from their oil resources. As Page pointed out, however, the companies did not formally negotiate with OPEC. After 1968, OPEC gradually increased its influence in petroleum negotiations. When the producing nations decided, in 1969 and 1970, to begin implementing the 1968 "Declaratory Statement of Petroleum Policy," Iran lost its top priority negotiating position to OPEC.

The 1968 OPEC "Declaratory Statement" had declared that both new and existing contracts (or concession agreements) should be open to revision "at predetermined intervals" subject to changing circumstances. Certainly circumstances changed after September 1, 1969, when Colonel Muammer al-Qadhdhafi [Qaddafi] and other Libyan Army officers overthrew their government. Qaddafi emerged as leader of the new governing body, the Revolutionary Command Council, and relations between the oil companies and the new government [discussed in more detail later in this section] quickly deteriorated. In 1970, Libya stimulated sharp changes in OPEC, and in other producing states as well. In a short time, OPEC had assumed the dominant role in oil pricing.

X   X   X

In May, 1970, the Jersey Management Committee reviewed the demands made by the Shah on the consortium. They agreed that the company interests would take no exception to acceding to the Shah's request for an increase in tax and royalty income, to provide Iran with $1,030 million for the year beginning April 1, 1970, provided the other participants agreed. Under the same condition, they saw no objection to a 45 day advance payment for the same year. This would "have the effect of paying for oil as it is lifted and . . . building 400,000 b/d spare capacity by the end of 1972." The committee recommended to C.J. Hedlund, president of Esso Middle East [created in 1968] that he carry these positions for discussion with other members of the consortium. Later in the year when news came of a Libyan settlement, the Management Committee concurred in a proposal from Esso Middle East to increase the consortium tax rate up to 55 percent and to raise the price of Iranian crude by nine cents per barrel. The consortium agreed with the Jersey position. The companies quickly offered the same terms to other producing states who, while they accepted them, were not pleased. By the time of the next OPEC meeting, Venezuela had seized the initiative by raising the income tax on oil companies to a flat 60 percent, which made the company-government profit split there 20 percent-80 percent respectively.

This began the price ratcheting between the Mediterranean state producers and the Persian Gulf states. Iraq and Saudi Arabia, with their pipelines to the Mediterranean, in effect benefitted at both ends of the ratchet. Jersey leaders had for some years anticipated such a chain reaction to price increases among the producing nations, and tried to avoid being placed in a position where the price of oil could be whipsawed among the producing states—in some circles this was referred to as leapfrogging. The Libyan revolutionary movement had achieved a larger price increase in a few months than had any negotiating session in the previous decade, largely because Libya threatened to—and the companies knew it could afford to— either reduce production or shut it down. This made the future of oil prices uncertain. And while those companies in the consortium negotiated with the Libyans, the Shah continued to make threats about expropriating their property.

During 1970, the representatives of the oil companies met in New York and London, discussing how next to react to Libyan demands and how to deal with them. They judged the Shah fully as ready to take over operations in Iran as Qaddafi had been in Libya.

It was interesting for this author to observe that Iranian pressures on the consortium to increase the volume of oil lifted began almost as soon as the companies and the Iranian government ratified the Page-Amini agreement in 1954. As the Shah's plans to develop Iran unfolded, oil income necessarily provided almost all of the funds. Too, the United States' and United Nations' loans and grants, which provided other monies for these programs, imposed many requirements for specific actions by the Iranian government. The U.S. aid programs sought to keep Iran among the anti-Communist nations; further, as planned they were intended to help it become a showcase democratic nation in the Middle East and the principal ally of the Western World there. To achieve this, the Shah needed large sums of money for the economic changes necessary to transform the Iranian economy from its relatively primitive state to one more modern—funds for education, industrial development and agricultural reform—and defense. The State and other departments also encouraged the Shah in this direction. Again, because Iran had the largest population of any Middle East nation, the Shah also advocated allocation of production according to that statistic.* Finally, as we have seen, his pro-Western stance in the first and second Arab-Israeli crises and on other occasions, in his opinion, merited special consideration from the consortium companies.

In the end, many of the Shah's reforms ran counter to the teachings of several of the more rigid Moslem sects. Hence, the more he changed Iran, the more opposition he encountered. Later events established that many of his improvements were superficial.

Despite this constant pressure from outside forces to change Iran, and despite the fact that the Shah remained confident that the consortium could, if it would, ease his problems in developing Iran simply by drilling more wells and increasing production, relations between the consortium and the Shah, his ministers, and NIOC never reached the breaking point. The consortium partners sought to reconcile Iranian demands with both the existing world oil markets and the conflicting interests of the other nations in which they held concessions. The Iranians, too, proved somewhat accommodating, though in fairness it must be said that throughout the troubled years of negotiations, Iran moved ever closer to full control of its natural resources.

---

*While Saudi Arabia has only 2 percent of the total OPEC population, across the years it has produced as much as 45 percent of total OPEC oil and in recent years, never less than 25 percent. Clearly, the Shah saw this as unfair. Of course, Saudi Arabia had 40 percent of the crude oil of OPEC and about three times as much in crude oil reserves as did Iran.

TABLE 1.

GROSS EXXON CRUDE LIFTINGS FROM
MIDDLE EAST ARRANGEMENTS
(Thousands of barrels daily)

	1975	1970	1965	1960	1955	1952
Crude Oil Offtake Under						
Equity Arrangements (1)						
Saudi Arabia	848	1,079	611	365	278	248
Iraq, Qatar and Abu Dhabi	89	258	201	225	89	50
Libya	104	631	520	–	–	–
Total Middle East and Libya	1,041	1,968	1,332	490	367	298
Crude Oil Offtake Under Special						
Arrangements (2)						
Saudi Arabia (Govt. Participation Crude Buyback) (i)	1,142	–	–	–	–	–
Iran (NIOC) Purchase) (ii)	241	245	128	67	21	–
Purchases under other special arrangements (iii)	344	663	220	164	100	53
Total Middle East	1,727	908	348	231	121	53

(1)   Includes 100% of Exxon and majority owned affiliate production and Exxon ownership percentage of the production of companies owned 50% or less.

(2)   Includes—

    (i)   Buyback of parts of the governments' share of production under terms of participation agreements effective January 1, 1973, and as amended thereafter.

    (ii)   Offtake from Iran under consortium arrangements initiated in 1954 and revised as set out in the Purchase and Sale Agreement of March 21 , 1973.

    (iii)   Major purchases from BP (Kuwait) and others under long term special arrangements.

N.B.   This table used by permission of Exxon Corporation. See *letter*, J.P. O'Halloran (Esso Middle East) to George H. Lewis (Public Affairs), September 1, 1982, (Copy), Files, C.B.H.S.

Chapter 19

Jersey and the Iraq Petroleum Company

IN 1928, STANDARD OIL COMPANY (NEW JERSEY) [Jersey] and Standard Oil Company of New York [Socony],[1] along with several other American companies,[2] created the Near East Development Corporation [NEDC], which purchased 23³/₄ percent of the stock in the Iraq concession of the Turkish Petroleum Company, Limited* [TPC].[3] With the approval of the U.S. Department of State [State], NEDC also accepted the terms of the Red Line Agreement—which bound all the partners, termed Groups, in an agreement not to seek separate oil concessions in most of the territory of the old Turkish empire, denoted at that time by a red line drawn on a map.[4]

By the advent of World War II, however, the U.S. Department of Justice [Justice], the Department of State, and other federal departments and agencies had begun to look askance at the restrictive Red Line arrangement. London counsel of both Jersey and Socony-Vacuum Oil Company, Inc. [Socony-Vacuum]** advised those companies that, because Compagnie Française des Pétroles [CFP] and Participations & Investments, Ltd. [P & I] operated in territory overrun by the Germans during the war, the 1928 Agreement had been terminated and the Red Line restrictions no longer applied to the Groups in TPC—now the Iraq Petroleum Company, Limited [IPC]. Jersey sought additional advice from other British lawyers, and all of them confirmed the opinion of company counsel.

Yet Jersey was suspicious of any new agreement that might raise problems with the U.S. State and Justice Departments because it imposed territorial restrictions on the Groups, or because it would violate the traditional "Open Door" foreign policy of the United States as, in fact, the original Red Line pact did. About the same time, Jersey and Socony-Vacuum learned that CFP and P & I would fight in court to protect the Red Line Agreement and to recover their share of the income from IPC oil lifted during the war.

*See Note 3 for details of company name changes.
**See Note 1 for details of company name changes.

637

Among other reasons advanced for Jersey's effort to end the 1928 Agreement were: the desire of Jersey and Socony-Vacuum to obtain a larger supply of oil from the Middle East—from Arabian American Oil Company [Aramco], in this case—for their own crude-hungry companies; and the pressures of the U.S. Departments of War and Navy* on both companies to secure additional sources of oil in the Middle East, for reasons of national security.

But whatever combination of motives may have led to the decision, the Red Line and other partner agreements were considered to be in effect until they were changed in 1948. On November 3 of that year, new agreements replaced them.

These 1948 Agreements resulted from discussions among the partners and they contained a number of unusual compromises, not seen in earlier agreements and contracts. As Sir Maurice Bridgeman, of British Petroleum Company, Ltd. [BP], remarked, "we have now succeeded in making the Agreement completely unintelligible to anybody."[5] Jersey's London barrister, Gerald Gardiner (later Lord Gardiner and Lord High Chancellor), reviewed the agreement made in 1948, and in a letter to Jersey characterized it as being "in such a state that it has, we feel become impossible for anyone to be reasonably certain what these arrangements are. These contractual relations," he went on to say, "are now contained in at least 69 documents consisting of:

22 Agreements
38 Letters (about half of which are, and half of which are not, legally
             scheduled to Agreements)
6 Cables
2 Undertakings
1 List of Assurance
___
69"

Gardiner also continued:

> From any lawyer's point of view the position has now got out of hand, with the result that it is becoming increasingly impossible to say with any confidence what rights or obligations may not *be* being *created*, or what burdens may not be being assumed of a nature which no one intended because of the increasing difficulty of construing 69 or more different documents together when they are nearly all inter-connected, and the wording of No. 3 may well affect the construction of No. 6, or vice versa.[6]

*The Department of Defense was created in 1947.

Clearly, the agreement of 1948, called the "48 Documents," contained ample justification for ongoing conflict.

Within the agreement there were specific points that, despite the most careful calculations, seemed bound to engender additional unintentional disputes. CFP, backed by the French government and P & I, privately owned by Calouste Sarkis Gulbenkian, sought ways to insure that crude production not wanted by the partners could be bought or overlifted, as well as devices to insure that the other Group members who had oil supplies elsewhere did not restrict production or fail to increase productive capacity.

Guaranteeing this required that the 48 Documents include provisions to establish productive capacity. Since the exploration costs for the oil discovered in Iraq, and the production of that oil, had been financed by all the partners according to the equity position of each, the agreement makers felt that some of this expense should be recoverable in case any member sold to another its own share of excess production. In an effort to achieve this flexibility, for the first time, a producing company agreement included a provision permitting partners in joint production to take crude oil out of proportion to their equity share. Too, since the production would be taken either from the proven reserve of all partners or from the share of the selling partner, the makers believed that some profit would be due on it. So they provided a limit on the amount of overlift that could be taken, and then arranged to cover the profit due the seller. Smith D. Turner, Jersey's Middle East regional coordinator, described this: "And they were told, 'Since you are taking some of our share of the oil reserves, we must have some of the profit.' There was just no basis for splitting such profit, so it was 'split down the middle,' and the 'half-way price' ... was born" giving the overlifter a share of the profit.[7]

Originally the basis for the price-profit arrangement was the "average market value of oil" but the partners quickly tied the new sharing plan to the posted price, with limiting effects later on Iraq production. After 1959, the market price of oil fell below the posted price, thereby reducing the share of the profit of the overlifter. Overlifted oil purchased from a partner by another partner—usually by CFP—also carried with it the added-on halfway price. Of course this higher price influenced sales and may have helped to keep production down; nevertheless, it should be noted that agreements (made after 1948) involving partners usually included provisions of this type—aimed at achieving overlift and profit-sharing results, but with varying mechanisms to bring them about.[8]

Numerous other factors served to complicate the operation of IPC between 1950 and 1975. Although the 1948 Agreements included no production schedule for the years 1948-1951 (since it was expected to be the maximum the pipelines could carry), its makers did plan one for the five years, 1952-1956. The first five years after the war were spent getting the company underway again, a formidable task. For example, only two twelve-inch pipelines had been laid before and during World War II. After the war, with only sixteen-inch pipe available in the sterling area,* IPC began construction of two new lines paralleling the earlier two—one to Tripoli in Lebanon and the other to Haifa in Palestine. Work was finished on the Tripoli line in 1950, while that to Haifa was never completed due to Arab-Israeli hostility. When Jersey helped to arrange an IPC purchase of thirty and thirty-two inch pipe in the United States, construction began on a line from the Kirkuk oil fields to Banias, Syria.[9] With the completion of this line in 1952, delivery of 500,000 barrels of oil per day [b/d] could be assured.[10]

New issues and questions began to arise between IPC and Iraq after 1948. Some of this discord and enmity derived from international politics; to the Iraqis, IPC, regardless of real ownership, remained largely a British company. Iraqi Nationalists and ethnocentrists alike detested the British for their influence over—some of them said control of—their government. Then, after 1947, the same group focused new hatred on the U.S. because of its pro-Israel policy. In fact, as was true of most Arab peoples, the Iraqis distrusted all foreigners.

Other problems between IPC and the government stemmed from the concession itself and the manner in which IPC managed it. For example, the question of sharing income from the concession surfaced while the government, at the same time, became concerned over the totality of the original (still unchanged) concession. While such problems increasingly demanded the attention of IPC directors, several of the younger Jersey executives involved in IPC also became concerned over internal management problems. David A. Shepard, who had been in London during most of World War II and who had taken an important part in the negotiations leading to the 1948 Agreements, conveyed his dissatisfaction with the entire IPC operations in 1948. W.L. Butte echoed Shepard's discontent, and pointed to the necessity of keeping Iraq satisfied with the operation. Howard W. Page, who replaced Shepard as IPC director in October, 1949, agreed with much that his colleagues had said about the IPC situation. A number of these Jersey representatives had served on the IPC board of directors or represented the company in NEDC, and they deplored the fact that virtually every consequential deci-

---

*The area of the group of countries which hold their international reserves in pounds sterling deposited in banks in the United Kingdom.

sion had to be made by the Groups in London. Monies appropriated for IPC improvements on occasion went unspent, or worse, were expended in a haphazard way rather than being coordinated. The Jersey men believed that IPC needed better management in the field—management with the freedom to make many of the decisions now so often routed to London. Coming from a company with a tradition of developing managers through a decentralized administrative structure, they sensed that IPC, despite its excellent prospects, actually functioned less effectively than it should. Soon they found other reasons to become disenchanted with the IPC operation.

Page initially thought the major cause of IPC's failure to become a better operating company could be traced to Gulbenkian, the owner of the 5 percent share. Early in 1951, Page wrote Jersey Vice President Orville Harden, stating exactly that, and he also suggested three courses the other Group members might pursue to buy out the Gulbenkian interests. In Page's view, such a move—or any step that removed that Group "entirely out of all operations. . . would be a great step forward." Page sent a copy of this letter along with a personal note to Middle East Advisor, Paul J. Anderson. Anderson in reply agreed with Page's assessment of IPC's operational problems, but he doubted whether even with the Gulbenkian interest removed, the friction among the Groups would disappear. He endorsed the Page idea of putting "responsible management in the field," along with more company officials resident in Baghdad; "I wonder if the objective could not be accomplished more easily by retaining the English corporate structure and putting 'mind and management' outside of England." Anderson questioned whether the British government would agree to permit IPC to leave that country for "Iraqi jurisdiction." He told Page there was no question that many changes would be required to achieve "your broader objective of a corporate reorganization."[11]

Interestingly enough, while Page singled out Gulbenkian as the major barrier to more effective IPC management, Gulbenkian himself had written Harden earlier to complain of "divergencies" among the IPC Groups. He agreed that management should be "on the spot and not in London," and indicated strongly that he believed IPC to be poorly managed.[12]

In January, 1950, shortly after he had visited Iraq, Page wrote representatives of the other Groups that Jersey had across the years invested "more money into this venture than they have taken out by a wide margin."[13] He added that "IPC today is not competitive in this sense. There are far more attractive investments, not only in the Middle East but elsewhere." Two years later, he expressed renewed doubts that IPC would ever be a good operating company, and that while all the Group representatives talked about economy, none of them *"will do much about it when it comes to a showdown."* Yet even then, he believed that IPC had all the components of "one of the best oil companies anywhere."[14]

Interestingly but not surprisingly, it is evident from the correspondence cited above (as well as from other records available) that Prime Minister Nuri-al-Said-Pasha and other prominent Iraqis knew much about the Groups' disputes, as well as about the company's administrative problems. Even Gulbenkian warned Harden that the Iraqis most assuredly "took note of these divergencies, and bided their time."[15] Part of the IPC problem, as the Jersey representatives indicated from time-to-time, was this constant internal friction, which Iraq officials always sought to use to their own advantage.

During the immediate post-World War II period, political events, and especially Iraq's involvement in them, served to complicate relations between the companies and Iraq. Both the Regent, Abdul Ilah, and the young King-to-be, Faisal II, because of long-standing obligations to Great Britain, were expected to influence their country toward Arab-British and other forms of Western accord [in 1942, Iraq had become a recipient of U.S. lend-lease aid]. But when the United Nations General Assembly voted, in 1947, for a resolution to partition Palestine, a wave of anti-American, anti-British riots swept the country. Iraq Nationalists then began to press for an end to all ties with Britain, and their numbers swiftly grew.

Even so, Prime Minister Nuri Pasha braved this storm in an effort to keep Iraq on its pro-Western course. In 1953, Faisal II ascended the throne and for a time, he helped to blunt some of the opposition. Nuri Pasha then took a leading role in creating the anti-Russian alliance, the Baghdad Pact, uniting the "Northern Tier" of Middle Eastern nations bordering Russia. This alliance crumbled during the Suez Crisis (1956-1957), and Iraq next joined Saudi Arabia in an effort to counter the influence of Egypt's General Gamal Abdul Nasser, but to no avail. Meanwhile, the Cairo radio constantly attacked Iraq, its monarch and its pro-Western leaders. The powerful influence of the propaganda emanating from that source helped to create the revolution that occurred in Iraq on July 14, 1958.

X   X   X

When Venezuela successfully concluded the first 50-50 profit-sharing agreement between a producing country and producing companies in 1948, it quickly sought to export the details of the arrangement to Middle Eastern nations. The Venezuelans believed 50-50 agreements there would lead to an increase in the cost of oil from the Middle East, and the higher price would more nearly equal the cost of their own crude. At the time Venezuela suffered from the competition of low-cost oil from the Persian Gulf area. Officials in Iran and Saudi Arabia refused to see the Venezuelan diplomatic delegation bearing the news, but in Baghdad and Basrah, Iraqi officials did talk to the Venezuelan delegates. *[16] This information added to the weight of demand for more favorable terms from IPC. Then anti-British, anti-American and anti-foreign riots occurred in Baghdad in 1949, just as earlier, this ethnocentrism had flamed violently during the 1948 Arab-Israeli War.

In 1950, Arabian American Oil Company and Saudi Arabia also concluded a 50-50 profit-sharing agreement. This development further increased the pressures on the Iraq government to demand revision of the company agreement. Negotiations between IPC and Iraq remained in prospect for some months.

On May 1, 1951, neighboring Iran implemented nationalization of the Anglo-Iranian Oil Company, Limited [AIOC] concession. Again, the crisis in Iraq heightened. While moderates pushed for a 50-50 agreement with IPC, radicals demanded that the company be nationalized.[17] Rioting continued, and anti-foreign demonstrations added to Nuri Pasha's problems.

Iraq also contended that IPC had restricted all top management positions in the company to foreigners; that it had neither trained nor arranged facilities to train Iraqis in necessary technical or management skills. Iraq alleged that IPC had made no genuine exploration or development efforts, and asserted that the company books were so kept as to absorb many costs and charges that more properly belonged as expenses on the ledgers of the Groups or their affiliates. In all, the list of unresolved questions and charges grew too extensive to detail there. Not the least among these was the old (since 1932 at least) argument about Iraq's four shilling in gold per ton royalty, and the meaning of the phrase "current price of gold."[18] Iraq took the latter issue to court.[19] One new issue developed slowly; the flaring of natural gas in oil fields. But gradually the government acquired more control of this product, ultimately demanding ownership. [In 1958, it was reported that IPC had agreed to furnish Iraq's chemical and petrochemical industries with reduced-price gas, provided any products made therefrom were sold only in local markets.][20]

*An account may be found in chapter 15.

The IPC five-year program planned for production five years in advance. This proved to be too inelastic to permit Iraq's production to benefit fully from the post-World War II consumer demand. Productive capacity increases required new capital investments, so the apparently rigidly scheduled production quotas actually served to limit installation of the equipment that would have enabled IPC to increase production quickly. The Iraqis alleged that the use of a complicated rule governing production kept production down.[21] [But it is also true that IPC and Iraq did change these five-year schedules both before and after they took effect.]

Howard Page played an important role in the negotiations that began in the spring of 1951 and were completed in early 1952. He had worked at understanding IPC for two years, and later he related the details of how he became involved in these negotiations. After making a quick tour of the Middle East, he sought to get the Groups to reach some agreement as to "the position IPC should take when Iraq demanded a revision of the original concession agreement, which appeared to be inevitable." His associates turned down that proposal. But as he put it, "when all hell broke loose out there with Abadan [Iran] being taken over, we finally decided we'd better get prepared for a negotiation. . . inasmuch as I'd studied this [50-50 profit-sharing] very carefully, I was selected as one of the team to go out there."

For a first-time Middle East negotiator, Page did remarkably well. He had a hand in shaping about 90 percent of the articles in the eventual 1952 Agreement: "I worked them out with the back room man for Nuri Pasha, the Prime Minister, and his name was Nadim Pachachi [Dr. Nadim al-Pachachi, later Iraq's Minister of National Economy]." Page added that, after he and Pachachi agreed, they presented their reports to Horace S. Gibson, IPC managing director, and Nuri Pasha, who in turn referred the agreements to the other principals involved.[22]

In May, 1951, Nuri Pasha announced that IPC had agreed to a 50-50 profit-sharing arrangement and that the negotiators were near full agreement. The American State Department had been kept fully informed of the negotiations. On December 4, 1951, a spokesman for that department stated, "Iraq now has the best terms in the Middle East; they are embarrassingly good. They get 75 cents a barrel approximately. It is the same 50-50 agreement." Drafting the final documents required several additional months. Then, everything had to be put in proper legal terms; the Groups had to approve this version and so did Nuri Pasha. Finally, when the agreements reached Parliament, it acted quickly to approve them, amid cries from the political opposition that the government had truckled to the Imperialists and was pro-British. Had Nuri Pasha been less skillful and vigorous in suppressing the street demonstrations and aborting a proposed general strike, his government might have fallen. As it was, he resigned some months later, only after some measure of calm had been restored.[23]

The 1952 Agreement between Iraq and IPC included the three oil companies operating in Iraq, all of them owned by the Groups. These were: IPC; Mosul Petroleum Company, Ltd. [MPC]; and Basrah Petroleum Company, Ltd. [BPC].* The last two companies named had developed some productive capacity by the time of the 1952 Agreement; however, as later improvements were made, production from both companies increased—and so did their problems.

The 1952 Agreement included a clear statement that "any doubt, difference or dispute" between the company and the government concerning the interpretation or execution of any provision of the contract should be settled by arbitration. It also detailed the procedures for this action, a provision that many scholars believe served to irritate the Iraqis for years afterwards. The Iraqis contended their national sovereignty was infringed upon—IPC possessed no sovereign powers. To the Iraqis, the relationship was that of company-government, not government-government.

As another result of the agreement, IPC established an industrial training center at Kirkuk, with a five-year training course for clerical and technical personnel. Students enrolled there received monthly stipends. In the same year, the company also sent, at its expense, fifty students to England to continue their studies. These students pledged themselves to work for either the company or the Iraq government for a period twice as long as they spent abroad furthering their education. The 1952 Agreement regularized this arrangement, which did serve to increase the number of trained Iraqis in managerial and technical positions. [By 1957, there were several Iraqis in junior positions moving up the managerial ladder. Some of them suffered hardships during and after the 1958 revolution because of their British educations, and their company positions.]

Briefly summarized, other portions of the 1952 Iraq-IPC Agreement provided for 50-50 profit-sharing; Iraq could take its $12^1/2$ percent royalty in oil, for sale in the open market; Iraq was to receive an annual royalty-profit guarantee of 25 percent of the posted price value of total production; its annual minimum production guarantee was set at 28 million metric tons [7.46 barrels each]; the company assured the government of 20 million

---

*For the sake of convenience, IPC or "the company" will be used to refer to all three companies collectively, while their names or acronyms will be used in other cases.

British pounds sterling annual income; and local refinery needs were to be supplied at 5¹/₂ shillings [British] per metric ton. Further, IPC agreed that Iraq could request royalty increases comparable to those received by its neighbors, and that two Iraqis would be appointed to each of the boards of the three IPC companies.[24]

During the early 1950s several forces led to significant increases in Iraqi production: the additional military needs of the U.S. (because of the Korean War); completion of the BPC pipeline from Basrah to the Iraqi port of Fao (near the mouth of Shatt-al-Arab) on the Persian Gulf; and the 1952 Agreement. Iraq's output grew fourfold between 1950 and 1954.[25] Yet despite these increases, the government of Iraq consistently pressed IPC for additional production.

The 1952 Agreement included some clauses—believed to be precise—that defined and governed the cost accounting practices to be used by the oil companies. Despite precautions, however, the agreement had scarcely been approved by the Iraq Senate before questions arose over the very points thought to be so clearly covered. In 1954, Iraq raised a question regarding the price to be used in determining income taxes due. In 1955, Page told the Jersey Executive Committee that a lump sum payment had apparently settled the dispute on the oil lifted after September 16, 1954, but that the parties had not agreed on the "retroactivity" claim. He reported that a partial agreement had been reached on the "crude oil price to be used when determining taxes payable to that Government in accordance with the 1952 Agreement." Yet the government still reserved the right to demand upward adjustment in prices; nor did it agree that the payments received settled the back taxes issue.[26] In all, the price question never ceased to trouble negotiators for IPC.

In 1953, V.C. Georgescu, serving as assistant coordinator of Jersey's worldwide producing activities, wrote M.L. Haider, then deputy coordinator for that function. Georgescu, after conversations with Anderson and Page, advocated assigning a competent technician, either with a producing background or one who knew producing operations, to study IPC production reports. All three men agreed that the Near East Development Corporation would be of greater assistance to IPC if it had a link to its management at the analysis, planning and working levels. Georgescu, Page, and Anderson believed that IPC now looked toward NEDC (the Americans wrote) for assistance—a circumstance new to the company, and one they believed to be worth acting upon. Georgescu noted:

In London, the principal Jersey representative, Mr. P. J. Anderson, spends most of his time in discussions, negotiations and meeting with other Group representatives in formulating policy for IPC management's guidance. This leaves little time for close contact with IPC on technical matters, and IPC, a producing company, could benefit by consulting with an American producing man in London.

Such a person might also serve as an alternate for Jersey's NEDC director, he concluded. Haider probably discussed this request with the Socony-Vacuum representative, and then put the question to the other Groups, some of whom opposed it. But it is interesting to note that six Jersey officials involved in Middle Eastern affairs essentially agreed that IPC could use advice in planning, coordination, and especially in the Producing Department. Despite this Jersey view, no change in IPC resulted.

X X X

Between the 1952 Agreement and Iraq's nationalization of some IPC properties in 1972, a major problem relating to IPC derived from pipeline transit rights from the giant Kirkuk and the Mosul oil fields in Northern Iraq, across Syria, to Banias. Certainly, the Syrians knew of the IPC-Iraq negotiations and of the agreement reached in 1952, which contained better terms for the latter. So Syria demanded, and received, an amended convention governing oil transit and loading—one also including an agreement by IPC to build a refinery in Syria, to handle product requirements. Pressure from the Iraq government to increase production led IPC to announce in May, 1955, that it would make the investments necessary to increase IPC's productive capacity and then build a new 24-inch pipeline. This project had scarcely been approved when the Suez Crisis began, in 1956, and Syria cut the existing pipelines to Banias and Tripoli. Instead of increased revenue, Iraq received no revenue from the Kirkuk and Mosul oil fields until service through these lines commenced again in the spring of 1957. More important, because of the Syrian action, both IPC and the United States now had serious reservations about making new investments or building pipelines through territory controlled by hostile regimes.[27] This pipeline problem remained to trouble IPC and Iraq for many years.

The 1952 Agreement specifically stated that, in lieu of free market prices for "individual commercial sales," the posted price should mean "fair prices fixed by agreement between the Government and the Companies" or agreed upon by arbitration that paid proper attention to comparable posted prices elsewhere. From 1952 until 1955, the parties argued over this formula. In the latter year, Iraq agreed on a "Memorandum" concerning IPC posted prices. Jersey's IPC director at the time, Harold W. Fisher, later remarked on his understanding of the 1955 Memorandum: "it was agreed that posted prices should be the prices posted by the seller of the oil!"—the traditional posted price. Nevertheless, some months later, when the companies reduced the posted price by as much as 5 cents per barrel without consulting the government, Iraq protested sharply. Nadim Pachachi, the Minister of National Economy, urged that the government not agree to this principle because unilateral price-setting established a dangerous precedent. This argument, once begun, continued for months.[28]

Also affecting Iraq's intransigency on prices, expenses, and profits were the actions of Iran and Saudi Arabia. In July, 1957, Iran's national oil company began to let oil contracts to concerns other than those in the Iranian Consortium; while the Saudis, in December, 1957, contracted with the Japanese-owned Arabian Oil Company to develop some offshore concessions. News of these negotiations in neighboring states served to remind the Iraq government that the IPC concession included virtually all of Iraq's land area, and that despite conversations, no portion of this territory had been relinquished. Then rumors of the supposed 56 percent-44 percent Saudi-Japanese profit-sharing arrangement arrived. This caused special concern because the 1952 Agreement had specifically provided that the company would re-examine the 50-50 agreement with Iraq if neighboring states began to receive more income from their oil.

Additionally, the Suez Crisis of 1956-1957 seriously injured Iraq's economy. The "willful cutting" of the pipelines by Syria caused a stoppage of oil to the Mediterranean, and the canal closure disrupted tanker movement from Fao. Again, the pressure mounted on IPC to increase production when the pipeline resumed operation early in 1957, and then intensified when Suez reopened. In the Iraq Parliament, loud demands for nationalization were heard. Clearly bargaining between Iraq and IPC appeared to be in order.

Some talks apparently went on at a lower level in 1958, although Fisher, who (in 1957) had been loaned to IPC to serve as joint managing director,* does not recall them. Nor does he believe that Jersey President Monroe J. Rathbone ever participated "in any of the formal Group meetings," as has

---

*With Geoffrey H. Herridge as the new managing director.

been stated in numerous accounts of IPC-Iraq government negotiations. Rather Fisher states that Rathbone and his family, in company with the David Shepards, were touring the Middle East, and during February and early March, 1958, they visited Iraq. Fisher accompanied the party in Iraq and on to Beirut. And while Jersey's chief executive officer [CEO] learned a good deal about the Middle East and Iraq on the tour, he never applied it in a negotiating session over Middle East oil.

Herridge, with legal advisors, did arrive in Baghdad on July 4, 1958, to renew formal talks. Prime Minister Nuri Pasha while in London earlier that year had indicated that the "Iraqi demand for better terms would not be long delayed." The talks (according to the Iraq press) covered re-appraisal of 50-50 profit sharing, the acreage relinquishment question and an increase in IPC production. The question of accounting continued to emerge, as did the demand for IPC nationals to have higher managerial positions in the company. After four days (July 12), the negotiators agreed to a recess, probably in order to let the Herridge team consult with its principals. At the time of the recess, press accounts indicated near agreement, with the companies agreeing to relinquish part of the concession and to double oil production in the next two years.[29]

On July 14, the Iraq revolution led by Brigadier-General Abdul Karim Kassem [Qasim] occurred. It ended the Hashemite regime in Iraq—as the rebels executed the entire royal family and Nuri Pasha. Qasim, who headed the new anti-Western government, informed the oil companies that he had no intention of nationalizing the concession or their property. In retrospect, the revolution that he led seems to have represented no real unifying principle other than the obvious objective of realigning Iraq with the other Arab nations in opposition to Western "imperialism." Yet rumors spread concerning Iraq's intention to nationalize the petroleum industry. At the same time, IPC continued to discuss future investment plans for virtually doubling its productive capacity. Qasim, despite his professions against nationalization, indicated that increases in oil revenue would be necessary to reach the goals of the revolution. Still, for at least a year, apparent harmony existed between the two principals. IPC indicated a general willingness to satisfy Iraq's demands even if some surrender of concessions was required.

For the record, as early as 1951, Page and the other IPC negotiators had offered to give up half the concession territory, provided only that the company could select the portion to be relinquished. Three months after the Iraq Parliament approved the 1952 Agreement, this issue returned and created a further problem. Jersey then indicated that it would agree, should its partners be willing, to relinquish half the concession territory with two provisos:

that the time period for the transaction be sufficient to allow IPC to agree on its location; and provided all three companies "concerned be treated as a unit for relinquishment purposes."[30]

Again, in 1958, two weeks after the Qasim revolution, Page called the attention of the Jersey Executive Committee to the IPC problem:

> he pointed out that although a number of concessions in the Middle East provide no legal basis upon which the governments concerned can insist on acreage relinquishment by the concessionaires, increasing pressures along these lines may be expected from the governments of such countries as well as other countries in the Middle East.

Next, he recommended, and the Committee approved, that a proposal be made to other shareholders participating with the company in such concessions: where a government had officially requested that territory be given up, the companies would voluntarily return acreage that had been explored or on which no further exploration was planned. For such returns, the companies should receive "suitable quid pro quos." Page also wanted relinquishment of certain schedules already negotiated, a proposal that the IPC board finally rejected.

Some years later Page, reflecting on this situation and other negotiations like it, claimed that the partnership arrangements Jersey had with other companies handicapped it, and in the last analysis cost it dearly. Page questioned whether Western companies should argue for the sanctity of contracts, especially in view of what was happening in places like the Middle East. To him "a lot of these [earlier] agreements no longer made sense," especially in view of the changes in the contracts made necessary by altered social and economic circumstances. As he said:

> But there were, and still are, a lot of people in our industry who insist that a contract is a contract, and that the signatories have to live up to it.... So when we got into negotiations with Iraq, some companies didn't agree to any changes, on the basis that the contracts were unbreakable.

Page added, "I was looked on as a dangerous liberal for agreeing to study situations, ... and say, 'Well if you were starting all over again, this is what you would have to give!' I was looked on as a dangerous character for suggesting that we should." Fisher, in general, shared the Page views on contracts. But Jersey had only an 11⅞ percent share in IPC and did not prevail.

Some months after his election to the Jersey Board in May, 1959, Fisher returned to New York. He was assigned contact responsibility for part of the Middle East, with Howard Page handling the other share. Fisher had supervision over Jersey interests in Iraq and the Arab Emirates. With Fisher gone, IPC Managing Director Herridge took over negotiations with the Iraq government. The new Qasim regime pressed to have IPC production increased and also demanded greater participation in company management. By April, 1959, it appeared that some of the problems might be settled. The government announced that IPC had promised to double oil production during the next three years. Further, at the year's end, Qasim reported that he had rejected a company offer to relinquish 90,000 square kilometers of the concession; while he demanded instead 60 percent of the whole.* [Apparently the Page-Jersey-NEDC proposal of 1958 had not been enthusiastically received by the other Groups in IPC.][32] Unfortunately for IPC-Iraq relations in 1959-1960, the market price of oil dropped sharply below the posted price. British Petroleum Company, Ltd. cut the posted price in 1959 *without consulting* the producing countries. Jersey followed with another reduction in 1960. Earlier, Iraq had bitterly protested the 1956 price cut of 5 cents per barrel for crude exported from Fao. Now, it quickly offered its capital city, Baghdad, as the site for the scheduled September conference with four other oil exporting nations (Iran, Saudi Arabia, Venezuela, Kuwait). The IPC-Iraq negotiations that began on August 15, 1960, over equity participation, revenue sharing, and the return of concession acreage became quite bitter before adjournment on August 31. So Iraq moved to improve its relations with other producing states by supporting vigorous action against the companies. Fisher reported to the Jersey Executive Committee on these negotiations and on the problems pending in September. In the meantime, at their meeting in Baghdad, the oil producing states formed the Organization of Petroleum Exporting Countries [OPEC].[33]

In March, 1961, Iraq reorganized its oil ministry and announced a new oil law which superseded all existing statutes and decrees. Under this law the ministry exercised control of oil policy and responsibility for the development of both petroleum and natural gas resources. In the following month the new ministry asserted that twelve issues remained to be settled with IPC. These included all of the old demands plus such relatively new ones as active Iraqi participation on the board of directors; Iraq control—to a degree—of company expenditures; use of Iraqi tankers to carry the oil; and 20 percent government participation in the companies. Heated discussions with IPC representatives ensued. By late April, 1961, negotiations between IPC and

*When the company agreed to that figure, Qasim demanded 75 percent.

the government were suspended. Yet the text of these discussions had been broadcast in Arabic and English and printed versions had appeared.

The Jersey Executive Committee discussed the text of Herridge-Qasim negotiations, and then decided that Fisher should inform Jersey IPC Director Paul J. Anderson of its opinion on the issues. The Committee recommended that Anderson be told that Jersey favored "prompt submission by IPC of formal proposals to the Iraqi Government" providing for immediate relinquishment of 75 percent of the concession acreage with 15 percent more to follow within seven years. IPC would have the right to select the acreage subject to the government's views regarding properties to be retained.[34] Further, the Committee supported suspension of the issue of "dead rent" payments pending the outcome of arbitration on that subject.[35] Then, the Committee struck sharply at a major internal problem which IPC had confronted during negotiations with Iraq: the failure to grant its representatives enough flexibility to negotiate properly. Jersey wanted "top-level management representatives of the shareholder companies ... available in Baghdad, Iraq, during all future negotiations," where they could advise, provide "policy guidance," and participate, when invited. Fisher liked this idea; Group representatives would "be available for consultation and to experience the ambience of the talks."

One reason for the fragile nature of all IPC-Iraq discussions derived from the limited degree of authority granted to the participants; all too often, IPC negotiators sent from London possessed little freedom of decision. Except for the terms of reference with which they came, they could not respond to changes with offers of real flexibility. Instead, when confronted by new proposals, they had to ask for a recess and return to London to consult their principals. On the other hand, and especially after Nadim Pachachi published for the Iraqi public the records of his negotiations, Iraq's negotiators operated in a political wind; they feared to endorse IPC proposals because of the savage and uncertain nature of their country's politics. So this Jersey proposal aimed at putting at the table persons empowered to resolve questions and to act promptly. However, by authorizing negotiators to make decisions, Jersey also lost some support from P & I, CFP, and perhaps others in the Groups.

When the Executive Committee met in May, 1961, Fisher reported on the status of the recessed IPC-Iraq negotiations and on the London IPC meeting, which had sought to reach agreement on the issues as well as on plans to resume the talks. The Committee noted that during the course of subsequent talks, "it would be inadvisable to undertake at this time the negotiating proposal recommended by the Company on April 25"—obviously the other

IPC Groups had rejected the Jersey-NEDC gambit to immediately surrender 75 percent of the concession, with 15 percent to follow within seven years. The Committee also noted that future talks with Iraq would be led by Herridge, Fred Stephens, chairman of the Shell Transport & Trading Company, Limited [Shell] and Fisher, "with the latter serving as chief spokesman." And while the Groups had rejected the Jersey relinquishment idea, they did agree to have "top level management representatives" available in Baghdad. Fisher then told the Committee that arrangements had been effected to keep the State Department fully informed about developments in Iraq.

Fisher next went to San Francisco with Page, to "just sit in" at an Arabian American Oil Company meeting. Smith Turner, Jersey's Middle East coordinator, telephoned Fisher there to tell him that the IPC Groups had chosen him to head its negotiators. Fisher later observed, "I didn't particularly relish the project." So he flew back to New York and then on to London, where he told the IPC shareholders that he would take the assignment only if they would agree on his fall-back positions; "I just didn't want to go over there and tell the old boy [Qasim] 'No,' on everything. That would flop." In retrospect he added, "We didn't get very much bargaining power after all those meetings we had in London, but we had some."

Fisher did not recall precisely how many times he went to Baghdad during the summer of 1961. He said:

> So the whole summer of 1961, I represented all of the shareholders of the Iraq Petroleum Company and I guess Iraq Petroleum itself. I was shuttling back and forth to London [from New York] and on to Baghdad.... We had, I guess, about four or five meetings with Qasim, under, I would say, quite difficult circumstances, including martial law.

On one of his stops in New York, Fisher requested that the Executive Committee accept as answer to Iraq's demand for all natural gas produced on the IPC concession, that the negotiating team be authorized to agree that the gas would be provided to the government "free of charge at the point of production or in close proximity thereto." Gas not taken by the government would become the property of the company. The Committee approved the Fisher proposal, and he indicated that he would try to get the other IPC shareholders to accept the plan.

Fisher always remembered details of the place where the parties met—in the office of the Minister of Defense, which position Qasim held, along with that of Prime Minister. Contrary to what the Groups' spokesmen told him,

Qasim, rather than impartially presiding, did most of the talking for Iraq, "with the Oil Minister saying hardly anything."

In the meeting room the blood-stained uniform Qasim had worn during the attempt to assassinate him (1959) was displayed in a large glass case. Yet Qasim remained absolutely convinced of the loyalty of the people of Baghdad, and he once took Fisher touring the city in a bulletproof Cadillac.*

Qasim took a hard-line approach for several reasons: because Iraq Nationalists demanded nationalization; because British influence in the Middle East had declined and the power vacuum thus created had been filled only at intervals by the United States—Western influence was at a low point; because Soviet Russia had indicated its readiness to assist Iraq; and because many Iraqis considered IPC a foreign, exploiting, imperialistic company. Educated Iraqis, however, knew that their fall-back position—nationalization—would leave them in a situation similar to that of Iran in 1950; there were not enough trained Iraqis to operate the company. So they blamed IPC for that too, although as Fisher pointed out, educated Iraqis, especially those trained in England, were frequently persecuted by the revolutionaries.

In all the talks, whether with Fisher alone or with the team, Qasim emphasized Iraq's need for more income. After that, he hammered at IPC's failure to accommodate his demand for relinquishment of part of the concession area, and next at "cargo and port-loading fees at Basrah." [Despite a 1955 agreement, Iraq raised these charges several times; in 1960, "the new imposition amounted to a twelvefold increase."] The IPC team secured Qasim's agreement not to push this issue at the scheduled September meetings, and the companies paid additional charges under protest.

On September 12, 1961, Fisher again met with the Executive Committee. The Committee outlined the Jersey position on the major issues. It reiterated its earlier position on giving up territory and "dead rents" and on several less significant points. On the question of "equity participation" that had again entered the lists, Jersey took the position that it would find "unacceptable" any proposal dealing with "IPC's present proved productive acreage (approximately 3.7% of its total concession)" to which the company should retain title. It would support, however, the establishment of a new company, with IPC owning an 80 percent equity share and Iraq 20 percent, to explore and develop as much as 15 percent of the concession. Jersey indicated its willingness to have this latter acreage selected in some joint fashion by the

---

*Qasim had an account of this trip (and others like it) published.

parties. Finally, the Jersey Committee indicated that the question of port dues at Fao should be settled by securing an agreement to have them reduced, with any future increases based on actual costs for services rendered. This September 12, 1961 discussion consumed an entire day in the Executive Committee. After this meeting Fisher left for London, where he conferred with Paul Anderson and representatives of the Groups. Then Fisher, Herridge, and Stephens—the negotiating team—accompanied by a considerable body of experts from the companies, left for Baghdad, where further sessions had been scheduled.

Qasim and his oil minister again assumed the primary task of representing the government in the negotiations. This time all of the earlier issues between the parties were trotted out. IPC apparently did not haggle over the lesser ones, preferring to center its attention on the matter of relinquishments, and on who would decide questions of acreage, equity participation, price, and profits.

Fisher, as head of the IPC team, sympathized with the needs and aspirations of Iraq, especially in a time of declining oil prices and income. But as he pointed out, the purpose of the meeting was to arrive at "a commercial arrangement between two parties." The vague, frequently changing Iraqi demands led to a recess, since even with "top level management representatives" present some of the Qasim proposals still lay outside delegated authority. After consulting with their home offices, the IPC representatives returned to Baghdad on October 6. After five additional days of discussion, the parties appeared no nearer agreement. Qasim then made what he termed his "final offer," which Fisher rejected on the ground that it was a "sudden proposal." Since in three years of fairly steady conferring, the parties had come to no accord, they decided to adjourn. Qasim then told the IPC delegates that "we will take the other areas according to legislation we have prepared." He dismissed the IPC team on October 11, 1961. *[36]

On December 11, 1961, the Iraqi government proclaimed Law 80, which nationalized 99.5 percent of the IPC concession. This nationalization did not affect the Mosul or Basrah Petroleum Companies. IPC protested the law and asked for arbitration, as the 1928 and 1952 Agreements had provided. As its representative, IPC selected for arbitrator the Right Honorable Lord McNair, former president of the International Court of Justice and an outstanding international lawyer. But for some months, Qasim simply chose to ignore IPC protests.

*The discussions of 1961 were recorded by the Iraqis and broadcast (radio) in Arabic. When the meeting ended, they were translated into English, and the British Broadcasting Corporation broadcast them.

Jersey and Socony Mobil Oil Company [Socony Mobil] protested Iraq's unilateral reclaiming of 99.5 percent of the IPC concession [without compensation] to the U.S. Department of State, while BP involved the British Foreign Office. Qasim ignored notes from both governments and charged the British with imperialistic motives. Iraq invited other companies to apply for concessions in the newly nationalized territory, and when some expressed interest, the government furnished them with geological data (provided by IPC) for their examination. But the sudden action of Iraq in taking over the principal portion of one of the earliest oil concessions in the Middle East— one that belonged to outstanding companies sponsored by powerful governments—did not encourage other firms to seek concessions immediately. Further, at this time a large worldwide surplus producing capacity existed, and crude oil prices remained low.

In February, 1963, another revolution took place in Iraq. Qasim was executed, and his one-time associate, General Abdul Salem Aref, emerged to head the army group that joined with the Baath Socialist party to form a new government. In the spring of 1963, oil negotiations began again, with Iraq represented by its University of Texas-trained petroleum engineer, Abd al-'Aziz al-Wattari, and an IPC team headed by Executive Director L.F. Murphy. Aziz al-Wattari skillfully untangled the issues, and agreement appeared near on several of the less important ones, but disagreement continued over Law 80 and other substantive matters.

Beginning in 1964, some independent oil companies moved to assess the opportunities available to them in the territory recaptured for Iraq by Law 80. Oil-short Sinclair Oil & Refining Company [Sinclair], after being assured by the Iraq government that Law 80 would remain "on the books without change," made inquiries of Baghdad. But Jersey, Socony Mobil, and British Petroleum had let it be known in 1961 [date of Law 80] that they would institute legal action against any company moving into the IPC concession or buying oil therefrom, until after IPC-Iraq arbitration [negotiations] had been concluded. This threat of possible court action, coupled with the negative attitude of the U.S. State Department served as a powerful restraint on those oil companies now becoming more interested in Iraq.

In a discussion with a representative of one such oil company, W. Averell Harriman, Under Secretary of State For Political Affairs, emphasized to E.L. Steiniger, Sinclair board chairman, that the U.S. maintained a firm policy of non-interference in oil company negotiations, but he added that the U.S. did not like arbitrary actions by host nations; problems should be settled through bargaining. Harriman pointed out that "expropriation is one of the things which the U.S. Government does not condone." The Sinclair executive then expressed the view that his company had no interest in any oil property still subject to any form of litigation, and that he and his company fully understood the dangers of unilateral contract abrogation. Harriman declared that if Iraq successfully stripped IPC of its concession, the public in other Middle Eastern producing states would demand equally arbitrary actions by their governments. The Sinclair spokesman explained that the problem did not lie with Western firms becoming impatient with the slow pace of negotiations. Most U.S. oil companies, he believed, would honor the IPC determination to arbitrate; but while they waited, "foreign firms" might well get concessions in what many oil experts believed to be prime territory.[37]

During the next several weeks Standard Oil Company (Indiana) [Indiana], Union Oil Company of California [Union], Paul Getty of Getty Oil Company, and Iricon Agency, Ltd., plus others all expressed varying degrees of interest and enthusiasm about possible concessions in the former IPC territory.[38] The State Department interceded with them "to deter them from making offers to GOI [Government of Iraq] while ... critical negotiations [were] in progress." Yet Union appeared quite persistent, Sinclair sent some personnel to Baghdad, and as the negotiations dragged on, the interest of the independent oil companies increased, until Harriman requested a legal opinion on Law 80 and on the IPC claim from a State Department attorney.[39]

The counsel drafting the legal brief noted that "all of the principal issues raised are at the frontier between international law and politics where no firm answer can be given by a lawyer alone." After succinctly tracing the "long and complicated" history of IPC, the attorney stated that, as he saw the matter, the principal loss the company had suffered was from breach of contract, since IPC still operated the same oil fields that it had when Law 80 took effect. He did not think arbitrating the case would benefit the major Western countries concerned because of the nature of some of the evidence, which could be easily misunderstood. The attorney concluded by stating that he foresaw no effective remedy in law for IPC; however, he advised, "There is nothing improper in United States support of IPC, and doubtless our political interests, as well as worldwide respect for concessions and negotiations, would be furthered by a settlement...." As for the State Department, "we have no firm legal basis for telling independent American companies—let alone foreign companies—to stay out of Iraq."[40]

The State Department then called for a conference with Jersey and Socony Mobil on the Iraq oil situation. On October 26, 1964, Paul J. Anderson, now a Jersey director, accompanied by Earl Neal, also of Jersey, attended.

George W. Ball, Under Secretary of State, presided at the meeting and gave the department's view of the situation to the Jersey and Socony Mobil representatives. State, he said, had actively discouraged many companies from making offers to the Iraqis because of the ongoing negotiations, but now it needed to know "whether progress was being made." He also told the company officials that State "had no way of putting a complete quarantine on the situation." Anderson indicated that Aziz al-Wattari and the Iraq negotiators appeared willing to settle, but he believed that the boiling political situation directly affected them because their proposals changed frequently, as the Iraq political situation sputtered. Hence, the conversations went on and on. The oil men indicated their appreciation of the role of State in dissuading other companies from making proposals to Iraq while the talks continued. Every offer, they pointed out, tended to be mentioned at the conference table.[41]

The Iraq situation began to seethe in November, as messages poured into Washington indicating that German, French, Japanese, and Italian firms had become interested in the nationalized IPC territory. Ente Nazionale Idrocarburi [ENI] sought a contract to operate in Iraq, but change in the Iraqi cabinet forestalled the conversations. State cautioned the Japanese Foreign Office about the risks involved for any concern contracting for either the oil or a concession in Iraq. The failure of IPC and Iraq to reach agreement, State believed, served to make the U.S. independents "increasingly nervous." Despite such fretting, the concerned companies remained interested, and they indicated they would inform State of any discussions that might lead to a signing of documents.[42]

On February 8, 1964, Law 11 established the Iraq National Oil Company [INOC], with power to regulate many activities relating to the oil business. IPC protested this action, and then suggested further negotiations. These did begin. In 1965, the Iraq government and IPC initialed a new agreement. Aziz al-Wattari made the main features public: the company would be permitted to keep the producing fields (as in Law 80), to which Iraq now added the controversial North Rumaila field discovered by Basrah Petroleum Company, Ltd., but not in full production in 1961; the companies were to pay Iraq an agreed upon amount to settle some of the unresolved questions; and they agreed upon large production increases annually. The Iraq Parliament never ratified this new agreement, due to political upheavals and many cabinet changes.

All the IPC Groups except Jersey approved a joint IPC (minus Jersey)-INOC venture to explore a small acreage of the original concession, with the old company owning 66²/₃ percent of the equity share. This agreement engendered sharp public controversy in Iraq, more especially since Aziz al-Wattari had conducted the negotiations in private.

While at the time relations between IPC and Iraq appeared to be improving, those with Syria sharply declined. In 1955, IPC had reached agreement with that country covering transit charges to be levied on the pipeline to Banias and export facilities there. Late in August, 1966, Syria "claimed back payments" of $100 million pipeline charges due it, and it also increased by around $14 million the tariff on throughput in each future year. IPC contended that these claims were outside the 1955 Agreement, that they derived from entirely different factors and formulae. Negotiations to resolve this difference began on September 10, 1966. But on November 23, the Syrian government representatives left the bargaining table.

Subsequently, on December 8, 1966, the Syrian government by decree attached the pipeline and other properties of IPC. Before it would permit tankers to load at Banias, it demanded payment of $10.5 million allegedly due (the new annual rate) for 1966. A few days later, the government closed the loading facilities at Banias, and within a week, the IPC line to Tripoli in Lebanon closed down. This ended the 950,000 b/d flow of oil from IPC fields to the Mediterranean.

To compensate for this suspension of supply caused by the pipeline closure, Jersey utilized alternative sources of producing capacity. As a result, no affiliate or contract customer failed to receive a scheduled shipment. Yet efforts to settle the dispute dragged on for months, with Egypt and other Arab states supporting Syria and encouraging Iraq to nationalize IPC, on one hand, while that country pressed IPC to settle with Syria and eventually, to get the oil moving. Despite the injection of political issues into what began largely as a financial matter, the parties agreed on the amount of pipeline royalties due Syria in early March, 1967, "on terms close to those that IPC had offered at an earlier stage of the struggle." The Arab-Israeli conflict of June, 1967, occurred however, before the full flow of oil resumed.[43]

By 1967, after another revolution plus a governmental change occasioned by the death of General Aref, Iraq's policy shifted. The general's brother, also a general, Abd al-Rahman 'Aref, had become head of the government on April 16, 1966, but before his new government established its policies, Israel defeated the Arabs in the 1967 Six-Day War. Coupled with the Syrian closure of the pipeline to the Mediterranean, this event fueled new Iraq antagonisms and heightened hostility to the West. On August 4, 1967, Iraq announced Law 97, which effectively barred any return of the North Rumaila oil fields to IPC. Of equal importance, Law 123 reorganized INOC and made it more responsive to political pressures. Now IPC became the obvious target upon which Iraq could focus its anger toward the United States and Great Britain, in return for their support of Israel. So Iraq readily joined other Arab states in embargoing oil shipments to those countries and to West Germany.

France was not subject to this proscription, and President Charles De-Gaulle moved on October 23, 1967, to establish state-to-state relations with Iraq, which he hoped would lead to a new arrangement for the French state-owned oil company, Enterprise de Recherches et d'Activités Pétrolières [ERAP]. Of course, Jersey and Mobil Oil Corporation [Mobil] firmly resisted this French move. The U.S. Department of State supported their position and on its own initiative, pointed out to concerned companies and governments that, by applying for concessions, CFP and other French companies were in effect acknowledging the legality of Law 80 and Law 97. Charles Boyer represented Jersey at some of the early conferences held to discuss means by which the French moves could be blocked.[44] Jersey and State Department sources indicate that the French government actually ordered CFP to join with ERAP in negotiations with Iraq. On October 13, 1967, Jersey Executive Vice President Howard Page and Mobil Executive Vice President H. L. Schmidt met with Under Secretary of State Nicholas deB. Katzenbach to discuss the French attempt to take over a portion of the IPC concession. Subsequently, Secretary of State Dean Rusk, in a strongly worded telegram, told the U.S. Ambassador to France that Katzenbach had sent a vigorous protest to the French government over the ERAP venture. Some French officials showed considerable irritation at the actions of State and of the American companies. They evinced concern as to how far Jersey and Mobil would push their threatened legal actions against companies that purchased oil from the former IPC concession.

One spokesman said that if the two American companies did take legal action, the French government "would consider this a declaration of hostilities and take appropriate action against the companies in France." Victor De Metz and Jean Duroc Danner of CFP travelled to New York to see Jersey Board Chairman Michael L. Haider (as well as the Mobil chairman) and explain the CFP position. Two years later, they were still explaining, and according to a Mobil spokesman, they remained "considerably embarrassed" over the French action, contending that the move to obtain an IPC concession had been made by the French government "without prior consultation with CFP." The CFP leaders believed that U.S. government support of IPC would influence the French government.[45]

Yet despite all the protests, meetings, telegrams and other efforts aimed at blocking the French move toward an oil agreement with Iraq, the two parties concluded one on November 23, 1967. Ironically Iraq made its agreement with one group of French companies, while CFP and its associated Groups still contended that the nationalizing action of Law 80 was illegal, and while CFP still officially advocated arbitration. Equally ironic was the fact that CFP had made a proposal for the North Rumaila field in 1967, only to be turned down. Ten years earlier, immediately after the 1958 Qasim revolution, when pressures in Iraq mounted to nationalize the French section of IPC as reprisal for French actions in Algeria, the Groups "informed the Iraqi Government that IPC must be considered a unit and a move against one part would be considered a move against the whole."[46] Iraq took no action at that time. But the French move in 1967, as Secretary Rusk later pointed out, enhanced "French oil interests in Iraq at the expense of the other shareholders in the IPC." Subsequently, in July, 1968,* Iraq also concluded an agreement with Soviet and Czech firms, and began to develop the North Rumaila oil fields on its own. The Russians agreed to provide technical assistance, lay pipelines, and perform other petroleum-related services.

After 1967, IPC continued to move some oil from the port of Fao, though port charges there remained a continual problem, leading to much protest and more discussion. Among other demands, Iraq asked for direct participation in the company, as well as for increased production. The company held out for arbitration of the Iraqi expropriation of its concession, including the North Rumaila field.

Some progress was made in negotiating several questions at issue. Basrah Petroleum Company agreed to pay higher port fees in 1969, while awaiting final adjudication of that dispute. The parties reached agreement on production levels for 1970 and 1971 (and presumably thereafter), and in 1971, the company placed several Iraqis on its board of directors. But by the end of that year, a dramatic demonstration of OPEC's new power altered the Middle East oil situation.

Iraq, on January 15, 1972, presented IPC with seventeen proposals. Principally, it sought a greater increase in production, retroactive royalty payments, 20 percent government participation in the company, and more Iraqis on the IPC board. Negotiations began in January. While these went on, Aramco and Saudi Arabia averted open confrontation by agreeing on 20 percent government participation in the company. This came after a two-

*Some sources say 1967.

day OPEC meeting. Aramco forced the participation offer—agreement in principle—in a letter to Saudi Oil Minister Sheik Ahmed Zaki Yamani. The four companies that owned Aramco tendered the same degree of participation in their other companies operating in the Persian Gulf to the other five countries there. This agreement also provided for 20 percent participation in the capital and the concessions.

Aramco's action forced the hand of IPC, and on March 19, it reached agreement in principle with Iraq for 20 percent participation. But the negotiations over compensation and the concession continued into May. Iraq gave the company two weeks to settle or face legislative action on not only the 20 percent participation but the other sixteen proposals as well. The company made an offer on May 31, which Iraq immediately rejected.

On June 1, 1972, Iraq nationalized the remaining oil fields of IPC (but not those of the other two companies). Then, Iraq and CFP announced on June 18, that CFP would lift 23.75 percent of the oil which had formerly belonged to IPC. In this manner the French company again deserted its partners. The other Groups refused to recognize the nationalization action.

Subsequent negotiations between IPC and Iraq led to an agreement on February 28, 1973; it provided company payment of $348 million in royalty arrears (1964-1972), and called for the surrender of the Mosul oil field (uneconomic anyway). Iraq "agreed to deliver one million tons of crude per month for 15 months to Banias and Tripoli" to compensate IPC for the loss of the Kirkuk field. It did not nationalize the Basrah Petroleum Company. It kept its bargain and delivered the agreed quantity, even though the price increases in 1973 quadrupled the value of the oil to the companies.

On October 6, 1973, the Yom Kippur War broke out, and the next day Iraq passed a law purporting to selectively nationalize the interests in BPC (a United Kingdom corporation) of both Exxon Corporation and Mobil. Not until December 8, 1975, did Iraq finally nationalize the other 57 percent interest of the remaining owners of Basrah Petroleum Company.[47]

X X X

The IPC-Iraq oil venture for Jersey ended as it had begun, in a blaze of controversy. Operations prior to 1950 provided materials upon which several Justice Department and Federal Trade Commission attorneys based their recommendations for antitrust suits, while subpoenaed documents made headlines for congressional committees. The fact is, however, that from 1950-1975, Iraq's political instability made that country a poor place for investments of any sort.

At least one authority on Middle East oil identifies the outstanding characteristic of Iraq over the almost fifty years of the IPC oil concession there as "political and general instability." Certainly in the years, 1950-1975, the government constantly changed. The Communist party, the oldest political party in Iraq, exerted great influence, although some of its leaders at times served as cabinet ministers and at other times fled for their lives. The Iraq revolutions, whatever their alleged purposes, seemed to derive their force mainly from anti-imperialistic rhetoric and the anti-Western posture of their leaders. They did not serve positive goals, and they became more punitive than progressive. In all, governmental instability occasioned by revolutions, coups, attempted coups, bloody military actions against portions of the population, public trials, purges, and comparable events saddled Iraq with serious handicaps and obvious disadvantages as a place for foreign companies to invest and to operate. During one particularly gory civil struggle in 1959, Allen W. Dulles, director of the U.S. Central Intelligence Agency, told the Senate Foreign Relations Committee that the Iraq situation was, "the most dangerous in the world today."* As one State Department expert later noted, "Since 1962 we have advised American companies to stay out of Iraq."[48]

In fact, during the years of the IPC concession, Iraq confirmed the wisdom of the adage: in Arab politics, nothing is permanent. After signing the 1948 Heads Agreements, some members of IPC felt that operations were being hedged by rules, other agreements, and the obstinacy of at least two other Group members, who consistently fought against every proposed innovation in administration. P & I and CFP Groups seemed to pursue their own interests only, and they remained ready to prevent proposed changes in the agreements. Thus, authority remained in London with the Groups; it did not transfer to the field, where realistic negotiation seemed possible. Meanwhile, Jersey's Executive Committee, Board of Director's minutes—and Page's interviews—tell a story of that company's apparent willingness to be flexible—to be more realistic—and to anticipate each Iraqi demand.

Howard Page, who served as Middle East contact director for most of these years, often did anticipate demands and trends, and he used what he knew in recommendations to the Jersey Board. Fisher also played an important role in Iraq. He enjoyed the confidence of Nuri Pasha before the 1958

---

*However, Fisher pointed out regarding this: "During the 1958 revolution there was no exodus of [expatriate] men or women.... [Some small children were sent to England.] The women stayed, which was particularly tough in Mosul, which was the site of several counterrevolutions, with opposing teams 'dragging' their enemies through town, a favorite means of execution." IPC also followed the policy of unrestricted housing; there was no company compound for foreigners.

revolution and he gained new respect from several of Qasim's ministers after that event. Obviously, the Groups trusted his judgment although Fisher often belittled his accomplishments by claiming that Page-engineered agreements lasted, while those he negotiated simply dissolved within a few years. Other experts, including Herridge and Stephens disagree.

Together Page and Fisher, supported by Shepard, Anderson, and later, George T. Piercy—all Jersey directors—provided a talented negotiating team for discussions of Middle East policy. And on the next administrative level, Smith D. Turner, who like Shepard had been involved in Middle East affairs since World War II, offered not only experience but also shrewd analysis of Middle East problems. Many others at Jersey supported the work of these more celebrated Middle East oil diplomats.

Still, given the complexity of IPC's management structure, and of the relationships between Shell and the Dutch and British governments, BP and Great Britain, and CFP and France, it is easy to see why the Iraqis often identified the oil companies with the policies of national governments—and with their pre-World War II practices. Even if Jersey had been entirely altruistic and devoted a full effort to improving relations with Iraq, it seems improbable that the company would have been able to effect significant changes in IPC actions.

Until 1959, IPC negotiators were able to bargain meaningfully with Iraq. But once Qasim began to use the media as a way of influencing the Iraq public (and in turn, the IPC negotiators), bargaining sessions became increasingly meaningless. One scholar remarked:

> Qasim was able to create public support for his negotiations, tactics and aims so solid and insistent that it eventually made politically impossible any substantial retreat from the stand that he had taken.

By August, 1961, the Iraq government made it clear before negotiations resumed that:

> cancellation of the companies' concessions will be the natural outcome of the oil talks if the companies persist in their tyranny and disregard of Iraqi rights.

This writer concludes, "Qasim had clearly gone beyond the point of no return."

By 1961, Iraq had settled on a set of basic demands: territorial relinquishment, equity participation, increased production and income, and full rights to natural gas. Another close student of Middle East affairs says of this, "The [Iraqi] Government refused to come to agreement on any individual issue, insisting on a 'package deal' which would include settlement of all issues." She goes ahead to declare what appeared to be clear: that Iraq's purpose in holding out for full settlement of all the issues "was to break the existing pattern of profit-sharing and concession arrangements that had up till then prevailed in the Middle East...."[49]

After 1961, IPC continued to press its claim for arbitration of Law 80 (and later Law 97 as well). One legal expert has indicated his opinion that, after Law 80 reclaimed 99.5 percent of the concession, "All subsequent discussions have shattered on the selection of this crucial 0.5 percent—the company insisting that it include North Rumaila and the government insisting that it not." Despite the importance of the other issues, this question remained central to all the negotiations conducted after 1961, and, in turn, left little room in which negotiators could maneuver. Finally, Iraq nationalized IPC.

During the decade of the 1960s, political uncertainties prevented IPC from increasing production in Iraq at a rate equal to that achieved in Saudi Arabia or Iran. And, Libyan production increased more than any of these, which as Page told the Senate Subcommittee on Multinational Corporations, all occurred "with Iraq down." If IPC had increased Iraq production, smaller increases would have been necessary in the other three countries. At the same committee session, both Page and Piercy categorically denied as false the assertion that IPC ever plugged wells or kept them hidden from the Iraq government. Both men also stated that large quantities of oil had been found in the North Rumaila field, but that under Law 80, it was seized from IPC before it could be produced.

Troubles with Syria and with the IPC pipelines, for which the Iraq government blamed the company, coupled with Law 80, together put Iraq in a unique position—at least in the view of Jersey and other major companies. Certainly after 1956, any plans for another Mediterranean pipeline deserved the most serious study before approval of such a massive capital outlay. Later, the passage of Law 80 provoked the companies into even more careful consideration of the entire investment climate in Iraq. What appeared far safer was an increasing investment in alternate producing capacity in Saudi Arabia or Iran or another Middle East state, especially since every major oil company held concessions elsewhere. Into those places went available capital.

Meanwhile, in Iraq, because of Law 80, the dream of many Arabs, "Arab Oil for Arabs," became a reality. Yet IPC remained for eleven years after

Law 80, still producing oil, and still negotiating, although without reaching a satisfactory agreement. Finally, when Iraq did invite the French and Russians in, the results were neither satisfying to the host country nor conclusive as a demonstration of how oil resources in the Middle East should be managed. The new arrangements, in the opinions of many oil experts at least, brought the Iraqis less total value for their oil than IPC had offered earlier in the negotiations. For all parties to the quarrels in Iraq, the price of participation was high.

Chapter 20

Jersey in Libya

SHORTLY AFTER WORLD WAR II, the Standard Oil Company (New Jersey) [Jersey] began to show interest in exploring for oil in Libya. In 1947, two company geologists made a brief survey of that country, and six years later, a Jersey executive secured from the Libyan government several surface exploration permits.

In December, 1951, Libya declared its political independence. As "a constitutional and hereditary monarchy," it became the first nation to place itself under the "auspices of the United Nations." Its King, Sayid Mohammed Idris al-Mahdi es Senussi I [Idris], was a Moslem religious leader who allegedly had supported Great Britain during World War II, when he led local resistance against both the Italian and German occupiers of Libya. His kingdom comprised 680,000 square miles, of which only about six percent had proven habitable for the 1,125,000 population, most of whom lived along the more than 1,200 miles of Mediterranean Coast.* For many centuries, agriculture represented the principal way of Libyan life.

On December 24, 1951, when Idris became King, the Libyan constitution required the monarchy to maintain two capital cities—Tripoli in the West and Benghazi in the East—with the government residing two years alternately in each. When the government moved to Tripoli, the King lived on a nearby farm instead of in the palace; when it moved to Benghazi, he resided in Baida, a pleasant mountain retreat that eventually became the true seat of government. In fact, Idris cared very little about being King. At one point he tried to abdicate, and actually wrote a letter of resignation; but when the desert tribesmen heard about it, they refused to let him step down. In an emotional, almost melodramatic incident, tribal leaders forced Idris to return to Baida and to his throne. Since he had no direct heirs, however (and given his lack of interest in running the country), it now seems almost inevi-

---

*In 1977, the Libyan population was estimated to be 2,600,000.

table that plots to depose him (and to guarantee a particular successor) would develop. In time, of course, they did.

X   X   X

Meanwhile, before Jersey's Libyan exploration crews could begin their survey and data gathering, some of the three million land mines that remained from World War II had to be removed. Their presence was not surprising since Libya had been the locale where, in wartime, German Field Marshal Rommel and his Afrika Korps dueled with Generals Auchinleck, Montgomery, and the British Eighth Army, as well as with U.S. troops. All the armies planted mines as they fought; and later, some of the same Germans who were involved laying the mines actually helped to clear them, so that Jersey geologists who began to arrive in March, 1954, could move across the desert—as one of them described it "up the jebels, down the wadis." Those intrepid geologists used "vehicles especially designed for travel over both desert sand and desert rocks," while their more fortunate fellows mapped the overall concession areas from the air, far beyond the reach of any mines.

Already for many years, Jersey had marketed in Libya, principally through its chief affiliate in the area, Esso Standard (Near East), Inc. Until 1953, however, Libya had no local law that dealt with oil exploration. Thus, the first oil seekers simply operated under the sanction of the Libyan Mining Department. When several other companies also applied for exploration permits in 1953, however, Libya took steps to draft a proper law. As M.A. "Mike" Wright, then coordinator of Jersey's worldwide producing activities recalls, "We helped them write a petroleum law. Our lawyers went over and worked" with the Libyans. Another Jersey employee, Zeb Mayhew, a geologist by training but by experience a very knowledgeable student of oil leases, then assigned to Producing Coordination, also went to Libya during the time when the 1955 law was being written. While he remained there for only a short while, he later remembered lawyers from several companies became involved in writing the final version, and that he and others were consulted about some of the provisions.

The new law took effect in June, 1955, after all previously issued permits for surface exploration had expired. By then, Jersey's geological teams had partially appraised their findings, and Jersey was ready to apply for Libyan concessions.[1]

Still, in 1955, despite some exploration, Libya had little real evidence on the basis of which to claim significant oil reserves. Simply, no one knew how much oil might be there. Given this situation, the Libyan legislature approved the new law, which offered very generous terms as a means of encouraging foreign-based companies to explore and invest in Libya. Wright characterized the new act as "a model petroleum law"; under its terms all areas would be open for application without further announcement by the government that the process for selection would be on a "first come" basis. Royalties would be paid on the realized price of the crude oil.

While a Jersey publication, *The Lamp*, endorsed this view, calling the 1955 act an "excellent new petroleum law," other observers were not so certain. Jersey's Middle East Contact Director Howard W. Page, for example, knew all too much about the company's problems with the Persian Gulf states over royalties and taxes; therefore, he sought to have the Libyan law written so as to provide for taxes to be collected on the posted price of oil, rather than on the realized price. Walter Levy, a noted oil consultant, represented Libya in discussions, and he agreed with Page. Both men felt that this arrangement would prove to be more equitable for the Libyans and less trouble for everyone in the future. Page and Levy believed that, by using the realized price as a tax base, the lawyers were borrowing trouble for the future. Their wisdom, which seemed more like pessimism at the time, would be proved by later events. For, despite their urging that royalties be collected on posted prices, the 1955 Libyan petroleum law called for the payment of royalties on realized prices. Almost as bad, it failed to define "income for tax purposes" precisely.[2] Together, these two problems opened the way for widespread price-cutting by independent companies that were unable to absorb their share of Libyan oil. As their production increased, these companies disposed of their oil at the best obtainable price, but usually far below the posted price; and they paid taxes on only the price they realized. Naturally, Libya became disturbed by such applications of the 1955 pact.

On the other hand, however, the 1955 law did achieve one national objective: attracting new investors to Libya. Whereas in 1953—when Jersey received its first exploration permits—only nine companies were active in Libya; in 1955, the Libyan government granted eighty-four concessions to seventeen companies, "in the first round of bidding."[3] This popularity demonstrated that at least some part of Libyan economic policy benefited from the pioneering legislation.

Back home in 1956, the Jersey Executive Committee heard Director David A. Shepard's report of a conversation with the head of the French government's bureau for handling petroleum affairs. The French had requested that the "Company submit some form of statement disclaiming any interest in oil resources in the vicinity of the discovery made by a non-affiliated company in the Libyan/Algerian border area." Also, France wanted Jersey to make available a 20 percent interest in Esso Libya's "large acreage block" along the Algerian border.* While the French might drop the former request, they seemed determined to achieve participation in Libya. Shepard told the Executive Committee that he had informed the French government that Jersey would not acquiesce to either request, and the Committee approved his actions, asking that the U.S. Department of State [State] be kept informed of the Jersey decision.[4]

On its part, State did exhibit concern about the presence of American oil companies in Libya. Shortly after the first Suez Crisis (1956-1957), and while Great Britain continued its political retreat from North African and Middle Eastern responsibilities, Jersey Director Arthur T. Proudfit, accompanied by W.R. Carlisle of the Law Department, met with State's Director of Northern Africa Affairs. Already the Libyan government of King Idris appeared to be less secure than it had seemed before the Suez Crisis. Now, State assured Proudfit and Carlisle that the U.S. government would continue to support Libya "even if there were an internal change in the Libyan government," and that as Britain withdrew economic aid, the U.S. would increase its support. State representatives thought that Britain would continue "its present military policy towards all the North African countries," and certainly the U.S. would continue military aid to Libya, subject only to congressional action. Should oil be found in Libya, State said, "internal problems will undoubtedly be present but they will be met as they arise." Proudfit then informed the State Department representatives that the Jersey budget for exploration and development in Libya had been increased from the 1956 amount of $4 million to $7 million for 1957, and that the company expected to spend $13 million in 1958.[5]

Before 1955, Jersey had not supported exploration ventures in the Middle East or North Africa because the company was compelled to adapt its marketing facilities to the oil acquired through the 1946 Agreement—which soon brought it a 30 percent share in Arabian American Oil Company [Aramco]. Also, in 1946, Vice President Orville Harden negotiated another contract with Anglo-Iranian Oil Company, Limited [AIOC]** for the pur-

---

*Being desert, the border between Libya and Algeria, until the early 1960s, was reckoned to be the Westernmost camel track from Ghat to Ghadames.

**In 1954, the company name was changed to British Petroleum Company, Ltd. [BP].

chase of 800 million barrels over a twenty year period. * Page later called the Aramco concession "the best concession in the world without any doubt," while the move into Libya seemed to him much more risky: ". . . we took the chance on it and . . . everybody . . . essentially left it up to me just saying, 'Are you willing to take the risk?' And I said 'yes.' I thought the Libyan thing was valuable enough to take a risk on."

Page's decision at first looked good, for by 1956, Jersey became very interested in the potential for oil in Libya. Not only had Director Proudfit begun to monitor local reports, but Wright himself made several field visits to Libya. As he later explained, "one of our purposes in going into Libya was to try to find an oil that would compete with the Middle East, . . . therefore, we would have an improved position with the Saudis [Saudi Arabia] by having another source of crude." Too, Jersey President Monroe J. Rathbone continually sought to improve the company's reserve position—although some company officials thought that he took unnecessary risks in the search—as Wright said, "This was something that Rathbone dared to spearhead as far as I was concerned. . . ."

Under Libya's 1955 petroleum law, Jersey received nine concessions, covering more than 70,000 square miles. To carry out local exploration, Jersey on April 11 of that year incorporated Esso Standard (Libya) Inc. [Esso Libya].** Soon, Jersey and Esso Libya made a decision to drill in Western Libya—the wrong place, as things turned out—in its Concession 1, near the Algerian border, where the French had found gas at Hassi Massoud on the Algerian side in 1956. Here, during June, 1957, in their third relatively shallow hole, Atshan Number 2, drill crews struck oil—the first oil found in Libya, although it was not present in commercial quantities.

In fact, further drilling there proved no significant shows of oil or gas; and early in 1958, Esso Libya moved its crews to Cyrenaica in Eastern Libya. Still, Esso Libya maintained an interest in the West and later returned there to resume drilling.

The move from Western Libya to Cyrenaica pleased Pete Collins, Jersey's head of Libyan drilling, for he (and his associates) had steadily contended that oil could be located in the western Sirte Basin, where a reefal show existed. Drilling began on the 10,000 square mile Concession 6, about one

*In 1959, Page signed another agreement with BP for 215 million barrels of oil over a twelve-year period. Just before the Harden contract expired, Smith D. Turner negotiated a new BP contract for 2.1 billion barrels over a fifteen-year period. This contract terminated in 1979, after BP had delivered 81 percent (app.) of the oil.

**The SONJ *Annual Report* first referred to it in 1955 but not again until 1959. The parentheses were dropped in 1961.

hundred miles from the coast; as Wright put it, while "everybody else was drilling over on the Western side [of Cyrenaica] ... we discovered the first oil there," on April 18, 1959, at Zelten. Wright had become a Jersey director in 1958, so when the coded cable reporting the strike reached him, he immediately passed the information to the Executive Committee. As the story ran, "We had this well and we debated for a day or so as to what we should do." The question was whether to open it up, let it flow long enough to prove its volume and size, or to stall for a time; "so we finally decided that because of the effect on other nations where we had our oil, that it probably would be advantageous to get a big test on it. The well tested 17,500 barrels per day [b/d]." Then Wright added: "And the price of Jersey stock dropped $2.00."

The significance and importance of the Zelten strike cannot be overemphasized. Twice the Western nations had been victimized by Suez blockages and pipeline shutdowns. Western oil companies with supplies from Venezuela and the United States prevented any real shortage in Europe and Japan in 1956-1957, but the interruption to the pipeline lasted longer after the Arab-Israeli War of 1967. But in both incidents, the closing of Suez served to frighten some Europeans, and to heighten awareness of their vulnerability to closure of the canal and the pipelines.

Only Jersey persisted in drilling in Libya when other companies indicated their intention to end exploration activity. The French oil companies had expended up to the limit of their budgets; Gulf Oil Corporation, according to reports, had spent $30 million in its search. British Petroleum reputedly had even larger funds involved. Allegedly, BP had begun to dispose of its warehouse stocks, leases, and villas preparatory to leaving. Independents proved to be equally discouraged. Zelten completely changed every company's outlook. The rush was on again, especially since Libyan oil could easily be shipped to Europe. In volume, it would alter the international oil business.

A second well drilled came in at 15,000 b/d, and within 28 months Jersey had 34 producing wells.[6] Planning for a pipeline to take the crude oil to the Mediterranean quickly began. No port facilities existed there, so Jersey and Esso Libya arranged for a hydrographic survey of the Marsa el Brega area only a few months after the Zelten strike. Brega possessed a mini-peninsula jutting out to sea that they believed would offer a measure of protection to a harbor. In July, 1960, ingenuity prevailed; Esso Libya engineers took 30-inch diameter pipes in 40-foot lengths, delivered offshore by freighters, and plugged them at each end, jettisoned them into water, and had them towed ashore by launches powered with heavy duty outboard motors. They next constructed a jetty built of 14 huge caissons and fitted them into holes from which the sand had been dredged.

More dredging deepened the harbor for ocean-going vessels. About the same time Esso Libya started laying a pipeline, one segment from the Zelten field went up an escarpment 20 miles away, and the second 90 miles from that point to Brega. Once the oil was pumped over the ridge, it flowed by gravity to the port. Pipelines were wrapped in tape and asphalt felt, and then buried. To ease construction problems, as well as for future use, Esso Libya built a 20-foot wide hard surface road alongside the pipeline route. All these enterprises and especially the speed with which they were completed attested to the importance with which Jersey treated its Libyan oil. And the drilling of additional wells went on simultaneously.

Marsa el Brega proved to be about the worst place for a port on the entire North Coast of Africa. Storms beginning in the Pyrenees often roared southeasterly, across the Mediterranean, toward Marsa el Brega. Oil installations located there would be very hard to protect, and to operate. When the harbor was completed, next came the erection of a loading dock to receive materials. Plans also called for construction of a tank farm with some million barrels of capacity. Within two years, more than fifteen Libyan and foreign contractors, employing at least 2,000 foreign and local employees, had begun working on these various projects.

As early as 1959, Esso Libya undertook a broad-gauged, yet intensive program of training Libyans for company jobs demanding everything from the various necessary laborer's skills to administrative and professional competence. The Libyans had to be taught enough English, at the same time, to enable them to identify and recognize supplies, tools, and equipment. Despite a lack of formal education, many of them did well. Even during early exploration days, the company sought to employ Libyans; as one foreign driller remarked, "They have a real aptitude for handling machinery." Eventually native Libyans came to constitute a large percentage of Esso Libya geological and seismic crews. But at the beginning, Esso Libya borrowed expert welders from The Carter Oil Company and from the Creole Petroleum Corporation [Creole] and also started a welding school. Ten Libyans joined the first class. Training was intensive; five months were required to complete the course. Yet the company took the long view. Looking to the future of management, Esso Libya also instituted programs to aid the "advancement of students at the university level."[7] A formal training center was established in 1965.

When Esso Libya and Jersey agreed on refinery plans, the latter had a scale model made in New York and sent to Libya. Trainees could study the model as they learned English and their refining jobs. From Imperial Oil

Limited of Canada, Jersey borrowed Training Coordinator Colm Bradley, who had experience in the refining of petroleum, to supervise this program and prepare an adequate labor force for the completed refinery.

In itself, Libyan oil possessed little sulphur, but a considerable amount of wax; still, it compared favorably with the best Arabian light crude. Combined with Libya's proximity to Europe, this high quality created an attractive economic proposition—one that encouraged additional exploration activity by interested companies. Meanwhile, the Libyan government, now assured of a sizeable oil resource, insisted that a refinery be built "to take care of the local gas and oil situation." Simply put, the Libyans demanded a refinery, and to get it, they were willing to assign to Esso Libya certain concessions. Eventually, on one of these special concessions, Esso Libya discovered its rich Raguba oil field. Since operating the refinery proved to be both expensive and onerous, Raguba oil could be said to be Esso Libya's pay for serving as operator.

Preliminary studies by Esso Standard (Near East), Inc. and Esso Libya indicated that a refinery of 8,000 b/d capacity would be sufficient to meet local demand for some years ahead. But Esso, the only American marketer in Libya, also hoped to secure the contract to supply petroleum products to Wheelus Air Force Base, the largest base outside the U.S. This would require a larger refinery, so a 15,000 b/d refinery was designed.* Another Libyan law required that foreign contractors have Libyan partners or Libyan capital, and Jersey experts anticipated that the number of bidders interested in building a refinery at Marsa el Brega would be few. Their prediction proved to be accurate, so the unattractive situation led them to study alternatives to on-site construction. The same problem also prevailed with respect to the construction of another plant—this one designed to generate the electric power needed because of Marsa el Brega's isolated location.

After considering both problems, management found a single solution. They decided to have both plants built on barges away from the final sites in places where materials and skilled labor were readily available. Then, the barges would be towed to Marsa el Brega, for final siting of the refinery and power plant. Thus, an oil refinery was begun in Belgium on a barge; later, the completed plant was towed to Brega, where a site had been dredged; and the refinery on its barge moved into place. In a similar manner, an electric power plant was constructed on a barge in Spain and towed to Brega , where it was "plugged in just like you plug in an electric fan."[8] Of course, during the time used to build both plants, pipeline and tank farm construction went forward.

---

*At the time of installation, the capacity was only 9,000 b/d.

Back at headquarters, Jersey leaders Proudfit, Wright, and others knew that they needed to build a creditable organization to run the new company. Wright suggested Robert A. Eeds, who had considerable experience with Creole and later with Aramco, where he had served as vice president for Government Relations. Wright recalls, "The big work was yet to come. And so I met Eeds and his wife in Switzerland and took them down to Tripoli, ... introduced them to the people, the government, and from then on he ran it for quite a while." Eeds, Wright said, "had an acceptance in that country that was extremely high."[9]

At Esso Libya, Eeds quickly took control, thereby validating Wright's estimate of his managerial ability. The company needed trained personnel, especially managers and supervisors. Simply put, once the oil was found, rationalizing a support structure remained to be completed. Esso Libya employees of the early 1960s recall this operation as remarkable, for Eeds brought in a whole range of talented people, and in "Operation Sandwich," he reorganized the company. Persons who had been in Libya for years suddenly reported to different managers or departments, as informal procedures became regularized. To his credit, Eeds accomplished this evolution of a major producing operation quite smoothly, provoking little employee resentment.

H.W. "Haywire" Brown headed Esso Libya briefly in 1959-1960, but became executive vice president when Eeds took over. Pete Collins moved in to become a director. Eeds found a number of competent persons already in place. Once staffed properly, Esso Libya ran well—just as Eeds intended.

By March, 1961, Jersey employed 130 expatriate personnel in Libya, including 35 on loan from affiliates. The Carter Oil Company and Creole together supplied more than one-half of these, who were mainly technicians and engineers. Yet Eeds continued to be concerned by the problem of obtaining enough skilled labor and staff. In fact, Jersey affiliates encountered this problem frequently, everywhere except in North America and Europe. Yet recruiting expatriate engineers, skilled workmen, technicians, and managers proved to be especially expensive for Esso Libya, which also had to bear the costs of recruiting local laborers, and providing education (including training schools) for them. Not all Americans who matured after World War II cared to spend their lives outside the U.S., at least in places like North Africa or the Middle East. One of the results of this American reluctance to live there was that Esso Libya hired many skilled nationals from other countries—Canadians, German, Italians, Spaniards, Norwegians, Britishers, Danes, Dutchmen among others. This labor proved quite effective and much less expensive. Judged solely by the homelands of the persons working for it, Esso Libya represented Jersey's first truly international affiliate, if not the first international company.

As a matter of policy, Jersey always gave "additional credit in its retirement plan"* for overseas service, and in some countries, the company paid employees an additional "hardship premium," which was adjusted regularly. But even so, the company found that:

> it is difficult to attract and retain overseas employees . . . in competition with service in other areas, especially as compared to South America or Europe. Furthermore, they point out that social and cultural conditions in Libya, with its Moslem environment, as well as circumstances pertaining to isolation and inadequate community facilities, are factors which point up hardship conditions in Libya.

Some expatriates had trouble becoming accustomed to the Libyan weather, especially to conditions at Benghazi, and in the oil fields of Zelten and Raguba, where employees were housed on bachelor status. Esso Libya, by 1963, had begun to supplement salaries by 5 percent of base pay up to a maximum of $1,000 for foreign workers at Tripoli; and 10 percent, up to $2,000 to those employed in Benghazi, Brega, and the oil fields.[10] But the shortage of skilled workers remained a problem.

Jersey Board member Wright was given contact responsibilities for Libya, and Siro Vázquez, a Venezuelan from Creole, became coordinator for production. In particular, he pursued Wright's interest in developing and maintaining good labor relations in Libya. About a year after the Zelten oil field came in, Vázquez and Proudfit visited there and noted that, despite all discussions, little had actually been done to provide proper housing for Esso Libya employees. Proudfit and Vázquez began to review the entire problem, when Proudfit remarked, "Hell, Larry Birney (Creole's manager of Industrial Relations) knows more about community integration and the development of communities near oil fields than anybody else. Why don't you get him from Creole?"

Birney would later join Esso Libya as a director (1961), but at that time he sent Ralph Dale, then head of Creole's community integration program, to survey Libyan labor housing needs. Thus when Birney arrived in Libya, he had the results of this basic study to work with. On that basis, he recommended that an open community (where company and non-company people would live together, like several that Creole operated in Venezuela) be developed at Brega. Briefly, Jersey believed that unrestricted housing would

*To Americans.

contribute to good relations with all the workers.[11]

Eeds, Birney, Dale, and George McCammon (who supervised major drilling operations) represented only the first contingent of Creole employees who would either move to Libya on loan or join Esso Libya permanently. Proudfit, Wright, and Vázquez all possessed Creole backgrounds; not surprisingly, Esso Libya acquired a strong Creole flavor, which quickly began to pay dividends.[12]

Hugh de N. Wynne, another Creole veteran with wide South American experience, joined Esso Libya in 1963, as vice president and director.* Later, he explained the company policy on worker housing: "I think you have to go back to [Eugene] Holman and his predecessors.** They were the ones that said, 'All right, we have to house them, so let's house them properly.'" Thus Esso Libya built housing for both Libyan and foreign staff, and for many of their families. Wynne recalled:

> We built family housing at Brega for those employees who were primarily supervisors or technicians. The laborers we put up in bachelor accommodations. But in the fields everyone was housed on bachelor status. Because of the experience we had, particularly in Venezuela, we wanted to make sure that what we did in Libya, we did properly. We wanted to build good housing right from the start. ... We were severely criticized by other members of the industry for the manner in which we followed that practice. In Tripoli and Benghazi our employees rented existing facilities.[13]

Within a few years, the housing at Brega was regarded as a model of what could be done; Libya's rulers were impressed.

X   X   X

When the oil pipeline was completed and, as Wright put it, they had "built a harbor of sorts," Esso Libya decided to open the new line with a proper ceremony at Brega, at the time still largely a desert. To this celebration on October 25, 1961 came the Libyan King and Queen, Jersey President and Mrs. Rathbone, plus an assortment of other dignitaries. Wright and others first believed that only a few hundred persons would attend the ceremony, but Eeds insisted that a very large turnout was likely. More than two thousand persons appeared. As Wright told the story: "So out in the desert, hundreds of miles from any city, we had this event. And you know, we had big tents, we had barbequed lamb, principally. I don't remember any

*Wynne became president of Esso Libya in 1965.

**Jersey chief executive officer [CEO], 1944-1960.

camel." He continued, "These people came out of the desert ... like you wouldn't believe and you had no idea how the word got around, ... but this was a big feast." Rathbone and his wife joined the King on the platform, and the King "opened the valve for the pipelines."[14]

During the ceremonies Libyan officials, Esso Libya President Robert Eeds, and President Rathbone of Jersey made speeches. All the speakers emphasized the role of oil in the future of Libya. Eeds announced a 1962 shipping target from Port Brega of 125,000 b/d. The celebration went off well, and Jersey's relations with the government of King Idris remained pleasant for so long as he ruled. Rathbone later wrote a note for *The Lamp*, "Exporting Technology," in which he related that he had attended "the colorful ceremonies." He pointed out that in a mere two and one-half years, oil from Libya had become a commercial reality. Terming that achievement "something of a record in the petroleum industry," he gave credit to the Libyans for their impressive accomplishments in learning about modern technology.[15]

On August 25, 1961, Siro Vázquez received a cablegram from J.F. Moore, Jr. in Libya: "Oil reached Brega tankage 0500 hours this morning twenty fifth. Have produced 440,000 [barrels] and currently producing and shipping at 34,000 barrels per day." Moore also stated that the first tanker would arrive within two weeks, and that by that time two of the Brega storage tanks would be filled.[16] Once it began, Libyan production soon soared; Page's gamble was paying off. Meanwhile, Jersey created Esso Sirte Inc. [Esso Sirte], in 1959, to serve as owner of properties it might acquire in Tripolitania and Cyrenaica. In 1960, the new company acquired one-half undivided interest in concessions in those districts* from Libyan American Oil Company [Libam] and W.R. Grace & Company. Esso Sirte began to explore these concessions and discovered new oil at Raguba in January, 1961. So later, a branch pipeline from Raguba was constructed to join that from Zelten.

X X X

Both Esso Libya and Esso Sirte (although somewhat later) encountered the problem of what to do with the ever increasing quantities of gas that came to the surface with the oil—"wet gas." Given the limited market for natural gas within the country, some method of selling it abroad seemed desirable; otherwise, the gas had to be flared—burned. Former Creole staff members and managers had become thoroughly familiar with this type of

*Concessions 16 and 17 in the former, and 20 in the latter.

problem in Venezuela's rich oil fields, where gas had been reinjected into the reservoirs to maintain pressures. Wynne recalled, "Thoughts got started in Libya about how to use the gas. .... We didn't have to be strapped by tradition. .... There are some practices that have been built up by the oil industry elsewhere ... that are unnecessary or impractical or maybe even detrimental." The usual flaring of gas in fields located great distances from industrial centers, Wynne and his associates thought to be a traditional but wasteful practice. Instead, they hoped to put this energy to better use, and perhaps even to make it profitable for both Libya and their company.

In fact, Esso Libya began a study of the gas problem in 1962, soon after oil production started. As it increased, concern about flaring so much gas mounted as well. But markets had to be found before gas could be sold. At Jersey, a study of how such quantities of Libyan gas could be moved to market, and of where such markets existed, involved Directors William R. Stott, Siro Vázquez,* and Mike A. Wright, as well as Robert H. Milbrath, president of Esso International Inc. In a memorandum to Wright, M.A. Matheson reported: "We plan first to look at the big, expensive and likely less feasible solutions"; first [pumping the gas via] a pipeline to the European mainland—likely to Naples; and second, liquefied natural gas [LNG] and shipping it to market by tanker. Matheson went on to say that "the odds are definitely against the two solutions based on present evidence," and he suggested possible use in fertilizer production and in making cement.

During an interview several years later, Wynne explained the difficult problem:

> Of course with ever increasing quantities of gas becoming available as production increased, ... [the problem] was to sell large quantities of gas for long periods of time. We weren't interested in some guy who was prepared to take a million [cubic] feet of gas [per day] for two years. What we needed was someone who was prepared to take a hundred million cubic feet [per day] for 15 or 20 years.

Referring to a gas line to Europe, Wynne said:

> Engineering-wise it was possible** to lay a pipeline across the Mediterranean; however, you sort of lock yourself in [if you do]. If

*Vázquez became a director in May, 1965.

**Wynne knew what he was talking about. As early as 1944, the British built a pipeline under the English Channel (Pipeline Under The Ocean) termed PLUTO, through which 1 million gallons of gasoline per day reached allied armies. Other lines tapped this system; see Wallace E. Pratt and Dorothy Good, *World Geography of Petroleum* (Princeton, 1950), 346-347.

the line went to Italy and your Italian customer suddenly stops buying from you and you find another customer, say in the United States, that pipeline to Italy isn't going to do you much good. So there's a certain amount of flexibility in using tankers even if it means a tremendous investment in liquefaction facilities and liquefied natural gas tankers.[17]

The study of what could be done with the gas continued for several years. In the meantime, "plumes of black smoke" continued "going up in the air from the flares at Zelten and Raguba" as the surplus Libyan gas was burned.

Yet this wasted gas represented only one part of the larger problem of conservation at Zelten and other fields. In October, 1961, the Jersey Executive Committee listened to reports concerning "optimizing the development and operation" of Zelten. They were told that studies indicated "that early pressure maintenance of the Zelten reservoir would result in a tripling of the fields' ultimate recoverable reserves and would permit a substantial increase in its maximum efficient producing rate." The Committee "expressed no objection" to Esso Libya's plans to spend $34 million on "a 500,000 b/d capacity water injection pressure maintenance system" with beach wells and a pump station at Brega, plus a thirty-six inch water line from Brega to Zelten. Power for this system would come from the $11 million electric power generating plant at Brega.[18]

This problem of maximizing available oil and gas reserves continued to confront Esso Libya and Esso Sirte for several years. In June, 1962, W. J. Barnett, assistant coordinator of Jersey's producing activities, sent Wright (and others) a note saying that the new water injection project had created a "controversy in Libyan politics," and that the Libyan prime minister needed a statement for presentation to Parliament concerning the company's reason for using water injection. The answering memorandum pointed out that the concession agreement required Esso Libya to operate the oil field in a "workmanlike manner," in "accordance with good oil field practices" and with "appropriate scientific methods." Water injection met these obligations, and it would:

  "1. Prolong the life of the field,
   2. Increase its production,
   3. Permit the extraction of the greatest possible quantities from field reservoirs."[19]

By November, 1962, all components of the "Libyan LNG Project" had been brought together for discussion in a report presented to the Jersey Board Advisory Committee [BAC]. The report indicated that Libya's proximity to Europe would make a LNG plant economically feasible; potential markets existed in Italy, Spain, and the United Kingdom. On the other hand, the Libyan government could not be expected to condone waste, even "if Jersey decides to take no action regarding flaring." Clearly, the days of flaring gas were almost over.

In 1962, neither Jersey nor any other oil company had much real experience in liquefying gas in quantity. Jersey interests did share in a vast gas field in Groningen, Netherlands, but due to the field's location, the gas could be pumped by pipeline to major Western European industrial centers. No liquefaction was involved here, or with Jersey gas within the U.S. In the case of Libyan gas, however, Jersey believed that, with liquefication, it could be moved and marketed economically in those parts of Europe bordering on the Mediterranean. At the time only Algeria had a plant (actually still under construction) designed to liquefy large quantities of gas. However, Jersey researchers believed that they could find a more efficient method of liquefying the gas than the one to be used in the Algerian plant, despite the fact that the Algerian process had already been "successfully demonstrated."[20]

Esso Research and Engineering Company [ERE] immediately embarked on a full scale program for developing a more efficient, less expensive gas liquefaction process. While ERE experimented, Wynne "felt a need to go over all this with the petroleum minister in Libya and make sure we got what we felt was necessary to justify going forward with a $300 million LNG project." In effect, Wynne sought to clear the way for construction of a liquefication plant even before research was completed. By so doing, he hoped to avoid bureaucratic delays that might keep the liquid gas from reaching market.

After becoming Jersey chief executive officer in May, 1960, Monroe J. Rathbone pushed the Libyan LNG project at Marsa el Brega. He enjoyed a close working relationship with Vice President, Director, and marketing specialist William R. Stott, who felt confident that all the gas available could be marketed in Spain and Italy. Proposals to build the plant, however, ran afoul of Libyan authorities even before ERE had finished experimenting. Finally, in 1964, the Libyan government approved the long-term sale of LNG, and this move accelerated planning, construction of facilities, and especially the search for markets. The Jersey *Annual Report* (1965) predicted that the plant would be ready in time for shipments to begin during 1968.

On July 7, 1964, Esso International Inc., the marketing agent for Jersey and Esso Libya, signed a letter of intent with an Italian company, SNAM

S.p.A.* Subsequently in November, 1964, the parties signed a formal agreement. In 1965, Jersey announced that Italian and Spanish companies would purchase 345 million cubic feet of gas per day. While Director Stott participated in these negotiations, generally D.M. Latimer or R.A. Longmire of Jersey's Gas Coordination Department conducted them.

Jersey assigned a top ERE expert to supervise the LNG plant construction. Liquefied gas requires special handling and facilities, so plants for that purpose also had to be built in Italy and Spain. And since the parent company had affiliates in both countries, it became the affiliates' responsibility to supervise and control construction in these.

The Italian plant quickly turned into a problem. Esso Standard Italiana [ESI] joined SNAM in obtaining a permit to build a plant at Panigaglia. Across the next four years, 1965-1969, ENI, Esso Europe Inc. [Esso Europe], and Jersey surmounted a number of difficulties in their effort to have the Italian plant ready for early deliveries from Libya. Complications such as that concerned with proper security arose, as ERE questioned whether Italian construction companies using Communist labor unions should have free access to designs and processes. But in the end, the Panigaglia plant was ready before the Libyan gas could be processed.

Among the reasons for Jersey's interest in developing a major market for LNG in Italy was that of forestalling the entry of Russian gas into that nation. For some ten years, Jersey had been active in opposing the sale of Russian oil and natural gas in Western Europe—firm in its belief that among the major Soviet objectives were:

> increased foreign exchange to Russia, allowing the purchase of the large diameter pipe and compressors badly needed by the Russians for their own natural gas distribution network.

Jersey, the Executive Committee said, had a vested interest in keeping the Russian gas out of Italy. So Jersey's Gas Coordination proposed a master plan to achieve that purpose: Use natural gas from Groningen, pipelined across France into Northern Italy, and then market Libyan LNG over the rest of the peninsula. At company headquarters in New York, the Libyan LNG project received almost constant study from several departments. By 1966, Gas Coordination Manager D. M. Latimer had taken over as Libyan LNG coordinator. While Jersey and Esso Libya studied and planned the LNG plant and examined potential markets, other U.S. companies operating in Libya approached

---

*A wholly-owned subsidiary of Ente Nazionale Idrocarburi [ENI], and a gas transmission company.

Wynne to express an interest in selling their gas. Jersey authorized Wynne to explore in some detail these prospects for gas delivered at Brega.

When J. Kenneth Jamieson joined the Jersey Board in September, 1964, CEO Rathbone asked him to review progress at the plant. Already, construction costs had run beyond all estimates, and still the plant was behind schedule. Jamieson recalled:

> I got involved in the Libyan LNG thing. I don't know how many people agreed but it was Rathbone's idea. He had a plan that he was going to make this a model of how to develop a producing property in a foreign country where you conserve the gas. ... So they conceived this idea of putting in the LNG plant to sell the gas. And we charged ahead into that LNG thing and made some pretty damned sorry contracts.

Jersey had persuaded the customer-marketing companies in Spain and Italy to finance almost half the entire cost of the building and apparatus—pipelines, liquefaction plant, harbor improvements, etc.—representing a joint investment of almost $300 million. Four specially constructed LNG ships were to be built by other companies and leased to Jersey. These shipbuilding companies, in turn, also invested ashore.

Then suddenly, the entire Libyan LNG program began to sour. The expected 1969 start-up was delayed for months; then the Libyan government prohibited shipments until Esso Libya agreed to a higher tax. The government also demanded an option to buy 50 percent equity in the plant. While intermittently company-government negotiations went on, delays resulted in penalty clauses being enforced, and idle tankers brought on demurrage charges. Of course, the interest on investment represented an added cost, and the 1969 revolution also created problems.

The *Middle East Economic Survey* reported that Libya was using the gas price-tax issue as still another means of forcing Esso Libya to raise the posted price for oil. Yet by refusing to discuss the issues, the Libyans actually were hurting themselves, at least in Wynne's view:

> They were the greatest self-mutilators you'll ever find. ... We had been burning that gas for nine years and now we had spent over $300 million to send it to the market for the benefit of Libya and for the benefit of Esso and they weren't going to let us do it. So ... we continued to burn it.

Finally, in 1971, Esso Libya received authorization to ship LNG, but new technical difficulties forced plant shutdowns and thus made deliveries uncertain. So costs continued to mount. Full operation was achieved during 1972, yet in that year deliveries averaged only 50 percent of the scheduled annual production. By 1973, the plant reached 85 percent of its productive capacity although that rate did not fulfill the early promise envisioned by Jersey planners.

In fact, the Libyan LNG project proved to be a continuous source of expense and irritation to both Esso Libya and the Exxon Corporation.* When the Libyans demanded and received tax increases, the company had to renegotiate contracts with Ente Nazionale Idrocarburi (SNAM) of Italy and Catalana de Gas Natural of Spain. Certainly the American technology used did not, at that time, prove to be as far advanced as did the French process used in Algeria. One critic has labeled the entire project "one of the most ill-fated deals in the history of the oil business," and whether it would have ever paid out still remains a question. All together, Exxon poured considerably more than the publicized figure of $125 million into the Brega LNG plant, about which the company at one time boasted: "it is the largest of its kind in the world."[21] Surely the miscalculation was great.

But to Wright and Vázquez—and to Exxon—the contretemps with Libya about taxes and pricing natural gas could not be termed unique; for earlier, Esso Libya had experienced the same governmental displeasure when the Brega oil refinery was completed. In that case, the Libyan government attempted to dictate product prices to Esso Libya even before allowing the refinery to operate. Again, the company negotiated a settlement. But after the two such incidents, it became apparent to Exxon's management that the Libyan government treated agreements as temporary and subject to change.

X   X   X

As Esso Libya developed, teams of functional experts from Jersey headquarters visited the company. One such team arrived in early 1962; it included Wright and George A. Lawrence of Public Relations. Larry Birney had been given overall responsibility, and he asked Lawrence to provide "basic guidelines" for building and maintaining good public relations, as well as good employee relations at Esso Libya. Wright forwarded these guidelines to Eeds with a reminder that, since Birney had no experience in the field, "it is going to be necessary for you to take a personal hand in the public relations situation for a period of time." In fact, Birney's relations with Libyan officials and others proved effective for the company—until

---

*Standard Oil Company (New Jersey) became Exxon Corporation on November 1, 1972.

larger forces, such as political nationalism, acted to reshape the entire Arab world.

Initially, however, both Wright and Lawrence acknowledged that good public relations were essential to Esso Libya's success. Wright asked that he be sent, as part of the "bi-weekly report, a section dealing with your activities and programs in this [public relations] area." Lawrence interpreted company policy to Birney in this way:

> The actions of Esso Libya must be the actions of a company which represents the interests of the Libyan people and which expects, permanently, to be a respected member of the Libyan community.
> . . .

More than a decade later, Wynne gave his view of this initial phase of the company's operations:

> Here was an outfit, Esso [Libya], that played the game according to the rules, that always acted like a responsible, permanent member of the community, that responded to the educational, social and financial needs of the country, that cooperated with the government across a broad front. This all gave Esso Libya a certain amount of respect in certain quarters.

Then Wynne added:

> To the degree we could, we tried to identify ourselves with the social aspirations of the country. We built and operated top-quality schools, clinics, training centers, mess halls and clubs. We provided first-class housing for our technicians regardless of their nationality and bachelor quarters for the laborers, and everywhere we tried to minimize the difference between national and foreigner.

In the technical area, Esso Libya introduced many new developments in Libya that benefitted the company and the country alike:

> Esso adopted the use of flow meters to measure volumes of oil instead of strapping its tanks; it built tank farms with ponding areas instead of fire walls; it located chokes at collection stations rather than at the well heads; it developed bow and single point mooring devices for loading tankers rather than adopt the traditional procedures; it engineered a whole new concept for liquefying natural gas; it developed a new technique for sand dune stabilization.

In the financial area, Wynne believed that Esso Libya "paid its royalty on its posted price, not its realized price as required by law, because it was in

Libya's best interest. It cost us more but we wanted to be a good citizen."[22] Evidence to back up that assertion is certainly not conclusive, although basically the statement is correct.

Whatever their motivation, however, Jersey affiliates had to take seriously their role in changing Libya and life for the people who lived there. [One difficulty, as it now appears, was that few Libyans either knew or cared about the things that concerned Jersey.] Esso Libya and Esso Sirte's determination to use appropriate scientific and engineering methods in developing their oil fields has been mentioned, as have been its training programs for potential employees and its attempt to provide proper housing. Moreover, Esso Libya planned to operate for some time; its every installation was substantially built, to convey a message that the company had not come to exploit, make quick profits, and then run away. Again, it was the only foreign company to undertake large scale experiments in sand dune control and reforestation, both in the best interest of the host country.

In 1961, working with the Forestry Department of the Libyan government, Esso began an experimental program designed to stabilize sand dunes, especially those dunes encroaching on the once arable land along the coast. Workmen sprayed the dunes with a light coat of crude oil and subsequently planted seedlings. The oil kept the sand from being blown away, which would have exposed the trees' young roots to the killing sun. By 1963, the "Sand Dune Stabilization Program" appeared to need "something dramatic to give it a major stimulus." So Luke W. Finlay, president of Esso Mediterranean, Inc., recommended that Libya be selected for "a dramatic development" both because the encroaching sand dunes had to be checked and because agricultural land already ruined by sand intrusions (as well as by wind erosion) badly needed restoration. Finlay believed that Jersey would be assured of "top-level interest [from the Libyan government] in any proposal," especially since the Libyan prime minister seemed friendly to Jersey. Producing Coordinator Vázquez and Directors Wright and Lloyd W. Elliott reviewed Finlay's letter and decided "not to press this matter but to continue experimental programs as proposed by Bob Eeds." So research on sand dune stabilization continued for several years, and it enjoyed considerable success. But after Exxon left Libya for good, on December 1, 1981, the desert sand once again began to invade usable farm land.[23]

Meanwhile, company relations with the Libyan government of King Idris, while generally smooth, nevertheless did provide anxious moments for the managements of both Esso Libya and Jersey. In July, 1961, for example, the government issued a circular letter to the oil industry barring two U.S. construction companies "from any further petroleum contracts without the

Petroleum Commission's express written approval." The Americans were "blacklisted" for doing work in Israel even though one of them, the Bechtel Corporation, had already contracted to build Esso Libya's oil separation plant at Zelten. Wright noted on the cablegram, "Squeeze on Bechtel." The Libyan government instigated a newspaper campaign against Bechtel, but when that company added a local firm as partner, that campaign ended and Bechtel was dropped from the blacklist.[24]

Another incident illustrative of early company-government relations occurred in 1961, when Radio Libya broadcast, and the Tripoli daily Arabic newspaper copied the official Libyan Petroleum Commission's objection "to the price determined by Esso Standard Libya Inc. for the sale of Libyan crude petroleum." The media alleged "that such price is damaging to the rights and interests of Libya" and announced that the commission proposed to fix "an appropriate price." But instead, the commission chose to seek the price change in an amicable manner, until as will be shown, a fair accord could be reached.[25]

In its early stages, Libyan oil production grew rapidly. As early as November, 1962, in fact, the crude available* exceeded Jersey's European refining capacity and gave the parent company a new set of concerns. For as they studied this problem, Jersey experts realized that "even if this refining limitation is removed, our ability to absorb the Libyan crude would still be limited by politically acceptable minimums from other Jersey crude sources." The same writer went on to point out that "substantial volumes of Libyan crude ... above our ability to refine and/or absorb in the total crude demand picture" would almost certainly exist from 1963 to 1965. Still, discounting to customers, exchange, and sales to third parties in the United States "via purchase of import tickets**" all seemed to be possible solutions, although in the latter case, "This could result in a Humble [Humble Oil & Refining Company] refinery running Jersey Libyan crude imported under a third party's quota." A cablegram explained the situation succinctly:

LIBYAN Availability In 1963-1965 Exceeds Our Capability to Refine in Europe. Even If Refining Limitation Is Removed And Libyan Maximized, it would be necessary to Restrict Crude Growth In Other Producing Areas.[26]

In its hunt for crude, Jersey had achieved so much success as to present a new

*Libyan plus other Middle East crude.

**Required for imported oil under the mandatory quota system imposed by the U.S. government in 1959.

set of difficulties. Of course, there was no easy way to solve the balancing problems created by booming production in Libya.

In March, 1974, Howard W. Page, Jersey's foremost Middle East expert, now retired, explained to a subcommittee of the U.S. Senate Committee on Foreign Relations that, at the moment when Libyan oil development occurred, Iraq's production had sharply declined. This reduction enabled Jersey (and other companies) to utilize high Libyan production as a means of offsetting the loss of Iraq's oil and of meeting new increases in demand. Page admitted that he had managed or "orchestrated" Jersey's Middle East production, but he also confessed to facing "some tough problems" in doing so.

In truth, Jersey never found a satisfactory solution to the oil glut in the 1960s; nor could it have solved the problem. Iran and Saudi Arabia, among many host countries, constantly urged the companies operating in their fields to produce more oil. Iraq was the exception, as has been shown; there, production of the Iraq Petroleum Company dropped off because of politics, those well recorded disagreements between the oil company and the Iraq government. Libya filled that gap and more; and before long the era of shortage had ended. By 1970, demand was beginning to absorb surplus producing capacity, and the new oil age opened with the ill-fated [for the West] Tehran and Tripoli Agreements of 1971.[27]

X   X   X

Among the attractive features in the 1955 Libyan petroleum law was the fact that tax provisions (as stated earlier) were based on realized, rather than on posted, prices—an arrangement that proved especially advantageous to independent companies.[28] For although Esso Libya discovered the first oil field in Libya, soon afterward a consortium of independent companies known as Oasis Oil Company* began to bring in oil wells on its concessions. Oasis paid taxes only on realized prices. Once Libya began to study its situation, the host government realized that its oil was bringing in far less revenue than did the oil produced in Persian Gulf states. So in a Royal Decree of July 3, 1961, by acts of Parliament in October, and by Petroleum Commission regulation in December of that year, Libya sought to alter the method of computing taxes, and thus to increase its income. These revisions required the posting of prices, but they also allowed a deduction of expenses for marketing crude oil. Thus, in 1961, royalty became based on posted price, but income tax continued to be based on realizations.

---

*Oasis consisted of Marathon Oil Company, Continental Oil Company and Amerada Petroleum Corporation [Amerada]. Later, Royal Dutch-Shell Group [Shell] bought a share of Amerada's interest. Already, Amerada had been selling all its crude to Shell.

As usual, the Jersey Executive Committee followed these Libyan actions closely; then it requested Esso Libya (and Esso Sirte) to make the parent company's position known to the Libyan Petroleum Commission. Finally, in December, 1961, the Committee "expressed no objection" to Esso Libya and Esso Sirte indicating their willingness "to adapt their existing concession interests to the terms of the amended law," when the Parliament approved it.[29] Thus, a 1963 company pamphlet, "Esso In Libya" (printed in both Arabic and English), could declare that Esso Libya had been, "Among The First To Agree To The Petroleum Law Amendments of 1961."[30] Yet, despite such cooperation, all the rulings, decrees, and laws together did not provide as much income as Libya expected. One writer described the situation this way:

> As Libyan output rapidly increased, some disturbing anomalies appeared. In 1964 Esso International sold most of its oil to affiliates at a posted price of $2.21-$2.22 a barrel and paid taxes to the government averaging about 90 cents a barrel on an average realized price of about $2.16. In contrast, the Oasis group sold its oil at an average of $1.55 (a discount of 67 cents a barrel) and paid to the government an average of less than 30 cents a barrel—only slightly more than bare royalty and rental charges.[31]

Not surprisingly, the continuing shortfall in revenues increasingly alarmed the Libyan government.

Wynne has described how this marketing differential came about:

> Based on Exxon's experience in the Persian Gulf, Howard Page was very anxious from the outset that it be written into the petroleum law that taxes be paid on the posted price [of crude oil]. Other companies, mostly independents, wanted taxes paid on the realized price and this was the way it turned out. Fuad Kabazi, the Petroleum Minister, kept telling us that exploration in Libya would never have been so rapid or extensive had taxes been paid on posting. Nevertheless, Esso, in the interest of being a good citizen, paid its taxes, on all sales to its affiliates—about 97% of its exports—on posted price, and on sales to third parties—about 3% of its exports—on realized prices as called for by the law.

Wynne went on to point out how the companies' paying taxes on realized prices actually worked to Libya's disadvantage—as well as to the detriment of those major international companies with large downstream investments:

What was happening was you would have 2 tankers loading in the Gulf of Sirte, one at Esso's terminal at Marsa al Brega [sic] and the other at Oasis's terminal at El Sider. Once loaded these 2 tankers would track each other out of the Mediterranean to Germany, let's say, where they would discharge their cargoes at refineries literally situated side-by-side at Hamburg. Here the crude would be chewed up into gasoline to be sold at company owned service stations on the same autobahn. In the one case, Esso was paying taxes to the Libyan Government on the basis of its $2.21 posted price while the Oasis partners were paying the Libyan Government on the basis of their $1.60 realized price. [32]

This inequity in tax payments naturally created considerable ill-will in Tripoli, because the Libyan government understood that it was being deprived of substantial revenue. Yet, ironical as it seems, although paying higher taxes Esso Libya gained no favor with the government. On the other hand, many of the independent companies that found oil in Libya had no substantial marketing operation outside the United States; thus, the imposition of mandatory U.S. quotas on oil imports (1959) served to prevent them from marketing their surplus there. So they began to look for other markets, especially in Europe. To compete in Europe, these companies found it necessary to discount prices on crude oil, and then to augment total income by increasing Libyan production. As Wynne put it, "Many of these other companies ... didn't give a damn if they were a good citizen or whether they employed good oil field practices. Their attitude was: 'let's get the oil and get out.'" The contrasting behavior did not protect Esso Libya's position, however.

In June, 1962, Libya joined the Organization of Petroleum Exporting Countries [OPEC]. Libyan officials attended the fourth OPEC conference held that year in Geneva, where members established their first program to deal with prices, royalty expensing, and marketing allowances. OPEC members also agreed to negotiate with the oil companies on matters of: a linking of oil prices to the price index of goods which they imported; discontinuing the companies' practice of deducting royalty payments from income tax liability; and an end to all contributions by producing countries to help pay marketing expenses.

Although this last issue was largely settled by 1964, the price and royalty payments problem still remained for subsequent negotiations. When the sixth OPEC conference accepted a proposal to expense royalties with a graduated reduction (over a three-year period) of the discount from posted prices for tax purposes, the Libyans moved to force oil companies to comply. Esso

Libya received a letter on November 7, 1964, giving it ten days to agree to the OPEC—now the Libyan—plan. The company acted swiftly and agreed. In 1965, both royalty and income tax became based upon posted price.[33]

X   X   X

While Esso Libya and the other majors had agreed to an arrangement earlier, the question of deducting royalties as an expense emerged anew and almost wrecked the Djakarta meeting of OPEC in November, 1964. Independent oil companies refused to go along with the majors on this matter, just as they continued to pay taxes on the realized price of oil, rather than on posted prices. Until January, 1966, Libya did not act to force these independents into compliance, although the Libyan Parliament in December, 1965, had authorized the Council of Ministers to stop all oil production and exports, and to "expropriate the assets of companies which did not comply" with the regulation that taxes be based on posted prices and "royalties expensed in line with the majors' offer." Further, the government refused to decide on bids for new concessions, thus providing additional time for amending existing agreements as a means to safeguard its interests. Both taxes and deductions made the independents special targets for government action.[34]

Unlike legislation in other Arab nations, Libyan law included special bonus clauses for companies in compliance. For example, the government could forgive outstanding claims incurred before 1964. An additional clause provided in extreme cases for "arbitration under international law."[35] Yet neither possibility satisfied the independent companies; they continued to oppose the 1964 decree on the grounds that any increase in taxes would force them out of Libya. Such a withdrawal, they contended, would mean that Libya must become part of the oil majors network, where production would be balanced with that of the Persian Gulf states. In this view, at issue was Libya's independence. Finally, in December, 1966, OPEC came out in support of Libya by passing a "resolution that all member countries refuse any oil rights of any nature whatsoever 'to any company or the subsidiary of any company refusing to comply with Libya's new oil policies.' "[36] Of course this OPEC action helped to persuade most of the reluctant companies to comply with the Libyan decree, and the December law, with its additional authorization for expropriation, provided enough of a threat to force compliance from the remaining independents (not, however, before their actions had aroused much hostility in Libyan government circles).

The three-year posted price tax discount agreement was to remain in effect through 1966, while the other problem remained unresolved.

Negotiations on royalty expensing dragged on, into June 1967, when the Six Day Arab-Israeli War interrupted. That struggle simply barred lesser considerations, for the war radically altered the entire world situation in oil.

X   X   X

In February, 1966, after months of evaluating bids submitted by twenty or more American and European oil companies for new concessions offered by the Libyan government under the 1961-1965 revisions to the petroleum law, awards were announced. Esso Libya, which bid on the two best blocks, received none. Wynne remembers the feeling of "surprise, shock and hurt" he experienced when he heard the news. He knew that Esso had bid competitively, and he questioned why "after all Esso had done for Libya since receiving its first exploration permits in 1953" it deserved such treatment. No two government officials gave him the same explanation. Subsequently, no fewer than six of the companies to which Libya did award concession acreage approached Esso Libya with farm-out* or acquisition proposals.

Among the independent oil companies that obtained concessions was Occidental Petroleum Company [Occidental].** Under the leadership of Dr. Armand Hammer, in 1966, Occidental obtained two surrendered blocks of the Mobil Oil Corporation's [Mobil]*** concession released to Libya under the terms of the 1955 petroleum law. Both of these blocks of acreage lay in the Sirte Basin, not far from Esso's Zelten field. When Occidental began drilling on its newly acquired properties with two contracted rigs in 1967, Siro Vázquez reported to Jamieson that both tracts appeared to be likely oil properties. So some time later the Jersey president explored with Hammer the possibility of purchasing them. Wynne, on the scene, reported to New York that a "potentially extensive reefal play situated in the trough"—roughly comparable to that in which Jersey had brought in the first Zelten well—existed in the two tracts that Occidental now owned. Jamieson's later efforts to buy them failed, however.[37]

Earlier in March, 1967, Wynne learned from the local Occidental manager in Tripoli that Dr. Hammer was anxious to see him. Knowing that Occidental had neither drilling nor producing experience, and that it lacked marketing facilities, Wynne suspected that Hammer would either make a

*A farm-out is a sharing of oil or gas exploration activities and costs.

**This company was acquired by Dr. Armand Hammer in 1956. Prior to that time, it claimed a book value of $34,000 and was located principally in California, where it had no oil operations.

***In 1966, Socony Mobil Oil Company became Mobil Oil Corporation.

proposal for use of Esso's pipeline and terminal facilities or offer to sell outright. Wynne flew to New York where, after consulting with Jersey, he went on to Los Angeles for a meeting with Hammer and four of his aides. As Wynne remembers:

> Hammer opened the discussion with the comment 'Mr. Wynne, I don't understand Esso ... ever since I got these two concessions in Libya everybody has been knocking on my door with a proposition ... but I never heard boo from Esso. ...

Wynne replied, "Well, Dr. Hammer, we bid on one of those concessions and didn't get it. Some you win, some you lose. We can take our knocks. We're producing over 600,000 barrels a day in Libya. We're doing all right."

Hammer, seeming to forget that he had invited Wynne to come and listen, then asked: "Well, what is your proposition?" Wynne replied, "My proposition? You invited me here to see you! I thought you had a proposition to make Esso." Finally, Hammer excused himself and, with his senior aide, left the room. About 20 minutes later, they returned and offered Esso an undivided 50 percent interest in Occidental's two concessions* for $100 million, and Occidental's share of the production at $1.55 per barrel with a 3 percent override. After discussions during the next 30 days, Wynne and Hammer reached substantial agreement, and Wynne left for New York to review the deal with Siro Vázquez, alternate contact director for Libya.

Like other Board members, Jamieson wanted to buy the Occidental property outright. So after Wynne had left for Libya, Jamieson asked Vázquez to go see Hammer and try to buy the two leases. Vázquez did so, and then he reported on the meeting to Jamieson. For some reason, Jersey dropped its offer to buy the property. The result, as Wynne recalls, left room to reconsider conventional oil wisdom:

> Well I am sure there were a lot of Jersey directors who had second thoughts about having told Hammer to 'forget it' when they watched Occidental not only find oil, but get their production up to over 800,000 b/d, overtaking Esso Libya as the largest producer in the country.

After the find at Intisar in 1967, Occidental production did reach 800,000 b/d by 1970.[38] In retrospect, Jersey lost a golden opportunity, in part because of the unusual course of negotiations with Hammer, which included harsh terms for a farm-in, but also because of the Department of Justice's then

*Then numbered Blocks 102 and 103. Previously when Esso Libya checked them, they were numbered 42B and 44

abiding interest in U.S. oil companies abroad. Jersey avoided both the prize and the problem.

X   X   X

By June 1, 1967, the Arab-Israeli confrontation in the Middle East had simply eclipsed every other regional concern; it was a major crisis. In a letter to Siro Vázquez, Wynne described the Libyan situation as "tense" and the atmosphere as "brittle." A Libyan political group demonstrated in front of "the U.S. and U.K. embassies," and the newspapers announced that both the King and the prime minister were making "personal donations" to Egypt in support of the Arab cause. Yet Libyan troops sent to Egypt by King Idris "were turned back at the Egyptian border as unacceptable," because they were little more than policemen. In fact, they were needed at home because of some rioting in Tripoli. In Libya, attacks on Westerners found alone became common, and Americans especially felt the strain.

On Monday, June 5, 1967, war between the Arabs and Israel broke out. Israeli military planes from Tel Aviv, flying well out over the Mediterranean, past Alexandria, suddenly turned back and attacked Cairo from the northwest, destroying on the ground every plane of the Egyptian Air Force. Gamal Abdul Nasser's response was immediate and predictable. Using "Radio Cairo," Nasser planted the "Big Lie." He informed the Arab world that Cairo had been attacked by U.S. planes from the Sixth Fleet and from Wheelus Air Force Base in Tripoli, Libya. The "Six Day War" had started and Libyan tenseness turned first into bitterness, and then into violent reprisals against foreigners—particularly Americans. In Libya, frustration over Arab impotency in the war ran so deep that—even after the bombing story had been proven false—Arabs still insisted that the U.S. had promoted Israeli aggression by failing to assist the Arabs, "which tacitly guaranteed that an Israeli defeat would not be permitted. ..."[39]

When the news of Israel's attack on Egypt reached Libya, fresh violence broke out, including demonstrations in front of the U.S. Embassy, which led Esso Libya's Controller's Department to abandon their nearby building. Another group of demonstrators visited the Oasis Oil Company and forced it to shut down all operations for several hours. Then, the Libyan government declared a state of emergency, including a daily curfew in Tripoli from 7 p.m. to 6 a.m. Nonetheless, burning, pillaging, and looting continued throughout the city.

On June 6, U.S. Ambassador David Newsom, recalled from vacation by order of the State Department, "categorically denied U.S. involvement in the conflict" and "advised [the heads of U.S. concerns] that an emergency airlift of military dependents living off the [Wheelus] Base had commenced." He urged all U.S. dependents to join the evacuation, and Wynne quickly sent the Tripoli-based dependents of Esso Libya to Wheelus. Next, he telephoned the heads of the Italian, Spanish, and German affiliates of Jersey, and he told them:

> We're having a problem over here and our people are going to be evacuated by the U.S. Air Force. I'd like you to contact the U.S. Embassy to find out if any of those planes will be landing in your country and if so make arrangements to contact our people and see that they are taken care of.

Later, he pointed out, "That's the advantage of being a member of a multinational like Esso." Wynne also noted that, although the evacuation was designed for dependents, some officials of other companies operating in Libya—even their top executives—"jumped on the planes with their dependents." In one case, a departing CEO called Wynne and asked him to "watch out for their operations." All together some 8,000 dependents of American and British oil and construction company personnel were evacuated from Wheelus by the U.S. Air Force; while another 6,000 dependents flew from the Tripoli airport on commercial airlines.

At first, the Libyan government demanded that oil operations be continued, and Esso Libya employees remained on the job. A day later, however, the oil ministry reversed its position and ordered Esso Libya along with the rest of the industry, to cease producing and loading oil. The company virtually closed down all field operations. Yet enough oil was produced to feed the local refinery, which the government required Esso Libya to operate. Since normal communication with the outside world was shut down by the government, Esso Libya had to establish a new line to Jersey headquarters in New York, using the tug Cyclops, at anchor in Brega harbor.

On June 7, bombs thrown over the wall around Wheelus caused little real damage, but they did heighten the tension among dependents awaiting evacuation. Shell held a contract with the Algerian government to supply aviation gas, and Algerian MIGs "enroute to the combat zone" in Egypt began to land at Tripoli for refueling, adding to the general war fever. When Shell's refueler broke down, Esso Libya was compelled to refuel the Algerian planes, bringing the company directly into the conflict. In all, some 600 dependents of expatriate Esso Libya employees were evacuated to Naples or Rome, where Esso Standard Italiana took care of them. A few others were flown to Spain, and the evacuation stood complete.

In Libya, despite the efforts of the petroleum minister to start up production and exports, the labor situation dictated a continuing shutdown. Many Libyans had joined the Petroleum Workers Union, an organization led by an American-educated, Syrian-born lawyer of Palestinian descent (who had been working for Esso), Mahmud Sulayman al-Maghribi. He, along with other Arab Nationalists, felt that labor should display strong support for the Arab cause; so this "fledgling but vociferous group" issued bulletins denouncing "Western Imperialism," called for the elimination of all zionism, and proposed the boycotting of all U.S., British, and West German planes and ships. In a less strident way, the Petroleum Workers Union also requested a study to determine how the lot of the Libyan laborers could be improved. With this agenda in view, a series of walkouts began on June 16, and led to a city-wide strike in Tripoli on July 5. Finally, on July 7, a general strike ensued, which forced the government to impose a curfew. By July 9, calm had been restored, the curfew lifted, and workers began to trickle back to their jobs.

During this time all attempts to resume oil production had been effectively thwarted by the labor organization. Now, the government took steps to educate the public concerning the dire results to Libya's economy that would occur if oil did not soon begin to flow. Gradually, however, the situation became less tense, and by August, oil once again was moving into world trade. According to Wynne's estimate, however, the shutdown cost Esso some $11.5 million in lost profits.

Meanwhile, within the Libyan government, the Six Day War was having serious political consequences. Arab activists became more vocal in their demands for ever stronger Libyan support, and a cabinet crisis developed on June 28, which resulted in the resignation of the prime minister, on July 1. Aside from the oil question, this minister had opposed the continued operation by the U.S. of the huge Wheelus Air Force Base. King Idris, on the other hand, insisted that the U.S. base remain open, and this conflict may well have led the King to ask for his minister's resignation. In a compromise move, the new prime minister permitted oil to be exported to West Germany, but none could be shipped to the U. S. and the U. K.

During the shutdown and the disturbances, Esso Libya's dependents remained abroad, but finally they were able to return on July 19. Yet the Libyan government decided not to recognize (for tax purposes) the costs to the company of sending these dependents out of their country for their own safety. A more positive evaluation of these Libyan affairs is possible, of course; for the evacuation provides one important endorsement of the overall effectiveness of Jersey's philosophy of decentralized management, which encourages managers to deal with contingencies at the scene of the action.[40] Libya made that policy look good to the world, while Wynne explained it from the manager's perspective:

> You were in charge, you were given guidelines and goals, and as long as you could operate within these guidelines and achieve the goals, ... [you] ... were left alone. There wasn't someone looking down your neck all the time saying 'Do it my way.' ... You had the right to go ahead and do it your way. But ... if you did it and didn't achieve those goals, then you were looking for another job.[41]

Even during difficult days, Wynne and his people made this philosophy work. With cooperation from other affiliates, a local company protected with great success both property and people. In retrospect, Wynne, on the scene, could handle each situation as it arose, secure in the knowledge that this was the Jersey policy.

X   X   X

For 1967, Esso Libya had targeted a production goal of 582,000 b/d; however, the Six Day War and the twenty-seven day ban on exports in June and July defeated that estimate. Of course, other factors also operated to curtail production elsewhere, including the closure of the Suez Canal and the Iraq Petroleum Company [IPC] pipeline; a slowdown in European economic activity; and an unseasonably warm winter. Seeing profits drop, Wynne grew restive and began to question whether Esso International coordinators and worldwide logistics specialists were paying adequate attention to the availability of Esso Libya's oil. C.O. Peyton, vice president and director of Esso International, defended his staff by taking issue with Wynne's evaluation:

> It is disconcerting to receive cables from Esso Libya which appear to not recognize the efforts of a number of dedicated people, and which frequently in tone and language indicate a lack of appreciation of physical limitations which can prevent the achievement of crude export targets.

Peyton went on to exonerate his company's supply representatives, who "are thoroughly familiar with the current and short term Brega crude outlook. ..." As the year wore on, however, Libyan crude did come to occupy a more important place in the Esso International supply picture—as tanker freight rates escalated and the longer haul from the Persian Gulf via the Cape of Good Hope took additional time.[42]

Perhaps predictably, Occidental Petroleum Company continued to make life difficult for the oil majors in Libya. Once the export of Libyan oil resumed in 1967, Occidental signed a contract making Signal Oil and Gas Company [Signal] Occidental's exclusive agent for the sale of crude in Europe. Before that time, Signal had purchased 70,000 to 80,000 b/d of crude from Esso Libya. Prior to publication of this contract, both Jersey and Gulf Oil had negotiated with Occidental, hoping to buy crude oil. But on the day the contract story appeared, Hammer telephoned a Jersey official to say that "he was no longer interested in discussing our [the Jersey-Gulf] proposal." Whatever the advantage in Hammer's mind, his deal with Signal has been explained this way:

> The Signal arrangement will give Occidental the freedom to get Libyan production up to a maximum potential in the shortest possible time whereas the Jersey/Gulf deal would have 'bottled him up.'[43]

Libya also kept busy looking after its growing oil interests. Shortly after the 1967 War ended, a "permanent" price commission was established to study "the possibility of raising posted prices during the current conditions of international supply dislocations."[44] This commission immediately began to press the oil companies to increase prices, which would also increase Libyan income. At the same time the Libyans demanded that oil companies phase out their percentage allowance off the posted price—a discount that was, by prior agreement, subject to negotiation in 1967.[45] Wynne defined the government position in this way: "Satisfaction to Libya ... boiled down almost entirely to what can be gotten now during present extraordinary circumstances resulting from Suez closure." Every other concern, such as tax allowances, adjustments based on the better gravity Libyan oil, and similar matters, he wrote, "pales by comparison with that [tax money] available through an increase in posted price. ..." But the problems remained complex, and finally, the Libyan oil minister indicated that he might have to turn the matter over to the Council of Ministers, which action, Wynne believed, would result in Parliament becoming involved.

Wynne himself held the view that the oil companies should not yield to demands for an increase in posted price. He pointed out to the minister that, because of the war, Esso Libya already had increased production substantially. Since the minister stated that he had to act quickly, Wynne urged full consultation with all the managers of oil companies in Libya before the minister acted.[46] Libya remained adamant, however, and in January, 1968, the companies agreed to remove the allowance off the posted price until Suez reopened. In the meantime Siro Vázquez, aware of other negotiations with Aramco, cabled Wynne and told him not to agree to any terms of settlement with Libya that might "in case of *failure* of negotiations in Persian Gulf[,] result in our being subject to *upward whipsaw* between *Libya* and *Persian Gulf*."[47] Later, in May, 1968, Occidental claimed from Libya the 6 1/2 percent allowance off the posted price permitted for 1966 (and subsequent years) under the revised concession agreements.[48] Responding for his government, the oil minister called in the Occidental manager and told him to eliminate that allowance, citing the Suez closure, and the company agreed. In some ways at least, the Libyans seemed to be gaining control over the pricing of resources.

Yet overall, Libya remained dissatisfied with its oil income, so the government began to question the yearly tax returns of oil companies, and to disallow certain of their depreciation and other expense claims. Esso Libya's 1965 return, for example, was challenged at five places, among them Esso Research and Engineering Company charges for gas liquefaction study and research. The government also contended that Esso Sirte owed an additional $145,000 on its 1965 tax. Many of these claims were adjusted amicably, and finally, most settlement costs paid by Jersey affiliates were recovered under Libyan tax laws.[49] But immediate income still dominated the list of national priorities in oil.

Again, the government asked for, and received from the seventeen companies producing oil in Libya, a "phase-in" of all tax payments on a quarterly basis,* with all royalty payments to be tendered on a monthly basis, and all monies due thirty days after the end of each month. In a letter to the companies, the oil minister advanced reasons for this step, and then he added a note of warning: Libya would be "unable to give clearance on accounts for 1966 and 1967, until it has an opportunity to review these accounts." Additionally, he predicted that when the 1965 accounts had been cleared, there would be fewer difficulties. When it agreed to all requests, Esso Libya secured clearance for its 1965 taxes.[50] As Wynne recalls:

> Esso took a lead in some of this, in trying to speed up some of these payments for the benefit of Libya. Or if we didn't take the initiative, at least we didn't discourage them when they suggested this, provided their suggestions were reasonable. The other companies, most of them fought it like a stuck pig. ....[51]

*The concession agreement specified annual payment.

But on other issues, the companies enjoyed common interests, so Esso Libya continued to explore every opportunity that might lead to profitable new investments (either with other companies or in wholly owned concessions) in Libya.[52] Drilling also continued in Esso's western concessions, but little of promise appeared.

By 1968, the Libyan government seemed to be trying several new ways of harassing the oil companies. First, it announced cancellation of the concession of four American oil companies, citing their failure to begin operations within the required eight months after receiving their grants, and ignoring the companies' protests that they had complied with the law. Next, the government brought additional pressure against Esso Libya and the two other companies that marketed petroleum products in Libya, by alleging that their terminals in Libya were unsafe. In fact, the minister of municipal affairs simply notified all three to relocate their marketing terminals (without compensation). Esso Libya denied that its terminals were unsafe, but agreed to review with the minister any "safety aspects" in question. Esso Libya received no reply. Next, the Libyan government refused to renew the Benghazi operating permits (and other licenses) of Esso Libya and the two other companies involved in the terminal safety order. Then, all three companies were refused permits to build new retail outlets in Tripoli. Marketing plans were crippled, but at this point, Esso Libya determined to maintain a low profile and to await developments, especially since rumors persisted that the government-owned Libyan Petroleum Company [Lipetco] now planned to expand into local marketing.

By this time in 1968, Wynne and his associates at Esso Libya understood that the Libyan government had altered its attitude toward the major oil companies. It was recruiting engineering, geological, and other trained personnel to work in a newly created Technical Affairs Department within the petroleum ministry, and that ministry was also creating a Planning Bureau for as yet unexplained purposes. So far, at least, the company remained sympathetic; in its "Monthly Corporate Review," Esso Libya urged both the U.S. government and Jersey to assist the Libyans in hiring trained specialists, "Because of the favorable influence and attitude U.S. technologists might bring to the government sector of the petroleum industry in Libya." Yet no clear directions had been signalled by the Libyan moves, although it seemed likely that Libya proposed to study its domestic oil industry, with the intention of making significant changes therein.[53]

Now in retrospect, we can see these moves for what they were: strengthening Libya's position in OPEC. So much was not clear at the time, although Wynne has stated: "We could see the handwriting on the wall." Yet, on the surface at least, Libya appeared to settle down: laborers returned to work, violence ended, and the political situation became less tense. Perhaps most misleading of all, and as if to confirm the state of normalcy, King Idris took a long vacation in late 1969. What he left behind was a pot ready to boil.

X   X   X

King Idris faithfully served his people as spiritual leader and political ruler, even while relegating much of his temporal power to his ministers. But he was helpless in the flux of Arab nationalism and a victim of alleged corruption in high places. Popular criticism of local politicians soon became criticism of the King from whom these ministers had received their authority. Yet little reform ensued. Instead, after the Arab defeat in the short 1967 struggle with Israel, political intrigue seemed to become the principal order of court business, especially within the Libyan Army. Now and again, King Idris tried to establish better order, but he did not succeed in quieting the popular clamor against his government.

Thus, beneath the seeming calm, at least two groups of plotters continued their intrigues. The first, composed of ranking army officers and members of the royal circle, became fairly well known. This group, fearing that King Idris would abdicate (as he had once attempted) or die, made preparation to take control in either instance. Essentially, they proposed to maintain the monarchy and their own style of easy living. In part, they were moved by rumors, especially the well-circulated one that Libya would become a democracy, and the fear that King Idris might try to achieve that end.[54] When, in the summer of 1969, King Idris did announce his visit to Turkey, almost everyone in Libya expected something dramatic to happen in his absence. Of course, something did, although not what most people had expected.

Instead, a second group of dissident army officers (unknown to the first), consisting of a dozen or so captains and lieutenants, took control. The most prominent member of this circle, Muammer al-Qadhdhafi [Qaddafi], had taken advanced military training in Great Britain. He was intrigued by the power of radio in the largely illiterate Arab world, and he made himself expert in radio transmission and reception. Qaddafi often spent his spare time in camp listening to Gamal Abdul Nasser on Radio Cairo. All of the officers around him were younger than thirty, and they shared a fundamentalist, fanatical Islamic faith, while Qaddafi in his own puritanism approached a religious mystic. This faction was composed of Arab, anti-Zionist, anti-Western, anti-Imperialist, xenophobic, and desert-lean

Socialists.[55] Yet within the social and political order then existing in Libya, these men operated at a level so far down as to be largely unobserved and unknown. They quickly proved, however, to be a potent force in Middle East politics.

These rebels struck on September 1, 1969, while the King was in Turkey. First, they gained the support of their fellows at the army base, and then they captured the radio stations. So swiftly did they move—no blood was shed—that some of the colonels in the other cabal believed that *their* revolution had started. When they rushed to join in, they were seized and imprisoned. King Idris himself never returned to Libya but went instead to Egypt, where Nasser provided him with a palace in Alexandria.*

Whether Qaddafi masterminded the plot is still not clear, but he and his cohorts quickly formed a Revolutionary Command Council [RCC], of which he became the head. On September 2, he announced that Libya would honor all existing agreements with the oil companies, thus reassuring both the Libyan petroleum affiliates and their Western parents. Less comforting, however, was the new foreign minister's statement: "Oil," he said, "must be used against our enemies; it does not make sense to sell oil to America which supplies Phantom planes to Israel, which are aimed to kill Arabs."[56] Meanwhile, the Organization of Arab Petroleum Exporting Countries [OAPEC]—uncertain about the revolutionary government's attitude on oil matters and viewing Qaddafi as a radical Socialist—postponed its next meeting, scheduled for Tripoli during September.

On September 8, 1969, the RCC appointed Maghribi as prime minister. A Syrian-born Palestinian with a Ph.D. from George Washington University, he had once been employed in the Esso Libya Law Department. Later, Maghribi was imprisoned by the government of King Idris for encouraging strikes against that regime, so he looked attractive to Qaddafi's revolutionaries.

Qaddafi himself adopted a policy of continuous confrontation with foreign oil companies (and some foreign governments), coupled with open hostility toward their Libyan representatives. This policy allowed him simply to ignore all previous contracts and agreements. Nor was he concerned about OPEC, for instead of stability of agreements he sought an immediate increase in Libya's oil income. Henceforth, the objective of the new government was to secure control of all production. Such control, Qaddafi and the RCC believed, would demonstrate to those Arab nations still controlled by moderates, that the Libyan revolution must be regarded as genuine. By

---

*He lived in Alexandria for 14 years, until his death in 1983, at the age of 93.

increasing production, the RCC—especially Qaddafi—could obtain more oil money to use in "the battle to liberate Palestine."[57] From his point of view, this policy provided for everything.

In 1970, after securing his hold on Libya, by getting rid of the U.S. and the U.K. air bases, Qaddafi removed Maghribi and assumed the position of prime minister himself. Major Abdesselam Jallud became his deputy and Izzidin al-Mabrouk took over as petroleum minister. To help with the oil programs, the Libyans employed two oil consultants, Dr. Nicholas Sarkis and Abdullah Tariki. Sarkis, a Beirut oil merchant and publisher, had been advocating for some years that the Arab states nationalize their oil industries. Tariki, once Director General of Petroleum and Mineral Resources and later, Minister of Petroleum Resources, for Saudi Arabia (1957-1962) had already played a leading role in the founding of OPEC. With Sarkis and Tariki as advisors, the Libyan Revolutionary Command Council had two bitter anti-Western, anti-Imperialists who promoted an even more radical position on oil matters.

Qaddafi also appeared determined to force the oil companies to submit to all of Libya's demands. As he said, "We must show we are masters here." New government-company negotiations over the posted price of crude oil began on January 20, 1970, and again on January 29, Qaddafi told the representatives of twenty-one companies: "People who lived without oil for 5,000 years can live without it again for a few years in order to attain their legitimate rights." Then he declared that Libyan oil had been "priced too low in relation to its production cost, its high quality and its nearness to markets," and added that "Libyan oil workers were not being treated fairly; and that Libya could live without oil revenues while it trained its own oil technicians."[58]

Even before the oil companies had assessed the full import of the Qaddafi speech, the Libyan Parliament amended the 1955 petroleum laws by vesting full control of all oil and concession areas in the Council of Ministers (thus removing control from the Ministry of Petroleum). Yet Minister Mabrouk began to call in the heads of oil companies for pointed conversations about the posted price. Next, the new government ordered that all "street signs, tickets, and letterheads were to be in Arabic only." At Esso Libya service stations, the oval red, white and blue signs had long identified Esso, with English letters on one side and Arabic on the other. The English side, although a registered trademark, now became unacceptable; it had to be changed.

In retrospect, it is clear that the Libyan revolutionary leaders enjoyed playing their new roles: For example, after President Tito of Yugoslavia—no

great oil market—visited Libya, he solemnly announced that he intended to
cooperate in oil development there. Then, Mabrouk went to visit Moscow
and returned with the message that the Soviets too would help explore and
develop Libyan oil. Soon the Soviet ambassador began to talk about cooperation in oil matters, and the RCC announced the arrival of a Soviet team
assigned to prepare a geological map for Libya [they came in May]. Finally,
Libya's own oil company, Lipetco, was renamed the Libyan National Oil
Corporation [Linoco], perhaps to announce that it was under new management.

The new Libyan Oil Commission put considerable pressure on Esso Libya
as [at that time] "the biggest producer, the biggest investor and ... the
biggest contributor to the revenue of that country. ..." Commission members told Wynne "that the government wanted the posted price to be
increased 43 cents," and as he recalls his reaction: "43 cents in those days,
Good God! That was out of this world. So we talked about five cents, possibly with an extension to ten cents. ..." After Esso Libya failed to agree to
any substantial increase, the commission "started working on some of the
other companies, particularly those companies that had all their eggs in the
Libyan basket and no other place to turn." Most prominent among those
companies was Occidental, for more than half of its crude supply came from
Libya. On behalf of Esso Libya, Wynne continued to offer a moderate
increase (around ten cents) in the posted price, which the Libyans in turn
rejected. Qaddafi repeated his stern warnings, and rumors spread that
Libya would nationalize all oil companies; Mabrouk denied any such intention "at this time" (April 11, 1970).

The oil companies especially disliked the Libyan demand that any posted
price increase agreed upon had to be retroactive to the time each company
had first begun to produce Libyan oil (in the case of Esso Libya, 1961).
Wynne said that no such claim could be justified, but that Esso Libya would
negotiate a settlement if possible. He repeated his earlier offer of an increase
of ten cents in the per barrel posted price. Other sources report that Esso
Libya offered to raise the per barrel price from thirteen to twenty-three
cents. In any case, the talks were stalled by April 22, 1970: but in May, Esso
Libya resumed negotiations with the host government. By that time the
whole situation was so muddled that both Jersey President Milo M. Brisco
and Board Chairman J. Kenneth Jamieson publicly expressed their opinion
that Libya would not expropriate Jersey's operation. After a few days, the
May negotiations again recessed.

As if he had been waiting for that cue, Qaddafi immediately began to
threaten American interests, claiming that they were biased in favoring

Israel over the Arabs. At the same time, Libya resumed harassment of Esso Libya and the other oil companies: port charges for tankers went up, and the government questioned the tax reference price of the liquefied natural gas soon to be exported from the Brega plant. Clearly, the screw had been turned on Esso Libya, and, in different ways, on the other oil companies as well. In June, 1970, the Libyan government directed Occidental to cut production back from the April figure of 800,000 b/d to 500,000 b/d. This decrease, Libya justified under Regulation 8 of the petroleum law, based on the OPEC Resolution for the Conservation of Natural Resources. Libya claimed that the Occidental field was being over-produced; and shortly afterward, similar reductions were ordered for two other independent companies, again in the interest of conservation.[59]

When the Trans-Arabian Pipeline [Tapline] was closed on May 30, 1969, because of sabotage,* this loss deprived Aramco and their Western European customers of some 500,000 b/d of crude. In all, the worldwide combination of an unusable Suez Canal, a closed Tapline, and new Libyan "conservation" added up to a transportation shortage, rather than an oil shortage. Because of Libyan proximity to Europe, this transportation crisis immediately strengthened Qaddafi's hand at the same time that it increased pressure on the oil companies to effect a settlement. Even some U.S. Department of State officials considered the Libyan demands reasonable—given the existing conditions.

On July 4, 1970, Wynne was attending a holiday reception at the U.S. Embassy when the Libyan petroleum minister telephoned to request a meeting with him and two other representatives of companies marketing in Libya. After waiting "an hour or so," Wynne remembers, "we went into the minister's office and he proceeded to tell us that they were nationalizing our marketing operations." The head of Esso Libya did not question Libya's right to do so, "provided that there was adequate compensation." He then asked what had occasioned the step, and one of the ministers remarked, "You've enjoyed the profit from marketing for a long time. Now it's our turn."

Aside from symbolism, the threat of nationalization chiefly provided one means of keeping the "squeeze" on the companies; Libya's real goal always remained to force the companies to accept a higher posted price, so as to increase Libyan income. Of course, Esso Libya and Jersey had already decided to make additional efforts to satisfy Libya's aspirations, but they did not wish to formalize in any new agreement the inflated freight rates

---

*This sabotage occurred in the Golan Heights, which was Syrian territory until the Israelis occupied it during the Six Day War.

occasioned by the closure of both Suez and Tapline. Freight rates, Jersey contended, must be left flexible enough to change with the times. On the other side, Libya indicated a willingness "to sacrifice current revenues to achieve changes in concession agreements." For the moment, this position seemed justified, as George T. Piercy, Jersey contact director for the Middle East,* later testified:

> Libya was in a unique position to do this. Its population was small, its monetary reserves high, and its oil increasingly valuable because of the transportation shortage. The new revolutionary government was determined to pursue its nationalistic aims which included ever-increasing control over its oil.[60]

Moving to increase that control, the Libyans began by selecting target companies for their new price changes. Jersey was chosen as the oil major with the largest production there, and Occidental as the independent company most dependent on Libyan oil—the one that could not afford a prolonged interruption in supplies.

Already, Hammer had grown increasingly worried because his company's production had been cut back. His European refineries needed crude oil and he had contracted to sell much of his production. So on July 10, 1970, he flew to New York to see Jersey Board Chairman Jamieson, the company's chief executive officer. Hammer's intention was to persuade Jamieson that Occidental could not stand against the Libyans without receiving crude from some other source; Jersey could be that source. As Jamieson later recalled:

> Hammer came to me. . . . He didn't agree with the new terms and conditions Qaddafi was demanding. . . . [He] tried to get me to sell him oil at cost. He said, 'If you would agree to do that, why I'll stand firm in Libya.' I said, 'Armand, we can't sell you oil at cost. We'll sell you oil at our lowest so-called third party price'—because if we sold him oil at cost, we'd have to tear up all our contracts with everybody else. So we offered him that oil . . . and . . . , I think I told him I could probably get Shell to sell him some oil at that price too, and I sent him over to talk to somebody at Esso International, which I'm not sure he did or did not do. . . . The next I heard about a week or ten days later, he was over in Iran trying to make his own deal with the Shah. That's the last I heard of him. But what he said later was a batch of fiction—that the sale [of Jersey oil to him] would have stopped the whole movement in the Middle East— that's just nonsense.

---

*Defined as the area from Libya to Afghanistan. In 1966, Piercy succeeded Howard W. Page as Middle East contact director.

Here, Jamieson refers to Hammer's public statements that, had Jersey provided him with crude oil, the whole Libyan movement to gain control of the pricing mechanism would have collapsed.[61] Whatever the reason for the Hammer mission to Iran, it obviously failed.

X   X   X

Wynne performed well as manager of Esso Libya. In Libya, he developed a proud philosophy of responsible operations and he demonstrated an abiding loyalty to Jersey. But Siro Vázquez, who knew Wynne well, thought that he was needed to run Jersey's Spanish companies, just then getting underway. So Wynne moved on to Spain, leaving behind him a considerable reputation. For Esso Libya, he had provided sound leadership—clear direction signalled with a great personal flair. Tough when he needed to be (when the occasion required), he always maintained a sense of humor. Further, Wynne habitually expressed himself exceptionally well, and many diplomats and executives found him to be quite charming. Even as he passed from the Libyan scene, his real achievement was clear to all.

Meanwhile George T. Piercy went to Tripoli during August, 1970, to negotiate with the Libyans, who now demanded a minimum increase of 21 cents in the posted price. Piercy countered by offering a fixed increase of ten cents, with "a variable freight premium of eleven cents which would fluctuate as freight rates went up or down." Next, the Libyans insisted on 21 cents as the fixed rate, plus a variable element. Piercy, who believed that the "higher posted price [had to] be fair and defensible" when finally arrived at, also thought the Libyan demand to be "above that justified on the basis of normal freight rates." Accepting their demands, he insisted, would create a "spillover" to which the Persian Gulf states would respond with new demands of their own. In turn, this reaction would undo any Libyan settlement and create "even further leapfrogging demands." So Piercy offered instead the variable freight amount of 11 cents per barrel. He also read the Libyans' reaction to his offer as positive enough to allow that "we would ultimately be successful in negotiating an adjustment in posted prices that would contain a variable freight element."[62] At that moment, however, the Libyans simply broke off negotiations with Esso Libya and concentrated their attention on Occidental.

Once again, Hammer had been ordered by the Libyans to reduce oil production; and now, with the first anniversary of the revolution due to be celebrated on September 1, he sensed that it was urgent to complete his negotiations with them. Every day, he commuted between Paris and Tripoli, until finally, on September 2, an agreement between Occidental and

Libya was announced. It called for a 30 cents increase in the posted price (without provision for a flexible freight rate) with a fixed-price future escalation of two cents per year for five years. Hammer also agreed to pay a supplemental tax of from 4 to 8 percent above the established 50-50 profit sharing split. As an immediate reward, Libya provisionally increased the Occidental production ceiling by almost 300,000 b/d.*

By its terms, this Occidental agreement created problems for other oil companies operating in Libya. The tax increase (4-8 percent) represented the first major break in the 50-50 profit-sharing agreements—the standard operative throughout the Middle East at that time. Understandably, on September 10 and 11, 1970, the Libyan producers met in New York to consider means of resisting the new increase in posted price. Again, John J. McCloy served as counsel to all of these companies, and he cleared this meeting in advance with the U.S. Department of Justice.** When Occidental refused to divulge details of its settlement, little could be done, except to speculate that, since the RCC had unilaterally cancelled all arbitration clauses in concession agreements, the Libyans might move against each of the other companies in turn, just as it had against Occidental.

This speculation proved to be justified. One week later, the Oasis partners were forced to agree to the same posted price increase and essentially the same tax-share increase that Occidental had accepted. Immediately, Libya rewarded Oasis with an increased allowable production, to 900,000 b/d. Together, the Occidental and Oasis settlements represented more than half of Libya's 3,000,000 b/d production (April, 1970), and more importantly, they largely increased government income. Now that control had been established, Libya gave the remaining oil companies one week to accept comparable terms or to face shutdowns (and possible expropriation). By October, most remaining members of the industry in Libya—all except Jersey, Mobil, and British Petroleum—had agreed to the new terms; and before the end of October, the three holdouts also signed. Clearly, the Libyans held a winning hand, and the Arab world watched their success.

Qaddafi and the RCC proved that oil could be translated into political power. Fascinated, OPEC followed developments and drew lessons: they saw the United States government do little or nothing to support American business interests abroad. Sir David Barran, head of Shell,*** also studied the Libyan situation with great interest; he remembers realizing that "sooner or

---

*On September 2, 1970, Esso Libya's production was cut by 110,000 b/d.

**For more details of McCloy's role, see the chapter, "Confrontation in Tehran and Tripoli."

***For the Barran role, see the chapter, "Confrontation in Tehran and Tripoli."

later we, both oil company and consumer, would have to face an avalanche of escalating demands from the Producer Governments and that we should at least try to stem the avalanche." He recalls, as well, that U.S. government officials did not understand, much less agree with, Shell's assessment of the oil situation.

Armand Hammer did agree with the Barran (Shell) view, and on October 9, 1970, he visited Barran. Hammer pointed out that the only way independents such as Occidental could resist further Libyan demands would be to have the major oil companies guarantee to provide substitute oil to any independent that would not capitulate to Libya (and thereby lose its Libyan production). From these Barran-Hammer conversations came the Libyan Producers Agreement—the famous "Safety Net"—which later would play so large a role in the next rounds of oil company-producing country negotiations. In effect, as Barran noted, the Hammer plan provided a mutual support mechanism for all Libyan producers, major and independents alike.[63]

Among the OPEC countries also, the Libyan agreements triggered swift responses. First, Iraq increased by 20 cents per barrel the posted price of its oil delivered by pipeline to the Mediterranean. Then, Iran declared a 16 cents per barrel increase, plus an increase in income tax to 55 percent; and Kuwait demanded an increase in the posted price of its oil. The companies—although certain that the "avalanche" had begun and that they were being whipsawed—nevertheless agreed to these increases. Further, in most cases profit-sharing agreements were changed to 55 (host)-45 percent; and Venezuela, simply by legislative action, changed its contract ratio to 60-40 percent in its favor. Some other tax arrangements were placed on the agenda for the next meeting of OPEC, scheduled in Caracas, Venezuela, during December, 1970.

Back in Libya, as soon as the Libyans heard of the new oil company agreements with the Persian Gulf states, they complained that their September-October contracts had not achieved their full purpose. So they proposed a completely new set of demands. By this time, of course, the power of the oil-producing nations to ratchet oil prices upward, level after level, had been clearly demonstrated to the world. Petroleum Minister Mabrouk told the companies that, since an OPEC resolution called for all member countries to raise their tax rate to 55 percent, "Libya deserved a higher tax rate than the 45/55% prevailing." A few days later, Deputy Prime Minister Jallud announced that on January 12, 1971, new tax negotiations would begin. More price ratcheting appeared to be inevitable.

X   X   X

"A watershed for international oil," one thoughtful writer has termed the triumph of Libya over the oil companies. For the companies did lose control—in Libya and elsewhere—of the power represented by oil (and natural gas) in countries where they were guests. Among the Libyan leaders, Qaddafi stands out: He cleverly targeted his peculiar combination of warnings, demands, and threats at the companies that most relied on Libyan oil for their survival. Once they had yielded, the industry front had broken, and no other single company could stand against the Libyan government. Elsewhere, no national government, either American or European, became interested enough to aid the oil companies in their plight. Majors and independents alike were left to capitulate. And, as a result, since that day of surrender in 1970, international business built on oil and gas has been far different

A decade after the majors' defeat, Jamieson, who headed Jersey during those years of painful negotiation with Libya (and other nations), was asked whether, in retrospect, these (and later) crises could have been avoided. He replied:

> I don't think so. It was just one of those—probably—inevitable things in history. . . . You had a whole changing set of conditions in the oil industry . . . supplies were very, very tight. You certainly had nationalistic tendencies to take over the oil industry. . . . I suppose you could have said that if they'd ever been able to effect a peaceful arrangement between the Israelis and the Arabs, the wars wouldn't have happened. But nothing had been done, . . . so to me it was just one of those kinds of flows of history that was almost inevitable.[64]

For Qaddafi, a fortuitous series of largely unpredictable events made him a dictator to oil companies, great and small. Briefly, the most compelling of these circumstances were: cuts in the oil pipeline (Tapline); blockage of the Suez Canal; a drop in worldwide spare crude oil producing capacity, combined with a rise in demand; an increase in tanker freight rates; and the inconsistencies of U.S. foreign policies in the Middle East, which seemed to center on a single point of interest: continued support for Israel. The last of these circumstances, in particular, made Jersey (and other American companies) vulnerable. Even so moderate an Arab as Saudi Arabia's King Faisal pointed out that this U.S. policy must drive all the Arab nations, including the Communist-hating Saudis, into the arms of Russia.

When Qaddafi and his RCC took control in Libya, they also accepted leadership of the Arab extremists. Certainly, no other Arab leader indulged himself more fully in incendiary, anti-Western (especially, anti-U.S.) rhetoric than did Qaddafi. His respect in the Arab world grew from an impression that he worked hard to create— that he did not care to rebuild or reshape Libya along Western lines. Rather than embracing what oil companies represented in Libyan life, the RCC seemed to stand boldly alone. As one observer saw them: "The Libyans were competent men in a strong position: they played their hand straight, and found it a winning one."

For Jersey, the Libyan venture proved to be yet another learning experience. When Libya began to signal trouble, especially in the early 1970s, Jersey executives at home understood their need to reconsider the future direction of the oil business. They knew that they could not ignore the forces of change, in the Middle East or elsewhere. As had become traditional at Jersey, the Management Committee in 1970, acted to create another working committee—this one with the title, "Worldwide Business Environment Policies Committee." To chair it, they selected the veteran Siro Vázquez. Almost from the beginning, Vázquez had been active in the Esso Libya venture; he had served as that company's contact director until 1969, when (largely due to his advice) the Jersey Board of Directors merged Libya into larger Middle East territory. In all, Vázquez was as well qualified as anyone to foresee the future. Yet his "Worldwide Business Environment Policies Committee" reached no dramatic conclusion and proposed no new strategy; at most it may have helped both Esso Libya and its parent company to anticipate some future demands by host countries. A witness to the fortunes of Jersey in Libya, Vázquez later smiled wryly as he summarized his experience: "The oil industry as we had known it would not exist much longer."[65] For Jersey, the lesson of Libya remained sad but clear.

Yet Libya also served as another example of Jersey's resourcefulness. For Europeans, the Libyan oil strike proved one of the most important events in post-World War II history. It emancipated Europe from bondage to the Persian Gulf oil states, and permitted economic growth undreamed of prior to the Zelten strike of 1959. The fact that Jersey hung on and found oil after all other companies had dropped out provided for its Exploration Department and, in fact, for its entire organization, confirmation of their traditional motto, "If there's oil there, we'll find it."

With its record in Libya—a proved record of discovery, creativity, and planning—many persons in Esso Libya failed to comprehend just how the Libyans could single their company out for attack, even after the flames of nationalism had begun to glow. But few positive distinctions that might favor some outsiders over others are made in such movements in history. And Esso Libya, like the other oil companies, became state property on December 1, 1981, ending one of Exxon's most successful stories.

Chapter 21

The Cazzaniga Affair

BEGINNING IN 1974, A NUMBER of American corporations received severely critical attention in the news media and in Congress for "improper" political payments and for alleged bribery in the U.S. and foreign countries. For Exxon Corporation,* its Italian affiliate, Esso Standard Italiana [ESI], and its managing director, Dr. Vincenzo Cazzaniga, provided the focus. Cazzaniga had been involved with political contributions in Italy for a number of years. Although corporate political contributions were legal and commonplace there, the American public, which had been stunned by Watergate and other scandals, did not regard them favorably. The circumstances in Italy, while eventually demonstrated to be altogether unrepresentative within Exxon, came to be a painful embarrassment to the corporation's management.

One of the strengths of Exxon over many years had been its decentralized operating structure: managers of affiliates could draw on the varied resources and expertise of the holding company, while their capital budgets received careful scrutiny in the New York office. But local managers made most of the routine decisions. In unusual situations, when they encountered problems on which they believed advice would be beneficial, they checked with the appropriate Board contact director. Except for the capital budget review, the system of control remained almost federal—yet it worked. Additionally, many of Exxon's employees gained valuable experience with the various affiliates, by working in foreign countries and by establishing company-wide contacts that would prove useful in broadening perspectives and nourishing morale and teamwork.

In Italy there was a particular need for a strong executive to cope with the intricacies of local political and economic problems. Traditional Italian deference to the person of the chief executive served to enhance Cazzaniga's

---

* Standard Oil Company (New Jersey) [Jersey] became Exxon Corporation on November 1, 1972.

discretionary powers. A Price Waterhouse & Co. report to the Jersey Audit Committee in 1972 noted:

> In Italy, much emphasis is put on the prestige of top-level executives of first-class organizations, the requisites of whose positions may include access to senior government officials. It appears that any indications that the Board did not have full confidence in the managing director could be detrimental, both to him and to his business relationships. In other words, on the Italian scene, the strong man vested with status deriving from wide powers may be in a better position to serve the Corporation.

This national attitude converged with Exxon's philosophy of management, and the company followed the accepted Italian practice of giving its managing director substantial powers, including "sole signatory powers on corporation bank accounts or the power to take down loans."[1]

In 1975, in testimony before the Subcommittee on Multinational Corporations of the Committee on Foreign Relations of the United States Senate headed by Senator Frank Church of Idaho [the Church Committee], Archie L. Monroe, Exxon controller, succinctly described Cazzaniga: "As chief executive of Esso Italiana for about 20 years, he was admired and respected not only by Exxon but throughout Italian industry and government."[2] Some months later, a Special Committee on Litigation of the Exxon Corporation Board of Directors elaborated on the Monroe description:

> Cazzaniga was an imposing man in his role as Managing Director [of ESI]. Of great ability, energy and charm, he stood out on the Italian business and political scene. He twice headed the Italian national petroleum industry association. He was on intimate terms with the leading political figures in Italy. He appeared to handle the affairs of Esso Italiana effectively, in a difficult environment.... Highly honored by Church and State, Cazzaniga was thought to be of impeccable integrity. Indeed, the high regard in which he was held within Exxon may well have delayed the ultimate exposure.[3]

These appraisals reflect a consensus among those Exxon employees who knew Cazzaniga. Numerous former company employees and executives recall that the ESI manager was exceptionally thoughtful, always kind, a pleasing personality and a "workhorse." He remembered birthdays, anniversaries, graduations, christenings and like events by sending cards, telegrams and on occasion, by personal telephone calls. His urbane manner and solicitous attention to the families of executives earned good will from most business acquaintances. He had worked for the Exxon affiliate in Italy for almost 40 years.

Until Esso Europe Inc. began to investigate Cazzaniga's activities in 1971, despite reservations about known contributions to political parties which ESI had made, no one in the parent company seems to have questioned his integrity.[4] Some felt that he tended to "push his authority to its limits, and perhaps beyond," but it was generally believed that he sought prompt approval when he did. As M.W. Johnson, vice president of Esso Europe, told L.B. Shore when the latter became an Esso Italiana director in 1968: "You'll find Cazzaniga does things in a way you may find unusual, but he's held in high esteem: don't try to interfere with his management style."[5] And his loyalty was unquestioned.

The parent company maintained the standard forms of financial control over ESI. These included those normally exercised by the Controller's Department, internal and outside audit functions and others. Internal procedures required that Cazzaniga consult with or report to the Esso Italiana Board of Directors and/or Management Committee and with shareholder representatives, depending on the nature and importance of the particular matter. As it turned out, he succeeded for a number of years in circumventing these and other controls.

Ultimately the investigation of Cazzaniga centered on his use of company funds ostensibly for political payments. Within Italy, political parties depended upon contributions, and petroleum was one of the industries expected to contribute heavily. It did. It is also apparent that even state-owned companies made substantial contributions to the political parties. Cazzaniga said contributions were a regular and necessary part of the Italian business scene, and in Italy, no prohibition on political contributions by companies existed.

Thus, within the oil industry, even Ente Nazionale Idrocarburi [ENI], the company through which the Italian government participated in the oil business, took part in the process of political contributions. Enrico Mattei, a dynamic and ambitious executive with a strong desire to compete successfully with the multinational oil companies directed ENI. Prior to his death in October, 1962, he had developed considerable popular support for his viewpoints and operating techniques, and, indeed, ENI had become ESI's principal competitor in Italy.

The original motive for investigating Cazzaniga and his management of the Italian affiliate did not relate to political payments. Rather, ESI Director Shore called Esso Europe's attention to vague rumors of conflicts of interest involving Cazzaniga in late 1971, and acting on these hints of misconduct, an investigation was begun in December.[6] With the discovery of secret agreements involving the sale of liquefied natural gas [LNG], its scope greatly expanded and soon included the area of contributions to political parties.

A report by the general auditor of Esso Europe in 1972 indicated that Cazzaniga had told the auditors that payments had been made by ESI to either political parties or their newspapers as early as 1948, in which year Cazzaniga rejoined the company. [Before that time, Cazzaniga had served with an Esso organization, S.I.L. Bedford, in Italy until World War II.]

By 1961, the Jersey audit staff had become concerned about sloppy accounting for payments of a confidential nature, and efforts were made to tighten these procedures. Three years later, on December 3, 1964, at a meeting attended by Cazzaniga, Jersey auditors "laid ground rules for payments of this nature to third parties"—rules which were not fully implemented by the ESI president. In April, 1966, a Jersey auditor pointed out continuing deficiencies in documentation and accounting for confidential payments. But Cazzaniga offered assurances that the various persons in the hierarchy considered sufficient.[7]

With the formation of Esso Europe Inc. in 1966, the new company assumed oversight of Jersey's European companies. In the case of ESI's political contributions, the execution of this responsibility was limited by the necessity, in Cazzaniga's view, of keeping the contributions confidential. He explained that the parties and political figures insisted on confidentiality. Moreover from ESI's point of view confidentiality was thought desirable to avoid comparisons and the matching expectations that would follow.[8] While payments to newspapers and public relations firms did appear on the books, the identity of the ultimate recipients was not revealed. Thus, the methods used to keep confidential the contributions in Italy also served to protect them from Esso Europe's scrutiny, for there was no way to verify who, in the final analysis, received what.[9]

In the 1960's, the level of authorized political contributions grew steadily. When the outlays reached $5.8 million in 1968,[10] New York and London demanded an explanation for the burgeoning requests. Cazzaniga responded by insisting "The amount had to rise with the expectations of political parties in the light of the expanding activities of Esso Italiana and its subsidiaries." Even so, this bothered Nicholas J. Campbell, Jr., president of Esso Europe, who learned of the contributions in the late summer of 1968. The fact that Cazzaniga reported all these contributions in a single annual total also caused him concern. So in August, 1968, Campbell sent D.J. Thompson, controller of Esso Europe, to Rome "to get a handle on the problem of political contributions." Thompson and Cazzaniga agreed on a "Special Budget" with a "format" to include references to "special problems or matters of concern, coupled with monetary figures allocated to each." This arrangement required Cazzaniga "to agree each year in advance with his contact executive in Esso Europe the payments necessary in the Company's general interest by broad categories."[11] Additional payments could not be made during the year without approval. Thompson later testified that the budget format "was of no value for control or audit purposes, since he could not say as fact whom the payments were made to or for what purpose." He did believe, however, that he had devised a mechanism that would enable H.T. Cruikshank, director and senior vice president of Esso Europe, to obtain sufficient information to enforce effectively a ceiling on the contributions.[12] Thompson reported that he questioned whether the payments being made were bribes, but Cazzaniga assured him he expended them only as political contributions.

Campbell remained determined to get the Special Budget "reduced to the $30,000/year budget level of the early 1960s." Cazzaniga, however, pressed him not to abruptly terminate the expenditures for fear of "hurtful repercussions in Italy." Campbell, who never saw a breakdown of the payments, chose not to risk political retaliation by ending them suddenly.

In the last half of 1968, E.B. Paust, general counsel for Jersey, and A.O. Savage, Jersey controller, informed J. Kenneth Jamieson, then president of the company, that political contributions were being made in Italy. They assured Jamieson that such donations were legal and also that they were being phased down by Esso Europe. After the death of Savage in 1968, R.E. Mays was appointed Jersey controller. In the early summer of 1969, Mays visited Europe where he talked with Campbell and expressed concern about the absence of controls on the Italian political payments. He urged the elimination of all contributions. Campbell explained to him Esso Europe's determination to control the practice, as well as his reasons for permitting them to be gradually decreased.[13] Later in 1969, while visiting New York, Campbell asked Paust about the legality of the payments. Paust sought an opinion from an Italian attorney which reaffirmed an understanding that contributions to political parties by companies like Esso Italiana were indeed legal.[14] Yet, despite these assurances and the possibility of undermining Cazzaniga's standing, Campbell continued to insist on the reduction of the contributions. The 1971 budget dropped the amount of authorized contributions to $3.7 million, the lowest figure since 1966.[15] That amount was still very much larger than the contributions budgets of the early 1960s,[16] but by the time Campbell left Esso Europe at the end of 1970, he at least had reason to expect that Esso Italiana's political contributions were being reduced.

Unfortunately, that expectation rested upon an illusion. The Special Budget did not then—and probably never did—represent an accurate accounting of Esso Italiana's political donations. Nor did anyone know of the additional funds that Cazzaniga diverted, allegedly for political purposes. Cazzaniga apparently found it possible to circumvent control procedures established in 1963-1968. A report by Esso Europe's general auditor commented: "In practice, Jersey and Esso Europe failed to maintain their appreciation that these control dangers applied equally to Dr. Cazzaniga." The same report also stated that it was virtually impossible to reconcile "the payments reported by Dr. Cazzaniga with those actually made since they bear so little relationship to each other." Cazzaniga's reporting was largely "at variance with what actually took place."[17]

In August, 1968, Howard C. Kauffmann had become executive vice president of Esso Europe. While he had no specific responsibility for Esso Standard Italiana, during Mays' visit to Europe in July, 1969, he learned that political contributions were being made in Italy by that company. Campbell informed Kauffmann that the payments were legal and that he need not become involved. Kauffmann became Esso Europe president in January, 1971, and received the figures on the total contributions made for 1970, as well as those authorized for 1971. Kauffmann was unhappy about the situation even though assured of the legality of the contributions. In mid-1971, after reluctantly acceding to an addition to the 1971 budget because he believed Cazzaniga had already committed the funds, he ordered the end of all political contributions and stated that none would again be approved by Esso Europe.[18]

Cazzaniga blatantly ignored the Kauffmann order. On March 17, 1972, Cazzaniga sent to ESI Controller E. Consigliere invoices totalling $500,000 from a Rome newspaper, "for services rendered." Consigliere, who by then had learned of the investigation of Cazzaniga, told Cruikshank that he had received the invoices, whereupon the latter told him that the "Special Budget" had ended. Consigliere "was shocked to learn ... that Cazzaniga was capable of such deception." Consigliere managed to intercept this payment, but Cazzaniga had already disbursed over $1 million without authorization in 1972.[19]

It was in early March, 1972, that Jersey learned that in the mid-1960s Cazzaniga had entered into four side-agreements with SNAM S.p.A., an arm of ENI which bought Libyan LNG from Esso International Inc. The discovery of these agreements, along with the revelation of the attempt to make an unauthorized political contribution, led the parent company to widen the

Cazzaniga investigation and to place it under the direction of Emilio G. Collado, a Jersey director and executive vice president. In all, the investigation produced a series of twenty-three internal audit reports issued between summer, 1972, and spring, 1973. Cazzaniga was removed as ESI managing director in March, 1972. It was later discovered that before being severed totally from Esso in May, he had made a last unauthorized political payment of $685,000 from an affiliated company.[20] The twenty-three audit reports, plus further investigations and analyses, conducted during 1973-1975, finally led to court proceedings against Cazzaniga by ESI.

Not until the special audit reports of 1972-1973, did the character and extent of the problems in Italy become clear; and even then the confidential nature of the payments from newspapers and public relations firms to others precluded a full accounting. The newspapers and agencies were understood to be conduits to the respective political parties. However, the special audits did render, at last, a somewhat more complete picture of Cazzaniga's activities.

Most of the controversial expenditures for which Cazzaniga had authority could be identified on Esso Italiana's books as payments to newspapers and public relations firms. Others were disbursements from an off-the-books bank account which was known to ESI personnel and funded from rebates and income from other sources. Cazzaniga could make cash withdrawals from this account either on his own authority, or for larger amounts, with the concurrence of one other ESI director.[21] In either case, the ultimate recipient of the funds could not be traced or controlled.

But that wasn't all. Out of Exxon's internal audit investigation, company executives learned of forty *secret* bank accounts, previously known only to Cazzaniga and a few trusted personal assistants. At the time of their discovery, most of the secret accounts still in existence were overdrawn; the amount of the overdraft indebtedness totalled $19.2 million. None of the financial officers of Esso Italiana, Esso Europe or Exxon learned of these accounts before 1972. Not even Price Waterhouse & Co., Exxon's external auditors, ever contemplated that banks might not respond truthfully to regular audit inquiries. In these cases, Italian bank officials had cooperated with Cazzaniga to keep the accounts secret.

In the years before the 1972 investigation, Cazzaniga, with the involvement of these Italian bank managers, had used the forty secret bank accounts as receptacles, "funded with overdrafts and rebates from third parties, including rebates of $13.5 million of monies which were purportedly expended under the 'Special Budgets' as political contributions."[22] These accounts he personally controlled, operating them in such a fashion that it was later difficult to establish the total cash flow through them. In all, it was eventually established that $39.3 million had passed through these accounts during the period 1963-72. Another $6 million was disbursed from regular Esso Italiana accounts by Cazzaniga without proper authorization.[23] Exxon auditors verified that Cazzaniga paid $10 million from the secret accounts to SNAM. They could not verify his allegations that he expended the rest for political contributions. ESI's eventual suit against Cazzaniga demanded among other things a full accounting for these funds. The company, however, paid his unauthorized commitments—such as the overdrafts—when in the opinion of counsel they could not be successfully contested. [The 1972 accounts properly reflected these payments.]

The unauthorized payments to SNAM developed from Jersey's efforts to sell Libyan natural gas to European customers. Protracted negotiations between Jersey's Gas Coordination Department and SNAM finally resulted in a formal letter of intent signed on July 7, 1964, between Esso International, as seller, and SNAM. Although Esso Italiana was not a party in either the negotiations or the agreement, Cazzaniga had been brought in because of his knowledge of the Italian situation and his prior relationships with the parties involved.[24] Then, in the months after the parties had reached agreement, rumors began to surface about a secret letter given by Cazzaniga to SNAM that "undercut the terms of the executed letter of intent." Cazzaniga denied the existence of such a secret letter, but the 1972 investigations would prove his denial to have been false.[25]

As the investigation showed, a few weeks before Esso International and SNAM signed the formal agreements in November, 1965, Cazzaniga executed four agreements with SNAM, "ostensibly on behalf of Esso Italiana," but in fact without any authorization. When these secret agreements came to light in 1972, Cazzaniga at first claimed that he had been authorized to execute the side agreements by William R. Stott, Jersey's contact director for Europe during the period. Later, however, when informed that Stott categorically denied any such authorization for these agreements or knowing of them, Cazzaniga "expressly disavowed any claim that Stott had authorized ... the agreements in question."[26] Apparently, the only indication of the existence of these agreements prior to 1972 came in the form of rumors that could have reached New York in late 1966.[27]

The $10 million payment from secret bank accounts during 1971 satisfied two of the secret SNAM agreements. In June, 1972, ESI commenced arbitration proceedings against SNAM seeking to recover the funds already paid by Cazzaniga, and a declaration that the agreements were null and void. SNAM filed a counterclaim, seeking a declaration that all four agreements were binding. The two unfulfilled agreements involved the granting to SNAM of discounts and extended credit terms, beyond those already granted by Esso International. The cost of honoring these agreements over their lifetimes was estimated in late 1975 to be $40 million.[28] Later, in June, 1976, the parties agreed to settle the matter rather than continue the arbitration to conclusion.

For whatever reasons the situation at Esso Italiana endured, the evidence tends to support one executive's opinion that, given Cazzaniga's long tenure at ESI, his assumption of authority and the skill with which he manipulated the political fund, irregularities would have occurred whatever steps had been taken to try to govern his behavior.

Nonetheless, beginning in 1972, Jersey took action to prevent any recurrence of such financial irregularities as those revealed in its investigation of Cazzaniga and the "Special Budget." The management of Esso Italiana was changed "to a more collegial approach ... vesting the full executive power in three managing directors"; auditing and control procedures were tightened; and more rigid accounting procedures adopted.[29] Realizing that the potential for abuse of a decentralized management system existed throughout its worldwide operations, the parent corporation took a number of corrective steps applicable to its entire organization.

Jamieson, who had become Jersey Board chairman and chief executive officer [CEO] in 1969, issued "a restatement of the Corporation's views on 'Morality in Management,' " in October, 1972. He called attention to Exxon's longstanding policy of complete integrity and honesty and "strict observance of all laws applicable to its business." Jamieson noted:

> Our policy does not stop there. Even where the law is permissive, Jersey chooses the course of the highest integrity ... honesty is not subject to criticism in any culture. Shades of dishonesty simply invite demoralizing and reprehensible judgements. A well-founded reputation for scrupulous dealing is itself a priceless company asset.[30]

Jamieson directed that "all transactions ... be properly booked and that no information of any kind ... be withheld from the Corporation's auditors or senior management." He also informed all regional chief executives that compliance with this policy statement would be part of their responsibilities since it would become "a standard part" of the parent company's annual review of regional management. Exxon's general auditor also received instructions to monitor compliance by affiliates. In 1975, the general auditor and the company's independent accountants, Price Waterhouse & Co., were ordered "to report all deviations to the Exxon Board Audit Committee" with copies of their report sent to every director.[31]

At the same time the company's worldwide internal audit staff increased substantially, and all internal auditors were given the right to report any deviation directly to the Board Audit Committee. Beginning in 1974, the Board Audit Committee shifted from having a majority of outside [non-employee] directors to being composed entirely of non-employee directors. * Other measures followed. These included "more restrictive standards for controlling bank accounts and banking transactions" and a requirement that each affiliate's chief executive and controller certify that their annual financial statements "were prepared in accordance with prescribed practice, including proper booking of all transactions and recording of all cash and other assets." Finally, early in 1976, Exxon banned corporate contributions to political candidates or parties anywhere in the world, without regard to legality.[32]

X   X   X

Even in the early stages of the Cazzaniga investigation, Jersey's management had to make a decision about disclosing its findings. Corporate obligations in the matter centered upon the concept of materiality, "which is defined as including those matters as to which an average prudent investor ought reasonably to be informed." Historically, the courts and the Securities and Exchange Commission [SEC] had determined "materiality in nearly all cases on a quantitative basis; i.e., the significance of the amounts in question to the total operations of the enterprise, particularly its consolidated balance sheet and income statement, or the effect on these operations."[33] Since the question of materiality and thus of public disclosure was necessarily technical, it required expert opinion. Here, Jersey's management depended upon advice from the company controller, its Law Department, its outside accounting firm, Price Waterhouse & Co., and its outside counsel. Prior to the March 29, 1972 meeting of the Board, most of these parties as well as Jersey's financial vice president and at least some members of the Management Committee had learned of the side agreements made by Cazzaniga with SNAM. Since both the controller and Price Waterhouse were convinced that the corporation's obligations under these agreements did not materially affect the full year 1971 financial statements, which had already been

*This move was not related to the ESI problem but rather was part of a trend in the U.S. business community.

approved, no one raised the question of disclosure at that meeting. Thus, in approving various registration statements and amendments to previous registration statements to be filed with the SEC, the Board acted on "the unarticulated judgments of the Controller and Price Waterhouse that the financial statements remained correct," despite new information.[34]

The registration statements were filed with the SEC on March 31, 1972, and became effective on April 13, 1972, by which time Jersey had delivered to Price Waterhouse & Co. a letter signed by CEO Jamieson and Controller Mays. The letter referred to both the SNAM agreements and the unauthorized overdrafts made by Cazzaniga; it noted that Price Waterhouse & Co. had recently been advised of the discovery of "certain unbooked alleged obligations of Esso Italiana as of year-end 1971"; it expressed the opinion of the writers that these obligations even if valid would not be material "in relation to Exxon's total assets."[35] Apparently Price Waterhouse & Co. agreed with this judgment, for it allowed the registration statements to become effective without objection or comment. Exxon had put the question of disclosure of the SNAM agreements to outside counsel, Sullivan & Cromwell. On May 11, 1972, that counsel confirmed earlier advice that no further disclosure was required. Jersey's Law Department advised the controller and other officers of their concurrence in that opinion.[36]

The Exxon investigations continued for some months more. In February, 1974, wire service reports summarizing Italian press articles, and printed in U.S. newspapers, disclosed that an official investigation of alleged political bribery by Italian oil companies had begun in Italy, and that an arrest warrant had been issued for Cazzaniga who was president of the Italian petroleum trade association, Unione Petrolifera. On February 22, the *New York Journal of Commerce* reported that the Italian investigation had already resulted in the jailing of one of Cazzaniga's closest aides and that the former managing director himself could not be found.[37] Allegedly, he was in the United States. Other newspapers also commented on the Italian government's investigation of oil companies. These press notices served to destroy any effective confidentiality, since at least some of the information was now in the public domain. Yet the newspaper accounts did not appear to motivate reporters or Exxon shareholders to question Exxon about either Cazzaniga or the "Special Budget." Nothing which occurred between the 1972 decision against disclosure until the spring of 1975 caused reconsideration of the 1972 disclosure decision. However, in 1975, the revelation of political contributions and alleged bribery by other American firms, coupled with the public outcry against such practices, as well as a change in the SEC's ideas about materiality, obliged Exxon to re-examine the matter.

Noting that the SEC in challenging "questionable" payments was taking the position that a corporation's duty of disclosure can arise out of the nature of certain acts without regard to whether the amounts involved are quantitatively material, Exxon's Law Department recommended that the Audit Committee and the Board review the decision of 1972 and the entire Italian affair.[38] The review, which took place on May 14-15, 1975, included participation by the controller, the Law Department, Price Waterhouse & Co., and outside counsel from Sullivan & Cromwell. Looking at the whole subject again, the experts reaffirmed their collective opinion that the corporation had no obligation to make further disclosures. The Sullivan & Cromwell attorneys, however, did advise the Board that it had "the discretion to decide in the good faith exercise of its judgment whether it was in the interest of the Corporation..." to make such disclosures. The Board voted not to restate prior accounts, but deferred a decision on any further disclosures pending a report by the Board Audit Committee.[39] The Board never had to make the decision. External events, including a question at the New Orleans shareholders meeting on May 15, 1975, made the issue moot.

For Jamieson, the 1975 gathering was the last Exxon shareholder meeting over which he presided. He handled it well, fielding questions from the floor with good humor and patience. Among the questions printed on the proxy statement and distributed prior to the meeting, however, was one requiring the company to publish in newspaper advertisements details of any political contributions. Obviously, Jamieson expected questions on this subject, and in response to one specific question about political contributions, he stated that contributions had been made in Canada and Italy, and that such contributions were legal in both countries. Those in Italy, he said, had ended in 1971 or 1972.[40] Subsequently, Jamieson and Exxon Company, U.S.A. Chairman M.A. Wright answered questions about these contributions from representatives of the communications media. But the discussion did not end there.

In addition to publicity in the press about an Italian parliamentary investigation of political contributions by all oil companies operating there, the SEC had begun probing "questionable payments" by U.S. businesses worldwide. And the U.S. Senate Subcommittee on Multinational Corporations, headed by Senator Frank Church, was investigating political contributions made abroad by American companies.

The Church Committee in July, 1975, added Exxon to the companies asked to answer questions concerning the extent of political contributions by U.S.-owned firms in Italy and elsewhere.[41] [Representatives of Mobil Oil Corporation and Gulf Oil Corporation were among others who also appeared.] Archie L. Monroe, who had become Exxon controller in 1973, testified that when Exxon first learned of the unauthorized financial practices of Cazzaniga, it had moved to end them, and that the company had engaged in a period of extremely intensive internal investigations. After Monroe's opening remarks, Senator Church questioned why the full Exxon Board had not been informed earlier of the activities of Cazzaniga. The subcommittee had been supplied with a number of documents, including internal audit reports made by Esso Europe General Auditor G.N. Martin and his staff, and an evaluation made to the Board Audit Committee by Jersey General Auditor D. Schersten. In his commentary, Monroe stated that the full Board is not normally asked to review expenditures of such a comparatively small magnitude but that, in retrospect, perhaps it should have been advised in this case.[42]

The Church Committee disclosures were linked by the communications media to the growing energy problems of the United States, Western Europe and Japan. In Italy, the press quickly publicized the hearings, and some of these newspaper stories suggested that corrupt practices by the international companies somehow had a role in oil shortages and price increases.

Subsequent to the Church Committee hearings, three shareholders filed separate derivative suits against Exxon directors, and a fourth "threatened the institution of a shareholder derivative suit if the Board of Directors did not accede to (a) demand" that the corporation sue its Board of Directors "with respect to certain sizable political contributions made by Esso Italiana to support political parties in Italy." The shareholder derivative suits alleged that the defendants:

> permitted or acquiesced in the waste and misuse of more than $59 million as bribes or for other illegal, improper or ultra vires purposes, including misappropriation for the benefit of unidentified persons; that they permitted or acquiesced in the concealment of the purpose of the payments and the identity of the recipients; that they permitted or acquiesced in the inaccurate recording of these payments thereby enabling Exxon unlawfully to reduce Italian taxes; and that they failed to impose adequate controls on the chief executive of Esso Italiana.[43]

The complaints went on to accuse the Exxon Board of improperly failing to disclose these facts to the SEC, the national stock exchanges and to company shareholders. They alleged a conspiracy among the Board members to conceal such information from the proper persons and authorities. And they argued that the company had filed false tax returns in both Italy and the United States.[44]

On September 24, 1975, the Board Audit Committee submitted its findings in a three-volume report.[45] The Exxon Board of Directors on the same day appointed a "Special Committee on Litigation," and delegated to it the full authority of the Board for the purpose of determining:

> whether Exxon should take legal action against any present or former Exxon Corporation Directors or officers in connection with these Italian matters. Its decision is final and not subject to review or approval by the Board.

The Board appointed three members to serve on the committee and empowered them to conduct "such review, analysis and further investigation" as it deemed necessary. When the committee made its determination about litigation, it was to "undertake and supervise any action necessary or appropriate to implement ... " its decision.[46] None of the Special Committee members had any connection with the matters under investigation, and all had joined the Board only in 1975. Non-employee Directors Edward G. Harness, chief executive officer and chairman of the board of The Procter & Gamble Company, and Sir Richard Dobson, chairman and chief executive of British American Tobacco Company, Limited, had been elected in May. They served the Special Committee as chairman and vice-chairman respectively. The third member, Jack F. Bennett, had held a number of posts at Exxon before leaving the company to join the U.S. Treasury Department prior to the investigations of Esso Italiana. He returned to Exxon in August, as a senior vice president and director. The Special Committee had its own staff composed of several Exxon employees, plus as special counsel, Joseph Weintraub, a retired Chief Justice of the New Jersey Supreme Court.[47]

In addition to studying the Internal Audit Reports prepared in 1972-1973, the Board Audit Committee Report of September, 1975, and other documents, the Special Committee conducted formal interviews with 105 individuals both inside and outside the company. After satisfying itself that its investigation, in conjunction with the earlier ones, provided "as complete a record as reasonably can be made," the Special Committee submitted its report on January 23, 1976.

After tracing the history of the Cazzaniga investigations and the Board's response to those revelations, the Special Committee determined that Exxon's management had acted legally throughout the period under investigation. After a careful examination of the issues, the committee concluded that Exxon shareholders' best interests would not be served by litigation against any Exxon officer or director. Among reasons for this conclusion, it advanced the low probability of success and the high cost, as well as possible damage to the company's operations and reputation both at home and abroad. But the committee went beyond that point. It said the events must be viewed "not only in the context of the fact that the same management in Exxon was responsible for the great accomplishments over these same years in many other parts of the world, but also in the light of other aspects of the company's Italian experience." The committee said that what occurred came as "the result of errors of business judgment, not the result of lack of devotion to duty or of breach of faith." They found no evidence that any director of Exxon "ever knowingly approved or condoned an illegal act." Exxon had acted promptly, "long before Watergate and long before current public interest in corporate political contributions." It ended all such contributions knowing that "substantial commercial risk was involved," since in Italy they were expected as a matter of course. Further, no one had suggested that Exxon had made illegal political contributions in the United States. The committee expressed regret that a "small number of officials" had concurred in the political contributions and had not referred the matter to higher executives in the corporation. Exxon management teams had successively misjudged Cazzaniga, and in doing so had ignored certain warning signals. In the committee's opinion, the audit controls and procedures in place in ESI had been inadequate.[48]

The report pointed out that Exxon had acted in 1971 to prohibit political contributions in Italy [a prohibition made worldwide in January, 1976] and that it had tightened financial controls on its own initiative. On the question of possible negligence in failing to disclose the results of earlier investigations, the committee again took some care to note that the officers and directors relied in good faith upon experts in the area, and that "any error of judgment, if there was error, could not be characterized as negligent."[49]

The corporation filed a copy of the committee's report with the United States District Court for the Southern District of New York. In January, 1977, the court granted summary judgment in Exxon's favor in one of the shareholder's [Joan L. Gall] actions, noting that the Special Committee on Litigation had "acted independently and in good faith." The New York State Court subsequently dismissed two other suits [Abraham Wechsler and Kathleen E. Shaw] on the ground that the issues had been decided by the Federal Court.[50]

In the meantime, in November, 1975, ESI filed suit against Cazzaniga in Italy, seeking a full accounting of his handling of company funds both in the secret bank accounts and through other unauthorized disbursements.

By the time the Special Committee reported in 1976, Exxon (and Cazzaniga) had become involved in another investigation, this one initiated by the SEC in July, 1975. In September of that year, Exxon filed, in connection with an offering of debentures, a registration statement disclosing in detail the Italian expenditures, and referring to incidents of improper payments and improper record keeping in other foreign countries. Similar information was revealed in an amended 8-K statement filed with the SEC in the same month, and in the corporation's 1976 proxy statement.[51]

In September, 1977, following extended investigation and negotiation, the SEC filed suit in the U.S. District Court for the District of Columbia seeking a permanent injunction against Exxon and Cazzaniga. The complaint charged Exxon and Cazzaniga with failing to make proper disclosure of material events allegedly consisting of payments totalling $55.25 million in Italy between 1963 and 1972, some of which was allegedly used to influence political parties and government officials. It also cited the improper accounting associated with the payments, including the use of unrecorded bank accounts. The complaint also alleged that Exxon failed to disclose additional payments totalling $1.25 million in fifteen other foreign countries, and masked the true nature of these with improper accounting practices.[52]

The Board of Directors determined it would be in the best interests of Exxon and its shareholders to avoid the burden and disruption of protracted litigation with the SEC. Accordingly, and without admitting or denying the allegations of the complaint, the corporation consented to the entry of an order, simultaneously with the filing of the complaint, enjoining it and its affiliates from disseminating in the future annual or other reports which fail to disclose material* information concerning unlawful payments to foreign government officials, political parties or politicians, or false bookkeeping regarding such payments. The order also enjoined the company from engaging in materially false bookkeeping with respect to such payments.[53]

Settlement of the SEC suit ended the story of the "Cazzaniga Affair" in the United States; however, political repercussions continued in Italy. There, Exxon's woes were part of a larger ordeal affecting the entire foreign oil industry and encompassing much more than the question of political payments. As early as the mid-1960s, ENI had complained of the fact that while Italian affiliates of foreign oil companies chronically failed to show a profit, their parent companies continued to prosper. ENI charged

*The order does not define what constitutes "material" information.

that the multinationals manipulated accounts in order to avoid Italian taxes.[54] Such charges created bad publicity for the oil industry, but they had little serious effect until after the 1973 Organization of Petroleum Exporting Countries [OPEC] oil embargo. In the wake of that action, crude oil prices shot upward at an unprecedented rate. Yet, the Italian government refused to allow the retail price increases necessary for oil companies to realize a sufficient return. [In 1975, Esso Italiana's inability to increase its product prices caused its annual operating loss to reach $100 million.[55]] At the same time, a government investigation into charges of hoarding petroleum products revealed evidence of political contributions by some oil companies. Although corporate political contributions were legal and of long-standing, their disclosure at this time increased already intense political opposition to product price relief, while crude oil prices continued to rise. Esso Italiana found itself in a situation in which the oil industry became for some the chief villain of the energy trauma.[56]

At this point, the ESI management responded with a campaign to counter what it considered an unfair attack by the press. ESI emphasized that it had acted to end Cazzaniga's wrongdoing and to prevent any further abuses well before publicity of its problems. The company urged that the private oil industry had an important and necessary part to play in Italy's future, and that continued attacks on the industry would only impede its functioning in that role.[57] Yet hostile press attacks continued for a time, and the company continued to lose money. [Both British Petroleum Company and Royal Dutch-Shell disposed of their Italian affiliates.] By the end of 1976, the climate of opinion had moderated, but government oil pricing policy remained unresolved and continued to hamper the industry's development.[58]

Esso Italiana had weathered the storm generated by the "Cazzaniga Affair." For the Italian affiliate, the departure of Cazzaniga marked the end of a turbulent era and the beginning of a new and different history. In the case of Exxon, the disturbing developments in Italy acted as a catalyst to speed the establishment of strengthened financial controls. Exxon had already established its regional management companies with the aim of more effective oversight and coordination of its far-flung operations.

Out of this unsettling experience, Exxon's standards of business ethics now came to be more rigorously applied to all its foreign operations. Of course, the tendency to hold headquarters managers accountable for company actions at the operational level did not originate during "the Cazzaniga Affair." More than any single event, however, that episode underscored the fact that a growing capacity to guide operations worldwide must inevitably be matched by heightened accountability of top managers. In responding to these changes, Exxon not only met the challenge of the Italian crisis but also prepared itself for the uncertainties of the future.

Chapter 22

The Company Changes Names— Jersey to Exxon

"I T'S [JERSEY] A PRODUCING COMPANY that tolerates marketers," said Emile E. Soubry, a retired director of that company and fifty-year veteran marketer. He went on to remark that he never knew a good company salesman who did not decry the confusion attendant upon marketing the products of the Standard Oil Company (New Jersey) [Jersey] nor one who did not long for a single name and trademark under which to sell—like the Shell—"a marketing company that tolerated producers." Many persons contend that Soubry correctly emphasized one of the major problems confronting the huge company for at least sixty of the first seventy-two years of the twentieth century. In truth, in its management, Jersey had largely utilized strong-minded producers and refiners. But in the 1960s, that situation was altered.

Between the Standard Oil Company (New Jersey) of 1882 and the Exxon Corporation of 1973, there were many differences in both company structure and operating strategy; no single change, however, can be compared to the court-ordered dissolution of 1911. This externally forced divestiture of thirty-three companies formerly owned and controlled by Jersey served to engrave in the public consciousness the notion of a Standard Oil monopoly; despite the fact that, as a result of the court's decree, on December 1, 1911, the company dissolved into thirty-four free, separate, and independent corporation entities. Eight of the new companies bore the "Standard Oil" name, and in the period of uncertainty that followed, none dared give up the commercial value associated with that familiar appellation. So the public remained convinced that the company still lived; in fact, two generations of journalists, politicians and government employees fattened on this notion. As a result, this lack of a separate, easily realized identity injured Standard Oil Company (New Jersey) for many years because, as the parent and the largest company—and the best known—all things bad in any Standard Oil Company came as a foundling to the stoop of Jersey.[1]

The almost continuous adjustments made in response to evolutionary changes in Jersey's economic, political, and social environment are far more significant than those resulting from outside pressure, even though in most cases they are imperceptible to the general public. In the spring of 1972, however, Jersey began a series of internal changes that attracted unprecedented public interest, for they affected the basic links between the company and the public.

Humble Oil & Refining Company [Humble] announced the first actions in this process of change on May 9. On that date Jersey's principal domestic affiliate made public its adoption of a new primary trademark for its operations throughout the United States. Beginning July 1, the Esso, Enco, and Humble trademarks, under which the affiliate operated in various parts of the country, started to be replaced as the primary marketing identification by a single name—Exxon—a completely new word. One company executive remembered that:

> Exxon was selected after tedious study of old lists of names discussed in committees over the years proved fruitless as did lists of computer-generated letter combinations; and it was only after a lengthy and labored 'brainstorming' session of the Committee with its outside consultants that the long-sought nugget EXXON finally emerged.

Nevertheless, the adoption of the new trademark did not prove the most surprising part of the company's statement, for industry observers had been speculating about the possibility of a trademark change ever since Humble began field tests of Exxon the previous October. Although the importance of a new primary trademark for one of the nation's largest oil companies could not be minimized, the second part of the Humble statement evoked even more comment.

In order to realize the full benefit of its new trademark, Humble announced it would change its corporate name to Exxon Company effective January 1, 1973.[2] That announcement gave impetus to even more intriguing speculation that Humble's parent company, Standard Oil Company (New Jersey), might also change its corporate name. A few weeks later John Kenneth Jamieson, Jersey's chairman and chief executive officer [CEO], confirmed that the Board of Directors had been considering such a step, and on June 21, the Board passed a resolution recommending shareholder approval for changing the parent company's name to Exxon Corporation.[3] Around one thousand shareholders of Jersey Standard's 780,000 did attend a meeting on October 24, at the Roosevelt Hotel in New York, to debate the corporate name change. Most of the absentees had sent in proxies, and more than 96 percent of the shareholders approved the change to Exxon. A week later the great corporation officially ended its ninety years as the Standard Oil Company (New Jersey); it became the Exxon Corporation. This change symbolized a momentous step for the company. A well-established trademark has incalculable value as a corporate asset, so the simultaneous replacement of three [Esso, Enco, Humble] of them as the company's primary trademarks constituted a remarkable move. Since Esso was considered one of the best-known trademarks in the world, replacing it as a principal identification in the domestic market had particular significance.

So, in a few short months the company had not only replaced one of the world's best-known trademarks in its largest market; it also had prepared to supplant its even more venerable corporate name with the newly created non-word. Behind the apparent hastiness of these acts lay a set of historical circumstances dating back over more than half a century, circumstances that finally made the change almost inevitable.

The conditions underlying the name change of 1972 had their origin in the earlier United States court-ordered dissolution of Standard Oil in 1911. The complete omission in the decision of any provision regarding brand names or trademarks led each of the companies to assume that it retained the right to use the "Standard Oil" trademark as well as the old brand name in its territory; nor, contrary to common belief, did the Supreme Court delineate marketing areas. As one attorney explained:

> All the 1911 decree said was that you [S.O.(N.J.)] have to divest yourself of your shares in ... all of these Standard Oil Companies. ... After the divestiture, these companies continued to do business in their respective states, their territories, if they had done business there before the breakup.

World War I arrived and few companies considered expanding. So they continued to operate in the old fashion.

But this *ad hoc* solution to the trademark problem would eventually become a source of constant public confusion and intercompany contention. After the war, the various one-time Standard Oil affiliates resumed operations. The automobile boom created a strong demand for their products. Few conflicts developed, largely because the companies did not enter each others' marketing territories. Thus, the rapid expansion of the overall market during the 1920s permitted each company to grow without expanding geographically. A company generally avoided commercial conflict with other "Standard Oil" companies in their strongholds.[4] Where the companies did move out of their traditional areas, they, of necessity, used corporate names and trademarks which avoided legal conflict but which also left them at some competitive disadvantage. In fact, as George S. Gibb and Evelyn H. Knowlton pointed out in *The Resurgent Years*, "after 1911 the once mighty Standard Oil name was, in effect, a liability."[5]

By the 1930s, however, conditions had changed markedly. The onset of the Depression coincided with the opening of those East Texas fields that precipitated a sudden and dramatic increase in the quantity of crude oil produced in the United States. In the absence of effective conservation legislation, this capacity translated into a glut that drove prices down, virtually eliminating profit margins, driving several major oil companies to near bankruptcy. Under these adverse conditions, competitive collision between the various "Standard Oil" companies could hardly be avoided.

The basis for the ensuing conflict had been laid early in the 1920s, when Jersey began to seek a trademark that could be used nationally without forfeiting the valuable connection with the "Standard Oil" name. In 1920, a change in the U.S. trademark laws permitted Jersey's registration of the "Standard" trademark with the Patent Office. However, the company's legal department stopped further registration on the ground that establishing the sole claim to the mark would be impossible. This led Jersey men "to experiment with names to which the cobwebs of history still clung." On February 3, 1923, the name Esso—a phonetic rendition of the old Standard Oil monogram—occurred to C. H. Straw, director of the patent and trademark division. Jersey quickly registered the mark and after altering it to accommodate objections of the Standard Oil Company of New York [now Mobil Oil Corporation] began in April, 1926, to market Esso gasoline as a premium grade.[6] Earlier that same year, the company received approval of the word Esso as a brand name for all of its products both at home and abroad. By the 1930s, Jersey was using Esso throughout its traditional domestic marketing areas as well as outside the United States, and it had prepared for wider use by registering the trademark in all of the other states.

With the spread of Esso as a brand name and the sharpening of competition, the stage was set for the first round of conflicts over the common law trademark rights enjoyed by the several "Standard Oil" companies. Ever vigilant to protect its territory, Standard Oil Company (Indiana) gave Jersey prior warning that it considered its rights to "Standard Oil" in its marketing area preemptive to any use by others of words connoting that name.

Nevertheless, in 1935, Jersey invaded the Indiana company's territory with Esso service stations in St. Louis, Missouri. According to the historian of Standard of Indiana, "Esso's invasion was interpreted by some in the oil trade as retaliation for the invasion of Jersey's territory by Indiana."[7] A subsidiary of Standard Oil Company (Indiana) had indeed moved into the Jersey domain, using other brand names for its gasoline.

Although Jersey took out large newspaper ads in St. Louis, to inform the public that Esso had no link to Standard Oil Company (Indiana), it used the same red, white, and blue color combination identified with that company, and even some of the same brand names. When protests against this encroachment brought no response from Jersey, Indiana filed suit on May 15, 1935, in the court of the Eastern Judicial District of Missouri, at St. Louis. Charging trademark infringement and unfair competition, the complaint asked for relief in the form of a permanent injunction barring Jersey and its affiliates from using Esso or any other name connoting "Standard Oil" in any part of Indiana's marketing territory.

Two years later, on July 8, 1937, Judge George H. Moore rendered his decision in the case. That opinion, while addressed solely to the facts of the case at hand, had far-reaching consequences. Judge Moore agreed with the Indiana Company's argument that the millions of dollars it had spent in promoting the "Standard Oil" name throughout its territory insured its historic claim to that trademark and to all others derived from or connoting it. As a result, the judge granted a perpetual injunction barring Esso, Incorporated, a marketing subsidiary of the Standard Oil Company of New Jersey, from using the term Esso in the fourteen states of Indiana's marketing territory.* Although the decree applied to that midwestern area only, its effect as a clear legal precedent served to recognize the common law trademark rights of each of the "Standard Oil" Companies within its own area and to solidify the trademark practices that had evolved since the dissolution.[8]

In the wake of the 1937 decree and its affirmation by the U.S. Eighth Circuit Court of Appeals the following year, Jersey adapted its marketing practices to the legal constraints on the Esso brand. Continuing its marketing in the Midwest, the company used a variety of trademarks at different times: HUMBLE, PENOLA, CARTER, PATE, and OKLAHOMA. In Texas, its own historic marketing area, it sought to protect Esso, suing Standard of Texas, a subsidiary of Standard Oil Company of California [SOCAL], to block confusing advertisements that stressed the letters S. O.

---

*A fifteenth state was later added by Indiana Standard to its "Standard" territory.

While it adjusted to the conditions imposed by the decree, Jersey never lost sight of the advantages inherent in a single brand name, one that could be used nationally. Nor could it afford to do so. More than twenty years after the court-ordered dissolution of the company, Jersey hired the famed pollster, Elmo Roper, to survey the public reaction to the Standard Oil Company name. He found that only 22 percent of the public "thought of the several Standard Oil Companies as individual concerns." Jersey then embarked on an intensive two-year advertising and promotion effort. Roper later found that 27 percent of those he polled knew "that [only] Standard Oil (New Jersey) and Standard Oil of California, for example, are not one and the same concern."

Even in advertising, the name-identity problem bothered officials. Late in 1949, Jersey Chairman Frank W. Abrams and President Eugene Holman met with Robert E. Wilson and A. W. Peake of the Standard Oil Company (Indiana) to consider Jersey's sponsorship of the "New York Philharmonic Symphony broadcast through stations located in the areas in which the Standard Oil Company (Indiana) operates." Abrams reported to the Executive Committee that Indiana considered any such action an encroachment "upon Indiana's legal rights." So he and Holman had general counsel draw up a memorandum outlining all advertising and public relations policies "that should govern Jersey with respect to the use of the Standard Oil name or the Esso trademark." When the Committee had approved the memorandum they sent it to all affiliates.[9]

However, Jersey's management continued to encounter an identity problem. In 1949, the company hired a public opinion polling firm to survey gasoline dealer opinion in several states. They sought to learn whether dealers understood the relationship between Esso Standard Oil Company and the Standard Oil Company (New Jersey). Too, they wanted to find out whether dealers understood the difference among the Pan American Petroleum Corporation, the Standard Oil Company (New Jersey), and the Standard Oil Company (Indiana). The findings "indicated confusion in the minds of dealers concerning the relationship of these various companies."[10]

The Executive Committee appointed a special committee on September 22, 1949, to consider all matters relating to the company and the principal brand names of its affiliates. Later, on June 20, 1950, they listened to a report from the company's General Counsel, Edward F. Johnson, on aspects of the 1937 St. Louis decision that restrained Jersey from using the brand name, Esso, within the territory of the Standard Oil Company (Indiana). Johnson advised the Committee that he saw "no feasible method of having the provisions of the St. Louis decree set aside."[11]

So between 1940 and 1965, Jersey and its affiliates had at least six study groups assigned to recommend a trademark for nationwide use in place of Esso. One of these research teams worked early in the era of President Monroe J. Rathbone (1960-1965). For it, he selected two trusted aides, turned over to them the lists of names and other materials accumulated by earlier name and trademark seekers, and assured them of complete secrecy and privacy while they sought an acceptable replacement for "Esso." Some weeks later, he called on the pair to present their findings to the Executive Committee. This they did, one offering reasons for and the other arguing against the name they had chosen. Rathbone listened intently for the full hour. Then he said, "Boys, I've never heard a better presentation in all my years with Jersey. You have touched all the bases in your study and I commend you for the diligence with which you worked on this—but your name stinks."[12] Rathbone, one of them remembered, really liked Esso. The recurrence of these studies and their increasing frequency throughout the 1950s testified to the persistence of the search for a suitable national trademark, but the results of these efforts indicate the growing reluctance to depart from the familiar Esso. In each case the recommended substitute bore a resemblance to the existing trademark. Each had four letters, began with an *E* and ended with *O*. Yet Jersey adopted none of these new trademarks until 1959.

The 1960 merger of Jersey's domestic affiliates into the new Humble Oil & Refining Company seemed to clear the way for nationwide marketing under a single trademark, but Enco, the brand name developed in 1959, became a primary trademark only in areas forbidden to Esso. The new trademark did, however, permit a substantial simplification of the situation. Continuing to use Esso where it could do so, Humble spread the new Enco brand to all of the other states where it marketed, except Ohio. There, the company marketed under the Humble trademark. Thus, the problem of a split identity remained unresolved. One consulting firm reported that, "the dual system is creating confusion or something less than a clear impact on the customer." In a letter to Director Wolf Greeven, they pointed out that a continuation of this problem could result in loss or diffusion of "some of the tremendous equity that has been built up over the years in the Esso name."[13] However, the goal of nationwide marketing under a single trademark was not to be suppressed for long.

In 1961, events launched Humble on a protracted effort to free Esso for nationwide use. The first thrust of this drive came in response to the Standard Oil Company of California's acquisition of Standard Oil Company (Kentucky). Because Humble and its predecessor companies had for decades been the principal suppliers of the Kentucky Company's products, Jersey affiliates had never marketed directly in the states that constituted Kentucky's territory: Mississippi, Alabama, Florida, Georgia, and Kentucky. With the loss of that supply contract, Humble's managers decided to protect the outlet by opening their own stations in the area. Maintaining that a trademark licensing agreement between Humble and Kentucky had recognized legitimate ownership of the Esso trademark in the region, and that the provisions of the contract between the two companies had permitted Esso to use its brand names, they expanded rapidly under the Esso sign. In 1961, when Kentucky challenged Humble's interpretation of the trademark license, the latter went into U.S. District Court in Biloxi, Mississippi, seeking a declaratory judgment affirming their exclusive right to use Esso throughout the area. Kentucky's countersuit to block that use pending the outcome of the case was soon set aside, and by July, 1963, when the case came to trial, Humble had 650 Esso stations in the contested territory, with hundreds more due to open by the end of the year.[14]

On the day after the Biloxi trial began, Humble took an even bolder step toward its goal of removing the restraints on Esso. Making a direct assault on the source of the legal restrictions, in July of 1963, it filed suit in the U.S. District Court for the Eastern District of Missouri asking the court to revise "the injunction ... entered against Esso in 14 midwestern states in 1937" [later enlarged to cover 15 states]. They asked the court to review that decision in the light of such changed conditions as population growth and better communications. Humble argued that the 1937 decree had created a geographical wall around the various Standard Oil Companies, and that the injunction had proved to be "anti-competitive," contrary to the "spirit and intent" of being an antitrust law. It "had the effect of creating monopolies in the market for Standard Oil products in the respective territories of the different Standard Oil Companies." Humble testified that it had no objections to other Standard Oil Companies marketing in "their territory," provided they properly identified themselves; rather, they welcomed such competition. Humble desired to use Esso throughout the area in which the Indiana Company enjoyed exclusive rights to the "Standard" name.

Humble also maintained, Esso "had achieved great acceptance as a unique and distinctive identifying symbol"[15] exclusive of its being a spelling out of the letters "S.O." Thus it no longer simply symbolized "Standard Oil." Moreover, the Jersey affiliate held that the good will attached to the "Standard Oil" name could be attributed to the actions of all the Standard Companies and not just of Standard of Indiana. Finally, Humble's brief maintained that because of a general public confusion as to the meaning of Standard Oil in the midwest, "Standard" used on its stations there no longer served the true function of a trademark—to identify the source of its products. On this basis, Humble argued that while "Standard" could no longer be considered a trademark deserving exclusive court protection, "Esso" could. To Jersey, the Indiana Company's insistence on its exclusive rights to "Standard" and to all words connoting it simply served as a way of avoiding effective competition in its original marketing territory,[16] when, in fact, the market conditions that had developed since 1937 made the whole concept of geographically allocating the good will attached to "Standard" both anachronistic and detrimental to free and fair competition.

The first fruit of its legal campaign to lift the geographic limits on "Esso" proved heartening for Humble. In 1964, the Federal District Court in Mississippi issued a declaratory judgment in favor of the company, but the tide soon began to flow in the other direction. In July, 1966, the Court of Appeals for the Fifth Circuit unanimously reversed the District Court ruling and, with the refusal of the U.S. Supreme Court to issue a writ of certiorari, Humble prepared to withdraw the Esso brand name from the southeastern states and to replace it with Enco.

Soon after the Court of Appeals ruling in *Humble vs. Standard (Kentucky)*, the company received more bad news. On September 29, 1966, the U.S. District Court for the Eastern District of Missouri ruled in favor of the Indiana Company, finding that the facts governing the 1937 decree had not changed; therefore, the earlier injunction was continued in full force. Humble, Jersey, and Esso Standard Oil Company appealed the decision to the U.S. Court of Appeals, Eighth Circuit, but on January 9, 1969, that court upheld the earlier decision. In May, the U.S. Supreme Court refused to hear the appeal, and the legal challenge to the 1937 decree ended.[17]

While the Kentucky case went through various appeals, both Jersey and Humble took steps leading to alternate strategies, to be used if they lost in court. The initiative in this countervailing course began at Humble. President Charles F. Jones recommended the creation of a joint Humble-Jersey task force "charged with developing alternate approaches to the solution of [Humble's] trademark problems, taking into account both Humble's overall corporate interest and the appropriate implications for Jersey on a worldwide basis."[18]

Headed by Richard P. Ryan, Humble's associate general counsel, this task force approached its undertaking by considering the range of possible alternatives. If legal clearances to use Esso could be obtained either through litigation or negotiation with other "Standard" Companies, the preferred course would be the use of Esso throughout the United States, but the committee also considered the possibility of continuing to use Esso and Enco in different parts of the country, and the use of Enco or some other existing Jersey trademark nationwide. Finally, it investigated the possibility of developing a new trademark for the domestic market. In addition to these alternatives, the task force studied the question of changing corporate names to correspond with the trademark chosen. On the basis of an intensive study, the Ryan Committee concluded in November, 1966, that Humble needed a single trademark that could be used nationwide, and that the most practical choice for that name would be Enco. It also recommended that with the adoption of a new nationwide trademark, Humble's corporate name be changed to one that allowed close identification of the company with its products.[19]

Despite the breadth of its charge, several components limited the Ryan Committee's work. Although it had Jersey representation, Humble executives made up the bulk of the committee's membership, and its analysis pertained almost exclusively to the domestic situation. In fact, the whole study took place in an atmosphere pervaded by the pressing need to establish a trademark policy covering those southeastern states where the elimination of Esso seemed imminent. As a result, the committee focused on short-term needs; it did not fully and freely explore all the alternative courses of action.

On November 15, 1966, the Ryan Committee made its report to Jersey's Operations Review Committee, a body consisting of all employee directors. This committee lacked the authority to approve the report but it did refer "Humble's recommendation" to the Executive Committee for decision later that same morning. That Committee, with its five regular members in attendance, did not approve the Ryan Committee report, "even tacitly." They did agree, however, "that Humble could use Enco in the five southeastern states" and could institute various steps to use that trademark "in the rest of the country." But they went no further in considering the report.[20]

During the discussion of the Humble plan, Chairman M. L. Haider "expressed concern" that should Jersey adopt Enco, it might resolve the pressing domestic problem, while at the same time it could "create an international one (and a U.S. chemical one) in its stead." Haider did not desire to "internationalize all trademark problems"; instead he sought to prevent them. When the discussion ended, Haider then gave Director Greeven [not a member of the Executive Committee] the assignment of surveying the entire trademark situation and recommending the course Jersey should pursue in charting a strategy suited to its worldwide interests.[21]

Greeven, who was contact director for Marketing, had spent almost thirty years with the marketing affiliates of Jersey. He knew all about Humble's trademark problems and he wasted no time in pressing for a definitive solution. Busy with the budget and anticipating difficulty in scheduling meetings during the holiday season, Greeven set a two-man team, Paul E. Morgan of Public Relations and Richard W. Kimball of Secretary's, to work in secrecy at 30 Rockefeller Plaza. They received instructions to sift through the various lists of names gleaned by the six previous name-study committees plus a computer list compiled for the Ryan Committee. They sought to "determine whether anything of importance had been overlooked," a brand name that might be used as a corporate name for the domestic affiliates, the parent company or both. Kimball and Morgan for some time had few contacts with the oil company people other than the Jersey Law Department. The two men continued their research for several weeks.

While this name study took place, Greeven recommended to the Executive Committee the appointment of a steering group to consider the use of outside consultants to advise the group and Committee. Acting on Greeven's recommendation, the steering group appointed consisted of Greeven as chairman; Director George T. Piercy, vice chairman; John O. Larson, secretary of the company; Henry B. Wilson, head of Jersey's Public Relations Department; and G. W. Butler, marketing coordinator and head of one of the earlier trademark study groups. With Kimball as secretary and Frank X. Clair, a trademark expert from Jersey's legal department as counsel, the initial steering group was completed.[22]

In Wolf Greeven, the group had a leader with well-developed ideas about the company's trademark needs. The operational problems arising from the inability to market nationally under a single trademark were readily apparent. But the redundant costs of multiple advertising programs and sales administration, and the confusion experienced by customers moving from one area to another were only the most superficial of the disadvantages arising from the existing situation. To Greeven, the basic problem lay elsewhere. In his years as a marketer of petroleum products, he had come to believe that despite the importance of the quality of the products, consumers did not decide primarily on that basis. The reasons buyers chose different brands had more to do with the qualities they associated with the companies behind the products than with the products themselves. From that analysis, the manifold disadvantages hampering Jersey's domestic marketing, because of its trademark problems, could be easily seen. Only in Ohio did the company use a trademark that could be clearly associated with its principal domestic affiliate. Esso might be connected with the parent company, but the confusion among the various "Standard Oil" Companies prohibited clear perception. The fact that their corporate names did not clearly reveal the relationship between Humble and Jersey completed a confused picture.

Under existing conditions, any positive values that might be attached to Humble or Jersey could never be fully translated into the sale of products.[23] This confusion also hampered the company in other areas, such as recruiting and public relations. Thus, Greeven began his work as head of the steering group with the firm conviction that the company had to use a single trademark nationwide, and that the corporate names of both the domestic affiliate and the parent company had to convey that trademark directly to consumers.

Kimball and Morgan reported in January, highlighting a number of possible names including two, "ECCO" and "ARCON" which seemed the best to them.[24] The reaction to this report was not enthusiastic, but the Committee, continuing to use the assistance of Kimball, persisted in its efforts.

Despite Greeven's certitude, all members of the steering group in its early days did not agree on the need for a name change. The group did, however, quickly agree on the necessity of using outside consultants in seeking a definitive trademark policy. Too, it accepted Greeven's suggestion to label the search "Project Nugget" because of his belief that in the mass of material reviewed by Kimball and Morgan, a piece of "gold had been overlooked."[25] Its first major task became the selection of an industrial design firm, since all previous trademark studies had been conducted by internal committees, and Greeven's group had no experience upon which to base a choice. Moreover, because their whole undertaking had to be cloaked in the strictest secrecy, any efforts to seek information would have to be made with the utmost caution. Finally, however, they struck upon a solution to their problem.

For assistance, the group turned to Arthur Newmyer, Jersey's long-time government relations consultant. Newmyer not only had some familiarity with the design field but could investigate it without attracting attention to Jersey's plans. The Committee asked him to conduct a confidential survey of the appropriate firms and to recommend those he thought most suitable for the project. To preserve secrecy, Newmyer did not even inform his own organization of his assignment. During his alleged Christmas vacation, he toured the country, conducting interviews with a dozen leading organizations in the field of trademark creation and design. On February 20, 1967, he presented the Greeven committee with a secret report that evaluated a dozen firms and recommended three for final consideration.[26]

The steering group then recommended the employment of Raymond Loewy-William Snaith, Inc., despite initial misgivings on the part of some members about Snaith's casual style. The Jersey Executive Committee then approved the retention of that firm.[27] As Project Nugget developed, all of those persons who came in contact with members of that organization, especially Snaith, came to appreciate the general personal expertise as well as the excellence of its employees.

Loewy-Snaith immediately set to work. They ordered new computer printouts of words and word combinations. After they reduced the list, ten thousand names were deemed worthy of consideration. They then secured linguistic experts to check the pronunciation and meaning of a number of these names in virtually every known language. They checked hundreds of telephone directories for both corporate and family names. This research impressed the steering group.

The Loewy-Snaith team working closely with the Project Nugget group finally narrowed the list down. Late in the summer of 1967, they had a particularly long and exhaustive session during which they reviewed the eight names most seriously considered. Subsequently, one of the group hit on Exxon. After "kicking the name around," they decided to keep that name confidential, "sleep on it" for several nights and then meet again. This they did. And all of them liked Exxon.

On September 13, 1967, the Project Nugget steering group met with the Executive Committee. During the meeting the eight words and non-words including Enco and Exxon were projected on a screen. So far as is known, up to this time, Jersey's top management had not been presented with these names. Jamieson, who attended that session, later recalled that he "shuddered and shook for a moment or two," upon first seeing the combination of letters spelling Exxon.

Later sessions of the Project Nugget group changed the initial eight names. Enjay, Chairman Haider's favorite, soon adorned the list. Snaith's experts quickly pointed out that many languages did not have in them the letters "J" and "Y." However, the final coup against the Haider choice came at a meeting during which the Executive Committee listened to a recording on which various names were pronounced as they would be in different languages. Some of the pronunciations of Enjay evoked chuckles among the members.[28]

Although the Greeven committee appreciated the continued preference for the nationwide use of Esso (if it could legally be done), the steering group continued the search for another trademark. The group's charge from the Executive Committee had dictated the criteria to be used. As a marketing tool, the trademark had to be distinctive and have no negative connotations. It also had to be easily recognized and remembered. Finally, the possible international use of the name meant that it had to be readily pronounceable in all major languages and yet have no negative associations in any of them.

These standards strongly suggested that the alternative trademark should have no actual meaning in any language, so that its only significance would be that derived from its association with the company.

Project Nugget seemed to be moving rapidly toward the selection of a trademark to replace Esso, but that did not mean Esso would, in fact, be replaced. The final decision clearly turned on the success of efforts to remove the restraints on Esso. Among Jersey's top management, commitment to the existing trademark remained strong; few, if any, executives wanted to abandon it without making every reasonable effort to avoid that step. In 1968, the Jersey Board had initiated negotiations with several oil companies, in an attempt to buy the right to use Esso in their territories, contingent on full court approval of any action taken. Even after 1969, when the suit to overturn the 1937 decree failed, those negotiations continued,[29] indicating that the Board still hoped to bargain for what it had failed to gain through litigation. Thus, while Project Nugget proceeded, it did so under a tacit understanding that any new trademark would be adopted only after every possibility for using Esso had been exhausted.

The pace of work of the steering group quickened as the final court decision on Esso neared. On June 9, 1969, Jersey's Management Committee* approved an eighteen-week testing program to evaluate further the remaining potential trademarks. The tests included nearly seven thousand interviews at home and abroad designed to gauge public reaction to various names.[30]

Loewy-Snaith came out strongly for Exxon in presentations to the Management Committee, and the results of the tests verified their judgment.[31] When measured against the criteria of ease of recognition and recall and lack of unfavorable connotations in any language, Exxon led all other names. But the test results did not guarantee that Exxon—or any new trademark—would be adopted. Negotiations for the right to use Esso across a wide area continued well into 1970.[32] Nevertheless, by early 1970, the possibility of Exxon being adopted as both a trademark and as a corporate name, had increased substantially.

Exxon's improved prospects derived from a number of sources. The final court decision upholding the 1937 decree eliminated the possibility of removing restraints on Esso through litigation, and generally undermined the

*The Management Committee is composed of the chairman of the board, president and all senior vice presidents. Constituted in May 1969, the committee was established by the chief executive officer to assist him in carrying out his responsibilities for the general care and supervision of the business and affairs of the corporation. The committee has final review authority for all business matters that have not been delegated or that have not been reserved to the Board of Directors by statute, by-laws or resolutions of the Board.

whole idea of nationwide use of that trademark. Meanwhile, support for Exxon with Jersey had been growing. By the time he retired, in 1969, Wolf Greeven had seen the steering group of Project Nugget come to favor the change unanimously.[33] Moreover, the group now included, in addition to Robert H. Milbrath, who had replaced Greeven as its head, a third director, Clifton C. Garvin, Jr. Perhaps more importantly, the company's new Chairman and CEO, J. Kenneth Jamieson, absolutely convinced that Jersey had no other option, saw a change of trademark and corporate name as a necessary investment in the future. He soon became the leading advocate of adopting Exxon.[34] Still many directors scoffed at the idea of changing the company name to what some termed "the double cross."

In February, 1970, Jersey's Management Committee approved in principle the conversion of all the company's domestic petroleum operations and of Humble's corporate name to the new trademark. It also raised the possibility of extending the new name to its domestic chemical business and to the company's other worldwide operations, but deferred action on those steps. Although the committee kept open the possibility of using Esso, if that course became feasible, it authorized the steering group to take further steps toward the adoption of Exxon. The committee asked the group to oversee a worldwide campaign designed to avoid challenges to the new name by negotiating settlements with all firms or individuals holding similar and potentially conflicting trademarks. When that program had proceeded sufficiently, a definite timetable for conversion was to be developed.[35] However, before that could be done, Exxon had to overcome one final challenge.

By October, 1970, the drive to develop a new trademark and corporate name had reached a critical juncture. The negotiations for the extended use of Esso had failed, and a final decision on Exxon had to be made. When the Management Committee met to review the situation, however, it soon became evident that a change to Exxon was not a foregone conclusion. While earlier resistance to Exxon had centered on a widespread, somewhat emotional attachment to Esso, the new challenge rested upon objective grounds. This objection served to focus Management Committee attention on the fundamental nature of the problem they sought to resolve; on the one hand, they desired the resolution of a United States marketing problem in the most expeditious and economic manner; on the other hand, they wished to create a valuable new asset for the corporation, i.e., a name that could be used worldwide by all future managers. The agenda included a discussion of whether or not the marketing benefits of using one name for corporate entities, brands, and products could warrant the cost of securing that special opportunity. W. M. McCardell, Jersey's vice president for Marketing and a member of the steering group, told the Management Committee that "the *immediate, quantifiable marketing benefits*" associated with the adoption of a single trademark and corporate name would not justify the cost of a worldwide conversion to a single corporate and brand name. With the concurrence of Humble's top management, he recommended a much more limited change, a switch to Enco that would be confined to petroleum product marketing in the United States and would not include chemicals, TBA, or any foreign operations. Since Enco had long been established in a large portion of Humble's marketing area, the switch represented the least costly way to solve the company's trademark problems. While endorsing the recommendation, Humble's managers considered even the change to Enco a matter of relatively low priority when compared to the major effort underway to improve profitability; they saw it as retaining "a lower order of priority for at least 4-5 years." As for the long range marketing benefits of a single worldwide trademark and corporate name, McCardell did not envision events within the next twenty to thirty years "which would cause the ability to use one-name worldwide to generate sufficient additional market profits" to justify a large scale outlay in the early 1970s.[36]

At the same meeting where McCardell presented the case for a lower cost solution to the marketing problem via a limited change to Enco, Milbrath, head of the steering group, pointed out "only one proposed name, 'EXXON,' provided the means to accomplish in full the directive with which the group was charged." The Management Committee then agreed to postpone the worldwide clearance program for Exxon while the steering group reexamined in detail the alternatives to the proposed name. In doing so, they directed the group to focus on corporate names and trademarks that might be used solely in North America if Esso continued to be used throughout the rest of the world. As an aid in its consideration of the matter, the Management Committee asked the steering group to include, in its reexamination, estimates of the cost differentials associated with the alternative courses of action.[37]

Despite the last minute agonizing reappraisal over the choice of names and the extent of the change, the adoption of some new trademark and corporate name seemed assured. While sending the steering group back to reconsider alternatives, the Management Committee also asked it to consider establishing a timetable for changing the parent company's name and suggesting similar changes at Humble and possibly at other affiliates in North America. Finally the committee asked for the group's advice on "any particular timing relationship" that might be necessary between the corporate name changes and conversion to a new trademark.[38]

Reexamination of the alternative courses of action had little effect on the steering group's evaluation of Exxon or on its view of the long-term benefits that could be realized by having a unified worldwide identity for the company, but it did result in a compromise on the immediate changes to be effected. On November 13, 1970, the steering group returned to the Management Committee with a set of recommendations for the adoption of a new corporate name and trademark. Believing that a corporate name change for the parent company would not be feasible before the regular 1972 annual shareholders' meeting, the steering group recommended that in the meantime the company clear Exxon promptly, and conduct marketing tests in the United States to test its acceptance and to establish firmly all rights to the trademark.

The legal work involved in clearing the Exxon name and trademark in all of the states and territories of the United States proved to be time-consuming and expensive. Only in Nebraska, however, did company lawyers run into a problem. There, they learned that a company already existed under the name, "Exon's Inc.," an office-furniture and equipment business owned by the Governor, J. J. Exon [later U.S. Senator]. Jersey persuaded him to change the name of his concern to The J. J. Exon Company.

The steering group also recommended that the parent company use the same period to develop with Humble a strategy for using the new word as a brand name. As soon as practicable (after favorable consideration by the shareholders), the group proposed that Jersey and Humble change their corporate names to reflect the new trademark. Other North American affiliates would then consider the possibility of similar changes. Although these recommendations for action assumed that immediate changes would be limited to North America, the group stressed that Exxon should be cleared worldwide and that a study should be undertaken on the question of eliminating the word "Standard" from the corporate names of affiliates around the world. Concurring with this plan, the Management Committee cleared the way for the Board of Directors to consider the question.[39]

When the Jersey Board met on November 25, 1970, Milbrath began by reviewing the history of the Esso trademark, the problems associated with its use in large parts of the United States, and the actions taken over the years to remove the restrictions on its use. Recalling the unsuccessful results of those actions, he referred to the Executive Committee's initiative of 1966 that had launched the drive toward "a definitive trademark policy compatible with the interests of the Company and its affiliates worldwide." With that background, he proposed the development of a plan and timetable for converting the names Jersey, Humble, and other appropriate domestic affiliates to new names incorporating the word Exxon. When they approved the proposal, the Board also endorsed plans to clear Exxon for worldwide use as a trademark and company name, with the understanding that Humble and other North American affiliates would use it "soon." Once they had obtained clearance and made detailed plans, the Board would again review the matter, to decide on implementation and prepare a proposal for consideration by the shareholders.[40]

In recognition of its changing functions as it approached the actual adoption of the new trademark, the Management Committee reconstituted the steering group. Milbrath, the chairman, remained the only director as the group planned for the implementation of the name change. Other members from the parent company included the vice presidents for Finance, Marketing, and Public Affairs, the secretary, and the general counsel. In addition, they asked Humble, Imperial, and Esso Chemical Company, Inc. to designate representatives. In testing the new trademark, the group would also work with other Jersey and Humble departments where appropriate. The Management Committee also appointed a Design Platform Committee with its representation composed of persons comparable to that of the steering group. This committee had the task of developing a suitable design for the name chosen. Vice President McCardell chaired this committee.[41]

The Board's approval of Exxon as the choice for the new primary trademark increased both the level of Project Nugget activity and the need for secrecy. The process of developing a final trademark design and preparing for field tests of the new brand name required the participation of greater numbers of people. In addition to expanding the activity at Jersey, Humble, and Loewy-Snaith, the new stage of development meant that other affiliates which might use the trademark had to be consulted and kept informed. The changing nature of the work itself also made security more difficult. In choosing a word to be used as a trademark, much of the communication could be limited to oral exchanges or handwritten notes that could be promptly destroyed.[42] While those forms could continue to be used in some cases, the development of an actual trademark design, logos, and various identifying signs all required full-scale models and examples in a variety of colors, type styles, and designs, all of which had to be retained and made available for comparison by a relatively large number of individuals. Until the company made the official announcement and cemented its claim by active use of the name, Exxon remained highly vulnerable. Moreover, as the date of implementation neared, the potential cost of a security leak increased constantly.

While the risks applied to all aspects of the process, the selection of a design for the main service station identification sign presented a special problem. Small versions could be used for screening dozens of variations, but in order to make the final decision correctly, conditions approximating those under which a motorist would encounter the sign had to be created. That meant building full-scale mockups and having company executives drive past at varying speeds both during the day and at night. To minimize the risk of exposure during these tests, Loewy-Snaith rented an abandoned air strip near the town of North, South Carolina. There they erected models with six different signs like those that would be used at service stations. All these actions required such secrecy that they hired Pinkerton guards to prevent accidental discovery of the plan by persons living nearby. Few, if any, people at Humble knew about this proposed "trial run." Members of the Design Platform Committee, together with representatives of Loewy-Snaith and the company's advertising agencies, then flew down to South Carolina where automobiles simulating actual driving conditions took them by the signs in daylight and after dark.[43] Movies were made which were later reviewed by members of the Management Committee. After the Board discussed their reactions to the signs, they reduced the designs to two: a red oval with white letters and a white rectangle with red letters underlined by a broad blue accent bar.[44] Both emphasized the distinctive double X's by interlocking them.

After thousands of additional interviews to test Exxon and the two designs, Jersey's Board of Directors approved a pilot service station testing program to be conducted by Humble. Early in October, 1971, thirty-four stations in six relatively small communities spread across the United States were converted to Exxon stations under one of the two designs.[45] The results of the program confirmed the faith of the Project Nugget participants. With the added stimulus of a substantial communication program, sales under the Exxon sign increased by 5 percent over previous levels, exceeding the projections made by company marketing experts. Tests also showed quick acceptance of the new name. Since the rectangular design proved to have greater appeal, that type of sign became standard by the end of the testing period, and with only minor alterations, that version became the new trademark design.[46]

As the "complete success" of the market tests became evident, Jersey's Management Committee began in January, 1972, to discuss seriously plans for changing the corporate name of the parent company as well as those of its largest domestic affiliates.[47] The name change would make the restrictions of the 1937 decree inapplicable anywhere. Now, the new circumstances would allow the renamed Humble and Esso chemical companies to be merged into the parent company as operating divisions. Although this step would not endanger the principle of decentralized management, it would create a simpler and more flexible structure.[48]

By April, Jersey's top management was completing plans for consummating the trademark and name changes and the organizational realignment. Despite the ready acceptance of Exxon in the test markets, the planners favored a gradual conversion that would allow customers and dealers to make the transition with as little confusion as possible. Thus, while the brand names would be changed on July 1, 1972, the process of changing station identification signs would not begin until November, when customers had become accustomed to the Exxon brand name. The parent company's name would be changed as soon as possible after stockholders approved the move, but changes in affiliate corporate names would not take effect until January, 1973.[49]

Although they had approved all of the program in principle before the Humble announcement in May, the Jersey Board did not take final action on changing the parent company's name and corporate structure until the morning of June 21, 1972. Having done so, they next informed the company's employees. While its customers had readily accepted the new name, the internal response to a new corporate identity could not be easily predicted. That afternoon most of Jersey's New York employees left the company's offices to convene in the Grand Ballroom of the nearby New York Hilton Hotel, where they heard Chairman Jamieson announce the changes that would take place if the shareholders approved. Answering questions for an hour after the announcement, Jamieson and Jersey President Milo Martin Brisco stressed the advantages of the new unified corporate identity and reassured employees that working relationships within the Jersey system would not be altered by the change.

Although the "first reactions were, at best, neutral," Jamieson assured workers that the name would soon grow on them. Despite a few *EXXIT* signs that appeared around the company's building, the weeks that followed seemed to bear out his prediction.[50] The resistance that did develop followed a predictable pattern: older employees seemed less amenable to the new name than their younger co-workers.[51] But due to strenuous efforts on the part of the Employee Relations and Public Affairs Departments, the problem never became serious.

By the end of the summer, the first stage of the changeover was well underway. Gasoline pumps across the country had been changed to the new brands, and Humble had launched an "intensive changeover advertising campaign" to promote the switch. To star in the new drive, the company recalled its familiar cartoon tiger, the most successful advertising symbol in its history, out of semiretirement. Stressing the continuity of the company's emphasis on product quality and service, his smiling face quickly appeared everywhere it seemed, also with his new slogan: "We're changing our name, but not our stripes." The owners and operators of Exxon service stations got the same message at a series of regional dealer conventions designed to promote the change and help dealers explain it to their customers. At company headquarters, specially assigned members of the Public Affairs Department kept busy answering some two thousand letters generated by the change.

These letters included a number from Jersey's shareholders. Since, as the company's owners, they still had to approve the name change, their written and spoken opinions received careful attention. Some stockholders agreed with Rathbone and other former executives who opposed the plan. They strongly objected to the idea of dropping a corporate name with such a long and distinguished history. Among the most vocal in the group were those who considered themselves "Jersey men" and who resented a step they thought would destroy some of the value of that beloved identification. However, Jamieson had listened enough. Now his mind was fixed. The name change would go through. Rumors spread, telephone bills soared, meetings took place, tempers shortened. At one point, three former Jersey Board chairmen, Leo D. Welch, Haider and Rathbone met for lunch to have "it all out." Welch backed the name change proposal as did Haider. Some persons say that "Haider handled it very well," and that afterwards Rathbone "kind of slowed back off." Certainly speculation soon ended about a possible "palace" revolt.[52] And, despite the intensity of their feelings, the objecting shareholders remained only a small minority of the whole.

On October 24, 1972, one thousand Jersey stockholders attended a special shareholders' meeting at the Roosevelt Hotel, to vote on the name change. Even the most outspoken opponents of the name change held out little hope of blocking the move, however. With 96 percent of the outstanding shares being voted in favor of the change, management moved quickly. On November 1, 1972, Standard Oil Company (New Jersey) became Exxon Corporation, and the familiar "J" symbol on the stock ticker gave way to a new symbol, "XON."[53]

The parent company's name change coincided with the beginning of a massive physical conversion of company property. Gasoline pumps and product packages had already been changed, but now the company had to begin the work of changing identification signs and insignia everywhere in the United States. For the company's twenty-five thousand service stations, that meant replacing not only the main identifying signs, but also approximately fifty smaller signs at each location. Even so the service stations represented only a portion of the task. Signs on each of the company's twenty-two thousand oil wells and eighteen thousand other buildings and storage tanks all had to be replaced or repainted. Each of its domestic tankers carried some two hundred pieces of equipment required to bear the company's name, and five hundred private roads needed new Exxon signs. The list of miscellaneous equipment to be changed ranged from identification badges and embroidered emblems to mudflaps for tank trucks. In addition, printed material, from sales slips to job application forms, all had to be revised.[54] At the same time, the company undertook a crash program to replace its eleven million credit cards with new Exxon cards by the end of 1972.

Industry observers estimated the cost of the changeover, including advertising to promote the new name, at one hundred million dollars. Considering the information a competitive secret, company spokesmen refused to divulge the actual cost or even to give out estimates. It is doubtful that the Accounting Department ever undertook to itemize and total the cost of the name change. When officials did discuss the subject, they took pains to place it in perspective by pointing out that some of the outlay replaced normal maintenance costs and that the final advertising expenditure depended upon the reaction of competitors. They also carefully noted that since much of the conversion cost represented capital investments that could be amortized, the impact on the company's earnings in any single year would not be substantial.[55] While recognizing that the change necessarily was very expensive, they viewed it as an investment in the future which would be amply repaid by the benefits flowing from a stronger, more unified corporate identity.

On January 1, 1973, the process of changing the corporate names of domestic affiliates began when Humble Oil & Refining Company became Exxon Company, U.S.A. and the Esso Chemical Company, Inc., Exxon Chemical Company. The two affiliates then merged with the parent company, to become its operating divisions. At the same time, Enjay Company, Inc. was renamed Exxon Chemical Company U.S.A. and, as the domestic division of Exxon Chemical Company, also merged with the parent. Finally, Jersey Enterprises, Inc., took a new name, Exxon Enterprises, Inc. A few months later Esso International became Exxon International, and the next year Esso Research and Engineering Company took the Exxon name.

Although the company registered the Exxon trademark in every country where it could do so, and even placed the Exxon sign over a few of its European service stations, it planned to retain Esso as its principal trademark outside the United States, at least for the foreseeable future. Abroad, no compelling reason existed for the abandonment of the Esso name and trademark. In similar fashion, it took steps to protect the "Esso" and "Standard" names in its domestic marketing area. However, those steps were intended to defend the good will in the old names rather than to keep open options for future moves.

When the name change had been legally and completely accomplished and Exxon had replaced the once cherished company names and trademarks, Director Milbrath, as chairman of the steering group, presented small mementos to those directly involved. To Jamieson he awarded "a symbolic Purple Heart ... for having confronted internal and external perils in leading the company to its new trademark and name."

By early 1973, the positive impact of the Exxon program had become evident. Not only had the new trademark been readily accepted in the marketplace, but the unified identity that had been the ultimate goal of the change seemed to be developing rapidly. "Exxon" had ceased to be simply a combination of letters and had become instead a word with definite meaning, derived from its association with the company, and solely from that source.

Out in Tucson, Arizona, former Jersey vice president and world famed geologist, Wallace E. Pratt, then eighty-seven years old, had followed the details of the name change struggle through the press and by corresponding with Chairman Jamieson. Pratt wrote him on August 9, 1972: "This is a fan letter. *The Lamp* has come with your little essay on Exxon and we think it is splendid. Our Warm Commendation and Thanks." When the change became final, he wrote Jamieson, "Exxon is more to be applauded every day."[56]

With the passage of time, the original critics of the change—even those who maintained their nostalgia for the old name—increasingly came to admit the success of the new one. Rathbone, who fought the change as hard as he could, later admitted it was "the best damn thing that ever happened to the company."[57]   And product sales in January and February, rose by 9 percent.

Chapter 23

Confrontation in Tehran and Tripoli, 1971

ORLD EVENTS IN 1969 AND EARLY 1970 fulfilled even the most dire predictions of the major oil companies. In Libya, for example, the Revolutionary Command Council [RCC], led by Colonel Muammer al-Qadhdhafi [Qaddafi] succeeded in redefining permanently all existing procedures for negotiating oil prices throughout the world. Moreover the Libyans accomplished this in less than a year.

The *New York Times* on January 12, 1970, printed a long story under the heading, "A Gusher of Trouble In The Oil Fields," which quoted an oil executive: "For the first time in my 20-year involvement with the Arab world, I am afraid for the lives and property of our employees." Another article, by William D. Smith, "Oil: A World of Deepening Strife," carried the subtitle: "Arabs Hold Trumps," and traced the shift in bargaining power away from the international oil companies to the producing nations. Later that year, to make certain the oil companies knew that they were serious, the Libyans ordered production cutbacks.

Not long afterwards (in the fall of 1970) the Libyans began to leapfrog* the price of oil—the hard-nosed tactic—one that shocked the oil companies almost as much as the new high price for oil. Also operating in Libya at that time were a number of foreign and U.S. companies relatively new in foreign producing operations, companies referred to collectively as "independents." One of these, Occidental Petroleum Company [Occidental], with large Libyan operations, "caved in" to government demands and symbolized the new state of negotiations. Since such independents had virtually all their production in Libya and could not compensate by increasing production elsewhere, most of them had little appetite for prolonged struggles with the Libyans. Hence, shutdowns were out of the question. The important Washington

---

*Leapfrogging as used here refers to the practice of one oil producing country taking as a base agreed-upon contract prices of another country, and then jumping its prices above those of the second country; then, the second country repeats that operation.

753

Special Actions Group* also met to consider a report that indicated Libyan oil and gas would be put in jeopardy only if the United States antagonized Qaddafi. "Libyan oil," the group agreed, "is literally the only 'irreplaceable' oil in the world, from the point of view both of quality and geographic location." So the United States government assumed a passive role, while developments in Libya continued. Once begun, the action initiated by Libya led the Persian Gulf producing nations to demand that each of their contracts be renegotiated. Iran, for example, started new conversations with the Iranian Oil Participants, Limited [IOP]—known as the consortium**— that lasted for six weeks. On November 16, 1970, the consortium agreed to increase Iran's tax share from 50 to 55 percent of net earnings—the amount earlier agreed to as the Libyan tax rate. And naturally such price leapfrogging spread to other producing countries. A domino effect prevailed.

What became obvious to all oil companies was that the Persian Gulf oil producing nations would continue to make demands; and as each new round of demands was settled, Libya would leap ahead to demand an even higher price. Although the strategy of every oil producing nation may have seemed the same, the pressure tactics of the Libyans emerged as by far the most disturbing—because they proved effective. Libyan crudeness and rudeness did not make their success easy to accept. No real protection against such tactics could be found, even though oil companies all felt certain that negotiations with Middle East producing nations would determine the stability of all Western economies. Equally obvious was the need of the oil industry to find some way of stabilizing Persian Gulf conditions. So Standard Oil Company (New Jersey) [Jersey] and other large companies began to search for some device that would enable their industry to confront Libya's new-found political power.

For Jersey, the changing circumstances affecting Middle East oil supplies presented especially serious problems. That company was the largest producer in Libya; it owned 30 percent of Arabian American Oil Company [Aramco], 11 7/8 percent of Iraq Petroleum Company, Limited [IPC], and a 7 percent share in the Iranian Consortium. Jersey also held contracts with

---

*This group, composed of members from several departments, was responsible for coordinating policy and reviewing contingency plans during crises. The U.S. President's National Security Advisor chaired it. It included persons of comparable or near equal rank from the Departments of State [State] and Defense [Defense], and from the Central Intelligence Agency [CIA].

**An operating group of oil companies in Iran.

other companies for Middle East oil. In fact, probably no other oil company held a greater stake in the successful resolution of Middle East conditions.

X   X   X

Among the oil majors operating throughout the Middle East and in Libya (as well as among other companies operating in Libya) everyone understood that oil demand had increased beyond productive capacity, and that the result would be marketing problems in both Europe and Japan. For some time, in fact, Jersey executives in both Esso Europe Inc. [Esso Europe] and New York had sought to persuade members of the European Economic Community [EEC] to take steps that would insure a supply of oil and products in the event of an emergency. Many oil spokesmen indicated their certainty that such supplies would be needed. In October, 1970, after a survey of the potential problem that included production in the Middle East and Africa, Esso Europe recommended to the EEC that each member country permit storage of a supply of oil sufficient for ninety days. The involved oil companies should also be allowed to add the storage price when they marketed this oil. Esso Europe also urged that each country develop an adequate rationing program. Whether the EEC ever recommended such steps, however, is not clear.[1]

X   X   X

In the years after World War II, Jersey Director Howard W. Page replaced Orville Harden as principal negotiator for the company in the Middle East.* But Page became overburdened, his health declined; and while he continued to have some responsibilities for the Middle East and to negotiate there, George T. Piercy, who served briefly as Middle Eastern representative in London, assumed an increasing role in on-the-spot negotiations. Already, Piercy had received highly favorable reviews of his work at Jersey; Cecil Morgan, vice president of the Standard Oil Company of Louisiana, once said about him: "George was so advanced in his thinking that he almost got himself fired ... in Baton Rouge ... George had an awful lot of capacity." But in his case, the real question concerned timing; Piercy himself phrased it:

> I think here the COED [Compensation and Executive Development] system wasn't working to perfection. ... Jamieson was presi-

*Jersey Vice President Orville Harden served as Middle East contact director until 1954.

dent ... and he asked me to go to London to head up our Middle
East office there [1965]. ... Page was on the Board and head of
Middle East [contact director] and they [the Board] had decided at
that time that there was no successor for Page. There was none in
... the Middle East pipeline. ...

Thus Piercy, unchallenged, emerged as Jersey's Middle Eastern man. He
had a fine background in economics and in supply; which, he believed,
constituted the main reason for this selection. His principal task, he
thought, would be "in managing that oil that was there, in getting it out and
making a profit on it."

Piercy went on the Board in May, 1966, and became a vice president in
May, 1967, with responsibility for the Middle East. Yet the entire process
bothered him even before he joined the Board. So he went to President
Jamieson to discuss the Middle East assignment: "I said, 'Ok, what alternatives do I have if I won't take it?' ... He [Jamieson] rolled his eyes back and
he said, 'I don't think you've got any.' So I said, 'I take the job.' " In that
dramatic manner, Jamieson confirmed a key, and, as it turned out, a very
wise decision concerning Howard Page's replacement. In 1970, Jersey sent
Piercy to Libya to negotiate a price contract. Piercy had already been involved in discussions with the Persian Gulf states. Thus, he had vastly more
negotiating experience than did ranking officials of other oil companies; and
increasingly, he assumed a leading role in the effort to resolve the industry's
problems with the Middle East governments.

X   X   X

The Royal Dutch-Shell Group [Shell] reported its version of events during
1970-1971, in a letter written August 16, 1974, by its chief executive, Sir
David Barran, to U.S. Senator Frank Church. Church, at the time, chaired
the Subcommittee on Multinational Corporations of the Senate Foreign
Relations Committee, and his subcommittee already had been probing
(among other things), reasons for the dramatic increase in the per barrel
price of oil in the years after 1970. Barran's letter pointed out to Church that
the oil companies—twenty-four of them—had agreed to present a united
front for negotiations with all producing countries, as the industry mounted
a collective effort designed to end price ratcheting between producing nations on the Mediterranean and those in the Persian Gulf. Oil companies
viewed this step as a calculated gamble, but to them it represented the last
hope of maintaining any semblance of control over price increases. Both the
producing governments and the oil companies knew that, inexorably, the

volume of demand was approaching surplus producing capacity. This trend added persuasive strength to the producing countries' arguments. While this united effort by the oil companies did ultimately fail, this failure did not occur for lack of trying every possible approach to the producing countries.[2] Finally, nothing availed the worldwide industry in its attempts to solve a localized problem.

Among the companies, Shell, in particular, had concluded early that only a concerted action by the industry promised any chance of success. In fact, Shell took the lead, becoming the first company openly to advocate a united front. As an oilman explained: "Shell had some very forward thinking people in an office that followed OPEC [Organization of Petroleum Exporting Countries] affairs, a quasi-political office, I guess you would call it. ..." Jersey and many other U.S. companies might have thought, considered, and even believed in industry action; but sixty years of antitrust confrontation had already taught them that, in discussing such subjects, words must be carefully chosen.

Barran later reminded Church that nothing had established the correctness of the Shell view, nor could anyone say that either the consuming countries or the companies would have been better off had they adopted the Shell position. Barran went on:

> At this time [1970] our Shell view was that the avalanche had begun and that our best hope of withstanding the pressures being exerted by the members of OPEC would lie in the companies refusing to be picked off one by one in any country and by declining to deal with the procedures except on a total, globe basis.[3]

Now, it seems clear that many of the economic problems of non-Communist nations during the 1970s began with the failure of Western governments and international oil companies to acknowledge the new reality of the bargaining strength exercised by oil producing countries; and then to take effective countermeasures. Both a sharp increase in international inflationary pressures and a discernible economic slowdown in the West can be dated from the oil crisis of 1971.

Yet Barran and his Shell advisors, while perhaps ahead of others in the industry, were slow to assess OPEC's potential as a price-controlling, or at least a price-influencing body. As late as 1960, after OPEC had emerged, Jersey's Piercy remembers conversations about whether "there would be any possibility of the West combining and bargaining what OPEC wanted versus what they had, oil ... something that could oppose OPEC." Piercy also recalls that when he and other executives broached this subject to French and British oil representatives, their attitude was different; the French "would always stand aloof on this sort of thing." Apparently the Americans discovered quite early that not all members of the European Economic Community regarded the formation of such a counterforce to OPEC as the best available strategy.[4]

Some years earlier when Jersey and other major oil companies had first encountered Russian oil in European markets, their attorney (for seven of these companies*) John J. McCloy of Milbank, Tweed, Hadley, and McCloy, was the same man who had served as consultant to President John F. Kennedy. In the course of one of their conversations, McCloy had indicated to the President that at some future date, national security interests might require some joining together of the oil companies against the Russian oil threat and later, against OPEC. When Kennedy agreed, McCloy acted swiftly. He went to see Attorney General Kennedy and Deputy Attorney General Byron White. Later he visited each successive Attorney General and repeated this conversation for their benefit, just in case the proper occasion for such a united company stand arose. At this time (1970), McCloy represented twenty-six oil companies as counsel.**[5]

McCloy does not remember the exact date on which he decided to meet with U.S. Attorney General John N. Mitchell,*** to tell him, "I thought the time had come for some form of collective bargaining on the part of the oil companies with OPEC and with the producing governments." The most likely time would have been between a meeting called by the U.S. State Department with McCloy and representatives of the oil companies on September 25, 1970, and a meeting of OPEC members held in Caracas, Venezuela, December 9-12; or perhaps soon after that session.[6]

On September 24, 1970, Shell's Barran lunched with the British Foreign Secretary, Sir Alec Douglas-Home, in New York. Barran sought to persuade the minister that Shell should refuse to comply with Libyan demands even if such opposition meant a shutdown of oil production—as he put it, "of the need to stand firm against Libyan demands." However, Douglas-Home already knew that member nations of the EEC took a dim view of any possible interruption in their oil supplies, so he did not endorse Barran's proposal.

*These were Gulf Oil Corporation [Gulf], Socony Mobil Oil Company [Socony Mobil], Jersey, Standard Oil Company of California [Socal], Texaco Inc. [before 1959, The Texas Company], Shell, and British Petroleum Company, Ltd. [BP].

**In a letter dated January 29, 1974, to Senator Frank Church, McCloy stated that he represented these during the OPEC negotiations (*MNC*, Part 6 [Appendix to Part 5] p. 289).

***1969-1972.

The September 25 meeting in Washington, called by the U.S. Department of State, included leading executives not only of U.S. companies but also those of several foreign based companies. Under Secretary Alexis Johnson, James E. Akins, of the Office of Fuels and Energy (a major State Department advisor on petroleum affairs), and other State officials attended this session. Barran represented Shell, and he remembers that his proposal for a united company position "to stem the avalanche" did not persuade the State officials. As for the oil company people present, "Some were less impressed than others and some not at all." Obviously on September 25, only Barran, a few other oil company executives, and McCloy saw real advantage in the companies uniting. Given the advantage of hindsight, it now seems difficult to understand why so few representatives were convinced. Viewed in the same perspective, a united front did have much chance of succeeding. Yet U.S. oil companies always had to consider—lived in its shadow, in fact—the U.S. system of antitrust. They had to guard continuously against remarks that would attract attention from the Antitrust Division of the Department of Justice [Justice].

While Barran sought support in the United States, Libya had stopped Shell from lifting oil, on the grounds that the company had refused to change profit sharing from 50-50 to 54-46 percent. As Barran claimed, these developments gave his advocacy of a united front an "especially keen cutting edge." Yet he did not succeed, and after the meeting as he flew back to New York with Jersey's Chief Executive Officer [CEO] J. Kenneth Jamieson and McCloy, he told them that without real cooperation "the game was up."

Akins had just returned from Europe, where the Libyan oil question was receiving much attention. He knew that McCloy and the company executives agreed with State that "any shutdown of Arab oil would clearly have had most serious economic consequences." So when State representatives expressed the particular fear of a Libyan shutdown, Akins understood why McCloy and the oil men insisted that a shutdown there would create a crisis much more serious than either of the two Suez Crises (1957 and 1967). Under Secretary Johnson told the group that U.S. relations with Libya were in a poor state; therefore any appeal to the government must prove "ineffective at best." The oil men explained that, while many OPEC members disagreed with Libyan tactics and behavior, no other nation would "stand still if the Libyan Government received a greater government 'take' " than it did. The probable result, therefore, would be leapfrogging price escalations throughout OPEC; and these increases would mean additional problems for the U.S. and the West. Something, they pointed out to the State Department, had to be done, and soon.

State remained troubled by the delicate Libyan situation. During the conversation, McCloy noted, almost everyone agreed that Libyan demands were outrageous, "and in the interests of all the need for combined resistance to the demands was apparent." No party present expressed any fear that the companies, by combining, would be risking antitrust action; rather, McCloy recalled:

> It was my view that action to combat inordinate government demands would not be within the 'mischief' of the antitrust laws ... though I believed it would be a wise precaution to keep the Department of Justice fully advised of any joint action taken.

McCloy remembered that, along with most of the oil companies, he maintained some hope of resolving the outstanding issues until the December, 1970 OPEC meeting, when "a whole new set of demands emanated" from the Caracas conference. Then, he and many company* leaders knew for certain that they had to "join together to discuss and formulate the means of offsetting this concerted onrush of government demands," in an effort to stabilize oil agreements and supplies. McCloy and a spokesman for each company began in December, 1970, to meet regularly with officials of the Justice and State Departments. Both departments agreed that a united front of oil companies to bargain collectively with OPEC now seemed desirable.

Yet not every State Department official approved of this course. James Akins believed that the Libyans deserved to receive the price increases that they had already demanded (from January 20, 1970 through January 3, 1971) because, according to his calculations, their oil had been underpriced. Akins stated that he and others in his office checked the Libyan numbers and found them to be correct; during 1970, "we took this up with the oil companies and asked them to show us where we were in error.... . And they were not able to point out to our satisfaction any errors in our calculations."

Years later, Akins told the Church Committee that he offered his opinion to representatives of every oil company operating in Libya. When asked about the identities of specific representatives, he said that, among others, he had spoken to Charles J. Hedlund, president of Esso Middle East. Several of the companies, Akins claimed—especially the independents—appeared willing to accept the Libyans' proposed price increase. But most of the oil majors told him that they preferred to hold out and bargain for a better deal. Since many independents obtained most of their oil from Libya, clearly they

---

*Ente Nazionale Idrocarburi [ENI] and Enterprise de Recherches et d'Activités Petrolières [ERAP] did not join with the other companies.

did not fear any price escalation elsewhere. The major companies did have other oil production and they had to fight to keep prices down in Libya—or watch increases spread to their other holdings. Akins obviously ignored the fact that although the posted price was relatively uniform, the prices on which government revenues were based had no such uniformity; some of the companies used the posted price as a basis for calculating tax while others used the much lower net price on sales. This and other comparable points might well have weakened Akins' argument.

Major oil company leaders believed that they must either join together or watch Occidental and other independent companies be nationalized, after which they would confront the Libyan government selling "their" oil. In Akins' view, the position taken by these companies seemed too narrow; it ignored the dilemma of the Europeans and Japanese. Akins remembered that he made the following argument to the companies: "That if the companies are nationalized in Libya and they tried to block the sale of Libyan oil, they will not succeed because Europe under no circumstances is going to do without Libyan oil. . . ."[7]

Later, some informed persons remarked that pressure from Akins and his associates at State opened the way for the oil companies' submission to Libyan demands within a few days of this September 25 meeting. The U.S. State Department, these persons insisted, acted as advocate for a uniform policy of "cooperation not confrontation" in the Middle East and especially in Libya. Whatever can be said in support of Akins' case, to the oil company executives, his proposal represented total capitulation. At that time, few of them were ready to give in, at least not without a struggle even though they recognized Akins' influence. He had, after all, become known as "Mr. Oil" in the State Department.[8]

To the U.S. international oil companies (and in particular those with Middle Eastern operations) Akins' views seemed symptomatic of an attitude in the State Department which they found somewhat disconcerting; they believed that State always pushed the oil companies to settle differences with OPEC by yielding to national demands. This seemed especially true of differences that involved Iran and its Shah [Muhammad Riza Pahlevi]. In the companies' view, since Akins had come to State as a Middle East expert, with much experience in that part of the world, what he had found there was an atmosphere congenial to his personal position. Interestingly enough, Henry Kissinger, at that time President Richard M. Nixon's Assistant for National Security Affairs, expressed a similar opinion of certain State Department officials. He wrote:

The working level of our goverment, especially in the State Department, operated on the romantic view that Third World radicalism was really frustrated Western liberalism. Third World leaders, they believed, had become extremist because the West had backed conservative regimes, because we did not understand their reformist aspirations, because their societies were backward and eager for change—for every reason, in fact, other than the most likely: ideological commitment to the implacable anti-Western doctrines they were espousing.

Much earlier, during the Cold War years, Iran had become the prized ally of the U.S. in the Middle East, and many persons in the American government continued to believe that nothing should be allowed to endanger that special relationship.

Yet rarely in modern times has the United States and its Department of State played a role of so little importance in the Middle East. During the last year of President Lyndon B. Johnson's administration, a virtual vacuum existed in many U.S. Embassies in the area. In fact, not all Middle Eastern ambassadorial and ministerial positions were filled after the 1967 Arab-Israeli conflict and during the ensuing oil boycott of the U.S. by several producing states. Too, though Egypt's Gamal Abdul Nasser stated that he erred in charging that U.S. war planes had aided the Israelis, he refused to make the formal apology upon which Johnson insisted; so at least some Arabs continued to believe Nasser's inaccurate charge.

Small wonder that, early in 1971 (January 17) the *Washington Post* could report as accurate its unidentified source: "The big international companies have always made a big impact on the State Department. That is justified—the companies provide information that the government can't get any other way, especially from the Middle East." If Secretary of State William P. Rogers had any Middle East goals in mind, they seemed to be submerged in higher priorities, such as Vietnam. As for President Nixon, Kissinger wrote: "Nixon thought Middle East diplomacy was a loser from the domestic point of view and [he] sought to deflect its risks from himself." As a general rule, except for crisis situations, Washington devoted little attention to the problems of the Middle East.

Meanwhile, oil companies big and small could not—or would not—understand precisely how the public viewed them. In particular, the European and American public appeared to accept as fact that oil companies were not concerned about crude oil prices paid the producing countries, holding that the companies merely passed on increases to the consumers. Neither the oil companies nor their governments ever successfully explained just how the companies had managed to provide the world with a steady supply of relatively inexpensive energy over many years. Nor did the public understand that arbitrary and capricious crude oil price increases accelerated world inflation and, in turn, produced economic slowdown. OPEC leaders, on the other hand, understood these matters, as did oil company economists. Again, some of the OPEC leaders appreciated the danger of unrestricted price movements; others did not care.

In 1969, President Nixon selected Henry Kissinger as his advisor on national security affairs. To some oil company leaders, as well as others, the choice was puzzling. But Kissinger operated largely in a political and military world—just where the Shah remained a stout friend of the U.S.

The Shah, whether from innate shrewdness or by sheer good luck, came out of the 1967 Arab-Israeli War as a ruler courted by both the Organization of Arab Petroleum Exporting Countries [OAPEC]* and the Western powers, especially the United States. During that brief war, he had permitted Iran's oil production to increase by 20 percent; while the unsuccessful oil boycott demonstrated to the Arab oil ministers that they could exercise real power over oil supplies—only if they could persuade Iran (and Venezuela) to join them. So the Shah enjoyed flattering attention from both his Arab neighbors and his American friends. For it was the United States, acting with assistance from Great Britain, that restored the Shah to the peacock throne in 1953. At that time also, Jersey's Howard W. Page, persisting after all others had quit, negotiated with Iran's Dr. Ali Amini a consortium agreement with the National Iranian Oil Company [NIOC]. Of course, the Shah could not forget his American partners.

After his restoration, the U.S. encouraged the Shah to institute political and economic reforms. Although no one would accuse the Shah of being a liberal reformer, he did manage, especially during the 1960s, to achieve important reforms, based on his use of NIOC oil money. Meanwhile under the direction of Presidents Johnson and Nixon, the United States continued to befriend and sponsor Iran. That country, throughout those uncomfortable years, represented in the minds of many American thinkers the major Middle Eastern barrier to Russian expansion.

Certainly the Shah, from 1954 until his downfall, must have resented the role of the U.S. at some moments. But he also proved to be a political realist. So he followed a careful policy of conscious drift—by diplomatic zigzagging between Western and Eastern suitors. To Middle Eastern media, he fre-

*OAPEC was formed in January, 1968. The non-Arab oil countries, such as Iran and Venezuela, did not belong to it.

quently revealed his favorite dream of a modern, up-to-date Iran joining the club of world powers. This dream, he said, would come true as soon as Iran began to control the destiny of its oil—from the field, through refining, to the market. Without listening to the details of his dream, the U.S. heard what they hoped he was saying; and Americans continued to court the Shah.

X   X   X

In Caracas, on December 9-12, 1970, OPEC quickly reached agreement to demand increases in the posted price of crude oil and in each government's share of oil profits [Resolution XXI, 120].* Now, for the first time, OPEC found willing enforcers; both King Faisal of Saudi Arabia and the Shah of Iran agreed (in advance) that oil companies which refused to comply with OPEC demands would not obtain oil. In general, all the producing states agreed to eliminate the discounts previously allowed to companies; and to demand from these companies a new system of pricing adjustments for variations in the specific gravity of oil. At the insistence of Iran and Saudi Arabia, OPEC tentatively decided to allow a committee of ministers from the Persian Gulf countries (Iran, Iraq, Saudi Arabia, Kuwait, Abu Dhabi, and Qatar) to begin negotiations with the companies lifting there. Then, they scheduled the first negotiating sessions to begin on January 12, 1971, in Tehran. Should these negotiations not achieve the planned results, an extraordinary session of OPEC would be convened three weeks later. As a final alternative, OPEC planned for concerted action by all the member states to deny oil to the involved companies.

As head of the Gulf negotiating team which would bargain for all of them, the Persian Gulf states chose Dr. Jamshid Amouzegar, finance minister of Iran. Amouzegar had been educated at Cornell University, and he had married a German woman. Later, he acquired a reputation as a linguist and as a diplomat, two qualities that would be useful to the OPEC team, which would be meeting the company spokesmen in his native country, where the Shah was expected to exert special influence on the deliberations. Also on the Gulf Committee was Sheik Ahmed Zaki Yamani, oil minister of Saudi Arabia and formerly a student of Arab law at the University of Cairo, and of international law at both New York University and Harvard University. Even though Yamani was a "commoner," King Faisal had elevated him to oil

---

*Results were announced on December 28. OPEC Resolution XXI, 120 sought for all OPEC countries the establishment of a 55 percent minimum income tax, a uniform price increase, elimination of all discounts and adjustments of the gravity price differential, among other changes.

minister when he was only thirty-one years old. Jersey's Howard W. Page later remarked that Yamani proved to be one of the toughest opponents with whom he ever negotiated. Another committee member, Dr. Saadoun Hamadi, president of the Iraq National Oil Company, had earned his Ph.D. in agricultural economics at the University of Wisconsin, and at times Abdulrehman Salim Al-Ateeqy, Kuwait's minister of finance, joined the talks. Together, the Gulf states' ministers constituted a resourceful and formidable team.

At the time of the OPEC discussions, the oil market was tight; with demand pressing spare producing capacity. The general posted price for Gulf oil hovered around $1.80 per barrel, with certain variations for quality. The average government take per barrel amounted to slightly more than 90 cents. The Suez Canal remained closed; the Trans-Arabian Pipeline [Tapline] had been cut (and Syria refused to let it be repaired), and marine freight rates were simply astronomical.

Earlier, in the 1950s and 1960s, plentiful and cheap supplies coupled with significant new production served to stimulate the use of oil as a source of energy. As could have been predicted, however, the producing countries continued to demand more production for the income needed to support their national economic plans. Spare producing capacity, plentiful in the 1950s and adequate in the 1960s, by 1971 had disappeared.

Even the U.S., which had plenty of crude oil spare producing capacity through the 1967 Arab-Israeli War, by 1971, was utilizing all of its surplus capacity. What world surplus producing capacity remained was located largely in the Middle East, especially in Saudi Arabia and Iran. This fact undoubtedly influenced the attitudes of Western European nations and of Japan, both areas that were more concerned with the continuity of oil supplies than with the price that companies had to pay for oil. Certainly the U.S. Department of State and President Nixon made it clear that the U.S., too, was more concerned over a stable supply of oil than the per barrel cost of it. Naturally OPEC leaders knew this.

For the actual loss of spare capacity, a number of reasons can be offered: both the companies and the U.S. government had failed to estimate correctly the real growth in demand for oil as an energy source; the Prudhoe Bay Alaskan discovery, which oil men once believed would make 2 million barrels per day [b/d] available by 1971, became snarled in litigation and slowed coming on stream; inside the U.S., spare producing capacity had declined rapidly as discoveries of new oil failed to add significantly to the supply; Libya, and later Kuwait, cut back on existing production while several producing countries placed new limits below capacity on production; offshore drilling in some parts of the U.S. had been suspended; and the list could be extended. In effect, as a final result, oil companies lost most of the influence they had exercised over prices, although it took three years for them to acknowledge this new situation. Yet the OPEC ministers at the meeting were already aware of many of these facts. They knew that while the U.S. began the 1950s depending on the Middle East for only a small quantity of its crude oil, that amount had grown steadily as time passed. Given the numerous factors proportionately reducing production elsewhere, along with soaring demand, U.S. dependence on imported oil simply had to increase. Oil specialists in OPEC and elsewhere read and understood this trend as well as did oil company experts.

In 1971, as an opening shot, Libyan Deputy Prime Minister Abdesselam Jallud called together all the local representatives of oil companies on January 3, 1971.* He told them that the Revolutionary Command Council of Libya intended to make them all suffer until the United States changed its Middle East policy of aid to Israel. To make its point, Libya proposed to implement the resolutions recently adopted at Caracas, except that companies producing in Libya had to pay a much higher tax rate—50 cents per barrel—and also sell "cost oil" to the host country. Clearly the Libyans intended to "cash in" as quickly as possible on their proximity to Europe, especially since world events had dictated freight rate increases. The long haul around Africa, made necessary by Suez closing (1967), took time and proved to be expensive. Since the Persian Gulf price plus freight established the price of oil generally, the Libyans wanted to reap extra profit from the freight cost increases by charging more for their oil than did other producing states. In addition, the RCC would require oil companies to invest in exploration in Libya 25 cents for each barrel of oil exported. All together the proposed increases threatened to double—almost—the October, 1970 prices of Libyan oil.[9]

In response to these Libyan demands, the U.S. Department of State "established an inter-agency task force and began consultations, with England, France, and the Netherlands; ..." in which nations, along with the U.S., all the companies being threatened were domiciled. Later consultations included other consuming countries, plus leading oil companies. The participants quickly agreed that another spiral of price increases should be avoided, if at all possible. Only Great Britain and the Netherlands, however, strongly backed united action by the companies.

Further complicating the situation, the Libyans called in both the Nelson

*In some references, this date is given as January 2.

Bunker Hunt Oil Company [Hunt] and Occidental representatives, on January 9, 1971, and told them that they had one week to agree to the demands—without negotiating—just accept them or else!

Back in the New York offices of Mobil Oil Corporation [Mobil],* the Chase Manhattan Bank, and in the University Club, oil company spokesmen began to meet with McCloy on an "informal and company basis." Initial conversations concentrated on a possible union of the oil majors. But Libya's obvious intention of picking off the independents separately, as it had done during the previous September, presented a special risk for the remaining independents. Moreover, the Justice Department discouraged any plan of action that did not include the independents. Realizing that they had "even worse problems" ahead than did the majors, these companies soon decided to join the union movement. As Barran wrote Church, "The first plans for a united industry front did not exclude the independents so much as recognize as a fact of life that some of them might wish to exclude themselves." In fact, several of these companies were almost wholly dependent on Libyan oil.[10]

Naturally this high level of activity by so many prominent oil executives led the news media to search for causes. The *New York Times* headlined its key story, "Fateful Meeting on Oil," and the writer, William D. Smith, predicted that the companies would soon journey to Tehran to meet with OPEC. After describing what had already happened in Caracas, Smith continued:

> Some well-informed observers find Washington's lack of concern and inactivity less understandable. One said, 'When you consider what is at stake may be at the very least our worldwide balance of payments position, it is difficult to understand the apparent lack of attention.'

Among oil executives, many of whom had been enthusiastic Nixon supporters, a feeling of bafflement now seemed general.

Interesting enough because of its own importance, but doubly important because of later events, Jersey's expert Middle East negotiator, Howard W. Page, retired on November 30, 1970.** Page's old friend, "Billy Fraser," of

---

*In 1966, Socony Mobil Oil Company became Mobil Oil Corporation.

**In all, Page served Jersey for more than forty years. After this Tehran meeting, he did not again participate in negotiations with the producing states. In 1976, an Exxon publication referred to Page "as prime architect of the key international agreements between the major oil companies and the producing countries during the 1950s and 1960s," and went on to state: "Page more than any other man shaped the complex pattern of international agreements around the world that now in the 1970s, are being phased out." That compliment was justified, as these chapters on the Middle East reveal.

BP—now Lord Strathalmond—seized that opportunity to invite Page to the Tehran negotiations. Page accepted on the condition that he represent no oil company. He realized that Lord Strathalmond would participate in the discussions, and that this BP executive was not known to OPEC ministers. Page, on the other hand, knew virtually all of the OPEC delegation, and he was a close friend of several senior ministers, with whom he had bargained across the years. Lord Strathalmond, while an excellent attorney, lacked that valuable experience. So Page prepared to go to London, and then to Tehran.

On January 7, 1971, McCloy and a number of company officials again called upon Under Secretary of State Alexis Johnson. In his office, they discussed recent statements from OPEC, and McCloy noted that the new demands of the producing countries "ignored pre-existing agreements." He also pointed out that these demands were "couched in terms which permit no fair or good faith negotiation between the parties"; he added:

> Heavy investment values are in jeopardy, balance of payment relationship[s] are seriously threatened and a widespread disruption of the principle of access on reasonable terms to oil supplies on the part of the Free World consumers is impending. I think I need [not] to point out the political and economic consequences of a continuance of such tactics. The Free world economy could become seriously impaired, particularly since coal as a main source of energy has largely disappeared and for the time being, at least, the shortages of transportation facilities have contributed to the problem.

Any concerted action by the companies seemed certain to raise questions concerning American antitrust laws, but McCloy assured Secretary Johnson that he had secured clearance from the Department of Justice for joint company action. Finally, he told Johnson that representatives of the companies had been working out some plan for the equitable sharing of any production cutbacks in Libya (or any other country) should that country carry out its "ugly threats"—a plan that McCloy referred to as the "Safety Net,"* and one suggested earlier by Occidental's Armand Hammer.[11]

Having secured adequate assurances from the Justice Department, McCloy and the oil executives—now meeting together instead of company-by-company—began to compose a joint message to be sent to OPEC, inform-

---

*It should be noted that in May, 1970, Libya, Iraq, and Algeria established a "joint co-operative fund"—their safety net—to provide reciprocal financial support should any one of them lose income as a result of a shutdown occasioned by failure in negotiations with the oil companies. Armand Hammer probably knew about this. In 1972, OPEC established a comparable fund.

ing them about unified negotiations. Because of the importance of this matter to the United States (as well as to other oil consuming nations), the Justice Department sent its own representative, Dudley Chapman, to advise the drafters, while State sent Akins to act as an observer. Neither McCloy nor the companies requested this step; however, the two government representatives did prove helpful in preparing the final message. The actual drafting took place in both Washington and New York. Jamieson, Piercy, and Hedlund represented Jersey at these sessions.

On January 13, McCloy took the draft message to Washington for Justice to review. He incorporated the department's suggestions, and Justice approved the text on that day, by having an Assistant Attorney General initial each page. Then, McCloy took the message to State, where it again received approval. Armed also with the Business Review Letter from Justice that cleared the combination of oil companies to negotiate as a unit, McCloy then returned to New York.

While the message was being prepared, some of the independents expressed fears concerning production cuts or even nationalization by Libya. In such cases the original "Safety Net" discussion called for pro-rata sharing among producers of any cutback during six months, or up to December 31, 1971. On January 13, 1971, the companies decided to confine the agreement to Libya, and to arrange for backup oil from the Persian Gulf. Next, McCloy took the additional precaution of securing a second letter from the Justice Department covering that second agreement—now formally termed the "Libyan Producers Agreement." Several independents supported the plan because it provided an alternative to full capitulation; but also and more important, it symbolized the solidarity of the companies—"a willingness to share the burden equitably."

In all, thirteen oil companies signed the message to the OPEC countries which was sent to Tehran on January 15.* This message made a proposal:

> that an all-embracing negotiation should be commenced between representatives of ourselves, together with such other oil companies as wish to be associated with this proposal, on the one hand, and OPEC as representing all its member countries on the other hand, under which an overall and durable settlement would be achieved.

The message went on to offer revisions in existing posted prices, which were to be adjusted annually against a yardstick of "Worldwide Inflation" and continued along with other adjustments. Copies of the message were sent to

---

*Eleven additional companies signed later.

every OPEC member at the same time, and also to other oil companies that might wish to join the united company front. Later, the message was released to the public by the companies—"we also feel it right to make this proposal public knowledge promptly," and they stated, "we are not able to conclude negotiations with individual member countries." Replies could be sent to either Jersey or British Petroleum.[12]

In the meantime, the gulf between OPEC and the oil companies had grown so wide and OPEC demands so outlandish that the situation appeared impossible for the companies. Then while John McCloy and the companies planned a single-front negotiation, word came of newly increased OPEC demands, coupled with a 24-hour ultimatum to be served on the companies and on the Western powers: accept these or take a total shutdown. Now everything related to the oil company-OPEC negotiations stopped. At this critical moment, company leaders turned to Jersey's one-time Middle East diplomat Howard W. Page.

Page's first task required a high order of diplomacy. OPEC's head, Nadim al-Pachachi of Iraq, was an old friend with whom Page had first negotiated in 1951. Twenty years later, as Page recalled, his approach required even greater care: "I briefed myself on the issues involved and then called Pachachi and asked if I could come to see him [in Vienna]." When Pachachi expressed his pleasure, Page went to confront the OPEC head alone. Pachachi insisted that OPEC demands were inflexible, while Page contended that a share of the responsibility for the impasse rested on some OPEC members. Finally, Pachachi agreed to ask the OPEC negotiators to continue with earlier plans for continued negotiations. Once the immediate crisis had ended, Page went back to London.

Meanwhile, the State Department had asked McCloy and several company executives to come to Washington on January 15, 1971, to discuss the oil crisis in the Middle East. Once there, McCloy sought to persuade State to take a more direct role by bringing pressure on the OPEC governments to "moderate their demands" and to engage in "fair bargaining practices." Secretary of State William P. Rogers and Under Secretary Johnson replied that Under Secretary John N. Irwin II would soon visit several Middle East countries for that purpose. McCloy remembers that the reason given for the Irwin trip "was to urge the producer states to moderate their demand[s] in view of the Free World's dependence on a continuity of the flow of oil at reasonable prices." McCloy and the oil men emphasized the point that only collective bargaining could help the companies, and that if the united front broke, price ratcheting among the producing countries would surely follow.

Irwin and the oil representatives also discussed the united front approach.

Yet Irwin was selected for a Middle East mission not because McCloy or any oil company chieftain suggested him to State, but because President Richard Nixon wanted the trip to start immediately. Akins suggested Irwin because he was available at that time, and the two began their trip on January 16, 1971—the same day that the Iranians received the oil companies' message.[13] On that day also, the Occidental and Hunt representatives met with the Libyan oil ministry in response to its January 9 ultimatum. Before Irwin and Akins left Washington, however, Secretary Rogers, speaking for the President, reiterated that, in talking with the Middle East rulers, Irwin should stress the importance of continuity in oil supplies to the economies of all consuming countries—particularly those in Europe (and Japan). In the limited time he had before leaving, Irwin did not receive a full briefing concerning the importance of the united front to the oil companies. Later events would demonstrate just how crucial that omission was.

Throughout the entire trip, in fact, Irwin seemed to consider his mission to be one of mollifying the leaders of producing countries, as he assured them of "the concern the United States would feel if oil production were cut and halted." As he later testified, Irwin explained to the heads of state in Iran, Iraq, and Saudi Arabia that the U.S. "did not intend to become involved in the details of the producing countries' negotiations with the oil companies." The U.S. State Department, he told them, "had urged the companies to be cooperative and reasonable"; the "Message to OPEC" Irwin explained as the companies' declaration of their willingness to negotiate OPEC's Caracas resolutions.[14]

Iranian Consortium representatives met with Iranian government officials on January 12, 1971. This date had been set by OPEC at the Caracas meeting and this session merely conformed to the OPEC schedule.[15]

In the meantime, the oil companies had chosen George T. Piercy and Lord Strathalmond of BP, to head the team formed to discuss the "message" and then to establish ground rules for later negotiations. The two men left immediately for Tehran. Piercy later advanced, as the reason compelling this abrupt departure, that the Iranian government and Finance Minister Amouzegar "were very upset with what they thought were delaying tactics of the oil companies. . . . The two of us were sent overnight to see if we could calm things a bit."[16] Piercy and Lord Strathalmond left for Tehran on January 16, 1971.

Both Piercy and Lord Strathalmond realized (as did McCloy, Barran, Jamieson and some others) that their mission had an unusual importance— beyond the value of oil price or concession negotiations. If they could persuade the oil producing states to agree to a program that would guarantee reasonable stability of oil supplies and prices, the worldwide economic situation would be improved. Such a meaningful agreement would benefit not only the industrial nations but also the oil-short, newly emerging Third World countries, where every oil price increase threatened to create new deficits in their balance of payments. In effect, the Piercy-Strathalmond mission could become an important step toward a healthy world economy.

The *Washington Post* on February 6, 1971 suggested that the producing countries now appeared determined to exploit their power; for them the whole affair "has turned into a crusade" for higher prices. Against such sovereign powers, the companies must unite for self protection. As one executive phrased it: *"Producing Company Solidarity* was seen to be entirely reasonable and the only effective response to firm *producing country* solidarity,"—meaning OPEC.[17] By this time, the basic issue confronting the companies (as well as most oil consuming nations) they believed to be clear: either they would maintain a united front during collective bargaining (and, after time, expect to stabilize oil production) or they could break ranks and expect to be intimidated one by one, and made to accept arbitrary price increases—that is, to be whipsawed or ratcheted by producing countries. Learning from the example of Qaddafi, the companies knew that no prior agreement could now be considered binding, and that no written agreement then in force would protect any single oil company from retroactive tax assessments, or other *ex post facto* adjustments. Instead, OPEC members had gained their hold on the companies by employing the Libyan strategy.

X   X   X

On January 16, 1971, in Iran, Jan Van Reeven, senior representative of Iranian Oil Participants, hand-delivered a copy of the companies' "Message to OPEC" to the Iranian Finance Minister, Dr. Amouzegar. After reading it, Amouzegar observed that while he could understand the desire of the companies to stabilize prices and end leapfrogging, OPEC could not stop Libya from making "crazy demands." Then, he told Van Reeven that, if the companies did not settle with the Persian Gulf states, a scheduled extraordinary meeting of OPEC would take place, and the companies would very likely confront the most radical demands as a "common denominator." The Iranian minister also appeared to be disturbed by news that the "high level team" (Piercy and Lord Strathalmond) would be empowered only to explore implications of the "Message to OPEC," by trying merely to arrange for later negotiations rather than by entering directly into a substantial discussion of the problems. Van Reeven went on to tell Amouzegar that the "high level team" would arrive in Tehran on January 19.

Under Secretary Irwin and James Akins showed up in Tehran on January 17, 1971, having made the first leg of their mission to the Middle East. Together with the U.S. Ambassador to Iran, Douglas MacArthur II (a nephew of the famous general), they joined Amouzegar for a two hour audience with the Shah on January 18. Afterwards, Irwin and MacArthur talked with Amouzegar for an hour. Then, Irwin, Akins, and MacArthur conferred together, after which Irwin cabled the State Department. State summarized the contents of the cable in this way:

> In that report, the Under Secretary [Irwin] stated his belief, based on all of his conversations in Tehran, that in view of the assurances which the Persian Gulf producers were prepared to give [the Shah held the proxies of Saudi Arabia and Kuwait] on future price ratcheting, the companies should be urged to negotiate in Tehran with the Gulf producers.

Irwin also requested that Secretary of State Rogers communicate these views to McCloy (or to the chief executives of the oil companies). State did review Irwin's proposal, and Rogers called McCloy to suggest parallel negotiations with the Gulf countries and Libya. "The companies would be well advised to do this," Rogers said. Next, State cabled Ambassador MacArthur "to brief the companies and the representatives of certain European governments" concerning Irwin's position. State also told MacArthur that McCloy had been informed.

While the cables flew, Piercy and Lord Strathalmond continued on their trip, unaware of any change in strategy. On January 19, they arrived in Tehran, where MacArthur reported to them that the Shah and Amouzegar seemed friendly and reasonable—but that Iran could accept the companies' proposal only if the companies did not increase consumer prices. Iran would accept a five-year agreement because it sought price stability, and because it found the companies' general position to accord with the Caracas OPEC Resolution XXI, 120. More important, the Shah thought that united front negotiations would be a "monumental error"; rather, the companies should deal with the Gulf states first. He believed that no other Gulf state would object to any settlement which he had approved. According to MacArthur, the Shah assured those present that the Gulf countries would stand by all agreements. What seemed evident was that the Shah expected to dominate negotiations, and very likely, to dictate the final settlement.

When MacArthur stated that the Shah wanted any agreement back-dated to January 1, 1971, Piercy responded that "discussion of an effective date certainly at this early stage was a trap pure and simple and he hoped we could avoid this." MacArthur continued: The Shah had stated that if the companies "tried any tricks the entire Gulf would be shut down and no oil would flow"—if the companies wanted to avoid serious problems, then, they would agree to negotiate with the Gulf group. Finally came the clincher: this information had already been sent to Secretary of State Rogers, who in turn informed "Jack McCloy urging [that] he designate 'negotiators' to deal with the Gulf group." Now, MacArthur urged that the companies deal with Gulf nations first and with Libya later.

Piercy quickly responded to the ambassador by saying that any splitting of the negotiations after the companies had agreed on a unified front would violate both the "Message to OPEC" and the Business Review Letter from Justice. He asked MacArthur and the others to present any evidence that had led them to believe that the Gulf nations would keep their word—assuming that a "preposterous" deal would be made later with Libya.[18] To say that Piercy became upset is an understatement; certainly he presented the case for collective bargaining by the companies with striking force. Some years later, with regard to Irwin, Akins, and MacArthur, Piercy said:

> The thing that I was critical of was that they had agreed, I don't [yet] know whether they got any clearance on it, but they agreed with the Shah that we would negotiate on the Persian Gulf . . . and that came within hours almost of [our] signing a letter to OPEC saying we're going to deal with you collectively. And since they separated Libya from these, I said to MacArthur . . . 'Well, you've put in jeopardy every independent in Libya with this action because there's no constraining force now on Libya.' Before we'd hoped that the entire OPEC would hold Libya in line.

Then he added: "Jack [John Irwin] really felt that was the only thing, that was the only way they could hold it together to make that agreement with the Shah."

To Irwin, the Shah indicated that the only way he would consider any unified negotiations was to separate the Persian Gulf from the others. Piercy said: "To me it made the whole exercise look silly as hell." Still, he and Lord Strathalmond together proved unable to convince the U.S. and British ambassadors about the complex nature of the company agreement involving majors and independents.* On the other hand, the Dutch Ambassador, Mr.

---

*For example, although it had no Libyan production, Gulf Oil Corporation joined in the Safety Net Agreement.

Jonker, quickly grasped the crux of the matter—what was being suggested was in clear violation of the "Message to OPEC" submitted on Saturday, January 16, 1971.[19]

On January 18, the other two OPEC ministers arrived: Saudi Arabia's Sheik Ahmed Zaki Yamani and Iraq's Dr. Saadoun Hamadi.* The next day, Piercy and Lord Strathalmond met with them and Amouzegar. Piercy opened the session by explaining that the companies had joined forces to negotiate because they needed five years of stable oil supplies, and while the companies wanted price agreement, they also sought an end to extreme demands and price ratcheting. Further, Piercy noted that some of the OPEC resolutions at Caracas seemed to support "the Libyans and their demands." Finally, he asked the three ministers, who after all represented only the Gulf states, what they could do.

Amouzegar replied for his colleagues. He pointed out that they would discuss only the OPEC resolution concerning price [No. XXI, 120] and that they had found the companies' "Message to OPEC" not much different from that resolution—"one of approach and procedure and not of substance." Then Amouzegar returned to the points he had made earlier: vast differences existed among the producing countries, so the three ministers could not negotiate "with other industry groups." Negotiations on a global basis, he said, were neither logical nor practical. Amouzegar told Piercy and Lord Strathalmond that OPEC had considered overall negotiation at Caracas; some states (Iran and Saudi Arabia) had objected so strenuously that the conference finally approved a different approach.

Next, Lord Strathalmond secured from the three ministers a statement that they did not support the Libyan demands; and he went on to ask whether OPEC would set up a Mediterranean Committee to work with their (the Gulf) Committee. To this, Amouzegar replied, "No." OPEC did not have supranational powers.

Yamani offered some advice to the Westerners: "They should conclude in the gulf [sic], then let it flow to the rest of the world resulting in a new stability." Lord Strathalmond referred to Libya's demands, and pointed out that he and Piercy spoke for many independents who obtained oil solely from Libya; for all the companies, but especially those, Libya's demands were outrageous. Again, the Gulf representatives spoke up, saying essentially that they had no power over the Libyans.

After reviewing Libya's position, Piercy returned the conversation to the

*During the later negotiations, they were sometimes joined by Kuwait's Abdulrehman Salim Al-Ateeqy.

"Message to OPEC." Again he made the point that before the companies sent it, they had submitted the text to their governments, who approved it. Since he felt committed by that agreement to a comprehensive negotiation, he did not believe a Gulf only approach would work. At that point, Amouzegar interjected a remark: "If you think you have a problem with your governments I am quite confident that they will agree to a regional or Gulf approach."

His display of confidence on that point—although at the time Piercy and Lord Strathalmond could have interpreted it as largely bluster—should have given this audience pause. Neither of the company men knew of any change in any government's position regarding the negotiations. Nor were they aware that someone must have revealed the contents of the State Department cable of January 18 (the day before they arrived) to some person— either Amouzegar directly, or someone else—who told the Iranian minister that State had recommended parallel negotiations.

The January 19, 1971 conference ended soon after this exchange, but not before the three oil ministers had agreed to delay the OPEC conference two days (from January 23 to January 25, 1971). This extra time gave the companies a chance to reconsider, and to decide whether they should pursue the collective approach or change plans and begin negotiating with the Gulf states first. Making the latter possibility attractive, the ministers announced that when OPEC met on February 3, they would attempt to have a Mediterranean negotiating committee appointed. If that move succeeded, they would endeavor in turn to get Iraq and Saudi Arabia represented on it. Finally, Hamadi raised strong objections to giving the companies any additional time before beginning to negotiate; and on that sour note the Tehran meeting ended.*[20]

After this session, Yamani asked Piercy to accompany him to his hotel and then to the airport. Piercy seized the opportunity for a private conversation with the oil minister, for whom he had great respect. Piercy believed in head-to-head talks, even though, as he recalls, "Yamani would rub it [Libya] in all the time." In fact, "It was a lesson to him on how to treat us." But, as Piercy continued, "You don't settle the issues with six nations opposite you. You settle them by some private conversations and see if you can get some movement." So the two men left together.

During the trip to the airport, Piercy told Yamani that the Libyan

---

*At virtually every session, Hamadi accused the companies of stalling. His protests followed the tenor of the agreement among Libya, Iraq, and Algeria of May, 1970, which established time limits for "lengthy and fruitless negotiations."

demands had reached such a crisis state that the companies operating there must either accept them or face a shutdown. Yamani replied, "George, you know the supply situation better than I. You know you cannot take a shutdown." After that remark, Yamani boarded his plane to Kuwait, picked up Al-Ateeqy, and returned to Tehran for an airport conference with Amouzegar.

Later, Yamani revealed to Piercy what he (Yamani) had told Amouzegar: that until the oil crisis ended, the Gulf states would not demand new terms comparable to those demanded by the Libyans. And after the crisis ended, adjustments would then be made only to achieve price harmony concerning the Libyan freight allowance. This assurance, Yamani told Piercy, had already been approved by the Shah, who added the caveat that the companies should not increase the price to consumers. When, in answer, Piercy first argued and then demonstrated from financial data, that such action would tie the companies to the low profit years of the 1960s, Yamani replied that something might be done during the negotiations to improve company profits.

Yamani also verified as fact the rumor that an oil embargo designed to strengthen OPEC demands had been considered by OPEC at Caracas, and that King Faisal had supported it. This shocked Piercy, who pointed out that the Saudis had never embargoed oil except in wartime. When he pressed Yamani not to consider that step, Yamani replied, "I don't think you realize the problem in OPEC. I must go along." To this Piercy answered, "I don't think you realize the impact to the rest of the world and what it will do to the prestige of these producing countries." For emphasis, Piercy told Yamani that if Venezuela joined the boycott, that action would make a mockery of its effort to obtain preferential treatment from the U.S. When Piercy questioned Yamani as to "whether he believed that any agreement would last for any length of time," the Saudi minister left the impression with Piercy that he did not believe that the companies would get the five years of stability they were seeking. As Piercy later cabled the U.S. State Department, "the meeting was cordial but as usual Frank [sic], specific and blunt." Indeed it was![21]

On January 20, Piercy told MacArthur that he had received a cable stating that Secretary Rogers had asked McCloy to have representatives designated for "parallel negotiations," one with the Gulf producers and another with Libya. Since their meetings would require some definition of terms of reference, Van Reeven of IOP, who was with Piercy, suggested to MacArthur that he help explain the situation to Amouzegar and then encourage him to postpone the February OPEC session.

Back in London, the receipt of so many cables set in motion another meeting of oil company executives. From New York, Jamieson sent Associate General Counsel Jack G. Clarke and Charles J. Hedlund, and from London, Elliot R. Cattarulla to help plan strategy.

In their report to the company chief executives on the meeting of January 19, Piercy and Lord Strathalmond summarized what had happened. They also added new weight to Amouzegar's statement about Gulf first negotiations:

> It was perfectly clear that Amouzegar believes he and H.I.M. [His Imperial Majesty] have convinced American Government in recent discussion of the correctness of their position on a Gulf negotiation coming first with the result that our negotiating stand on procedure to be adopted is by no means an easy one.

The company spokesmen concluded:

> It is not easy to advise on what should be done. If we commence with Gulf negotiations we must have very firm assurances that stupidities in the Mediterranean will not be reflected here. On the other hand, if we stick firm on the global approach we cannot help but think, in view of the meetings reported separately, that there will be a complete muddle for many months to come. Somehow we feel the former [latter?] will in the end be inevitable.[22]

The companies now confronted a dilemma, since actual events had already defeated plans so carefully made. Now, they required some new course of action—and one that did not seem obvious—since State Department concern for Iran and the Shah had overbalanced the oil companies' common interest. The American government had moved, and now Piercy and Lord Strathalmond felt the pressures that accompany international political diplomacy. Both men knew that they must doubt whether any plan formulated by the companies could be effective.

X   X   X

If, as he remarked years later, Piercy was "disappointed" by the actions taken by Irwin and MacArthur and by the advice they cabled to the State Department, so were many other oil company executives who agreed with his view. Certainly it must have irritated him to learn *first* from Amouzegar that the U.S. Department of State had withdrawn its support for united front negotiations. Again based on remarks made some years later, John McCloy seems to have been even more indignant. State's reversal caught

McCloy by surprise. The company committee apparatus, planned to assist Piercy and Lord Strathalmond, had not yet been established when their cables arrived from Tehran asking the companies for a decision about sticking to the original plan or accepting the substitute program of the Shah, Irwin, MacArthur, and the U.S. Department for parallel negotiations.

Later, McCloy told the Church Committee that the final decision on whether to "buck" the Shah and Libya rested with the oil companies; but, he added: "We weren't too much impressed, if I may say so, by the attitude of the U.S. Government." He went on: "I will tell you why, because usually the man that is on the ground is usually inclined to be quite sympathetic to the local desires." When asked to explain that statement, McCloy singled out the U.S. Ambassador to Iran, Douglas MacArthur: "Yes. He is more Persian than the Persians sometimes." Yet once again he conceded that the U.S. government's attitude had not been the deciding factor, and that he always "took for granted that Mr. MacArthur would side with the Shah."[23]

George Henry Mayer Schuler of the Hunt Company told the Church Committee, in essence, that the collapse of the united front and the substitute plan to use parallel negotiations was an "unmitigated disaster." He continued, "I opposed it at every step of the way. . . . I had consistently taken a view that what we are doing was playing into the hands of OPEC, destroying what we had set out to do, which was to have solidarity." The companies hoped "to somehow find a countervailing power to the cartel of the producing states," he said, and "it [the united front] was the only thing we could do to meet it. We had had enough experience trying to do it individually, so we did it this way. . . . I am afraid that the oil producing states realized that there was no way we could stand against them at that time. . . . "[24]

Thus the carefully formulated plan for uniting the oil companies into a single force potent enough to counter OPEC demands had failed. Of course, there is no certainty that any united front of the companies would have succeeded; but many experts felt that, at the least, it had a good chance. Meanwhile, critics of the major companies have found it easy to explain the companies' failure in the Middle East.[25] Yet the situation remained paradoxical, as McCloy wrote Senator Church: "The United States has benefitted perhaps more than any other nation from the presence and activities of its companies in the Middle East." There, the U.S. had secured a "vitally important non-domestic source of oil," which lessened the effects of the heavy drain on domestic supplies occasioned by two world wars. Further, this oil had made possible three decades of progress—improvements in the standard of living, and increased productivity in the U.S., in Western economies, and in the Middle East—supplies of relatively cheap energy at relatively stable prices. In McCloy's view, some U.S. government support of the companies that had made all this possible appeared to be the natural and proper American policy. Instead, the U.S. government failed to encourage oil companies to cooperate among themselves and with foreign companies, to keep the "oil flowing at a reasonable low price." This failure, in McCloy's opinion, had proven ruinous, especially in the face of "a large well-organized international government cartel whose avowed purpose was to keep and maintain the price of their oil high."[26] Certainly Church recognized the strength of McCloy's charges.

Today, it appears certain that the Irwin-Akins visit—and especially Irwin's report to the State Department—triggered the swift reversal of policy. As Senator Charles H. Percy remarked when Irwin testified some three years after his trip, "From a practical standpoint when the No. 2 man [Irwin] makes a recommendation, there is only one man [the Secretary of State or, of course, the President] who can change it."[27] The Irwin cable led the State Department to recommend separate negotiations with the Persian Gulf producing states, and with those in the Mediterranean. If the companies followed that suggestion—and they had little choice—there could be no united front against OPEC.

X   X   X

The chief executives of the large oil companies now undertook to establish various advisory committees designed to assist them in conducting global negotiations, as well as to furnish useful information to Piercy and Lord Strathalmond. On January 20, in response to the January 19 and 20 cables from Piercy and Lord Strathalmond in Tehran, the companies organized a new joint office in London, despite the fact that many of the major corporate offices were in New York. For Middle East negotiations, however, London was closer to the scene; so the signers of the "Message to OPEC" wished "to establish there some kind of central organization along functional lines." In this organization, each oil company was to be represented by one senior executive (or his deputy), although more than a single representative from each company could attend the sessions. To chair this group, the representatives selected Joseph Addison, then retiring as general manager of IOP. He presided over what soon became known as the London Policy Group, meeting in British Petroleum headquarters building, Brittanic House.

Since discussions between the united front and OPEC were scheduled to open first in Tehran, company spokesmen agreed that all communications there, including press releases, would be cleared with Jan Van Reeven, who had been selected to replace Addison at IOP. When requested to do so, Van Reeven also joined Piercy and Lord Strathalmond in their conversations with Dr. Amouzegar and with the Gulf ministers.

The London Group held the responsibility for selecting members of the negotiating team; but its more substantive function was to support their work by formulating terms of reference and fallback positions for them to use. At all times its office staff acted to keep members informed about the status of the negotiations, and its London office soon became a message center for the companies. Additionally, the staff kept minutes of all meetings and prepared reports for McCloy to use in briefing the U.S. Departments of State and Justice. As a backup for the London Group, the companies decided to create a special New York office, headed by William Lindenmuth, general manager of Mobil's Middle East Department. For a time at least, his team, composed largely of senior executives, met in Mobil corporate offices. *

Both support groups were expected to develop questions that the negotiators would seek to answer, and they jointly selected the technical aides to assist the negotiating team. Together, the groups hoped to prepare for all possible modifications of the terms of reference, as well as for all questions that the negotiators might ask. The executives in New York were specially designated to review the work of the London Group, in particular all policy decisions. More generally, the New York office had the advantage of ready access to the wealth of economic information and technical expertise which the negotiators might need.

Despite these joint preparations, each company remained free to act alone. Group meetings, once started, were kept relatively open, but strict secrecy applied to communications with the negotiators in the field, who sometimes used diplomatic cables for very important messages. The London and New York Groups met often, and they kept in touch with each other during early 1971. Meanwhile, John McCloy arranged for a partner from his law firm to be present at every session of each group, thus assuring that all discussions remained consistent with the terms of the "Message to OPEC, the Libyan Producers Agreement and the pertinent business review letters from the Antitrust Division." [All told, legal counsel missed only three meetings during the entire negotiations.] McCloy himself attended many of the New

---

*While the Tehran negotiations went on, these representatives met at Mobil headquarters. Later, when negotiations with Libya began at Tripoli, William J. Greenwald, executive vice president of Esso Middle East, replaced Lindenmuth and the New York Group then met in the Esso building at 15 W. 51st Street.

York meetings, while his partner, William E. Jackson, ordinarily attended sessions in London.*

By January 20, 1971, the New York office had established four working committees, later to be paralleled by the London Group. One committee handled all public relations; a second tackled the thorny transportation (freight) problem; the third studied the oil supply problem; and the fourth, named the "Economic Evaluation Committee," gathered information for the entire group.[28] During the Tehran negotiations, all offers and counteroffers had to be analyzed in terms of their effects not only on taxes due to the producing countries, but also on taxes that might be levied in the United States. Such analysis proved to be tedious, and finally it served to slow down the entire negotiating process.

Meanwhile, the London Group began the principal task—to formulate tactics. Basically, they concluded these alternatives: either risk a complete shutdown by OPEC members (and thus chance economic havoc in Western Europe and Japan) by holding fast to a policy of global price negotiations; or follow the State Department's suggestions for parallel negotiations which might well destroy the united company front. This second alternative, of course, meant yielding to the pressure and threats of the Shah and of the Gulf oil ministers, and it practically guaranteed new problems with Libya. For, even if the companies decided to conduct two separate but concurrent negotiations—one with the Persian Gulf states and another with the Mediterranean states, neither negotiating team could agree to final terms without consulting with the other or the collective bargaining front would be shattered. Moreover, given these circumstances, the companies could not be certain that the Gulf states would agree to any such procedure.

For OPEC, sensing that the momentum of oil decision-making had shifted toward the producing states, which already possessed the power to dictate their own terms, no easy or simple solution could be found. On January 20, McCloy cabled Jackson: "State Department takes the view that parallel negotiations with Gulf and Libya might be favorable provided that they were linked in such a way as to ensure that they were within the principles and framework of our OPEC Message" (of January 16). But he foresaw a problem: how to link parallel negotiations in form and manner so as to come within "the principles and framework" of Justice's Business Review Letter as well as the OPEC message. McCloy also echoed George Piercy's earlier remark to Ambassador MacArthur: "I would imagine the companies would repose little faith in assurances by Iran and others that they would stand still

*Mr. Isaac Shapiro of the same firm also attended some London sessions.

even if more favorable terms were granted the Libyans." In fact, McCloy actually anticipated the danger that "substantially unrelated but parallel negotiations" might go beyond the Justice clearance letter, and impair "the rationale of the objectives of the entire operations from the industry's point of view." Finally, he told Jackson that Justice and State were "exploring" this problem.

Later on the same day, Greenwald of Esso Middle East reported to Jackson that the Justice Department had cleared only the "Message to OPEC." If the companies deviated from the procedure detailed in that letter, some additional support from Justice should be sought. And, as if to complicate the already heavy task before the London Group (and also stymie McCloy for a time), the Justice spokesman now claimed that his department could take no action "without knowing the facts and the specifics of the negotiating procedure and/or plans."[29]

Akins of State left Tehran on January 18, and went to London, where he remained for "about a month." He recalled that he saw members of the London Group several times every day. Meanwhile, Lord Strathalmond met privately with Dr. Amouzegar, who continued to paint a dark picture: an extraordinary meeting of OPEC would have to be called, so that the Venezuelan terms* could be approved by the delegates. Afterwards, however, Dr. Amouzegar expressed his personal distaste for reconvening such a meeting. He went on to mention several approaches that Piercy and Lord Strathalmond might use to provide him adequate cause for delay. After cabling details of their meeting to London, they concluded the message with this warning: "Vital that you progress proposals. It may be difficult for us to hold things open even a week."[30]

X   X   X

On January 20, in its first meeting, the London Group debated responses to the demands of OPEC and the Gulf ministers. Led by Hunt representatives, several independents argued steadily against changing the "global approach," contending that the threats of Dr. Amouzegar, the Shah, and the Gulf states' ministers were actually meaningless, since all Libyan producers would get the "highest common denominator" in any case. Finally, however, the London Group unanimously endorsed separate but connected negotiations, and approved a letter to be sent to Tehran on January 21, 1971.** This letter was sent to Piercy and Lord Strathalmond for delivery,

*60 percent tax, 16.67 percent royalty, and unilaterally established tax-reference prices.

**A copy of this letter was delivered to Libya on January 23.

along with an additional explanatory statement and "fallback" proposals. It stated that the companies were prepared to follow the January 16 "Message to OPEC" with specific offers within a week; in the meantime, OPEC members should agree on guarantees that would eliminate leapfrogging. The letter continued:

> We should prefer, and we should have thought that it would be mutually beneficial in the interests of time, that the negotiations should be with a group representing all the OPEC members. Nevertheless we should not exclude that separate (but necessarily connected) discussions could be held initially with groups comprising fewer than all OPEC members. However if we are to embark on such separate but connected negotiations it is important that it be realized that any negotiated settlement must be acceptable overall and will not lead to further leapfrogging. It would be our intention to table one comprehensive document embodying the proposals referred to above.[31]

After reading this letter, Piercy and Lord Strathalmond cabled London with objections and opinions: to follow the steps outlined would "only put us back to square one in the Gulf," so they suggested holding the letter and sending it to the Secretary General of OPEC at some later time.

But the London Group insisted that the letter be immediately delivered to the Amouzegar committee, as originally requested, "and hopefully through them to the other OPEC members (particularly Libya). ..." London also urged that the men in the field not explore any assurances against leapfrogging that the Gulf ministers might offer, nor did London agree that the presence of one or two representatives of independent companies would help in Tehran, as had been suggested by Piercy, Lord Strathalmond, and McCloy.

Back in Tehran, the negotiators did deliver the letter. First, however, they met privately on the morning of January 21 with Dr. Amouzegar, and persuaded him to send the letter to all OPEC members. In their cablegram relating the details of this meeting, Piercy and Lord Strathalmond characterized the Iranian minister as interested in postponing the extraordinary meeting of OPEC, based on his notion of what Algeria, Libya, and Iraq might propose. Dr. Amouzegar, they said, feared that in a full meeting these nations would gain control. In turn, Piercy and Lord Strathalmond declared that they would "do our best not to blow it nor [in] any way compromise our global approach," nor would they discuss assurances, while they would explain the absence of spokesmen from the independents.

The meeting scheduled for January 21 began at 4:00 p.m. Yamani had not yet arrived, but Al-Ateeqy was there, as the meeting quickly turned into a rough negotiating session. First, Hamadi asked the company negotiators whether there could be any agreement with the Gulf states if Algeria and Libya held out. Their negative answer provoked more comment, and Amouzegar suggested that settlement be made with the Gulf states before February 3. He then sought agreement to discuss OPEC Caracas Resolution XXI, 120, but Piercy objected, saying such discussion could involve Libya, a topic that would be inappropriate. After a further exchange of opinion, Piercy interjected: "What I understand you to be saying to us Dr. Amouzegar is that either we move ahead and discuss one of these items* or we break off negotiations." To the categorical "yes" of Amouzegar, Piercy replied, "We most assuredly do not want to break on procedural matters. We will discuss this matter as long as it is appreciated that such discussion does not compromise our position given in the letter that was just handed to you."

With that new understanding, the discussion ranged over many topics, including assurances against leapfrogging, gravity allowances, and a number of other questions—despite assertions by Piercy and Lord Strathalmond that they were not prepared to agree to anything. Piercy again made the point that the companies remained committed to a global negotiating approach, and Dr. Amouzegar insisted that he would tell OPEC that, since negotiations had started, no meeting would be necessary on Monday, January 24. Hamadi again contended that the companies were stalling, which led Piercy and Lord Strathalmond to caucus outside the room.

When they returned to the negotiating table, their attitude had hardened; they would not meet on Saturday and Sunday. Further, when discussions did start, they would demand a package deal. Lord Strathalmond told Dr. Amouzegar that a proposal from the companies would be sent on Thursday, January 28; and that if the ministers felt compelled to call an OPEC meeting, he and Piercy would consider the move unfortunate. In any event, Piercy planned to leave Thursday evening for New York.

Obviously the Iranian did not want to hold an OPEC session at that time, so he recommended to the other two ministers that they await further proposals. Finally, the session broke up, as company spokesmen were assured that, if they really wanted a global approach, OPEC would aim to match the Venezuelan terms. Piercy and Lord Strathalmond agreed to have company experts immediately begin technical discussions with the Gulf teams on those portions of OPEC, XXI, 120 which affected only the Persian Gulf

*In OPEC Caracas Resolution XXI, 120.

states. Further, the Arab ministers refused to disengage the Gulf settlement from any other, although they did provide plentiful verbal assurances against leapfrogging. They also wanted proposals from the companies by January 28, and they extended the deadline for settlement to February 1.

Summing up the meeting in a cablegram, Lord Strathalmond and Piercy wrote:

> We have not yet brokeen [sic] off and the Monday OPEC meeting is
> cancelled. However we are far from home particularly in regard to
> a Gulf settlement not being dependent on a settlement elsewhere.
> We are sure this will not repeat not be agreed to here and if we push
> for a global settlement this will only lead to a demand for Venezue-
> lan terms everywhere.

In all, the Thursday negotiating session lasted four and one-half hours. That night Piercy left Tehran, while Strathalmond made plans to leave for London the next day. Both men had agreed to return on Thursday, January 28, to try to clear up several procedural points.[32] Their earlier cablegram had recommended that the London Group send a copy of the January 21 letter to the Secretary General of OPEC, in advance of its regular meeting on February 3.

These cablegrams led to a reply from London urging that Piercy and Lord Strathalmond discourage Dr. Amouzegar from replying to the January 21 letter. If he sent a reply, the London Group claimed they would have to reassert the position already taken in the "Message to OPEC," and also to make an issue of the unilaterally set deadline for an end to negotiations. Strathalmond responded by declaring that he had talked with both Sir Denis Wright, and MacArthur, who agreed with him that "to attempt to do as you all ask can only finish everything here and now." Then, he pointed out to London that the proposed letter represented a ruse by Amouzegar, designed merely to postpone the forthcoming OPEC sessions. Moreover since the companies were already accused of stalling until a Libyan settlement could be achieved, any attempt to stop the letter would fuel that suspicion. In fact the letter, Lord Strathalmond said, had already been drafted and signed by all the ministers except Yamani, who had left Riyadh for Tehran, to sign it. Finally, the Englishman noted: "In any event we must face position set out your letter which Amouzegar described as a poor lawyers effort. I too am a bit lost on it." In the morning, Strathalmond carried a copy of the Amouzegar letter, dated January 22 and signed by all four ministers, with him to London.

The State Department had reversed its position. Originally it indicated support for the companies' uniting to deal with all the oil producing nations; now, on the basis of a survey of the Western consuming nations and Japan, it seems to have informed oil company chieftains that the consumers had uniformly answered, "Take a price increase and do nothing to cause them [OPEC] to shut down." The results of that poll plus Irwin's report to State caused the department to withdraw its endorsement of industrial unity. Naturally this change added to the existing confusion in the oil company camp.[33]

X X X

On January 22, 1971, Strathalmond returned to London, and the London Policy Group released a statement to the press:

> The oil companies objective remains the achievement of an overall settlement covering the producing countries concerned. The companies reaffirm that any settlement with the oil producing governments concerned must be reached simultaneously.

So far, the company front remained unbroken, as a new round of events began. First, the Gulf countries sought to strengthen their position by persuading the Shah, who was then very popular in the West, to have a press audience on January 24. In this interview the Shah took a firm position against global negotiations—based perhaps on his conversations with MacArthur and Irwin. Again, he threatened that if the companies continued to insist on global terms, OPEC would demand parity with Venezuela. As a lever, he reintroduced the possibility of a total Persian Gulf shutdown. In summary, the Shah insisted:

> What we are trying to tell the companies is this: That the front you have formed to say that you want to talk to all the oil producing countries at once, on one day, and to sign a single contract with all of them can be taken either as a joke or as an intent to waste time.

Then, he concluded by asserting the ultimate threat: after existing concession agreements had ended, the producing countries should simply take over all oil operations, from field to customer.[34]

After a meeting of CEOs, both the New York and London Groups convened to study the information from Tehran and to prepare a reply to the oil ministers' letter. On virtually every point the ministers had rejected the terms of the "Message to OPEC" of January 16, as well as those in the letter of January 21. Further, the Gulf ministers flatly stated that, once negotiations with them had produced an agreement, its terms should not depend on "settlement with any other country."[35]

Acting with the concurrence of all the companies that joined the united front, on January 25, the London Group selected the Tehran negotiators. Lord Strathalmond headed the team. [While Piercy's name appeared on the original Tehran list, in fact he headed the Tripoli team.*] Page, who went to Tehran as an observer, remembered that the group included a number of well-informed oil company executives, including three persons of top rank. Two of these represented major oil companies and the other headed an independent company. A six-person back-up group included Cattarulla and Clarke of Jersey. The Tripoli team that Piercy headed included, among five top executives of independent companies, Schuler of Hunt and Enno Schubert of Gelsenberg A.G. of Germany.[36]

When Lord Strathalmond and his team left for Tehran on January 26, they carried new terms as well as different fall-back positions. For the first time, they also possessed authority to proceed with the Gulf negotiations; they could complete an agreement provided that "suitable assurances are secured." In general, the new letter was more diplomatic in approach and tone than would have been the reply that the companies instructed Lord Strathalmond to write on January 22—but which did not get written. And the January 28 letter stated explicitly: "Simultaneous negotiations are starting in Libya by other members of the industry team, led by Mr. Piercy."[37]

In Libya, meanwhile, executives of the oil companies had been faced with threats of nationalization, as the price for retaining the united front. On January 18, however, Libyan Petroleum Minister Izzidin al-Mabrouk appeared uncertain and indecisive, when he met with representatives of Hunt. On January 19, the minister called in Esso Standard (Libya) Inc. [Esso Libya] executives, and they listened to Major Jallud, who assured them that "Libya would do its duty" before he extended the companies' deadline for abandoning the united front to January 24. When that day came, however, Libya's under secretary of the oil ministry talked to Hunt spokesmen, but without making new threats.

Yet within days, just after Libyan Oil Minister Mabrouk and other ministers received news from Amouzegar about the hard line already established in overall discussions, he reverted to his familiar threats of Libyan power. First he flatly rejected the companies' global approach—whether separate or not—and announced that at the February 3 OPEC conference a shutdown

---

*After the effort to conduct separate but connected negotiations with Libya failed on January 30, Piercy doubled back to the Tehran team.

would be considered—provided that no agreement with the companies had yet been reached. Nor was Mabrouk alone in speaking out. The Venezuelan oil minister joined in and talked tough: "each country is sovereign and if they want to demand higher prices, that is their right." Clearly, the Venezuelan also regarded his words as a warning to the oil companies.

Piercy cabled the Libyan oil ministry on January 27, to request an appointment with Major Jallud for the next day. Before Piercy could see Jallud, however, Mabrouk announced that he would meet Piercy as a representative of Esso alone. When Piercy saw Mabrouk, the latter denounced the company "cartel" and refused to read the joint proposal of the companies. Later on January 28, Piercy sent Mabrouk a note with an enclosed "Aide Memoire," that spelled out the companies' proposals in some detail. Moreover, all the members of the Libyan negotiating team had signed this portion of the Piercy letter—so it expressed the united front. Yet the Libyans certainly rejected it for the original was returned to the files of Esso Libya.

By now the Piercy team had determined that the Libyans would not negotiate, but rather await results from the Tehran negotiations. Accordingly, on January 30, Piercy sent Mabrouk a new note stating that all the members of Piercy's team were leaving for London, to remain there until the Libyan minister chose to resume discussions. The Libyan end of the joint negotiations had reached stalemate.[38]

X   X   X

In Tehran, Lord Strathalmond's team met with Yamani, Hamadi, and Amouzegar for some six hours on January 28; they made no progress. In fact, the freight differential issue loomed larger and larger, as the Gulf states demanded in any agreement a clause that would guarantee an extension of the temporary Libyan freight differential to the entire Gulf area, in case the companies could not reduce that differential as planned. Assuming that the companies agreed to a 25 cents per barrel temporary freight premium on approximately two million b/d of Libyan oil (and that Libya refused to surrender the premium after normal trade channels reopened), this meant that the six Gulf states would expect the same premium per barrel on some twenty million b/d of Gulf production, as a way of eliminating any price disparity. Such a guarantee, the Strathalmond team could not grant. Small wonder that in his cabled report of the session Lord Strathalmond stated: "We felt we might be facing a break up within the next two days since a reasonable area of compromise has not yet presented itself."[39]

John McCloy reported to the Justice Department that the initial demands of the six Gulf producing states amounted to an increase "of approximately $20.5 billion over five years." These demands included a 55 percent tax rate, a posted price increase of 54 cents per barrel, plus an inflation factor to provide an additional increase of 4.5 percent per year, as well as increases for gravity and other quality differences. For two days, as McCloy later phrased it, "intensive and arduous" bargaining continued, and the Gulf producing nations proved adamant. In all, the OPEC Committee "made threats of embargoes, legislation and expropriation. It became clear that no settlement was likely to be reached on the basis of the industry's initial offer."[40]

At this point, the Tehran team caucused, and then cabled possible new terms to London, where they were received in a "moderately doleful session," and finally passed on to New York, along with Strathalmond's request that "one or two senior members of industry come here for final session on Sunday [January 31]." From London, Addison added: "Our reaction is to doubt whether principals should join in the negotiating fracas in Tehran." Meanwhile, several members of the Strathalmond team returned to London on January 30, met with the group, and headed back to Tehran on the same day. Unfortunately, just as the Tehran negotiators left, some of the Piercy-Libyan team arrived in London, where they learned of the decision to unhinge the negotiations—that is, to end the global approach and to separate the Gulf discussions from the agenda in Libya.

This important decision had been approved by the "chiefs" in New York, even though no representatives from Hunt had been present. In London, Schuler objected strenuously to the lack of a tie-in agreement on the Eastern Mediterranean postings (the pipeline oil), and the London Group reported his views to New York. Soon, on a conference call, Jersey's Ken Jamieson came on the speaker phone and asked Schuler to state his views again for the CEOs in New York. When he had finished, Jamieson thanked Schuler and assured him that the "chiefs" would seriously consider his views. Yet despite his best efforts, Schuler could only get one other company, Gelsenberg A.G., to join him in condemning the abandonment of the global approach. To Jack Clarke of Jersey, Schuler later revealed his misgivings:

> I said that this was a bad mistake and that ... this was going to be, the end. There was no salvaging anything, but I said I wanted it clearly understood that it was not an effort of the major oil companies to steamroller us, but rather it was the failure of the independents to stand up for what, in my eyes, was their obvious protection.

Despite Schuler's objections, the change went through, and the united company front was abandoned. Along with it went the "separate but necessarily related" approach, so that when negotiations began once again in Tehran, they assumed a completely different form.

As a beginning, Amouzegar quickly agreed to have his advisors work with the legal experts on Lord Strathalmond's team, to formulate the assurances against leapfrogging on which the companies insisted. But at the same time, the Iranian minister refused to postpone the next OPEC conference scheduled for February 3. He did, however, extend the deadline for an oil settlement until the morning of that day. So Strathalmond cabled London to ask whether, if by "mischance" they did arrive at an agreement by that time, the Business Review Letter would cover those American members initialling the document. As he noted: "To tell the world that U.S. laws hindered an international agreement would not be very palatable anywhere."[41] Even in Iran, a Britisher had to reckon with the long shadow of U.S. antitrust policy. If the assurances offered by the oil ministers still appeared to be inadequate, they were the best that the negotiators could obtain. In effect, they constituted a promise: Even if terms on Iraq or Saudi oil pipelined to the Mediterranean turned out to be better than those accepted by the six Gulf states, none of the latter would seek higher prices. Further, the Gulf nations agreed not to take any action collectively or individually against any oil company that refused to accept higher terms (from Libya) resulting in production cutbacks; nor would they in any manner restrict the flow of Gulf oil.* Finally, the ministers agreed to a proportional increase in income if Libya should refuse to reduce or terminate the special freight allowance. These were the terms which Van Reeven cabled to Addison in London.

Within a few hours Van Reeven had a response from the London Group; it must have seemed frustrating:

> We are greatly disturbed by the reports on assurances and by the implication that these are virtually agreed. We feel that the assurances set out in your VR0067 fall far short of the outline of assurances set out in our last night's message to you through west end friends [British Foreign Office]. ... We are following with a further message containing a detailed commentary on your restatement but in essence we feel that you must revert to all the basic protections ... as set out in our last night's message.

In short, the London Group insisted that the assurances against leapfrogging should include guarantees against cutbacks by Iraq and Saudi Arabia in

---

*For Iraq and Saudi Arabia, this applied to Gulf crude only.

pipeline oil sent to the Mediterranean. Also required was a guarantee "that tax and royalty rates applicable to them in the gulf" would "also apply to their crude exported from the Mediterranean."

The Tehran team pointed out that neither Yamani nor Hamadi would discuss those matters, or even meet to discuss them. In his cablegram, Van Reeven stated that the two countries expected to obtain for their pipelined oil the same treatment that Libya received. Then he added: "Basically the most serious part of the problem is that they will not now guarantee against a future oil embargo as part of Mediterranean claims, even their own." This meant, ironically, that they were willing to increase production in the Gulf to offset any embargo that might result from a failure in Libyan negotiations.

Now the U.S. Department of State sent John McCloy a message "that something should be released by the companies before any break ensues." This, he interpreted as a request for a public statement that the oil companies hoped to continue negotiations, but they were having problems getting assurances concerning a stable flow of oil at reasonable prices. Before any public statement was released, both the London and New York Groups (including the "chiefs") convened to study particulars for a "firm offer" to OPEC to be made on January 31. McCloy later estimated that this offer would have added $6.29 billion over five years to the Gulf producing countries' income.[42]

As it developed, the "firm offer" involved hard work by both groups, the CEOs, and all the negotiators. The London Group especially feared any appointment of an OPEC Mediterranean Committee "of four, in which 3 are radicals," and it ceased to encourage any movement in that direction. When London tried to get Iraq and Saudi Arabia to agree on a price structure for their Mediterranean oil, it suggested that Howard Page or George L. Parkhurst* (both of whom were with the Tehran team and knew Yamani) be asked to contact the Saudi minister—"an informal approach." But the Tehran team did not follow up, preferring instead to continue seeking some price arrangement that would end the OPEC threat of legislation, and also prevent the Gulf countries from leapfrogging.[43]

On February 1, Piercy left London for Tehran. He carried a message to be used as a last resort.** The Tehran team, so the message claimed, had not yet offered OPEC ministers enough money, nor did these negotiators have authority to offer enough. On their part, the OPEC ministers claimed not to

*Of Standard Oil Company of California.

**Piercy remained in Iran until February 3, working with the Strathalmond team.

have sufficient authority to settle the East Mediterranean position, nor could they give the companies adequate assurances against leapfrogging. As the last resort, the Tehran team should ask them to go and get additional authority to cover those matters: then the ministers could return to bargain with the companies and obtain better financial terms. However, this draft letter was to be used *only* if a breakdown occurred in the current negotiations.

Later that day, the Strathalmond team met with the ministers. Their cable to London described the best available statement as an understanding that, if Libya followed the course that all parties assumed it would, then the companies would increase the amounts due Persian Gulf countries according to some formula.[44] But the negotiations did continue, and new misunderstandings developed. During a closed meeting, for example, the Iraq and Saudi ministers denied that they had agreed earlier in the day to apply the Persian Gulf standard of reasonableness to the Mediterranean. In particular, this allowed the ministers to retreat on their assurance against leapfrogging (except among Gulf countries and within the Gulf area). Now, it was clear that the OPEC members knew they held an advantage.

As this stalemate continued, the companies suffered a serious blow. A key cablegram from the London Group that disclosed the full terms of reference somehow reached the Tehran press corps. The press, in turn, bitterly attacked these terms of reference, on grounds that they offered far less than the Iranians had been led to expect. By approving the text for publication, Amouzegar showed his shrewdness, as he correctly judged that the newspaper account would serve to create public pressure, and thus place the Strathalmond team under a handicap. Not yet content, Amouzegar harangued the negotiators. After much discussion, the Strathalmond team finally asked for the ministers' "rock bottom" offers, since companies had to decide whether to accept or reject them before the next OPEC meeting opened at 10 a.m. on the following day. Each of the ministers then provided a separate set of figures, which were discussed for some time.[45]

When the "chiefs" moved to London on February 2, the group offices became a scene of confusion, with both cables and "phone calls coming in from Teheran to particular 'chiefs' superseding cables which arrived later in time. . . ." Yet no final offer from the ministers received approval. Once the "chiefs" took control in London, they cabled the remaining members of the Strathalmond committee in Tehran that neither a one-year nor a five-year term of agreement would be acceptable. To this point, the "chiefs" persisted in believing that they still possessed real bargaining power, especially with regard to Mediterranean pricing. So they insisted on strong assurances on that matter. At the same time, the chiefs failed to understand why Yamani and Hamadi would not discuss any possibility of tying their pipeline oil in the Mediterranean to Gulf prices (plus a short haul freight premium).

When Piercy and Lord Strathalmond told Amouzegar that neither offer—the one-year nor the five-year deal—would be acceptable, the Iranian warned them flatly that the Shah would soon announce new oil legislation. In the interim, the terms of the one-year contract would apply; the consortium must either accept them or risk a shutdown. Negotiations were ended for the time. The final letter from the Strathalmond team to the ministers followed the essential outlines of the London letter that Piercy had brought on the preceding day. Among its terms, Piercy assured Amouzegar, the companies had offered "the makings of a good deal and if we [the negotiators] only had time we could work it out."[46]

When Piercy, Lord Strathalmond, and most of the team reached London, a *New York Times* report described the top negotiators as "rather dispirited." The *Times* traced their "gloom" to two causes: first, "the international oil business has been transformed from a buyers' to a sellers' market over the past year"; and second, "is a new found aggressiveness on the part of the producing countries to press their advantage." The rest of this story treated the Tehran proceedings as a three-cornered struggle among the producing countries, the consuming countries, and the international oil companies. In summing up the Tehran meeting, the *Times* observed that the balance of oil power had shifted to the producing countries.

In his own statement to the newspapers, Dr. Amouzegar cited the Gulf Committee's limited grant of authority, and went on to emphasize the Shah's willingness to stabilize oil prices—provided only that the companies agree not to increase product prices to customers. Amouzegar then explained that the price increase which the Gulf producing countries wanted for the crude oil actually amounted to merely one-half "the amount the oil companies have already added to the price of oil products."[47]

Each of the negotiating parties, taking its case to the press, sought to shift the burden of failure to the other. In this campaign for public support the London Group went further, issuing to the press its own long statement, which detailed the history of the dispute, beginning with the Libyan actions in the fall of 1970 and concluding with the negotiations of January and February, 1971. The London Group contended that if the companies had accepted the OPEC Committee's demands, the resulting annual increase in revenue for the six Gulf states "when generalized to include non-Gulf countries" would equal two-thirds of all the companies' total earnings everywhere except in the Western Hemisphere. Such earnings would include producing, refining, and marketing. In short, the London release refuted the notion that oil companies could absorb such increases as OPEC demanded. The oil companies reminded OPEC that they and the public still required precise assurances against leapfrogging of prices.[48] There, they rested their case.

X   X   X

On February 3, the formal OPEC meeting began. It lasted only one day. Naturally, the OPEC ministers were concerned because serious negotiations had broken off; and that concern dominated the Shah's welcoming address. First, the Shah briefly summarized the history of OPEC, and then turned to his real topic: the world price of oil. Essentially, he covered the same ground that had been reported after the press conference of January 25; now— contrary to Amouzegar's earlier interpretations—the Shah recommended a moderate course for OPEC:

> I now suggest that the countries of this region should adopt a system which would be rational and reasonable; a system in accord with the resolutions of the United Nations safeguarding the sovereign rights and independence of the countries; a system that has precedents in other areas, and ensures the stability and confidence which is the objective of consuming countries. Of course any legal or legislative measures necessary will be taken by the member countries at the same time.

Following the Shah, each head of a delegation responded in turn. Most praised the Shah's speech and announced their willingness to follow his advice. Mabrouk of Libya, however, used only a few words to alter the tone of this opening session, hinting at more ominous possibilities:

> You have indeed stressed in detail what the committee has faced in dealing with the oil companies. We do indeed seek an agreement with oil companies, we always try not to act unilaterally, we always try to give them a chance to express their own views, but in case we fail in that direction our rights are so obvious that we don't hesitate to take action. What Your Imperial Majesty suggested, legislation, is the exercise of our own sovereignty, and I believe the first action in the right direction for achieving our own aims.

Afterwards, when a reporter questioned the Shah concerning rumors that Western governments had begun to make "new moves" aimed at mediation between the companies and the producing countries, His Majesty replied that, after his earlier warning to the consuming countries to stay out of the negotiating process, they had done exactly that—"[they] have not shown the slightest sign of any interference or support of the companies." Yet the Shah must have known otherwise; for as Lord Strathalmond told Schuler, virtually every ambassador in town [Tehran] had pressed the OPEC ministers to delay legislation. Given the strategic circumstances, it would have been most unusual if the U.S., through Ambassador MacArthur, had not engaged in conversations with the Shah on that subject. In answer to another question, the Shah also suggested new regional negotiations. But he would not speculate about whether legislation might follow the Venezuelan pattern.[49] Already, the Shah's leadership of the Persian Gulf states had reached the limits of its usefulness.

On February 4, the OPEC delegates adopted two resolutions: the first called for each member country to introduce legislation on February 15, to implement OPEC Caracas Resolution XXI, 120. Should the oil companies fail to accept terms within seven days of that date, the new laws would take effect. The second resolution expressed full support for the Libyan people and against any collective act by the oil companies.* This second resolution might easily have been interpreted as rendering ineffective the "Safety Net" already devised by the companies. Together, the resolutions carried with them an implied threat of shutdowns; the first in all OPEC countries, and the second, in Libya.[50]

The Tehran team met with the CEOs in London on February 5, to assess the situation. Their discussion centered on alternate strategies: whether to yield, or to hold out and risk legislation. In support of the latter, some company executives contended that OPEC members would not legislate price increases because of uncertainties about the reaction of consumer countries. Since OPEC had already sought to place on the companies the full responsibility for price increases, if higher prices were now legislated, much credibility of the producing countries would be destroyed. Also, if the producing countries legislated price increases, such action would bring home to some influential consuming nations (such as France) their strategic error in relying on government-to-government negotiations to secure better prices. This realization would work to the disadvantage of OPEC, although it appeared doubtful that its more radical members cared. In essence, however, the theme of the London debate could be expressed in a simple ques-

*These resolutions were numbered XXII, 131 and 132.

tion: How much could the companies yield without appearing to capitulate completely?

When the "chiefs" and the groups had covered all possible answers, they drafted new terms of reference, memoranda of agreements, and other documents. These would guide negotiations in future discussions. A legal team provided the texts of several specific assurances which, they believed, would "preclude the Gulf supporting unreasonable Libyan claims." Schuler objected strongly to a loosely worded "No embargo" guideline concerning the Gulf producing states, and Bunker Hunt cabled his support of Schuler's objection. Hunt reaffirmed the position of oil company solidarity in all bargaining with the producing countries. Yet the elements of harmony that had produced a united company front now simply eroded under joint pressure from the oil producing governments and the consuming countries. Certainly some wanted oil at any price. Victor de Metz of Compagnie Française des Pétroles [CFP], for example, argued that legislation regarding petroleum must inspire comparable legislation concerning other resources, and he correctly pointed out that the haggling over ironclad assurances and watertight legal documents appeared a bit foolish when "they have it already in their concessionary agreements."[51]

On February 9, Dr. Amouzegar met with Piercy and Lord Strathalmond in Paris. The Iranian—now away from the other Gulf negotiators— suddenly seemed more flexible. He agreed that, should Libya not keep its pledge to lower the special freight rate when the Suez Canal opened, the Gulf states would then be entitled to only 10 percent of what the companies would have paid on a barrel-for-barrel basis. (The oil companies operating in the Gulf had ten times as much production there as in Libya.) This sum was estimated at an acceptable—to the companies—$77 million, as opposed to $770 million.

Back in the U.S., however, more complications appeared. When Akins returned from London, he went to New York to talk with McCloy on February 9. Judging from McCloy's memorandum of the conversation, either the Shah and other heads of state in the Persian Gulf had furnished State with a version of their negotiating position unlike that given the Strathalmond team by their oil ministers, or the Strathalmond team's interpretation of the Gulf ministers' proposals differed almost totally from State's interpretation.

Akins seemed at a loss to understand what assurances the companies sought. He said:

> The Shah had from time to time indicated that he was prepared to
> go a long way in respect of assurances, that he was out of sympathy

with some of the arbitrary positions taken by the Libyan Company [sic] and that he was prepared to enter into agreements against leapfrogging.

Akins also told McCloy about a rumor that OPEC "would not wish to shut down," a view some of the company negotiators shared. In neither Europe nor the United States did any responsible public official explain that full day-to-day publicity could only hamper oil negotiations by making compromise more difficult. Moreover, no one in high government office pointed out that both the U.S. Departments of State and Justice were actually monitoring the entire negotiating procedure. Not surprisingly, in this vacuum of authoritative information, rumors abounded and most of these rumors were filled with strong anti-big business prejudice. No wonder the general public remained confused.

In part because of the failure in public understanding, the steadily increasing pressure from consuming countries had become more and more powerful as an influence in all discussions. When Lord Strathalmond and his team left once again for Tehran (February 10), the Gulf producers brought their most effective public weapon—the Shah—to speak for the consumer (and against the companies) on the same day. The Shah allowed an interview to be broadcast on the British Broadcasting Corporation [BBC] show "24 Hours," in which His Majesty pointed out that the companies had already raised oil product prices. Further, he said, what the Gulf states asked "was only 50 % of what they [the companies] had already augmented." He also promised that the Gulf states would accept a five-year agreement and not ask for changes unless the world price index rose. (That event would require readjustment of world oil prices.) In all, the Shah's interview accomplished two important things: first, its moderate tone gave the companies some hope that a compromise could be reached; and second, his support of OPEC encouraged the Arab ministers.[52] Of course, it directed media attention to the oil problem.

Aside from these purposes, his interview followed the strategy then being used by all spokesmen for OPEC—that of trying to drive a wedge between the oil companies and the consumer governments. As the Shah and other OPEC spokesmen described the situation—should negotiations break down—all the consumer countries would have to do would be deal directly with the producers. For the gullible, this strategy made the matter seem simple. At no point did OPEC spokesmen mention the logistics involved in moving oil from the well to the consumer, or of making crude into refined products which the consumer could use.

Negotiations in Tehran resumed on February 12. As yet, Amouzegar had not been able to persuade Hamadi and Yamani to honor the temporary understandings reached in Paris with Piercy and Lord Strathalmond, so these disagreements threatened an end to all efforts at collective bargaining, although this fact did not quickly become apparent. At the opening, old issues reappeared, and some progress seemed to be made. While the London Group was considering the cost of a new Amouzegar proposal on freight rates, the Tehran Group met with Amouzegar and Hamadi, and heard the former declare that an agreement had to be reached quickly, and that all parties must sign it before the February 15 deadline. The Gulf ministers proposed a five-year contract, with a no-leapfrogging clause for the Gulf only, plus an assurance of no support in the Gulf for any OPEC member demanding any increase in government take "above the terms now agreed." Eventually, the London Group carefully studied every line of the ministers' proposal as it was cabled from Tehran. Company experts scrutinized every clause for hidden meanings as much as for obvious implications. As Addison remarked in a cable to Van Reeven, "Fuzzy agreements usually work to our detriment." This time, London intended to take every precaution.

While the London and New York Groups studied the proposed final agreement, Middle East expert Howard Page said to Lord Strathalmond's team: "I definitely and unequivocally recommend grabbing the agreement." He then went on to say, "It's the best thing you can get." Page emphasized that the amounts involved now represented about one-tenth of those demanded at the time he saw Pachachi, earlier in the year.

At first, it appeared that a final draft would be ready for signing as early as February 14, but suddenly Hamadi announced that he could not sign this agreement because several companies sought to include a quit-claim concerning various disputes with Iraq.* In response, the team prepared for Hamadi a special codicil in a side letter, and late on February 13, cleared it with the companies involved. On February 14, they secured Hamadi's agreement to this letter, which exempted certain disputes and claims, and went on to state that the main agreement applied only to Iraq oil in the Gulf.** With that objection overcome, and with details of the principal document settled, on February 14, 1971, all parties signed the Tehran Agreement. It was designed to last for five years. Speaking for the Gulf Committee and himself, Amouzegar summed up: "I was so happy I had tears in my eyes."

X   X   X

*For this controversy, see the chapter on Iraq.

**This letter was signed on February 15, 1971.

Perhaps the clearest gain for the companies came from the clause that ended leapfrogging or price ratcheting. No longer could the Persian Gulf countries expect that their price for Gulf oil would be kept equal to that of any other country. And the further agreement of the countries not "to take any action in the Gulf to support any OPEC member which may demand either any increase in government take above the terms now agreed, or any increase in government take or any other matter not covered by [OPEC] Resolution XXI, 120"—spelled an end to embargoes by the Gulf states. Again, the tax share of the producing countries advanced to a uniform 55 percent, although specific adjustments on gravity and freight differentials were also included. In general, as Lord Strathalmond and Piercy pointed out in a statement to the press, "The Agreement establishes security of supply and stability in financial arrangements for the 5-year period of 1971-1975." At least the companies hoped for five years of stability.

For the producing nations, the terms of the Tehran Agreement provided an increase in total income of $1.2 billion in 1971. And through an escalator clause, the amount would rise to $3 billion in 1975.[53] The per barrel price of oil agreed upon for 1971 (effective February 15), rose from $1.80 to $2.18; while the producing states' take per barrel increased by approximately 36 cents. The five-year agreement provided that these numbers would increase on January 1, 1975, to a posted price of $2.615, of which the governments would receive $1.525. Clearly, oil and oil products would become much more expensive to the consumer.

For the companies, the Tehran Agreement of 1971 became the second major revision in the status of concessions within six months—the first being the Libyan agreements of September and October, 1970. For Jersey's George Piercy, who had been at the bargaining table, Tehran represented not what he wanted, but rather an arrangement his company "could live with." Henry Schuler, whom John McCloy once termed "one of the hardnosed people," reacted quite differently: After Tehran, Schuler said, "it was never quite the same. It wasn't going to be; nobody was pretending it was going to be. ..." At Tehran, on February 14, 1971, Schuler believed, the "momentum" passed from the oil companies to the producing countries. Elsewhere, the Shah of Iran expressed at the time his complete satisfaction with the Tehran Agreement, to the surprise of no one.[54]

Yet few experts assumed that the Tehran Agreement would finally settle the question of long-term crude oil prices. Nor did everyone believe that the Libyans, for example, might well consider Tehran prices as more than a point of departure for their own future demands. For Piercy, nothing in the record of Tehran supported an optimistic view. In the British House of Commons, however, the Foreign Secretary, Sir Alec Douglas-Home, stood up to declare that "maintenance of supplies of oil at reasonable and stable prices had been achieved." This, he said, "is also important to the economies of all consumer States, including a great many developing countries." Around the world many persons hoped that events would prove him right.

Elsewhere in London, "the Coordinating Committee of the oil companies" termed the agreement: "fair, appropriate and [the] final settlement of all matters related to the bases of taxation and the levels of posted prices." Largely overlooked in their press release was a statement that the oil companies were unable to absorb the new cost increases—meaning that consumers would have to pay. For consumers, then, the results of the Tehran Agreement would be quickly realized—oil and products would be more expensive.

In the U.S., after the February 3, 1971 OPEC meeting, oil matters received increased media attention. The *New York Times*, as early as February 8, termed the sessions "confrontation in Tehran" and reported that the ten OPEC members were "putting the squeeze on ... for a huge price increase." In an editorial on February 10, the *Christian Science Monitor* pointed out that the companies had no alternative but to accept higher prices, and then wisely added, "optimistic expectations that the companies would win their case with the diplomatic backing of their governments have proven ill-founded. ..." When news of the signed agreement reached New York, the *Times* called it "Surrender in Tehran," and stated that it represented "a dramatic demonstration of the power that an effective cartel of raw material producers can exert upon the highly industrialized nations of Western Europe and Japan. ..." But of all accounts and reports, that in the *Financial Times* (London) seemed to explain the proceedings at Tehran with greatest accuracy and clarity:

> In the long run the countries of Europe have little choice but to work towards the day when such pressures are impossible ... concentrate upon developing Europe's own sources of energy. ... With every new success in the North Sea the pressure on prices outside Europe will be that much relieved.

In the business of choosing winners, however, the press made no mistake; for the Shah was first on every list. He "strengthened his position as the dominant political figure in the Persian Gulf through his active and successful role in the Teheran oil negotiations," according to the *Financial Times*. And the *Washington Post* noted that his gain in "political status ... should serve him well in his American-supported effort to become the premier power—

replacing the retiring British—in the Persian Gulf."[55] Everyone appeared to agree that the Shah had reached a new apogee of power.

X   X   X

One day before the parties signed the Tehran Agreement (February 13), Hamadi told Strathalmond that OPEC had already formed a Mediterranean Committee, including oil ministers from Iraq, Algeria, Saudi Arabia, and Libya. He also said that OPEC would schedule their first meeting for February 16 in Tripoli. Yamani later confirmed this date, and added that both he and Hamadi would attend.[56] Dr. Amouzegar was less certain; first he announced that he planned to go to Tripoli, but for some reason, the Shah later asked him to take a ten-day holiday in Germany. Then, a rumor spread that Yamani would not personally go to Libya; and finally, when the Tehran team returned to London, they indicated that, although Hamadi and Yamani had been invited to Libya, no one appeared certain that OPEC would form its Mediterranean Committee.

Once again the presence of Yamani—already a central figure in the Tehran negotiations—seemed to hold special significance for the companies. An able negotiator, Yamani, up to the time of Tripoli, appeared to play a moderate role, at least in contrast with some other OPEC members. Yet, later reconsideration of Tehran, especially in light of the testimony before the Church Committee, suggests that Yamani's role in the Tehran negotiations may have been seriously misunderstood. After Tehran, clearly Yamani showed little taste for service on OPEC's Mediterranean Committee alongside representatives of the most radical members—Algeria, Libya, and Iraq. Yet, for the companies, at least, Yamani remained the key to the events that transpired in Tripoli.

Yamani also faced a difficult problem at Tripoli. While he may not have enjoyed radical company, he could not forget that the Libyans, in particular, might be successful. Libya, with a disregard for all other interests and a loud threat to shut down all oil companies, also was in a key position; that power might be translated into a better price for Libyan oil, or even for all Mediterranean oil. The Gulf and its ministers would be left behind unless they cooperated. Could Yamani take such a chance? Could he leave Saudi Arabia behind, outside the range of such a deal? The price for being moderate might come high at Tripoli, even though that had not proved to be the case at Tehran. Once he realized that difference, he graciously accepted a lesser role at Tripoli and played it with style, remaining always offstage and apart from the radicals—letting the Libyan extremists make every new demand.

Although they mattered little, some attempts were made to restrain Libyan extremism. On February 16, for example, the Shah cautioned them to be moderate: "It is better to get a reasonable price that guarantees stability to the producer and to the consumer." And the vice president of the European Economic Community suggested that any Mediterranean settlement should take account not only of the "special situation of the producing countries, but also of the great importance of Mediterranean oil for the [European] community."[57]

The oil companies' negotiating team knew that Libya's pricing would be the crux of the forthcoming negotiations. Already, as has been noted, after the fall 1970 agreements, the Libyans enjoyed a price differential in crude over the Gulf states. A portion of this differential could be termed a temporary short-haul freight premium, but with allowances for the Suez Canal closure added to freight premiums, Libyan crude oil commanded $2.55 per barrel, as opposed to the Gulf price of $1.80.* When word of the Tehran settlement reached them, some spokesmen for independent oil companies immediately decried the Tripoli negotiations as a cause lost in advance.

The Libyans, Deputy Prime Minister Jallud and Petroleum Minister Mabrouk, in all their conversations with company spokesmen, insisted that the familiar united front approach by the companies represented a clear challenge to Libyan sovereignty. Consistently at every session after January 25, Mabrouk raised the sovereignty issue almost as prefatory to any discussion. Yet, while he denigrated the company collective bargaining approach, he also made it clear that the Tripoli negotiations should provide a single settlement that would define the worldwide pattern for all the companies. On February 18, Major Abdesselam Jallud established the Libyan position in a speech that interpreted the meaning of Tehran:

> The Tehran oil agreement did not fulfill the aspirations of the Gulf peoples, but the joint stand of the producing countries against the industrialized countries and their monopolistic companies was a victory. ... The Tehran agreement will allow us to increase our income considerably, but we shall not be satisfied with what was obtained in that agreement.[58]

Next, Oil Minister Mabrouk handed to the local British Petroleum Company representative a list of Libyan demands. Finally, Mabrouk made clear that he expected his answer to be endorsed by the entire petroleum industry, and

---

*In the end Libya gained a 90 cents per barrel increase, of which 65 cents was permanent and 25 cents temporary.

that he wanted a complete response by February 23.

When the London Group considered the Libyan demands, the range of choices appeared rather limited. London believed that the Libyans would try to protect the price advantage resulting from agreements with Occidental Petroleum Company (and others) made in September-October, 1970. This increase, Libya would use as a basing point on which to build new price demands for increases, to reach a "permanent" posted price. This meant demands for a series of small increases rather than a single large jump. So, in London's view, the companies obviously had to *fight* against each of these increases, or watch their control over the oil pricing structure erode further. This strategy underlay the initial moves of their team, made one day before the Libyan deadline.

In Tripoli, Major Jallud welcomed the Gulf ministers with a speech at a mass meeting in which he declared that Libya expected better terms from the companies than those agreed upon at Tehran. From the ministers' private session came a press release threatening a shutdown if the oil companies failed to meet Libya's demands.[59] From the moment they arrived, the oil companies' group, headed by Piercy, operated in an atmosphere designed to intimidate them. According to Schuler's analysis, the chairman of the company experts told his team: "contrary to earlier hopes that Yamani was normally committed to Gulf terms for the Mediterranean, it was now certain that both Saudi Arabia and Iraq would support the [threatened] embargo." In these circumstances, the chairman questioned whether a hard-line approach could work, or whether his team should simply wait and plan to fight another day.

Later on the evening of February 24, Major Jallud announced a meeting of representatives from all the oil companies operating in Libya, at 7:30 p.m., in the oil ministry. There, after some two hours, Jallud began to speak to a group that included H.H. Goerner and M.S. Mahmud of Esso Libya. Mahmud understood Arabic and kept an aide memoire which showed the Major to have said: "I personally and all of us in Libya laughed a lot when we received the oil companies cable and though[t?] the companies did not understand the situation with us for about a year and a half [after the RCC took over]." And later, "Negotiations would be for a very short period. Demands are precise. . . . If we reached a conclusion with one company the other companies would be accordingly notified and they will have to say yes or no." Jallud added that negotiations had already started, and that three days would be allowed the companies to get ready for them. In two weeks, he said, negotiations would end. Then, Jallud's aides distributed to each company officer two sheets of paper with the Libyan demands written on them. One sheet, each company representative signed and returned to establish receipt.

Following Jallud's speech, the team decided against offering to Libya any of the terms formulated in London after the Tehran Agreement. Instead, the team left for London immediately; while they were in transit, on February 24, Algeria, in order to emphasize the producing countries' willingness to act dramatically, announced to the world that it had unilaterally nationalized 51 percent of all its oil production, as well as 100 percent of all local pipelines and gas production. Algeria agreed to compensate oil companies for the properties nationalized but at Algerian valuations, which included retroactive claims against the companies—not at company estimates of value.

When the Libyan team had returned, the London Group met again to consider further action. Jallud's demands called for a posted price of $3.75 per barrel, including a short-haul freight premium of 41 cents per barrel, retroactive to the 1967 Suez Canal closure. Also included were increases in the tax rate up to 55 percent, and on each barrel of oil exported, a 25 cents reinvestment-in-Libya tax. Except for the tax rate figures, all of these sums were considerably higher than the comparable numbers in the Tehran Agreement. Libya held power and used it.

By February 25, the oil companies' thin pretense of union had all but collapsed, in fact as much as theory. Even though some independent companies were very much dependent on Libyan production, they pressed the London Group to oppose the Jallud ultimatum. At the same time, the larger companies with Persian Gulf production showed that they were far more willing to continue the struggle. Here, it should also be noted that most representatives of independent oil companies had little experience in dealing with the Gulf countries or with their representatives. This crisis produced heated exchanges between spokesmen for big and little oil companies. Up to this time there had been relative harmony between oil majors and independents; the common objective they pursued had helped. Too, John McCloy possessed exceptional administrative ability and he had used it to keep his clients together. The tacit understanding of the companies' purpose by the Justice Department had also been important in this first major union of the oil companies. Finally, however, rather than orderly consideration of the problems, confusion characterized the London meetings.

Even so, the Libyan demands still required that new terms of reference be drafted for the Tripoli negotiating team. By now, all the oil companies operating in Libya recognized that no unified approach would be acceptable there. After many hours, new terms were drafted, and the Piercy team again left for Tripoli on February 28. On March 3, Jallud called in executives of seven companies, but in the end he spoke only to the BP representative, offering him a five-hour sermon. In it, Jallud threatened that Libya would legislate even higher posted prices, as well as demand a Mediterranean embargo—unless the companies accepted Libyan terms within four days. This ultimatum led the Gelsenberg and BP representatives to fly to New York in an attempt to persuade the "chiefs" to accept a one-year agreement. But the "chiefs" wanted a five-year package to avoid any negative reaction by the Gulf countries. After additional discussion, however, they confirmed the terms prepared in London, which met some, but not all, of the Libyan demands.

Back in Tripoli, when Occidental attempted to present the new price schedule agreed to by the companies, Jallud refused to review it. Instead, he repeated that Libya might well nationalize its oil. Again, he threatened an embargo and added a new message: Libya did not care what happened in Europe! Jallud's deadline for acceptance was moved back to March 9, but even that delay offered little hope for the company negotiators who had observed Jallud and Mabrouk in their intemperate behavior. What went on at Tripoli could hardly be termed negotiation.

When the oil industry made its offer on March 9, Jallud and Mabrouk quickly rejected the proposals as insufficient. By now, the lengthy negotiations had attracted new support for Libya from other oil producing nations and from Syria also, for reasons which still remain unclear. Among the principals, Hamadi and Yamani presented themselves in Tripoli on March 15, voicing new support for Jallud, even though the companies had little stomach for more fighting. When George Piercy met with the two Libyan spokesmen on March 14, Jallud again proved to be "extremely rude" as he threatened the chairman by saying, "You've tied the hands of the independents." When Yamani arrived, he and Hamadi quickly fell in line with the Libyan proposals, adding some new ideas, which were later described as "leapfrogging in the biggest way." Yamani showed no power or inclination to moderate Libyan demands, at least in the meetings of the ministers. On March 16, he assured an oil company spokesman in private that Jallud was first determined to nationalize 51 percent of all Libyan oil companies; only by consulting the ill Colonel Qaddafi had Yamani been able to prevent this action. So the oil companies barely escaped expropriation.[60]

Sensing victory in the offing, Jallud extended the deadline, this time until March 18, after which an embargo would begin. But in other ways, Libya appeared intractable. When Piercy's team met on March 17, to consider the Libyan demands, they reviewed the chief reasons in favor of a quick settlement: maintaining the oil industry as middleman, preventing nationalization, and moving the oil. Another undebated reason for settling, perhaps the most important of all, was the companies' failure in public relations; for the companies felt that they might not be able to convince European consumers that a break in negotiations with Libya was really necessary. Equally, the unsettling effect of the recent agreement made at Tehran may have been a compelling force in advancing a Libyan settlement at the time.

On March 18, 19, and 20, negotiations with Jallud and Mabrouk did take place. On March 20, after a seven-hour meeting with Mobil's Andrew Ensor and Gelsenberg's Enno Schubert, Jallud conceded that agreement was near; even so he continued to display a studied lack of common civility [the meeting was tape recorded] by repeatedly ignoring every document that Schubert and Ensor sought to introduce. This meeting recessed to allow the companies time to review the detailed proposals for "temporary" versus regular freight split and then, more days passed while the Libyans argued over a five-year term and some form of Suez retroactivity tax. Eventually Jallud had his way; and the companies yielded more ground, until on April 2, fifteen oil companies signed the final Tripoli Agreement.[61] Here, its extraordinary terms require review.

This final settlement covered a five year period. Counting the first-year escalation, the posted price agreed upon amounted to a 35 percent increase, while the take of the government increased by 46 percent. In dollars, Libyan income from oil rose in 1971 by $700 million.* In all, the agreement actually gave Libya about 20 cents per barrel more than the amount received by the Persian Gulf states under the Tehran Agreement.

Yet the Gulf states kept their promise: no leapfrogging or price ratcheting occurred. In the Gulf there was, however, some unhappiness over the Tripoli terms, and reportedly, the Shah of Iran became agitated when he heard about the terms accorded Libya. At that time, Dr. Amouzegar told a petroleum consortium executive that His Majesty had vowed to support the radical members of OPEC in the future.

After Tripoli, agreements with Algeria, Iraq, and Saudi Arabia still remained to be concluded. But once it became understood that Tripoli terms had superseded those of Tehran, Iraq and Saudi Arabia soon came to terms with the companies on pipelined oil (June, 1971). But the Algerians went ahead to set new terms for the French government—which were more radical than those of Libya.

In justice to the Piercy team, it now seems clear that they were over-

---

*$1.6 billion estimated before April to $2.3 billion after, including the retroactivity tax.

matched at Tripoli. They had come to negotiate in the traditional way and according to customary practices. But even before Piercy reached the table, Libya had changed the rules of the game, and utilized the public press to announce more changes. Confronted by overwhelming evidence of Libyan recklessness, Piercy had no alternative except to yield. The Libyans capitalized on their advantage and altered forever the basic strategy of oil diplomacy.[62]

X   X   X

Several years after Tripoli, Piercy and Schuler testified before the Subcommittee on Multinational Corporations of the Senate Foreign Relations Committee, which was seeking to fix responsibility for the high price levels of OPEC oil. Piercy thoughtfully evaluated the Tehran team effort; then he said:

> I think, as I look back at it, the negotiations were about as good as you could expect. If you look at what the prices that we were able to come out of Teheran and Tripoli with and you look at the 10 to 15 years of a flat price before that, and you recognize the problems that these nations had, I think it was a reasonably good job. So the disaster, and I would even use Henry's [Schuler] words, 'unmitigated disaster,' is October 16 [1973] when they quit negotiating. ...
> I believed them when they threatened us with what they call the Venezuelan system and that is, 'We will post the price, we will do it and we will do it whenever we darn please.'

Schuler, in his testimony, agreed with Piercy that the Tehran-Tripoli negotiations represented a tragedy for the companies as well as for Japan and the nations of the West. At Tehran and Tripoli, "the momentum" in oil negotiations passed over to OPEC and the producing states. It had not yet been recaptured. Piercy freely admitted: "there were some errors made when I arrived in Teheran." But then he went on to a more telling point: "After we had been through the wringer a few times in Libya ... I must say on reflecting back I don't know how we could have brought Libya to the same table." Schuler, in his turn, traced some of the difficulties of Tehran and Tripoli to human error. Then he added: "I agree 100 per cent it would have been extremely difficult to bring the Libyans to the table."

While supporters of a strong U.S. antitrust policy may have chuckled at the public discomfort displayed by American oil companies from January to April, 1971, thoughtful observers pondered the more important decline in

U.S. prestige and the gains in OPEC's power. Here, the major cause must be found in the absence of any settled Middle East policy within the U.S. government.

While oil supplies had been tight, no major shortages existed. But a transportation [tanker] shortage did exist. This made the surplus producing capacity, almost all concentrated in the Persian Gulf, difficult of access; and at the same time, the proximity of Libyan oil to European markets gave strength to the Libyan negotiators. With production of almost two million b/d, Libya alone could create a major energy crisis for Europe. The Libyans knew it, OPEC knew it, and oil company negotiators could not ignore the shared fact of that knowledge.

Given these circumstances, John McCloy and some leaders of the oil industry believed that the time for some union of the oil companies had come. Whether they actually persuaded some (or many) State Department officials to adopt that position, is still not known at this time. But two things are clear: The U.S. Department of Justice Antitrust Division did endorse without penalty such a combination of companies; and second, John McCloy would not have arranged such a combined front had he not believed the idea had support from the U.S. State Department. So the State Department's quick reversal today remains unanswered.

Of course, the Irwin cablegram from Iran helped to precipitate this striking change in national policy; and the polls of western consuming countries (and Japan) during the Tehran conference also played a part, as respondents answered, "Accept a price increase and do nothing to cause them [OPEC] to shut down production." Finally, an American interest in strengthening Iran added its weight, and the combination rendered nearly impossible the task of Piercy and the other company representatives who met the producing nations at the negotiating table. In fact, if they had a chance at all, it lay in the concerted action of the companies, supported by the consuming nations. As it was, the negotiators bought almost two and one-half years' time for the West and Japan—time largely frittered away by the U.S. operating under the questionable leadership of a President deeply mired in Watergate.[63]

Writing about the Tehran and Tripoli Agreements, Henry Kissinger remarks: "Our hands-off policy ordained the result; the companies yielded." Then he adds: "Both producing groups agreed to maintain this level for five years—a solemn promise that must hold a world record in the scale and speed of its violation." But in reality, for a few months some measure of stability did appear in world oil markets—thus reassuring consumer governments during that time. Viewed in the perspective of hindsight, however, it now seems doubtful that any united front could have held together for long.

Yet, at the time, Jersey entered into negotiations with real optimism. For years the company had permitted its negotiators great freedom to make decisions. (Other oil companies within the united front did not delegate authority so freely.) But, as the documents subpoenaed by the Church Committee clearly show, here the negotiators exercised little real authority; their hands were tied by instructions, provided through a bureaucratic apparatus which proved to be cumbersome and ineffective. Finally, tangled lines of authority led to too much conversation and too little authentic negotiation.

For example, although it was never completely clear that a single oil company had the power to block a decision, each of the companies seemed determined to protect its particular interests—thus, the potential influence of a united front suffered. In theory, the idea of collective action by the companies was sound; but to be effective, it required a grant of authority from each of the participants. That any one company would have yielded sufficient authority remains doubtful. And even had Jersey done so, it probably would have stood alone. So far as available records show, no executive or set of executives really took charge for the companies.

For any individuals to have been effective, however, this delegation of power would have also required enlightened support of the consumer governments. Yet the oil companies did not make this clear enough at the time, so a united oil company front never combined with powerful consumer interest—or at least not until the cause of the consumer had already been lost in the intimidating rhetoric of OPEC spokesmen at Tehran and Tripoli. Years later, Piercy and others involved recalled their sense of disillusionment and disbelief: the United States government had supported foreign producers against the American companies.[64]

But viewed in another light, given the competitive spirit in the industry, the real miracle is that even so much unity was achieved. Oil companies did pull together reasonably well, considering their history of cooperation only under government sanction in periods of crisis.

As to what caused the final collapse, oil men will ever differ. Fourteen years after Tehran, Howard W. Page, who was there, reflected on what had happened and doubted that the Irwin mission made as much difference as other people thought. He questioned that any better agreement would have resulted had Irwin *not* made his trip. Interestingly, Jersey's Piercy, certainly the key industry negotiator, agreed with Page. Both concluded that, in the end, the Irwin mission did not much matter. In time, both Jersey leaders came to view the OPEC success as inevitable.

On the other hand, Jersey CEO Jamieson, who was an able negotiator himself, pointed to that mission as crucial. Jamieson believed that the State

Department representatives, involved before the Tehran negotiations began, undercut any possible success of the oil companies. Several years after he retired from Exxon, George Piercy reflected on his tribulations during 1971. He expressed doubt that any head-on confrontation with the producing nations would have worked, because he still believed that the State Department had not supported the companies' position; rather, it continually pressed to have the matter settled—to "Pay the OPEC price."[65]

Chapter 24

"The Wheels Come Off"

TEHRAN AND TRIPOLI, in 1971, represented for the international oil companies a practical end to their influence over oil prices. At that time, no way seemed to exist for the companies to develop quickly enough the additional producing capacity needed to match the soaring world demand for oil. Public statements about this simple fact—the dramatic shift in control of international oil from the companies to the producing countries—appeared daily in the news media. This popular attention almost guaranteed that, in the future, negotiations could not be conducted as they had been in earlier years.

For a time after the five-year agreements signed at Tehran and Tripoli, peace ensued. The Persian Gulf states did not either leapfrog or ratchet prices. But the fact that Shah Muhammad Riza Pahlevi believed Iran had been short-changed in the Tehran Agreement was widely discussed, especially among correspondents and editorial writers in the United States and Great Britain, who called attention to the fragility of the 1971 oil agreements. Several openly predicted that the next step of the Organization of Petroleum Exporting Countries [OPEC] would be to demand participation in the companies.[*1]

While the oil producing countries did not win every point in the Tehran and Tripoli Agreements, they did establish primacy in the negotiations. It soon became apparent that a principal part of the burden of maintaining the oil companies' position would devolve on several executives of Standard Oil Company (New Jersey) [Jersey], especially on Middle East Contact Director George T. Piercy, and on the company Chief Executive Officer [CEO] J. Kenneth Jamieson. Jersey had more oil interests in the Middle East than did any other U.S. company, and Director Piercy had more Middle East negotiating experience than did any other company executive. Jamieson, who later became the principal negotiator with Iran, followed as CEO the prac-

*As used here, participation means ownership by an oil producing country in an oil company operation.

tice of delegating authority to competent people. Then, he left his chosen lieutenants alone.

Both Jamieson and Piercy were familiar with possible OPEC demands, so the idea of producing country participation in the companies did not surprise them. Yet, in all fairness, it should be said that neither Jersey executive, nor any oil company leader, was prepared for the swiftness with which change came. Nor, in truth, did the OPEC leader anticipate the fortuitous series of events that brought their participation about, as participation, indeed, did become the next important goal of all major producing countries.

The question of participation had surfaced as early as June, 1968, during the OPEC meeting at which the oil ministers drafted their "Declaratory Statement of Petroleum Policy in Member Countries." That declaration included a clause stipulating: Where no provision for governmental participation in the ownership of the concession-holding company had been included in a contract, the host government "may acquire a reasonable participation on the grounds of the principle of changing circumstances." Something new was being proposed for future oil contracts.

One important petroleum economist, the Englishman Paul Frankel, immediately grasped the implications of this new principle—that it would destroy the old safeguards in international law and business. As he pointed out:

> The center of the problem of the last few years stems from the doctrine that a sovereign state can override all commercial commitments. This has downgraded international intercourse to the level of the jungle.[2]

Whether oil negotiations merited the designation "jungle" may be debatable, but everyone recognized the severity of the changes that had taken place. After 1971, oil agreements seemed to last far less time than it took producing countries to approve them formally. Almost before the ink had dried, the agreement had become a worthless document. And oil men stood by, almost helpless. As Jersey's retired Middle East specialist, Smith D. Turner, described these years: "[They] were the one[s] in which the wheels came off—not four at once, but one at a time, ending with a situation little like that at the start."

X   X   X

Foremost among the problems pressing OPEC during the late spring and summer of 1971 was inflation, or as the ministers termed it, "monetary parity." Caracas Resolution XXI, 122 stated that whenever the real value of the monies paid to OPEC nations dropped, "posted or tax-reference prices should be adjusted so as to reflect such changes." And after August, 1971, the value of the U.S. dollar in international exchange steadily fell. At the same time, the radical Arab states within OPEC became more and more determined to use their oil as a political weapon, aimed at the U.S. in an attempt to force that nation to modify policies that seemed to the Arabs more pro-Israel than even-handed. Such radical pressure to alter U.S. Middle East policy ultimately forced Saudi Arabia to choose between extremist pro-Arab policies and its friendship with the United States. But as late as the summer of 1973, King Faisal still held to a conservative course, as he refused to permit any Saudi use of oil to put political pressure on the U.S.[3]

For their part, the oil companies contended that the Tehran and Tripoli Agreements clearly provided adequate provision and compensation for both inflation and currency fluctuation (a 2.5 percent annual guaranteed price increase and a five cents per barrel bonus allowance). The oil ministers answered that those numbers were insufficient to take care of "freight escalation and/or inflation," and that nothing in the original agreements covered changes in monetary values. Moreover, when OPEC leaders reviewed the record of negotiations at the Tehran and Tripoli meetings,* they could not identify a single essential issue on which the companies had not yielded to some extent, nor any evidence that consuming countries had supported the oil companies. In all, OPEC's victories during 1971 encouraged the oil ministers afterward to demand greater participation and additional compensation, as well as increases in the value of currency payments for oil. Once these issues were introduced, OPEC continued to pursue them.

In defending the companies from new assaults, John J. McCloy, the veteran legal counselor, again took a leading role. On July 14, 1971, he and his associate, William E. Jackson, accompanied by Jersey's Jamieson and four other company chief executive officers met with U.S. Department of State [State] personnel, including Joseph J. Sisco, Assistant Secretary of State (Near East-South Asia), to discuss the new OPEC demands and, in particular, problems with Libya. Next, McCloy's group met with Deputy Under Secretary Nathaniel Samuels on July 19, when the oil delegation received assurances that State would try to prevent any publication of the Libyan Producers Agreement. But more important, the oil "chiefs," McCloy, and

*See preceding chapter.

State representatives also discussed "at some length" the need for planning "an overall Government Policy in regard to the production of oil in the Middle Eastern area." For the State Department, this show of concern may have represented too little action too late; yet the atmosphere seemed encouraging. The U.S. government did appear concerned. *

On the companies' behalf, McCloy next wrote a cautionary letter to Richard W. McLaren, Assistant Attorney General, Antitrust Division, U.S. Department of Justice [Justice], on July 23, 1971. McCloy noted the fragile nature of the Tehran and Tripoli Agreements; and then he added an opinion that, because of the Shah's irritation over the Tripoli settlement, Iran might well support the radical Libyans in future negotiations. As McCloy explained:

Recent private pronouncements by Iranian Finance Minister [Dr. Jamshid] Amouzegar to the effect that Iran's dissatisfaction with the Libyan and Eastern Mediterranean agreements will lead it to support the more radical elements in OPEC unless additional concessions are granted to Iran have a particularly ominous ring, carrying on an echo of the events which led to the original Caracas resolutions of December 1970.

Finally McCloy remarked: "Recent disturbing events just at the time of OPEC's first meeting since the negotiations are a cause of deep anxiety both for the companies and for the consuming countries to which stabilization represents such an important objective." To this legal counselor, the situation seemed threatening.

During the next stage of negotiations, the participation issue suddenly moved onstage, as a follow-up to the Tehran-Tripoli Agreements of early 1971. In fact, one key participant, Sheik Ahmed Zaki Yamani, the Saudi Arabian oil minister, told the world press that in his view the five-year pacts made at Tehran and Tripoli did not extend to cover either the decline in the value of the U.S. dollar or participation by the host country. If they had done so, Yamani contended, he would not have signed them.

George Piercy, along with other oil company spokesmen, pronounced this interpretation of the Tehran and Tripoli contracts incorrect; Piercy said that the agreements explicitly provided for fluctuations in the value of the U.S. dollar, as well as for host country participation.

Nevertheless, OPEC, meeting as a whole, in Vienna during July, 1971,

---

*The original Libyan Producers Agreement or Safety Net Agreement provided for pro-rata sharing among producers in case of any government ordered cutback over a six-months period. This term agreement was later extended. Backup oil was to be provided from the Persian Gulf.

called on member states to implement immediately participation agreements with the oil companies. This represented a new strategy, and to prepare for it, an extraordinary OPEC meeting was set for September in Beirut. Participation meant more profit, or so the ministers reasoned.

On August 13, 1971, OPEC published the resolutions of its July meeting. These included two specific demands: first, for an adjustment in currency parity; and second, for "immediate implementation of rights" to participate. In response, McCloy, on October 5, advised McLaren that the London Administrative Group* would meet on October 11. At the extraordinary OPEC meeting in Beirut, on September 22, 1971, participation negotiations received priority. Since Algeria had already achieved 51 percent—participation by means of nationalization, OPEC members saw a clear advantage in moving toward majority participation. By September, in fact a majority of the five major Persian Gulf producing states had indicated their preference for initial 20 percent participation—gradually escalating to 51 percent while other OPEC members appeared to be seeking some percentage between those two numbers.

At Beirut, two other issues also engaged the OPEC oil ministers, who felt that both were likely to prove difficult to resolve in forthcoming negotiations with the companies: first, as compensation, the ministers stated that host governments would pay the companies no more than the book value of the "net fixed assets" for shares purchased. This standard meant that the companies would not receive fair market price (even if such could be determined); second, to provide continuity in oil supplies while the national companies acquired experience as marketers, a crude oil "buy-back" price would have to be established. Thus, the concessionaires (companies) would guarantee to buy this oil back at a figure half-way between tax-paid costs and the posted price.[4] Both of these issues, the OPEC ministers agreed, could be expected to appear on the agenda for a future meeting.

At this time, McCloy continued to assert to the U.S. government the companies' view that OPEC seemed determined to settle the currency issue, on which the oil ministers believed the companies to be united against any change. The Iranians especially, McCloy thought, were pushing the issue of adjusting posted prices to compensate for inflation. On September 26, for another example, Libya's oil minister had sent a letter telling all companies to adjust upward their next tax payments—to compensate for inflation. The Libyan instruction, McCloy noted, violated the terms of the Tripoli Agree-

---

*The London Policy Group became the London Administrative Group when the former ceased to function as a support unit for the Tehran-Tripoli negotiations.

ment. He went on to express a personal opinion, based on years of Middle East experience: the gravity of the present situation was so great, McCloy told McLaren, that all oil consuming countries should be notified. Already a company "study group"* had been formed to assess the need for reinstituting the Libyan Producers Agreement, and to set a date to begin new negotiations concerning monetary values.

In fact, this study group first met on October 18, 1971, when it produced two documents: a "Memorandum of Intent," and a "Memorandum of Confirmation." Twenty oil companies signed the first, and seventeen, the second. Yet neither agreement could take effect until McCloy advised the companies that the U.S. Department of Justice:

> has stated in writing that it has no present intention to take any action under the Antitrust laws with respect to the making of either of the Memoranda or the activities contemplated therein.

Acting for the companies, McCloy sought a Business Review Letter designed to minimize the possibility that Justice would object when the companies met together to formulate a joint response to the producing countries. McLaren replied on October 22, that Justice "does not presently intend to institute any proceedings under the antitrust laws" with regard to either memorandum, but Justice reserved the right to do so if later information warranted antitrust action.

Meanwhile, a team of company spokesmen headed by Jersey's George Piercy flew to Tehran, to discuss fiscal demands with Dr. Jamshid Amouzegar. As support for these discussions, a small group of financial specialists was formed in Vienna to consider the entire range of matters relating to inflation and purchasing power, which impacted on both the companies and OPEC. After arranging for that meeting, Piercy returned to New York, where the oil company "chiefs" were busy considering this new evidence of OPEC dissatisfaction with standing agreements.

In the *Wall Street Journal* for November 5, 1971, Jersey's Chairman and CEO Jamieson explained the company position.** When asked about the role that Jersey would take in the forthcoming negotiations on currency fluctuation, Jamieson indicated that the company would hold firm. Since "The ink was barely dry" on the agreements concluded at Tehran and Tripoli, Jersey saw no reason to reassess them. Indeed, Jersey believed that "the

---

*This group consisted of representatives of thirteen oil companies, including Jersey.

**Jamieson and President Milo M. Brisco spoke to the New York Society of Security Analysts on November 4.

effects upon them (OPEC) of currency changes to date ... [do not] establish a case in equity for any relief." In response to later questions about host country participation, the CEO replied that his company believed "it would be extremely difficult to implement ... in a manner that would serve the best interests of the producing and consuming countries and the petroleum industry." Jamieson did indicate that he expected an amicable settlement to be reached, that Jersey had always tried to be fair, and that in the past the company had found most of the host countries to be reasonable.[5]

Late in November, Piercy joined the team of financial experts in Vienna, where they conferred with producing government representatives for two weeks. At that time the financial meetings adjourned so that the ministers of producing countries could attend the December 7 OPEC meeting in Abu Dhabi.

Meanwhile in Libya, all the oil companies protested against the petroleum minister's September 26 demand for increased payments to offset a decline in the dollar's value. Simply, the companies refused to comply with the exchange directives expressed in the government letter. By now, it had become clear that the Tripoli Agreement of April 2, 1971, had settled little of importance. Within a month, Libya's Deputy Prime Minister, Major Abdesselam Jallud, resumed his campaign of harassing oil companies.* Significant revisions to the basic petroleum law of 1955 now required that the companies engage in expensive gas conservation projects—despite the fact that, for gas, only a limited market existed. Companies found in violation could be summarily fined by the Libyan government. Next, the government began to use the Bank of Libya to exact additional forms of tribute. Thus, the state fixed the value of the Libyan "dinar" by decree and then required that all company payments be made in that unit of currency. Already for some years, the Bank of Libya had routinely decreased the value of the dinar when payments fell due, and then later increased its value. Since dinars had no circulation outside Libya, when the bank arbitrarily fixed their value, it generally cost the companies more to pay taxes and royalties. This clever practice began in 1969. In 1971, to avoid Libya's unilateral abrogation of provisions covering payment supposedly made a few weeks earlier, the companies met in New York and agreed (again) to stand together to protect their rights.**

*Apparently, individual Libyan ministers negotiated under the special pressure of being penalized for failure to obtain better terms than those offered to other OPEC ministers. Jallud mentioned this threat frequently during the first Tripoli negotiations.

**The *Middle East Economic Survey*, several American newspapers, and some public figures termed these meetings secret. Yet the companies made no effort to hide the fact that they were meeting, or the subjects being discussed.

The Libyan letter of September 26 changed the dollar-dinar ratio from $2.94-1 dinar to $2.80-1, creating a premium of about 5 percent for Libya. When tax and royalty charges came due in October, 1971, the companies paid Libya at the old dollar-dinar exchange rate, but they also provided letters explaining their actions. For reasons yet unknown, Libya accepted all payments except that of Esso Standard (Libya) Inc. [Esso Libya]; to that company, the government sent a new letter which threatened to attach company property as a penalty for underpayment. Next, Nelson Bunker Hunt Oil Company [Hunt] representatives were told that, if they continued to refuse to pay, their entire concession might be cancelled. Esso then sent to Libya a detailed explanation of its action, and an executive of each of the other oil companies signed the Esso text. Still, Libyan harassment continued, as the host government in early November, 1971, attached Esso Libya's funds—by "administrative sequestration." Libyan banks then refused to honor Esso checks, but the company quickly arranged for payroll checks to be cashed; and within a week all company accounts were released. Next, Libya peremptorily deducted the amount it claimed that Esso Libya owed on taxes and royalties under the new exchange rate—$900,000. This direct government action represented a new Libyan tactic in dealing with oil companies.[6]

At the Abu Dhabi meeting on December 7, OPEC called on the companies to meet with the Persian Gulf ministers (led by Dr. Amouzegar) on January 10, 1972, to settle the currency problem within ten days. After that settlement, negotiators could consider the question of joint participation. But the money decision had to come first, as once again the Shah and the Iranians forced their priorities on OPEC. For them, inflation provided a convenient excuse to press for an increase in crude oil posted prices, as well as a new opportunity to express their exasperation with the Tripoli Agreement. Strangest of all, the Shah could demand changes in Tripoli terms and still remain within the boundaries set by the Tehran Agreement—at least according to the OPEC view.

On December 7, 1971, in an unrelated but significant political action, Libya acted to nationalize the properties of the British Petroleum Company, Ltd. [BP]. This move was explained as a retaliation for alleged plotting by the government of the United Kingdom, with the "puppet government of Iran," aimed at allowing Iran to take possession of three small islands* in the Persian Gulf, which Libya claimed actually belonged to Iraq. Since BP had jointly owned Concession Number 65 with Hunt, Libya now established a

*The Greater and Lesser Tumbs, plus Abu Musa.

new national oil company to operate with Hunt, while continuing to press for increased production and marketing of Libyan oil.

This nationalization brought a quick response from BP, which formally requested that the Libyan Producers Agreement go into effect. Meanwhile, other oil producers agreed that the safety net should apply, so the principle of company unity faced a test. When McCloy related these incidents to Justice, he indicated that every Libyan producer feared that it might also suffer either production cutbacks or nationalization. To protect themselves, the twenty-two foreign companies operating in Libya agreed to a "Further Memorandum of Confirmation," dated December 16, 1971, which added to the original Libyan Producers Agreement supplementary coverage or protection against partial or total nationalization. Then, McCloy asked that Justice issue a Business Review Letter to cover this new memorandum, adding that he represented all of the signers; Justice agreed to his request.

The public, too, was becoming conditioned to learning of new Middle East oil problems as the January, 1972 meeting between OPEC representatives and oil executives drew advanced notice in the newspapers. "Oil Replay," the *New York Times* headlined, reminding its readers that the optimistic statements by company spokesmen after the Tehran and Tripoli Agreements had proved no more accurate than Neville Chamberlain's promise of "Peace in Our Time," made after Munich in 1938. The *Times* account made the point that the result of new negotiations would be identical with the old in Tehran and Tripoli—"a rout of the oil companies and an increase in the price consumers will have to pay. ... " At Jersey, *Times* readers wondered.

On January 4, 1972, Ken Jamieson, George Piercy, G.F. Cox, and Charles J. Hedlund—all of Jersey—and John J. McCloy flew to London, to meet with the London Group over the next several days.* On January 10, 1972, Piercy and eleven other negotiators, representing twenty-four oil companies, gathered in Geneva, Switzerland, in response to OPEC's demands that company representatives engage OPEC's Persian Gulf Committee for discussions of monetary parities.

At many times these conversations seemed unfriendly, for the question of the relative value of currencies proved complex and it aroused strong emotions. The companies offered "to determine the exact impact of the changes on each nation through use of a quarterly index compiled from statistics of the International Monetary Fund [IMF]." OPEC spokesmen met this proposal by demanding a flat 12 percent increase in taxes and royalties; they

*Jamieson returned to New York on January 7.

used against the proposed linking of money values to an inflation index the same argument they had used earlier at Tehran. (Basically, that the proposed index did not include increases in costs in such sectors of their economy as "services, investments, (and) military equipment.") However, a *Wall Street Journal* reporter probably came closer to the substance of OPEC's response when he wrote, "It seemed clear that the biggest objection was that the index wouldn't provide enough money." In any event, the OPEC Committee rejected the company offer and continued to demand a flat 12 percent increase. After two days, talks broke off and the meeting recessed.

When talks resumed, the Gulf states' representatives dropped their demand for a 12 percent increase and indicated a willingness to settle for 8.57 percent, the estimated decline in the value of the U.S. dollar. In turn, the conferees agreed to an increase in tax reference prices of 8.49 percent, a figure just below the dollar drop in value. Meanwhile, observers estimated that the combined increase would cost consumers more than $700 million in 1972, in funds paid to the Gulf host countries. Geneva I, as this January, 1972 meeting became known, took "into account the rise in the price of gold vis-a-vis the U.S. dollar." More important, gold here became established as the reference point for future discussions. Posted prices were to be changed to accord with an index of the currency movements in nine countries.* After the parties present had signed this agreement, other producing countries made corresponding agreements with the oil companies. In this manner all price adjustments were arranged.

Before adjourning, company negotiators and the Gulf states' spokesmen also agreed to begin new talks on participation. Although these conversations represented an aboutface in the companies' position, the explanation was simple: the oil ministers had threatened the companies with "combined action" (meaning a complete shutdown) if no discussions of share ownership were to take place.[7] Later, Howard W. Page, now retired as a Jersey executive vice president and director, in answer to a question from the U.S. Senate Subcommittee on Multinational Corporations of the Committee on Foreign Relations, headed by Frank Church, provided a simple explanation for the price capitulation by the oil companies:

> Once you have got complete control of the major markets by a cartel of producing countries, then you really have had it so far as

---

*These countries were: Belgium, France, Great Britain, Holland, Italy, Japan, Sweden, Switzerland, and West Germany.

price is concerned, and I don't know exactly what you do about it except to go without oil if you don't like the price.[8]

X   X   X

Exactly when Yamani first conceived of participation as an alternate to (or temporary substitute for) nationalization is not known; however, several sources confirm that as early as 1965, he was talking about host countries playing such a direct role. To him, participation provided a way for host countries to maintain more control over their natural resources, but without either interrupting the inward flow of foreign capital needed for their economic well-being or creating a sudden crisis in world petroleum markets—as nationalization surely would. Thus, oil companies could provide needed capital, and also give to producing nations both markets and operating experience. Participation seemed to Yamani, and eventually to others, an idea whose time had come. By 1968, Sheik Yamani had become its leading advocate within OPEC. Even earlier, OPEC had sought equity shares in oil companies, and also during the 1960s, several major international concerns did agree to joint participation ventures. But more than any other single event, the Arab defeat in the 1967 war with Israel thrust the participation issue into the foreground; and more than anyone else, Sheik Yamani kept it there.

Despite Western claims of bringing progress to the Organization of Arab Petroleum Exporting Countries [OAPEC] states, many nationals of the host countries considered the oil companies as an alien—foreign—Western presence. Naturally, when Arab public opinion flamed against the United States just after the Israeli victory in 1967, much Arab frustration was vented on the oil companies, and especially on their Western employees. In that moment Sheik Yamani saw the appropriate opportunity for the commencement of an equity participation debate with the companies. He saw that an ownership share in operations would at least partially placate those who demanded more OPEC oil for OPEC, and that any gains in participation could be the opening wedge that would lead eventually to majority control of the most profitable [producing] operations. In political terms, participation was equally attractive, because a role in ownership would ultimately lead to the host countries having more control of their petroleum resources. So in 1968, at the conference in Beirut, by introducing the participation issue, Yamani displayed his complex logic for the benefit of the oil companies and of the world press.

Already, Yamani had proved himself to be a strong leader. As Minister of

Petroleum and Mineral Resources of Saudi Arabia, he commanded the respect appropriate to a soft-spoken oil diplomat who represented his country effectively. As Standard Oil Company of California [Socal] Vice Chairman and Arabian American Oil Company [Aramco] Director, George M. Keller admitted: "Yamani is a superb negotiator." Since the country he represented, Saudi Arabia, was regarded as being the most politically moderate of the oil producing states, Yamani, as we have shown earlier, did not symbolize the radical extremes of OPEC.

On his part, Yamani brought a shrewd intelligence to his job. He realized that Saudi Arabia, at that time, had too few trained nationals to operate the vast Aramco installations.* Hence, his country would be best served by a gradual inclusion in company operations—rather than by any sudden nationalization. Yamani watched Libya gain power and wealth even while it continued to antagonize the oil companies, the consuming countries, and even the more moderate OPEC states. As he revised his earlier estimates of necessary delays before the Algerians and Libyans could expect to gain majority control of oil production, Yamani knew that he must accept the increasing tempo of the participation movement, as much as its growing force. Yet even as he evaluated events, Yamani still held the view that participation would slow down the overall move toward nationalization, while at the same time providing the oil companies with a relatively stable tie to the producing countries—a relationship useful to both sides.[9]

Now, with the advantage of hindsight, it is easy for us to see that the companies misunderstood both the symbolic and the actual importance of participation. Or, perhaps oil company negotiators understood only too clearly what shared ownership would mean to the producing states—encouragement to take over everything. In either case, it now appears unlikely that any counter force could long have forestalled OPEC's potent drive to gain political power, beginning with participation, and moving on to majority control. By 1971, the world oil situation had simply become explosive; at best, perhaps another year or two might have been bought by negotiations—but not without more active participation by the oil consuming countries.

In 1972, George T. Piercy, Jersey's Middle East negotiator, met at least twelve times with Yamani. Later, Piercy remarked that, given what actually happened, it really seemed silly to have haggled for so long over a few pennies per barrel of oil, in the belief that those additional cents would

---

*The four owners of Aramco were: Jersey, Socal, and Texaco Inc. [before 1959, The Texas Company], each with 30 percent, and Mobil with 10 percent.

create havoc in world oil markets. Yet Piercy and other oil company representatives did haggle; in turn, the OPEC leaders became accustomed to arguing over every fractional increase. Meanwhile, the participation issue was largely ignored.

Not surprisingly, company leaders, like most Westerners, held a view of participation directly opposed to that of Yamani. Counselor McCloy, as their spokesman, interpreted participation as a euphemism for expropriation: the *New York Times* stated that participation would enable the producing countries to keep "the oil companies as a viable hostage." President Richard M. Nixon's National Security Advisor, Henry Kissinger, labeled "equity participation in the companies" by the producing countries a form of "creeping nationalization."[10] George Henry Mayer Schuler, the one-time U.S. diplomat whom Bunker Hunt had persuaded to serve as vice president for his Libyan oil company, said that moderates coined the word "participation" as a way of softening demands for outright nationalization by more radical producing nations. And Schuler added, "both are 'confiscation' to a greater or lesser degree." Sir David Barran, executive and director of the Royal Dutch-Shell Group [Shell], pronounced 51 percent participation "intolerable" and indicated that his company would prefer nationalization.

As early as 1969, Yamani made clear that the Gulf countries planned to use participation as a holding device—a way of helping stabilize oil price levels while national oil companies around the Gulf gained training and experience. The politics of gradualism, he believed, would encourage development without creating crises in world oil. Participation to him, however, remained important because it represented the requisite first step in any meaningful change.[11]

Piercy later attempted to explain participation to the Church Committee, on February 1, 1974:

> Participating ceded them [producing nations] equity ownership, ceded them the right to the oil, and they made more money out of it. ... And the fact that the general point that we have lost our bargaining strength applies to the participation bargaining as well as posted prices.

In Piercy's view, the struggles over posted price (and taxes) at Tehran and Tripoli represented battles fought over issues entirely different from that of participation; "one did not precipitate the other." Yet at Tehran and Tripoli, and in later discussions, oil companies continued to sacrifice their control over the future; this loss of power itself made inevitable their defeat on the participation issue. As one writer put it: "From the very beginning of the [participation] negotiations it was obvious that the countries would get what they wanted."[12]

X X X

Formal talks on participation began on January 15, 1972. The OPEC Gulf ministers demanded 20 percent participation from the oil companies effective immediately, climbing to 51 percent equity ownership in all producing properties within a decade. Payment would be determined by net book value. The companies were not receptive to that proposal; and when no progress in the talks appeared likely, the meeting recessed. Informally, the other Gulf ministers agreed to let Yamani negotiate Saudi participation in Aramco in advance of any resumption of talks on the larger question.

Of course, these Aramco negotiations with Yamani held major significance for every oil company that operated in any of the OPEC states. The four U.S. companies in Aramco were ranked among the ten largest oil companies in the world, the Aramco concession was the largest in the world, and oil production there far outstripped the output anywhere else. Thus, any agreement on participation reached between Saudi Arabia and Aramco would likely provide the basis for similar negotiations with other nations. So the Yamani-Aramco talks came to have both a real and a symbolic meaning in the world oil business.

Twice in February, 1972, Yamani met with Frank Jungers, Aramco's president, and four other Aramco officials. But the parent companies did not authorize any movement on the basic participation issue. On February 15, 1972, Aramco did offer a counter-proposal to the Saudis: 50 percent participation in several proven oil properties, none of which had yet been developed. This, Yamani summarily rejected before leaving to attend another OPEC meeting on March 1. When he returned, Yamani asked the Aramco officials:

> Who are the so-called moderates? There is only Saudi Arabia, Kuwait, and a handful of small Gulf states. There is a worldwide trend toward nationalization and Saudis cannot stand against it alone. The industry should realize this and come to terms so that they can save as much as possible under the circumstances.[13]

It seems clear now that once King Faisal had decreed Saudi equity participation in Aramco, the question became one of whether the companies would simply yield or fight and risk nationalization. And Faisal did move. When at the Yamani-Jungers talks in February, 1972, Aramco proved unresponsive, the Saudis increased their pressure, and the U.S. companies appealed to Washington for help. In response, President Nixon wrote to King Faisal and urged moderation of Saudi demands. But his letter angered both the King and Minister Yamani, and, in turn, the King released to the press both the text of the Nixon letter and the royal reply dated February 18, 1972. Then, he sent to Aramco a strong personal note—one that soon led it to agree to 20 percent host participation on March 10.

During this time, both John McCloy and the U.S. State Department were kept informed. When James E. Akins, head of State's Fuels and Energy Division, telephoned McCloy, late in February, 1972, William E. Jackson took the call in McCloy's absence. The two men discussed "the hardening situation in Saudi Arabia ... which had led some of the companies to the view that State Department intervention and possibly other steps vis-a-vis the King would be essential to save the day." Akins reported that the U.S. Ambassador had already seen the King, and had been encouraged by the King's attitude. Yet the U.S. Ambassador almost certainly misunderstood King Faisal's position on the subject of participation; for on March 7, in a meeting with Aramco officials, the King bluntly told them that the Saudis wanted a "minimum of 20% participation" immediately. Before Yamani left for the OPEC meeting in Beirut on March 11, he also warned the Aramco executives that another resolution already prepared would be adopted there. It provided that, in case the companies balked, one or two companies would be selected and individually—in the Libyan style—forced to accept participation, or face an oil cutoff.

Put under this threat, on March 10, Aramco yielded. It's letter to Yamani read: "[I]n response to His Majesty's request, Aramco and its shareholders accept the principle enunciated by His Majesty of 20% [equity] participation in Aramco by the Saudi Government." In addition, the letter suggested that other details of the transaction should be negotiated; but the Aramco surrender seemed clear. Yet even this swift reversal of positions did not result in cancellation of the extraordinary Beirut session of OPEC. The seeds of change were sprouting.

In Beirut on March 11 and 12, observers noted that the Aramco letter "defused a tense situation," and helped to avert confrontation. But Aramco's appeasement was not the end. OPEC delegates agreed on a "toughly worded" resolution which stated that all oil companies operating in member countries should adopt a participation plan—or face concerted OPEC action, meaning a cutoff of oil supplies. Before adjourning, Yamani and other OPEC ministers made clear their timetable: they expected all details of participation to be settled in 1972.[14]

X   X   X

Back in Houston, meanwhile, the future seemed less clear. On May 18, 1972, at the Jersey shareholders meeting, Board Chairman Jamieson reported on the sharing agreement reached earlier in March. He went on to say that the company had no illusions about the difficulty of working out participation. Then, he added:

> The weight of power is at present with the producing countries. We believe that by agreeing to the principle of participation we have taken a long step toward meeting their aspirations. We now expect the governments to demonstrate their willingness to continue our mutually advantageous relationship.

Jamieson's optimism was not entirely shared by George Piercy, however. Earlier Piercy pointed out that "wide differences exist between the two groups [OPEC and the companies] as to the proper price for the asset [participation]"; yet Piercy expressed hope for "a mutually profitable settlement ... with OPEC countries." Taking up that theme in Pittsburgh, on May 25, responding to a question, Nicholas J. Campbell, Jr., Jersey senior vice president and director, pointed out that, when the producing countries achieved equity participation, they should not only pay a fair price for that share in the company; but they should "also have to pay a fair share of development costs for future production—in short, be full paying partners. ..." This statement presented the most complete definition of participation so far considered—but one that had not yet entered the discussions with OPEC. Piercy's skepticism would prove prophetic.

At the time, neither Western governments nor Japan responded dramatically to the events of 1972. The obvious power of OPEC, demonstrated in 1970 and after, largely served to stimulate dissension among the major consuming countries, so that no joint action became possible for several years. In 1972, the United States seemed uninterested in the oil scene; or, as National Security Advisor Henry Kissinger related in *Years of Upheaval*, national policy was guided by President "Nixon's desire to avoid a domestic blowup over the Middle East in an election year." Whether or not any concerted action by consumers would have arrested the rise of OPEC's power over the consuming nations remains today a moot question. Certainly this volume will not provide the answer.

On June 12-15, 1972, the oil companies of the European Organization for Economic Cooperation and Development [OECD] met in Paris to discuss an emergency plan for oil apportionment, and the United States, Canada, Australia, and Japan received invitations to join in the "scheme." Gatherings of a "General Working Group" preceded sessions of the "High Level Group meeting (government representatives only)," to which the Office of Oil and Gas of the U.S. Department of the Interior sent delegates.

What happened in Paris that June illustrates the fragile nature of the Western Alliance when it confronted a vital question—which the assurance of oil supplies surely represented. All delegations present wanted to maintain a common front, but quietly, so as not to offend OPEC and thus endanger supplies. Delegates made clear that they wanted the governments to downplay the oil crisis, and to maintain an illusion of "a quiet political and economic atmosphere" in which OPEC would benefit from continuing investment. Europeans accused the U.S. delegates of doing a disservice by spreading "doom and gloom" forecasts, which later proved to be correct. The delegations from France and Holland said that Europe had been concerned about its increasing dependence on oil imports for many years, and now that U.S. domestic supplies had dwindled, it had become a "doomsayer." France said it had "learned to keep its cool," while the Dutch "indicated that Europe had learned to 'walk on the razor's edge.'" The OECD secretary termed this exchange "the frankest ... ever to take place between member delegations." Finally, after listening carefully, the United States responded to the OECD invitation to joint action by saying that an answer would be forthcoming at some later date.

In all, five Jersey executives* served on the general OECD Petroleum Advisory Committee. To them, despite speeches at the meeting by Secretary Rogers C.B. Morton and other Americans, nothing that took place in Paris could have seemed reassuring. For what appeared obvious was that OPEC held the real power in oil and that no present plan for rationing or allocation would soon change that reality.[15]

X   X   X

Even while Yamani talked with Piercy and other representatives of the Aramco parent companies, the Iraq Petroleum Company, Limited [IPC] began to feel special pressures. Late in May, 1972, the Iraq government issued an ultimatum to the company: Increase oil exports or face legislative

---

*They were: C.O. Peyton, F.A. Holloway, Stephen Stamas, J.W. Hanson, and J.B. Meredith.

action. Already, because of political uncertainties, IPC had allowed production to decline in the northern fields. So pipelines to Banias (Syria), and Tripoli (Lebanon) operated at less than capacity. Now, the Iraqi government demanded full production, 20 percent participation in the company's operation, and a guaranteed annual production increase of 10 percent. One Iraqi official even vowed to bring IPC "down on its knees" unless it agreed to government demands, while the Arab Oil Congress expressed full support for Iraq. During this confrontation process, the Iraqi Oil Minister, Dr. Saadoun Hamadi, sent to IPC yet another ultimatum: Either boost production and let the government market any surplus oil, or abandon the northern fields.*

Hamadi's threats still failed to produce a satisfactory solution, so Iraq nationalized the remaining IPC and related concession areas except those of the Basrah Petroleum Company, Ltd. [BPC]. These consisted almost wholly of producing properties.** Syria soon followed the Iraqi action by expropriating the IPC pipeline through its territory. In February, 1973, IPC and Iraq negotiated a settlement for all expropriated properties. The IPC companies received compensation for their property and at the time gave up their remaining MPC properties due to the inability of the reservoir to produce the minimum production called for under the terms of the concession.

The failure of Western governments to act decisively reaffirmed the view of producing states that the West's (and Japan's) need for oil overrode all other considerations.*** Naturally, Iraq's overall success added momentum to the trend towards greater participation in—and possible nationalization of—foreign properties.

Yet a real question remained unanswered: To what extent would OPEC back these actions in Iraq and elsewhere? At the time, both Iran and Saudi Arabia had strong reasons for not supporting Iraq, and no member seemed greatly concerned with Libya. Still, everyone interested realized that any overt failure to endorse Iraq might well destroy—and certainly must seriously damage—the power of OPEC. Since both Iran and Saudi Arabia were strongly anti-Communist, they had no cause to favor the revolutionary

---

*Interestingly, by this time, one of Iraq's earlier demands had been dropped: the requirement that nationals receive training at IPC expense. The company already had made a genuine effort to train Iraqis as operators. By May, 1972, there were only fifty foreigners in Iraq working for IPC.

**In December, 1961, Iraq had nationalized 99.5 percent of the total concession area held by IPC and its associated company, the Mosul Petroleum Company, Ltd. [MPC].

***For a time, however, many potential customers refused to buy the expropriated oil because they feared expensive legal action.

government of Iraq, especially after two Communist members had joined its ruling council. Already, Iran and Iraq had for many years engaged in bitter territorial disputes, which would ultimately develop into a bloody war. Some Western observers—perhaps more full of hope than of reason—saw the Iraqi action as damaging to OPEC because it would result in dissension that could only undermine OPEC's role in future bargaining. A more pragmatic observer pointed out, however, that the only certain winner in the oil confrontation would be the Soviet Union; and that the one certain loser must be the Western consumer.[16]

When news of the Iraqi expropriation of IPC reached the Eighth Annual Arab-Petroleum Congress meeting in Algeria, on June 1, 1972, the delegates seemed delighted. As one outspoken Nationalist, Sheik Abdullah Tariki of Beirut, told a reporter: "The present [short] world crude supply has made the time ripe for takeover." Talk of participation and nationalization was on everyone's tongue, and the speech delivered by the U.S. Department of State's James Akins did nothing to dampen Arab support for Iraq. Akins claimed that the Iraqi actions "needn't necessarily be an unmitigated disaster," and he went on to predict optimistically that they would "galvanize Washington and other countries to seek other sources of energy." Another colorful statement attributed to Akins reminded his listeners: "The old concessions weren't written by divine will on tablets of stone, and I am not saying they shouldn't or couldn't be changed." Then he added:

> The companies themselves have been remarkably flexible ... [they] have yielded under the sledge-hammer blows of the producing countries. All the changes have been in the favor of the producing countries. I don't think you have any reason to be ashamed of your success so far. I'm just suggesting that it might continue to be of benefit to the producing governments to continue to bend, not to break, this steel rod [i.e. the companies].

Akins went on to assure this Arab audience that higher prices for their oil were appropriate: "on this question the consumers have had their heads in the sands, like a collection of especially stupid ostriches. ..." The Soviet Union approved the Iraqi action; and it signed a trade and economic pact with Iraq on June 6, 1972.

At the time of the Iraqi action, Jersey's share of IPC production amounted to 11 7/8 percent—around 125,000 b/d. Six weeks after the final nationalization of its northern fields, IPC indicated that, for the next three months, a moratorium on further legal action regarding the purchase of its oil would be in effect.* This delay would provide time for mediation concerning terms of settlement. While some sources indicate that IPC took this action because of pressure from Compagnie Française des Pétroles [CFP], analysts of U.S. foreign policy assert that the U.S. Department of State also influenced the decision.[17]

More important, on June 28, the Shah pulled Iran out of the participation talks, and indicated that his country had no further interest in them. Symbolic support for participation had shattered, as this Iranian decision to withdraw caused OPEC "concern and embarrassment." In all, the Iranian action marked a turn in the Middle East story. It will be discussed in detail later in this chapter.[18]

During mid-July, company officials and spokesmen for Western governments received at least some encouragement when the Libyan government announced that Muammer al-Qadhdhafi [Qaddafi] had stepped aside. After a power struggle in the Revolutionary Command Council [RCC], Major Abdesselam Jallud became prime minister in his own right, and he issued reports suggesting that Libya might slow its drive to nationalize foreign oil companies. Now Western optimists claimed that "Jalloud ... is considered a pragmatist in contrast to his more idealistic and impulsive colleague Qaddafi." But Piercy and his Jersey colleagues were not so easily swayed by the hopeful statements of Middle East specialists at State and elsewhere. Already Jersey had dealt with Jallud; the company knew better. And despite the shift in offices, no change occurred in Libyan relations with other nations.[19]

X   X   X

Yamani continued to enjoy the confidence of the Saudi government, so he naturally represented the key to an agreement on participation. The King had become convinced that it was necessary for the U.S.-owned companies in Aramco to yield on the participation issue, as a means of helping Saudi Arabia demonstrate to other Arab countries that it was truly a sovereign nation. Earlier, Algeria and Libya had leveled charges of American domination that rankled the Saudi monarch. Only participation would allow him to save face and to keep the leadership in OPEC, which his oil had given him. In all fairness, it must be said that King Faisal made every effort to inform the Nixon administration about the royal views. Any American politician— indeed, any interested observer—should have been able to understand the

---

*The moratorium was later extended.

special plight of the Saudis. King Faisal did respect and depend on the United States, but also he could not fail to demonstrate his commitment to the Arab cause. Yamani, as King Faisal's principal spokesman to the West, had to carry through that message.

So while other Aramco representatives talked about increasing productive capacity, Piercy discussed participation, and especially, specific terms for agreement with Yamani. But their quiet conversations were temporarily interrupted by dramatic events—first by Iraq with its June, 1972 nationalization of the IPC properties, and then by the Shah's withdrawing Iran from further participation talks. Finally, on July 14, the Saudi Royal Cabinet became impatient. It issued a statement that threatened "some form of government takeover" if Yamani's talks did not prove successful. The message was blunt.

In August, Saud al Faisal, one of King Faisal's sons and a deputy to Yamani, travelled to Washington to tell the Nixon administration that the companies—not the consumers—would suffer if no agreement was reached. Kissinger recalls that at the request of the oil companies, he met with the Prince, who asked for "some indication ... what terms the United States considered fair" with respect to compensating them for the participation share. When he reported this interview to the oil companies, Kissinger contends that they objected to his suggesting anything of substance to the Saudis. Among the reasons for this company attitude, Kissinger cites the American oil industry's dissatisfaction with the State Department mission in 1971, led by John Irwin II. Aside from Kissinger, other sources indicate that Saud al Faisal reiterated to State officials that Saudi Arabia planned to order a production cut unless the companies took some early action on the participation issue.

Late in September, 1972, Sheik Yamani came to the U.S. to resume formal negotiations on the terms of participation with Piercy and the oil companies' team. Before arriving in New York for this session, Yamani stopped in Washington, D.C., where he addressed the Middle East Institute in Georgetown on September 28. After estimating that Saudi Arabian oil production would reach 20 million b/d by 1980, he predicted that the national oil companies in the producing states would soon be performing "exactly the same role played at present by the international companies." His remarks indicated clearly that he expected producing countries to have an increased involvement in the oil business.

In New York, before entering discussions with Piercy, Yamani gave an interview on September 29, in which he made a proposal accurately termed "bold and imaginative" by one oil executive. In fact, Yamani virtually offered the U.S. an oil partnership with Saudi Arabia. He said that, in view of projected oil imports of 8 million b/d by the U.S. in 1980, Saudi Arabia was prepared to deal directly with the U.S. on a most favored nation basis, and to satisfy American requirements. In return, Yamani asked that the oil be exempted from the levies of the U.S. quota system (which still governed imports) and that the U.S. give the Saudis "privileged status" for economic investments. At no time in the interview, by word or tone, did Yamani indicate that the Saudis might use their oil as a political weapon, to influence the U.S. to take an active role in the troubled Arab-Israeli negotiations. King Faisal, Yamani said, in fact had "ruled out any such policy" of political pressure. Later Yamani delivered to President Nixon a formal letter from the King that contained the essentials of the Saudi proposal.

The Nixon administration found procrastination easier than formulation of a definite reply. First, it indicated to Yamani that the President's energy policy speech—not yet drafted—would have to precede any formal answer. Finally, not until February, 1973, did Nixon send his reply to the September, 1972 letter from King Faisal. Tardiness, in the Saudi view, indicated a lack of any American concern as well as simple bad manners. At a lower diplomatic level, however, a significant "reply" had been offered by Akins of State, who immediately welcomed Yamani's speech and termed the Saudi proposal "important." President Nixon and many of his principal aides were engaged in Watergate cover-up attempts. Everywhere, the nation suffered but in no place so grievously as the Middle East. Akins' reply, though courteous, should never have been assumed to reflect any views but his own.[20]

Participation talks with Piercy and the companies' team began in New York on October 1, 1972. After intensive bargaining sessions, the parties reached a general agreement on October 5, 1972—a pact signed by twenty-three companies on one side, and by five Persian Gulf states on the other. For Yamani, this settlement represented a personal triumph; he remarked: "I believe that participation will be proven to be the instrument available in the oil trade that will provide prosperity and stability for posterity." His prediction guaranteed the Saudi minister no greater reputation as a prophet than those predictions earlier made by the many company spokesmen, who insisted on viewing the historical record of the Middle East as grounds for optimism.

Yet there was agreement: Under its terms, each government would immediately acquire a 25 percent equity in each producing oil company concession within its domain. This percentage share would increase by steps to reach 51 percent by 1983, where it would remain constant until the concession agreement expired. Instead of net book value, which OPEC had wanted as a basis for payment, Piercy and the company negotiators persuaded the Gulf states to accept a different plan.* Piercy once remarked that it represented more of an adjustment than did book value. The latter was original cost less depreciation; taxes were about 50 percent. In the new agreement, the companies were paid by the government one-half of depreciation, and the updated book value was calculated by adding back the other one-half. Piercy said:

> The concept that was proposed, updated book [value], was [that] the companies would receive the current value of their investment less any portion previously absorbed by the government granting them depreciation.

On the point of value of the property, neither OPEC nor the companies obtained exactly what they sought. The final agreement also included procedures by which the companies could buy back the Gulf countries' share of oil.

On October 19, 1972, Sheik Yamani asked George Piercy to meet with him in Riyadh to go over the terms of agreement. The Kuwait legislature had objected to some of the provisions, and called for a complete review of the agreement. Next, Iraq balked, as that country claimed a need to study the effect of the new provisions on its already nationalized IPC property. In this situation, even though it appeared that Saudi Arabia, Qatar, and Abu Dhabi were ready to sign, the oil ministers scheduled an OPEC meeting for October 25. In Riyadh, several other members of the companies' negotiating team joined Piercy, and they began working out details of the general New York agreement—but without being certain that it would be signed. "Conceivably, it could still come unglued," said one oil executive who knew that the Persian Gulf ministers planned to assess again the final pact in advance of the full OPEC meeting. Now to add to the uncertainty, another factor emerged during these negotiations—one that had remained dormant for some years, but which assumed real significance after 1972: the fluctuating quantity of crude oil—or as the companies preferred to call it, "the quest for assured supply," upon which every company entered. In fact, so great was the demand for the producing countries' shares of crude oil that price offers for it soared. This competition superseded earlier plans of the companies to purchase most of this buyback oil for their downstream requirements. That

---

*Kuwait never signed the Participation Agreement, so it demanded and obtained net book value for its properties. The other countries then demanded the same. Iraq also balked at signing the Participation Agreement because it feared it would have to pay updated book value for the IPC properties it had nationalized on June 1.

assumption had collided head-on with demand from outsiders. So market disruption was in the wind. One newspaper reported: "The Arab states are being swamped with direct offers from elsewhere for the oil." This market pressure whetted the desire of OPEC members for greater participation in the companies—and for more participation oil to capitalize on the rising prices. In early 1973, Sheik Mana Saeed al-Otaiba of Abu Dhabi told reporters that, at then-current prices, his country would receive over $200 million additional profit from the Participation Agreement, over the next three years.[21]

At the conclusion of the OPEC meeting in Riyadh (October 25-27, 1972), the *four* Gulf states planning to sign the "General Agreement on Participation," as the Riyadh agreement was termed, received congratulations from all OPEC members except Libya and Algeria. In the uncertain oil markets of late 1972 and early 1973, however, not all of the provisions in this General Agreement were implemented, as the trend toward nationalization accelerated. Yet the General Agreement did affect prices—increasing the cost of Gulf oil by nine cents per barrel. Of course the terms of the Participation Agreement also rendered the Tehran Agreement an almost forgotten casualty of the times. It no longer mattered.

X   X   X

Elsewhere, Iran also continued to play a leading role in Middle Eastern oil, as the Shah arranged meetings with the negotiating team representing the fourteen companies in Iranian Oil Participants, Limited [IOP].* Since the Shah spent most of each February skiing at St. Moritz, the consortium arranged to meet him at his Swiss villa, on Monday, February 14, 1972. The IOP negotiating team was headed by Jersey's Jamieson, and David Steel (vice chairman of BP) was selected as vice chairman.**

---

*The original companies in the consortium were British Petroleum Company, Ltd., 40 percent; Royal Dutch-Shell Group, 14 percent; Compagnie Française des Pétroles, 6 percent; Standard Oil Company (New Jersey), 8 percent; Standard Oil Company of California, 8 percent; Gulf Oil Corporation, 8 percent; Texaco Inc., 8 percent; and Mobil Oil Corporation, 8 percent. In 1955, each of the American companies assigned an eighth of its holdings to the Iricon Agency, which gave Iricon 5 percent and reduced the share of each of the large U.S. companies to 7 percent. Iricon Agency was composed of eight independent American oil companies: Richfield Oil, Signal Oil and Gas, Tidewater, Getty Oil, American Independent Oil, San Jacinto Petroleum, Standard Oil (Ohio), and Atlantic Refining.

**The logical choice to head the team to negotiate with the Shah was Sir Eric Drake, chairman of BP; however, he had become *persona non grata* to Iran because of his years with Anglo-Iranian Oil Company, Limited [AIOC]. Sir David Barran of the Royal Dutch-Shell Group stepped aside for personal reasons. Since the team wanted a chief executive officer to lead them, they chose Jamieson.

Jamieson had never met the Shah, and on February 14, the Shah's prompt-
ness impressed him: Unlike "most of those characters ... he wouldn't keep
you sitting around for an hour." Jamieson recalls that the Shah,

> took us into a kind of study, and it had an L-shaped table along the
> wall where we sat down and the Shah sat here [pointing to one end]
> and I sat here [pointing to the opposite end]. It's what I call eyeball
> to eyeball negotiating ... he had only one man with him but I had
> about four.

Smiling, Jamieson described the scene in detail:

> He [the Shah] had on what I describe as kind of a velveteen uni-
> form that was black and kind of velvet-buttoned, with a high col-
> lar and brocade down the front. Well, we got to going at it pretty
> strong—negotiating [the general terms of a new IOP contract]. It
> was a real rough-rough session, and he was just as cold as they come
> ... he was a little paunchy at that time—and he'd make some
> profound statement and lean back in his chair; and this suit was
> fastened with dome fasteners and one of the damn fasteners would
> fly open.

Humorous as it seems in retrospect, Jamieson did not dare to laugh as the
Shah "kind of fumbled around" and re-fastened the errant snap. Then, for
fifteen minutes or so, the Shah and the IOP negotiators would go at each
other again. At last the Shah would sound-off, lean back, and the same
fastener would give way. Jamieson quickly saw that the fasten-unfasten
routine embarrassed His Majesty. Nevertheless, the discussions continued
for two hours, leaving Jamieson with an unforgettable first impression of the
Shah.

Before the IOP team's next meeting with the Shah, scheduled for Wednesday, February 16, BP's Steel breakfasted with Jamieson and appeared quite
upset. He said, "Ken, I am afraid we've had it. I really think we've had it in
Iran." Jamieson agreed that might be the case, but he added, "At least we've
got another session with the man. I think I have a little bit of bargaining
advantage because of that damned dome fastener of his." Certainly it
relieved some of their tension.

When the team met with the Shah on Wednesday afternoon, they saw
that His Majesty had tumbled down a ski slope and his swollen wrist looked
very painful. Jamieson detected another difference in the Shah: "He was a
completely changed personality. He was friendly and affable. So, obviously
what I had been going through was a testing period. .... From then on we got

along very well together." Apparently Jamieson's self-control during the initial negotiating session had convinced the Shah that the Jersey executive was a formidable opponent. Certainly, human relations had improved, and negotiations benefitted. For during the next two years, Jamieson led IOP negotiators in at least twelve separate meetings with the Shah. A regular pattern emerged, as the men got to know each other:

> We'd kind of agree on the broad principles with the Shah. Then we had negotiators on both sides try and work out the details ... every once in a while they'd get stalemated over something ... and I'd have to go see the Shah to see if we could agree on something to give these fellows guidance. ...

The Jersey CEO developed great respect for Dr. Reza Fallah,* who did much detail work for the Shah.

Reviewing his extensive experience with Iran and the Shah, Jamieson insisted on playing down his own role; however, when pressed he did relate a conversation he had around 1975 with Sir Peter Ramsbotham. Ramsbotham, who had been British Ambassador to Iran and, later, to the United States, insisted that Jamieson represented the best oil industry negotiator ever to deal with the Shah of Iran—a great compliment indeed.[22]

Over time, talks between the Shah and the IOP team provided a new basis of understanding; in fact, although press reports indicated an Iranian presence at the participation talks on June 24, 1972, it soon became obvious that the Shah had set a different course for his country. Now, the Shah clearly believed that he had erred in permitting his country to be a party to the Tehran Agreement, and that he lost part of his bargaining edge by joining with the other Persian Gulf states. The United States, by courting Iran with aid programs and military equipment, made it obvious that the Shah enjoyed some advantages over other Gulf states. So it is conceivable that the Shah believed that, if he broke away from his Arab neighbors—played his hand properly—and entered individual negotiations, he and his country stood to benefit.

On June 28, the Shah announced that Iran would withdraw from the talks and work out its own agreement with IOP. At the time, Dr. Jamshid Amouzegar told delegates at the regular OPEC conference that, by its independent action, Iran did not intend to disrupt the ongoing participation

---

*Earlier, Reza Fallah had served as deputy managing director of the National Iranian Oil Company [NIOC].

talks. More important and equally independent, Amouzegar pledged that Iran would honor the general 1954 Consortium Agreement until its expiration in 1994.* Already, the consortium companies [IOP] had agreed to increase Iranian production and to furnish the National Iranian Oil Company with more crude oil, which it was free to market. While the Shah's withdrawal angered other OPEC ministers, oil company leaders proclaimed his action realistic. They especially liked that portion of his statement which noted that, when the Tehran Agreement expired, Iran would agree to tax reference prices adjusted to an "inflation index based on an 'international basket for prices.' "[23]

For the member companies in IOP, Iran's independent actions—while welcome—created new problems that required additional consultation. Eventually, the topics discussed included: turning over the Abadan refinery to Iran; completing additional construction at Kharg Island to make it one of the world's largest terminals; increasing by 12 to 20 percent the share of Iranian oil production; constructing deep water ports for supertankers; assisting Iran in setting up a marketing network for its oil; helping to finance these and other improvements and additional construction. In all, the consortium proposed to increase Iranian production by one million b/d in 1972, and by an additional 10 percent per year afterwards, until it reached eight million b/d in 1980. This entire agreement, the Shah referred to as a company-state partnership. For the oil companies, it spelled out the price of Iranian participation—a price not without its uncertainties.

As a first step, extending the oil agreement to 1979 required some two months of work, during which new terms satisfactory to both Iran and IOP were reached. The results, in fact, provided Western oil executives a rare feeling of successful achievement, since the participation talks with all five of the other Persian Gulf countries appeared to be stalled. As one company vice president remarked, "Iran has seen sooner than the other governments the need not to discourage private investment." To him and to many others, it seemed foolish for national oil companies to spend their own revenues on "ventures [that] private capital is willing to undertake for them."[24] Iran seemed reconciled to this view.

Yet the need for oil diplomacy and the vital roles of Jamieson and others did not end when the Shah tentatively agreed on proposals. Instead, each month brought the companies some new problems with Iran, as various

---

*The original term of the Iranian Oil Participants Agreement was from 1954 to 1979; however, a clause permitted renewal for three terms of five years each. This accounts for the date "1994" and for the confusion attendant using both that date and "1979." Both as used are correct.

experts sought to work out specialized details. But for six months, negotiations accomplished little, even though Jamieson and his team flew to Iran for sessions with the Shah at least once during each of the last four months of 1972. Then in January, 1973, after a mid-month meeting, the Shah declared his intentions in a major speech: When the Consortium Agreement ended in 1979, he would not propose to extend it to 1994, despite his earlier promises to do so. After ridiculing the IOP Consortium for alleged mistakes, the Shah announced new terms under which it must operate in the future: IOP should increase production from the 1972 figure of slightly more than five million b/d to eight million b/d, and do so quickly; otherwise, IOP must leave Iran when the contract expired. Finally, the Shah also warned his listeners that Iran might assume control of all oil production, leaving the companies to become merely customers.

Most IOP principals believed that the Shah used his speech simply to justify his refusal to renew the agreement in 1979, since it would take many months to bring new fields into production and to install recovery equipment. IOP also knew that the requested increased production of three million b/d would require at least $1 billion capital outlay—while satisfactory guarantees of new production sufficient to amortize that figure simply could not be had, even from the Shah. Adding everything together, it became clear to IOP executives that, while Iran expected IOP to supply the capital needed for expanding the entire oil production system, it would not or could not furnish responsible guarantees to insure recovery of their financial investment.

Once again, in February, 1973, Jamieson and his IOP team went to St. Moritz, where on the 22nd and 24th they resumed discussions with the Shah. A "general understanding" was reached, as the press quoted several conferees who confirmed that only details remained to be worked out before Iran and the companies could claim full agreement.

A few days later, parts of that agreement became known. Essentially, it provided that all of the IOP companies would become Iran's "long-term customers," beginning in 1974. In the future, Iran would make all decisions relating to volumes of production, refining, distribution, and investment. An earlier difficulty had been created because IOP wanted to extend the 1954 agreement (in amended form); while the Shah consistently sought to replace that old agreement with a new one. Now, it appeared that the companies had yielded, although, as the Shah remarked, the consortium would receive oil "at a fair price with a discount that anyone grants to its good customers."[25]

Confirmation of a new twenty-year agreement (completed on March 5) came on May 24, when news stories indicated that Iran and IOP members were prepared to sign it. One source also predicted that Iran would "receive revenues not less than those which accrue to other Persian Gulf producing countries," and this report also stated that the new arrangement would both insure a continued flow of Iran's oil to markets and, by agreed-upon steps, give Iran full control of its oil. Nevertheless, the Shah's deputy, Reza Fallah, who headed the NIOC team, soon became a target for sharp criticism from other Gulf ministers and from some members of his own group, "for granting the companies unnecessary concessions over prices, compensation and investment." Even so, by late July (1973), the IOP-Iranian contract had received all necessary signatures. Now, the Shah considered that the new agreement had vested 100 percent control in Iran, and under its terms, the National Iranian Oil Company received 200,000 b/d of crude oil for sale—which amount was to be increased by steps to 1.5 million b/d by 1981.[26] The agreement between Iran, the National Iranian Oil Company, and each member company of Iranian Oil Participants, Limited was structured differently from any of the participation agreements, although the net financial result could not have been too dissimilar. The 1973 agreement provided that Iran would sell the oil and the consortium companies would purchase all of Iran's oil, except the amount needed for domestic consumption and some that NIOC needed to fulfill export contracts. But changes occurred so rapidly in the international oil system that by 1975, Iran sought to begin conversations leading to modification of the 1973 agreement.

Jamieson and his IOP team had agreed that member companies could buy the remainder of Iran's more than three million b/d production [1973], while they would continue to provide operating and technical services for the Iranian national company. The agreement also guaranteed that the price which Iran received for its oil would not fall below that of comparable Persian Gulf oil, while the charges for technical services would not exceed those for comparable assistance elsewhere. The terms were comprehensive—so much so that the Shah could later claim that his effort to have NIOC obtain full control of Iran's oil had consumed fully fifteen years of his time (1958-1973). As he concluded, "Finally I won out—72 years of foreign control of the operations of our oil industry was ended on July 31, 1973."[27] Even as the completeness of his victory remained in question, events thereafter overwhelmed these terms before either the Shah or Iran could realize the full benefit of this drastic change.

X   X   X

Another crack appeared in what little remained of the agreements of 1970 and 1971, as Piercy negotiated with Yamani on the price of buyback oil. Piercy said:

> Yamani has made the decision to break the agreement made in New York on prices. Individual country negotiations, firming markets, various attacks on him have all contributed. But perhaps it is all part of a plan. He has decided to grab for big price increases.[28]

Piercy sensed that things had changed completely; so he urged his group to make a "firm" and final offer on buyback oil. But the Aramco owners would not accept the Exxon plan.* Simply, they needed Saudi oil and continued access to it took priority over price. As a result, the final agreements reached with Saudi Arabia and Abu Dhabi (both signed the December 20 Participation Agreements) moved forward the date on which the host countries would acquire 51 percent participation from 1983 to 1982. Also, and more important, for the first four years of the agreement, the companies agreed to buy back the bulk of the governments' equity oil at the market price or at a small discount. Oil at any price had won the battle for participation, but oil companies would learn to regret their lack of restraint and their failure to recognize the wisdom of Piercy's views.

X   X   X

As usual, the Libyans watched events in Saudi Arabia with great interest. When on October 5, 1972, Piercy and Yamani agreed on preliminary terms for Saudi participation in Aramco, Libya quickly called in representatives of the Hunt Oil Company and demanded immediate 50 percent participation. To other companies Libya indicated that 50 percent participation would be the standard of sharing participation and that shares would be paid at net book value only. Hunt Oil refused to comply; then, Libyan Prime Minister Jallud issued an ultimatum giving Hunt until November 23 to agree. Hunt, in turn, asked other Libyan oil companies to join with it in extending once again the Libyan Producers Agreement—the safety net—until 1975. This, the companies agreed to do; so John J. McCloy drew up a supplement to the Producers Agreement and secured a Business Review Letter covering their cooperation. The companies signed it just two days before the Libyan ultimatum expired.[29]

When Hunt simply continued to resist Libyan pressures, that

---

*Standard Oil Company (New Jersey) became the Exxon Corporation on November 1, 1972.

government, in retaliation, refused to allow Hunt to export any oil—leaving Hunt to institute arbitration proceedings on December 20, 1972. Now, sensing Hunt's toughness, Libya pushed hard against the smaller independent U.S. companies, while considering new actions against Hunt. Finally, on June 11, 1973—the third anniversary of the expulsion of the U.S. Air Force from Wheelus Air Base—Libya made the grand gesture of expropriating Hunt. Qaddafi was pictured with Idi Amin Dada, the Ugandan dictator, and Egypt's Anwar Sadat, as he told the world that he had given the United States "a big hard blow . . . on its cold insolent face." To his Libyan followers, Qaddafi proudly exclaimed:

> The nationalization of the American company is only a warning to the oil companies to respond to the demands of the Libyan Arab Republic (and) . . . also a warning to the United States to end its recklessness and hostility to the Arab nation.[30]

In fact, the Hunt seizure represented the first actual expropriation of a U.S. producing company by an Arab nation. The U.S. Department of State issued a strong protest, claiming that Libya had violated established principles of international law. When Qaddafi ignored this protest, the U.S. government took no further action. In the *New York Times*, a story about the expropriation stated that the united company front remained firm; and that Libya would have difficulty in selling its oil directly because Europe needed Libyan oil less than it had two years earlier. Incredibly, the *Times* managed to ignore the larger climate of spot shortages and virtual panic among consumers. But the *Times* story did point out correctly that Libya continued to menace three additional oil companies. Mr. Bunker Hunt reacted predictably: He threatened to "take court action against anybody dealing in or with oil extracted from the Sarir field." But again, world demand for oil swept aside Hunt's protest and ignored his attorney's legal actions. Instead of being intimidated, other companies flocked to Libya to buy one-time Hunt-British Petroleum oil. Greed triumphed over State principles. Qaddafi appeared to be riding the wave of the future.

On August 11, 1973, Libya nationalized 51 percent of Occidental Petroleum Company [Occidental] and established a three-man board to supervise its operations, of which two members were Libyans. Other independent companies also signed 51 percent nationalization agreements "on terms similar to but in some respects more onerous" than those accorded Occidental. Then on August 21, Esso Libya, Mobil, Socal, Texaco, and Shell executives received instructions to come to Abdesselam Jallud's office, where he told them that Libya was nationalizing 51 percent of each company's holdings.

While some of the company representatives refused to accept this procedure, the earlier actions of the independents in acquiescing to the Libyan nationalization seemed to undercut the opposition of the major companies. Then too, they feared a domino effect—especially that the Libyan action would spread to the Persian Gulf states. So for ten days they talked of resisting, while they sought in vain for effective support from the governments of the U.S., Great Britain or Holland; finally on September 1, several oil companies acceded under protest to Libyan demands because no other course seemed open to them. *

What should be noted here, as it was seven months after that September day when Hunt Vice President Schuler testified before the U.S. Senate Subcommittee on Multinational Corporations of the Committee on Foreign Relations, is this: The collapse of concession agreements in Libya did not result from actions by the major oil companies. Rather, as Schuler claimed:

> I think I must say in fairness that the major oil companies have been willing to live up to their obligation[s] far more readily than the independent oil companies. ... The major oil companies until they were 51 percent nationalized in Libya last fall, continued to meet their obligation to us [under the Producers Agreement] and we had no complaints.

Schuler mentioned Exxon, BP, Socal, and Texaco, among the majors, as companies that had lived up to the terms of the Libyan Producers Agreement and thus helped Hunt to survive.

Libya required that "all employees of the companies regardless of nationality" continue working—involuntary servitude. This meant that employees could not resign their positions without approval from the new management committees. Nationalization was effected by decree—despite protests, arbitration proceedings, and threats of litigation by the companies—and the Libyans also moved to make things rougher on the companies by alternately cutting and then raising allowable production levels. All Libyan actions were accompanied by intemperate threats and general harassment of the companies.[31]

For Western governments, the adage about history teaching lessons seemed to be numbered among the many casualties of Libyan action, for the U.S. and other powers took no meaningful actions. They neglected to protest the Libyans' open contempt for judicial process, and for international law;

---

*Exxon and Mobil fought their case with the Libyan government until April, 1974. Then, they accepted 51 percent nationalization.

and they failed to prevent physical abuse of their citizens, just as, forty years earlier, they had failed to protect their nationals against German actions. In retrospect, it now seems that the world's growing appetite for oil simply created Libya's wealth and influence. Unheard by their governments, a few realists, such as U.S. Senator Hubert H. Humphrey, pointed out that, by continually allowing its "giant industrial concerns" to yield to economic blackmail, the United States was permitting "the more militant Arab states to gain virtual control over the Middle East and the Persian Gulf." But Humphrey's voice was lost in the cries of the marketplace—the search for security of supply.

X   X   X

Before the signed Participation Agreements between the companies and the Persian Gulf countries could be implemented, world attention to the bothersome Middle East had already shifted; first to Iran, then to Libya. Yamani, always the center of concern, joined eight Aramco executives in a meeting that included Exxon's Aramco Director C.J. Hedlund, on May 23, 1973.* In Geneva, the group reviewed the January agreements before Yamani declared that, although they might agree on implementation, he would sign nothing because of his "fear of critics using this against him with respect to what may transpire [in] Iran and Libya and possibly Nigeria." While he delayed, Saudi Arabia would tacitly agree to abide by any new pact that was signed, Yamani added. At this point, as if to assert its sovereignty for effect, Kuwait threatened to nationalize 51 percent of the Gulf Oil Corporation-British Petroleum concession in its territory. Then in Libya, as Qaddafi confiscated more oil company properties, he continued to undermine every possibility of real success in these Geneva negotiations. While all this evidence mounted around him, Sheik Yamani—astute though he was—must have raised his estimate of the place that oil held in the pantheon of Western values. Not surprisingly, the Saudis must share some responsibility for the absence of real progress in the Geneva talks; for conversations simply stalled before the discussions found answers to the twin questions of allowable increases in production levels, and of specific prices to be paid for participation oil. On this point, no better understanding emerged and no progress occurred.

Here, the Nixon administration provided little help to the companies. The American public, and with them the communications media, had—in the

---

*Hedlund also served as vice president of Exxon Corporation and president of Esso Middle East.

absence of leadership from the White House—become near hysterical on the subject of energy (to the public, energy translated into a single word: "gasoline"). Screaming headlines led into stories based on few facts but filled with many predictions. That combination sold papers; yet it provided little information. M.A. Adelman, a veteran petroleum economist at the Massachusetts Institute of Technology [MIT], did little to explain the problems of the companies in his essay, "Is The Oil Shortage Real? ...," which answered in the negative, while contending that the oil companies had become the tax-collecting agents of foreign powers. *

Yet oil company profits rose slightly, as sales volumes continued to increase and shadow markets in products began to develop. Earlier, in August, 1971, President Nixon had frozen oil prices. Now, on February 27, 1973, John Ehrlichman, one of the President's closest advisors, announced that the administration had no plans to suspend quotas on imported oil. Five weeks later, on April 12, Nixon ended all such controls. Small wonder the American people were confused.

Senators and congressmen, governors and legislators—all under intense pressure from that public—began to investigate, and predictably they soon found a convenient scapegoat—the multinational oil companies. On the national level, both the Senate Foreign Relations Committee and the House Foreign Affairs Committee held special hearings. Politics replaced negotiation. Foreign Relations Committee Chairman J. William Fulbright announced:

> The Arab oil producers are militarily insignificant—gazelles ... in a world of lions. ... They should take account of the pressures and temptations to which the powerful industrial nations would be subjected if their economies should be threatened by severe and protracted energy crises. [32]

In the marketplace, however, Fulbright's words changed nothing, as the ripple effect from spot gasoline and heating oil shortages began to threaten the costs of heating and transportation.

In April, 1973, Secretary of the Treasury George P. Shultz, several high-level government officials, and a delegation of senators and congressmen flew to Bermuda for an off-the-record conference with eighty leading executives from energy companies. Shultz was expected to provide insight into the

---

*Adelman also asserted that the failure of the U.S. to act positively to support the oil companies during the 1971 Tehran negotiations had given the Arabs a political victory and a false sense of power.

forthcoming "President's long-awaited message on energy policy." Instead, the energy people were astonished to find that Shultz had come not to teach but to "learn." Already, Nixon's first energy message had provided "only a Band-Aid for the problem," in the words of one specialist. Now, the public and the industry expected something more, but disillusion followed at Bermuda and after—as the President continued to procrastinate on the energy problem.

Feeling more pressure from the American public, the Senate completed hearings in the Commerce Committee and recommended a crash program on energy policy, to be administered by a three-man council. Sensing its loss of prestige on that issue during the summer of 1973, the administration, led by Secretary of State William P. Rogers, repeatedly expressed support for the old idea of union among consuming nations.[33] This sentiment proved meaningless, however, because the time to unite had passed.

While Exxon awaited some signal from the President, the company recommended that he put an immediate end to all oil quotas. This reversed Exxon's position of fifteen years standing, and it allied the company with Texaco (among the majors). Both companies now held that an end to quotas on imported oil would permit greater access to foreign supplies while greater domestic reserves were developed and new refineries constructed. So when newspapers indicated that the President's as-yet-unspoken energy speech would recommend new tariffs on imported oil, Exxon Corporation also announced its opposition to all new taxes on oil (or products) and called for a reduction of existing taxes.[34] But the rumor mill was not slowed by these positive statements.

When the *New York Times* appeared on April 16, 1973, it jarred the entire oil industry. According to one story, the U.S. State Department was planning to form "an international organization of oil-importing countries." Akins, now State's senior oil expert, explained to the *Times* that this consumers' group would determine the "international allocation of the available oil if chronic shortages occur." Moreover, he predicted that, in the future, the role of the international oil companies "will be more limited than it is today," since allocation of oil supplies by other groups appeared reasonable. President Nixon, the *Times* account continued, would tell the public more about the program in his forthcoming energy speech.[35]

Finally, on April 18, 1973, President Nixon did deliver his long-awaited energy message. It contained little of substance, being composed largely of Nixon rhetoric. Certainly, the speech sounded no trumpet for American (or united) action to discourage OPEC. Although the President proposed no new agreements or different plans, the oil companies still endorsed the President's words, perhaps because none of his statements seemed to constitute an attack on them. Of the international companies, Exxon appeared the "most vociferous—possibly the most complimentary" toward the President's remarks. A company press release stated: "We are impressed by the comprehensiveness of the message and the administration's grasp of the energy problems that the nation faces." Certainly, if business support of the Presidential address could help to ease the public mind, Exxon wished to do its share.[36]

Yet, Exxon did not endorse all of Nixon's energy initiatives, objecting in particular to suggestions by the State Department that oil consuming countries join in forming some users' pact (or cartel) designed to offset OPEC's growing power. In Tokyo in April, 1973, Exxon CEO Jamieson spoke out against the Nixon administration proposal for the creation of a bloc of oil consuming nations. He pronounced the notion of dividing the world into two power blocs "dangerous" because it could interrupt "the smooth flow of energy." He added: "Such a confrontation could create a hostile atmosphere and harden the opposition between the two camps rather than open up lines of mutual accommodation." Then, he listed five objectives for improving relations among all nations, one of which, "a firm adherence to contracts" represented his chief hope. In the whole speech, Jamieson emphasized his belief that privately owned oil companies should continue to play a vital role in achieving real cooperation between consuming and producing countries. In fact, his Tokyo address represented an exercise in corporate diplomacy, and it seemed statesmanlike to his listeners and readers. Moreover, it attracted worldwide attention, most of it approving. Jamieson could claim a rare success in Exxon public policy.[37]

X   X   X

When in February, 1973, the United States devalued its dollar by 11.11 percent, a connection between the dollar's decline and the inflationary effect created by the price increase in OPEC oil (resulting from the 1971 Tehran and Tripoli Agreements) seemed clear. The *Washington Post* underscored the point: "There is a clear relationship between energy supplies and inflation," and currency specialists contended that the Arab nations had used their surplus funds to speculate against the dollar in foreign money markets. Piercy and his eleven-man team had arranged to deal with possible inflation in drafting the terms of the Geneva I Agreement, written during January, 1972. But OPEC saw larger problems, including a need to convene in an extraordinary session, apparently to discuss these inflationary circumstances, in March, 1973. At this time, Sheik Yamani used his increasing influence to silence angry demands for an immediate 10 percent oil price increase. But the pressure grew intense, and in Beirut later that month, the oil ministers met again to establish a three-man team (with ministers from three of the radical states, Iraq, Libya, and Kuwait) to negotiate a price increase with the companies. The OPEC Committee also announced that the 6.5 percent increase already due producing countries (under the Geneva I Agreement) had been rendered inadequate by inflation. Negotiation would be required.

As a first step, Piercy and a group of company experts met with the OPEC Committee in April, only to have their offer of a 7.2 percent increase rejected outright. Later, another offer of more than 8 percent received a like rejection, after which the OPEC Committee turned its attention to watch events in Iran and Libya. Stalling for time, the OPEC Committee finally ignored specific terms and threatened new sanctions unless the companies increased their pricing offer.[38] By May, negotiations had resumed, and the parties appeared to be separated by only 1 or 2 percent. One source stated: "The difference amounts to hundreds of millions of dollars that eventually would have to be financed by Western consumers." Piercy and John Sutcliffe of British Petroleum jointly headed the company team which held out for revision of the Geneva I Agreement. The OPEC representatives argued that it no longer applied.

The impasse finally ended on June 1, 1973, with the capitulation of the companies, who actually agreed to 11.9 percent, an increase greater than the original OPEC demand. This figure amounted to almost twice as much as Geneva I called for, but it also took into account the real shrinkage in the value of the dollar after the February devaluation. Geneva II, the revised arrangement, called for monthly price adjustments and also added Canada and Australia to the list of countries on the price index of currency movements. In Geneva II, deflationary clauses (from Geneva I) were eliminated because of OPEC objections, and the per barrel price of oil increased 15-16 cents. For example, Persian Gulf light oil was priced at $2.59 per barrel by the Geneva I Agreement; under the pricing mechanism made there, this price would have risen to $2.74 [roughly, by 6 + percent]. Under the Geneva II terms, it went to $2.90.* The revised agreement affected only the Persian Gulf, Libya, and Nigeria.

During the actual negotiations, it soon became apparent that OPEC representatives knew they would succeed; yet they remained fractious, even

*The original $2.59 figure increased by 11.9 percent.

after the addition to the OPEC Committee of Iran's Jamshid Amouzegar and Saudi Arabia's Yamani—both regarded as moderates. The Piercy-Sutcliffe company team held out against all sorts of threats and intimidation— ranging from a possible freeze on production to a complete shutdown. At the end, however, several observers noted a more pragmatic, businesslike character at the sessions, as company spokesmen seemed resigned to results that favored OPEC. Several persons also predicted that Geneva II would be the last major negotiation between oil companies and producing countries. George Piercy, for one, hoped that this would prove to be true. As John McCloy said of Piercy: "He has been out there under the gun ... and I think he has a great sense of frustration."[39]

X   X   X

Elsewhere during 1973, Saudi Arabia's King Faisal agonized over the political predicament in which he and his country found themselves. He held strong pro-Muslim-anti-Zionist convictions; yet he considered himself a personal friend of President Nixon and he tried to keep his country sympathetic to the United States. After visiting the President in 1971, Faisal wrote several letters to Nixon which outlined the King's personal struggle to balance his loyalties and solve his problems. Increasingly, as he grew older, Faisal brooded over the obvious lack of real progress in implementing United Nations Resolution 242*—especially the fact that Israel had occupied, and still controlled, all of the Old City of Jerusalem. As a devout follower of Mohammed, Faisal often expressed a desire to pray again in the Mosque of Omar there.**

The King was often visited by an Aramco Committee representing the four owners.***[40] During virtually every call, Faisal reminded listeners of the Saudi's growing isolation from their neighbors—the price of cordial relations with the United States, at a time when other Arab countries were growing increasingly hostile to the U.S.—and to the Saudis. Although they

---

*Resolution 242 called for withdrawal of Israeli forces from occupied Arab areas; an end to the state of belligerency between the Arab nations and Israel; and a general acceptance of the territorial integrity, sovereignty, and independence of every nation in the area. Secure national boundaries had to be recognized, as did the right of all nations to use the area's international waterways. The refugee problem was to be settled as well.

**Interestingly on March 14, 1973, Prime Minister Golda Meir of Israel offered Jordan guardianship of the Islamic shrines in the Old City of Jerusalem.

***Usually the members were: C.J. Hedlund of Exxon, A.C. DeCrane, Jr. of Texaco, H.C. Moses of Mobil, and W.J. McQuinn of Socal. Aramco President Frank Jungers also accompanied them.

understood Faisal's dilemma, Hedlund and the others took a larger view: Aramco considered U.S. foreign policy in the Middle East to be based on preservation of the status quo and continued regional stability. As practiced, however, that policy had led to a "steady and continuing deterioration of American influence in the Arab World," and to threats of unpredictable changes. U.S. support for Israel had long been too uncritical, the Aramco Committee believed, and this blind support had aroused Arab propagandists, who continuously attacked the U.S. and its policies, especially in newspapers and on the radio. As Aramco executives viewed the situation then, Soviet Russia alone had profited from the failure of U.S. policies and the decline of U.S. influence in the Middle East. Of course, Saudi Arabia had suffered too, so much so that Aramco leaders marvelled at the political courage of Sheik Yamani (backed by King Faisal) in making his September, 1972 proposal for "a special economic relationship between Saudi Arabia and the United States." This proposal was repeated several times, thus attracting even more critical attention from Arab neighbors.

Nevertheless, Aramco President Jungers noted in January, 1973, that, despite Arab pressure on King Faisal he probably would continue his pro-U.S. policy "unless there is renewal of Arab-Israeli War or unless there is another instance of over-reaction by Israelis followed by some blatantly pro-Israeli action by the United States." In either case, Jungers felt that "Saudi Arabia might be forced to decide to cut off oil exports to U.S.A. and Western Europe." Other oil company officials whose companies operated in the Middle East agreed with this view.

Jungers cabled the headquarters of the four Aramco owners (on May 2, 1973) after holding a discussion with Kamal Adham, the Saudi advisor. Adham had expressed his surprise that the U.S. did not begin to use its influence to bring about a Middle East peace. He recalled that when Egyptian President Sadat had expelled 15,000 military advisors and other Soviet personnel on July 18, 1972, the Saudis "as well as other Arab countries were amazed by the fact that Sadat's moving the Russians out failed to produce results in the Israeli situation." By now, according to Adham, it seemed clear that only the U.S. could relieve that impasse. Then, as he concluded this conversation, Adham gave Jungers a jolt: "He stated that he therefore was quite sure that Sadat would embark on some sort of hostilities even though the Egyptians themselves considered it hopeless."* So, in the Saudi view,

---

*Asked by a *Wall Street Journal* reporter, two days after the war began, whether or not Egypt could win, the Egyptian Foreign Minister, Muhammed Hassan el-Zayyat, replied: "Frankly, no. But you don't struggle because you are assured of success. You struggle because you are right."

Sadat took this dramatic risk with hope of forcing the U.S. into some action that would end the Middle East stalemate—or so Jungers understood. Adham also told Joseph J. Johnston, Aramco senior vice president, that the Saudis "were deeply concerned that the tide of events would force the Americans out of the area [f] or good, even in Saudi Arabia."[41]

In later conversation with both King Faisal and Kamal Adham, Johnston received the same message: the Saudis already stood alone. When the expected armed hostilities broke out, Saudi Arabia would no longer be able to hold that solitary position. Help from friends abroad would be required.

Later that month (May 29), Hedlund and representatives of the other three owners joined Aramco Vice President R.W. Powers, Jungers, and Johnston in Geneva, where they held talks with Yamani on implementing participation in Aramco. To their surprise, Yamani offered to arrange an audience with Saudi King Faisal. Already the King had visited Egypt, where Sadat gave him "a bad time" about his failure to support Arab causes. Now, Faisal had come to Geneva for rest; but he did meet with the Aramco group to tell them that "time is running out with respect to U.S. interest[s] in Middle East, as well as Saudi position in the Arab world." Although "Saudi Arabia is in danger of being isolated among its Arab friends," King Faisal made clear that he would not let this happen. "You will lose everything," he finally warned, and he suggested that the companies make the U.S. government and the consuming public aware of their "true interests" in the Gulf area.

Hedlund and his associates went to Washington May 30, to see representatives of the State and Defense Departments and the White House. The King's message was delivered to all of them, including Joseph Sisco, Assistant Secretary of State for Middle Eastern Affairs, and Acting Secretary of Defense William Clements. But the five company men were not prepared for their reception by the government "experts." As Johnston recalled:

> The general atmosphere encountered was attentiveness to the message and acknowledgement by all that a problem did exist but a large degree of disbelief that any drastic action was imminent or that any measures other than those already underway were needed to prevent such from happening. It was pointed out by several from the Government that Saudi Arabia had faced much greater pressures from Nasser than they apparently face now and had handled such successfully then and should be equally successful now.
>
> The impression was given that some believe H.M. [His Majesty] is calling wolf when no wolf exists except in his imagination. Also,

> there is little or nothing the U.S. Government can do or will do on
> an urgent basis to affect the Arab/Israeli issue. . . .

In short, the U.S. government remained skeptical; and it obviously took no action. That attitude, Johnston judged accurately.[42]

Henry Kissinger's personal account of feelings in the Nixon administration has since added to the puzzle. His description of what happened before October, 1973, reveals the disabling indecision that appeared chronic to Nixon's foreign policy in the Middle East:

> For three years, a new infrastructure had been elaborated by the oil producing nations built on the weakness and irresolution of the consumers. Free-market theology had kept the consumer governments, and especially the United States out of negotiations as the companies were rendered defenseless. Political demands had become mingled with economics.

Even better than Kissinger, *Washington Post* columnist Hobart Rowan caught the flavor of these years in his trenchant analysis, "The Energy Crisis: Who's In Charge?" This question, Rowan answered with a simple declaration: "There probably is no knottier dilemma on the horizon, but in Watergated Washington it's not getting the attention it needs." No one was in charge. And no one felt any responsibility, Rowan added.

Here, should be noted another important article, this one written by James Akins and published in *Foreign Affairs* (April, 1973) under the title, "The Oil Crisis: This Time The Wolf is Here." Akins gave comfort to OPEC, while making the difficult position of the companies even more arduous. He used the pages of a prestigious international journal to proclaim that consuming countries *expected* to pay more for oil. As Akins predicted, they soon would; and the price might reach $10 per barrel—which one critic later identified as "a classic self-fulfilling prophecy." Akins—still serving in the U.S. State Department—advanced the case so often made by the group of professional Arabists in the government. Then, he defended the role of State between 1970-1973, and concluded that, in case of a boycott of the consuming countries by OPEC, the U.S. would have to choose among three alternatives: economic collapse, war, or concession to OPEC's demands. In his analysis, Akins declared, in substance, that the companies must pay higher prices to the producing countries.

Among the governments in consuming countries, the U.S. government was not alone in its inability to set a firm policy. Other governments also failed to support the petroleum industry, despite frequent invitations to consider the consequences. In September, 1973, Jamieson and other Exxon company spokesmen (who, like Secretary of State Rogers, anticipated some form of embargo by Middle East producing nations) urged the European Common Market countries to prepare some international allocation plan— to meet the possible emergency. But instead, individual actions by the European Economic Community [EEC] nations included separate attempts to court Arab favor, by Great Britain, Italy, Spain—and especially France. Thus, any unity of purpose disappeared. In fact, OPEC leaders saw little evidence to warrant practical concern about any union of consuming countries. So the oil companies, including Exxon, could only conclude that Middle Eastern affairs ranked far down on the list of government priorities worldwide. And each company felt compelled to respond to King Faisal's request in its own way.

Exxon made no overt move, although some persons believe that former Jersey (Exxon) Vice President Howard W. Page voiced a response of sorts in a speech critical of the U.S. government, which Page made in New York City, to the alumni of the American University of Beirut, an institution that he had served as a trustee.[43] Socal and Mobil, on the other hand, attempted to draw public attention to the Saudi and Arab position in letters to stockholders, and in print advertisements. The success of these efforts remains hard to measure, but during this time the Aramco office in Washington kept communication channels open in the hope of government action by the Nixon administration. Before anything occurred in Washington, however, political events in the Middle East suddenly changed the situation.[44]

X X X

During the summer of 1973, Sheik Yamani repeatedly told Aramco executives that the Tehran Agreement no longer applied, and that OPEC would soon double the posted price of crude oil.*[45] Yet King Faisal also continued to believe that some gradual shifting of U.S. policy might well occur, so that he would not be forced to use his oil as a weapon. In fact, as Faisal once said to Jungers, "a simple disavowal of Israeli policies and actions by the USG [United States Government] would go a long way toward quieting the current anti-Arab feeling." Sheik Yamani told U.S. officials in April that the King had virtually decided to involve more key advisors in the search for a solution to the problem of the proper diplomatic position of Saudi Arabia. Some members of the royal family obviously believed King Faisal's policy of

*Persian Gulf light then sold for $3.01 per barrel.

courting the U.S. strained their relations with other Arab nations. Many of them felt that Saudi Arabia should not continue to increase oil production; some, in fact, thought that production should be reduced, to demonstrate Saudi unity with the other Arab states. While King Faisal did deem the latter course wise (and while Yamani opposed it) pressures kept growing within the Saudi government for a more active pro-Arab course.

In May, 1973, disturbed by these divisions among his advisors, King Faisal appointed a Supreme Petroleum Council, with his half brother Prince Fahd as chairman. He gave that body the assignment of devising a Saudi oil policy. As the King's patience wore thin, he wrote Nixon a letter on June 1, stating clearly that the message he had sent with the oil men in late May conveyed his intentions exactly. Perhaps sensing failure, Faisal then decided to change his tactics.[46]

In July, King Faisal granted interviews to newspaper correspondents from the *Washington Post* and the *Christian Science Monitor*. In these, he emphasized that the oil weapon would be used only if the U.S. remained inactive; and he claimed to be seeking a more "even handed and just Middle East policy."[47] In August, Faisal permitted the National Broadcasting Corporation to videotape an interview during which he elaborated on the same theme; ten days after this, *Newsweek* published an interview in which the King reiterated his opposition to the use of oil as a political weapon.[48] In every medium, King Faisal made himself crystal clear. Faisal had spent several years as Saudi Foreign Minister (which title he still retained), and now he proved his ability as an effective communicator.

This concern with the effort to influence American foreign policy had pushed other problems aside, at least temporarily. Aramco, for example, had never reached agreement with the Saudis on buyback oil; so in San Francisco early in September, 1973, Piercy and Yamani began to discuss both the price and the quantity of oil that the Saudis would sell. On September 13, Aramco owners met in New York to approve terms of reference and amounts of these purchases. Here, Jamieson gave fresh evidence of his characteristic good judgment. Speaking for Exxon, he indicated its belief that Aramco should allow Yamani flexibility and avoid provoking him into unilaterally taking additional oil for Petromin [the Saudi national company]. By offering some latitude now, Aramco would probably be assured of receiving more oil when the final amount was actually determined. Socal's Otto N. Miller and Texaco's M.F. Granville supported Jamieson, while Mobil's Rawleigh Warner expressed his own concern that the Saudis were encroaching on Aramco's rights to stable supply. Mobil, always short of crude oil, questioned the readiness of the other three companies to test Yamani on the price and volume issues. After all was said in New York, the four companies could not reach agreement; Mobil reserved the option of making its own case directly to the Saudi government. So again, any cloak of company unity disappeared, and as a result, Piercy's position as a negotiator was not strengthened.[49]

OPEC met on September 15 and 16, 1973, in Vienna, where the oil ministers agreed that serious talks with the companies on the subject of prices should begin in Geneva on Monday, October 8. This action accorded with the negotiating strategy proposed by the OPEC Gulf Committee (Yamani, Amouzegar, and Hamadi), and the tone of this call to meeting left little doubt that OPEC now felt ready to dictate an increase in prices, or perhaps that both OPEC and the companies realized by now that something had to be done about oil prices. The market price for crude was increasing daily; and the margin between posted price and market price narrowed considerably in 1973. So the companies stood to make most of the profit from equity crude.* Outside the Middle East, Venezuela set an example by unilaterally raising prices several times during the year. In fact, many events combined to encourage OPEC members and make them more confident. In particular, OPEC oil ministers felt certain that they could still use the threat of curbs on oil production as an ultimate weapon against the companies.[50]

While they waited for the important October 8 talks, Piercy and Yamani concluded a tentative agreement on Aramco prices and sale procedures. Upon approval, it would begin immediately to provide substantial new income for Saudi Arabia. Future crude oil prices were to be based on the price Petromin obtained for crude in the open market, so Aramco must meet this price in order to buy its share of that oil. But Petromin also was entitled to a larger share of oil than that provided in the 1972 agreement.[51] Now, the Saudis seemed to be well protected; yet this tentative Aramco agreement soon attracted bitter criticism, especially from other OPEC members, who objected to potential profits for the companies.

In consuming countries, the fact that the oil companies' profit margins had been quite low for several years did not temper this criticism, nor moderate public indignation over what came to be called "exorbitant" profits. Everywhere politicians responded to the public ferment, to attack the oil companies as enemies of the people, until in this process, the oil companies learned that their relations with the public had reached a disastrous state. The general public simply did not grasp the rudimentary economics

---

*Equity crude is the oil owned by the producing company as a result of the Participation Agreements.

involved in the oil crisis.

News stories relating high company (second quarter) profits appeared in late July and early August, while world-wide inflation continued. This combination led Sheik Yamani, on September 7, to restate his position: The Tehran Agreement had either to be amended or considered dead. Further, if the oil companies chose not to cooperate, OPEC would be left free to act unilaterally, in accordance with the Statement of Petroleum Policy, passed in 1968.* Already, Yamani had held conversations with the Iraqis, and he and Amouzegar (of Iran) had also agreed that some increase in the posted price was necessary. They also agreed that the producing countries needed a more reliable inflation indicator. So after their cordial meeting, it seemed obvious that the Gulf Committee would demand a major increase in posted prices from oil company negotiators.

A parallel reason for OPEC's good fortune was a new crisis in the Nixon administration during September and October, 1973, which kept attention turned away from Middle Eastern diplomacy. When he replaced Rogers as Secretary of State on September 22, Kissinger arrived fresh from dealing with the Soviet Union on a global basis. For the moment, he had little time to study the Middle East, as normally he would have done; and when he did turn to problems in that area, Kissinger himself admits that he misinterpreted vital information. At the same moment, Vice President Spiro Agnew resigned and Gerald Ford was sworn in to replace him—as the jaws of the Watergate vise were closing on the President himself. Clearly, under these circumstances, the crisis in Washington meant little government help for the Middle East.

There, for Exxon and its three partners in Aramco, they continued to believe that King Faisal would no longer hesitate to use oil as a weapon. In September, 1973, Piercy made a tentative offer of a price increase, but Yamani quickly rejected it as inadequate. In fact, no ground was gained in these negotiations, which dragged on until it became unmistakably clear that something dramatic was about to happen in the Middle East, involving Egypt, Syria, and Israel. By the summer of 1973, the political situation had become tense enough to force the Israelis to mobilize their army twice. Yet, no one in American public office seemed to accept as real the Saudi threat to cut back oil production—despite warnings from virtually every oil company that operated in the Middle East. Simply, no one wanted to listen to them.[52]

---

*The 1968 statement affirmed the right of members to determine "posted or tax reference prices in such a manner as to prevent any deterioration in their relationship to the prices of manufactured goods traded internationally."

Then, Piercy, Andre J.P. Benard (of Shell), and other negotiators* met in Vienna on October 7; the long awaited blow was struck. Even before they arrived, President Anwar Sadat's Egyptian Army hit Israel in the Sinai, on Sunday, October 6, 1973—the Jewish high holy day, Yom Kippur, and during the Muslim fasting month of Ramadan. At the same time, Syrian forces also moved against Israel, and both the Egyptian and Syrian Armies achieved early victories. The Middle East again had burst into flames.

For obvious reasons, the United States expressed surprise at the Arab action; although after the fact, it seemed clear that the U.S. should have anticipated an attack on Israel. King Hussein of Jordan, for example, had provided information earlier in May, and at the Summit Conference in June, the Russians passed along several warnings to the American delegation. Perhaps Secretary of State Kissinger explained this American failure most accurately:

> It resulted from the misinterpretation of facts available for all to see, unbeclouded by any conflicting information. Sadat boldly all but told what he was going to do and we did not believe him. ...
> We knew everything but understood too little. And for that the highest officials—including me—must assume responsibility.[53]

On October 7, the government of Iraq nationalized all property of Exxon and Mobil. This sudden move naturally led the other two owner companies of Aramco, Socal and Texaco, to express their mutual concern for the security of Aramco. As a further step, the other three owner companies delegated to Exxon the task of briefing the U.S. State Department on the oil supply situation. According to testimony before the Senate Subcommittee on Multinational Corporations, in 1973, the Exxon representatives often conferred with State throughout the period of war crisis.

In London, the Policy Group had reconvened to establish terms of reference for Piercy's team. Joint actions by the companies still were covered by earlier Business Review Letters issued by the Justice Department. Now with the outbreak of war, the situation in Vienna on October 8 no longer seemed to encourage effective discussion. Instead, a constant flow of radio reports from the Arab world claiming that the U.S. was flying armaments for Israel directly to the front served to keep the atmosphere strained. Yet, these Vienna negotiations did open on schedule, when the war was only two days old. First, Piercy made the companies' initial offer; then Yamani replied

---

*Others in the group were: J.D. Bonney, vice president of Socal; R.F. Cox, vice president of Arco; P.I. Walters, managing director, British Petroleum.

with a call for a posted price figure of $6.00 per barrel, roughly a 100 percent increase. When Piercy asserted that the company spokesmen were empowered to offer no more than a 25 percent increase, Sheik Yamani and Dr. Amouzegar dropped their request to $5 per barrel, still representing a 60 percent increase. Privately, Yamani urged Piercy to agree, because Yamani feared that the war and a possible embargo would serve only to confuse the price issue.* Certainly Faisal's oil minister appeared to try to avoid a breakdown in the talks.

After the initial discussions, the oil men asked for a two-day recess, in which to consult their principals. This, the Gulf ministerial committee granted. Shortly, the OPEC proposal was under study in both London and New York. In London, David Drummond, an Esso Europe Inc. employee, speaking for the oil industry, declared that the increases "would have exceptionally serious and wide-ranging implications, not only for the companies but also for the world economy at large." In Egypt, President Sadat called on King Faisal to cut off all oil to the U.S. if that country acted to re-arm Israel, while Arab sources in Beirut were quoted widely as saying that King Faisal had already sent a blunt warning to President Nixon—no oil if Israel is re-supplied.

Through John J. McCloy, the companies also persuaded the U.S. Department of State to sound out several North Atlantic Treaty Organization [NATO] allies regarding some united political action. They answered in the negative, perhaps fearing the loss of oil supplies. As a London journalist observed, Europe appeared "too divided, and too nervous of confrontation with the producers to sort out any agreed upon response among themselves." Any hint of disrupted lines of supply could produce panic.

When Piercy and Benard received from London the companies' negative answer to the Gulf Committee offer, the hour was after midnight. Without waiting for morning, they went to see Yamani, and in "a dramatic 1 A.M. encounter" they asked him to persuade the other Gulf ministers to accept a two-week recess, so that the companies could consult with their home governments without breaking off the talks. Yamani agreed to telephone the Kuwait Oil and Finance Minister, Abdulrehman Salim Al-Ateeqy, who came to Yamani's room in his pajamas, and showing signs of anger. The Kuwait minister absolutely refused to agree to any delay in reaching the price decision. After he had left, Yamani told Piercy that, in any event, Aramco could expect a production cutback of around 14 percent, from 8.3

*Yamani had evidence to support his view. The Kuwait legislature in January, 1973, for example, had passed a resolution stating that in case of a war with Israel, the oil weapon should be used against the U.S.

million b/d to 7 million b/d. Finally, Yamani declared that the Gulf states would not change the terms of their offer; and Piercy replied that the company representatives could not accept them. After the famed "1 A.M." meeting ended, Yamani left for Saudi Arabia and Piercy returned to London. Just before they parted, Piercy asked the Saudi minister what would happen next. According to one version, Yamani replied: "You can hear it on the radio."[54]

In the meantime, Kuwait's call for an OAPEC meeting had been greeted with concern in the headquarters of Aramco's four parent companies. For each of them, of course, the Aramco concession represented a valuable piece of property.* Jamieson, Warner, and Granville felt this anxiety and joined in drafting a memorandum to President Nixon. It called his attention to oil shortages that might result from Arab actions. This document was sent to Miller (Socal) for a fourth signature, while Jamieson telephoned John McCloy and asked him to pass their message to President Nixon. Instead, McCloy suggested sending it to General Alexander M. Haig, Nixon's chief-of-staff, to whom the counselor addressed a covering note. Also at McCloy's suggestion, a copy was added for Secretary Kissinger. Of course, McCloy himself read the Aramco message, about which he later wrote: "The statement seems to have been quite prophetic."

In the memorandum itself, the oil men noted that no spare producing capacity presently existed anywhere in the oil industry, which they said "is now operating wide-open." Market conditions had already led to some price increases. OPEC demands, they termed "unacceptable" and stated that they proposed to resist OPEC pressures. The CEOs went on to indicate that their own information had led them to believe that King Faisal would order some cutback in Saudi crude oil production as a reaction against U.S. policy toward the Israelis. "A more substantive move," they believed to be in store against oil companies if there should be "further evidence of increased U.S. support of the Israeli position." In all, the Middle East, they wrote, was "highly charged," and a "single action taken by one producer" against the U.S. government might snowball into a major crisis. Since Europe and Japan needed Middle East oil to a greater degree than did the U.S., our allies might, in the event of a crisis, consider their own interests alone. Finally, the memorandum pointed out that "much more than our commercial interests in the area is now at hazard." U.S. primacy in the Middle East could well be supplanted by the Japanese, the Europeans, or even the Russians "to the

---

*Exxon's total offtake of Aramco oil, in 1971, represented 533 million barrels; in 1972, it was 633 million barrels, and in 1973, 784 million barrels.

detriment of both our economy and our security." Yet, aside from sending these words of warning, there appeared to be nothing else that the four executives (or their companies) could do at that time.

Whether or not this memorandum ever reached President Nixon himself remains unclear. General Haig replied to McCloy some three days (October 15) after receiving the Presidential copy, promising that the message would be read by both the President and Secretary Kissinger.[55] Kissinger, in his memoirs, attaches no special significance to the memorandum; obviously, if he read it, his diplomacy showed no signs of its influence. Unrelated special messages from President Nixon and Secretary Kissinger to King Faisal elicited formal diplomatic replies, but the King continued to insist that peace must be predicated on Israel's withdrawing to her 1967 borders. Even as the diplomatic cables crossed, the U.S. began to send military supplies to Israel. Predictably, this American aid to the Israelis disturbed the Arabs.

Not without some reason. As Nixon told Kissinger: "The Israelis must not be allowed to lose," and both men wished to establish a public record of aid to Israel. As they reasoned, such action would enable them to play a decisive role in ending the war and defining the peace. On the other hand, Secretary of Defense James Schlesinger and Under Secretary Kenneth Rush questioned this aid, arguing that Israel did not need more weapons to defeat Syria and Egypt, and that significant help to the Israelis so early in the conflict might mean the loss of Middle East oil for the U.S. But on October 10, confirmation of reports that Russian planes were supplying Syria settled the issue. Immediately, the U.S. began preparations for a massive airlift to Israel.[56]

Even then, King Faisal still hoped that the U.S. would avoid the worst consequences by keeping quiet—at least, not calling the world's attention to new American aid for Israel. In fact, he sent his Minister of State for Foreign Affairs, Omar Saqqaf, to see both President Nixon and Secretary Kissinger, and to deliver Faisal's personal letter to Nixon. Saqqaf was instructed to learn whether the U.S. proposed to reinforce Israel, and then to ask that American policy in the Middle East be kept even-handed. Accompanied by the foreign ministers of Morocco, Algeria, and Kuwait, Saqqaf met with Kissinger on the morning of October 17, when he provided the delegation with little comfort as he explained U.S. policy: to end all fighting in the Middle East, and thus avoid a "great-power confrontation there." Before they left, the ministers urged Kissinger to encourage Israel to accept Sadat's offer of a cease-fire based on an Israeli return to its pre-1967 boundaries.

Next, the ministers went to see President Nixon, who proved to be in his best form. As Kissinger recalls: "Nixon with his mastery of intangibles knew exactly how to strike the right note: to promise a major diplomatic effort without committing himself to a particular outcome." Nixon did promise Saqqaf that Kissinger would personally handle Middle East negotiations, and the Secretary of State himself later urged the foreign ministers to counsel their governments in favor of caution. Saqqaf apparently liked both Nixon and Kissinger, and he was impressed by the President's appeal, but nothing in his report of the American response managed to alter the determination of the Saudi King.[57] Faisal had been disappointed by the American friends he trusted.

X   X   X

One day before Saqqaf presented Faisal's letter to the American statesmen, OAPEC representatives gathered in the Kuwait City Sheraton Hotel.* There, on October 16 (17), 1973, six representatives of Persian Gulf nations took a series of steps that shook the foundations of world economy. First, they unilaterally established a new market price for crude oil: $3.65 per barrel. Then, they decided that the posted (tax reference) price should be 40 percent above the market price; so they added $1.46 to the new market price for a total of $5.11. Finally, these OAPEC representatives announced that the 40 percent price differential would apply in the future—but without showing precisely how it would be figured.

In all, the $5.11 posted price represented a 70 percent increase over the existing price. Then following the precedent established at this meeting, OPEC countries could appropriate for themselves the right to determine the full posted price that the companies must pay for oil. So the future had been placed in the hands of producing countries, and reverberations would last for many years. Although the companies protested this appropriation of power, at the time they had no option other than to accept.**

George Piercy remembers that the October 16 date actually marked the very debacle that George Henry Mayer Schuler had predicted earlier, when the united company front was first broken during negotiations with OPEC. In testimony before the Church Committee, Piercy described what had happened in Vienna and again at Kuwait:

They ... called us to Vienna to patch up the Teheran Agreement,

*The Organization of Arab Petroleum Exporting Countries frequently met, at times with observers from OPEC. Several meetings occurred a few days before regularly scheduled OPEC meetings.

**Because of the tax rate on the posted price, there was no way the companies could pass the cost increase along to customers without raising market prices even more.

but that was not what they were after when we got there. They hit us with a demand of a doubling of the posted price, which was more than a patch-up; it was a reflection that things were in their control.

It was such a huge demand that I asked for 2 weeks' delay so we could consult with consuming governments on this and they did not give me that. They went to Kuwait and acted unilaterally.

To OPEC, the new price increase meant a logical adjustment of posted prices to then-current spot market prices. Already the market price had passed the posted price, thus limiting the governments' tax and royalty income. Yet, whether or not the argument offered by OPEC to justify increases could be regarded as valid remained long in dispute, especially among the consuming countries.[58] As always, one nation's idea of justice seemed to another an example of extortion—"highway robbery" was the term most often used.

X   X   X

Unlike oil problems, the Russian airlift to Syria produced a quick response from the U.S. government. On October 13, 1973, U.S. Air Force planes began the airlift to Israel, and on October 15, the American intention to re-supply Israel was publicly announced. Two days later, the OAPEC ministers, meeting in Kuwait, agreed to retaliate by cutting their oil production by 5 percent in October, and again in each month thereafter, until such time as world opinion might force Israel to evacuate all the territory it had occupied during the 1967 war—and until "the legitimate rights of the Palestinian people are restored." This initial cut amounted to about 1 million b/d; but more than lost production, it symbolized "a new phase of post-war history"—a demonstration of power by the producing states. Yet even in this heady atmosphere, Sheik Yamani carefully followed King Faisal's instruc-tions, and while the orders came from the King, the gradual Saudi policy shift had been influenced by the Supreme Petroleum Council. Sheik Yamani—and King Faisal himself—were moderates when compared to some members of this body. Yamani faithfully did the King's bidding as he fought off all attempts to cut production more. His moderating skill pre-vailed, while his threatened cut remained before the oil men.*

Now, having some evidence of the seriousness of the King's intentions, Jamieson and other company executives felt compelled to call on Under

---

*Some evidence indicates that the letter from King Faisal to Nixon stated an ultimatum: Unless the U.S. ended all aid to Israel within forty-eight hours, an embargo on Saudi oil would follow.

Secretary of State Sisco (and other officials) to discuss recent developments. Once in the room, Jamieson quickly came to the point: "Look, there's going to be an oil embargo [against the U.S.]." In disbelief, Sisco replied, "Oh no, no no! You're wrong, our source of information is much better than yours." Hearing that, Jamieson knew that additional conversation would only prove pointless; so he terminated the discussion. Clearly, the U.S. government was determined to deny all possibility of an oil embargo.

By October 18, Abu Dhabi had embargoed all oil destined for the United States. This example was quickly followed by Libya and Qatar. Still, Washington remained unmoved, and on October 20, the President sent to Congress his request for $2.2 billion to finance aid for Israel. Both Nixon and Kissinger seemed to believe that this grant was necessary to counter Soviet assistance to Syria, but to King Faisal, the American move ended all hopes of an even-handed U.S. policy in the area. Once these hopes were dashed, Faisal was left to feel rebuffed for his patient efforts; so on that same day, the King ordered Sheik Yamani to begin an embargo of oil to the U.S. Yamani was prepared, and now the loyal minister proved himself as zealous and able in administering the embargo as he had been earlier in opposing it. Kuwait, Bahrain, and Dubai soon followed the Saudi lead, and Bahrain rescinded its agreement with the U.S. for use of the port there.[59]

Only hours before Faisal gave his final order—on October 19—Frank Jungers and Yamani had quietly discussed the implications of Saudi participation in Aramco. Then, on October 20, the bottom fell out, and Jungers and other Aramco managers were summoned to meet with Sheik Yamani, at 1 p.m. on October 21. Also present were a number of ranking Saudi officials, plus Princes from the large royal family. An unsmiling Yamani told the Aramco executives that the meeting had been called to lay out the ground rules for implementing the Saudi embargo. Aramco's production had to be cut back 10 percent from the September total, and all direct or indirect shipments to the U.S. (including military) were to be eliminated. Yamani then provided a list of approved countries to which Aramco oil could be shipped. Despite their preparations, the Saudis seemed to be aware that the embargo would be difficult to administer; so they looked "to Aramco to police it." Finally, Yamani told Jungers that if the cuts failed to achieve results, the next step would be nationalization.[60] After listening to the minister, Jungers could not doubt the Saudi's serious intent or that the King had approved the embargo.

X X X

At the time the embargo took effect, the oil world already was troubled; with the embargo in place, the results amounted to chaos. George Piercy summarized the oil story in testimony before the Subcommittee on Multinational Corporations on March 28, 1974:

> No, there was ample oil, but not a glut up until October, 1973. It was touch and go but we had oil. ... There was a cutback instituted on October 19, the meeting of the Arab ministers in Kuwait City ... not only embargoed oil to the United States and Netherlands, but cut production. ... That cutback got as high as about 25 percent or about 4 million barrels a day. Now that created ... chaos in the market that we are living through today because ... there was a real shortage of oil at that time.[61]

This time, the Saudi threat finally did materialize. It did little to achieve substantive political change in the Middle East; but it did both revolutionize the oil industry and restructure the world economy. Even the United States—long thought invulnerable—suddenly became energy conscious and energy thrifty. In this revolution among energy consumers, none of King Faisal's original goals was achieved. In effect, Faisal had changed the world without changing the U.S. policy toward Israel.

Aramco had little choice except to carry out the Saudi orders with respect to Saudi oil. The United States used only 650,000 b/d of Saudi oil, but the embargo also extended to U.S. armed forces stationed abroad. So Exxon and the other owners of Aramco "fully alerted and advised" both the U.S. Department of State and the Department of Defense. As Exxon's Jamieson explained:

> They concurred with what we were forced to do. We were, however, able to move non-Saudi supplies around, and I think I can say that the U.S. Armed Forces did not get short of products.[62]

In fact, until the embargo against the U.S. ended, on March 18, 1974, the entire American petroleum industry met this crisis with creativity, and in general effectively, by rerouting oil and other supplies to ease shortages as they appeared. In retrospect, despite the companies' efforts, the picture was confusion itself; for example, in spite of the Saudi Arabian production cuts executed by Aramco, 3 percent more oil was actually produced in that country during the last quarter of 1973 than in the comparable period of 1972. So, that oil became available for use by someone somewhere. Yet all together, Europe operated for the period on 11 percent less oil than it had during the same time in 1972, while the oil available to the U.S. dropped by 6.9 percent. Oddly, during the cutback and the embargo, the amount of oil available to Canada and Japan actually increased.[63]

Even more puzzling, OAPEC's decision to use the oil embargo went unchallenged by the consuming countries, although it violated two United Nations declarations (1965 and 1970) which seven OAPEC nations had enthusiastically supported.* Sensitive to the issue, the Arab nations advanced as legal justification for their action the doctrine that oil concessions had been awarded as political favors, and not as matters of contract. Of course, for international oil companies such as Exxon to accept this doctrine would have required the company to alter traditional principles for doing business. In the main, these business principles had been adhered to in Exxon's foreign operations, especially the principle of not becoming involved in the political affairs of a host country. Since the company had always relied on contracts (rather than political favors) the Arab justification proved unacceptable to Exxon.

When Libya joined the embargo and cut its share of production, it also arbitrarily raised the market price of Libyan crude to $4.90 and the posted price to $8.92 per barrel.** Iraq, however, refused to follow the same path, asserting instead that the time had come for more drastic action—expropriation. When their motion to expropriate failed, the Iraqis left Kuwait; but later, they did embargo oil to the U.S. Iran, a principal supplier of oil to Israel, simply did not support the Kuwait actions, and Iranian production rose.

Finally, the entire oil shortage proved to be less severe than the public believed, as rumors and hysterical predictions refused to give way before facts. Libyan and Iraqi oil continued to be marketed outside the U.S., but Howard Page and George Piercy later testified that so far as they knew, very little—if any—of that oil ever reached the United States. In fact, the embargo worked pretty much as the Arabs had planned it. When Senator Edmund S. Muskie cited Treasury Secretary William E. Simon's allegations to the Church Committee that leakage from the embargo amounted to 400,000 barrels per day, Piercy replied that "Exxon had no direct evidence that there was this leakage."[64]

*In December, 1973, the U.N. also adopted another resolution concerning permanent sovereignty over natural resources, which decried any use of economic coercion in the settlement of disputes. Of the OAPEC members, only Kuwait signed this one.

**In December, 1973, Libyan production was 1,800,000 b/d which represented a 5 percent cut from the September, 1972 production level. While Colonel Qaddafi continued to fulminate against the U.S., Libya sold oil (as did Iraq) to all comers—except the U.S.—for whatever price it could get.

But as the records of these hearings also suggest, the greatest impact of the embargo was not in the market so much as on the public. In the U.S., every report of production cutbacks led media persons and politicians alike to register new concern about the adequacy of oil supplies, especially for the approaching winter heating season. For one, Secretary of the Interior Rogers C.B. Morton refused to be reassured or to reassure his listeners. Instead, he claimed that everything depended on "old man Winter and where he strikes." Other public officials were less poetic but no less discomforting.[65] And the lack of any real leadership from President Nixon played a part in magnetizing the public's fears—especially since the media preferred to amplify the psychological impact of the embargo, rather than to search out and publish harder facts.

Although key oil executives like Jamieson had often—long before the October War—announced the existence of a "real and serious" energy crisis, none of these statements seemed important enough to prepare the general public for reality. Understanding the depth of its problems, Exxon tried to allay public fears by using its domestic affiliate, Exxon Company, U.S.A. [Exxon USA], to run a newspaper advertisement headed, "Will there be enough heating oil to keep us warm this winter?" The text below this title stated that 75 percent of the required oil would be refined during the winter, 15 percent was already in storage, and 10 percent was left to be imported. As the explanation continued: with U.S. production declining annually, increasing amounts of oil would have to be imported in the future. Since world supplies already were tight, the needed oil would probably come from Caribbean and European refineries. So the company was asking that the U.S. government—to help in this situation—modify (temporarily) its sulphur restrictions on oil and coal. Finally, the advertisement called on the public to conserve supplies.[66]

At least this part of Exxon's message found the public responsive. As the *Wall Street Journal* editorialized: "The only thing that stands a chance of working is a no-nonsense organized program of conservation, one that emphatically impresses on Americans the need to save fuel." Obviously Americans had begun to learn.

Yet the emergency continued. On November 1, 1973, the United States instituted a mandatory heating oil allocation program, designed to enable the country to use available supplies efficiently. Suppliers were authorized to sell to regular customers only as much oil as they had used in 1972; but if stocks became short, dealers would make proportional cuts "among all their customers." Confusion greeted consumer questions about how the plan should be implemented.

On the same day, Exxon announced that it would resume allocating gasoline in the Gulf and East Coast markets, despite the fact that it expected to have available a greater volume of that product than it had in November, 1972. Everywhere, it seemed, at least one effect of the oil situation appeared as a threat of crisis or an impending disaster. The American public had surely awakened to the importance of oil—if not yet to the potency of oil politics.

Ironically, it now appears that King Faisal's idea of using oil as an instrument to influence U.S. foreign policy became something of a boomerang; once oil became an issue, the more often Americans heard the Arab threat, the more anti-Arab they became. Even after the October War, when OAPEC embargoed oil to the U.S., many Americans termed that action "blackmail"; as eight distinguished American economists, including four Nobel Prize winners, urged the U.S. to resist this "oil blackmail." Perhaps diplomat George Kennan caught the national spirit best when he declared that, to yield to such pressure "would represent a humiliation ... for which future generations of Americans would be unlikely to forgive us."

Meanwhile, the oil ministers of the six Persian Gulf states resumed discussions in Beirut on November, 1973, where they decided to postpone consideration of a new market price mechanism. They did, however, agree to increase the posted price of their oil by six cents per barrel, to $5.17, as a way of allowing for recent currency fluctuations. The new posted price was apparently based on spot sales of crude—at prices well above any established market price. New discussions with oil company representatives were announced for November 17.

Also in November, Exxon held its own energy outlook press conference. President Clifton C. Garvin, Jr.* told reporters that, even if all the proposals of President Nixon were implemented, it would still be impossible during 1973 to recover the losses caused by the Arab oil embargo on the U.S. Exxon's supplies alone had been curtailed by 17 percent. "The sad fact," he continued, "is that as far as we can see there isn't any spare crude capacity." Next, Exxon USA President Mike A. Wright questioned whether the U.S. needed a "crash" program to resolve the oil problem. In his view, few substantial measures had as yet been given thoughtful consideration, and many possible solutions remained untried. Among the possibilities he listed were: coastal exploration, coal gasification, and the unbuilt Alaska pipeline.[67] Wright showed that now as so often before, politicians were confronting problems with words rather than with usable ideas. In all, Exxon's leaders

---

*Garvin became president of Exxon on November 1, 1972.

made clear that promising solutions to the nation's energy "crisis" possibly were already available for the trying.

X   X   X

On October 16, 1973—the same day that the Persian Gulf states increased by 70 percent the posted price of their oil—the following words of Jamieson appeared in Sylvia Porter's column in the *New York Post*:

> We [the U.S.] are short of the fundamental resources, such as oil and natural gas [,] which create energy. And it is energy that ... makes our daily lives possible as we live them today. ... In our desire to achieve rapid economic growth and high standards of living, we Americans have been prodigal with resources that once seemed limitless. But energy resources as well as air, water and usable space are finite and the war crisis in the Middle East is only part of the problem.

Jamieson went on to suggest that conflicts between energy goals and environmental objectives had to be reconciled "through appropriate government processes." His was a voice of experience, and he viewed the Middle East for what it truly represented—a complicated set of circumstances, which included the larger problem of energy consumption. That larger problem, as Jamieson knew and often pointed out, existed in the public mind, but only so long as it was not pushed aside by something more dramatic—such as public concern over oil company profits.

On October 20, along with the news that Exxon's earnings had reached an all-time high, Jamieson offered his own careful explanation, pointing out that the company had gained profits from both increases in demand and higher prices. His words were lost in the clamor, as press accounts emphasized Exxon's 80 percent 1973 third quarter increase in profits (measured against the same period in 1972) to the exclusion of everything else. In fact, reporters seemed unwilling to understand what Jamieson meant when he said that the Exxon organization had delivered a capacity performance during the past quarter. Moreover, the sensationalism with which the media surrounded the numbers did little to offset the popular belief that oil companies were simply passing all OPEC price increases along to the customers while increasing their own profits.[68]

Certainly, Exxon did make profits during 1973. But uncertainties about both supplies and prices also shook the structure of the corporation and created strained relationships between the parent company and the affiliates. R.C. Darcey, assistant secretary to the Management Committee, reported on one such strain to C. O. Peyton, president of Exxon International Company; Darcey related precisely what the committee felt about a wholly-owned affiliate's charges for oil.* Already the committee had met four times in five months, to confront the need for drafting guidelines "to affiliate crude prices and the pass-through to third-party crude oil customers of changes in government take." Traditionally, Exxon always sought to keep its oil prices "market oriented," especially when Middle East governments or their national companies established prices that could be used as "indicators of market value." But, while those governments frequently made price increases retroactive, the Management Committee believed that all company changes should (where practical) be made on a current, rather than a retroactive, basis. Thus, for example, Exxon International did not recover the infamous 70 percent government increase of October 16, since the committee declined to increase company prices above prevailing world market levels, even though the new price did not fully reflect the additional government share. Nor did the original pricing end the Management Committee's work; for it still expected to make additional adjustments in price based on the final prices of buyback oil from OPEC countries. Briefly then, the spiral of crude oil prices not only tested the corporation's executive management in negotiations and elsewhere; it also established a need for shrewd executive guessers—who filled a void whenever no specific answers could be found. Exxon had both types of managers.

X   X   X

For Exxon's management overall, 1973 proved to be a memorable year, even though one of continuous trial. On November 17, as previously scheduled, George Piercy led an oil company team in negotiations with the OPEC Persian Gulf Committee in Vienna. These conversations achieved nothing, as the companies sought some form of price stabilization, while the oil ministers wished to establish some "mechanism that would allow prices to rise from time to time." After two days of disagreement, the meeting

---

*The Management Committee is composed of the chairman of the Board, president and all senior vice presidents. Constituted in its present form in May, 1969, the committee was established by the chief executive officer to assist him in carrying out his responsibilities for the general care and supervision of the business and affairs of the corporation. The committee has final review authority for all business matters that have not been delegated or that have not been reserved to the Board of Directors by statute, bylaw, or resolutions of the Board.

adjourned, leaving the OPEC members to disagree among themselves on posted prices.

OPEC itself met in Vienna on November 19 and 20. Here, Iran's Amouzegar took a prominent role as he ranted about the lack of company suggestions for new posted prices. Meanwhile, the OAPEC representatives had met on November 19, and approved an exemption of oil supplies for all members of the European Economic Community from the forthcoming 5 percent production cut, scheduled to go into effect in December. This action rewarded those sometime-friends of the U.S. who had refused to follow American leadership, and who enacted instead (on November 6), a set of resolutions which met with OAPEC's approval.[69]

Of all the oil states, Iran appeared most desperate in its need for additional funds. Even the recent price increases of October 16, coupled with production increases, had not provided enough income to reduce Iran's enormous foreign debt. Now, the Shah wanted more money to finance a new five-year development plan—which he claimed to be a political necessity. Anxious to show sympathy, the other Gulf ministers agreed to hold their extraordinary December meeting in Tehran, Iran's capital.

A majority of OAPEC ministers agreed in November to watch spot auctions as possible indicators of oil price, but Yamani and King Faisal were opposed. They regarded high auction prices merely as artificial reflections of recent production cutbacks and of the embargo. They contended that, when the political question of Israel was settled, oil production would increase and the price must fall.[70] Other oil problems also could be traced to the atmosphere of uncertainty, as world demand for oil continued to increase while unpredictable oil supplies led some companies to defer building new refineries or upgrading older ones—steps that might help meet the demand.

In Houston, when the American Petroleum Institute met on November 13, 1973, some oil company executives asserted that new refinery construction would have to wait for firm guarantees of more adequate supplies. Earlier in 1973, Exxon had announced a $400 million domestic refinery expansion program; now, at least in the view of Mike Wright, CEO of Exxon USA, the situation had changed. Before spending to expand, "We'll take a second look," he said; and as if to underline his words, on the same day, Exxon USA began to allocate supplies of certain chemical feedstocks to customers. In Houston both Jamieson and Wright once again deplored the failure of the U.S. government to speed construction of the Alaskan pipeline, and to auction new offshore leases for exploration. While the government delayed, they urged that a "coherent long-term federal policy on imports" be put into place. [71] Then, Jamieson returned to his earlier themes, as he questioned how fast oil stocks could be increased so long as no priorities on consumption existed. Soberly, both executives agreed that the current oil crisis would continue for a decade.

X   X   X

Early in November, 1973, Saudi Arabia moved to acquire 51 percent of Aramco—years in advance of the previously agreed upon date, 1982. Yet, since the Saudis already held effective control over both production levels and pricing structure, Aramco's owners evidenced little concern about this accelerated takeover. In fact, all the oil companies had other things to think about. By December, 1973, spot auctions brought Iran's prices up to $17.34 per barrel, and the Shah clearly expected something close to this record figure for every pint of his country's oil. Elsewhere, the OPEC nations also raised posted prices of their oil, despite some feeble efforts by consumer governments to oppose these moves. Here, the Shah led the way, as he later recalled: "Earlier that year during the Arab oil embargo, we had sold oil on the spot market for $35 a barrel.* That told us something. Demand for oil was so strong, price was no object. Oil had been underpriced for too long."[72]

When the OPEC Gulf ministers met in Tehran on December 22 and 23, four other member nations sent delegations, and Venezuela sent an observer. The stage was set for dramatic events, especially since one week earlier the Shah had told the Arabs that prices should be related to the spot auction prices which Iran was receiving for oil. Now, much to the discomfiture of Sheik Yamani and of the Iraqi delegation, the Shah took a leading role in the conference. As he reported later:

> In response to my call the ministers of OPEC met in Teheran on December 22 and 23, 1973. This assembly decided to raise the price of each barrel of oil from 5.032 to 11.651 dollars. ... I was also convinced that in the long run the world economy would be healthier when oil sold at a price which would foster exploration of other forms of energy.[73]

In the Shah's simplified account, however, too much has been omitted. During the meetings, Yamani and Hamadi—a strange union founded on common distrust of the Shah and suspicion concerning his dominant role in the conference—joined in urging caution. When Iranian Finance Minister

*For 1973, no such spot market price as $35 is listed in any available source.

Amouzegar asked for a market price of $12-14 per barrel, with a posted price of $17-20, Yamani and Hamadi opposed this increase. Eventually, with approval from the Shah, Amouzegar agreed to lower the price figure. Despite Yamani's efforts, however, the other OPEC ministers agreed to accept a posted price of $11.65 per barrel. Speaking for Hamadi and himself, Yamani told the *Middle East Economic Survey*: "In our opinion a lower posted price would have been more equitable and reasonable. However, we went along with the majority." King Faisal expressed his royal anger, for he wanted no such price increase and had so instructed Yamani. But he did not censure his oil minister for joining in support of it.

What happened at Tehran did heighten the rift between Saudi Arabia and Iran, however. And that division would grow. In other ways too, the conference uncovered new grounds for disagreement, such as the problem of defining a satisfactory mechanism to compensate producing nations for the effects of inflation. So, when the OPEC meetings adjourned in Tehran, the Arab ministers moved to Kuwait City, where they decided (December 24, 25) to restore 10 percent of the 25 percent production cut from the September, 1973 production level. At that same time they also cancelled the additional 5 percent production cut scheduled for January, 1974.[74]

Many discussions of this December, 1973 Tehran conference have pointed to the fact that, during the summer and fall of 1973, market prices moved above the posted price. This price disparity failed to increase the governments' take; thus the new increase actually represented an OPEC effort to restore the previous 40 percent margin of posted price over the market price. By making the change, OPEC hoped to squeeze profits which, these discussants contended, the oil companies had enjoyed when the market price rose sharply. So the December price, while it satisfied neither the "hawks" (the radical OPEC states such as Libya) nor the "doves" (the moderate nations), did increase government take and reduce company profits. On the surface, this explanation may seem attractive; but it pays too little attention to the real reasons for the escalation in oil market prices beginning in the summer of 1973.

The overall effect of the Tehran meeting was to raise the posted price of oil 130 percent above the October 16 figure, as well as to boost the market price almost that much. Moreover, the new prices were scheduled to take effect shortly, on January 1, 1974. No wonder then that the news from Tehran provided a shock to the world. Yet—alone of the heads of government in consuming countries—President Nixon protested to the Shah with real vigor. Today, we know that Kissinger actually wrote that protest message for Nixon, which called attention to the destabilizing effect of the new prices on the world economy, and to the "catastrophic problems it could pose for the international monetary system." Pointing out that the price increases could produce a worldwide recession, the U.S. message urged that the Tehran price scale be reconsidered, but it had no effect on the final decision, which was approved unanimously by OPEC. *[75]

For the consuming world, the sudden escalation of oil prices did bring swift changes. Before the February 14, 1971 Tehran Agreement, the highest posted price for any Persian Gulf crude was $1.93 per barrel for Qatar's fine Dukhan oil; while Saudi Arabian light sold for $1.80. According to the first Tehran Agreement (1971), Qatar's oil rose to $2.28 and Saudi oil to $2.18; in June, the figure for Qatar's oil had risen to $2.387 and for Saudi to $2.285. However, on October 16, 1973, Qatar oil rose to $5.834, Saudi light to $5.119, and the price of Abu Dhabi's Murban oil rose to $6.045. After the second Tehran session (1973) the new prices were: Qatar, $12.414; Saudi, $11.651; and Abu Dhabi, $12.636. For the greatest quantity of oil sold to consuming countries, that from Saudi Arabia, the price rose from $1.80 per barrel to $11.651.

Yet the true impact of the 1973 Tehran Agreement—with the turmoil it caused—can only be measured against the price consistency of the preceding decade. At that time, stability of energy prices had been achieved by companies like Exxon successfully negotiating lower prices. As George Piercy asserted: "If you look at ... the prices that we were able to come out of Tehran and Tripoli (1971) with and you look at the 10 to 15 years of a flat price before that, and you recognize the problem that those nations had, I think that it was a reasonably good job."[76] In the end, however, it was less OPEC's success in gaining control over pricing that defeated the oil companies than the defection or abandonment of the companies by the consumer governments, including that of the U.S., for the U.S. Department of State insisted that oil companies come to terms, first with Iran and the Persian Gulf states, and later with Libya. That tacit assistance to OPEC—more than any specific terms in any of the agreements—finally encouraged OPEC to enact its will by taking control of pricing, and thus by punishing consumer countries.

Today, it seems clear that from the beginning, without the support of the

*At the same time, President Nixon had Kissinger write for him a letter to King Faisal. James Akins, who received the letter to be passed on to the King, had become ambassador to Saudi Arabia in November, 1973. Akins showed the letter to Omar Saqqaf, who advised him not to give it to the King. Akins returned it to Kissinger and asked that it be rewritten. It was rewritten and sent to the King. Kissinger does not mention this letter in *Years of Upheaval*, 887-893. Also see Holden and Johns, *The House of Saud*, 352-353.

consuming countries, the only chance for price stability resided in a united company front—in joint negotiations and joint decisions. Every division among the companies vitiated their influence and weakened their power. With the loss of company unity went the last hope of oil market stability. Instead, oil prices shot up to historic levels in less than three years. For U.S. consumers, the total costs of imported oil, for example, rose from $5 to $21 billion. And that 300 + percent increase in the U.S. was more than equalled by the increase in import costs to Europe and Japan. Energy bills soared, and as the industrialized free world suffered unprecedented trade deficits, oil revenues flowed into the oil producing countries at such a rate that these nations literally could not spend their wealth quickly enough.

Any success of a united oil company front must have been based on support from the governments of the consuming countries. In theory, the combination of company and country could have succeeded; but for all practical purposes, few nations possessed leaders with vision enough to foresee the dangers caused by wildly escalating oil prices. Those far sighted persons who did anticipate the economic chaos resulting from 1,000 percent oil price increases lacked the influence needed to persuade their own people of imminent danger. Both in the U.S. and abroad, a combination of political and economic ignorance with a disabling fear, so great as to produce paralysis instead of action, destroyed even hope of effective opposition to the OPEC program.

X   X   X

As a consequence of change, the Tehran price levels prevailed for less than one month. Bolivia raised its posted price to $16, thus surpassing the previous high, a Venezuelan price of $14.08. Next Libya announced on January 2, a posted price of $15.768, setting a new high for that country. More ominously, on New Year's Day, 1974, retiring Venezuelan President Rafael Caldera, in a nationwide television appearance, recommended early nationalization for oil (and other key industrial) operations in Venezuela— wherever a high percentage of foreign ownership existed. Exxon, for example, owned more than 95 percent of Venezuela's Creole Petroleum Corporation.* So, although the Venezuelan President had only little more than two months to serve, Exxon listened to his message with some concern, for it appeared likely that the incoming President, Carlos Andrés Pérez, and the new Congress might take steps to speed nationalization ahead of its

---

*In 1975, Exxon acquired 100 percent ownership of Creole.

scheduled date in 1983. But suddenly, even before Andrés Pérez took office, the Caldera government, on January 7, simply took over seven Creole concessions. And coincidentally, on the same day, the OPEC oil ministers failed again to find any basis for agreement on an inflation adjustment device. All of these changes (and even more price increases) took effect within a mere two weeks—just after the December 23-24 meeting in Tehran.[77] In the world of oil, the pace of change had clearly accelerated.

X   X   X

Escalating crude oil prices, coupled with persistent rumors of overlarge 1973 company profits, caused J. Kenneth Jamieson, Clifton C. Garvin, and other Exxon leaders to decide to join with the entire industry in "Operation Candor"—a program planned by the companies to provide full information about their earnings—and, perhaps, to improve the public's assessment of the companies at the same time. Once committed, under Jamieson's leadership, Exxon plunged ahead, hoping to act decisively to improve the company's public relations, which now seemed to have utterly failed. Already, for more than twenty years, Exxon had commissioned the Roper Organization, Inc. to make surveys of public attitudes on all manner of questions. During the 1970s, that firm took opinion polls for Exxon (and other companies) to assess the public's attitude and response to the oil crisis.

Using the data collected between 1971 and 1976, Roper described the public as cynical, bitter, and uninformed. The first two conditions, Roper believed, could be explained by the fact that people felt that they had been lied to about Vietnam, completely deceived by a President about Watergate, and more recently betrayed by both the CIA and the FBI. So they were not inclined to believe anything told them by "authorities." A majority of these people believed that for almost three years oil companies had lied to them, had created an oil crisis, and then inflated prices for profit. As for the companies, the public's distrust of oil spokesmen became evident in early polls, and it was confirmed by later results. Obviously, oil companies had failed to get their message across to the public. So Jamieson, Garvin, and Wright, along with their industrial colleagues, faced a tough assignment.

Jamieson himself took the lead. Henry B. Wilson, one-time head of the Exxon Public Relations Department, rated him as one of the top communicators to serve the company over a twenty-five year period. Wilson recalls that Jamieson would always come to him after a speech and say, "Henry, how did I do?" Then he adds, "Jamieson, if he didn't want to do it, knew how to do it. ... He could put on a damn good show. ..." And Jamieson did it often; he spent much of his last three years in office trying to improve the public's opinion of Exxon and the oil business.

His subjects ranged from the OPEC assumption of control of prices and production to the energy crisis and company profits. Exxon arranged conferences with the news media in such places as New Orleans, San Francisco, Philadelphia, and New York. The blitz proved quite intensive; for example, on January 11, 1974, Exxon executives held three sessions in Washington, D.C. Both Jamieson and Garvin attended all of these, and they were accompanied by other executives and key resource persons.* At virtually every appearance, questions were entertained and answered. Yet it should be noted here that later analysis of conference tapes seems to reveal a gulf between questioner and responder that rarely was breached; the Exxon leader frequently appeared to be at a disadvantage, perhaps because only a few of the media representatives indicated that they had a sound basis for comprehending the answers. Granted that several of the Exxon resource persons proved unable to eliminate technical jargon from their explanations, which helped widen the chasm; but Jamieson, Garvin, Piercy, Wright, and Collado made clear explanations and frequently used simple analogies to help make their answers understandable.

At one point, President Garvin moved in to assist Jamieson by emphasizing the fact that "a great deal of misunderstanding" existed about oil companies and their profits. Garvin termed the business "tremendously complex and complicated," and remarked at the Washington luncheon session:

> I'll be very candid with you, the people who've got to understand it best in terms of background is the media, whatever form they're in, because you are the instrument that people listen to and so I've got a candid impression it was not to change you today for tomorrow. What we're after is to enhance your understanding over a period of time.

Concerning the company profits, he stated somewhat facetiously: "You know I'm not a guy that goes sneaking off in the corner with a lot of money in his pocket—it's not my money. I'm just a paid manager." Most of that cash flow, plus borrowed money, Garvin said, would be used to fund the "stepped

---

*Some of these were: Emilio G. [Pete] Collado, executive vice president and director; David J. Jones, vice president for Finance; Stephen Stamas, vice president for Public Affairs; Robert Anderson, secretary; Archie L. Monroe, controller; George Weed, assistant controller; George T. Piercy, executive vice president and director and contact director for the Middle East; Howard W. Page, former director and contact director for the Middle East [retired in 1970]; M.A. Wright, chairman and CEO of Exxon Company, U.S.A.; William T. Slick, Jr., vice president for Public Affairs at Exxon USA; and Jim Hanson, chief economist at Exxon.

up ... expenditure pattern for the next four to five years." Arab control of production and prices limited Exxon's planning for any longer period of time.

President Garvin repeatedly stated that he had no idea how much Exxon might have to pay for Middle East oil in the future. As he remarked, "I don't know how high is up." But "up" remained the only direction in sight. Garvin, Piercy, and others mentioned that Middle East governments often insisted that agreements be backdated, and Garvin pointed out that every government he knew about, except Germany, had an oil products price control mechanism. Then he said:

> I've never known of one yet that made a retroactive adjustment and I don't expect them to in this case. Now you think you've got a tiger by the tail—we have—because we're talking about literally hundreds of millions of dollars. ... It's a very bad situation and don't misunderstand me—but our friend Yamani's got bigger fish to fry right now, and he knows he's got us in that sense and he's just not pushing.

To assist in telling the story to the public, Jamieson brought in Wright, who was a polished communicator, plus Exxon USA's vice president for Public Affairs, W.T.[Bill] Slick. Wright explained the U.S. supply and demand situation to persons who attended these sessions. He, Slick, and other executives answered questions about oil-finding, storage, pipelines and refining; but in the last analysis, somewhat wryly, Wright doubted that his answers made much of an impression on the media. As he consistently pointed out to them—as a percentage of average total assets, oil company profits could not be considered exorbitant; however, and more sadly he continued, "I don't know how to convince the man on the street about that." For his part, Garvin admitted there was little specific information about the details of the Arab oil embargo. As for the company profits, he declared simply that the public believes, "anybody who makes a $1.6 billion profit [as Exxon did in 1972] is suspect."

After the January 11 conferences, one important participant, the *Wall Street Journal*, did its best to tell the Exxon story. Jamieson and Garvin pronounced the corporation's profit "record shattering," since it indicated that the per share profit almost equalled the price per barrel "of some domestic crude oil these days." Yet, unlike less thoughtful stories about Exxon profits, the *Wall Street Journal* account also mentioned that before 1972, oil profits had not been adequate; and it even found space to include a

measurement of earnings and profits against the shareholder's equity.[78]
Here, this oil story related background information, and if hardly laudatory,
at least it made sense of the profit question.

At another well-attended news conference, held on January 23, 1974,
Exxon brought together a large number of key executives who could be
expected to provide specialized information to the media. Jamieson
explained when the meeting began, "They will provide me with the answers
which I may not have." Then he laid out the basic figures for 1973; Exxon
earnings totalled $2,440 million or $10.89 per share, an increase of 59 percent over the 1972 earnings of $1,532 million and earnings of $6.83 per share.
Next, he broke down 1973 earnings by geographical region: U.S. earnings
showed profit margins of 12.4 percent (up from the 1972 figure of 11.4 return
on average total assets); Western Hemisphere earnings outside the U.S. represented profits of 13.2 percent in 1973. Finally, the CEO explained the large
increase in Eastern Hemisphere earnings: Product prices had been low for
several years, and after 1971, they were raised in the face of increased
demand; and devaluation of the dollar also resulted in increased earnings in
other currencies. Overall, Jamieson said that Exxon averaged 32 cents profit
per barrel of oil in 1972, as opposed to 25 cents profit in 1973—simply put,
the volume sold made most of the difference in total profits, although chemical and other related earnings also rose significantly in 1973. Here, his formal presentation ended.

Now answering questions, Jamieson advanced as one reason for holding
the meeting, the fact that "the industry has been under such attack for
secrecy for not trying to inform the public." In this session, Exxon was trying
to meet the public's need to know. After that, Jamieson went to the particular: the industry had to keep inventory stocks higher than usual because of
the Arab embargo; he also pointed out that the energy industry needed high
profits for exploration, development, and research. Other questioners
wanted to know why the U.S. government had been forced to come to the
industry to obtain accurate statistics on production, etc.; and here, the
Exxon leader had to confess that he did not know the answer, especially since
Exxon filed with the various governmental agencies more than 150 reports
each year. But the Exxon CEO gave a commonsensical answer:

> In some cases, those reports are not too well coordinated within the
> government itself. But you have a new situation where we have this
> energy problem on our hands, and I think the government right-
> fully so is saying 'We need more information than we had in the
> past so that we have a better understanding of the problem.' And in
> the past, they didn't need all the information.

Later, Jamieson defended Exxon's record profits—"I am not embarrassed." But he restated what was painfully obvious when he admitted: "To be quite frank, our public image is at a low ebb."

President Garvin also explained Exxon efforts to contain inflation; he stated that the amount of oil required from producing countries constantly increased, and that the company's choice was to buy or not. Usually Exxon bought, although the consuming governments (not the company) regulated what portion of the cost could be passed on to the consumers. At the next stage, Exxon's allocation of oil (85 percent of total 1972 volume allocated for 1973) to dealers put severe pressure on many of them, Garvin remarked, and this cut into dealers' profits thus endangering good relations with the company. To another questioner asking whether costs of crude oil might soon go down, Jamieson replied that Exxon saw no indications that they would. Here, Garvin digressed to point out that Middle East oil represented only a small part of the entire oil price problem, saying, "There is a lot more Venezuelan oil coming into the United States than there is Mideast oil."

During the conference, the media representatives seemed to be gaining a larger understanding, although they remained skeptical, as press reports demonstrate. The *New York Times* acknowledged that Jamieson "conducted the news conference with an easy self-assurance and was evidently well-prepared, seldom shunting questions to other officers present," but otherwise that paper showed little appreciation for Exxon's position. Perhaps the *Financial Times* (London) showed the most complete understanding of what the company's CEO had undertaken to do on that January morning in 1974—and from the beginning of his tenure:

> Ken Jamieson has constantly attempted to project the contribution and role of the major oil companies to the public through speeches, press conferences, advertisements and private contacts .... he has established an unusual reputation among his staff for fairness, patience and decision and a reputation outside the industry, particularly with Governments, for integrity and international outlook.[79]

Aside from Jamieson's personal achievement, the press proved less sympathetic or even understanding on the matter of earnings and profits. At least one item of planned expenditure—the huge $6,100 million set for exploration which would increase every budget through 1977, did find a place in most accounts: but not much more of this Exxon story reached the general public through the media. For the latter, the company effort to level with them must have been taken as a non-event. So Exxon began 1974 by defending itself against charges of unusual financial success, while actually this temporary increase in income had resulted from the company's limited success at the bargaining table. The allegations, rumors, and stories in the media simply were not offset by all the efforts of Exxon executives to furnish accurate information; public clamor increased.

At the time Exxon sought to set the record straight, it was competing for public attention with the televised and more fully reported hearings of the Senate Permanent Subcommittee on Investigations of the Committee on Government Operations, chaired by Henry M. Jackson. Here, in a theatrical forum, reputations were being made and lost, as senators competed for the committee limelight. As an obvious result of the sudden glut of publicity, public indignation focused on the oil industry, and oil companies became identified as villains—self-seeking creators of an artificial oil crisis. The bright lights of free publicity helped the senators to identify their target; while no defense based on logic, fact or reason could guarantee fair judgment in such a forum.* The kangaroo court met to show the public that the oil shortage had been occasioned by neither OPEC nor the Arabs; rather, it was created for profit by American and British oil executives. Senator Jackson's remarks to Exxon USA Vice President Roy A. Baze exposed these proceedings at their nadir. As Baze vainly tried to explain the complex nature of Exxon's operation, Jackson interrupted: "Those are just childish responses Mr. Baze."

Senator Jackson also threatened to subpoena Exxon records, because Baze could not supply detailed information about the parent company's last quarter, 1973 earnings.[80] Overall, television proved its potency as the medium of drama rather than news. For Exxon and the oil industry, nothing in the press or on the screen suggested that Americans better understood the real problems that the company faced in the Middle East.

X   X   X

While Exxon and other oil companies were seeking to restore their tarnished credibility with the public, OPEC also kept busy. On January 7-9, OPEC held its first 1974 meeting in Vienna, where the oil ministers began by ignoring an increase in the value of the dollar. Instead, they decided to eliminate the previously operative principle that posted prices of oil were to remain 40 percent above the market price. By this time, the confusion

---

*After hearing the responses by Senators Jackson and others to testimony by oil executives, representatives of the industry charged that Jackson ran the hearings like a "criminal trial."

attendant constantly changing prices in the spot market, production cutbacks, and the embargo had served to make that statutory differential virtually impossible to maintain. After January, 1974, oil market prices generally
approximated 93 percent of the posted price.

Despite an agreement reached with three Persian Gulf states that they
should have 25 percent equity in the companies operating in their domain, it
soon became clear (by January 1, 1974) that all producer nations were determined to press for at least 60 percent participation.* This response represented one proof that the newly acquired power of OPEC had not generated
sufficient confidence to impress at least some of the oil ministers. So, beginning in January, 1974, they continued to demand ownership shares in the
operating companies. Kuwait led the way in January, 1974, with a 60 percent takeover of the joint concession previously owned by British Petroleum
and Gulf Oil Corporation. Then, Saudi Arabia, Qatar, and Abu Dhabi
moved in quick succession to increase their participation in companies
within their domains. This trend of accelerated takeovers quickly resulted in
government ownership of majority shares in every producing company—an
ownership achieved without prior planning or much preparation on the
part of OPEC governments. Once begun, the participation movement developed its own special momentum; and no resistance proved useful in stopping
it.

Perhaps because of the success of the participation movement, the January, 1974 OPEC conference did not enact additional price increases. During
the same month, U.S. Defense Secretary James Schlesinger warned (as Senator Fulbright had earlier) that if the Arabs persisted in their oil embargo,
they "risked" military confrontation. Finally, on January 9, President Nixon
announced that he had invited leading consumer nations to attend an
energy conference on February 11. Like the Schlesinger threats, this American initiative served only to increase Arab anger; and even today, it remains
unclear precisely what Nixon hoped to achieve, although he seemed to be
announcing a new role for the U.S. as leader of the consuming countries. At
Exxon, at least one observer pinpointed the reasons for Arab hostility to any
union of consuming countries. Arabs were certain to be opposed:

> on the grounds that it will lead to a confrontation between pro-
> ducers and consumers. Clearly most do not want to see their lever-
> age reduced; and agreement among consumers on such issues as

*Saudi Arabia, Abu Dhabi, and Qatar signed the October, 1972 agreement. It was never
implemented.

supply sharing, price bidding ceilings, and limitations of unilateral government to government deals could reduce producer power
to some extent.

Henry Kissinger also irritated the Arabs by terming their continued
embargo "pressure tactics." He cited the peaceful role played by the U.S., as
well as his own personal efforts to achieve a Middle East peace. For Kissinger, and for the entire Nixon administration, it appeared obvious that
their concern centered on the embargo rather than on the escalated price of
crude oil. Kissinger said: "If now pressure tactics are continued, it can only
be construed as blackmail.... ," and it seemed clear that the Secretary of
State spoke for several members of the Nixon cabinet, even while the Arabs
boiled over the Kissinger remarks.

Meanwhile, Nixon's own problems, especially those connected with
Watergate, served largely to undercut American effectiveness as leader of
the consuming countries. For despite their acceptance of his invitation
(except for France), consuming nations continued their "intense scramble"
for "special relationships" with OPEC. Between their self-interest and his
notoriety, Nixon's union of consumers was doomed to be a hypothetical goal
only. Among other countries affected, Britain and France offered to barter
arms, technology, and industrial goods for oil. In fact, so anxious were the
French to curry Arab favor that they entered into a contract with the Saudis
for oil at a price higher than Aramco would have charged. (France was not
under embargo.) Britain (and other nations) signed similar contracts.[81]
Across the globe, it appeared that one theme was constant: A threat of
shortage—and in every Western country and Japan, the need for guaranteed
supplies of oil determined the course of practical diplomacy.

X   X   X

Aware as always of power politics among nations, Jamieson continued
nevertheless to present the case for Exxon to what he believed was a poorly
informed American public. In a speech on January 28, 1974, before the
Economic Club of Detroit, which proved to be, for some reason, a hostile
audience, he warned that energy problems would persist far beyond the time
when the Arab embargo ended. In the same talk, he admitted that Exxon
had underestimated energy consumption during the 1970s by at least by 5
percent per year; while, conversely, the company had overestimated coal
production and utilization. Oil had become much more important, as

Jamieson pointed out, and only a few persons in either the industry or the U.S. government had anticipated the problem of supply posed by Arab unity—especially in the form of an oil embargo. In the U.S., oil companies had failed to build new refineries because the quota system prevented their bringing in the necessary foreign oil, and also because of uncertainties occasioned by irregular crude supplies, high construction costs, and the opposition of environmentalists.

Already, Jamieson had noted changes in his personal life style resulting from the growing credibility gap between Exxon and the public. Even so early in 1974, rumors and allegations that loaded Exxon tankers hovered offshore, waiting for the next price increase, filled the news media. So Jamieson remarked, "Cocktail parties used to be fun. Now they're inquisitions all presided over by some fellow who wants to know why we have all that oil loaded on tankers offshore waiting for prices to go up."

The oil shortage affected Exxon's chief executive officer in other ways too. When he drove to his favorite service station for gasoline, he took a place in a long line. One Saturday during February, 1974, when Jamieson drove into the station, the manager told him he had no gasoline but a truck was expected quite soon. The Exxon CEO decided to wait, and he kept his place in the line of cars while he listened to the conversation of other drivers. One customer talked to the manager, and then turned away to say, "I'm going out and picket Ken Jamieson." The manager quickly replied: "There's no need to go to his house to picket him. He's standing right over there." What happened next, the newspaper did not report; however, the writer did include some of Jamieson's opinions about the oil shortage.

Naturally enough, Jamieson felt that something should be done about lines at gasoline stations. This, he believed to be the task of the Federal Energy Office [FEO] which allocated gasoline to the states. Jamieson did not advocate rationing, however, as he had so often remarked, he felt that the oil shortage would last five or more years, and nothing had happened to change that opinion. Exxon's CEO advocated a program of conservation to help ease the shortage. He said people should walk every place they could, and where possible, use public transportation. These represented an easy beginning for conservation, and it would save gasoline. In an aside, he suggested that it was not inconceivable that cities might "bloom" again as people moved back into them. But even for Jamieson, no extra gasoline emerged from the pump. In public and private addresses, the Exxon chief tried to restore some balance in public opinion. The popular distrust of all oil companies, and the public's lack of information about oil deeply bothered him. Jamieson and his Exxon associates viewed this widening credibility gap with alarm.

On June 1, 1973, he promoted Stephen Stamas to vice president of Public Affairs, replacing Charles O. Peyton, who became head of Exxon International Company. Stamas, a one-time Rhodes scholar, held the Harvard doctorate in economics, and he had served for years in the U.S. government. Earlier, he had worked in the Exxon Financial and Planning Departments, as well as in Public Affairs. In 1974, Exxon needed such diverse talents in public relations; as Jamieson told an interviewer:

> We got everybody involved in the problem, I would say, that we could possibly think of. Of course, that was when we really started beefing up our whole P.R. effort. Certainly, as an individual, I was forced to play a much more public role than I had ever in the past, probably than any other chief executive in the history of the company had ever. We were forced into it, and we recognized that we had a terrible information gap, and we started trying to do everything that we possibly could about it.

Jamieson also hosted a luncheon for representatives of all the major television networks. Walter Cronkite, Harry Reasoner, and John Chancellor came with staff people. After a brief talk, the CEO told them he'd like to hear what they had to say about Exxon's failures—or successes—in supplying them with needed information and answers. He recorded their reaction: "Well, look, we tried to get in touch with you fellows, and we can't get any response." On the basis of that reply, Jamieson moved quickly: "We set up a little television studio on the lower level of the building so that when they'd call up about two o'clock in the afternoon and say they'd like a statement from Jamieson [or some other Exxon executive] for the seven o'clock news, we'd say 'Come on over.'"

Soon, he and other company officials found that certain requests from the press, radio, and television had not been properly handled; "... in a sense, our lower level P.R. people were shielding us from these inquiries and had built up a lot of resentment in the press, and in the other media also. So we started doing things about that." Jamieson, Garvin, Wright, among others, worked at resolving the information problem; they became available to the news media, and this helped to change some preconceptions about Exxon. But no sudden shift in the public's attitude toward the company occurred. In September, 1975, Roper Associates found attitudes toward Exxon (and other oil companies) still remained about what they had been across two decades. Much of the public's suspicion concerning big oil can be traced to the industry's relations with the Arab nations, profit levels of Exxon and other companies, and their role during the oil crisis. For the press, at least, these charges became a national indictment.

X   X   X

At almost the same time that Jamieson sought to get the company position before the public, Senator Frank Church, quoting freely from testimony before his committee and from documents it had subpoenaed, placed full responsibility for the energy crisis on the oil companies and the U.S. government. Although his charges for the most part were dramatic, many of the senator's conclusions could not be verified. For example, little evidence from either testimony or documents actually supported Church's conclusion that a major division had existed between big and little oil companies in the days shortly before the first Tehran negotiations (1971). Following Church, Senator Jackson charged that Exxon in particular had stopped supplying products to the armed forces during the embargo—an action he imputed to the company's lack of patriotism.* The press took up these accusations and added them to the indictment. As one small town newspaper editorialized: "The oil embargo is a spin-off from bungling within the United States." Neither Exxon nor the U.S. government was spared in the press campaign to denounce real and imagined examples of domestic ineptitude. In this atmosphere, few voices spoke kindly about the oil industry.[82]

Jackson's subcommittee focused its investigation on Exxon, especially on the company's role as part owner in Aramco before and during the Saudi Arabian embargo. Now Exxon International President Charles O. Peyton took his turn before Jackson's group, telling them that the cutbacks had combined with the embargo to shorten Exxon's supplies by one million b/d (or about 17 percent of total demand). His company, he said, had notified the U.S. State Department by cable of the Saudi boycott on October 21, 1973, the same day that notice was received by Exxon. In fact, Exxon communicated every Saudi order to senior officials of the U.S. government Defense Department (as well as to other departments), and company officials also went to Washington to discuss the situation with prominent government officials. So, the company did keep the government constantly informed about its plans, including deliveries to the U.S. armed forces of some 45,000 b/d, which, after the Saudi cutbacks, derived from alternate

---

*As was explained earlier in this chapter and is reiterated in the next paragraph, this charge was erroneous.

sources. In summary, Peyton said, "At no time have we been told by the U.S. Government that national security interests warranted any change in our actions."[83] Thus, patriotism was not a real issue, despite the fervor.

When the shareholders met on May 16, 1974, in Los Angeles, Jamieson again defended Exxon's actions regarding deliveries to the military. He pointed out that the U.S. Department of Defense (and others in the U.S. government) knew at all times exactly what the company was doing and what the company planned to do. Responsibilities for supporting U.S. military forces, both at home and overseas, were never ignored. Then, turning to the future, Jamieson laid out several principles that the nation should follow: first, the U.S. had to be made more energy efficient, especially through effective conservation; second, new proposals to tax energy companies should be considered as justifiable only within the context of larger national goals, to keep taxes from becoming counter-productive; third, the U.S. and other consuming countries should "pursue the objective of reducing their growing dependence on foreign oil imports." Finally, Jamieson declared that, despite the economic revolution of 1971-1974, the large oil companies would continue for many years to play a vital role in providing consuming countries with energy. As usual the speaker insisted that stockholders and executives must share a long-range view.

A pointed question to President Garvin brought the meeting back to the present, when a shareholder insisted that Exxon was doing a poor job of keeping the public informed. In reply, Garvin first recounted everything that the company had done and was doing; then he concluded: "All I can say is that we will just have to triple and quadruple this type of effort. Economic education in this country has got to become a must." No oil executive in the room would have disagreed. Yet all available evidence indicates that Exxon shareholders received management's answers with confidence and good will, perhaps because the company spokesmen had done their homework, as they filled the air with statistics. Naturally, shareholders were pleased to hear that first quarter, 1974 earnings showed an increase of 39 percent over the comparable period in 1973.

The shareholders also were pleased to learn that the December, 1973 *Dun's Review* had picked Exxon as one of the five best managed U.S. corporations, terming it "The Nimble Giant" because of "a unique ability to see the future." Certainly Jamieson and the Exxon shareholders had derived some measure of comfort from the statement made by Jerome Levinson, staff counsel to the Church Committee, quoted in the *New York Times* a few days earlier, when the newspaper ran a feature story on Exxon. As Levinson described the company:

Perhaps most notable, its [Exxon's] grasp of global politics and of enlightened self-interest enabled it during the 1973-74 Arab oil embargo to shuffle oil supplies around like checkers. No nation was denied oil, but the companies nevertheless honored the letter of the embargo. Exxon had the clearest vision of balancing all the interests and the need not to push anyone to the wall. ...[84]

But no one at the stockholders' meeting could feel entirely comfortable with the unspoken implications of the gathering. Although the point was never raised, in 1974, everyone who followed oil knew that the real power over the world's supply rested with the producing countries. No increase in corporate profits could alter that fact.

X   X   X

In the Middle East, OPEC's structure already appeared to be flawed: Kuwait, after assuming 60 percent participation in the BP-Gulf concession, still could not force the two companies to meet its price demands for its share of crude. After negotiations in May, 1974, other Gulf states joined with Kuwait in setting a buyback price schedule which, during the following October, would reach 93 percent of the posted price—a slight drop from the 94 percent that Kuwait had demanded for oil sold from January to March. But the more radical states could achieve no greater gains, largely because, on the one hand, Saudi Arabia favored reduction in per barrel prices; while on the other side, the United States pressed all producers for a substantial price rollback.[85]

And prices did drop. Later in 1974, this decline seems to have resulted from a worldwide economic recession, brought about in part at least by the record-breaking oil prices, set by producers late in 1973—when high energy costs produced a ripple effect throughout the world economy. At the same time, overall oil demand decreased, due to both the mild 1973-1974 winter and the higher prices being charged for oil. As one measure of the changes, spot prices for quality Persian Gulf crude declined from a December, 1973 quoted high of around $17 to less than $10 quoted per barrel in September, 1974. In brief, instability ruled the market.

Aramco, battered earlier in the same Jackson subcommittee hearings where Exxon was singled out for the largest share of blame, now confronted (during 1974) a further threat of seizure by the Saudi government. Conversations on that subject started in February, when Yamani found the owner companies "uncooperative," and before the end of that month, he informed Liston F. Hills, Aramco board chairman and chief executive officer, that

Saudi Arabia now considered invalid the October, 1972 agreement that provided for 51 percent Saudi participation by 1982. Responding to a question about Saudi plans for Aramco, Yamani replied, "I can say two things. ... The present arrangement is not satisfactory and a 60% deal isn't satisfactory." In fact, Aramco was never told what would be satisfactory, before it was simply taken over, later on.

For the moment, the Saudi government continued talking while George T. Piercy testified about the sharp drop in Exxon's Aramco profits for 1973. Piercy contended that his company, despite this showing, remained committed to lower world oil prices. As he saw Exxon's struggle:

> I can assure you as one who has been in this thing continually since 1970 that if we have not been fighting then I don't know what fighting is. We have taken a lot of personal abuse. Our record is clear, it is defensible, it is honorable.

In an earlier press conference, Piercy had made the same points, but with much more emphasis.[86]

Yet public opinion remained skeptical, especially when it was revealed that in its first quarter, 1974 financial report, Exxon had reduced profit totals to accommodate future taxes on stored oil, as well as in anticipation of another increase in the price of Mideast crude oil. While securities analysts disagreed on whether or not this handling of quarterly profit figures accorded with accepted accounting practices, they did agree that quarterly reports did not have to meet the same strict specifications of annual reports. In an effort to clear the air, Jamieson pointed out that company accounts had been reviewed by the firm's auditors, Price Waterhouse & Co.

Jamieson surely must have chuckled when he first read about himself in *Current Biography* (June, 1974), especially a line about his being "The single most powerful man in the American oil industry." While that sort of assertion remained difficult to prove, it seemed more obvious that none of his power influenced in any favorable way the general consensus that the great international oil companies had been, and were, "ripping the public off." Jamieson's role continued to be a lonely one.

Elsewhere, President Andrés Pérez of Venezuela moved to nationalize all foreign oil companies on May 17, 1974, by appointing a national commission to plan the actual takeover. The President promised to pay adequate compensation—an amount not to be based on the value of assets however— and he predicted that this reversion of ownership would take place within two years. Since Exxon owned more than 95 percent of Creole, Venezuela's largest producing oil company, dismay was general among Exxon executives. But what could they do? Pérez made the point aptly: "Now is not the time for judgment or evaluation of the international oil companies now managing the oil business. They are our partners and should be treated as such."[87] Yet as Venezuela's partner, Exxon was being retired—with or without its consent.

In fact, the Venezuelan decision helped to shape Exxon's fate in the Middle East: It paved the way for acceptance of a Libyan nationalization of 51 percent of Esso Libya, in the spring of 1974; and both Libya and Venezuela together encouraged Saudi Arabia to come to agreement with those OPEC nations that favored increased participation. Noting these actions, and Exxon's reaction— and while many Saudis, including Yamani, continued to express admiration for the high quality and performance of the Aramco personnel—the Saudi government moved ahead with its plan to take over ownership of Aramco. One news story, issued on June 10, 1974, suggested Saudi Arabia would acquire 60 percent participation. That account appeared, significantly enough, just before the OPEC meeting in Quito— just in time to give Sheik Yamani additional leverage when he argued for lower prices and lower taxes.

OPEC did meet at Quito, and a struggle for control ensued. Yamani and the Saudis threatened to set their own production and price levels. By now, the other OPEC members realized that the Saudis—with their vast reserves—could develop enough spare producing capacity to establish the world oil price; so the Quito meeting became a standoff between OPEC factions. They did agree, however, that royalty rates were to be raised by 2 percent, thereby increasing each government's take.

In September, when OPEC met again, all members (except Saudi Arabia) agreed to raise royalty and tax rates, effective October 1, 1974. The new royalty rate rose by 2.17 percent, and the tax rate by 10.75 percent, thus increasing the per barrel price by 33 cents. The OPEC producers cited excess profits by multinational oil companies as justification for the raises, while almost everyone who watched OPEC agreed that no end was yet in sight. Rather than a goal, participation now seemed to represent a mere holding action, sufficient until the world oil situation settled a bit. No final resolution on prices or ownership appeared in view.[88]

At home, Exxon confronted other problems. A company request for an injunction against the mandatory crude allocation program of the Federal Energy Office brought a speedy hearing in the U.S. District Court, District of Columbia. FEO had ruled that the act of Congress that created it also required Exxon (and other companies) to sell crude oil to independent refiners. In Exxon's case, the amount involved was 210,000 b/d. On its part,

Exxon contended that, since the company had only 11 percent of the total U.S. refining capacity, it should be required to provide no more than 11 percent of all crude used by U.S. refiners. But FEO ruled that Exxon should furnish 20 percent of refinery capacity, at a price claimed by the company to be below replacement cost. While this case was before the court, Federal Energy Administrator John Sawhill publicly charged the big oil companies with "foot dragging," and with endangering the success of an entire government program designed to make independent refiners competitive. On July 2, 1974, the court refused to grant the injunction, and Exxon bowed to FEO.[89]

When negotiations for the Saudi takeover of Aramco appeared to be stalled, King Faisal called a meeting of the Persian Gulf ministers, for November, 1974, in Abu Dhabi. At the same time, he charged that the oil companies had sought to involve both the U.S. State Department and the news media in a worldwide campaign of opposition to the Aramco takeover. Then, at the Abu Dhabi meeting, Yamani secured the Gulf ministers' approval for new higher tax and royalty rates that reduced company per barrel profit margins. Crude oil posted prices, all agreed, should be reduced by 40 cents per barrel. The overall effect of the changes was that the average government take rose by 4 percent, or 40 cents per barrel—from $9.728 to $10.125. The new price of buyback oil, at 93 percent of posted price, was $10.46. This made the difference between the tax-paid cost of equity oil and the new market price extremely small; the new tax and royalty system had made equity ownership a potential liability for the companies, rather than an asset.

Further study of the Saudi oil fields indicated that large amounts of capital as well as advanced technology and perhaps, ingenuity would be necessary before major increases in production would be possible. The necessary capital would have to come from the four owner companies. Yet, with the new high price of buyback oil, as well as a soft oil market, this combination could result in negative cash flows for the equity owners. So confronted by the possibility of a severe squeeze, the corporate owners yielded to pressure and agreed in principle to a 100 percent Saudi takeover of Aramco.

Later in December, at the OPEC conference in Vienna, the ministers endorsed the Abu Dhabi decision, specifying in the announcement of the OPEC action that "the new average government take figure of $10.12/barrel for marker crude"* would continue. This communique indicated that the posted price had been phased out of use (except in a few cases) for tax

---

*The benchmark price.

reference purposes, as Iran had suggested doing earlier. The ministers also agreed that the $10.12 price would hold until the last of September, 1975.[90] This action, in effect, closed a long chapter in the history of Exxon in the Middle East.

X   X   X

In 1970, gross production of crude oil and natural gas liquids from the Middle East and Africa—principally Libya—was more than double Jersey's production from all U.S. sources; this in turn was more than ten times the production of the Jersey affiliate in Canada, and almost 25 percent more than that from Jersey's other Western Hemisphere sources (largely Venezuela). In Jersey's total sales of petroleum products, Europe bought 20 percent more than did the United States. These figures supported the idea that major changes loomed ahead for the Jersey organization—changes that would require a full reassessment of the entire organization. Even allowing for transfers of types of crude oil to maintain the correct quality needed within these sales regions, in 1970, it was perfectly clear that the Middle East and Libya contributed almost 40 percent of the oil refined within the complete Jersey organization. Thus, that region became the major source of Jersey crude and natural gas liquids, replacing Latin America—Venezuela largely—just as earlier, Latin America had surpassed U.S. production.

Yet long before 1975, Jersey's Middle East and Libyan organizations had ceased to have more than minimal control over production or management of properties there; while changes in the flow of oil began to show a marked effect on all the countries and peoples in the Free World. Of course, this economic revolution (for so it was) did not affect Exxon alone, nor did it occur by simple default. Rather, this global change involved in the same bitter struggle every oil company with operations in the Middle East and Africa—and , in the end, all oil companies everywhere. But just as had been the case in other years and in earlier crises, Jersey executives were often selected as industrial leaders to head committees and lead teams of their fellow oil men. Of equal importance, among its top executives, Jersey numbered persons who possessed solid experience and true comprehension of the issues confronting the company during these years of transition. Jamieson and Piercy for example, proved as able as any representatives of the West. Yet their stories, like those of their colleagues, remain tales of loss.

Just why the oil companies failed to retain their Middle East properties is neither clear nor simple. At this time, at least, no single explanation provides a satisfactory answer. But in time, perhaps, the continuing study of company records as they become available* and of the documents of several governments (especially those of the United States) will supply enough additional information so that better judgments can be made. At present, many U.S. oil men believe that the revolution of the 1970s resulted directly from the failure of their government to pursue any consistent overall policy in the Middle East. Certainly, the U.S. government did not support U.S. business abroad, as had been traditional, and this failure proved a major factor in the success of the OPEC nations.

To students of world affairs, and especially of the Middle East, the success of OPEC (1970-1973) marked the end of an epoch. Whatever else the oil companies may have represented, never again would they exercise influence and power in the world petroleum business comparable to what had been theirs before 1970. Nor did the governments in which oil companies were domiciled effectively replace them at the bargaining table; rather, after 1970 some combination of producing nations and natural economic forces worked together to determine oil prices. Individual oil companies could not compete. Yet Western experts familiar with the internal problems of several Arab nations did not agree that the OPEC cartel would work indefinitely. Howard W. Page expressed his doubt as early as 1974 that the OPEC nations could long hold their cartel together. He foresaw internal economic and social pressures within the member nations which would increasingly disrupt cartelized production and pricing.

On his part, Sheik Ahmed Zaki Yamani often warned other oil ministers that, when they broke ranks, OPEC would be doomed. Yamani feared, as had King Faisal, that if OPEC priced its oil too high, consumer nations would reduce consumption, thus creating a surplus of crude that would cause prices to drop and require production cutbacks. So Saudi Arabia, in 1973, counseled its OPEC allies to keep prices down. Even the Shah of Iran seemed to sense the need for continued producing country unity when, in 1971, he told the press that one major defeat would crush OPEC. As the decade of the 1970s drew to a close, no solution to OPEC's power had yet made its appearance.

X   X   X

For Exxon, a memorable era ended with Jamieson's retirement as chairman and chief executive officer. To the end, his personal schedule remained

---

*The records of Atlantic Richfield Company [Arco] and British Petroleum are open to scholars now. The Royal Dutch-Shell Group is gradually opening its records also.

demanding—filled with speeches, media appearances, and virtually incessant travel. In the months before retirement, Jamieson kept telling the Exxon story to the American public.

On January 27, 1975, for example, he appeared on the television show "Washington Straight Talk," hosted by Paul Duke. Jamieson answered Duke's questions with confidence and a mastery of the world energy problem. As he viewed it, while there would be no significant reduction in oil prices for some years, OPEC could be expected to stabilize prices, and then to control production in the effort to prevent prices falling again. Oil, Jamieson believed, had become a political weapon; hence, every decision regarding OPEC was also a decision that involved American national security. Yet he opposed any military action against Arab nations, in part because of the ease with which oil fields, pipelines and loading terminals could be sabotaged.

Jamieson also told Duke that President Gerald M. Ford's* recent proposal to conserve energy by adding new taxes was sound, although Jamieson questioned whether anyone had yet devised an effective means for reducing consumption. Altogether, the Exxon leader defended his company, even though he freely admitted that its long-range planning had not been adequate. But neither had the U.S. government planned for what had happened, as he said. Finally, Jamieson agreed with Duke that the public's attitude toward oil companies had changed—and not for the better. The situation, in fact, now seemed almost adversarial—American government versus American business. But a remedy would be found, and the energy problem would be solved.[91] The final note was optimistic. Three days later, in Miami, Jamieson covered almost the same ground in an address to the American Bankers Association. For this more specialized audience, he added new signs of future improvement—changes could be expected when the Alaskan pipeline began to move oil south, when North Sea production entered the market, and when new offshore leases in the U.S. began to produce oil in commercial quantities.

Of course, these two instances are only samples of the many speeches, interviews, and appearances that made Jamieson a public figure in all seasons. One thing is certain: Jamieson made a major contribution toward restoring industry credibility, and to the Western World's information about petroleum, by his straightforward efforts to explain the oil crisis.

The week before he stepped aside as CEO, *Exxon Manhattan* featured an interview, "Ken Jamieson Looks Back," and on a television show, "Exxon

*Ford became President when Nixon resigned, August 8, 1974.

Conversations," he, with host Andy Purcell questioning him, provided employees a glimpse of some of the events in his career: "It's a Picture of Ken Jamieson that few employees had the opportunity to see." Characteristically Jamieson minimized the significance of his achievements, while he found time to include things he had disliked about the job: "Something I won't miss at all are those numerous black tie stag dinners. New York City is the only place in the world where I could wear out two dinner jackets in one year."

Having reached retirement age, Jamieson stepped aside on July 31, 1975, passing his responsibilities to Clifton C. Garvin, Jr., who had been marked for the top post much earlier. The Exxon Board asked Jamieson to continue as "outside director."* As usual at Exxon, succession was orderly, even during a turbulent time. Jamieson moved to Houston, Texas, where he had lived earlier, while serving as president of the Humble Oil & Refining Company. The Texans, he once remarked in jest to an interviewer, readily accepted him because they assumed his birthplace of Medicine Hat (Canada) could be found somewhere in West Texas.[92]

*"Outside" means not a company employee.

Epilogue

Exxon after 1975

TODAY, IN EVALUATING the substantial attempts of Frank W. Abrams, Eugene Holman, Wallace E. Pratt, Orville Harden, Robert T. Haslam, and others to prepare Standard Oil Company (New Jersey) for the second half of the 20th century, it is interesting to note that many of the changes they initiated came to seem less important during the years, 1950-1975. Jersey constantly had to adapt to changed world conditions—circumstances that no wise leader had been able to predict.

The corporate personnel selection process—certainly one of the finest in the world—remained in place throughout these 25 years. While some of its features dated back to the administrations of Walter C. Teagle and Ralph W. Gallagher, under Holman and his successors, the Compensation and Executive Development Committee became a powerful management tool as well as a supplier of top executives. M. A. Wright, chairman of Exxon Company, U.S.A., once remarked that the committee always had several well qualified prospects for every major opening. Exxon achieved fame for developing good managers.

As the third quarter of the century ended, it seemed clear that Exxon's Middle Eastern ventures had turned sour; of its national partners in the region, only Saudi Arabia continued to be a relatively reliable source of crude oil. Uncertainty attended every effort to obtain oil from Iraq, Iran, and Libya, and virtually all Exxon property in OPEC nations, including Saudi Arabia, had been nationalized—much of it without adequate compensation. Once this situation became clear, the company moved as rapidly as possible to increase supplies from non-OPEC sources, and with significant success increased its program of exploration.

Herbert Stein, a noted economist, writing "Commentary" in *Newsweek* on November 19, 1979, effectively summarized the American public's view of the organization of Petroleum Exporting Countries. Already OPEC, which began life on October 16, 1973, has assumed full control over their oil. Now Stein wrote:

Americans are confused and angry. They face rising energy prices and disruptions, which they suspect are unnecessary. They feel they are victims of a vast conspiracy among OPEC the oil companies and the government, and that the plot would be unmasked if the facts were known.

Earlier, in 1975, Exxon Corporation's CEO, Clifton C. Garvin, Jr.,* had written in the company *Annual Report* to explain why this feeling about the oil industry persisted:

After decades of readily available low-cost energy, it has been hard to accept the fact that energy is not limitless in any practical sense and that the only incremental supplies still readily available come at high cost, established as a matter of national policy by the member nations of OPEC.**

Naturally, Exxon did not welcome this shift in control to the producing nations any more than did the buying public. Yet the company had no choice except to recognize OPEC's influence as a fact of oil life—a condition that Exxon must adjust to.

Garvin as CEO for eleven years directed Exxon in a series of bold steps that served, even in this troubled era, to stabilize the corporation in its role as the top-ranked U.S. petroleum company. Writing in 1976, he remarked that despite the "nationalization of certain affiliates abroad . . . it is essential that Exxon find ways of working constructively with both producing and consuming countries even as we attempt to find new sources of oil and develop other forms of energy which will lessen that dependence . . ." on OPEC.

Finally, the Arab oil embargo and the price escalations of and after 1973 accomplished something larger than merely forcing change upon the oil industry; as a result, the entire structure of the world's economy was altered. Over five years oil prices rose from approximately $2 per barrel to about $40. A disproportionate oil price, as Saudi Arabia's Sheik Ahmed Zaki Yamani had predicted before the embargo, actually served a dual portion of grief: while less developed nations plunged more deeply into debt, their more prosperous neighbors frantically searched for some escape from their habitual dependence on OPEC. In turn, both responses sparked a counter-reaction—energy conservation—which eventually acted to reduce world demand for oil. Then, the resulting surplus dropped price levels to around $28 per barrel before 1980. In 1986, the power of OPEC had been severely

*Garvin replaced J. Kenneth Jamieson as chairman and CEO, July 31, 1975.

**Howard C. Kauffmann, president, also signed this report, dated March 3, 1976.

tested, and some prices had actually fallen below $15 per barrel.

Over the years, much damage had been done; and, even so, just what lessons (if any) the public had learned during that turbulent decade remained unclear. For Exxon, however, some responses seemed appropriate. First, the company brought on stream crude oil from geographic areas where, ten years earlier, production costs would have been prohibitive. Given the special problems of the North Sea and those of the North Slope in Alaska, getting oil out represented major scientific and technological breakthroughs.

Although, in these years, many companies in which Exxon held an equity interest were nationalized, this did not in every case mean loss of oil; service agreements became operative with both Saudi Arabia and Venezuela, for example. Yet even with 100 percent company-owned oil, equity shares of production in other companies, long-term agreements with governments, and additional supplies available under other special arrangement, the total production available for Exxon refineries dropped.

Even then, cash flow—influenced by both demand and inflation—remained quite strong. Beginning when the shortage perceptibly eased in 1981-1982, the company inaugurated a new policy of austerity, with the goal of increasing operating efficiency. Overall, Exxon's profit level continued to increase: from $3 billion in 1974 to $5.5 billion in 1985. Dividend payments also increased.

In 1978, Garvin said that his principal task was to keep Exxon "in the best shape possible for [the] men who come after me." That job proved to be interesting, and it required of Garvin a commitment to non-traditional management techniques. Of course, public confusion and governmental ineptness in the field of energy did not lighten his burden. Although Exxon's CEO loudly proclaimed that "Energy conservation must become a way of life," few listeners seemed to heed his remarks. Government agencies, in particular, gave little shrift to expert opinion.

Perhaps Garvin had this in mind when he wrote, "no effective energy policies can be built on a foundation of self-deception." Certainly as Exxon's leader, Garvin did not fool himself.

Simply, the pre-1974 world of oil had come apart. As the Exxon chieftain wrote in 1978, "consuming nations remain dangerously vulnerable to actions of others beyond their control." All nations suffered because of the revolution among the world's oil suppliers, and the results broadly affected modern civilization.

For a time Exxon pursued diversification in areas other than chemicals, and at least some of these efforts achieved a degree of success. But when the price of oil declined, Exxon felt it wise to sell some of its non-oil properties, an electric motor company, for example. In retrospect, none of these investments altered the basic interests of the Exxon Corporation.

Garvin, Kauffmann, and the Exxon Board of Directors did anticipate a decline in the quantity of OPEC oil produced (and a corresponding drop in per barrel prices).* However Garvin reported to the shareholders in 1981 that the so-called oil glut would not be a permanent feature of the energy world. Thus, "Exxon's future will depend on technological innovation and the management of technological change." At the same time that it was altering its existing corporate structure, the company also continued to seek attractive investment opportunities in all areas of the petroleum business.

In fact, the economic recession of 1982, combined with a declining demand for oil, led Exxon leaders to accelerate their planned changes. Still, adapting operations to lower levels of demand was not an easy or an instantaneous process. Policy statements defined their objective: "to maintain a substantial resource base, trim and maintain flexible operating capability and maintain financial strength," sufficient to enable the company to handle "the unexpected and the unforeseen." By 1983, the company had begun to reduce some of its less profitable operations and the size of its work force. as management saw it, the situation in the petroleum industry left them little choice. To survive, Exxon had to achieve the most efficient use of personnel, plants, and resources. So modifications took place in many well established procedures.

In May, 1985, when President Kauffmann retired, to be replaced by Lawrence G. Rawl, personnel changes, which had been moving slowly during 1981-1982, began to speed up. By December 31, 1986, when Garvin's retirement from the company became effective, significant reductions had been made in personnel and operations, and a number of refineries and tankers had been sold. Then, a Japanese firm became the owner of the Exxon building in New York City, where the corporation remained a tenant. In fact, the leaner company proved financially stronger, as Rawl and others in charge sought to make it more capable of adapting to unforeseen world events.

Earlier, some years before Garvin left Exxon, a writer had borrowed a book title from the American author John Reed, to label the years 1974-1984, "Ten Years That Shook The World." In reality for Exxon, however, there were twelve traumatic years, ending with 1986. For by January, 1987, the

---

*The *New York Times*, May 28, 1987, pointed out that while Saudi Arabia ranked first as an oil supplier, no other Persian Gulf country ranked among the top ten.

new Exxon differed substantially from the old Jersey with the changes being deep and profound. Jersey had been reborn as Exxon, but the new firm was almost as a stranger to many of those who had known the old.

Appendix 1

Officers and Directors of the Exxon Corporation, 1950-1975*

Chief Executive Officers	From	To
Eugene Holman	June 12, 1944	April 30, 1960
Monroe J. Rathbone	May 1, 1960	February 28, 1965
Michael L. Haider	March 1, 1965	September 30, 1969
J. Kenneth Jamieson	October 1, 1969	July 31, 1975
Clifton C. Garvin, Jr.	August 1, 1975	December 31, 1986

Chairmen of the Board		
Frank W. Abrams	January 1, 1946	December 31, 1953
Eugene Holman	January 1, 1954	April 30, 1960
Leo D. Welch	May 1, 1960	March 31, 1963
Monroe J. Rathbone	April 1, 1963	February 28, 1965
Michael L. Haider	March 1, 1965	September 30, 1969
J. Kenneth Jamieson	October 1, 1969	July 31, 1975
Clifton C. Garvin, Jr.	August 1, 1975	December 31, 1986

Presidents		
Eugene Holman	June 12, 1944	December 31, 1953
Monroe J. Rathbone	January 1, 1954	March 31, 1963
Michael L. Haider	April 1, 1963	February 28, 1965
J. Kenneth Jamieson	March 1, 1965	September 30, 1969
Milo M. Brisco	October 1, 1969	October 31, 1972
Clifton C. Garvin, Jr.	November 1, 1972	July 31, 1975
Howard C. Kauffmann	August 1, 1975	May 16, 1985

Vice Presidents		
Orville Harden	April 4, 1935	December 31, 1953
John R. Suman	February 1, 1945	March 31, 1955
Robert T. Haslam	January 1, 1946	October 1, 1950
Chester F. Smith	December 1, 1946	August 11, 1955
Jay E. Crane	October 19, 1950	September 6, 1956
Emile E. Soubry	December 1, 1951	February 28, 1961
Lloyd W. Elliott	January 1, 1954	March 31, 1965
Stewart P. Coleman	June 2, 1955	March 31, 1961
Henry H. Hewetson	August 18, 1955	May 27, 1959
Leo D. Welch	September 6, 1956	April 30, 1960

*The Standard Oil Company (New Jersey) became the Exxon Corporation, January 1, 1973.

900

Hines H. Baker	May 23, 1957	September 4, 1958
David A. Shepard	January 8, 1959	January 31, 1966
Myron A. Wright	May 1, 1960	May 18, 1966
	also January 1, 1973	March 31, 1976
Marion W. Boyer	May 1, 1960	May 18, 1966
Michael L. Haider	May 1, 1960	March 31, 1963
Peter T. Lamont	May 1, 1960	May 25, 1961
Howard W. Page	May 1, 1960	November 30, 1970
William R. Stott	May 1, 1960	January 31, 1968
Emilio G. Collado	May 3, 1962	November 30, 1975
Harold W. Fisher	May 3, 1962	May 14, 1969
John R. White	May 3, 1962	May 12, 1971
Eger V. Murphree	July 5, 1962	October 29, 1962
J. Kenneth Jamieson	September 10, 1964	February 28, 1965
Cecil L. Burrill	May 18, 1966	May 17, 1967
Milo M. Brisco	May 17, 1967	September 30, 1969
Wolf Greeven	May 17, 1967	May 14, 1969
George T. Piercy	May 17, 1967	November 30, 1980
Leroy D. Stinebower	May 17, 1967	November 30, 1969
Siro Vázquez	May 17, 1967	November 1, 1974
Clifton C. Garvin, Jr.	February 1, 1968	October 31, 1972
Charles J. Hedlund	February 1, 1968	October 31, 1979
Robert H. Milbrath	May 14, 1969	October 31, 1973
Frederic A. L. Holloway	May 13, 1970	October 31, 1978
David J. Jones	May 13, 1970	December 8, 1976
Robert T. Bonn	May 13, 1970	June 30, 1976
Thomas W. Moore	May 13, 1970	September 30, 1972
Roy A. Baze	May 13, 1970	October 1, 1973
Charles O. Peyton	May 13, 1970	December 15, 1976
William M. McCardell	July 1, 1970	September 30, 1975
	also August 1, 1977	August 31, 1980
Nicholas J. Campbell, Jr.	January 1, 1971	November 30, 1976
Donald M. Cox	May 12, 1971	May 16, 1985
James F. Dean	August 1, 1971	May 31, 1973
	also September 1, 1978	August 31, 1983
Thomas D. Barrow	May 18, 1972	November 15, 1978
Hugh H. Goerner	January 1, 1973	June 30, 1974
Randall Meyer	January 1, 1973	To date
Donald O. Swan	January 1, 1973	April 26, 1976
Edward C. Holmer	January 1, 1973	February 28, 1986
Norton Belknap	April 1, 1973	May 31, 1979
Stephen Stamas	June 1, 1973	June 30, 1986
Russel H. Herman, Jr.	October 1, 1973	May 17, 1979
Howard C. Kauffmann	January 1, 1974	July 31, 1975
Carl R. Patterson	July 1, 1974	July 31, 1977
Robert A. Winslow	August 28, 1974	May 14, 1979
	also January 1, 1981	March 31, 1985
Jack F. Bennett	August 1, 1975	To date
Terry A. Kirkley	October 1, 1975	July 14, 1978
	also October 1, 1979	February 28, 1981
Jack G. Clarke	December 1, 1975	To date

902

Directors

Frederick H. Bedford, Jr.	October 14, 1927	December 3, 1952
Orville Harden	April 12, 1929	December 31, 1953
Frank W. Abrams	June 4, 1940	December 31, 1953
Eugene Holman	June 4, 1940	May 25, 1960
Robert T. Haslam	December 1, 1942	June 7, 1950
Frank W. Pierce	December 1, 1942	March 1, 1953
Jay E. Crane	June 6, 1944	May 22, 1957
Chester F. Smith	June 6, 1944	May 23, 1956
John R. Suman	February 1, 1945	March 31, 1955
Bushrod B. Howard	August 31, 1945	November 1, 1954
Stewart P. Coleman	January 1, 1946	March 31, 1961
Emile E. Soubry	January 1, 1949	May 24, 1961
John W. Brice	November 1, 1949	June 8, 1951
Monroe J. Rathbone*	November 1, 1949	April 30, 1965
Henry H. Hewetson	June 7, 1950	May 27, 1959
Lloyd W. Elliott	June 8, 1951	March 31, 1965
David A. Shepard	December 1, 1951	January 31, 1966
Leo D. Welch	May 27, 1953	March 31, 1963
Peter T. Lamont	January 1, 1954	May 24, 1961
Howard W. Page	January 1, 1954	November 30, 1970
Arthur T. Proudfit	November 4, 1954	February 5, 1959
Marion W. Boyer	May 25, 1955	May 18, 1966
William R. Stott	May 23, 1956	January 31, 1968
Hines H. Baker	May 22, 1957	December 31, 1958
Myron A. Wright	May 28, 1958	May 18, 1966
	also January 1, 1973	March 31, 1976
Michael L. Haider	January 1, 1959	September 30, 1969
Cecil L. Burrill	May 27, 1959	May 17, 1967
Harold W. Fisher	May 27, 1959	May 14, 1969
Emilio G. Collado	May 25, 1960	November 30, 1975
John R. White	May 25, 1960	May 12, 1971
Paul J. Anderson	May 24, 1961	May 19, 1965
Wolf Greeven	May 24, 1961	May 14, 1969
Leroy D. Stinebower	May 22, 1963	November 30, 1969
J. Kenneth Jamieson**	September 10, 1964	May 14, 1981
Nicholas J. Campbell, Jr.	May 19, 1965	May 18, 1966
	also January 1, 1971	November 30, 1976
Siro Vázquez	May 19, 1965	November 1, 1974
Milo M. Brisco	February 1, 1966	November 1, 1972
Frederick R. Kappel	May 18, 1966	May 13, 1970
George T. Piercy	May 18, 1966	November 30, 1980
Julius A. Stratton	May 18, 1966	May 13, 1970
Bert S. Cross	May 17, 1967	May 16, 1974
Howard J. Morgens	May 17, 1967	May 14, 1969
Clifton C. Garvin, Jr.	February 1, 1968	December 31, 1986
William H. Franklin	May 14, 1969	May 16, 1979
Robert H. Milbrath	May 14, 1969	October 31, 1973

*Non-employee Director from February 25, 1965-April 30, 1965.
**Non-employee Director beginning August 1, 1975.

T. Vincent Learson	May 14, 1969	December 15, 1975
Franklin A. Long	October 30, 1969	May 14, 1981
Donald S. MacNaughton	May 13, 1970	To date
Donald M. Cox	May 12, 1971	May 16, 1985
Guido Colonna di Paliano	October 27, 1971	May 15, 1975
Otto Wolff von Amerongen	October 27, 1971	To date
Thomas D. Barrow	May 18, 1972	November 15, 1978
Howard C. Kauffmann	January 1, 1974	May 16, 1985
Martha Peterson	May 16, 1974	May 21, 1987
Sir Richard Dobson	May 15, 1975	May 17, 1984
Edward G. Harness	May 15, 1975	November 15, 1984
Jack F. Bennett	August 1, 1975	To date
Jack G. Clarke	December 1, 1975	To date

Non-Employee Directors		Firm
Frederick R. Kappel	1966-1970	American Telephone & Telegraph
Julius A. Stratton	1966-1970	Ford Foundation & M. I. T.
Bert S. Cross	1967-1974	3M
Howard J. Morgens	1967-1969	Proctor & Gamble
William H. Franklin	1969-1979	Caterpillar
T. Vincent Learson	1969-1975	International Business Machines
Franklin A. Long	1969-1981	Cornell University
Donald S. MacNaughton	1970-	Prudential Life Insurance Co.
Guido Colonna Di Paliano	1971-1975	Rinascente
Otto Wolff von Amerongen	1971-	Otto Wolff Group
Martha Peterson	1974-1986	Beloit College
Sir Richard Dobson	1975-1984	British American Tobacco
Edward G. Harness	1975-1984	Proctor & Gamble
J. Kenneth Jamieson*	1975-1981	Exxon (retired)

Secretaries	From	To
Adrian C. Minton	June 6, 1933	February 23, 1955
John O. Larson	February 24, 1955	November 30, 1969
Robert B. Acker	December 1, 1969	October 31, 1972
Robert E. Anderson	November 1, 1972	September 30, 1977
Richard E. Faggioli	September 30, 1977	August 31, 1985
Elliott R. Cattarulla	September 1, 1985	To date

Controllers		
James C. Anderson	June 12, 1944	March 31, 1957
Alfred O. Savage	April 1, 1957	November 1, 1968

Employee Director from September 10, 1964-July 31, 1975.

Robert E. Mays	December 16, 1968	June 30, 1973
Archie L. Monroe	July 1, 1973	May 31, 1977

Treasurers

Leo D. Welch	October 16, 1944	May 26, 1954
Emilio G. Collado	May 27, 1954	May 25, 1960
Leroy D. Stinebower	May 26, 1960	May 22, 1963
Lester B. Johnson	May 23, 1963	August 4, 1965
David J. Jones	August 5, 1965	May 12, 1970
Allan C. Hamilton	May 13, 1970	May 31, 1983

Appendix 2

	Net Production Crude Oil and Natural Gas Liquids	Proved Reserves Crude Oil and Natural Gas Liquids	Sales of Natural Gas	Proved Reserves Natural Gas	Petroleum Product Sales	Refinery Runs
	ESSO EASTERN INC.					
	SELECTED OPERATING DATA*					
Year	1,000 Barrels Daily	Million Barrels Daily	Million Cubic Feet Daily	Billion Cubic Feet Year	1,000 Barrels Daily	1,000 Barrels Daily
1962	36	138	2	2,274	312	162
1963	35	130	3	2,260	342	182
1964	28	122	7	2,253	382	235
1965	29	161	7	3,026	417	254
1966	29	280	9	3,447	510	259
1967	26	252	9	3,587	548	280
1968	25	874	11	3,562	575	303
1969	24	860	21	4,374	602	319
1970	92	833	57	4,382	671	321
1971	172	796	109	4,429	712	413
1972	186	802	239	4,464	765	507
1973	219	914	333	3,694	845	569
1974	212	1,002	342	4,190	648	426
1975	227	947	418	3,793	604	392

*This table, from the Standard Oil Company (New Jersey) Financial and Statistical Supplement to the 1971 Annual Report, covers the period 1962-1971. The figures for 1972-1975 were obtained from each separate Exxon Corporation Annual Report.

Notes

PROLOGUE

1. Harry S. Truman, *Memoirs*, 2 Volumes (New York, 1955-1956), II, p. 332.

2. Executive Committee Minutes [hereafter cited as E.C. Minutes], July 24, 1950, Secretary's Office, Exxon.

3. Copy in Exxon Corporation, Administrative Services Department, Information Center [hereafter cited as Exxon Information Center], XVIII, 75.

4. *Ibid.*

5. *Interview*, Henrietta M. Larson, Evelyn H. Knowlton, and Frank M. Surface with Earl Newsom, June 2, 1959. Courtesy, Miss Larson. Seventeen Justice Department lawyers searched the Jersey files. As one observer remembered, these files had not been subject to court order since 1911, and the government searchers operated with an "amazing amount of freedom" from all company interference.

6. *Letter*, Edward F. Johnson to Bennett H. Wall, July 31, 1978, Files, Center for Business History Studies [hereafter cited as C.B.H.S.]. Johnson became general counsel of Standard Oil Company (New Jersey) [Jersey] on January 1, 1945. On January 6, 26, February 6, 24, and March 11, 1942, Johnson discussed the case with the Executive Committee. See E.C. Minutes for those dates.

7. E.C. Minutes, March 25, 1942. Note: There is no record of the Washington meeting in the Board Minutes. The Board decided that:

> to obtain vindication by trying the issues in court would involve considerable expense and months of time of the Officers and Directors and many employees of the Company and its subsidiaries; that their work is of greater importance to the stockholders than court vindication and that the Company ought not to remain in a position which the Department of Justice considers to be questionable.

8. The evidence submitted to the Truman Committee supported the company position. The *New York Times* editorialized: "In the light of this evidence Mr. Arnold's charges that the Standard Oil Company is responsible for the shortage of synthetic rubber evaporate. It is apparent that he did not have the facts." The committee found: "No question of moral turpitude or of subjective unpatriotic motivation on the part of Standard or any of its officials." *New York Times*, April 2, 1942, quoted in L.L.L. Golden, *Only By Public Consent* (New York, 1968), p. 166. See the editorial in the *New York Times*, April 5, 1942. This story is well covered in Henrietta M. Larson, Evelyn H. Knowlton, and Charles S. Popple, *New Horizons, 1927-1950: History of Standard Oil Company (New Jersey)*, (New York, 1971), 428-443. Edward Johnson's account of the informal Board meeting in Washington received corroboration from Margery Porter, assistant company secretary to Adrian C. Minton, company secretary. Courtesy, Dorothy A. Windels, October 9, 1978. See *interview*, Wall and C. Gerald Carpenter with Johnson, May 3, 1977, Files, C.B.H.S.

9. Ralph W. Hidy (Foreward), *Oil's First Century* (Boston, 1960), statement by Wallace E. Pratt concerning this investigation, p. 58 ff.

10. *New York, Times*, March 27, 1942; several sources related in confidence that prior to

Jersey officials testifying before the Truman Committee, Thurman Arnold suggested settling the I.G. Farben affair, stating that he would tell the committee that the government had been satisfied with its examination of company papers; or, he could read damaging letters to the committee. Arnold wanted a quick answer. President W.S. Farish recommended that Jersey attorneys stall to give him time to effect a higher level settlement. But this proved unsuccessful. Arnold read the letters, with disastrous results for Jersey. *Interview*, Wall with Pratt, June 19, 1974, Files, C.B.H.S.; *Interview*, Wall and Carpenter with Johnson, May 3, 1977, *ibid*.

11. Ralph W. Gallagher had spent many years with the East Ohio Gas Company. In the parlance of the Standard Oil Company (New Jersey), this phrase denoted such a background, in contradistinction to the self-explanatory "oil man."

12. A full account of this period is provided in Larson, Knowlton, and Popple, *New Horizons, 1927-1950*; see especially 431-448.

13. Charles S. Popple, *Standard Oil Company (New Jersey) in World War II* (New York, 1952), passim. See also Larson, Knowlton, and Popple, *New Horizons, 1927-1950*; John W. Frey and H. Chandler Ide (editors), *A History of the Petroleum Administration for War, 1941-1945* (Washington, 1946).

14. Confidential *memorandum*, Larson, March 25, 1945. Courtesy, Miss Larson.

15. These characterizations of Frank W. Abrams and the following Board members are based on *Interviews* with retired directors and other employees of Jersey who served with them in 1950 and earlier, and on the specific *Interviews* noted: *Interviews*, Larson with Pratt, November 16, 1959; Johnson, July 19, 1956; Henry E. Bedford, February 23, 1954, Courtesy, Miss Larson; *Interview*, Larson and Knowlton with Abrams, December 8, 1954, *ibid.; Interview*, Larson, Knowlton, and Surface with Newsom, June 2, 1959, *ibid.; Interviews*, Wall with Pratt, November 24-29, 1974, June 15-18, 1975, May 31-June 4, 1976, July 1-5, 1977; Johnson, February 10, 1975, May 7, July 28, 1976; George H. Freyermuth, December 18, 1975, April 26-27, 1976, December 15, 1977, Files, C.B.H.S.; *Interviews*, Wall and Carpenter with Hines H. Baker, March 1, September 15, 1976; Johnson, May 3-4, 1977; Abrams, May 8, 1975;

Courtney C. Brown, June 4, 1975; J.O. Larson, October 15-16, 1975; William P. Headden, June 5, 1975; and others, *ibid*. See also Larson, Knowlton, and Popple, *New Horizons, 1927-1950, passim*. Each time a new director was elected, *The Lamp* carried a sketch of his career; e.g., see Vol. XXVII, No. 2 (April, 1945), p. 22.

16. Carl Maas, "Jersey and the Arts," July 19, 1972; see Appendix, p. 1 [a mimeographed report for the Standard Oil Company (New Jersey)]; *Interview*, Wall with Maas, August 1, 1975, Files, C.B.H.S.; *Interview*, Wall with Freyermuth, April 26-27, 1976, *ibid*.

17. *Ibid*. Golden in his *Only By Public Consent*, p. 169, supports the statements regarding John D. Rockefeller, Jr., and Robert T. Haslam. He interviewed many of the participants.

18. David A. Shepard, "Receiving the History of Standard Oil Company (New Jersey)," 157-160, Lewis P. Cain (ed.), *Proceedings of the Business History Conference*, Second Series, Vol. I (Indiana University, Bloomington, Indiana, 1973).

19. Larson, Knowlton, and Popple, *New Horizons, 1927-1950*, p. 445.

20. *Interviews*, Wall with Freyermuth, April 26-27, 1976, Files, C.B.H.S.

21. *Ibid.*, December 18, 19, 1975.

22. *Ibid.*

23. *Ibid.*

24. Edward L. Bernays, *Public Relations* (Norman: University of Oklahoma Press, 1952), p. 339.

25. *Letter*, Haslam to Newsom, May 15, 1942, quoted in Larson, Knowlton, and Popple, *New Horizons, 1927-1950*, p. 444.

26. *Interviews*, Wall with Freyermuth, December 18-19, 1975, Files, C.B.H.S.

27. *Ibid.*, April 26-27, 1976.

28. *Ibid.*

29. "Summary of conclusions from Elmo Roper's Survey of employees' attitudes at Baton Rouge Refinery of the Standard Oil Company of Louisiana, May 29, 1945," Courtesy, Burns Roper.

30. *Interview*, Larson with Frank G. Clark (Baton Rouge, LA), March 17, 1945. Courtesy, Miss Larson.

31. *Interview*, Larson with F.R. McGrew (Shreveport, LA), March 13, 1945, *ibid*.

32. Most Jersey employees applauded the Board's decision to improve public relations. Elmo Roper, in an article, "The Public Looks at Big Business," printed in *The Lamp*, December, 1943, left his readers with a strong conviction that the Jersey Company leaders had operated responsibly, honestly, and morally. He based the article on answers to his various surveys. Jersey employees shared that conviction. Maas, "Jersey and and the Arts," p. 2.

33. *Interviews*, Wall with Freyermuth, December 18-19, 1975, Files, C.B.H.S.

34. *Interview*, Wall with Maas, August 1, 1975, *ibid*.

35. Maas, "Jersey and the Arts," p. 12.

36. Actually, the film opened in Abbeville, LA.

37. Shepard, "Receiving the History of Standard Oil Company (New Jersey)," p. 159.

38. *Letter*, N.S.B. Gras to N. D'Arcy Drake, October 7, 1947, Public Affairs File, Administration Organization, History-Company-Business, History Foundation, 1947-1949 [hereafter cited as Public Affairs Files, Company History, Exxon Records Center (E.R.C.)].

39. *Interviews*, Wall with Freyermuth, April 26-27, 1976, Files, C.B.H.S.

40. Pratt, "The Value of Business History in the Search for Oil," in Hidy (Foreward), *Oil's First Century*, p. 58 ff.

41. Gras and his associates formed the Business History Foundation, a tax exempt educational organization chartered in 1947, in New York State. Freyermuth and other Jersey leaders negotiated terms of a grant with this group. Of course Jersey depended on its outside counsel, Davis, Polk, Wardwell, Sunderland & Kiendl for legal advice. "Memorandum of a presentation before the Board of Directors of the Delaware Corporation at 26 Broadway, February 9, 1945," by Freyermuth and Gras. Courtesy, Miss Larson. E.C. Minutes, March 3, 1947, Secretary's Office, Exxon; *Telegram*, Eugene Holman to John R. Suman, Haslam, Chester F. Smith, Frank W. Pierce, Bedford, and Abrams, March 3, 1947, Public Affairs Files, Company History, E.R.C.

42. Ralph W. Hidy and Muriel E. Hidy, *Pio-neering in Big Business, 1882-1911: History of Standard Oil Company (New Jersey)*, (New York, 1955); George S. Gibb and Evelyn H. Knowlton, *The Resurgent Years, 1911-1927: History of Standard Oil Company (New Jersey)*, (New York, 1956); Larson, Knowlton, and Popple, *New Horizons, 1927-1950*.

43. *Interview*, Wall with Cecil Morgan, July 18, 1978, Files, C.B.H.S.; *New York Times*, August 20, 1946; *Interview*, Wall with Pratt, June 19, 1974, Files, C.B.H.S.; Dr. Courtney C. Brown joined Jersey in 1946. Prior to that time, Brown had taught in the Columbia University School of Business, from which he took leave of absence in 1941. He worked for Chase National Bank until 1942, when he joined the U.S. Department of State and the War Food Administration. In the State Department he became chairman of the Inter-Agency Committee on wartime trade controls. *Interview*, Wall and Carpenter with Brown, June 4, 1975, Files, C.B.H.S.; *Interview*, Wall and Carpenter with Siro Vázquez, May 5, 1975, *ibid*.

44. *Interview*, Wall with M.A. Wright, August 26, 1982, *ibid*.; *Interview*, Wall with J. Larson, August 18, 1975, *ibid*.

45. Larson, Knowlton, and Popple, *New Horizons, 1927-1950, p. 479.*

46. *Ibid.*, p. 482.

47. *Interview*, Wall with Pratt, June 15, 1975, Files, C.B.H.S.

48. Larson, note on trip, early 1944 (Venezuela); *Interview*, Larson with Dr. Guillermo Zuloaga, February 4, 1949, Courtesy, Miss Larson.

49. Vice President Howard W. Page, in *The Economic Journal* (London), September 1960, quoted in George W. Stocking, *Middle East Oil* (Nashville, 1970), p. 150, stated:

> that the fixed royalty payment [agreed upon] no longer gave the equitable division originally intended and that additional royalty payments were neither economically practical nor a permanent method of maintaining equity between the parties. The 'rapid spread' of 50/50 was the result of this recognition of the need to restore the equity which had been frustrated by drastically changed conditions beyond the control of either party and unforeseen at the time the agreements were negotiated.

50. In the traditional sense, the agreement reached in Venezuela in 1943 did not provide for 50/50 profit sharing; still it provided the base upon which such agreements developed. In 1945 and in 1946, special taxes passed by the Venezuelan government did bring the payments up to 50 percent. In 1945, Dr. Juan Pablo Pérez Alfonso stated that he considered the "equal sharing of profits by the oil companies and the Venezuelan government as 'acceptable and equitable.'. . . The country which provides oil deposits and the industry which provides capital and technique should share half and half," *New York Times*, January 9, 1946.

51. Larson, Knowlton, and Popple, *New Horizons, 1927-1950*, p. 484.

52. *Notes* from Creole Annuitants' Meeting, May 11-15, 1976, Files, C.B.H.S.; *Interviews*, Wall with Joe L. Biffle, Mrs. Helen Campbell, Karl "Dit" Dallmus, and William S. Nancarrow, *ibid*.

53. Stocking, *Middle East Oil*, 43, 57; Benjamin Shwadran, *The Middle East, Oil and the Great Powers* (New York, 1955), p. 246, FN 7.

54. These companies were: Standard Oil Company (New Jersey), Socony-Vacuum Oil Company, Inc., Gulf Oil Corporation, Atlantic Refining Company, Pan-American Petroleum & Transport Company (Standard of Indiana), Sinclair Consolidated Oil Corporation, and The Texas Company. Ultimately, all of the above companies except Jersey and Socony dropped out.

55. For example, see the *New York Times*, July 15, September 22, December 16, 1943, January 23, 1944, ff. See also John A. DeNovo, "The Quest for a National Oil Policy During World War II," a paper prepared for the SHAFR session on "Energy and American Foreign Policy," Georgetown University, August 15, 1975, Courtesy, Dr. DeNovo.

56. *Interview*, Larson, Knowlton, and Surface with Newsom, June 2, 1959, Courtesy, Miss Larson.

57. For the story of aid to Ibn Saud, see Shwadran, *The Middle East, Oil and the Great Powers*, 301-317; Stocking, *Middle East Oil*, 89-103.

58. *Memorandum* of conversations at meeting called by Fred M. Vinson, June 22, 1945, Decimal File 890F. 51/6-2245, Record Group 59, National Archives, Washington, D.C.

59. *Memorandum* [not sent], James V. Forrestal to James F. Byrnes, August 1, 1945, Folder 36-1-30, Forrestal Files, Record Group 80, General Records of the Department of the Navy, *ibid*. It should be noted that the reference to Gulf appeared because Gulf had a concession in Kuwait.

60. *Letter*, C. Stuart Morgan to Wallace Murray, March 3, 1941, Decimal File 890G. 6363 T84-653, Record Group 59, General Records of the State Department, National Archives, Washington, D.C., Courtesy, Irvine H. Anderson, Jr.; *Opinion*, A. Andrewes Uthwatt, January 14, 1941, *ibid*.; *Letter*, R.W. Sellers to H.F. Sheets, Jr., January 21, 1941, *ibid*.

61. Some years later, Pratt wrote:

> After I had left New York, Gene [Holman] got in touch with me on my ranch in West Texas and asked if I would object if Jersey moved the date of my retirement back to the first of January 1945 so the record would show that John [Suman] was elected to fill my place at the stockholders' meeting in May 1945. I agreed to this.

Letter, Pratt to Wall, December 29, 1978, Files, C.B.H.S.

62. *Letter*, Page to Paul E. Morgan, April 24, 1978, *ibid*.

63. *Memorandum* of conference between Senator Joseph C. O'Mahoney and Mr. Holman, February 5, 1947, E.R.C., 164-5-C.

64. *New York Times*, February 23, 1944.

65. *Ibid*., February 22, 1945.

66. *Ibid*., January 3, 1945.

67. *Ibid*., August 1, 1945.

68. *Ibid*., August 21, 1945.

69. *Ibid*., December 9, 1946.

70. *Ibid*., December 13, 1946.

71. *Interview* (telephone), Wall with Woodfin L. Butte, February 9, 1977, Files, C.B.H.S.

72. *Interview* (incomplete), Wall with Shepard, Johnson, and Smith D. Turner, May 6, 1976, *ibid*.

73. E.C. Minutes, December 26, 1946, Secretary's Office, Exxon.

74. *Interview* (telephone), Wall with Butte, February 9, 1977, Files, C.B.H.S.

75. E.C. Minutes, May 2, 1947, Secretary's Office, Exxon.

76. *Interview* (telephone), Wall with Butte, February 9, 1977, Files, C.B.H.S.; *Interview* (incomplete), Wall with Shepard, Johnson, and Turner, May 6, 1976, *ibid.*

77. *Interview* (telephone), Wall with Butte, February 9, 1977, *ibid.*

78. *Portrait in Oil, The Autobiography of Nubar Gulbenkian* (New York, 1965), 224-226.

79. Exxon executives and attorneys consulted the following government officials, the following domestic attorneys and judges, and the following public relations consultants regarding termination of the Group Agreement of 1928 and purchase of a share of Aramco: Harold L. Ickes, Secretary of the Interior, and Ralph K. Davies, Acting Head of the Oil and Gas Division, Department of the Interior (1/3/45); Robert Nathan, Deputy to Director, Office of War Mobilization and Reconversion (9/27/45); James V. Forrestal, Secretary of the Navy, Donald S. Russell, Assistant Secretary of State, and Ralph K. Davies (11/21/46); Tom Clark, Attorney General (12/5/46, 1/16/47, and 12/23/48); Julius Krug, Secretary of the Interior, Robert P. Patterson, Secretary of War (12/9/46); Judge Manley O. Hudson (12/18/46, 3/11/47, and 6/21/48); Ambassador James Wadsworth, Department of State (1/7/47); W.L. Clayton and Loy W. Henderson, Department of State (1/10/47); John W. Davis and Charles Evans Hughes, Jr. (3/11/47); W.L. Clayton (7/6/48); John A. Loftis, Department of State (1/16/47); John W. Davis (1/27/47); Senators Tom Connally and E.H. Moore (2/19/47); John Foster Dulles, Attorney (8/25/47); Representatives of the United States Navy (3/23/48); Milo Perkins, Consultant, Foreign Operations (4/16/48); Sullivan and Cromwell, attorneys, and Charles Evans Hughes, Jr. (12/23/46 and 7/2/48); Paul G. Hoffman, United States Economic Cooperation Administration (9/20/48); Paul H. Nitze, Joseph C. Satterthwaite (and others) of the Department of State (12/23/48).

These citations are from the E.C. Minutes, Secretary's Office, Exxon, which do not indicate the precise day Jersey executives saw the above individuals; rather, they represent the days the Committee heard reports on their advice and opinions. The Board also secured an opinion from Josiah Stryker (3/11/47),

Board of Directors Minutes [hereafter cited as Bd. Dir. Mins.], Secretary's Office, Exxon. Some evidence indicates that Charles Evans Hughes, Jr., John W. Davis, and attorneys at Sullivan and Cromwell were consulted much more often than the records show.

80. *Interview*, Wall with Pratt, June 18, 1975, Files, C.B.H.S. Shepard and others confirmed that such a trip did occur.

81. E.C. Minutes, November 18, 1948, Secretary's Office, Exxon.

82. *Ibid.*, December 2, 1948.

83. Larson, Knowlton, and Popple, *New Horizons, 1927-1950*, 656-660. Surface paper dated March 5, 1957.

84. E.C. Minutes, May 23, 1947, Secretary's Office, Exxon.

85. Larson, Knowlton, and Popple, *New Horizons, 1927-1950*, 656-660; *The Lamp*, Vol. XXVIII, No. 5 (November, 1947), 20, 22.

86. *New York Times*, August 3, 1948.

87. *Ibid.*, September 18, October 21, 1945, December 6, 12, 1946.

88. *Ibid.*, July 16, September 24, 1948. In July, Jersey ordered eight tankers of 228,000 tons each.

89. *Ibid.*, September 17, 24, 1948. It should be noted that the statements of Holman and other leading oil company spokesman sometimes appear to be contradictory, depending on how one reads their caveats. See the report of Holman's speech, *ibid.*, December 27, 1946.

90. The first offshore well report appeared in the *New York Times* on March 9, 1947. The Humble Company well came in on August 12, 1948. *Ibid.*, August 13, 1948.

91. *Ibid.*, May 28, 1947.

92. *Ibid.*, February 11, 1949, November 18, 1946. It should be noted that in December, 1946, the U.S. government dropped the so-called "Mother Hubbard" case against the American Petroleum Institute [API] and 225 oil companies, including Jersey. This case, originally involving 366 companies charged with conspiracy, had been held "in abeyance" during the war by agreement of the President, the Attorney General, and other cabinet officers. Jersey understood that this agreement did not signal any lessening of interest on the part of the U.S. government.

CHAPTER 1

1. Henrietta M. Larson, Evelyn H. Knowlton and Charles S. Popple, *New Horizons, 1927-1950: History of Standard Oil Company (New Jersey)*, (New York, 1971), 786-88.

2. Standard Oil Company (New Jersey) [hereafter cited as SONJ], *Annual Report*, 1950, p. 5.

3. Larson, Knowlton and Popple, *New Horizons*, p. 807.

4. Gilbert Burck, "The Jersey Company," *Fortune*, Vol. XLIV, No. 10 (October, 1951), p. 99.

5. Larson, Knowlton and Popple, *New Horizons*, 806-807.

6. *Ibid.*, 8-9.

7. *Memorandum*, July 1, 1944, quoted in "Standard Oil Company (New Jersey)," *The Corporate Director* (Special Audit No. 102, February, 1952), p. 2.

8. Burck, "The Jersey Company," p. 176.

9. SONJ, *The Annual Meeting of the Stockholders of Standard Oil Company (New Jersey)*, (June 7, 1950), p. 17.

10. SONJ, Employee Relations Department, *This Is Jersey Standard* [c. 1962], Files, Center for Business History Studies [hereafter cited as C.B.H.S.].

11. *Interview*, Larson and Knowlton with Frank W. Abrams, December 8, 1954. Courtesy, Miss Larson.

12. Quoted in Courtney R. Hall, *History of American Industrial Science* (New York, 1954), p. 414.

13. "GDM" to G. Clark Thompson, January 25, 1955, Exxon Records Center [hereafter cited as E.R.C.], 192-7-B.

14. Executive Committee Minutes [hereafter cited as E.C. Minutes], March 24, 1950, Secretary's Office, Exxon; (See also Board of Directors Minutes [hereafter cited as Bd. Dir. Mins.], December 22, 1953 and November 4, 1954, *ibid.*).

15. Bd. Dir. Mins., December 7, 1950, May 10, June 12, 1951, *ibid.*

16. *Interview*, Bennett H. Wall with Leo D. Welch, May 21, 1975, Files, C.B.H.S.

17. Larson, Knowlton and Popple, *New Horizons*, 587-588, 601.

18. *Ibid.*, p. 589.

19. SONJ, *Annual Meeting* (1950), p. 17.

20. Burck, "The Jersey Company," p. 103.

21. Ralph W. Hidy and Muriel E. Hidy, *Pioneering in Big Business, History of the Standard Oil Company (New Jersey), 1882-1911* (New York, 1955), p. 333.

22. *Interviews*, N. S. B. Gras with W. R. Finney, April 25, 1945, Courtesy, Miss Larson; H. Larson, Knowlton and Frank M. Surface with Edward F. Johnson, July 19, 1956, *ibid.*; H. Larson and Knowlton with Wallace E. Pratt, February, 1956, *ibid.*; *Interviews*, Wall with John O. Larson, November 19, 1976, Files, C.B.H.S.; Wall and C. Gerald Carpenter with Lloyd W. Elliott, May 5, 1975, *ibid.*; Wall with Hines H. Baker, September 15, 1976, *ibid.*; Wall with Pratt, June 19, November 25, 1974, June 15, 16, 18, 1975, April 24, 1976, July 3, 1977, *ibid.*

23. These characterizations of Abrams and the following Board members are based on interviews with retired directors and other employees of Jersey who served with them in 1950 and earlier and on the specific interviews noted: *Interviews*, H. Larson with Pratt, November 16, 1959, Courtesy, Miss Larson; H. Larson with Johnson, July 19, 1956, *ibid.*; H. Larson with Henry E. Bedford, February 23, 1954, *ibid.*; *Interview*, H. Larson and Knowlton with Abrams, December 8, 1954, *ibid.*, *Interview*, H. Larson, Knowlton, and Surface with Earl Newsom, June 2, 1959, *ibid.*; *Interviews*, Wall with Pratt, November 24-29, 1974, June 15-18, 1975, May 31-June 4, 1976, and July 1-5, 1977, Files C.B.H.S.; Wall with Johnson, February 10, 1975, May 7, 1976, July 28, 1976, and May 3-4, 1977, *ibid.*; Wall with H. Baker, September 15, 1976, *ibid.*; Wall with George H. Freyermuth, December 18, 1975, April 26-27, 1976, December 15, 1977, *ibid.*; Wall with H. Larson, October 15-16, 1975, *ibid.*; Wall with William P. Headden, June 5, 1975, *ibid.*; *Interviews*, Wall and Carpenter with Abrams, May 8, 1975, *ibid.*; Wall and Carpenter with Baker, March 1, 1976, *ibid.*; Wall and Carpenter with Courtney C. Brown, June 4, 1975, *ibid.*; and others, *ibid.*, Files, C.B.H.S. See also Larson, Knowlton and Popple, *New Horizons, passim.* Each time a new director is elected, *The Lamp* carries a sketch of his career, *e.g.*, see Vol. XXVII, No. 2 (April, 1945), p. 22 for that of John R. Suman.

24. *Interviews*, Wall with J. Larson, August 18, 1975, November 19, 1976, Files, C.B.H.S.

25. *Letter*, Howard W. Page to Paul E. Mor-

gan, April 24, 1978, Morgan Files, C.B.H.S. The title of Suman's book was *Petroleum Production Methods* (Houston, 1921).

26. *Interview,* H. Larson with Bedford, February 23, 1954, Courtesy, Miss Larson. *Interview,* Knowlton, H. Larson and Surface with Johnson, July 19, 1956, *ibid.*

27. "General Policy: Working Relations with Standard Oil Company (New Jersey)," International Petroleum Company, Limited, September 16, 1953, E.R.C., 192-7-B.

28. Henry Ozanne, "The Jersey Standard Plan: Decentralization For Efficiency," *World Oil* (March, 1949).

29. Alfred D. Chandler, Jr., *Strategy and Structure: Chapters in the History of the American Industrial Enterprise* (Cambridge, Mass., 1963), p. 221.

30. Quoted in Burck, "The Jersey Company," p. 101.

31. *Interview,* H. Larson with H. L. Shoemaker, November 12, 1945, Courtesy, Miss Larson.

32. E.C. Minutes, February 20, 1950, Secretary's Office, Exxon.

33. *Interview,* Wall with M. W. Boyer, August 7, 1975, Files, C.B.H.S.

34. E.C. Minutes, August 5, 1952, Secretary's Office, Exxon.

35. *Interview,* Wall and Carpenter with Page, June 3, 1975, Files, C.B.H.S.

36. SONJ, *Annual Report,* 1950, p. 34.

37. Burck, "The Jersey Company," p. 104.

38. *Letter,* Page to C. F. Smith, April 4, 1947, E.R.C., 192-7-B; Burck, "The Foreign Policy of the Jersey Company," *Fortune,* Vol. XLIV, No. 11 (November, 1951), p. 198. (Mediterranean Standard Oil Company, also based in New York, served a similar purpose on a more limited basis, handling the purchase and sale of Middle Eastern crude oil.)

39. SONJ, *Annual Report,* 1950, p. 28; Imperial Oil Limited, *Annual Report, 1950,* 3-6; Burck, "Jersey Company," p. 104.

40. Burck, "Foreign Policy," 81-83, 182.

41. *Ibid.,* 81-83.

42. *Interview,* Wall and Carpenter with H. W. Fisher, June 25, 1975, Files, C.B.H.S.

43. *Letter,* Emile Soubry to Morgan, May 3, 1975, Morgan Files, C.B.H.S.

44. The 1942 Consent Decree stated that Jersey should not engage in practices restraining trade or act in violation of the antitrust laws of the United States.

45. George Bull, "The History of Esso Petroleum Company," [1888-1963], (unpublished MS), Files, C.B.H.S., 204-05.

46. Burck, "Foreign Policy," p. 83.

47. SONJ, *Annual Report,* 1951, p. 44.

48. The standard work on Stanvac up to 1941 is Irvine S. Anderson, Jr., *The Standard-Vacuum Oil Company and United States East Asian Policy, 1933-1941* (Princeton, 1975).

49. *Letter,* Page to Smith, April 4, 1947, E.R.C., 192-7-B.

CHAPTER 2

1. *Letter,* Cecil L. Burrill to Paul E. Morgan, August 15, 1979 [Burrill wrote this after reviewing a rough draft of this section], Morgan Files, Center for Business History Studies [hereafter cited as C.B.H.S.].

2. Secretary's Office, Exxon.

3. *Interview,* Bennett H. Wall and Earl N. Harbert with David J. Jones, December 7, 1979 [hereafter cited as Jones *Interview*], Files, C.B.H.S.

4. *Letter,* M. A. Wright to Morgan, August 21, 1979 [Wright, a director in 1960, wrote this in response to a rough draft of this section], Morgan Files, C.B.H.S.

5. Jones, "Basic Assumptions in Preparation for Investment Proposals," [hereafter cited as Jones Report. This report has a section "Conclusions" separately numbered, and that portion will be cited as "Conclusions" to avoid confusion]. Exxon Records Center [hereafter cited as E.R.C.], 397-2-B.

6. *Ibid.,* p. 2.

7. *Ibid.,* p. 4.

8. *Ibid.,* 3-4. Later Jones blamed himself for using the term "fail-safe" for he believed that to some directors it indicated that he did not advocate risk-taking. He explained that he advocated, "Take risks consciously . . . if profits [were] worthwhile," Jones *Interview,* Files, C.B.H.S.

9. Jones Report, 49-50.

10. "Conclusions," 1-2.

11. *Ibid.,* p. 2.

12. Jones Report, p. 20.

13. *Ibid.*, p. 48.

14. "Conclusions," 1-2.

15. Jones Report, 50-52.

16. *Ibid.*, p. 47.

17. *Ibid.*, 50-51; "Conclusions," 4-5.

18. Jones Report, p. 51.

19. "Conclusions," p. 6.

20. *Ibid.*, 4-5.

21. *Ibid.*, 5-6.

22. *Ibid.*, p. 5.

23. *Ibid.*, p. 6.

24. *Memorandum*, M. J. Rathbone to M. W. Boyer, *et al.*, September 22, 1960, E.R.C., 397-2-B.

25. *Letters*, E. G. Collado to Rathbone, October 10, 1960; M. L. Haider to Rathbone, October 7, 1960; L. W. Elliott to Rathbone, October 25, 1960; M. W. Johnson to Rathbone, October 6, 1960; C. H. Carpenter to Rathbone, October 13, 1960; H. W. Fisher to Rathbone, October 10, 1960; L. W. Welch to Rathbone, November 30, 1960; *ibid.*

26. *Letter*, Carpenter to Rathbone, October 13, 1960, *ibid.*

27. *Letters*, H. G. Burks, Jr. to Rathbone, October 10, 1960; A. O. Savage to Rathbone, October 7, 1960; L. D. Stinebower to Rathbone, October 10, 1960; W. A. M. Greeven to Rathbone, October 5, 1960; *ibid.*

28. *Letters*, Stinebower to Rathbone, October 10, 1960; L. F. Kahle to Rathbone, October 6, 1960; Wright to Rathbone, October 6, 1960; Johnson to Rathbone, October 6, 1960; D. A. Shepard to Rathbone, October 6, 1960; Siro Vázquez to Rathbone, October 6, 1960; Greeven to Rathbone, October 5, 1960; G. W. Butler to Rathbone, October 4, 1960; E. E. Soubry to Rathbone, November 4, 1960; Carpenter to Rathbone, October 13, 1960; Haider to Rathbone, October 7, 1960; *ibid.*

29. *Letters*, Stinebower to Rathbone, October 10, 1960; Collado to Rathbone, October 10, 1960; W. R. Stott to Rathbone, November 23, 1960; Elliott to Rathbone, October 25, 1960; *ibid.*

30. *Letters*, Kahle to Rathbone, October 6, 1960; Stott to Rathbone, November 23, 1960; Soubry to Rathbone, November 4, 1960; Carpenter to Rathbone, October 13, 1960; Fisher

to Rathbone, October 10, 1960; J. R. White to Rathbone, October 10, 1960; Boyer to Rathbone, October 7, 1960; Savage to Rathbone, October 7, 1960; Stinebower to Rathbone, October 10, 1960; Johnson to Rathbone, October 6, 1960; Wright to Rathbone, October 6, 1960; Shepard to Rathbone, October 6, 1960; Vázquez to Rathbone, October 6, 1960; *ibid.*

31. *Letters*, Wright to Rathbone, October 6, 1960; White to Rathbone, October 10, 1960; Haider to Rathbone, October 7, 1960; Boyer to Rathbone, October 7, 1960; Stinebower to Rathbone, October 10, 1960; S. P. Coleman to Rathbone, October 7, 1960; *ibid.*

32. *Letter*, Soubry to Rathbone, November 4, 1960, *ibid.*

33. *Interview*, Wall and Carpenter with H. W. Page, May 17, 1978, Files, C.B.H.S.

34. *Letter*, Page to Rathbone, September 29, 1960, E.R.C., 397-2-B.

35. *Ibid.*

36. *Letter*, Wright to Rathbone, October 6, 1960, *ibid.*

37. *Letter*, Vázquez to Rathbone, October 6, 1960, *ibid.*

38. *Letter*, Coleman to Rathbone, October 7, 1960, *ibid.*

39. *Letters*, Carpenter to Rathbone, October 13, 1960; Burrill to Rathbone, October 7, 1960; Savage to Rathbone, October 7, 1960; Butler to Rathbone, October 4, 1960; Vázquez to Rathbone, October 6, 1960; *ibid.*

40. *Letter*, Greeven to Rathbone, October 5, 1960, *ibid.*

41. *Letter*, Fisher to Rathbone, October 10, 1960, *ibid.*

42. *Letter*, Greeven to Rathbone, October 5, 1960, *ibid.*

43. *Letter*, Welch to Rathbone, November 30, 1960, *ibid.*

44. *Letters*, Fisher to Rathbone, October 10, 1960; Boyer to Rathbone, October 7, 1960; G. T. Piercy to Rathbone, October 7, 1960; *ibid.*

45. *Letters*, Stott to Rathbone, November 23, 1960; Haider to Rathbone, October 7, 1960; Fisher to Rathbone, October 10, 1960; Coleman to Rathbone, October 7, 1960; Vázquez to Rathbone, October 6, 1960; Wright to Rathbone, October 6, 1960; Soubry to Rathbone, November 4, 1960; Collado to Rathbone, October 10, 1960; Burks to Rathbone, October 10, 1960; Burrill to Rathbone, Octo-

ber 7, 1960; Stinebower to Rathbone, October 10, 1960; Kahle to Rathbone, October 6, 1960; Butler to Rathbone, October 4, 1960; Carpenter to Rathbone, October 13, 1960; Boyer to Rathbone, October 7, 1960; Elliott to Rathbone, October 25, 1960; *ibid.*

46. *Letters*, Piercy to Rathbone, October 7, 1960; Burrill to Rathbone, October 7, 1960; Savage to Rathbone, October 7, 1960; Carpenter to Rathbone, October 13, 1960; Greeven to Rathbone, October 5, 1960; *ibid.*

47. *Memorandum*, by Rathbone, December 13, 1960, *ibid.*

48. *Ibid.*

49. Executive Committee Minutes [hereafter cited as E.C. Minutes], December 14, 1960, Secretary's Office, Exxon.

50. *Ibid.*

51. *Ibid.*

52. *Letter*, Boyer to G. D. MacConnachie, September 29, 1961, *ibid.*

53. *Ibid.*

54. "Draft Statements of Investment Policy," September, 1961, Secretary's Office, Exxon.

55. *Ibid.*

56. *Ibid.*

57. *Ibid.*

58. *Ibid.*

59. *Ibid.*

60. *Ibid.*

61. *Ibid.* [Disinvestment proved to be one area in which long debate ensued, for not all the directors agreed that such outlets for oil as service stations should be profitable. Later the marketers came to accept that view, leading to much disinvestment of unprofitable service stations.]

62. "Draft Statements of Investment Policy," September, 1961, Secretary's Office, Exxon.

63. *Ibid.*

64. *Ibid.*

65. *Ibid.*

66. *Ibid.*

67. *Ibid.*

68. *Ibid.*

69. *Ibid.*

70. *Ibid.*

71. *Letter*, Boyer to MacConnachie, September 29, 1961, *ibid.*

72. "Statement of Policy for Chemical Activities," March 23, 1962, *ibid.*

73. E.C. Minutes, April 11, May 31, June 13, 1962, *ibid.*

74. *Press Release*, "Jersey Standard Adopts New Management Approach for Worldwide Research Activities," September 24, 1962, Files, C.B.H.S.; "Organization of Research Activities," June 12, 1962, Secretary's Office, Exxon.

75. E.C. Minutes, April 22, 1963, *ibid.; Letter*, Stott to V. Cazzaniga, October 17, 1962, E.R.C., 281-18-A.

76. Organizational Chart, September, 1963. Board of Directors Minutes [hereafter cited as Bd. Dir. Mins.], November 8, 1963, Secretary's Office, Exxon.

77. Standard Oil Company (New Jersey) [hereafter cited as SONJ], *Annual Report*, 1961, 9, 11.

78. E.C. Minutes, September 11, 1959, January 12, 1961, Secretary's Office, Exxon; "Jersey's Giant Strides in Chemicals," *Chemical Week* (July 15, 1961), p. 110; SONJ, *Annual Report*, 1960, p. 18; *ibid.*, 1961, p. 9; "More Food For More People," *The Lamp*, Vol. XLVII, No. 4 (Winter, 1965), p. 24; *A History of Imperial Oil* (Imperial Oil Limited [c. 1967]), p. 28.

79. E.C. Minutes, February 4, 1957, March 9, 1960, January 16, February 17, 1961, Secretary's Office, Exxon.

80. SONJ, *Annual Report*, 1961, 7, 9; *Letter*, Carl E. Reistle, Jr. to Wright, September 3, 1963, E.R.C., 82-10-A; Bd. Dir. Mins., December 1, 1965, Secretary's Office, Exxon.

81. E.C. Minutes, December 20, 1963, *ibid.*; Bd. Dir. Mins., January 30, 1964, *ibid.*; E.C. Minutes, February 26, 1965, *ibid.*; Bd. Dir. Mins., July 30, 1964, *ibid.*

82. E.C. Minutes, January 29, June 22, 1965, *ibid.*; Bd. Dir. Mins., August 5, 1965, *ibid.*

83. *Ibid.*, January 30, April 16, 1964, *ibid.*; E.C. Minutes, December 18, 1968, *ibid.* Burrill and Jones both remarked later that, despite the guidelines, the haste to diversify led to many investments without proper study, and they cited several fertilizer ventures as examples. Of the purchase of American Cryogenics, Inc., Burrill stated that since he was the contact director involved in that purchase,

he should have stopped it. See *letter*, Burrill to Morgan, August 14, 1979, Morgan Files, C.B.H.S., and Jones *Interview*, Files, C.B.H.S.

84. *Memorandum*, W. O. Twaits "To All [Imperial] Management," April 26, 1963, E.R.C., 281-18-A.

85. Bd. Dir. Mins., May 14, 1964, Secretary's Office, Exxon.

86. *Ibid.*, October 1, 1964.

87. *Ibid.*, February 18, March 18, 1965.

88. *Letter*, J. Kenneth Jamieson to Morgan, September 2, 1981, Morgan Files, C.B.H.S. (Italics Mine.)

CHAPTER 3

1. Quoted in "Standard of Jersey's New Plan for Realignment," *Business Week* (August 6, 1960) [hereafter cited as "Jersey's New Plan for Realignment"], p. 44. See also Walter Guzzardi, Jr., "How Rathbone Runs Jersey Standard," *Fortune*, Vol. LIII, No. 7 (January, 1963), p. 172.

2. The conditions facing the company were essentially those described by Alfred D. Chandler, Jr., in *Strategy and Structure: Chapters in the History of American Industrial Enterprise* (Cambridge, Mass., 1962), p. 299. In his study of the early evolution of the decentralized, multidivisional corporation Chandler found: 'the administrative load on the senior executive officers increased to such an extent that they were unable to handle their entrepreneurial responsibilities efficiently.' In the 1920s when Jersey first adopted a decentralized structure, the excessive burden on its top executives consisted of operating responsibilities. The solution at that time was to shift all operations to affiliates and make Jersey exclusively a holding company. As a result the principal executives were freed to concentrate on broader concerns.

3. Humble Oil & Refining Company, *Annual Report*, 1954, p. 4.

4. "Legal and Political Conditions Which Affect Humble's Public Relations," March 26, 1954, Exxon Records Center [hereafter cited as E.R.C.], 192-07-B; *Interview*, Henrietta M. Larson with Hines H. Baker, October 10, 1945, Courtesy, Miss Larson.

5. The Waters-Pierce Oil Company *vs.* The State of Texas (Court of Civil Appeals of Texas, March 9, 1898), 44 S.W. 936.

6. "Legal and Political Conditions Which Affect Humble's Public Relations," March 26, 1954, E.R.C., 192-07-B; See also 103 S.W. 836; 105 S.W. 851; 106 S.W. 326; 106 S.W. 919; and 53 L.Ed. 417.

7. "Memorandum Relating to the Position of the Standard Oil Company (New Jersey) Under the Laws of Texas," November 28, 1955, E.R.C., 192-07-B.

8. *Ibid.*, See also 263 S.W. 319; *Interview*, Larson with Baker, October 10, 1945, Courtesy, Miss Larson.

9. "Legal and Political Conditions Which Affect Humble's Public Relations," March 26, 1954, E.R.C., 192-07-B.

10. *Interview*, Bennett H. Wall with Baker, September 15, 1976, Files, Center for Business History Studies [hereafter cited as C.B.H.S.].

11. The Socony Mobil Oil Company had reorganized its domestic operations before Jersey did, and the Standard Oil Company (Indiana) was known to be planning a similar change. "Jersey's New Plan for Realignment," p. 45.

12. Humble Oil & Refining Company, *Annual Report*, 1954, p. 4.

13. Standard Oil Company (New Jersey) [hereafter cited as SONJ], Executive Committee Minutes [hereafter cited as E.C. Minutes], April 18, 1955, and July 8, 1955, Secretary's Office, Exxon.

14. *Interiew*, Wall with Edward Johnson, December 14, 1977, Files, C.B.H.S.

15. *Interiew*, Wall with H. Baker, September 15, 1976, *ibid.*

16. *Interview*, Wall with Johnson, December 14, 1977, *ibid.*

17. The case referred to by the Bakers was the U.S. *vs.* E. I. Dupont de Nemours & Company; *Interview*, Wall with Rex G. Baker, March 2, 1976, Files, C.B.H.S.

18. Chandler, *Strategy and Structure*, p. 299.

19. *Interview*, Wall with H. Baker, September 15, 1976, Files, C.B.H.S.

20. *Ibid.; Interview*, Wall with John O. Larson, November 19, 1976, *ibid.; Interview*,

Wall with Marion W. Boyer, August 5, 1976, *ibid.*

21. SONJ, Board of Directors Minutes [hereafter cited as Bd. Dir. Mins.], September 2, 1959, Secretary's Office, Exxon.

22. *Ibid.*; "The Proposed Merger," *The Humble Way* (September-October, 1959), p. 1; *Letter*, Eugene Holman and M. J. Rathbone to Shareholders, September 2, 1959, Files, C.B.H.S.

23. "Brief History of Humble Oil & Refining Company," Humble Public Relations Department, 1959, *ibid.*

24. Guzzardi, "How Rathbone Runs Jersey," 171, 179.

25. *Speech*, Rathbone, "The Challenge of Change," before the National Petroleum Assciation, Atlantic City, New Jersey, September 10, 1958, Files, C.B.H.S.

26. *Interview*, Wall with H. Baker, September 15, 1976, *ibid.*

27. *Interview*, Wall with J. Larson, August 18, 1975, *ibid.*; "Lessons in Leadership, Part I: Deciding the Tough Ones," *Nation's Business* (June, 1965) [hereafter cited as "Lessons in Leadership"], 35, 50.

28. *Memorandum*, Rathbone to Board Members, February 25, 1958, and attached Plan of Organization, Secretary's Office, Exxon.

29. *Ibid.*; Bd. Dir. Mins., March 20, and May 29, 1958, *ibid.*

30. E.C. Minutes, August 12, 1958, *ibid.*

31. "Jersey's New Plan for Realignment," p. 46.

32. One other change that decentralized responsibilities formerly carried by directors was the abolition of the Advisory Committee on Human Relations. Never able to make concrete recommendations, the committee had been headed for dissolution as early as 1956 when staff studies recommended the discontinuance of having directors serve as chairman and vice chairman. After experiments with other forms, the Board ended the committee's work as part of the 1960 reorganization, E.C. Minutes, January 9, 1956, October 21, 1958, Secretary's Office, Exxon; "Jersey's New Plan for Realignment," p. 51.

33. Guzzardi, "How Rathbone Runs Jersey," p. 171.

34. "Professional Orientation Program,"

Humble Oil & Refining Company (Bayway refinery), 1960, p. 3, Files, C.B.H.S.

35. Bd. Dir. Mins., June 8, 1950, Secretary's Office, Exxon.

36. *Interview*, Wall and Carpenter with Howard W. Page, June 3, 1975, Files, C.B.H.S.; *Interview*, Wall and Carpenter with David A. Shepard, June 25, 1975, *ibid.*

37. E.C. Minutes, September 9, 1958, Secretary's Office, Exxon.

38. *Ibid.*, September 9, 1959.

39. *Ibid.*, December 7, 1959.

40. *Letter*, M. A. Wright to Rathbone, June 15, 1959, E.R.C., 192-07-B.

41. *Memorandum*, Rathbone, "Memorandum on Organization," January 25, 1960, Secretary's Office, Exxon; *Interview*, Wall with J. Larson, November 19, 1976, Files, C.B.H.S.

42. *Memorandum*, Rathbone, "Memorandum on Organization," January 25, 1960, Secretary's Office, Exxon; The "Far East, East, and South Africa" region was to get a coordinator also, but his duties were not specified in the memorandum.

43. *Ibid.*

44. SONJ, Organization Chart, May 25, 1960, Files, C.B.H.S.

45. *Memorandum*, Rathbone, "Memorandum on Organization," January 25, 1960, Secretary's Office, Exxon.

46. *Ibid.*

47. *Interview*, Wall with J. Larson, November 19, 1976, Files, C.B.H.S.; Bd. Dir. Mins., February 4, 25, 1960, Secretary's Office, Exxon.

48. SONJ, Organization Chart, May 25, 1960, Files, C.B.H.S.

49. *Interview*, Wall with Boyer, August 5, 1976, *ibid.*

50. *Memorandum*, Rathbone, "Memorandum on Organization: Investment Planning and Review," March 7, 1960, *ibid.*

51. "Jersey's New Plan for Realignment," p. 52.

52. *Letter*, Wright to Morgan J. Davis, May 31, 1961, E.R.C., 82-10-A.

53. *Memorandum*, Rathbone, "Memorandum on Organization: Investment Planning

and Review," March 7, 1960, Files, C.B.H.S.; see also the attached *memoranda,* "Organization for Investment Planning," and "Duties and Responsibilities of Investment Analysis Department" and, *Interview,* Wall and Carpenter with Page, June 26, 1975, *ibid.*

54. *Interview,* Wall with Boyer, August 7, 1975, *ibid.;* SONJ, Annual Meeting of Shareholders Minutes, May 28, 1958; "The Changing Face of Jersey Standard," *Forbes,* (November 1, 1960), p. 19; *Letter,* Cecil L. Burrill to Wall, July 30, 1977, Files, C.B.H.S.

55. *Memorandum,* Rathbone, "Memorandum on Organization: Investment Planning and Review," March 7, 1960, *ibid.*

56. *Interview,* Wall with Boyer, August 5, 7, 1975, *ibid.; Interview,* Wall with Leo D. Welch, May 21, 1975, *ibid.; Interview,* H. Larson with H. L. Shoemaker, November 12, 1945, Courtesy, Miss Larson.

57. "This is Jersey Standard," SONJ, Employee Relations Department (*c.* 1962), Files, C.B.H.S.; *Letter,* Burrill to Wall, July 30, 1977, *ibid.*

58. "Jersey's New Plan for Realignment," p. 47; see also "Lessons in Leadership, Part I," p. 50.

59. *Interview,* Wall with Boyer, August 5, 1976, Files, C.B.H.S.

60. "Jersey's New Plan for Realignment," p. 52.

61. Bd. Dir. Mins., April 28, 1960, Secretary's Office, Exxon.

62. "The Changing Face of Jersey Standard," *Forbes,* November 1, 1960; The *Financial Times,* March 6, 1958; "Jersey Standard's Corporate Longevity," *Dun's Review and Modern Industry* (March, 1963).

63. E.C. Minutes, January 18, 1961; February 1, 4, 20, April 22, September 4, 1963, Secretary's Office, Exxon; Bd. Dir. Mins., April 6, 12, July 5, 1962; February 21, May 29, June 13, November 21, 1963; February 27, May 11, 1964, *ibid.; Pamphlet,* "Standard Oil Company (New Jersey): What It Is and How It Helps Meet the World's Expanding Energy Needs," Standard Oil Company (New Jersey), *c.* 1962; *Pamphlet,* "Standard Oil Company (New Jersey)," *ibid., c.* 1964; "Jersey Standard's Corporate Longevity," *Dun's Review and Modern Industry* (March, 1963).

CHAPTER 4

1. Humble Oil & Refining Company [Humble], *Annual Reports,* 1947-1955; Henrietta M. Larson and Kenneth Wiggins Porter, *History of Humble Oil & Refining Company* (New York, 1959), 634-647.

2. *Interview,* Bennett H. Wall with Hines H. Baker, September 15, 1976, Files, Center for Business History Studies [hereafter cited as C.B.H.S.]; *Interviews,* Wall with Rex G. Baker, March 2, 1976, August 25, 1982, *ibid.; Interview,* Wall and Carpenter with Richard J. Gonzalez, November 30, 1977, *ibid.;* Gilbert Burck, "The Jersey Company," *Fortune,* Vol. XLIV, No. 10 (October, 1951), 98-112, 186-200, 106; Larson and Porter, *History of Humble Oil & Refining Company.* Ms. Larson related to this author on a number of occasions the open-handed, warm reception she received at Humble while completing this volume. Such was not the case with all the companies, nor all the executives. *Letter,* Wallace E. Pratt to Wall, December 29, 1978, Wall Personal Files; see also "Remarks," Morgan J. Davis, May 27, 1957, Courtesy, Jay Rose.

3. When Everit E. Sadler of the Standard Oil Company (New Jersey) [hereafter cited as Jersey or SONJ] Board of Directors visited W. S. Farish, Humble president, saying he needed Eugene Holman, Farish sent for Wallace E. Pratt, who headed exploration, and asked him what he thought—did Humble have a replacement for Holman? Pratt told Farish that he knew where to find a better geologist than Holman. He hired L. T. Barrow. *Interview,* Wall with Pratt, June 19, 1974, Files, C.B.H.S.; *Letter,* Pratt to Wall, August 30, 1979, Wall Personal Files.

4. Ernest R. Bartley, *The Tidelands Oil Controversy* (Austin, 1953), 4-5; United States *vs.* California (1947), 332 U.S. 19, p. 38; United States *vs.* Texas (1950), 339 U.S., 707, 719-720; United States *vs.* Louisiana (1950), 339 U.S., 699, 705-706; Humble, *Annual Reports,* 1948, 1949.

5. Burck, "The Jersey Company," p. 106; Humble, *Annual Reports,* 1947-1958; SONJ, *Annual Reports,* 1950-1958; *Interview,* Wall and Carpenter with Davis, November 29, 1977, Files, C.B.H.S. For offshore oil see: "Deep-Water Oil," *The Lamp,* Vol. 31, No. 3 (September, 1949), 8-14; "Oil From Under the Gulf," *ibid.,* Vol. 37, No. 3 (September, 1955), 10-15; Norman Ritter, "Oil and The

Undersea World," *ibid.*, Vol. 48, No. 2 (Summer, 1966), 9-13; Richard Rutter, "Oil from the Deep Sea," *ibid.*, Vol. 53, No. 2 (Fall, 1971), 26-28.

6. Humble, *Annual Report*, 1956.

7. Humble, *Annual Reports*, 1948-1959. For what happened to Golden Esso, see: *Memorandum*, "Review of Humble Gasoline Program," [no person] July 6, 1962, Exxon Records Center [hereafter cited as E.R.C.], 82-10-A; Fred C. Allvine and James M. Patterson, *Competition LTD.: The Marketing of Gasoline* (Bloomington and London, 1972), p. 142 discusses the "third-grade" gasoline price war that followed Esso's introduction of Golden Esso Extra.

8. Larson and Porter, *History of Humble Oil & Refining Company*, p. 618 explains the changes made in the late 1940s in the Annuity & Thrift Plan. In 1960, Humble, Esso Standard Oil Company [Esso], and The Carter Oil Company merged their retirement and benefit programs. Generally this resulted in increases; see Executive Committee Minutes [hereafter cited as E.C. Minutes], July 1, 1960, Secretary's Office, Exxon; Humble, *Annual Reports*, 1948-1959.

9. "Depth Interviews About Jersey Standard," prepared by Roper Associates, September, 1975 (obviously 1957).

10. Humble, *Annual Report*, 1956; Burck, *The Jersey Company, p. 107.*

11. "Depth Interviews About Jersey Standard," Roper Associates, September, 1957; Burck, *The Jersey Company*, p. 107; Humble, *Annual Reports*, 1950-1958; *Interview*, Wall and Carpenter with Carl E. Reistle, Jr., September 14, 1976, Files, C.B.H.S.; *Interview*, Wall with Cecil Morgan, July 18, 1978, *ibid.*

12. Burck, *The Jersey Company*, p. 106; Humble, *Annual Reports*, 1947-1958; *Interviews*, Wall and Carpenter with Reistle, September 14, 1976, Files, C.B.H.S. *Interviews*, Wall with Reistle, March 3, 1976, May 16, 1980, February 17, 1983, March 25, 1985, *ibid.*; *Interview*, Wall and Carpenter with Davis, November 29, 1977, *ibid.*; *Interview*, Wall with confidential source, November 30, 1977, *ibid.*; E.C. Minutes, May 2, 1955, Secretary's Office, Exxon.

13. *Interview*, Wall with Reistle, February 17, 1983, Files, C.B.H.S.; *Interview*, Wall with M. A. Wright, August 26, 1982, *ibid.*; *Ibid.*, August 27, 1982.

14. *Ibid.*, August 26, 27, 1982; *Memorandum*, (Unknown), "Legal and Political Conditions Which Affect Humble's Public Relations," March 26, 1954, E.R.C., 192-07-B; *Memorandum*, (Unknown) to D. A. Shepard, November 18, 1955, containing "Memorandum Relating to the Position of the Standard Oil Company (New Jersey) Under the Laws of Texas," *ibid.*

15. *Interview*, Wall with Reistle, March 3, 1976, Files, C.B.H.S.; *Interview*, Wall with Wright, August 26, 1982, *ibid.*; Dan Cordtz, "They're Holding Feet to the Fire at Jersey Standard," *Fortune*, Vol. LXXXII, No. 1 (July, 1970), p. 80.

16. W. S. Farish approved Humble employees moving on to Jersey. Eugene Holman believed that such shifting of personnel helped both companies. Wallace Pratt liked the idea of sending good men to Jersey. He believed this kept Humble management flexible and dynamic. Harry C. Wiess, Barrow, and Davis disliked such exchanges. Carl Reistle described his position:

> I was always hiring so many good men and I would get two or three that were on a par. . . . I could see that I was going to have some problems, so I could unload them into the Jersey organization and know their future would be assured. It gave me an opportunity to feed in some more bright young men. You helped Jersey and you helped yourself.

He added:

> In our exploration and in the rest of the oil company, Mr. Wiess and Mr. Barrow didn't like to see those good men go.

Interview, Wall and Carpenter with Reistle, September 14, 1976, Files, C.B.H.S.; see also, "Humble Policy on Interchange of Personnel with Jersey and Its Affiliates," E.R.C., 192-7-B.

17. *Interview*, Wall with Henry B. Wilson, May 17, 1977, Files, C.B.H.S.; *Interview*, Wall with Wright, August 26, 1982, *ibid.*; *Interview*, Wall and Carpenter with Gonzalez, November 30, 1977, *ibid.*

18. *Interview*, Wall with Wright, March 26, 1985, *ibid.*

19. For example, see E.C. Minutes, September 20, 1948, stock purchased from Wiess Estate, Secretary's Office, Exxon; see also *ibid.*, April 18, 1955, *ibid.*; *Interview*, Wall with John O. Larson, November 19, 1976, Files, C.B.H.S. As late as the 1960s and 1970s, Pratt

corresponded with Wright and J. Kenneth Jamieson, Wall Personal Files.

20. *Interview*, Wall with Johnson, December 14, 1977, Files, C.B.H.S.; *Interviews*, Wall with Rex Baker, March 2, 1976, August 25, 1982, *ibid.*

21. Larson and Porter, *History of Humble Oil & Refining Company*, p. 425.

22. Hines Baker found in his files his longhand *Memorandum*, "Steps," in which he outlined the details of the merger. He gave Wall a copy in 1976. See also the chapter entitled, "1960 Reorganization" for other details of this merger. See also SONJ, "Notice of Special Stockholders Meeting," at New York, NY, December 1, 1959 (dated October 7, 1959); *Interview*, Wall with Hines Baker, September 15, 1976, Files, C.B.H.S.; *Interview*, Wall and Carpenter with Davis, November 29, 1977, *ibid.*; *Interviews*, Wall with Rex Baker, March 2, 1976, August 25, 1982, *ibid.*; *Letter*, Pratt to Monroe J. Rathbone, May 9, 1957, E.R.C., 192-07-B; *Interview*, Wall with Reistle, March 25, 1985, Files, C.B.H.S.; *Interview*, Wall with Larson, November 19, 1976, *ibid.*

23. *Letter*, Pratt to Rathbone, May 9, 1957, E.R.C., 192-07-B.

24. "U.S. Government V. E. I. DuPont de Nemours & Co. et al.," *U.S. Reports*, Vol. 353, Oct. Term 1956, 586-656 (Wash., 1957).

25. *Interview*, Wall with Hines Baker, September 15, 1976, Files, C.B.H.S.; *Interviews*, Wall with Rex Baker, March 2, 1976, August 25, 1982, *ibid.*; Hines Baker *Memorandum*, "Steps," 1959. Note: Rex Baker in one *interview* said he wrote to his brother, Wall with Rex Baker, March 2, 1976, Files, C.B.H.S.

26. SONJ, "Notice of Special Stockholders Meeting," (dated October 7, 1959); *Interview*, Wall and Carpenter with Davis, November 29, 1977, Files, C.B.H.S.

27. Technically, the new Humble Company began operations only after the Jersey shareholders approved the merger, on December 1, 1959; however, some sources indicate that papers had to be filed according to New Jersey law. Hence, the December 31 date. *Letter*, Rathbone to Presidents of Affiliated Companies, E.R.C., 192-7-B.

28. "Consolidation, U.S. Affiliates," *The Lamp*, Vol. 41, No. 3 (Fall, 1959), 22-25; "The Proposed Merger," *The Humble Way*, Vol. XV, No. 3 (September-October, 1959), Reference No. 2; "A Look at the New Humble Company," *ibid.*, Vol. XV, No. 6 (May-June, 1960), *ibid.*

29. *Ibid.*; *Interview*, Wall and Carpenter with Davis, November 29, 1977, Files, C.B.H.S.; *Business Week*, April 27, 1963; E.C. Minutes, December 8, 1959, Secretary's Office, Exxon; Ramsey was quoted in Allvine and Patterson, *Competition, LTD.: The Marketing of Gasoline*, p. 32 quoting "New Humble's No. 1 Marketer Looks at Oil Marketing Today," *National Petroleum News* (February, 1960), p. 122.

30. "Humble's New Organization," *The Humble Way*, Vol. XVI, No. 2 (October, November, December, 1960), Reference No. 5.

31. *Interviews*, Wall and Carpenter with Reistle, September 14, 1976, Files, C.B.H.S.; *Interview*, Wall with Reistle, March 3, 1976, February 17, 1983, *ibid.*; *Interview*, Wall with Hines Baker, September 15, 1976, *ibid.*; *Interview*, Wall with Wright, August 26, 1982, *ibid.*; *Interview*, Wall with Jamieson, May 16, 1980, *ibid.*; *Interview*, Wall and Carpenter with Davis, November 29, 1977, *ibid.*; *Interview*, Wall with confidential source, May 16, 1980, Wall Personal Files; Humble, *Annual Reports*, 1957, 1958; *Business Week*, April 27, 1963.

32. As late as 1968, Jersey remained concerned about this. See *letter*, Jersey Director Clifton C. Garvin to Wright, June 13, 1968, E.R.C., 208-5-A. Garvin said, "We question whether marginal, onshore wildcatting operations are being reduced at a rapid enough pace."

33. "Humble's New Organization," *The Humble Way*, Vol. XVI, No. 2 (October, November, December, 1960), Reference No. 5; *Manuscript*, "Brief History of Humble Oil & Refining Company"; *Interview*, Wall and Carpenter with Reistle, September 14, 1976, Files, C.B.H.S.; *Interview*, Wall with Reistle, February 17, 1983, *ibid.*

34. *Interview*, Wall and Carpenter with Davis, November 29, 1977, *ibid.*; *Interview*, Wall with Jamieson, May 16, 1980, *ibid.*; *Business Week*, April 27, 1963; *Letter*, Davis to William R. Stott, August 25, 1961, E.R.C., 82-10-A; *The Lamp*, Vol. 51, No. 1 (Spring, 1969), p. 21.

35. *Interview*, Wall with Wright, August 26,

1982, Files, C.B.H.S.; *Interview*, Wall with Reistle, March 3, 1976, *ibid.*; *Interview*, Wall and Carpenter with Reistle, September 14, 1976, *ibid.*

36. Larson and Porter, *History of Humble Oil & Refining Company*, p. 243; Henrietta M. Larson, Evelyn H. Knowlton and Charles S. Popple, *New Horizons, 1927-1950, History of Standard Oil Company (New Jersey)* (New York, 1971), 276, 278, 298-299, 403, 535, 546, 669, 775; *Interview*, Wall with Reistle, February 17, 1983, Files, C.B.H.S.; *Interview*, Wall and Carpenter with Davis, November 29, 1977, *ibid.*; *Interview*, Wall with Wright, August 26, 1982, *ibid.*; *Interview*, Wall with Jamieson, August 27, 1982, *ibid.*

37. The U.S. District Court in St. Louis had issued in 1937 a permanent injunction against Jersey marketing in Standard Oil Company (Indiana) territory. It should be noted that the 1911 decree did not divide the Standard Oil Trust territory. That division came as a result of the separate companies continuing to do business in their respective areas and in states where they had engaged in business before 1911.

38. On May 8, 1960, the Jersey Executive Committee approved the cash purchase of 27 service stations in Arizona, California, and Nevada, E.C. Minutes, Secretary's Office, Exxon. In some cases, Jersey stock was used in payment. These properties were then given to Humble Oil & Refining Company as a capital contribution. For example, see E.C. Minutes, April 29, June 15, 1960, September 27, 1961, March 16, 1962, *ibid.*

39. *Ibid.*, May 27, June 15, 1960; *Time*, January 27, 1961; *Business Week*, April 22, June 10, 1961; *U.S. News and World Report*, June 19, 1961; *Marketing*, November 4, 1961; *Fortune*, Vol. LXIV, No. 1 (July, 1961).

40. *Interview*, Wall with confidential source, July 16, 1978, Wall Personal Files; *Business Week*, June 10, 1961; *U.S. News and World Report*, June 19, 1961.

41. E.C. Minutes, June 5, 6, 8, July 5, 7, August 16, 1961, Secretary's Office, Exxon; "The Big Change to ENCO," *The Humble Way*, Vol. XVI, No. 3 (Spring, 1961); *Manuscript*, "Brief History of Humble Oil & Refining Company"; *Barron's*, January 29, 1962; *Business Week*, July 3, 1965; *Interview*, Wall and Carpenter with Davis, November 29, 1977, Files, C.B.H.S.; *Interview*, Wall and

Carpenter with Reistle, September 14, 1976, *ibid.*

42. E.C. Minutes, January 11, May 17, June 8, 1962, Secretary's Office, Exxon.

43. *Ibid.*, November 22, 1963. On June 13, 1962, Humble purchased Olin Oil and Gas Corporation with production in Arkansas, Louisiana, and Texas. Olin and Monterey were the only two purchases that possessed reserves of oil and gas.

44. *Ibid.*, October 28, November 22, 1963; *Marketing*, February 22, 1964; *Interview*, Wall with Wright, August 26, 1982, Files, C.B.H.S.; *Interview*, Wall and Carpenter with Reistle, September 14, 1976, *ibid.*

45. *Press Release*, November 27, 1963, E.R.C., 82-10-A; *Interview*, Wall with Reistle, February 17, 1983, Files, C.B.H.S.

46. *Ibid.*

47. *Business Week*, April 18, May 9, May 25, 1964; *New York Times*, April 3, 16, 1964; *Interview*, Wall with Jamieson, August 27, 1982, Files, C.B.H.S.; Notes on "Remarks," Garvin at Conference, September 4, 1986, *ibid.*

48. *Interview*, Wall with Reistle, February 17, 1983, *ibid.*; *Interview*, Wall with Wright, August 26, 1982, *ibid.*

49. *Ibid.*

50. *Interview*, Wall with Reistle, February 17, 1983, *ibid.*; *Interview*, Wall with Wright, August 26, 1982, *ibid.*; *Business Week*, May 9, 1964; *Interview*, Wall with Jamieson, August 27, 1982, Files, C.B.H.S. Jamieson agreed with Wright that the Board believed Jersey's size was the factor.

51. *Interview*, Wall with Reistle, February 17, 1983, Files, C.B.H.S.

52. Harold Gilliam, "Battle of the Bay," *The Lamp*, Vol. 51, No. 3 (Fall, 1969), 3-9.

53. *Interview*, Wall and Carpenter with Davis, November 29, 1977, Files, C.B.H.S.; *The Lamp, Vol. 54, No. 4 (Winter, 1972)*, p. 26; *Manuscript*, "A Brief History of the Humble Oil & Refining Company"; "That Traveling Esso Tiger," *The Lamp*, Vol. 47, No. 4 (Winter, 1965), 8-9. The tiger was drawn by a young artist who had once worked for Walt Disney in California. While employed by another agency, he also did some work for McCann Erickson. *Letter*, Paul E. Morgan to

Wall, April 8, 1986, March 24, 1987, Files, C.B.H.S. During this period, Jersey affiliates in the Netherlands, Australia, Belgium, England (a second time), and the Philippines all used the tiger at different intervals.

54. *Business Week*, February 18, 1967.

55. Pratt, "Oil Fields in the Arctic," *Harper's*, Vol. 188, No. 1124 (January, 1944), 107-112, 107, 108; Pratt, "Petroleum in the North," in Hans W. Weigert and Vilhjalmur Stefansson, *Compass of the World* (New York, 1945), 336-347; see also Pratt, "Petroleum on the Continental Shelves," *Bulletin*, American Association of Petroleum Geologists, Vol. 31, No. 4 (April, 1947), 657-671; Pratt, "Distribution of Petroleum in the Earth's Crust," *ibid.*, Vol. 28, No. 7 (July, 1944), 924-952; George Sweet Gibb and Evelyn H. Knowlton, *The Resurgent Years, 1911-1927, History of the Standard Oil Company (New Jersey)*, (New York, 1956); Larson and Porter, in their *History of Humble Oil & Refining Company*, did not mention any search for oil in Alaska before 1948. Earlier, *Popular Mechanics* had run two brief articles on Arctic oil: "Hunting for Oil Beneath Polar Snows," 42: 597-599, October, 1924; and "Hunting Oil in the Frozen North," 46: 451-454, September, 1926; while the *Saturday Evening Post* presented an article by G. Meyers, "Black Gold in the White North," Vol. 217, No. 6 (December 30, 1944). This latter article appeared almost eleven months after that of Pratt, cited above. Bern Keating, "Through The Northwest Passage," *The Lamp*, Vol. 52, No. 1 (Spring, 1970), 2-9.

56. Pratt, "Oil Fields in the Arctic," *Harper's*, 107-112; W. L. Copithorne, "The Worlds of Wallace Pratt," *The Lamp*, Vol. 53, No. 3 (Fall, 1971), 10-14.

57. Humble, *Annual Reports*, 1956, 1957; Charles S. Jones, *From the Rio Grande to the Arctic, The Story of the Richfield Oil Corporation* (Norman, 1972) relates the story of that firm's oil exploration in Alaska and its merger with Humble. Details in this volume differ from those found in several Humble publications. Varying in some places also are versions related by several Humble interviewees who occupied important positions in the company while Alaskan exploration and development took place.

58. *Interview*, Wall and Carpenter with Reistle, September 14, 1976, Files, C.B.H.S.;

Interview, Wall with Reistle, February 17, 1983, *ibid.*; *Search* [A magazine for Exxon USA Exploration], Vol. 2, No. 1 (September-October, 1986), 6-13.

59. *Interview*, Wall and Carpenter with Reistle, September 14, 1976, Files, C.B.H.S.; *Interview*, Wall with Reistle, February 17, 1983, *ibid.*; Humble, *Annual Reports*, 1957, 1958; *Interview*, Wall and Carpenter with John L. Loftis, Jr., December 1, 1977, Files, C.B.H.S.

60. Humble, *Annual Reports*, 1951-1958; *Search*, 6-13; Jones, *From the Rio Grande to the Arctic*, 283-331.

61. *Interview*, Wall and Carpenter with Reistle, September 14, 1976, Files, C.B.H.S.; *Interview*, Wall with Wright, August 26, 1982, *ibid.*; Jones, *From the Rio Grande to the Arctic*, 283-331; *Search*, 6-13; *Interview*, Wall and Carpenter with Loftis, December 1, 1977, Files, C.B.H.S.

62. Jeremy Main, "The Hot Oil Rush in Arctic Alaska," *Fortune*, Vol. LXXIX, No. 4 (April, 1969), 120-125, 136-142, p. 140; Jones, *From the Rio Grande to the Arctic*, 316-321; *Interview*, Wall with Wright, August 26, 1982, Files, C.B.H.S.; Mary Clay Berry, *The Alaska Pipeline* (Bloomington and London, 1975), 263-264.

63. The British Petroleum Company Limited, *Our Industry Petroleum* (London, 1976), p. 158; the merger was approved by the shareholders of the two companies on December 30, 1965, and became effective January 3, 1966.

64. Jones, *From the Rio Grande to the Arctic*, 327-331; *New York Times*, July 19, 28, December 17, 1968; Daniel Jack Chasan, *Klondike '70. The Alaskan Oil Boom* (New York, et al., 1971), 3-22, 103; Board of Directors Minutes [hereafter cited as Bd. Dir. Mins.], January 30, 1969, Secretary's Office, Exxon. North Slope oil is medium gravity crude, American Petroleum Institute rated 27-28°, with less than 1 percent sulphur by weight. The best light crude is rated 40°. Downs Matthews, "Protecting Prudhoe Bay," *Exxon USA*, Vol. XVIII, No. 4 (Fourth Quarter, 1979), 16-22.

65. E.C. Minutes, December 4, 1968, Secretary's Office, Exxon; *Letter*, C. F. Lindsley to Wright, December 10, 1968, E.R.C., 208-5-A; *Journal of Commerce* (New York), February 11, 1969; *The Lamp* carried a number of articles on the North Slope discovery and pro-

duction. See: "Oil Hunt In The North Coun-
try," Vol. 35, No. 1 (March, 1953), p. 24; Jack
Long, "Alaska," Vol. 51, No. 1 (Spring, 1969),
7-14; "Alaska's North Slope," Vol. 51, No. 4
(Winter, 1969), p. 16; British Petroleum, *Our
Industry Petroleum*, 41-42.

66. *Manuscript*, "Brief History of Humble
Oil & Refining Company"; *U.S. News &
World Report*, March 3, 1969, 66-67; Main,
"The Hot Oil Rush in Arctic Alaska," p. 124;
New York Times, December 17, 1968, June 4,
1969; Bern Keating, *The Northwest Passage*
(Chicago, *et al.*, 1970), p. 140. Keating states
that the vessel carried only 126 persons (p.
141) but he obviously did not include the crew.
Washington Post, August 11, 12, 1969; Ms.
Bentley had been Maritime reporter for the
Baltimore Sun. Critique, "From Humble Oil
& Refining Company to Exxon Company,
U.S.A.," by Wright, May 7, 1986, Files,
C.B.H.S.

67. *Interview*, Wall with Wright, August 26,
1982, *ibid.*; *Interview*, Wall and Carpenter
with Reistle, September 14, 1976, *ibid.*; Bd.
Dir. Mins., January 30, 1969, Secretary's Office, Exxon; E.C. Minutes, December 4, 1968,
January 31, 1969, *ibid.*; *New York Times*,
May 1, 1969.

68. *Interview*, Wall with Wright, August 26,
1982, Files, C.B.H.S.; E.C. Minutes, April 22,
1969, *ibid.*

69. *New York Times*, June 20, September 26,
28, December 20, 1969.

70. *San Francisco Examiner*, May 14, 1969;
Journal of Commerce (New York), July 22,
23, 1969.

71. Chasan, *Klondike '70*, 11, 149; *New York
Times*, September 1, 12, 1969; *Interview*,
Wall with Wright, August 26, 1982, Files,
C.B.H.S. Note: Wright's specific comments
on the main objections were not tape recorded
at his request; Berry, *The Alaska Pipeline*, 98-
101. Chasan quotes the Sierra Club *Bulletin*,
but this writer could not find it in the place
cited. The December, 1969 *Bulletin* strongly
urged a three-year moratorium on Alaskan oil
exploration, p. 2.

72. Chasan, *Klondike '70*, 33-34, 142-143;
Berry, *The Alaska Pipeline*, 98-99.

73. The eight companies in Alyeska were:
Amerada Hess Corp., Atlantic Richfield
Company [Arco], Standard Oil Company
(Ohio) [Sohio], Exxon, Mobil Oil Corpora-

tion, Phillips Petroleum Company, Union Oil
Company of California, and British Petroleum; *Telephone Conversation*, Wall with
Leslie C. Rogers, February 26, 1986, Files,
C.B.H.S.; Bd. Dir. Mins., June 23, 1970, Secretary's Office, Exxon. Precisely what percentage Jersey (Exxon) actually owned in
Alyeska is difficult to determine. In the Jersey
Annual Report for 1970, the statement is
made, "Humble Pipe Line Company has a 25
percent interest in Alyeska." In the 1973 *Annual Report*, the figure is "the long-delayed
trans-Alaska pipeline, in which Exxon has a
25.52 interest. . . ." The reports for 1974 and
1975 state that Exxon Pipeline Company has a
20 percent interest; *Press Release*, Alyeska
Pipeline Service Company, June 8, 1976,
Files, C.B.H.S.; *Note*, Rogers to Wall, February 11, 1986, *ibid.*

74. Chasan, *Klondike '70*, p. 175; Cordtz,
"They're Holding Feet to the Fire at Jersey
Standard," *Fortune* (July, 1970), p. 80; *New
York Times*, June 4, 1969; Keating, *The
Northwest Passage*, p. 140.

75. The two voyages are described in: *Wash-
ington Post*, August 11, 12, September 16,
1969, February 21, 1970; *New York Times*,
August 31, November 10, 1969, March 18, 29,
April 1, 2, July 31, 1970; *Journal of Commerce*
(New York), January 7, February 25, April 3,
May 21, 1970; *Christian Science Monitor*,
September 16, 1969; Keating, *Northwest Pas-
sage*, describes only the first voyage as is indi-
cated by the title.

76. *New York Times*, March 29, August 10,
October 22, 1970; *Wall Street Journal*, April
2, August 12, October 22, 1970; *Journal of
Commerce* (New York), April 3, 1970; *Interview*, Wall with Wright, August 26, 1982,
Files, C.B.H.S. Wright's figure does not include interest and some other charges. The
probable cost was near $10 billion.

77. *New York Times*, August 10, 1970, February 2, 1972.

78. *Telephone Conversation*, Wall with Rogers, February 7, 1986, Files, C.B.H.S.; Chasan, *Klondike '70*, 23-26; *Wall Street Journal*,
March 10, 1969, states that no drilling reports
were made before March; Robert Easton,
*Black Tide. The Santa Barbara Oil Spill and
its Consequences* (New York, 1972).

79. *Wall Street Journal*, February 4, 22,
1971; *New York Times*, March 5, 1971; *Christian Science Monitor*, February 13, 1971;

Washington Post, February 15, 16, 22, 1971; *Journal of Commerce* (New York), February 22, 1971; Berry, *The Alaska Pipeline*, 254-278; *Newsweek*, April 16, 1973; *U.S. News & World Report*, July 30, 1973.

80. Berry, *The Alaska Pipeline*, 254-282; *New York Times*, November 18, 1973; The *Financial Times* (London), January 20, 1973; *Business Week*, November 10, 1973; *Letter*, B. W. Massey to R. T. Bonn, April 4, 1972, Secretary's Office, Exxon; Bd. Dir. Mins., September 25, 1974; March 26, April 30, 1975, Secretary's Office, Exxon.

81. The *Financial Times* (London), January 20, 1973; Berry, *The Alaska Pipeline*, p. 262; *Multinational Corporations and United States Foreign Policy, Hearings Before The Subcommittee on Multinational Corporations of the Committee on Foreign Relations, United States Senate, Ninety-Third Congress, Second Session*, 17 Parts (Washington, 1973-1976) [hereafter cited as *MNC*], Part IV, p. 192 states that there are at least 50 billion barrels of crude oil and 80 trillion cubic feet of natural gas in Pet Four.

82. *Time*, February 18, 1974, June 27, 1977; *U.S. News & World Report*, June 10, 1974. For additional details on the pipeline, *New York Times*, September 26, 1975; The *Times-Picayune* (New Orleans), June 29, 1975; *Christian Science Monitor*, November 30, 1977; *Miami Herald*, June 21, 1977; *Wall Street Journal*, October 25, 1976, June 20, 1977; *States-Item* (New Orleans), December 1, 1976.

83. Larson and Porter, *History of Humble Oil & Refining Company*, 390-425; *Interviews*, Wall with Pratt, June 15, 16, 1975, Files, C.B.H.S. *Interview*, Wall and Carpenter with Reistle, September 14, 1976, *ibid.*; *Interview*, Wall with Reistle, March 25, 1985, *ibid.*

84. E.C. Minutes, April 12, 1950, Secretary's Office, Exxon; "Humble Helps Build A City," *The Lamp*, Vol. 46, No. 2 (Summer, 1964), 2-5; "Biography of an Oil Well," *ibid.*, Vol. 42, No. 3 (Fall, 1960), 20-21; *Interview*, Wall and Carpenter with Reistle, September 14, 1976, Files, C.B.H.S.

85. *Telephone Conversation*, Wall with John B. Boles, September 10, 1985, Files, C.B.H.S.; *Interview*, Wall and Carpenter with Reistle, September 14, 1976, *ibid.*; *Interview*, Wall with Reistle, March 25, 1985, *ibid.*; *Interviews*, Wall with Rex Baker, March 2, 1976, August 25, 1982, *ibid.*

86. Stephen B. Oates, "NASA's Manned Spacecraft Center at Houston, Texas," *Southwestern Historical Quarterly*, Vol. LXVII, No. 1 (January, 1964), p. 352.

87. William D. Angel, Jr., "Politics of Space," *Houston Review*, Vol. VI, No. 2 (1984), p. 84.

88. *Ibid.*, p. 72.

89. E.C. Minutes, September 20, 1961, Secretary's Office, Exxon; *Interview*, Wall and Carpenter with Reistle, September 14, 1976, Files, C.B.H.S.; *Interview*, Wall with Reistle, March 25, 1985, *ibid.*

90. "Humble Helps Build A City," *The Lamp*, 2-5; E.C. Minutes, September 20, 28, 1961, Secretary's Office, Exxon; *Letter*, L. R. Moore to L. D. Stinebower, December 28, 1961, E.R.C., 82-10-A; "Jersey Standard Explores Non-Oil Fields," *Investor's Reader*, Vol. 48, No. 11 (June 7, 1967), 13-16; *Interview*, Wall and Carpenter with Reistle, September 14, 1976, Files, C.B.H.S.; *Interview*, Wall with Reistle, March 25, 1985, *ibid.*

91. *Manuscript*, "Brief History of Humble Oil & Refining Company"; *Interview*, Wall and Carpenter with Reistle, September 14, 1976, Files, C.B.H.S.; *Interview*, Wall with Reistle, March 25, 1985, *ibid.*

92. *Interviews*, Wall with Wright, February 17, 1983, March 25, 1985, *ibid.*; *Interview*, Wall and Carpenter with Reistle, September 14, 1976, *ibid.*; *Interviews*, Wall with Reistle, February 17, 1983, March 25, 1985, *ibid.* On October 10, 1966, Humble purchased 1,500 acres adjoining Bayport, E.C. Minutes, October 10, 1966, Secretary's Office, Exxon.

93. E.C. Minutes, September 23, 1969, *ibid.*; SONJ, *Annual Reports*, 1969, 1970, 1971; "Humble Helps Build A City," *The Lamp*, 2-5.

94. *Letter*, Davis to Wright, September 29, 1961, E.R.C., 82-10-A; *Interview*, Wall and Carpenter with Reistle, September 14, 1976, Files, C.B.H.S.; *Interviews*, Wall with Reistle, March 3, 1976, March 25, 1985, *ibid.*; *Business Week*, July 3, 1965; "A Look at the New Humble Company," Reference No. 2; *Manuscript*, "Brief History of the Humble Oil & Refining Company."

95. *Interview*, Wall and Carpenter with Jamieson, November 30, 1977, Files, C.B.H.S.; *Interview*, Wall with Wright, August 26, 1982, *ibid.*; *Interview*, Wall and Carpenter with Reistle, September 14, 1976, *ibid.*; *Interview*, Wall with Reistle, March 3, 1976, *ibid.*;

Notes, Wright Career, *ibid.*; *Telephone Conversation*, Wall with Wright, May 5, 1986, *Notes, ibid.*

96. SONJ, *Annual Reports*, 1963, 1967, 1970, 1971; E.C. Minutes, January 13, 24, 1967, Secretary's Office, Exxon; *Telephone Conversation*, Wall with Wright, May 5, 1986, Notes, Files, C.B.H.S.; *Notes*, Wright Career, *ibid.*; Wright, *The Business of Business: Private Enterprise and Public Affairs* (New York, 1967), p. 34.

97. E.C. Minutes, January 13, 24, 1967, Secretary's Office, Exxon; SONJ, *Annual Reports*, 1963, 1967, 1968, 1970; *Memorandum*, A. A. Draeger to Members of the Board, May 28, 1968, E.R.C., 208-5-A.

98. E.C. Minutes, September 20, October 15, 1962; June 14, September 22, 23, 1966; September 24, 1968, Secretary's Office, Exxon; SONJ, *Annual Reports*, 1962-1975; "A New Coal Company," *The Lamp*, Vol. 56, No. 4 (Winter, 1974), p. 21; "Carter Oil Subsidiary to Open Illinois Coal Mine," *The Humble Way*, Reference No. 10 (no issue, no date).

99. E.C. Minutes, December 13, 1967, April 23, 1968, Secretary's Office, Exxon; SONJ [Exxon], *Annual Reports*, 1967-1975; *New York Times*, September 8, November 10, 1968, December 6, 1970; see Easton, *Black Tide*, for a biased account of the Santa Barbara spill.

100. *New York Times*, November 14, 1970, March 5, 1971; "The Pollution Problem," *The Lamp*, Vol. 49, No. 1 (Spring, 1967), p. 1; S. Dillon Ripley, "The Fragile Sea," *ibid.*, Vol. 55, No. 1 (Spring, 1973), 2-5; J. K. Jamieson, "Exxon and the Environment," *ibid.*, Vol. 56, No. 4 (Winter, 1974), inside cover and p. 1; "Cleaning the Air," *ibid.*, Vol. 57, No. 2 (Fall, 1975), p. 27; Jamieson, "Energy and The Environment: Striking A Balance," *ibid.*, Vol. 54, No. 1 (Spring, 1972), inside cover and p. 1

101. *Letter*, Wright to Garvin, July 15, 1968, E.R.C., 208-5-A; Humble, *Annual Report*, 1967; *Interview*, Wall with Wright, February 17, 1983, Files, C.B.H.S.

CHAPTER 5

1. *Speech*, O. V. Tracy, "How Jersey Got Into The Chemical Business," May 21, 1956, p. 3. Courtesy, Tracy.

2. In this account the author relied largely on George S. Gibb and Evelyn H. Knowlton, *The Resurgent Years, History of Standard Oil Company (New Jersey), 1911-1927* (New York, 1956), 521-526; Frank A. Howard, *Buna Rubber, The Birth of An Industry* (New York, 1947), 1-9; *Interviews;* and company literature.

3. *Speech*, Tracy, May 21, 1956, p. 2. For another account of the beginnings of chemicals at Jersey see "In on the Beginning," *Chemsphere*, July, 1971, Esso Chemical Company.

4. *Interview*, Bennett H. Wall with George H. Freyermuth, December 18, 1975, Files, Center for Business History Studies [hereafter cited as C.B.H.S.] *Letter*, Freyermuth to Wall, October 4, 1978, *ibid.* A number of these persons achieved great success in the Jersey corporation: Harold W. Fisher, David A. Shepard, Marion W. Boyer, and Robert P. Russell. See also *letter*, F. H. Kant to Wall, May 18, 1981, Files, C.B.H.S.

5. A. Donald Green, "An Engineer Putting Chemistry to Work," (n.p., 1977), 6-7.

6. See glossary.

7. See glossary.

8. *Interview*, Wall with George T. Piercy, September 20, 1978, Files, C.B.H.S.

9. *Speech*, Tracy, May 21, 1956, p. 8.

10. See glossary.

11. See glossary.

12. From the perspective of the 1960s, M. J. Rathbone characterized the period 1940-1945 as the "Coming of Age" of the petrochemical industry; ["Petrochemicals No Longer a Stepchild," *Chemical Engineering Progress* (May, 1969), p. 24]; *Speech*, Tracy, "Developments in Petroleum Chemicals," January 5, 1953. Courtesy, Tracy.

13. *Memorandum*, M. B. Hopkins, "Oil-Chemical Raw Materials," March 31, 1944, Exxon Records Center [hereafter cited as E.R.C.], 192-7-B.

14. *Memorandum*, Howard, "Oil-Chemical Business of Standard Oil Co. (N.J.) Interests," July 22, 1943, E.R.C., 192-7-B. It should be noted that in addition to being a brilliant patent lawyer, ninety patents were issued on Howard's inventions, most of them relating to the oil business.

15. *Ibid.*

16. On this point, see Green, "An Engineer . . . ," 18-22.

17. *Ibid.*, p. 22.

18. E. C. Holmer, "Notes on Exxon Chemical," February 21, 1980, Files, C.B.H.S.

19. *Letter,* F. W. Abrams to Eugene Holman, April 4, 1944, E.R.C., 192-7-B.

20. *Memorandum,* Hopkins, "Oil-Chemical . . . ," March 31, 1944, *ibid.*

21. *Memorandum,* Howard, "Main Considerations Involved in the General Planning of the Jersey Chemical Business," March 1, 1944, *ibid.*

22. Rathbone, "Petrochemicals No Longer a Stepchild," *Chemical Engineering Progress* (May, 1969), p. 24.

23. *Memorandum,* Howard, "Oil-Chemical . . . ," E.R.C., 192-7-B.

24. *Memorandum,* Orville Harden, October 19, 1943, *ibid.*; "Memorandum on Meeting in Board Room," October 19, 1943, *ibid.* [Significantly, Haslam was not on this committee.]

25. *Ibid.*

26. "Form New Enjay Company," *The Esso Marketer,* January, 1947.

27. *Interview,* Wall and C. Gerald Carpenter with Fisher, June 25, 1975, Files, C.B.H.S.

28. Executive Committee Minutes [hereafter cited as E.C. Minutes], January 2, August 20, 1946; January 10, 1947; March 15, 1948, Secretary's Office, Exxon.

29. *Letter,* Howard to Holman, January 6, 1949, E.R.C., 192-7-B.

30. Some observers contend that Howard wrote *Buna Rubber* as an answer to his critics. See Howard, *Buna Rubber.*

31. *Letter,* Howard to Abrams and Holman, May 12, 1950. Courtesy, Tracy; see also E.R.C., 192-7-B.

32. Rathbone's college degree was in chemical engineering. In the 1920s and 1930s he had managed the Louisiana Company, where the chemical research developments occurred.

33. *Letter,* Abrams to Rathbone, July 10, 1950. Courtesy, Tracy.

34. *Letter,* Rathbone to Abrams, January 10, 1951, E.R.C., 192-7-B.

35. *Letter,* E. V. Murphree to Rathbone, January 22, 1951, *ibid.*

36. *Letter,* Howard to Abrams, January 29, 1951, *ibid.*

37. *Letter,* Holman to S. P. Coleman and C. F. Smith, April 11, 1949, *ibid.*

38. E.C. Minutes, April 25, 1951; February 13, March 11, 1952; Secretary's Office, Exxon; "General Statement with Respect to Petroleum-Chemical Operations of Jersey Affiliated Companies," Rathbone, November 9, 1951, E.R.C., 192-7-B; *Letter,* Stanley C. Hope to Rathbone, January 10, 1952, *ibid.*

39. *Letter,* Howard to Abrams, January 29, 1951, *ibid.*

40. "General Statement with Respect to Petroleum-Chemical Operations of Jersey Affiliated Companies," Rathbone, February 13, 1952, *ibid.*

41. *Interview,* Wall and Carpenter with Fisher, June 25, 1975, Files, C.B.H.S.

42. *Letter,* Murphree to W. J. Edmonds, October 21, 1954, E.R.C., 192-7-B.

43. "General Statement . . . ," February 13, 1952, *ibid.*

44. *Letter,* William J. Haley to Smith, February 15, 1951, *ibid.*

45. Enjay Chemical Co. 1950-1970 Financial Summary, Files, C.B.H.S.

46. Standard Oil Company (New Jersey) [hereafter cited as SONJ], *72nd Annual Meeting [of the Shareholders],* Linden, New Jersey, May 26, 1954, p. 13.

47. "The Chemical Business at Mid-Year," July 16, 1954, E.R.C., 192-7-B; SONJ, *Annual Report,* 1955.

48. *Memorandum,* Piercy, "Report on the Chemical Organization," May 8, 1961, Courtesy, Piercy; "General Statement . . . ," February 13, 1952, E.R.C., 192-7-B; *Letter,* C. H. Carpenter to Coleman and H. W. Page, *ibid.*; E.C. Minutes, May 24, 1955, Secretary's Office, Exxon.

49. *Letter,* Coleman to Smith, July 18, 1955, E.R.C., 192-7-B.

50. *Letter,* H. G. Burks to Jackson, August 31, 1955; *Memorandum,* Burks, September 16, 1968, Courtesy, Ted A. White.

51. Holmer, "Notes on Exxon Chemical," February 21, 1980, Files, C.B.H.S.

52. *The Lamp,* Vol. 37, No. 2 (June, 1955), p. 24.

53. SONJ, *Annual Report,* 1955, 31-32, 38-39; *Interview,* Wall and Carpenter with Carl E. Reistle, Jr., September 14, 1976, Files, C.B.H.S.

54. Rathbone, "Jersey Standard Looks Ahead," a *speech* before the New York Society of Securities Analysts, December 5, 1955, 17-20; Rathbone, "Executive Sees Markets Despite a Promise," *Christian Science Monitor*, February 12, 1958; Rathbone, "The Challenge of Change," a *speech* before the National Petroleum Association, September 10, 1958, 5, 10, Files, C.B.H.S.

55. SONJ, *73rd Annual Meeting*, p. 9; Board of Directors Minutes [hereafter cited as Bd. Dir. Mins.], December 20, 1951, Secretary's Office, Exxon; E.C. Minutes, March 24, 1955, *ibid*.

56. *Ibid*., March 24, April 26, October 14, 1955, *ibid*.

57. *Interview*, Carpenter with J. L. Ernst, July 19, 1977, Files, C.B.H.S.; Robert A. Gilbert, "Resourceful Giant: A Giant Company is Meeting the Challenge of Adversity," *Barron's* (July 4, 1960), p. 10.

58. *Speech*, Rathbone, "Jersey Standard . . . ," December 5, 1955, 6-7; *Letter*, Howard to Rathbone, September 6, 1957, E.R.C., 192-7-B.

59. E.C. Minutes, December 9, 1954, Secretary's Office, Exxon.

60. SONJ, *74th Annual Meeting*, 1956, p. 15.

61. P. T. Lamont, "Memorandum of Discussion with Sir Leonard Sinclair at Burgenstock, Switzerland," June 22, 26, 1956, E.R.C., 192-7-B.

62. "Petrochemical Organization for Affiliates," July 29, 1957, *ibid*.

63. E.C. Minutes, March 24, 1955, Secretary's Office, Exxon.

64. *Letter*, Jackson to Tracy, August 29, 1958, E.R.C., 192-7-B.

65. SONJ, *Annual Report*, 1956, p. 43.

66. *Ibid*., 1957, 27, 32.

67. *Interviews*, Wall and Carpenter with J. K. Jamieson, November 30, 1977; Wall with Jamieson, May 12, 1980, Files, C.B.H.S.; *Letter*, Howard to Rathbone, September 6, 1957, E.R.C., 192-7-B.

68. SONJ, *Annual Report*, 1959, p. 19.

69. E.C. Minutes, December 16, 1959, Secretary's Office, Exxon.

70. Donald P. Burke, "Esso Chemical—Today the World; What's for Tomorrow?," *Chemical Week* (September 21, 1968), p. 50; SONJ, *Annual Report*, 1959, p. 16.

71. *Letter*, Jackson to Tracy, August 29, 1958, E.R.C., 192-7-B; "Chemicals: An Expanding Frontier," *The Lamp*, Vol. 50, No. 1 (Spring, 1968), p. 2.

72. SONJ, Report of Special Meeting of Shareholders, December 1, 1959, p. 7.

73. Gilbert, "Resourceful Giant . . . ," *Barron's* (July 4, 1960), p. 10.

74. SONJ, *Annual Report*, 1960, p. 18.

75. *Interview*, Wall and Carpenter with Fisher, June 25, 1975, Files, C.B.H.S.; Shepard, "Recent Trends in the U.S. Chemical Industry," [translated from French], presented to "La Conference Internationale des Arts Cliniques," Parts, April 26, 1962, *ibid*.

76. John Mellecker, "Jersey Standard: Chemical Success with a Management Structure That Simply Couldn't Work," *Chemical Processing* (July, 1961), p. 5; Rathbone, "The President's Letter: The Promise of Petrochemicals," *The Lamp*, Vol. 42, No. 3 (Fall, 1960), p. 1.

77. *Ibid*., p. 2.

78. *Ibid*., p. 1; SONJ, *Annual Meeting*, 1960, p. 10; SONJ, *Annual Report*, 1960, p. 18.

79. *Letter*, Rathbone to Abrams, January 10, 1951, E.R.C., 192-7-B.

80. *Interviews*, Wall and Carpenter with Fisher, June 25, 1975; Wall with Fisher, September 9, 1975, Files, C.B.H.S.

81. "The Changing Face of Jersey Standard," *Forbes* (November 1, 1960), 19-20.

82. "Jersey's Giant Strides in Chemicals," *Chemical Week* (July 15, 1961), p. 109; Rathbone, "Memorandum on Organization," March 7, 1960, Secretary's Office, Exxon.

83. Mellecker, "Chemical Success . . . ," *Chemical Processing* (July, 1961), p. 2.

84. "Jersey's Giant Strides in Chemicals," *Chemical Week* (July 15, 1961), p. 109.

85. Mellecker, "Chemical Success . . . ," *Chemical Processing* (July, 1961), p. 3.

86. *Ibid*., p. 6.

87. *Interview*, Carpenter with J. B. Smith, October 11, 1977, Files, C.B.H.S.

88. SONJ, *Annual Report*, 1960, p. 18; *ibid*., 1961, p. 9.

89. Gilbert, "Resourceful Giant . . . ," *Barron's* (July 4, 1960), p. 9; "Plastic from Oil," *The Lamp*, Vol. 42, No. 3 (Fall, 1960), p. 8.

90. Gilbert, "Resourceful Giant . . . ," *Barron's* (July 4, 1960), p. 10.

91. "Jersey's Giant Strides . . . ," *Chemical Week* (July 15, 1961), p. 109; Mellecker, "Chemical Success . . . ," *Chemical Processing* (July, 1961), 3, 6; SONJ, *Annual Report*, 1960, p. 18; *ibid.*, 1961, p.9; *ibid.*, 1962, 7-8; Shepard, "Recent Trends . . . ," April 26, 1962, Files, C.B.H.S.

92. E.C. Minutes, May 4, 1961, Secretary's Office, Exxon; SONJ, *Annual Report*, 1961, p. 9.

93. *Ibid.*, 9, 11.

94. "Petrochemicals," SONJ [For Company Use Only], (May, 1963), p. 5.

95. "No Sleeping Giant," *Barron's* (January 29, 1962), p. 15; SONJ, *Annual Report*, 1961, p. 9.

96. "Jersey's Giant Strides . . . ," *Chemical Week* (July 15, 1961), 109-110; SONJ, *Annual Report*, 1961, 9-11; *ibid.*, 1962, p. 13; SONJ, *Report of Annual Meeting*, 1961, p. 7.

97. SONJ, *Annual Report*, 1961, p. 9; "Jersey's Giant Strides . . . ," *Chemical Week* (July 15, 1961), p. 109.

98. SONJ, *Annual Report*, 1961, p. 9; E.C. Minutes, December 16, 1959; January 10, 1961, Secretary's Office, Exxon.

99. "Jersey's Giant Strides . . . ," *Chemical Week* (July 15, 1961), p. 110.

100. *Ibid.*, 109-110; Mellecker, "Chemical Success . . . ," *Chemical Processing* (July, 1961), 2, 6.

101. "Jersey's Giant Strides . . . ," *Chemical Week* (July 15, 1961), p. 110; *Interview*, Wall with Fisher, September 9, 1975, Files, C.B.H.S.

102. Fisher quoted in "Jersey's Giant Strides . . . ," *Chemical Week* (July 15, 1961), p. 111.

103. *Ibid.*

104. *Interview*, James B. Quinn with Fisher, May 3, 1977, Files, C.B.H.S.; Mellecker, "Chemical Success . . . ," *Chemical Processing* (July, 1961), p. 6.

105. *Ibid.*, 5-6; "Jersey's Giant Strides . . . ," *Chemical Week* (July 15, 1961), p. 115.

106. *Interview*, Wall and Carpenter with Fisher, June 25, 1975, Files, C.B.H.S.

107. *Ibid.*; "Jersey's Giant Strides . . . ," *Chemical Week* (July 15, 1961), p. 113.

108. *Ibid.*; *Memorandum*, Piercy, "Report on the Chemical Organization," Courtesy, Piercy; "Statement of Policy for Chemical Activities," March 23, 1962, E.R.C., 192-7-B; see also E.C. Minutes, March 23, 1962, Secretary's Office, Exxon.

109. *Ibid.*; *Interviews*, Wall with Piercy, September 20, 1978; January 26, 1981; Wall with Jamieson, May 16, 1980, Files, C.B.H.S.

110. *Ibid.*; SONJ, *Press Release*, June 4, 1962, Exxon Central Library; *Interview*, Quinn with Fisher, May 3, 1977, Files, C.B.H.S.

111. E.C. Minutes, October 4, 1962, Secretary's Office, Exxon.

112. SONJ, *Press Release*, April 29, 1963, Public Affairs Department, Exxon.

113. *Letters*, J. O. Lofstrom to J. Harkins, November 9, 1963, italics mine; Harkins to Wolf Greeven, November 15, 1963, E.R.C., 306-2-B.

114. *Letter*, Rathbone to Board Members, October 3, 1963, *ibid.* (Italics mine.)

115. Esso Chemical Company Inc., Chemical's Financial and Operating Review [hereafter cited as F & O Review], April 9, 1964. Courtesy, Holmer.

116. *Ibid.*; E.C. Minutes, April 17, August 14, 1963, Secretary's Office, Exxon.

117. *The Lamp*, Vol. 44, No. 4 (Spring, 1963), 2-9.

118. SONJ, *Annual Report*, 1962, 4-5; John R. White, "Private Industry in a Growing Colombia," Address before the National Association of Industrialists, Cartagena, Colombia, January 24, 1963, Files, C.B.H.S.; see *The Lamp*, Vol. 48, No. 3 (Fall, 1966), 11-15.

119. E.C. Minutes, September 11, 1959; January 12, 1961, Secretary's Office, Exxon; "Jersey's Giant Strides . . . ," *Chemical Week* (July 15, 1961), p. 110; SONJ, *Annual Report*, 1960, p. 18; *ibid.*, 1961, p. 9.

120. E.C. Minutes, January 12, 1961, Secretary's Office, Exxon.

121. "Jersey's Giant Strides . . . ," *Chemical Week* (July 15, 1961), p. 110; E.C. Minutes, February 8, and May 23, 1961, Secretary's Office, Exxon.

122. *Ibid.*, March 26, April 4, 18, June 10, July 23, 1963; September 16, 1964; May 25,

1966, *ibid.*; Bd. Dir. Mins., August 19, October 8, 22, 29, November 19, 1964, *ibid.*

123. *Memorandum*, D. O. Swan to R. L. Brown, Jr., "Proposed 1966 Chemical Investments," October 8, 1965, Courtesy, Holmer.

124. Esso Chemical Company, Inc., "Executive Committee Presentation," October 29, 1965, *ibid.*; "Notes on Exxon Chemical," February 21, 1980, Files, C.B.H.S.

125. Copet was a joint company owned by Esso Argentina and Manufactura Forti S.A. [Forti] that manufactured polyester fibers. The company never realized the glowing expectations of its sponsors. E.C. Presentation, October 29, 1965, Courtesy, Holmer.

126. *Memorandum*, Swan to Brown, October 8, 1965, *ibid.*

127. *Ibid.*

128. Esso Chemical Company Inc. F & O Review, May 2, 1966, *ibid.*

129. E.C. Minutes, December 7, 1965, Secretary's Office, Exxon.

130. Bd. Dir. Mins., February 27, June 11, and October 15, 1964, *ibid.*

131. *Letter*, T. A. White to Wall, February 21, 1979, quoting E. C. Holmer, remarks on an early draft of this chapter, Files, C.B.H.S.

132. *Letter*, Holmer to Wall, February 21, 1980, *ibid.*; E.C. Minutes, June 19, September 26, 1963; June 25, October 8, 1965, Secretary's Office, Exxon.

133. *A History of Imperial Oil*, Imperial Oil Limited (c. 1967), 28-29; E.C. Minutes, June 16, 1964, Secretary's Office, Exxon; Bd. Dir. Mins., June 10, November 4, 1965, *ibid.*; *Memorandum*, "Chemical Organization," February 9, 1965, E.R.C., 76-12-A.

134. E.C. Minutes, August 8, September 26, 1963, Secretary's Office, Exxon; SONJ, *Annual Report*, 1965, 12, 14.

135. E.C. Minutes, July 10, 1964, January 20, December 15, 1965, Secretary's Office, Exxon; SONJ, *Annual Report*, 1965, 3, 11.

136. *Ibid.*, 1963, 7, 8; *ibid.*, 1964, 9, 11; *ibid.*, 1965, p. 12; *Chemical Week*, October 31, 1964.

137. Burke, "Esso Chemical . . . ," *Chemical Week* (September 21, 1968), p. 51.

138. "Guidelines for Chemicals Organization," January 4, 1965, E.R.C., 76-12-A.

139. "Chemical Organization," February 9, 1965, *ibid.*

140. E.C. Minutes, September 21, 1965, Secretary's Office, Exxon.

141. *Ibid.*; *ibid.*, May 25, 1966, *ibid.*; *Interview*, Wall and Carpenter with Fisher, March 17, 1976, Files, C.B.H.S.

142. "Chemical Organization," February 9, 1965, E.R.C., 76-12-A; "Summary of Recommendations for Corporate Framework of Jersey's European Chemical Business," September 13, 1965, *ibid.*

143. "Chemical Organization," February 9, 1965, *ibid.*

144. "Standard of Jersey—A Revitalized Giant," *Moody's Stock Survey* (January 27, 1964), p. 760.

145. SONJ, *Annual Report*, 1963, p. 2; E.C. Minutes, March 16, 1962, Secretary's Office, Exxon.

146. "More Food For More People," *The Lamp*, Vol. 47, No. 4 (Winter, 1965), p. 24; "The Earth Can Feed Us All," *ibid.*, Vol. 45, No. 1 (Spring, 1963), 2, 5; "The Jersey Story," *Oil and Coal News* (July, 1965), p. 26; SONJ, *Annual Report*, 1965, p. 7.

147. E.C. Minutes, August 26, 1963; April 10, 1964, Secretary's Office, Exxon; "The Earth Can Feed . . . ," *The Lamp*, Vol. 45, No. 1 (Spring, 1963), 2, 5; SONJ, *Annual Report*, 1964, p. 12; *ibid.*, 1965, p. 6.

148. E.C. Minutes, January 14, 1964, Secretary's Office, Exxon.

149. Bd. Dir. Mins, May 2, 1962; July 2, 1964, *ibid.*; E.C. Minutes, November 28, 1962; December 17, 1965, *ibid.*; "More Fertilizer Means More Food," *The Lamp*, Vol. 48, No. 3 (Fall, 1966), p. 12.

150. SONJ, *Annual Report*, 1964, p. 12; E.C. Minutes, April 9, May 28, 1963; May 27, 1964, Secretary's Office, Exxon; Bd. Dir. Mins., March 26, 1964; February 18, August 26, 1965, *ibid.*

151. E.C. Minutes, March 26, 1962; April 4, 1963; June 18, 1965, *ibid.*; Bd. Dir. Mins., January 23, 1964, *ibid.*

152. *A History of Imperial Oil*, Imperial Oil Limited (c. 1967), p. 28.

153. "Jersey Standard—Diversified Giant," *Financial World* (June 8, 1966), p. 10.

154. "Oil Is No Longer Enough," *Business Week* (July 3, 1965), 48-57.

155. "Confidential—Proposed Jersey Corporate Investment Objectives," E.C. Minutes, February 2, 1966, Secretary's Office, Exxon.

156. Bd. Dir. Mins., December 7, 1964, *ibid.*

157. E.C. Minutes, October 29, 1965, *ibid.*

158. SONJ, *Annual Meeting of the Shareholders*, 1966, 6, 9-10.

159. *Ibid.*, 9-10; "Oil Is No Longer Enough," *Business Week* (July 3, 1965), 48-57.

160. "Jersey Affiliates' Chemical Business," 1964 F & O Review, April 15, 1965. Courtesy, Holmer.

161. [Esso Chemical Company], Executive Committee Presentation, October 29, 1965, *ibid.*

162. *Letter*, C. C. Garvin, Jr. to Esso Chemical Employees, February 17, 1966, *ibid.*

163. *Ibid.*; See also SONJ, *Annual Meeting of Shareholders*, 1966, p. 6; "Confidential—Proposed Jersey Corporate Investment Objectives," E.C. Minutes, February 2, 1966, Secretary's Office, Exxon; Elkins Oliphant II, "Europe in the Age of Chemicals," *The Lamp*, Vol. 53, No. 1 (Spring, 1971), p. 48.; "Esso Chemical S.A.—Corporate Objectives," April 5, 1967, E.R.C., 25-13-A.

164. E.C. Minutes, May 25, 1966, Secretary's Office, Exxon.

165. *Interview*, Wall and Carpenter with Fisher, June 25, 1976, Files, C.B.H.S.; E.C. Minutes, June 7, August 26, 1966, Secretary's Office, Exxon; *Letters*, Garvin to John F. Wright, October 4, 1967, R. H. Milbrath to Fisher, November 28, 1967, E.R.C., 25-13-A.

166. Burke, "Esso Chemical . . . ," *Chemical Week* (September 21, 1968), p. 51.

167. *Ibid.*, p. 50.

168. *Ibid.*, 53, 57; "Confidential—Proposed Jersey Corporate Investment Objectives," E.C. Minutes, February 2, 1966, Secretary's Office, Exxon; Bd. Dir. Mins., November 30, 1967, *ibid.*; "Esso Chemical S.A.—Corporate Objectives," April 5, 1967, E.R.C., 25-13-A.

169. Burke, "Esso Chemical . . . ," *Chemical Week* (September 21, 1968), p. 55.

170. "Confidential—Proposed Jersey Corporate Investment Objectives," E.C. Minutes, February 2, 1966, Secretary's Office, Exxon; Burke, "Esso Chemical . . . ," *Chemical Week* (September 21, 1968), 50, 55; SONJ, *Annual Report*, 1966, p. 20.

171. *Ibid.*; *A History of Imperial Oil*, 28-29; Burke, "Esso Chemical . . . ," *Chemical Week* (September 21, 1968), p. 50; E.C. Minutes, January 25, June 3, July 27, August 9, September 6, October 18, October 25, 1966; February 10, 1967; January 10, March 8, July 16, September 3, 1968, Secretary's Office, Exxon.

172. *Ibid.*, April 12, November 2, 1966; February 10, 1967; January 24, July 16, 1968, *ibid.*; "Chemicals: An Expanding Frontier," *The Lamp*, Vol. 50, No. 1 (Spring, 1967), p. 5.

173. *A History of Imperial Oil*, p. 25; SONJ, *Annual Report*, 1966, p. 20; E.C. Minutes, September 6, 1966, Secretary's Office, Exxon.

174. *Ibid.*, March 28, 1966; February 17, July 5, 1967; May 8, 1968, *ibid.*

175. "Confidential—Proposed Jersey Corporate Investment Objectives," *ibid.*, February 2, 1966, *ibid.*

176. *Ibid.*, April 22, June 7, December 7, 1966; March 7, 1967, *ibid.*; *Letter*, Edoardo de Pedys to George A. Lawrence, September 22, 1967, E.R.C., 25-13-A.

177. *A History of Imperial Oil*, 28-29; "Confidential—Proposed Jersey Corporate Investment Objectives," E.C. Minutes, February 2, 1966, Secretary's Office, Exxon; SONJ, *Annual Report*, 1966, p. 20; E.C. Minutes, July 27, September 6, December 7, 1966; July 13, 1967, Secretary's Office, Exxon.

178. Bd. Dir. Mins., March 31, 1966, *ibid.*; E.C. Minutes, August 9, 1966; January 10, September 3, 1968, *ibid.*

179. F & O Review, May 12, 1966, Courtesy, Holmer.

180. "Notes on Exxon Chemical," Holmer, February 21, 1980, Files, C.B.H.S.

181. *Ibid.*; Esso Chemical Company Inc. Management Committee Minutes, September 28, 1966, Courtesy, Holmer.

182. Esso Chemical Company Inc. Five Year Business Plan and 1967 Investment Plan, October 10, 1966, *ibid.*; "Notes on Exxon Chemical," by Holmer, February 21, 1980, Files, C.B.H.S.

183. "The Management of Jersey's Chemical Organization," February 1, 1966, *ibid.*

184. *Ibid.*, p. 51; "Confidential—Proposed Jersey Corporate Investment Objectives," E.C. Minutes, February 2, 1966, Secretary's

Office, Exxon.

185. *Letter,* Garvin to Esso Chemical Employees, February 17, 1966, Courtesy, Holmer.

186. "Notes on Exxon Chemical," Holmer, February 21, 1980, Files, C.B.H.S.

187. *Ibid.*

188. SONJ, *Report of the 85th Annual Meeting of Shareholders,* May 17, 1967, p. 6.

189. Dan Cordtz, "They're Holding Feet to the Fire at Jersey Standard," *Fortune,* LXXXII, No. 1 (July, 1970), p. 127; E.C. Minutes, May 17, 1968, Secretary's Office, Exxon; *Interview,* Wall with Reistle, March 3, 1976, Files, C.B.H.S.

190. "Esso Chemical S.A.—Corporate Objectives," April 5, 1967, E.R.C., 25-13-A.

191. Burke, "Esso Chemical . . . ," *Chemical Week* (September 21, 1968), p. 54; *Interview,* Wall with Fisher, September 9, 1975, Files, C.B.H.S.; SONJ, *Annual Report,* 1967, p. 20; "More Fertilizer Means More Food," *The Lamp,* Vol. 48, No. 3 (Fall, 1966), p. 12.

192. E.C. Minutes, June 18, 1965, Secretary's Office, Exxon.

193. *Letter,* C. F. Van Berg to Garvin, June 1, 1967, E.R.C., 25-13-A; Notation by Karl H. Karlson on C. J. Carven to W. R. Stott, February 3, 1967, ibid.; E.C. Minutes, September 8, 1966, Secretary's Office, Exxon.

194. *Letter,* Cecil Burrill to Morgan, August 14, 1979, Files, C.B.H.S.; Esso Chemical Accomplishment and Objective Review [hereafter cited as A & O Review], May 2, 1967, Courtesy, Holmer.

195. *Interview,* Wall with Fisher, September 9, 1975, Files, C.B.H.S.

196. *Letter,* Swan to R. K. Dix, *et al.,* January 9, 1967, Courtesy, Holmer.

197. Esso Chemical A & O Review, May 2, 1967, *ibid.*

198. Holmer, Presentation to Jersey IAC, November 1, 1967, *ibid.*

199. Holmer, Presentation to Jersey Executive Committee, December 5, 1967, *ibid.*

200. "Esso Chemical S.A.—Corporate Objectives," April 5, 1967, E.R.C., 25-13-A; E.C. Minutes, May 31, 1967, Secretary's Office, Exxon.

201. Cordtz, "They're Holding Feet to the Fire . . . ," *Fortune* (July, 1970), p. 130.

202. SONJ, *Annual Report,* 1968, p. 21; E.C. Minutes, December 6, 1967, Secretary's Office, Exxon.

203. "Notes on Exxon Chemical," by Holmer, February 21, 1980, Files, C.B.H.S.

204. *Ibid.*

205. *Ibid.* On October 11, 1970, Swan wrote the IAC that by 1973 the chemical company would have divested $536 million; *Performance and Outlook Letter,* Swan to Garvin, April 11, 1969, Courtesy, Holmer.

206. *Letter,* A. R. Martin, Jr. to J. F. Dean, January 30, 1970, Secretary's Office, Exxon.

207. SONJ, *Annual Report,* 1969, 18-19; E.C. Minutes, December 11, 1966; December 6, 1967, Secretary's Office, Exxon.

208. Burke, "Esso Chemical . . . ," *Chemical Week* (September 21, 1968), p. 54.

209. SONJ, *Annual Meeting of Shareholders,* 1970, p. 12.

210. *Letter,* Holmer to Swan, May 11, 1970, Files, C.B.H.S.; *Interview,* Carpenter with Ray B. Nesbitt, October 11, 1977, *ibid.*

211. *Letter,* E. A. Herberich to A. O. Savage, October 10, 1968, Courtesy, Holmer.

212. *Letter,* Holmer to R. K. Dix, *et al.,* December 17, 1968, *ibid.*

213. *Letter,* Swan to Garvin, April 11, 1969, *ibid.*

214. *Letter,* Swan to Garvin, April 20, 1970, *ibid.*

215. *Pamphlet,* Holmer, "Presentation to the Jersey Management Committee, May 7, 1970," *ibid.*

216. Bd. Dir. Mins., November 25, 1970, Secretary's Office, Exxon; See also "Introducing Exxon Chemicals," *Exxon Chemicals,* Vol. V, No. 4 (Winter, 1972-73), p. 3; *Chemsphere,* July, 1972, p. 3; *Ibid.,* February 1973, inside cover.

217. John Brooks, "The Marts of Trade: It Will Grow on You," *The New Yorker* (March 10, 1973).

218. SONJ, *Annual Report,* 1972, p. 22.

219. *Letter,* Swan to Garvin, October 4, 1971, Courtesy, Holmer.

220. SONJ, *Annual Report*, 1973, 2, 18.

221. *Ibid.*, p. 18.

222. *Memorandum*, "Technology Study Exxon Chemical Company," February 16, 1973. Courtesy, Holmer.

223. *Letter*, Swan to Howard C. Kauffmann, March 29, 1974, *ibid.; Chemsphere* (July, 1971), p. 11.

224. *Letter*, Swan to Kauffmann, March 29, 1974, Courtesy, Holmer.

225. Exxon Corporation, *Annual Report*, 1973, p. 1.

226. *Wall Street Journal*, November 14, 1973.

227. *Letter*, Swan to Kauffmann, May 1, 1974, E.R.C., 752-5-A.

228. *Ibid.*

229. *Ibid.; Pioneers in Petrochemicals*, Exxon Chemical Company U.S.A., 1975, p. 7.

230. *Letter*, Swan to Kauffmann, May 1, 1974, E.R.C., 752-5-A; SONJ, *Annual Report*, 1972, p. 22; *Letter*, R. C. Darcey to Swan, July 12, 1973, Management Committee Files, Secretary's Office, Exxon; *Letter*, Richard W. Kimball to Swan, September 18, 1973, *ibid.*

231. *Letter*, Donald M. Cox to Swan, June 10, 1974, E.R.C., 752-5-A.

232. *Letter*, Swan to Kauffmann, May 1, 1974, *ibid.*

233. *ChemReport '74*, cover; *Letter*, Swan to Kauffmann, March 24, 1975. Courtesy, Holmer.

234. *Chemsphere*, July-August, 1974, p. 20. On May 3, 1976, Holmer returned to Exxon Chemical as president, replacing Swan, who died the week before, Exxon Chemical Company, *Press Release*, May 3, 1976.

235. Exxon Corporation, *Annual Report*, 1975, p. 18.

CHAPTER 7

1. In 1975, Jersey's claim for "War Damage and Nationalization" against Rumania amounted to $20,000,000 (app.), while that against Hungary was $27,000,000; *Memorandum* (with enclosures), E. G. O'Neill to D. G. Gill, June 6, 1976, Files, Center for Business History Studies [hereafter cited as C.B.H.S.].

2. *Interview*, Bennett H. Wall with M. A. Wright, February 17, 1983, *ibid.*

3. For discussions of the search for oil in Europe, see *ibid.*, and "Oil-Hungry Europeans Seek Added Production," *The Lamp*, Vol. 30, No. 1 (January, 1948); "Ten Years of Progress in Free Europe," *ibid.*, 75th Anniversary Issue, 1957; W. L. Copithorne, "Europe's Energy Revolution," *ibid.*, Vol. 53, No. 1 (Spring, 1971); Don Kliewer and Robert E. Spann, "Private Enterprise Scores Again," *World Oil*, July, 1954. Wallace E. Pratt, noted Jersey executive and geologist, discussed the search for oil in Western Europe in numerous *interviews* with the author, Files, C.B.H.S.; *Interview*, Wall with Zeb Mayhew, January 30, 1984, *ibid.*

4. Paul Ruedemann, an American of German birth, before World War II had managed Jersey's Hungarian affiliate, Magyar Amerikai Olajipari Részvénytársaság [MAORT]. He returned to Hungary and on September 18, 1948, he was arrested by the Russian Communists in Hungary and tortured until he signed a phony confession. On the day (September 25, 1948) the Russians released him, they nationalized MAORT. To Director Wright and many others who knew him, he was a strong figure. It is not inconceivable that Ruedemann learned of the Shell discoveries in Western Europe in 1948 after his release. He worked for Jersey for several years after the war. Wright insists that Ruedemann provided him with the first information he had heard about the Shell-Jersey agreement. In that sense Jersey's interest in the Groningen gas field and other oil and gas fields started with him.

5. *Interview*, Wall with Wright, August 27, 1982, Files, C.B.H.S. For the material upon which this account of Groningen is based, see the following: *Letter*, C. R. Smit to W. R. Stott, May 8, 1961, Exxon Records Center [hereafter cited as E.R.C.], 397-2-B; *Memorandum*, Wolf Greeven to Stott, May 16, 1961, *ibid.; Memorandum*, R. H. Milbrath to Greeven, June 5, 1961, *ibid.; Memorandum*, "Dutch Gas Situation," by Smit, September 21, 1961, *ibid.; Cablegram*, Greeven to Unknown (Canesso), October 25, 1961, *ibid.; Cablegram*, Greeven to Smit and [Murray] Matheson, October 31, 1961, *ibid.; Cablegram*, Smit to Coordination Department, January 31, 1963, E.R.C., 281-18-A; *Cablegram*, Smit to P. J. Anderson, February 7, 1963, *ibid.; Cablegram*, Coordination Department to Smit, February 7, 1963, *ibid.; Memorandum*, "Presentation on Dutch Gas,"

by M. W. Johnson, March 6, 1963, *ibid.*; *Letter*, G. W. Butler to H. W. Haight, May 22, 1963, *ibid.*; *Letter*, D. M. Latimer to D. M. Cox, October 18, 1963, *ibid.*; *Memorandum*, Latimer to Anderson, *et al.*, October 31, 1963, *ibid.*; *Memorandum*, Latimer to Stott, November 1, 1963, *ibid.*; *Memorandum*, Latimer to Greeven, November 26, 1963, *ibid.*. In addition, see Executive Committee Minutes [hereafter cited as E.C. Minutes], Secretary's Office, Exxon, for the following dates: 1961—January 5, 18; February 10; June 21; July 24; October 12, 26; December 4; 1962—September 19; November 1, 8; December 14; 1963—March 7, 18; April 1; June 17; November 20; December 6; 1964—March 3, 17; September 16; November 11; See also, *Letter*, Milbrath to Wall, August 31, 1983, Files, C.B.H.S.; *Letter* (with enclosures), E. de Pedys to Paul Morgan, July 4, 1980, including "A Brief History of the Dutch Gas Venture," Morgan Files, *ibid.*; See also Standard Oil Company (New Jersey) [hereafter cited as SONJ], *Annual Reports*, 1959, 1960, 1961, 1962; Other details may be found in: Bryan Cooper and T. F. Gaskill, *North Sea Oil—The Great Gamble* (London, 1966); Report: "The Technology Assessment Group Science and Public Policy Program" (Irvin L. Kash, *et al.*), *North Sea Oil and Gas* (Norman, Oklahoma, 1973), [hereafter cited as *North Sea Oil and Gas*]; *Interview*, Wall with Mayhew, January 30, 1984, Files, C.B.H.S.

6. For this section of the European exploration story, rather than repetitiously cite SONJ *Annual Reports* and *Annual Meeting of Shareholders Minutes*, the writer consolidated them. See SONJ, *Annual Reports*, 1959-1975, *Annual Meetings of Shareholders Minutes*, 1959-1975; the following articles from *The Lamp* were found useful: "Opening Holland's Treasure," Vol. 46, No. 1 (Spring, 1964); "Zero Hour in the North Sea," Vol. 46, No. 4 (Winter, 1964); "The Boom That Has Europe on the Move," Vol. 47, No. 1 (Spring, 1965); Charles W. Frey, "New Growth For Natural Gas," Vol. 48, No. 1 (Spring, 1966); "The Sea Around Them," Vol. 50, No. 4 (Winter, 1968); Norman Richards, "Oil From the Fiercest Sea," Vol. 55, No. 2 (Summer, 1973); Sandford Brown, "North Sea Venture," Vol. 57, No. 4 (Winter, 1975). See also *Letter* (with enclosures), de Pedys to Morgan, July 4, 1980, including "A Brief History of the Dutch Gas Venture," Morgan Files, C.B.H.S.; E.C. Minutes, October 20, 1960, January 5, 1961; December 6, 1963, Secre-

tary's Office, Exxon; See also *Letter*, Smit to Stott, May 8, 1961, E.R.C., 397-2-B; *Memorandum*, Milbrath to Greeven, June 5, 1961, *ibid.*; *Memorandum*, "Dutch Gas Situation," September 25, 1961, *ibid.*, and others; *Interview*, Wall with Mayhew, January 30, 1984, Files, C.B.H.S.; *New York Times*, August 8, September 14, 1972.

CHAPTER 8

1. *Speech*, George Catlett Marshall, "Harvard Commencement Address," June 5, 1947, *Vital Speeches of the Day*, Vol. XIII, No. 18 (July 1, 1947), 553-554.

2. "Up From the Ashes," *The Lamp*, Vol. 33, No. 3 (September, 1951), 2-5.

3. *Ibid.*

4. *Ibid.*; "Oil's Role in Britain's Struggle," *The Lamp*, Vol. 35, No. 3 (September, 1953), 2-5; *Interview*, Bennett H. Wall with Emile E. Soubry, October 14, 1977, Files, Center for Business History Studies [hereafter cited as C.B.H.S.].

5. "Up From the Ashes," *The Lamp*, 2-5.

6. *Ibid.*; "Oil's Role in Britain's Struggle," *ibid.*, 2-5; *Interview*, Wall with Soubry, October 14, 1977, Files, C.B.H.S.; "Refinery Going Up," *The Lamp*, Vol. 32, No. 3 (September, 1950), 2-5; Executive Committee Minutes [hereafter cited as E.C. Minutes], July 30, October 31, November 22, 1946, January 13, 15, 17, May 21, 23, August 14, 21, September 5, November 14, 17, October 24, 1947, Secretary's Office, Exxon.

7. "Up From the Ashes," *The Lamp*, 2-5.

8. *Letter*, George Koegler to David E. Longanecker, April 10, 1947, Exxon Records Center [hereafter cited as E.R.C.], 646-1-A; *Letter*, Koegler to Longanecker, December 10, 1947, *ibid.*; *New York Times*, May 11, 1950; *Memorandum*, "Stanic Industria Petrolifera," September 12, 1956, E.R.C., 192-7-B.

9. "Up From the Ashes," *The Lamp*, 2-5.

10. *Ibid.*

11. "Energy in Europe's Economic Renaissance," *ibid.*, Vol. 37, No. 3 (September, 1955), 2-7; "Oil's Role in Britain's Struggle," *ibid.*, 2-5; W. L. Copithorne, "Europe's Energy Revolution," *ibid.*, Vol. 53, No. 1 (Spring, 1971), 16-21.

12. *Letter*, P. T. Lamont to Gerhard Geyer,

December 21, 1954, E.R.C., 726-5-B. See also *Memorandum*, "J. C. Anderson's Notes re Italy," August 21, 1956, *ibid.*, 192-7-B.

13. *Letter*, Lamont to Geyer, December 21, 1954, *ibid.*, 726-5-B. Other supply agreements are mentioned in E.C. Minutes, March 17, October 29, November 5, 1954, Secretary's Office, Exxon. The company failed to seek a refinery in Finland, apparently because a supply agreement could not be included, *ibid.*, August 23, 1954.

14. *Ibid.*, January 20, 1953, November 28, 1955, August 15, 1956.

15. "Up From the Ashes," *The Lamp*, 2-5; Refinery histories, Files, C.B.H.S.; "New Refineries for Europe," *The Lamp*, Vol. 40, No. 4 (Winter, 1958), p. 19.

16. Refinery histories, Files, C.B.H.S.; "Up From the Ashes," *The Lamp*, 2-5; E.C. Minutes, February 2, April 20, October 26, December 6, 1954, October 2, 30, 1956, November 13, December 22, 1958, Secretary's Office, Exxon; "Blueprints For People," *The Lamp*, Vol. 35, No. 4 (November, 1953), 16-17.

17. Refinery histories, Files, C.B.H.S.; E.C. Minutes, January 20, 1953, March 26, 1954, Secretary's Office, Exxon; *Memorandum*, "Stanic Industria Petrolifera," September 12, 1954, E.R.C., 192-7-B.

18. Refinery histories, Files, C.B.H.S.; E.C. Minutes, May 5, August 20, 1954, November 28, 1955, August 20, 1956, Secretary's Office, Exxon.

19. *Letter*, Lamont to D. L. Wright, E.R.C., 726-5-B; E.C. Minutes, January 20, June 3, 1953, March 18, 1954, Secretary's Office, Exxon.

20. *Ibid.*, November 28, 1955, *ibid.*; *Memorandum*, "Stanic Industria Petrolifera," September 12, 1956, E.R.C., 192-7-B; *Memorandum*, Lamont, "A Visit to the State Department," January 7, 1955, *ibid.*, 191-13-A; *Memorandum*, Lamont, Conversation with Ambassador [Clare Booth] Luce, January 7, 1955, *ibid.*; *Memorandum*, "J. C. Anderson's Notes re Italy," August 21, 1956, *ibid.*

21. *Speech*, R. C. Kiddoo, "Remarks to Libyan Press Tour Group," August 8, 1963, *ibid.*, 281-18-A; *Letter*, M. W. Johnson to unknown?, February 2, 1963, *ibid.*, 306-2-B; E.C. Minutes, March 20, 1964, Secretary's Office, Exxon.

22. Refinery histories, Files, C.B.H.S.; E.C. Minutes, January 20, 1953, February 2, March 16, November 29, 1956, Secretary's Office, Exxon; *Memorandum*, "Norway Report," May 1, 1958, E.R.C., 646-1-A.

23. Refinery histories, Files, C.B.H.S.

24. *Ibid.*; E.C. Minutes, November 28, 1955, May 31, July 23, 1957, Secretary's Office, Exxon.

25. Speech, Kiddoo, "Remarks to Libyan Press Tour Group," August 8, 1963, E.R.C., 281-18-A; Refinery histories, Files, C.B.H.S.; *Letter*, Johnson to unknown, February 2, 1963, E.R.C., 306-2-B.

26. Refinery histories, Files, C.B.H.S.; *Memorandum*, "Dansk Veedol Kalundborg Refinery," September 20, 1962, E.R.C., 281-18-A.

27. Refinery histories, Files, C.B.H.S.; E.C. Minutes, October 11, 1963, Secretary's Office, Exxon.

28. Refinery histories, Files, C.B.H.S.; *Memorandum*, Johnson to W. R. Stott, January 25, 1962, E.R.C., 281-18-A; *Memorandum*, Johnson to Stott, April 13, 1962, *ibid.*; *Memorandum*, Vincenzo Cazzaniga to Stott, August 31, 1962, *ibid.*; *Letter*, M. J. Rathbone to Stott, September 10, 1962, *ibid.*; *Letter*, H. G. Burks to Stott, October 23, 1962, *ibid.*

29. Refinery histories, Files, C.B.H.S.; E.C. Minutes, October 11, 1963, Secretary's Office, Exxon.

30. *Memorandum*, "Possible Major Expansion of Esso Engineering Operations in Europe," July 9, 1965, E.R.C., 76-12-A.

31. *Letter*, C. R. Egeler to Johnson, June 6, 1963, E.R.C., 306-2-B. The company also considered building a small refinery on the Cape Verde Island as a way of entering the Portuguese market if that was the only means of doing so; *Memorandum*, W. H. Relien, "Clean Products Refinery at Cape Verde," May 13, 1963, E.R.C., 306-2-B; E.C. Minutes, March 14, 1966, Secretary's Office, Exxon; Refinery histories, Files, C.B.H.S.

32. *Letter*, Stott to Gregorio Lopez Bravo, March 26, 1963, E.R.C., 306-2-B; *Memorandum*, R. B. Carder to Stott, September 20, 1963, *ibid.*; *Memorandum*, "Esso Investment Proposal for Spain," October 2, 1963, *ibid.*, 281-18-A; *Memorandum*, Johnson to W. A. M. Greeven, August 19, 1965, *ibid.*, 306-2-B.

33. *Letter*, M. L. Coppen to Johnson, August

15, 1963, *ibid.*; *Letter*, E. G. Muldoon to Johnson, November 5, 1963, *ibid.*; *Letter*, Serge Scheer to Johnson, July 5, 1963, *ibid.*; *Memorandum*, "Regulations Concerning Petroleum Products of French Origin," October 23, 1961, *ibid.*, 397-2-B.

34. E.C. Minutes, March 24, 1966, December 11, 1968, Secretary's Office, Exxon; *Memorandum*, "Summary of ESSAF Stewardship Review," [1968], E.R.C., 208-5-A.

35. E.C. Minutes, March 22, 1968, Secretary's Office, Exxon; Refinery histories, Files, C.B.H.S.; *Letter*, F. O. Canfield to Stott, May 9, 1962, E.R.C., 281-18-A.

36. Refinery histories, Files, C.B.H.S.; E.C. Minutes, September 23, 1966, July 18, 1967, Secretary's Office, Exxon.

37. Refinery histories, Files, C.B.H.S.

38. *Ibid.*

39. E.C. Minutes, August 9, 1966, Secretary's Office, Exxon; Refinery histories, Files, C.B.H.S.

40. *Ibid.*

41. *Ibid.*; E.C. Minutes, January 25, 1966, August 30, 1967, Secretary's Office, Exxon.

42. Refinery histories, Files, C.B.H.S.

43. E.C. Minutes, March 14, 15, July 20, 1966, Secretary's Office, Exxon; *Cable*, Gassenheimer to J. K. Jamieson, June 12, 1968, E.R.C., 47-1-B; *Letter*, H. T. Cruikshank to Jamieson, July 2, 1968, *ibid.*; *Letter*, F. N. Breekland to Jamieson, July 25, 1968, *ibid.*; *Memorandum*, H. F. Stevenson, "Status Report: Greece," [December, 1968], *ibid.*

44. *Letter*, N. J. Campbell to M. L. Haider, February 1, 1967, *ibid.*, 25-13-A; *Letter*, Richard W. Kimball to Campbell, April 17, 1967, *ibid.*; *Cable*, [R. T.] Bonn to Greeven, March 5, 1968, *ibid.*, 208-5-A; E.C. Minutes, April 6, 1967, June 12, 1968, Secretary's Office, Exxon.

45. Copithorne, "Europe's Energy Revolution," *The Lamp*, 16-21.

46. *Ibid.*; *Letter*, J. G. Finley to Howard W. Page and Siro Vázquez, July 28, 1970, E.R.C., 306-2-B.

47. *Letter*, C. F. Lindsley to C. C. Garvin, July 31, 1970, Files, C.B.H.S.

48. Refinery histories, *ibid.*

49. *Ibid.*; E.C. Minutes, June 26, 1968, Secretary's Office, Exxon.

50. Theodor Swanson, "Europe Cleans House," *The Lamp*, Vol. 53, No. 1 (Spring, 1971), 36-41.

51. *Letter*, Finley to Page and Vázquez, July 28, 1970, E.R.C., 306-2-B; *Memorandum*, R. H. Milbrath to Haider, April 7, 1969, *ibid.*, 164-5-A; *Letter*, Milbrath to Campbell, June 22, 1970, Secretary's Office, Exxon.

52. *Memorandum*, Milbrath to Haider, April 7, 1969, E.R.C., 164-5-A; *Letter*, Finley to Page and Vázquez, July 28, 1970, *ibid.*, 306-2-B; *Memorandum*, Finley to Milbrath, August 11, 1970, *ibid.*; *Letter*, Milbrath to Campbell, August 14, 1970, *ibid.*

53. Refinery histories, Files, C.B.H.S.; *Letter*, B. W. Massey to H. C. Kauffmann, March 7, 1973, Secretary's Office, Exxon.

54. Refinery histories, Files, C.B.H.S.

55. *Letter*, J. F. Dean to Campbell, May 6, 1974, E.R.C., 752-5-A.

56. Refinery histories, Files, C.B.H.S.

CHAPTER 9

1. Esso Standard Italiana S.p.A., "Statistical Data for Period, 1950-1975," Corporate Planning Department (Rome), June 20, 1980, Files, Center for Business History Studies [hereafter cited as C.B.H.S.]; Esso A.G., "Development of Profitability, 1950-1975," June, 1980, *ibid.*; Esso Petroleum Company, "Operating Results," June 13, 1980, *ibid.*; *Telex*, Esso France to Esso Europe, Inc., "Operating Results and Sales Patterns, Net Income—Financial Statements," June [1980], *ibid.*; *Interview*, James B. Quinn with Michael L. Haider, May 6, 1977, Courtesy, Mr. Quinn; *Interview*, Bennett H. Wall with John O. Larson, November 19, 1976, Files, C.B.H.S.

2. Standard Oil Company (New Jersey), [hereafter cited as SONJ], *Annual Reports*, 1955-1963.

3. *Interview*, Wall and C. Gerald Carpenter with Nicholas J. Campbell, Jr., May 5, 1977, Files C.B.H.S.; *Interview*, Wall and Carpenter with Haider, June 6, 1975, *ibid.*

4. "Jersey Geography," *Forbes*, December 15, 1967, p. 19.

5. *Interview*, Wall and Carpenter with Haider, May 6, 1975, Files, C.B.H.S.; *Inter-*

view, Wall and Carpenter with Campbell, May 5, 1977, *ibid.*

6. *Ibid.*

7. *Interview*, Wall and Carpenter with Haider, June 6, 1975, *ibid.*

8. Henrietta M. Larson, Evelyn H. Knowlton and Charles S. Popple, *New Horizons, 1927-1950: History of Standard Oil Company (New Jersey)* (New York, 1971), 321-322.

9. Executive Committee Minutes [hereafter cited as E.C. Minutes], September 9, 1958, Secretary's Office, Exxon; *Interview*, Wall and Carpenter with Campbell, August 31, 1977, Files, C.B.H.S.

10. E.C. Minutes, December 7, 1959, Secretary's Office, Exxon.

11. Monroe Jackson Rathbone, "Memorandum on Organization," January 25, 1960, Files, C.B.H.S.; SONJ, "Organization Chart," May 25, 1960, *ibid.*

12. Rathbone, "Memorandum on Organization," January 25, 1960, *ibid.*; *Interview*, Wall with J. Larson, August 18, 1975, *ibid.*

13. *Ibid.*; *Interview*, Wall and Carpenter with Campbell, August 31, 1977, *ibid.*

14. *Interview*, Wall and Carpenter with Campbell, May 5, 1977, *ibid.*

15. *Interview*, Quinn with Haider, May 6, 1977, Courtesy, Mr. Quinn; *Interview*, Wall and Carpenter with Haider, May 6, 1975, Files, C.B.H.S.

16. *Interview*, Wall and Carpenter with Campbell, May 5, 1977, *ibid.*

17. Board of Directors Minutes [hereafter cited as Bd. Dir. Mins.], February 14, 1963, Secretary's Office, Exxon.

18. *Interview*, Wall and Carpenter with Campbell, May 5, 1977, Files, C.B.H.S.

19. *Interview*, Wall and Carpenter with Haider, May 6, 1975, *ibid.*

20. *Interview*, Wall and Carpenter with Campbell, May 5, 1977, *ibid.*

21. *Interview*, Wall and Carpenter with Campbell, August 31, 1977, *ibid.*; *Interview*, Quinn with Haider, May 6, 1977, Courtesy, Mr. Quinn; *Interviews*, Wall and Carpenter with Haider, May 6, 1975, May 18, 1978, Files, C.B.H.S. *Interviews*, Wall with J. Larson, August 18, 1975, November 19, 1976, *ibid.*

22. *Memorandum*, J. Larson, "Jersey Organization Principles and Structure," June 14, 1965, Secretary's Office, Exxon.

23. *Ibid.*

24. *Ibid.*; *Interview*, Wall and Carpenter with Haider, May 6, 1975, Files, C.B.H.S.; *Interview*, Quinn with Haider, May 6, 1977, Courtesy, Mr. Quinn.

25. *Interview*, Wall and Carpenter with Campbell, May 5, 1977, Files, C.B.H.S.; *Memorandum*, Larson, "Jersey Organization Principles and Structure," June 14, 1965, Secretary's Office, Exxon.

26. *Ibid.*

27. Bd. Dir. Mins., May 19, 1965, *ibid.*

28. *Memorandum*, Larson, "Jersey Organization Principles and Structure," June 14, 1965, *ibid.*

29. Bd. Dir. Mins., June 28, 1965, *ibid.*

30. *Interview*, Wall and Carpenter with Campbell, August 31, 1977, Files, C.B.H.S.

31. *Letter*, George W. Butler to Campbell, September 14, 1965, *ibid.*

32. *Interview*, Wall and Carpenter with Campbell, August 31, 1977, *ibid.*; Bd. Dir. Mins., June 28, 1965, Secretary's Office, Exxon.

33. Compensation and Executive Development Committee Minutes [hereafter cited as COED Minutes], November 3, 1965, Secretary's Office, Exxon.

34. *Interview*, Wall and Carpenter with Haider, May 6, 1975, Files, C.B.H.S.; *Interview*, Quinn with Haider, May 6, 1977, Courtesy, Mr. Quinn.

35. *Ibid.*; *Interview*, Wall and Carpenter with Haider, May 6, 1975, Files, C.B.H.S.

36. Bd. Dir. Mins., November 29 and December 2, 1965, Secretary's Office, Exxon.

37. *Interview*, Wall and Carpenter with Campbell, August 31, 1977, Files, C.B.H.S.

38. Bd. Dir. Mins., November 29 and December 2, 1965, Secretary's Office, Exxon; COED Minutes, November 29, 1965, *ibid.*; *Interview*, Wall and Carpenter with Campbell, August 31, 1977, Files, C.B.H.S.

39. Bd. Dir. Mins., December 16, 1965, Secretary's Office, Exxon.

40. *Press Release*, February 25, 1966, Files, C.B.H.S.

41. *Interview*, Wall and Carpenter with Campbell, August 31, 1977, *ibid*.

42. *Ibid*.

43. *Ibid*.

44. *Ibid*.

45. *Ibid*.

46. *Interview*, Quinn with Haider, May 6, 1977, Courtesy, Mr. Quinn; *Interview*, Wall and Carpenter with Campbell, August 31, 1977, Files, C.B.H.S.

47. *Ibid*.

48. *Interview*, Wall and Carpenter with Campbell, May 5, 1977, *ibid*.

CHAPTER 10

1. This story is partially covered in the Prologue to this volume. So far as this author knows, no full account has been written.

2. *Letter*, Smith D. Turner to Bennett H. Wall, September 6, 1982, Files, Center for Business History Studies [hereafter cited as C.B.H.S.]; Richard Funkhouser, "Middle East Oil (Used as a Background Paper for September 11, 1950 Meeting with Oil Executives) State Department Policy Paper on September 10, 1950," *Multinational Corporations and United States Foreign Policy, Hearings Before The Subcommittee On Multinational Corporations of the Committee on Foreign Relations, United States Senate, Ninety-Third Congress, Second Session*, 17 Parts (Washington, 1973-1976) [hereafter cited as *MNC*], Part 7, 122-134, 125.

3. The Board of Directors directed Counsel Edward F. Johnson to monitor compliance with the Antitrust Consent Decree. Each year after 1942, Johnson reported to the company secretary with copies to the directors on his findings. After he retired in 1957, other counsel continued to check on each affiliate's operations to insure compliance. After World War II, Jersey shareholder representatives in Europe spent much time explaining the importance of the decree to executives of the affiliates; examples are: *Letter*, Johnson to John O. Larson, August 1, 1956, Exxon Records Center [hereafter cited as E.R.C.], 192-7-B; *Memorandum*, J. K. Jamieson to Various Departments of Jersey, June 28, 1967, *ibid.*, 25-13-A.

4. Henrietta M. Larson, Evelyn H. Knowlton, and Charles S. Popple, *New Hori-zons, 1927-1950: History of Standard Oil Company (New Jersey)*, (New York, 1971) 694-713.

5. *Interview*, Wall and C. Gerald Carpenter with Howard W. Page, June 3, 1975, Files, C.B.H.S.; Paul E. Morgan, "Emilio G. Collado, A Sketch," Morgan Files, *ibid*.

6. Larson, Knowlton, and Popple, *New Horizons*, p. 713.

7. "Up From the Ashes," *The Lamp*, Vol. 33, No. 3 (September, 1951), 2-5; *Memorandum*, Dansk Veedol Kalundborg Refinery, September 20, 1962, E.R.C., 281-18-A; Executive Committee Minutes [hereafter cited as E.C. Minutes], October 4, 1954, Secretary's Office, Exxon; *Ibid.*, February 2, March 16, May 22, July 12, November 29, 1956, *ibid.*; Standard Oil Company (New Jersey) [hereafter cited as SONJ], *Annual Report*, 1957.

8. *Foreign Claims Settlement Commission of the United States Claim No. W-7625, Decision No. W. 21483*, Standard Oil Company (Incorporated in New Jersey), May 17, 1967.

9. Contact directors for marketing were: Emile E. Soubry, 1949-1954; Peter T. Lamont, 1954-1961; William R. Stott, 1956-1962; Wolf Greeven, 1962-1969; Robert H. Milbrath, 1969-1973; Howard C. Kauffmann, 1974-1975; *Letter*, W. M. Date to Stott, May 21, 1956, E.R.C., 646-1-A.

10. *Letter*, J. C. Anderson to A. O. Savage, July 23, 1956, *ibid.*; see also E.C. Minutes, December 8, 1955, Secretary's Office, Exxon.

11. *Letter*, Gerhard Geyer to Lamont, November 30, 1954, E.R.C., 726-5-B; *Letter*, Lamont to Geyer, December 21, 1954, *ibid*.

12. See Anderson's Notes *re* Germany, August 22, 1956, E.R.C., 646-1-A.

13. *Memorandum*, "Visit to State Department in Washington," February 3, 1955, *ibid.*, 191-13-A; Anderson, "Notes *re* Italy," (app. 1956) *ibid.*, 192-07-B; "White Paper," RTB [R. T. Bonn], September 26, 1961, *ibid.*, 306-2-B.

14. "Esso In Ireland," author and date unknown (1978), Morgan Files, C.B.H.S.; *Interview*, Wall with Soubry, October 14, 1977, Files C.B.H.S.

15. "Ten Years of Progress in Free Europe," *The Lamp*, 75th Anniversary Issue, 1957; "Energy in Europe's Economic Renaissance," *ibid.*, Vol. 37, No. 3 (September, 1955), 2-7;

"France: A Land in Need of Oil," *ibid.*, Vol. 30, No. 1 (January, 1948), 14-19; "Oil's Role in Britain's Struggle," *ibid.*, Vol. 35, No. 3 (September, 1953), 2-5; "Esso in Ireland," Morgan Files, C.B.H.S.

16. *Interview*, Wall with Soubry, October 14, 1977, Files, C.B.H.S.; "Ten Years of Progress in Free Europe," *The Lamp*, 75th Anniversary Issue, 1957.

17. "Energy in Europe's Economic Renaissance," *The Lamp*, Vol. 37, No. 3 (September, 1955), 2-7.

18. *Letter*, Lamont to C. R. Smit, December 11, 1957, E.R.C., 192-07-B; E.C. Minutes, November 3, 1954, February 4, October 11, 1957, Secretary's Office, Exxon; SONJ, *Annual Report*, 1958.

19. *Letter*, Lamont to Geyer, December 21, 1954, E.R.C., 726-5-B.

20. *Interview*, Wall with Soubry, October 14, 1977, Files, C.B.H.S.

21. "This Company and the Emergency," *Esso* [U.K.] *Magazine*, 1957.

22. *Letter*, J. G. Finley to Page and Siro Vázquez, July 28, 1970, E.R.C., 306-2-B.

23. E.C. Minutes, October 8, 1953; June 25, July 2, 8, 29, August 3, 1954, Secretary's Office, Exxon; For 1954 Denmark, see Telegram, I. Doughten, III to Stott, July 7, 1954, E.R.C., 726-5-B; *Memorandum*, Doughten, III to Lamont and Stott, July 9, 1954, *ibid.*; *Letter*, Stott to Directors, Dansk Esso A/S, July 23, 1954, *ibid.*; *Letter*, Robert Murphy to M. J. Rathbone, July 28, 1954, *ibid.*; *Letter*, Rathbone to John Foster Dulles, July 22, 1954, *ibid.*; *Letter*, Loren F. Kahle to Directors, Dansk Esso A/S, September 9, 1958, *ibid.*, 306-2-B.

24. *Letter*, D. J. Thompson to M. W. Johnson, October 6, 1961, *ibid.*; *Memorandum*, Swiss Company, October 6, 1961, *ibid.*; *Memorandum*, L. J. Bourgeois to Greeven, November 22, 1961, *ibid.*, 397-2-B; *Memorandum*, Swiss Company, November 22, 1961, *ibid.*

25. *Letter*, Geyer to Greeven, July 7, 1961, *ibid.*

26. *Letter*, Vincenzo Cazzaniga to Johnson, April 14, 1961, *ibid.*, 306-2-B (with enclosure); *Note*, B. G. (unknown) to Greeven, October 9, 1961, *ibid.*; (enclosing copy, ESI White Paper, prepared by Bonn, September 26, 1961); *Letter*, Bonn to Johnson, October

20, 1961, *ibid.*; *Memorandum*, Ente Nazionale Idrocarburi [ENI], November (n.d.), 1961, *ibid.*; Report on Ente Nazionale Idrocarburi, November (n.d.), 1961, *ibid.*; *Letter*, Cazzaniga to Johnson, April 13, 1967 (containing untitled enclosure, March 10, 1967) *ibid.*, 25-13-A.

27. E.C. Minutes, March 24, May 15, 1964, Secretary's Office, Exxon; "Oil is No Longer Enough," reprinted from *Business Week*, July 3, 1965, Files, C.B.H.S.

28. E.C. Minutes, February 19, 26, August 13, 1963; February 5, December 7, 1964; April 12, May 19, 1966, Secretary's Office, Exxon; *Letter*, Greeven to William F. Turner, December 8, 1965, E.R.C., 76-12-A; *Letter*, H. T. Cruikshank to Cazzaniga, July 17, 1967, *ibid.*, 25-13-A; *Letter*, Cruikshank to Greeven, August 3, 1967, *ibid.*

29. E.C. Minutes, February 5, 1964, May 19, 1966, May 10, 1967, Secretary's Office, Exxon; *Letter*, C. F. Lindsley to Nicholas J. Campbell, May 18, 1967, E.R.C., 25-13-A; *Letter*, Cruikshank to Greeven, August 3, 1967, *ibid.*; "Europe's New Motor Hotels," *The Lamp*, Vol. 49, No. 1 (Spring, 1967), 23-25; *Letter*, Milbrath to Campbell, June 22, 1970, Secretary's Office, Exxon; SONJ and Exxon, *Annual Reports*, 1966, 1968, 1971, 1972, 1974.

30. Affiliate compilations on sales, market shares, and profitability, Files, C.B.H.S.; "Oil is No Longer Enough," reprinted from *Business Week*, July 3, 1965, *ibid.*; "Summary of ESSAF Stewardship Review," [1968], E.R.C., 208-5-A; "Minutes of the 18th Scandinavian Committee Meeting," April 22-23, 1963, *ibid.*, 281-18-A; *Letter*, Hugh C. Tett to Greeven, April 21, 1961 (with enclosures), *ibid.*, 397-2-B.

31. "Minutes of the 18th Scandinavian Committee Meeting," April 22-23, 1963, *ibid.*, 281-18-A.

32. *Letter*, Campbell to Stott, September 7, 1967 (with enclosures), *ibid.*, 564-3-B.

33. *Letter*, Milbrath to Cruikshank, October 22, 1969, *ibid.*, 82-10-A.

34. Theodor Swanson, "Europe Cleans House," *The Lamp*, Vol. 53, No. 1 (Spring, 1971), 36-41; *Letter*, Milbrath to M. L. Haider, April 7, 1969, E.R.C., 164-5-A.

35. E.C. Minutes, January 21, February 26, 1963; March 14, 15, July 20, October 5, No-

vember 17, December 16, 1966, Secretary's Office, Exxon; SONJ, *Annual Reports*, 1959, 1963, 1964, 1965.

36. *Letter*, Cruikshank to Jamieson, July 2, 1968, E.R.C., 47-1-B; *Letter*, F. N. Breckland to Jamieson, July 25, 1968, *ibid.*; *Letter*, Jamieson to Campbell, July 30, 1968, *ibid.*; *Letter*, F. D. Dennstedt to H. F. Stevenson, July 31, 1968, *ibid.*; *Letter*, Stevenson to Jamieson, December 10, 1968 (with enclosure), *ibid.*; *Letter*, Lindsley to Campbell, September 10, 1969, Secretary's Office, Exxon; *Cablegram*, W. B. Cook to Kauffmann, January 28, 1974, E.R.C., 752-5-A.

37. *Memorandum*, by Y. Klidjian, "Spanish Study," no date (probably July, 1962), E.R.C., 306-2-B; *Letter*, Luke W. Finlay to Stott, July 13, 1963, *ibid.*; *Letter*, Stott to Gregorio Lopez Bravo, March 26, 1963, *ibid.*; *Letter*, A. C. Reiners to Stott, April 5, 1963, *ibid.*; *Memorandum*, R. B. Carder to Greeven, August 13, 1963, *ibid.*, 281-18-A; *Memorandum*, Carder to Stott, September 20, 1963, *ibid.*, 306-2-B; *Memorandum*, "Esso Investment Proposal for Spain," October 2, 1963, *ibid.*, 281-18-A; *Memorandum*, Johnson to Greeven, August 19, 1965, *ibid.*, 306-2-B; *Letter*, Watt H. Denison, Jr. to Johnson, August 17, 1965, *ibid.*; *Letter*, Campbell to Haider, February 1, 1967, *ibid.*, 25-13-A; *Letter*, R. W. Kimball to Campbell, April 17, 1967, *ibid.*; *Cablegram*, Bonn to Greeven, March 5, 1968, *ibid.*, 208-5-A; *Note*, Hugh de N. Wynne to Kauffmann (no date), enclosing translation of principal editorial, *ABC*, June 16, 1974, *ibid.*, 752-5-A; *Letter*, Clifton C. Garvin, Jr. to Campbell, May 6, 1974, *ibid.*

38. *Letter*, Joseph F. Moore, Jr. to M. A. Wright (with enclosures), May 18, 1961, E.R.C., 397-2-B; *Letter*, Clark R. Egeler to Johnson, June 6, 1963, *ibid.*, 306-2-B; *Letter*, F. B. de Castro Neves to Egeler, July 16, 1963, *ibid.*

39. SONJ, *Annual Reports*, 1968, 1969, 1970.

40. SONJ and Exxon, *ibid.*, 1970-1975.

41. *Memorandum*, Finley to Milbrath, August 11, 1970, E.R.C., 306-2-B; *Letter*, Lindsley to Garvin, July 31, 1970, Secretary's Office, Exxon; *Letter*, Milbrath to Campbell, June 22, 1970, *ibid.*; SONJ, *Annual Report*, 1972.

42. *Letter*, Vázquez to Johnson, April 28, 1967, E.R.C., 727-5-B; *Letter*, Jamieson to Campbell, September 23, 1967 [N.B. This let-

ter, p. 2, is dated 1968], *ibid.*, 25-13-A; *Letter*, Milbrath to Cruikshank, October 22, 1969, *ibid.*, 82-10-A.

CHAPTER 11

1. Material in the paragraphs preceding this number is drawn from: *Memorandum*, "Soviet Competition In The Free World Oil Market," March-May, 1960 [authors unknown but a portion of the study was written by Jersey's General Economics Department, other sections by the Coordination and Petroleum Economics Department, and at least one section by I. S. Salnikov, an economist from Producing Coordination], Exxon Records Center [hereafter cited as E.R.C.], 726-5-B. Other sources used were: *U.S. Trade and Investment in the Soviet Union and Eastern Europe; A Staff Report Prepared For the Use of the Subcommittee on Multi-National Corporations of the Committee on Foreign Relations. United States Senate* (Washington, 1974); John P. Hardt, George D. Holliday and Young C. Kim, *Western Investment in Communist Economies*, *ibid.*, (Washington, 1974); George Lenczowski, *Soviet Advances in the Middle East* (Washington, 1972); Benjamin Shwadran, *The Middle East, Oil and The Great Powers* (New York and Toronto, 1973). See the section entitled "The Middle East, 1957-1970," reference note number 38. See also R. W. Campbell, *The Economics of Soviet Oil and Gas* (Baltimore, 1968); George Sweet Gibb and Evelyn H. Knowlton, *The Resurgent Years, 1911-1927: History of Standard Oil Company (New Jersey)*, (New York, 1956); Henrietta M. Larson, Evelyn H. Knowlton and Charles S. Popple, *New Horizons, 1927-1950: History of Standard Oil Company (New Jersey)*, (New York, 1971); Ralph W. Hidy and Muriel E. Hidy, *Pioneering in Big Business, 1882-1911: History of Standard Oil Company (New Jersey)*, (New York, 1955).

2. *Memorandum*, "Status of [Exxon] War Damage and Nationalization Claims, 1947-1976," Files, Center for Business History Studies [hereafter cited as C.B.H.S.]. The Rumanian War damage claim was prepared by Joseph T. Mantsch, who had worked for Romano-Americana S.A., assisted by V. C. Georgescu, who served as president of that company, and whose rescue from the Communists represents an unusual story, even in that day of many like episodes. Mantsch secured the records of war damage from Ro-

mano company files, while the nationalization process took place. See *Letter*, Joseph T. Mantsch to David A. Shepard, July 20, 1977, *ibid.*; *Letter*, Mantsch to T. H. Tiedemann, Jr., June 3, 1977, *ibid.*; *Manuscript*, "Fifty years with Jersey," by Mantsch, *ibid.* It should be noted that Jersey attorneys on October 27, 1944 instituted a series of memoranda to guide the company in preparing international claims. See *Memorandum* (No. 1), "International Claims," October 27, 1944, by George Koegler, *ibid.*; *Letter*, H. W. Kull to W. R. Stott and L. D. Welch, October 1, 1956, E.R.C., 192-07-B; *Memorandum*, "War Damage and Nationalization Claims Filed with Foreign Claims Settlement Concession," December 31, 1957, Files, C.B.H.S. The Piercy incident is described in George T. Piercy, "An Oilman Looks at Russia," *The Lamp*, Vol. 42, No. 4 (Winter, 1960); and in, *Interview*, Bennett H. Wall with Piercy, July 30, 1982, Files, C.B.H.S. The discussion of the evolution of Jersey policy toward Soviet bloc trade is found in numerous reports, memoranda and letters. The material cited came from: *Letter*, Welch to David Rockefeller, October 20, 1959, E.R.C., 192-7-B; *Letter*, Rockefeller to Welch, October 15, 1959, *ibid.*; *Pamphlet*, "Statement of Position on Oil Trade With the Soviet Bloc," June 26, 1963, E.R.C., 726-5-B; *Petroleum Intelligence Weekly*, October 7, 1963; *Letter*, W. C. Asbury to H. W. Fisher, October 22, 1963, E.R.C., 726-5-B; *Letter*, S. B. Sweetser to Fisher, October 24, 1963, *ibid.*; *Report*, "The Effect of Selling Refineries to the Soviet Union," November 18, 1963, *ibid.*; *Memorandum*, "Soviet Bloc General Trade and Oil Trade: A Review and Recommended Jersey Policies," December 20, 1963 [revised January 15, 1964], by R. M. Lilly, *ibid.*; *Letter*, Luke W. Finlay to Wolf Greeven, March 23, 1964, *ibid.*

3. J. H. Carmical writing in the *New York Times*, February 8, 1959; Robert E. Ebel, *The Petroleum Industry of the Soviet Union* (n.p., 1961), p. 154; *Letter*, Chester F. Smith to Frank W. Abrams, September 1, 1953, E.R.C., 726-5-B; "Draft Letter to All Affiliates," April 2, 1954, *ibid.*, 192-7-B.

4. C. L. Bauer also cited 10,000 b/d going to Argentina, an unknown amount to India, and possible sales to Israel and Japan. *Memorandum*, Bauer, "Soviet Controlled Oil in World Markets," December 16, 1953, E.R.C., 726-5-B. The Soviet oil movement was worldwide,

but since the bulk of the oil flowed to Europe and that was Jersey's main concern, in this study, non-European exports are discussed only incidentally.

5. *Ibid.*

6. The Russian-dominated Rumanian government had nationalized without compensation Jersey's operations and properties in that country.

7. *Memorandum*, "Denmark," July 1, 1954, E.R.C., 726-5-B.

8. *Ibid.*

9. Peter R. Odell, *Oil and World Power: A Geographic Interpretation* (New York, 1970), p. 50.

10. *Memorandum*, "Denmark," July 1, 1954; *Memorandum*, "Russian Oil Products to Denmark," May 8, 1954, E.R.C., 726-5-B.

11. *Memorandum*, I. Doughten, III to P. T. Lamont, June 16, 1954, *ibid.*

12. *Ibid.*; *Letter*, D. L. Wright to Stott, June 16, 1954, *ibid.*

13. *Memorandum*, Doughten to Lamont, June 22, 1954, *ibid.*

14. *Memorandum*, Doughten to Stott, July 7, 1954, *ibid.*; *Memorandum*, Doughten to Lamont, July 13, 1954, *ibid.*; *Memorandum*, July 16, 1954, *ibid.*; *Letter*, Robert Murphy to M. J. Rathbone, July 28, 1954, *ibid.*

15. *Letter*, Stott to Directors, Dansk Esso A/S, July 23, 1954, *ibid.*

16. *Letter*, Erik Frandsen to Stott, August 3, 1954, *ibid.*

17. *Ibid.*

18. *Memorandum*, Stott to Lamont, September 28, 1954, *ibid.*

19. *New York Times*, December 6, 8, 1959.

20. Robert E. Ebel, *Communist Trade in Oil and Gas: An Evaluation of the Future Export Capability of the Soviet Bloc* (New York, 1970), 83, 60.

21. *Memorandum*, "Oil Development in the Soviet Bloc, 1948-1955" [May, 1956], E.R.C., 726-5-B; *Memorandum*, "Recent Developments in Russian Petroleum Industry," April 15, 1957, *ibid.*; *Memorandum*, "U.S.S.R.: Analysis of 1959-1965 Oil Industry Targets," July, 1959, *ibid.*; Richard E. Welch, Jr., *Response to Revolution. The United States and*

The Cuban Revolution, 1959-1961 (Chapel Hill, 1985), 6-7, 52; *Time*, July 11, 1960; Mira Wilkins, *The Maturing of Multinational Enterprise: American Business Abroad from 1914 to 1970* (Cambridge, Mass., 1974), p. 355; *Memorandum*, "Iron Curtain Crude Oil and Financial Products...." (No Author), February 3, 1959, E.R.C., 192-7-B; M. W. Johnson, "Notes on Visits to Denmark, Sweden, Norway and Belgium, September 6 to 23rd, 1959," *ibid.*

22. Wilkins, *The Maturing of Multinational Enterprise*, p. 355.

23. "The world now has a far greater supply of known oil reserves in relation to demand than at any time in recent history." [L. F. McCollum to the Fifth World Petroleum Congress, quoted in *New York Times*, June 2, 1959.]

24. *Memorandum*, "U.S.S.R.: Analysis of 1959-1965 Oil Industry Targets," July, 1959, E.R.C., 726-5-B.

25. *Memorandum*, "Soviet Competition in the Free World Oil Market," February 8, 1960, *ibid.*

26. *Ibid.*

27. *Ibid.*

28. *Ibid.*

29. *Ibid.*

30. *Memorandum*, J. S. Baldwin, Jr. to M. W. Boyer, July 21, 1960, *ibid.*

31. *Letter*, Rathbone to Emile Soubry, July 25, 1960, *ibid.*

32. *Letter*, Soubry to R. W. Adams, G. W. Butler, Greeven, G. M. Parker, and L. D. Stinebower, October 4, 1960, *ibid.*

33. *Letter*, Baldwin to Soubry, August 16, 1960, *ibid.*; *Letter*, Piercy to Gunter Last, November 11, 1960, *ibid.* For example, see Executive Committee Minutes [hereafter cited as E. C. Minutes], June 16, 1961, Secretary's Office, Exxon.

34. *New York Times*, August 14, 1960.

35. *Memorandum*, November 3, 1960, E.R.C., 726-5-B; *New York Times*, November 11, 1960.

36. *Memorandum*, November 3, 1960, E.R.C., 726-5-B.

37. *New York Times*, March 12, 1959.

38. *Memorandum*, November 3, 1960, E.R.C., 726-5-B.

39. *New York Times*, November 15, 1960.

40. *Letter*, Greeven to Stott, December 14, 1960, E.R.C., 726-5-B; *Letter*, Vincenzo Cazzaniga to F. O. Canfield, December 9, 1960, *ibid.*; "Excerpts from an article published on 'PAESE SERA,' December 6/7, 1960," and "Article published on the 'CARLINO SERA,' December 1, 1960," *ibid.*

41. *Memorandum*, R. Ihrie, "Strategy for Meeting the Threat of Soviet Oil," February, 1961, *ibid.*

42. *Ibid.*; *New York Times*, October 8, 1960.

43. *Memorandum*, Ihrie, "Strategy for Meeting the Threat of Soviet Oil," February, 1961, E.R.C., 726-5-B.

44. *Ibid.*

45. *Ibid.*

46. *Ibid.*

47. *Memorandum*, "Use of Company Facilities for Handling Soviet and Satellite Oil," June 26, 1961, *ibid.*

48. *Letter*, Rathbone to Allen W. Dulles, March 13, 1961, *ibid.*

49. *Letter*, Dulles to Rathbone, April 8, 1961, *ibid.*

50. *Memorandum*, Greeven to "All Heads of Common Market Affiliates," February 23, 1961, *ibid.*; *Memorandum*, "East Bloc Oil and Common Market," February 24, 1961, *ibid.*

51. *Memorandum*, "Growth of Soviet Oil Exports," March 30, 1961, *ibid.*

52. *Memorandum*, W. R. Carlisle to Rathbone, June 27, 1961, *ibid.*

53. Letter, Rathbone to Dean Rusk, July 14, 1961, *ibid.*; *Memorandum*, July 13, 1961, *ibid.* See also *Memorandum*, Piercy, "The Soviet Oil Offensive," July 12, 1961, *ibid.*

54. *Memorandum*, July 13, 1961, *ibid.*

55. *Ibid.*

56. *New York Times*, May 10, 1961, quoting C. L. Sulzberger; *New York Times*, May 14, 1961.

57. *Ibid.*, June 9, 1961.

58. *Ibid.*, June 21, 1961.

59. *Ibid.*, June 22, 1961.

60. *World Oil*, September, 1961, p. 78.

61. *New York Times*, October 30, November 1, 1961.

62. *Ibid.*, November 14, 1961; Wilkins, *The Maturing of Multinational Enterprise*, 71-72.

63. *Multinational Corporations and United States Foreign Policy, Hearings Before The Subcommittee on Multinational Corporations of the Committee on Foreign Relations, United States Senate, Ninety-Third Congress, Second Session*, 17 Parts (Washington, 1973-1976) [hereafter cited as *MNC*], Part 5, "Testimony of John J. McCloy," 255-258; *Memorandum*, Rathbone, July 17, 1961, E.R.C., 726-5-B; *Letter*, W. D. Bayles to Carlisle, July 24, 1961, *ibid.*; *New York Times*, July 29, 1961; *Interview*, Wall with J. Kenneth Jamieson, August 27, 1982, Files. C.B.H.S.

64. *Letter*, Baldwin to Greeven, August 17, 1961, E.R.C., 726-5-B; *Letter*, Greeven to Rathbone, August 22, 1961, *ibid.*

65. *Letter*, Rathbone to Greeven and Shepard, July 6, 1961, *ibid.*

66. *Minutes*, Russian Oil Committee, August 11, October 12, 1961,*ibid.*

67. *Memorandum*, Greeven to R. J. Yoder, November 21, 1961, *ibid.*; *Memorandum*, Greeven, November 21, 1961, *ibid.*; *Minutes*, Russian Oil Committee, November 24, 1961, *ibid.*; *Memorandum*, "Statement of Position on the Threat of Communist Trade," December 26, 1961, *ibid.*

68. *Ibid.*

69. *Ibid.*

70. *Minutes*, Russian Oil Committee, January 11, 1962, *ibid.*

71. *New York Times*, January 21, June 26, 1962. Pamela Hartland Thunberg, "The Soviet Union in the World Economy," in *Dimensions of Soviet Economic Power*, prepared for the Joint Economic Committee, Congress of the United States, 87th Congress, 2nd Session, 1962, p. 415.

72. *New York Times*, April 5, May 9, 1962; August 8, 1960.

73. *Ibid.*, March 4, September 20, October 28, 1962.

74. Ebel, *Petroleum Industry of the Soviet Union*, 147-150, 155-157.

75. *New York Times*, October 27, 1962; February 14, 1963.

76. *Letter*, Parker to Johnson, February 16, 1962, E.R.C., 726-5-B; *New York Times*, December 28, 31, 1962.

77. *Ibid.*, October 21, 1960; May 14, December 28, 1961; January 7, July 8, 29, September 30, 1962. *Letter*, Yoder to Greeven, October 4, 1962, E.R.C., 726-5-B.

78. *New York Times*, March 23, April 27, 28, 1963; *Christian Science Monitor*, May 3, 1963; *Cablegram*, Hugh C. Tett (Esso Petroleum Company, Limited) to Jersey (probably Rathbone), October 30, 1962, E.R.C., 281-18-A.

79. *New York Times*, November 15, 1963.

80. *Ibid.*, September 28, 1963; January 17, 1964. *Letter*, Asbury to Fisher, October 22, 1963, E.R.C., 726-5-B; *Letter*, Sweetser to Fisher, October 24, 1963, *ibid.*; *Memorandum*, "Effect of Selling Refineries to the Soviet Bloc," November 18, 1963, *ibid.*

81. *Memorandum*, Lilly, "Soviet Bloc General Trade and Oil Trade: A Review and Recommended Jersey Policies," January 15, 1964, *ibid.*

82. *New York Times*, January 10, February 13, 14, 16, 25, March 1, 7, 8, 11, 12, 14, 17, 26, April 1, 9, 19, 26, 30, June 2, 10, September 13, 16, October 24, November 19, December 18, 1964.

83. *Memorandum*, "Soviet Bloc Oil and Gas Industries' Effect on the West" [April, 1965], E.R.C., 726-5-B; *Letter*, J. W. Hansen to C. L. Burrill, Greeven, and Shepard, April 27, 1965, *ibid.*; *Letter*, Finlay to Greeven, March 23, 1964, *ibid.*

84. *New York Times*, April 30, 1964.

85. *Ibid.*, December 9, 1967, May 15, 1968; Ebel, *Communist Trade in Oil and Gas*, 123-125; Odell, *Oil and World Power*, p. 56.

86. *New York Times*, November 11, 18, 1966, April 22, 1967; Ebel, *Communist Trade in Oil and Gas*, 135, 140-142; Odell, *Oil and World Power*, p. 56; *Letter*, Greeven to D. M. Cox, November 16, 1966, E.R.C., 726-5-B.

87. *Letter*, Greeven to Cox, November 16, 1966, *ibid.*; *Memorandum*, "Common Market Energy Policy," Finlay to M. L. Haider, *et al.*, January 3, 1969, *ibid.*, 63-6-A.

88. Ebel, *Communist Trade in Oil and Gas*, p. 83; *New York Times*, January 19, March 10,

1964, May 5, 1968.

89. Ebel, *Communist Trade in Oil and Gas*, p. 123.

90. *New York Times*, January 19, August 23, 1965, June 5, 1966.

91. *Ibid.*, March 10, 1964.

92. *Ibid.*, December 9, 1967.

93. *Ibid.*, January 11, 1969.

94. Ebel, *Communist Trade in Oil and Gas*, p. 104.

95. Robert W. Campbell, *Trends in the Soviet Oil and Gas Industry* (Washington, 1976), p. 75.

CHAPTER 12

1. *Imperial Oil Limited, Annual Report*, 1950 [hereafter cited as *Annual Report* by year].

2. *The Lamp*, Vol. 62, No. 4 (Winter, 1980); see also *Interviews*, Bennett H. Wall and C. Gerald Carpenter with M.L. Haider, May 6, 1975, May 18, 1978, Files, Center for Business History Studies [hereafter cited as C.B.H.S.]; *Interviews*, Wall with Wallace E. Pratt, June 19, August 4, November 25, 1974; June 15, 16, 18, 1975; April 24, July 20, 1976; July 2, 1977; *ibid.* Haider related many interesting details about the Leduc strike during these interviews. Pratt detailed the role in this strike of his friend, Ted Link of Imperial, as well as furnishing details of Imperial's exploration ventures, 1937-1942. From his files, he extracted and read paragraphs of interesting contemporary letters from Link, Louis Weeks and others.

3. *The Lamp*, Vol. 62, No. 4 (Winter, 1980).

4. Earle Gray, *The Great Canadian Oil-Patch* (Toronto, 1970), 136-140.

5. *Annual Report*, 1950.

6. *Esso Mariners, A History of Imperial Oil's Fleet Operations From 1899-1980* (Toronto, 1980), p. 62 [hereafter cited as *Esso Mariners*].

7. *Memories, The Story of Imperial's First Century as Told by Its Employees and Annuitants* (Toronto, 1980), p. 56 [hereafter cited as *Memories*]. One factor that caused Imperial to agonize over this decision was the continuing decline in production of the Turner Valley field discovered in 1914. Across the years, production from this field had fluctuated widely.

In 1936, a new strike occurred there and within a few years, production reached 27,000 barrels per day. By 1945, production had dropped again.

8. *Fortune*, XLI, No. 1 (January, 1950), 71-79, 138-142.

9. "Jersey Press Release," no date [pencilled, April 18], 1952, Files, C.B.H.S.; in 1945, Canada had fewer than one and one-half million vehicles on the highways. By 1960, there were more than five million, *The Lamp*, Vol. 62, No. 4 (Winter, 1980).

10. *Imperial Oil Review*, Vol. 44, No. 4 (August, 1960).

11. Gray, *The Great Canadian Oil-Patch*, 143-150; James B. Hilborn (consulting editor), *Dusters and Gushers. The Canadian Oil and Gas Industry* (Toronto, 1968), 125-132 [hereafter cited as Hilborn, *Dusters and Gushers*].

12. Gray, *The Great Canadian Oil-Patch*, p. 142.

13. *Annual Reports*, 1951, 1960, 1975.

14. *Ibid.*, 1956; see also *Interviews*, Wall with John Kenneth Jamieson, March 1, 1976, October 31, 1979, May 16, 1980, October 13, 1981, Files, C.B.H.S.; *Interview*, Wall and Carpenter with Jamieson November 30, 1977, *ibid.*; *Interview*, Andy Purcell with Jamieson, July 15, 1975, *ibid.*

15. *Interview*, Wall and Carpenter with Jamieson, November 30, 1977, *ibid.*

16. *Annual Report*, 1964; *A History of Imperial Oil* (Toronto, circa, 1967), p. 28.

17. *Annual Report*, 1967; *A History of Imperial Oil*, p. 28.

18. *Annual Reports*, 1972, 1973, 1974.

19. *Ibid.*, 1969.

20. *Ibid.*, 1974.

21. *Ibid.*, 1975.

22. *Ibid.*, 1954.

23. *Ibid.*, 1959.

24. Hilborn, *Dusters and Gushers*, 21-22.

25. *Ibid.*, p. 20.

26. *Ibid.*, p. 22

27. *Ibid.*, p. 201.

28. *Imperial Oil Review*, Vol. 42, No. 5 (October, 1958).

29. In September, 1982, in "Imperial Oil Limited, Second Submission to the Restrictive Trade Practices Commission on the state of competition in the Canadian petroleum industry," the company stated that Jersey had given up an East Coast market that its affiliate, Creole Petroleum Corporation, had been supplying to assist Imperial expand its market for Canadian crude in the Pacific Northwest area of the United States; Part VII, 20-22, Courtesy, W.G. Charlton.

30. Hilborn, *Dusters and Gushers*, p. 201.

31. *The Lamp*, Vol. 62, No. 4 (Winter, 1980).

32. *Annual Reports*, 1960-1961.

33. *Ibid.*, 1970.

34. *Ibid.*

35. *Memories*, 104-105; *The Story of Imperial Oil* (circa, 1970), p. 21.

36. *PACE*, press release, Toronto, December 2, 1970.

37. *The Lamp*, Vol. 62, No. 4 (Winter, 1980).

38. *Annual Report*, 1950.

39. *Interview*, Charlton with R.A. Wilson, December 22, 1981, Courtesy, Charlton.

40. *Annual Reports*, 1950-1975.

41. *Ibid.*, 1955.

42. *Ibid.*, 1952.

43. *Ibid.*, 1962.

44. Henrietta M. Larson, Evelyn H. Knowlton and Charles S. Popple, *New Horizons, 1927-1950: History of Standard Oil Company (New Jersey)*, (New York, 1971), p. 632.

45. *Letter*, Charlton to Wall, September 30, 1981, Files, C.B.H.S.

46. *Letter*, W.O. Twaits to J.R. White, January 3, 1963, Exxon Records Center [hereafter cited as E.R.C.], 281-18-A.

47. E.G., see *Imperial Oil Review*, Vol. 59, No. 5 (Issue No. 325, 1975).

48. *Annual Reports*, 1950-1975; *Imperial Oil Review*, 1950-1975.

49. *Hearings Before the Subcommittee on Multinational Corporations of the Committee on Foreign Relations, United States Senate, Ninety-Third Congress*, 17 Parts Washington (1973-1976), Part 12, p. 255.

50. *Letter*, Twaits to All Management, April 26, 1963, E.R.C., 281-18-A.

51. *Speech*, White to Canadian Petroleum Association, quoted in *Imperial Oil Review*, Vol. 41, No. 3 (June, 1957).

52. *Annual Reports*, 1958, 1959. Both contain a ten year summary.

53. *Memorandum*, Nicholas J. Campbell, Jr. to H.H. Hewetson, July 10, 1958, E.R.C., 176-3-B; *Telegram*, Newmyer Associates to Campbell, July 10, 1958, *ibid.*

54. *Imperial Oil Review*, Vol. 49, No. 5 (October, 1965).

55. *Annual Reports*, 1964, 1968, 1970, 1971.

56. *Speech*, J.A. Armstrong to Junior Investment Dealers of Toronto, Ontario, April 6, 1970, Courtesy, Charlton.

57. *Annual Report*, 1970.

58. *Speech*, Armstrong to Petroleum Seminar, Toronto, November 22, 1977, Courtesy, Charlton.

59. *Speech*, Armstrong to Canadian Club of Ottawa, December 8, 1976, *ibid.*

60. *Canadian Petroleum Association Statistical Handbook* (1981), Section III, Table 1.

61. *Annual Report*, 1969.

62. *Speech*, R.G. Reid to Financial Executives Institute, Winnipeg, Manitoba, February 14, 1975, Courtesy, Charlton.

63. Imperial's efforts to get research started on the Athabasca oil sands were discussed fully in the Jersey Executive Committee. See Executive Committee Minutes, March 5, September 16, 1959, and later [hereafter cited as E.C. Minutes], Secretary's Office, Exxon. For Syncrude, see *Imperial Oil Review*, Vol. 65, No. 2 (Issue No. 358, 1981).

64. *Speech*, D.D. Lougheed to "open house," Cold Lake, Alberta, September 11, 1975, Courtesy, Charlton.

65. *Letter*, D.H. Cooper to White, September 27, 1963, E.R.C., 281-18-A.

66. *Annual Reports*, 1960-1975.

67. *Ibid.*

68. *Memorandum*, W.R. Robinson to Wolf Greeven, February 9, 1965, E.R.C. 218-18-A.

69. *Speech*, Armstrong, April 6, 1970, Courtesy, Charlton.

70. *Annual Report*, 1969.

71. *Ibid.*

72. *Speech*, R.E. Landry to the Canadian

Chamber of Commerce, Saskatoon, Saskatchewan, September 24, 1975; Armstrong to Annual Meeting, April 25, 1975, Courtesy, Charlton.

73. Marie-Josée Drouin and Harald B. Malmgren, "Canada, The United States And The World Economy," *Foreign Affairs*, Vol. 60, No. 2 (Winter, 1981-1982), 393-413. Some Canadian governmental officials remarked on this publicly, *e.g.*, see *Speech*, J.J. Greene (Minister of Energy, Mines and Resources) to Pacific Gas Association, Portland, Oregon, September, 1970, Courtesy, Charlton.

74. *Speech*, Pierre Trudeau to National Press Club, Washington, D.C., March 26, 1969, *ibid.*; see also Drouin and Malmgren, "Canada, The United States And The World Economy," p. 397.

75. *Speech*, Twaits to Imperial Shareholders, Toronto, April 22, 1974, Courtesy, Charlton.

CHAPTER 13

1. For this story see *The New York Times*, December 7, 9, 13, 14, 29, 1973; February 15, March, 11, 13, 14,15, April 4, 11, 30, May 1, 3, 1974.

2. For details of this transaction see Henrietta M. Larson, Evelyn H. Knowlton and Charles S. Popple, *New Horizons, 1927-1950, History of Standard Oil Company (New Jersey)*, (New York, 1971), p. 47.

3. *Internal Memorandum*, Esso Brasileira, Coordination Planning and Economics Department, April 8, 1980, Files, Esso Inter-America.

4. *Ibid.*

5. *Ibid.*

6. *Ibid.*

7. *New York Times*, October 25, 1974.

8. George S. Gibb and Evelyn H. Knowlton, *The Resurgent Years, 1911-1927, History of Standard Oil Company (New Jersey)*, (New York, 1956), 369-383.

9. *Colombia: Profile of a Nation*, a publication sponsored by INTERCOR, Bogotá, Colombia, 1978.

10. *Ibid.* On September 9, 1980, Exxon Corporation announced that it would participate in the development of the coal resources in this region. Exxon's wholly owned Colombian affiliate, INTERCOR, will operate the concession. Carbones De Colombia (Carbocol) and

Exxon participate in this project on a 50-50 basis. *Exxon Press Release*, September 9, 1980, Files, Center for Business History Studies [hereafter cited as C.B.H.S.].

11. International Petroleum Company—Essosa orientation, December 7, 1962, Employee Relations Department IPC. See also *Internal Memorandum*, presentation of G. E. Golden, assistant manager, Esso Caribbean, February 4, 1974, Files Esso Inter-America.

12. *Ibid.*

13. *Discussions*, Paul E. Morgan with the Staff, Exxon Secretary's Department, 1974-1980, Files, Esso Inter-America.

14. *Internal Document*, "Jersey in Latin America," Presentation by Latin American Coordination Office, January 21, 1971, Files, C.B.H.S.

15. *Ibid.*

16. *Ibid.*

CHAPTER 13-CREOLE

1. *Manuscript*, "History of Creole," Private Collection, Files, C.B.H.S.

2. *Ibid.*, 33, 84-91; Henrietta M. Larson, Evelyn H. Knowlton, Charles S. Popple, *New Horizons, 1927-1950* (New York, 1971), 134-517; Wyne C. Taylor, John Lindeman, Victor R. Lopez, *The Creole Petroleum Corporation in Venezuela*, National Planning Association Series, "United States Business Performance Abroad," 1955, 87-88.

3. Larson, Knowlton and Popple, *New Horizons*, 134-157; 485; *Manuscript*, "History," p. 254; *Interviews*, Gene S. Yeager with Nicanor Garcia, November 28, 1975; Yeager with George Breffeilh, May 14, 1976; Yeager with Joseph Kosewickz, December 2, 1975, Files, C.B.H.S.

4. E. E. Barberii, "Petroleum History of Eastern Venezuela," *World Oil*, 131:6 (November, 1950), p. 290; Edwin Lieuwen, *Petroleum in Venezuela* (Berkeley, 1954), p. 108; Larson, Knowlton and Popple, *New Horizons*, 479-481.

5. The most complete discussion of this affair is found in *New Horizons*, 480-485. See also *Interview*, Larson with Wallace E. Pratt, November 16, 1958, Courtesy, Miss Larson; also, *Manuscript*, "History," 246-248; and *Interview*, Frank M. Surface, Larson and Knowlton with Arthur J. Proudfit, June 5, 1956, Courtesy, Miss Larson.

6. "Creole Petroleum: Business Embassy," *Fortune*, Vol. LIX (February, 1959), p. 177.

7. Lieuwen, *Petroleum in Venezuela*, 108-109; Executive Committee Minutes [hereafter cited as E. C. Minutes], January 17, August 4, 1957, Secretary's Office, Exxon; *Hispanic World Report*, April, 1949; Glen L. Kolb, *Democracy and Dictatorship in Venezuela, 1945-1958* (New London, 1974), p. 30.

8. Rómulo Betancourt, *Venezuela: política y petróleo* (Mexico, 1956), 236-237; Robert J. Alexander, *Venezuelan Democratic Revolution* (New Brunswick, 1964), p. 23; E. C. Minutes, October 23, 25, 30, 1945, Secretary's Office, Exxon.

9. E. C. Minutes, March 18, 1946, February 17, 1948, Secretary's Office, Exxon: *The Daily Journal* (Caracas), June 15, 1975; "Creole Petroleum: Business Embassy," p. 178.

10. E. C. Minutes, January 2, 10, February 5, 1946; Secretary's Office, Exxon.

11. E. C. Minutes, December 17, 19, 1946, January 17, November 25, 1947, April 22, May 19, 1948, *ibid*. Critics maintained that it created powerful disincentives to exploration and production. Joseph E. Pogue, *Oil in Venezuela: An Economic Study* (New York, 1949); Jersey's advocacy of 50-50 is clear in M. J. Rathbone's "Fair Shares in Oil," *The Financial Times*, March 6, 1958, and in W. L. Butte, "Memorandum re: '50-50,' " January 25, 1952, Exxon Records Center [hereafter cited as E. R. C.], 192-7-B.

12. Kolb, *Dictatorship and Democracy*, p. 51; Pedro E. Mejía Alarcón, *La industria del petróleo en Venezuela* (Caracas, 1977), p. 117; Lieuwen, *Petroleum*, 110-114; "It's Hot in Venezuela," *Fortune*, XXXIX (May, 1949), 101-108, 150-164; E. C. Minutes, December 6, 1958, Secretary's Office, Exxon.

13. "Creole Petroleum: Business Embassy," p. 179.

14. *Ibid*.

15. *Ibid*.; "La Creole convertida en la mayor productora de petróleo en el mundo," *Farol*, CXXXII (1951), 8-9; Robert Spann, "Boom in Venezuela," *World Oil*, 133:1 (July 1, 1951), 239-241.

16. "Creole Petroleum: Business Embassy," p. 95.

17. *Interview*, Yeager with George Culp, May 13, 1976, Files, C.B.H.S.

18. "Creole Petroleum: Business Embassy," p. 95.

19. "En buena compañia," *El Farol*, CXXV (October, 1949), 2-7.

20. "El paludismo en derrota," *El Farol*, CIX (June, 1948), 2-5; "Hospitales, baluarte de salud," *ibid*., CVII (April, 1948), 2-5.

21. "Mujeres en la Creole," *Farol*, CXVII (March, 1949), 10-15.

22. "El entrenamiento de trabajadores criollos," *ibid*., CXXIX (1950), 2-5; "Becas para nuestra juventud," *ibid*., CIV (January, 1948), 28-29; *Interview*, Yeager with Federico Baptista, December 4, 1975, Files, C.B.H.S.

23. Burck, "The Foreign Policy," p. 23.

24. *Manuscript*, "History," Files, C.B.H.S.

25. *Interview*, Yeager with Dr. Guillermo Zuloaga, December 5, 1975, *ibid*.

26. E. C. Minutes, May 3, 7, December 5, 1946, Secretary's Office Exxon; *Interview*, Yeager with Muguerza López, December 3, 1975, Files, C.B.H.S.; *Bienvenido a Amuay* (Caracas, n.d.); Juan Toro Martínez, *Amuay: 25 años en la historia de una refinería* (Caracas, 1975), 16-22.

27. "Creole Petroleum: Business Embassy," p. 182. See also "Tubos en vez de Tanqueros," *Farol*, CXV (December, 1948), 2-7, and "Donde hasta los segundos cuestan dinero," *ibid*, CXXVI (January, 1950), 2-7.

28. Toro Martínez, *Amuay*, p. 98: Aquiles Rojas, "La derrota de le Sed," *Farol*, XXII: 190 (September-October, 1960), 9-14; Agua para el desierto," *ibid*., CXIX (April, 1949), 2-7; "Llegó el agua en Amuay," *ibid*., 128 (1950), 12-15; "Siburua: Culminación de una batalla por agua," *Nosotros*, 145 (November, 1960), 2-4.

29. E. C. Minutes, May 7, 24, 1946, September 13, 1948, October 10, 1949, Secretary's Office, Exxon; E. C. Minutes, Creole Budgets and Appropriations, October 14, 1947, *ibid*.

30. *Interview*, Yeager with Hugh Jencks, January 6, 1976, Files C.B.H.S.; E. C. Minutes, April 6, 1949, Secretary's Office, Exxon.

31. *Ibid*., August 29, September 25, November 1, 1950; Board of Directors Minutes [hereafter cited as Bd. Dir. Mins.], December 8, 1949, p. 1550, June 22, 1950, p. 1606, July 6, 1950, p. 1608, August 24, 1950, p. 1624, Sectretary's Office, Exxon.

32. Jenkin Lloyd Jones, "I Saw American Enterprise and Oil Helping Build a Nation," *World Oil* (March, 1953), 242-246.

33. "Judibana, Ciudad futura," *Farol*, 141 (September, 1952), 18-21; Reuel Denney, "Oil Town: New Style," *The Lamp*, 40:3 (Fall, 1958), 6-17; "Judibana: Realidad presente proyectada al futuro," *Nosotros* 124 (February, 1959), 5-9.

34. Julián Ferris, "La vivienda en las áreas industriales," *Farol*, XVII:160 (October, 1955), 12-13; Warren L. Baker, "Creole Undertaking Community Intergration," *World Oil*, 145:6 (November, 1957), 176-178.

35. Interviews, Yeager with García, November 28, 1975; with J. R. Demori, November 27, 1975; with Federico Baptista, December 4, 1975, Files, C.B.H.S. Also, Jones, "I Saw American Enterprise," p. 246.

36. Denney, "Oil Town," 6-17; "When Nations Sow the Petroleum," *The Lamp*, Vol. 35, No. 1 (March, 1953), 2-5; Clark, the "Case Study," 35-36.

37. E. C. Minutes, May 3, 1951, March 1, 1956, Secretary's Office Exxon; "New Deep-Water Channel in Venezuela to be Dedicated," *World Oil*, 143:7 (December, 1956), 208-210.

CHAPTER 13-PERU

1. Alberto J. Pinelo, *The Multinational Corporation as a Force in Latin American Politics: a Case Study of the International Petroleum Company in Perú* (New York, 1973), 4-14.

2. *Ibid.*, 31-34; *Peruvian Times*, July 28, 1967. The text of the Laudo is reproduced with related documents in Rieck B. Hannifin, *Expropriation by Perú of the International Petroleum Company: a Background Study of the Legal Issues, Political Consequences of the Controversy* (Washington, D.C.; Library of Congress Legislative Reference Service, 1969), 106-126 and in *The La Brea y Pariñas Controversy* (n.p.: International Petroleum Company, 1969), Vol. I, 2-11.

3. Henrietta M. Larson, Evelyn H. Knowlton, Charles S. Popple, *New Horizons, 1927-1950* (New York, 1971), p. 730; Board of Directors Minutes [hereafter cited as Bd. Dir. Mins.], May 18, 1950, Secretary's Office, Exxon.

4. "The Underground Puzzle of Perú," *The Lamp*, Vol. 38, No. 1 (Spring, 1956), 6-7. For a general description of the difficulties of marketing petroleum products in Perú, see "Oil in the High Andes," *The Lamp* Vol. 37, No. 3 (September, 1955), 21-25.

5. Eduardo Bustamante Ramos, "El futuro de la industria del petróleo," *Fanal*, XXI:78 (1966), 23-27; *Noticias del Petróleo*, May, 1964; April, 1968; Benjamin C. Craft and M. F. Hawkins, *Applied Petroleum Reservoir Engineering* (Englewood Cliffs, N.J., 1959), p. 369.

6. George M. Ingram, *Expropriation of U.S. Property in South America: Nationalization of Oil and Copper Companies in Perú, Bolivia and Chile* (New York, 1974), p. 38.

7. *Ibid.*, 39-41; Pinelo, *The Multinational Corporation*, 65, 71.

8. *Petroleum Press Service*, June, 1952, 211-214; March 1953, 107-108.

9. Executive Committee Minutes [hereafter cited as E. C. Minutes], January 15, February 6, 18, March 10, 11, 16, August 13, 1948 and February 2, 1949, Secretary's Office, Exxon; See also Bd. Dir. Mins., May 27, 1958, *ibid.*

10. E. C. Minutes, April 4, 1951; December 5, 1955, *ibid.*

11. Larson, *et al.*, *New Horizons*, p. 373.

12. Charles T. Goodsell, *American Corporations and Peruvian Politics* (Cambridge, Mass., 1974), 60-61.

13. Pinelo, *The Multinational Corporation*, p. 55; Benjamín Núñez Bravo, "La Talara que yo ví," *Fanal*, XX:73 (1965), 25-31; Gilbert Burck, "The Foreign Policy of the Jersey Corporation," *Fortune*, XLIV (November, 1951), 182-185.

14. Pinelo, *The Multinational Corporation*, p. 53.

15. Hannifin, *Expropriation*, p. 24.

16. G. W. Potts and Paul E. Morgan, "An Examination of the La Brea y Pariñas Problem and a Review of International Petroleum Company's Public Relations in Perú, with Resumés of Private Conversations, May 3-May 19, 1961," *Internal Report*, Exxon Corporation, Morgan Files, C.B.H.S.; *Interview*, Bennett H. Wall with Morgan, April 30, 1980, Files, C.B.H.S.

17. Pinelo, *The Multinational Corporation*, 110-114; *Peruvian Times*, July 28, 1967. For the Exxon version of this story, see "The La Brea y Pariñas Controversy—A Resumé," Standard Oil Company (New Jersey), 1962.

18. Pinelo, *The Multinational Corporation*, p. 115.

19. *Ibid.*, 118-119.

20. *Ibid.*, 119-123; Bd. Dir. Mins., June 11, 1964, Secretary's Office, Exxon.

21. Hannifin, *Expropriation*, p. 38.

22. Ingram, *Expropriation*, p. 53.

23. Among these are Augusto Zimmerman Zavala, *La Historia secreta del petróleo* (Lima: Editorial Gráfica Labor, 1968); and Frente Nacional de Defensa de Petróleo, *Declaración de principios y exposición de motivos* (Lima, 1960). Unique for its authors' insistence that the case be decided on a rational, economic basis was Harries-Clichy Peterson and Tomas Unger, *Petróleo: hora cero* (Lima, 1964). On this point see Robert E. Kingsley, "Two Studies of Anti-Americanism in Perú," September 15, 1960, *Internal Memorandum*, Exxon, Jarvis M. Freymann Files, C.B.H.S.

24. C. T. Goodsell, "Diplomatic Protection of U.S. Business in Perú," in Daniel A. Sharp, ed., *U.S. Foreign Policy and Perú* (Austin, 1972), p. 247.

25. This position was later adopted by the government. Perú, Oficina Nacional de Información, *Petroleum in Perú, For the World to Judge: The History of a Unique Case* (Lima, 1969).

26. Pinelo, *The Multinational Corporation*, 140-145; Goodsell, *American Corporations*, p. 54; *Wall Street Journal*, March 20, 1968.

27. *Ibid.*, October 11, 1968.

28. Goodsell, *American Corporations*, p. 100; Pinelo, *The Multinational Corporation*, p. 140. The conflicting IPC and government arguments are explained by Richard N. Goodwin, "Letter from Perú," *The New Yorker*, May 17, 1969; *New York Times*, December 14, 1968.

29. Goodsell, "Diplomatic Protection," p. 251; *Air Mail Latin America*, October 11, 1968; *Wall Street Journal*, December 18, 1968; *New York Times*, April 5, 1969; Roderick T. Groves, "Expropriation in Latin America: Some Observations," *Inter-American Economic Affairs*, XXIII:3 (Winter, 1969), p. 56.

30. "Nelson Rockefeller Begins Mission to Latin America," *The Department of State Bulletin*, LX:1562 (June 2, 1969), 470-472; Richard M. Nixon, "A New Approach to Pan American Problems," *ibid.*, LX:1558 (May 5, 1969), 384-386; *Andean Air Mail and Peruvian Times*, August 8, 1969.

31. Peter R. Odell, "Oil and the State in Latin America," *International Affairs*, X:4 (October, 1964), demonstrates that nationalistic considerations often have priority over the advice of economists and planners.

32. Goodsell, "Diplomatic Protection," p. 129.

33. *General Memorandum*, Freymann to Messrs. S. Stamas, E. J. Hess, J. B. Meredith, J. A. Morakis, December 19, 1974, Freymann Files, C.B.H.S.; see also *Press Release*, same date, *ibid.*; see also David A. Gantz, "The United States-Peruvian Claims Agreement of February 19, 1974," *International Lawyer*, Volume X, No. 3, 389-399, *ibid.*

CHAPTER 15

1. Henrietta M. Larson, Evelyn H. Knowlton and Charles S. Popple, *New Horizons, 1927-1950: History of Standard Oil Company (New Jersey)*, (New York, 1971), 382-555 contains an excellent account of the Jersey Company's wartime activities. See also, Charles S. Popple, *Standard Oil Company (New Jersey) in World War II* (New York, 1952).

2. *Interview*, Bennett H. Wall with Edward F. Johnson, September 21, 1972, Files, Center for Business History Studies [hereafter cited as C.B.H.S.].

3. *Memorandum* For the Attorney General (Francis Biddle), by Wendell Berge, December 24, 1943, Department of Justice, Records, Case 60-012-20 [hereafter cited as D.O.J.], Courtesy, Burton I. Kaufman [all such records cited D.O.J. are courtesy of Kaufman]. See also *letters*, John M. Coffee to Berge, Assistant Attorney General, October 20, 1943; Berge to Coffee, October 23, 1943, D.O.J. 60-57-140. For a slashing rejoinder to accusations that some oilmen served PAW as agents, rather than as patriotic citizens, see Max W. Thornberg to Senator Owen Brewster, February 10, 1948, copy, courtesy, Gerald T. White. The only extensive study on the oil cartel case is that of Burton I. Kaufman, *The Oil Cartel Case* (Westport, Connecticut, 1978).

4. D.O.J., 60-57-140, *Memoranda*, 1943-

1950.

5. *Letter*, Johnson to James P. McGranery, March 4, 1944, Exxon Records Center [hereafter cited as E.R.C.], 164-5-A.

6. *Letter*, Frank W. Gaines, Jr. to DiGirolamo, April 1, 1946, D.O.J., 60-122-32.

7. *Letter*, George V. Holton [Vice President and General Counsel of the Socony-Vacuum Oil Company] to Berge, October 23, 1944, *ibid.*, 60-122-48.

8. *Letter*, Gaines to Herbert A. Berman, January 18, 1946, *ibid.*, 60-122-32.

9. *Letter*, Gaines to Robert A. Nitschke, September 16, 1948, *ibid.*, 60-122-48.

10. *Memorandum*, Gaines to Nitschke, October 4, 1946, *ibid.*, 60-122-48.

11. *Memorandum*, W. B. Watson Snyder to Nitschke, October 4, 1946, *ibid.*

12. *Memorandum*. For the Attorney General (Tom C. Clark), by Berge, November 7, 1946, *ibid.*, 60-57-140.

13. *Letter*, Thomas W. Palmer to M. C. Pollak, November 19, 1946, *ibid.*, 60-122-48.

14. *Letter*, R. T. Haslam to Clark, December 11, 1946, *ibid.*, 60-57-140.

15. *Memorandum*, Gaines to Nitschke, September 18, 1946, *ibid.*, 60-122-48.

16. In 1948, Intava merged with the Standard Oil Export Corporation.

17. Larson, Knowlton and Popple, *New Horizons*, p. 701.

18. *Memorandum*, Watson Snyder to Berge, January 8, 1947, D.O.J., 60-57-140. For Berge's view on business combinations, see Wendell Berge, *Cartels. Challenge To A Free World* (Washington, 1944).

19. *Memorandum*, Berge to Watson Snyder, January 10, 1947, D.O.J., 60-57-140; *Memorandum*, Watson Snyder to Berge, January 10, 1947, *ibid.*

20. *Memorandum*, For the Files, January 24, 1947, by Watson Snyder, *ibid.*

21. *Letter*, Harry T. Klein to Clark, March 12, 1947, *ibid.*; *Memorandum*, Watson Snyder to Nitschke, March 31, 1947, *ibid.*

22. *Memorandum*, For the Files, August 18, 1952, by Leonard J. Emmerglick, *ibid.* Emmerglick stated *re* Jersey, Socony and California, "It is also noted that these three companies are commonly related in that the Rockefeller interests are the dominant stockholders in all three."

23. *Memorandum*, Nitschke to H. Graham Morison, January 31, 1947, D.O.J., 60-57-140.

24. *Memorandum*, Gaines to Nitschke, May 22, 1947, *ibid.*; (Italics mine). *Letter*, Haslam to Clark, August 8, 1947, E.R.C., 164-5-A.

25. *Memorandum*, George P. Comer to John Ford Baecher, June 7, 1948, D.O.J., 60-57-140; *Memorandum*, Gaines to Baecher, May 27, 1948, *ibid.*; *Memorandum*, Watson Snyder to [Baecher?], May 26, 1948, *ibid.*

26. *Memorandum*, Rodolfo A. Correa to Herbert A. Bergson, January 10, 1947, *ibid.*

27. *Memorandum*, Watson Snyder to Bergson, March 10, 1949, *ibid.*

28. *Memorandum*, For the Files, June 7, 1949, by Watson Snyder, *ibid.*

29. Senate Select Committee on Small Business, Commission Print No. 6, *The International Petroleum Cartel: Staff Report to the Federal Trade Commission* (Washington, D.C., 1952), p. 1. [Hereafter cited as *FTC Staff Report.*]

30. Statement of John M. Blair, *Hearings Before the Subcommittee on Multinational Corporations of the Committee on Foreign Relations, United States Senate, Ninety-Third Congress, Second Session*, 17 Parts, Washington, 1973-1976 Part 9, p. 192 [hereafter cited as *MNC*]. Blair was assisted by one-time Cornell University economist and commission staff member Roy Pruitt, and J. W. Adams, another long-time staff member. Some informed persons state that Corwin Edwards also played a prominent role in drafting the report.

31. *Memorandum*, For the Files, February 24, 1950, by Holmes Baldridge, D.O.J., 60-57-140.

32. *New York Times*, February 17, 1950.

33. *Letter*, Nitschke to Morrison [sic], January 31, 1947, cited as *Letter*, Nitschke to John F. Sonnett, June 4, 1947, in Kaufman, "Oil and Antitrust: The Oil Cartel Case and the Cold War," *Business History Review*, Vol. LI, No. 1 (Spring, 1977), 38-39 [hereafter cited as Kaufman, "Oil and Antitrust"]. Note: Morison's comment made in longhand on the bottom of the Nitschke letter replied to Nitschke's recommendation that the New Red Line and

other oil company agreements in the Middle East be investigated, whether connected to the "AS IS" agreements or not. Morison wrote:

No!! Definitely no, until it appears affirmatively that restraint of trade and enjoining to competition exists. This approach completely ignores the fact that distribution of oil both foreign & [word illegible] domestic is in fact a quasi-public utility business although not legally recognized as such & thus is inately [sic] by nature of a monopolistic character.

Morison misspelled "innately" in several other letters. This author did not locate the Sonnett letter mentioned but he has no doubt that it exists, D.O.J., 60-57-140.

34. Standard Oil Company (New Jersey) [hereafter cited as SONJ], *Annual Meeting of the Shareholders*, Linden, New Jersey, June 7, 1950.

35. *Ibid.*, June 8, 1951.

36. Bruce K. Brown, *Oil Men in Washington* (The Evanil Press, 1965), 11-33; *Telegram*, Eugene Holman to Oscar L. Chapman, July 2, 1951, E.R.C., 303-14-B; *Letter*, Peter T. Lamont to Foreign Affiliates, July 27, 1951, *ibid.*; *Letter*, Chapman to Manly Fleishman, February 15, 1952, *ibid.*; *Letter*, Thomas E. Monaghan to Howard W. Page, January 5, 1954, *ibid.*; *Memorandum*, "Prepared Statement by Stewart P. Coleman, MEEC," Before Senate Subcommittees on March 5, 1957, *ibid.*

37. Wright Patman, veteran Texas Congressman of well known populist sentiments and cousin of Elmer Patman, chaired the House Committee, while John Sparkman of Alabama filled the comparable position on the Senate Committee.

38. *New York Times*, December 30, 1951.

39. *Letter*, Watson Snyder to Morison, December 17, 1951, D.O.J., 60-57-140; *Letter, ibid.*, January 23, 1952, *ibid.; Memorandum*, Report of John Edgar Hoover (Director, Federal Bureau of Investigation) to the Acting Attorney General relative to the Report of the Federal Trade Commission on the International Petroleum Cartel, May 7, 1952, *ibid.*

40. *Letter*, Thomas C. Hennings, Jr., to FTC Chairman James M. Mead, *ibid.*

41. *New York Times*, April 18, June 22, July 11, 1952.

42. *FTC Staff Report*, V-VI.

43. Section 708 of the 1950 Act allowed private oil companies exemption from the antitrust laws and from FTC rulings "so that they could enter upon voluntary agreements and supply pools to provide oil in periods of emergency."

44. *E.g.*, "Middle East Oil," by Richard Funkhouser (used as background paper for September 11, 1950, meeting with oil executives), State Department Policy Paper, September 10, 1950, *MNC*, Part 7, 122-134. See also "Testimony of Richard Funkhouser," *MNC*, , Part 7, 121-171, including documents submitted, especially p. 166. See also "Testimony of and documents submitted by George C. McGhee," *MNC*, Part 4, 83-102.

45. *MNC*, Part 7, p. 99.

46. John M. Blair, *The Control of Oil (New York, 1976), p. 72.*

47. *New York Times*, July 11, 1952.

48. Board of Directors Minutes [hereafter cited as Bd. Dir. Mins.], Standard Oil Company (New Jersey), Secretary's Office, Exxon, July 17, 24, August 14, September 4, 1952. The Special Committee, chaired by Director David A. Shepard, included Vice President Orville Harden, Directors Jay E. Crane and Monroe J. Rathbone, General Counsel Johnson, Counsel Monaghan, Secretary Adrian W. Minton, Public Relations Director George H. Freyermuth, outside Counsel Taggart Whipple of Davis, Polk, Wardwell and Kiendl, and Public Relations Consultants Earl Newsom and Arthur Newmyer.

49. Ultimately, the court set January 1, 1941, as the beginning year; *New York Times*, November 11, 1952.

50. *Letter*, Holman and Frank W. Abrams, "To the Shareholders and Employees of Standard Oil Company (New Jersey), September 19, 1952," Files, C.B.H.S.; Executive Committee Minutes [hereafter cited as E.C. Minutes], September 16, 1952, Secretary's Office, Exxon.

51. *Ibid.; Letter*, Holman "To Heads of Affiliated Companies, September 26, 1952," Files, C.B.H.S.; E.C. Minutes, September 18, 1952, Secretary's Office, Exxon.

52. *Memorandum*, Watson Snyder to Morison, January 23, 1952, D.O.J., 60-57-140; Brown, *Oil Men*, 252-258.

53. E.C. Minutes, September 11, 1952, Secretary's Office, Exxon.

54. *MNC*, Part 8 (Appendix to Part 7), 13-30.

55. *Memorandum*, For the Files, by Emmerglick, August 15, 18, 21, 1952, D.O.J., 60-57-140.

56. For example see, *Letter*, L. A. Donker and J. M. A. Luns to Caltex Petroleum Maatschappij, October 2, 1952, *ibid.; Letter*, Geoffrey Lloyd to the Anglo-Iranian Oil Company, Ltd. [AIOC], October 2, 1952, *ibid.*

57. *Memorandum*, For the Files, by Philip Welsh and Watson Snyder, December 29, 1952, *ibid.; Letter*, Stephen J. Spingarn to McGranery, January 14, 1953, *ibid.*

58. Kaufman, "Oil and Antitrust," p. 38.

59. George W. Stocking, *Middle East Oil* (Nashville, 1970), 152-154; Dates given for the Russian troop withdrawal vary; e.g., see George Lenczowski, *The Middle East in World Affairs* (Ithaca, 1952), p. 174. It should be noted that Stocking, Leonard Mosley, Lenczowski, and other authors fail to agree on the dates of key Iranian events. Sunil Kanti Ghosh, *The Anglo-Iranian Oil Dispute* (Calcutta, India, 1960), provides many correct dates; however, its value as a source is not consistent. For example, compare Ghosh, 48-49 and Lenczowski, p. 184.

60. See also *Speech*, "The Problem of Near Eastern Oil," Remarks by Ambassador Funkhouser Before the National War College, December 4, 1951, *MNC*, Part 7, 160-171; Page, "Nationalism in the Middle East" [a paper presented at The Jersey Roundtable], June 19, 1958, Files, C.B.H.S.

61. Jersey's Lloyd W. Elliott had W. L. "Woody" Butte, a sometime member of the Near East Committee and an accomplished international lawyer draw up a definition which he circulated in the company. *Letter*, Elliott to John O. Larson, March 11, 1952; *Memorandum*, *re* "50-50," January 25, 1952, by Butte, E.R.C., 192-7-B. Butte pointed out that the phrase 50-50 "standing alone" had no precise meaning. His analysis of four such contracts with Jersey or with those in which the company had interests (Venezuela, Saudi Arabia, Iraq, and Kuwait) noted that they all differed significantly (e.g., whether the sharing occurred before or after foreign taxation). Therefore, the financial results differed. Equal contributions provided the theoretical

basis for 50-50 sharing; from the governments, oil resources, protection and services; and from the companies, capital and skill. In Butte's opinion, "political leaders feel no compulsion to attack the relationship." He cautioned against making or supporting "any deal anywhere . . . in which the government's share is greater than 50 percent." In cases where Jersey could "live with such," the ground would be cut from under the company; all countries would rush to get more income.

Page stated regarding the so-called 50-50 agreements,

> In the old days, the four shilling gold [payment] per ton was, in effect, a 50-50 profit at the time the agreements were made. . . . Later the U.S. and British Governments froze the price of gold, so the producing countries all felt cheated and I feel so myself. . . . They had agreed to gold payment per ton, for the simple reason that gold normally followed market values in the world. And inasmuch as it became completely artificial during and after the war [World War II], . . . the price in the open market in Baghdad for gold was two or three times what the U.S. Government would pay for it. . . . The original 50-50 no longer existed. In the early days there was no reason for an income tax because the royalty already paid the full 50-50.

Interview, Wall with Page, November 1, 1979, Files, C.B.H.S.; Mana Saeed al-Otaiba, *OPEC and the Petroleum Industry* (New York, 1975), who served on occasion as Minister of Petroleum and Mineral Resources of the United Arab Emirates, essentially agrees with Page. He states that oil prices escalated because the purchasing power of oil production income to host countries constantly declined, "eroded by inflation."

62. Larson, Knowlton and Popple, *New Horizons*, p. 737; *FTC Staff Report*, p. 146.

63. *Ibid.*, 145-160.

64. *Letter*, B. B. Howard to George McGhee, May 18, 1951, E.R.C., 194-21-B; E.C. Minutes, May 15, 17, June 29, 1951, Secretary's Office, Exxon; *Letter*, Larson to Members of the Board, May 21, 1951, E.R.C., 194-21-B; "For the Press," Department of State, "United States Position on Iranian Oil Situation, May 18, 1951 (No. 412)," *ibid.; Memorandum*, Fred L. Palmer to E.[arl] N.[ewsom], *et al.*, June 27, 1951, *ibid.; Memorandum*,

Rathbone to the Board, July 23, 1951, *ibid.*; *Memorandum*, Freyermuth to Rathbone, July 30, 1951, *ibid.*; [The Executive Committee Minutes state specifically that Wilson made the request, but the Holman telegram confirmed to Chapman the "advice given June 29, 1951," which is the date the Committee considered the request]; *Letter*; Lamont to Foreign Affiliates, July 27, 1951, E.R.C., 303-14-B; *Letter*, Lamont to E. E. Soubry, July 30, 1951, *ibid.*; *Interview*, Wall with Page, November 1, 1979, Files, C.B.H.S.

65. *Memorandum*, "Internal State Department Memoranda—Analysis of Middle Eastern Oil Concessions and Their Implications for U.S. Foreign Policy," by Funkhouser, July 3, 1950 (revised September 10, 1953), *MNC*, Part 7, 5-139.

66. *Letter*, Shepard to Howard Quinlan, March 24, 1948, *ibid.*, Part 8 (Appendix to Part 7), 259-260.

67. *Letter*, Butte to Quinlan, no date (certainly March, 1948), *ibid.*, 260-262.

68. *Letter*, Page to Neville Gass, *et al.*, January 3, 1950, *ibid.*; *Memorandum*, "I.P.C. Functions, Performance and Suggested Guiding Financial Principles," by Page, same date, *ibid.*, 287-290.

69. *Interview*, Wall with Page, November 1, 1979, Files, C.B.H.S. The bracketed portion represents an explanatory comment made by Page after reading this chapter.

70. *Letter*, Page to John W. Brice, January 26, 1950, *MNC*, Part 8 (Appendix to Part 7), 292-293. This letter was marked: Attention F. O. Canfield.

71. *Letter*, Calouste S. Gulbenkian to Harden, March 21, 1949, *ibid.*, 282-284.

72. Stocking, *Middle East Oil*, p. 151.

73. *Interview*, Wall with Page, November 1, 1979, Files, C.B.H.S.; *Letter*, Page to Wall, October 4, 1981, *ibid.*

74. *MNC*, Part 8 (Appendix to Part 7), 347-349.

75. Leonard Mosley, *Power Play, Oil in the Middle East* (Baltimore, 1973), 194-195.

76. *MNC*, Part 8 (Appendix to Part 7), 372-378 contains the text of these proclamations; *Interview*, Wall with Page, November 1, 1979, Files, C.B.H.S.

77. *MNC*, Part 8 (Appendix to Part 7), 372-374.

78. Dean Acheson, *Present At The Creation* (New York, 1969), p. 506.

79. *MNC*, Part 7, 168-172.

80. Dwight D. Eisenhower, *Mandate For Change, 1953-1956* (Garden City, 1963), p. 159; Benjamin Shwadran, *The Middle East, Oil and The Great Powers* (New York, 1955), 103-152.

81. *Ibid.*, 106-107; Lenczowski, *The Middle East in World Affairs*, 184-185; Robert H. Ferrell (ed.), *The Eisenhower Diaries* (New York, 1981), p. 192.

82. *Memorandum, Foreign Service of the United States* (from London), March 20, 1951, "British Oil Interests in Iran," D.O.J., 60-57-140; N.B. On July 5, 1951, the International Court at the Hague handed down a 10-2 decision in the oil dispute case. This decision called for Iran to reinstate the AIOC to full control of its assets and operations. This, Iran refused to do and continued to deny that the court had jurisdiction in this affair. On July 22, 1952, the court reversed the earlier opinion on the ground that it did, indeed, lack jurisdiction.

83. Acheson, *Present At The Creation*, 501-503; Shwadran, *The Middle East*, 110-132; Stocking, *Middle East Oil*, 155-158; Lenczowski, *The Middle East in World Affairs*, 184-189; *MNC*, Part 7, 122-139, 160-171.

84. *Memorandum*, Page to Wall, November 1, 1979, Files, C.B.H.S.

85. *Interview*, Wall with Page, November 1, 1979, *ibid.*

86. Henry Longhurst, *Adventure in Oil. The Story of British Petroleum* (London, 1959), p. 143.

87. *Memorandum*, Page to Wall, November 1, 1979, Files, C.B.H.S.; *Interview*, Wall with Page, November 1, 1979, *ibid.*

88. Acheson, *Present At The Creation*, 682-683.

89. *Ibid.*, p. 600.

90. *New York Times*, August 19, 1952.

91. E.C. Minutes, August 26, September 3, 10, 12, 16, 18, 23, 1952, Secretary's Office, Exxon; *Memorandum*, "For the Attorney General, October 11, 1952," *The International Petroleum Cartel, The Iranian Consortium and U.S. National Security* (Washington, 1974), 20-21 [hereafter cited as *The IPC . . . and U.S. National Security*].

92. *Memorandum*, "For the Attorney General, October 13, 1952," *The IPC . . . and U.S. National Security*, 22-23.

93. Acheson, *Present At The Creation*, 681-685; *The IPC . . . and U.S. National Security*, 26-28; *Letter*, Adrian S. Fisher to McGranery, December 17, 1952, *ibid.*

94. *New York Times*, November 9, 1952; *Fortune*, Vol. XLVI, No. 10 (October, 1952), p. 113.

95. *New York Times*, December 11, 16, 1952.

96. *Ibid.*, December 19, 1952.

97. Acheson, *Present At The Creation*, p. 682.

98. *Ibid.*

99. *Memorandum*, "Material Relevant to the International Petroleum Cartel Case," *MNC*, Part 8 (Appendix to Part 7), 1-29; See also, *The IPC . . . and U.S. National Security*, 33-34.

100. "Testimony of Leonard J. Emmerglick" [hereafter cited as "Emmerglick Testimony"], *MNC*, Part 7, 102-103; See also, *The IPC . . . and U.S. National Security*, p. 33.

101. *New York Times*, January 12, 1953.

102. *Interview*, Wall and C. Gerald Carpenter with Johnson, May 3, 1977, Files, C.B.H.S.; *New York Times*, January 13, 1953.

103. *Ibid.*

104. "Emmerglick Testimony," *MNC*, Part 7, p. 118.

105. *New York Times*, January 13, 1953; *Memorandum: re* Pending Criminal Proceeding Against Oil Companies, D.O.J., 60-57-140; *Letter*, Chapman to McGranery, January 17, 1953, *ibid.*

106. Norman A. Graebner, *The Age of Global Power* (New York, *et al.*, 1979), p. 119; Paul Y. Hammond, *Cold War and Detente* (New York, *et al.*, 1975); Carl Solberg, *Oil Power* (New York, 1976), chapters 7 and 8; "Minutes of Telephone Calls of John Foster Dulles and of Christian Herter (1953-1961)" [hereafter cited as Dulles Telephone Calls], Microfilm, University Publications of America, Inc. (Washington, D.C. 20015), April 1, 6, 7, 1953; "Minutes and Documents of the Cabinet Meetings of President Eisenhower (1953-1961)" [hereafter cited as Eisenhower Cabinet Meetings], Microfilm, University Publications of America, Inc. (Washington, D.C. 20015), January 23, 1952; William R. Polk,

The United States and the Arab World (Cambridge, Massachusetts and London, England, 1975), p. 373.

107. Dulles Telephone Calls, April 1, 6, 7, 1953.

108. *The IPC . . . and U.S. National Security*, p. 34.

109. Dulles Telephone Calls, April 20, 1953.

110. *The IPC . . . and U.S. National Security*, 35-47.

111. Kermit Roosevelt, *Counter Coup: The Struggle for the Control of Iran* (New York, 1979), p. 21. Acheson, *Present At The Creation*, p. 681 indicates that he had been aware of these discussions earlier; See also Eisenhower, *Mandate For Change*, 159-166.

112. SONJ, *Annual Meeting of the Shareholders*, 1953, p. 14.

113. Dulles Telephone Calls, August 25, 1953; Eisenhower Cabinet Meetings, August 27, 1953.

114. *Letter*, Walter B. Smith to Herbert Brownell, Jr., April 27, 1953, D.O.J., 60-57-140. Several of these companies were later reinvolved in the case. Judge John Cashin of the U.S. District Court for the Southern District of New York ordered the oil companies to produce the foreign documents in the fall of 1959. On August 6, 1953, the NSC directed the Secretary of State "to assume direction of steps being taken to protect the [oil] interests of the free world in the Near East . . . consulting with the Attorney General on legal aspects of the subject." *The IPC . . . and U.S. National Security*, p. 55.

115. Dulles Telephone Calls, August 31, 1953; See also, *The IPC . . . and U.S. National Security*, p. 48.

116. Larson, Knowlton and Popple, *New Horizons*, p. 484.

117. Dulles Telephone Calls, September 4, 8, 1953.

118. His title was Special Representative of the United States Government within the Department of State, to deal with problems related to an Anglo-Iranian oil settlement.

119. *The IPC . . . and U.S. National Security*, 54-55.

120. Shwadran, *The Middle East*, 180-181.

121. *Letter*, Brownell to Herbert Hoover, Jr., December 8, 1953, D.O.J., 60-57-140.

122. *Memorandum,* For the Files, by Stanley N. Barnes, September 24, 1953; *Memorandum,* For the Files, by Emmerglick, October 15, 1953; *Letter,* Harden to Dulles, December 4, 1953; *Letter,* Smith to Harden, December 10, 1953; *Memorandum,* Barnes to Brownell, December 10, 1953, *The IPC . . . and U.S. National Security,* 49-50, 54-55, 58-60.

123. Shwadran, *The Middle East,* p. 181.

124. *Memorandum,* Barnes to Brownell, December 10, 1953; *Letter,* Herman Phleger to Brownell, January 8, 1954, *The IPC . . . and U.S. National Security,* 58,60-61.

125. *Memorandum,* by Dulles, January 7, 1954, *ibid.,* p. 61; See also, Dulles Telephone Calls, December 29, 1953.

126. *Memorandum,* To the National Security Council, by Brownell, January 13, 1954, *The IPC . . . and U.S. National Security,* p. 66.

127. "Testimony of Howard W. Page" [hereafter cited as "Page Testimony"], *MNC,* Part 7, p. 304. It should be noted that Page gave this testimony as "hearsay but . . . good hearsay."

128. *Interview,* Wall with Page, November 1, 1979, Files, C.B.H.S.

129. *The IPC . . . and U.S. National Security,* p. 52.

130. *Ibid.,* 73-74.

131. *Ibid.,* 75-76.

132. *New York Times,* February 7, 1954.

133. "Page Testimony," *MNC,* Part 7, p. 296.

134. The companies in the consortium were British Petroleum Company (formerly Anglo-Iranian Oil Company), 40 percent; Royal Dutch-Shell Group, 14 percent; Compagnie Française des Pétroles, 6 percent; Standard Oil (New Jersey), 8 percent; Standard Oil of California, 8 percent; Gulf Oil Corporation, 8 percent; Texaco, 8 percent; Socony-Vacuum Oil, 8 percent. In 1955, each of the American companies assigned an eighth of its holdings to the Iricon Agency, which gave Iricon 5 percent and reduced the share of each of the large U.S. companies to 7 percent. Iricon Agency was composed of eight independent American oil companies: Richfield Oil, Signal Oil and Gas, Tidewater, Getty Oil, American Independent Oil, San Jacinto Petroleum, Standard Oil (Ohio), and Atlantic Refining.

135. The best and most accurate version of the Harden and Page team negotiations is found in L. P. Elwell-Sutton, *Persian Oil* (London, 1955), though Elwell-Sutton presses the "cartel" power theme; Shwadran, *The Middle East,* 162-163; *Memorandum,* Smith D. Turner to Paul E. Morgan, October 13, 1981, Morgan Files, C.B.H.S. Turner gave details of the Harden team negotiations; *Letter,* Page to Wall, December 15, 1981, *ibid.*

136. *Memorandum,* "Brief Summary of Iranian Consortium Negotiations," by Page, *ibid.,* no date, probably written from notes in 1981, in response to this author's request; *Letter,* Page to Wall, December 15, 1981, *ibid.* For example, see Paul H. Frankel, *Mattei: Oil and Power Politics* (New York and Washington, 1966), p. 95.

137. *Interview,* Wall with Page, November 1, 1979, Files, C.B.H.S.; *Memorandum,* Page notes on Middle East chapter, October 11, 1982 [hereafter cited as "Page Notes"], *ibid.*

138. *Ibid.; Memorandum* (no place, no date), *ibid.*

139. Interestingly, Torkild Rieber, one-time Chairman of the Board of the Texas Company, then representing the International Bank for Reconstruction and Development, was engaged as special oil advisor for the Iranian government, see Shwadran, *The Middle East,* p. 183; see footnote 137; "Page Notes," Files, C.B.H.S.

140. *Memorandum,* "History of Iranian . . . Negotiations," by Page, Files, C.B.H.S. In writing about these events, the author also used the cited *interview* with Page plus his memorandum about the negotiations, *ibid.;* "Page Notes," *ibid.*

141. *E.g.,* Gulf; *Memorandum,* For the Files, by Barbara J. Svedberg, May 20, 1957, *MNC,* Part 8 (Appendix to Part 7), 560-561.

142. Longhurst, *British Petroleum,* p. 159.

143. "Page Testimony," *MNC,* Part 7, p. 297; Turner Critique of Middle East Chapter [hereafter cited as Turner, "Critique"], November 13, 1981, Files, C.B.H.S.

144. *Interview,* Wall with Page, November 1, 1979, *ibid.*

145. Mira Wilkins, *The Maturing of Multinational Enterprise: American Business Abroad from 1914 to 1970* (Cambridge, 1974), [hereafter cited as *The Maturing*], p. 323.

146. In the Church Committee hearings, *MNC,* Part 7, "Page Testimony," p. 306, Page

stated, "We, all of the companies, said they would have nothing whatever to do with the selection of the independents." Then the Committee Chief Counsel, Jerome Levinson commented, "Based on our investigation, I think Mr. Page is absolutely right, that the independents were selected without any input from the major oil companies"; in a *Memorandum*, however, entitled "Memorandum of Conversation," September 27, 1954, by a Department of State participant, Assistant Attorney General Barnes, told State Department representatives "that he thought it would be unwise for the Department to become involved in handling applications by the independents," *MNC*, Part 8 (Appendix to Part 7), p. 557; Herbert Hoover, Jr., in a *Memorandum* entitled "Department of State—For the Record," states that American Independent Oil Company had failed to provide Price Waterhouse & Company with a "satisfactory statement of financial resources" and that the former had been told they would not be allowed to participate in Iricon. Hoover said that any decision to exclude that company would have to be made by the participants. Hoover had discussed this with George Koegler of SONJ who told him that American Independent Oil Company would not be allowed to join because of that company's failure to satisfy Price Waterhouse, *ibid.*, 558-559. That company was later admitted. Conceivably, Price Waterhouse & Company through analysis of the financial statements, made the key decisions.

147. Wilkins, *The Maturing*, p. 322.

148. "Page Testimony," *MNC*, Part 7, p. 301.

149. *Interview*, Wall with Johnson, December 14, 1977, Files, C.B.H.S.; See also "Testimony of John J. McCloy," *MNC*, Part 5, 59-75.

150. *New York Times*, April 22, 1953.

151. "Attorney General's Conference Re International Oil Cartel, U.S. Department of Justice, Washington, April 14, 1953," 5, 8, 9, 13-14, quoted in Robert Engler, *The Politics of Oil* (New York, 1961), 216-217; *Interview*, Wall with Nicholas J. Campbell, Jr., August 31, 1978, Files, C.B.H.S.

152. Dulles Telephone Calls, August 31, September 4, 1953; it should be noted that Arthur Dean and Dulles had been members of the same law firm (Sullivan & Cromwell) and that they were close personal friends. Dean frequently talked to Dulles but not about the "cartel" case, according to the transcripts.

153. "Testimony of Mrs. Barbara J. Svedberg" *MNC*, Part 7, 56-85, [hereafter cited as "Svedberg Testimony"]; *Memorandum*, Kenneth R. Harkins to Worth Rowley, March 31, 1955, D.O.J., 60-57-140; *Memorandum*, Watson Snyder to Wilbur L. Fugate, June 30, 1955, *ibid.*

154. *Letter*, Fugate to Victor R. Hansen, January 19, 1959, *ibid.* Fugate ultimately headed the Justice staff for the "cartel" case.

155. "Statement of David I. Haberman" [hereafter cited as "Haberman Statement"], *MNC*, Part 7, p. 55d; *Memorandum*, Fugate to Barnes, November 3, 1955, D.O.J., 60-57-140; *Memorandum*, For the Files, by Svedberg, June 28, 1956, *ibid.*; *Memorandum*, Fugate and Charles L. Whittinghill to Barnes, November 7, 1955, *ibid.*, charged that they had proof of the original and later "AS IS" agreements.

156. *Interview*, Wall and Carpenter with Elliott, June 26, 1975, Files, C.B.H.S.

157. *Interview*, Wall with Johnson, December 14, 1977, *ibid.*

158. *The Lamp*, Vol. 43, No. 1 (Spring, 1961), p. 2.

159. Bd. Dir. Mins., December 26, 1957, Secretary's Office, Exxon; *Memorandum*, Fugate to Barnes, January 24, 1956, D.O.J., 60-57-140; *Memorandum*, For the Files, by Svedberg, June 28, 1956, *ibid.*; *Memorandum*, For the Files, by Roy S. Kulby, May 15, 1957, *ibid.*; *Memorandum*, Fugate to Hansen, June 3, 1957, *ibid.*; *Memorandum*, "For the Attorney General," by Hansen, June 20, 1957, *ibid.*; *Letter*, Hansen to Monaghan, September 23, 1957, *ibid.*; Shepard and Monaghan reported frequently to the Jersey Executive Committee on their discussions with Justice attorneys; for example see E.C. Minutes, July 18, 1957, August 23, 30, 1957, September 19, 25, 1957, November 21, 1957, Secretary's Office, Exxon.

160. *Interview*, Wall with Page, November 1, 1979, Files, C.B.H.S.

161. Bd. Dir. Mins., December 26, 1957; October 29, November 5, 1959; April 7, July 21, September 30, 1960, Secretary's Office, Exxon; *Letter*, Joseph C. O'Mahoney to William P. Rogers, November 20, 1957, D.O.J., 60-57-140; *Memorandum*, For the Files, by Fugate, December 3, 1957, *ibid.*; *Memorandum*, For the Files, by Robert A. Bicks, December 6, 1957, *ibid.*; *Letter*, Estes Kefauver to Hansen,

December 27, 1957, *ibid.; Memorandum*, For the Files, by Fugate, November 13, 1959, *ibid.; Letter*, Fugate to Bicks, December 10, 1959, *ibid.; Memorandum*, For the Files, by Fugate, December 15, 1959, *ibid.*

162. "Stanvac Documents," Secretary's Office, Exxon.

163. Final Judgment, U.S. *vs.* Standard Oil Company (New Jersey), *et al.*, November 14, 1960, Files, C.B.H.S.

164. *New York Times*, November 15, 1970; *The Lamp*, Vol. 41, No. 1 (Spring, 1961), p. 2.

165. *Interview*, Wall and Carpenter with Campbell, May 5, 1977, Files, C.B.H.S.

166. *Interview*, Wall with Page, November 1, 1979, *ibid.*

167. *New York Times*, November 15, 1960; Wilkins, *The Maturing*, p. 338, note.

168. *The Lamp*, Vol. 43, No. 1 (Spring, 1961), p. 2.

169. *Ibid.*, 3-4; The young man referred to, M. Yashiro, soon became president of Esso Standard Sekiyu K. K., the Esso Eastern affiliate in Japan. Page believed him to be "the youngest president [at that time] of any important Japanese company." Yashiro ultimately became executive vice president of Esso Eastern in Houston; *Letter*, Page to Wall, December 2, 1981, Files, C.B.H.S.

170. *Interview*, Wall with Page, November 1, 1979, *ibid.*

171. "Haberman Statement," *MNC*, Part 7, p. 44.

172. "A Report to the National Security Council by the Departments of State, Defense and the Interior," *MNC*, Part 8 (Appendix to Part 7), 1-9.

CHAPTER 16

1. "From Africa to the Pacific," *The Lamp*, Vol. 43, No. 1 (Spring, 1961), 2-3.

2. Much of this has been detailed in the chapter entitled, "Jersey, Antitrust, Iran and Stanvac." The court entered its Superseding Final Judgement on May 29, 1968. Until that time the judgment dated November 14, 1960, remained in effect; see U.S. District Court, Southern District of New York, Civil No. 86-27, U.S. vs. Standard Oil Company (New Jersey) *et al.*, May 29, 1968.

3. Standard Oil Company (New Jersey) [hereafter cited as SONJ], *Annual Report*, 1961.

4. *Letter*, H. W. McCobb to Gentlemen, December 12, 1961, Exxon Records Center [hereafter cited as E.R.C.], 82-10-A.

5. "My Impressions As To The Development of Standard-Vacuum," by W. B. Cleveland, June 10, 1983, Files, Center for Business History Studies [hereafter cited as C.B.H.S.]; *Interview*, Bennett H. Wall with Jack Wexler, August 25, 1982, *ibid.; Letter*, Wexler to Wall, January 15, 1984, *ibid.*

6. "My Impressions . . . of Standard-Vacuum," by Cleveland, June 10, 1983, *ibid.; Interview*, Wall and C. Gerald Carpenter with L. W. Elliott, June 26, 1975, *ibid.; Letter*, Grosvenor Blair to Wall, September 7, 1982, *ibid.; Interview*, Wall with Richard W. Barham and Wexler, August 25, 1982, *ibid.; Interviews*, Wall with Wexler, August 25 and 26, 1982, *ibid.; Interview*, Wall with Blair, August 25, 1982, *ibid.*; Executive Committee Minutes [hereafter cited as E.C. Minutes], December 7, 20, 1961, Secretary's Office, Exxon; see also, *Letter*, Howard W. Page to Wall, September 1, 1983, Files, C.B.H.S.

7. "My Joining Stanvac in 1959 and Subsequent Developments," by Cleveland, June 10, 1983, *ibid.*

8. "Report on Trip to Southeast Asia, June 2-June 28, 1962," September 4, 1962, *ibid.*; D. V. Schworer of Esso Standard Eastern joined the party in the Philippines; E.C. Minutes, September 17, 1962, Secretary's Office, Exxon.

9. *Letter*, Walter F. Spath to Elliott, *et al.*, May 7, 1962, E.R.C., 281-18-B.

10. SONJ, *Annual Report*, 1962; *Memorandum*, H. W. Howze to Wolf Greeven, May 29, 1962, E.R.C., 281-18-A; *Memorandum*, "Aide Memoire, South Korean Refinery," Cleveland to Elliott, June 27, 1962, *ibid.*, 306-2-B; *Memorandum*, "Tanganyika Refinery," September 27, 1962, *ibid.*, 281-18-B; *Letter*, Greeven to Spath, September 7, 1962, *ibid.*

11. *Memorandum*, "Aide Memoire, South Korean Refinery," Cleveland to Elliott, June 27, 1962, *ibid.*, 306-2-B; "My Joining Stanvac . . . ," by Cleveland, June 10, 1983, Files, C.B.H.S.; *Hearings Before the Subcommittee on Multinational Corporations of the Com-*

mittee on Foreign Relations, United States Senate, Ninety-Fourth Congress, First Session, 17 Parts, Washington, 1973-1976 [hereafter cited as *MNC*] Part 12, 4-58, 9.

12. *Memorandum*, M E. Jones to David A. Shepard, *et al.*, September 27, 1962, E.R.C., 281-18-B; *Memorandum*, Telephone Conversation From Claus W. Ruser to Jones, September 26, 1962, *ibid.*; *New York Times*, May 13, 1961.

13. *Memorandum*, "ESE Position Paper on India," September 27, 1962, E.R.C., 281-18-B; *Memorandum*, "Summary of Position at Time of Break-Off of Negotiations for Exploration Rights in Rajasthan, India," October 4, 1962, *ibid.*

14. *Memorandum*, D. E. Stines to Leo D. Welch, April 24, 1962, *ibid.*

15. *Letter*, J. W. Sinclair (signed by C. B. Thomas) to R. T. Burton, January 3, 1962, *ibid.*

16. *Memorandum*, Telephone Conversation from Ruser to Jones, September 26, 1962, *ibid.*; *New York Times*, November 14, 21, December 9, 1961, March 31, August 30, 1962.

17. *Interview*, Wall with Wexler, August 26, 1982, Files, C.B.H.S.

18. *Letter*, Thomas J. Cory to Elliott, *et al.*, November 1, 1962, E.R.C., 281-18-B.

19. "My Joining Stanvac . . . ," by Cleveland, June 10, 1983, Files, C.B.H.S.

20. *Memorandum*, Spath to Greeven, October 5, 1962, E.R.C., 281-18-B.

21. "My Joining Stanvac . . . ," by Cleveland, June 10, 1983, Files, C.B.H.S.

22. SONJ, *Annual Report*, 1963.

23. *Letter*, Burton to McCobb, January 29, 1963, E.R.C., 306-2-B; "ESE's Attitude Regarding Minority Participation," January 29, 1963, *ibid.*; "Historical Summary of Local Government/Private Participation in the ESE Area Oil Industry (Excluding India)," January 29, 1963, *ibid.*

24. SONJ, *Annual Report*, 1964.

25. "My Joining Stanvac . . . ," by Cleveland, June 10, 1983, Files, C.B.H.S.; *Letter*, Wexler to Wall, January 15, 1984, *ibid.*

26. SONJ, *Annual Report*, 1966; *Wall Street Journal*, "Exxon to Spend $450 Million in

Hong Kong," May 10, 1983.

27. "A History of Oil and Gas Development in Bass Strait, Australia," [by Jim Davis], March, 1970, Files, C.B.H.S.; *Discovery In Bass Strait*, by the Broken Hill Proprietary Company Limited and Standard Oil Company (New Jersey), December, 1971 [Courtesy, M. A. Wright]; "My Joining Stanvac . . . ," by Cleveland, June 10, 1983, Files, C.B.H.S.; *Interviews*, Wall with Wright, August 26, 27, 1982, February 17, 1983, *ibid.*; *Letter*, J. R. Hasulak to Comptroller's Department of All Affiliates, January 10, 1967, E.R.C., 25-13-A; SONJ, *Annual Reports*, 1964, 1967, 1968, 1971, 1972; *Notes*, Telephone Conversation, Wall with Zeb Mayhew, November 29, 1983, Files, C.B.H.S.; *Interview*, Wall and Carpenter with Siro Vázquez, May 16, 1978, *ibid.*

28. "My Joining Stanvac . . . ," by Cleveland, June 10, 1983, Files, C.B.H.S.; SONJ, *Annual Reports*, 1965, 1966; "A History of Oil and Gas Development . . . ," [by Davis], Files, C.B.H.S.; *Interviews*, Wall with Mayhew, January 6, 30, 1984, *ibid.*

29. See *Letter*, Greeven to R. H. Milbrath, June 18, 1962, E.R.C., 281-18-A; *Memorandum*, Milbrath to Greeven, July 5, 1962, *ibid.*; *Letter*, R. E. Howe to Greeven, August 5, 1962, *ibid.*, 76-12-A; *Memorandum*, "An Estimate of the Situation in Thailand," Luke W. Finlay to Cleveland and T. H. Tonnessen, July 2, 1968, *ibid.*, 1208-2-B; *Letter*, A. F. Evans to A. B. Hecker, September 11, 1968, *ibid.*; *Letter*, E. H. Weinman to [Evans?], August 29, 1968, *ibid.*; *Memorandum*, Charles O. Peyton to Department Heads, May 28, 1970, *ibid.*, 163-6-A; SONJ, *Annual Reports*, 1965, 1966, 1967, 1968.

30. *Letter*, H. W. Coxon to Spath, May 15, 1968, E.R.C., 1208-2-B.

31. *Letter*, J. L. Spivey to Gentlemen, April 8, 1968, *ibid.*, 164-5-A.

32. *Note*, M. M. Brisco to Employee Directors, January 8, 1969, containing copy, *Letter*, John H. Hamlin to Cleveland, December 31, 1968, *ibid.*, 163-6-A.

33. *Letter*, M. L. Haider to A. M. Satterthwaite, January 17, 1968, *ibid.*, 208-5-A. On the copy of this letter, Haider states that he cleared this with Brisco and that Cleveland suggested the letter.

34. *Memorandum* (confidential), "For M. A. Wright In Anticipation of His Visit With Sir Ian McLennan," June 19, 1968, *ibid.*, 208-5-A.

35. *Memorandum*, F. O. Dennstedt to Brisco and H. W. Fisher, August 9, 1968, *ibid.*, 1208-2-B.

36. SONJ, *Annual Report*, 1971; *Letter*, Richard W. Kimball to Cleveland, November 13, 1973, Secretary's Office, Exxon; *Letter*, C. R. Patterson to R. C. Curtis, February 1, 1974, E.R.C., 752-5-A; *Letter*, Curtis to H. H. Goerner, February 8, 1974, *ibid.*; *Letter*, Patterson to M. E. J. O'Loughlin, March 11, 1974, *ibid.*

37. *Letter*, (Unknown) to E. C. Wells, Jr., June 28, 1974, *ibid.*

38. *Memorandum*, H. M. Hartzband to W. M. McCardell and Coxon, August 8, 1974, *ibid.*

39. *Trip Memorandum*, to W. A. Konrad, "W. E. Meiers' Visit to Singapore and Japan," February 20, 1967, E.R.C., 25-13-A; SONJ, *Annual Reports*, 1969, 1970, 1971, 1972; *Letter*, Cleveland to Nicholas J. Campbell, Jr., December 14, 1973, E.R.C., 752-5-A.

40. *Ibid.*; *Interview*, Wall with Wexler, August 25, 1982, Files, C.B.H.S.; *Letter*, J. Kenneth Jamieson to David Packard, May 22, 1970, E.R.C., 163-6-A. *Note:* The contract also called for Esso Eastern to supply all the civil administration's petroleum requirements for three years. See also E.C. Minutes, May 28, 1968, Secretary's Office, Exxon.

41. *Letter*, Cleveland to Campbell, December 4, 1973, E.R.C., 752-5-A.

42. *Interview*, Wall with Wexler, August 26, 1982, Files, C.B.H.S.; *Interview*, Wall with Wright, August 26, 1982, *ibid.*; SONJ and Exxon, *Annual Reports*, 1965-1975.

43. *Ibid.*; *Letter*, A. C. Reiners to Directors, *et al.*, April 9, 1974, E.R.C., 752-5-A; *Memorandum*, Goerner to Vázquez, *et al.*, April 12, 1974, *ibid.*

44. "My Joining Stanvac . . . ," by Cleveland, June 10, 1983, Files, C.B.H.S. Esso Eastern's operating loss for the Philippines (excluding chemicals) in 1973, reached $3,800,000 but after divestment the net loss for that year was only $2,200,000. See *letter*, H. J. Keith, Jr. (for G. V. Sherman) to Directors, *et al.*, February 27, 1974, E.R.C., 752-5-A; SONJ, *Annual Report*, 1972.

45. *Ibid.*, 1971, 1972, 1973.

46. "My Joining Stanvac . . . ," by Cleveland, June 10, 1983, Files, C.B.H.S. (italics his); *Letter*, Keith, Jr. (for Sherman) to Directors, *et al.*, February 27, 1974, E.R.C., 752-5-A.

47. "My Joining Stanvac . . . ," by Cleveland, June 10, 1983, Files, C.B.H.S.; Exxon, *Annual Report*, 1974.

48. *Ibid.*, 1970-1975; *Memorandum* (no author) to H. C. Kauffmann, January 18, 1974, E.R.C., 752-5-A; *Telegram*, Esso Eastern to D. L. Snook, January 22, 1974, *ibid.*; Exxon *Daily Bulletin*, September 16, 1976, Files, C.B.H.S.; *Letter*, B. W. Massey to Kimball, April 9, 1974, E.R.C., 752-5-A; *Memorandum*, Sherman to O'Loughlin, July 24, 1974, *ibid.*; *Letter*, Dale E. Owens to Kauffmann, June 25, 1974, *ibid.*; *Memorandum*, Coxon to McCardell, August 1, 1974, *ibid.*; *Wall Street Journal*, January 21, March 15, 1974. The price quoted in the last dated story was $30 million. The Foreign Claims Settlement Commission, in August, 1985, awarded Esso Eastern $30,199,453.30 with interest at the rate of 6 percent per annum from May 1, 1975 (date of loss) to date of settlement. An additional $4,853,707.00 was awarded to Esso Exploration on its separate claim against Vietnam.

Letter, J. V. Pickering to Elliott, September 26, 1963, E.R.C. 306-2-B; *Letter*, George P. Case to Harold Midtbo, December 7, 1966, *ibid.*, 76-12-A; *Notes*, Conversation with Dilip Mukerjee in New York, by Marc Pardue (for W. M. Craig of Esso Eastern), April 19, 1967, *ibid.*, 25-13-A; *Letter*, David G. Gill to Paul E. Morgan, March 18, 1986, Files, C.B.H.S.

49. *Interview*, Wall with Wexler and Barham, August 25, 1982, Files, C.B.H.S.; *Letter*, Brisco to Cleveland, June 13, 1968, E.R.C., 1208-2-B. "A" Areas included Japan, Australia, Singapore, Malaysia, Hong Kong, Thailand and Indonesia (producing). "B" Areas included Philippines, India, Pakistan, Bangladesh, Laos, Cambodia and Vietnam. See *Letter*, Cleveland to Campbell, December 14, 1973, E.R.C., 752-5-A.

50. Exxon, *Annual Report*, 1973; *Telegram*, Esso Eastern to Snook, January 22, 1974, E.R.C., 752-5-A; *Letter*, Cleveland to Campbell, December 14, 1973, *ibid.*; *Letter*, Brisco to Cleveland, January 15, 1968, *ibid.*, 163-6-A; *Memorandum*, M. J. Callahan to At-

tached List (not attached), March 29, 1974, *ibid.*, 752-5-A.

51. *Interview*, Wall with Wexler, August 25, 26, 1982, Files, C.B.H.S.; *Interview*, Wall with Wexler and Barham, August 25, 1982, *ibid.*; "My Joining Stanvac . . . ," by Cleveland, June 10, 1983, *ibid.*

52. *Ibid.*; *Interview*, Wall with Wexler, August 26, 1982, *ibid.*

53. *Letter*, Brisco to Cleveland, June 13, 1968, E.R.C., 1208-2-B; *Memorandum*, Sherman to O'Loughlin, July 24, 1974, *ibid.*, 752-5-A.

54. *Cablegram*, [?] Cohen to O'Loughlin, November 7, 1974, *ibid.* (with copies to Case, J. S. Baldwin, A. M. Natkin at Esso Eastern and to Jamieson and Kauffmann of Exxon Corporation and Peyton of Exxon International); *Memorandum*, Peyton to Jamieson, November 14, 1974, *ibid.*

55. *Letter*, (K. F. Huff?) to Wells, Jr., June 28, 1974, *ibid.*; *Letter*, T. D. Barrow to Kauffmann, November 22, 1974, *ibid.*; *Letter*, Campbell to O'Loughlin, November 14, 1974, *ibid.*; see also *letter*, Patterson to Curtis, February 1, 1974, *ibid.*; *Letter*, Curtis to Goerner, February 8, 1974, *ibid.*; *Letter*, Patterson to O'Loughlin, March 11, 1974, *ibid.*

CHAPTER 17

1. The literature on the Suez Crisis is extensive and highly contradictory. The roles of Egypt, Israel, Great Britain, France, Russia and the United States receive differing interpretations from the various commentators. This author relied on a number of volumes and articles for background information, especially M. A. Fitzsimmons, "The Suez Crisis and the Containment Policy," *The Review of Politics*, Vol. 19, No. 4(October, 1957), p. 432, 438-439; Executive Committee Minutes [hereafter cited as E.C. Minutes], April 10, 1956, July 31, 1956, Secretary's Office, Exxon; *The Middle East: U.S. Policy, Israel, Oil and the Arabs* (third edition), Congressional Quarterly, Inc. (Washington, 1977); Dean Acheson, *Power and Diplomacy* (Cambridge, Massachusetts, 1959); Tarun Chandra Bose, *The Superpowers and The Middle East* (Bombay, 1972); J. Stanley Clark, *The Oil Century. From the Drake Well to the Conservation Era* (Norman, 1955); Robert A. Divine, *Eisenhower and the Cold War* (New York and Oxford, 1981); Roscoe Drummond and Gaston

Coblentz, *Duel at the Brink* (New York, 1960); Dwight D. Eisenhower, *Waging Peace, 1956-1961* (New York, 1961); Robert H. Ferrell (ed.), *The Eisenhower Diaries* (New York and London, 1981); Herman Finer, *Dulles Over Suez* (Chicago, 1964); George Lenczowski, *Soviet Advances in the Middle East* (Washington, 1971); Anthony Nutting, *I Saw For Myself* (Garden City, 1958); Herbert S. Parmet, *Eisenhower and the American Crusades* (New York and London, 1972); William R. Polk, *The United States and the Arab World* (Cambridge, Massachusetts and London, England, 1975); Benjamin Shwadran, *The Middle East, Oil and the Great Powers* (New York and Toronto, 1973); *Memorandum*, Thomas E. Monaghan to Stewart P. Coleman, February 10, 1956, Exxon Records Center [hereafter cited as E.R.C.], 303-14-B; Aaron S. Klieman, *Soviet Russia and the Middle East* (Baltimore and London, 1970).

2. Fitzsimmons, "The Suez Crisis and the Containment Policy," 419-445.

3. *Cablegrams*, August 8, 9, 1956, E.R.C., 303-14-B.

4. *Memorandum*, August 10, 1956, *ibid.*

5. Robert Engler, *The Politics of Oil* (New York, 1961), p. 261; Marguerite Higgins, "Colonialism and Mr. Dulles," *New York Herald Tribune*, October 8, 1956, quoted in Fitzsimmons, "The Suez Crisis and the Containment Policy," p. 442.

6. *Memorandum*, August 9, 1956, E.R.C., 303-14-B; See also E.C. Minutes, July 27, 30, 31; August 9, 13, 16, 23; September 28; October 23, 24; December 21, 1956, Secretary's Office, Exxon; Eisenhower, *Waging Peace*, p. 34.

7. *Memorandum*, August 10, 1956, E.R.C., 303-14-B; *Hearings Before the Subcommittee on Multinational Corporations of the Committee on Foreign Relations, United States Senate, Ninety-Third Congress, Second Session*, 17 Parts, Washington, 1973-1976 [hereafter cited as *MNC*], Part 8 (Appendix to Part 7), p. 561.

8. *Memorandum*, August 15, 1956, E.R.C., 303-14-B.

9. *Letters*, Felix Wormser to Coleman, August 21, 22, 1956, *ibid.*

10. *Memorandum*, August 15, 1956, *ibid.*

11. *Telegram*, Wormser to Coleman, August 21, 1956, *ibid.*; *Letter*, Wormser to Coleman, August 21, 1956, *ibid.*; *Letter*, Wormser to

Coleman, August 22, 1956, *ibid.;* Draft, *Letter,* Coleman to Wormser, August 23, 1956, *ibid.; Memorandum,* "Background Memorandum Concerning the Nature and Activities of the Middle East Emergency Committee, August 10 through January 31, 1957," by Coleman, *ibid.*

12. *Letter,* Wormser to Coleman, August 22, 1956, *ibid.*

13. *Letter,* Fred G. Andahl, Acting Secretary of the Department of the Interior, to H. Wilkinson, August 23, 1956, *ibid.; Letter,* T. W. Moore to Coleman, August 27, 1956, *ibid.*

14. *Letter,* Coleman to H. A. Stewart, August 27, 1956, *ibid.*

15. *Memorandum,* August 27, 1956, *ibid.*

16. *Letter,* Coleman to Fred A. Seaton, September 7, 1956, *ibid.;* Coleman again offered to resign but Stewart refused to accept it. See *Letter,* Coleman to Wormser, September 11, 1956, *ibid.*

17. *New York Times,* February 8, 1957, quoting Coleman statement of September 19, 1956.

18. *Letter,* C. R. Smit (Esso Nederland N.V.) to J. J. Winterbottom, September 18, 1956, E.R.C., 303-14-B.

19. *Letter,* N. J. Campbell, Jr. to Representatives and Alternates of Standard Oil Company (New Jersey) on the Sub-Committee of the MEEC, September 20, 1956, *ibid.* Italics his.

20. *Letter,* A. J. May to Coleman, September 24, 1956, *ibid.*

21. *Letter,* Harold W. Fisher to Coleman, October 2, 1956, *ibid.; Letter,* Campbell to Fisher, October 9, 1956, *ibid.*

22. *Memorandum,* October 10, 1956, *ibid.; Interview,* M. J. Rathbone, "Will Suez Force Gas Rationing?" *U.S. News and World Report,* September 28, 1956.

23. *Letter,* Coleman to Wormser, October 18, 1956, E.R.C., 303-14-B.

24. *Memorandum,* Rathbone to D. A. Shepard, October 24, 1956, *ibid.*

25. Engler, *The Politics of Oil,* 144, 238; *New York Times,* July 1, 1956; See also the *Statement* of Hines H. Baker, *Joint Hearing Before Subcommittees of the Committee on the Judiciary and the Committee on Interior and Insular Affairs, United States Senate,* *Eighty-Fifth Congress, First Session,* Pursuant to Senate Resolution 57 (4 parts), I, 764-830 [hereafter cited as *Joint Hearings on the Oil Lift*]; *Letter,* P. F. Lineau to Austin P. Foster, November 13, 1956, Department of Justice Files [hereafter cited as D.O.J.], 60-57-140, Courtesy, Burton I. Kaufman; *Letter,* John P. Tolbert to Willard A. Colton, January 15, 1957, *ibid.; Memorandum,* "Background Information on MEEC for Meeting, February 11, 1957," states that MEEC was inactive because its actions might compromise U.S. (or U.N.) policy or actions, *ibid.*

26. Colonel Ernest O. Thompson had powerful political allies. When he testified before a committee of the House of Representatives, early in 1957, Speaker Sam Rayburn introduced him and then Senate Majority Leader Lyndon B. Johnson observed the proceedings, *New York Times,* February 16, 1957; Engler, *Politics of Oil,* p. 400. On Thompson, see James A. Clark, *Three Stars For the Colonel. The Biography of Ernest O. Thompson* (New York, 1954).

27. *Letter,* W. D. Bayles to W. R. Carlisle, December 14, 1956, E.R.C., 303-14-B; For the view of the U.S. President, see Eisenhower, *Waging Peace,* 20-99.

28. *New York Times,* November 11, 1956.

29. *Ibid.*

30. *Memorandum,* November 20, 1956, E.R.C., 303-14-B.

31. Recommendation of the OEEC Council, November 26, 1957, *ibid.; Cablegram,* Bayles to Carlisle, November 21, 1956, *ibid.; Letter,* Campbell to Fisher, November 23, 1956, *ibid.*

32. Certainly no action was taken, nor does the correspondence indicate that Flemming considered any.

33. Coleman Testimony, *Joint Hearings on the Oil Lift,* II, p. 1038.

34. Rathbone Testimony, *ibid.,* II, p. 1040.

35. *Memorandum,* December 3, 1956, E.R.C., 303-14-B. Some sources indicate that the Director of Defense waited until December 5, 1956, to reactivate the committee. However, in Coleman's files, a *Memorandum,* dictated by Arthur Newmyer, dated November 30, 1956, states that the President, after conferring with John F. Dulles and Walter Bedell Smith requested Seaton to authorize the fifteen U.S. companies to coordinate their ef-

forts to resolve the oil supply problem, *ibid.*

36. *Memorandum* (no author, probably George H. Freyermuth), December 3, 1956, *ibid.*

37. Rathbone Testimony, *Joint Hearings on the Oil Lift*, II, 1089-1094.

38. *Letter,* R. S. Fowler, Director of the Voluntary Agreement, to Coleman, December 5, 1956, E.R.C., 303-14-B.

39. *Telegram,* Campbell to Fisher, December 5, 1956, *ibid.*

40. *Memorandum,* December 5, 1956, *ibid.*

41. *Letter,* Wormser to Coleman, December 10, 1956, *ibid.*

42. *Letter,* Campbell to C. J. Hedlund, December 12, 1956, *ibid.*

43. *Letter,* Bayles to Carlisle, December 14, 1956, *ibid.*

44. *Letter,* Campbell to Hedlund, December 12, 1956, *ibid.; Letter,* Campbell to Fisher, December 17, 1956, *ibid.*

45. *Letter,* Freyermuth to George M. Parker, December 18, 1956, *ibid.*

46. *Letter,* Fisher to Shepard, December 19, 1956, *ibid.* Italics his.

47. *Letter,* Cogan to Fisher, December 21, 1956, *ibid.*

48. *Letter,* Fisher to P. T. Lamont, December 24, 1956, *ibid.*

49. Engler, *The Politics of Oil,* p. 238.

50. *Letter,* Fowler to Coleman, December 28, 1956, E.R.C., 303-14-B.

51. *Memorandum,* W. P. Headden to Stewart Schackne, December 28, 1956, *ibid.*

52. *Cablegram,* Fisher to Coleman, December 31, 1956, *ibid.*

53. *Letter,* Shepard to Carlisle, December 31, 1956, *ibid.;* For a report of what Schackne and Theodor Swanson learned, see *Letter,* Schackne to Information Subcommittee of MEEC, February 4, 1957, D.O.J., 60-57-140.

54. Standard Oil Company (New Jersey) [hereafter cited as SONJ], *Annual Report,* 1956.

55. *New York Times,* January 4, 1957; *Memorandum,* "Statement on Petroleum Prices," January 28, 1957, E.R.C., 176-3-B [obviously a Jersey statement explaining the Humble

action]; *Memorandum,* "Statement on Crude Oil Prices," by Baker, February 4, 1957, *ibid.*

56. *Memorandum,* James Crayhon to Shepard, December 21, 1956, E.R.C., 303-14-B. Italics his.

57. *Telegram,* Newmyer Associates, Inc. to Headden, January 8, 1957, *ibid.*

58. *Cablegram,* F. O. Canfield to Shepard, January 9, 1957, *ibid.; Report,* "Political and Economic Effects of Middle East Developments and the European Fuel Crisis," January 11, 1957, by Bayles, D.O.J., 60-57-140.

59. *Telegram,* Coleman to Wormser, January 15, 1957, E.R.C., 303-14-B.

60. *Letter,* Cogan to Fisher, January 15, 1957, *ibid.*

61. *Telegram,* Coleman to J. A. Neath, January 17, 1957, *ibid.*

62. *Letter,* Cogan to Fisher, January 15, 1957, *ibid.*

63. "Allocation of European Oil Supplies. Principles of Equalization," January 22, 1957, *ibid.*

64. "Report Concerning Estimated Petroleum Product Normal Requirements and Availability in the West of Suez Area for the First Quarter, 1957," (two reports), January 24, 1957, *ibid.*

65. *Cablegram,* Coleman to Fisher, January 29, 1957, *ibid.*

66. *New York Times,* February 10, February 16, February 18, 1957.

67. *Joint Hearings on the Oil Lift,* I, p. 1.

68. *Ibid.,* I, p. 2.

69. *Ibid.,* I, p. 9. Senator Matthew M. Neely represented a state dominated by powerful coal interests and the United Mine Workers.

70. "Statement of Victor R. Hansen," Assistant Attorney General, Antitrust Division, *ibid.,* I, 650-728.

71. Baker Testimony, *ibid.,* I, 764, 792-794; Rathbone and Coleman Testimony, *ibid.,* II, 1026-1040, 1089-1107. See also *Memorandum,* "Statement on Crude Oil Prices," by Baker, February 4, 1957, E.R.C., 176-3-B. For a garbled, but often quoted version of the Baker statement, see statement of Senator Estes Kefauver, *Joint Hearings on the Oil Lift,* I, p. 703.

72. *New York Times*, March 2, 1957; Hines H. Baker, who served for years as Humble's attorney, board member, and president, and who became a director of the Standard Oil Company (New Jersey), once pointed out that because of "a feeling and understanding in the State of Texas that by virtue of the statutes and by virtue, possibly, of a preliminary injunction, the Jersey [Company] had been 'run out of Texas.' " While this was not true, he added that Jersey purchased "the Humble Company as an investment and not as a subterfuge or front; care was exercised from the beginning not only to avoid any exercise of control or domination by the Jersey Company over the conduct of Humble's affairs." "Jersey-Humble Relations. Notes on conversation of Baker with Henrietta Larson, October 10, 1945," Courtesy, Miss Larson, Files, C.B.H.S.

73. Testimony of Seth MacDonald [Federal Trade Commission], *Joint Hearings on the Oil Lift*, II, 1232-1234. MacDonald made this statement in response to a question from Campbell, Jersey Associate General Counsel.

74. *Letter*, Interior Secretary Fred A. Seaton to Attorney General Herbert Brownell, February 16, 1957, E.R.C., 303-14-B; *Letter*, Brownell to Seaton, March 13, 1957, *ibid.*

75. *Joint Hearings on the Oil Lift*, II, p. 1245.

76. *Ibid.*, II, 1245-1247.

77. *Ibid.*, II, p. 1245.

78. Coleman Testimony, *ibid.*, II, p. 1026.

79. *Ibid.*

80. *New York Times*, February 21, 1957.

81. *Joint Hearings on the Oil Lift*, III, 2024-2028.

82. "Verbatum De La Conference De Presse," February 11, 1957, p. 5, E.R.C., 303-14-B.

83. *Ibid.*, 16-17.

84. *New York Times*, February 14, 1957.

85. *Letter*, J. A. Cogan to Fisher, February 15, 1957, E.R.C., 303-14-B.

86. *Letter*, Edward G. Moline to Coleman, February 20, 1957, *ibid.*

87. "Memorandum concerning the Implementation of Schedule No. 2, Item 3," February 25, 1957, *ibid.*

88. "Department of the Interior [Press] Release, March 1, 1957," *ibid.*

89. *Cablegram*, unidentified, March 6, 1957, *ibid.*

90. *New York Times*, March 10, 1957.

91. *Letter*, Cogan to Seaton, March 7, 1957, E.R.C., 303-14-B.

92. *Letter*, Cogan to Fisher, March 7, 1957, *ibid.*

93. "Department of the Interior [Press] Release, March 7, 1957," *ibid.*

94. *New York Times*, March 26, 1957.

95. "Associated Press Dispatch," March 6, 1957, E.R.C., 303-14-B.

96. *Letter*, Coleman to White, March 13, 1957, *ibid.*; *Letter*, Coleman to Cogan, March 13, 1957, *ibid.*

97. *Pamphlet*, "To the Shareholders of Standard Oil Company (New Jersey)," March 15, 1957 [From Eugene Holman], Files, C.B.H.S.; "Full Text, Open Hearing over WABC TV (N.Y.) and ABC Television Network," *Radio Reports, Inc.*, March 17, 1957, E.R.C., 303-14-B.

98. *Letter*, Coleman to *Journal of Commerce* (New York), March 21, 1957, *ibid.*

99. *Letter*, Wormser to Coleman, March 22, 1957, *ibid.*

100. *Memorandum*, C. J. Bauer to A. C. Reiners, April 3, 1957, *ibid.*

101. *Letter*, L. A. Astley-Bell to Coleman, April 10, 1957, *ibid.*

102. *Cablegram*, unidentified, April 10, 1957, *ibid.*; *Letter*, Astley-Bell to Coleman, April 10, 1957, *ibid.*; *Letter*, Rathbone to C. D. Dillon [Under Secretary of State], April 10, 1957, *ibid.*

103. *New York Times*, March 30, 1957.

104. *Letter*, Rathbone to Dillon, April 10, 1957, E.R.C., 303-14-B; Rathbone discussed this planned pipeline thoroughly with the Jersey Executive Committee and kept them informed about the reaction to it, from government officials in Washington and London; for example, see E. C. Minutes, January 31, April 3, 1957, Secretary's Office, Exxon.

105. *Letter*, Coleman to Wormser, April 15, 1957, E.R.C., 303-14-B.

106. *Letter*, May to Coleman, May 1, 1957, *ibid.*

107. *Cablegrams*, May 3, 6, 1957, *ibid.*

108. *E.g., Letter,* Astley-Bell to Coleman, May 9, 1957, *ibid.;* Draft, *Letter,* Coleman to Astley-Bell, May 10, 1957, *ibid.*

109. *New York Times,* May 14, 1957.

110. "Department of the Interior [Press] Release," May 14, 1957, E.R.C., 303-14-B.

111. *Letter,* Fowler to Coleman, May 16, 1957, *ibid.; Letter,* Seaton to Coleman, May 23, 1957 [?], *ibid.*

112. *Memorandum,* June 5, 1957, *ibid.*

113. *Letter,* M. E. Jones to Bayles, July 21, 1957, *ibid.*

114. *Letter,* Joseph C. O'Mahoney to Coleman, July 24, 1957, *ibid.*

115. *Letter,* Jones to Coleman, July 29, 1957, *ibid.*

116. SONJ, *Annual Report,* 1957.

117. See George W. Stocking, *Middle East Oil* (Nashville, 1970), 214-269; Engler, *The Politics of Oil,* p. 309; Leonard Mosley, *Power Play* (Baltimore, 1973), 278-288.

118. *United States vs. Arkansas Fuel Oil Corporation, et al.,* February 13, 1960 (U.S. District Court for the Northern District of Oklahoma), Judge Savage delivered the opinion orally from the bench. It is published in *CCH 1960 Trade Cases* [Para. 69, 619], 76, 473-76, 496; *United States vs. Arkansas Fuel Oil Corporation, et al.,* Criminal Number 3456 (U.S. District Court For the Eastern District of Virginia), May 29, 1958; *New York Times,* February 1-21, 1960; See John C. McLean and Robert W. Haigh, *The Growth of Integrated Oil Companies* (Cambridge, Massachusetts, 1954).

119. *Ibid.;* Ruth Sheldon Knowles, "Oil, Vaccine and Mr. Bicks," *Fortune,* Vol. LXI, No. 6 (June, 1960), 168-170.

120. *Memorandum,* G. T. Piercy to J. H. Doores and others, January 9, 1961, E.R.C., 303-14-B.

CHAPTER 18

1. For example, see *Letter,* Henry H. Hewetson to John W. Brice, September 21, 1950, Exxon Records Center [hereafter cited as E.R.C.], 192-07-B; *Letter,* George Koegler to Edward F. Johnson, January 14, 1952, *ibid.,* 194-21-B; *Letter,* Milo Perkins to Eugene Holman and Monroe J. Rathbone, December 26, 1956 (tentative draft), *ibid.,* 192-07-B;

Letter, Hewetson to Perkins, January 4, 1957, *ibid.; Memorandum,* Summary of ACHR [Advisory Committee on Human Relations], "Recommendations on the Perkins Memorandum," April 1, 1957, *ibid.; Pamphlet,* "Nationalism In the Middle East," by Howard W. Page, June 19, 1958 [A Jersey Roundtable Presentation], Files, Center for Business History Studies [hereafter cited as C.B.H.S.].

2. Page served as contact director for the Middle East until 1966 when he was replaced by George T. Piercy. Smith D. Turner retired in 1967. Harold W. Fisher joined the Board in 1959. David A. Shepard became a vice president that year. *Multinational Corporations and United States Foreign Policy, Hearings Before The Subcommittee on Multinational Corporations of the Committee on Foreign Relations, United States Senate, Ninety-Third Congress, Second Session,* 17 Parts (Washington, 1973-1976) [hereafter cited as *MNC*], Part 7, "Remarks by Ambassador Richard Funkhouser Before the National War College, December 4, 1951, — The Problem of Near Eastern Oil," 160-171, p. 171; State Department Internal *Memoranda,* "Analysis of Middle Eastern Oil Concessions and Their Implication For U.S. Foreign Policy," by Funkhouser, July 3, 1953 (revised September 10, 1953), *ibid.,* p. 137; *Interview,* Bennett H. Wall and C. Gerald Carpenter with Page, June 3, 1975, Files, C.B.H.S.

3. *Letter,* Turner to Wall, December 26, 1981, *ibid; New York Times,* February 17, 1970.

4. *Current Antitrust Problems. Hearings Before Antitrust Subcommittee* (Subcommittee No. 5) *of the Committee on the Judiciary, House of Representatives, Eighty-Fourth Congress, First Session,* 3 Parts (Washington, 1955) [hereafter cited as *Current Antitrust Problems*], Part 2, 1563-1651, has the full contract.

5. Executive Committee Minutes, August 27, 1954 [hereafter cited as E.C. Minutes], Secretary's Office, Exxon.

6. *Pamphlet,* "Report by Stewart Schackne and [Theodor] Ted Swanson on Inspection Trip to Middle East, March 7-April 15, 1958," E.R.C., 192-7-B; *Current Antitrust Problems,* Part 2, p. 1573; see also *Pamphlet,* "Nationalism in the Middle East," a Presentation by Page at the Jersey Roundtable, June 19, 1958, Files, C.B.H.S.

7. E.C. Minutes, May 5, 1955, August 15, 1957, Secretary's Office, Exxon.

8. *Ibid.*, October 11, 1955, April 23, June 25, 1956, June 27, 1957, *ibid.*

9. *Current Antitrust Problems*, "Testimony of Stanley N. Barnes" [hereafter cited as Barnes Testimony], Part 1, 203-285; see pages 268-272; *Press Release*, Standard Oil Company (New Jersey) [hereafter cited as SONJ], August 5, 1954, Files, C.B.H.S.

10. *Current Antitrust Problems*, Barnes Testimony, Part 1, 268-272.

11. *Pamphlet*, "Standard Oil Company (New Jersey) and Middle East Oil Production" (Summer, 1954), Files, C.B.H.S.; No copy of the supplement could be located. That used appears in *Current Antitrust Problems*, Part 1, 276-278.

12. *Ibid.*, Part 1, p. 277; Herbert Hoover, Jr., former Under Secretary of State testified in 1957, "It was not a case in which these companies wanted to go in. It was a case where we believed it was to our national security interests that they should come in. ... The companies accepted this consideration," *The President's Proposal on the Middle East, Hearings Before the Committee on Foreign Relations and the Committee on Armed Services, United States Senate, Eighty-Fifth Congress, First Session*, 2 Parts (Washington, 1957) [hereafter cited as *The President's Proposal on the Middle East*], Part 2, p. 625. Hoover stated earlier "one of the greatest difficulties we had ... was to get the companies who did participate to do so," *ibid.*

13. *Current Antitrust Problems*, Part 2, 1556-1559.

14. *Ibid.*, Part 2, 1561-1562 (*Letter*, Emanuel Celler to John Foster Dulles, July 20, 1955; *Letter*, Dulles to Celler, August 16, 1955); Versions of these agreements appeared in numerous publications other than those cited; e.g., *Wall Street Journal* and the *New York Times*; a copy of the participants agreement (without day or month) is published in *The International Petroleum Cartel, The Iranian Consortium and U.S. National Security, Prepared For The Use of Subcommittee on Multinational Corporations of the Committee on Foreign Relations, United States Senate, Ninety-Third Congress, Second Session*, (Washington, 1974) [hereafter cited as *The International Petroleum Cartel ... and U.S. National Security*], 95-116.

15. These figures are given in cubic meters (article 20) in the Iranian government-oil company agreement, *Current Antitrust Problems*, Part 2, p. 1600 and in barrels per day in "The Iranian Oil Agreement," *ibid.*, I, 276-278.

16. *New York Times*, August 11, October 14, 1954. Additional information regarding U.S. aid to Iran as well as the Seven Year Plans may be found in *U.S. Aid Operations in Iran, Hearings Before a Subcommittee of the Committee on Government Operations, Eighty-Fourth Congress, Second Session* (Washington, 1956) [hereafter cited as *U.S. Aid Operations in Iran*]. For example, see pages 896-898, 997-998. N.B. In *U.S. Aid Operations in Iran*, p. 84, U.S. aid to Iran is given as: 1951, $1.3 million; 1952, $23.5 million; 1953, $22 million; 1954, $84 million; and 1955, $76 million (all figures are rounded off). This author cannot account for the discrepancy between the 1953 figure above and that in the text. Conceivably more than one aid program existed. The $45 million figure is given in The *Middle East. U.S. Policy, Israel, Oil and the Arabs* (Congressional Quarterly, Inc., Washington, 1977) p. 155, which indicates this was made by President Eisenhower.

17. E.C. Minutes, January 23, February 16, 1956, Secretary's Office, Exxon. The minutes clearly state "about 475,000 b/d" while "The Iranian Oil Agreement," *Current Antitrust Problems*, I, 276-278 states with equal clarity "450,000 barrels a day in the second year [1955]." Both were published by Jersey.

18. E.C. Minutes, August 9, 21, 1956, Secretary's Office, Exxon.

19. *Ibid.*, October 19, 1956.

20. *Ibid.*; See also *ibid.*, November 23, 26, 1956.

21. *Ibid.*, August 23, 1956, January 24, February 4, 1957; See *ibid.*, March 14, 20, May 21, June 18, July 17, August 12, September 23, 1958.

22. *Ibid.*, February 6, May 22, September 21, 28, November 7, 8, 12, 13, 1956; August 15, 1957.

23. *Interview*, Page with (unknown), September 17, 1957, Files, C.B.H.S. (This interview was typed on Aramco Government Relations Department letterhead, obviously for radio news, carrying the date, August 29, 1957); *Interview*, Page with (unknown), no

date (obviously September, 1957), *ibid*; *Interview*, Wall and Carpenter with Page, June 3, 1975, *ibid*; George W. Stocking, *Middle East Oil* (Nashville, 1970), 165-167.

24. Benjamin Shwadran, *The Middle East, Oil and the Great Powers* (New York, Toronto and Jerusalem, 1973), 157-170; Paul H. Frankel, *Mattei: Oil and Power Politics* (New York and Washington, 1966); Edith T. Penrose, *The Growth of Firms, Middle East Oil and Other Essays* (London, 1971). Page questioned some of the facts and conclusions in her essay on "Profit Sharing Between Producing Countries and Oil Companies in The Middle East," originally published in *The Economic Journal*, Vol. LXIX, June, 1959 and reprinted in this volume, 151-167. Ms. Penrose printed Page's reply, 167-171, and her rejoinder, 171-174.

25. *MNC*, Part 7, p. 282; State Department *Telegram* (outgoing), April 20, 1966, *MNC*, Part 8 (Appendix to Part 7), 561-562.

26. On Soviet oil, see "Western Investment In Communist Economies," prepared for the *MNC* Subcommittee, August 5, 1974 (Washington, 1974), V-VII, 1-37; George Lenczowski, *Soviet Advances in the Middle East* (Washington, 1972); Aaron S. Klieman, *Soviet Russia and the Middle East* (Baltimore and London, 1970); *Memorandum*, Page, "Notes on this chapter," September 19, 20, 1982 [hereafter cited as "Page Notes"], Files, C.B.H.S. Page wrote, "I think that there was a real possibility that if we hadn't saved Iran by the agreement, Russia could have gained control of the country through the Tudeh party. However, I know that Herb Hoover, Jr. and Dulles saw a Commie under every bed and greatly exaggerated the threat—just as rightwing groups are doing today." A *Memorandum to File*, March 28, 1968, provides the classic example of the U.S. Department of State's efforts to influence the oil companies to accede to the requests of a Middle Eastern producing state for increased crude oil offtake—in this case Iran. Eugene Rostow, Under Secretary of State and other representatives of that department told representatives of five American participants in the Iranian Consortium, "Iran wants increased offtakes to increase its revenue." The *memorandum* indicates that they urged such action; *MNC*, Part 7, 274-275; See statement in *Letter*, Turner to Wall, December 26, 1981, "Don't lose the point that Jersey was *pushed*

by the U.S. government." (Italics his), Files, C.B.H.S.; *Interview*, Wall with Dean Rusk, February 11, 1983, *ibid*.

27. See table at end of this chapter.

28. *Letter*, Turner to Paul E. Morgan, November 16, 1981, Morgan Files, C.B.H.S.; *Letter*, Turner to Wall, December 26, 1981, August 29, 1982, Files, C.B.H.S.; *Letter*, Page to Wall, February 9, 1982, *ibid*.; "Page Notes," *ibid*.; *Interview*, Wall and Carpenter with Page, June 3, 1975, *ibid*.

29. *MNC*, Part 7, p. 288; *Letter*, Wall to Page, February 1, 1982, Files, C.B.H.S.; *Letter*, Page to Wall, February 9, 1982, *ibid*.; "Page Notes," *ibid*.; *Letter*, Turner to Wall, December 26, 1981, *ibid*.; *Interview*, Wall and Carpenter with Page, June 3, 1975, *ibid*.; [Anthony Sampson, *The Seven Sisters* (New York, 1975), states that Page received strong support in his opposition to the cut from Director William R. Stott. He cites no source for this remark.] The idea of creating an organization of producing countries had occurred in many individuals in the Middle East, Venezuela and elsewhere. Arab oil spokesmen realized that the success of such a body would depend a great deal on the attitude toward it of such non-Arab nations as Iran and Venezuela. By 1949, the latter country had become concerned over the relative production cost of oil there and in the Middle East, and feared the loss of its crude oil markets. In that year Venezuela sent a three-man delegation to the Middle East to expound on the 50-50 principle operative in that country and to promote the coordination of oil policies toward consuming nations. Venezuela also sent a delegation to the Arab Petroleum Congress in 1959 (as did Iran). Resolutions passed and agreements made there, and resolutions passed by a subcommittee of the Arab League Economic Council expressed the concrete idea of cooperation of the member states with Iran and Venezuela. These provided the groundwork for OPEC. In addition to the five named founding countries, the following states have since joined OPEC: Abu Dhabi, 1967; Algeria, 1969; Ecuador, 1973; Gabon, 1975; Indonesia, 1962; Libya, 1962; Nigeria, 1971 and Qatar, 1961.

30. Ian Seymour, *OPEC/Instrument of Change* (New York, 1981), p. 35; It should be noted here that Seymour, who is an exceptionally competent student of Middle Eastern affairs, describes his book as "made possible

through the generous sponsorship of OPEC." The Shah was quoted in *Petroleum Week*, December 9, 1960, Seymour, *OPEC*, 35-36; on OPEC, see also Mana Saeed al-Otaiba, *OPEC and the Petroleum Industry* (New York, 1975). Henry Kissinger, *Years of Upheaval* (Boston and Toronto, 1982), p. 668. It should be noted that Kissinger also states that he paid little attention to the economics of the Iranian situation. He concerned himself only with the diplomatic and security aspects of the problems.

31. Seymour, *OPEC*, p. 37; on Juan Pablo Pérez Alfonso, see *New Orleans Times-Picayune*, October 1, 1979, Files, C.B.H.S.

32. *Interview*, Wall and Carpenter with Page, June 3, 1975, Files, C.B.H.S.; E.C. Minutes, October 10, 1960, Secretary's Office, Exxon; See also *ibid.*, November 25, 1959.

33. Seymour, OPEC, p. 37; See also *OPEC, The Statute of the Organization of Petroleum Exporting Countries* (no place, July, 1980).

34. *Memorandum*: For the Files (by Wilbur L. Fugate), September 8, 1961, Department of Justice Records [hereafter cited as D.O.J.], 60-57-140, (Italics mine), Courtesy, Burton I. Kaufman.

35. Shwadran, *The Middle East, Oil and the Great Powers*, p. 159; In some Middle Eastern countries, the 50-50 profit-sharing arrangement consisted of a royalty payment of 12 1/2 percent of posted price and taxes at 50 percent of net income (with royalty payments credited against taxes), for example:

Posted price	$1.800
Royalty payment	
@12 1/2%	0.225
Cost of production	0.250
Taxable income	1.550 (1.800 –
	0.250 ÷ 2)
Income tax @ 50%	0.775
Total Government take	
(with royalty credited	
against tax)	0.775

The OPEC recommended plan to expense royalty payments—provided no offset against taxes, for example:

Posted price	$1.800
Royalty payment	
@ 12 1/2%	0.225
Cost of production	0.250
Taxable income (excluding royalty)	1.325 (1.800 –
	0.475 ÷ 2)

Tax receipts @ 50%	0.6625
Total Government take	
(tax plus royalty)	0.8775

36. E.C. Minutes, September 21, 1962, Secretary's Office, Exxon.

37. *Memorandum*, Turner to Morgan, November 16, 1981, Morgan Files, C.B.H.S.; See also *Letter*, Turner to Wall, December 26, 1981, Files, C.B.H.S.; Seymour, *OPEC*, p. 46.

38. *Letter*, Turner to Wall, August 29, 1982, Files, C.B.H.S.; "Page Notes," *ibid.*; *Letter*, Turner to Wall, December 26, 1981, *ibid.*; Seymour, *OPEC*, p. 46, citing *Middle East Economic Survey*, October 26, 1952; in a *Letter*, Page to Wall, February 9, 1982, the writer stated "while the formation of an OPEC was inevitable, I seriously doubt if it would have occurred before 1971 at the earliest and 1973 at the latest when demand equalled total producing capacity. Throughout the entire 1965-1971 period the payments to the governments were 6 1/2¢, falling to 1 1/2¢ less than if the Posted Price had not been changed. Therefore in a narrow sense the cutting of Posted Price was a very successful action." Files, C.B.H.S. This change from realized price to posted price occurred in Libya in 1965.

39. E.C. Minutes, May 16, 20, June 28, 1963, October 1, 1964, Secretary's Office, Exxon. It should be noted most of the oil companies operating in the Middle East did place bids. Only Jersey and Continental Oil Company did not.

40. *Ibid.*, July 29, August 15, 19, September 4, 23, October 31, 1963; Seymour, *OPEC*, 46-47, has this conference taking place in August; however, the source he cites is for January 8, 1964, which fact, plus the failure of the Jersey Executive Committee to refer to it, leads this author to conclude that it *did* occur in September.

41. E.C. Minutes, October 31, 1963, Secretary's Office, Exxon.

42. *Interview*, James B. Quinn with Page, April 22, 1977, Files, C.B.H.S., Courtesy, Mr. Quinn; *Letter*, Page to Wall, January 1, 1982, *ibid.*; *Letters*, Turner to Wall, November 16, December 26, 1981, *ibid.*; *Interview*, Wall with Piercy, July 30, 1982, *ibid.*; "Page Notes," *ibid.*; *Letter*, Turner to Wall, August 29, 1982 (figures supplied August 30, 1982), *ibid.*

43. *Middle East Economic Survey*, September 16, 1960, quoted in Seymour, *OPEC*, p. 38.

44. *Letter*, Turner to Armin H. Meyer, April 18, 1966, *MNC*, Part 8 (Appendix to Part 7), 567-568. No author is given to this letter but *Telegram* (outgoing), [George W.] Ball to [probably U.S. Embassy in Tehran], April 20, 1966, refers to "details of this complex question set forth in April 18 letter from Turner to Ambassador Meyer," *ibid.*, 561-562. Further, the author of the April 18 letter refers to Piercy as "My Colleague." When asked whether he wrote this letter Turner replied, "I think probably I did, though I can't document this and do not specifically recall it. But I had the material, and knew A.M. [Armin H. Meyer] (he called on me in New York soon after this). *Letter*, Turner to Morgan, February 16, 1983, Morgan Files, C.B.H.S. Note: The relevant section of the participants' agreement may be found in *MNC*, Part 8 (Appendix to Part 7), p. 570; Page sought without success to explain this procedure to the U.S. Senate Subcommittee on Multinational Corporations when he testified. See *MNC*, Part 7, 280-345.

45. *Telegram* (outgoing), Ball (Acting) to U.S. Embassy, Tehran, April 20, 1966, *ibid.*, Part 8 (Appendix to Part 7), 561-562. No embassy is specified but the evidence indicates Tehran.

46. *Telegram* (incoming), U.S. Embassy in Iran to U.S. Department of State, April 25, 1966, *ibid.*, 562-563. Neither the author nor addressee of this telegram is identified; however, the contents indicate both.

47. Stocking, *Middle East Oil*, 184-185, quoting *Kayhan International*, a Tehran English-language daily paper.

48. Much of this information came from interviews and correspondence with Page, Piercy, Turner and others, plus copies of Jersey and Exxon press releases. The names of the individuals and the dates of the press releases concerning them are: Paul J. Anderson, June 28, 1957, February 27, 1961, August 10, 1964; Charles M. Boyer, November 22, 1965; Elliot R. Cattarulla, July 30, 1976; Jack G. Clarke, March, 1976, September 24, 1976; Page, August 5, September 21, 1954, May 25, 1966, August 8, 1968, August 28, 1970; Piercy, May 6, 1955, June 5, 1958, June 14, 1960, June 29, 1961, July 2, November 22, 1965, February 3, 1966, May 19, 1967, May 15, 1970, February, 1977; Turner, May 3, 1960. It should be noted here that so far as this writer can learn, there has been no systematic effort made at Exxon to preserve old press releases. Obviously great

numbers of them have been destroyed. These copies are located in Files, C.B.H.S.

49. *Interview*, Wall with J. Kenneth Jamieson, March 1, 1976, *ibid.*

50. *Memorandum*, Fugate to Donald F. Turner, June 15, 1966, D.O.J., 60-57-140.

51. *Letter*, Page to Wall, January 16, 1983, Files, C.B.H.S.; *Memorandum*, "Iranian Demands on Oil Consortium," Department of State Memorandum of Conversation, October 24, 1966, *MNC*, Part 8 (Appendix to Part 7), 571-574.

52. *Ibid.*

53. *Ibid.*, 568-569.

54. *Memorandum*, "Discussions With United Kingdom on Iran — November 2-3, Briefing Paper . . . ," *MNC*, Part 8 (Appendix to Part 7), 564-567; *Memorandum*, "Walter Levy's Concern About Consortium Overlifting Arrangements," [no date], *ibid.*, 574-575.

55. *New York Times*, November 7, 1966.

56. *Ibid.*; Stocking, *Middle East Oil*, 177-180. There is reason to doubt that this was a revolutionary contract. See *ibid.*, p. 180.

57. *Ibid.*, p. 193; *Letter*, Page to Wall, January 16, 1983, Files, C.B.H.S.

58. *Interview*, Wall with Rusk, February 11, 1983, *ibid.*; *Interview*, Wall with Jamieson, May 16, 1980, *ibid.*

59. Some of the books used in preparing this particular section were: Fred J. Khouri, "United Nations Peace Efforts," Malcolm H. Kerr (ed.), *The Elusive Peace in The Middle East* (Albany, 1975), 19-102, p. 53; Ibrahim Abu-Lughod (ed.), *The Arab-Israeli Confrontation of June 1967: An Arab Perspective* (Evanston, 1970); Shlomo Aronson, *Conflict & Bargaining In The Middle East/An Israeli Perspective* (Baltimore and London, 1978); Surendra Bhutani, *The United Nations And The Arab-Israeli Conflict* (Gurgaon, India, 1977); Tarun Chandra Bose, *The Superpowers and the Middle East* (Bombay, India, 1972); Randolph S. Churchill and Winston S. Churchill, *The Six Day War* (Boston, 1967); Moshe Davis (ed.), *The Yom Kippur War* (New York, 1974); C.H. Dodd and M.E. Sales, *Israel and the Arab World* (New York, 1970); Robert Engler, *The Brotherhood of Oil* (Chicago, 1977); Tim Hewat (ed.), *War File* (London, 1967); *Keesing's Research Report*, *The Arab-Israeli Conflict* (New York, 1968); John Norton Moore (ed.), *The Arab-Israeli*

Conflict, Readings and Documents (New Jersey, 1977); Edgar O'Ballance, *The Third Arab-Israeli War* (Hamden, 1972); Mira Wilkins, *The Maturing of Multinational Enterprise: American Business Abroad from 1914 to 1970* (Cambridge and London, 1974); Stocking, *Middle East Oil* (Nashville, 1970); *Articles:* El Hassan Bin Talal, "Jordan's Quest for Peace," *Foreign Affairs*, Vol. 60, No. 3 (Spring, 1982), 802-813, p. 807; Claudia Wright, "Iraq — New Power in the Middle East," *ibid.*, Vol. 58, No. 2 (Winter, 1979-1980), 257-278.

60. Jordan joined later. Klieman, *Soviet Russia and the Middle East*, p. 74; Lyndon Baines Johnson, *The Vantage Point* (New York, *et al.*, 1971), see chapter 13, 287-304.

61. *Letter*, Leroy Marceau (counsel) to Piercy, June 7, 1967, E.R.C., 565-3-B; *Copy*, Proposed Plan of Action, Office of Oil and Gas, June 7, 1967, *ibid.*; United States Department of the Interior *Press Release*, June 10, 1967, *ibid.*, *Copy*, Foreign Petroleum Supply Committee, Proposed Plan of Action, June 15, 1967, *ibid.*; *Letter*, Farris Bryant to Standard Oil Company (New Jersey), June 30, 1957, *ibid.*, enclosing copy, *Letter*, O.P. Lattu, Chairman, Foreign Petroleum Supply Committee, submitting a Plan of Action under the Voluntary Agreement, dated June 20, 1967, *ibid.*; *Letter*, Lattu to John Ricca, July 5, 1967, *ibid.*; *Letter*, Richard Borwick to Henry B. Wilson, July 13, 1967, *ibid.*; Report of the Supply and Distribution Subcommittee of the Emergency Petroleum Supply Committee, August 15, 1967, *ibid.*; Stocking, *Middle East Oil*, 458-459; Noel Mostert, *Superships* (New York, 1974).

62. Zuhayr M. Mikdashi, *The Community of Oil Exporting Countries* (London, 1972); *Middle East Economic Survey* quoted in Stocking, *Middle East Oil*, p. 459; *Telegram* (outgoing), [Dean] Rush[k] to Unspecified [probably U.S. Embassy, Tehran], no date [between November 15-22, 1967], *MNC*, Part 8 (Appendix to Part 7), p. 576; El Hassan Bin Talal, "Jordan's Quest for Peace," *Foreign Affairs*, Vol. 60, No. 4 (Spring, 1982), 802-813, p. 807.

63. *Letter*, Lattu to Jamieson (no date, but obviously around June 19, 1967), E.R.C., 725-6-B; *Memorandum*, R.H. Herman to Page, February 12, 1970, E.R.C., 82-10-A; *Memorandum*, Leslie Pincott to Jamieson, January 20, 1970, *ibid.*; *Telegram* (outgoing),

[Dean] Rush[k] to Unspecified [probably U.S. Embassy, Tehran], no date [between November 15-22, 1967], *MNC*, Part 8 (Appendix to Part 7), p. 576. In testimony before the Church Committee, E.L. Shafer of Continental Oil Company, in answer to a direct question from Committee Associate Counsel Jack Blum, stated that rumors had identified the French company, CFP, as the one who gave the Shah a copy of the overlift arrangements, "Testimony of E.L. Shafer," *ibid.*, Part 7, 244-274.

64. *Letter*, Robert L. Dowell, Jr. to Walter M. McLelland, December 4, 1967, enclosing "Reflections and Observations On Consortium/NIOC Relations," by R.H. Harlan, *ibid.*, Part 8 (Appendix to Part 7), 583-584, italics theirs; *Airgram*, U.S. Embassy, Tehran to Department of State, December 5, 1967 [signed Armin H. Meyer], *ibid.*, 582-583; *Telegram*, no person [probably Averell Harriman] to Secretary of State, December, 1967, *ibid.*, 576-577; *Interview*, Wall with Jamieson, May 16, 1980, Files, C.B.H.S.

65. *Telegram*, Rusk to Unknown [probably U.S. Embassy London or Tehran], no date [obviously December, 1967], *MNC*, Part 8 (Appendix to Part 7), p. 577; *Telegram*, David Bruce to Rusk, December, 1967, *ibid.*, 577-578.

66. *Memorandum*, "Department of State Memorandum of Conversation," December 13, 1967, *ibid.*, 578-579.

67. *Letter*, Anthony M. Solomon to [Eugene V.] Rostow, December 11, 1967, *ibid.*, 580-581, italics theirs; *Telegram*, American Embassy, London to Secretary of State, December [no date], 1967, *ibid.*, 579-580; Bruce to Secretary of State, December [no date], 1967, *ibid.*, p. 583.

68. *Ibid.*

69. *Letter*, John B. McGrath to James E. Akins, February 21, 1968, *ibid.*, p. 585;*Letter*, Akins to McGrath, March 5, 1968, *ibid.*,585-586.

70. *Telegram*, [Dean] Rusk to [Unknown], February 29, 1968, *ibid.*, p. 590.

71. Shwadran, *The Middle East . . .*, p. 173.

72. *Ibid.*, p. 174.

73. *Memorandum* to File, "Meeting in Washington with State Department," March 28, 1968, *MNC*, Part 7, 274-275.

74. al-Otaiba, *OPEC and the Petroleum Industry*, 115-118; Stocking, *Middle East Oil*, 437-439; Seymour, *OPEC: Instrument of Change*, 63-66.

75. *Letter*, Robert L. Dowell, Jr. to Akins, November 27, 1968, *MNC*, Part 8 (Appendix to Part 7), p. 588.

76. *Memorandum*, "Iricon's Views on the Iranian Oil Situation," February 12, 1969, *ibid.*, 587-588.

CHAPTER 19

The full story of the Iraq Petroleum Company is told in the following volumes: George Sweet Gibb and Evelyn H. Knowlton, *The Resurgent Years, 1911-1927: History of Standard Oil Company (New Jersey)*, (New York, 1956); Henrietta M. Larson, Evelyn H. Knowlton and Charles S. Popple, *New Horizons, 1927-1950: History of Standard Oil Company (New Jersey)*, (New York, 1971).

1. In 1934, Standard Oil Company of New York [Socony] merged with Vacuum Oil Company to become Socony-Vacuum Oil Company, Inc. [Socony-Vacuum]; in 1955, it became Socony Mobil Oil Company [Socony Mobil] and in 1966, it became Mobil Oil Corporation [Mobil].

2. All the companies except Standard Oil Company (New Jersey) [Jersey] and Socony either sold out or withdrew.

3. Turkish Petroleum Company, Limited [TPC] was renamed Iraq Petroleum Company, Limited [IPC] in 1929. Its concession was for "the whole of Iraq, covering the provinces of Mosul and Baghdad, except Transferred Territory and the Basra area." In May, 1932, Iraq granted all of the territory west of the Tigris and north of the 33-degree line to the British Oil Development Company, Limited [BOD] which later expanded to include Mosul Oil Fields, LTD [MOF]. Several years later, IPC created Mosul Oil Holdings, Ltd. to acquire MOF. When in control of MOF, IPC renamed the company Mosul Petroleum Company, Ltd. [MPC] which also took over the BOD concession in 1938. Both MOF and BOD were liquidated.

In 1938, Iraq granted IPC an additional concession covering the Basrah area for its wholly owned subsidiary, Basrah Petroleum Company, Ltd. [BPC]. In this manner IPC gained the concession for all of Iraq except a small area along the Saudi Arabian neutral zone.

The IPC holding company, Petroleum Concessions, Ltd., had numerous subsidiary companies that acquired concessions, explored, and produced oil in the small sheikdoms in the Persian Gulf area. Among a number of others these included Aden, Qatar, Abu Dhabi, Trucial Coast, Oman, and Dhofar.

The five IPC owners were: Participations and Investments, Ltd. [P & I] (Gulbenkian), 5 percent, and 23.75 each to Anglo-Persian Oil Company, Limited, Royal Dutch-Shell, Compagnie Française des Pétroles [CFP], and Near East Development Corporation [NEDC]—(Standard Oil Company (New Jersey) and Standard Oil Company of New York). In 1935, Anglo-Persian became Anglo-Iranian Oil Company, Limited [AIOC], and in 1954, that company became British Petroleum Company, Ltd. [BP]. On November 1, 1972, Standard Oil Company (New Jersey) changed its name to Exxon Corporation [Exxon].

For the sake of convenience and clarity, IPC or "the company" will be used to refer to that company as well as the Mosul and Basrah companies, or to all three collectively.

When the U.S. Department of State permitted NEDC to sign the restrictive Red Line Agreement, it sanctioned a violation of the U.S. Open Door policy. This exception became a major issue between the companies and the Justice Department.

4. For the story see, Gibb and Knowlton, *The Resurgent Years, 1911-1927: History of Standard Oil Company (New Jersey)*; Larson, Knowlton and Popple, *New Horizons, 1927-1950: History of Standard Oil Company (New Jersey)*; See also, *Multinational Corporations and United States Foreign Policy, Hearings Before The Subcommittee on Multinational Corporations of the Committee on Foreign Relations, United States Senate, Ninety-Third Congress, Second Session*, 17 Parts (Washington, 1973-1976) [hereafter cited as *MNC*]; Senate Select Committee on Small Business, Commission Print No. 6, *The International Petroleum Cartel: Staff Report to the Federal Trade Commission* (Washington, D.C., 1952), [hereafter cited as *FTC Staff Report*].

5. Nubar Gulbenkian, *Portrait in Oil. The*

Autobiography of Nubar Gulbenkian (New York, 1965), p. 277.

6. *Memorandum*, Smith D. Turner to Paul Morgan, November 16, 1981 (enclosing page of Gerald Gardiner *letter* to Standard Oil Company (New Jersey), no date), Morgan Files, Center for Business History Studies [hereafter cited as C.B.H.S.].

7. *Ibid.*

8. *FTC Staff Report*, 104-106; Edith Penrose, *The Large International Firm in Developing Countries* (Cambridge, Mass., 1968), 157-160, see especially note 1, p. 158.

9. Executive Committee Minutes [hereafter cited as E.C. Minutes], March 11, 1949, Secretary's Office, Exxon.

10. *Standard Oil Company (New Jersey) and Middle East Production* (revised edition, New York, 1954).

11. *Letter*, Howard W. Page to Orville Harden, November 3, 1951, *MNC*, Part 8 (Appendix to Part 7), 314-315; *Letter*, Paul J. Anderson to Page, March 13, 1952, *ibid.*, 333-335; *Letter*, Page to Neville Gass, *et al.*, January 3, 1950, *ibid.*, 287-290.

12. *Letter*, C.S. Gulbenkian to Harden, March 21, 1949, *ibid.*, 282-284.

13. This view was often expressed. For a contrary opinion, see "State Department Memoranda of Conversation Concerning 1950 Tax Decision," *ibid.*, 341-345, especially p. 343.

14. *Letter*, Page to Gass, *et al.*, January 3, 1950, *ibid.*, 287-290; *Letter*, Page to J.W. Brice, January 26, 1950, *ibid.*, 292-293 (italics his).

15. *Letter*, Gulbenkian to Harden, March 21, 1949, *ibid.*

16. "Remarks of Ambassador Richard Funkhouser Before the National War College, December 4, 1951—The Problem of Near Eastern Oil" [hereafter cited as "Funkhouser Remarks, December 4, 1951"], *MNC*, Part 7, p. 168.

17. George Lenczowski, *The Middle East in World Affairs* (Ithaca and London, 1980), p. 279.

18. Great Britain went off the gold standard in 1932.

19. The Jersey Executive Committee had asked Page to take the position, in any contract negotiations with Iraq, that the company would agree to "a dollar clause with payment in sterling for a gold clause in order to stabilize the purchasing power of payments to the Iraq Government," E.C. Minutes, July 17, 1950, Secretary's Office, Exxon. Two years later Middle East Coordinator Anderson told the Executive Committee that a proposal had been made [in the negotiations?] to settle the case by a lump sum payment plus IPC relinquishing 50 percent of its concession acreage "over a four to five-year period" with IPC determining which acreage, *ibid.*, May 16, 1952. On July 18, 1952, Director Bushrod B. Howard reported to the Committee that the gold clause suit had been settled by a cash payment, *ibid.*, July 18, 1952.

20. Penrose, *The Large International Firm in Developing Countries*, p. 250 quoting Nadim al-Pachachi, *Iraq Oil Policy August 1954—December 1957* (Beirut, 1958).

21. Ms. Penrose explained the 5/7 rule in this way:

> To illustrate the nature of the restrictive effect of the '5/7' rule if one or two groups wanted to use it to keep down total output, consider the following example: (I ignore for simplicity the 5 per cent interest of Gulbenkian and assume equal shares for the four major groups). Suppose that the two low groups asked for 150 and 200 'units' of oil; the highest amount any group could get would then be restricted to 5/7 × 350 = 250. Imagine, then, that the following set of demands existed: group A, 150; B, 200; C, 220 and D, 350—a total of 920. D's demand would be reduced to 250 and total output to 820. Group A, by further reducing its demand to 80 units could restrict all others to 200 units and total output to 680 units. Group A could conceivably consider this a useful way of restricting the competitive power of the others, while at the same time selling its own excess at a higher price to others since, out of a total output of 680, A would be entitled to 1/4 or 170 units. If A took only 80, having plenty of oil elsewhere, it could dispose of 90 to the others; but even if D took all the excess oil, it would be unable to satisfy its original requirements.

Penrose, *The Large International Firm in Developing Countries*, p. 158, footnote 1.

22. *Interview*, Bennett H. Wall and C. Gerald Carpenter with Page, June 3, 1975, Files, C.B.H.S.; *Letter*, Page to Wall, September 1, 1983, *ibid.*

23. Lenczowski, *The Middle East in World Affairs*, p. 279; "Funkhouser Remarks, December 4, 1951," *MNC*, Part 7, 160-171, 167.

24. *Letter*, H.W. Fisher to Wall, November 29, 1983, Files, C.B.H.S.; *MNC*, Part 8 (Appendix to Part 7), p. 502.

25. Penrose, *The Large International Firm in Developing Countries*, p. 69.

26. E.C. Minutes, February 14, March 18, 1955, Secretary's Office, Exxon.

27. *Letter*, V.C. Georgescu to M.L. Haider, June 15, 1953, Exxon Records Center [hereafter cited as E.R.C.], 192-7-B; see also E.C. Minutes, October 28, 1952; August 12, September 30, November 3, 4, December 1, 1955, Secretary's Office, Exxon; Benjaman Shwadran, *The Middle East, Oil and the Great Powers* (New York, Toronto and Jerusalem, 1973), p. 270.

28. George W. Stocking, *Middle East Oil* (Nashville, 1970), 209-210, p. 255, quoting "Minutes of 1961 Meetings," *Iraq Ministry of Oil*, p. 49.

29. Shwadran, *The Middle East, Oil and the Great Powers*, 269-272; Stocking, *Middle East Oil*, 210-214; Lenczowski, *The Middle East in World Affairs*, 286-289. Stocking states in regard to the 56-44 profit-sharing arrangement:

> The companies correctly pointed out that revenue per ton depended as much on the size of the profit margin as on the proportions in which it was divided; that the profit margin depended both on costs and selling price; that costs varied with numerous factors—character of the reserves, rate of flow, the distance of the well from the terminal, the quality of the oil, and the like—and that crude prices varied from terminal to terminal.

Iraq received at the time (1960) more revenue per barrel than any other country. Stocking, *Middle East Oil*, 225-226; see *Letter*, Fisher to Wall, November 29, 1983, Files, C.B.H.S.; Note that the usually reliable Stocking, *Middle East Oil*, p. 213, states that Monroe J. Rathbone did participate in the talks while Fisher who was in Baghdad at the time, states

categorically that he did not.

30. E.C. Minutes, May 16, 1952, Secretary's Office, Exxon.

31. *Ibid.*, July 29, 1958.

32. *Ibid.*; see also September 15, 1960; the Page quotation is from Leonard Mosley, *Power Play* (Baltimore, 1974), 278, 283-284. ·

33. Letter, Fisher to Wall, November 29, 1983, Files, C.B.H.S.; Ian Seymour, *OPEC/Instrument of Change* (New York, 1981), 1-38; Mana Saeed al-Otaiba, *OPEC and the Petroleum Industry* (New York, 1975), 57-60.

34. E.C. Minutes, September 9, 15, November 9, 1960; April 25, May 16, 22, 26, June 29, September 5, 12, October 20, 1961; Secretary's Office, Exxon; Shwadran, *The Middle East, Oil and the Great Powers*, 267-273; *Interview*, Wall and Carpenter with Fisher, March 17, 1976, Files, C.B.H.S.; *Letter*, Fisher to Wall, November 29, 1983, *ibid.*

35. "Dead rents" are the rents that companies agree to make between the signing of agreements and the discovery and production of minerals; in this case, oil.

36. *Interview*, Wall and Carpenter with Fisher, March 17, 1976, Files C.B.H.S.; *Letter*, Fisher to Wall, November 29, 1983, *ibid.*; E.C. Minutes, April 25, May 16, June 29, September 5, 12, October 20, 1961, Secretary's Office, Exxon; Stocking, *Middle East Oil*, 215-249. For information on Iraq after the 1958 revolution, see *The Middle East. U.S. Policy, Israel, Oil and the Arabs* (Congressional Quarterly Inc., Washington, 1977); Seymour, *OPEC*; Lenczowski, *The Middle East in World Affairs*, 288-313; Shwadran, *The Middle East, Oil and the Great Powers*, 267-297; Humphrey Trevelyan, *The Middle East In Revolution* (Boston, 1970).

37. *Memorandum*, "Sinclair Interest in Iraq Oil Concession," May 6, 1964, *MNC*, Part 8 (Appendix to Part 7), 532-534.

38. The others were: Sinclair Oil & Refining Company, Union Oil Company of California, Standard Oil Company (Indiana), Continental Oil Company, Marathon Oil Company, Pauley Petroleum Incorporated, and Phillips Petroleum Company.

39. *Memorandum*, "Standard of Indiana's Interest in Iraq," May 19, 1964, *MNC*, Part 8 (Appendix to Part 7), p. 534; *Memorandum of Conversation*, "Interest of Union Oil Company in Iraq," May 20, 1964, *ibid.*, 534-535;

Ibid., June 10, 15, 16, and 18, 1964, *ibid.*, 535-536; *Cablegram*, to American Embassy, London, July 8, 1964, *ibid.*, 536-537.

40. *Memorandum*, For the Under Secretary, October 24, 1964, *ibid.*, 537-541.

41. *Memorandum of Conversation*, "Iraq Oil," October 26, 1964, *ibid.*, 542-543.

42. *Telegram*, Rome to Washington, May 13, 1967, *ibid.*; *Telegram*, Tokyo to Washington, November 14, 1967, *ibid.*, p. 544; *Telegram*, Washington to Tokyo (no date, obviously 1964), *ibid.*, 544-545.

43. *Memorandum*, To Presidents of Affiliates, by Henry B. Wilson, December 19, 1963, Courtesy, Morgan; *Ibid.*, January 25, 1967, *ibid.*; Stocking, *Middle East Oil*, 270-299; Shwadran, *The Middle East, Oil and the Great Powers*, 275-277.

44. *Memorandum of Conversation*, "Petroleum: Iraqi Law 97: Company and Government Protests," August 17, 1967, *MNC*, Part 8 (Appendix to Part 7), 547-548.

45. *Ibid.*; *Memorandum*, Anthony M. Soloman to Nicholas deB. Katzenbach, October 13, 1967, *ibid.*, 550-552; *Telegram*, Dean Rusk to American Embassy, Paris, October 17, 1967, *ibid.*, 552-553; *Telegram*, Paris to Washington, October (no date), 1967, *ibid.*, 553-554; *Letter*, Robert G. Barnes to James E. Akins, August 8, 1969 (including enclosure, in part deleted), *ibid.*, 554-555; Concerning the French behavior in this period, U.S. Senator Stuart Symington remarked:

> The French already, ever since I have been around, have consistently done exactly what they please when it came to NATO, to oil, to the gold standard, to getting off the gold standard and floating the franc, and so forth. They always go unilaterally.

Ibid., p. 283.

46. *Memorandum*, Soloman to Katzenbach, October 13, 1967, *ibid.*, 550-552.

47. *Telegram*, Rusk to American Embassy, Paris, October 17, 1967, *ibid.*, 552-553; "Documents and Materials Relating to Iraq and the Iraq Petroleum Company," I. A Chronology of Events Relating to Developments in Oil in Iraq, Prepared For *MNC* by Clyde R. Mark, Analyst in Middle Eastern Affairs, The Library of Congress, *ibid.*, 495-508; *The Middle East. U.S. Policy, Israel, Oil and the Arabs* (Congressional Quarterly Inc., Washington,

1977), Middle East Developments Between 1945 and 1977, 151-183; *New York Times*, March 13, 20, 1972.

48. *Memorandum*, Soloman to Katzenbach, October 13, 1967, MNC, Part 8 (Appendix to Part 7), 550-552; John F. Dulles quoted in Lenczowski, *The Middle East in World Affairs*, p. 293.

49. Stocking, *Middle East Oil*, 240-244 (quoting *Middle East Economic Survey*, August 25, 1961); Penrose, *The Large International Firm in Developing Countries*, p. 70.

CHAPTER 20

1. Information for the above came from: *The Middle East/U.S. Policy, Israel, Oil and the Arabs* (Congressional Quarterly, Inc., Washington, 1977), 31-33; "Esso In Libya" (Tripoli, Libya, 1963), Files, Center for Business History Studies [hereafter cited as Files, C.B.H.S.]; George Rodger, "Desert Search," *The Lamp*, 1957 (Vol. 39, No. 3); "Oil Strike At Zelten," *ibid.*, Fall, 1959 (Vol. 41, No. 3); *Interview*, Bennett H. Wall with M.A. Wright, August 27, 1982, Files, C.B.H.S.; *Telephone Conversation*, Wall with Zeb Mayhew, November 29, 1983, *ibid.* See also Executive Committee Minutes [hereafter cited as E.C. Minutes], February 25, March 30, 1955, Secretary's Office, Exxon.

2. *Interview*, Wall and C. Gerald Carpenter with Hugh de N. Wynne, August 30, 1977, Files, C.B.H.S.; Edith Tilton Penrose, *The Large International Firm in Developing Countries* (Cambridge, MA, 1968), 202-206; *Letter*, Smith D. Turner to Wall, June 10, 1984, Files, C.B.H.S.

3. "New Horizons in Foreign Oil," *The Lamp*, Summer, 1956 (Vol. 28, No.2).

4. E. C. Minutes, April 23, 1956, Secretary's Office, Exxon; see also *Memorandum*, C.H. Carpenter to Wright, May 7, 1956, Exxon Records Center [hereafter cited as E.R.C.], 191-13-A.

5. *Memorandum* re Libya, "Conference in Washington, D.C., on October 28, 1957," by W.R. Carlisle, *ibid.*; *Interview*, James B. Quinn with Howard W. Page, November 29, 1977, Courtesy, Mr. Quinn.

6. *Ibid.*; "Oil Strike at Zelten," *The Lamp*, Fall, 1957 (Vol. 39, No. 3); "Esso In Libya," Files, C.B.H.S.; *Interview*, Wall with Wright, August 27, 1982, *ibid.*; Standard Oil

Company (New Jersey) [hereafter cited as SONJ], *Press Release*, October 26, 1961, *ibid.*

7. "Esso In Libya," Files, C.B.H.S.

8. *Interview*, Wall and Carpenter with Wynne, August 30, 1977, *ibid.*; *Interview*, Wall with Wright, August 27, 1982, *ibid.*; *Memorandum*, Libyan Progress Report No. 2, by M.H. Clapp, February 1, 1961, E.R.C., 397-2-B; SONJ, *Press Release*, October 26, 1961, Files, C.B.H.S.; *Ibid.*, no date probably 1963 (two photo-feature stories released by Robert H. Scholl of the Public Relations Department), E.R.C., 1209-3-B.

9. *Interview*, Wall with Wright, August 27, 1982, *ibid.*

10. *Memorandum*, H.W. Brown to Wright, July 23, 1963, E.R.C., 306-2-B.

11. *Interview*, Wall with Larry Birney, June 21, 1977, Files, C.B.H.S.; *Memorandum*, Joseph F. Moore to M.L. Haider, February 2, 1961, E.R.C. 397-2-B; *Interview*, Wall and Carpenter with Siro Vázquez, May 5, 1975, Files, C.B.H.S.

12. *Memorandum*, Libyan Refinery, Progress Report No. 2, February 1, 1961, *ibid.*; for example, see E.C. Minutes, August 29, 1963, Secretary's Office, Exxon; *Interview*, Wall and Carpenter with Vázquez, May 16, 1978, Files, C.B.H.S.

13. *Interview*, Wall and Carpenter with Wynne, August 30, 1977, *ibid.*

14. *Interview*, Wall with Wright, August 27, 1982, *ibid.*

15. *The Lamp*, Spring, 1962 (Vol. 44, No.1).

16. *Cablegram*, Moore to Vázquez, August 25, 1961, E.R.C., 397-2-B.

17. *Interview*, Wall and Carpenter with Wynne, August 30, 1977, Files, C.B.H.S.; *Memorandum*, M.A. Matheson to Wright, December 8, 1961, E.R.C., 397-2-B.

18. E.C. Minutes, October 3, 1961, Secretary's Office, Exxon.

19. *Memorandum*, W.J. Barnett to Wright, *et al.*, June 12, 1962, E.R.C., 281-18-A.

20. *Memorandum*, Esso Standard Libya, Inc., "Libyan LNG Project," from Marketing Coordination, Gas Coordination Division, by D.M. Latimer, *et al.*, *ibid.*

21. *Interview*, Wall and Carpenter with Wynne, August 30, 1977, Files, C.B.H.S.; Standard Oil Company (New Jersey) and Exxon Corporation, *Annual Reports*, 1964-1974; The *Financial Times* (London), June 16, 1972; *Interview* (telephone), Wall with George H. Lewis, September 14, 1984, Files C.B.H.S.; E.C. Minutes, June 25, July 26, 1965, Secretary's Office, Exxon; SONJ, Board of Directors Minutes [hereafter cited as Bd. Dir. Mins.], July 29, 1965, *ibid.*; *Interview*, Wall with J. Kenneth Jamieson, August 27, 1982, Files, C.B.H.S.

22. *Letter*, Unknown [GAL] to D.A. Shepard, April 16, 1962, E.R.C., 281-18-A; *Letter*, Wright to R.A. Eeds, April 23,1962, *ibid.*; *Interview*, Wall and Carpenter with Wynne, August 30, 1977, Files, C.B.H.S

23. "Esso In Libya," *ibid.*; *Memorandum*, Luke W. Finlay to Wright, August 12, 1963, E.R.C., 306-2-B.

24. *Cablegram*, Esso Libya to [W.J] Barnett, July 6, 1961, *ibid.*, 397-2-B.

25. *Cablegram*,[Robert] Eeds to ? [Fully coded], August 15, 1961, *ibid.*

26. "Libyan Crude," by GHL, November 19, 1962, *ibid.*, 281-18-A; Italics theirs.

27. *Hearings Before the Subcommittee on Multinational Corporations of the Committee on Foreign Relations, United States Senate, Ninety-Third Congress, Second Session*, 17 Parts, Washington (1973-1976) [hereafter cited as *MNC*], Part 7, 282-284, 307-310.

28. George W. Stocking, *Middle East Oil* (Nashville, 1970), p. 374.

29. E.C. Minutes, October 16, November 10, December 4, 1961, Secretary's Office, Exxon.

30. *Ibid.*; "Esso In Libya," Files, C.B.H.S.

31. Stocking, *Middle East Oil*, p. 375.

32. *Interview*, Wall and Carpenter with Wynne, August 30, 1977, Files, C.B.H.S.

33. *Ibid.*; See also Penrose, *The Large International Firm in Developing Countries*, 203-206.

34. Ian Seymour, *OPEC/Instrument of Change* (New York, 1981), 48-54. The major oil companies in July, 1964, made an offer on royalty dispensing based on the formula, "8.5% allowance off posted prices for tax purposes, ... from January 1, 1964 ... reduced to 7.5% in 1965 and 6.5% in 1966 (subject to a gravity adjustment in favour of the heavier crudes). ..." This offer was made to Reza Fallah, Deputy Managing Director of National

Iranian Oil Company; *MNC*, Part 8 (Appendix to Part 7), p. 767.

35. Stocking, *Middle East Oil*, p. 377.

36. *Ibid.*, 378, quoting the *Middle East Economic Survey*, December 31, 1965.

37. Robert Engler, *The Brotherhood of Oil* (Chicago, 1977), p. 18; The *Christian Science Monitor*, October 7, 1983; *Interview*, Wall with Jamieson, May 16, 1980, Files, C.B.H.S.; *Letter*, Wynne to Vázquez, November 28, 1967, E.R.C., 546-3-B; *Interview*, Wall and Carpenter with Wynne, August 30, 1977, Files, C.B.H.S. For information on the Libyan Occidental Petroleum Corporation, see No. 67 Civil 4011, U.S. District Court, Southern District New York, August 30, 1974. *Federal Supplement*, Vol. 382, p. 1052. See also Allen & Company, Plaintiff-Appellant, vs. Occidental Petroleum Corporation, Defendant-Appellee, No. 824, Docket 74-2340. United States Circuit Court of Appeals, Second Circuit. Argued May 15, 1975; Decided June 23, 1975. *Federal Reporter* (Second Series), Volume 519, F.2d, p. 788.

38. *Ibid.*; *Interview*, Wall with Jamieson, May 16, 1980, *ibid.*; *MNC*, Part 4, 160-179.

39. *Letter*, G.E. Golden to Vázquez, October 17, 1967, with enclosure, "Esso Libya Environmental Outlook - Fall 1967 Supplement," E.R.C, 546-3-B; *Letter*, Wynne to Vázquez, November 28, 1967, *ibid.*

40. *Interview*, Wall and Carpenter with Wynne, August 30, 1977, Files, C.B.H.S.; *Memorandum*, J.N. Gorringe to M.M. Brisco, June 20, 1967 (with enclosed *Memorandum* by J.N.G. on subsidies for Libyan personnel evacuated), E.R.C., 546-3-B; *Letter*, Wynne to V. Cazzaniga, June 28, 1967, *ibid.*; *Letter*, Wynne to Vázquez, July 11, 1967, *ibid.* It should be noted that Jersey had an established policy covering such contingencies (so did Mobil Oil Company and Oasis Oil Company). However, the previous Jersey Emergency Policy had been issued on May 1, 1962, and it did not fully cover all aspects of the Libyan crisis, so Wynne on his own initiative improvised as he thought proper. Later Jersey altered its policy to provide affiliate management more flexibility in deciding what to do. Gorringe in his *Memorandum* to Brisco (cited above) indicated that Wynne's actions had been reviewed and stated, "We believe the Esso Libya action to be appropriate."

41. *Interview*, Wall and Carpenter with

Wynne, August 30, 1977, Files, C.B.H.S.; Stocking, *Middle East Oil*, p. 459.

42. *Letter*, C.O. Peyton to Wynne, May 4, 1967, E.R.C., 546-3-B.

43. *Memorandum*, L.A. Smith to Brisco, August 9, 1967, *ibid.*; See also *New York Times*, August 9, 1967.

44. "Esso Libya Environmental Outlook, Fall, 1967," E.R.C., 546-3-B, italics theirs.

45. See the chapter entitled, "The Middle East, 1957-1970," and note 42.

46. *Cablegram*, Wynne to Vázquez, October 14, 1967, E.R.C., 546-3-B.

47. *Cablegram*, Vázquez to Wynne, October 31, 1967, *ibid.*, italics theirs.

48. See the chapter entitled, "The Middle East, 1957-1970," and note 42.

49. *Letter*, Wynne to Vázquez, November 28, 1967, E.R.C., 546-3-B; E.C. Minutes, January 19, September 25, October 15, 1968, Secretary's Office, Exxon; "Esso Libya Monthly Corporate Review, June 1, 1968," E.R.C., 1208-2-B.

50. *Letter*, Kingdom of Libya, Ministry of Petroleum Affairs to Amerada Petroleum Corporation of Libya, *et al.*, September 16, 1968, *ibid.*; *Memorandum*, prepared at the request of Jamieson and Haider for a meeting with Shell executives, October 23, 1968, *ibid.*

51. *Interview*, Wall and Carpenter with Wynne, August 30, 1977, Files, C.B.H.S.

52. E.C. Minutes, January 19, September 25, October 15, 1968, Secretary's Office, Exxon; "Esso Libya Monthly Corporate Review, June 1, 1968," E.R.C., 1208-2-B.

53. *Ibid.*

54. *Interview*, Wall and Carpenter with Wynne, August 30, 1977, Files, C.B.H.S.

55. Patrick Seale and Maureen McConville, *The Hilton Assignment (London, 1973)*; *The Middle East. U.S. Policy, Israel, Oil and the Arabs*, 31-33; *Interview*, Wall and Carpenter with Wynne, August 30, 1977, Files, C.B.H.S.; *Interview*, Wall with Wright, August 27, 1982, *ibid.*; George Lenczowski, *The Middle East in World Affairs* (Ithaca and London, 1980), 759-764.

56. Albert E. Danielson, *The Evolution of OPEC* (New York, *et al.*, 1982), p. 154.

57. "Statement of George Henry Mayer

Schuler ..." [hereafter cited as "Schuler Statement"], *MNC*, Part 6 (Appendix to Part 5), p. 2.

58. Leonard Mosley, *Power Play* (Baltimore, 1973), p. 356; *MNC*, Part 4, p. 162; *Ibid.*, Part 6 (Appendix to Part 5), p. 2; Seymour, *OPEC/ Instrument of Change*, 66-68; Lenczowski, *The Middle East*, 759-764; *Re* Tariki, see for example Stocking, *Middle East Oil*, p. 317, Mosley, *Power Play*, p. 274. He left Saudi Arabia for Lebanon in 1962. Some knowledgeable scholars have tagged Tariki a brilliant speaker, but "more effective as an agitator than administrator." Yet there can be no doubt that he vigorously advocated complete Arab control of Arab resources. Certainly, these two influential men advocated extreme anti-oil company measures, including expropriation—a fact well known to company representatives who negotiated with the Libyans.

59. Seymour, *OPEC/Instrument of Change*, 66-68; *MNC*, Part 4, 160-179; *Ibid.*, Part 5, p. 3; *Ibid.*, Part 6 (Appendix to Part 5), 164-167; *Interview*, Wall and Carpenter with Wynne, August 30, 1977, Files, C.B.H.S.

60. *Ibid.*; *Interview*, Wall with Wright, August 27, 1982, *ibid.*; *The Middle East. U.S. Policy, Israel, Oil and the Arabs*, 32-33; *New York Times*, August 9, 1969; January 26, June 16, 21, July 12, 26, August 30, 1970; Mosley, *Power Play*, 374-377; "Prepared Statement of Mr. G. T. Piercy, etc." [hereafter cited as "Piercy Statement"], *MNC*, Part 5, 213-215; "Schuler Statement," *ibid.*, Part 6 (Appendix to Part 5), 1-5; Seymour, *OPEC/Instrument of Change*, 65-70.

61. *Interviews*, Wall with Jamieson, May 16, 1980; August 27, 1982, Files, C.B.H.S.; Seymour, *OPEC/Instrument of Change*, 67-74; "Piercy Statement," *MNC*, Part 5, 211-217.

62. Seymour, *OPEC/Instrument of Change*, 67-74; "Piercy Statement," *MNC*, Part 5, 211-217, p. 214; "Schuler Statement," *MNC*, Part 6 (Appendix to Part 5), p. 205; *Letter*, C.F. Lindsley to C.J. Hedlund, July 7, 1970, Secretary's Office, Exxon.

63. Barran made these statements in a *letter* to Senator Frank Church on August 16, 1974, *MNC*, Part 8 (Appendix to Part 7), 771-773.

64. *Interview*, Wall with Jamieson, May 16, 1980, Files, C.B.H.S.

65. *Interviews*, Wall and Carpenter with

Vázquez, May 5, 1975, May 16, 1978, Files, C.B.H.S.

CHAPTER 21

1. *Document*, "Determination and Report of the Special Committee on Litigation," January 23, 1976, Secretary's Office, Exxon, 12-13, (hereafter cited as *Determination and Report*).

2. *Hearings Before the Subcommittee on Multinational Corporations of the Committee on Foreign Relations, United States Senate, Ninety-Third Congress, Second Session,* 17 Parts, Washington, 1973-1976 [hereafter cited as *MNC*], "Archie L. Monroe Testimony," Part 12, 241-267 [hereafter cited as "Monroe Testimony"].

3. *Determination and Report*, p. 7.

4. *Interview*, Bennett H. Wall and C. Gerald Carpenter with Nicholas J. Campbell, Jr., August 31, 1977, Files, Center for Business History Studies [hereafter cited as C.B.H.S.]; N.B. The Securities and Exchange Commission [SEC], *Form 8-K*, Exhibit A, filed by Exxon Corporation, September 27, 1977, p. 3, alleged that the payments made were in cash, without receipts, and channeled through intermediaries such as newspapers, journals, public relations firms, consultants and schools, to "certain political parties, government officials and employees." [For the sake of brevity, this author has referred only to "newspapers and public relations firms"], Files, C.B.H.S.

5. *Determination and Report*, p. 62.

6. *Ibid.*

7. *MNC*, Part 12, p. 287.

8. *Determination and Report*, 20-22.

9. *Ibid.*

10. *Ibid.*, p. 46.

11. *Ibid.*, 20, 22-24, 47; Thompson did not remember whether he or Cazzaniga devised the "format" idea; *MNC*, Part 12, p. 288.

12. *Determination and Report*, p. 23.

13. *Ibid.*, 25-26.

14. *Ibid.*, p. 26.

15. *Ibid.*, 46, 27.

16. *Ibid.*, p. 22.

17. *MNC*, Part 12, 270, 271, 288.

18. *Determination and Report*, 26-28; SEC, *Form 8-K* , p. 2.

19. *Ibid.*, p. 3.

20. *MNC*, Part 12, p. 276.

21. *Determination and Report*, p. 21.

22. *Ibid.*, p. 56.

23. *Ibid.*

24. *Ibid.*, p. 53.

25. *Ibid.*

26. *Ibid.*, 53-54.

27. *Ibid.*, 54-56.

28. *Ibid.*

29. *Ibid.*, 13-14.

30. *Ibid.*, Attachment 7; Chief Executive Officer Clifton C. Garvin, Jr., in September, 1975, asked the Exxon Board of Directors to reaffirm this policy with only minor changes. On September 24, he reissued it with the title "Policy Statement on Business Ethics," and pointed out that "it continues to be the policy of Exxon." The Jamieson statement referred to earlier is misquoted in the Church Committee hearings. See *MNC*, Part 12, p. 263.

31. *SEC, Form 8-K* , p. 3.

32. *Ibid.*; The ban, however, does not apply to situations where it is lawful to expend corporate funds to set forth the corporation's views and position regarding propositions and referenda submitted to voters, which would affect materially the company business, or for administrative expenses on behalf of employee or shareholder political action committees.

33. *Determination and Report*, p. 63.

34. *Ibid.*, p. 67.

35. *Ibid.*

36. *Ibid.*, p. 68.

37. *Journal of Commerce* (New York), February 12, 22, 1974.

38. *Determination and Report*, p. 76.

39. *Ibid.*

40. Exxon Corporation, *"Report of the 93rd Annual Meeting of Shareholders, May 15, 1975"* (New York, 1975).

41. *Determination and Report*, p. 76.

42. *MNC*, Part 12, p. 266. For the "Monroe Testimony," see *ibid.*, 241-267.

43. Shareholder derivative suits were filed by Abraham Wechsler, Joan Levin Gall, and Kathleen E. Shaw. The demand threatening a suit came from Stephen W. Lane. *Determination and Report*, p. 1.

44. *Ibid.*, p. 2.

45. *Ibid.*, p. 4.

46. Exxon Corporation, *Press Release*, January 30, 1976, Files, C.B.H.S.; *Determination and Report*, Attachment 1.

47. *Ibid.*, Attachments 2 & 4.

48. *Ibid.*, 79-81.

49. *Ibid.*, 78-81; Exxon Corporation, *Press Release*, January 30, 1976, Files, C.B.H.S.

50. *SEC, Form 8-K* , p. 2.

51. *Ibid.*, 4-7.

52. *Ibid.*, 1-7; "Complaint for Permanent Injunction and Certain Ancillary Relief," *Securities and Exchange Commission vs. Exxon Corporation and Vincenzo Cazzaniga*, United States District Court for the District of Columbia, September 27, 1977, Civil Action No. 77-1681, Files, C.B.H.S.; Later in 1977, the Federal Court, on the SEC's motion, entered a default judgment against Cazzaniga, who had informed the SEC that he did not intend to defend himself; "Default Judgment," *Securities and Exchange Commission vs. Exxon Corporation and Vincenzo Cazzaniga*, December 1, 1977, Files, C.B.H.S.

53. *SEC, Form 8-K* , p. 1; See also Exhibit B.

54. *Memorandum*, ENI, Cazzaniga to M.W. Johnson, April 13, 1967, Exxon Records Center, 25-13-A.

55. *Pamphlet*, "Esso in Italy, 1974/1977: A Public Affairs Analysis," 3-4.

56. *Ibid.*

57. *Ibid.*

58. *Ibid.*, 9-10.

CHAPTER 22

1. The quotation in paragraph one, page one, is from an *interview*, Bennett H. Wall with Emile E. Soubry, October 14, 1977, Files, Center for Business History Studies [hereafter cited as C.B.H.S.]. Soubry's opinion runs counter to that of many Jersey men. Most of the former executives remembered the company as one that specialized in marketing. Some said that the influence of Charles Pratt

and his son, Charles M. Pratt, early Board members and great marketers, persisted well into the 1930s. Concerning singling out Jersey, the company customarily received top billing in all court cases. The newspapers then (usually) referred to court proceedings as involving the Standard Oil Company (New Jersey) and "other major oil companies." See the Department of Justice files on the oil cartel case where in all correspondence with lawyers representing the involved companies, the reference was to *United States vs. Standard Oil Company (N.J.), et al.*, Civil 86-27, S.D.N.Y. [Justice Department Files Numbers 60-012-20 and 60-57-140, Courtesy, Burton Kaufman].

2. *New York Times*, May 10, 1972.

3. J. K. Jamieson, *letter*, "To Shareholders of Standard Oil Company (New Jersey)," September 11, 1972.

4. *Interview*, Wall with Frank X. Clair, May 18, 1977, Files, C.B.H.S. This account of the historical reasons for the name change relies heavily on Paul H. Giddens, "Historical Origins of the Adoption of the Exxon Name and Trademark," *Business History Review* (Autumn, 1973), 353-66. For details on the background of the dissolution itself, see Ralph W. Hidy and Muriel E. Hidy, *Pioneering in Big Business, 1882-1911: History of the Standard Oil Company (New Jersey)*, (New York, 1955), 671-718.

5. George Sweet Gibb and Evelyn H. Knowlton, *The Resurgent Years, 1911-1927: History of the Standard Oil Company (New Jersey)*, (New York, 1956), p. 493.

6. *Ibid.*, 494-495.

7. Giddens, "Historical Origins. . . ," *Business History Review*, Vol. XLVII, No. 3 (Autumn, 1973), p. 360. Study of the Jersey Executive Committee Minutes [hereafter cited as E.C. Minutes] indicates constant alertness on the part of Indiana against invasion of its territory by Jersey. For example, see E.C. Minutes, February 3, 1950, Secretary's Office, Exxon.

8. Giddens, "Historical Origins. . . ," *Business History Review*, Vol. XLVII, No. 3 (Autumn, 1973), 355-57.

9. *Business Week*, September 8, 1945, p. 5; E.C. Minutes, September 14, October 21, 1949, Secretary's Office, Exxon.

10. *Ibid.*, August 9, 1949.

11. *Ibid.*, September 22, 1949, June 20, 1950.

12. *Ibid.*, September 9, 1960; *Letter*, Paul E. Morgan Files, C.B.H.S.; *Interview*, Wall with Clair, September 20, 1978, Files, C.B.H.S.; See also *letter*, Richard W. Kimball to Morgan, April 15, 1981, Morgan Files, *ibid.*

13. *Memorandum* to W. A. M. Greeven from Lippincott and Margulies, Inc., "Effective Implementation of Esso's New Identity System," November 17, 1965, Exxon Records Center [hereafter cited as E.R.C.], 164-2-A.

14. Giddens, "Historical Origins. . . ," *Business History Review*, Vol. XLVII, No. 3 (Autumn, 1973), p. 365; *Business Week*, July 20, 1963, p. 64.

15. *Interview*, Wall with Clair, May 18, 1977, Files, C.B.H.S.; Kimball and Morgan, "Report of Task Force," January 16, 1967, Morgan Files, C.B.H.S.; Board of Directors Minutes [hereafter cited as Bd. Dir. Mins.], April 6, 1950, Secretary's Office, Exxon; *Humble vs. American Oil Company, et al.*, Amended complaint, quoted in John R. White to W. O. Twaits, May 24, 1965, E.R.C., 82-10-A.

16. *Business Week* (July 20, 1963), p. 64; *Interview*, Wall with Clair, May 18, 1977, Files, C.B.H.S.

17. Giddens, "Historical Origins . . . ," *Business History Review*, Vol. XLVII, No. 3 (Autumn, 1973), p. 365.

18. *Letter*, Charles F. Jones to M. A. Wright, September 2, 1966, E.R.C., 164-2-A.

19. E.C. Minutes, November 15, 1966, Secretary's Office, Exxon.

20. *Ibid.*

21. *Letters*, Kimball to Morgan, July 31, 1978, April 15, 1981, Morgan Files, C.B.H.S.; E.C. Minutes, November 15, 1966, Secretary's Office, Exxon.

22. *Letters*, Kimball to Morgan, July 31, 1978, April 15, 1981, Morgan Files, C.B.H.S.; *Interview*, C. Gerald Carpenter with Henry B. Wilson, May 4, 1977, Files, C.B.H.S.

23. *Interview*, Wall and Carpenter with Greeven, May 17, 1978, *ibid.* See also "Effective Implementation. . . ," *Memorandum* to Greeven from Lippincott and Margulies, Inc., November 17, 1965, E.R.C., 164-2-A.

24. Kimball and Morgan, "Report of Task Force," January 16, 1967, Morgan Files, C.B.H.S.

25. *Letter*, Kimball to Morgan, July 31, 1978, *ibid.*

26. *Interview,* Wall and Carpenter with Wilson, May 17, 1977, Files, C.B.H.S.

27. E.C. Minutes, February 20, 1967, Secretary's Office, Exxon.

28. *Ibid.,* September 13, 1967; *Interview,* Wall and Carpenter with Greeven, May 17, 1978, Files, C.B.H.S.; John Brooks, "The Marts of Trade: It Will Grow on You," *The New Yorker* (March 10, 1973), 106, 108; *Letters,* Kimball to Morgan, July 31, 1978, April 15, 1981, Morgan Files, C.B.H.S.; *Interview,* Wall with Morgan, February 6, 1981, Files, C.B.H.S.; *Interview,* Andy Purcell with Jamieson, July 15, 1975, *Exxon Conversations; Letter,* Michael L. Haider to Morgan, December 30, 1981, Morgan Files, C.B.H.S.

29. Management Committee Minutes, May 23, July 3, 1968, February 10, 1970, Secretary's Office, Exxon.

30. *Interview,* Wall with George T. Piercy, March 19, 1981, Files, C.B.H.S.; Brooks, "The Marts of Trade . . . ," p. 108.

31. *Letter,* Kimball to Morgan, July 31, 1978, Morgan Files, C.B.H.S.

32. Management Committee Minutes, February 10, April 2, October 23, 1970, Secretary's Office, Exxon.

33. *Interview,* Wall and Carpenter with Greeven, May 17, 1978, Files, C.B.H.S.

34. *Interview,* Purcell with Jamieson, July 15, 1975, *Exxon Conversations.*

35. *Letters,* C. F. Lindsley to R. H. Milbrath, March 3, 1970, Management Committee Files, Secretary's Office, Exxon. W. M. McCardell made several suggestions clarifying his presentation to the Management Committee. He also provided a copy of the "Handout" that he distributed at the meeting of the Management Committee on October 23, 1970. *Letter,* McCardell to Morgan, August 4, 1978, Morgan Files, C.B.H.S. (Emphasis in original).

36. McCardell, "Comments at Management Committee" [sic], October 23, 1970, Management Committee Files, Secretary's Office, Exxon. (Emphasis in original.)

37. *Letter,* B. W. Massey to Milbrath, November 5, 1970, *ibid.*

38. *Ibid.*

39. *Letter,* Massey to Milbrath, November 19, 1970, *ibid.; Wall Street Journal,* January 9, 1973.

40. Bd. Dir. Mins., November 25, 1970, Secretary's Office, Exxon.

41. *Letter,* Massey to Milbrath, January 22, 1971, Management Committee Files, Secretary's Office, Exxon; *Letter,* T. H. Tonnessen to Morgan, March 12, 1981, Morgan Files, C.B.H.S.

42. Brooks, "The Marts of Trade . . . ," p. 109.

43. *Interview,* Wall with Clark R. Egeler, June 25, 1978, Files, C.B.H.S.

44. "A New Trademark," *The Lamp,* Vol. 54, No. 3 (Fall, 1972), 3-4; "Background Information—The Search for EXXON," n.d., Files, C.B.H.S.

45. Bd. Dir. Mins., August 25, 1971, Secretary's Office, Exxon. (The cities involved were Athens, Georgia; Zanesville, Ohio; Manchester, New Hampshire; Battle Creek, Michigan; Nacogdoches, Texas; and San Luis Obispo, California.)

46. "A New Trademark," 3-4; *Business Week* (May 13, 1972), 53-54.

47. D. F. Dickey, quoted in "Humble To Use EXXON As National Trademark And Company Name," The *Humble Extra* (May-June, 1972), p. 1.

48. *Letter,* Lindsley to Milbrath, May 8, 1972, Management Committee Files, Secretary's Office, Exxon.

49. Bd. Dir. Mins., April 26, 1972, *ibid.*

50. Brooks, "The Marts of Trade . . . ," p. 110.

51. *Interview,* Purcell with Jamieson, July 15, 1975, *Exxon Conversations.*

52. "Tiger on a Pole," *The Lamp,* Vol. 54, No. 4 (Winter, 1972), p. 26; *The Humble Extra* (May-June, 1972), 12-13; Brooks, "The Marts of Trade . . . ," p. 111; *Interview,* Morgan with Milo M. Brisco, March, 1976 [n.d.], Files, C.B.H.S. Another version of this meeting is that Jamieson invited the four (Rathbone, Welch, Haider and Brisco) to lunch and that after hearing him out, all of them except Rathbone approved changing the company name to Exxon. Thus convinced, Jamieson went ahead with the plans; *Letter,* Haider to Morgan, December 30, 1981, Morgan Files, *ibid.*

53. Brooks, "The Marts of Trade . . . ," p. 110-11.

54. James C. Tanner, "Name Change Brings Excedrin Headaches and Costs Approximately $100 Million," *Wall Street Journal*, January 9, 1973; Piercy, Exxon Director, testifying before the Church Committee, stated in response to the question of the cost of the name change to the company that "it is something approaching $100 million before these offsetting costs," *Hearings Before the Subcommittee on Multinational Corporations of the Committee on Foreign Relations, United States Senate, Ninety-Third Congress, Second Session*, 17 Parts, Washington, 1973-1976, Part 5, p. 205.

55. "Master List of Jersey-Exxon Questions and Answers" [June 15, 1972], 10-11, Files, C.B.H.S.

56. *Letter*, Milbrath to Morgan, March 31, 1981, Morgan Files, C.B.H.S.; *Interview*, Morgan with Brisco, March, 1976 [n.d.], Files, C.B.H.S.; *Letters*, Wallace E. Pratt to Jamieson, November 26, 1972, May 5, 1973, Courtesy, Jamieson.

57. *Interview*, Wall with Morgan, February 4, 1976, Files, C.B.H.S. See also *Letter*, Haider to Morgan, December 30, 1981, Morgan Files, *ibid.*

CHAPTER 23

General Note:

Much of this chapter is based on *Hearings Before the Subcommittee on Multinational Corporations of the Committee on Foreign Relations, United States Senate, Ninety-Fourth Congress, First Session*, 17 Parts, Washington, 1973-1976 [hereafter cited as *MNC*], Part 5 as indicated: "Testimony of James E. Akins, October 11, 1973" [hereafter cited as "Akins Testimony"], 1-28; "Testimony of John J. McCloy, January 24, 1974 (accompanied by George T. Piercy). . . ." [hereafter cited as "McCloy Testimony"], 59-75, 247-287; "Testimony of George Henry Mayer Schuler, January 28, 1974" [hereafter cited as "Schuler Testimony"], 75-133, 237-247; "Testimony of Norman Rooney, January 28, 1974," [hereafter cited as "Rooney Testimony"], 133-145; "Testimony of John N. Irwin, II, January 31, 1974" [hereafter cited as "Irwin Testimony"], 145-173; "Testimony of George Piercy, February 1, 1974" [hereafter cited as "Piercy Testimony"], 175-237.

From *MNC*, Part 6 (Appendix to Part 5): "Statement of George Henery [sic] Mayer Schuler" [hereafter cited as "Schuler State-

ment"], 1-60; "Exhibits referred to in *ibid.* [hereafter cited as "Schuler Exhibits"], 60-223; "Correspondence between John J. McCloy . . . and . . . Department of Justice" [hereafter cited as "McCloy-Justice Correspondence"], 223-271; "Analysis by Norman Rooney" [hereafter cited as "Rooney Analysis"], 271-289; "Exchange of correspondence between Senator Frank Church and John J. McCloy" [hereafter cited as "Church-McCloy Correspondence"], 289-297; "McCloy Memorandums and Notes of Conversations with State Department" [hereafter referred to as "McCloy Memoranda and Notes"], 297-311; "Statement of John M. [N.] Mitchell" [hereafter cited as "Mitchell Statement"], 311-315; "Additional Material on the NEPCO Case," 315-342.

Some of the books used for this chapter are: The Cabinet Task Force on Oil Import Control, *The Oil Import Question* (Washington, 1970); Albert L. Danielsen, *The Evolution of OPEC* (New York, 1982); James McGovern, *The Oil Game* (New York, 1981); Zuhayr Mikdashi, *The International Politics of Natural Resources* (New York, 1976); Edith T. Penrose, *The Large International Firm in Developing Countries* (Cambridge, 1968); Ian Seymour, *OPEC/Instrument of Change* (New York, 1981); Amin Saikal, *The Rise and Fall of the Shah* (Princeton, 1980); Marvin Zonis, *The Political Elite of Iran* (Princeton, 1971); Henry Kissinger, *Years of Upheaval* (Boston, 1982); Yusif A. Sayigh, *Arab Oil Policies in the 1970s* (Baltimore, 1983).

It should be noted that there are numerous typographical errors to be found in the *MNC* material. Some of the testifiers gave differing dates for events also.

X   X   X

Beginning around 1965, Jersey consistently included Libya in all references to the Middle East. Siro Vázquez once explained how that came about. First, Jersey combined Libya with the Middle East and then, later added Algeria and Egypt to that territory. He said:

They made me contact director for Libya right after I moved up here [to the Board of Directors in 1965] . . . . And there was a growing concern about the same types of problems that we are having in the Middle East today. . . . I said, 'Look, it doesn't make any sense to have two contact directors; one for the Middle East, one for

Libya and then we have to be coordinated by the Executive Committee.
...'

George T. Piercy had become Middle East
contact director in 1966, when Howard W.
Page became executive vice president of Jersey. At that time Vázquez became alternate
contact director for that area. After the 1969
revolution, Libyan problems required negotiation in the same way and for the same reasons that other Middle East problems did. So
the ongoing Vázquez suggestion led in 1969, to
Libya's being included in the Middle East by
Jersey. Piercy, like Vázquez, approved the
idea. *Interview*, Bennett H. Wall and C.
Gerald Carpenter with Vázquez, May 5,
1975, Files, Center for Business History Studies [hereafter cited as C.B.H.S.].

X X X

1. *New York Times*, January 10, 11, 17, 26;
February 22; March 22; April 13; May 7; June
12, 21; July 12, 26, 1970; *Letter*, Fred A.L.
Holloway to Page, October 22, 1970, enclosing copy, *Letter*, Holloway to Clifton C. Garvin, Jr., October 2, 1970, entitled "Esso
Europe Security of Supply Study," Exxon
Records Center [hereafter cited as E.R.C.],
163-6-A.

2. *Letter*, Sir David Barran to Church,
August 16, 1974, *MNC*, Part 8 (Appendix to
Part 7), 771-773.

3. *Ibid.*; "Schuler Testimony," *ibid.*, Part 5,
p. 80; Kissinger, *Years of Upheaval*, p. 860.

4. *Interview*, Wall with Piercy, July 30, 1982,
Files, C.B.H.S.; "McCloy Testimony," *MNC*,
Part 5, 255-258; McCloy, in 1957, headed a
United Nations mission sent to negotiate the
reopening of the Suez Canal with Colonel
Gamal Abdul Nasser. He won the confidence
of the Egyptian leader at that time. Later,
McCloy became a friend of Shah Muhammad
Riza Pahlevi of Iran and of King Hussein of
Jordan, while also establishing warm rela-
tions with the House of Saud; Alan Brinkley,
"Minister Without Portfolio," *Harper's*, Feb-
ruary, 1983, 31-46.

5. "McCloy Testimony," *MNC*, Part 5, p.
257; the list of firms Milbank, Tweed, Hadley
and McCloy represented may be found in
MNC, Part 6 (Appendix to Part 5), p. 289.

6. "McCloy Testimony," *ibid.*, Part 5, 257-
262.

7. *Ibid.*, 247-287, 255-256; "Mitchell Statement," *ibid.*, Part 6 (Appendix to Part 5), p.
311; *Letter*, McCloy to Jerome I. Levinson
(Church Committee Counsel), July 12, 1974,
ibid., Part 8 (Appendix to Part 7), 767-769;
"Akins Testimony," *ibid.*, Part 5, 1-28. For
another Arabist point-of-view within the
State Department, see J. Rives Childs, *Let the
Credit Go: An Autobiography* (New York,
1983).

8. "Schuler Testimony," *MNC*, Part 5, p. 127;
Kissinger, *Years of Upheaval*, 196, 859;
Stephen R. Graubard, *Kissinger, Portrait of a
Mind* (New York, 1973); Marvin Kalb and
Bernard Kalb, *Kissinger* (Boston, 1974); John
G. Stoessinger, *Henry Kissinger: The Anguish
of Power* (New York, 1976); Roger Morris,
Henry Kissinger & American Foreign Policy
(New York, 1977).

9. Seymour, *OPEC/Instrument of Change*,
74-77; "Schuler Statement," *MNC*, Part 6
(Appendix to Part 5), 1-60, see especially p. 7;
"Akins Testimony," *ibid.*, Part 5, 1-28.

10. "McCloy Testimony," *ibid.*, 247-287;
Letter, Barran to Church, August 16, 1974,
ibid., Part 8 (Appendix to Part 7), 771-777;
"Schuler Statement," *ibid.*, Part 6 (Appendix
to Part 5), 7-8.

11. *Interview*, Wall with Page, March 2,
1985, Files, C.B.H.S.; "McCloy Memoranda
and Notes," *MNC*, Part 6 (Appendix to Part
5), 297-298; it should be noted that when
Mitchell stated that McCloy sought and
secured a Business Review Letter from the
Justice Department [Justice], McCloy agreed.
He answered Church by saying that, "I neither sought nor obtained a waiver by the
Department of Justice of the antitrust laws of
the United States or any exemption therefrom." Clearly McCloy resented the "misunderstanding" based upon versions of the story
appearing in the press. He went on to say that
in his judgement, the Justice Department
could not waive the application of the antitrust laws, *MNC*, Part 5, 257-262; *New York
Times*, January 10, 1971.

12. "Schuler Statement," *MNC*, Part 6
(Appendix to Part 5), 7-12. Note: Schuler
states that twenty-three companies signed the
letter but "Schuler Exhibits," *ibid.* (Exhibit
1), 60-61, lists only sixteen. In the "Piercy Testimony," *ibid.*, Part 5, p. 215, the figure given
is thirteen companies signing with eleven others adding their signatures later; "McCloy

Testimony," *ibid.*, 247-287; *Letter*, J. Kenneth Jamieson to Wall, January 23, 1984, Files, C.B.H.S. Note: Armand Hammer made the suggestion for a safety net to Barran on October 7, 1970; see *Letter*, Barran to Church, August 16, 1974, *MNC*, Part 8 (Appendix to Part 7), 771-773; *New York Times*, January 10, 1971.

13. *Interview*, Wall With Page, March 2, 1985, Files, C.B.H.S.; "McCloy Testimony," *MNC*, Part 5, 259-263; "Irwin Testimony," *ibid.*, 145-162. Note that Irwin stated (p. 155) that he believed McCloy to have suggested that Irwin be sent on the trip, while McCloy (p. 263), in answer to a direct question as to whether he suggested Irwin, gave as his answer, "no." McCloy went on to say that he wanted sent "a person of dignity, with clout. . . ."; *Letter*, McCloy to Levinson, July 12, 1974, *ibid.*, Part 8 (Appendix to Part 7), 767-769.

14. "Irwin Testimony," *ibid.*, Part 5, 145-173; "McCloy Testimony," *ibid.*, 262-265.

15. "Piercy Testimony," *ibid.*, p. 215; "Schuler Testimony," *ibid.*, 75-133, 237-247; "Schuler Statement," *ibid.*, Part 6 (Appendix to Part 5), 1-60; "Schuler Exhibits," *ibid.*, 60-223; *Interview*, Wall with Piercy, July 30, 1982, Files, C.B.H.S.

16. *Ibid.*; "Irwin Testimony," *MNC*, Part 5, p. 148; "Piercy Testimony," *ibid.*, p. 220.

17. "Schuler Statement," *ibid.*, Part 6 (Appendix to Part 5), 7-12, italics his; Seymour, *OPEC*, 295-296, Note 1, Chapter VI, citing *Events* Magazine, January 14, 1977; *Interview*, Wall with Piercy, July 30, 1982, Files, C.B.H.S.

18. "Irwin Testimony," *ibid.*, Part 5, 145-162; "McCloy Testimony," *ibid.*, 259-263; "Schuler Statement," *ibid.*, Part 6 (Appendix to Part 5), 1-60; "Schuler Exhibits," *ibid.*, 60-223.

19. *Ibid.; Interview*, Wall with Piercy, July 30, 1982, Files, C.B.H.S.

20. "Schuler Testimony," *MNC*, Part 5, 75-133, 237-247; "Piercy Testimony," *ibid.*, 175-237; "Schuler Statement," *ibid.*, Part 6 (Appendix to Part 5), 1-60; "Schuler Exhibits," *ibid.*, 60-223; *Interview*, Wall with Piercy, July 30, 1982, Files C.B.H.S.

21. "Schuler Testimony," *MNC*, Part 5, 75-133, 237-247; "Piercy Testimony," *ibid.*, 175-237; *Interview*, Wall with Piercy, July 30, 1982, Files, C.B.H.S.; "Schuler Statement,"

MNC, Part 6 (Appendix to Part 5), 1-60; "Schuler Exhibits," *ibid.*, 60-223.

22. "McCloy Testimony," *ibid.*, Part 5, 264-268; "Schuler Exhibits," *ibid.*, Part 6 (Appendix to Part 5), Exhibit 7, 69-70.

23. "Piercy Testimony," *ibid.*, Part 5, p. 221; "McCloy Testimony," *ibid.*, 266-267.

24. "Schuler Testimony," *ibid.*, p. 122.

25. *Interview*, Wall with Jamieson, August 27, 1982, Files, C.B.H.S.

26. *Letter*, McCloy to Church, March 25, 1974, *MNC*, Part 6 (Appendix to Part 5), 290-293.

27. "Irwin Testimony," *ibid.*, Part 5, p. 168.

28. *Letter*, McCloy to Richard W. McLaren, July 23, 1971, *ibid.*, Part 6 (Appendix to Part 5), 231-245; "Schuler Exhibits," *ibid.* (Exhibit 8), 71-72.

29. *Ibid.* (Exhibits 9-10), 72-73.

30. *Ibid.* (Exhibit 12), p. 74; "Akins Testimony," *ibid.*, Part 5, 1-28.

31. "Schuler Exhibits," *ibid.*, Part 6 (Appendix to Part 5), (Exhibit 2), p. 61; "Schuler Statement," *ibid.*, 13-14.

32. "Schuler Exhibits," *ibid.* (Exhibits 14-16), 75-78.

33. *Ibid.* (Exhibit 14), 75-76; *Interview*, Wall with Page, March 2, 1985, Files, C.B.H.S.; *Memorandum on Negotiating* by Page, *ibid.*

34. *Ibid.; Interview*, Wall with Page, March 2, 1985, *ibid.*; "Schuler Statement," *MNC*, Part 6 (Appendix to Part 5), p. 17; "Schuler Exhibits," *ibid.* (Exhibit 23), 82-87.

35. *Ibid.* (Exhibit 22), 80-82.

36. *Interview*, Wall with Page, March 2, 1985, Files, C.B.H.S.; *Letter*, McCloy to McLaren, July 23, 1971, *MNC*, Part 6 (Appendix to Part 5), 231-245. The Libyan backup group of thirty-seven men also included Clarke (who had been listed on the Tehran team). Jamieson, the Jersey CEO, assigned G.F. Cox, G. Demakis, Hugh H. Goerner, W.D. Kruger, and Frank Mefferd to assist in the discussion with the Libyans. Jersey's affiliate Esso Libya was the largest of the major international companies operating in Libya at this time, and this fact as well as the known competence of the individuals whom Jamieson sent may have explained the large

number of Jersey men on the list.

37. "Schuler Exhibits," *ibid.* (Exhibit 25), 88-89.

38. "Schuler Statement," *ibid.*, 7-19; "Schuler Exhibits," *ibid.* (Exhibit 24), p. 87.

39. *Ibid.* (Exhibit 27-29), 89-93.

40. *Letter*, McCloy to McLaren, July 23, 1971, *ibid.*, 231-245.

41. "Schuler Testimony," *ibid.*, Part 5, 75-133, 237-247; "Rooney Testimony," *ibid.*, 133-145; "Schuler Statement," *ibid.*, Part 6 (Appendix to Part 5), 15-23; "Schuler Exhibits," *ibid.* (Exhibits 37-45), 96-103; *Letter*, McCloy to McLaren, July 23, 1971, *ibid.*, 231-245.

42. "Schuler Exhibits," *ibid.* (Exhibits 41-44), 99-103; *Letter*, McCloy to McLaren, July 23, 1971, *ibid.*, 231-245.

43. "Schuler Exhibits," *ibid.* (Exhibits 40-48), 98-104.

44. *Ibid.* (Exhibits 40-53), 98-108; "Schuler Statement," *ibid.*, 23-29.

45. *Ibid.*, 23-25; "Schuler Exhibits," *ibid.* (Exhibits 47-62), 104-112.

46. *Ibid.* (Exhibits 52-63), 107-112.

47. *Ibid.* (Exhibits 62-63), 112-113; *New York Times*, February 7, 1971.

48. "Schuler Exhibits," *MNC*, Part 6 (Appendix to Part 5), (Exhibit 64), 113-117.

49. *Ibid.* (Exhibit 65, quoting Katan International, February 4, 1971), 117-123; "Schuler Statement," *ibid.*, 25-29. At a press conference after the speeches, the Shah revised his figures. According to this new version, the oil companies got $13 [in products] from each barrel of oil after paying Iran taxes. The actual figure, according to some estimates, was nearer $9.

50. "Schuler Exhibits," *MNC*, Part 6 (Appendix to Part 5), (Exhibit 66), 125-126.

51. *Ibid.* (Exhibits 67-73), 124-150; "Schuler Statement," *ibid.*, 28-33; *Letter*, McCloy to McLaren, July 23, 1971, *ibid.*, 231-245.

52. "Schuler Exhibits," *ibid.* (Exhibits 74-75), 150-162.

53. *Ibid.* (Exhibits 73, 74, 76-83), 137-158, 162-174; "Schuler Statement," *ibid.*, 27-35; "Piercy Testimony," *ibid.*, Part 5, 175-237; "McCloy Testimony," *ibid.*, 247-287; "Schuler Testimony," *ibid.*, 75-133, 237-247; *Inter-*

view, Wall with Page, March 2, 1985, Files, C.B.H.S.

54. "Schuler Testimony," *MNC*, Part 5, 75-143; "Schuler Exhibits," *ibid.*, Part 6 (Appendix to Part 5), (Exhibits 82-84), 169-175; "Schuler Statement," *ibid.*, 31-35.

55. "Schuler Exhibits," *ibid.* (Exhibits 83-86), 169-176; *New York Times*, February 8, 16, 1971; *Christian Science Monitor*, February 10, 1971; *Financial Times* (London), January 18, 1971; *Washington Post*, February 17, 1971.

56. "Schuler Exhibits," *MNC*, Part 6 (Appendix to Part 5), (Exhibits 84-85), 174-175; "Schuler Statement," *ibid.*, 33-42. For some reason, Yamani and Hamadi refused to give Piercy this date. But Hamadi did give it to Cattarulla of Jersey, R. DeMontaigu of CFP, and B.A. Carlisle of Shell.

57. "Schuler Exhibits," *MNC*, Part 6 (Appendix to Part 5), (Exhibit 86), 175-176.

58. *Ibid.* (Exhibits 24, 88-89), 87, 176-184; "Schuler Statement," *ibid.*, 23-42; "McCloy Testimony," *ibid.*, Part 5, 247-287; Seymour, *OPEC*, 78-90.

59. "Schuler Exhibits," *MNC*, Part 6 (Appendix to Part 5), (Exhibits 89-92), 179-191; "Schuler Statement," *ibid.*, 35-42; "Schuler Testimony," *ibid.*, Part 5, 237-247; "McCloy Testimony," *ibid.*, 269-287.

60. "Schuler Exhibits," *ibid.*, Part 6 (Appendix to Part 5), (Exhibits 93-96), 191-204; "Schuler Statement," *ibid.*, 35-42.

61. *Ibid.*; "Schuler Exhibits," *ibid.* (Exhibits 98-100), 212-222; Seymour, *OPEC*, 90-96.

62. *Ibid.*, p. 92 states that "the result in posted price terms fell somewhere in between the position of the two sides before the March 15 ultimatum" of the Libyans.

63. *Interview*, Wall with Page, March 2, 1985, Files, C.B.H.S.; "Piercy and Schuler Testimonies," *MNC*, Part 5, 236-239; The Church Committee subpoenaed some of the documents exchanged between the London and New York Groups and the Tehran and Tripoli teams, *ibid.*, Part 4, 109-113; *Ibid.*, Part 6 (Appendix to Part 5), 60-223. These clearly reveal that the groups virtually dictated everything except the dialogue between the negotiating teams and both the Persian Gulf ministers and the Libyan ministry. They also reveal that the oil producing countries exhibited no real concern for company problems.

64. *Interview*, Wall with Page, March 2, 1985, Files, C.B.H.S.; *Ibid.*, Wall with Piercy, July 30, 1982; see also Seymour, *OPEC*, 70-109, especially p. 72; Kissinger, *Years of Upheaval*, p. 865.

65. "Piercy and Schuler Testimonies," *MNC*, Part 5, 236-239; "Piercy Testimony," *ibid.*, 175-237; *Interview*, Wall with Page, March 2, 1985, Files, C.B.H.S.; *Ibid.*, Wall with Jamieson, August 27, 1982, *ibid.*

CHAPTER 24

1. *New York Times*, January 10, June 27, 1971.

2. *Ibid.*, June 27, 1971; Mana Saeed al-Otaiba, *OPEC and the Petroleum Industry* (New York, 1975), 163-164, citing a lecture by Saudi Arabian Oil Minister Ahmed Zaki Yamani, American University of Beirut, Spring, 1969; Ian Seymour, *OPEC/Instrument of Change* (New York, 1981), Chapter X, 216-235.

3. For example, *ibid.*, p. 110, quoting Cairo's *al-Musawar*, August 4, 1972; *Middle East Economic Survey*, August 4, 1972; King Faisal said:

> I recall that such a suggestion was made by some at the Rabat (Arab Summit) Conference, but it was opposed by Gamal Abdel-Nasser on the grounds that it would affect the economies of the Arab countries and interfere with their ability to support Arab staying power; at the same time such a measure would not affect America because America does not need any of our oil or other Arab Gulf oil before 1985. Therefore my opinion is that this proposal should be ruled out, and I see no benefit in reviving its discussions at this time.

4. *Multinational Corporations and United States Foreign Policy, Hearings Before The Subcommittee on Multinational Corporations of the Committee on Foreign Relations, United States Senate, Ninety-Third Congress, Second Session*, 17 Parts (Washington, 1973-1976) [hereafter cited as *MNC*], John J. McCloy, "Memorandum to Files," July 15, 1971, Part 6 (Appendix to Part 5), p. 302;

Ibid., July 19, 1971, p. 302; *Letter*, McCloy to Richard W. McLaren, July 23, 1971, *ibid.*, 231-245; *Ibid.*, August 5, 1971, p. 245; Seymour, *OPEC/Instrument of Change*, 216-221; al-Otaiba, *OPEC and the Petroleum Industry*, 161-166.

5. *Letter*, McCloy to McLaren, October 15, 1971, *MNC*, Part 6 (Appendix to Part 5), 245-246; *Ibid.*, October 19, 1971, 246-247; Exhibit A, "Memorandum of Intent," October 18, 1971, *ibid.*, 247-248; Exhibit B, "Memorandum of Confirmation," October 18, 1971, *ibid.*, p. 248; *Letter*, McLaren to McCloy, October 22, 1971, *ibid.*, 248-249; "Notes of Meeting to McLaren and Walker B. Comegys (Deputy Assistant Attorney General)," October 28, 1971, *ibid.*, 249-250; *Letter*, McCloy to Comegys, December 16, 1971, *ibid.*, 250-253; "Further Memorandum of Confirmation," December 16, 1971, *ibid.*, 253-255; Seymour, *OPEC/Instrument of Change*, 96-97; al-Otaiba, *OPEC and the Petroleum Industry*, 153-156; *Wall Street Journal*, November 5, 1971; Standard Oil Company (New Jersey) [hereafter cited as SONJ], *Press Release*, Files, Center for Business History Studies [hereafter cited as C.B.H.S.].

6. *Letters*, McCloy to Comegys, October 28, 1971, *MNC*, Part 6 (Appendix to Part 5), 250-253; "Statement of George Henry Mayer Schuler" [hereafter cited as "Schuler Statement"], *ibid.*, 44-49; *Wall Street Journal*, November 5, 1971.

7. *Ibid.*, January 12, 24, 1972; *New York Times*, January 9, 1972; "Schuler Statement," *MNC*, Part 6 (Appendix to Part 5), p. 50; *Letter*, McCloy to Comegys, December 16, 1971, *ibid.*, 250-255.

8. *New York Times*, June 27, 1971; Leonard Mosley, *Power Play: Oil in the Middle East* (New York, 1973), 395-396; "Testimony of Howard W. Page" [hereafter cited as "Page Testimony"], *MNC*, Part 7, p. 293.

9. Mosley, *Power Play*, 395-396; *New York Times*, June 27, 1971; "Testimony of George M. Keller" [hereafter cited as "Keller Testimony"], *MNC*, Part 7, 405-439, p. 433.

10. *Ibid.*; *Wall Street Journal*, January 24, 1972; Henry Kissinger, *Years of Upheaval* (Boston, 1982), p. 866.

11. Concerning the crash development of systems to manage Arab oil, see Yusif A. Sayigh, *Arab Oil Policies in the 1970s* (Baltimore, 1983), p. 7 and throughout; Mosley, *Power*

Play, 395-396, quoting his *interview* with Yamani; J.B. Kelly, *Arabia, The Gulf and The West* (New York, 1980), p. 368.

12. "Testimony of George T. Piercy" [hereafter cited as "Piercy Testimony"], *MNC*, Part 5, p. 231; *New York Times*, October 6, 1972.

13. Seymour, *OPEC*, 221-222; Robert Engler, *The Brotherhood of Oil* (Chicago, 1977), 35-41; Anthony Sampson, *The Seven Sisters* (New York, 1975), 235-237; Robert Sherrill, *The Oil Follies of 1970-1980* (New York, 1983), 132-133; *New York Times*, February 20, 1972; "Schuler Statement," *MNC*, Part 6 (Appendix to Part 5), p. 50; *Financial Times* (London), March 3, 1972; "Memorandum to Mr. McCloy," by William E. Jackson, *MNC*, Part 6 (Appendix to Part 5), p. 305; a participations chronology may be found in *ibid.*, Part 5, 229-231. This is a basic reference.

14. "Prepared Statement of J.J. Johnston" (Secretary and Director, Aramco) [hereafter cited as "Johnston Statement"], *MNC*, Part 7, p. 220; *Wall Street Journal*, March 13, 1972; *Financial Times* (London), March 13, 1972; *New York Times*, March 13, 25, 1972; *Washington Post*, March 15, 1972.

15. SONJ, *90th Annual Meeting of Shareholders*, May 18, 1972; *Houston Business Journal*, May 22, 1972; *Boston Herald Traveler*, April 25, 1972; *Oil Daily*, April 27, 1972; *Pittsburgh Post-Gazette*, May 25, 1972; *Letter* (with enclosure), A.W. Jessup to C.O. Peyton, *et al.*, July 10, 1972, containing *Memorandum* to OECD [Organization for Economic Cooperation and Development] Advisory Group, June 20, 1972, from Robert E. Ebel, Files, C.B.H.S.; *Middle East Oil and Gas* (Exxon Background Series), New York, December, 1984.

16. *Wall Street Journal*, June 2, 5, 1972; *New York Times*, May 18, June 2, 3, 30, 1972; *Washington Post*, June 2, 1972; *New York Post*, June 26, 29, 1972.

17. *New York Times*, May 18, 31, June 1, 1972; *Wall Street Journal*, April 10, May 30, June 2, 1972; *Christian Science Monitor*, May 27, 1972; "Schuler Statement," *MNC*, Part 6 (Appendix to Part 5), 50-52.

18. *New York Times*, June 27, 1972; *Wall Street Journal*, June 27, 28, 1972; *Journal of Commerce* (New York), June 28, 1972; Kelly, *Arabia, The Gulf and The West*, p. 370.

19. *Washington Post*, July 24, 1972.

20. "Testimony of Dillard P. Spriggs, Executive Vice President, Baker-Weeks & Co. Inc." [hereafter cited as "Spriggs Testimony"], *MNC*, Vol. 4, 54-83; *New York Times*, October 6, 15, 1972; Kelly, *Arabia, The Gulf and The West*, 370-373; David Holden & Richard Johns, *The House of Saud* (London, 1981), 316-329; Kissinger, *Years of Upheaval*, p. 867.

21. "Piercy Notes on this Chapter," April 10, 1985, Files, C.B.H.S.; *New York Times*, October 6, 15, 22, December 22, 1972; *Journal of Commerce* (New York), October 17, 1972; *Wall Street Journal*, October 19, 22, 30, November 30, December 18, 22, 1972; Seymour, *OPEC*, 55-98; Sayigh, *Arab Oil Policies in the 1970s*, 52-54.

22. *Interview*, Bennett H. Wall with J. Kenneth Jamieson, May 16, 1980, Files, C.B.H.S.

23. *Financial Times* (London), June 27, 1972.

24. *Washington Post*, August 31, 1972.

25. *Wall Street Journal*, January 24, February 27, 1973; *New York Times*, February 27, 1973; *Washington Post*, March 3, 1973; *Letter* (with enclosure), Jamieson to Wall, January 23, 1983, Files, C.B.H.S.

26. *Cablegram*, Dow, Jones & Co., May 24, 1973; *New York Times*, May 25, 1973; *Wall Street Journal*, May 25, 1973.

27. Mohammed Reza Pahlevi, *Answer to History* (New York, 1980), p. 96.

28. Kelly, *Arabia, The Gulf and The West*, p. 372.

29. "Schuler Statement," *MNC*, Part 6 (Appendix to Part 5), 52-55; *Letter*, McCloy to Thomas E. Kauper, December 5, 1972, *ibid.*, 256-260; *Letter*, Kauper to McCloy, December 20, 1972, *ibid.*, p. 260; Kelly, *Arabia, The Gulf and The West*, 373-374.

30. "Schuler Statement," *MNC*, Part 6 (Appendix to Part 5), 54-57; Kelly, *Arabia, The Gulf and The West*, 384-385; *The Middle East. U.S. Policy, Israel, Oil and the Arabs* (Congressional Quarterly, Inc., Washington, 1977), 31-33, 136-138.

31. *New York Times*, June 18 August 23, 1973; *Wall Street Journal*, July 5, 6, 1973; *Washington Post*, August 22, 1973; *Letter*, McCloy to Kauper, September 21, 1973, *MNC*, Part 6 (Appendix to Part 5), 261-263; *Ibid.*, September 28, 1973, *ibid.*, 265-269; "Schuler Statement," *ibid.*, 56-59; Kelly, *Arabia, The Gulf and The West*, 384-387.

32. Kissinger, *Years of Upheaval*, p. XIV; M.A. Adelman, "Is The Oil Shortage Real? Oil Companies as OPEC Tax-Collectors," *Foreign Policy*, No. 9 (Winter, 1973-1974), 69-107. [It should be noted that Jersey had employed Adelman earlier. The Executive Committee Minutes of October 27, 1952, state that Director David A. Shepard moved and the Committee approved the retaining of Professor Morris A. Adelman as consultant to the Law Department at the request of company counsel, to prepare "a summary of, and commentary on," the Federal Trade Commission Staff Report. Adelman received $200 per day plus expenses.] Other citations referred to above are: Kelly, *Arabia, The Gulf and The West*, 379-390; Holden and Johns, *The House of Saud*, 320-337; *New York Times*, January 26, April 1, 13, 16, 17, June 10, 1973; *Wall Street Journal*, April 10, 1973; *Philadelphia Inquirer*, July 22, 1973; *Journal of Commerce* (New York), March 26, 27, April 16, May 10, 1973; *New York Post*, March 27, 1973; *Financial Times* (London), March 29, July 14, 1973; *Washington Post*, April 11, 13, 16, June 28, 1973; *Christian Science Monitor*, April 25, 1973; "Memorandum (Confidential), Meeting with Yamani, May 23, 1973," Material Pertinent to the January 28 [1974] and June 20 Hearing Days, *MNC*, Part 7, 470-567, Exhibit 7, 504-505 [hereafter cited as "Material Pertinent to the ... Hearing"].

33. *Wall Street Journal*, April 10, 1973; *Christian Science Monitor*, April 10, 1973; *New York Times*, April 14, 1973.

34. *Journal of Commerce* (New York), April 16, 1973; *New York Times*, April 14, 1973; *Chicago Today*, April 16, 1973; *Forth Worth Star-Telegram*, April 16, 1973; *Cincinnati Enquirer*, April 16, 1973; *Kansas City Star*, April 16, 1973; *Youngstown Vindicator*, April 16, 1973. It should be noted that, contrary to the simplified explanations in the newspapers, Exxon's policy toward import quotas had been complex and, at times, unclear.

35. *Washington Post*, March 10, April 20, 1973; *Financial Times* (London), March 29, 1973; *New York Times*, April 16, 17, 1973; *Christian Science Monitor*, April 20, 1973.

36. *Ibid.*; *Oklahoma World* (Tulsa), April 22, 1973; Sherrill, *The Oil Follies of 1970-1980*, 167-168.

37. *Daily Oklahoman* (Oklahoma City), April 24, 1973; *New York Times*, April 25, 1973; *Chicago Tribune*, April 24, 1973; *Bos-*

ton Globe April 25, 1973; *Herald-Examiner* (Los Angeles), April 25, 1973; *Natchez Democrat* (Mississippi), April 25, 1973; *News Tribune* (Tacoma), April 25, 1973; Jamieson, "Cooperation In Energy," Remarks by ... at the Symposium on New Order In the World Economy, ... Tokyo, Japan, April 24, 1973.

38. Kelly, *Arabia, The Gulf and The West*, 380-387.

39. *Ibid.*; *Washington Post*, March 9, 1973; *Journal of Commerce* (New York), March 26, 1973; *New York Times*, May 28, June 4, 1973; *Financial Times* (London), March 26, 1973; "Testimony of John J. McCloy" [hereafter cited as "McCloy Testimony"], *MNC*, Part 5, p. 281.

40. *The Middle East/U.S. Policy* (Congressional Quarterly), 88-89; "Material Pertinent to the ... Hearing," *MNC*, Part 7, 504-544; Kelly, *Arabia, The Gulf and The West*, 407-413.

41. "Material Pertinent to the ... Hearing," Exhibit 7 [*Memoir* (Confidential), Meetings in Geneva, May, 1973, dated May 29, 1973, by W.J. McQuinn; *Cablegram*, Johnston to A.C. DeCrane, Jr., *et al.*, May 2, 1973; *Cablegram*, Frank Jungers to DeCrane, *et al.*, May 2, 1973; *Cablegram*, Jungers to M.H. Ameen [Washington representative], DeCrane, *et al.*, May 3, 1973], *MNC*, Part 7, 504-507; *Ibid.*, Exhibit 14 [Briefing Paper, Enclosure in *Letter*, Jungers to Johnston, July 24, 1973], *ibid.*, 517-528; *Ibid.*, Exhibit 18 [*Cablegram*, Jungers to McQuinn, January 16, 1973], *ibid.*, 534-536.

42. *Ibid.*, Exhibit 7 [*Memoir* (Confidential), Meetings in Geneva, May 1973, dated May 29, 1973, by McQuinn], *ibid.*, Part 7, 504-505; *Ibid.*, Exhibit 8 [*Letter*, Johnston to DeCrane, *et al.*, June 1, 1973], *ibid.*, p. 509.

43. In the speech, Page expressed amazement that the U.S. government had permitted OPEC to force U.S. businesses to sign such agreements as they had in Tehran when oil was so vital to national security. *Washington Post*, June 28, 1973; Kissinger, *Years of Upheaval*, p. 871; Sampson, *The Seven Sisters*, 167-168; James Akins, "The Oil Crisis: This Time The Wolf is Here," *Foreign Affairs*, Vol. 51, No. 3 (April, 1973), 462-490.

44. "Material Pertinent to the ... Hearing," Exhibit 9 [*Cablegram*, McQuinn to Keller, June 26, 1973], *MNC*, Part 7, p. 510; *Ibid.*, Exhibit 10 [*Letter*. O.N. Miller (Board Chairman, Socal) to Stockholders, July 26, 1973;

Letter, McQuinn to Johnston, August 7, 1973; *Cablegram* (Confidential), R.W. Powers to Johnston, August 13, 1973; *Ibid.*, August 18, 1973; *Memorandum* with enclosure (Confidential), Johnston to DeCrane, *et al.*, August 28, 1973], *ibid.*, 510-513.

45. "Testimony of William P. Tavoulareas" [hereafter cited as "Tavoulareas Testimony"], *ibid.*, Part 9, p. 107.

46. "Material Pertinent to the … Hearing," Exhibit 19 [*Cablegram* with confidential enclosure, Ellas (Majed Elass) to Johnston, dated August 28, 1974], *ibid.*, Part 7, 541-542.

47. Kelly, *Arabia, The Gulf and The West*, 390-400; Holden and Johns, *The House of Saud*, 328-337; *Washington Post*, July 6, 1973; *Christian Science Monitor*, July 6, 10, 14, 1973.

48. "Material Pertinent to the … Hearing," Exhibit 19 [*Cablegram* with confidential enclosure, dated August 28, 1973] *MNC*, Part 7, 541-542.

49. *Ibid.*, Exhibit 15 [Notes of Meeting of Aramco CEOs on September 12, 1973], *ibid.*, 528-529; Kelly, *Arabia, The Gulf and The West*, p. 392.

50. *Wall Street Journal*, July 26, 1973; *New York Times*, July 21, 26, 1973; Kelly, *Arabia, The Gulf and The West*, p. 393; Robert Lacey, *The Kingdom* (New York, 1981), 396-404; Holden and Johns, *The House of Saud*, p. 336; Seymour, *OPEC*, p. 108 quoting the *Middle East Economic Survey* (supplement), September 7, 1973.

51. *New York Times*, September 30, 1973; *Financial Times* (London), October 1, 1973.

52. Kissinger, *Years of Upheaval*, 459-599; Kelly, *Arabia, The Gulf and The West*, 392-402; Holden and Johns, *The House of Saud*, 330-336; Lacey, *The Kingdom*, 388-422; Seymour, *OPEC*, 105-125.

53. Kissinger, *Years of Upheaval*, 459, 467; Kelly, *Arabia, The Gulf and The West*, 388-405; "Keller Testimony," *MNC*, Part 7, 405-439; Lawrence L. Whetten, *The Canal War: Four Power Conflict in the Middle East* (Cambridge, MA, 1974) p. 284; Malcolm H. Kerr (ed.), *The Elusive Peace in the Middle East* (Albany, 1975), 249-310; Moshe Davis (ed.), *The Yom Kippur War* (New York, 1974), 81-93.

54. "Keller Testimony," *MNC*, Part 7, 311-340; "Piercy Testimony," *ibid.*, Part 5, 175-

237, 216-217; *Washington Post*, October 8, 1973; *Journal of Commerce* (New York), October 9, 13, 16, 1973; *New York Times*, October 9, 13, 16, 17, 1973; *Financial Times* (London), October 17, 18, 1973; *Wall Street Journal*, October 16, 17, 18, 1973; *New York Post*, October 16, 1973; *Washington Post*, October 11, 1973. It is interesting to note that the versions of the October 8 meeting in Sampson, *The Seven Sisters*, 14-15, 249-251 (published in 1975); Kelly, *Arabia, The Gulf and The West*, 394-396 (published in 1980); and Lacey, *The Kingdom*, 404-406 (published in 1981) are quite similar.

55. A year later, former Socal Board Chairman and Chief Executive Officer Miller did not recall very much about the letter. See "Testimony of Otto N. Miller" [hereafter cited as "Miller Testimony"], *MNC*, Part 7, 446-457; "Piercy Testimony," *ibid.*, 325-326; Holden and Johns, *The House of Saud*, 266-267; "Material Pertinent to the … Hearing," Exhibit 25 [*Letter*, Richard E. Keresey to Senator Frank Church, July 1, 1974; *Letter*, McCloy to General Alexander M. Haig, October 12, 1974 [1973]; *Memorandum to the President*, (signed by) Jamieson, Rawleigh Warner, Jr., M.F. Granville, Miller; *Letter*, Haig to McCloy, October 15, 1973], *MNC*, Part 7, 546-547; *Letter*, McCloy to Jerome I. Levinson (Church Committee counsel), July 12, 1974, *ibid.*, Part 8 (Appendix to Part 7), 767-769.

56. Kissinger, *Years of Upheaval*, 450-544.

57. Sources disagree on the date of the Omar Saqqaf-Kissinger and Nixon meetings. Kissinger, *Years of Upheaval*, p. 534, says Wednesday, which was October 17; *Ibid.*, 534-536; *The Middle East/U.S. Policy* (Congressional Quarterly), 171-172; *New York Times*, October 17, 1973; *Wall Street Journal*, October 17, 18, 1973.

58. "Piercy Testimony," *MNC*, Part 5, 175-237, p. 236; "Testimony of George Henry Mayer Schuler" [hereafter cited as "Schuler Testimony"], *ibid.*, 237-247.

59. *Interview*, Wall and C. Gerald Carpenter with Jamieson, November 30, 1977, Files, C.B.H.S.; *The Middle East/U.S. Policy* (Congressional Quarterly), 171-172; Holden and Johns, *The House of Saud*, 334-342; Kelly, *Arabia, The Gulf and The West*, 396-399; Holden and Johns, *The House of Saud*, 337-342; Lacey, *The Kingdom*, 406-412.

60. "Material Pertinent to the ... Hearing," Exhibit 17 [*Cablegram* to McQuinn, December 19, 1973], *MNC*, Part 7, 532-533; *Ibid.*, Exhibit 12 [*Cablegram*, Jungers to DeCrane, *et al.*, October 21, 1973], *ibid.*, 516-517; *The Middle East/U.S. Policy* (Congressional Quarterly), 89-92, 171-172; Kissinger, *Years of Upheaval*, 450-545.

61. "Piercy Testimony," *MNC*, Part 7, p. 327.

62. Exxon Corporation, *Report of the 92nd Annual Meeting of Shareholders, May 16, 1974. Los Angeles, California.*

63. Kelly, *Arabia, The Gulf and The West*, 405-415; Seymour, *OPEC*, 98-126; "Testimony of Howard W. Page; ... [and] George T. Piercy...." [hereafter cited as "Page and Piercy Testimony"], *MNC*, Part 7, 280-345.

64. *Ibid.*, 314-316.

65. *Wall Street Journal*, October 17, 1973; *Current Biography*, April, 1973, p. 24.

66. Exxon Company, U.S.A., *advertisement*, Files, C.B.H.S.

67. *Wall Street Journal*, October 25, November 1, 9, 28, 1973; *New York Times*, November 1, 9, December 2, 1973; *Journal of Commerce* (New York), November 6, 1973.

68. *New York Post*, October 16, 1973; *New York Times*, October 20, 1973; *New York News*, October 20, 1973; *Washington Post*, October 20, 1973; *Wall Street Journal*, October 22, 1973.

69. L. Scott Miller, "Recent Oil Price Developments, Chronology With Comments," enclosure in *Memorandum*, Donald L. Snook to S. Stamas, *et al.*, February 18, 1974 (Public Affairs Department), [hereafter cited as "Recent Oil Price Developments"], Files, C.B.H.S.; *The Middle East/U.S. Policy* (Congressional Quarterly), 137-140; Kelly, *Arabia, The Gulf and The West*, 407-411; Seymour, *OPEC*, 121-125.

70. *Ibid.*, 120-125; Kelly, *Arabia, The Gulf and The West*, 412-417.

71. *Wall Street Journal*, November 14, 1973; *Washington Star-News*, November 14, 1973.

72. *Washington Post*, November 17, 1973; *New York Times*, November 23, 1973 (quoting the *Middle East Economic Survey*, November 20, 1973); Kelly, *Arabia, The Gulf and The West*, 415-417; Pahlevi, *Answer to History*, p. 97.

73. *Washington Post*, November 23, 1973; Holden and Johns, *The House of Saud*, 349-353; Seymour, *OPEC*, 120-125; Pahlevi, *Answer to History*, p. 97; Kissinger, *Years of Upheaval*, 885-891; Kelly, *Arabia, The Gulf and The West*, 414-418.

74. Holden and Johns, *The House of Saud*, p. 351, quoting the *Middle East Economic Survey*, December 28, 1973; Kelly, *Arabia, The Gulf and The West*, 413-417.

75. Kissinger, *Years of Upheaval*, 885-891; see J.E. Hartshorn, "Two Prices Compared: OPEC Pricing in 1973-1975 and 1978-1980," in Mallakh, Ragaei El, *OPEC/Twenty Years and Beyond* (Boulder, Colorado and London, England, 1982), 17-33.

76. Miller, "Recent Oil Price Developments," Files, C.B.H.S.; "Piercy Testimony," *MNC*, Part 5, 175-236, p. 236.

77. Miller, "Recent Oil Price Developments," Files, C.B.H.S.; *New York Times*, January 2, 6, 1974; *Wall Street Journal*, January 7, 1974.

78. *Washington Star News*, January 12, 1974; *Jersey City Journal*, January 14, 1974; *Houston Post*, January 12, 1974; *Wall Street Journal*, January 14, 21, 1974; *Interview*, Wall with Henry B. Wilson, May 17, 1977, Files, C.B.H.S.; *Address*, Burns W. Roper to Atomic Industrial Forum, February 9, 1976, Courtesy Mr. Roper, *ibid.*; *Press Conference* (Audio Tapes), January 11 (3 sessions), January 23, March 29, November 14, 1974, *ibid.*

79. *Ibid.*; "Earnings Press Conference," Jamieson and Garvin, January 23, 1974 (copy), Files, C.B.H.S.; "Excerpts from J.K. Jamieson's Remarks at a Press Conference, January 23, 1974," *ibid.*; "The Oil Shortage: Real or Contrived," by Jamieson, before Economic Club of Detroit, Michigan, January 28, 1974, *ibid.*; Exxon, *Annual Report*, 1972; *New York Times*, January 24, 1974; *Financial Times* (London), January 26, 1974.

80. *Press Conference* (Audio Tapes), January 11 (3 sessions), January 23, March 29, November 14, 1974, Files, C.B.H.S.; *Oklahoma City Times*, January 22, 1973; *New York Post*, January 23, 1974; *New York Times*, January 22, 24, 25, 1974; *Wall Street Journal*, January 23, 24, 25, 1974; *New York News*, January 24, 1974; *Washington Post*, January 24, 1974; *Times-Picayune* (New Orleans), January 24, 1974; "Summary of Testimony of Roy A. Baze, Senior Vice President of Exxon Company, U.S.A. before the Committee on Government

Operations, Senate Permanent Subcommittee on Investigations, January 21, 1974," Files, C.B.H.S.

81. Kelly, *Arabia, The Gulf and The West*, 416-419; Seymour, *OPEC*, 126-133, 216-230; Miller, "Recent Oil Price Developments," Files, C.B.H.S.; *Wall Street Journal*, February 7, 1974; *Christian Science Monitor*, February 8, 1974; *New York Times* (Magazine), March 24, 1974.

82. "The Oil Shortage: Real or Contrived," by Jamieson, January 28, 1974, Files, C.B.H.S.; *New York Times*, January 29, February 1, 1974; *Wall Street Journal*, January 29, 1974; *Journal of Commerce* (New York), January 29, 1974. Jamieson's remarks received wide coverage and many editorials, e.g., *State Journal* (Springfield, Illinois), January 29, 1974. A "canned" pro-industry editorial appeared in such widely separated newspapers as the *New York Herald* (Red Creek, NY) on February 28, 1974; the *News-Topic* of Lenoir, N.C. on February 19, 1974; the Americus (Georgia) *Times-Record* of February 9, 1974; the *Industrial News-Review* of Hillsboro, Oregon, February 14, 1974; the *Cedar Creek Lake Beacon* of Kemp, Texas on April 4, 1974; and *Shreveport Journal* (Louisiana), February 1, 1974; *Interview*, Wall with Jamieson, November 30, 1977, Files, C.B.H.S.; "Depth Interviews about Jersey Standard," Roper Associates, September, 1975, Courtesy, Mr. Burns Roper, *ibid.*; The *Reporter-Dispatch* (White Plains, NY), February 18, 1974.

83. "Statement of Charles O. Peyton, President Exxon International Company (a division of Exxon Corporation) Before The Permanent Subcommittee on Investigations of the Senate Committee on Government Operations, April 22, 1974," Files, C.B.H.S.

84. Exxon Corporation, *Report of the 92nd Annual Meeting of Shareholders, May 16, 1974. Los Angeles, California*; Douglas Martin, "The Singular Power of A Giant Called Exxon," *New York Times*, May 9, 1982; *Dun's Review*, December, 1973.

85. Seymour, *OPEC*, 126-147; "Statement Submitted By The Standard Oil Company of California in Response to the Hearings of the Subcommittee on Multinational Corporations," *MNC*, Part 8 (Appendix to Part 7), 653-701; Holden and Johns, *The House of Saud*, 352-362.

86. *New York Times*, December 27, 1973; *New York Post*, January 23, 1974; *Shreveport Journal* (Louisiana), February 1, 1974; *Washington Post*, February 3, 1974; *Christian Science Monitor*, February 12, 1974; *New York Times*, February 13, March 30, 1974; *Wall Street Journal*, February 15, March 5, 1974; *Journal of Commerce* (New York), February 19, 1974; *Press Conference*, George T. Piercy, March 29, 1974 (New York), Files, C.B.H.S.

87. *New York Post*, April 26, 27, 1974; *New York News*, April 27, 1974; *New York Times*, April 17, 28, May 17, 18, 1974; *Wall Street Journal*, April 17, 1974; *Current Biography*, June, 1974, 22-25.

88. *Washington Post*, May 26, 1974; *New York Times*, June 11, 12, 1974; Seymour, *OPEC*, 132-135.

89. *Journal of Commerce* (New York), June 18, 1974; *Wall Street Journal*, July 1, 1974; Seymour, *OPEC*, 132-138, 141.

90. *Ibid.*; Kelly, *Arabia, The Gulf and The West*, 446-450; *Letter*, Frank S. Pramuk to Paul E. Morgan, June 13, 1985, Files, C.B.H.S.; however, Seymour, *OPEC*, 131-141, gives a different account of this event.

91. *Transcript*, Jamieson's Appearance, "Washington Straight Talk," January 27, 1975, Files, C.B.H.S.

92. Jamieson, "The Oil Industry," *Vital Speeches of The Day*, Vol. XXXXI, No. 9 (February 15, 1975), 278-280, Files, C.B.H.S.; Paul E. Morgan (unpublished manuscript), "The Man From Medicine Hat," April 23, 1975, *ibid.*; *Exxon Manhattan*, July 25, 1975, *ibid.*

Acknowledgements

O VER THE DECADE AND MORE of research and writing represented by this volume, the principal author and his associates have accumulated innumerable debts, most of them to individuals who assisted in this work. Listed below are the names of current or former employees of Exxon or one of its affiliates. All quotations attributed to these people carry dates of their interviews, or of their correspondence.

Robert E. Anderson
Frank W. Abrams
Hines H. Baker
Rex G. Baker
Federico Baptista
Richard W. Barham
L.T. Barrow
Stewart R. Bengston
Joe L. Biffle
Larry Birney

Grosvenor Blair
Marion W. Boyer
George A. Breffeilh
M.M. Brisco
Robert F. Brocksmidt
Courtney Brown
Cecil L. Burrill
Woodfin L. Butte
Helen Campbell
Nicholas J. Campbell, Jr.

Peter Campfield
Cecil H. Cass
Elliot R. Cattarulla

Laverne Chandler
Roger Chandler
W.G. Charlton
Frank X. Clair
Jack G. Clarke
William B. Cleveland
Edward H. Clevely

Emilio G. Collado
W.L. Copithorne
Donald M. Cox
Jim Crayhon
George Culp
Karl "Dit" Dallmus
Ed Davis
Morgan Davis
Stanley Davis
Reinaldo Demori

Harry O. Diamond
Duval Dickey
R.K. Dix
Richard J. Dolan
Robert A. Eeds
C.R. Egeler

Eduardo Elejalde
L.W. Elliott
J.L. Ernst
R.E. Faggioli

J. Faris
Harold W. Fisher
William H. Franklin
Ruth Frey
George H. Freyermuth
Jarvis M. Freymann
J.C. Gamble
Nicanor Garcia
Clifton C. Garvin, Jr.
Sue Geriak
David G. Gill

Richard J. Gonzalez
A. Donald Green
Wolf Greeven
Merrill Haas
M.L. Haider
George Hall
William P. Headden
Ed Holmer
Frank Hooke
J. Kenneth Jamieson

Hugh Jencks
Edward F. Johnson
Princetta Johnson
David J. Jones
Nelson Jones
Hugh Jordan
Dudley F. Judd
Howard C. Kauffmann
Richard W. Kimball
Joseph Kosewicz

John O. Larson
Henry Lartigue
George A. Lawrence

Rodolfo Ledgard
George H. Lewis
John L. Loftis, Jr.
David E. Lombardi
F.A. Long
Muguerza Lopez
Alexander Lorenz

Leo Lowry
Carl Maas
Albert M. Maney
Zeb Mayhew
Robert Mays
Donald S. MacNaughton
W.M. McCardell
Ralph R. McCoy
Eleanor T. McNamara
William E. McTurk

Frank H. Mefferd
James B. Meredith
Robert H. Milbrath
Isalou Moody
J.A. Morakis
Matt Moran
Cecil Morgan
Paul E. Morgan
William S. Nancarrow
R.F. Neblett

R.H. Nesbitt
Robert F. Neu
Helen G. Nurge
M.E. O'Loughlin
John K. Oldfield
Howard W. Page
Carl Patterson
Stanley J. Pelc
Martha Peterson
George T. Piercy

David Powell

Wallace E. Pratt
Andy Purcell
Kurt Quick
Monroe J. Rathbone
Carl E. Reistle, Jr.
Werner Renberg
Hector Requezes
Helen P. Roche
Leslie C. Rogers

Jay Rose
Edward R. Sammis
Charles L. Scarlott
David A. Shepard
T.M. Sheriff
James B. Smith
Don Snook
Emile E. Soubry
Steve Stamas
C.M. Standahl

Edward Stanley

Ozzie C. Stroud
John F. Sturgeon
Ernesto Sugar
Blaine F. Townsend
Osgood V. Tracy
Lee W. Traven
Smith D. Turner
Siro Vázquez
Otto Wolff von Amerongen

Dr. W.E. Wakeley, Jr.
Leo D. Welch
Jack Wexler
T.A. White
Henry B. Wilson
Dorothy A. Windels
M.A. Wright
Hugh de N. Wynne
Stephen L. Wythe
Guillermo Zuloaga

Hines Baker, Cecil Burrill, Bill Cleveland, Jarvis M. Freymann, Dave Gill, Donald Green, Ed Holmer, Ken Jamieson, Ed Johnson, Carl Maas, W.M. McCardell, Howard Page, Wallace Pratt, Otz Tracy, Jack Wexler, Hugh Wynne, David Winans, and Paul Morgan, among others, loaned memoirs or like materials of their years with the company or its affiliates.
A number of historians provided timely advice and help. Their names are listed below:

Irvine Anderson
John B. Boles
William N. Greene
Burton I. Kaufman
Henrietta M. Larson

Stephen J. Randall
Joseph Clarke Robert
Charles B. Wallace
Gerald T. White
Mira Wilkins

Among those persons neither connected with Exxon nor professed historians, many furnished ideas and/or necessary materials: In particular, Bruce K. Brown, former president of PAN-AM Southern Corporation, shared his vast knowledge of oil and politics in the 1950s; Thomas E. Ward's book *Negotiations for Oil Concessions in Bahrain, El Hasa (Saudi Arabia), the Neutral Zone, Qatar and Kuwait...* (New York, 1965), was provided by his son Thomas E. Ward, Jr, as well as information about the Middle East; James B. Quinn of Dartmouth University loaned tapes of interviews with Jersey executives; Elmo Burns Roper made available typescripts of his many polls conducted for Exxon; Don Dedera contributed the tape of a fascinating interview with Wallace E. Pratt; Dean Rusk graciously granted me interviews covering his years in the State Department; and Stan Swinton permitted use of the Associated Press files; interviews with Osgood "Otz" V. Tracy furnished an informal overview of the company and a number of the men who served as executives during the period covered. Tracy proved to be well informed and very interesting.

Other individuals may also deserve a place in these listings. But, like other errors in this volume, such omissions are the sole responsibility of the principal author, who has attempted to do justice to all concerned with this ambitious project.

Bennett H. Wall
Athens, Georgia
1987

Index